Development of Bio-Based Materials: Synthesis, Characterization and Applications

Development of Bio-Based Materials: Synthesis, Characterization and Applications

Editors

Antonio M. Borrero-López
Concepción Valencia-Barragán
Esperanza Cortés Triviño
Adrián Tenorio-Alfonso
Clara Delgado-Sánchez

MDPI • Basel • Beijing • Wuhan • Barcelona • Belgrade • Manchester • Tokyo • Cluj • Tianjin

Editors
Antonio M. Borrero-López
Pro2Tecs
Universidad de Huelva
Huelva
Spain

Concepción Valencia-Barragán
Pro2Tecs
Universidad de Huelva
Huelva
Spain

Esperanza Cortés Triviño
Pro2Tecs
Universidad de Huelva
Huelva
Spain

Adrián Tenorio-Alfonso
Pro2Tecs
Universidad de Huelva
Huelva
Spain

Clara Delgado-Sánchez
Pro2Tecs
Universidad de Huelva
Huelva
Spain

Editorial Office
MDPI
St. Alban-Anlage 66
4052 Basel, Switzerland

This is a reprint of articles from the Special Issue published online in the open access journal *Polymers* (ISSN 2073-4360) (available at: www.mdpi.com/journal/polymers/special_issues/dev_bio_mater).

For citation purposes, cite each article independently as indicated on the article page online and as indicated below:

LastName, A.A.; LastName, B.B.; LastName, C.C. Article Title. *Journal Name* **Year**, *Volume Number*, Page Range.

ISBN 978-3-0365-6506-4 (Hbk)
ISBN 978-3-0365-6505-7 (PDF)

Contents

About the Editors

Antonio M. Borrero-López

Antonio María Borrero-López completed a degree in Chemical Engineering at the University of Huelva (Spain) in 2013 and later obtained a Master's degree in Product Formulation and Design from the International University of Andalusia (Spain). Afterwards, he received his PhD degree from the University of Huelva, with a thesis entitled "Development of new lignocellulosic-based thickening agents for biodegradable oleogel formulations with several industrial applications". This thesis was awarded the Best 2021–2022 Doctoral Thesis in the branch of Engineering and Architecture of the University of Huelva and received an honourable mention at the Annual European Rheology Conference 2022 (AERC 2022). Since his PhD, he has been working as a postdoc at the Institut Jean Lamour, University of Lorraine (France), researching new greener approaches to improve the performance of bio-based materials. He has participated in a total of eight research projects, including the European Project UCGWATERplus, which aims to remediate waters polluted with organic and inorganic contaminants through the formulation of different products via the valorisation of the residues from underground coal gasification and other processes. He has published a total of 22 articles and contributed to more than 30 international and national congresses.

Concepción Valencia-Barragán

Concepción Valencia Barragán received her PhD degree from the University of Extremadura. She is a Full Professor at the University of Huelva. She has carried out extensive research for more than 23 years, mainly focused on Chemical Product Engineering, specifically concerning the rheology of complex materials, the processing of non-Newtonian fluids, lubricants, adhesives and coatings, polymers and biopolymers, food colloids and emulsion technologies. In total, she has participated in more than 30 research projects with public funding, with some of them as a lead researcher and more than 40 contracts with private sector companies. Her main scientific and technical achievements include the modification of the rheological properties of lubricating greases by means of reactive and recycled polymeric additives as well as the development of biodegradable oleogels capable of replacing traditional lubricating greases formulated from non-renewable resources. As a result of this research activity, she has authored more than 118 papers in peer-reviewed journals and is the inventor of six patents. She has also presented more than 130 communications at national and international scientific conferences.

Esperanza Cortés Triviño

Esperanza Cortés-Triviño graduated from Industrial Technical engineering with a specialty in Industrial Chemistry in 2012 and became licensed in Chemical Engineering in 2014 at the University of Huelva. She completed her official Master's in Chemical Engineering in 2016, where she started her research career working with bioplastics based on wheat gluten. Meanwhile, she was hired by the University of Huelva as a researcher to conduct several tasks within the MINECO CTQ2014-56038-C3-1-R project funded by "Ministerio de Economía y Competitividad", also obtaining a grant from "Junta de Andalucía"and accomplishing a Doctoral Thesis in the Industrial and Environmental Science and Technology program (CyTIA), associated with the project TEP-1499, which is related to the development of new thickening agents of vegetable oil from different chemically modified lignocellulosic fractions. She obtained her PhD degree in 2019 with a Cum Laude Mention, unanimously, which was accompanied by a stay at the University of Applied Sciences of Hamburg to receive an International Mention. Her Doctoral Thesis was awarded the best 2019 Doctoral Thesis in the branch of Engineering of the University of Huelva and received an honourable mention at the Iberian Meeting on Rheology 2019 (IBEREO 2019). In her research career, she has attended multiple conferences and collaborated in some of their management activities. She is a member of The Complex Fluid Engineering Research Group (TEP-185) and The Research Center of Chemical Product and Process Technology (Pro2TecS), as well as The Spanish Rheology Group (GER) belonging to the Spanish Royal Society of Chemistry (RSEQ). Currently, she works as a postdoctoral researcher at the University of Huelva within an international project for the Procter & Gamble Company.

Adrián Tenorio-Alfonso

After graduating in Chemical Engineering in 2013 and upon the completion of his Master's in Product Formulation and Design with Applications in the Chemical, Food and Pharmaceutical Industry in 2014, both at the University of Huelva, Adrián Tenorio-Alfonso started his research activity with a final research project on the thermo-rheological characterization of foams and adhesives based on polyurethanes and biopolyurethanes. Afterwards, he was given a national grant for a PhD contract "Ayuda a la Formación de Profesorado Universitario (FPU13/01114)"by el Ministerio de Educación, Cultura y Deporte, carrying out his PhD thesis entitled Development of polyurethane formulations based on cellulose acetate and castor oil, affiliated with the Doctoral Programme in Industrial and Environmental Science and Technology at the Chemical Process and Product Technology Research Centre (Pro2TecS) from the University of Huelva. He has been a member of several national and regional research projects (MINECO CTQ2014-56038-C3-1-R and TIC 1499 funded by El Ministerio de Economía y Competitividad and La Conserjería de Innovación, Ciencia y Empresa de la Junta de Andalucía, respectively). Additionally, he has taken part in a research project for the development of synthetic ice rinks in collaboration with the Xtraice Rinks company. More recently, he has participated in a research project to study thermorheologically advanced suspensions for heat transport applications (CTQ2017-89792-R). In relation to this, he has been awarded a postdoctoral fellowship in the development of novel phase change materials, and is currently working on the research project "Development of phase change oil-in-oil emulsions with enhanced rheological, heat storage and heat transfer properties"at the Pro2TecS research centre.

Clara Delgado-Sánchez

Clara Delgado-Sánchez received her MSc in Chemical Engineering in 2013 and her Master's in Product Formulation and Technology in 2014, both at the University of Huelva. Afterwards, she received her PhD degree from the University of Lorraine (France) in 2017. During her thesis, she focused her research on the development of "New methods of optimization and characterization of tannin foams for thermal insulation of buildings"in the research group Biosourced Materials of the Institut Jean Lamour (—UMR CNRS 7198). Since her PhD, she has been a postdoc at the Chemical Process and Product Research Centre (Pro2Tec) at the University of Huelva, working on the project "Study of thermo-advanced dispersions for heat transport applications", a multidisciplinary project that led not only to her participation in the research and working groups of five other projects related to the same topic on a national and regional competitive basis and several publications, but also to her obtaining a postdoctoral fellowship that she currently enjoys. Currently, her line of research focuses on novel phase-change materials to obtain thermal-energy-storage systems as alternatives for efficient energy use and conservation. Moreover, in November 2020, she was awarded her first project as the principal researcher: "Valorisation of phosphogypsums in the development of foamed bitumen", financed by the "Cátedra de la Provincia"(University of Huelva). The results of her research efforts carried out between 2015 and 2022 were recorded in 22 articles in international journals and several contributions to national and international congresses.

polymers

MDPI

Article

Fe(III)-Rhamnoxylan—A Novel High Spin Fe(III) Octahedral Complex Having Versatile Physical and Biological Properties

Anum Hayat [1,2], Mohammad Saeed Iqbal [1,*], Naveed Ahmad [3,*], Nabil K. Alruwaili [3] and Atta ur Rehman [4]

[1] Department of Chemistry, Forman Christian College, Lahore 54600, Pakistan
[2] Department of Chemistry, Kinnaird College for women, Lahore 54000, Pakistan
[3] Department of Pharmaceutics, College of Pharmacy, Jouf University, Sakaka 72388, Al-Jouf, Saudi Arabia
[4] Department of Pharmacy, Forman Christian College, Lahore 54600, Pakistan
* Correspondence: saeediqbal@fccollege.edu.pk (M.S.I.); nakahmad@ju.edu.sa (N.A.)

Abstract: An iron (III) complex with rhamnoxylan, a hemicellulose from *Salvia plebeia* seeds, was synthesized and characterized by elemental analysis, spectroscopic and magnetic susceptibility measurements, thermal analysis and scanning electron microscopy. The rhamnoxylan was found to be a branched hemicellulose consisting of β-1,4-linked xylose main chain and rhamnose attached to the chain at β-1,3 positions. The complex was found to contain 18.8% *w/w* iron. A high-spin octahedral geometry of Fe^{3+} was indicated by the electronic absorption spectrum of the complex. In other experiments, the complex exhibited good electrical and magnetic properties. In vivo efficacy, as hematinic, of the complex in induced anemia was demonstrated equivalent to that of iron protein succinylate (taken as standard) as evidenced by raised red blood cell count, hemoglobin, hematocrit and total iron in rabbit. The complex was found to be non-toxic with LD50 > 5000 mg kg^{-1} body weight in rabbit. Thus, iron(III)-rhamnoxylan hold the potential for application as hematinic for treatment of iron deficiency anemia.

Keywords: Iron(III)-hemicellulose complex; rhamnoxylan; iron supplement; magnetic polymer; hematinic

Citation: Hayat, A.; Iqbal, M.S.; Ahmad, N.; Alruwaili, N.K.; Rehman, A.u. Fe(III)-Rhamnoxylan—A Novel High Spin Fe(III) Octahedral Complex Having Versatile Physical and Biological Properties. *Polymers* **2022**, *14*, 4290. https://doi.org/10.3390/polym14204290

Academic Editors: Antonio M. Borrero-López, Concepción Valencia-Barragán, Esperanza Cortés Triviño, Adrián Tenorio-Alfonso and Clara Delgado-Sánchez

Received: 18 August 2022
Accepted: 10 October 2022
Published: 12 October 2022

1. Introduction

The iron deficiency anemia is of concern in developed countries and a serious problem in underdeveloped countries. It is caused by low iron diet in developed countries and scarcity of nutritional diet in underdeveloped countries [1]. Iron deficiency causes a drop in hemoglobin level in blood. Typical hemoglobin levels in humans are >13 g dL^{-1} (males) and >12 g dL^{-1} (females); these levels may drop to 3 g dL^{-1} or lower in case of extensive blood loss due to hemorrhage, menstrual flow and bleeding ulcer [2]. In order to treat such conditions iron compounds are administered as therapeutics or included in dietary supplements. The iron salts currently used as therapeutics include ferrous sulphate, ferric chloride, ferric ammonium citrate, ferrous fumarate and complex compounds like ferrous gluconate, iron dextran, iron sorbitex, ferric carboxymaltose, ferric hydroxide polymaltose complex and iron protein succinylate (IPS) [3]. Among these, polymeric iron complexes are considered more useful as simple iron salts, when administered orally, interact with food and other medications resulting in reduced bioavailability [4]. On the other hand, the polymeric complexes are relatively stable and release iron in a sustained manner through ligand exchange processes in the body. Therefore, the complexes are considered beneficial for oral administration. Large molecule polysaccharide metal complexes exhibit higher safety as compared with traditional small molecule metal complexes. Therefore, polysaccharide metal complexes from single polysaccharides can be used as promising novel drugs.

Fe^{3+} forms both low-spin and high-spin octahedral complexes with different ligands. High-spin complexes are formed with weak-field ligands such Cl^- or OH^- ions. Low-spin and high-spin complexes play important role in spin-cross-over reactions in the biological system. Normally, the low-spin state is more favorable for Fe^{3+} complexes. The proteins having octahedral Fe^{3+} environment with all the six sites occupied include cytochromes and ferredoxins, which are involved in electron transfer processes. The title complex being reported here would mimic the role of these proteins. To the best of our knowledge no high-spin octahedral complex with a hemicellulosic ligand has been reported.

The problem with the currently available iron pharmaceuticals and nutritional supplements is that they cause gastric irritation and low efficacy. Ferric hydroxide polymaltose complex and IPS, considered to be safe [5], also exhibit side effects to varying degrees. Treatment with ferric hydroxide polymaltose is usually associated with diarrhea and lower onset effect [4,5], whereas the IPS may not be tolerated by the patients showing casein allergy. So, it is desirable to design and synthesize new iron complexes having lower toxicity, better tolerance and bioavailability. It has been reported that hemicelluloses have the potential to be used as drug carriers because of being biocompatible, biodegradable and digestible by human [6–8]. The present study was designed to investigate the potential of hemicelluloses to bind enough quantity of iron for use as hematinic. For this purpose, rhamnoxylan, a hemicellulosic material isolated from *Salvia plebeian* seeds, commonly known as sage and regionally recognized as 'Samundarsok' and 'Kamarkas', was used as a model [9].

S. plebeia is a herb found along sides the streams and rivers in different parts of the world. The plant's classification is: Plantae (kingdom), Magnoliophyta (division), Magnoliopsida (class), Lamiales (order), Lamiaceae (family), Salvia (genus) and plebeia (species). The extracts of seeds and leaves of this plant have significant therapeutic value; the popular uses include sore throat, headache, asthma and inflammatory diseases [10]. When seeds of *S. plebeia* are soaked in water, they produce a mucilage mainly consisting of hemicelluloses identified as rhamnoxylans having weighted-average molar mass (Mw) of 2.47×10^6 Daltons containing Xylp 99.32% and Rhap 0.68%. In the present work, it has been hypothesized that rhamnoxylan forms an iron complex having better efficacy, bioavailability and toxicity profile for treating iron deficiency.

2. Results and Discussion

2.1. Synthesis of Fe(III)-Rhamnoxylan

A dark brown complex was obtained by the experimental procedure as discussed in the methods section. The role of NaOH is to facilitate dissolution of rhamnoxylan during complex formation. In order to remove $Fe(OH)_3$, that may have formed due to a reaction of $FeCl_3$ with NaOH, the complex was thoroughly washed with dilute HCl. Further washings with methanol and ether afforded a significantly dry product, which was further dried at 60 °C in an oven to a constant weight and ground to obtain a free-flowing powder in good yield (~90%). The complex was sparingly soluble (approx. 1 g in 100 mL) in dimethyl sulfoxide (DMSO) and dimethylformamide (DMF), and insoluble in other organic solvents and water.

2.2. Characterization

2.2.1. Elemental Analysis

Elemental analysis of the rhamnoxylan sample and the synthesized Fe(III)-rhamnoxylan complex (Table 1) shows the presence of C and H. The percentages in the complex of these elements (C: 31.1%; H: 5.15%) were lower than those of rhamnoxylan (C: 46.73%; H: 6.45%) as expected due to the presence of iron (18.40%) in the complex. However, the C:H molar ratio was similar in the two materials, which confirmed that iron is bonded to the rhamnoxylan. The $AgNO_3$ gave negative test for Cl^- ions. The water content in the complex (11.0%) was higher than that in the rhamnoxylan used (5.98%) indicating the hygroscopic nature of the complex.

Table 1. Elemental, moisture, monosaccharide and GPC analysis data (% *w/w* on dry substance basis).

Compound	Carbon	Hydrogen	Iron	Moisture	Rha*p*	Xyl*p*	M_w (Dalton)
Rhamnoxylan	46.73 ± 0.04	6.45 ± 0.02	-	5.98 ± 0.01	0.68 ± 0.02	99.32 ± 0.05	2.47×10^6
Fe(III)-rhamnoxylan	31.1 ± 0.03	5.15 ± 0.01	18.40 ± 0.02	11.0 ± 0.01	0.61 ± 0.01	89.12 ± 0.04	2.52×10^6

2.2.2. Monosaccharide Analysis

The monosaccharide analysis in Table 1 clearly indicates the presence of Xyl*p* and Rha*p* in the complex; their percentages are lower than those in pure rhamnoxylan due to the presence of iron as part of the complex. However, the Xyl*p*:Rha*p* ratio in the complex is similar to that in the rhamnoxylan. This effect is also noticeable in the relatively higher molar mass of the complex (2.52×10^6 D) compared to that of the rhamnoxylan (2.47×10^6 D) as determined by GPC (Table 1).

2.2.3. FT-IR and Raman Analysis

The FT-IR spectra of rhamnoxylan and the Fe(III)-rhamnoxylan complex are shown in Figure 1. All the absorption bands of the sugar moieties in rhamnoxylan were present in the spectrum of the iron complex. The characteristic bands due to ν(OH) at 3337–3371 cm^{-1}, ν(C–C) in rhamnonosyl side chain at ~1059 cm^{-1} and β-glycosidic C-H bending at ~910 cm^{-1} were observed in both the samples. A broad absorption at 3371 cm^{-1} in the spectrum of the complex appears to be shifted to a higher energy from 3337 cm^{-1} due to ν(O-H) [11] in the pure rhamnoxylan [12,13]. This clearly indicates formation of a coordinate covalent bond between the hydroxyl groups and iron. Slight shift of glycosidic linkage, C-O-C was observed at 1034 cm^{-1} [13–16]. A Fe-O absorptions below 450 cm^{-1} that falls in the far IR region is generally indicative of formation of new Fe-O bond; this could not be recorded due to instrumental limitation. For that Raman spectrum of the complex was compared with that of the rhamnoxylan shown in Figure 1b and ferric chloride reported in literature [17]. Prominent Raman shifts in the region 300–420 cm^{-1} were observed in the spectrum of the complex that confirmed the existence of Fe-O bond in the complex and the absorption at 255–258 cm^{-1} assigned to an A_{1g}(Fe-Cl) mode [17] was absent in our spectrum indicating that the complex was free from $FeCl_3$. The $AgNO_3$ test also confirmed the absence of Cl^- ion in the complex.

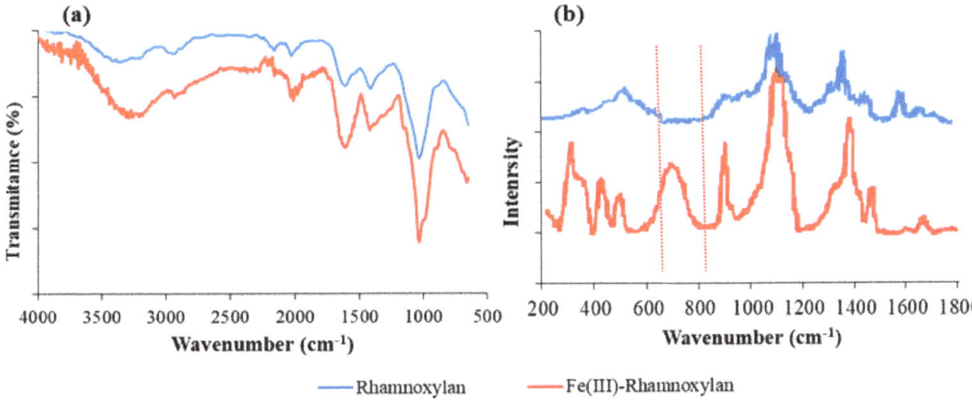

Figure 1. (**a**) FTIR and (**b**) Raman spectra of rhamnoxylan and Fe(III)–rhamnoxylan complex.

2.2.4. Structural Analysis of Rhamnoxylan by MALDI-ToF Mass Spectroscopy

The MALDI-ToF mass spectrum of the water soluble hemicellulose is shown in Figure 2. The assignments of fragment ions were done by mass interval analysis. The masses of residual ions of xylose (M = 150) and rhamnose (M = 164) after glycosidic cleavage (loss of 18 Da) were calculated as $(132)_n + 23$, $(132)_n + 23 + 18$ Da, $(146)_n + 23$ Da, $(146)_n + 23 + 18$ Da and $(146)_n + (132)_n + 23$ Da, where n is the number of monosaccharide units and 18 Da the mass of reducing end consisting of two hydrogen and one oxygen atoms. The mass difference between the daughter and the parent ions was used to identify the glycoside position in the monosaccharides. Such fragmentation patterns have been categorized into various types by Domon and Costello [18,19].

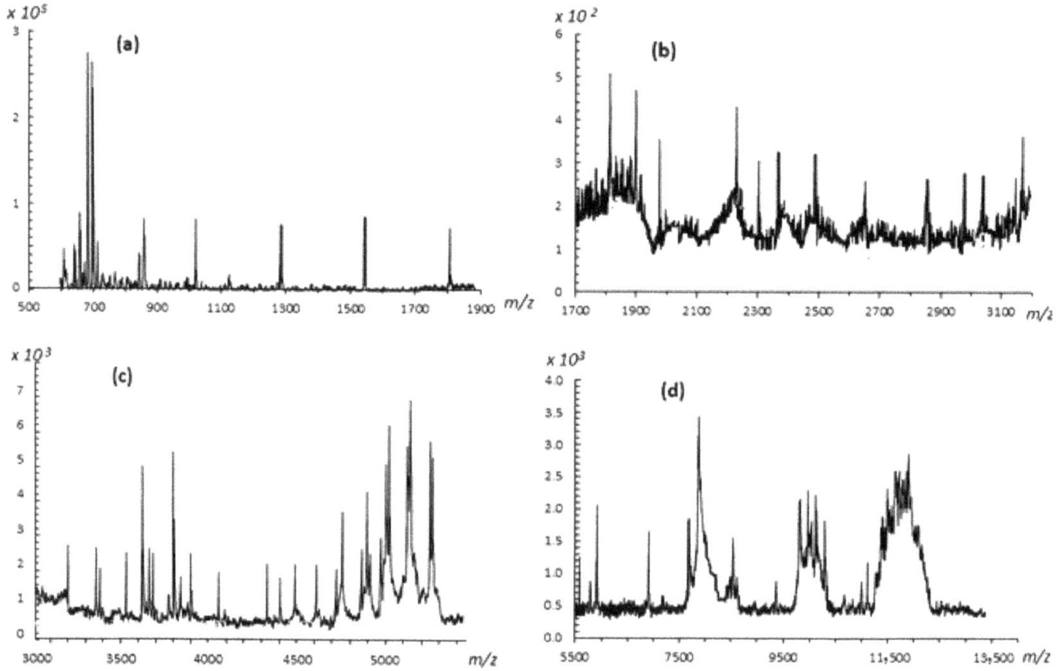

Figure 2. MALDI-TOF mass spectra of rhamnoxylan in (**a**) 500–1900, (**b**) 1700–3100, (**c**) 3000–6000 and (**d**) 5500–14,000 m/z ranges.

High intensity peaks were observed in the 600–900 m/z range. The intense peak at m/z 703 was assigned to a $[5\text{ Xyl} + \text{Na}]^+$ adduct, where Xyl is the xylose unit. The neighboring peaks at 643 and 626 m/z with a mass difference of 60 and 78 Da indicate β-1,4 linkages of xylosyl residues in the xylan backbone due to loss of $C_2H_4O_2$ and $C_2H_4O_2 + 18$ (cross-ring A/X-type cleavage). A peak at m/z~850 represents a $[\text{Xyl-Xyl-Xyl-Xyl-Xyl-Rh} + \text{Na}]^+$ residue, where Rh is the rhamnose unit. Two low intensity signals at 834 and 730 m/z having mass differences of 16 and 120 Da, respectively, suggest that rhamnose is also linked to xylose through β-1,4-linkage. Relatively higher intensities of these signals indicate that these groups are present as branches at the xylan chain. This is because bulky groups are more vulnerable to fragmentation. A low intensity peak at m/z 661 was assigned to $[\text{Xyl-Xyl-Xyl-Xyl-Xyl} + \text{H}]^+$ xylosyl residues. The lower intensity signal of the fragment is understandable because the carbohydrates are not easily protonated [20]. Very low intensity peaks were observed in the 900–3500 m/z range. There is a series of weak signals spaced by 132 Da; around m/z 1028, 1157, 1290, 1554 (dimer) and 1818.237 (dimer). Above m/z 1800, the rhamnosyl residues produced peaks frequently.

Some lower intensity peaks at m/z 1906, m/z 1894 and m/z 1866, having mass difference of 78, 90 and 118 Da from m/z 1984, due to cross-ring fragmentation (A/X-type) suggested that xylosyl and rhamnosyl residues are also linked through β-1,3 positions along with β-1,4. The lower intensities may be indicative of a rigid main chain difficult to undergo fragmentation. In the *m*/*z* 8000–13,000 range bunches of peaks were observed. These bunches most probably consist of a series of signals separated by 132 and 146 Da corresponding to xylosyl and rhamnosyl residues, respectively. Several signals with a difference of 16 and 17 Da were observed due to loss of O and OH, which are abundantly present in the polysaccharide structure. There was no evidence of the presence of uronic acids because of the absence of peaks having a mass difference of 178 Da.

This analysis shows that the main chain of polysaccharide consists of β-1, 4-D-xylose linkages branched with β-1, 3-linked rhamnose units. Thus, a block of the polymeric structure of rhamnoxylan can be proposed as shown in Figure 3. According to this, a mole of rhamnoxylan (2.47×10^6 Da) would consist of around 850 such blocks and 61,000 monosaccharide units.

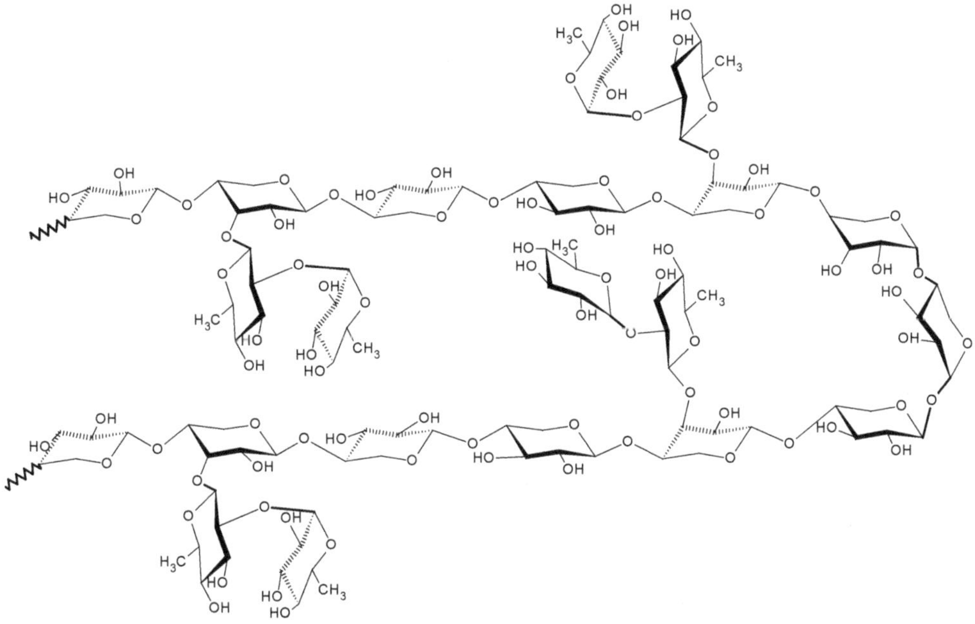

Figure 3. Structure of rhamnoxylan based on MALDI-ToF mass spectral analysis.

2.2.5. Electronic Absorption Spectrophotometry

Electronic absorption spectra of iron complexes in the UV-Visible range provide useful information about the oxidation state of the metal ion, chromophores in the ligand, charge transfer transitions and geometry of the complex ion. High-spin Fe^{3+} complexes contain five unpaired electrons in the d orbital, so all d-d transitions are spin forbidden, hence, should be weak. Most of the Fe^{3+} complexes exhibit weak d-d absorption bands, while some have been reported to show quite intense absorptions. The spectrum of the Fe(III)-rhamnoxylan was characteristic of a complex formation and entirely different from that of $FeCl_3{\cdot}6H_2O$ shown in Figure 4. The spectrum of $FeCl_3{\cdot}6H_2O$ consisted of a single absorption at about 300 nm [21] whereas, that of the complex exhibited three absorptions in the visible region centered at 400 (very broad), 570 (broad), 645 and 690 nm (sharp). This pattern resembles those of six- coordinate Fe^{3+} complexes.

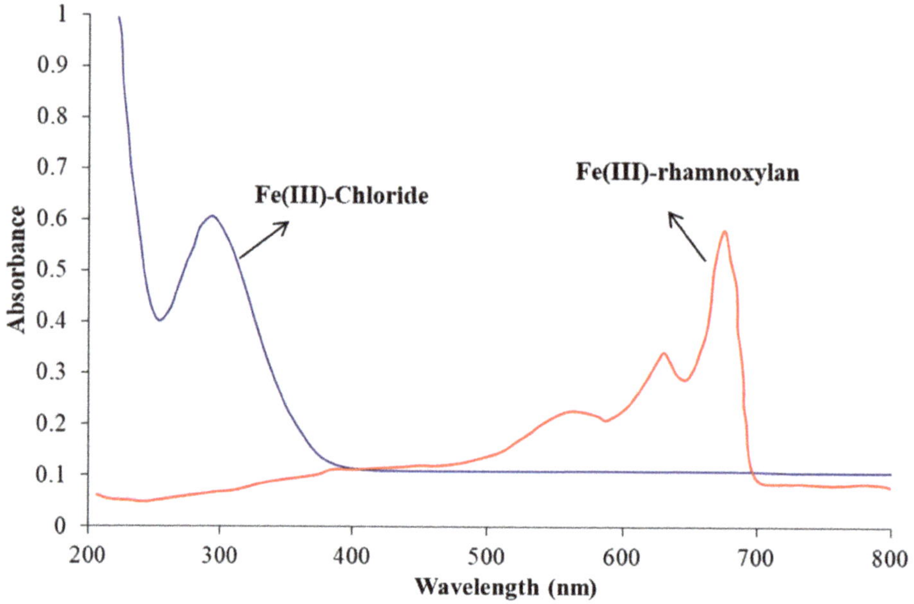

Figure 4. UV-Vis spectra of $FeCl_3 \cdot 6H_2O$ (in water) and Fe(III)-rhamnoxylan (diffuse reflectance).

The absence of any absorption in the regions 420–450 nm and 800–900 nm rules out the presence of Fe^{2+} ion in the complex [22]. The assignment of the observed absorptions is straightforward as for the octahedral complexes [23–26] and can be assigned as: ~400–$^6A_{1g} \rightarrow (^4A_{1g}, ^4E_g)$; ~570 nm–$^6A_{1g} \rightarrow {}^4T_{2g}$; ~645 nm (charge transfer, O-Fe LMCT); ~690 nm–$^6A_{1g} \rightarrow {}^4T_{1g}$. The sharpness of low-energy bands in the spectrum of the complex may be attributed to combination vibrational bands of the ligand.

2.2.6. Mössbauer Specroscopy

The Mössbauer spectrum recorded at room temperature is shown in Figure 5. The spectrum was fitted to a single doublet to compare it with a reported iron polysaccharide complex [27]. The ferrous doublet was absent that confirmed the absence of Fe^{2+} ion in the sample. The isomeric shift and quadrupole splitting were found to be 0.35 and 0.71 mm s^{-1}, respectively. These values correspond to ferrihydrite iron core [28].

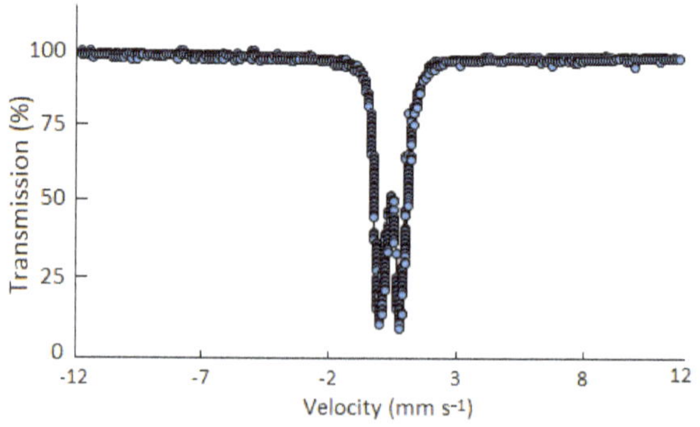

Figure 5. Mössbauer spectrum of Fe(III)–rhamnoxylan complex at room temperature (25 °C).

Based on the analytical (elemental and monosaccharide analyses) and spectroscopic (FT-IR/Raman and electronic absorption) evidence presented so far, the coordination sphere of Fe^{3+} ion in the complex appears to have an octahedral geometry, where the macromolecule acts as an anionic ligand. A structural block of the Fe(III)-rhamnoxylan complex can be shown as in Figure 6. Accordingly, a mole of the complex (2.52×10^6 Da) would consist of around 578 such blocks. This indicates that the degree of polymerization has been reduced to ~68% on complex formation. The salient features of the proposed structure are: (i) the rhamnoxylan acts as an anionic ligand under highly alkaline condition, (ii) coordinated water (1–3 molecules per iron atom) is present, (iii) varying number (1–2) of free OH^- ions in the alkaline medium are also found in the coordination sphere. The structure represents a polynuclear ferrihydrite iron core similar to the *Enteromorpha* polysaccharide-iron (III) complex [29]. The Fe(III)-rhamnoxylan complex pre-dominantly appeared to be hydrophobic (lipophilic) as shown by the solubility data.

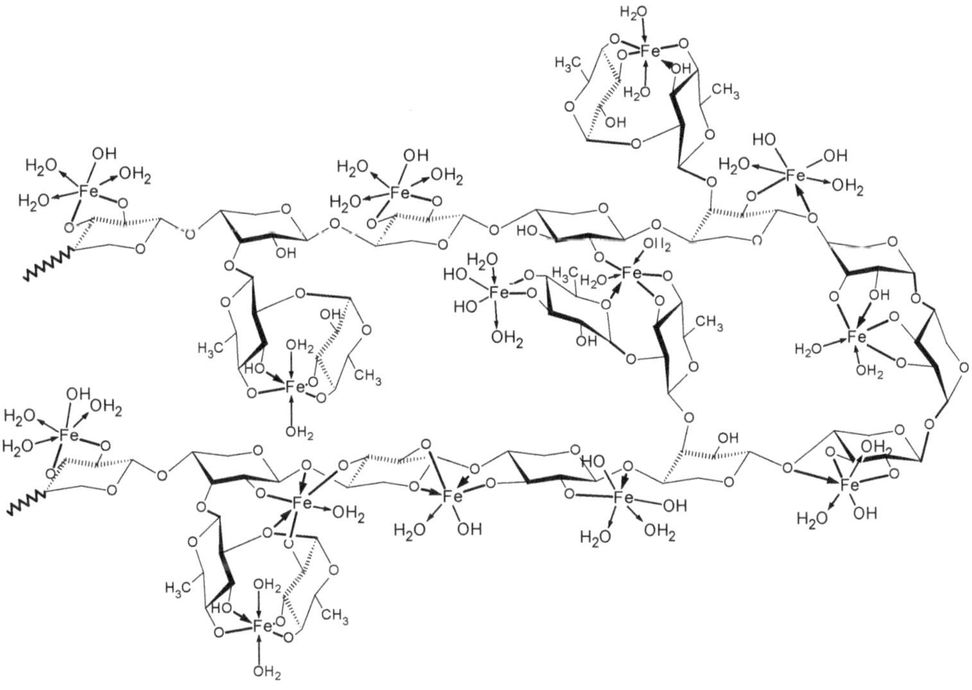

Figure 6. Proposed structure of the Fe(III)–rhamnoxylan complex.

2.2.7. Powder X-ray Diffraction Analysis

The p-XRD spectra of rhamnoxylan and the Fe(III)-rhamnoxylan complex are shown in Figure 7. Rhamnoxylan appeared to be mainly in amorphous state, whereas the complex exhibited four peaks at 2θ values 33°, 45°, 57° and 75° indicating a crystalline nature of the complex [30–34] and favoring octahedral geometry. The sharp peak around 33° corresponds to that of iron oxide nanoparticle [31,32,35], whereas at 45° is related to the iron crystal [32]. The appearance of peaks represents different crystal sizes ranging from nano- to micro-meter [23]. The significant difference in the spectra suggests a phase change due to complex formation. These peaks represent different crystal sizes ranging from nano- to micro-meter scale. Attempts to grow a single crystal for further structure elucidation were not successful.

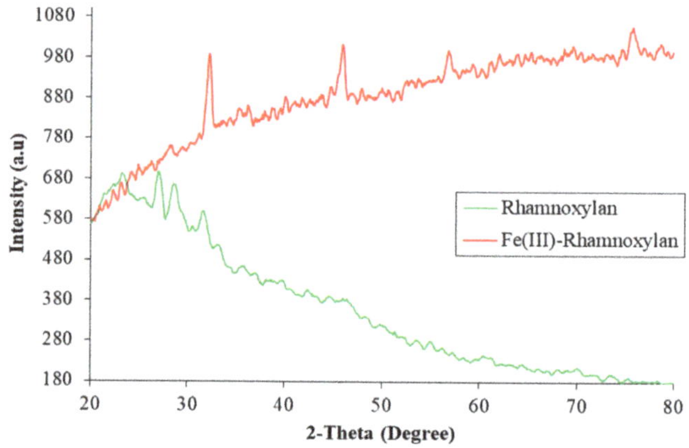

Figure 7. X-ray diffraction patterns of rhamnoxylan and Fe(III)-rhamnoxylan complex.

2.2.8. Magnetic Properties

The magnetic moment was found to be 5.9 B.M. that corresponds to the spin only value of 5.92 B.M for Fe(III) complexes. This result confirms the high spin configuration of Fe(III) ion in the complex. Thus, the complex exhibits paramagnetism [33,35,36]; the property that may be exploited in biomedical applications including cell separation, immunoassay, magnetic resonance imaging and drug delivery.

2.2.9. Scanning Electron Microscopy

The SEM images of rhamnoxylan and the Fe(III)-rhamnoxylan complex are shown in Figure 8. A comparison of the two micrographs clearly shows the formation of a new material with change in the morphology after complexation.

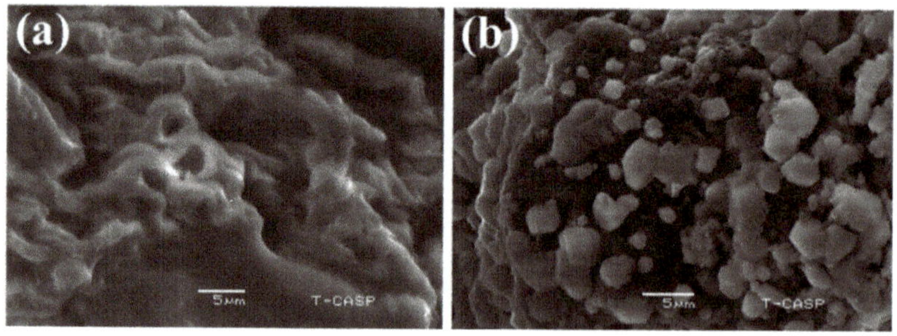

Figure 8. SEM images of (**a**) rhamnoxylan and (**b**) Fe(III)-rhamnoxylan complex.

2.2.10. Thermal Analysis

TGA was used to determine thermal stability of the complex and to further verify the presence of organic component in it. Thermograms of rhamnoxylan and the Fe(III)-rhamnoxylan are shown in Figure 9. A three-stage decomposition of the complex was observed. The stage I at ambient to ~120 °C corresponds to a loss of absorbed moisture (~11.3%) in the complex; whereas in case of rhamnoxylan this loss was ~5.8% due to absorbed moisture at ~100 °C. It was noted that decomposition of the complex starts at a higher temperature (~271 °C) compared to that of the rhamnoxylan (~225 °C); this suggests that the complex is thermally more stable than the rhamnoxylan. Two major losses occurred at stages II and III, which account for the decomposition of the polymeric ligand. A relatively lower weight loss (~20%) at stage II probably represents side chain decomposition compared with a major loss (~40%) at stage III due to decomposition of the main polymer chain. The residue of the complex was ~26% that corresponds to Fe_2O_3.

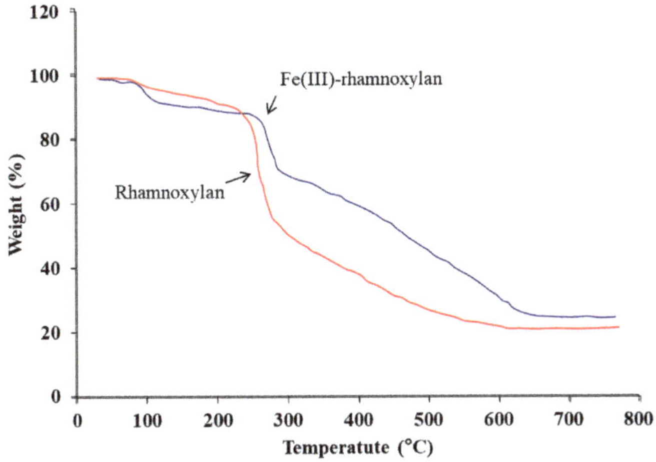

Figure 9. TGA curves of rhamnoxylan and Fe(III)-rhamnoxylan recorded at 10 °C min^{-1} under nitrogen flow (30 mL min^{-1}).

2.2.11. Electrical Properties

The impedance spectra of the samples are shown in Figure 10. The resistance and ionic conductivity data at different temperatures are presented in Table 2. It can be seen that resistance decreases and conductivity increases with a rise in temperature in accordance with the kinetic theory. These data show that the complex can be used as a solid polymer electrolyte.

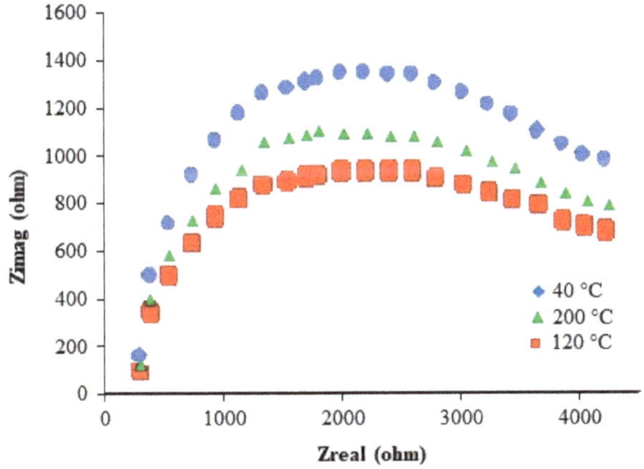

Figure 10. Impedance spectra of the iron complex recorded at 40, 120 and 200 °C.

Table 2. Resistance and conductivity data of the iron complex at different temperatures.

Temperature (°C)	R_g (Ω)	R_{gb} (Ω)	R_t (Ω)	σ (S cm^{-1})
40	205.35	1650.46	1845.64	2.51×10^{-4}
120	100.86	456.24	507.21	1.01×10^{-3}
200	36.32	65.89	107.21	4.99×10^{-3}

2.3. Stability Study

The results of the six-month accelerated stability study are depicted in Figure 11. It was observed that the Fe^{3+} content remained unchanged throughout the study period and there was no Fe^{2+} impurity detectable in the product. This is understandable because Fe^{2+}, the only expected impurity, may be produced under reducing environment and cannot creep up in the presence of air and humidity. Moreover, no significant change in the FT-IR and electronic absorption spectra was observed during this period. So, based on this study and TGA data, it can be concluded that the product is highly stable.

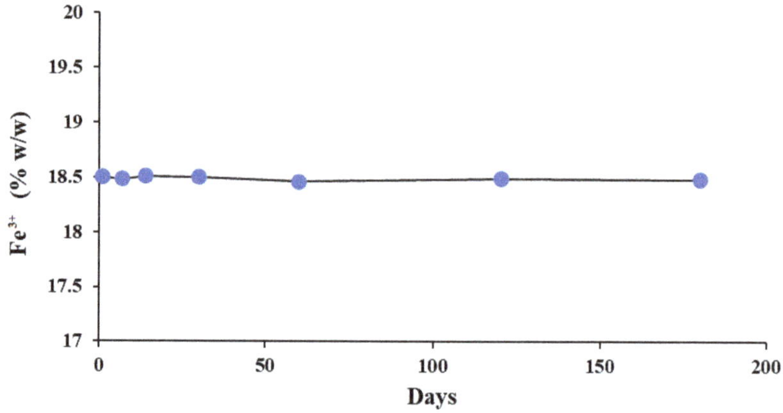

Figure 11. Fe^{3+} content in the Fe(III)-rhamnoxylan complex over a period of six months stored at 50 ± 1 °C and 75% relative humidity.

2.4. Biological Evaluation

2.4.1. Acute Oral Toxicity

According to OECD guidelines, 24 h after administration of the test dose (5000 mg kg^{-1} body weight) none of the animals was found to be dead and the animals showed no significant change in eye color, hair fall and weight for next 48 h. During that time, they remained active with normal heart and breathing rate and no irritation. The eye reflexes were also positive as assessed by a standard method [37]. No redness or ulceration was found in the stomach walls, indicating a good tolerability needed for long term treatment. The isolated stomach, spleen, liver, intestine, heart and kidneys were found to be normal as examined microscopically. The results of this test suggested that the synthesized Fe(III)-rhamnoxylan complex is non-toxic and may be categorized as 'unclassified' according to GHS categorization [38].

2.4.2. In Vivo Absorption and Excretion of Iron

After oral administration of 5000 mg kg^{-1} body weight of Fe(III)-rhamnoxylan in the acute toxicity test, the stomach material was microscopically found to contain folds of the hemicellulose (Figure 12a) indicating that the complex remains intact in the stomach and reaches intestine slowly releasing iron there, after dissolution of rhamnoxylan at the alkaline pH in the duodenum. The complex is predominantly lipophilic and slowly dissolves in alkaline medium. Some amount of iron in feces and significant amount in blood were detected indicating that iron from the complex is released and distributed throughout the body. Stomach was almost clear of the complex after 4 days (Figure 12b).

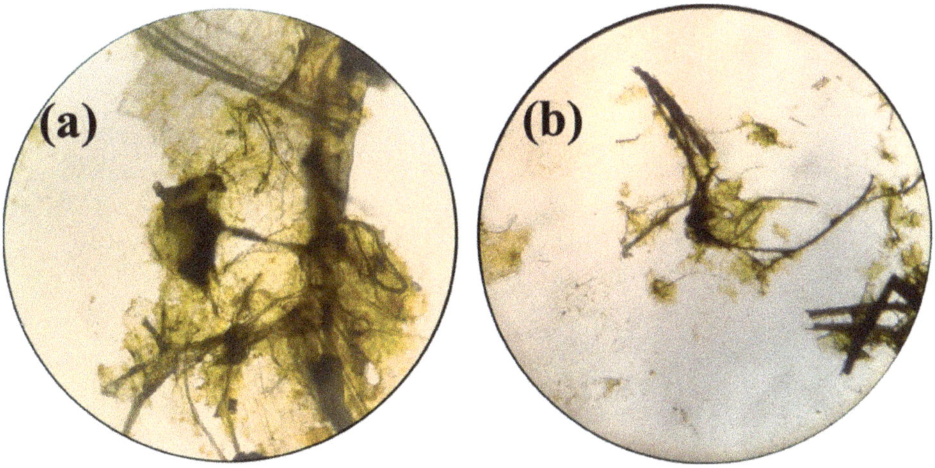

Figure 12. Microscopic image of (**a**) Fe(III)-rhamnoxylan in stomach after 24 h and (**b**) stomach material after 14 days.

As the storage of iron in the body depends upon several factors including that in stomach, spleen, liver, macrophages, absorption from GIT and simultaneous release through bile [39], the bioavailability of iron cannot be accurately determined from the AUC. So, the bioavailability was assessed as the % increase of iron in blood after 14 days (Table 3) and it was estimated to be ~13.29% that was slightly higher than that of the IPS standard (~9.78%).

Table 3. Bioavailability data of Fe(III)-rhamnoxylan and IPS.

Treatment	Amount of Fe(III) in Control Animals after 14 Days (mg)	Amount of Fe(III) in Blood after 14 Days (mg)	Bioavailability (% Increase)
IPS (33 mg equivalent of iron)	6.4	7.02 ± 0.3	~9.78
Fe(III)-rhamnoxylan (33 mg equivalent of iron)	6.4	7.26 ± 0.2	~13.29

2.4.3. Therapeutic Evaluation

The Figure 13a shows the % change in Hb level for 14 days experiment. The minimum levels of Fe, Hb, HCT and RBCs count were observed on the 6th day. However, the minimum levels of all these parameters were better in the treatment groups compared to those for the untreated control (Table 3). A comparison of the Hb levels on the day 14 indicates that recoveries were: 89% (Fe(III)-rhamnoxylan), 87% (IPS) and 70% (negative control/anemic), suggesting that the Fe(III)-rhamnoxylan complex is equally or slightly more effective in raising the Hb level than IPS (the standard drug). Similar trends were observed in RBCs (Figure 13b), HCT (Figure 13c) and blood iron (Figure 13d). Overall, the in vivo study demonstrates better ($p < 0.05$) efficacy, bioavailability and toxicity profile of Fe(III)-rhamnoxylan, in the animal model under investigation, compared with a commercially available standard (IPS). Fe(III)-rhamnoxylan can be considered as a potential candidate for clinical trials.

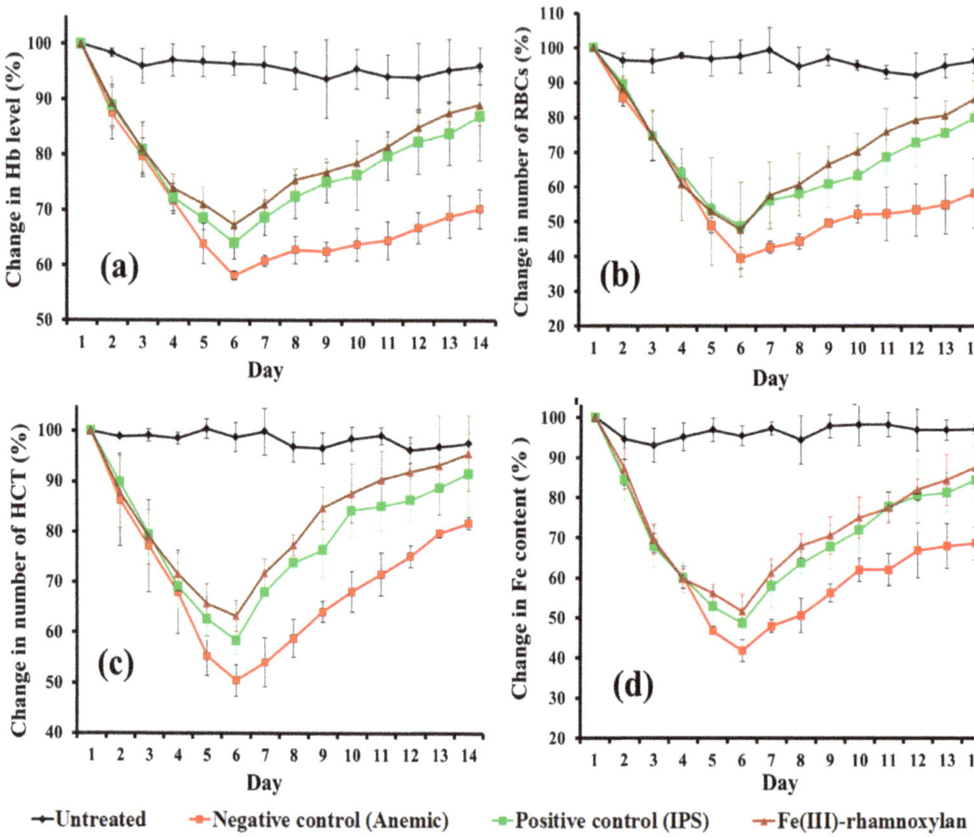

Figure 13. Effect of Fe(III)-rhamnoxylan on (**a**) % Hb, (**b**) % RBCs, (**c**) % HCT and (**d**) Fe content.

3. Materials and Methods

3.1. Materials

Rhamnoxylan from *S. plebeia* (Mw 2.47 × 10^6 Daltons containing 99.32% Xylp and 0.68% Rhap) was used from a sample previously prepared and characterized in our laboratory [40]. Ferric chloride hexahydrate ($FeCl_3 \cdot 6H_2O$), silver nitrate ($AgNO_3$) sodium hydroxide (NaOH), hydrochloric acid (HCl, 37%) and 2,5-dihydroxybenzoic acid (DHB), all were from Sigma-Aldrich, St. Louis, MO, USA. IPS (Italfarmaco S.p.A. Milan, Italy) was used as the standard drug and manganese chloride ($MnCl_2$, QQ 153, Sherwood Scientific Ltd., Cambridge, UK) was used as the calibrating standard for magnetic measurements. All the chemicals were of analytical grade and used without further purification. Distilled water was used throughout the experimental work.

3.2. Synthesis of Fe(III)-Rhamnoxylan Complex

Rhamnoxylan (2 g) was suspended in hot water (at 75–80 °C) and allowed to soak for about an hour under constant stirring by a magnetic stirrer. To the stirred suspension, $FeCl_3 \cdot 6H_2O$ (5.35 mL of 75% aqueous solution) was added followed by dropwise addition of NaOH (40% aqueous solution) to adjust the pH to 11.5. The mixture was stirred for further 60 min at 80 °C. The brown colored complex formed as precipitate was isolated by filtration under vacuum. The complex was washed with dilute HCl to remove any ferric hydroxide that may have formed followed by several washings with methanol. The complex was finally washed with ether and dried at 60 °C in an oven and ground into a powder (yield 1.8 g; 90% of the rhamnoxylan used). The complex formation occurred in the pH range 8–11.5 with better yield at 11.5. Similar pH has been used in the synthesis of an iron-saccharide complex [41]. The purity of the complex was ascertained by the single peak appearing in the gel permeation chromatogram.

3.3. Characterization

The elemental (carbon and hydrogen) analysis of the synthesized complex was carried by Leco 4201 elemental analyzer (Leco, St Joseph, MI, USA). Iron was determined by atomic absorption spectrophotometery. For this, the complex (100 mg) was digested in aqua regia (15 mL) followed by dissolution of the iron salt in water to make 100 mL of the solution. The solution was filtered and analyzed by flame atomic absorption spectrometer AAS FS240 (Varian, Palo Alto, CA, USA) using iron hollow cathode lamp set at 248.3 nm and an air-acetylene flame. The standards (prepared from 1.0 mg mL^{-1} stock solution, lot# D2-FE03121 in 2% HNO_3 *v/v*, from Inorganic Ventures, Christiansburg, VA, USA Cat# AAFE1-5) and samples were analyzed in triplicate. The R^2 value of the six-point calibration curve was 0.9998. The $AgNO_3$ test was performed to detect Cl^- ions if any. Moisture content of the complex was determined by Karl-Fischer titration using Titrino 701KF (Metrohm, Herisau, Switzerland) auto titrator.

Monosaccharide analysis was carried out after hydrolysis of the rhamnoxylan and the iron complex with sulfuric acid and electro chemical detector according to a reported method [42]. Rhamnose and xylose (Sigma Aldrich, St. Louis, MO, USA) were used as standards. The chromatographic conditions were: isocratic elution using 95% water and 5% 0.2 M NaOH at room temperature (25 ± 1 °C); flow rate 0.5 mL min^{-1}; injection volume 50 µL.

Molar masses were determined by gel permeation chromatography (GPC) using Agilent 1200 series (Agilent, Waldbronn, Germany) system equipped with PL aquagel-OH mixed column and refractive index detector (G1362A) using DMSO containing 0.50% LiBr as eluent (flow rate 0.5 mL min^{-1} at 70 °C) and injection volume of 10 µL. Pullulan (96351 Supelco standard set Mp ~350–700,000) and dextran (31430 Supelco standard set Mp 1000–400,000) were used as calibration standards.

The FT-IR spectra were recorded at 2 cm^{-1} resolution, in transmission mode, using a Carry 630 FTIR spectrophotometer (Agilent Technologies, Santa Clara, CA, USA). The powdered sample was placed on the diamond crystal and the spectrum was recorded in the 4000–630 cm^{-1} range. Raman spectra were recorded by use of a reflex microscope (inVia™ confocal Raman microscope, Renishaw). A near-infrared diode laser was used for excitation at 785 nm. Electronic absorption spectra were recorded in diffuse reflectance mode by UV-1800 spectrophotometer (Shimazdu, Kyoto, Japan) in the 200–800 nm range.

MALDI-ToF mass spectrometric analysis was carried out to determine structural features of the rhamnoxylan used for complex formation. This technique is appropriate as it produces singly charged ions; this has been demonstrated in our recent work [20]. The spectra were recorded in the 500–14,000 m/z range by MALDI-ToF/ToF (Bruker Daltonics Inc, Fremont, CA, USA) spectrometer using single ToF option in the reflectron mode. The desalted sample (1.5 mL) was spotted in triplicate on MTP 384 polished steel BC targets (Bruker Daltonics) and the samples were overlaid with 2,5-DHB (1.5 mL) and dried. The rest of the procedure was same as reported earlier [20].

The Mössbauer spectrum was recorded at room temperature using a 25 mCi ^{57}Co(Rh) source. The velocity was calibrated by laser interferometry. The powdered sample (~125 mg) was pressed in a brass ring with area of 1.75 cm^2 and sealed with benzene-styrofoam. Aluminum foil was used as the backing. The spectrum was fitted to single doublet assuming a Lorentzian line shape.

Powder X-ray diffraction (p-XRD) spectra were recorded by D2 PHASER diffractrometer (Bruker, Bremen, Germany). The samples were tightly packed in the sample holder and diffractions were recorded using Cu radiation (1.54184 Å, 30 kV, 10 mA) and Lynxeye array detector over the 19°–80° 2θ range.

Magnetic susceptibility measurements were performed by using MSB MK 1 magnetic susceptibility balance (Sherwood Scientific, Cambridge, UK). The balance was calibrated by using $MnCl_2$, standard (QQ153, Sherwood Scientific Ltd., UK). The effective magnetic moment (μ_{eff}) was calculated by the formula: μ_{eff} (B.M.) = $2.828(\chi AT)^{1/2}$, where χA is the atomic susceptibility of the paramagnetic ion (Fe^{3+} in this case) corrected for the ligands and T is the absolute temperature.

Surface morphology was studied by scanning electron microscopy (SEM) using JEOL JSM-6480LV microscope after sputter coating with gold (in case of rhamnoxylan) and without sputter coating (in case of Fe(III)-rhamnoxylan as the complex was self-conducting due to the presence of iron in it). For this, samples were fixed on the stub with a double-sided adhesive tape and images were recorded at different magnifications.

Thermal analysis was carried out in TGA mode by SDT Q600 analyzer (TA Company, La Vergne, TN, USA). Samples (approx. 6 mg) were heated in aluminum sample pans from 10 °C to 700 °C at a heating rate of 10 °C min^{-1} under nitrogen flow (30 mL min^{-1}).

To determine electrical properties, sample was cold-pressed into cylindrical pellets (10 mm dia) at 40 MPa. The pellets were painted with silver paste and impedance spectra were recorded in air at 40–200 °C (heating rate: 15 °C) using Gammry Interface 1000 potentiostat. The total resistance was determined by the formula: $R_t = R_g + R_{gb}$, where R_g and R_{gb} are the resistance of grain interior and grain boundary, respectively. The ionic conductivity (σ) was calculated using the formula: $\sigma = l/RA$, where l is the thickness of the sample and A the cross-section area.

A six-month accelerated stability study was performed, according to WHO Q1F Stability Guideline [43] as follows. Three different batches of the complex contained in plastic PP Securitainers® (VWR International Ltd., Lutterworth, UK) were stored at 50 ± 1 °C and 75% relative humidity and assayed for Fe^{3+} according to a validated method [44]. In addition to this the FT-IR and electronic absorption spectra were also recorded before and after six months.

3.4. Biological Evaluation

The synthesized complex was subjected to acute toxicity test and therapeutic evaluation in vivo in standard Dutch rabbit. These studies were approved by the institutional review board (vide IRB-124/12-2018) of the Forman Christian College (A chartered University), Lahore, Pakistan. All animal experiments complied with the ARRIVE guidelines in accordance with the U.K. Animals (Scientific Procedures) Act, 1986 and associated guidelines (EU Directive 2010/63/EU for animal experiments).

3.4.1. Acute Oral Toxicity Test

This test was performed according to OECD guidelines for determination of oral toxicity [38]. Ten male rabbits (800–1000 g) were fed with fodder and free water adaptively for a week. Five of the animals were orally fed with the synthesized complex at a dose of 5000 mg kg^{-1} body weight, while the other group of five was kept as control. Both the groups were monitored for 48 h for any physical and behavioral changes including eye color, eye reflexes, hair fall, weight change, heart rate, breathing rate, irritability and activity. Eye reflexes were examined by flashing light into the side of the rabbit's eye. Corneal reflexes were examined by smoothly touching the cornea with a clean cotton swab to check for any induced blink reflex [37].

3.4.2. Therapeutic Evaluation

Twenty male rabbits (6–8 weeks old, weighing 1000 ± 20 g) were purchased from Tollinton Market, Lahore, Pakistan and housed in the animal storage facility at Forman Christian College. They were fed with plain water and a low iron diet, adaptively for seven days. The animals were divided equally into four groups as: A (untreated), B (negative control/anemic), C (positive control/IPS) and D (Fe(III)-rhamnoxylan). Iron-deficiency anemia was induced as reported earlier [45] by oral administration of 11 µg kg^{-1} per day phenylhydrazine for two days. The treatment was started after 4th day of the induction of anemia. The groups C and D were administered, orally, a dose of the standard (IPS) and the test sample equivalent to Fe 33 mg kg^{-1} body weight, respectively [46]. Blood samples (1mL) were collected over 14 days with an interval of 24 h from the marginal ear vein, using a 1 mL disposable syringe (with the needle 26G × 5/8 in.) containing K_2-EDTA, The effect of phenylhydrazine on Hb, RBCs, HCT and Fe levels was continuously monitored separately. The percent change in the values was plotted against time (days), where % change was calculated by the equation:

$$\% \; change = \left[\frac{100 - 100 \times (Initial\ value\ of\ the\ parameter - Value\ on\ the\ particular\ day)}{Initial\ value} \right]$$

3.5. Statistical Analysis

The student t test was applied to observe the significant difference between the negative control (anemic) and the Fe(III)-rhamnoxylan group; the significance was determined at $p < 0.05$.

4. Conclusions

A molecular complex of Fe^{3+} with rhamnoxylan from *Salvia plebeia* seeds having the octahedral coordination environment and high magnetic susceptibility was successfully prepared. The complex exhibited good electrolytic properties and accelerated stability. The Fe(III)-rhamnoxylan was non-toxic and exhibited better hematinic property as compared with iron protein succinylate (a commercially available product) with no gastric irritation. This provides for a safer potential candidate as hematinic for clinical trials.

Author Contributions: Conceptualization, M.S.I. and N.A.; methodology, M.S.I., A.u.R. and A.H.; formal analysis, M.S.I., A.u.R. and A.H.; investigation, M.S.I., A.u.R., N.A. and A.H.; resources, M.S.I., N.K.A. and N.A.; data curation, N.A. and M.S.I.; writing—original draft preparation, A.H., A.u.R. and M.S.I.; writing—review and editing, N.A. and N.K.A.; supervision, M.S.I. and N.A.; project administration, M.S.I. and N.A.; funding acquisition, N.A. and N.K.A. All authors have read and agreed to the published version of the manuscript.

Funding: This work was supported through grant number "375213500" from the Deputyship for Research and Innovation, Ministry of Education in Saudi Arabia.

Institutional Review Board Statement: The animal study protocol was approved by the Institutional Review Board of the Forman Christian College (A chartered University), Lahore, Pakistan (vide IRB-124/12-2018).

Acknowledgments: The authors extend their appreciation to the Deputyship for Research and Innovation, Ministry of Education in Saudi Arabia supporting this study. The authors thank the Department of Pharmacy, Forman Christian College (A chartered University), Lahore, Pakistan for providing access to their animal and laboratory facilities.

Conflicts of Interest: The authors declare no conflict of interest.

References

1. Haas, J.D.; Brownlie, T. Iron deficiency and reduced work capacity: A critical review of the research to determine a causal relationship. *J. Nutr.* **2001**, *131*, 676S–690S. [CrossRef] [PubMed]
2. Lim, W.H.; Choi, E.K.; Han, K.D.; Lee, S.R.; Cha, M.J.; Oh, S. Impact of hemoglobin levels and their dynamic changes on the risk of atrial fibrillation: A nationwide population-based study. *Sci. Rep.* **2020**, *10*, 1–8. [CrossRef] [PubMed]
3. Cancelo-Hidalgo, M.J.; Castelo-Branco, C.; Palacios, S.; Haya-Palazuelos, J.; Ciria-Recasens, M.; Manasanch, J.; Pérez-Edo, L. Tolerability of different oral iron supplements: A systematic review. *Curr. Med. Res. Opin.* **2013**, *29*, 291–303. [CrossRef] [PubMed]
4. Funk, F.; Canclini, C.; Geisser, P. Interactions between Iron(III)-hydroxide polymaltose complex and commonly used medications laboratory studies in rats. *Arzneim.-Forsch. Drug Res.* **2007**, *57*, 370–375. [CrossRef]
5. Shilpashree, B.G.; Arora, S.; Sharma, V.; Singh, A.K. Preparation of succinylated sodium caseinate-iron complex by adopting ultrafiltration technology: A novel food fortificant. *Innov. Food Sci. Emerg. Technol.* **2015**, *32*, 165–171. [CrossRef]
6. Iqbal, M.S.; Akbar, J.; Hussain, M.A.; Saghir, S.; Sher, M. Evaluation of hot-water extracted arabinoxylans from ispaghula seeds as drug carriers. *Carbohydr. Polym.* **2011**, *83*, 1218–1225. [CrossRef]

7. Ahmad, N.; Tayyeb, D.; Ali, I.; Alruwaili, N.K.; Ahmad, W.; ur Rehman, A.; Khan, A.H.; Iqbal, M.S. Development and characterization of hemicellulose-based films for antibacterial wound-dressing application. *Polymers* **2020**, *12*, 548. [CrossRef] [PubMed]
8. Holloway, W.D.; Tasman-Jones, C.; Lee, S.P. Digestion of certain fractions of dietary fiber in humans. *Am. J. Clin. Nutr.* **1978**, *31*, 927–930. [CrossRef] [PubMed]
9. Liang, Y.-Y.; Wan, X.-H.; Niu, F.-J.; Xie, S.-M.; Guo, H.; Yang, Y.-Y.; Guo, L.-Y.; Zhou, C.-Z. Salvia plebeia R. Br.: An overview about its traditional uses, chemical constituents, pharmacology and modern applications. *Biomed. Pharmacother.* **2020**, *121*, 1–15. [CrossRef] [PubMed]
10. Jung, H.; Seon, Y.; Lim, C.; Park, E. Anti-Inflammatory, anti-angiogenic and anti-nociceptive activities of an ethanol extract of *Salvia Plebeia* R. Brown. *J. Ethnopharmacol.* **2009**, *126*, 355–360. [CrossRef] [PubMed]
11. Gao, W.; Huang, Y.; He, R.; Zeng, X. Synthesis and characterization of a new soluble soybean polysaccharide-iron(iii) complex using ion exchange column. *Int. J. Biol. Macromol.* **2018**, *108*, 1242–1247. [CrossRef] [PubMed]
12. Zhao, W.; Odelius, K.; Edlund, U.; Zhao, C.; Albertsson, A.C. In situ synthesis of magnetic field-responsive hemicellulose hydrogels for drug delivery. *Biomacromolecules* **2015**, *16*, 2522–2528. [CrossRef] [PubMed]
13. Kang, M.; Oderinde, O.; Liu, S.; Huang, Q.; Ma, W.; Yao, F.; Fu, G. Characterization of xanthan gum-based hydrogel with Fe^{3+} ions coordination and its reversible sol-gel conversion. *Carbohydr. Polym.* **2019**, *203*, 139–147. [CrossRef] [PubMed]
14. Kačuráková, M.; Wellner, N.; Ebringerová, A.; Hromádková, Z.; Wilson, R.H.; Belton, P.S. Characterisation of xylan-type polysaccharides and associated cell wall components by FT-IR and FT-Raman spectroscopies. *Food Hydrocoll.* **1999**, *13*, 35–41. [CrossRef]
15. Peng, P.; She, D. Isolation, Structural characterization, and potential applications of hemicelluloses from bamboo: A Review. *Carbohydr. Polym.* **2014**, *112*, 701–720. [CrossRef] [PubMed]
16. Geetha, K.; Raghavan, M.S.S.; Kulshreshtha, S.K.; Sasikala, R.; Rao, C.P. Transition-metal saccharide chemistry: Synthesis, spectroscopy, electrochemistry and magnetic susceptibility studies of iron(iii) complexes of mono- and disaccharides. *Carbohydr. Res.* **1995**, *271*, 163–175. [CrossRef]
17. Sharma, S.K. Raman Study of ferric chloride hexahydrate and ferric chloride hexadeutrate in crystalline, molten and glassy states. *J. Non. Cryst. Solids* **1974**, *15*, 83–95. [CrossRef]
18. Domon, B.; Costello, C.E. A systematic nomenclature for carbohydrate fragmentations in FAB-MS/MS spectra of Glycoconjugates. *Glycoconj. J.* **1988**, *5*, 397–409. [CrossRef]
19. Viseux, N.; De Hoffmann, E.; Domon, B. Structural assignment of permethylated oligosaccharide subunits using sequential tandem mass spectrometry. *Anal. Chem.* **1998**, *70*, 4951–4959. [CrossRef] [PubMed]
20. Iram, F.; Massey, S.; Iqbal, M.S.; Ward, D.G. Structural investigation of hemicelluloses from plantago ovata, mimosa pudica and lallemantia royleana by MALDI-ToF mass spectrometry. *J. Carbohydr. Chem.* **2018**, *37*, 285–301. [CrossRef]
21. Abderrazak, H.; Dachraoui, M.; Lendl, B. A novel flow injection procedure for determination of phosphate in industrial raw phosphoric acid. *Analyst* **2000**, *125*, 1211–1213. [CrossRef]
22. Kudasheva, D.S.; Lai, J.; Ulman, A.; Cowman, M.K. Structure of carbohydrate-bound polynuclear iron oxyhydroxide nanoparticles in parenteral formulations. *J. Inorg. Biochem.* **2004**, *98*, 1757–1769. [CrossRef] [PubMed]
23. Rossman, G.R. The optical spectroscopic comparison of the ferric iron tetrameric clusters in amarantite and leucophosphite. *Am. Mineral.* **1976**, *61*, 933–938.
24. Rossman, G.R. Spectrosopic and magnetic studies of ferric iron hydroxy sulfates: Intensification of color in ferric iron clusters bridged by a single hydroxide ion. *Am. Mineral.* **1975**, *60*, 698–704.
25. Fontana, I.; Lauria, A.; Spinolo, G. Optical absorption spectra of Fe^{2+} and Fe^{3+} in aqueous solutions and hydrated crystals. *Phys. Status Solidi Basic Res.* **2007**, *244*, 4669–4677. [CrossRef]
26. Pansuriya, P.; Patel, M.N. Iron(III) Complexes: Preparation, characterization, antibacterial activity and DNA-Binding. *J. Enzyme Inhib. Med. Chem.* **2008**, *23*, 230–239. [CrossRef] [PubMed]
27. Coe, E.M.; Bowen, L.H.; Speer, J.A.; Wang, Z.; Sayers, D.E.; Bereman, R.D. The recharacterization of a polysaccharide iron complex (Niferex). *J. Inorg. Biochem.* **1995**, *58*, 269–278. [CrossRef]
28. Murad, E.; Schwertmann, U. The Mössbauer spectrum of ferrihydrite and its relations to those of other iron oxides. *Am. Mineral.* **1980**, *65*, 1044–1049.
29. Cui, J.; Li, Y.; Yu, P.; Zhan, Q.; Wang, J.; Chi, Y.; Wang, P. A novel low molecular weight enteromorpha polysaccharide-iron (iii) complex and its effect on rats with Iron Deficiency Anemia (IDA). *Int. J. Biol. Macromol.* **2018**, *108*, 412–418. [CrossRef] [PubMed]
30. Bhosale, M.A.; Ummineni, D.; Sasaki, T.; Nishio-Hamane, D.; Bhanage, B.M. Magnetically separable γ-Fe_2O_3 nanoparticles: An efficient catalyst for acylation of alcohols, phenols and amines using sonication energy under solvent free condition. *J. Mol. Catal. A Chem.* **2015**, *404*, 8–17. [CrossRef]
31. Balasubramanian, C.; Joseph, B.; Gupta, P.; Saini, N.L.; Mukherjee, S.; Di Gioacchino, D.; Marcelli, A. X-ray absorption spectroscopy characterization of iron-oxide nanoparticles synthesized by high temperature plasma processing. *J. Electron Spectros. Relat. Phenomena* **2014**, *196*, 125–129. [CrossRef]
32. Dadashi, S.; Poursalehi, R.; Delavari, H. Structural and optical properties of pure iron and iron oxide nanoparticles prepared via pulsed Nd:YAG laser ablation in liquid. *Procedia Mater. Sci.* **2015**, *11*, 722–726. [CrossRef]

33. Rafi, M.M.; Ahmed, S.Z.; PremNazeer, K.; Kumar, D. Antibacterial Activity of iron oxide nanoparticles on polysaccharide templates: Synthesis, characterization and magnetic Studies. *Malays. Polym. J.* **2015**, *10*, 16–22.
34. Shinkai, M. Functional magnetic particles for medical application. *J. Biosci. Bioeng.* **2002**, *94*, 606–613. [CrossRef]
35. Chang, P.R.; Yu, J.; Ma, X.; Anderson, D.P. Polysaccharides as stabilizers for the synthesis of magnetic nanoparticles. *Carbohydr. Polym.* **2011**, *83*, 640–644. [CrossRef]
36. Yagi, K.; Tokuda, M.; Kobayashi, T.; Kishita, T. Design and synthesis of magnetic ultra fine polysaccharide particles. In Proceedings of the International Symposium on Micro Machine and Human Science; IEEE: Piscataway, NJ, USA, 1999; pp. 157–162.
37. Maciel, C.B.; Youn, T.S.; Barden, M.M.; Dhakar, M.B.; Zhou, S.E.; Pontes-Neto, O.M.; Silva, G.S.; Theriot, J.J.; Greer, D.M. Corneal Reflex testing in the evaluation of a comatose patient: An ode to precise semiology and examination skills. *Neurocrit. Care* **2020**, *33*, 399–404. [CrossRef]
38. *Acute Oral Toxicity-Fixed Dose Procedure*; OECD guidelines for the testing of chemicals, Section 4; OECD: Paris, France, 2002; ISBN 9789264070943.
39. Lancker, J.L.V. Iron and bile pigment metabolism. In *Molecular and Cellular Mechanisms in Disease*; Springer: Berlin/Heidelberg, Germany, 1976; pp. 361–395.
40. Massey, S.; Iqbal, M.S.; Wolf, B.; Mariam, I.; Rao, S. Comparative Drug loading and release study on some carbohydrate polymers. *Lat. Am. J. Pharm.* **2016**, *35*, 146–155.
41. Montgomery, K.O.; Jhaveri, C.R. Process for preparing an iron-saccharide complex. U.S. Patent 3821192, 28 June 1974.
42. Weitzhandler, M.; Barreto, V.; Pohl, C.; Jandik, P.; Cheng, J.; Avdalovic, N. CarboPac k PA20: A new monosaccharide separator column with electrochemical detection with disposable gold electrodes. *J. Biochem. Biophys. Methods* **2004**, *60*, 309–317. [CrossRef]
43. Annex 2. In *Stability Testing of Active Pharmaceutical Ingredients and Finished Pharmaceutical Products*; WHO Technical Report Series, No. 953; WHO: Geneva, Switzerland, 2009; Available online: https://www.qlaboratories.com/wp-content/uploads/2019/10/World-Health-Org-WHO-Stability-testing-guidelines.pdf.
44. Krishnamurti, G.S.R.; Huang, P.M. Spectrophotometric Determination of Fe(II) with 2,4,6-Tri(2′-Pyridyl)-1,3,5-Triazine in the presence of large quantities of Fe(III) and complexing ions. *Talanta* **1990**, *37*, 745–748. [CrossRef]
45. Nawaz, H.; Shad, M.A.; Iqbal, M.S. Optimization of phenylhydrazine induced hyperbilirubinemia in experimental rabbit. *Exp. Anim.* **2016**, *65*, 363–372. [CrossRef]
46. Kheiri, R.; Koohi, M.K.; Sadeghi-Hashjin, G.; Nouri, H.; Khezli, N.; Hassan, M.A.; Hoomani, F.; Shams, G.; Rasouli, A.; Motaghinejad, M. Comparison of the Effects of Iron Oxide, as a New Form of Iron Supplement, and Ferrous Sulfate on the Blood Levels of Iron and Total Iron-Binding Globulin in the Rabbit. *Iran. J. Med. Sci.* **2017**, *42*, 79–84. [PubMed]

polymers

MDPI

Review

A Review of Rigid Polymeric Cellular Foams and Their Greener Tannin-Based Alternatives

Antonio M. Borrero-López [1], Vincent Nicolas [1,*], Zelie Marie [1], Alain Celzard [1,2] and Vanessa Fierro [1,*]

[1] Université de Lorraine, CNRS, IJL, F-88000 Epinal, France
[2] Institut Universitaire de France (IUF), France
* Correspondence: vincent.nicolas@univ-lorraine.fr (V.N.); vanessa.fierro@univ-lorraine.fr (V.F.)

Abstract: This review focuses on the description of the main processes and materials used for the formulation of rigid polymer foams. Polyurethanes and their derivatives, as well as phenolic systems, are described, and their main components, foaming routes, end of life, and recycling are considered. Due to environmental concerns and the need to find bio-based alternatives for these products, special attention is given to a recent class of polymeric foams: tannin-based foams. In addition to their formulation and foaming procedures, their main structural, thermal, mechanical, and fire resistance properties are described in detail, with emphasis on their advanced applications and recycling routes. These systems have been shown to possess very interesting properties that allow them to be considered as potential substitutes for non-renewable rigid polymeric cellular foams.

Keywords: rigid foams; tannin-based foams; polyurethane foams; phenolic foams

Citation: Borrero-López, A.M.; Nicolas, V.; Marie, Z.; Celzard, A.; Fierro, V. A Review of Rigid Polymeric Cellular Foams and Their Greener Tannin-Based Alternatives. *Polymers* **2022**, *14*, 3974. https://doi.org/10.3390/polym14193974

Academic Editor: José Ignacio Velasco

Received: 31 August 2022
Accepted: 13 September 2022
Published: 23 September 2022

Publisher's Note: MDPI stays neutral with regard to jurisdictional claims in published maps and institutional affiliations.

1. Introduction

It is necessary for industry to adapt to new policies, consumer opinions, and economic demands, and to propose alternatives aimed at the reduction of material costs, the minimization of resource consumption, and the development of more environmentally friendly resources. In order to compensate for the scarcity of petroleum derivatives, raw materials, industrial processes, and production infrastructures are evolving towards the incorporation of bio-based materials and renewable and/or recyclable energy. A large number of political, industrial and academic institutions have therefore embarked on the search for more environmentally friendly energy, chemical, and physical extraction processes.

Among these initiatives, this review is focused on recent efforts in the area of the thermal insulation of buildings. Indeed, policies implemented to mitigate the effects of climate change and pollution follow transportation and energy consumption trends with respect to buildings. Numerous incentives to renovate buildings have been launched in order to curb fuel poverty and provide access to healthy, comfortable, and less energy-consuming structures. Thus, new regulations and government subsidies have been implemented, aiming to reduce the energy needs of buildings by improving their bioclimatic design, reducing heat loss (reinforced insulation), promoting solar and internal energy contributions, and installing more efficient systems (heating, mechanical ventilation, etc.). Insulation is therefore a critical point in the design of a new building or the rehabilitation of an old building. The objective is to design an envelope that reduces heat transfer between the interior and the exterior, thus minimizing heat loss. Increasing the overall insulation capacity of the building can be achieved by selecting the appropriate wall, roof, and opening materials, the optimal installation techniques for reducing infiltration losses, and the best design methods for limiting thermal bridges (Figure 1).

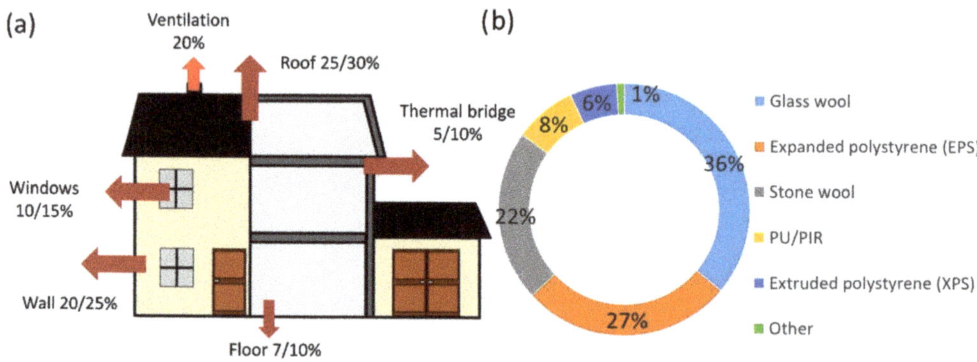

Figure 1. (**a**) Heat losses for a standard building in France [1], (**b**) European insulation market in 2019 [2].

Thus, insulating a building involves the use of materials with very low thermal conductivity, as well as the load-bearing structure of the walls. These materials must also meet strict standards with respect to their structural properties (e.g., compressive strength, lack of dimensional shrinkage, fire resistance, durability, etc.) and comply with toxicity regulations. Today, the insulation market includes a wide range of materials. These can be formed of different structures (cellular, fibrous, or loose) and can be composed of various materials (bio-based, synthetic, or inorganic). Each solution has its own facilities and costs of implementation and dismantling that need to be taken into account when making the final choice, depending on the intrinsic performance of the material. Thus, some formulations with a very good thermal resistance have higher installation costs, while less efficient insulators necessitate greater thicknesses, therefore requiring higher volumes to be processed at the end of the product's life. Glass wool and rock wool are widely used in Europe due to their good performance and ease of installation, as shown in Figure 1. Other solutions also exist, but remain marginal when comparing the volumes used; renewable insulation materials accounted for around 1.4% of the European market in 2019 [2]. Cellulose wadding, straw, and natural fibers such as hemp, flax, wood, and wool can also be mentioned [3,4]. In this respect, the development of a bio-based material with good thermal performance offers an alternative to petrochemical materials.

Inspired by the cellular phenolic and polyurethane materials commonly used in insulation, tannin-based biosourced foams have been designed and produced in recent years at the laboratory scale, with promising performance for insulation and other applications [5–9]. In this review, foams developed for insulation are carefully evaluated through the analysis of the different formulations, foaming methods, and end-of-life processes available for these materials. Finally, special attention is paid to tannin-based foams as an emerging alternative to the use of non-renewable precursors.

2. Rigid Cellular Materials

Rigid cellular materials constitute a very broad family. Depending on the application, different kinds of foams can be produced, which possess very different structures covering a wide range of mechanical properties:

- *Metallic foams*, the skeleton of which can, for instance, be made of aluminum, nickel, steel, titanium, copper, etc., offer very good resistance to chemical attacks and high temperature [10].
- *Ceramic foams* based on alumina, silicon carbide, zirconia, etc., are also used for their thermal and chemical resistance.
- *Carbon foams* can be produced from a matrix that is either a polymeric carbon precursor or a support for the deposition of carbon prior to pyrolysis, or can be obtained by

exfoliation and compaction of graphite. Their high but narrow porosity, electrical conductivity, and good resistance to high temperatures and fire make them good candidates in applications such as thermal energy storage, electrodes, gas adsorption/desorption, electromagnetic insulation, and fire protection [11].

- *Polymeric foams*, flexible or rigid, are composed of a wide range of macromolecules (polyurethane, polypropylene, polyethylene, polyvinyl chloride, phenol–formaldehyde, melamine–formaldehyde, polyimide, ethylene-vinyl acetate, polystyrene, etc.), and can cover a wide variety of applications depending on the targeted properties and the constraints of price, weight, physical and chemical compatibility.

As a summary, Figure 2 provides an overview of the usual range of values for the main properties typically used to characterize foams.

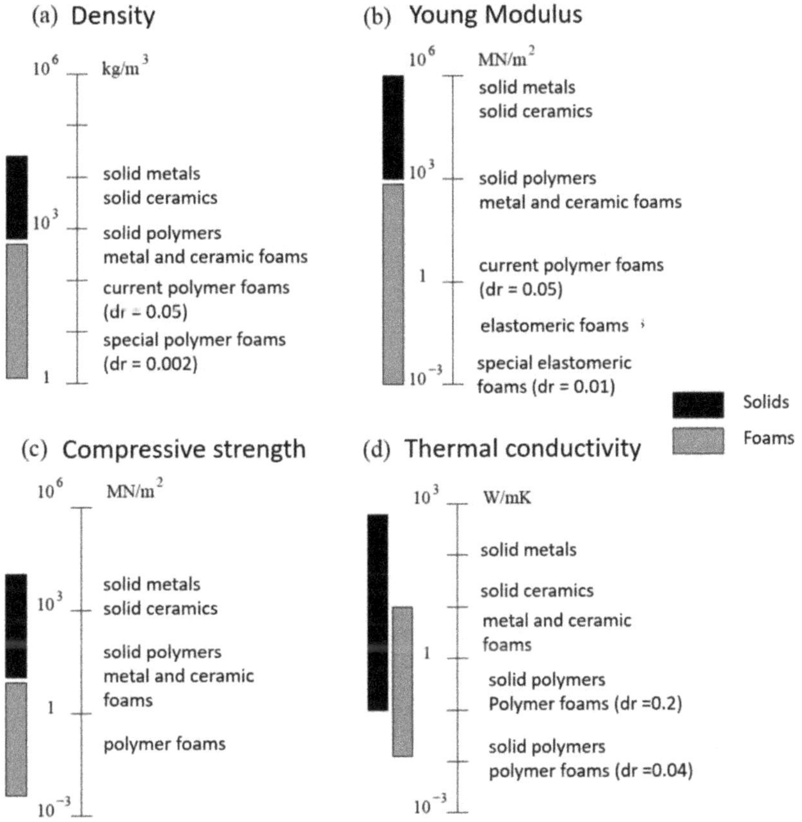

Figure 2. Typical range of values for some properties, i.e., (**a**) density, (**b**) Young Modulus, (**c**) compressive strength and (**d**) thermal conductivity, of foams depending on their nature (adapted from [12]).

The first polymeric cellular materials were designed by incorporating or directly producing gas bubbles within the polymer matrix, resulting in cells. Because of their own polymerization, thermoset foams can in some cases spontaneously produce the gas necessary for the foaming process, whereas when using the thermoplastic method, it is necessary to insert gas bubbles in order to obtain polymeric foams. The final cellular materials can therefore be defined as combining at least two different phases: a porous polymer matrix derived from a liquid reaction mixture or from the cooling of a molten phase, forming a structural skeleton, and a gaseous phase, which is entrapped and forms cavities dispersed therein [13].

The interface between these two phases is called the cell wall, the thickness of which depends on the chemical and physical properties of the matrix, and the proportion of gas introduced [14]. During foam formation, the gas first diffuses into the liquid phase until it forms pockets. Then, the bubbles grow, fed by the continuous diffusion and/or expansion of the gas as the internal pressure and/or temperature in the liquid increases. Finally, the cells can be isolated or merged by creating windows in the walls, an event that opens the pores. Materials are then defined as being "closed cell" when there are no windows in the cell walls. On the contrary, open-cell foams can undergo different degrees of pore opening, from the merging of a few windows to the formation of a structure only based on struts. In this extreme latter case, they are called reticulated foams. The properties of the material, and in particular its structure, depend greatly on the density and distribution of the gas bubbles. Foams can therefore be characterized by considering three different scales of study: microscopic, mesoscopic, and macroscopic.

- At the *microscopic scale*, pores with a diameter of less than 2 nm, i.e., micropores can be present in the solid walls, in response to mechanical and thermal stresses, and may bear different surface functions depending on the chemical nature of the foam (Figure 3a).
- At the *mesoscopic scale* (2 nm–500 μm), one can observe the arrangement of cells according to the theories of ideal cubic [12], spherical [15], tetrakaidecahedral [16] or mixed [17] structures (Figure 3b–e). At this level, it is possible to determine the size and shape of the cells and to evaluate their orientation, spatial distribution, and possible connectivity by scanning electron microscopy or microtomography (Figure 3a).

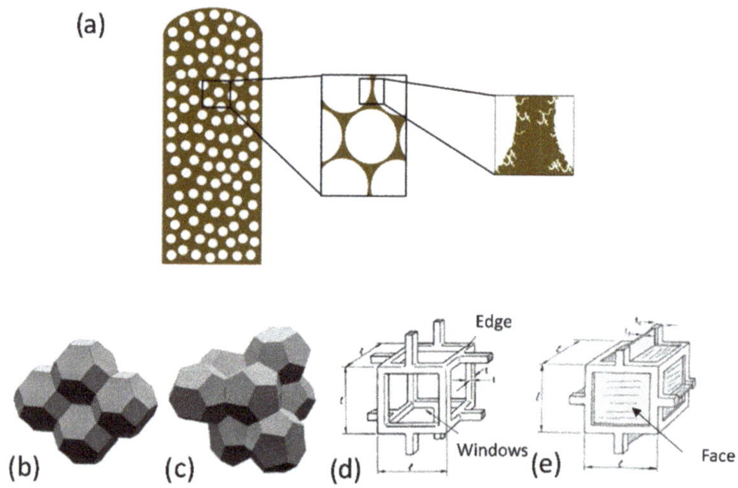

Figure 3. (**a**) Cross-section of a cellular material plane at macroscopic, mesoscopic, and microscopic scales (from left to right). 3D representation of model cell structures according to: (**b**) Kelvin (adapted from [18]); (**c**) Wheaire and Phelan (adapted from [18]); (**d**) Ashby and Gibson (open cells, adapted from [19]); (**e**) Ashby and Gibson (closed cells, adapted from [12]).

- At the *macroscopic scale* (>0.5 mm), the porous characteristics can be distinguished by the naked eye (Figure 3a). Thus, relative density ρ can be used to describe the material, defined as the ratio of the apparent (or bulk) density ρ_a (determined by the weight divided by the total volume: solid and cells) to the skeletal density ρ_s (i.e., not considering cells). The porosity, $\phi = 1 - \rho$, is also often referred to as a complement of the relative density. To complete the picture, a cellular material can be characterized by two additional quantities. The tortuosity θ is defined as the ratio of the actual length of the flow path when a fluid passes through it to the actual distance between the two ends; and the permeability k accounts for the difficulty for that fluid to pass through the medium under consideration when subjected to a pressure gradient. These four parameters are widely used to characterize foams.

2.1. Formulations

Throughout history, the discovery of new families of polymeric materials has led to profound changes in the production of goods. As a result, new polymeric materials are lighter, easy and inexpensive to manufacture, and offer a wide variety of shapes and colors. These are employed in industry in order to satisfy the needs of mass production. The global polymer foam market was worth about USD 100 billion in 2015, and was expected to have reached about USD 122 billion by 2021 [20]. More recently, a new report established foam production growth of about 5%/year from 2021 to 2026 [21]. The ubiquity of foams has stimulated studies aiming to improve their chemical, physical, or mechanical properties, while simultaneously achieving greater lightness. Like polymeric resins, foams can be classified into different categories based on:

1. *The chemical mechanisms of polymerization*: polyaddition (production of a polymer chain without by-products); polycondensation (production of a polymer condensate and by-products, such as CO_2 or H_2O); cyclotrimerization (production of a polymer composed of cyclic units, obtained from three monomers); ring-opening (production of a polymer by opening the ring from which their monomers are composed); and free radical polymerization (production of a polymer chain by the reaction of radical monomers).
2. *The physical mechanisms for their shaping*: irreversible hardening under the action of heat to obtain thermosetting materials [22], or reversible shaping by softening under the action of heat to shape thermoplastic foams [23], or by crosslinking in elastomers [24].
3. *The mechanical behavior under a light load*: rigid and brittle foams, or flexible and deformable foams, without rupture under load (rubber-like behavior), or semi-rigid foams with an intermediate behavior [25].

To understand the mechanisms involved in the expansion of foams, a literature review was carried out with respect to cellular materials derived from commonly used thermosetting polymers available on the market, as summarized in Table 1. Many thermosetting resins have been developed that form foams when a gas fraction is introduced during synthesis. However, the current market is mainly focused on two large families of materials, which can be characterized on the basis of their main crosslinking agent: (i) foams formed from isocyanates, i.e., comprising N=C=O functional groups; and (ii) foams that instead incorporate phenol derivatives, i.e., composed of an aromatic ring and a hydroxyl group. The former are the main materials used in this sector, and include polyurethanes, polyureas, and polyisocyanurates, while the latter are less widespread and are known as phenolic foams. Polyester foams also exist, but their use is marginal compared to the foams mentions above.

Table 1. Examples of rigid thermosetting cellular materials.

	Poly Addition	**Poly Condensation**	**Cyclo-Trimerization**	**Ring-Opening**	**Radical Polymerization**
		Isocyanates			
Polyurethane PUR (*)	X				
Polyurea PUA (*)	X				
Polyisocyanurate PIR (*)			X		
Polycarbodiimide		X			
Polyimide		X			
		Phenolic			
Phenol–formaldehyde PF (*)		X			
Epoxy EP				X	
Tannin–furanic foams TFF (*)		X			
		Polyesters			
Unsaturated polyester					X

(*) especially considered in this review article.

2.1.1. Isocyanate Chemistry

The production of rigid cellular materials, as mentioned above, is largely dominated by polymers derived from isocyanate chemistry, based on the N=C=O functional group. The most emblematic foam in this category is polyurethane foam, from which the other polymers in this family of materials are derived.

Polyurethanes

Polyurethanes (PURs) comprise a family of polymers obtained by polyaddition of urethane units, i.e., the production of urethane units without the formation of by-products during the main reaction, as defined by Carothers [26] (Figure 4a). Nevertheless, careful protocols must be followed if only PURs are being targeted, because through the reaction of isocyanate groups with water or moisture, polyureas can also be formed, as presented in Figure 4a and discussed in the next section.

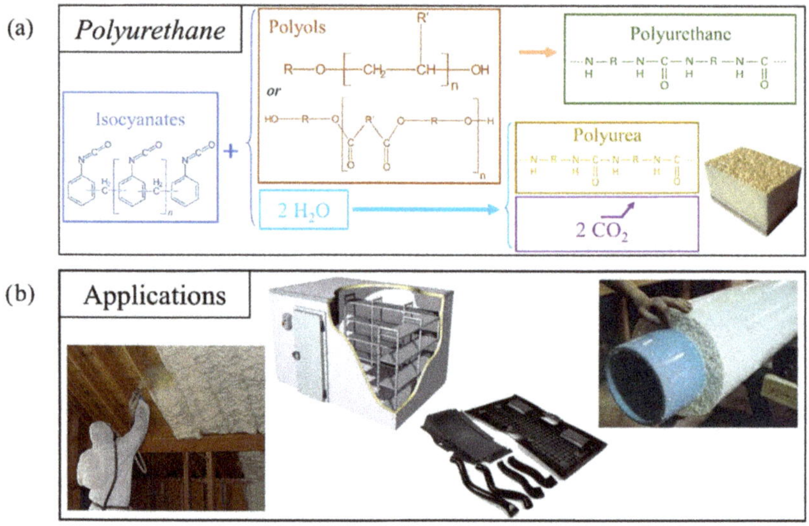

Figure 4. (**a**) Reaction diagram for PUR and PUA systems, containing (from left to right) the isocyanate precursors, some examples of polyols for the formation of PUR, and, in the presence of water, the formation of PUA and the consequent liberation of CO_2; (**b**) examples of PUR foam applications, such as the thermal insulation of buildings, different polymeric devices, and thermal insulation of pipes.

Developed in the 20th century as a result of the research of Bayer, published in 1937 [27], PURs emerged in German industry as an alternative to hard-to-produce synthetic fibers during World War II, and have become key polymers in the plastics market due to their versatility and their vast number of applications (see Figure 4b) [28,29]. Within this family of PURs, rigid foams account for about 25% of all polyurethanes produced—a market valued at USD 60.5 billion in 2017 [30]. According to a more recent study by Polaris Market Research, growth to USD 92.26 billion is projected for 2026 [31]. The name *polyurethane* does not refer to a unified family of macromolecules like most polymers. In fact, naming conventions set polymer names such that *PolyAAA* is a chain of *AAA* monomer. However, PURs are actually polymeric chains that end in urethane groups, but are not necessarily composed of urethane monomers, as the chains often have other units. The incorporation of additional reagents, with the selected proportions and reaction kinetics, allows the development of myriad polyurethane types, leading to the coverage of a wide range of properties, and thus applications [32]. It is, therefore, difficult to provide an exhaustive list of possible reactions and reactants, but the most notable are mentioned below. However, there are examples of the addition of amines to produce urea units, a possible route for crosslinking through the formation of biurets, carboxylic acids, epoxies, etc. [29]. Modified and hybrid PUR foams based on various proportions of these non-PUR units for mild or advanced modifications, respectively, can be found in the literature [33–35].

The polymerization reaction of PUR takes place through a main reaction between isocyanates and polyols and, more generally, compounds rich in oxygen or nitrogen groups such as hydroxyls, amines, amides, acids, epoxides, etc. The main components of foam formulations can be obtained by using foaming agents to allow expansion, catalysts for the different reactions, additives such as surfactants and flame retardants, etc. These different constituents are presented in the following sections.

1. Isocyanates

Polyisocyanates comprise a family of chemicals that are highly reactive in the presence of a free hydroxyl or amine-rich component, and under appropriate thermal conditions. Thus, they react readily to the addition of alcohols, amines, carboxylic acids, water, etc. Most of the polyisocyanates used in polyurethane synthesis are aromatic compounds, such as toluene diisocyanate (TDI) (Figure 5a), methylene diphenyl diisocyanate (MDI) (Figure 5b), pure monomeric MDI (mMDI), and especially polymeric MDI (pMDI) (Figure 5c), which is the main isocyanate in rigid foams [29,36–51]. Other isocyanates can be considered for specific uses [29,52–55].

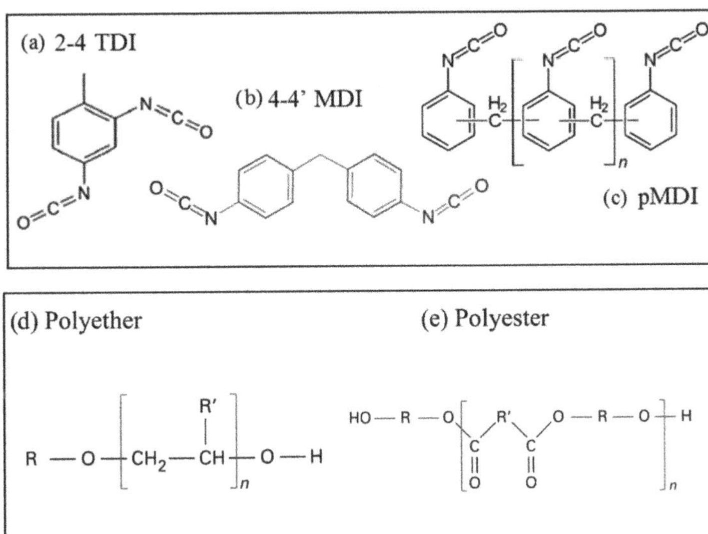

Figure 5. (**a**–**c**) Main isocyanates used in the formulation of PUR. General schematic formulations of main polyols used for PUR production: (**d**) polyether and (**e**) polyester.

As mentioned earlier, the synthesis of polyurethanes is based on the reaction between an isocyanate and a polyol generating a chain of urethane units [38]. For foam production, the polyols used in a standard way are often in aqueous medium. When mixed with isocyanates, the main reaction between polyols and isocyanates forms the polymer, while a second reaction between water and isocyanates takes place and produces CO_2 and urea. As a result, the self-reaction of the isocyanates and the release of CO_2 become the engine of expansion, thus avoiding the use of flammable or fluorinated foaming agents. In addition, the formed urea units offer improved fire and compression resistance [32].

2. Polyols

The PUR polymerization reaction takes place between any chemical with at least two primary hydroxyl groups and diisocyanates. Polyols are commonly used because of the diversity of their molecules, making it possible to achieve a wide range of properties [29]. The majority of the molecules are oligo-polyols, i.e., polyols with units other than -OH groups in their terminations or in their branches. They generally have a low molecular weight and high functionality, which leads to dense polymerization through crosslinking reactions and thus to rigid matrices [38]. Two main types of polyols can be used for polyurethane formulation: polyethers and polyesters (Figure 5).

Polyether polyols account for 80–90% of the polyols used in the PUR foam market due to their thermal stability, high compressive strength, and the possibility of degradation by hydrolysis of their C-O-C bond (Figure 5d). Their low viscosity and low cost are also advantageous during processing. In addition, their high solubility in many solvents makes it possible to use high-viscosity polyols [29].

Polyesters account for most of the other polyols, acting as aromatic molecules for rigid foams, and aliphatic constituents for other types of polyurethane products [56,57] (Figure 5e), such as polyurethanes alone. Their susceptibility to degradation by hydrolysis and heat has been a huge obstacle in the past, but new environmental and recycling requirements make them ideal new candidates for foam production. Also worth mentioning is the incorporation of halogenated and phosphorated polyols or antimony oxides in polymer chains with flame-retardant properties [29].

3. Additives

The reactions of isocyanates with compounds rich in -OH or -NH groups do not necessarily require a chemical catalyst. However, a catalyst is frequently used when it is desirable to control certain foaming properties (curing, expansion kinetics, etc.), to boost less reactive reactions, or to retard the reaction for specific applications (such as spraying). As an example, the formation of isocyanurates requires a synergistic catalyst composed of a tertiary amine and a metal catalyst [29]. In fact, the two main categories of catalysts used in PUR foaming are: (i) tertiary amines to control the foaming parameters (triethylamine, diazabicycloundecene, etc.); and (ii) metal salts to adjust the reaction kinetics [58]. These can be combined to make the transformation protocol as precise as possible, thus obtaining the optimal desired properties. Other catalysts can also be incorporated, such as buffered amines, or acids like hydrogen chloride, to promote and control reactions to specific products (polyurethane, polyurea, polyisocyanurate, etc.) [58–61].

On the other hand, foaming agents are also frequently used. The addition of these chemicals to the reaction mixture allows the production of gases that cause the expansion of the polymer matrix. They act by increasing the quantity and pressure of the gas produced. Foaming can occur either by chemical expansion, due to the formation of internal gases, such as the generation of CO_2 during the reaction between isocyanates and water, or by physical expansion, due to the phase change of a liquid with a low boiling point present in the mixture. Exothermic reactions, or heat input, vaporize the liquid, which is the foaming agent, to generate gas. The most popular foaming agents for rigid polyurethanes include n-pentane and hydrofluorocarbons [62].

Other additives commonly used in PUR formulations include surfactants, which are added to control cell size and degree of opening, flame retardants, and crosslinking agents. In addition, specific applications require the addition of UV stabilizers, dyes, pigments, preservatives, fillers such as fibers to increase matrix strength and dimensional stability, etc.

- Surfactants are generally copolymers chosen to modify the surface tension of the reactive mixture to achieve specific properties, e.g., to help solubilize or disperse a compound, increase the wetting of the mixture in the mold, plasticize the polymer, facilitate the nucleation of gas bubbles, increase the porosity, etc. [63]. To achieve this last objective, a surfactant is commonly added to polymer precursors in order to control the cell size and the type of porosity. Indeed, depending on the properties of the surfactant, the polymer film at the edge of the gas bubbles will be more or less resistant to the gas pressure generated by the bubbles, and will ensure a given dispersion of the latter. The gas bubbles trapped in the polymer mixture grow and create the cells of the matrix. As the material and gas expand, the cell wall becomes thinner and must resist the internal pressure. The ideal surfactant must provide the polymer with enough elasticity to resist deformation during expansion and harden when crosslinking, and likewise during cooling to prevent shrinkage [29]. Therefore, the use of surfactants and the amount added have an impact on the structural, thermal, and mechanical properties of the developed foams [52].

- Flame retardants are chemical reagents added to reduce the thermal degradation and flammability of PUR. Strict regulation with respect to furniture, transportation and construction have spurred research into flame-retardant components, as PUR foams burn easily and can spread incandescent materials. Fire-resistant components work by slowing combustion, incorporating halogens that react with the radicals responsible for increased combustion, and phosphorus that helps to form a carbon barrier and reduce the production of flammable gases. The fire behavior of PURs is the subject of many studies [33,50,64–69], including the incorporation of other types of resin in the formulation, such as urea–formaldehyde [70].
- Crosslinking agents and curing agents are chemical compounds that can react with the linear chain of the polymer and link several of its branches. These bonds stiffen the main chain and establish the polymer network with other precursors and additives. Various less common crosslinking reactions may be involved in PUR chemistry, occurring in the final stage of its formation. High temperatures are usually reached in the reactor as a result of the exothermic reactions of urethane formation and CO_2 generation. These elevated temperatures promote the reaction of isocyanates with other urethane and urea units in the chain or with themselves, forming dimers or trimers. These branches stiffen the matrix and improve the mechanical properties of conventional PUR resins. In addition, PUR foams can be reinforced by adding fibers or other fillers to improve compressive strength [71].

4. Properties and Applications

PUR foams have been used in many technical applications, such as in pressure sensors and bacterial scaffolds, and for oil/water separation [72,73], as well as in insulating materials for buildings, heaters, refrigeration systems, automotive parts, structural components, etc. [29,74,75] (Figure 4b). The diversity of formulations allows a very wide range of mechanical, thermal, water absorption, and fire resistance properties to be obtained. For example, an increase in the amount of surfactant affects polymerization kinetics by increasing the creaming and/or gelling time, which also affects the structural properties by decreasing the density and cell size. These changes have a direct impact on the mechanical properties, resulting in increased compressive strength, and thermal properties, resulting in decreased thermal conductivity [76].

The measured properties and potential applications depend mainly on the formulation and manufacturing conditions (reagent content, additives, water content, temperature, etc.) [76–82]. For example, rigid foams have a low density (30–58 $kg \cdot m^{-3}$) [83] and a high content of closed cells, which gives them a low thermal conductivity (0.0267–0.034 $W\ m^{-1} \cdot K^{-1}$) [84]. These properties make them good candidates for exterior insulation, heat retention, and, ultimately, energy savings [85]. In addition, polyurethane padding panels based on open-celled foams also offer good acoustic insulation [86] and shock absorption properties [29]. To comply with building regulations, acceptable flame resistance must be guaranteed; therefore, flame retardants can be incorporated to achieve better combustion behavior. Improvements in fire resistance properties have also been achieved using chemical modifications of the polyurethane network, such as via the introduction of urea, isocyanurate, or carbodiimide bonds into the polyurethane chains [87]. Moreover, their cellular organization provides a lightweight and rigid polymeric structure that can also be used for seat frames, vehicle and engine interior components, and decorative structures [88]. Some PUR foams are also ideal for water filtration systems, as they possess a hydrophilic character and an open structure capable of absorbing water [73,89].

Polyureas

Polyurea foams (PUAs) have been developed as a less expensive alternative to PUR and polyisocyanurate (PIR) foams. PUAs are cellular materials consisting of a polymeric structure composed of urea units and incorporating optional units of urethanes, isocyanurates, amides, imides, and carbonates [90]. They exist as flexible, semi-flexible and rigid foams, depending on the nature of the cell wall material. PUAs are sprayed or applied in blocks for building insulation or molded into specific and complex shapes in car interiors [91].

PUA foams can also be based on isocyanate chemistry to produce the polymer matrix (Figure 6). Polyureas generally consist of MDI [92–94] or pMDI [95,96], which, when coming into contact with water, first produce amines, and then polyureas (see Figure 6). Therefore, to maximize the amount of urea formed by the reaction with isocyanates, some studies incorporate amines directly into the polymer blend, such as aniline [97] or oligomeric diamines [98].

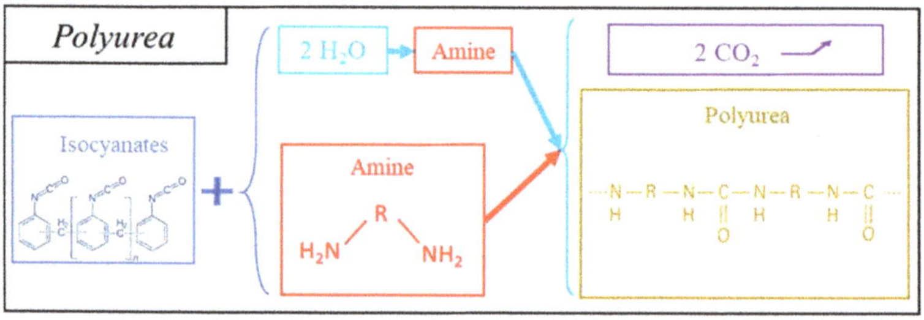

Figure 6. Reaction scheme for the formation of polyurea by contact with water or by direct combination with amines.

More specifically, the polymerization of urea follows a two-component process through two consecutive reactions. The first is the foaming reaction, which leads to the generation of an unstable carbamic acid that rapidly converts to amine and CO_2. There has been some research taking advantage of the addition of metal salts such as copper or zinc salts to increase gas generation, thus maximizing expansion [99]. The second step is the polymerization of urea units catalyzed by tertiary amines. Both routes of PUA formation have been described as producing new materials. Water-based formulations have been developed for semi-rigid or rigid building insulation boards and sprayed insulation devices. This is less expensive, because it involves the use of water, which also produces gas for foaming without introducing a foaming agent. However, some foams can be expanded with a fluorocarbon foaming agent [95].

PUA foams have a very low density (~8 $kg \cdot m^{-3}$) and good insulating properties, even though their thermal conductivity is higher than that of PUR foams (0.050 $W\ m^{-1} \cdot K^{-1}$) [93]. As with PUR, some formulations incorporate additives that can also be added to the polymer blend to achieve the desired properties, such as fibrous fillers, which stiffen the polymer network and compensate for poor compression characteristics [94]. There are also surfactants, plasticizers and stabilizers, which provide a specific cell structure in the form of open or closed cells, fine or coarse, or fire retardants added to the mixture to ensure fire resistance and adequate smoke emission.

Polyisocyanurates

Polyisocyanurate, or PIR, foams are formed by a cyclotrimerization reaction, in which three molecules react together in a ring configuration (Figure 7). PIR foams are obtained by reacting high-index isocyanates, where an index is defined as the ratio of the equivalent amount of isocyanate used to the theoretical equivalent amount, multiplied by 100, and in large proportion in a PUR formulation [38,100]. Therefore, PIR foams form a heterogeneous polymer containing an isocyanurate ring linked to polyols through urethane and urea linkages. The foaming agents used for polyurethane chemistry can be applied in the same way to the formulation of PIR foams [101,102]. In addition, four types of catalyst can be used individually or as mixtures: tertiary amines, quaternary ammonium salts, metal salts and potassium-based fatty acids. Other components can be incorporated into PIR formulations, such as surfactants, flame retardants, and fillers.

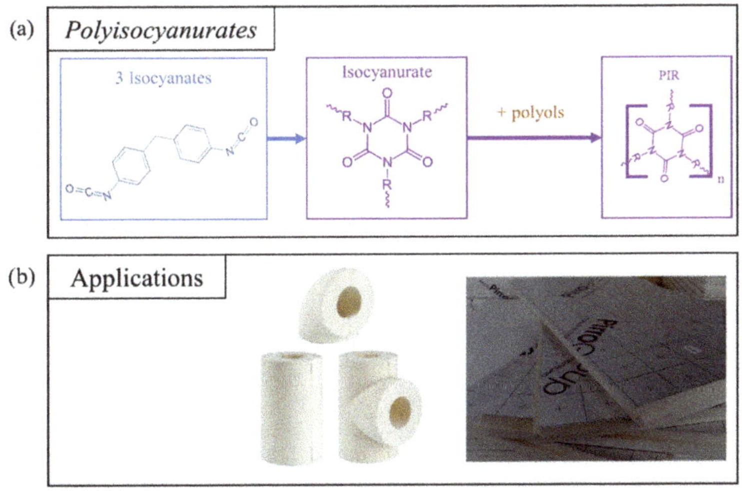

Figure 7. (**a**) Reaction diagram for obtaining isocyanurate from isocyanate and then formulating PIR by adding polyols; (**b**) pipe insulation and insulation panels for covering walls, ceilings, roofs, etc., as examples of PIR foam applications.

PIR foams have been designed that retain the advantages of PUR foams while improving their thermal behavior and fire resistance [29]. As a result, PIR foams can be used for construction, thermal insulation, and heating systems, either in panel form or directly foamed on site, [32,103]. Although their fire performance is superior to that of PUR, PIR foams are dangerous when burned, because they are able to release hydrogen cyanide, a compound that prevents the cells from using oxygen [104]. The rigidity and strength of this lightweight material also make it attractive for use in the automotive industry [74,101].

2.1.2. Phenolic-Based Formulations

Polyphenolic foams or phenol–formaldehyde (PF) foams refer to a family of foams resulting from the polymerization of phenolic monomers with aldehydes, in particular formaldehyde (Figure 8a) [105,106]. The process can be described in two simultaneous steps: the crosslinking reaction of phenolic compounds and aldehydes; and the foaming of the polymer mixture.

The polycondensation reaction takes place between phenolic compounds and aldehydes. To activate the crosslinking mechanisms, a minimum temperature of 70 °C [107] is typically required, which is achieved either by secondary exothermic reactions, as in the example of tannin–furanic foams, or by external heat input. The thermal conditions applied for each formulation are variable, and depend on the proportion of reactants, catalysts and surfactants used. Some formulation protocols require premixing before being placed in an oven at a temperature between 50 and 140 °C [108–111], while others subject the material to high temperatures only after expansion for curing [112], which gives the foams different properties.

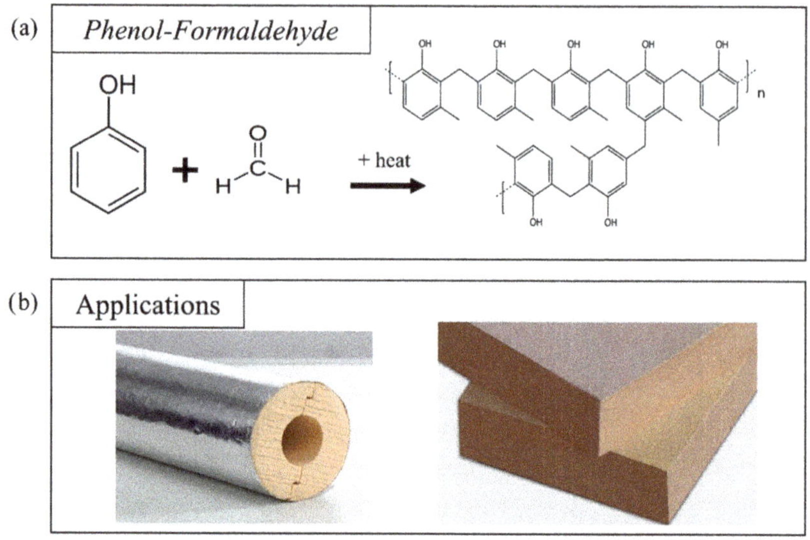

Figure 8. (**a**) General reaction of PF chemistry, showing an example of phenolic precursor, aldehyde crosslinker, and the resulting polymer; (**b**) examples of tubular insulation and panels made of PF foam.

Phenols are aromatic compounds with a hydroxyl function bonded to the ring and have three reactive sites (one in the *para* position and two in the *ortho* positions). As a result, phenols can share three bonds with aldehyde molecules [105], resulting in dense crosslinking, producing the characteristic hardness of these rigid foams. Other phenolic derivatives in which one or more sites have been substituted can also be used for the development of PF foams. Some examples are given below (Figure 9a). These are mainly petrochemical derivatives that can also be found in tars and mineral oils from coal or charcoal [109,113].

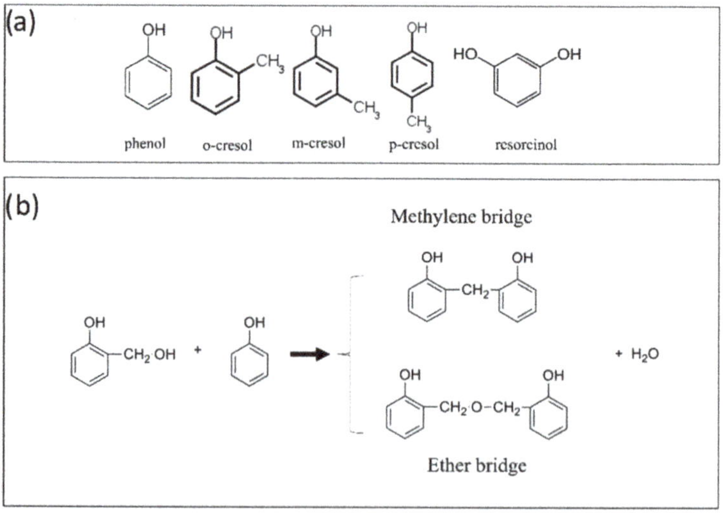

Figure 9. (**a**) Phenol and derived phenolic compounds used in PF foams (based on [113]); (**b**) reaction scheme for the formation of hydroxymethyl phenols (adapted from [113]).

The other main reagent of PF foam production is formaldehyde, which is the simplest of the aldehydes. Due to its low price and wide availability, its use is suitable for large industrial volumes. However, other types of aldehydes can be considered to optimize specific properties. Thus, the use of glyoxal, paraformaldehyde, acetaldehyde, or furfural has been reported in the literature [109,113]. Hexamethylenetetramine can also be used for its ability to decompose by the thermal conditions imposed, providing, on the one hand, aldehydes for the PF reaction, and on the other hand, amino constituents that can catalyze it [109].

After mixing, formaldehyde reacts by binding to the C_2, C_4 or C_6 sites of the phenol ring, forming a hydroxymethylphenol. This compound can then polymerize via an ether bond or a methylene bridge, formed by a subsequent reaction with another previously reacted hydroxymethylphenol or a simple phenol, respectively, leading in both cases to the subsequent release of water (Figure 9b). The latter configuration produces bisphenol F, a precursor of polycarbonate plastics. Compared to foams made from other precursors, these generally exhibit less brittleness and fragility, which is the main weakness of PF foams [106]. Therefore, many studies have attempted to improve the mechanical properties of PF foams by adding epoxy resins (EP) or bioresources to the formulation [106,108,110,112]. Another disadvantage that is worth mentioning is that PF foams release acidic vapors over time, which corrode nearby metals, limiting their applications [114]. The phenol–formaldehyde reaction requires the use of an acidic or basic catalyst, characterizing the reaction pathways known as the Novolac and Resole pathways, respectively. Metal salts can also be used to catalyze these two types of reaction [113].

The Novolac route follows a two-step process. In the first step, excess phenol or cresol is mixed with an acid, which acts as a catalyst. Common examples include sulfuric acid, oxalic acid, hydrochloric acid, sulfonic acid, and p-toluenesulfonic acid [111,115,116]. In a second step, the formed Novolac prepolymer can be molded in the presence of formaldehyde at temperatures between 70 and 140 °C. It behaves like a thermoplastic resin that cures slowly, and therefore requires the addition of a hardener to fix the geometry and obtain a thermoset network. This process favors the formation of methylene bonds, although ether groups can also be formed [113] (Figure 9b). Instead, the Resole route is a one-step reaction process, where excess formaldehyde and phenols react in an alkaline synthesis medium (pH around 10). The process begins by mixing phenol and formaldehyde (with the usual formaldehyde/phenol molar ratio being higher than 1:1) with water at room temperature. The hydroxymethylphenol-rich mixture deforms and hardens, with or without a crosslinking agent, after being placed in an oven at the appropriate temperature [113].

PF foams are obtained by means of physical expansion, i.e., by adding a component that is vaporized during the change of state and becomes gaseous. The list of possible compounds includes diisopropyl ether [117], methyl formate [118], diglycidyl ether [108], pentane [111,116], CFC-113 [107], water [107], etc. Other formulations have been investigated by introducing a compound that degrades upon heating and is released into the matrix, producing a gas that drives expansion [109,110,112]. Finally, the use of microwaves has also been considered [119,120].

PF foams are mainly used as insulators in refrigeration and heating installations or in buildings (Figure 8b). Indeed, they possess good thermal conductivity, with values between 0.029 and 0.060 $W \cdot m^{-1} \cdot K^{-1}$ for densities between 60 and 130 $kg \cdot m^{-3}$, values slightly higher than those measured for PUR foams [116,117,119–121]. Rickle and Denslow also showed the effect of the presence of water in the formulation on the final conductivity, where decreasing amounts of water resulted in a decrease in conductivity [122]. In addition, PF foams have excellent fire resistance [100,108], making them good candidates for use in building insulation. This performance can be further improved by adding phosphate compounds to the formulation [111]. Nevertheless, PFs have been reported to be brittle materials, and need to be strengthened in order to improve their mechanical properties. Several studies have attempted to improve their mechanical performance by adding different fillers, such as fibers [106,116,121,123–125], or by chemical modification. In addition, the chemical similarity of epoxy (EP) to that of PF foams allows the ease of reaction and good mechanical properties of both systems to be combined. Thus, Auad et al. obtained a compressive strength between 620 and 2870 kPa, depending on the PF:EP ratio under consideration, in parallel with good expansion [108], while the literature reports values around 138 and 620 kPa [107]. Yang et al. also obtained foams with higher tensile strength and elongation at break by modification with epoxy resins of a mainly PUR backbone [126]. Other authors, on the other hand, focused their work on more environmentally friendly modifications. For example, De Carvalho et al. [105] achieved improved mechanical properties by combining synthetic phenols and lignin-derived phenols, while compromising the characteristic lightness of polymeric foams. In more recent studies, improvements to the properties of PFs have also been reported through the addition of microcrystalline cellulose, clays, graphene, nano-silica or (nano)-lignin [106,127]. Following this strategy, some alternatives to petroleum-based reagents for foam manufacturing are examined in the next section.

2.1.3. Alternatives Incorporating Bio-Based Reagents

To introduce bio-based materials, much work has been performed on replacing petroleum-based reagents. Thus, a wide range of biosourced PUR foams have been developed, which are known in the literature as bio polyurethanes (bPUR). A first route is to add vegetable oils as alternative polyols, while the use of lignin-derived compounds is also very common. Examples include coconut oil, castor oil, linseed oil, palm oil, soybean oil, sunflower oil, evening primrose oil, and polyols derived from lignin [63,83,86,128–139]. There are also fillers and fibers of plant origin such as lignin particles [89]. Studies have also been conducted on other materials in the PUR family using alternatives to bio-based polyols. In addition, new materials are emerging through the removal of isocyanates, the mainstay of PUR chemistry, from PUR foams. Thus, isocyanate-free PURs can be developed by reacting polyols with cyclic and bicyclic diamines and carbonates [140–144].

Phenolic compound chemistry has also considered renewable sources of phenols. The first method was to synthesize phenolic molecules from bio-based materials. Lignin is a natural three-dimensional polymer that constitutes and provides rigidity and impermeability to the cell wall of plants, of which it is also one of the most abundant components. It is always composed of a combination of the following units: p-hydroxyphenyl, called unit H, derived from coumaryl alcohol; guaiacyl, called unit G, derived from coniferyl alcohol; and syringyl, called unit S, derived from sinapyl alcohol [145] (Figure 10a). Nevertheless, many other phenolic and non-phenolic units are frequently present [146,147]. These three monolignols are derived from phenol, and have been studied to obtain bio-based alternatives to these molecules. Despite the spatial configuration of lignin and the loss of reactivity of the phenolic groups thus extracted, this process has been considered for the chemical valorization of the by-products of the wood and paper industries [107,148]. However, the structure of lignin is complex and difficult to describe as a whole. Its properties depend on the extraction methods used, which partially degrade it. Nevertheless, the model proposed by Adler provides a first approximation of this molecule [149] (Figure 10b). The processes employed in the paper industry and biorefineries aim to separate the carbohydrate components from the lignin, and to valorize this waste, which is very often burned for heat production. However, the most recent research has tended towards finding alternative uses for lignin [147,150–152]. Indeed, lignin seems to be promising as a precursor for foams with good thermal properties, low density, and sufficient mechanical strength, and for this reason, there are a number of studies in which phenol is completely replaced by lignin, leading to lignin–formaldehyde foams [153,154]. Lignin has also been considered as a source of phenol and polyol, usually by means of chemical modifications, and can thus be incorporated in PUR or PF blends [155,156].

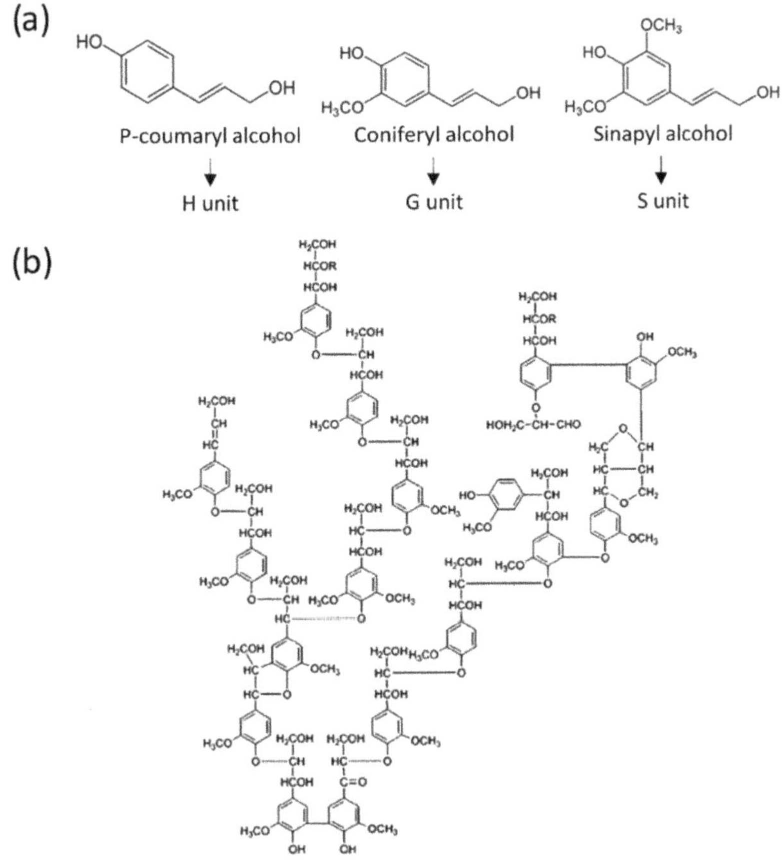

Figure 10. (**a**) Lignin monolignols; (**b**) representation of lignin proposed by Adler (based on [149]). Reprinted from Dobado et al. [149], with permission from Wiley.

Among other plant-based alternatives to phenol in PF foams, cardanol has interesting properties, since it is involved in crosslinking with aldehydes. In addition, when it contains phosphorus, it improves fire resistance [157,158]. Finally, tannins, another phenolic extract from plants, are an important area of research as substitutes for PF foams and resins, and are discussed in detail in Section 3.

2.2. Foams from Cradle to Grave

The quality of a final foam depends on its thermal, rheological, mechanical, and other properties, which are based on the selection of the precursor, crosslinker and other components. Nonetheless, the selection of an appropriate gas, foaming method and processing method is also necessary to obtain a satisfactory final material. Although very different chemical mechanisms exist, the generation of cellular foams follows the same general steps: mixing of components, processing, expansion, curing, and commissioning of the material before its end of life, as explained in the following sections with reference only to thermoset foams.

Foaming Pathways of Thermoset Foams

The cellular structure of such polymeric foams is achieved by a gaseous component whose internal pressure is the driving force for expansion. This gas component, as mentioned earlier, is called the foaming (or blowing) agent, and it depends on the foaming mechanism. In the literature, three types of expansion are generally possible for rigid thermoset foams: chemical expansion using a chemical blowing agent (CBA); physical expansion using a physical blowing agent (PBA); and mechanical expansion (mechanically blown foam, MBF).

Physical Foaming

The physical blowing process aims to create gaseous cavities within the polymer by a physical process, hence the name. For example, it can be achieved by introducing a low-boiling liquid PBA into the mixture, which changes phase to become gaseous when the mixture is heated by exothermic reaction or by external input. In contrast, some specific production processes inject the liquid into the mixing head and cause vaporization by depressurizing the liquid at the nozzle [159]. In both cases, the matrix polymerizes and hardens during or shortly after the phase change to trap the gas and stabilize the cell structure.

A promising family was studied and used during the second half of the 20th century: fluorocarbon compounds such as those of the Freon family. They can be distinguished into four subcategories: chlorofluorocarbons (CFCs), hydrochlorofluorocarbons (HCFCs), hydrofluorocarbons (HFCs), and perfluorocarbons (PFCs). The best-known and most widely used fluorocarbon is CFC-11 [160]. Fluorocarbons are carbon chains fully or partially substituted with fluorine and chlorine atoms, and can be handled in liquid form at room temperature. Their low boiling point is often between 20 and 40 °C. This type of blowing agent improves the final properties of foams, providing increased mechanical properties and a closed-cell structure. In addition, they have remarkable thermal properties, with a very low value of thermal conductivity, λ [32]. In addition, they have excellent non-corrosive and non-flammable properties. For these reasons, they quickly became the most suitable PBAs for foams in building insulation and refrigeration systems. However, the commercialization of these components was compromised in the late 1980s by the discovery of their involvement in environmental problems. Studies have indeed shown that CFCs and their relatives are responsible for ozone depletion in the stratosphere [161] and contribute to greenhouse effects. As a result, global regulation was adopted by the Montreal Protocol in 1987 [162]. Industries gradually replaced CFCs in their products, until their final ban in 1995, with HCFCs and HFCs, which are less aggressive than CFCs. However, this is a temporary solution, because HCFCs also contribute to the destruction of atmospheric ozone and have a significant global warming impact. Their gradual withdrawal from the market was planned to take place between 1996 and 2030 for developed countries, and 2040 for the rest of the world in the 1992 Copenhagen amendment [163]. The prerogatives of the Montreal Protocol were extended to also include a ban on HFC gases by 2050 by the Kyoto Protocol (COP03) in 1999 [164] and the Kigali agreement of 2016 [165].

Today, PBAs are less frequently used, due to the progressive withdrawal of fluorocarbons from the market. HFCs and PFCs, albeit to a lesser extent because of their chemical incompatibility with polyurethanes, are still used for specific industrial applications. However, companies are looking for alternatives, because these PBAs are more expensive and less efficient than CFCs. Non-fluorinated PBAs have been investigated in the form of hydrocarbon ethers. For example, some research has been conducted with diethyl ether (DE) in a natural PF formulation [9,166,167]. In addition, the use of alkanes such as n-pentane, which has no ozone-depletion potential, and has supplanted fluorinated hydrocarbons in PUR and PF insulating foams, despite its flammability, can also be mentioned [159,168,169]. Its application process and gas content are highly regulated for building insulation. In addition, the spray-applied thickness must be greater to compensate for the difference in thermal properties between alkanes and CFCs. Therefore, research has been conducted on alkanes as PBAs for PUR foam applications to explore new options that meet new economic requirements. Therefore, they must have a low boiling point and be environmentally friendly, non-flammable and cheap. Al-Moameri et al. investigated the cell morphology of

PUR foams produced with n-pentane, cyclopentane, isopentane, n-hexane, cyclohexane, a mixture of n-pentane and methylene chloride, and methyl formate in order to estimate the mass transfer coefficients and cell size for each of the considered PBAs by numerical simulation [168,170]. The introduction of CO_2 as an inexpensive and safe blowing agent has also been investigated recently, with the possibility of creating microcellular epoxy foams with cell densities of about 100 cells/cm^3 [171]. However, corrugation and surface defects have often been reported due to the use of CO_2, which is why Breuer et al. tested mixtures of CO_2 and organic solvents such as acetone, ethyl acetate or ethanol [172].

Finally, the example of physical foaming using microwaves to allow the expansion of phenolic foams can be cited [119,120,173]. The reagents are mixed mechanically to ensure good homogenization and the incorporation of a large fraction of air in the reactive preparation. The latter is subjected to microwave heating between 150 and 250 °C, allowing a more uniform heating of the different constituents than by conduction. However, with this method, foam deformations on the order of 5 to 10 times the initial volume have been obtained. The latest possibilities and main drawbacks of this technique, including the formulation of high-density foams, have been reported elsewhere [174].

Chemical Foaming

The use of a CBA is the first foaming method historically used in the early days of PUR foam research. This technique involves the production of gas as a by-product of the polymerization reaction or as a side reaction. The most notable example is the production of CO_2 in PUR foams [32,85]. Water can be introduced into the formulation by the incorporation of free water, or simply by the natural presence of moisture. As shown previously, isocyanates readily react with water to form a carbamic acid (Figure 11). The unstable nature of this component causes it to react instantaneously to produce urea and carbon dioxide. Primarily used for the production of flexible PUR foams, it is also an important means of creating rigid PUR or tannin-based foams [175]. This reaction has been known since the beginning of PUR chemistry. However, investigations into this pathway were slowed down in the 1950s because it was considered to inhibit the main polymerization reaction that produces the PUR chain. Controlling the moisture content of the mixture then became a key parameter in the PUR foam manufacturing process. On the other hand, at the same time, fluorinated PBAs were booming and were used extensively. Nevertheless, their ban prompted companies to reconsider the use of chemical foaming.

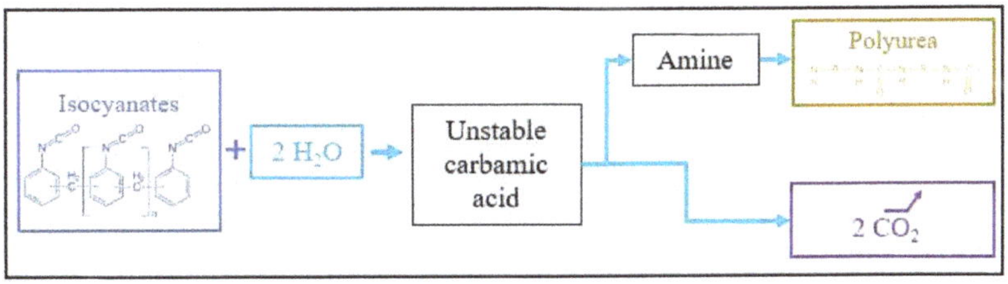

Figure 11. Chemical foaming reaction between isocyanates and water in polyurethane chemistry.

Nevertheless, there are also several drawbacks to the reaction of isocyanates with water to produce CO_2 as CBA, since it requires a larger amount of this component. This is because two units of isocyanate are needed to produce one unit of urea, instead of one single isocyanate producing a unit of urethane. This increases the demand for TDI, and has had a significant impact on the cost of recent products. In addition, the thermal properties of CO_2 are inferior to those of CFCs, as similar insulating performance would be achieved by increasing the thickness (twice that required with CFCs), which would be achieved with a greater number of reactants [32]. Through chemical foaming, hardening takes place at the same time as the release of the CBA. It stiffens the cell structure during expansion, and it must be controlled to optimize expansion and avoid collapse. The materials produced by this route are therefore slightly different from those physically expanded. They are also generally harder: the heat produced by both reactions without induction of phase change promotes polymerization and hardening, and the cells are generally smaller. These variations affect the structure and properties of the material, and determine the manufacturing process to obtain the desired foam.

Mechanical Foaming

A third foaming mechanism has recently been investigated for the removal of hazardous, expensive, and environmentally damaging chemicals from reagent mixtures. In this case, the foaming agent is air, incorporated by mechanical agitation, and trapped in the monomeric solution. Two techniques have been presented for producing MBFs: stiff liquid foam resembling beaten egg whites, for which air is incorporated by whipping with a propeller [176,177]; and foams produced by injecting air into a pressurized nozzle containing the reactive liquid [178]. To optimize the properties of foams, studies have presented ways of combining the effects of two foaming agents. Many examples can be found in the literature on polymer foams blown by a chemical agent and a PBA [168,179,180], as well as studies combining the latter mechanical techniques with chemical routes [176].

2.3. *End of Life and Recycling*

The polymer foam market was estimated to be worth USD 83.9 billion (25.3 million tons produced) in 2019 for all resins, and is expected to reach USD 134 billion by 2027 [181]. Thus, given this production in recent years, the volume of polymers reaching end of life represents a large amount of waste requiring disposal [182–186]. Such wastes can be sent to traditional generic treatment methods such as incineration or landfill. However, these methods raise environmental concerns regarding groundwater pollution, soil pollution, and greenhouse gas production. The increasing use of polymeric materials brings this issue to the forefront of environmental struggles [187–189]. Therefore, alternative methods for managing the end of life of these materials have been studied; in particular, special attention has been paid to the biodegradation mechanism in order to optimize this step, or to the removal of the polymer matrix during recycling for second life applications (Figure 12).

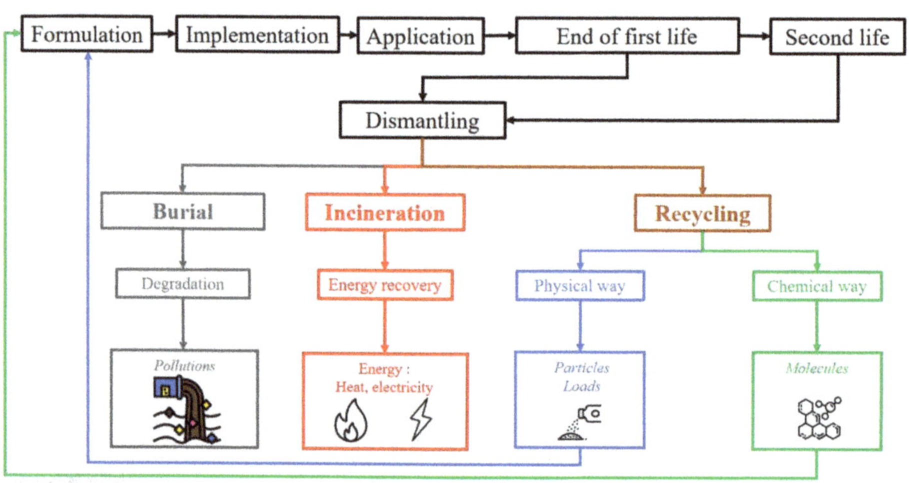

Figure 12. Life cycle of rigid cellular materials.

2.3.1. Incineration and Landfill

The two main waste management methods used around the world are incineration and landfill. The incineration process involves the complete oxidation of waste by combustion [85]. The heat produced can be used to contribute to the heating of buildings or to transform this energy into electricity. This combustion also produces ash containing inorganic and metallic elements, smoke, and fine particles. The fumes released by incineration are mainly composed of CO_2, H_2O, NO_x, SO_x, N_2, O_2, HCl, HF, etc., and fine particles, rich in heavy metals, dioxins and furans. Regulations have been imposed requiring these fumes to be filtered before rejecting them into the atmosphere by trapping the particles and dangerous gases. Thus, while this method emits non-recoverable greenhouse gases, which is an obvious drawback, it also produces energy and ash, which can be used as filler in cement, concrete, or asphalt.

Burying waste in landfills is another common method that allows the natural mechanisms of material degradation and decomposition to occur. Microorganisms, bacteria, and enzymes break down polymer chains into smaller molecules that can leach into the soil and water. Most studies on foam degradation have focused on PUR foams [190]. The decomposition of PUR has been studied in order to track the stages of polymer degradation [191]. This study showed that polyester-based PUR was more susceptible to being attacked by aerobic microorganisms compared to polyether-based PUR, due to the more decomposition-resistant chemical structure of polyether. However, PUR resists biodegradation in anaerobic environments and under low-temperature and -humidity conditions [192]. In addition, the stability is highly dependent on the polymer structure and the degree of crosslinking of the network [193]. However, the depolymerization steps are similar between the two types: free isocyanates still trapped in the polymer network are attacked first. Then, the urea and amide bonds are targeted before the degradation of the urethane groups takes place. Finally, the isocyanuric acid rings are the most resistant to microbial activity [194]. Although this is a natural process, possible soil pollution and infiltration into groundwater have led some countries to ban this method. In addition, it wastes land resources for agriculture. It is also worth mentioning that PUR foams degrade faster than other PUR formulations due to their high porosity, which makes them more accessible to potential degradation agents [195]. Further information can be found elsewhere [196,197].

2.3.2. Recycling and Reuse

The previous methods have limited efficiency for waste recovery. Long-term methods are based on physical and chemical recycling, and were compared by Yang et al. [185]. Physical recycling involves treating materials by mechanical or thermal processes. Chemical recycling is the most efficient recycling pathway, but the technology is complex to implement on an industrial scale. Nevertheless, some studies have attempted to obtain compounds by chemical treatment of polyurethanes in order to valorize them for other polymer formulations.

- Physical recycling is based on thermal or mechanical treatments to cause the polymer matrix to break down into smaller segments such as fillers, powders, etc. These solid particles can then be used with a resin to produce composites or to reinforce polymers or other materials such as concrete [198,199]. Gaseous compounds from the thermal recycling of PUR and PIR have also been studied as potential compounds for high value-added uses [200]. They can also be more easily used for further chemical processing, such as hydrolysis, acidolysis or glycolysis [201].
- Chemical recycling is based on the transformation, through chemical reactions, of polymers to produce smaller molecules that can be used for appropriate reactions. Most studies have focused on PUR polymers, one of the main polymers on the market, and are detailed in the literature [185,202–209]. This kind of recovery protocol would be the most comprehensive, giving a second life to polymer waste. However, some steps are complex, requiring a lot of energy, purification, and processing in order to be incorporated into new polymeric structures. Five main reactions have been studied: the alcoholysis, glycolysis, hydrolysis, amine, and phosphate methods [28].

- Alcoholysis consists of the depolymerization of PUR by combining it with a short-chain alcohol and a catalyst to obtain a new polyol, aromatic molecules and amines. When the alcohol used is a glycol, it is called glycolysis.
- Hydrolysis instead involves treating PUR with steam and a metal hydroxide catalyst to produce diamines, polyols, and carbon dioxide.
- The amine process uses a fatty amine to break down the polymer into polyols, aromatic constituents, and amines.
- The phosphate ester process treats the material to produce halogenated components, depending on the initial reactant. The resulting products can be added as flame retardant additives to polymer blends.

Other more recent strategies, such as the catalytic hydrogenation of PUR, have emerged thanks to new catalysts developed recently [210], and acidic processes have also been evaluated [211]. PUR is a very heterogeneous polymer with a variety of formulations, chemical structures, and polymeric structures. Therefore, finding a single route for chemical recovery is extremely complex. Studies on landfill biodegradation have led to the idea of reusing polymers and applying this process in a controlled industrial environment to recover depolymerized chemicals [201]. They focus on biodegradation by genetic and biochemical pathways. The molecules resulting from these depolymerization processes can then be reused as polyols in the formulation of new materials with similar performance to foams produced from virgin compounds [209,212]. The interest in reusing PUR is such that studies comparing reference PUR foams with recycled PUR foams have been carried out, showing minimal changes in mechanical properties [213]. Elastomeric materials based on PUR and ethylene vinyl acetate foams with compositions ranging from 0 to 100% recycled resources have also been evaluated, with 100% recycled formulations exhibiting properties suitable for shoe soles [214]. PUR coatings based on acid-treated PUR waste foams have also been produced, with excellent hydrophobic and mechanical properties [211].

3. Tannin–Furanic Foams (TFFs)

Tannin–furanic foams are bio-based formulations in which the initial mixture reacts, self-organizing into a three-dimensional network of polymer chains via crosslinking. The final foam has its final rigid form following a curing step that completes the polymerization.

3.1. Formulations

Environmental concerns have led to the search for alternatives to current sources of energy and raw materials. Thus, many studies are trying to develop new polymers based on biosourced materials. The result is the substitution of current non-renewable reagents of existing materials with more environmentally friendly solutions. Various methods of upgrade have emerged to create a wide range of bio-based substitutes [215]. Some improvements have contributed to the creation of new resins, such as tannin–furanic foams. In the 1980s, bio-based alternatives to traditional wood adhesives were presented [216], and these have continued to be studied to the present day [55,150,217]. The polycondensation reaction has been adapted for the design of tannin-based resins and foams that have properties similar to those of products produced from non-renewable resources [167]. While the main research on rigid adhesives and foams is reviewed in the next section, flexible foams and elastomers also open new horizons for the generation of phenol-based polymers [180] and other bio-based structures [152,218,219]. The manufacturing process for tannin-based foams follows the following general formulation: wood tannins, furfuryl alcohol, formaldehyde, catalyst, and foaming agent [167,220].

3.1.1. Tannins

Tannins are phenolic compounds found in different plant tissues such as leaves, fruits, and shells, and especially in the wood and bark of trees [221]. They are involved in the defense mechanisms of plants against parasitic organisms such as insects, fungi, and bacteria [222]. Tannins can be classified into three types of molecules: hydrolyzable, complex, and condensed [223]. The first category refers to a huge proportion of molecules in plants and corresponds to simple phenolic compounds such as gallotannins, gallic and digallic acids, and ellagitannins based on ellagic acid. Tannins from chestnut (*Castanea sativa*) and tara (*Caesalpinia tinctoria*) are among them, and are used in the production of leather goods. However, their low reactivity limits the development of green products based on these molecules. Nevertheless, some studies have attempted to incorporate them as a replacement for phenolic constituents in PF resins or isocyanates in PUR resins [142,224,225].

Condensed tannins are the most studied tannins because they can be recovered from many different trees, including mimosa (*Acacia mearnsii*) (Figure 13a), spruce (*Picea abies*), pine (*Pinus radiata* and *Pinus pinaster*), quebracho (*Schinopsis balansae* and *Schinopsi lorentzii*), and larch (*Larix gmelinii*) [166,226–229]. Condensed tannins are primarily extracted from barks (Figure 13a) and heartwood, which is achieved by first grinding the original material and then extracting them with solvents, usually hot water with sulfites. The final liquid is filtered and spray dried to concentrate the tannins into a fine brown powder (Figure 13a). Low manufacturing costs, wide availability, and ease of management have enabled the use of these resources in the development of new materials capable of competing with synthetic foams [230].

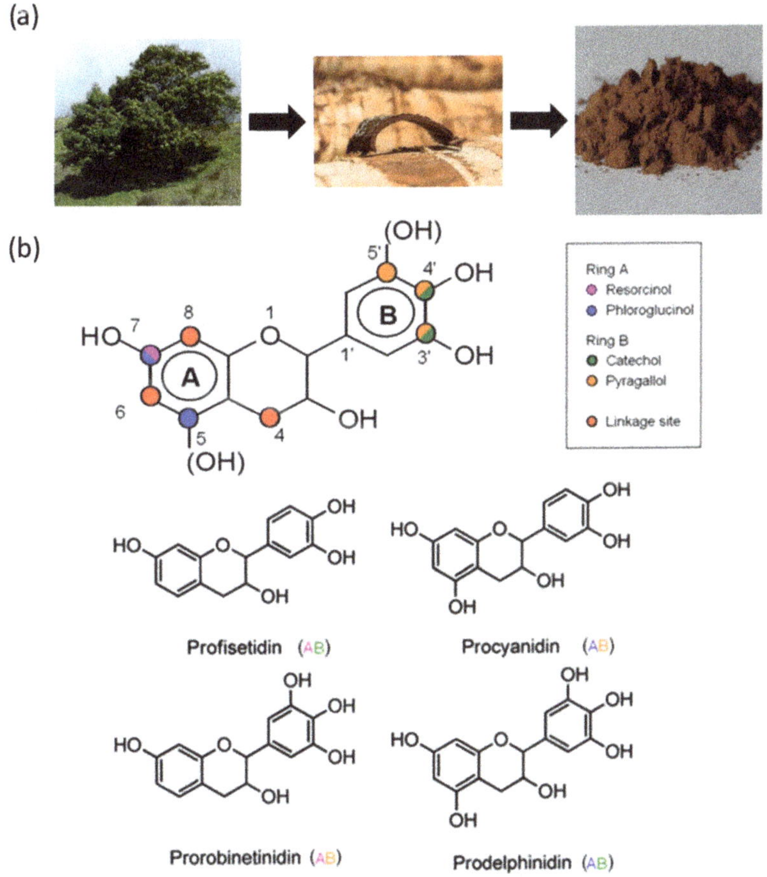

Figure 13. (**a**) Mimosa tree (Acacia mearnsii), mimosa wood after debarking, and tannin powder; (**b**) general structure of flavonoids followed by the main four types of flavonoids (based on [231]). Reproduced from Ref. [231] with permission from the Royal Society of Chemistry.

Condensed tannins are polyflavonoid compounds consisting of 3 to 10 flavonoid units, usually derived from flavan-3-ol and/or flavan-3,4-diol. Each unit is characterized by the combination of an A-ring (resorcinol or phloroglucinol type) and a B-ring (catechol or pyrogallol), as illustrated in Figure 13b. These combinations lead to four main categories of polyflavonoids: procyanidin, prodelphinidin, prorobinetinidin and profisetidin (Figure 13b). Among the most widely used species, mimosa tannins consist of prorobinetinidin units linked together by C_4-C_6 bonds. Quebracho tannins, on the other hand, result from profisetidin units linked by C_4-C_6 bonds. Pine tannins are mainly composed of procyanidin units, branched by C_4-C_8 bonds [232]. Finally, complex tannins are molecules resulting from the bonding of hydrolyzable tannins (ellagitannins) and flavonoid units (flavan-3-ol) [223].

The good reactivity of condensed tannins is a result of their structure, which is characterized by two aromatic rings and nucleophilic OH sites of interest. Previous research has mainly been focused on ring A, which is the most reactive and resorcinol-like, in the case of mimosa tannins, which do indeed resemble the resorcinol used in adhesives. Therefore, the chemistry of these new materials is based on well-known reaction mechanisms. The main reactive mechanism relies on the use of aldehydes to form a polymer under appropriate pH conditions or through catalysts. Under acidic or basic conditions, tannins can self-condense, thus producing more complex polymer networks [233,234]. In addition, tannins also react with added furfuryl alcohol to produce heat for the physical blowing process. Finally, tannins react with hexamine, a compound obtained from the reaction between formaldehyde and ammonia. This component is soluble in water and can be decomposed either in acidic medium into its precursors or in alkaline medium into formaldehyde and trimethylamine. The tannins react with the aldehyde formed and create benzylamine bonds to extend the polymer network [235].

3.1.2. Furfuryl Alcohol

Furfuryl alcohol (FA) can be derived from furfural, a molecule of great value in the biofuel and green chemistry industries that is obtained by the hydrolysis of C5 sugars contained in hemicelluloses from biomass waste [236,237] (Figure 14(a1)). FA has a good capacity to react with tannins [238] (Figure 14(a2)) and also to self-condense [239,240] (Figure 14(a3)). These two exothermic reactions generate a large amount of heat, triggering the vaporization of the physical foaming agent introduced in the formulation in the case of the physical blowing of the foam. It also participates in the formation of the polymer network. In addition, FA can also be added to the mixture in order to form a highly reactive intermediate with formaldehyde (Figure 14(a4)), which also extends the polymer network.

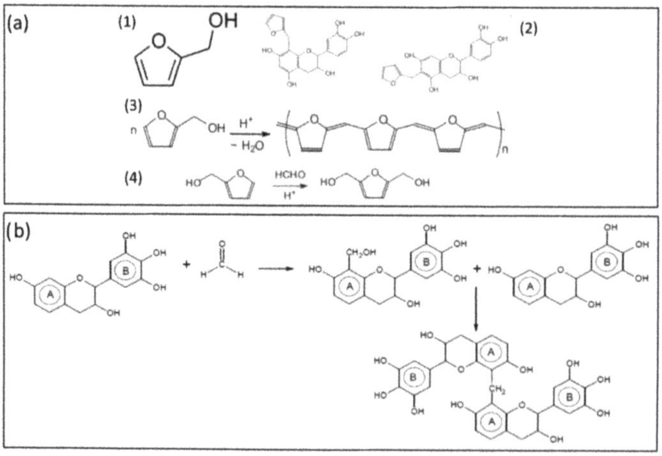

Figure 14. *Cont.*

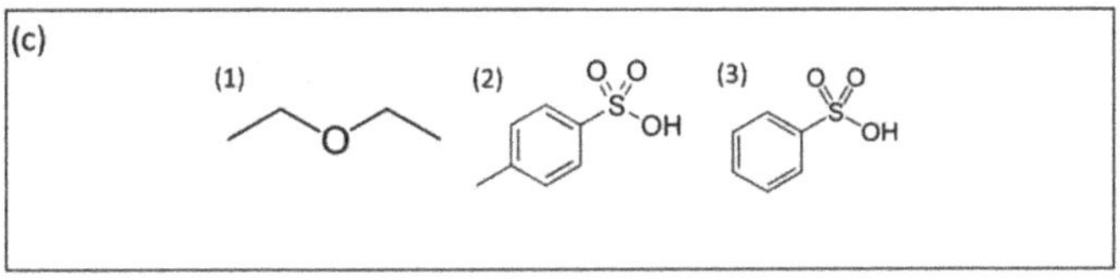

Figure 14. (**a1**) Chemical formula of furfuryl alcohol; (**a2**) products of the reaction of furfuryl alcohol with catechin units of tannin; (**a3**) self-condensation reaction of furfuryl alcohol; (**a4**) reaction between furfuryl alcohol and formaldehyde; (**b**) reaction of formaldehyde with prorobinetinidin (adapted from [11,241]); chemical formula of (**c1**) diethyl ether; (**c2**) para-toluenesulfonic acid; (**c3**) and phenolsulfonic acid.

3.1.3. Formaldehyde

Aldehydes, typically formaldehyde, are added to the mixture as curing agents for the crosslinking reactions. The reactions that take place are similar to those that occur in industrial PF foams between aldehydes and resorcinol. They occur between the aldehyde and ring A, e.g., the resorcinol-type ring in the case of mimosa tannins. It should be noted that under alkaline conditions (pH greater than 10), the reaction with the aldehyde also affects ring B. Following its listing as a human carcinogen by the U.S. National Toxicology Program in 2011 [242], growing concern about the release of formaldehyde has spurred research into possible replacements. Previous work on tannin-based adhesives for wood applications presents several alternatives, considering different aldehydes [243]. For example, formaldehyde-free tannin-based foams have also been studied in the last decade [5,173]. Despite the lower reactivity of alternatives aldehydes, this can be overcome by increasing the amount of FA [175].

3.1.4. Diethyl Ether

Diethyl ether (DE) is used as PBA for "standard" tannin-based foam formulations (Figure 14(**c1**)) due to its low boiling point (34 °C) and absence of reactivity with the other components. Thus, as the temperature of the mixture increases due to the polymerization reactions, DE vaporizes, easily causing the entire mixture to expand.

3.1.5. Para-Toluenesulfonic Acid

A variety of different catalysts can be added to facilitate the desired polymerization reaction under the appropriate pH conditions, depending on the formulation. Two main routes have been investigated: the use of acidic catalysts and the use of alkaline catalysts. Most methods are based on acidic conditions, achieved by the addition of para-toluenesulfonic acid (Figure 14(c2)) or phenolsulfonic acid (Figure 14(c3)) [227,244,245]. The reaction between tannins and formaldehyde is the easiest, but side reactions with isocyanates to produce urethane– and urea–formaldehyde resins have also been considered. For alkaline processes, a sodium hydroxide solution is usually added to the tannin–water–formaldehyde mixture.

3.1.6. Additives

The spectrum of possible applications for these new materials has been broadened through the optimization of the initial formulation and the introduction of new reagents. These additives can be of different types, each of which can be used to improve specific properties. The first example is the use of plasticizers to improve the elasticity of the materials, which are usually rigid, by introducing glycerol [226,246]. The plasticizer promotes the sliding of macromolecular chains relative to each other. Thus, the polymer network gains flexibility and exhibits a more elastic behavior. Conversely, the rigidity of the foam can be enhanced by supplementing the formulation with a filler such as hexamine [235,247], or recent more environmentally friendly alternatives, such as biochar [248,249] or lignin [250]. The addition of isocyanates to create urethane bridges can also be carried out to serve this purpose [180,220,251]. In addition, a good way of combining the effects of physical and chemical expansion in order to obtain a lighter foam is by maximizing cell formation [251]. In addition, surfactants can be included in the mixture to better control the size of the cells, while obtaining a more homogeneous distribution.

3.2. Expansion Steps

The expansion of tannin foams follows the same sequence of steps as those presented for other thermoset cellular polymers. Thus, the first step is to mix the polymer fraction without catalyst (tannin, water, formaldehyde, furfuryl alcohol) using a propeller agitator (Figure 15a), resulting in homogeneous and lump-free tannin-based mixtures. The DE is then incorporated into the mixture using a stirrer. Finally, the catalyst is added to the preparation under vigorous stirring; this marks the beginning of the FA polymerization and condensation reaction (Figure 15b). The latter reaction is highly exothermic, and therefore heats the polymer mixture, which maintains constant volume, but exceeds the boiling point of the foaming agent. After a few seconds, the DE changes phase and is vaporized. Then, the internal pressure and gas fraction increase, causing the matrix to expand. The foam reaches its maximum height when the blowing agent has been completely vaporized. Polymerization continues, and the polymer backbone sets and stiffens, forming a 3D polymer network. The material slowly cools following the depletion of the furfuryl alcohol, and completes its curing and drying, which can result in shrinkage in some formulations (Figure 15c). A general overview of the compounds used in the foam formulation and a photo of the final product are shown in Figure 15d.

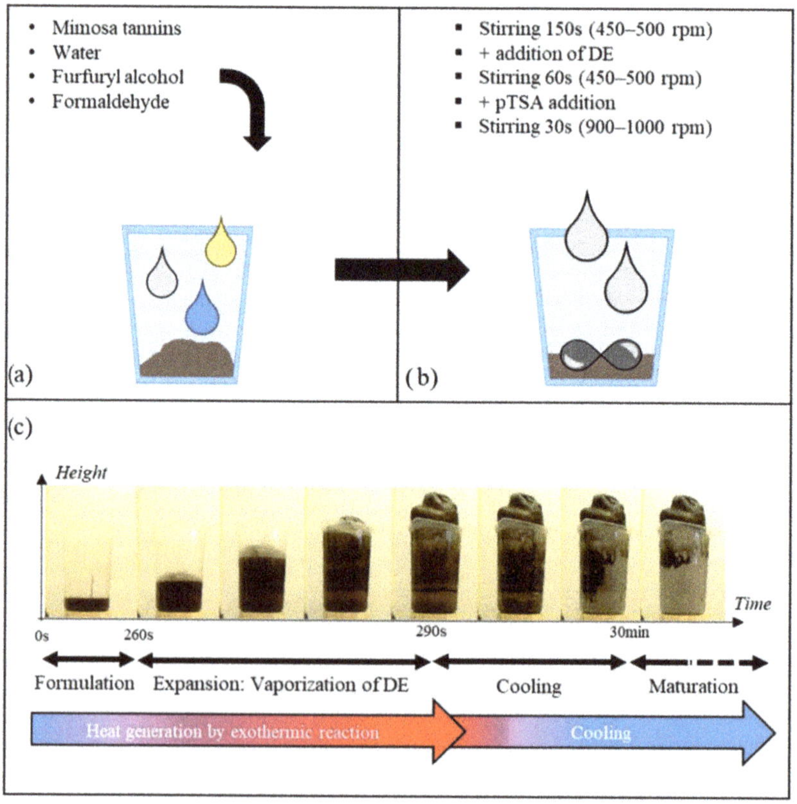

Figure 15. *Cont.*

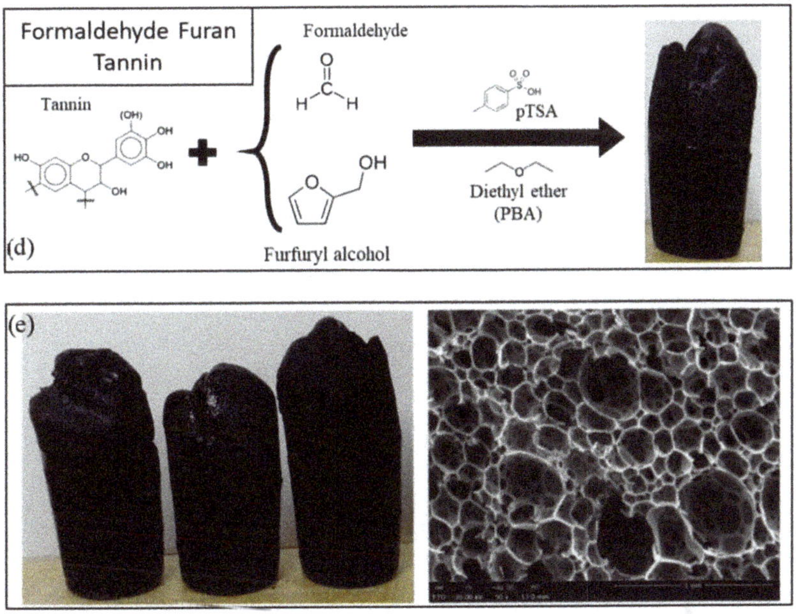

Figure 15. Steps for obtaining TFF foam: (**a**) addition of reagents; (**b**) mixing and addition of PBA and catalyst; (**c**) processing, expansion, and maturation of the material (with an example of time evolution; (**d**) reaction scheme for formulating TFF, (**e**) example of TFs (left) and TFF observed using scanning electron microscopy (×50, right, adapted from Letellier, 2015 [11]).

3.3. Physical Properties

3.3.1. Structural Properties

Figure 15e shows the partially open cell structure of a standard TFF. First, this opening of the walls may be a result of the high pressure within the gas bubbles during expansion. When this pressure is too high for the surface tension of the material, windows appear. This phenomenon can also occur at a later stage, during the drying and crosslinking of the polymer matrix, causing the thin walls to shrink, which can ultimately lead to the formation of new windows. The preparation of these lab-made materials in the laboratory is achieved using molds that are taller than they are wide. Expansion is therefore guided so that it only occurs along the vertical axis, creating a slight anisotropy in terms of cell size and shape. Nevertheless, an inherent heterogeneity in rigid tannin–furanic foams has generally been considered, as confirmed in a recent study [252].

A wide range of densities can be covered by adapting the formulation of tannin foams. The choice of foaming method, the proportions of reagents, the use of additives, etc., influence the structure of the foam with respect to cell size, wall thickness, pore opening, and thus the density, which also has implications on the foam's thermal and mechanical properties. For standard proportions, as presented elsewhere [166], the density of the foam ranges from 0.055 to 0.075 g·cm^{-3}, while the density of the solid fraction reaches around 1.59 g·cm^{-3} [220,253,254]. Thus, Zhao et al. implemented TFFs of the same formulation in molds of different diameters (6 and 12 cm) [254]. The materials produced did not show the variations in density or differences in cell shape reported in previous studies [255,256]. It was also highlighted in [254] that limitation of vertical expansion, a recurrent phenomenon for foaming in closed molds or in situ, also disturbed the structure and created anisotropy. The simplest way to control the density is to vary the amount of foaming agent added to the preparation. By introducing different amounts of DE, within safety limits [257], a direct relationship between the mass of DE used and the inverse of the density was demonstrated

(Figure 16a,c). Similarly, a relationship was established between the inverse of the bulk density—and thus the mass of DE—and cell size (Figure 16b,d).

However, adjusting the density by varying the other reactants is more complex due to the different chemical reactions that take place during polymerization, making predictions more difficult. For example, increasing the amount of FA may intensify the exothermic self-condensation reaction, generating more heat and facilitating the evaporation of DE. Nevertheless, this also promotes early matrix hardening, thus stabilizing the structure and inhibiting its expansion.

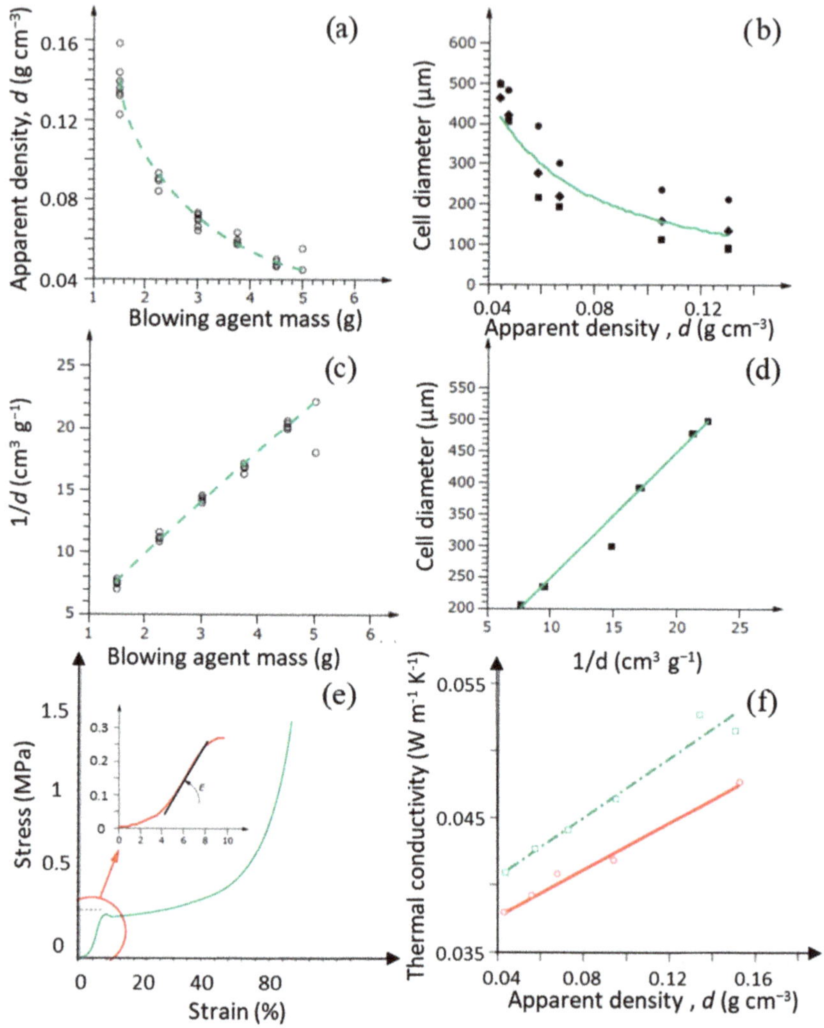

Figure 16. (**a**) Apparent density of TFFs as a function of the mass of introduced PBA; (**b**) cell diameter of TFFs as a function of apparent density; (**c**) inverse of the apparent density of TFFs as a function of the mass of introduced PBA; (**d**) cell diameter of TFFs as a function of apparent density [254]; (**e**) typical compression curve of a TFF (ρ = 0.068kg.m^{-3}) (adapted from [7]); (**f**) thermal conductivity of TFFs measured perpendicular to (green dashed line) and along (red solid line) their direction of

expansion (adapted from [258]). Figure 16a–d, reprinted from Materials Chemistry and Physics, Zhao et al., Effect of composition and processing parameters on the characteristics of tannin-based rigid foams. Part I: Cell structure, 175–182, Copyright (2010), with permission from Elsevier. Figure 16e, reprinted from Materials Science and Engineering A, Celzard et al., Mechanical properties of tannin-based rigid foams undergoing compression, 4438–4446, Copyright (2010), with permission from Elsevier. Figure 16f, reprinted from Materials and Design, Martinez de Yuso et al., Structure and properties of rigid foams derived from quebracho tannin, 208–212, Copyright (2014), with permission from Elsevier.

3.3.2. Thermal Properties

Good insulators are materials with low total thermal conductivity, which depends on the contribution of conduction in the solid and gas phases, convection in the cells, and radiation. However, the cell size is generally too small to allow convective motion, so this quantity is neglected [11]. To obtain suitable performance for insulation, the material used must therefore have high porosity, with closed cells, capable of trapping gas of low thermal conductivity. The contribution of the gas fraction depends on the intrinsic conductivity of the gas and the porosity. Many models of porous materials exist for defining the contribution of the solid phase, typically based on the conductivity of the solid fraction and on porosity [12,77]. Finally, the role of radiation must also be taken into account, since it can contribute from 20 to 30% of the total conductivity [11,77].

Like other rigid cellular materials, tannin foams exhibit very good insulating qualities with very a low thermal conductivity (0.032–0.050 $W \cdot m^{-1} \cdot K^{-1}$) [8]. However, these results depend on the homogeneity of the structure: in the case of anisotropy, measurements parallel and perpendicular to the expansion axis must be performed, because the conductivities in these directions may differ. In addition, thermal conductivities also depend significantly on the porosity of the material, as mentioned above. In other words, bulk density and total thermal conductivity are correlated, as shown in Figure 16f. Tannin–furanic foams can therefore compete with phenolic or PUR-based foams, and can be considered as a serious alternative for building insulation.

3.3.3. Mechanical Properties

The mechanical behavior of the foam depends on the internal organization of the cells and the degree of polymerization of the matrix. The mechanical behavior is usually characterized by compression both along and perpendicular to the foaming axis in the case of anisotropy. This behavior can be divided into three different phases: the first phase is linear elastic; the second phase, during which stress is maintained while the strain increases significantly, is a serrated plateau; and the third phase corresponds to densification. On the basis of this global behavior, the compressive strength and modulus of elasticity can be obtained as the stress at the end of the linear part and as the slope of the linear part, respectively (Figure 16e). Although the elastic phase of these materials is quite limited, they are capable of absorbing large amounts of mechanical energy by means of irreversible cell wall rupture.

Tannin-based foams are considered rigid materials, i.e., their compressive strength is greater than 0.08 MPa. However, the formulation and addition of hardeners or plasticizers changes the mechanical behavior of the material by altering the bonds between the polymer chains, so that they can be made stiffer or more elastic. With respect to compression, the literature reports values between 0.05 and 1.75 MPa [7,259], which is a good ratio of strength to relative density. These foams are therefore good candidates for use in shock and vibration absorption applications. On the other hand, the removal of formaldehyde from the formulation also increases the elasticity of the material, due to the absence of a rigid 3D polymer network created by the crosslinking agent [5].

3.3.4. Fire Resistance Properties

Tannin foams have very good fire resistance properties, even without the addition of flame retardants [260,261], as the energy input required to start combustion corresponds to severe conditions [220,253,260,261]. For instance, when comparing PF foams and TFFs by applying an incident heat flux of 50 $kW \cdot m^{-2}$, an ignition time of 6s was observed for the former [108], while no ignition was observed for the latter [190]. This unique behavior conferred by tannin allows the derived foams to stand out from other insulation materials, a property that is often a critical point, and is a major vulnerability of PUR foams, in particular. This makes tannins advantageous for applications in sensitive, fire-prone, or public buildings or systems. The numerous studies carried out on TFF foams in which fire resistance properties were evaluated have led to the development of a range of formulations for designing materials with similar or even better properties than the PUR foams they are intended to replace.

3.4. Applications and End of Life

The physical and chemical characterizations performed suggest a wide range of possible uses. First, TFFs are lightweight and have low thermal conductivity. Combined with excellent fire resistance, this makes them a good alternative for building insulation. Formulations with open-cell structures have typical acoustic properties for this type of material, even excellent for medium and high frequencies [262]. However, other more sophisticated applications have recently been reported, such as biosubstrates for Raman spectroscopy [263], or the adsorption of different pollutants, as described below. However, one of the main problems related to the use of TFFs for engineering applications is the high friability of these materials. This makes them difficult to handle and is a big obstacle at the industrialization stage. To overcome this disadvantage, they have been studied as core materials of sandwich panels (plywood, medium density fiberboard, wood). The composite materials thus obtained combine the lightness of TFFs and the bending strength of the other elements, without the problem of the friable TFF core crumbling [173,264].

Tannin-based foams are still new products, and are still under development. Therefore, the issue of their dismantling and recycling has not yet been fully explored. Some techniques generally used for cellular materials can be employed, however. Sánchez-Martín et al. developed another application for this material by testing it for wastewater treatment [265]. Tests carried out on dyes, surfactants, and pharmaceutical compounds have demonstrated the good capacity of these foams to depollute [266,267]. This adsorption capacity had already been demonstrated by Tondi et al. for the capture of metal ions (Cu^{2+} and Pb^{2+}) from wastewater [268]. This new perspective allows us to consider a method for recycling foams applied as insulation in buildings by transforming them into adsorbents for wastewater remediation. Moreover, other studies have used these materials as precursors to obtain carbon foams by pyrolysis. This step increases the porosity of the materials, in particular by creating micropores. These treatments introduce these foams in areas such as catalysis and absorption [269,270], and pave the way to end-of-life recovery of the foams.

4. Conclusions

This review illustrates the wide range of rigid polymeric foam solutions that exist today. However, this mature sector, with well-defined processes and facilities, is still highly dependent on petrochemical feedstocks, which are necessary for the formulation of these polymeric foams. In recent years, some initiatives have attempted to introduce increasing proportions of bio-based materials in order to replace petrochemical reagents. For example, porous materials such as PUR, PIR, PUA and PF foams are playing an important role in this revolution, because their lightweight but robust structure preserves material and energy. Their thermal, mechanical, and acoustic properties make them useful in many applications, the most common being saving energy in the thermal insulation of buildings or industrial systems (coolers, tanks, ducts, etc.). Moreover, these conventional solutions are being progressively challenged by the emergence of new materials such as

TFFs, which are positioning themselves as credible alternatives with similar or even better performances than the previous materials, while simultaneously incorporating a higher fraction of renewable raw materials. Although they seem to be promising candidates for replacing PUR or PF formulations, these new products still need to be developed and optimized before reaching the industrialization phase. To conclude this review, Table 2 gathers the main properties of the foams described here.

Table 2. Summary of bulk density, thermal conductivity, compressive strength, and fire resistance properties of the foams reported here.

Foam	Density (g/cm^3)	Thermal Conductivity ($W\cdot m^{-1}\cdot K^{-1}$)	Compressive Strength (MPa)	Fire Resistance	[Refs.]
PUR	0.026–0.058	0.010–0.057	0.45–4.8	Bad	[271–279]
PUA	0.008–0.400	0.013–0.050	0.03–2.83	Good	[93,280–283]
PIR	0.019–0.073	0.032–0.035	0.20–0.32	Good	[29,284–289]
PF	0.016–0.314	0.029–0.060	0.07–3.45	Bad	[116,117,119–121,220,229,290–292]
TFF	0.030–0.200	0.032–0.070	0.05–1.75	Very good	[8,220,253,260,261,293]

On the basis of this table, it can be concluded that TFFs have similar or even better properties than those observed for common foams based on non-renewable resources, and thus place them as potential substitutes for many applications.

Author Contributions: A.M.B.-L.; Writing—Review and Editing. V.N.: Writing—Review and Editing, Supervision. Z.M.; Writing—Original Draft Preparation. A.C.; Writing—Review and Editing, V.F.; Writing—Review and Editing, Supervision, Funding Acquisition, Project Administration. All authors have read and agreed to the published version of the manuscript.

Funding: This research was funded by the European Regional Development Fund (ERDF), grant number TALiSMAN project (2019-000215).

Institutional Review Board Statement: Not applicable.

Informed Consent Statement: Not applicable.

Data Availability Statement: Not applicable.

Acknowledgments: This study was made possible by the TALiSMAN project (2019-000215) funded by the European Regional Development Fund (ERDF).

Conflicts of Interest: The authors declare no conflict of interest.

Acronyms and Symbols

bPUR	biopolyurethane
CBA	chemical blowing agent
CFC	chlorofluorocarbon
DE	diethyl ether
EP	epoxy
FA	furfuryl alcohol
HCFC	hydrochlorofluorocarbon
HFC	hydrofluorocarbon
MBF	mechanically blown foam
MDI	methylene diphenyl diisocyanate
mMDI	pure monomeric MDI
PBA	physical blowing agent
PF	phenol–formaldehyde
PFC	perfluorocarbon

PIR	polyisocyanurate
pMDI	polymeric MDI
PUA	polyurea
PUR	polyurethane
TDI	toluene diisocyanate
TFF	tannin–furanic foams
θ	tortuosity
k	permeability
ρ	relative density
ρ_a	apparent density
ρ_s	skeletal density (i.e., of the solid fraction)
ϕ	porosity

References

1. ADEME. Pourquoi et Comment Isoler Son Logement? Available online: https://librairie.ademe.fr/urbanisme-et-batiment/976-pourquoi-et-comment-isoler-son-logement-.html (accessed on 23 August 2022).
2. White World Packaging srl, Insulation in EU. Available online: https://wgp4.com/en/insulation-in-eu/ (accessed on 23 August 2022).
3. Abu-Jdayil, B.; Mourad, A.-H.; Hittini, W.; Hassan, M.; Hameedi, S. Traditional, State-of-the-Art and Renewable Thermal Building Insulation Materials: An Overview. *Constr. Build. Mater.* **2019**, *214*, 709–735. [CrossRef]
4. Asdrubali, F.; D'Alessandro, F.; Schiavoni, S. A Review of Unconventional Sustainable Building Insulation Materials. *Sustain. Mater. Technol.* **2015**, *4*, 1–17. [CrossRef]
5. Basso, M.-C.; Li, X.; Fierro, V.; Pizzi, A.; Giovando, S.; Celzard, A. Green, Formaldehyde-Free, Foams for Thermal Insulation. *Adv. Mater. Lett.* **2011**, *2*, 378–382. [CrossRef]
6. Celzard, A.; Szczurek, A.; Jana, P.; Fierro, V.; Basso, M.-C.; Bourbigot, S.; Stauber, M.; Pizzi, A. Latest Progresses in the Preparation of Tannin-Based Cellular Solids. *J. Cell. Plast.* **2015**, *51*, 89–102. [CrossRef]
7. Celzard, A.; Zhao, W.; Pizzi, A.; Fierro, V. Mechanical Properties of Tannin-Based Rigid Foams Undergoing Compression. *Mater. Sci. Eng. A* **2010**, *527*, 4438–4446. [CrossRef]
8. Lacoste, C.; Basso, M.C.; Pizzi, A.; Laborie, M.-P.; Celzard, A.; Fierro, V. Pine Tannin-Based Rigid Foams: Mechanical and Thermal Properties. *Ind. Crops Prod.* **2013**, *43*, 245–250. [CrossRef]
9. Tondi, G.; Pizzi, A.; Olives, R. Natural Tannin-Based Rigid Foams as Insulation for Doors and Wall Panels. *Maderas Cienc. Tecnol.* **2008**, *10*, 219–227. [CrossRef]
10. Jiejun, W.; Chenggong, L.; Dianbin, W.; Manchang, G. Damping and Sound Absorption Properties of Particle Reinforced Al Matrix Composite Foams. *Compos. Sci. Technol.* **2003**, *63*, 569–574. [CrossRef]
11. Letellier, M. Optimisation de Mousses de Carbone Dérivées de Tannin Par l'étude et La Modélisation de Leurs Propriétés Physiques. Ph.D. Thesis, Université de Lorraine, Nancy, France, 2015.
12. Gibson, L.J.; Ashby, M.F. *Cellular Solids: Structure and Properties*; Cambridge University Press: Cambridge, UK, 1999; ISBN 1-316-02542-X.
13. IUPAC. Gold Book—Foam. Available online: https://goldbook.iupac.org/html/F/F02467.html (accessed on 13 July 2018).
14. Weaire, D.; Phelan, R. Cellular Structures in Three Dimensions. *Phil. Trans. R Soc. Lond. A* **1996**, *354*, 1989–1997. [CrossRef]
15. Kırca, M.; Gül, A.; Ekinci, E.; Yardım, F.; Mugan, A. Computational Modeling of Micro-Cellular Carbon Foams. *Finite Elem. Anal. Des.* **2007**, *44*, 45–52. [CrossRef]
16. Thomson, W. On the Division of Space with Minimum Partitional Area. *Acta Math.* **1887**, *11*, 121–134. [CrossRef]
17. Phelan, R.; Weaire, D.; Peters, E.A.J.F.; Verbist, G. The Conductivity of a Foam. *J. Phys. Condens. Matter* **1996**, *8*, L475–L482. [CrossRef]
18. Kusner, R.; Sullivan, J.M. Comparing the Weaire-Phelan Equal-Volume Foam to Kelvin's Foam. *Forma* **1996**, *11*, 233–242.
19. Fallet, A. Structure et Propriétés Mécaniques d'Empilements Aléatoires de Sphères Creuses: Caractérisation et Modélisation. Ph.D. Thesis, Institut National Polytechnique de Grenoble, Grenoble, France, 2008.
20. Smithers Rapra. High-Performance Polymer Foams Market Is Forecast to Reach $122.4 Billion by 2021. Available online: https://www.smithersrapra.com/resources/2017/february/high-performance-polymer-foams-market-forecast (accessed on 4 September 2019).
21. The Future of Polymer Foams to 2026. Market Reports and Research. Available online: https://www.smithers.com/services/market-reports/materials/the-future-of-polymer-foams-to-2025 (accessed on 3 February 2022).
22. IUPAC. Gold Book—Thermosetting Polymer. Available online: https://goldbook.iupac.org/html/T/TT07168.html (accessed on 13 July 2018).
23. IUPAC. Gold Book—Thermoplastic Elastomer. Available online: https://goldbook.iupac.org/html/T/TT07268.html (accessed on 3 September 2018).
24. IUPAC. Gold Book—Elastomer. Available online: https://goldbook.iupac.org/html/E/ET07547.html (accessed on 3 September 2018).

25. Imeokparia, D.D.; Suh, K.W.; Stobby, W.G. Cellular Materials. In *Encyclopedia of Polymer Science and Technology*; American Cancer Society: Atlanta, GA, USA, 2003; ISBN 978-0-471-44026-0.
26. Carothers, W.H. Studies on Polymerization and Ring Formation. I. An Introduction to the General Theory of Condensation Polymers. *J. Am. Chem. Soc.* **1929**, *51*, 2548–2559. [CrossRef]
27. Polyuréthanes—Société Chimique de France. Available online: http://www.societechimiquedefrance.fr/polyurethanes.html (accessed on 3 September 2019).
28. Gama, N.; Ferreira, A.; Barros-Timmons, A. Polyurethane Foams: Past, Present, and Future. *Materials* **2018**, *11*, 1841. [CrossRef] [PubMed]
29. Szycher, M. *Szycher's Handbook of Polyurethanes*, 2nd ed.; CRC Press Taylor & Francis Group, LLC: Boca Raton, FL, USA, 2013; ISBN 978-1-4398-3958-4.
30. Plastics Insight Polyurethane Production, Pricing and Market Demand. Available online: https://www.plasticsinsight.com/resin-intelligence/resin-prices/polyurethane/ (accessed on 4 September 2019).
31. Polyurethane (PU) Market 2021: Top Companies Report Covers, Industry Outlook, Opportunities and Showing Impressive Growth by 2030 with Dominant Regions and Countries Data. Available online: https://www.marketwatch.com/press-release/polyurethane-pu-market-2021-top-companies-report-covers-industry-outlook-opportunities-and-showing-impressive-growth-by-2030-with-dominant-regions-and-countries-data-2021-12-06 (accessed on 3 February 2022).
32. Landrock, A.H. *Handbook of Plastic Foams: Types, Properties, Manufacture and Applications*; Noyes Publication: Saddle River, NJ, USA, 1995; ISBN 0-8155-1357-7.
33. Zhang, S.; Chu, F.; Xu, Z.; Zhou, Y.; Qiu, Y.; Qian, L.; Hu, Y.; Wang, B.; Hu, W. The Improvement of Fire Safety Performance of Flexible Polyurethane Foam by Highly-Efficient P-N-S Elemental Hybrid Synergistic Flame Retardant. *J. Colloid Interface Sci.* **2022**, *606*, 768–783. [CrossRef]
34. Gu, X.; Gao, T.; Meng, X.; Zhu, Y.; Wang, G. Enhanced Hydrophobicity of Polyurethane with the Self-Assembly of Perfluoropolyether-Based Triblock Copolymers. *Prog. Org. Coat.* **2022**, *162*, 106561. [CrossRef]
35. Wu, M.; Li, T.; Wang, P.; Wu, S.; Wang, R.; Lin, J. Dual-Encapsulated Highly Conductive and Liquid-Free Phase Change Composites Enabled by Polyurethane/Graphite Nanoplatelets Hybrid Networks for Efficient Energy Storage and Thermal Management. *Small* **2022**, *18*, 2105647. [CrossRef]
36. Alavi Nikje, M.M.; Noruzian, M.; Tamaddoni Moghaddam, S. Investigation of Fe_3O_4/AEAP Supermagnetic Nanoparticles on the Morphological, Thermal and Magnetite Behavior of Polyurethane Rigid Foam Nanocomposites. *Polimery* **2015**, *60*, 26–32. [CrossRef]
37. Ferkl, P.; Karimi, M.; Marchisio, D.L.; Kosek, J. Multi-Scale Modelling of Expanding Polyurethane Foams: Coupling Macro- and Bubble-Scales. *Chem. Eng. Sci.* **2016**, *148*, 55–64. [CrossRef]
38. Ionescu, M. *Chemistry and Technology of Polyols for Polyurethanes*; Smithers Rapra Publishing: Shrewsbury, UK, 2005; ISBN 978-1-85957-501-7.
39. Javni, I.; Zhang, W.; Petrović, Z.S. Soybean-Oil-Based Polyisocyanurate Rigid Foams. *J. Polym. Environ.* **2004**, *12*, 123–129. [CrossRef]
40. Jin, J.F.; Chen, Y.L.; Wang, D.N.; Hu, C.P.; Zhu, S.; Vanoverloop, L.; Randall, D. Structures and Physical Properties of Rigid Polyurethane Foam Prepared with Rosin-Based Polyol. *J. Appl. Polym. Sci.* **2002**, *84*, 598–604. [CrossRef]
41. Karimi, M.; Droghetti, H.; Marchisio, D.L. Multiscale Modeling of Expanding Polyurethane Foams via Computational Fluid Dynamics and Population Balance Equation. *Macromol. Symp.* **2016**, *360*, 108–122. [CrossRef]
42. Lee, Y.; Baek, K.H.; Choe, K.; Han, C. Development of Mass Production Type Rigid Polyurethane Foam for LNG Carrier Using Ozone Depletion Free Blowing Agent. *Cryogenics* **2016**, *80*, 44–51. [CrossRef]
43. Pikhurov, D.V.; Sakhatskii, A.S.; Zuev, V.V. Rigid Polyurethane Foams with Infused Hydrophilic/Hydrophobic Nanoparticles: Relationship between Cellular Structure and Physical Properties. *Eur. Polym. J.* **2018**, *99*, 403–414. [CrossRef]
44. Pinto, M.L. Formulation, Preparation, and Characterization of Polyurethane Foams. *J. Chem. Educ.* **2010**, *87*, 212–215. [CrossRef]
45. Rao, R.R.; Mondy, L.A.; Long, K.N.; Celina, M.C.; Wyatt, N.; Roberts, C.C.; Soehnel, M.M.; Brunini, V.E. The Kinetics of Polyurethane Structural Foam Formation: Foaming and Polymerization. *AIChE J.* **2017**, *63*, 2945–2957. [CrossRef]
46. Sarier, N.; Onder, E. Thermal Characteristics of Polyurethane Foams Incorporated with Phase Change Materials. *Thermochim. Acta* **2007**, *454*, 90–98. [CrossRef]
47. Xu, Z.; Tang, X.; Gu, A.; Fang, Z. Novel Preparation and Mechanical Properties of Rigid Polyurethane Foam/Organoclay Nanocomposites. *J. Appl. Polym. Sci.* **2007**, *106*, 439–447. [CrossRef]
48. Yan, D.; Xu, L.; Chen, C.; Tang, J.; Ji, X.; Li, Z. Enhanced Mechanical and Thermal Properties of Rigid Polyurethane Foam Composites Containing Graphene Nanosheets and Carbon Nanotubes. *Polym. Int.* **2012**, *61*, 1107–1114. [CrossRef]
49. Zhai, T.; Li, D.; Fei, G.; Xia, H. Piezoresistive and Compression Resistance Relaxation Behavior of Water Blown Carbon Nanotube/Polyurethane Composite Foam. *Compos. Part Appl. Sci. Manuf.* **2015**, *72*, 108–114. [CrossRef]
50. Zhang, L.; Zhang, M.; Zhou, Y.; Hu, L. The Study of Mechanical Behavior and Flame Retardancy of Castor Oil Phosphate-Based Rigid Polyurethane Foam Composites Containing Expanded Graphite and Triethyl Phosphate. *Polym. Degrad. Stab.* **2013**, *98*, 2784–2794. [CrossRef]
51. Zhu, H.; Xu, S. Preparation and Fire Behavior of Rigid Polyurethane Foams Synthesized from Modified Urea–Melamine–Formaldehyde Resins. *RSC Adv.* **2018**, *8*, 17879–17887. [CrossRef]

52. Kim, S.H.; Kim, B.K.; Lim, H. Effect of Isocyanate Index on the Properties of Rigid Polyurethane Foams Blown by HFC 365mfc. *Macromol. Res.* **2008**, *16*, 467–472. [CrossRef]
53. Borrero-López, A.M.; Valencia, C.; Franco, J.M. Rheology of Lignin-Based Chemical Oleogels Prepared Using Diisocyanate Crosslinkers: Effect of the Diisocyanate and Curing Kinetics. *Eur. Polym. J.* **2017**, *89*, 311–323. [CrossRef]
54. Borrero-López, A.M.; Valencia, C.; Franco, J.M. Green and Facile Procedure for the Preparation of Liquid and Gel-like Polyurethanes Based on Castor Oil and Lignin: Effect of Processing Conditions on the Rheological Properties. *J. Clean. Prod.* **2020**, *277*, 123367. [CrossRef]
55. Tenorio-Alfonso, A.; Sánchez, M.C.; Franco, J.M. Impact of Moisture Curing Conditions on the Chemical Structure and Rheological and Ultimate Adhesion Properties of Polyurethane Adhesives Based on Castor Oil and Cellulose Acetate. *Prog. Org. Coat.* **2021**, *161*, 106547. [CrossRef]
56. Tenorio-Alfonso, A.; Sánchez, M.C.; Franco, J.M. Impact of the Processing Method on the Properties of Castor Oil/Cellulose Acetate Polyurethane Adhesives for Bonding Wood. *Int. J. Adhes. Adhes.* **2022**, *116*, 103153. [CrossRef]
57. Gallego, R.; Arteaga, J.F.; Valencia, C.; Franco, J.M. Thickening Properties of Several NCO-Functionalized Cellulose Derivatives in Castor Oil. *Chem. Eng. Sci.* **2015**, *134*, 260–268. [CrossRef]
58. Dworakowska, S.; Bogdał, D.; Zaccheria, F.; Ravasio, N. The Role of Catalysis in the Synthesis of Polyurethane Foams Based on Renewable Raw Materials. *Catal. Today* **2014**, *223*, 148–156. [CrossRef]
59. Sardon, H.; Engler, A.C.; Chan, J.M.W.; García, J.M.; Coady, D.J.; Pascual, A.; Mecerreyes, D.; Jones, G.O.; Rice, J.E.; Horn, H.W.; et al. Organic Acid-Catalyzed Polyurethane Formation via a Dual-Activated Mechanism: Unexpected Preference of N-Activation over O-Activation of Isocyanates. *J. Am. Chem. Soc.* **2013**, *135*, 16235–16241. [CrossRef]
60. Wegener, G.; Brandt, M.; Duda, L.; Hofmann, J.; Klesczewski, B.; Koch, D.; Kumpf, R.-J.; Orzesek, H.; Pirkl, H.-G.; Six, C.; et al. Trends in Industrial Catalysis in the Polyurethane Industry. *Appl. Catal. Gen.* **2001**, *221*, 303–335. [CrossRef]
61. Sridaeng, D.; Jitaree, W.; Thiampanya, P.; Chantarasiri, N. Preparation of Rigid Polyurethane Foams Using Low-Emission Catalysts Derived from Metal Acetates and Ethanolamine. *e-Polymers* **2016**, *16*, 265–275. [CrossRef]
62. Liu, P.S.; Chen, G.F. Chapter Five—Fabricating Porous Ceramics. In *Porous Materials*; Liu, P.S., Chen, G.F., Eds.; Butterworth-Heinemann: Oxford, UK, 2014; pp. 221–302. ISBN 978-0-12-407788-1.
63. Guo, A.; Javni, I.; Petrovic, Z. Rigid Polyurethane Foams Based on Soybean Oil. *J. Appl. Polym. Sci.* **2000**, *77*, 467–473. [CrossRef]
64. Akdogan, E.; Erdem, M.; Ureyen, M.E.; Kaya, M. Rigid Polyurethane Foams with Halogen-Free Flame Retardants: Thermal Insulation, Mechanical, and Flame Retardant Properties. *J. Appl. Polym. Sci.* **2019**, *137*, 47611. [CrossRef]
65. Bhoyate, S.; Ionescu, M.; Kahol, P.K.; Gupta, R.K. Sustainable Flame-Retardant Polyurethanes Using Renewable Resources. *Ind. Crops Prod.* **2018**, *123*, 480–488. [CrossRef]
66. Li, Q.; Wang, J.; Chen, L.; Shi, H.; Hao, J. Ammonium Polyphosphate Modified with β-Cyclodextrin Crosslinking Rigid Polyurethane Foam: Enhancing Thermal Stability and Suppressing Flame Spread. *Polym. Degrad. Stab.* **2019**, *161*, 166–174. [CrossRef]
67. Shi, L.; Li, Z.-M.; Xie, B.-H.; Wang, J.-H.; Tian, C.-R.; Yang, M.-B. Flame Retardancy of Different-Sized Expandable Graphite Particles for High-Density Rigid Polyurethane Foams. *Polym. Int.* **2006**, *55*, 862–871. [CrossRef]
68. Wang, S.-X.; Zhao, H.-B.; Rao, W.-H.; Huang, S.-C.; Wang, T.; Liao, W.; Wang, Y.-Z. Inherently Flame-Retardant Rigid Polyurethane Foams with Excellent Thermal Insulation and Mechanical Properties. *Polymer* **2018**, *153*, 616–625. [CrossRef]
69. Yang, R.; Hu, W.; Xu, L.; Song, Y.; Li, J. Synthesis, Mechanical Properties and Fire Behaviors of Rigid Polyurethane Foam with a Reactive Flame Retardant Containing Phosphazene and Phosphate. *Polym. Degrad. Stab.* **2015**, *122*, 102–109. [CrossRef]
70. Zhu, H.; Xu, S. Synthesis and Properties of Rigid Polyurethane Foams Synthesized from Modified Urea-Formaldehyde Resin. *Constr. Build. Mater.* **2019**, *202*, 718–726. [CrossRef]
71. Członka, S.; Strąkowska, A.; Strzelec, K.; Kairytė, A.; Vaitkus, S. Composites of Rigid Polyurethane Foams and Silica Powder Filler Enhanced with Ionic Liquid. *Polym. Test.* **2019**, *75*, 12–25. [CrossRef]
72. Liu, Q.; Liu, Y.; Shi, J.; Liu, Z.; Wang, Q.; Guo, C.F. High-Porosity Foam-Based Iontronic Pressure Sensor with Superhigh Sensitivity of 9280 KPa−1. *Nano-Micro Lett.* **2021**, *14*, 21. [CrossRef] [PubMed]
73. Zhou, M.; Zhao, S.; Zhou, K.; Mei, F.; Qian, X.; Shi, C. Flexible Polyurethane Foams Surface-Modified with FeOOH for Improved Oil-Water Separation and Flame Retardancy. *Mater. Chem. Phys.* **2022**, *276*, 125408. [CrossRef]
74. Billotto, F.V.; Mirdamadi, M.M.; Pearson, B.A. *Design, Application Development, and Launch of Polyurethane Foam Systems in Vehicle Structures*; SAE International: Warrendale, PA, USA, 2003.
75. Narine, S.S.; Kong, X.; Bouzidi, L.; Sporns, P. Physical Properties of Polyurethanes Produced from Polyols from Seed Oils: II. Foams. *J. Am. Oil Chem. Soc.* **2007**, *84*, 65–72. [CrossRef]
76. Lim, H.; Kim, S.H.; Kim, B.K. Effects of Silicon Surfactant in Rigid Polyurethane Foams. *Express Polym. Lett.* **2008**, *2*, 194–200. [CrossRef]
77. Glicksman, L.R. Heat Transfer in Foams. In *Low Density Cellular Plastics*; Springer: Berlin/Heidelberg, Germany, 1994; pp. 104–152.
78. Saha, M.C.; Barua, B.; Mohan, S. Study on the Cure Kinetic Behavior of Thermosetting Polyurethane Solids and Foams: Effect of Temperature, Density, and Carbon Nanofiber. *J. Eng. Mater. Technol.* **2011**, *133*, 011015. [CrossRef]
79. Sung, G.; Choe, H.; Choi, Y.; Kim, J.H. Morphological, Acoustical, and Physical Properties of Free-Rising Polyurethane Foams Depending on the Flow Directions. *Korean J. Chem. Eng.* **2018**, *35*, 1045–1052. [CrossRef]

80. Thirumal, M.; Khastgir, D.; Singha, N.K.; Manjunath, B.S.; Naik, Y.P. Effect of Foam Density on the Properties of Water Blown Rigid Polyurethane Foam. *J. Appl. Polym. Sci.* **2008**, *108*, 1810–1817. [CrossRef]
81. Traeger, R.K. Physical Properties of Rigid Polyurethane Foams. *J. Cell. Plast.* **1967**, *3*, 405–418. [CrossRef]
82. Wu, J.-W.; Sung, W.-F.; Chu, H.-S. Thermal Conductivity of Polyurethane Foams. *Int. J. Heat Mass Transf.* **1999**, *42*, 2211–2217. [CrossRef]
83. Calvo-Correas, T.; Mosiewicki, M.A.; Corcuera, M.A.; Eceiza, A.; Aranguren, M.I. Linseed Oil-Based Polyurethane Rigid Foams: Synthesis and Characterization. *J. Renew. Mater.* **2015**, *3*, 3–13. [CrossRef]
84. Jiang, T.; Wang, W.; Yu, D.; Huang, D.; Wei, N.; Hu, Y.; Huang, H. Synthesis and Characterization of Polyurethane Rigid Foams from Polyether Polyols with Isosorbide as the Bio-Based Starting Agent. *J. Polym. Res.* **2018**, *25*, 140. [CrossRef]
85. Lee, S.-T.; Scholz, D.P.K. (Eds.) Polymeric Foams. In *Polymeric Foams: Technology and Developments in Regulation*; CRC Press Taylor & Francis Group, LLC: Boca Raton, FL, USA, 2009; ISBN 978-1-4200-6125-3.
86. Riyapan, D.; Saetung, A.; Saetung, N. A Novel Rigid PU Foam Based on Modified Used Palm Oil as Sound Absorbing Material. *J. Polym. Environ.* **2019**, *27*, 1693–1708. [CrossRef]
87. Zatorski, W.; Brzozowski, Z.K.; Kolbrecki, A. New Developments in Chemical Modification of Fire-Safe Rigid Polyurethane Foams. *Polym. Degrad. Stab.* **2008**, *93*, 2071–2076. [CrossRef]
88. Oertel, G. *Polyurethane Handbook*; Carl Hanser Verlag: Munich, Germany, 1993.
89. Hatakeyama, H.; Kato, N.; Nanbo, T.; Hatakeyama, T. Water Absorbent Polyurethane Composites Derived from Molasses and Lignin Filled with Microcrystalline Cellulose. *J. Mater. Sci.* **2012**, *47*, 7254–7261. [CrossRef]
90. Chiriac, C.I.; Tanasă, F. Polyureas. In *Ullmann's Encyclopedia of Industrial Chemistry*; Wiley: Hoboken, NJ, USA, 2000; ISBN 978-3-527-30673-2.
91. Primeaux, I.D.J. Foamed Polyurea Elastomer-Rigid and Close-Celled. U.S. Patent #5,153,232, 8 February 1992.
92. Borsus, J.M.; Merckaert, P.; Jérôme, R.; Teyssié, P. Catalysis of the Reaction between Isocyanates and Protonic Substrates. II. Kinetic Study of the Polyurea Foaming Process Catalyzed by a Series of Amino Compounds. *J. Appl. Polym. Sci.* **1982**, *27*, 4029–4042. [CrossRef]
93. Creyf, H.S. Development and optimization of polyurea foam. In *New Ways to Save Energy*; Strub, A.S., Ehringer, H., Eds.; Springer: Dordrecht, Germany, 1980. [CrossRef]
94. Staendeke, H.; Lagoda, R. Flame-Resistant Polyurea Foam. Patent #ZA952401B, 30 July 1998.
95. Cartmell, M.J.; Brown, R.K. Polyurea Foam for Retrofit Insulation—An Update with Case Studies. *J. Therm. Insul.* **1983**, *6*, 250–262. [CrossRef]
96. Frisch, K.C.; Baumann, H. Polyurea Foams Prepared from Isocyanate Water and a Lower Alkanol. Patent #4,454,251, 12 June 1984.
97. Briody, J.M.; Narinesingh, D. Bifunctional Catalysis by Amides and Ureas in the Reaction of Amines with Phenylisocyanate. *Tetrahedron Lett.* **1971**, *12*, 4143–4146. [CrossRef]
98. Ramirez, B.J.; Kingstedt, O.T.; Crum, R.; Gamez, C.; Gupta, V. Tailoring the Rate-Sensitivity of Low Density Polyurea Foams through Cell Wall Aperture Size. *J. Appl. Phys.* **2017**, *121*, 225107. [CrossRef]
99. Borsus, J.M.; Jérôme, R.; Teyssié, P. Catalysis of the Reaction between Isocyanates and Protonic Substrates. I. Metal Salt–Amine Complexes as Catalysts in the Polyurea Foaming Process. *J. Appl. Polym. Sci.* **1981**, *26*, 3027–3043. [CrossRef]
100. Ashida, K. *Polyurethane and Related Foams: Chemistry and Technology*; CRC Press: Boca Raton, FL, USA, 2006; ISBN 978-1-58716-159-9.
101. Andersons, J.; Kirpluks, M.; Stiebra, L.; Cabulis, U. Anisotropy of the Stiffness and Strength of Rigid Low-Density Closed-Cell Polyisocyanurate Foams. *Mater. Des.* **2016**, *92*, 836–845. [CrossRef]
102. Naruse, A.; Nanno, H.; Kurita, M.; Inohara, H.; Fukami, T. Development of All Water-Blown Polyisocyanurate Foam System for Metal-Faced Continuous Sandwich Panels. *J. Cell. Plast.* **2002**, *38*, 385–401. [CrossRef]
103. Christian, J.E.; Courville, G.E.; Desjarlais, A.O.; Graves, R.S.; Linkous, R.L.; McElroy, D.L.; Weaver, F.J.; Wendt, R.L.; Yarbrough, D.W. *The Technical Viability of Alternative Blowing Agents in Polyisocyanurate Roof Insulation: A Cooperative Industry/Government Project*; Oak Ridge National Laboratoy: Oak Ridge, TN, USA, 1993.
104. Stec, A.A.; Hull, T.R. Assessment of the Fire Toxicity of Building Insulation Materials. *Energy Build.* **2011**, *43*, 498–506. [CrossRef]
105. Knop, A.; Pilato, L.A. *Phenolic Resins: Chemistry, Applications and Performance Future Directions*; Springer: Berlin, Germany, 1985; ISBN 978-3-662-02429-4.
106. Sandhya, P.K.; Sreekala, M.S.; Thomas, S. *Phenolic Based Foams: Preparation, Characterization, and Applications*; Springer: Berlin, Germany, 2022; ISBN 9789811652370.
107. De Carvalho, G.; Pimenta, J.A.; dos Santos, W.N.; Frollini, E. Phenolic and Lignophenolic Closed Cells Foams: Thermal Conductivity and Other Properties. *Polym. Plast. Technol. Eng.* **2003**, *42*, 605–626. [CrossRef]
108. Auad, M.L.; Zhao, L.; Shen, H.; Nutt, S.R.; Sorathia, U. Flammability Properties and Mechanical Performance of Epoxy Modified Phenolic Foams. *J. Appl. Polym. Sci.* **2007**, *104*, 1399–1407. [CrossRef]
109. Dos-Santos, C.G.; Costa, M.A.; Morais, W.A.D.; Pasa, V.M.D. Phenolic Foams from Wood Tar Resols. *J. Appl. Polym. Sci.* **2010**, *115*, 923–927. [CrossRef]
110. Stefani, P.M.; Barchi, A.T.; Sabugal, J.; Vazquez, A. Characterization of Epoxy Foams. *J. Appl. Polym. Sci.* **2003**, *90*, 2992–2996. [CrossRef]

111. Sui, X.; Wang, Z. Flame-Retardant and Mechanical Properties of Phenolic Foams Toughened with Polyethylene Glycol Phosphates. *Polym. Adv. Technol.* **2013**, *24*, 593–599. [CrossRef]
112. Alonso, M.V.; Auad, M.L.; Nutt, S. Short-Fiber-Reinforced Epoxy Foams. *Compos. Part Appl. Sci. Manuf.* **2006**, *37*, 1952–1960. [CrossRef]
113. Gardziella, A.; Pilato, L.A.; Knop, A. *Phenolic Resins: Chemistry, Applications, Standardization, Safety and Ecology*, 2nd ed.; Springer: Berlin, Germany, 2010; ISBN 978-3-642-08484-3.
114. Cornick, M. Foam. In *Phenolic Resins: A Century of Progress*; Pilato, L., Ed.; Springer: Berlin, Germany, 2010; ISBN 978-3-642-04714-5.
115. Dammel, R. Basic Chemistry of Novolaks. In *Diazonaphthoquinone-Base Resists Tutor*; SPIE: Bellingham, WA, USA, 1993.
116. Desai, A.; Auad, M.L.; Shen, H.; Nutt, S.R. Mechanical Behavior of Hybrid Composite Phenolic Foam. *J. Cell. Plast.* **2008**, *44*, 15–36. [CrossRef]
117. Lee, S.-H.; Teramoto, Y.; Shiraishi, N. Resol-Type Phenolic Resin from Liquefied Phenolated Wood and Its Application to Phenolic Foam. *J. Appl. Polym. Sci.* **2002**, *84*, 468–472. [CrossRef]
118. Papa, A.J. Methyl Formate as an Adjuvant in Phenolic Foam Formation. Patent #4,033,910, 5 July 1977.
119. Kim, B.G.; Lee, D.G. Development of Microwave Foaming Method for Phenolic Insulation Foams. *J. Mater. Process. Technol.* **2008**, *201*, 716–719. [CrossRef]
120. Song, S.A.; Oh, H.J.; Kim, B.G.; Kim, S.S. Novel Foaming Methods to Fabricate Activated Carbon Reinforced Microcellular Phenolic Foams. *Compos. Sci. Technol.* **2013**, *76*, 45–51. [CrossRef]
121. Shen, H.; Nutt, S. Mechanical Characterization of Short Fiber Reinforced Phenolic Foam. *Compos. Part Appl. Sci. Manuf.* **2003**, *34*, 899–906. [CrossRef]
122. Rickle, G.K.; Denslow, K.R. The Effect of Water on Phenolic Foam Cell Structure. *J. Cell. Plast.* **1988**, *24*, 70–78. [CrossRef]
123. Del Saz-Orozco, B.; Alonso, M.V.; Oliet, M.; Domínguez, J.C.; Rojo, E.; Rodriguez, F. Lignin Particle- and Wood Flour-Reinforced Phenolic Foams: Friability, Thermal Stability and Effect of Hygrothermal Aging on Mechanical Properties and Morphology. *Compos. Part B Eng.* **2015**, *80*, 154–161. [CrossRef]
124. Desai, A.; Nutt, S.R.; Alonso, M.V. Modeling of Fiber-Reinforced Phenolic Foam. *J. Cell. Plast.* **2008**, *44*, 391–413. [CrossRef]
125. Zhou, J.; Yao, Z.; Chen, Y.; Wei, D.; Wu, Y. Thermomechanical Analyses of Phenolic Foam Reinforced with Glass Fiber Mat. *Mater. Des.* **2013**, *51*, 131–135. [CrossRef]
126. Yang, Z.; Zhao, X.; Ding, Y.; Yang, Q.; Ye, L. Structure and Mechanical Stability of Epoxy Modified Polyurethane Foam Under Heat and Stress. *J. Macromol. Sci. Part B* **2019**, *58*, 113–127. [CrossRef]
127. Ma, Y.; Gong, X.; Jia, P. The Effects of DOPO-g-ITA Modified Microcrystalline Cellulose on the Properites of Composite Phenolic Foams. *J. Renew. Mater.* **2020**, *8*, 45–55. [CrossRef]
128. Chian, K.S.; Gan, L.H. Development of a Rigid Polyurethane Foam from Palm Oil. *J. Appl. Polym. Sci.* **1998**, *68*, 509–515. [CrossRef]
129. Cho, S.T.; So, J.I.; Jung, J.-Y.; Hwang, S.; Baeck, S.-H.; Shim, S.E. Polymerization Kinetics and Physical Properties of Polyurethanes Synthesized by Bio-Based Monomers. *Macromol. Res.* **2019**, *27*, 153–163. [CrossRef]
130. Herrán, R.; Amalvy, J.I.; Chiacchiarelli, L.M. Highly Functional Lactic Acid Ring-opened Soybean Polyols Applied to Rigid Polyurethane Foams. *J. Appl. Polym. Sci.* **2019**, *136*, 47959. [CrossRef]
131. Hu, Y.H.; Gao, Y.; Wang, D.N.; Hu, C.P.; Zu, S.; Vanoverloop, L.; Randall, D. Rigid Polyurethane Foam Prepared from a Rape Seed Oil Based Polyol. *J. Appl. Polym. Sci.* **2002**, *84*, 591–597. [CrossRef]
132. Kakroodi, A.R.; Khazabi, M.; Maynard, K.; Sain, M.; Kwon, O.-S. Soy-Based Polyurethane Spray Foam Insulations for Light Weight Wall Panels and Their Performances under Monotonic and Static Cyclic Shear Forces. *Ind. Crops Prod.* **2015**, *74*, 1–8. [CrossRef]
133. Kurańska, M.; Prociak, A. The Influence of Rapeseed Oil-Based Polyols on the Foaming Process of Rigid Polyurethane Foams. *Ind. Crops Prod.* **2016**, *89*, 182–187. [CrossRef]
134. Mahmood, N.; Yuan, Z.; Schmidt, J.; Xu, C. (Charles) Preparation of Bio-Based Rigid Polyurethane Foam Using Hydrolytically Depolymerized Kraft Lignin via Direct Replacement or Oxypropylation. *Eur. Polym. J.* **2015**, *68*, 1–9. [CrossRef]
135. Marcovich, N.E.; Kurańska, M.; Prociak, A.; Malewska, E.; Kulpa, K. Open Cell Semi-Rigid Polyurethane Foams Synthesized Using Palm Oil-Based Bio-Polyol. *Ind. Crops Prod.* **2017**, *102*, 88–96. [CrossRef]
136. Paciorek-Sadowska, J.; Borowicz, M.; Czupryński, B.; Isbrandt, M. Effect of Evening Primrose Oil-Based Polyol on the Properties of Rigid Polyurethane–Polyisocyanurate Foams for Thermal Insulation. *Polymers* **2018**, *10*, 1334. [CrossRef]
137. Petrović, Z.S. Polyurethanes from Vegetable Oils. *Polym. Rev.* **2008**, *48*, 109–155. [CrossRef]
138. Shirke, A.; Dholakiya, B.; Kuperkar, K. Novel Applications of Castor Oil Based Polyurethanes: A Short Review. *Polym. Sci. Ser. B* **2015**, *57*, 292–297. [CrossRef]
139. Yang, R.; Wang, B.; Li, M.; Zhang, X.; Li, J. Preparation, Characterization and Thermal Degradation Behavior of Rigid Polyurethane Foam Using a Malic Acid Based Polyols. *Ind. Crops Prod.* **2019**, *136*, 121–128. [CrossRef]
140. Bossion, A. New Challenges in the Synthesis of Non-Isocyanate Polyurethanes. Ph.D. Thesis, University of Bordeaux, Bordeaux, France, 2019.
141. Clark, J.; Farmer, T.J.; Ingram, I.; Lie, Y.; North, M. Renewable Self-Blowing Non-Isocyanate Polyurethane Foams from Lysine and Sorbitol. *Eur. J. Org. Chem.* **2018**, *2018*, 4265–4271. [CrossRef]
142. Thébault, M.; Pizzi, A.; Santiago-Medina, F.J.; Al-Marzouki, F.M.; Abdalla, S. Isocyanate-Free Polyurethanes by Coreaction of Condensed Tannins with Aminated Tannins. *J. Renew. Mater.* **2017**, *5*, 21–29. [CrossRef]

143. Thébault, M.; Pizzi, A.; Essawy, H.A.; Barhoum, A.; Van Assche, G. Isocyanate Free Condensed Tannin-Based Polyurethanes. *Eur. Polym. J.* **2015**, *67*, 513–526. [CrossRef]
144. Thébault, M.; Pizzi, A.; Dumarçay, S.; Gerardin, P.; Fredon, E.; Delmotte, L. Polyurethanes from Hydrolysable Tannins Obtained without Using Isocyanates. *Ind. Crops Prod.* **2014**, *59*, 329–336. [CrossRef]
145. Boerjan, W.; Ralph, J.; Baucher, M. Lignin Biosynthesis. *Annu. Rev. Plant Biol.* **2003**, *54*, 519–546. [CrossRef]
146. Borrero-López, A.M.; Martín-Sampedro, R.; Ibarra, D.; Valencia, C.; Eugenio, M.E.; Franco, J.M. Evaluation of Lignin-Enriched Side-Streams from Different Biomass Conversion Processes as Thickeners in Bio-Lubricant Formulations. *Int. J. Biol. Macromol.* **2020**, *162*, 1398–1413. [CrossRef] [PubMed]
147. Borrero-López, A.M.; Valencia, C.; Ibarra, D.; Ballesteros, I.; Franco, J.M. Lignin-Enriched Residues from Bioethanol Production: Chemical Characterization, Isocyanate Functionalization and Oil Structuring Properties. *Int. J. Biol. Macromol.* **2022**, *195*, 412–423. [CrossRef]
148. Alonso, M.V.; Oliet, M.; Pérez, J.M.; Rodríguez, F.; Echeverría, J. Determination of Curing Kinetic Parameters of Lignin–Phenol–Formaldehyde Resol Resins by Several Dynamic Differential Scanning Calorimetry Methods. *Thermochim. Acta* **2004**, *419*, 161–167. [CrossRef]
149. Calvo-Flores, F.G.; Dobado, J.A. Lignin as Renewable Raw Material. *ChemSusChem* **2010**, *3*, 1227–1235. [CrossRef] [PubMed]
150. Borrero-López, A.M.; Valencia, C.; Domínguez, G.; Eugenio, M.; Franco, J.M. Rheology and Adhesion Performance of Adhesives Formulated with Lignins from Agricultural Waste Straws Subjected to Solid-State Fermentation. *Ind. Crops Prod.* **2021**, *171*, 113876. [CrossRef]
151. Domínguez, G.; Blánquez, A.; Borrero-López, A.M.; Valencia, C.; Eugenio, M.E.; Arias, M.E.; Rodríguez, J.; Hernández, M. Eco-Friendly Oleogels from Functionalized Kraft Lignin with Laccase SilA from Streptomyces Ipomoeae: An Opportunity to Replace Commercial Lubricants. *ACS Sustain. Chem. Eng.* **2021**, *9*, 4611–4616. [CrossRef]
152. Borrero-López, A.M.; Wang, L.; Valencia, C.; Franco, J.M.; Rojas, O.J. Lignin Effect in Castor Oil-Based Elastomers: Reaching New Limits in Rheological and Cushioning Behaviors. *Compos. Sci. Technol.* **2021**, *203*, 108602. [CrossRef]
153. Fulcrand, H.; Rouméas, L.; Billerach, G.; Aouf, C.; Dubreucq, E. Advances in Bio-Based Thermosetting Polymers. In *Recent Advances in Polyphenol Research*; Halbwirth, H., Stich, K., Cheynier, V., Quideau, S., Eds.; John Wiley & Sons, Ltd.: Chichester, UK, 2019; pp. 285–334. ISBN 978-1-119-42789-6.
154. Mimini, V.; Kabrelian, V.; Fackler, K.; Hettegger, H.; Potthast, A.; Rosenau, T. Lignin-Based Foams as Insulation Materials: A Review. *Holzforschung* **2018**, *73*, 117–130. [CrossRef]
155. Wang, S.; Liu, W.; Yang, D.; Qiu, X. Highly Resilient Lignin-Containing Polyurethane Foam. *Ind. Eng. Chem. Res.* **2019**, *58*, 496–504. [CrossRef]
156. Zhang, X.; Kim, Y.; Eberhardt, T.L.; Shmulsky, R. Lab-Scale Structural Insulated Panels with Lignin-Incorporated Rigid Polyurethane Foams as Core. *Ind. Crops Prod.* **2019**, *132*, 292–300. [CrossRef]
157. Bo, C.; Hu, L.; Chen, Y.; Yang, X.; Zhang, M.; Zhou, Y. Synthesis of a Novel Cardanol-Based Compound and Environmentally Sustainable Production of Phenolic Foam. *J. Mater. Sci.* **2018**, *53*, 10784–10797. [CrossRef]
158. Bo, C.; Yang, X.; Hu, L.; Zhang, M.; Jia, P.; Zhou, Y. Enhancement of Flame-Retardant and Mechanical Performance of Phenolic Foam with the Incorporation of Cardanol-Based Siloxane. *Polym. Compos.* **2019**, *40*, 2539–2547. [CrossRef]
159. Berthier, J.-C. Polyuréthanes—PUR. In *Techniques de l'ingénieur*; Matériaux-Plastiques et Composites: Saint-Denis, France, 2009; p. 27. ISBN AM3425 v2.
160. Singh, S.N. *Blowing Agents for Polyurethane Foams*; Rapra Technology Ltd.: Shawbury, UK, 2002.
161. Fahey, D.W.; Hegglin, M.I. *Twenty Questions and Answers About the Ozone Layer: 2010 Update*; World Meteorological Organization: Geneva, Switzerland, 2010.
162. UNEP Montreal Protocol on Substances That Deplete the Ozone Layer. *Int. Negot.* **1987**, *1*, 49. [CrossRef]
163. UNEP. Copenhagen Amendment. 1992. Available online: https://unep.ch/ozone/french/Ratification_status/copenhagen_amendment.shtml (accessed on 31 May 2019).
164. UNEP. *Montreal Protocol on Substances That Deplete the Ozone Layer*; UNEP: Athens, Greece, 1999; p. 204.
165. UNEP. *Rapport de la Vingt-Huitième Réunion des Parties au Protocole de Montréal Relatif à des Substances qui Appauvrissent la Couche d'Ozone-28ème Réunion des Parties au Protocole de Montréal*; UNEP: Athens, Greece, 2016; p. 80.
166. Basso, M.C.; Pizzi, A.; Celzard, A. Dynamic Foaming Behaviour of Polyurethane vs. Tannin/Furanic Foams. *J. Renew. Mater.* **2013**, *1*, 273–278. [CrossRef]
167. Meikleham, N.E.; Pizzi, A. Acid-and Alkali-Catalyzed Tannin-Based Rigid Foams. *J. Appl. Polym. Sci.* **1994**, *53*, 1547–1556. [CrossRef]
168. Al-Moameri, H.; Zhao, Y.; Ghoreishi, R.; Suppes, G.J. Simulation Blowing Agent Performance, Cell Morphology, and Cell Pressure in Rigid Polyurethane Foams. *Ind. Eng. Chem. Res.* **2016**, *55*, 2336–2344. [CrossRef]
169. Modesti, M.; Adriani, V.; Simioni, F. Chemical and Physical Blowing Agents in Structural Polyurethane Foams: Simulation and Characterization. *Polym. Eng. Sci.* **2000**, *40*, 2046–2057. [CrossRef]
170. Al-Moameri, H.; Zhao, Y.; Ghoreishi, R.; Suppes, G.J. Simulation of Liquid Physical Blowing Agents for Forming Rigid Urethane Foams. *J. Appl. Polym. Sci.* **2015**, *132*, 42454. [CrossRef]
171. Ling, Y.; Yao, S.; Chen, Y.; Hu, D.; Xi, Z.; Zhao, L. Synergetic Effect between Curing Reaction and CO_2 Diffusion for Microcellular Epoxy Foam Preparation in Supercritical CO_2. *J. Supercrit. Fluids* **2022**, *180*, 105424. [CrossRef]

172. Breuer, R.; Hendriks, S.; Reinhardt, N.; Facklam, M.; Hopmann, C. Modeling Flow and Cell Formation in Foam Sheet Extrusion of Polystyrene with CO_2 and Co-Blowing Agents. Part I: Material Model. *Polym. Eng. Sci.* **2021**, *61*, 2799–2813. [CrossRef]
173. Link, M.; Kolbitsch, C.; Tondi, G.; Ebner, M.; Wieland, S.; Petutschnigg, A. Formaldehyde-Free Tannin-Based Foams and Their Use as Lightweight Panels. *Bioresources* **2011**, *6*, 4218–4228.
174. Zubair, M.; Ferrari, R.; Alagha, O.; Mu'azu, N.D.; Blaisi, N.I.; Ateeq, I.S.; Manzar, M.S. Microwave Foaming of Materials: An Emerging Field. *Polymers* **2020**, *12*, 2477. [CrossRef]
175. Basso, M.C.; Giovando, S.; Pizzi, A.; Celzard, A.; Fierro, V. Tannin/Furanic Foams without Blowing Agents and Formaldehyde. *Ind. Crops Prod.* **2013**, *49*, 17–22. [CrossRef]
176. Santiago-Medina, F.J.; Tenorio-Alfonso, A.; Delgado-Sánchez, C.; Basso, M.C.; Pizzi, A.; Celzard, A.; Fierro, V.; Sánchez, M.C.; Franco, J.M. Projectable Tannin Foams by Mechanical and Chemical Expansion. *Ind. Crops Prod.* **2018**, *120*, 90–96. [CrossRef]
177. Szczurek, A.; Fierro, V.; Pizzi, A.; Stauber, M.; Celzard, A. A New Method for Preparing Tannin-Based Foams. *Ind. Crops Prod.* **2014**, *54*, 40–53. [CrossRef]
178. Santiago-Medina, F.J.; Delgado-Sánchez, C.; Basso, M.C.; Pizzi, A.; Fierro, V.; Celzard, A. Mechanically Blown Wall-Projected Tannin-Based Foams. *Ind. Crops Prod.* **2018**, *113*, 316–323. [CrossRef]
179. Baser, S.A.; Khakhar, D.V. Modeling of the Dynamics of R-11 Blown Polyurethane Foam Formation. *Polym. Eng. Sci.* **1994**, *34*, 632–641. [CrossRef]
180. Li, X.; Basso, M.C.; Fierro, V.; Pizzi, A.; Celzard, A. Chemical Modification of Tannin/Furanic Rigid Foams by Isocyanates and Polyurethanes. *Maderas Cienc. Tecnol.* **2012**, *14*, 257–265. [CrossRef]
181. Smithers Rapra. Polymer Foams Market Forecast to 2019. Available online: https://www.smithersrapra.com/news/2014/may/polymer-foam-market-to-consume-25-3-million-tonnes (accessed on 4 September 2019).
182. Molero, C.; de Lucas, A.; Rodríguez, J.F. Recovery of Polyols from Flexible Polyurethane Foam by "Split-Phase" Glycolysis with New Catalysts. *Polym. Degrad. Stab.* **2006**, *91*, 894–901. [CrossRef]
183. Quadrini, F.; Bellisario, D.; Santo, L. Recycling of Thermoset Polyurethane Foams. *Polym. Eng. Sci.* **2013**, *53*, 1357–1363. [CrossRef]
184. Ulrich, H.; Odinak, A.; Tucker, B.; Sayigh, A.A.R. Recycling of Polyurethane and Polyisocyanurate Foam. *Polym. Eng. Sci.* **1978**, *18*, 844–848. [CrossRef]
185. Yang, W.; Dong, Q.; Liu, S.; Xie, H.; Liu, L.; Li, J. Recycling and Disposal Methods for Polyurethane Foam Wastes. *Procedia Environ. Sci.* **2012**, *16*, 167–175. [CrossRef]
186. Zia, K.M.; Bhatti, H.N.; Ahmad Bhatti, I. Methods for Polyurethane and Polyurethane Composites, Recycling and Recovery: A Review. *React. Funct. Polym.* **2007**, *67*, 675–692. [CrossRef]
187. Azzarello, M.Y.; Vleet, E.S.V. Marine Birds and Plastic Pollution. *Mar. Ecol. Prog. Ser.* **1987**, *37*, 295–303. [CrossRef]
188. Eriksen, M.; Lebreton, L.C.M.; Carson, H.S.; Thiel, M.; Moore, C.J.; Borerro, J.C.; Galgani, F.; Ryan, P.G.; Reisser, J. Plastic Pollution in the World's Oceans: More than 5 Trillion Plastic Pieces Weighing over 250,000 Tons Afloat at Sea. *PLoS ONE* **2014**, *9*, e111913. [CrossRef]
189. Howard, G.T. Biodegradation of Polyurethane: A Review. *Int. Biodeterior. Biodegrad.* **2002**, *49*, 245–252. [CrossRef]
190. Fajardo, C.; Blánquez, A.; Domínguez, G.; Borrero-López, A.M.; Valencia, C.; Hernández, M.; Arias, M.E.; Rodríguez, J. Assessment of Sustainability of Bio Treated Lignocellulose-Based Oleogels. *Polymers* **2021**, *13*, 267. [CrossRef]
191. Filip, Z. Decomposition of Polyurethane in a Garbage Landfill Leakage Water and by Soil Microorganisms. *Eur. J. Appl. Microbiol. Biotechnol.* **1978**, *5*, 225–231. [CrossRef]
192. Urgun-Demirtas, M.; Singh, D.; Pagilla, K. Laboratory Investigation of Biodegradability of a Polyurethane Foam under Anaerobic Conditions. *Polym. Degrad. Stab.* **2007**, *92*, 1599–1610. [CrossRef]
193. Kloss, J.R.; Pedrozo, T.H.; Dal Magro Follmann, H.; Peralta-Zamora, P.; Dionísio, J.A.; Akcelrud, L.; Zawadzki, S.F.; Ramos, L.P. Application of the Principal Component Analysis Method in the Biodegradation Polyurethanes Evaluation. *Mater. Sci. Eng. C* **2009**, *29*, 470–473. [CrossRef]
194. Borowicz, M.; Paciorek-Sadowska, J.; Lubczak, J.; Czupryński, B. Biodegradable, Flame-Retardant, and Bio-Based Rigid Polyurethane/Polyisocyanurate Foams for Thermal Insulation Application. *Polymers* **2019**, *11*, 1816. [CrossRef]
195. Lattuati-Derieux, A.; Thao-Heu, S.; Lavédrine, B. Assessment of the Degradation of Polyurethane Foams after Artificial and Natural Ageing by Using Pyrolysis-Gas Chromatography/Mass Spectrometry and Headspace-Solid Phase Microextraction-Gas Chromatography/Mass Spectrometry. *J. Chromatogr. A* **2011**, *1218*, 4498–4508. [CrossRef] [PubMed]
196. Magnin, A.; Pollet, E.; Phalip, V.; Avérous, L. Evaluation of Biological Degradation of Polyurethanes. *Biotechnol. Adv.* **2020**, *39*, 107457. [CrossRef]
197. Howard, G.T. Polyurethane Biodegradation. In *Microbial Degradation of Xenobiotics*; Environmental Science and Engineering; Singh, S.N., Ed.; Springer: Berlin, Germany, 2012; pp. 371–394. ISBN 978-3-642-23789-8.
198. Rebeiz, K.S.; Craft, A.P. Plastic Waste Management in Construction: Technological and Institutional Issues. *Resour. Conserv. Recycl.* **1995**, *15*, 245–257. [CrossRef]
199. Siddique, R.; Khatib, J.; Kaur, I. Use of Recycled Plastic in Concrete: A Review. *Waste Manag.* **2008**, *28*, 1835–1852. [CrossRef] [PubMed]
200. Eschenbacher, A.; Varghese, R.J.; Weng, J.; Van Geem, K.M. Fast Pyrolysis of Polyurethanes and Polyisocyanurate with and without Flame Retardant: Compounds of Interest for Chemical Recycling. *J. Anal. Appl. Pyrolysis* **2021**, *160*, 105374. [CrossRef]

201. Cregut, M.; Bedas, M.; Durand, M.-J.; Thouand, G. New Insights into Polyurethane Biodegradation and Realistic Prospects for the Development of a Sustainable Waste Recycling Process. *Biotechnol. Adv.* **2013**, *31*, 1634–1647. [CrossRef] [PubMed]
202. Aguado, A.; Martínez, L.; Moral, A.; Fermoso, J.; Irusta, R. Chemical Recycling of Polyurethane Foam Waste via Glycolysis. *Chem. Eng. Trans.* **2011**, *24*, 1069–1074.
203. Campbell, G.A.; Meluch, W.C. Polyurethane Foam Recycling. Superheated Steam Hydrolysis. *Environ. Sci. Technol.* **1976**, *10*, 182–185. [CrossRef]
204. Czupryński, B.; Paciorek-Sadowska, J.; Liszkowska, J. Properties of Rigid Polyurethane-Polyisocyanurate Foams Modified with the Selected Fillers: Properties of Rigid Polyurethane-Polyisocyanurate Foams. *J. Appl. Polym. Sci.* **2010**, *115*, 2460–2469. [CrossRef]
205. Gerlock, J.; Braslaw, J.; Zinbo, M. Polyurethane Waste Recycling. 1. Glycolysis and Hydroglycolysis of Water-Blown Foams. *Ind. Eng. Chem. Process Des. Dev.* **1984**, *23*, 545–552. [CrossRef]
206. Modesti, M.; Costantini, F.; dal Lago, E.; Piovesan, F.; Roso, M.; Boaretti, C.; Lorenzetti, A. Valuable Secondary Raw Material by Chemical Recycling of Polyisocyanurate Foams. *Polym. Degrad. Stab.* **2018**, *156*, 151–160. [CrossRef]
207. Nikje, M.M.A.; Mohammadi, F.H.A. Polyurethane Foam Wastes Recycling under Microwave Irradiation. *Polym. Plast. Technol. Eng.* **2010**, *49*, 818–821. [CrossRef]
208. Sendijarevic, V. Chemical Recycling of Mixed Polyurethane Foam Stream Recovered from Shredder Residue into Polyurethane Polyols. *J. Cell. Plast.* **2007**, *43*, 31–46. [CrossRef]
209. Shin, S.; Kim, H.; Liang, J.; Lee, S.; Lee, D. Sustainable Rigid Polyurethane Foams Based on Recycled Polyols from Chemical Recycling of Waste Polyurethane Foams. *J. Appl. Polym. Sci.* **2019**, *136*, 47916. [CrossRef]
210. Gausas, L.; Donslund, B.S.; Kristensen, S.K.; Skrydstrup, T. Evaluation of Manganese Catalysts for the Hydrogenative Deconstruction of Commercial and End-of-Life Polyurethane Samples. *ChemSusChem* **2022**, *15*, e202101705. [CrossRef] [PubMed]
211. Godinho, B.; Gama, N.; Barros-Timmons, A., Ferreira, A. Recycling of Different Types of Polyurethane Foam Wastes via Acidolysis to Produce Polyurethane Coatings. *Sustain. Mater. Technol.* **2021**, *29*, e00330. [CrossRef]
212. Nikje, M.M.A.; Garmarudi, A.B.; Idris, A.B. Polyurethane Waste Reduction and Recycling: From Bench to Pilot Scales. *Des. Monomers Polym.* **2011**, *14*, 395–421. [CrossRef]
213. Njuguna, J.K.; Muchiri, P.; Mwema, F.M.; Karuri, N.W.; Herzog, M.; Dimitrov, K. Determination of Thermo-Mechanical Properties of Recycled Polyurethane from Glycolysis Polyol. *Sci. Afr.* **2021**, *12*, e00755. [CrossRef]
214. Gama, N.; Godinho, B.; Barros-Timmons, A.; Ferreira, A. Insights into PU/EVA Blends Produced Using Industrial Residues Towards Eco-Efficient Materials. *J. Polym. Environ.* **2022**, *30*, 1451–1461. [CrossRef]
215. Kuo, P.-Y.; Yan, N.; Tratnik, N.; Luo, J. Applications of Bark for Bio-Based Adhesives and Foams. *Phys. Sci. Rev.* **2018**, *3*, 20170194. [CrossRef]
216. Pizzi, A. Tannin-Based Adhesives: New Theoretical Aspects. *Int. J. Adhes. Adhes.* **1980**, *1*, 13–16. [CrossRef]
217. Tenorio-Alfonso, A.; Sánchez, M.C.; Franco, J.M. A Review of the Sustainable Approaches in the Production of Bio-Based Polyurethanes and Their Applications in the Adhesive Field. *J. Polym. Environ.* **2020**, *28*, 749–774. [CrossRef]
218. Panja, S.; Siehr, A.; Sahoo, A.; Siegel, R.A.; Shen, W. Biodegradable Elastomers Enabling Thermoprocessing Below 100 °C. *Biomacromolecules* **2022**, *23*, 163–173. [CrossRef] [PubMed]
219. Kitphaitun, S.; Chaimongkolkunasin, S.; Manit, J.; Makino, R.; Kadota, J.; Hirano, H.; Nomura, K. Ethylene/Myrcene Copolymers as New Bio-Based Elastomers Prepared by Coordination Polymerization Using Titanium Catalysts. *Macromolecules* **2021**, *54*, 10049–10058. [CrossRef]
220. Tondi, G.; Zhao, W.; Pizzi, A.; Du, G.; Fierro, V.; Celzard, A. Tannin-Based Rigid Foams: A Survey of Chemical and Physical Properties. *Bioresour. Technol.* **2009**, *100*, 5162–5169. [CrossRef] [PubMed]
221. Sjöström, E. *Wood Chemistry: Fundamentals and Applications*, 2nd ed.; Academic Press: San Diego, CA, USA, 1993; ISBN 978-0-12-647481-7.
222. Pantoja-Castro, M.A.; González-Rodríguez, H. Study by Infrared Spectroscopy and Thermogravimetric Analysis of Tannins and Tannic Acid. *Rev. Latinoam. Quím.* **2011**, *39*, 107–112.
223. Khanbabaee, K.; van Ree, T. Tannins: Classification and Definition. *Nat. Prod. Rep.* **2001**, *18*, 641–649. [CrossRef] [PubMed]
224. Merle, J.; Birot, M.; Deleuze, H.; Mitterer, C.; Carré, H.; Bouhtoury, F.C.-E. New Biobased Foams from Wood Byproducts. *Mater. Des.* **2016**, *91*, 186–192. [CrossRef]
225. Chen, X.; Li, J.; Xi, X.; Pizzi, A.; Zhou, X.; Fredon, E.; Du, G.; Gérardin, C. Condensed Tannin-Glucose-Based NIPU Bio-Foams of Improved Fire Retardancy. *Polym. Degrad. Stab.* **2020**, *175*, 109121. [CrossRef]
226. Čop, M.; Laborie, M.P.; Pizzi, A.; Sernek, M. Curing Characterisation of Spruce Tannin-Based Foams Using the Advanced Isoconversional Method. *BioResources* **2014**, *9*, 4643–4655. [CrossRef]
227. Lacoste, C.; Čop, M.; Kemppainen, K.; Giovando, S.; Pizzi, A.; Laborie, M.-P.; Sernek, M.; Celzard, A. Biobased Foams from Condensed Tannin Extracts from Norway Spruce (*Picea Abies*) Bark. *Ind. Crops Prod.* **2015**, *73*, 144–153. [CrossRef]
228. Lacoste, C.; Pizzi, A.; Basso, M.-C.; Laborie, M.-P.; Celzard, A. Pinus Pinaster Tannin/Furanic Foams: PART I. Formulation. *Ind. Crops Prod.* **2014**, *52*, 450–456. [CrossRef]
229. Li, J.; Zhang, A.; Zhang, S.; Gao, Q.; Zhang, W.; Li, J. Larch Tannin-Based Rigid Phenolic Foam with High Compressive Strength, Low Friability, and Low Thermal Conductivity Reinforced by Cork Powder. *Compos. Part B Eng.* **2019**, *156*, 368–377. [CrossRef]

230. Sarika, P.R.; Nancarrow, P.; Khansaheb, A.; Ibrahim, T. Bio-Based Alternatives to Phenol and Formaldehyde for the Production of Resins. *Polymers* **2020**, *12*, 2237. [CrossRef] [PubMed]
231. Arbenz, A.; Avérous, L. Chemical Modification of Tannins to Elaborate Aromatic Biobased Macromolecular Architectures. *Green Chem.* **2015**, *17*, 2626–2646. [CrossRef]
232. Porter, L.J. Flavans and Proanthocyanidins. In *The Flavonoids*; Harborne, J.B., Ed.; Springer: Boston, MA, USA, 1988; pp. 21–62. ISBN 978-0-412-28770-1.
233. Masson, E.; Merlin, A.; Pizzi, A. Comparative Kinetics of Induced Radical Autocondensation of Polyflavonoid Tannins. I. Modified and Nonmodified Tannins. *J. Appl. Polym. Sci.* **1996**, *60*, 263–269. [CrossRef]
234. Merlin, A.; Pizzi, A. An ESR Study of the Silica-Induced Autocondensation of Polyflavonoid Tannins. *J. Appl. Polym. Sci.* **1996**, *59*, 945–952. [CrossRef]
235. Pichelin, F.; Kamoun, C.; Pizzi, A. Hexamine Hardener Behaviour: Effects on Wood Glueing, Tannin and Other Wood Adhesives. *Holz Als Roh Werkst.* **1999**, *57*, 305–317. [CrossRef]
236. Mariscal, R.; Maireles-Torres, P.; Ojeda, M.; Sádaba, I.; López Granados, M. Furfural: A Renewable and Versatile Platform Molecule for the Synthesis of Chemicals and Fuels. *Energy Environ. Sci.* **2016**, *9*, 1144–1189. [CrossRef]
237. Iroegbu, A.O.; Hlangothi, S.P. Furfuryl Alcohol a Versatile, Eco-Sustainable Compound in Perspective. *Chem. Afr.* **2019**, *2*, 223–229. [CrossRef]
238. Foo, L.Y.; Hemingway, R.W. Condensed Tannins: Reactions of Model Compounds with Furfuryl Alcohol and Furfuraldehyde. *J. Wood Chem. Technol.* **1985**, *5*, 135–158. [CrossRef]
239. Choura, M.; Belgacem, N.M.; Gandini, A. Acid-Catalyzed Polycondensation of Furfuryl Alcohol: Mechanisms of Chromophore Formation and Cross-Linking. *Macromolecules* **1996**, *29*, 3839–3850. [CrossRef]
240. Guigo, N.; Mija, A.; Vincent, L.; Sbirrazzuoli, N. Chemorheological Analysis and Model-Free Kinetics of Acid Catalysed Furfuryl Alcohol Polymerization. *Phys. Chem. Chem. Phys.* **2007**, *9*, 5359. [CrossRef] [PubMed]
241. Sánchez-Martín, J.; Beltrán-Heredia, J.; Carmona-Murillo, C. Adsorbents from Schinopsis Balansae: Optimisation of Significant Variables. *Ind. Crops Prod.* **2011**, *33*, 409–417. [CrossRef]
242. National Technical Reports Library. *Report on Carcinogens*, 12th ed.; Department of Health and Human Services, Public Health Service, National Toxicology Program: Durham, NC, USA, 2011; p. 507.
243. Rossouw, D. du T.; Pizzi, A.; McGillivray, G. The Kinetics of Condensation of Phenolic Polyflavonoid Tannins with Aldehydes. *J. Polym. Sci. Polym. Chem. Ed.* **1980**, *18*, 3323–3343. [CrossRef]
244. Basso, M.C.; Lagel, M.-C.; Pizzi, A.; Celzard, A.; Abdalla, S. First Tools for Tannin-Furanic Foams Design. *BioResources* **2015**, *10*, 5233–5241. [CrossRef]
245. Basso, M.C.; Giovando, S.; Pizzi, A.; Lagel, M.C.; Celzard, A. Alkaline Tannin Rigid Foams. *J. Renew. Mater.* **2014**, *2*, 182–185. [CrossRef]
246. Li, X.; Pizzi, A.; Cangemi, M.; Fierro, V.; Celzard, A. Flexible Natural Tannin-Based and Protein-Based Biosourced Foams. *Ind. Crops Prod.* **2012**, *37*, 389–393. [CrossRef]
247. Delgado-Sánchez, C.; Fierro, V.; Li, S.; Pasc, A.; Pizzi, A.; Celzard, A. Stability Analysis of Tannin-Based Foams Using Multiple Light-Scattering Measurements. *Eur. Polym. J.* **2017**, *87*, 318–330. [CrossRef]
248. Maia, L.S.; Zanini, N.C.; de Souza, A.G.; Barbosa, R.F.S.; Rosa, D.S.; da Barud, H.S.; Mulinari, D.R. Effective Oil Spill Cleaned up with Environmentally Friendly Foams Filled with Eucalyptus Charcoal Residue. *Iran. Polym. J.* **2021**, *31*, 383–398. [CrossRef]
249. Uram, K.; Kurańska, M.; Andrzejewski, J.; Prociak, A. Rigid Polyurethane Foams Modified with Biochar. *Materials* **2021**, *14*, 5616. [CrossRef]
250. Haridevan, H.; Evans, D.A.C.; Ragauskas, A.J.; Martin, D.J.; Annamalai, P.K. Valorisation of Technical Lignin in Rigid Polyurethane Foam: A Critical Evaluation on Trends, Guidelines and Future Perspectives. *Green Chem.* **2021**, *23*, 8725–8753. [CrossRef]
251. Basso, M.C.; Pizzi, A.; Lacoste, C.; Delmotte, L.; Al-Marzouki, F.; Abdalla, S.; Celzard, A. MALDI-TOF and 13C NMR Analysis of Tannin–Furanic–Polyurethane Foams Adapted for Industrial Continuous Lines Application. *Polymers* **2014**, *6*, 2985–3004. [CrossRef]
252. Cefarin, N.; Bedolla, D.E.; Surowka, A.; Donato, S.; Sepperer, T.; Tondi, G.; Dreossi, D.; Sodini, N.; Birarda, G.; Vaccari, L. Study of the Spatio-Chemical Heterogeneity of Tannin-Furanic Foams: From 1D FTIR Spectroscopy to 3D FTIR Micro-Computed Tomography. *Int. J. Mol. Sci.* **2021**, *22*, 12869. [CrossRef]
253. Tondi, G.; Pizzi, A. Tannin-Based Rigid Foams: Characterization and Modification. *Ind. Crops Prod.* **2009**, *29*, 356–363. [CrossRef]
254. Zhao, W.; Pizzi, A.; Fierro, V.; Du, G.; Celzard, A. Effect of Composition and Processing Parameters on the Characteristics of Tannin-Based Rigid Foams. Part I: Cell Structure. *Mater. Chem. Phys.* **2010**, *122*, 175–182. [CrossRef]
255. Hawkins, M.C.; O'Toole, B.; Jackovich, D. Cell Morphology and Mechanical Properties of Rigid Polyurethane Foam. *J. Cell. Plast.* **2005**, *41*, 267–285. [CrossRef]
256. Jackovich, D.; O'Toole, B.; Hawkins, M.C.; Sapochak, L. Temperature and Mold Size Effects on Physical and Mechanical Properties of a Polyurethane Foam. *J. Cell. Plast.* **2005**, *41*, 153–168. [CrossRef]
257. Celzard, A.; Fierro, V.; Pizzi, A. Nouveaux matériaux poreux multifonctionnems dérivés du bois. In *Techniques de l'Ingénieur*; Matériaux-Bois, Verre, Céramique et Textile: Saint-Denis, France, 2012; p. 32. ISBN N4203 v1.
258. De Yuso, A.M.; Lagel, M.C.; Pizzi, A.; Fierro, V.; Celzard, A. Structure and Properties of Rigid Foams Derived from Quebracho Tannin. *Mater. Des.* **2014**, *63*, 208–212. [CrossRef]

259. Lacoste, C.; Pizzi, A.; Laborie, M.-P.; Celzard, A. Pinus Pinaster Tannin/Furanic Foams: Part II. Physical Properties. *Ind. Crops Prod.* **2014**, *61*, 531–536. [CrossRef]
260. Celzard, A.; Fierro, V.; Amaral-Labat, G.; Pizzi, A.; Torero, J. Flammability Assessment of Tannin-Based Cellular Materials. *Polym. Degrad. Stab.* **2011**, *96*, 477–482. [CrossRef]
261. Delgado-Sánchez, C.; Sarazin, J.; Santiago-Medina, F.J.; Fierro, V.; Pizzi, A.; Bourbigot, S.; Celzard, A. Impact of the Formulation of Biosourced Phenolic Foams on Their Fire Properties. *Polym. Degrad. Stab.* **2018**, *153*, 1–14. [CrossRef]
262. Lacoste, C.; Basso, M.-C.; Pizzi, A.; Celzard, A.; Ella Ebang, E.; Gallon, N.; Charrier, B. Pine (*P. Pinaster*) and Quebracho (*S. Lorentzii*) Tannin-Based Foams as Green Acoustic Absorbers. *Ind. Crops Prod.* **2015**, *67*, 70–73. [CrossRef]
263. Sabathi, G.; Reyer, A.; Cefarin, N.; Sepperer, T.; Eckardt, J.; Neubauer, J.; Wendisch, F.J.; D'Amico, F.; Vaccari, L.; Tondi, G.; et al. Tannin-Furanic Foams Used as Biomaterial Substrates for SERS Sensing in Possible Wastewater Filter Applications. *Mater. Res. Express* **2021**, *8*, 115404. [CrossRef]
264. Zhou, X.; Pizzi, A.; Sauget, A.; Nicollin, A.; Li, X.; Celzard, A.; Rode, K.; Pasch, H. Lightweight Tannin Foam/Composites Sandwich Panels and the Coldset Tannin Adhesive to Assemble Them. *Ind. Crops Prod.* **2013**, *43*, 255–260. [CrossRef]
265. Sánchez-Martín, J.; Beltrán-Heredia, J.; Delgado-Regaña, A.; Rodríguez-González, M.A.; Rubio-Alonso, F. Optimization of Tannin Rigid Foam as Adsorbents for Wastewater Treatment. *Ind. Crops Prod.* **2013**, *49*, 507–514. [CrossRef]
266. Sánchez-Martín, J.; Beltrán-Heredia, J.; Delgado-Regaña, A.; Rodríguez-González, M.A.; Rubio-Alonso, F. Adsorbent Tannin Foams: New and Complementary Applications in Wastewater Treatment. *Chem. Eng. J.* **2013**, *228*, 575–582. [CrossRef]
267. Hao, B.; Wang, F.; Huang, H.-T.; Wu, Y.; Jia, S.; Liao, Y.; Mao, H. Tannin Foam Immobilized with Ferric Ions for Efficient Removal of Ciprofloxacin at Low Concentrations. *J. Hazard. Mater.* **2021**, *414*, 125567. [CrossRef]
268. Tondi, G.; Oo, C.W.; Pizzi, A.; Trosa, A.; Thevenon, M.F. Metal Adsorption of Tannin Based Rigid Foams. *Ind. Crops Prod.* **2009**, *29*, 336–340. [CrossRef]
269. Tondi, G.; Fierro, V.; Pizzi, A.; Celzard, A. Tannin-Based Carbon Foams. *Carbon* **2009**, *47*, 1480–1492. [CrossRef]
270. Varila, T.; Romar, H.; Luukkonen, T.; Lassi, U.; Varila, T.; Romar, H.; Luukkonen, T.; Lassi, U. Physical Activation and Characterization of Tannin-Based Foams Enforced with Boric Acid and Zinc Chloride. *AIMS Mater. Sci.* **2019**, *6*, 301–314. [CrossRef]
271. Zieleniewska, M.; Leszczyński, M.K.; Szczepkowski, L.; Bryśkiewicz, A.; Krzyżowska, M.; Bień, K.; Ryszkowska, J. Development and Applicational Evaluation of the Rigid Polyurethane Foam Composites with Egg Shell Waste. *Polym. Degrad. Stab.* **2016**, *132*, 78–86. [CrossRef]
272. Oushabi, A.; Sair, S.; Abboud, Y.; Tanane, O.; Bouari, A.E. An Experimental Investigation on Morphological, Mechanical and Thermal Properties of Date Palm Particles Reinforced Polyurethane Composites as New Ecological Insulating Materials in Building. *Case Stud. Constr. Mater.* **2017**, *7*, 128137. [CrossRef]
273. Bryśkiewicz, A.; Zieleniewska, M.; Przyjemska, K.; Chojnacki, P.; Ryszkowska, J. Modification of Flexible Polyurethane Foams by the Addition of Natural Origin Fillers. *Polym. Degrad. Stab.* **2016**, *132*, 32–40. [CrossRef]
274. Antunes, M.; Cano, Á.; Haurie, L.; Velasco, J.I. Esparto Wool as Reinforcement in Hybrid Polyurethane Composite Foams. *Ind. Crops Prod.* **2011**, *34*, 1641–1648. [CrossRef]
275. Yan, D.-X.; Dai, K.; Xiang, Z.-D.; Li, Z.-M.; Ji, X.; Zhang, W.-Q. Electrical Conductivity and Major Mechanical and Thermal Properties of Carbon Nanotube-Filled Polyurethane Foams. *J. Appl. Polym. Sci.* **2011**, *120*, 3014–3019. [CrossRef]
276. Dolomanova, V.; Rauhe, J.; Chr, M.; Jensen, L.R.; Pyrz, R.; Timmons, A.B. Mechanical Properties and Morphology of Nano-Reinforced Rigid PU Foam. *J. Cell. Plast.* **2011**, *47*, 81–93. [CrossRef]
277. Konig, A.; Fehrenbacher, U.; Hirth, T.; Kroke, E. Flexible Polyurethane Foam with the Flame-Retardant Melamine. *J. Cell. Plast.* **2008**, *44*, 469–480. [CrossRef]
278. Wong, E.H.H.; Fan, K.W.; Lei, L.; Wang, C.; Baena, J.C.; Okoye, H.; Fam, W.; Zhou, D.; Oliver, S.; Khalid, A.; et al. Fire-Resistant Flexible Polyurethane Foams via Nature-Inspired Chitosan-Expandable Graphite Coatings. *ACS Appl. Polym. Mater.* **2021**, *3*, 4079–4087. [CrossRef]
279. Furtwengler, P.; Avérous, L. Renewable Polyols for Advanced Polyurethane Foams from Diverse Biomass Resources. *Polym. Chem.* **2018**, *9*, 4258–4287. [CrossRef]
280. Priester, R.D.; Peffley, R.D.; Turner, R.B.; Herrington, R.M. High Resiliency Polyurea Foam-An Improved Flexible Foam Matrix. *J. Cell. Plast.* **1994**, *30*, 144–163. [CrossRef]
281. Ramirez, B.J.; Gupta, V. High Tear Strength Polyurea Foams with Low Compression Set and Shrinkage Properties at Elevated Temperatures. *Int. J. Mech. Sci.* **2019**, *150*, 29–34. [CrossRef]
282. Reed, N.; Huynh, N.U.; Rosenow, B.; Manlulu, K.; Youssef, G. Synthesis and Characterization of Elastomeric Polyurea Foam. *J. Appl. Polym. Sci.* **2020**, *137*, 48839. [CrossRef]
283. Lee, J.; Gould, G.; Rhine, W. Polyurea Based Aerogel for a High Performance Thermal Insulation Material. *J. Sol-Gel Sci. Technol.* **2009**, *49*, 209–220. [CrossRef]
284. Liszkowska, J.; Czupryński, B.; Paciorek-Sadowska, J.; Michałowski, S. Thermal and Flammable Properties of Rigid PUR–PIR Foams Obtained by Using New Compounds Based on 2-Hydroxypropane-1,2,3-Tricarboxylic Acid. *J. Cell. Plast.* **2016**, *52*, 321–341. [CrossRef]
285. Dominguez-Rosado, E.; Liggat, J.J.; Snape, C.E.; Eling, B.; Pichtel, J. Thermal Degradation of Urethane Modified Polyisocyanurate Foams Based on Aliphatic and Aromatic Polyester Polyol. *Polym. Degrad. Stab.* **2002**, *78*, 1–5. [CrossRef]

286. Levchik, S.V.; Weil, E.D. Recent Progress in Flame Retardancy of Polyurethane and Polyisocyanurate Foams. In *Fire and Polymers IV*; ACS Symposium Series; American Chemical Society: Washington, DC, USA, 2005; Volume 922, pp. 280–290. ISBN 978-0-8412-3948-7.
287. Ashida, K.; Saiki, K.; Goto, J.; Sasaki, K. Polyisocyanurate Foams Modified by Thermally Stable Linkages. In *Polymeric Foams*; ACS Symposium Series; American Chemical Society: Washington, DC, USA, 1997; Volume 669, pp. 81–100. ISBN 978-0-8412-3516-8.
288. Joanna, L.; Bogusław, C.; Joanna, P.-S. The Effect of Hydroxyalkyls, Derivatives of 2-Hydroxypropane-1.2.3-Tricarboxylic Acid, on Flammability and Thermal Properties of PUR–PIR Foams. *Polym. Bull.* **2018**, *75*, 3801–3823. [CrossRef]
289. Liszkowska, J. The Use of Citric Acid in the Production of Polyols for Rigid PUR-PIR Foams. *Polym. Bull.* **2017**, *74*, 283–305. [CrossRef]
290. Mougel, C.; Garnier, T.; Cassagnau, P.; Sintes-Zydowicz, N. Phenolic Foams: A Review of Mechanical Properties, Fire Resistance and New Trends in Phenol Substitution. *Polymer* **2019**, *164*, 86–117. [CrossRef]
291. Tseng, C.; Kuo, K. Thermal Properties of Phenolic Foam Insulation. *J. Chin. Inst. Eng.* **2002**, *25*, 753–758. [CrossRef]
292. Song, F.; Li, Z.; Jia, P.; Bo, C.; Zhang, M.; Hu, L.; Zhou, Y. Phosphorus-Containing Tung Oil-Based Siloxane Toughened Phenolic Foam with Good Mechanical Properties, Fire Performance and Low Thermal Conductivity. *Mater. Des.* **2020**, *192*, 108668. [CrossRef]
293. Sánchez, C.D. New Optimisation and Characterisation Methods of Tannin-Based Foams for Thermal Insulation of Buildings. Ph.D. Thesis, Université de Lorraine, Nancy, France, 2017.

polymers

MDPI

Editorial

Development of Bio-Based Materials: Synthesis, Characterization and Applications

Clara Delgado-Sánchez, Adrián Tenorio-Alfonso, Esperanza Cortés-Triviño, Antonio M. Borrero-López * and Concepción Valencia

Pro2TecS-Chemical Product and Process Technology Research Centre, Universidad de Huelva, 21071 Huelva, Spain

* Correspondence: am.borrero@diq.uhu.es

Citation: Delgado-Sánchez, C.; Tenorio-Alfonso, A.; Cortés-Triviño, E.; Borrero-López, A.M.; Valencia, C. Development of Bio-Based Materials: Synthesis, Characterization and Applications. *Polymers* **2022**, *14*, 3599. https://doi.org/10.3390/polym14173599

Received: 27 August 2022
Accepted: 29 August 2022
Published: 31 August 2022

The need to find suitable biomaterials and procedures from alternative products able to imitate or even enhance the performance of currently used products has become an important focus of research today due to the depletion of non-renewable resources and the increasing concern related to climate change, sustainability and environmental preservation. Thus, this book gathers different original articles and review manuscripts concerning the Special Issue **"Development of Bio-Based Materials: Synthesis, Characterization and Applications"**. The development of partial or fully bio-based materials has been included, with excellent outcomes in many different applications, as well as alternative procedures that can reduce the carbon footprint or optimize both production and energy consumption.

Given the interest in these materials, this book, including 27 articles and reviews written by research experts in their topics of interest, reports the most recent research on bio-based materials, with emphasis on the pharmaceutical and medical fields but covering a very extensive range of applications. Several novel and fascinating methods and studies related to thermo-responsive photopolymers (Figure 1A), biopolymer encapsulations, biopolymer-based films (Figure 1B), natural fibres production, biocompatible adhesives (Figure 1C), green composites (Figure 1D) and many other different materials have been introduced. The aim of this Special Issue, and now this book, is to provide a clear picture of the latest frontiers reached in the biomaterials field and their latest applications such as 3D bioprinting inks, immobilization of enzymes (Figure 1E), lubricating greases, elastomers and adhesives (Figure 1F), in which bio-based materials show great potential.

Emphasis on the biomaterial itself, the protocol followed, characterization and/or application have been reported, therefore contributing to a greener picture through the formulation of different more environmentally friendly products, but also to an understanding of the composition and structure of those systems, as well as applications thereof.

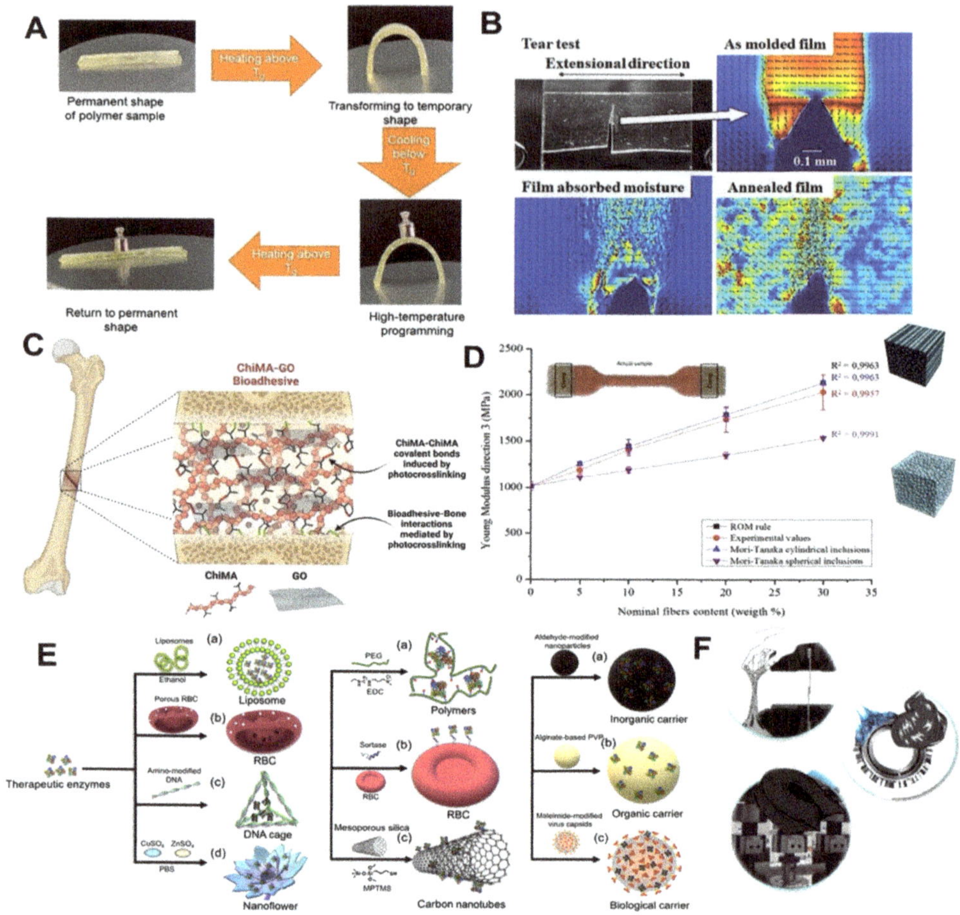

Figure 1. (**A**) Scheme of shape-memory characteristics of the bio-based polymers developed by Jaras et al. [1]. (**B**) Images of the in situ retardation measurements at the tip of the tear for the polylactic acid-based film developed in Yutaka et al. [2]. (**C**) Schematic overview of the bioadhesive formulation and application in Céspedes-Valenzuela et al. [3]. (**D**) Overview of the tensile properties of the green composites developed by Lemaire et al. [4]. (**E**) Different therapeutic enzymes studied in Zhu et al. [5]. (**F**) Adhesives, lubricating greases and elastomers as biomaterials studied in Borrero-López et al. [6].

Funding: This research received no external funding.

Conflicts of Interest: The authors declare no conflict of interest.

Short Biography of Authors

Clara Delgado-Sánchez received her MSc in Chemical Engineering in 2013 and her Master in Product Formulation and Technology in 2014, both at the University of Huelva. Afterwards, she received her PhD degree from the University of Lorraine (France) in 2017. During her thesis, she focused her research on the development of "New methods of optimization and characterization of tannin foams for thermal insulation of buildings" in the research group Biosourced Materials of the Institut Jean Lamour (IJL—UMR CNRS 7198). After her PhD, she has been a postdoc at the Chemical Process and Product Research Centre (Pro2Tec) at the University of Huelva, working on the project "Study of thermo-advanced dispersions for heat transport applications", a multidisciplinary project that led not only to her participation in the research and working groups of five other projects related to the same topic on a national and regional competitive basis and several publications but also to her obtaining a postdoctoral fellowship that she currently enjoys. Currently, her line of research focuses on novel phase-change materials to obtain thermal-energy-storage systems as alternatives for efficient energy use and conservation. Moreover, in November 2020, she was awarded her first project as the principal researcher: "Valorisation of phosphogypsums in the development of foamed bitumen", financed by the "Cátedra de la Provincia" (University of Huelva). The results of her research effort carried out between 2015 and 2022 were recorded in the publication of 22 articles in international journals and several contributions to national and international congresses.

Adrián Tenorio-Alfonso. After graduating from Chemical Engineering in 2013 and upon the completion of Master's in *Product Formulation and Design with Applications in the Chemical, Food and Pharmaceutical Industry* in 2014, both at the University of Huelva, he started his research activity with a final research project on the *Thermo-rheological characterization of foams and adhesives based on polyurethanes and biopolyurethanes.* Afterwards, he was given a national grant for a PhD contract *"Ayuda a la Formación de Profesorado Universitario (FPU13/01114)"* by el Ministerio de Educación, Cultura y Deporte, carrying out his PhD thesis entitled *Development of polyurethane formulations based on cellulose acetate and castor oil,* affiliated with the Doctoral Programme in Industrial and Environmental Science and Technology at the Chemical Process and Product Technology Research Centre (Pro^{2}TecS) from the University of Huelva. He has been a member of several national and regional research projects (MINECO CTQ2014-56038-C3-1-R and TIC 1499 funded by El Ministerio de Economía y Competitividad and La Conserjería de Innovación, Ciencia y Empresa de la Junta de Andalucía, respectively). Additionally, he has taken part in a research project for the development of synthetic ice rinks in collaboration with the Xtraice Rinks company. More recently, he has participated in a research project to study thermorheologically advanced suspensions for heat transport applications (CTQ2017-89792-R). In relation to this, he has been awarded a postdoctoral fellowship in the development of novel phase change materials, currently working on the research project *"Development of phase change oil-in-oil emulsions with enhanced rheological, heat storage and heat transfer properties"* at the Pro^{2}TecS research centre. As a result of his research experience, he has published nine research articles in international journals with high impact factors, accounting for more than 25 contributions to national and international congresses.

Esperanza Cortés-Triviño graduated from Industrial Technical engineering with a specialty in Industrial Chemistry in 2012 and became licensed in Chemical Engineering in 2014 at the University of Huelva. She completed her official Master's in Chemical Engineering in 2016, where she started her research career working with bioplastics based on wheat gluten. Meanwhile, she was hired by the University of Huelva as a researcher to conduct several tasks within the MINECO CTQ2014-56038-C3-1-R project funded by "Ministerio de Economía y Competitividad", also obtaining a grant from "Junta de Andalucía" and accomplishing a Doctoral Thesis in the Industrial and Environmental Science and Technology program (CyTIA), associated with the project TEP-1499, which is related to the development of new thickening agents of vegetable oil from different lignocellulosic fractions chemically modified. She obtained her PhD degree in 2019 with a Cum Laude Mention, unanimously, which was complemented by a stay at the University of Applied Sciences of Hamburg to obtain International Mention. Her Doctoral Thesis was awarded the best 2019 Doctoral Thesis in the branch of Engineering of the University of Huelva and received an honourable mention at the Iberian Meeting on Rheology 2019 (IBEREO 2019). Her research career has been completed by attending multiple conferences and by collaborating in some of their management activities. She also participated in other transfer activities such as "Café con Ciencia" and the European Researcher's Night and in more than 500 h of specialized training. She is a member of The Complex Fluid Engineering Research Group (TEP-185) and The Research Center of Chemical Product and Process Technology (Pro^{2}TecS), as well as The Spanish Rheology Group (GER) belonging to the Spanish Royal Society of Chemistry (RSEQ). Currently, she is working as a postdoctoral researcher at the University of Huelva within an international project for the Procter & Gamble Company. As a result of her research experience, she has published nine research articles in international journals with high impact factors, accounting with more than 25 contributions to national and international congresses.

Antonio María Borrero-López completed a degree in Chemical Engineering at the University of Huelva (Spain) in 2013 and later obtained a Master's degree in Product Formulation and Design from the International University of Andalusia (Spain). Afterwards, he received his PhD degree from the University of Huelva, with a thesis entitled *"Development of new lignocellulosic-based thickening agents for biodegradable oleogel formulations with several industrial applications"*. This thesis has been awarded the Best 2021–2022 Doctoral Thesis in the branch of Engineering and Architecture of the University of Huelva and received an honourable mention at the Annual European Rheology Conference 2022 (AERC 2022). After his PhD, he has been working as a postdoc at the Institut Jean Lamour, University of Lorraine (France), where new greener approaches for the performance of bio-based materials are being targeted. He has participated in a total of eight research projects, including the European Project UCGWATERplus, which aims to remediate waters polluted with organic and inorganic contaminants by the formulation of different products via the valorisation of the residues from underground coal gasification and other processes. As a result of his research, he has published a total of 22 articles and contributed to more than 30 international and national congresses.

Concepción Valencia Barragán received her PhD degree from the University of Extremadura. She is a Full Professor at the University of Huelva. She has carried out extensive research activity for more than 23 years, mainly focused on Chemical Product Engineering, specifically concerning the rheology of complex materials, the processing of non-Newtonian fluids, lubricants, adhesives and coatings, polymers and biopolymers, food colloids and emulsion technologies. In total, she has participated in more than 30 research projects with public funding, obtained in competitive calls, with some of them as a lead researcher and more than 40 contracts with private sector companies. Her main scientific and technical achievements have been focused on the modification of the rheological properties of lubricating greases by means of reactive and recycled polymeric additives as well as on the development of biodegradable oleogels capable of replacing traditional lubricating greases formulated from non-renewable resources. As a result of this research activity, she is the author of more than 118 papers in peer-reviewed journals and the inventor of six patents. She has also presented more than 130 communications at national and international scientific conferences.

References

1. Jaras, J.; Navaruckiene, A.; Skliutas, E.; Jersovaite, J.; Malinauskas, M.; Ostrauskaite, J. Thermo-Responsive Shape Memory Vanillin-Based Photopolymers for Microtransfer Molding. *Polymers* **2022**, *14*, 2460. [CrossRef] [PubMed]
2. Kobayashi, Y.; Ishigami, A.; Ito, H. Relating Amorphous Structure to the Tear Strength of Polylactic Acid Films. *Polymers* **2022**, *14*, 1965. [CrossRef] [PubMed]
3. Céspedes-Valenzuela, D.N.; Sánchez-Rentería, S.; Cifuentes, J.; Gantiva-Diaz, M.; Serna, J.A.; Reyes, L.H.; Ostos, C.; Cifuentes-De la Portilla, C.; Muñoz-Camargo, C.; Cruz, J.C. Preparation and Characterization of an Injectable and Photo-Responsive Chitosan Methacrylate/Graphene Oxide Hydrogel: Potential Applications in Bone Tissue Adhesion and Repair. *Polymers* **2022**, *14*, 126. [CrossRef] [PubMed]
4. Lemaire, T.; Rodi, E.G.; Langlois, V.; Renard, E.; Sansalone, V. Study of Mechanical Properties of PHBHV/Miscanthus Green Composites Using Combined Experimental and Micromechanical Approaches. *Polymers* **2021**, *13*, 2650. [CrossRef] [PubMed]
5. Zhu, C.-Y.; Li, F.-L.; Zhang, Y.-W.; Gupta, R.K.; Patel, S.K.S.; Lee, J.-K. Recent Strategies for the Immobilization of Therapeutic Enzymes. *Polymers* **2022**, *14*, 1409. [CrossRef] [PubMed]
6. Borrero-López, A.M.; Valencia, C.; Franco, J.M. Lignocellulosic Materials for the Production of Biofuels, Biochemicals and Biomaterials and Applications of Lignocellulose-Based Polyurethanes: A Review. *Polymers* **2022**, *14*, 881. [CrossRef] [PubMed]

polymers MDPI

Review

Potential Medical Applications of Chitooligosaccharides

Sukumaran Anil [1,2]

[1] Oral Health Institute, Department of Dentistry, Hamad Medical Corporation, Qatar University, Doha 3050, Qatar; asukumaran1@hamad.qa; Tel.: +974-50406670
[2] Pushpagiri Research Centre, Pushpagiri Institute of Medical Sciences and Research Centre (PIMS&RC), Thiruvalla, Pathanamthitta 689101, Kerala, India

Abstract: Chitooligosaccharides, also known as chitosan oligomers or chitooligomers, are made up of chitosan with a degree of polymerization (DP) that is less than 20 and an average molecular weight (MW) that is lower than 3.9 kDa. COS can be produced through enzymatic conversions using chitinases, physical and chemical applications, or a combination of these strategies. COS is of significant interest for pharmacological and medical applications due to its increased water solubility and non-toxicity, with a wide range of bioactivities, including antibacterial, anti-inflammatory, anti-obesity, neuroprotective, anticancer, and antioxidant effects. This review aims to outline the recent advances and potential applications of COS in various diseases and conditions based on the available literature, mainly from preclinical research. The prospects of further in vivo studies and translational research on COS in the medical field are highlighted.

Keywords: chitooligosaccharides; chitin; chitosan; anti-inflammatory; antioxidant; tissue engineering; wound healing; antimicrobial; drug delivery; antitumor

Citation: Anil, S. Potential Medical Applications of Chitooligosaccharides. *Polymers* **2022**, *14*, 3558. https://doi.org/10.3390/polym14173558

Academic Editor: Marek M. Kowalczuk

Received: 24 June 2022
Accepted: 23 August 2022
Published: 29 August 2022

1. Introduction

Chitin, a mucopolysaccharide, is produced by many living organisms and is usually present in a complex with other polysaccharides and proteins in insects, crustaceans, arachnids, myriapods, nematodes, algae, and fungi [1]. Chitin is a linear polysaccharide composed of (1 $\rightarrow$ 4) linked 2-acetamido-2-deoxy-β-d-glucopyranosyl units and occurs naturally in three polymorphic forms with different orientations of the microfibrils, known as α-, β-, and γ-chitin [2]. Chitin has been the focus of numerous therapeutic uses in addition to serving as a precursor for producing chitosan and chitooligosaccharides. Chitin has several uses in food, agriculture, wastewater treatment, textiles, microbiology, nanotechnology, chemistry, and material science [3]. It is biodegradable and is also considered as a promising biomaterial for tissue engineering and stem cell technologies [4].

Chitosan [poly-(β-1/4)-2-amino-2-deoxy-D-glucopyranose] is a natural nontoxic linear polysaccharide biopolymer produced through the deacetylation of chitin [5]. Commercial chitosan is made by deacetylating naturally occurring chitin and is used in dietary supplements, organic fertilizers, and cosmetics [6,7]. Chitin and chitosan can be distinguished based on the degree of acetylation of the D-glucosamine units. Chitin includes over 70% acetylated units, whereas chitosan contains less than 30% acetylation. In the presence of organic acids, including formic acid, acetic acid, and ascorbic acid, chitosan forms salt and becomes water soluble [8]. Chitosan possesses three reactive functional groups, including an amino- or N-acetamide group and two primary and secondary hydroxyl groups at the C-2, C-3, and C-6 positions. The amino- or N-acetamide groups distinguish the structure and physicochemical properties of various chitosans [9]. The fraction of N-acetylated residues (FA), degree of polymerization (DP), molecular weight (MW), MW distribution, and pattern or sequence of N-acetylation (PA) can be used to classify chitosan [10].

Chitosans with a DP of less than 20 and an average MW of less than 3.9 kDa are called chitooligosaccharides (COSs), chitosan oligomers, or chitooligomers [11,12]. Their characteristics, such as low molecular weight, low polymerization degree, and high water solubility, are superior to those of chitin and chitosan [11]. COS has a variety of biological activities and many potential uses in multiple fields, such as medicine, cosmetics, food, and agriculture [13]. In addition, chemical methods using acid, hydrogen peroxide (H_2O_2), or sodium nitrite ($NaNO_2$) are also used to extract COS. There are multiple ways to obtain chitosan oligomers. These methods are categorized as enzymatic, physical, or chemical depolymerizations [14]. Chemical methods using acid [15,16], H2O2 [17], or NaNO2 [18] physical methods, such as hydrothermal [19], microwave [20], ultra-sonication [21], and gamma rays [22]. Among the chemical methods for the hydrolysis of chitosan, acid hydrolysis is probably the best known. The enzymatic depolymerization of chitosan is characterized by the enzymes' selective cleavage of chitosan glycosidic bonds [23].

While hetero-chitooligosaccharides combine oligomers with varying DP, deacetylation degree (DD), and acetylation patterns (position of the N-acetyl residues in the chain), homo-chitooligosaccharides are oligomers made exclusively of either glucosamine (GlcN) or N-Acetylglucosamine (GlcNAc) units [3,24]. The attractive bioactive properties of COS make them suitable for various biological applications. A low molecular weight (1.5 kDa) results in aqueous solubility across a broad pH range and easy absorption through epithelial cells [10]. The physical and biological activities of chitosan and its oligomers are governed primarily by the DP, MW, and DD [3,25,26]. The most significant property of COS is its water solubility or solubility in physiological pH due to the freely accessible amino groups on the shorter chains [27]. All heterogeneous COS with a DP < 10 and DD between 50 and 100 % are completely soluble over the pH range extending from neutral to slightly alkaline, whereas commercially available chitosan with a higher DP precipitates out at such pH values [28].

COS exhibits superior solubility over a more extended pH range and at relatively higher concentrations than the corresponding chitosan, exhibiting exciting biological properties. Depending on their size, they are also soluble in other solvents, such as dimethyl sulfoxide, dimethylformamide, water, and alcohol. There has been a growing interest in modifying these oligomers to expand their applications. A low molecular weight (1.5 kDa) confers water solubility across a broad pH range and the capacity to be swiftly absorbed by epithelial cells. A wide variety of cells, including neuronal, embryonic, and bone-marrow-derived stem cells, can grow in COS with a degree of deacetylation (DD) > 85%, which is more compatible with cell growth than chitosan with lower DD levels [29]. The fabrication of compatible membrane materials, in which hydrogels are the most thoroughly investigated, is a key biomedical use of COS. Hydrogels are frequently used for therapeutic purposes, including tissue engineering and drug delivery, due to their propensity to expand in aqueous and biological fluids [30].

This review aims to provide a broad overview of chitooligosaccharides, their methods of preparation, and a comprehensive account of the preclinical and laboratory studies conducted to date on its application in the medical field. Based on the available preliminary information, future attention is being paid to translational studies to apply this preliminary information in clinical practice. The importance of focusing on translational studies in the future in order to apply the preliminary information in medical practice is highlighted.

2. Structure of Chitooligosaccharides

Chitosans with degrees of polymerization (DP) < 20 and an average molecular weight of 3900 Da are chitosan oligomers, chitooligomers, and chitooligosaccharides [12]. COS is synthesized by depolymerizing chitin or chitosan through acid hydrolysis, physical hydrolysis, and enzymatic degradation [11] (Figure 1). Chitooligosaccharides are composed of co-oligomers of *N*-acetyl-D glucosamine and D-glucosamine, usually prepared through the partial hydrolysis and deacetylation of chitin and chitosan [31]. The DD of COS, which is related to the molar units of GlcN on its backbone, depends on the DD of chitosan used as the starting material, determining the molecular weight distribution and N-acetylation pattern [32]. COS shares three common functional reactive groups that are directly responsible for inducing and enhancing its various biological effects. These reactive functional groups include the amino/acetamido group, and the primary and secondary hydroxyl groups, at the C-2, C-3, and C-6 positions, respectively, as well as the glycosidic bond at the β-(1→4) link, which is the link between the *N*-glucosamine units [33].

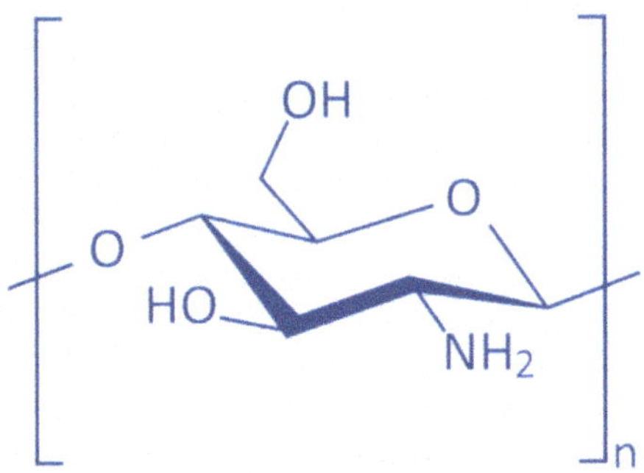

Figure 1. Basic chemical structure of N-deacetylated chitooligosaccharides. (Redrawn from the references: Liaqat and Eltem 2018 [3]; Naveed et al., 2019 [34]).

3. Preparation of Chitooligosaccharides

The depolymerization of chitin or chitosan and the production of chitooligosaccharides can be accomplished using physical (microwave treatment, ultraviolet radiation, and ultrasonic treatment), chemical (hydrogen peroxide oxidation and acid hydrolysis), enzymatic, electrochemical, and composite degradation methods [10]. Ultrasonication, hydrothermal treatment, hydrodynamic cavitation, electromagnetic, gamma rays, and microwave irradiation are the physical methods for chitin or chitosan hydrolysis [35–37]. Chemical methods for producing chitooligosaccharides include degradative methods, such as breaking down chitin or chitosan with a strong acid or oxidation, and synthesis methods [38]. Chemical approaches for synthesizing chitooligosaccharides produce low yields and necessitate protection/deprotection processes, the employment of hazardous chemicals, the production of hazardous waste, high costs, and the generation of a large quantity of monomers [39]. The chemical degradation process is easy to employ, but the relative molecular weight of the degradation products is widely dispersed, making separation and purification challenging. In addition, the amount of reagents used is considerable, and the post-treatment is complicated. Acid and oxidative degradation are the two main processes involved in the chemical decomposition of COS (Figure 2).

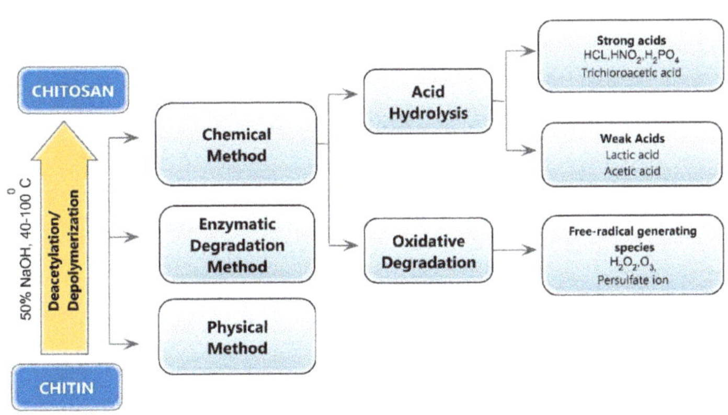

Figure 2. Various methods of chitooligosaccharide production.

3.1. *Acid Hydrolysis of Chitosan*

Chitosan can be hydrolyzed using hydrochloric acid, acid with electrolytes, nitrous acid, phosphoric acid, and hydrofluoric acid, as well as oxidative and reductive techniques involving hydrogen peroxide or persulfate [11,40]. The MW and DD of COS products are highly influenced by the acid concentration, temperature, and treatment time [11]. The most commonly used acid for obtaining COS is hydrochloric acid (HCl). At higher temperatures, concentrated hydrochloric acid degrades chitosan directly to form a chitooligomer and increases the product's DD. Jia and Shen [41] used 85% phosphoric acid to obtain water-soluble chitosan under optimal conditions at 60 degrees Celsius. Degradation by strong acids is difficult to control, the process is complex, and the process yields a lower quantity of CO. Weak acids, such as formic acid, acetic acid, and nitrous acid, can also be used to degrade chitosan. Nitrous acid also breaks down GlcN residues into 2,5-anhydromannose. A small amount of the amino group damage results from the GlcN ring being rearranged and released in this reaction, but GlcN is eventually repaired by sodium borohydride (NaBH4) [42].

3.2. *Enzymatic Hydrolysis*

Enzymatic chitosan hydrolysis is superior to chemical reagent-catalyzed chitosan hydrolysis regarding its predictability and controllability [11,43]. The enzymatic procedures are precise and involve minimal chemical alteration of the products (Figure 3). Additionally, enzyme-catalyzed reactions are controllable, producing COS with specific chemical properties by modifying production procedures, such as enzyme concentrations, pH, temperature, and reaction time [44]. Several glycoside hydrolases composed of specialized enzymes, such as chitinases and chitosanases, can catalyze the enzymatic hydrolysis of chitin and chitosan [45]. Enzymatic hydrolysis can also be performed using nonspecific enzymes, including cellulase, amylase, lipase, and pectinase, with higher yields and lower costs [46,47]. Many factors can affect the outcome of an enzymatic depolymerization reaction. In manufacturing chitooligosaccharides from chitin or chitosan, the choice of enzyme and substrate exert a significant effect [38]. COS exhibits diverse biological activities, which depend on their molecular weight and DD compared to chitosan with a high molecular weight. The most efficient method for obtaining longer chains of COS is to use enzymes, and this method has been effectively used to manufacture COS spanning from dimers (DP 2) to octamers (DP 8) [48]. Cellulose cellulase demonstrated the best hydrolytic ability among the six nonspecific enzymes employed to degrade chitosan [49]. The enzymatic hydrolysis method necessitates optimal conditions for a reaction, and the raw materials' structural properties may also impact the hydrolysis rate [50].

Several nonspecific enzymes, including proteolytic enzymes, such as papain, pepsin, and pronase, were used to produce COS from chitosan. Depolymerization was achieved using immobilized papain and chitosan. The advantage of immobilized papain was that it could be retrieved after the process, and the retrieval enzyme could be reused, despite the low yield and molecular weight of low-molar-mass chitosans (LMWCs) [51]. For chitosan depolymerization, nonspecific enzymes, such cellulase, hemicellulase, lysozyme, pectinase, and a-amylase, were also used. Due to the structural similarity of chitosan and cellulose, polymers of D-glucose linked by b-1,4-glycosidic linkages, many researchers have concentrated on cellulase to produce COS from chitosan. It appears that the enzyme does not strongly recognize the group at the C-2 position in glucose or glucosamine residues when the enzyme–substrate complex is formed [52].

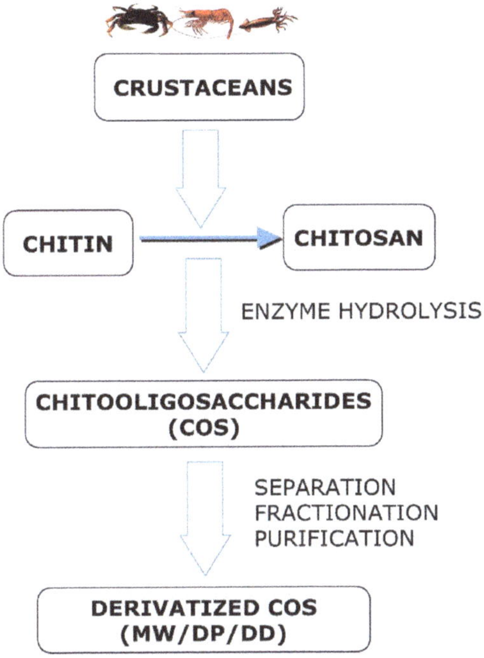

Figure 3. Schematic diagram showing the major steps in obtaining Chitooligosaccharides (enzymatic synthesis).

3.3. Oxidative Degradation

By using free radicals formed by oxidants in aqueous solutions, oxidative degradation attacks the chitosan's β -1,4 glycosidic bonds with active amino groups [53]. The response surface approach using 5.5% hydrogen peroxide chitosan can be converted into COS with a yield of 93.5% [54]. According to a first-order kinetics study, higher temperatures and concentrations result in prolonged reaction times and chitooligosaccharides with a higher mono-to-disaccharide-yield ratio [53]. The reduction in active amino groups and the browning of the product during the later stages of the degradation process are two drawbacks of the H_2O_2 degradation method. Two COSs with DDs of 69% and 95% can be separated from the same source material using a mixture of H_2O_2 and acetic acid in a ratio of 1:7 [55]

3.4. Electrochemical Degradation

The electrochemical degradation of COS is an innovative method for its preparation. This method is simple and contaminant-free but has significant drawbacks, such as a short electrode life and a high failure rate [56,57]. The protonated amino groups remained stable after electrochemical treatment, and the molecular weight of chitosan decreased with an increase in the current density of the Ti/TiO_2-RuO_2 electrode [56]. In addition, the Ti/Sb-SnO_2 electrode was observed to be more effective than the Ti/TiO_2-RuO_2 electrode due to the difference between electrode materials' characteristics [57].

4. Therapeutic Applications of Chitooligosaccharides

COS possesses many biological activities and promising applications in multiple fields, such as medicine, cosmetics, food, and agriculture (Figure 4). COS, which is recognized as low MW and water soluble, is in much greater demand than its precursor molecule chitosan. This section attempts to describe several biological activities of COS that can be applied in the field of medicine.

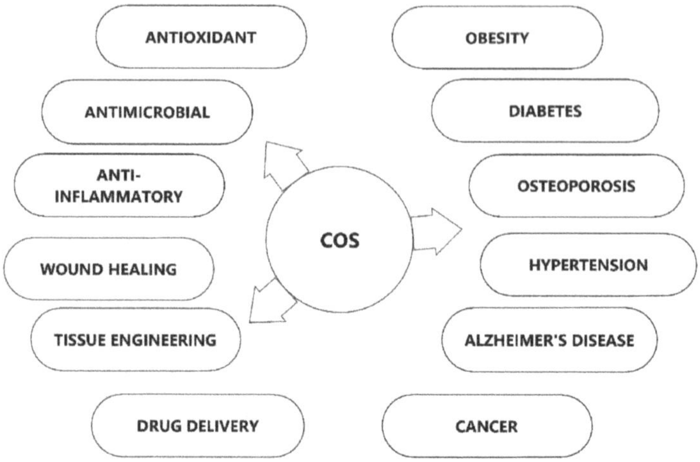

Figure 4. Therapeutic applications of chitooligosaccharides (COSs).

4.1. *Chitooligosaccharides as Antioxidant Agents*

Antioxidants can scavenge free radicals and protect the human body from highly toxic reactive oxygen species (ROS), slowing the progression of many chronic diseases. The ability of chitosan and its derivatives to scavenge free radicals and prevent oxidative damage by interrupting radial chain reactions is well established. Compared to chitosan, COS and its derivatives have better antioxidant properties [58,59]. COS antioxidant or radical-scavenging properties are primarily determined by their molecular weights and DD [46,58,60]. Several chronic diseases, such as cardiovascular diseases, atherogenesis, cancer, and Parkinson's disease, are all linked to oxidative stress [61]. The excessive generation of ROS damages proteins, lipids, and DNA, leading to inflammation, tissue degeneration, and cellular apoptosis. ROS plays a vital role in the wound healing process at low concentrations, and an excessive level of reactive oxygen species can hinder wound healing by stimulating processes, such as inflammation and fibrosis [62].

Though the molecular mechanism underlying COS scavenging activity is unknown, the presence of hydroxyl and free amino groups in COS is thought to be responsible for its antioxidant activity. Antioxidants are noted for their positive benefits on health by protecting cells from the harmful effects of oxidation. It was demonstrated that COS and its derivatives have a high overall reducing power and can effectively remove hydroxyl radicals and superoxide anions [63]. The COS scavenging mechanism is believed to be based on the reaction of hydroxyl and superoxide anion radicals with active hydrogen atoms in COS, producing stable macromolecule radicals [59,64]. As COS can provide positrons to free radicals and transform them into more stable products, it can interrupt the chain reaction caused by free radicals [65].

In human lung epithelial A549 cells, the gallic acid-conjugated COS was tested for its antioxidant and anti-inflammatory properties. Gallate-COS was discovered to have remarkably high DPPH radical scavenging capacity and protect against DNA damage by H_2O_2. This study observed that adding gallic acid to COS enhanced its anti-inflammatory and antioxidant capabilities [66]. The findings indicated that gallate-COS might be a potential preventive agent for lung inflammation and lung cancer brought on by free radicals. In a high-fat-diet mouse model, the ability of COS to scavenge free radicals was examined both in vitro and in vivo. It was discovered that COS had strong antioxidant properties that could shield animals from oxidative stress [67]. Catalase, superoxide dismutase, and glutathione peroxidase were significantly increased in the stomachs, livers, and sera of mice when COS was used in conjunction with a high-fat diet. This indicates that COS may possess some antioxidant properties and be able to restore the function of the enzymes that a high-fat diet compromises.

4.2. Chitooligosaccharides as Antimicrobial Agents

COS is an effective antibacterial agent that inhibits the growth of various microbes, such as bacteria, fungi, and viruses [68,69]. The antimicrobial activity of COS depends on the molecular weight (MW), degree of polymerization (DP), and pattern of acetylation (PA), as well as the type of organism. Though the actual mechanism of the antimicrobial activity is not clearly understood, low-molecular-weight chitosan can penetrate bacterial cell walls, bind with DNA, and inhibit DNA transcription and mRNA synthesis [70]. Conversely, chitosan with a high molecular weight can bind to the negatively charged elements of the bacterial cell wall. As a result, it modifies the permeability of the cell, creates an impermeable layer around it, and prevents transport into the cell [71,72].

The acetylated sequences in the COS structure are necessary for antibacterial action, and COS with a greater number of acetylated sequences and fewer free amino groups possesses improved antimicrobial activity [73]. It was observed that Gram-positive and Gram-negative bacteria have varied responses to the antibacterial action of COS. Gram-negative bacteria have a negatively charged cell surface; hence, the positively charged amine group in COS can severely impede their growth. However, COS does not prevent the growth of Gram-positive bacteria very effectively. The interaction between positively charged amino groups of COS and negatively charged carboxylic acid groups of bacterial cell surfaces paves the way for the formation of polyelectrolyte complexes, resulting in the formation of an impermeable coating around the bacterial cell and the suppression of metabolic activity [74]. As a result of COS antibacterial properties, a lower pH value and more polymerization have also been deemed advantageous [75]. Chitooligosaccharides significantly block *A. actinomycetemcomitan* growth due to its effect on cell membrane permeability. The release of cellular components and the unregulated entry of substances from the surrounding environment results in microbial cell death [69].

COS antimicrobial activities are influenced by a variety of factors, including deoxycholic acid (DA) or dicetyl phosphate (DP), as well as other physicochemical properties and microorganism types [76]. COS can change the permeability characteristics of microbial cell membranes, preventing materials from entering or triggering the cell component's leakage, ultimately leading to bacterial destruction. The bacterial envelope serves as the active site for COS, and membrane rupture may lead to the destruction of the microorganism. Chitosan penetration into bacterial DNA suppresses RNA transcription, an additional mechanism for killing microorganisms [77].

COS has antibacterial properties that promote tissue granulation and collagenase activity, two key factors in wound healing. Therefore, the antibacterial property of COS is responsible for its potential to promote wound healing. The antibacterial effects of COS are superior against both Gram-positive and Gram-negative microorganisms [78]. Antimicrobial action is enhanced with higher degrees of deacetylation than lower acetylation levels [79]. In a study conducted in relation to yeast, chitooligosaccharides with a degree of polymerization of 32 showed strong inhibitory efficacy [80].

Streptococcus mutans, a Gram-positive bacterium often observed in human dental caries, was susceptible to the antimicrobial action of chitooligosaccharide coupled with glycidyl trimethyl ammonium chloride after only two hours of exposure [81]. The antimicrobial activity of COS can also be enhanced by incorporating other nanomaterials. The impact of enhanced antibacterial potential was observed when the quaternary ammonium compounds of COS were added to zinc oxide and a palygorskite nanocomposite [82]. COS-coated silver nanoparticles showed efficient antimicrobial activity [83]. COS-coated silver nanoparticles with calcium hydroxide were tested against *Streptococcus mutans*, *Staphylococcus aureus*, and *Enterococcus faecalis* by Rajasekaran et al. [84]. The composite paste inhibited the investigated bacteria more efficiently, which could benefit pediatric pulp therapy.

An analysis of the antifungal activity of chitooligosaccharide with an average polymerization degree of 32 was observed to be effective against human fungal infections [80]. A study by Jang et al. [85] revealed that the low-molecular-weight chitooligosaccharide decreased the fluorescence intensity of the SARS-CoV-2 nucleocapsid protein of the virus-infected cells in a dose-dependent manner. They concluded that chitooligosaccharide can have an antiviral effect on SARS-CoV-2 and may be a potent natural treatment for COVID-19.

4.3. *Chitooligosaccharides as Anti-Inflammatory Agents*

Inflammation plays a crucial role in the pathology of various diseases, such as chronic asthma, rheumatoid arthritis, multiple sclerosis, inflammatory bowel disease, psoriasis, and cancer [86–89]. It has been demonstrated that the oral administration of COS inhibits the activation of myeloperoxidase, cyclooxygenase (COX)-2, and inducible nitric oxide synthase (iNOS), as well as the levels of proinflammatory cytokines, such as interleukin (IL)-6 and tumor necrosis factor (TNF)-α [90]. COS exposure may inhibit the generation of numerous proinflammatory cytokines associated with lipopolysaccharide (LPS)-induced inflammation without impairing cell viability [91]. NF-κB (nuclear factor kappa-light-chain-enhancer of activated B cells) nucleus translocation was decreased as a result of the COS suppression of LPS-induced inflammatory gene expression, which attenuated an LPS-induced inflammatory response in vascular endothelial cells [92]. Ma et al. [93] hypothesized that COS could reduce the phosphorylation levels of mitogen-activated protein kinases (MAPKs) and activate NF-κB and activator protein 1 (AP-1) to decrease the LPS-induced interleukin 6 (IL-6) and TNF-α generation in macrophages. In a study on the effects of COS on synoviocytes in rabbit knee joints, it was found that activating AMPK reduced the expression of the enzymes iNOS and COX-2, thereby reducing synovial inflammation [94].

COS supplementation in the diets of animals has been shown to have a potent immunoenhancing effect. The results obtained from an evaluation of the COS effect on cyclophosphamide-induced immunosuppression by Mei et al. [95] showed that in mice treated orally with COS and cyclophosphamide, the delayed-type hypersensitivity reaction, macrophage phagocytosis activities, and levels of cytokines IL-2, IL-12, and interferon were significantly increased, while the production of IL-10 was decreased. COS exhibited protective benefits against ovalbumin-induced lung inflammation in induced asthmatic mouse models at a maximal dose of 16 mg/kg per day, with a significant decrease in mRNA expression and protein levels of IL-4, IL-5, IL-13, and TNF-α in lung tissue and bronchoalveolar lavage fluid [96]. The serum levels of IL-1, IL-2, IL-6, immunoglobulin (Ig) A, IgG, and IgM were all raised in early weaned pigs when COS was reintroduced into their diets. Increased cell-mediated immunity in response to weaning stress may be possible if COS can modulate the levels of specific cytokines and antibodies [97]. COS inhibits the release of nitric oxide in LPS-induced RAW 264.7 cells and BV-2 microglia, diminishes the glycerol-induced inflammatory response in rat kidneys, and decreases organ failure in LPS-induced sepsis [98,99].

Glycoprotein YKL-40 promotes angiogenesis and functions as a negative regulator of the inflammasome. The high-affinity binding of N-acetylglucosamine to YKL-40 and subsequent stimulation of chondrocyte proliferation could be used as a novel therapeutic strategy for treating inflammatory rheumatoid diseases [100]. In treating rheumatoid arthritis, it may be beneficial to elucidate the proinflammatory mechanism of COS [101]. COS could be considered as a new functional diet for IBD patients based on the observations from the animal study. Orally administered COS decreased the shortening of the colon length and tissue injury in mice and controlled inflammation in the colonic mucosa [90].

These research findings based on the above-mentioned hypothesis indicate that COS can induce an anti-inflammatory effect by inhibiting cyclooxygenase and reducing prostaglandins. However, additional research is required to determine the effect on more extended inflammatory periods and the bioavailability of these substances.

4.4. Chitooligosaccharides and Their Anti-Obesity Activity

Obesity is a chronic trophic metabolic disorder primarily caused by an energy imbalance, resulting in the buildup of excess body fat. Obesity is associated with type 2 diabetes (T2D), hyperlipidemia, hypertension, cerebrovascular events, and cancer [102]. COS has excellent water solubility and lower viscosity than chitosan, and the intestine more easily absorbs it. In animal models, COS has also been shown to induce weight loss, lower triglyceride and cholesterol levels in serum, and prevent lipid buildup in hepatocytes and adipose tissues [103,104]. COS has demonstrated enhanced intestinal absorption, and studies have been performed to investigate its potential to reduce weight gain, blood triglyceride/cholesterol levels, and lipid buildup in the liver and adipose tissues [25,105,106]. COS anti-obesity activity has been the subject of numerous hypotheses, but its precise mechanism has not been fully elucidated. For instance, COS may block enzymes or interact with bile acids, resulting in decreased fat absorption and increased fecal fat excretion [107].

When obese rats were fed COS, higher amounts of high-density lipoproteins or cardioprotective lipid-containing particles were discovered than in control rats. These lipoproteins and particles are responsible for removing excess cholesterol from tissues and transporting them to the liver [103]. Low-MW COS appears to be more effective in increasing plasma and hepatic lipoprotein lipase activity [108]. The evidence obtained using mouse models suggested that COS inhibited the expression of apolipoprotein B, which reduced the amount of cholesterol found in the serum [109]. Comparing the anti-obesity activity of COS and resistant starch (RS) to their combination, COS–RS, in rat models of induced obesity and dyslipidemia, COS–RS displayed the most significant fat and lipid-lowering benefits, followed by COS and RS [110]. Their examination of RNA sequencing revealed an increased conversion of cholesterol to bile acids. COS reduces triglycerides, total cholesterol, low-density lipoprotein cholesterol serum levels, and the expression of endoplasmic reticulum stress pathway-related factors (GRP78, GRP94, ATF4, and CHOP), and increases oxidative lipid catabolism [111]. This effect may be mediated by metabolites or directly by the molecules themselves. These findings underline the significance of developing COS as a potential drug derived from natural products for the prevention and treatment of obesity.

4.5. Chitooligosaccharides as Antidiabetic Agents

COS has demonstrated the potential to protect β cells from excessive glucose by promoting pancreatic cell proliferation, resulting in enhanced insulin production to lower glucose levels [112,113]. Reducing blood glucose levels and restoring normal insulin sensitivity were among the documented benefits of COS therapy for diabetic rats [113]. COS possesses antidiabetic properties and may reduce diabetes mellitus (DM) incidence by affecting the glucose–lipid metabolic balance and glycemic control [114]. Additionally, COS therapy can potentially improve the general health of diabetic rats, alleviate diabetic symptoms, bring blood glucose levels back to normal, and restore normal insulin sensitivity.

In addition, chitooligosaccharides can stimulate the multiplication of beta cells and restore the functional capacity of injured beta cells. COS has been shown to promote the overgrowth of beta cells and isolated pancreatic islet cells, in addition to increasing insulin release from pancreatic cells [113]. Chitooligosaccharides have the potential to significantly increase the rate of proliferation of pancreatic islet cells. In diabetic mice, it was also found that dietary COS reduced hyperglycemia by activating hepatic glucokinase and increasing peripheral tissue glucose uptake, as well as by increasing pancreatic insulin secretion and improving skeletal muscle glucose uptake [115]. In a study using streptozotocin (STZ)-induced diabetic rats, COS was able to treat hyperglycemia at a dose of 1000 mg/kg by lowering fasting serum glucose and insulin levels, thus improving O-glycosyltransferase (OGT), enhancing the index of insulin sensitivity, and reducing insulin resistance [116]. Additionally, COS significantly increases the amount of glycogen in the liver by increasing glucokinase, which facilitates the transfer of blood glucose into liver glycogen. Low-molecular-weight COS also increases the plasma adiponectin levels in prediabetic subjects [117].

Significantly elevated serum glutamic pyruvic transaminase (SGOT) and serum glutamic pyruvic transaminase (SGPT) activities were observed in alloxan-induced diabetic mice. As blood levels of amino acids increase, gluconeogenesis and ketogenesis increase, which results in enhanced transaminase activity in the absence of insulin [118]. After receiving COS (10 mg/kg), the liver function indicators SGOT and SGPT returned to normal levels, indicating that the liver was functioning normally.

Chronic hyperglycemia may be acquired, consequent to the failure of peripheral tissues to utilize glucose properly. Glucose transporter type 4 (GLUT-4) is a vital glucose transporter and regulator of its metabolism found in skeletal muscles and adipocytes. The expression of GLUT-4 was decreased in adipose and skeletal tissues in a diabetic animal model [119]. The treatment of the diabetic animals with COS caused a significant increase in GLUT-4, which might upregulate GLUT-4 mRNA expression, thereby improving insulin resistance. COS may have therapeutic effects on type 2 diabetes due to its capacity to boost liver glycogen production and GLUT-4 gene expression in soleus muscle and adipose to ameliorate insulin resistance. COS, on the other hand, exhibited an antidiabetic impact by stimulating proliferation, increasing insulin release and GLUT-2 mRNA levels, and protecting against STZ-induced apoptosis in β cells. In addition, COS activities, such as enzyme inhibition, antioxidant activity, and immunomodulation, may assist non-obese diabetic mice in avoiding the development of type 2 diabetes mellitus, which promotes the preservation of pancreatic β cells and the standardization of the vital insulin secretion [120].

4.6. *Chitooligosaccharides in Osteoporosis*

Supplemental chitooligosaccharides are beneficial in calcium-deficient states, such as osteoporosis. Supplementing animals' diets with COS has been shown to improve calcium bioavailability in rat osteoporosis models induced by ovariectomy and concomitant low calcium intake [121]. A substantial reduction in the serum levels of inflammatory cytokines was observed after the oral administration of COS to older persons [122]. The mechanism behind the possible antiosteoporotic effect of COS was thought to be connected to its anti-inflammatory properties. Additionally, others have reported that COS suppressed the synthesis and expression of proinflammatory mediators in vitro [121,123].

The antiosteoporotic effect of COS may be partially explained by its anti-inflammatory effect via the downregulation of COX-2 expression levels, which provides additional evidence for the efficacy of selective COX-2 inhibition in preventing bone loss in estrogen-deficient animals and postmenopausal women [124]. The mineralization process and bone density depend on Ca^{2+}, which provides structural support. There is evidence that COS causes an increase in bone calcium deposition [121,125]. Jung et al. [121] discovered that COS effectively inhibited the development of insoluble calcium-phosphate salts, enhancing Ca^{2+} bioavailability and bone strength. COS (5 kDa) enhanced calcium retention and lowered bone turnover in a rat osteoporosis model, indicating that COS may have favorable effects as a calcium supplement in Ca^{2+} shortage, such as osteoporosis.

4.7. *The Antihypertensive Effect of COS*

COS can effectively control hypertension by inhibiting renin or angiotensin-converting enzyme activity. COS is an angiotensin-converting enzyme (ACE) inhibitor since the active binding site of ACE is positively charged and contains hydrogen-bond acceptors and zinc as a cofactor. Hong et al. [126] evaluated the ACE inhibitory actions of COS with varying degrees of polymerization from 1 to 10 and found that the chitotriose (DP = 3) derivative was the most potent. Studies have shown that the DD is inversely proportional to the ACE inhibitory activity of COS or its derivatives [127,128]. Carboxylated and sulfated COS have been synthesized, and it has been observed that they possess a significantly greater inhibitory action against ACE than unmodified COS [127,128]. It is believed that an increase in negative charges on the molecule is responsible for the improved binding of this modified COS to the integral active site of the enzyme, thereby boosting its ACE inhibitory action. The aminoethyl-conjugated COS exhibited enhanced ACE inhibitory action due to the formation of hydrogen bonds, which promotes COS binding [129].

4.8. Chitooligosaccharides and Alzheimer's Disease

The pathological changes in Alzheimer's disease (AD) are caused by neuronal apoptosis, and neuronal protection is crucial in treating it [130]. A study conducted to explore the neuroprotective effect of COS against AD showed that COS significantly decreased amyloid beta-induced cell apoptosis by reducing the expression of caspase 3 and Bax/Bcl-2 ratio activation [131]. COS significantly reduced the neuronal damage caused by oxidative stress and glucose deprivation, and these findings imply that COS has the potential to serve as a neuroprotective agent against neurodegenerative diseases, such as Alzheimer's disease [132]. The protein expression and acetylcholinesterase activity produced by amyloid β peptide in PC12 cells were inhibited by COS, which was revealed to be the first stage in the pathogenic cascade of AD [133]. A mouse model demonstrated that COS could be a valuable drug for treating sciatic nerve injury and reducing scar tissue formation [134]. Evidence shows that COS could have neuroprotective effects by reducing oxidative stress and neuroinflammatory responses [135].

4.9. Chitooligosaccharides as Antitumor Agents

Chitooligosaccharides and their derivatives have demonstrated potent antitumor activity against human cancer cells [136]. Though limited information is available on the antitumor mechanisms of COS, some hypotheses have been proposed (Figure 5). Initially, the mechanism of anticancer activity was linked to the cationic character of COS; however, it was later hypothesized that relative molecular weight was also a crucial factor [137]. The tumor-inhibiting effect of COS is likely a result of its ability to induce lymphocyte cytokines via promoting T-cell proliferation. The antitumor mechanism of COS is essentially enhanced by acquired immunity by increasing T-cell differentiation to boost cytotoxicity and preserve T-cell activity [138]. COSs are naturally occurring polysaccharides with a cationic charge; they can promote the apoptosis of numerous cancer cells, including liver cancer, breast cancer, cervical cancer, kidney cancer, lung cancer, leukemia, and colorectal cancer cells [25]. COS with a cationic charge can absorb tumor cells by modifying their permeability through electrostatic contact. The binding of COS to tumor cell membranes and subsequent changes in the ionic environment cause alterations in the permeability of tumor cells [139]. However, due to the mutual repulsion between COS and healthy cells with a uniform positive charge, only tumor cells are targeted, not healthy ones.

The antitumor activity of COS is linked to its ability to stimulate cytokine production through enhanced T-cell proliferation. It was established that COS could prevent tumor growth via immune-boosting actions [140]. Immunostimulatory activities of chitooligosaccharides indirectly contribute to tumor suppression by boosting the generation of lymphokines, inducing lymphocyte factors, and stimulating the proliferation of cytolytic T lymphocytes, hence producing an antitumor effect [95,141]. Furthermore, COS can inhibit tumor angiogenesis by inhibiting the expression of matrix metalloproteinase-9 (MMP-9), which in turn inhibits the expression of vascular endothelial growth factor (VEGF) [142]. MMP-9 is an essential component in cancer invasion and metastasis, and the upregulation of MMP-9 expression may lead to the release of VEGF and the formation of angiogenesis [143]. Additionally, by inhibiting the production of VEGF and urokinase-type plasminogen activator in vascular endothelial cells, COS increases its ability to inhibit tumor angiogenesis [144].

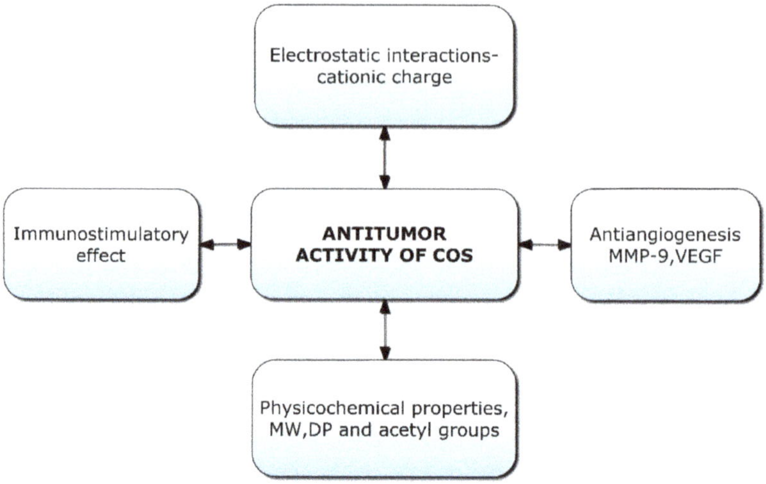

Figure 5. Potential factors responsible for antitumor activities of chitooligosaccharides (COSs). VEGF: Vascular endothelial growth factor; MW: Molecular weight; DP: Degree of polymerization.

Aminoethyl-chitooligosaccharides are made by replacing the hydroxyl group at C-6 with an aminoethyl group, and they were found to inhibit cell invasion and metastasis [145]. Necrosis was also found to be the principal mechanism by which a highly charged COS derivative inhibits the growth of cancer cells, according to a study analyzing DNA fragments [139]. However, low-molecular-weight COS and *N*-acetyl-d-glucosamine oligomers (NACOS) can potentially cause apoptosis in cancer cells [146,147]. COS also inhibits tumor progression by modulating key pathways in malignant growth, such as the suppression of β-catenin, mTOR, pyruvate kinase, and ornithine decarboxylase; the activation of caspase 3 in tumor cells; and the stimulation of interferon gamma (IFN-γ) and IL-12 secreted by natural killer (NK) cells [148].

Several potential mechanisms were explained based on the animal studies. COS-based drug delivery and therapy systems for anticancer applications have great potential if novel synthesis methodologies and a cost-effective strategy can be developed. Future research is required to investigate the molecular mechanism of COS on cancer cells and its effect on signaling pathways.

4.10. Chitooligosaccharides in Wound Healing

Hemostasis, inflammation, proliferation, and remodeling are some stages of the wound healing process that are influenced by the immunostimulating properties of materials that aid in healing, such as proinflammatory cytokines, inflammatory cells, and growth factors [149]. As a result, the demand for biomaterials with superior biological properties, such as antibacterial, antioxidant, and anti-inflammatory properties, has increased exponentially. The improved biological properties of COS, including cytocompatibility, antibacterial, antioxidant, anti-inflammatory, and immunostimulatory activities, may speed up the wound healing process. As a result of its beneficial biological actions, including antimicrobial protection, increased permeability to air and moisture, cellular adhesion, and cell proliferation, COS is a promising therapeutic agent in the treatment of tissue damage and wound healing.

The overexpression of miR-27a and the activation of the transforming growth factor-beta (TGF-β)-1-Smad2/3 pathway are two potential mechanisms of COS in the acceleration of wound healing and tissue regeneration [150]. COS is combined with other biopolymers for wound healing due to its low molecular weight. In addition to its biological properties,

COS can promote wound healing by enhancing a wound dressing's water absorption, flexibility, and mechanical strength.

However, the ability of COS to heal wounds is affected by its molecular weight and degree of acetylation [151]. The local injection of immunostimulatory COS into wounds could stimulate the production of growth factors and the migration of keratinocytes, both of which are essential for wound healing [152]. COS, which shares the same chain length as oligonucleotides, possesses higher immunostimulatory activity and enhances the activity of macrophages. In COS-dressed wounds, the more significant migration of macrophages to injured tissues stimulates the development of type-III collagen, expediting the healing process.

The wound healing activity of oxidized alginate-gelatin hydrogel was enhanced using COS and salicylic acid conjugates. Collagen proliferation at the wound sites was accelerated by the COS conjugate, leading to quicker healing [153]. Applying COS to the locations of sciatic nerve lesions demonstrated the drug's multifaceted impact on healing. The restoration of nerve function was accelerated by COS because it decreased the ratio of type I to III collagen, which boosted axon regeneration and slowed axonal demyelination. COS was discovered to slow the proliferation of fibroblasts, an essential component of scar tissues that prevents axonal regeneration [134,151]. COS has been found to induce the formation of type I and III collagen in wounds, repair injured nerves, expedite hemostasis, and protect against infections, making it an effective wound healing agent [153,154]. Combining medications, such as curcumin with COS, enhances their therapeutic efficacy [155]. COS is, therefore, superior to chitosan because it possesses a more potent biological activity without modification [153,154].

4.11. Chitooligosaccharides in Tissue Engineering

Cells, composite scaffolds, and signaling molecules are the three essential components for producing substituted tissues that repair, replace, or regenerate damaged tissues or organs. COS is comparable to glycosaminoglycans, a crucial part of the extracellular matrix of a cell (ECM). As it can create an environment that closely resembles the ECM, allowing for cell attachment and the preservation of growth factors, this makes it successful for scaffolds in tissue engineering applications [156]. In the creation of bone tissue engineering scaffolds, natural polymers, such as gelatin and COS, have been used. By regulating the genes that control osteoblast proliferation in bone tissues, COS has been shown to stimulate neuronal differentiation in PC-12 nerve cells [157]. Gelatin–COS scaffolds that were cultivated with mesenchymal stem cells (MSCs) obtained from bone marrow and capable of osteogenic differentiation demonstrated promising outcomes in the production of bone tissue [158]. Within two weeks of implantation, cell proliferation was observed inside the scaffolds; homogenous collagen distribution within the pores and calcium deposition on the scaffolds' surfaces indicated that the cells had successfully proliferated [159]. An increase in COS content caused the cross-linking density to decrease, which allowed the scaffolds to swell to more than twice their dry size and made them more permeable to cells. The COO−, C—O, and amino groups present in COS serve as nucleating sites for hydroxyapatite crystals [158].

Single-walled carbon nanotubes (SWCNTs) have been added to synthetic composite materials as alternatives to bone grafts. By enhancing SWCNT implants with organic polymers, such as chitosan and COS, it is possible to keep the graft in constant contact with the tissues, which speeds up bone tissue formation and eliminates the need for additional procedures [160]. A collagen the COS scaffold that was created to act as a synthetic skin tissue was discovered to have improved resistance to collagenase and other deteriorating enzymes. Collagenase would quickly degrade artificial collagen if used alone; hence, collagen was combined with COS to create densely linked microporous structures with enhanced mechanical characteristics [161]. In contrast to pure collagen scaffolds, scaffolds made with COS showed a higher rate of fibroblast growth.

4.12. Chitooligosaccharides in Drug Delivery

COS has excellent potential for usage in drug delivery systems (DDSs) due to its non-toxicity, biodegradability, and solubility in water [154]. Due to its solubility at physiological pH, the water solubility of COS is generally more effective for drug administration [162]. Recently, a novel alternative to red blood cells has been developed using a pectin-based COS–hydrogel microcapsule carrier designed to transport hemoglobin [163]. Oligosaccharides were used to extend the shelf life of these therapeutic drugs by several months. With COS of a 95% degree of deacetylation (DD) and an MW of 10 kDa, non-biodegradable polyethylene glycol (PEG) and cyclodextrin inclusion complexes are converted into hydrogels capable of transporting drugs.

MW and DD are influential parameters for modulating the actions of COS, offering a strategy to improve its efficacy in drug delivery systems (DDSs) [164]. COS is more significantly absorbed through negatively charged mucous membranes of tissues because it is more capable of binding under these situations [165]. DD also controls binding; at a low MW, the effect of DD on charge density is more substantial [28,166]. This may have a positive influence on the interaction of COS with oppositely charged copolymers or active medicines; nevertheless, the considerably stronger interactions can harm cell viability [167]. Therefore, when utilizing COS with a high DD in drug delivery, it is necessary to monitor and optimize the dosage rate to limit cytotoxic effects [164].

After combining it with the aqueous copolymer solution, COS is used as a hydrogel precursor to speed up the production of cyclodextrin hydrogel, which can be completed in as little as five minutes. Due to the synergistic interaction between hydrated COS and hydrophobic aggregated inclusion complexes of cyclodextrin with an MPEG-PCL graft and unbound PCL blocks, rapid gelation resulted in the formation of a stable supramolecular hydrogel. This was attributed to the hydrogel containing unbound PCL blocks [168]. Mahato et al. (2020) created hydrogel systems based on PVA and COS for drug delivery applications. In addition, the antibacterial agent lomefloxacin was added to the hydrogels, and sustained drug release was observed [169].

The decrease in chitosan's molecular weight increases its affinity for polyanionic complexes and mucosal membranes. This may increase the uptake rate of COS-based DDS through targeted tissues due to the substance's increased retention at the location [170]. Garaivo et al. [171] determined that chitosan oligomers are superior to chitosan for the intracellular delivery of genes because they are not taken up by endocytic vesicles coated with the clathrin protein, which transfers the polyplexes to the lysosome for degradation. This is the fate that linear chain chitosan-DNA nanoplexes typically face. The graft copolymer is created by graft–diblock copolymers of amphiphilic polyethylene glycol methyl ether and polycaprolactone onto hydrophilic COS.

5. Conclusions and Future Prospects

Chitooligosaccharides are derived from the deacetylation and depolymerization of chitin or chitosan through physical, chemical, or enzymatic hydrolysis. Though several methods can be used to convert chitin and chitosan, the lack of a standard method to obtain highly pure COS with a stable DP for large-scale applications is challenging. COS with a varying MW, DP, and concentrations has been shown to possess various therapeutic effects, including antimicrobial, antitumor, immunoregulatory, and antioxidant activities, which can potentially treat and prevent diseases, such as cancer, diabetes mellitus, cardiovascular disease, and infectious diseases. In addition, due to the cationic sphere on the more exposed shorter N-glucosamine units, COS is absorbable through the intestinal epithelium and displays a greater amount of cellular transduction. Despite its widespread use, its stability, safety, and purity have not been proved successful. At present, limited information is available on the definitive impacts of the physicochemical properties of COS on various biological functions and the molecular mechanisms underlying bioactivities. Examining the molecular structure of COS, which is accountable for its biological activity, is essential for comprehending its process. Several studies on the biological activities of

COS and its toxicities were performed in cell and animal models and using COS mixtures or purified COS with insufficiently defined chemical characteristics. Future investigations and translational research should investigate COS therapeutic effects and toxicity profiles with well-defined chemical characteristics in either large animal models or human subjects.

Funding: This research received no external funding.

Institutional Review Board Statement: Not applicable.

Informed Consent Statement: Not applicable.

Data Availability Statement: Not applicable.

Conflicts of Interest: The authors declare no conflict of interest.

Abbreviations

ACE	Angiotensin-converting enzyme
LPSs	Lipopolysaccharides
AD	Alzheimer's disease
MAPKs	Mitogen-activated protein kinases
AP-1	Activator protein 1
MMP-9	Matrix metalloproteinase 9
COS	Chitooligosaccharide
MSC	Mesenchymal stem cell
COX	Cyclooxygenase
MW	Molecular weight
DA	Deoxycholic acid
NaBH4	Sodium borohydride
DD	Deacetylation degree
NACOSs	N-acetyl-d-glucosamine oligomers
DD	Degree of deacetylation
$NaNO_2$	Sodium nitrite
DDSs	Drug delivery systems
NF-κB	Nuclear factor kappa-light-chain-enhancer of activated B cells
DM	Diabetes mellitus
NK	Natural killer
DP	Degree of polymerization
OGT	O-glycosyltransferase
ECM	Extracellular matrix
PEG	Polyethylene glycol
GlcN	Glucosamine
ROS	Reactive oxygen species
GlcNAc	N-Acetylglucosamine
RS	Resistant starch
GLUT-4	Glucose transporter type 4
SGOT	Serum glutamic pyruvic transaminase
H_2O_2,	Hydrogen peroxide
SGPT	Serum glutamic pyruvic transaminase
HCl	Hydrochloric acid
STZ	Streptozotocin
IFN-γ	Interferon gamma
SWCNTs	Single-walled carbon nanotubes
IL	Interleukin
TGF-β	Transforming growth factor beta
iNOS	Inducible nitric oxide synthase
TNF	Tumor necrosis factor
LMWCs	Low-molar-mass chitosans
VEGF	Vascular endothelial growth factor

References

1. Synowiecki, J.; Al-Khateeb, N.A. Production, properties, and some new applications of chitin and its derivatives. *Crit. Rev. Food Sci. Nutr.* **2003**, *43*, 145–171.
2. Rudall, K.M. The chitin/protein complexes of insect cuticles. In *Advances in Insect Physiology*; Beament, J.W.L., Treherne, J.E., Wigglesworth, V.B., Eds.; Academic Press: Cambridge, MA, USA, 1963; Volume 1, pp. 257–313.
3. Liaqat, F.; Eltem, R. Chitooligosaccharides and their biological activities: A comprehensive review. *Carbohydr. Polym.* **2018**, *184*, 243–259. [CrossRef]
4. Wan, A.C.; Tai, B.C. Chitin—A promising biomaterial for tissue engineering and stem cell technologies. *Biotechnol. Adv.* **2013**, *31*, 1776–1785. [CrossRef]
5. Lee, H.-W.; Choi, J.-W.; Han, D.P.; Lee, N.-W.; Park, S.; Yi, D.-H. Identification and production of constitutive chitosanase from bacillus sp. Hw-002. *J. Microbiol. Biotechnol.* **1996**, *6*, 12–18.
6. Roy, S.; Chakraborty, T.; Begum, J.; Hasnain, M.S.; Nayak, A.K. Chapter 1-chitosan: A versatile biopolymer. In *Chitosan in Biomedical Applications*; Hasnain, M.S., Beg, S., Nayak, A.K., Eds.; Academic Press: Cambridge, MA, USA, 2022; pp. 1–11.
7. Yeul, V.; Rayalu, S. Unprecedented chitin and chitosan: A chemical overview. *J. Polym. Environ.* **2012**, *21*, 606–614. [CrossRef]
8. Trombotto, S.; Ladaviere, C.; Delolme, F.; Domard, A. Chemical preparation and structural characterization of a homogeneous series of chitin/chitosan oligomers. *Biomacromolecules* **2008**, *9*, 1731–1738. [CrossRef]
9. Xia, W.; Liu, P.; Zhang, J.; Chen, J. Biological activities of chitosan and chitooligosaccharides. *Food Hydrocoll.* **2011**, *25*, 170–179. [CrossRef]
10. Aam, B.B.; Heggset, E.B.; Norberg, A.L.; Sorlie, M.; Varum, K.M.; Eijsink, V.G. Production of chitooligosaccharides and their potential applications in medicine. *Mar. Drugs* **2010**, *8*, 1482–1517. [CrossRef]
11. Lodhi, G.; Kim, Y.S.; Hwang, J.W.; Kim, S.K.; Jeon, Y.J.; Je, J.Y.; Ahn, C.B.; Moon, S.H.; Jeon, B.T.; Park, P.J. Chitooligosaccharide and its derivatives: Preparation and biological applications. *Biomed. Res. Int.* **2014**, *2014*, 654913.
12. Mourya, V.; Inamdar, N.; Choudhari, Y.M. Chitooligosaccharides: Synthesis, characterization and applications. *Polym. Sci. Ser. A* **2011**, *53*, 583–612. [CrossRef]
13. Dong, H.; Wang, Y.; Zhao, L.; Zhou, J.; Xia, Q.; Jiang, L.; Fan, L. Purification of dp 6 to 8 chitooligosaccharides by nanofiltration from the prepared chitooligosaccharides syrup. *Bioresour. Bioprocess.* **2014**, *1*, 20. [CrossRef]
14. Jung, W.J.; Park, R.D. Bioproduction of chitooligosaccharides: Present and perspectives. *Mar. Drugs* **2014**, *12*, 5328–5356. [CrossRef]
15. Sato, K.; Saimoto, H.; Morimoto, M.; Shigemasa, Y. Depolymerization of chitin and chitosan under hydrothermal conditions. *Sen'i Gakkaishi* **2003**, *59*, 104–109. [CrossRef]
16. Xing, R.; Liu, S.; Yu, H.; Guo, Z.; Wang, P.; Li, C.; Li, Z.; Li, P. Salt-assisted acid hydrolysis of chitosan to oligomers under microwave irradiation. *Carbohydr. Res.* **2005**, *340*, 2150–2153. [CrossRef]
17. Wu, T.; Zivanovic, S.; Hayes, D.G.; Weiss, J. Efficient reduction of chitosan molecular weight by high-intensity ultrasound: Underlying mechanism and effect of process parameters. *J. Agric. Food Chem.* **2008**, *56*, 5112–5119. [CrossRef]
18. Yoksan, R.; Akashi, M.; Miyata, M.; Chirachanchai, S. Optimal gamma-ray dose and irradiation conditions for producing low-molecular-weight chitosan that retains its chemical structure. *Radiat. Res.* **2004**, *161*, 471–480. [CrossRef]
19. Domard, A.; Cartier, N. Glucosamine oligomers: 4. Solid state-crystallization and sustained dissolution. *Int. J. Biol. Macromol.* **1992**, *14*, 100–106. [CrossRef]
20. Einbu, A.; Varum, K.M. Depolymerization and de-n-acetylation of chitin oligomers in hydrochloric acid. *Biomacromolecules* **2007**, *8*, 309–314. [CrossRef]
21. Lin, F.; Jia, X.-G.; Lei, W.-X.; Li, Z.-J.; Zhang, T.-Y. Spectra analyses of chitosans degraded by hydrogen peroxide under optimal conditions. *Spectrosc. Spectr. Anal.* **2009**, *29*, 43–47.
22. Morris, V.B.; Neethu, S.; Abraham, T.E.; Pillai, C.K.S.; Sharma, C.P. Studies on the condensation of depolymerized chitosans with DNA for preparing chitosan-DNA nanoparticles for gene delivery applications. *J. Biomed. Mater. Res. B Appl. Biomater.* **2009**, *89*, 282–292. [CrossRef]
23. Liang, S.; Sun, Y.; Dai, X. A review of the preparation, analysis and biological functions of chitooligosaccharide. *Int. J. Mol. Sci.* **2018**, *19*, 2197. [CrossRef]
24. Hamer, S.N.; Cord-Landwehr, S.; Biarnés, X.; Planas, A.; Waegeman, H.; Moerschbacher, B.M.; Kolkenbrock, S. Enzymatic production of defined chitosan oligomers with a specific pattern of acetylation using a combination of chitin oligosaccharide deacetylases. *Sci. Rep.* **2015**, *5*, 8716. [CrossRef]
25. Muanprasat, C.; Chatsudthipong, V. Chitosan oligosaccharide: Biological activities and potential therapeutic applications. *Pharmacol. Ther.* **2017**, *170*, 80–97. [CrossRef]
26. Naqvi, S.; Moerschbacher, B.M. The cell factory approach toward biotechnological production of high-value chitosan oligomers and their derivatives: An update. *Crit. Rev. Biotechnol.* **2017**, *37*, 11–25. [CrossRef]
27. Wang, Q.Z.; Chen, X.G.; Liu, N.; Wang, S.X.; Liu, C.S.; Meng, X.H.; Liu, C.G. Protonation constants of chitosan with different molecular weight and degree of deacetylation. *Carbohydr. Polym.* **2006**, *65*, 194–201. [CrossRef]
28. Maganti, N.; Venkat Surya, P.K.C.; Thein-Han, W.W.; Pesacreta, T.C.; Misra, R.D.K. Structure–process–property relationship of biomimetic chitosan-based nanocomposite scaffolds for tissue engineering: Biological, physico-chemical, and mechanical functions. *Adv. Eng. Mater.* **2011**, *13*, B108–B122. [CrossRef]

29. Ratanavaraporn, J.; Kanokpanont, S.; Tabata, Y.; Damrongsakkul, S. Growth and osteogenic differentiation of adipose-derived and bone marrow-derived stem cells on chitosan and chitooligosaccharide films. *Carbohydr. Polym.* **2009**, *78*, 873–878. [CrossRef]
30. Pella, M.C.G.; Lima-Tenorio, M.K.; Tenorio-Neto, E.T.; Guilherme, M.R.; Muniz, E.C.; Rubira, A.F. Chitosan-based hydrogels: From preparation to biomedical applications. *Carbohydr. Polym.* **2018**, *196*, 233–245. [CrossRef]
31. Riva, R.; Ragelle, H.; des Rieux, A.; Duhem, N.; Jérôme, C.; Préat, V. Chitosan and chitosan derivatives in drug delivery and tissue engineering. In *Chitosan for Biomaterials II*; Jayakumar, R., Prabaharan, M., Muzzarelli, R.A.A., Eds.; Springer: Berlin/Heidelberg, Germany, 2011; pp. 19–44.
32. Kaur, S.; Dhillon, G.S. The versatile biopolymer chitosan: Potential sources, evaluation of extraction methods and applications. *Crit. Rev. Microbiol.* **2014**, *40*, 155–175. [CrossRef]
33. Li, Z.; Yang, F.; Yang, R. Synthesis and characterization of chitosan derivatives with dual-antibacterial functional groups. *Int. J. Biol. Macromol.* **2015**, *75*, 378–387. [CrossRef]
34. Naveed, M.; Phil, L.; Sohail, M.; Hasnat, M.; Baig, M.; Ihsan, A.U.; Shumzaid, M.; Kakar, M.U.; Mehmood Khan, T.; Akabar, M.D.; et al. Chitosan oligosaccharide (cos): An overview. *Int. J. Biol. Macromol.* **2019**, *129*, 827–843. [CrossRef]
35. Baxter, S.; Zivanovic, S.; Weiss, J. Molecular weight and degree of acetylation of high-intensity ultrasonicated chitosan. *Food Hydrocoll.* **2005**, *19*, 821–830. [CrossRef]
36. Wu, Y.; Huang, Y.; Zhou, Y.; Ren, X.e.; Yang, F. Degradation of chitosan by swirling cavitation. *Innov. Food Sci. Emerg. Technol.* **2014**, *23*, 188–193. [CrossRef]
37. Zainol, I.; Akil, H.M.; Mastor, A. Effect of γ-irradiation on the physical and mechanical properties of chitosan powder. *Mater. Sci. Eng. C* **2009**, *29*, 292–297. [CrossRef]
38. Miguez, N.; Kidibule, P.; Santos-Moriano, P.; Ballesteros, A.O.; Fernandez-Lobato, M.; Plou, F.J. Enzymatic synthesis and characterization of different families of chitooligosaccharides and their bioactive properties. *Appl. Sci.* **2021**, *11*, 3212. [CrossRef]
39. Kaczmarek, M.B.; Struszczyk-Swita, K.; Li, X.; Szczesna-Antczak, M.; Daroch, M. Enzymatic modifications of chitin, chitosan, and chitooligosaccharides. *Front. Bioeng. Biotechnol.* **2019**, *7*, 243. [CrossRef]
40. Einbu, A.; Grasdalen, H.; Varum, K.M. Kinetics of hydrolysis of chitin/chitosan oligomers in concentrated hydrochloric acid. *Carbohydr. Res.* **2007**, *342*, 1055–1062. [CrossRef]
41. Jia, Z. Effect of reaction temperature and reaction time on the preparation of low-molecular-weight chitosan using phosphoric acid. *Carbohydr. Polym.* **2002**, *49*, 393–396. [CrossRef]
42. Tommeraas, K.; Varum, K.M.; Christensen, B.E.; Smidsrod, O. Preparation and characterisation of oligosaccharides produced by nitrous acid depolymerisation of chitosans. *Carbohydr. Res.* **2001**, *333*, 137–144. [CrossRef]
43. Jeon, Y.-J.; Kim, S.-K. Continuous production of chitooligosaccharides using a dual reactor system. *Process. Biochem.* **2000**, *35*, 623–632. [CrossRef]
44. Patil, R.S.; Ghormade, V.V.; Deshpande, M.V. Chitinolytic enzymes: An exploration. *Enzym. Microb. Technol.* **2000**, *26*, 473–483. [CrossRef]
45. Suresh, P. Enzymatic technologies of chitin and chitosan. In *Enzymatic Technologies for Marine Polysaccharides*; CRC Press: Boca Raton, FL, USA, 2019; pp. 449–466.
46. Laokuldilok, T.; Potivas, T.; Kanha, N.; Surawang, S.; Seesuriyachan, P.; Wangtueai, S.; Phimolsiripol, Y.; Regenstein, J.M. Physicochemical, antioxidant, and antimicrobial properties of chitooligosaccharides produced using three different enzyme treatments. *Food Biosci.* **2017**, *18*, 28–33. [CrossRef]
47. Montilla, A.; Ruiz-Matute, A.I.; Corzo, N.; Giacomini, C.; Irazoqui, G. Enzymatic generation of chitooligosaccharides from chitosan using soluble and immobilized glycosyltransferase (branchzyme). *J. Agric. Food Chem.* **2013**, *61*, 10360–10367. [CrossRef]
48. Choi, Y.J.; Kim, E.J.; Piao, Z.; Yun, Y.C.; Shin, Y.C. Purification and characterization of chitosanase from *bacillus* sp. Strain kctc 0377bp and its application for the production of chitosan oligosaccharides. *Appl. Environ. Microbiol.* **2004**, *70*, 4522–4531. [CrossRef]
49. Dong, H.; Wang, Y.; Zhao, L.; Zhou, J.; Xia, Q.; Qiu, Y. Key technologies of enzymatic preparation for dp 6-8 chitooligosaccharides. *J. Food Process. Eng.* **2015**, *38*, 336–344. [CrossRef]
50. Zhou, J.; Wen, B.; Xie, H.; Zhang, C.; Bai, Y.; Cao, H.; Che, Q.; Guo, J.; Su, Z. Advances in the preparation and assessment of the biological activities of chitosan oligosaccharides with different structural characteristics. *Food Funct.* **2021**, *12*, 926–951. [CrossRef]
51. Terbojevich, M. Molecular parameters of chitosans depolymerized with the aid of papain. *Carbohydr. Polym.* **1996**, *29*, 63–68. [CrossRef]
52. Xia, W.; Liu, P.; Liu, J. Advance in chitosan hydrolysis by non-specific cellulases. *Bioresour. Technol.* **2008**, *99*, 6751–6762. [CrossRef]
53. Chang, K.L.; Tai, M.C.; Cheng, F.H. Kinetics and products of the degradation of chitosan by hydrogen peroxide. *J. Agric. Food Chem.* **2001**, *49*, 4845–4851. [CrossRef]
54. Du, Y.; Zhao, Y.; Dai, S.; Yang, B. Preparation of water-soluble chitosan from shrimp shell and its antibacterial activity. *Innov. Food Sci. Emerg. Technol.* **2009**, *10*, 103–107. [CrossRef]
55. Tishchenko, G.; Šimůnek, J.; Brus, J.; Netopilík, M.; Pekárek, M.; Walterová, Z.; Koppová, I.; Lenfeld, J. Low-molecular-weight chitosans: Preparation and characterization. *Carbohydr. Polym.* **2011**, *86*, 1077–1081. [CrossRef]
56. Cai, Q.; Gu, Z.; Chen, Y.; Han, W.; Fu, T.; Song, H.; Li, F. Degradation of chitosan by an electrochemical process. *Carbohydr. Polym.* **2010**, *79*, 783–785. [CrossRef]

57. Gu, Z.; Cai, Q.; Liu, Y.; Li, F.-s. Electrochemical degradation of chitosan using ti/sb–sno2 electrode. *J. Polym. Environ.* **2012**, *21*, 479–486. [CrossRef]
58. Zhao, D.; Wang, J.; Tan, L.; Sun, C.; Dong, J. Synthesis of n-furoyl chitosan and chito-oligosaccharides and evaluation of their antioxidant activity in vitro. *Int. J. Biol. Macromol.* **2013**, *59*, 391–395. [CrossRef] [PubMed]
59. Yang, Y.; Shu, R.; Shao, J.; Xu, G.; Gu, X. Radical scavenging activity of chitooligosaccharide with different molecular weights. *Eur. Food Res. Technol.* **2005**, *222*, 36–40. [CrossRef]
60. Tomida, H.; Fujii, T.; Furutani, N.; Michihara, A.; Yasufuku, T.; Akasaki, K.; Maruyama, T.; Otagiri, M.; Gebicki, J.M.; Anraku, M. Antioxidant properties of some different molecular weight chitosans. *Carbohydr. Res.* **2009**, *344*, 1690–1696. [CrossRef]
61. Garcia-Sanchez, A.; Miranda-Diaz, A.G.; Cardona-Munoz, E.G. The role of oxidative stress in physiopathology and pharmacological treatment with pro- and antioxidant properties in chronic diseases. *Oxid. Med. Cell Longev.* **2020**, *2020*, 2082145. [CrossRef]
62. Steen, E.H.; Wang, X.; Balaji, S.; Butte, M.J.; Bollyky, P.L.; Keswani, S.G. The role of the anti-inflammatory cytokine interleukin-10 in tissue fibrosis. *Adv. Wound Care (New Rochelle)* **2020**, *9*, 184–198. [CrossRef]
63. Kim, K.W.; Thomas, R.L. Antioxidative activity of chitosans with varying molecular weights. *Food Chem.* **2007**, *101*, 308–313. [CrossRef]
64. El-Sayed, S.T.; Omar, N.I.; El-Sayed, E.-S.M.; Shousha, W.G. Evaluation antioxidant and cytotoxic activities of novel chitooligosaccharides prepared from chitosan via enzymatic hydrolysis and ultrafiltration. *J. Appl. Pharm. Sci.* **2017**, *7*, 50–55.
65. Feng, T.; Du, Y.; Li, J.; Wei, Y.; Yao, P. Antioxidant activity of half n-acetylated water-soluble chitosan in vitro. *Eur. Food Res. Technol.* **2006**, *225*, 133. [CrossRef]
66. Vo, T.-S.; Ngo, D.-H.; Bach, L.G.; Ngo, D.-N.; Kim, S.-K. The free radical scavenging and anti-inflammatory activities of gallate-chitooligosaccharides in human lung epithelial a549 cells. *Process. Biochem.* **2017**, *54*, 188–194. [CrossRef]
67. Qu, D.; Han, J. Investigation of the antioxidant activity of chitooligosaccharides on mice with high-fat diet. *Rev. Bras. Zootec.* **2016**, *45*, 661–666. [CrossRef]
68. Kendra, D.F.; Hadwiger, L.A. Characterization of the smallest chitosan oligomer that is maximally antifungal to *Fusarium solani* and elicits pisatin formation in *Pisum sativum*. *Exp. Mycol.* **1984**, *8*, 276–281. [CrossRef]
69. Choi, B.K.; Kim, K.Y.; Yoo, Y.J.; Oh, S.J.; Choi, J.H.; Kim, C.Y. In vitro antimicrobial activity of a chitooligosaccharide mixture against *Actinobacillus actinomycetemcomitans* and *Streptococcus mutans*. *Int. J. Antimicrob. Agents* **2001**, *18*, 553–557. [CrossRef]
70. Goy, R.C.; Britto, D.d.; Assis, O.B.G. A review of the antimicrobial activity of chitosan. *Polímeros* **2009**, *19*, 241–247. [CrossRef]
71. Goy, R.C.; Morais, S.T.B.; Assis, O.B.G. Evaluation of the antimicrobial activity of chitosan and its quaternized derivative on *E. Coli* and *S. Aureus* growth. *Rev. Bras. Farmacogn.* **2016**, *26*, 122–127. [CrossRef]
72. Zheng, L.-Y.; Zhu, J.-F. Study on antimicrobial activity of chitosan with different molecular weights. *Carbohydr. Polym.* **2003**, *54*, 527–530. [CrossRef]
73. Sanchez, A.; Mengibar, M.; Rivera-Rodriguez, G.; Moerchbacher, B.; Acosta, N.; Heras, A. The effect of preparation processes on the physicochemical characteristics and antibacterial activity of chitooligosaccharides. *Carbohydr. Polym.* **2017**, *157*, 251–257. [CrossRef]
74. Busscher, H.J.; Engels, E.; Dijkstra, R.J.; van der Mei, H.C. Influence of a chitosan on oral bacterial adhesion and growth in vitro. *Eur. J. Oral Sci.* **2008**, *116*, 493–495. [CrossRef]
75. Li, K.; Xing, R.; Liu, S.; Qin, Y.; Yu, H.; Li, P. Size and ph effects of chitooligomers on antibacterial activity against *Staphylococcus aureus*. *Int. J. Biol. Macromol.* **2014**, *64*, 302–305. [CrossRef] [PubMed]
76. Chung, Y.C.; Su, Y.P.; Chen, C.C.; Jia, G.; Wang, H.L.; Wu, J.C.; Lin, J.G. Relationship between antibacterial activity of chitosan and surface characteristics of cell wall. *Acta Pharm. Sin.* **2004**, *25*, 932–936.
77. Yan, D.; Li, Y.; Liu, Y.; Li, N.; Zhang, X.; Yan, C. Antimicrobial properties of chitosan and chitosan derivatives in the treatment of enteric infections. *Molecules* **2021**, *26*, 7136. [CrossRef]
78. Benchamas, G.; Huang, G.; Huang, S.; Huang, H. Preparation and biological activities of chitosan oligosaccharides. *Trends Food Sci. Technol.* **2021**, *107*, 38–44. [CrossRef]
79. Wang, Z.; Zheng, L.; Yang, S.; Niu, R.; Chu, E.; Lin, X. N-acetylchitooligosaccharide is a potent angiogenic inhibitor both in vivo and in vitro. *Biochem. Biophys. Res. Commun.* **2007**, *357*, 26–31. [CrossRef] [PubMed]
80. Ganan, M.; Lorentzen, S.B.; Agger, J.W.; Heyward, C.A.; Bakke, O.; Knutsen, S.H.; Aam, B.B.; Eijsink, V.G.H.; Gaustad, P.; Sorlie, M. Antifungal activity of well-defined chito-oligosaccharide preparations against medically relevant yeasts. *PLoS ONE* **2019**, *14*, e0210208.
81. Kim, J.Y.; Lee, J.K.; Lee, T.S.; Park, W.H. Synthesis of chitooligosaccharide derivative with quaternary ammonium group and its antimicrobial activity against *Streptococcus mutans*. *Int. J. Biol. Macromol.* **2003**, *32*, 23–27. [CrossRef]
82. Hui, A.; Yan, R.; Wang, W.; Wang, Q.; Zhou, Y.; Wang, A. Incorporation of quaternary ammonium chitooligosaccharides on zno/palygorskite nanocomposites for enhancing antibacterial activities. *Carbohydr. Polym.* **2020**, *247*, 116685. [CrossRef]
83. Dalavi, P.A.; Prabhu, A.; Shastry, R.P.; Venkatesan, J. Microspheres containing biosynthesized silver nanoparticles with alginate-nano hydroxyapatite for biomedical applications. *J. Biomater. Sci. Polym. Ed.* **2020**, *31*, 2025–2043. [CrossRef]
84. Rajasekaran, S.; Rao, S.S.; Dalavi, P.A.; Prabhu, A.; Anil, S.; Venkatesan, J.; Bhat, S.S. Rapid microwave-assisted biosynthesis of chitooligosaccharide coated silver nanoparticles: Assessments of antimicrobial activity for paediatric pulp therapy. *Adv. Nat. Sci. Nanosci. Nanotechnol.* **2021**, *11*, 045018. [CrossRef]

85. Jang, D.; Lee, D.; Shin, Y.C.; Lee, J.S.; Jung, J.; Ryoo, S. Low molecular weight chitooligosaccharide inhibits infection of SARS-CoV-2 in vitro. *J. Appl. Microbiol.* **2022**, *133*, 1089–1098. [CrossRef] [PubMed]
86. Anogeianaki, A.; Angelucci, D.; Cianchetti, E.; D'Alessandro, M.; Maccauro, G.; Saggini, A.; Salini, V.; Caraffa, A.; Tete, S.; Conti, F.; et al. Atherosclerosis: A classic inflammatory disease. *Int. J. Immunopathol. Pharm.* **2011**, *24*, 817–825. [CrossRef] [PubMed]
87. Mantovani, A. Cancer: Inflaming metastasis. *Nature* **2009**, *457*, 36–37. [CrossRef] [PubMed]
88. Ngo, D.-H.; Vo, T.-S.; Ngo, D.-N.; Kang, K.-H.; Je, J.-Y.; Pham, H.N.-D.; Byun, H.-G.; Kim, S.-K. Biological effects of chitosan and its derivatives. *Food Hydrocoll.* **2015**, *51*, 200–216. [CrossRef]
89. de Andrade, R.; de Araújo, N.K.; Torres-Rêgo, M.; Furtado, A.A.; Daniele-Silva, A.; de Souza Paiva, W.; de Medeiros Dantas, J.M.; da Silva, N.S.; da Silva-Júnior, A.A.; Ururahy, M.A.G.; et al. Production and characterization of chitooligosaccharides: Evaluation of acute toxicity, healing, and anti-inflammatory actions. *Int. J. Mol. Sci.* **2021**, *22*, 10631. [CrossRef]
90. Azuma, K.; Osaki, T.; Kurozumi, S.; Kiyose, M.; Tsuka, T.; Murahata, Y.; Imagawa, T.; Itoh, N.; Minami, S.; Sato, K.; et al. Anti-inflammatory effects of orally administered glucosamine oligomer in an experimental model of inflammatory bowel disease. *Carbohydr. Polym.* **2015**, *115*, 448–456. [CrossRef]
91. Jitprasertwong, P.; Khamphio, M.; Petsrichuang, P.; Eijsink, V.G.H.; Poolsri, W.; Muanprasat, C.; Rangnoi, K.; Yamabhai, M. Anti-inflammatory activity of soluble chito-oligosaccharides (chos) on vitd3-induced human thp-1 monocytes. *PLoS ONE* **2021**, *16*, e0246381. [CrossRef]
92. Li, Y.; Liu, H.; Xu, Q.S.; Du, Y.G.; Xu, J. Chitosan oligosaccharides block lps-induced o-glcnacylation of nf-kappab and endothelial inflammatory response. *Carbohydr. Polym.* **2014**, *99*, 568–578. [CrossRef]
93. Ma, P.; Liu, H.-T.; Wei, P.; Xu, Q.-S.; Bai, X.-F.; Du, Y.-G.; Yu, C. Chitosan oligosaccharides inhibit lps-induced over-expression of il-6 and tnf-α in raw264.7 macrophage cells through blockade of mitogen-activated protein kinase (mapk) and pi3k/akt signaling pathways. *Carbohydr. Polym.* **2011**, *84*, 1391–1398. [CrossRef]
94. Kunanusornchai, W.; Witoonpanich, B.; Tawonsawatruk, T.; Pichyangkura, R.; Chatsudthipong, V.; Muanprasat, C. Chitosan oligosaccharide suppresses synovial inflammation via ampk activation: An in vitro and in vivo study. *Pharm. Res.* **2016**, *113*, 458–467. [CrossRef]
95. Mei, Y.X.; Chen, H.X.; Zhang, J.; Zhang, X.D.; Liang, Y.X. Protective effect of chitooligosaccharides against cyclophosphamide-induced immunosuppression in mice. *Int. J. Biol. Macromol.* **2013**, *62*, 330–335. [CrossRef] [PubMed]
96. Chung, M.J.; Park, J.K.; Park, Y.I. Anti-inflammatory effects of low-molecular weight chitosan oligosaccharides in ige–antigen complex-stimulated rbl-2h3 cells and asthma model mice. *Int. Immunopharmacol.* **2012**, *12*, 453–459. [CrossRef] [PubMed]
97. Yin, Y.L.; Tang, Z.R.; Sun, Z.H.; Liu, Z.Q.; Li, T.J.; Huang, R.L.; Ruan, Z.; Deng, Z.Y.; Gao, B.; Chen, L.X.; et al. Effect of galacto-mannan-oligosaccharides or chitosan supplementation on cytoimmunity and humoral immunity in early weaned piglets. *Asian-Australas. J. Anim. Sci.* **2008**, *21*, 723–731. [CrossRef]
98. Yoon, H.J.; Moon, M.E.; Park, H.S.; Im, S.Y.; Kim, Y.H. Chitosan oligosaccharide (cos) inhibits lps-induced inflammatory effects in raw 264.7 macrophage cells. *Biochem. Biophys. Res. Commun.* **2007**, *358*, 954–959. [CrossRef]
99. Vo, T.S.; Ngo, D.H.; Ta, Q.V.; Wijesekara, I.; Kong, C.S.; Kim, S.K. Protective effect of chitin oligosaccharides against lipopolysaccharide-induced inflammatory response in bv-2 microglia. *Cell Immunol.* **2012**, *277*, 14–21.
100. Einarsson, J.M.; Bahrke, S.; Sigurdsson, B.T.; Ng, C.H.; Petersen, P.H.; Sigurjonsson, O.E.; Jonsson, H., Jr.; Gislason, J.; Thormodsson, F.R.; Peter, M.G. Partially acetylated chitooligosaccharides bind to ykl-40 and stimulate growth of human osteoarthritic chondrocytes. *Biochem. Biophys. Res. Commun.* **2013**, *434*, 298–304. [CrossRef]
101. Gudmundsdottir, S.; Lieder, R.; Sigurjonsson, O.E.; Petersen, P.H. Chitosan leads to downregulation of ykl-40 and inflammasome activation in human macrophages. *J. Biomed. Mater. Res. A* **2015**, *103*, 2778–2785. [CrossRef]
102. Franks, P.W.; McCarthy, M.I. Exposing the exposures responsible for type 2 diabetes and obesity. *Science* **2016**, *354*, 69–73. [CrossRef]
103. Huang, L.; Chen, J.; Cao, P.; Pan, H.; Ding, C.; Xiao, T.; Zhang, P.; Guo, J.; Su, Z. Anti-obese effect of glucosamine and chitosan oligosaccharide in high-fat diet-induced obese rats. *Mar. Drugs* **2015**, *13*, 2732–2756. [CrossRef]
104. Choi, E.H.; Yang, H.P.; Chun, H.S. Chitooligosaccharide ameliorates diet-induced obesity in mice and affects adipose gene expression involved in adipogenesis and inflammation. *Nutr. Res.* **2012**, *32*, 218–228. [CrossRef]
105. Chen, P.; Zhao, M.; Chen, Q.; Fan, L.; Gao, F.; Zhao, L. Absorption characteristics of chitobiose and chitopentaose in the human intestinal cell line caco-2 and everted gut sacs. *J. Agric. Food Chem.* **2019**, *67*, 4513–4523. [CrossRef] [PubMed]
106. Pan, H.; Fu, C.; Huang, L.; Jiang, Y.; Deng, X.; Guo, J.; Su, Z. Anti-obesity effect of chitosan oligosaccharide capsules (coscs) in obese rats by ameliorating leptin resistance and adipogenesis. *Mar. Drugs* **2018**, *16*, 198. [CrossRef] [PubMed]
107. Kang, N.H.; Lee, W.K.; Yi, B.R.; Park, M.A.; Lee, H.R.; Park, S.K.; Hwang, K.A.; Park, H.K.; Choi, K.C. Modulation of lipid metabolism by mixtures of protamine and chitooligosaccharide through pancreatic lipase inhibitory activity in a rat model. *Lab. Anim. Res.* **2012**, *28*, 31–38. [CrossRef] [PubMed]
108. Zhang, J.; Zhang, W.; Mamadouba, B.; Xia, W. A comparative study on hypolipidemic activities of high and low molecular weight chitosan in rats. *Int. J. Biol. Macromol.* **2012**, *51*, 504–508. [CrossRef]
109. Kao, C.H.; Hsiang, C.Y.; Ho, T.Y. Assessment of chitosan-affected metabolic response by peroxisome proliferator-activated receptor bioluminescent imaging-guided transcriptomic analysis. *PLoS ONE* **2012**, *7*, e34969. [CrossRef]

110. Shang, W.; Si, X.; Zhou, Z.; Wang, J.; Strappe, P.; Blanchard, C. Studies on the unique properties of resistant starch and chito-oligosaccharide complexes for reducing high-fat diet-induced obesity and dyslipidemia in rats. *J. Funct. Foods* **2017**, *38*, 20–27. [CrossRef]
111. Deng, X.; Ye, Z.; Cao, H.; Bai, Y.; Che, Q.; Guo, J.; Su, Z. Chitosan oligosaccharide ameliorated obesity by reducing endoplasmic reticulum stress in diet-induced obese rats. *Food Funct.* **2020**, *11*, 6285–6296. [CrossRef]
112. Karadeniz, F.; Kim, S.K. Antidiabetic activities of chitosan and its derivatives: A mini review. *Adv. Food Nutr. Res.* **2014**, *73*, 33–44.
113. Liu, B.; Liu, W.S.; Han, B.Q.; Sun, Y.Y. Antidiabetic effects of chitooligosaccharides on pancreatic islet cells in streptozotocin-induced diabetic rats. *World J. Gastroenterol.* **2007**, *13*, 725–731. [CrossRef]
114. Zhu, D.; Yan, Q.; Liu, J.; Wu, X.; Jiang, Z. Can functional oligosaccharides reduce the risk of diabetes mellitus? *FASEB J.* **2019**, *33*, 11655–11667. [CrossRef]
115. Yuan, W.P.; Liu, B.; Liu, C.H.; Wang, X.J.; Zhang, M.S.; Meng, X.M.; Xia, X.K. Antioxidant activity of chito-oligosaccharides on pancreatic islet cells in streptozotocin-induced diabetes in rats. *World J. Gastroenterol.* **2009**, *15*, 1339–1345. [CrossRef] [PubMed]
116. Ju, C.; Yue, W.; Yang, Z.; Zhang, Q.; Yang, X.; Liu, Z.; Zhang, F. Antidiabetic effect and mechanism of chitooligosaccharides. *Biol. Pharm. Bull.* **2010**, *33*, 1511–1516. [CrossRef] [PubMed]
117. Kim, H.J.; Ahn, H.Y.; Kwak, J.H.; Shin, D.Y.; Kwon, Y.I.; Oh, C.G.; Lee, J.H. The effects of chitosan oligosaccharide (go2ka1) supplementation on glucose control in subjects with prediabetes. *Food Funct.* **2014**, *5*, 2662–2669. [CrossRef]
118. Katiyar, D.; Singh, B.; Lall, A.; Haldar, C. Evaluation of antidiabetic and hypolipidemic activity of chitooligosaccharides in alloxan-induced diabetes mellitus in mice. *Int. J. Pharm. Bio. Sci.* **2011**, *2*, 407–416.
119. Zisman, A.; Peroni, O.D.; Abel, E.D.; Michael, M.D.; Mauvais-Jarvis, F.; Lowell, B.B.; Wojtaszewski, J.F.P.; Hirshman, M.F.; Virkamaki, A.; Goodyear, L.J.; et al. Targeted disruption of the glucose transporter 4 selectively in muscle causes insulin resistance and glucose intolerance. *Nat. Med.* **2000**, *6*, 924–928. [CrossRef] [PubMed]
120. Srinivasan, K.; Viswanad, B.; Asrat, L.; Kaul, C.L.; Ramarao, P. Combination of high-fat diet-fed and low-dose streptozotocin-treated rat: A model for type 2 diabetes and pharmacological screening. *Pharm. Res.* **2005**, *52*, 313–320. [CrossRef]
121. Jung, W.K.; Moon, S.H.; Kim, S.K. Effect of chitooligosaccharides on calcium bioavailability and bone strength in ovariectomized rats. *Life Sci.* **2006**, *78*, 970–976. [CrossRef]
122. Kim, H.M.; Hong, S.H.; Yoo, S.J.; Baek, K.S.; Jeon, Y.J.; Choung, S.Y. Differential effects of chitooligosaccharides on serum cytokine levels in aged subjects. *J. Med. Food* **2006**, *9*, 427–430. [CrossRef]
123. Pangestuti, R.; Bak, S.S.; Kim, S.K. Attenuation of pro-inflammatory mediators in lps-stimulated bv2 microglia by chitooligosaccharides via the mapk signaling pathway. *Int. J. Biol. Macromol.* **2011**, *49*, 599–606. [CrossRef]
124. Richards, J.B.; Joseph, L.; Schwartzman, K.; Kreiger, N.; Tenenhouse, A.; Goltzman, D. The effect of cyclooxygenase-2 inhibitors on bone mineral density: Results from the canadian multicentre osteoporosis study. *Osteoporos. Int.* **2006**, *17*, 1410–1419. [CrossRef]
125. Kim, S.; Park, P.; Jung, W.; Byun, H.; Mendis, E.; Cho, Y. Inhibitory activity of phosphorylated chitooligosaccharides on the formation of calcium phosphate. *Carbohydr. Polym.* **2005**, *60*, 483–487. [CrossRef]
126. Hong, S.-P.; Kim, M.-H.; Oh, S.-W.; Han, C.-K.; Kim, Y.-H. Ace inhibitory and antihypertensive effect of chitosan oligosaccharides in shr. *Korean J. Food Sci. Technol.* **1998**, *30*, 1476–1479.
127. Park, P.J.; Je, J.Y.; Kim, S.K. Angiotensin i converting enzyme (ace) inhibitory activity of hetero-chitooligosaccharides prepared from partially different deacetylated chitosans. *J. Agric. Food Chem.* **2003**, *51*, 4930–4934. [CrossRef] [PubMed]
128. Huang, R.; Mendis, E.; Kim, S.K. Improvement of ace inhibitory activity of chitooligosaccharides (cos) by carboxyl modification. *Bioorg. Med. Chem.* **2005**, *13*, 3649–3655. [CrossRef] [PubMed]
129. Ngo, D.-N.; Qian, Z.-J.; Je, J.-Y.; Kim, M.-M.; Kim, S.-K. Aminoethyl chitooligosaccharides inhibit the activity of angiotensin converting enzyme. *Process. Biochem.* **2008**, *43*, 119–123. [CrossRef]
130. Vila, M.; Przedborski, S. Targeting programmed cell death in neurodegenerative diseases. *Nat. Rev. Neurosci.* **2003**, *4*, 365–375. [CrossRef]
131. Dai, X.; Hou, W.; Sun, Y.; Gao, Z.; Zhu, S.; Jiang, Z. Chitosan oligosaccharides inhibit/disaggregate fibrils and attenuate amyloid beta-mediated neurotoxicity. *Int. J. Mol. Sci.* **2015**, *16*, 10526–10536. [CrossRef]
132. Xu, Y.; Zhang, Q.; Yu, S.; Yang, Y.; Ding, F. The protective effects of chitooligosaccharides against glucose deprivation-induced cell apoptosis in cultured cortical neurons through activation of pi3k/akt and mek/erk1/2 pathways. *Brain Res.* **2011**, *1375*, 49–58. [CrossRef]
133. Lee, S.H.; Park, J.S.; Kim, S.K.; Ahn, C.B.; Je, J.Y. Chitooligosaccharides suppress the level of protein expression and acetylcholinesterase activity induced by abeta25-35 in pc12 cells. *Bioorg. Med. Chem. Lett.* **2009**, *19*, 860–862. [CrossRef]
134. Hou, H.; Zhang, L.; Ye, Z.; Li, J.; Lian, Z.; Chen, C.; He, R.; Peng, B.; Xu, Q.; Zhang, G.; et al. Chitooligosaccharide inhibits scar formation and enhances functional recovery in a mouse model of sciatic nerve injury. *Mol. Neurobiol.* **2016**, *53*, 2249–2257. [CrossRef]
135. Xu, W.; Huang, H.C.; Lin, C.J.; Jiang, Z.F. Chitooligosaccharides protect rat cortical neurons against copper induced damage by attenuating intracellular level of reactive oxygen species. *Bioorg. Med. Chem. Lett.* **2010**, *20*, 3084–3088. [CrossRef] [PubMed]
136. Park, J.K.; Chung, M.J.; Choi, H.N.; Park, Y.I. Effects of the molecular weight and the degree of deacetylation of chitosan oligosaccharides on antitumor activity. *Int. J. Mol. Sci.* **2011**, *12*, 266–277. [CrossRef] [PubMed]
137. Shen, K.T.; Chen, M.H.; Chan, H.Y.; Jeng, J.H.; Wang, Y.J. Inhibitory effects of chitooligosaccharides on tumor growth and metastasis. *Food Chem. Toxicol.* **2009**, *47*, 1864–1871. [CrossRef]

138. Zou, P.; Yang, X.; Zhang, Y.; Du, P.; Yuan, S.; Yang, D.; Wang, J. Antitumor effects of orally and intraperitoneally administered chitosan oligosaccharides (coss) on s180-bearing/residual mouse. *J. Food Sci.* **2016**, *81*, H3035–H3042. [CrossRef]
139. Huang, R.; Mendis, E.; Rajapakse, N.; Kim, S.K. Strong electronic charge as an important factor for anticancer activity of chitooligosaccharides (cos). *Life Sci.* **2006**, *78*, 2399–2408. [CrossRef]
140. Suzuki, K.; Mikami, T.; Okawa, Y.; Tokoro, A.; Suzuki, S.; Suzuki, M. Antitumor effect of hexa-n-acetylchitohexaose and chitohexaose. *Carbohydr. Res.* **1986**, *151*, 403–408. [CrossRef]
141. Tokoro, A.; Tatewaki, N.; Suzuki, K.; Mikami, T.; Suzuki, S.; Suzuki, M. Growth-inhibitory effect of hexa-n-acetylchitohexaose and chitohexaose against meth-a solid tumor. *Chem. Pharm. Bull.* **1988**, *36*, 784–790. [CrossRef]
142. Xu, W.; Jiang, C.; Kong, X.; Liang, Y.; Rong, M.; Liu, W. Chitooligosaccharides and n-acetyl-d-glucosamine stimulate peripheral blood mononuclear cell-mediated antitumor immune responses. *Mol. Med. Rep.* **2012**, *6*, 385–390. [CrossRef]
143. Xiong, C.; Wu, H.; Wei, P.; Pan, M.; Tuo, Y.; Kusakabe, I.; Du, Y. Potent angiogenic inhibition effects of deacetylated chitohexaose separated from chitooligosaccharides and its mechanism of action in vitro. *Carbohydr. Res.* **2009**, *344*, 1975–1983. [CrossRef]
144. Xiao, J.; Duan, X.; Yin, Q.; Miao, Z.; Yu, H.; Chen, C.; Zhang, Z.; Wang, J.; Li, Y. The inhibition of metastasis and growth of breast cancer by blocking the nf-κb signaling pathway using bioreducible pei-based/p65 shrna complex nanoparticles. *Biomaterials* **2013**, *34*, 5381–5390. [CrossRef]
145. Hong, S.; Ngo, D.N.; Kim, M.M. Inhibitory effect of aminoethyl-chitooligosaccharides on invasion of human fibrosarcoma cells. *Environ. Toxicol. Pharm.* **2016**, *45*, 309–314. [CrossRef] [PubMed]
146. Zhai, X.; Yuan, S.; Yang, X.; Zou, P.; Li, L.; Li, G.; Shao, Y.; Abd El-Aty, A.M.; Hacimuftuoglu, A.; Wang, J. Chitosan oligosaccharides induce apoptosis in human renal carcinoma via reactive-oxygen-species-dependent endoplasmic reticulum stress. *J. Agric. Food Chem.* **2019**, *67*, 1691–1701. [CrossRef] [PubMed]
147. Zou, P.; Yuan, S.; Yang, X.; Zhai, X.; Wang, J. Chitosan oligosaccharides with degree of polymerization 2-6 induces apoptosis in human colon carcinoma hct116 cells. *Chem. Biol. Interact.* **2018**, *279*, 129–135. [CrossRef]
148. Nam, K.S.; Kim, M.K.; Shon, Y.H. Inhibition of proinflammatory cytokine-induced invasiveness of ht-29 cells by chitosan oligosaccharide. *J. Microbiol. Biotechnol.* **2007**, *17*, 2042–2045. [PubMed]
149. Yu, H.; Xu, X.; Chen, X.; Hao, J.; Jing, X. Medicated wound dressings based on poly(vinyl alcohol)/poly(n-vinyl pyrrolidone)/chitosan hydrogels. *J. Appl. Polym. Sci.* **2006**, *101*, 2453–2463. [CrossRef]
150. Okamoto, Y.; Watanabe, M.; Miyatake, K.; Morimoto, M.; Shigemasa, Y.; Minami, S. Effects of chitin/chitosan and their oligomers/monomers on migrations of fibroblasts and vascular endothelium. *Biomaterials* **2002**, *23*, 1975–1979. [CrossRef]
151. Jafari, H.; Bernaerts, K.V.; Dodi, G.; Shavandi, A. Chitooligosaccharides for wound healing biomaterials engineering. *Mater. Sci. Eng. C Mater. Biol. Appl.* **2020**, *117*, 111266. [CrossRef]
152. Yamamoto, M.; Sato, T.; Beren, J.; Verthelyi, D.; Klinman, D.M. The acceleration of wound healing in primates by the local administration of immunostimulatory cpg oligonucleotides. *Biomaterials* **2011**, *32*, 4238–4242. [CrossRef]
153. Oh, G.W.; Kim, S.C.; Kim, T.H.; Jung, W.K. Characterization of an oxidized alginate-gelatin hydrogel incorporating a cos-salicylic acid conjugate for wound healing. *Carbohydr. Polym.* **2021**, *252*, 117145. [CrossRef]
154. Park, H.H.; Ko, S.C.; Oh, G.W.; Jang, Y.M.; Kim, Y.M.; Park, W.S.; Choi, I.W.; Jung, W.K. Characterization and biological activity of pva hydrogel containing chitooligosaccharides conjugated with gallic acid. *Carbohydr. Polym.* **2018**, *198*, 197–205. [CrossRef]
155. Nguyen, M.H.; Lee, S.E.; Tran, T.T.; Bui, C.B.; Nguyen, T.H.; Vu, N.B.; Tran, T.T.; Nguyen, T.H.; Nguyen, T.T.; Hadinoto, K. A simple strategy to enhance the in vivo wound-healing activity of curcumin in the form of self-assembled nanoparticle complex of curcumin and oligochitosan. *Mater. Sci. Eng. C Mater. Biol. Appl.* **2019**, *98*, 54–64. [CrossRef] [PubMed]
156. Ragetly, G.R.; Griffon, D.J.; Lee, H.-B.; Fredericks, L.P.; Gordon-Evans, W.; Chung, Y.S. Effect of chitosan scaffold microstructure on mesenchymal stem cell chondrogenesis. *Acta Biomater.* **2010**, *6*, 1430–1436. [CrossRef] [PubMed]
157. Yang, Y.; Liu, M.; Gu, Y.; Lin, S.; Ding, F.; Gu, X. Effect of chitooligosaccharide on neuronal differentiation of pc-12 cells. *Cell Biol. Int.* **2009**, *33*, 352–356. [CrossRef] [PubMed]
158. Ratanavaraporn, J.; Damrongsakkul, S.; Kanokpanont, S.; Yamamoto, M.; Tabata, Y. Osteogenic differentiation of bone-marrow-derived stem cells cultured with mixed gelatin and chitooligosaccharide scaffolds. *J. Biomater. Sci. Polym. Ed.* **2011**, *22*, 1083–1098. [CrossRef] [PubMed]
159. Li, J.; Chen, Y.; Yin, Y.; Yao, F.; Yao, K. Modulation of nano-hydroxyapatite size via formation on chitosan–gelatin network film in situ. *Biomaterials* **2007**, *28*, 781–790. [CrossRef]
160. Pallela, R.; Venkatesan, J.; Janapala, V.R.; Kim, S.-K. Biophysicochemical evaluation of chitosan-hydroxyapatite-marine sponge collagen composite for bone tissue engineering. *J. Biomed. Mater. Res. Part A* **2012**, *100A*, 486–495. [CrossRef]
161. Chandika, P.; Ko, S.-C.; Oh, G.-W.; Heo, S.-Y.; Nguyen, V.-T.; Jeon, Y.-J.; Lee, B.; Jang, C.H.; Kim, G.; Park, W.S.; et al. Fish collagen/alginate/chitooligosaccharides integrated scaffold for skin tissue regeneration application. *Int. J. Biol. Macromol.* **2015**, *81*, 504–513. [CrossRef]
162. Sultankulov, B.; Berillo, D.; Sultankulova, K.; Tokay, T.; Saparov, A. Progress in the development of chitosan-based biomaterials for tissue engineering and regenerative medicine. *Biomolecules* **2019**, *9*, 470. [CrossRef]
163. Cherwin, A.; Namen, S.; Rapacz, J.; Kusik, G.; Anderson, A.; Wang, Y.; Kaltchev, M.; Schroeder, R.; O'Connell, K.; Stephens, S.; et al. Design of a novel oxygen therapeutic using polymeric hydrogel microcapsules mimicking red blood cells. *Pharmaceutics* **2019**, *11*, 583. [CrossRef]

164. Tabassum, N.; Ahmed, S.; Ali, M.A. Chitooligosaccharides and their structural-functional effect on hydrogels: A review. *Carbohydr. Polym.* **2021**, *261*, 117882. [CrossRef]
165. Thanou, M.; Florea, B.I.; Geldof, M.; Junginger, H.E.; Borchard, G. Quaternized chitosan oligomers as novel gene delivery vectors in epithelial cell lines. *Biomaterials* **2002**, *23*, 153–159. [CrossRef]
166. Sacco, P.; Cok, M.; Asaro, F.; Paoletti, S.; Donati, I. The role played by the molecular weight and acetylation degree in modulating the stiffness and elasticity of chitosan gels. *Carbohydr. Polym.* **2018**, *196*, 405–413. [CrossRef] [PubMed]
167. Chae, S.Y.; Jang, M.K.; Nah, J.W. Influence of molecular weight on oral absorption of water soluble chitosans. *J. Control. Release* **2005**, *102*, 383–394. [PubMed]
168. Zhao, S.; Lee, J.; Xu, W. Supramolecular hydrogels formed from biodegradable ternary cos-g-pcl-b-mpeg copolymer with alpha-cyclodextrin and their drug release. *Carbohydr. Res.* **2009**, *344*, 2201–2208. [CrossRef] [PubMed]
169. Mahato, K.K.; Sabbarwal, S.; Misra, N.; Kumar, M. Fabrication of polyvinyl alcohol/chitosan oligosaccharide hydrogel: Physicochemical characterizations and in vitro drug release study. *Int. J. Polym. Anal. Charact.* **2020**, *25*, 353–361. [CrossRef]
170. Shaikh, R.; Raj Singh, T.R.; Garland, M.J.; Woolfson, A.D.; Donnelly, R.F. Mucoadhesive drug delivery systems. *J. Pharm. Bioallied. Sci.* **2011**, *3*, 89–100.
171. Garaiova, Z.; Strand, S.P.; Reitan, N.K.; Lelu, S.; Storset, S.O.; Berg, K.; Malmo, J.; Folasire, O.; Bjorkoy, A.; Davies Cde, L. Cellular uptake of DNA-chitosan nanoparticles: The role of clathrin- and caveolae-mediated pathways. *Int. J. Biol. Macromol.* **2012**, *51*, 1043–1051. [CrossRef]

polymers

Article

Effect of Bleaching and Hot-Pressing Conditions on Mechanical Properties of Compressed Wood

Le Van Hai [1,2], Duc Hoa Pham [1] and Jaehwan Kim [1,*]

1 Creative Research Center for Nanocellulose Future Composites, Inha University, 100 Inha-ro, Michuhol-ku, Incheon 22212, Korea; levanhai121978@gmail.com (L.V.H.); phamduchoa.tdt@gmail.com (D.H.P.)
2 Pulp and Paper Dept, Phutho College of Industry and Trade, Phongchau, Phuninh, Phutho 290000, Vietnam
* Correspondence: jaehwan@inha.ac.kr; Tel.: +82-32-874-7325

Abstract: This paper reports on multiple stage bleaching and its effect on the mechanical and swelling properties of compressed wood (CW). The natural wood specimen was bleached with $NaClO_2$ in five steps and three hot-pressing conditions. Their effects were investigated in morphologies: lignin content, alpha-cellulose content, compression ratio, mechanical properties, swelling and, water contact angle. After compression, the wood specimens became dense and the most porous structures collapsed. The lignin content decreased as the bleaching steps progressed, and the highest alpha-cellulose content was observed at the third bleaching step. This CW showed the best mechanical properties: bending strength was 240.1 ± 35.7 MPa, and Young's modulus was 23.08 ± 0.89 Gpa. The CW swelling decreased as the bleaching step progressed, and was associated with the density decrease and the compression ratio increase with the bleaching step. The B3 is an optimum bleaching step that accounts for the best mechanical properties, which might be associated with the highest alpha-cellulose content.

Keywords: compressed wood; bleaching; lignin content; alpha-cellulose; swelling

Citation: Van Hai, L.; Pham, D.H.; Kim, J. Effect of Bleaching and Hot-Pressing Conditions on Mechanical Properties of Compressed Wood. *Polymers* **2022**, *14*, 2901. https://doi.org/10.3390/polym14142901

Academic Editors: Clara Delgado-Sánchez, Concepción Valencia-Barragán, Esperanza Cortés Triviño, Adrián Tenorio-Alfonso and Antonio M. Borrero-López

Received: 30 May 2022
Accepted: 15 July 2022
Published: 17 July 2022

1. Introduction

Wood is one of the polymer matrix composites nature produces. In total, 4–5 billion cubic meters of wood are harvested annually for various applications such as houses, furniture, construction, ships, bridges, pulp and papers, and many others [1]. Due to its organic cell-type structure, the specific strength and stiffness in certain cases of wood can reach notable values competitive with other construction materials, such as low carbon steel and aluminum alloys [2]. Compressed wood (CW), or densified wood, has been investigated for a long time, and there have been many reported applications of CW, including flooring products [3]. CW can reduce the porosity or void space of the wood and make highly compact wood, leading to better mechanical properties. There are many different approaches for fabricating CW [1,3–9]: thermo-hydro-mechanical (THM), thermo-mechanical, heat treatment, chemical, surface densification, steam treatment, non-steaming densification, and so on [10,11]. The thermo-mechanical method uses an open hot- press without controlling the humidity and moisture content. Densified wood showed great strength, low price, good stability, and an environmentally friendly character [11].

In CW, lignin removal and the compression condition are important for improving the mechanical properties. Yano's group bleached wood with sodium chloride and then infiltrated it with phenol-formaldehyde resins [12], such that the tensile strength was improved from 317.4 MPa to 453.7 MPa, and Young's modulus from 26 GPa to 39 GPa. They reported that the wood specimen's weight loss is from 14 to 24%, mainly from lignin and hemicellulose. However, they did not report the lignin and cellulose content after bleaching. Song et al., 2018 [13] densified the wood with sodium hydroxide and sodium sulfite bleaching followed by compression at 100 °C and 5 MPa for one day. The

CW exhibited a great improvement in the tensile strength from 100 MPa to more than 400 MPa. As the lignin content decreased from 20.8% to around 0.13% by cooking with sodium hydroxide and sodium silicate, the cellulose content reduced from 44.01% to 31.2%. Fang et al., 2012 [6] adopted different temperature and steaming conditions for preparing CW. Using different heating temperatures from 140 °C to 220 °C and the pressure from 4.5 MPa to 9.0 MPa, the tensile strength increased from 50–70 MPa to 150 MPa. Fang et al., 2020 [14] also reported that CW could gain super mechanical properties using sodium hydroxide and anthraquinone: 1.13 GPa of tensile strength and 50–60 GPa of Young's modulus. The lignin content reached around 1% after several hours.

There has not been any detailed research on the bleaching condition effect on the alpha-cellulose content that influences mechanical properties of CW, although the lignin content effect of CW has been reported. According to previous reports, the lignin content is very broad, from 0.13 to 14%, and the cellulose content of CW was not thoroughly investigated for improving the mechanical properties. Thus, this report focuses on multiple-stage bleaching and its effect on CW's mechanical and swelling properties. The influence of the lignin and cellulose contents was studied to improve the mechanical properties. Lignin is more hydrophobic than cellulose; thus, non-bleached wood absorbs less water than bleached wood. Due to lignin removal, bleached wood absorbs water easily and becomes unstable during humidity changes. Thus, the swelling properties are investigated with the lignin content and cellulose content. A bleached wood surface can easily absorb water as lignin was partially or mostly removed. However, the waterproof nature of CW is important for its applications. Thus, the water contact angle (WCA) of the prepared CW is measured. Lastly, the density changes and the hot-pressing conditions are also studied to monitor their effect on CW. In order to bleach wood, various bleaching agents or methods are used, such as NaOH, Na_2SO_3, $NaClO_2$, H_2O_2, or UV-assisted bleaching methods. Song et al., 2018 [13] used NaOH and Na_2SO_3 under a boiling condition to remove lignin. Xia et al., 2021 [15] used a UV-assisted bleaching method to decolorate wood. $NaClO_2$ is popularly used for bleaching wood [12,16,17].

In this research, $NaClO_2$ was selected for the bleaching process, and the detailed CW preparation process was explained. The specimens' morphologies were characterized using field-emission scanning electron microscopy (FESEM). The lignin content, alpha-cellulose content, compression ratio, and swelling ratio were examined, and the three-point-bending (TPB) test was performed to characterize the mechanical properties.

2. Materials and Methods

2.1. Materials

Pinewood (*Pinus koraiensis*) was purchased from a local market in Korea. The wood specimen size was 150 × 50 × 15 mm^3, and the grain direction was unidirectional along the longitudinal: 150 mm was the longitudinal direction, 50 mm was the tangential direction, and 15 mm was the radial direction. The wood sample was then dried in an oven drier at 103 °C for several days before the experiment. Sodium chloride ($NaClO_2$), sodium acetate (CH_3COONa), and acetic acid (CH_3COOH) were purchased from Sigma Aldrich. All chemicals used were analytical reagent grade (purity > 99%) and were used as received.

2.2. Methods

2.2.1. Bleaching Process

Figure 1 shows the CW preparation process consisting of bleaching and hot pressing. The bleaching condition adopted from Yano et al., 2001 [12] was modified by having multiple steps as follows:

- **Step 1:** 1% of $NaClO_2$ was adjusted pH to 4–5 with acetic acid-sodium acetate buffer, then the sample was bleached at 80 °C for 12 h. The bleached wood was washed under running water for around 2 h and kept in a water bath for a day. After washing, the sample was kept in an oven drier at 40 °C for 1–2 days before proceeding to the next bleaching step.

- **Step 2:** In total, 2% of $NaClO_2$ was adjusted to a pH of 4–5 with acetic acid-sodium acetate buffer, then the sample was bleached at 80 °C for 12 h. The bleached wood was washed under running water for 2 h and kept in a water bath for a day. After washing, the sample was kept in an oven drier at 40 °C for 1–2 days before proceeding to the next bleaching step.
- **Step 3:** Step 1 was repeated.
- **Step 4:** The sample was kept in 0.1% NaOH suspension for 1 day at room temperature, and then it was washed under running water for 2 h, followed by a water bath for 1 day. After washing, the sample was kept in an oven drier at 40 °C for 1–2 days before processing to the next step.
- **Step 5:** Step 1 was repeated.

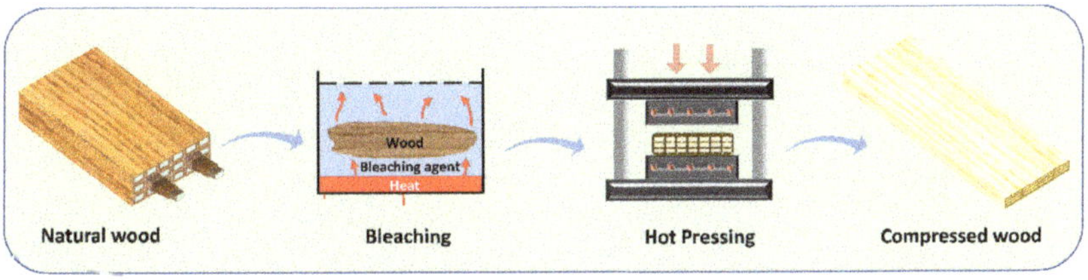

Figure 1. Compressed wood preparation process.

The pristine wood specimen was named "NW." After Step 1 finished, the specimen was called "B1". Likewise, the specimens were named B3, B4, and B5 after finishing Steps 3, 4, and 5. Since Step 2 was the same as Step 1, except for different $NaClO_2$ content, it was not much different, and the specimen was not prepared.

2.2.2. Compressed Wood Preparation

The bleached wood specimens were hot-pressed using a hot-pressing machine (SB3651, HY Industry, Incheon, South Korea) for preparing CW. The hot-pressing machine is a hydraulic-type machine, and the hot-press machine's maximum hydraulic pressure is limited to 10 MPa. The cylinder area is 0.0195 m^2. Generally, hot pressing is governed by temperature, pressure, and time. Two temperature conditions were tested for hot pressing: 100 and 150 °C. However, the 150 °C for 8 h was so high that the CW surfaces were always burned. Thus, the temperature was fixed to 100 °C.

The machine's hydraulic pressure was made only for 5 MPa and 10 MPa, and the hot-pressing time was varied to 4, 8, and 16 h. The pressure and the time variations have six combinations, which are too many cases. According to our preliminary observation, the pressure and time exhibited a similar effect: as the pressure and time increased, the compression increased. Thus, these hot-pressing conditions were accelerated by choosing the following conditions: 5 MPa and 4 h, 10 MPa and 8 h, and 10 MPa, and 16 h. Note that the hydraulic pressure values 5 MPa and 10 MPa correspond to the actual compression pressure of 13 MPa and 26 MPa on the wood specimens. To further increase the compression pressure, one wood specimen area was reduced to 150 × 25 mm^2 such that the compression pressure was made to 52 MPa. Consequently, the following hot-pressing conditions were made (in terms of actual compression pressure): HP-1: 13 MPa and 4 h; HP-2: 26 MPa and 8 h; and HP-3: 52 MPa and 16 h. The compression ratio was calculated by:

$$CR = (t_i - t_c)/t_i \tag{1}$$

where t_i is the initial thickness and t_c is the thickness after hot pressing. The prepared specimens were stored in a desiccator (VESD-350, GoodsGood Co., Ltd., Gyeryong-si, Chungcheongnam-do, South Korea) at 15% RH and 23 °C.

2.3. *Characterization*

2.3.1. Morphology

The cross-sectional morphologies of the wood specimens before and after compression were observed using the FE-SEM (H-4000, Hitachi, Tokyo, Japan). The specimens were cut by dipping into liquid nitrogen and breaking them. The specimens were coated with platinum in advance and observed with 10 kV.

2.3.2. Lignin and Alpha-Cellulose Contents

The Klason lignin in the wood was determined by following the TAPPI T 222 om-02 [18]. In total, 15 mL of 72 wt% sulfuric acid was added to a beaker containing 0.5 g of wood sample. The wood samples were stirred with a magnetic stirrer for 2 h at 20 °C. After 2 h reaction time, DI water was added to the 72 wt% sulfuric acids to dilute it to 3 wt% sulfuric acid (TAPPI T 222 om-02). The samples were then boiled in a 3-neck flask for 4 h, and the resultant samples were left to cool in a fume hood overnight and then filtered. After the reaction, the crucible cup 1G3 was used to collect and clean the remaining lignin. The collected samples were dried at 105 °C, and the weight of lignin content was determined by:

$$lignin\ (\%) = m/M \times 100 \tag{2}$$

where m is the weight of lignin and M is the oven-dried weight of the specimen.

The alpha-cellulose content was characterized according to Hai et al., 2013 [19]. In brief, a one-gram oven-dried weight of bleached cellulose was treated in 25 mL of 17.5% NaOH at room temperature and mixed for 30 min. Then, 25 mL of deionized (DI) water was added to stop the reaction and then washed through glass filers with DI water, after which point 10 mL of 10% acetic acid was added for 5 min and washed with DI water to neutralize it using a vacuum pump. The washed sample was kept in an oven drier at 105 °C overnight and its alpha-cellulose content was determined.

2.3.3. Mechanical Property

The TPB test was carried out using a tensile test machine (TEST ONE, South Korea) with the TPB test fixture. The distance between the two holding bars was 100 mm, the TPB specimen size was 150 × 10 mm^{2}', and the thickness ranged from 2.81 to 6.05 mm depending on different bleaching steps and hot-pressing conditions. The TPB test was carried out at 1 mm/min crosshead speed. The tensile test system automatically calculated the bending strength, strain-at-break, and Young's modulus from the given dimension from the stress-strain curves. At least five samples were tested for each case. All wood specimens were kept in a humidity-controlled chamber (30% RH and 25 °C). Samples were kept in a humidity and temperature-controlled box. The mechanical properties of the samples were characterized using a tensile test machine (TEST ONE, Siheung-si, Kyunggi-do, South Korea) with a humidity/temperature-controlled chamber.

2.3.4. Swelling Ratio

The swelling test was performed to address the stability of the materials under high moisture content. The CW specimens prepared under HP-3 with different bleaching conditions were placed in the humidity-controlled chamber set to 90% RH and 25 °C for 8 h. The swelling ratio was found by:

$$\text{Swelling Ratio}\ (\%) = 100 \times (t_s - t_i)/t_i \tag{3}$$

where t_s represents the thickness after swelling and t_i is the initial thickness.

3. Results and Discussion

3.1. *Morphology*

Figure 2 shows the cross-sectional FESEM images of the bleached wood specimensthat showed high porosity. As the bleaching steps progressed, more pore structures were observed. Figure 3A–D represents the cross-sectional FESEM images of the CW specimens with HP-3 condition. After compression, the wood specimens became dense and the most porous structures collapsed. The voids in the porous structure were retained in some regions, but most of the voids disappeared. Figure 3E–I shows the photographs of wood specimens for natural wood (E), compressed natural wood with different temperatures at 120 °C, 150 °C and 180 °C under 5 MPa pressure (F-H), and a B3 compressed specimen at 100 °C and 5 MPa. When the hot-pressing temperature was over 100 °C, the specimens were burnt. Thus, 100 °C was chosen for the hot-pressing temperature.

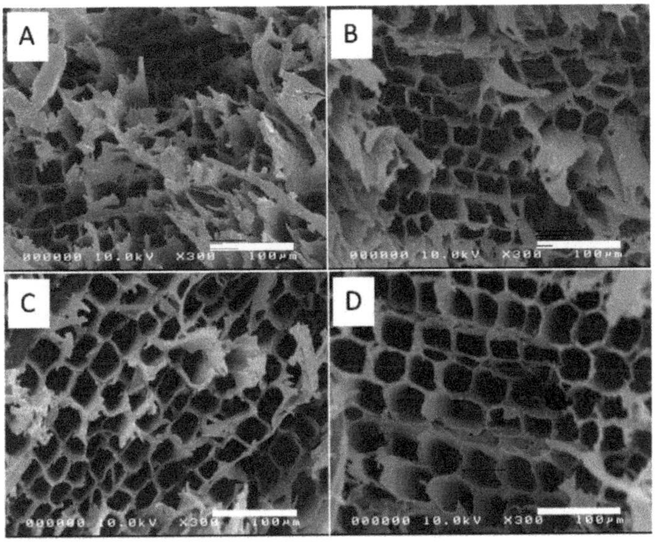

Figure 2. Cross-sectional FESEM images of bleached wood specimens before compression: (**A**) B1, (**B**) B3, (**C**) B4, and (**D**) B5.

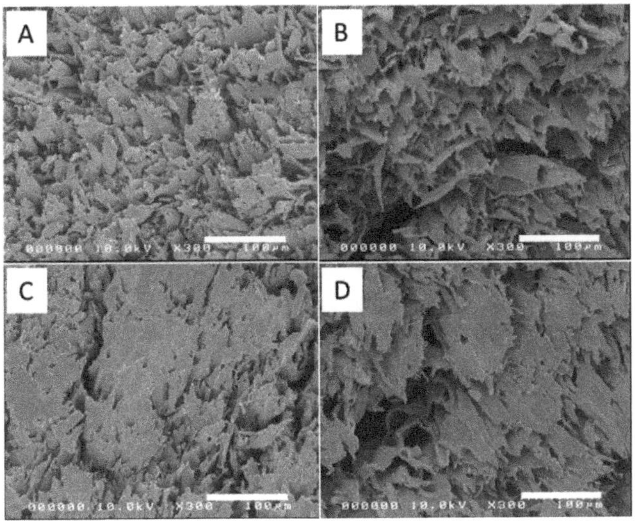

Figure 3. *Cont.*

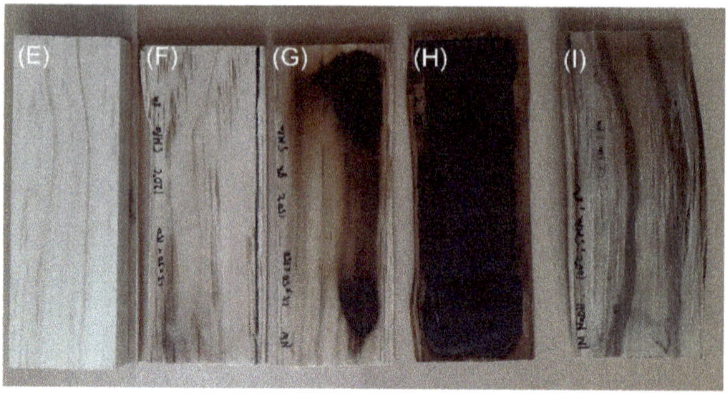

Figure 3. Cross-sectional FESEM images of compressed wood specimens (**A–D**): (**A**) B1, (**B**) B3, (**C**) B4, and (**D**) B5; Photographs of wood specimens (**E–I**): (**E**) natural wood, (**F–H**) compressed natural wood at 120 °C, 150 °C, and 180 °C under 5 MPa, (**G**) compressed B3 at 100 °C and 5 MPa.

3.2. Lignin and Alpha-Cellulose Contents

Lignin and alpha-cellulose contents were examined under different bleaching conditions, and the results are shown in Figure 4. The lignin content of the natural wood was initially 30.3%, but after bleaching, it decreased to 24.5%, 15.0%, 10.7%, and 5% as the bleaching steps progressed to B1, B3, B4, and B5, respectively. Decreasing the lignin content was natural as the bleaching steps progressed. At the same time, the alpha-cellulose content changed to 78.9%, 80.4%, 80%, and 76.7% for B1, B3, B4, and B5. The B3 and B4 specimens showed higher alpha-cellulose content than the B1. The B5 was the lowest alpha-cellulose content, attributed to over-bleaching. Wood degradation can happen when over-bleaching is done, and thus, the low molecular weight cellulose can be dissolved, leading to lower alpha-cellulose content. The alpha-cellulose content in this research is higher than the holocellulose content reported by several researchers [20,21]. It might be because alpha-cellulose content in this research relied on each bleaching stage using $NaClO_2$. One stage bleaching using $NaClO_2$ should remain hemicellulose and lignin contents, which were not possible by simple 17.5% NaOH use in the alpha-cellulose analysis.

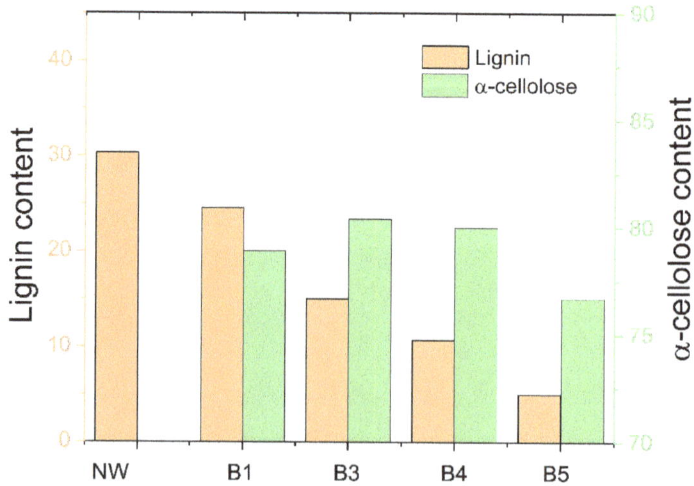

Figure 4. Lignin and alpha-cellulose content with the bleached steps.

3.3. The Compression Ratio

The bleached wood specimens were hot-pressured at 100 °C under HP-1, HP-2, and HP-3 conditions. Figure 5 represents the compression ratio to the bleaching and hot-pressing conditions. As the bleaching steps increased, the hot-pressing pressure, the time, and the compression ratio increased. The natural wood specimen had a 58% compression ratio under the HP-1 condition, termed "NW+P". It was elevated from 61.7% to 72.4% as the bleaching step progressed from B1 to B5. By increasing the hot-pressing pressure and time to HP-3, the compression ratio increased from its initial value of 68.2% to 77.5%. The compression ratio reached over 77% when the bleaching conditions were B3, B4, and B5 under the hot-pressing conditions. It revealed that under the HP-3 condition, the bleaching step over B3 gives the highest compression ratio.

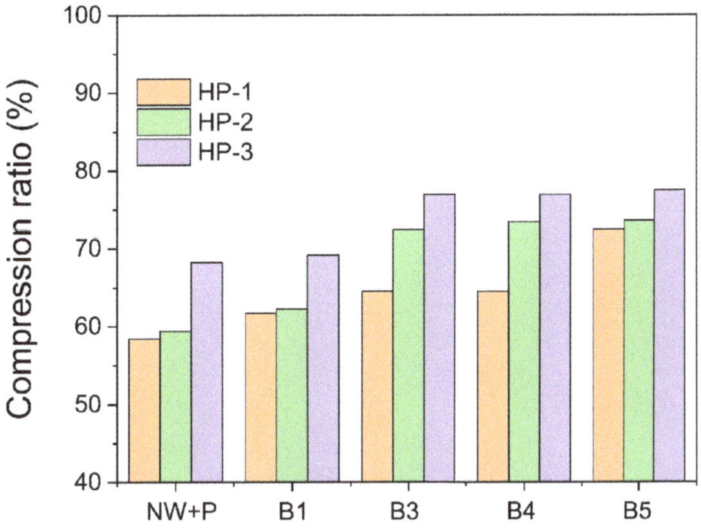

Figure 5. Compression ratios of the wood specimens under different hot-pressing conditions.

Note that the compression pressure of 52 MPa of the HP-3 condition is just over the compression yield strength of the wood in the transverse direction, 50 MPa [2]. The compression pressure should maintain the yield strength in the transverse direction of the wood to permanently acquire the collapsed deformation. In addition, the hot-pressing time should be long enough to fully collapse the cell wall and hydrogen bond it in the transverse direction. The longer the time, the better the CW's hydrogen bond might be formed. However, long compression time consumes much energy, which is not economical. Thus, 16 h of HP-3 might be enough for forming CW. As a consequence, HP-3 was an optimum condition for the hot pressing of CW.

3.4. Three-Point Bending Test

The mechanical properties of the wood specimens were investigated using the TPB test. Three hot-pressing conditions (HP-1, HP-2, HP-3) and four bleaching conditions (B1, B3, B4, B5) in addition to the natural wood pressing one (NW+P) were considered. Figure 6 shows the stress-strain curves of the wood specimens according to the bleaching conditions under HP-3 and 100 °C hot-pressing conditions. The original NW specimen was so weak that it was broken at low strength. Its bending strength and Young's modulus were 74.1 MPa and 3.7 GPa. The compressed NW (NW+P) was better than the NW: the first linear region gives it a Young's modulus of 12.15 GPa, and the second deformation region was observed near 1.5% strain, which is typical behavior in structural composites because of filament and matrix failures. The bending strength of NW+P was 142.0 MPa.

As the bleaching step progressed, the wood specimens increased their Young's modulus and bending strength. B3's bleaching condition exhibited 240.1 MPa bending strength and 23.08 GPa Young's modulus, which indicates that the CW became much tougher than the original NW.

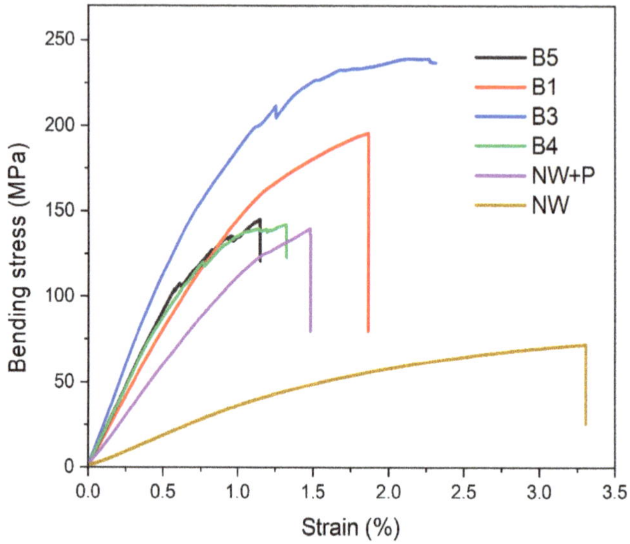

Figure 6. Stress-strain curves of wood specimens under HP-3 hot-pressing conditions (NW: natural wood; NW+P: natural wood hot-pressed; B1-B5: specimens after 1–5 bleaching steps).

Table 1 represents the bending strength, Young's modulus, and elongation-at-break for all hot-pressing and bleaching conditions in addition to the natural wood hot-pressed case. Figure 7 represents their bending strength trend. The bending strength increased with the bleaching condition up to B3 and decreased after over-bleaching. HP-3 condition exhibited better bending strength than other hot-pressing conditions: HP-1 and HP-2. The highest bending strength of 240.1 MPa was obtained from HP-3 hot-pressing and B3 bleaching conditions.

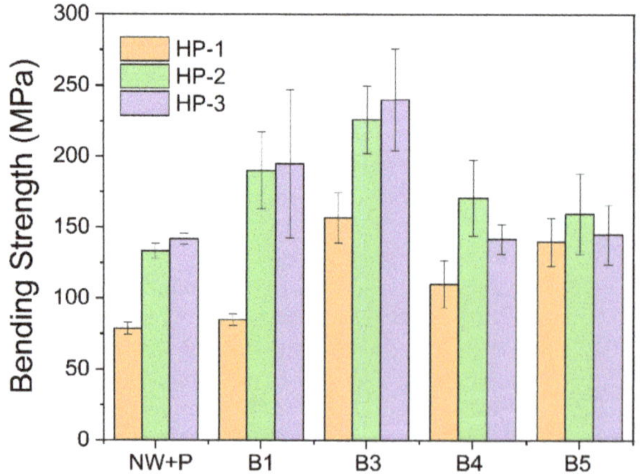

Figure 7. Bending strength of CW specimens prepared by various bleaching and hot-pressing conditions.

Table 1. Mechanical properties of CW specimens prepared by various bleaching and hot-pressing conditions.

Hot-Pressing Condition	Bleaching Condition	Bending Strength (MPa)	Young's Modulus (GPa)	Elongation-at-Break (%)
Natural wood		74.1 ± 2.9	3.7 ± 0.06	3.23± 0.05
HP-1 (13 MPa and 4 h)	NW±P	78.7 ± 4.3	4.76 ± 0.11	2.10± 0.10
	B1	85.0 ± 3.9	5.60 ± 0.19	1.90± 0.24
	B3	157.0 ± 17.6	10.60 ± 1.42	2.10 ± 0.24
	B4	110.0 ± 16.8	9.88 ± 2.58	1.30± 0.27
	B5	140.1± 16.9	11.50 ± 3.94	1.50± 0.16
HP-2 (26 MPa and 8 h)	NW±P	133.7 ± 5.1	9.76 ± 0.02	1.60 ± 0.01
	B1	190.1 ± 27.0	17.27 ± 1.90	1.48 ± 0.16
	B3	226.0 ± 23.9	24.49 ± 3.30	1.70 ± 0.17
	B4	171.1 ± 26.7	22.64 ± 5.8	1.18 ± 0.60
	B5	160.0 ± 38.4	18.90 ± 8.04	1.35 ± 0.05
HP-3 (52 MPa and 16 h)	NW±P	142.0 ± 3.9	12.15 ± 2.54	1.54 ± 0.01
	B1	195.0 ± 52.2	16.22 ± 4.78	1.62 ± 0.16
	B3	240.1 ± 35.7	23.08 ± 0.89	1.97 ± 0.17
	B4	142.0± 10.5	17.14 ± 3.48	1.28 ± 0.60
	B5	145.0 ± 21.0	17.20 ± 0.28	1.23 ± 0.05
Song et al. [13]	Partial lignin removal by NaOH, Na_2SO_3	549	42	1.3
Fang et al. [14]	alkaline sulfite-anthraquinone-methanol cooking	1000	60.2	3
Paril et al. [4]	chemical modification of gaseous ammonia	160	16	-

Figure 8 shows the Young's modulus trend of wood specimens. Young's modulus increased with the bleaching steps, and after B3, it decreased. The HP-2 condition exhibited a better Young's modulus than the other two conditions: HP-1 and HP-3. The maximum Young's modulus of 24.49 GPa was obtained at HP-2 hot-pressing and B3 bleaching conditions. Long hot-pressing time might result in an annealing effect such that it decreases Young's modulus and increases the strain-at-break and bending strength.

In this paper, the highest mechanical properties of CW were observed at the B3 bleaching condition, where the lignin content was 15.0%. Further proceeding with the bleaching step can increase the lignin removal, decreasing the mechanical properties due to decreased cellulose content. Thus, an optimum bleaching step is B3, and further bleaching is not recommended. Song et al., 2018, reported the best mechanical properties when the lignin content was 11.3% [13]. Fang et al., 2020 [14], made 3% lignin content to achieve good mechanical properties. The lignin content that achieves the best mechanical properties may differ depending on the chemicals and hot-pressing conditions. In terms of mechanical development, the bleached CW shows good mechanical properties. The strength of NW, 74 MPa reached over 240 MPa after compression, which is over three times high. According to Song et al. [13], the NW strength of 50 MPa reached over 500 MPa depending on the bleaching treatment and multilayers compression. Fang et al. [14] reported that the bending strength of NW, about 50 to 70 MPa, can be reached to 120–140 MPa, depending on the thermal treatment from 160 °C to 220 °C, whereas Yano et al., 2001, produced transparent wood with mechanical properties from 317 MPa to 453 MPa, which is combined by compression and filler reinforcement. Paril et al. [4] reported that the TPB strength of natural wood and compression wood was from 96 to 160 MPa. It shows that by bleaching and compression, the bending strength of bleached wood is much better than normal compression wood. This research presented that compression wood can reach 240 MPa with the three stages of bleaching treatment. As seen below, parts of Table 1 show mechanical properties of the previous reports for densified wood samples.

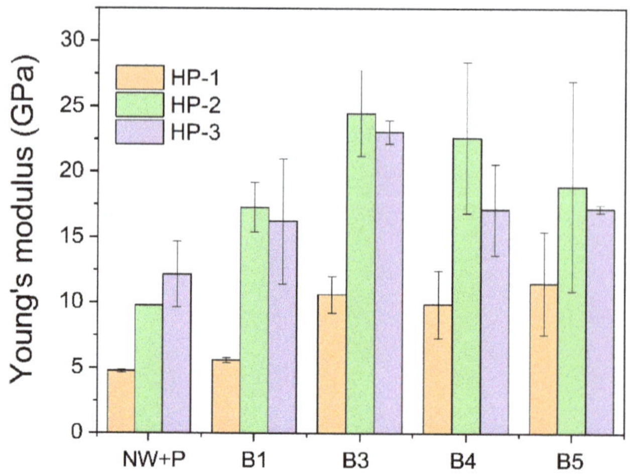

Figure 8. Young's modulus of CW specimens prepared by various bleaching and hot-pressing conditions.

3.5. Swelling

The swelling test was performed to address the stability of the materials under high moisture content. The CW specimens prepared under HP-3 with different bleaching conditions were placed in the humidity-controlled chamber set to 90% RH and 25 °C for 8 h. The CW swelled after bleaching step 1 (swelling ratio = 42%) and monotonically decreased to 20% as the bleaching step progressed, as shown in Figure 9. The swelling of woods varies and is closely related to the relative chemical content, morphological structures, temperature, and relative humidity [22]. The relationship between the carbohydrates and the other components (e.g., lignin and extractives) on the swelling characteristics of wood have already been well-documented. It has also been reported that the swelling of compressed wood increased as the density increased [23]. Figure 9 also shows that the density decreases as the bleaching step progresses. Note that the density was measured before compression. The longer the bleaching steps, the higher the lignin and hemicellulose removal, leading to a highly porous wood structure and low density of the bleached wood. The swelling decrease might be mainly associated with the density decrease, and the compression ratio increase as the bleaching step continued. Although B5 has the lowest lignin content (see Figure 4), it exhibited the lowest swelling ratio.

The weight and density changes of the specimens were measured before and after compression. Table 2 shows the results. As one can see, the weight significantly decreased as the bleaching step progressed, which resulted in the original density decrease. The weight decrease was associated with the lignin and hemicellulose removal by the continued process. Note that the hot pressing significantly increased the CW densities 3–4 times more than the original densities. As the hot-pressing condition changed to HP-1, HP-2, and HP-3, they increased. The CW density ranged between 1.0–1.2 under HP-1 but increased to 1.1–1.4 and 1.3–1.7 under HP-2 and HP-3. The CW density change by the bleaching steps increased as the hot-pressing condition changed from HP-1 to HP-2 and HP-3. The weight decrement with the bleaching steps mitigated the volume decrement by the compression as the bleaching steps progressed. Thus, the CW density change by the bleaching steps was not significant.

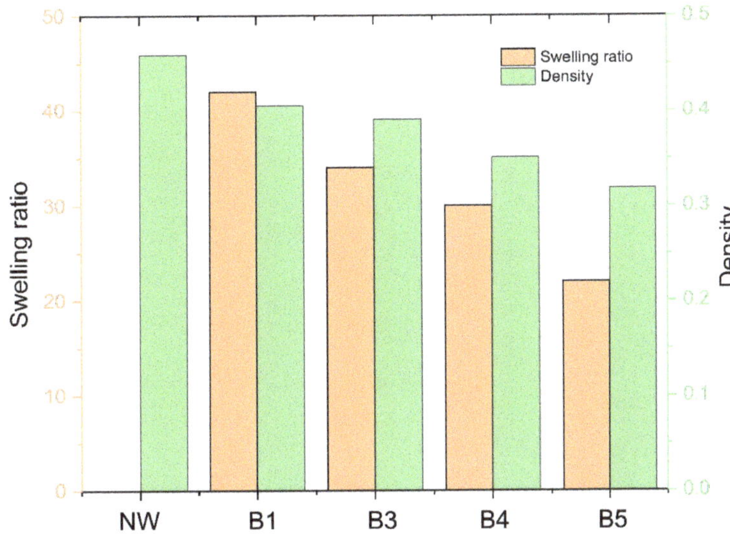

Figure 9. The swelling ratio of CW specimens and density change with bleaching steps.

Table 2. Density changes of specimens under various bleaching steps and hot-pressing conditions.

Specimen	Weight (g)	Original Density (g/cm^3)	Density after HP-1 (g/cm^3)	Density after HP-2 (g/cm^3)	Density after HP3 (g/cm^3)
NW	22.70	0.459	1.104	1.131	1.447
B1	18.05	0.405	1.059	1.074	1.314
B3	18.65	0.390	1.101	1.415	1.695
B4	15.90	0.350	0.986	1.317	1.518
B5	15.05	0.318	1.154	1.205	1.413

The water contact angle (WCA) of the heat-treated CW increases the hydrophobic properties [24,25]. The authors indicated the effects of different heat treatments and compacted surfaces of CW. However, in this research, only one heat-compression condition was carried out, limiting the heat treatment effects. Furthermore, different bleaching treatment leads to different effects on WCA of CW. Longer bleaching times and multiple bleaching stages raise the higher OH groups exposed to the surface of wood specimens. They lead to decreased hydrophobicity. The WCA of CW specimens was measured to investigate their water-proof behavior. Figure 10 shows the WCA results. As the bleaching process continued, the WCA decreased to 72.5° at B4 and bounced back at B5. Surface morphologies and lignin content can influence the WCA. Hemicellulose is the most hygroscopic polymer in wood, followed by lignin and crystalline cellulose. The WCA decreased as the lignin content decreased with the bleaching step (see Figure 4). The slightly bounced back WCA at B5 might be associated with the smooth surface morphology of B5, resulting in the highest compression ratio.

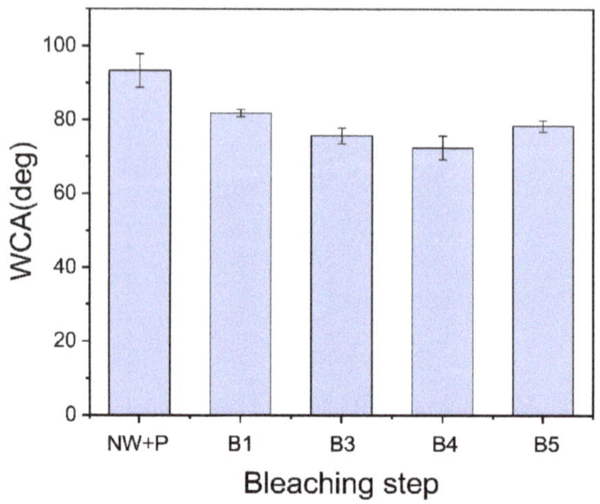

Figure 10. Water contact angles of CW specimens.

4. Conclusions

Compression wood was prepared by removing lignin with multiple steps of $NaClO_2$ bleaching, and the bleaching effects were investigated in terms of morphologies, lignin content, alpha-cellulose content, compression ratio, mechanical properties, swelling, and water contact angle. After compression, the CW densities increased, and the most porous structures collapsed. The lignin content decreased as the bleaching steps progressed, and the highest alpha-cellulose was observed at the B3 and B4 bleached specimens. The compression ratio reached over 77% when the bleaching condition was over B3 under the HP-3 hot-pressing condition, which was just over the compression yield strength of the wood in the transverse direction. The B3 CW showed the best mechanical properties: bending strength 240.1 ± 35.7 MPa and Young's modulus 23.08 ± 0.89 GPa. Over the B3 step, the mechanical properties decreased due to over-bleaching. The swelling of CW decreased as the bleaching step progressed, which was associated with the density decrease and the compression ratio increase with the bleaching step. The water contact angle decreased to 72.5° at B4 as the bleaching process continued. The B3 is an optimum bleaching step that accounts for high mechanical properties, which might be associated with high alpha-cellulose content.

Bleaching treatment with hot pressing can improve the mechanical properties of CW, which is applicable for environmentally friendly structures such as concrete reinforcement. However, the moisture absorption and water swelling behaviors need to be improved, and chemical inertness and thermal and sound absorbing behaviors need to be investigated.

Author Contributions: Conceptualization, L.V.H.; writing—original draft preparation, L.V.H. and D.H.P.; writing—review and editing, J.K. and D.H.P.; supervision, J.K. All authors have read and agreed to the published version of the manuscript.

Funding: This research was funded by Inha University Research Grant.

Data Availability Statement: Not applicable.

Conflicts of Interest: The authors declare no conflict of interest.

References

1. Navi, P.; Heger, F. Combined densification and thermo-hydro-mechanical processing of wood. *MRS Bull.* **2004**, *29*, 332–336. [CrossRef]
2. Kettunen, P.O. *Wood: Structure and Properties*; Trans Tech Publications, Ltd.: Uetikon am See, Switzerland, 2006.

3. Laine, K.; Segerholm, K.; Wålinder, M.; Rautkari, L.; Hughes, M. Wood densification and thermal modification: Hardness, set-recovery and micromorphology. *Wood Sci. Technol.* **2016**, *50*, 883–894. [CrossRef]
4. Pařil, P.; Brabec, M.; Maňák, O.; Rousek, R.; Rademacher, P.; Čermák, P.; Dejmal, A. Comparison of selected physical and mechanical properties of densified beech wood plasticized by ammonia and saturated steam. *Eur. J. Wood Wood Prod.* **2014**, *72*, 583–591. [CrossRef]
5. Hill, C.A.; Ramsay, J.; Keating, B.; Laine, K.; Rautkari, L.; Hughes, M.; Constant, B. The water vapour sorption properties of thermally modified and densified wood. *J. Mater. Sci.* **2012**, *47*, 3191–3197. [CrossRef]
6. Fang, C.-H.; Mariotti, N.; Cloutier, A.; Koubaa, A.; Blanchet, P. Densification of wood veneers by compression combined with heat and steam. *Eur. J. Wood Wood Prod.* **2012**, *70*, 155–163. [CrossRef]
7. Laine, K.; Belt, T.; Rautkari, L.; Ramsay, J.; Hill, C.A.; Hughes, M. Measuring the thickness swelling and set-recovery of densified and thermally modified Scots pine solid wood. *J. Mater. Sci.* **2013**, *48*, 8530–8538. [CrossRef]
8. Bekhta, P.; Proszyk, S.; Krystofiak, T. Colour in short-term thermo-mechanically densified veneer of various wood species. *Eur. J. Wood Wood Prod.* **2014**, *72*, 785–797. [CrossRef]
9. Diouf, P.N.; Stevanovic, T.; Cloutier, A.; Fang, C.-H.; Blanchet, P.; Koubaa, A.; Mariotti, N. Effects of thermo-hygro-mechanical densification on the surface characteristics of trembling aspen and hybrid poplar wood veneers. *Appl. Surf. Sci.* **2011**, *257*, 3558–3564. [CrossRef]
10. Cabral, J.P.; Kafle, B.; Subhani, M.; Reiner, J.; Ashraf, M. Densification of timber: A review on the process, material properties, and application. *J. Wood Sci.* **2022**, *68*, 20. [CrossRef]
11. Luan, Y.; Fang, C.H.; Ma, Y.F.; Fei, B.H. Wood mechanical densification: A review on processing. *Mater. Manuf. Process* **2022**, *37*, 359–371. [CrossRef]
12. Yano, H.; Hirose, A.; Collins, P.; Yazaki, Y. Effects of the removal of matrix substances as a pretreatment in the production of high strength resin impregnated wood based materials. *J. Mater. Sci. Lett.* **2001**, *20*, 1125–1126. [CrossRef]
13. Song, J.; Chen, C.; Zhu, S.; Zhu, M.; Dai, J.; Ray, U.; Li, Y.; Kuang, Y.; Li, Y.; Quispe, N. Processing bulk natural wood into a high-performance structural material. *Nature* **2018**, *554*, 224–228. [CrossRef] [PubMed]
14. Fang, Z.; Li, B.; Liu, Y.; Zhu, J.; Li, G.; Hou, G.; Zhou, J.; Qiu, X. Critical role of degree of polymerization of cellulose in super-strong nanocellulose films. *Matter* **2020**, *2*, 1000–1014. [CrossRef]
15. Xia, Q.; Chen, C.; Li, T.; He, S.; Gao, J.; Wang, X.; Hu, L. Solar-assisted fabrication of large-scale, patternable transparent wood. *Sci. Adv.* **2021**, *7*, eabd7342. [CrossRef]
16. Chen, H.; Baitenov, A.; Li, Y.; Vasileva, E.; Popov, S.; Sychugov, I.; Yan, M.; Berglund, L. Thickness Dependence of Optical Transmittance of Transparent Wood: Chemical Modification Effects. *ACS Appl. Mater. Interfaces* **2019**, *11*, 35451–35457. [CrossRef]
17. Hai, L.V.; Muthoka, R.M.; Panicker, P.S.; Agumba, D.O.; Pham, H.D.; Kim, J. All-biobased transparent wood: A new approach and its environmental-friendly packaging applications. *Carbohydr. Polym.* **2021**, *264*, 139–150.
18. TAPPI. *Acid-Insoluble Lignin in Wood and Pulp, Test Method T 222 om-21*; TAPPI: Atlanta, GA, USA, 2006.
19. Hai, L.V.; Park, H.J.; Seo, Y.B. Effect of PFI mill and Valley beater refining on cellulose degree of polymerization, alpha cellulose contents, and crystallinity of wood and cotton fibers. *J. Korea Tech. Assoc. Pulp Pap. Ind.* **2013**, *45*, 27–33. [CrossRef]
20. Gonultas, Z.; Candan, Z. Chemical characterization and FTIR spectroscopy of thermally compressed Eucalyptuswood panels. *Maderas Cienc. Tecnol.* **2018**, *20*, 431–442.
21. Rabemanolontsoa, H.; Saka, H. Holocellulose Determination in Biomass. In *Green Energy and Technology*; Springer: Cham, Switzerland, 2011. [CrossRef]
22. Sahin, H.T. Experimental determination of the anisotropic swelling and water sorption properties of chestnut wood. *Wood Res.* **2010**, *55*, 33–40.
23. Morsing, N. Densification of Wood: The Influence of Hygrothermal Treatment on Compression of Beech Perpendicular to Gain. Ph.D. Thesis, Technical University of Denmark, Kongens Lyngby, Denmark, 1998.
24. Candan, Z.; Gonultas, O.; Gorgun, H.V.; Unsal, O. Examining Parameters of Surface Quality Performance of Paulownia Wood Materials Modified by Thermal Compression Technique. *Drv. Ind.* **2021**, *72*, 231–236. [CrossRef]
25. Candan, Z.; Gorgun, H.V.; Korkut, S.; Ünsal, O. Surface roughness and wettability performance of thermally modified rowan wood as a fast-growing species. *Drewno* **2021**, *64*, 208.

polymers

Article

Thermo-Responsive Shape Memory Vanillin-Based Photopolymers for Microtransfer Molding

Justinas Jaras [1], Aukse Navaruckiene [1], Edvinas Skliutas [2], Jurga Jersovaite [2], Mangirdas Malinauskas [2] and Jolita Ostrauskaite [1,*]

[1] Department of Polymer Chemistry and Technology, Kaunas University of Technology, Radvilenu Rd. 19, LT-50254 Kaunas, Lithuania; justinas.jaras@ktu.edu (J.J.); aukse.navaruckiene@ktu.lt (A.N.)
[2] Laser Research Center, Faculty of Physics, Vilnius University, Sauletekis Ave. 10, LT-10223 Vilnius, Lithuania; edvinas.skliutas@ff.vu.lt (E.S.); jurga.jersovaite@ff.stud.vu.lt (J.J.); mangirdas.malinauskas@ff.vu.lt (M.M.)
* Correspondence: jolita.ostrauskaite@ktu.lt; Tel.: +370-37-300192

Abstract: Novel thermo-responsive shape-memory vanillin-based photopolymers have been developed for microtransfer molding. Different mixtures of vanillin dimethacrylate with tridecyl methacrylate and 1,3-benzenedithiol have been tested as photocurable resins. The combination of the different reaction mechanisms, thiol-acrylate photopolymerization, and acrylate homopolymerization, that were tuned by changing the ratio of monomers, resulted in a wide range of the thermal and mechanical properties of the photopolymers obtained. All polymers demonstrated great shape-memory properties and were able to return to their primary shape after the temperature programming and maintain their temporary shape. The selected compositions weretested by the microtransfer molding technique and showed promising results. The developed thermo-responsive shape-memory bio-based photopolymers have great potential for forming microtransfered structures and devices applicable on non-flat surfaces.

Keywords: bio-based polymers; dual curing; shape-memory; photocuring; microtransfer molding

Citation: Jaras, J.; Navaruckiene, A.; Skliutas, E.; Jersovaite, J.; Malinauskas, M.; Ostrauskaite, J. Thermo-Responsive Shape Memory Vanillin-Based Photopolymers for Microtransfer Molding. *Polymers* **2022**, *14*, 2460. https://doi.org/10.3390/polym14122460

Academic Editors: Antonio M. Borrero-López, Concepción Valencia-Barragán, Esperanza Cortés Triviño, Adrián Tenorio-Alfonso and Clara Delgado-Sánchez

Received: 16 May 2022
Accepted: 14 June 2022
Published: 16 June 2022

1. Introduction

Increased oil prices and environmental awareness have prompted the scientific community to seek alternative feedstocks for polymers, which nowadays are mainly derived from fossil fuels [1,2]. The main source of sustainable and environmentally friendly polymers are renewable raw materials, such as plants [3]. Currently, vanillin is one of a few bio-based and aromatic compounds that are industrially available [4]. Its derivatives have been successfully used in various polymerization techniques. For example, vanillin acrylate-based polymers, synthesized by thermal polymerization, possess high glass transition temperature and high mechanical strength [5]. Recently, vanillin-based polymers, synthesized by photopolymerization, demonstrated significant antimicrobial activity and were applied in optical 3D printing [6]. However, there are only a few examples of the usage of vanillin derivatives in dual curing systems [7].

Dual curing is a process that combines two curing reactions that occur simultaneously or sequentially [8]. In this process, the properties of the resulting materials can be easily changed by manipulating the composition of the resins by changing the ratio of the selected monomers [9]. Polymers with unique mechanical and thermal properties can be obtained during the dual curing process as the result of interpenetrating, or semi-interpenetrating, polymer networks [10]. Due to these extraordinary features, dual curing attracts huge attention among scientists as one of the possible ways to create smart materials, such as shape-memory polymers [11].

Shape-memory polymers are a part of intelligent polymeric systems, also known as smart materials [12]. Their unique property of changing their shape in response to a stimulus has significantly broadened their areas of application [13]. Bio-based shape-memory polymers attract huge interest in the scientific community, however, only a few examples of UV photocured shape-memory polymers are available [14]. Recently, novel shape-memory bio-based polymers have been successfully synthesized from castor oil with the introduction of dynamic pyrazole–urea bonds [15]. Bovine serum albumin-based shape-memory bioplastics have been produced and adapted for stereolithography [16]. Cholesterol was successfully used in the synthesis of thermo-responsive shape-memory polymers [17]. The novel poly(cholesteryl methacrylate) coatings were characterized as potential substrates for tissue engineering, showing the importance of shape-memory polymers [18]. In this work, we focused our attention on the development of novel bio-based shape-memory photopolymers suitable for the manufacturing of various structures and devices by nanoimprint lithography.

Photocuring requires a smaller amount of resources to produce the required part compared to other techniques [19]. It is a simple way to create unique products of complex shapes while also reducing manufacturing costs and carbon dioxide emission [20]. The final product can be successfully manufactured during the molding process and does not need further modification before usage [21]. Moreover, even the most complex parts can be duplicated easily if the original part was damaged during the operation process [22]. Microtransfer molding, or nanoimprint lithography, is a great way to reduce waste as it uses a minimal amount of material needed for the final product [23], and it is a versatile technique for nanofabrication of a wide range of micro and nanostructures and devices [24]. It offers the ability to fabricate nanostructures that exhibit vertically aligned nanoarrays and nanopatterns, and the ability to have precise control over the size and geometry of the nanostructures, and to form nanostructures attached to a bulk support. The main advantage of this technique is its fast, easy, and inexpensive, but accurate, reproduction of various 2D or 3D objects from the nanoscale to the macroscale [25,26].

In this work, different mixtures of bio-based monomer vanillin dimethacrylate with tridecyl methacrylate, derived from natural oil and 1,3-benzenedithiol, have been tested to find an efficient photocurable system for microtransfer molding. Ethyl (2,4,6-trimethylbenzoyl) phenylphosphinate was selected as the photoinitiator, due to its ability to cure the deep layers of the resin and its photobleaching effect [27]. This work is a continuation of our previous studies [7,28]. In the present work, the addition of tridecyl methacrylate to the resin composition has led to a reduction in the use of expensive vanillin monomer and, thus, to a reduction in the cost of the polymers, without sacrificing their properties. The polymers obtained demonstrated a wide range of thermal and mechanical properties as a result of the combination of the different reaction mechanisms, thiol-acrylate photopolymerization and acrylate homopolymerization, which were tuned by changing the ratio of monomers. The increase in tridecyl methacrylate content resulted in less rigid polymers, lower photocuring rate, and lower Young's modulus values. The reduction of thiol content increased the shrinkage of polymers, the photocuring rate, and the thermal and mechanical characteristics of the resulting polymers. All polymers demonstrated shape-memory properties. The most promising compositions have been tested by the microtransfer molding technique and showed promising results as photoresins for nanoimprint lithography. Microtransfered structures and devices formed from the developed thermo-responsive shape-memory bio-based photopolymers can be applied on non-flat surfaces, convex or concave, such as cylinders, tubes, etc.

2. Materials and Methods

2.1. Materials

Vanillin dimethacrylate (VDM, Specific Polymers, Castries, France), 1,3-benzenedithiol (BDT, Fluorochem, Glossop, United Kingdom), tridecyl methacrylate (C13-MA, VISIOMER® Terra C13-MA, Evonik, Essen, Deutschland), ethyl(2,4,6 trimethylbenzoyl)phenylphosphinate (TPOL, Fluorochem, Glossop, United Kingdom), (Figure 1) were used as received.

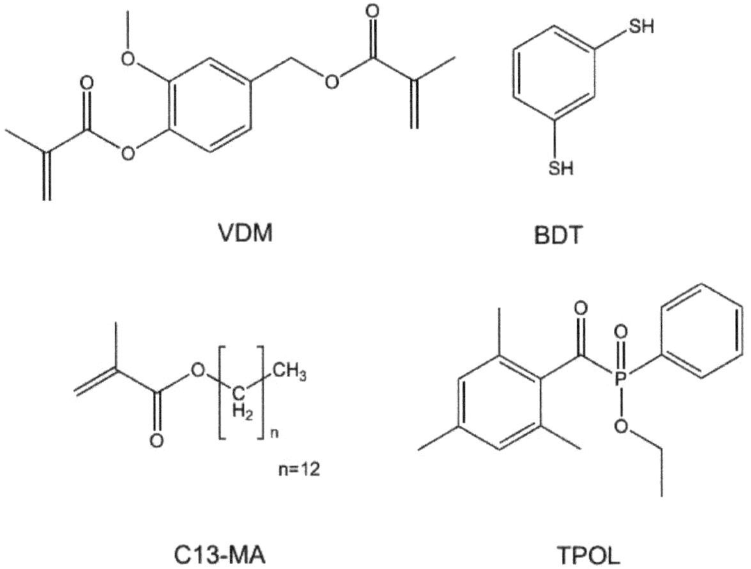

Figure 1. Chemical structures of vanillin dimethacrylate (VDM), 1,3-benzenedithiol (BDT), tridecyl methacrylate (C13-MA), and ethyl(2,4,6-trimethylbenzoyl)phenylphosphinate (TPOL).

2.2. Preparation of Cross-Linked Polymer Specimens

The mixtures containing 1 mol of VDM, 3 mol.% of TPOL, 1, 0.75, 0.5, or 0.25 mol of BDT and 1.5, 3, or 4.5 mol of C13-MA were stirred with a magnetic stirrer at room temperature (25 °C) for 1 min. When homogeneous mixtures were obtained, the resins were poured into a Teflon mold and cured for 5–7 min in the UV irradiation chamber BS-02 (Opsytec Dr. Grobel, Ettlinger, Germany) with an intensity of 30 mW/cm^2 and a wavelength range of 280–400 nm. The composition of the resins is presented in Table 1.

Table 1. Composition of resins **C1–C12**.

Resin	Amount of VDM, Mol	Amount of BDT, Mol	Amount of C13-MA, Mol	Amount of TPOL, Mol.%
C1	1	1	1.5	3
C2	1	0.75	1.5	3
C3	1	0.5	1.5	3
C4	1	0.25	1.5	3
C5	1	1	3	3
C6	1	0.75	3	3
C7	1	0.5	3	3
C8	1	0.25	3	3
C9	1	1	4.5	3
C10	1	0.75	4.5	3
C11	1	0.5	4.5	3
C12	1	0.25	4.5	3

2.3. Characterization Techniques

Fourier transformation infrared (FT-IR) spectroscopy spectra were recorded using a Spectrum BX II FT-IR spectrometer (Perkin Elmer, Llantrisant, UK). Reflection was measured during the test. The range of wavenumbers was (650–4000) cm^{-1}.

The Soxhlet extraction was used to determine the yield of the insoluble fraction. 0.2 g polymer samples were extracted with acetone for 24 h. After 24 h, the insoluble fractions were dried under vacuum until no weight changes were observed. The yield of insoluble fraction was calculated as the weight difference before and after extraction and drying.

Thermogravimetrical analysis (TGA) was performed on a TGA 4000 apparatus (Perkin Elmer, Llantrisant, UK). A heating rate of 20 °C/min was chosen in a nitrogen atmosphere (100 mL/min). The temperature range of (10−800) °C was used. Aluminium oxide pans were used.

Dynamical mechanical thermal analysis (DMTA) was performed on an MCR302 rheometer (Anton Paar, Graz, Austria) equipped with the SRF10-SN30777 measuring system. The Peltier-controlled temperature chamber was used. The temperature was increased from −80 °C to 100 °C, with a heating rate of 2.99 °C/min. The normal force was set at −0.1 N during the measurement. In all cases, the torsion mode was used with a frequency of 1 Hz and a torsion strain of 0.1%. The storage modulus (G′), the loss modulus (G″), and the loss factor (tanδ) were recorded as functions of temperature.

The mechanical properties of the synthesized polymers were determined by the tensile test. The tensile test was performed on a Testometric M500-50CT testing machine (Testometric Co. Ltd., Rochdale, UK) with flat-faced grips at room temperature (21.5 °C). The dimensions of the test specimens were 70 (±0.01) × 10 (±0.01) × 2 (±0.15) mm. The gap between the grips was set to 20 mm and the test was performed at a speed of 5 mm/min until the specimen broke. Young's modulus, tensile strength, and elongation at break were determined.

2.4. Real-Time Photorheometry

UV/Vis curing tests were performed with resins containing 1 mol of vanillin dimethacrylate, 3 mol.% of TPOL, 1, 0.75, 0.5, or 0.25 mol of BDT and 1.5, 3.0, or 4.5 mol of C13-MA on a MCR302 rheometer (Anton Paar, Graz, Austria) equipped with the plate/plate measuring system. The measuring gap was set to 0.1 mm and the samples were irradiated by UV/Vis light in a wavelength range of 250–450 nm through the glass plate using the OmniCure S2000 UV/Vis spot curing system (Lumen Dynamics Group Inc., Mississauga, ON, Canada). The temperature was 25 °C. The shear mode with a frequency of 10 Hz and a shear strain of 1% was used in all cases. The storage modulus (G′), the loss modulus (G″) and the complex viscosity (η*) were recorded as a function of the irradiation time and the values of each parameter taken after 300 s of photocuring are presented in Table 2. The gel point (t_{gel}) was calculated as the intersection point of the G′ and G″ curves. The shrinkage was calculated from the reduction of the height of the sample during the polymerization process. The normal force was set at 0 N during the measurement of the sample shrinkage. Five measurements of each resin were used to obtain the mean value and standard deviation.

Table 2. Rheological Characteristics of Resins.

Resin	Storage Modulus, G′, MPa	Loss Modulus, G″, MPa	Loss Factor, tanδ	Complex Viscosity η*, MPa·s	Gel Point t_{gel}, s	Induction Period, s	Shrinkage, %
C1	0.312	0.128	0.409	0.054	8.9	4.9	2
C2	3.030	2.061	0.680	0.635	6.8	2.6	7
C3	14.173	5.872	0.414	2.440	4.6	0.7	8
C4	17.343	5.661	0.327	2.900	3.4	0.3	9
C5	0.031	0.020	0.646	0.005	13.5	9.2	0
C6	0.721	0.333	0.461	0.013	10.6	7.3	2
C7	2.420	1.290	0.534	0.437	5.8	5.1	3
C8	10.148	4.470	0.442	1.760	3.7	0.9	4
C9	0.055	0.018	0.328	0.009	20.8	12.3	1
C10	0.542	0.230	0.425	0.094	14.4	11.9	3
C11	2.002	1.010	0.502	0.357	13.6	11.6	6
C12	2.930	1.450	0.497	0.520	9.8	8.8	10

2.5. Microtransfer Molding Technique

A microtransfer molding (μTM) technique, or nanoimprint lithography, was used to test **C1** and **C8** resins to make replicas [29]. First, a master structure was manufactured out of PlasGray material with the Asiga Pico2 39 UV table-top 3D printer (Asiga, Alexandria, Australia). Next, polydimethylsiloxane (PDMS) was poured over this structure and thermally cured at 100 C for 1 h, thus creating a soft mold (stamp). It was then used to make a replica of both resins. The printed structure was a 1951 USAF resolution target (25% of the downloaded vector file size), used as a sample in optical engineering for testing imaging systems. A UV diode emitting 365 nm wavelength light (CS2010, Thorlabs, Newton, NJ, USA) was used to cure the **C1** and **C8** resins and obtain the replicas. The curing time was set to 5 min for each sample. The soft mold and the replicas made were characterized using the Olympus IX73 optical microscope (Olympus Corporation, Tokyo, Japan).

3. Results

3.1. Monitoring of Photocuring Kinetics by Real-Time Photorheometry

The photocuring of the three-component resins was studied by real-time photorheometry. In this study, the most important parameters, the photocuring rate (characterized by t_{gel}, induction period and slope of the curves), rigidity (characterized by storage modulus G'), and shrinkage (described by volume loss during photopolymerization) have been monitored and examined [30]. The real-time photorheometry data of all resins are summarized in Table 2.

The dependence of the storage modulus G' on the irradiation time of all resins is presented in Figure 2. The storage modulus represents the stiffness and rigidity of the polymers that are the important parameters for the applications of the polymers. It was determined that the resin composition had a huge influence on the photocuring kinetics and on the properties of the resulting polymers. The most rigid polymer was **C4** prepared with the lowest amount of C13-MA and BDT. The highest amount of VDM in **C4** compared to other polymers caused the high rate of acrylate homopolymerization. After comparing polymer **C4** with **C8** and **C12**, it was determined that the rigidity of the polymers decreased with increasing volume of tridecyl methacrylate in the resins. The rigidity of the polymers was reduced from 17.343 (**C4**) to 2.930 (**C12**) MPa by increasing the content of C13-MA from 1.5 to 4.5 mol. The long carbon chain of tridecyl methacrylate was the reason for the formation of the soft and flexible polymer [31]. By increasing the amount of C13-MA, the amount of polymer to plasticize long flexible alkyl chains was increased, and, thus, the rigidity of the polymer was reduced, while the increase in the amount of VDM, having a relatively short and partially aromatic structure in comparison to C13-MA, resulted in higher rigidity of the polymers.

The amount of thiol also had a great impact on the polymer structure and, therefore, its characteristics. It was determined that the polymer rigidity was increased by reducing the amount of thiol in the composition. The high amount of thiol resulted in a higher rate of thiol-ene photopolymerization, which is a slower reaction in comparison to free radical polymerization. A high number of short polymer chains was formed as a result of competing free radical and thiol-ene photopolymerizations, and even increased with increasing amounts of thiol. However, according to the literature, thiols not only reduce the rigidity of polymers but also increase their flexibility due to the formation of flexible thioether linkages [7]. For example, the rigidity of polymers **C5**–**C8**, prepared with 3 mol of C13-MA, was increased from 0.031 (**C5**) to 10.148 (**C8**) MPa by decreasing the amount of thiol from 1 to 0.25 mol. This tendency was also visible in polymers **C1**–**C4** prepared with 1.5 mol of C13-MA, and polymers **C9**–**C12** prepared with 4.5 mol of C13-MA. It is important to note that different areas of application require different materials, and both, rigid and soft, materials are needed.

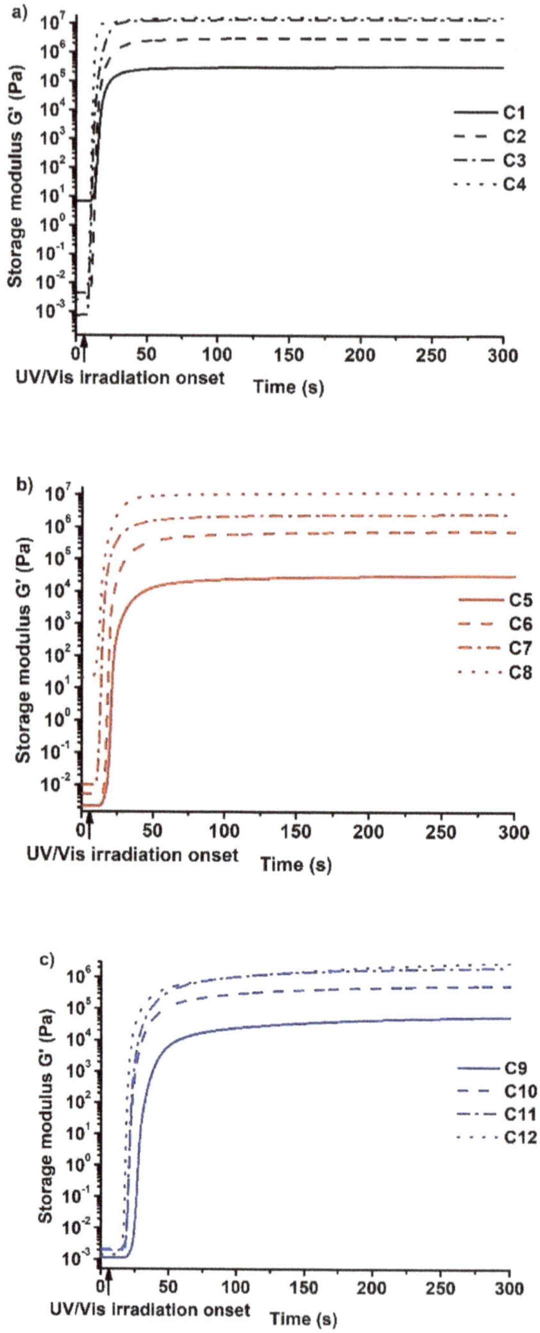

Figure 2. Dependency of storage modulus G′ of the resins **C1–C4** (**a**), **C5–C8** (**b**), and **C9–C12** (**c**) on the irradiation time.

The photocuring rate (described by the t_{gel}, induction period, and slope of the curve) is the other important parameter for the optimal curing process. The gel point is extremely important in microtransfer molding technology. The lowest values of t_{gel} and the shortest induction period were demonstrated by polymers prepared with the lowest amount of thiol. The curve of storage modulus of these polymers also reached the plateau faster than that of other vanillin-based polymers. As was stated earlier, free radical photopolymerization is a faster process in comparison to thiol-ene photopolymerization. However, not only the reaction mechanism, but also the structure of the monomer affects the gel point. C13-MA has a much longer structure compared to vanillin dimethacrylate and, as a result of that, the increase in C13-MA in compositions slowed the photopolymerization process, while the increase in VDM made it faster [28]. Subsequently, the lowest t_{gel} value was acquired by resin **C4** (t_{gel} = 3.4 s) prepared with 0.25 mol of thiol and 1.5 mol of tridecyl methacrylate. Resin **C12**, which was prepared with the same amount of thiol and a higher amount of C13-MA, had a much higher gel point (t_{gel} = 9.8 s) compared to resin **C4**, which was the result of an increased amount of flexible tridecyl methacrylate chains.

The shrinkage is also very important for the application of polymers. Low values of shrinkage are needed to form precise and unique structures, as high shrinkage could result in failed molding attempts. Values as low as 0–5% are mandatory in order to obtain high-quality products [32]. The lowest shrinkage values were demonstrated by polymers prepared with the highest amount of thiol. The shrinkage values of the polymers with 1 mol of thiol and different amounts of tridecyl methacrylate, **C1**, **C5**, and **C9**, were as low as 0–2%. However, resins with a lower amount of thiol and a higher amount of acrylate shrunk more. For example, the shrinkage increased from 1 to 10% as a result of the reduction of the amount of thiol from 1 to 0.25 mol in the resins **C9–C12**. This can be explained by the dominant reaction mechanism, as it is well known that acrylates have a high shrinkage rate during free radical polymerization [33], while thiol-ene photopolymerization results in a lower shrinkage volume [34]. That is because long-distance connections via weak Van der Waals force are replaced by short, strong covalent bonds between carbon atoms in monomer units during free radical acrylate polymerization, which results in shrinkage of the polymers [33].

After analyzing all of these results, the **C8** resin was selected as the most promising composition for the microtransfer molding technique. This composition demonstrated one of the lowest gel point values and relatively low shrinkage, which makes the process faster and allows the creation of complex and accurate shapes. Furthermore, the induction period of this resin was less than 1 s and the resultant polymer was a stiff material with a high value of G'. All these properties made the **C8** resin a suitable candidate for the microtransfer molding technique. For comparison, the resin **C1** that forms soft but non-brittle polymer was also selected for microtransfer molding.

3.2. *Characterization of Photocross-Linked Polymer Structure*

The chemical structure of the photocross-linked polymers was confirmed by FT-IR spectroscopy. Signals of C=O, which were present at 1714–1735 cm^{-1}, and those of the C=C group, which were present at 1606–1639 cm^{-1} in VDM and C13-MA spectra, were also reduced in polymer spectra. The signals of the S-H group, which were present at 2560 cm^{-1} in the BDT spectra, were not visible in the polymer spectra and the new signal of the C-S group was detected at 1120–1122 cm^{-1} in the polymer sample spectra. These changes in the polymer spectra indicated the formation of cross-linked structures in the polymers. As an example, the FT-IR spectra of VDM, C13-MA, BDT, and polymers **C5–C8** are presented in Figure 3.

The Soxhlet extraction was also performed in order to confirm the cross-linked structure of the polymers. The yield of the insoluble fraction of these polymers was in the range of 61–90% (Table 3). The high yield of the insoluble fraction indicated that the dense polymer network was formed during photopolymerization. However, polymers **C1, C5,** and **C9**, which were prepared with 1 mol of thiol, demonstrated a relatively low value of the yield of the insoluble fraction (61–70%). This was probably due to the large amount of soluble linear or branched polymer fragments, which were formed as the result of spatial hindrances as the polymer chains were unable to pass through each other during the photo polymerization process [35]. Acrylate free radical homopolymerization is faster than thiol-ene photopolymerization, and, as a result of that, the spatial hindrances were increased by increasing the amount of thiol.

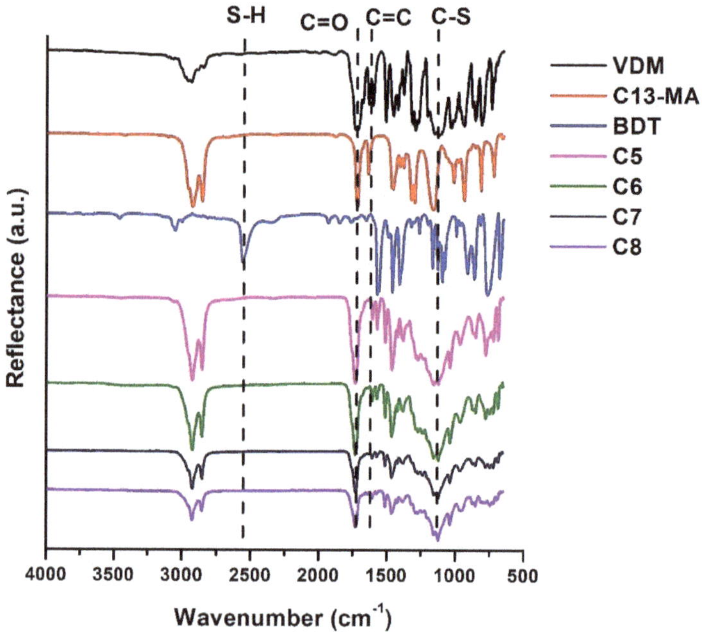

Figure 3. FT-IR spectra of VDM, C-13MA, BDT, and cross-linked polymers **C5–C8**.

Table 3. Yield of insoluble fraction, thermal, and thermomechanical characteristics of the polymers.

Resin	Yield of Insoluble Fraction, %	$T_{dec.-10\%}$, °C *	T_g, °C **	Storage Modulus G′ at −80 °C, MPa ***	Storage Modulus G′ at 100 °C, MPa ****
C1	69	308	−10	761.99	3.36
C2	84	328	10	1539.61	1.17
C3	87	329	28	1649.52	3.90
C4	90	332	54	2350.37	21.67
C5	61	328	−9	386.44	0.19
C6	82	338	0	499.35	0.88
C7	88	346	18	778.79	2.22
C8	85	344	28	902.59	4.84
C9	70	322	−7	19.08	0.58
C10	81	323	−1	911.49	3.15
C11	88	324	7	890.68	2.27
C12	79	331	17	661.37	2.69

* from TGA curves. ** from DMTA curves. *** before glass transition temperature. **** after glass transition temperature.

3.3. Thermal Properties of Cross-Linked Polymers

Dynamical mechanical thermal analysis (DMTA) and thermogravimetrical analysis (TGA) were used to study the thermal properties of the polymers. The results are summarized in Table 3. DMTA was used to determine the glass transition temperature (T_g) of the polymer samples. The DMTA curves are presented in Figure 4. The glass transition of the polymers was in the range of −10–54 °C. It was determined that the higher values of T_g were obtained when the lower volume of thiol was used. For example, T_g increased from −10 °C for polymer **C1** to 54 °C for polymer **C4** by reducing the amount of BDT from 1 to 0.25 mol. The low glass transition temperature was the result of softer polymer with a less dense structure and low yield of the insoluble fraction [36]. The amount of C13-MA also had a great influence on T_g. For example, the increase in the amount of C13-MA from 1.5 mol in the polymer **C2** to 4.5 mol in the polymer **C10** reduced the T_g from 10 to −1 °C. This was the result of a long carbon chain of C13-MA. Long chains made the polymer softer and flexible, and decreased the glass transition temperature of the polymers [37].

The thermal decomposition of the polymers occurred in one step (Figure 5). The temperature of 10% weight loss ($T_{dec.-10\%}$) was in the range of 308–346 °C. Consequently, to the glass transition temperature, the temperature of 10% weight loss was reduced by increasing the amount of BDT in the polymer. For example, $T_{dec.-10\%}$ increased from 308 °C to 332 °C by decreasing the amount of thiol from 1 to 0.25 mol. The results were correlated with the yield of the insoluble fraction of the polymers, as higher decomposition temperatures were shown by the polymers with the higher yield of the insoluble fraction and the denser inner polymer network. Different results were observed when the amount of C13-MA was increased. There was no consistent increase or decrease in $T_{dec.-10\%}$ when the amount of tridecyl methacrylate increased. Very similar values of the temperature of 10% weight loss were observed when 1.5 and 4.5 mol of C13-MA were used. However, polymers prepared with 3 mol of C13-MA demonstrated higher values of 10% weight loss. For example, polymers **C4** and **C12**, which were prepared with 1.5 and 4.5 mol of C13-MA, reached the temperature of 332 °C and 331 °C, respectively, while polymer **C8**, prepared with 3 mol of C13-MA, reached the temperature of 344 °C.

3.4. Thermomechanical Properties of Cross-Linked Polymers

Dynamical mechanical thermal analysis (DMTA) was performed to characterize the mechanical properties of the developed polymers. The dependence of the storage modulus G′ on temperature was measured (Figure 6) and the values of G′ before and after the glass transition temperature were compared (Table 3). For the compatible results, the storage modulus values were taken from the plateau of the curve as the plateau indicates that polymer is in the steady state and that no changes in its structure are happening. The chosen temperatures were −80 °C (before T_g) and 100 °C (after T_g). After the results were compared, it was determined that the storage modulus of these polymers decreases greatly as the temperature increases to values higher than those of their T_g. Because of this, the developed polymers are rigid materials below their glass transition temperature and become soft and flexible when the temperature increases above their T_g. For example, the storage modulus of polymer **C1** was reduced from 761.99 MPa to 3.36 MPa when the temperature increased from −80 °C to 100 °C. These results correlate to the rheological and thermal characteristics, as the same tendencies occur when these polymers are compared. For example, polymers **C4**, **C8**, and **C12**, prepared with different amounts of C13-MA, show that the storage modulus reduces from 2350.37 MPa to 661.37 MPa (at −80 °C) when the amount of C13-MA increases from 1.5 mol to 4.5 mol, which was the result of an increased amount of flexible tridecyl methacrylate chains, as mentioned above. Increased amounts of BDT in the polymer also reduced the rigidity of the polymers. For example, rigidity was reduced from 902.59 MPa (polymer **C8** at −80 °C) to 386.44 MPa (polymer **C5** at −80 °C) by increasing the amount of thiol from 0.25 mol to 1 mol, due to the formation of flexible thioether linkages, as also mentioned earlier.

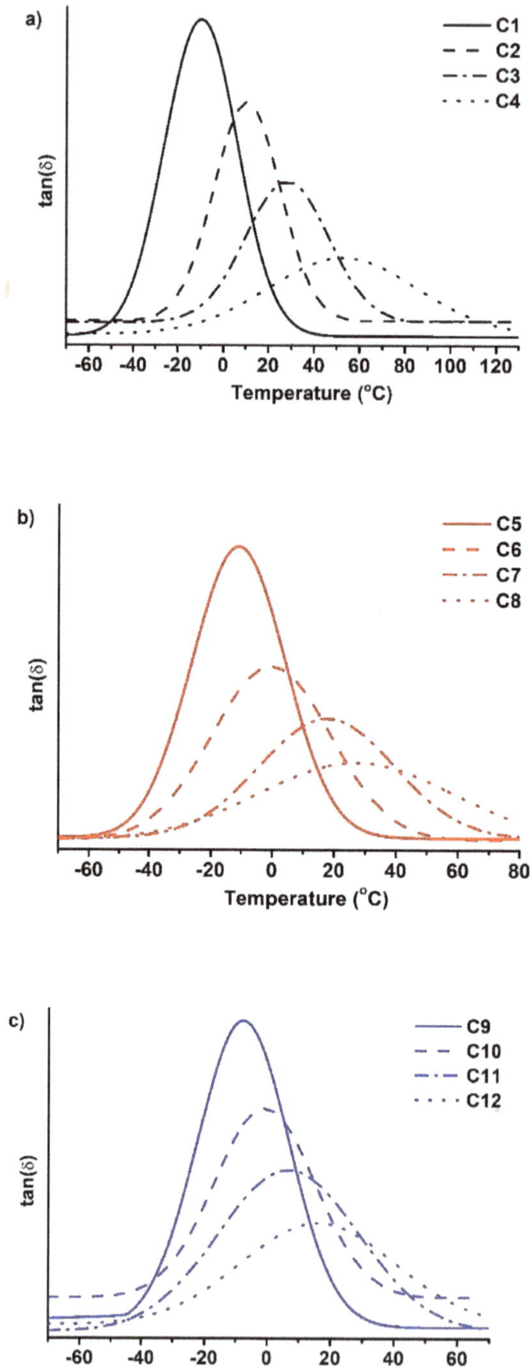

Figure 4. DMTA thermograms of cross-linked polymers **C1–C4** (**a**), **C5–C8** (**b**), and **C9–C12** (**c**).

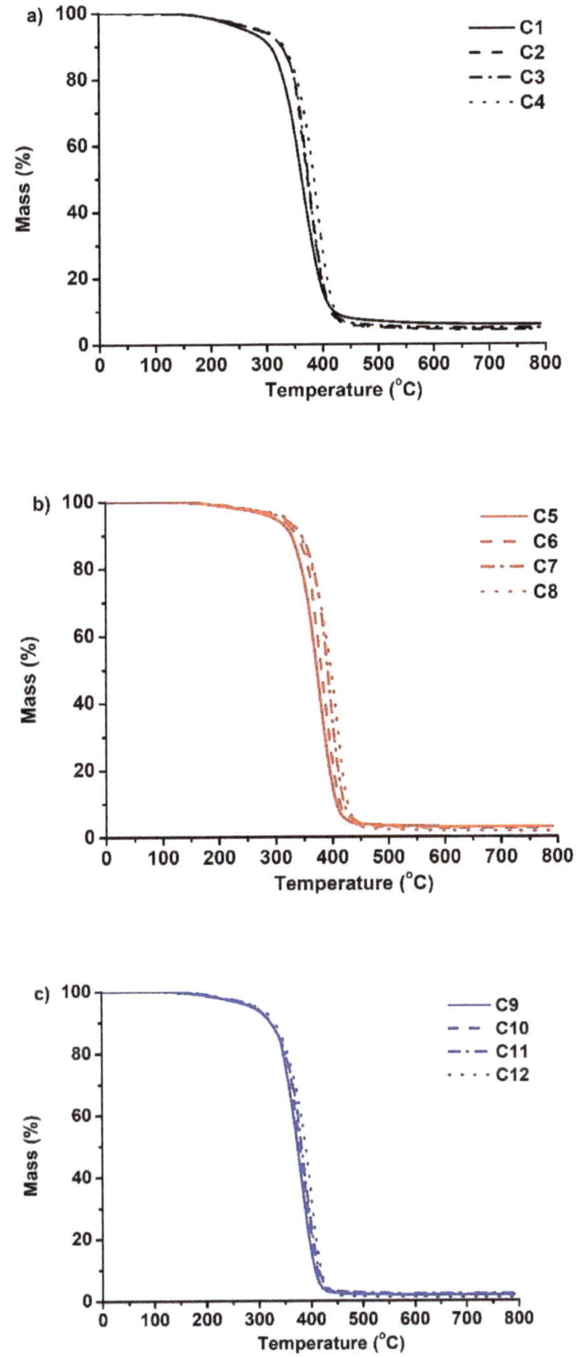

Figure 5. Thermogravimetric curves of cross-linked polymers **C1–C4** (**a**), **C5–C8** (**b**), and **C9–C12** (**c**).

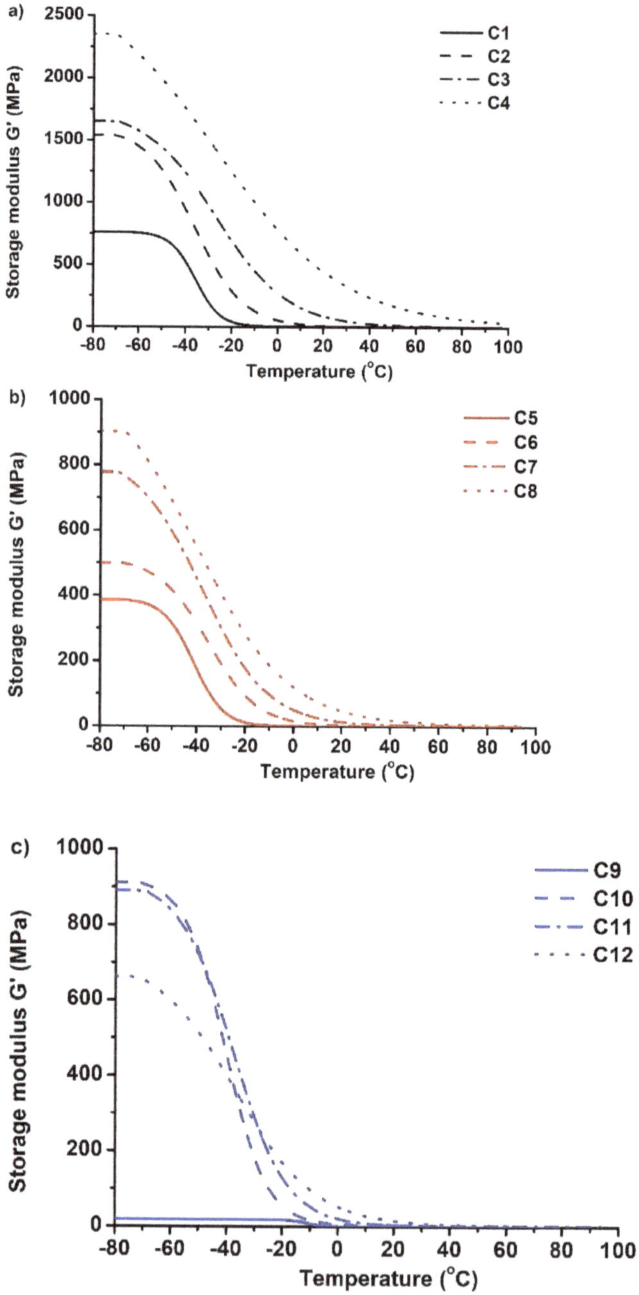

Figure 6. Dependency of the storage modulus G′ of polymers **C1–C4** (**a**), **C5–C8** (**b**), and **C9–C12** (**c**) on temperature.

3.5. Shape-Memory Properties of Cross-Linked Polymers

Thermo-responsive shape-memory polymers attract huge interest in the scientific community because of their unique ability to remember their shape. To show shape-memory properties, polymer samples were heated above their glass transition temperature, reformed to the temporary shape, and cooled down to the temperature lower than their T_g. All polymers were rigid materials and were able to maintain their temporary shape when the temperature was below their T_g. To return to a permanent shape, polymer samples were heated above their glass transition temperature, and all polymers were able to return to a permanent shape within seconds. The recovery to original shape conditions of these polymers is determined by their glass transition temperature [14].

Polymers were transformed to their temporary shape at 60 °C temperature and then placed in the refrigerator (−20 °C) to maintain their temporary shape. All developed polymers were able to maintain their permanent shape at a temperature below 0 °C, however, only a few of them were able to maintain it at room temperature (25 °C). Because of this, the application of these polymers is limited. The most promising results were demonstrated by the polymers **C3**, **C4,** and **C8**, because these polymers have relatively high glass transition temperature (from 28 to 54 °C) and can maintain their temporary shape at room temperature and above. The transformation of the **C8** sample is presented in Figure 7.

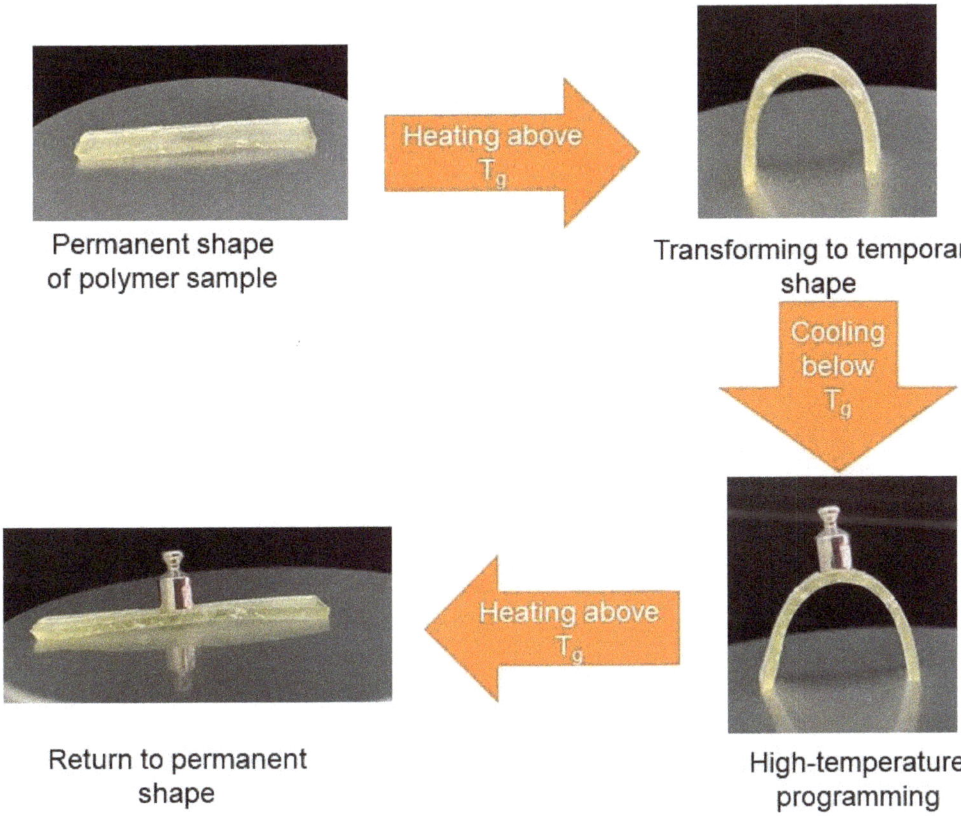

Figure 7. Scheme of shape-memory behaviour of polymer samples.

3.6. Mechanical Characteristics of Cross-Linked Polymers

The tensile test was performed to investigate the mechanical properties of the obtained polymers. The results are presented in Table 4. Polymers **C1**, **C2**, **C5**, and **C9** were too soft and/or brittle for testing machine, and it was impossible to detect the break during the measurement. The highest values of Young's modulus and the lowest values of elongation and break were demonstrated by polymer **C4**, which was prepared with the lowest amount of C13-MA and BDT (60.05 MPa and 5.47%). Polymer **C3**, which was prepared with the same amount of VDM and C13MA as polymer **C4**, but with a higher amount of thiol, demonstrated a lower Young's modulus value and greater elongation at break value (12.44 MPa and 17.3%). This behaviour was caused by the highly flexible thioether linkages. These linkages also explain why polymers with higher amounts of thiol stretched more during the tensile test and their elongation at break value was higher. Tridecyl methacrylate had a similar effect as thiol on the mechanical properties of the polymers. The increased amount of C13-MA in the composition led to the lower values of Young's modulus. For example, by increasing the amount of C13-MA from 1.5 to 3 and then to 4.5 mol the Young's modulus decreased as follows, from 60.05 MPa (polymer **C4**) to 23.033 MPa (polymer **C8**) and, finally, to 5.53 MPa (polymer **C12**). The elongation at break increased with increasing C13-MA in the resin from 5.47% (polymer **C4**) to 6.49% (polymer **C8**) and, finally, to 8.38% (polymer **C12**). This behaviour can be explained by the long carbon chain of tridecyl methacrylate. As a result, the polymer can stretch more, and is softer and more flexible [34].

Table 4. Mechanical characteristics of the polymers obtained by the tensile test.

Resin	Young's Modulus, MPa	Tensile Strength, MPa	Elongation at Break, %
C1	- *	-	-
C2	-	-	-
C3	12.436	1.256	17.30
C4	60.050	1.976	5.47
C5	-	-	-
C6	1.865	0.140	12.19
C7	3.991	0.309	11.10
C8	23.033	0.960	6.49
C9	-	-	-
C10	2.317	0.040	22.36
C11	3.054	0.129	10.71
C12	5.531	0.227	8.38

*—not determined due to excessively soft and/or brittle polymer.

3.7. Characterization of Microtransfered Structures

Two photocurable resins forming nonbrittle polymers, one rigid (**C8**) and the other soft (**C1**), have been tested in microtransfer molding (μTM). The results are depicted in Figure 8. Image (a) demonstrates the 3D printed USAF target. The thickest 3D printed lines were approximately 220–230 μm width, meanwhile the thinnest lines were about 68–69 μm width. Image (b) shows the PDMS stamp, which coincided well with the 3D printed target. The following images (c) and (d) depict replicas made of **C1** and **C8** resins. In both cases, the replicas corresponded to the PDMS mold, keeping its shape and all features (lines, letters, and numbers). Only minor drawbacks were observed on the replicas, which were voids (usually appearing due to air bubbles while dropcasting the resin) and loss of sharp edges due to peeling of the stamp. Despite this, both **C1** and **C8** resins showed great potential to be used for the μTM technique.

Figure 8. Results of μTM. (**a**) 3D printed 1951 USAF target (25% of the downloaded vector file size). (**b**) Soft PDMS mold. (**c**) and (**d**) are replicas made from **C1** and **C8** resins, respectively. Scale bars are the same in all the images and represent 5 mm.

4. Conclusions

Novel thermo-responsive shape-memory vanillin-based photopolymers have been developed and applied in microtransfer molding. Different mixtures of the two biobased monomers vanillin dimethacrylate and tridecylmethacrylate, to which 1,3-benzenedithiol has been added, have been tested as photocurable resins. The reduction of the thiol content was determined to increase the shrinkage of the polymers, the photocuring rate, and the values of the thermal and mechanical characteristics of the resulting polymers. The increase in tridecyl methacrylate content resulted in less rigid polymers, lower photocuring rate, and lower glass transition temperature. All polymers demonstrated great shape-memory properties and were able to return to their primary shape after the temperature programing. Furthermore, they were able to maintain their temporary shape when the temperature was lower than their glass transition temperature. Two photocurable resins forming nonbrittle polymers, one rigid and the other soft, have been tested in microtransfer molding and both resins demonstrated perfect replication, proving the novel bio-based photoresins to be suitable for microtransfer molding technique.

The developed thermo-responsive shape-memory bio-based photopolymers have great potential for forming microtransfered structures and devices that can be used on nonflat surfaces. This work offers a sustainable route for exploiting biorenewable materials to replace established commercial photoresins applied in stamping technologies that are already being scaled up for cost-effective everyday use.

Author Contributions: Conceptualization, J.O. and A.N.; methodology, J.O., A.N. and M.M.; formal analysis, J.J. (Justinas Jaras), J.J. (Jurga Jersovaite) and E.S.; investigation, J.J. (Justinas Jaras); writing—original draft preparation, A.N. and E.S.; writing—review and editing, J.O. and M.M.; visualization, A.N. and E.S.; supervision, J.O. and M.M.; project administration, J.O.; funding acquisition, J.O. All authors have read and agreed to the published version of the manuscript.

Funding: This research was funded by the Research Council of Lithuania (project No. S-MIP-20-17) and the Doctoral Fund of Kaunas University of Technology (No. A-410, approved 26 June 2019).

Institutional Review Board Statement: Not applicable.

Informed Consent Statement: Not applicable.

Data Availability Statement: Not applicable.

Acknowledgments: Ch. Delaite, A. S. Schuller, and Ch. Deguines from the University of Haute-Alsace are gratefully acknowledged for the opportunity to carry out DMTA measurements.

Conflicts of Interest: The authors declare no conflict of interest.

References

1. Poulopoulou, N.; Kasmi, N.; Bikiaris, D.N.; Papageorgiou, D.G.; Floudas, G.; Papageorgiou, G.Z. Sustainable Polymers from Renewable Resources: Polymer Blends of Furan-Based Polyesters. *Macromol. Mater. Eng.* **2018**, *303*, 1800153. [CrossRef]
2. Schneiderman, D.K.; Hillmyer, M.A. 50th anniversary perspective: There is a great future in sustainable polymers. *Macromolecules* **2017**, *50*, 3733–3749. [CrossRef]
3. Zhu, Y.; Romain, C.; Williams, C.K. Sustainable polymers from renewable resources. *Nature* **2016**, *540*, 354–362. [CrossRef] [PubMed]
4. Fache, M.; Boutevin, B.; Caillol, S. Vanillin, a key-intermediate of biobased polymers. *Eur. Polym. J.* **2015**, *68*, 481–487. [CrossRef]
5. Zhang, C.; Yan, M.; Cochran, E.W.; Kessler, M.R. Biorenewable polymers based on acrylated epoxidized soybean oil and methacrylated vanillin. *Mater. Today Commun.* **2015**, *5*, 18–22. [CrossRef]
6. Navaruckiene, A.; Skliutas, E.; Kasetaite, S.; Rekštytė, S.; Raudoniene, V.; Bridziuviene, D.; Malinauskas, M.; Ostrauskaite, J. Vanillin Acrylate-Based Resins for Optical 3D Printing. *Polymers* **2020**, *12*, 397. [CrossRef]
7. Navaruckiene, A.; Bridziuviene, D.; Raudoniene, V.; Rainosalo, E.; Ostrauskaite, J. Vanillin acrylate-based thermo-responsive shape memory antimicrobial photopolymers. *Express Polym. Lett.* **2022**, *16*, 279–295. [CrossRef]
8. Roig, A.; Ramis, X.; de la Flor, S.; Serra, À. Dual-cured thermosets from glycydil methacrylate obtained by epoxy-amine reaction and methacrylate homopolymerization. *React. Funct. Polym.* **2021**, *159*, 1381–5148. [CrossRef]
9. Ramis, X.; Fernández-Francos, X.; de la Flor, S.; Ferrando, F.; Serra, À. Click-based dual-curing thermosets and their applications. In *Thermosets Structure, Properties, and Applications*, 2nd ed.; Guo, Q., Ed.; Elsevier: Amsterdam, The Netherlands, 2017; Chapter 16.
10. Konuray, O.; Fernández-Francos, X.; Ramis, X.; Serra, À. State of the Art in Dual-Curing Acrylate Systems. *Polymers* **2018**, *10*, 178. [CrossRef]
11. Gamardella, F.; Sabatini, V.; Ramis, X.; Serra, À. Tailor-made thermosets obtained by sequential dual-curing combining isocyanate-thiol and epoxy-thiol click reactions. *Polymer* **2019**, *174*, 200–209. [CrossRef]
12. Zhao, Q.; Qi, H.J.; Xie, T. Recent progress in shape memory polymer: New behavior, enabling materials, and mechanistic understanding. *Prog. Polym. Sci.* **2015**, *50*, 79–120. [CrossRef]
13. Zare, M.; Prabhakaran, M.P.; Parvin, N.; Ramakrishna, S. Thermally-induced two-way shape memory polymers: Mechanisms, structures, and applications. *Chem. Eng. J.* **2019**, *374*, 706–720. [CrossRef]
14. Hager, M.D.; Bode, S.; Weber, C.; Schubert, U.C. Shape memory polymers: Past, present and future developments. *Prog. Polym. Sci.* **2015**, *50*, 3–33. [CrossRef]
15. Zhang, J.; Zhang, C.; Shang, Q.; Hu, Y.; Song, F.; Jia, P.; Zhu, G.; Huang, J.; Liu, C.; Hu, L.; et al. Mechanically robust, healable, shape memory, and reprocessable biobased polymers based on dynamic pyrazole-urea bonds. *Eur. Polym. J.* **2022**, *169*, 111133. [CrossRef]
16. Sanchez-Rexach, E.; Smith, P.T.; Gomez-Lopez, A.; Fernandez, M.; Cortajarena, A.L.; Sardon, H.; Nelson, A. 3D-printed bioplastics with shape-memory behavior based on native bovine serum albumin. *ACS Appl. Mater. Interfaces* **2021**, *13*, 19193–19199. [CrossRef]
17. Stetsyshyn, Y.; Raczkowska, J.; Budkowski, A.; Awsiuk, K.; Kostruba, A.; Nastyshyn, S.; Harhay, K.; Lychkovskyy, E.; Ohar, H.; Nastishin, Y. Cholesterol-Based Grafted Polymer Brushes as Alignment Coating with Temperature-Tuned Anchoring for Nematic Liquid Crystals. *ACS Langmuir* **2016**, *32*, 11029–11038. [CrossRef]
18. Raczkowska, J.; Stetsyshyn, Y.; Awsiuk, K.; Lekka, M.; Marzec, M.; Harhay, K.; Ohar, H.; Ostapiv, D.; Sharan, M.; Yaremchuk, I.; et al. Temperature-responsive grafted polymer brushes obtained from renewable sources with potential application as substrates for tissue engineering. *Appl. Surf. Sci.* **2017**, *407*, 546–554. [CrossRef]
19. Javaid, M.; Haleem, A.; Singh, R.P.; Suman, R.; Rab, S. Role of additive manufacturing applications towards environmental sustainability. *Adv. Ind. Eng. Polym. Res.* **2021**, *4*, 312–322. [CrossRef]
20. Haleem, A.; Javaid, M. Additive manufacturing applications in industry 4.0: A review 04. *J. Ind. Integr. Manag.* **2019**, *4*, 1930001. [CrossRef]
21. Gu, W.; Styger, E.; Warner, D.H. Assessment of additive manufacturing for increasing sustainability and productivity of smallholder agriculture. *3D Print. Addit. Manuf.* **2020**, *7*, 300–310. [CrossRef]
22. Rejeski, D.; Zhao, F.; Huang, Y. Research needs and recommendations on environmental implications of additive manufacturing. *Addit. Manuf.* **2018**, *19*, 21–28. [CrossRef]
23. Colorado, H.A.; Velásquez, E.I.; Monteiro, S.N. Sustainability of additive manufacturing: The circular economy of materials and environmental perspectives. *J. Mater. Res. Technol.* **2020**, *9*, 8221–8234. [CrossRef]
24. Zhang, X.; Cox, L.; Wen, Z.; Xi, W.; Ding, Y.; Bowman, C.N. Implementation of two distinct wavelengths to induce multistage polymerization in shape memory materials and nanoimprint lithography. *Polymer* **2018**, *156*, 162–168. [CrossRef] [PubMed]
25. Xu, Z.Y.; Li, L.; Du, L.; Wang, L.; Shi, L.Y.; Yang, K.K.; Wang, Y.Z. Multiscale shape-memory effects in a dynamic polymer network for synchronous changes in color and shape. *Appl. Mater. Today* **2022**, *26*, 101276. [CrossRef]
26. Lan, X.; Huang, W.; Leng, J. Shape Memory Effect in Micro-Sized Shape Memory Polymer Composite Chains. *Appl. Sci.* **2019**, *9*, 2919. [CrossRef]
27. Green, W.A. *Industrial Photoiniciators, A Technical Guide*, 2nd ed.; CRC Press: Boca Raton, FL, USA, 2010.

28. Navaruckiene, A.; Bridziuviene, D.; Raudoniene, V.; Rainosalo, E.; Ostrauskaite, J. Influence of Vanillin Acrylate-Based Resin Composition on Resin Photocuring Kinetics and Antimicrobial Properties of the Resulting Polymers. *Materials* **2021**, *14*, 653. [CrossRef]
29. Danilevičius, P.; Rekštytė, S.; Balčiūnas, E.; Kraniauskas, A.; Širmenis, R.; Baltriukienė, D.; Bukelskienė, V.; Gadonas, R.; Sirvydis, V.; Piskarskas, A.; et al. Laser 3D micro/nanofabrication of polymers for tissue engineering applications. *Opt. Laser Technol.* **2013**, *45*, 518–524. [CrossRef]
30. Mezger, T.G. *The Rheology Handbook*, 3rd ed.; Vincentz Network: Hanover, Germany, 2011.
31. Buruiana, T.; Melinte, V.; Stroea, L.; Buruiana, E. Urethane Dimethacrylates with Carboxylic Groups as Potential Dental Monomers. Synthesis and Properties. *Polym. J.* **2009**, *41*, 978–987. [CrossRef]
32. Gurr, M.; Mülhaupt, R. Rapid prototyping. In *Polymer Science: A Comprehensive Reference*, 1st ed.; Matyjaszewski, K., Möller, M., Eds.; Elsevier: Edinburgh, UK, 2012; Volume 1, pp. 77–99.
33. Jian, Y.; He, Y.; Jiang, T.; Li, C.; Yang, W.; Nie, J. Volume shrinkage of UV-curable coating formulation investigated by real-time laser reflection method. *J. Coat. Technol. Res.* **2013**, *10*, 231–237. [CrossRef]
34. Albeshir, E.G.; Alsahafi, R.; Albluwi, R.; Balhaddad, A.A.; Mitwalli, H.; Oates, T.W.; Hack, G.D.; Sun, J.; Weir, M.D.; Xu, H.H.K. Low-Shrinkage Resin Matrices in Restorative Dentistry-Narrative Review. *Materials* **2022**, *15*, 2951. [CrossRef]
35. Imakaev, M.V.; Tchourine, K.M.; Nechaev, S.K.; Mirny, L.A. Effect of topological constraints on globular polymers. *Soft Matter* **2015**, *11*, 665–671. [CrossRef] [PubMed]
36. Xie, R.; Weisen, A.R.; Lee, Y.; Aplan, M.A.; Fenton, A.M.; Masucci, A.E.; Kempe, F.; Sommer, M.; Pester, C.W.; Colby, R.H.; et al. Glass transition temperature from the chemical structure of conjugated polymers. *Nat. Commun.* **2020**, *11*, 893. [CrossRef] [PubMed]
37. Shrivastava, A. Introduction to plastics engineering. In *Plastics Design Library, Introduction to Plastics Engineering*, 1st ed.; Shrivastava, A., Ed.; William Andrew Publishing: New York, NY, USA, 2018; Volume 1, pp. 1–16.

polymers

MDPI

Review

Recent Advances in 3D Bioprinting: A Review of Cellulose-Based Biomaterials Ink

Wan Nazihah Liyana Wan Jusoh [1,2], Mohd Shaiful Sajab [1,2,*], Peer Mohamed Abdul [1,2] and Hatika Kaco [3]

1 Research Center for Sustainable Process Technology (CESPRO), Faculty of Engineering and Built Environment, Universiti Kebangsaan Malaysia, Bangi 43600, Selangor, Malaysia; nazihahliyana.j@gmail.com (W.N.L.W.J.); peer@ukm.edu.my (P.M.A.)
2 Department of Chemical and Process Engineering, Faculty of Engineering and Built Environment, Universiti Kebangsaan Malaysia, Bangi 43600, Selangor, Malaysia
3 Kolej GENIUS Insan, Universiti Sains Islam Malaysia, Bandar Baru Nilai, Nilai 71800, Negeri Sembilan, Malaysia; hatikakaco@usim.edu.my
* Correspondence: mohdshaiful@ukm.edu.my; Tel.: +60-3-8921-6425

Abstract: Cellulose-based biodegradable hydrogel proves to be excellently suitable for the medical and water treatment industry based on the expressed properties such as its flexible structure and broad compatibility. Moreover, their potential to provide excellent waste management from the unutilized plant has triggered further study on the advanced biomaterial applications. To extend the use of cellulose-based hydrogel, additive manufacturing is a suitable technique for hydrogel fabrication in complex designs. Cellulose-based biomaterial ink used in 3D bioprinting can be further used for tissue engineering, drug delivery, protein study, microalgae, bacteria, and cell immobilization. This review includes a discussion on the techniques available for additive manufacturing, bio-based material, and the formation of a cellulose-based hydrogel.

Keywords: 3D printing; additive manufacturing; biopolymer; cellulose; cell immobilization

Citation: Wan Jusoh, W.N.L.; Sajab, M.S.; Mohamed Abdul, P.; Kaco, H. Recent Advances in 3D Bioprinting: A Review of Cellulose-Based Biomaterials Ink. *Polymers* **2022**, *14*, 2260. https://doi.org/10.3390/polym14112260

Academic Editors: Antonio M. Borrero-López, Concepción Valencia-Barragán, Esperanza Cortés Triviño, Adrián Tenorio-Alfonso and Clara Delgado-Sánchez

Received: 18 April 2022
Accepted: 30 May 2022
Published: 31 May 2022

1. Introduction

The adaptation of 3D bioprinting with cell culturing and drug delivery has shown suitable trends over the past few years. One of the benefits of using additive manufacturing, such as 3D bioprinting, is its ability to work with living materials and cells during the printing process [1]. In 3D bioprinting, biopolymer hydrogel expresses perfect criteria to imitate tissue matrix and can be produced from biomaterial [2]. Hydrogel preparation commonly requires some integration with other materials to increase efficiency and performance. Based on studies conducted by some researchers, many successful findings have been concluded for the use of hydrogel, which led to its application in the human body. Moreover, culturing algae and bacteria on the hydrogel shows an excellent characteristic of hydrogel to support the growth of living cells.

Uses of natural resources from plants and microorganisms such as cellulose, lignin, alginate, chitosan, and algae have gained some interest in 3D bioprinting. All these resources can be obtained abundantly from specific parts of plants and animals but require some treatments during the extraction process. Cellulose, extracted from plants, is an attractive biopolymer in focus throughout this review. Pretreatment steps for 3D bioprinting are less complicated, and cellulose can survive during 3D bioprinting at specific process parameters [3]. Biopolymer is the main material in medical products as it shows a safe and convenient effect that is harmless to humans and animals.

In this review, 3D bioprinting involves several techniques, such as extrusion, laser-assisted bioprinting, and inkjet, which are discussed in detail. The need for biopolymer as the biomaterial ink for 3D bioprinting and biopolymer hydrogel was discussed, particularly regarding the properties and application. Then, the recent study on 3D bioprinting for uptake and release of protein, immobilization of algae, bacteria, and cell culturing of human and animal cells in the hydrogel are being reviewed. This review includes a thorough explanation of cellulose as a biopolymer for its application in additive manufacturing, including its characteristics, extraction techniques, and size reduction process. Furthermore, we emphasized the use of cellulose-based biomaterial ink and hydrogel preparation techniques. Three-dimensional bioprinting of cellulose-based biomaterial in other specific areas such as tissue engineering and drug delivery is reviewed. Finally, the visual bibliometric network and analysis are explained in the statistic of the recent study.

2. 3D Bioprinting Techniques

In the early 2000s, 3D printing was initially used in the medical industry before leading to 3D bioprinting. Thomas Boland's group invented the first bioprinter at Clemson University after using an inkjet printer for cell printing [4]. From the printing of cells, 3D bioprinting then developed into bones, organ, and tissue printing, which have comprehensive advantages in tissue engineering and organ transplant.

Most bioprinters use the same techniques as 3D printers, such as extrusion-based, inkjet, and laser-assisted. A primary step required before conducting 3D bioprinting is preparing the ink material and creating a 3D model using the software. The crucial part of 3D bioprinting is the formation of bioink made up of biocompatible material and living cells that typically need some culturing process. Bioink needs to show some criteria such as suitable viscosity, shear-thinning, and capability to be crosslinked [5]. Then, computer-aided design (CAD) is software used to model a 3D structure with internal and external details such as the size of pores [5]. The .stl file refers to standard triangle language and will be used for printing by adding the file to the 3D printing software [6]. After modeling the structure, the 3D bioprinting process can commence according to several specifications based on the required printed product.

3D bioprinting can overcome the shortage of organs and increase the possibility of organ transplants. It can produce a complex structure for medical purposes with high cell survival rates in shorter times and efficient costs [7]. The 3D bioprinting of the skin has a great possibility to be produced repeatedly with faster production [8]. The customization of specific tissues or organs is applicable with this 3D bioprinting. Direct deposition of the printing ink or cells on the targeted human body part is possible by using 3D bioprinting [9].

2.1. Extrusion/Liquid Deposition Modeling

Liquid deposition modeling (LDM) is a 3D printing method focusing on printing the viscous liquid made up of biomaterial. Liquid ink can be added and stored in the syringe for printing and released during the process with the help of external force. The composition and viability of the biomaterial can be maintained during the process as it is placed in a controlled temperature and pressure condition. The printing temperature and build plate temperature will influence the printing of each biomaterial. Furthermore, the printing process is controlled by specific parameters such as viscosity, the flow of fluid, shear-thinning, and gelation [10]. Other than that, the geometry of printed material is crucial for 3D printing. As the size becomes smaller with a channeled structure, the printing becomes more challenging and complicated [11]. Clogging and blockage are the main problems in the printing process, and as explained, smaller-sized print core leads to clogging.

There is a general procedure for extruding 3D bioprinting of bacterial nanocellulose (BNC) by dissolution in 1-ethyl-3-methylimidazolium acetate (EmimAC) [12]. The syringe is attached to the holder and connected to the cellulose solution, and the syringe pump will manage the syringe actuator. A needle for ink dispersion is attached to the syringe. A slight increase in the temperature of ink solution can cause dilution and ease the extrusion of bioinks. The increase in the concentration of biomaterial requires an increase in printing temperature to reduce the potential of blockage in the print head and nozzle [6].

Biomaterial must be in a printable solution state, which is in a paste-like form [13]. Additional chemicals such as isopropanol were used to clean the building plane and induce an easier transfer of printed material. For example, in extrusion bioprinting, the dual crosslinking technique is the recent finding where two different biomaterials, polyethylene glycol and cellulose nanofibrils (CNF), are required [14]. It needs to be crosslinked before and after bioprinting by placing it under the light for photo-crosslinking. The crosslinking of biomaterial improved the properties of the printed product and showed better potential to be used as a bioinks. LDM technique is suitable for printing liquid bioink due to the printer's design.

2.2. Stereolithography

Stereolithography, a compatible 3D printing technique, uses digital light processing where an ultraviolet lamp is used as the curing agent [15]. Moreover, the excess resin can be eliminated from the printed material by cleaning with a suitable alcohol solution. Recently, a new printing technique was discovered in which it uses fast hydrogel stereolithography printing [16]. This technique can produce a centimeter-sized printed product such as a hydrogel scaffold in several minutes by using photopolymerization. The printing setup uses digital light processor (DLP) devices to print the image on the photosensitive material in the liquid tank. Some parameters such as the condition of photopolymerization, force-driven, and velocity of hydrogel need to be controlled. Low printing duration gives advantages to this process as it lowers the chances of cell injury and deformation due to lesser printing time.

Besides that, it uses a liquid resin that can provide natural support in printing the hollow structure. Photopolymerization is an important process that affects the product. It can be managed by controlling the energy, printing speed, polymer concentration, and coefficient of photoinitiator absorption. In addition, there are studies on applying low force stereolithography using a Form 3B printer with seven printing resins in the formation of microfluidic devices for biological application [17]. Technically, low force stereolithography uses a combination of the galvanometer, laser beam, filter, fold, and parabolic mirror in the printing process. The maximum resolution is obtained by manipulating the channel and pillar using software and adjusting the orientation of the printing angle. This improved technique resulted in increased preciseness, accuracy, and less manufacturing stress on the printed product. Culturing human endothelial cells on the printed disc shows that the disc is biocompatible for cell seeding.

2.3. Inkjet

Inkjet printing is divided into two techniques, which are drop on demand (DOD) and continuous inkjet (CIJ). It is a contactless printer that does not cause any damage to the printed substrate during the process [18]. An exact small amount of bioinks droplet (picolitre) is released onto the substrate or culture dish [19]. Sterilization of the printer is required before adding the ink to minimize the risk of contamination. Sterilization can be performed by exposing the printer to ultraviolet light for at least overnight, spraying the cartridge with ethanol, and incubating the bio paper [20]. For inkjet printing of cellulose-based hydrogel, it is recommended to use low viscosity ink as it will ease the printing process [10]. This printing technique is able to print biomaterial ink directly on the body through an automated or hand-held device [21]. The printing is precisely based on the topography, and it is beneficial for an uneven wound area.

For 3D bioprinting of biomaterial using alginate and CNF, the centrifugation process and withdrawal of the unused supernatant enable improvement in the dispersion of CNF [22]. The accuracy of printing and excellent dispersion is determined by the micro-sized 300 μm diameters of the printer head valve. At the same time, the flow rate depends on the pressure of dispersion, time of valve opening, and distance of dosing. Additional printing speed parameter is required in managing the distance between each printed line. Besides that, stirring can lower the potential of ink from settling. CNF has better accuracy in terms of shape than alginates as it has low shear rates, higher viscosity, and shear-thinning, but mechanical force could destroy it.

2.4. Laser-Assisted Bioprinting

Laser-assisted bioprinting (LAB) is a bioprinting technique that uses the laser to deposit the bioink into the substrate surfaces. It applied the laser-induced forward-transfer (LIFT) effect with the involvement of three components, such as the pulse laser source, ribbon, or target, containing the bioink and substrate to accept the printed material [23,24]. The ribbon or target should be made of a non-absorbing material and coated with a thin layer of absorbing metal for easier deposition of printed material [23]. The printing process using LAB techniques takes almost 3 h for the whole printing. Moreover, it is required to have a laminar flow cabinet and incubator to reduce the chances of contamination and better printing efficiency, respectively [24]. The laser-assisted bioprinting workstation must include specifications such as the laser wavelength, pulse duration, repetition rate, beam quality, galvanometric mirror, CAD/CAM software, and printer cartridges [25].

For this technique, the bioink is in the liquid state, and its viscosity and cell composition will determine printed cell quality [25]. Furthermore, the ink solution needs to maintain its sterilization condition and viscoelastic properties during the printing process [24]. For efficient printing, analyzing the cartridges under the microscope before printing can help maintain the concentration and uniform distribution of cells to be entirely deposited on the slide [24]. Besides that, loss of water and stable viscoelastic properties can be controlled by locating the cartridge and holder onto the ice blocks before printing. Laser-assisted bioprinting is an advanced bioprinting technique that can specify the cell density and the 3D organization of cells [23]. Other than that, it can imitate the physiological performance of the actual component. This technique enables automation in processing and high reproducibility of the printed material.

3. Potential and Limitation of Hydrogel as Biomaterial Ink

3.1. Biomaterial Ink

Three-dimensional bioprinting is an example of additive manufacturing. It is an advanced 3D printing technique that can print living cells using biomaterial to support and enhance the cells. The biocompatible material will undergo modification based on the desired printed living cell before being added to the syringe. By using 3D bioprinting, it avoids a high-risk cell seeding process and promotes high distribution of cells at multiple locations as well as high cell density [26]. A significant comparison between 3D printing and bioprinting was illustrated in Figure 1. Many studies have been conducted on the cell and tissue for application in the medical field.

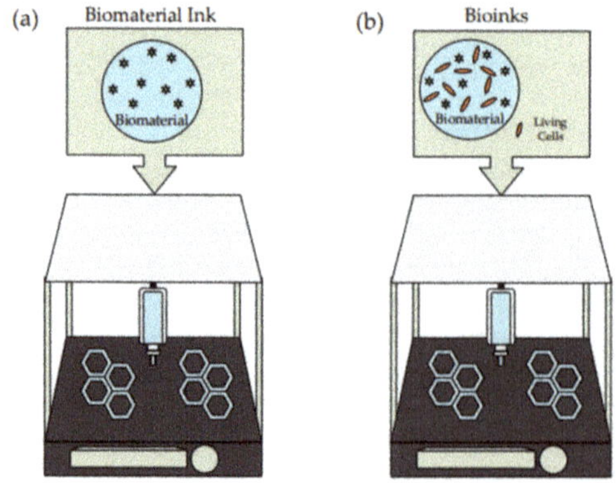

Figure 1. Difference between (**a**) 3D printing and (**b**) 3D bioprinting with the illustration using liquid deposition modeling (LDM) technique.

Significant differences in 3D printing and 3D bioprinting are based on the type of ink being extruded from the nozzle. Biomaterial ink is the typical type of ink used for 3D printing involving bio-based material in ink production. The biomaterial can interact with or respond to the living cells where it is widely applied in medical treatment [27,28]. In the late 1800s, the uses of biomaterials gained much attention and are being used widely for medical purposes [29]. The interaction of biomaterial with biological cells might have good or bad effects depending on the characteristics of the biomaterial [28]. At the early stage of its development, the biomaterial is relatively inert and cannot be totally accepted by the body [27]. The biomaterial should be biocompatible, non-toxic, bioactive, and biodegradable for the human body [27,30].

For 3D bioprinting, the ink consists of a combination of living cells with the biomaterial called bioinks [31]. Generally, it can be divided into two categories in which natural and synthetic bioinks [32]. Bioink should express similar criteria as the other ink, such as excellent mechanical properties and the ability to retain the cell throughout the printing process. As it involves biologically, the bioinks must be biocompatible and cytocompatible toward the living cell and extracellular matrix [31].

3.2. Potential of Biopolymer Hydrogel

Biopolymers are monomers attached by a covalent bond to form a larger molecule. The phrase bio refers to the material produced by living things such as plants or microbes. For medical application, most biopolymer is derived from microorganisms and are suitable for drug delivery and tissue engineering [33]. The increase in waste production from agriculture triggered the development of biopolymer material. Moreover, it can reduce the possibility of global warming as the waste disposed into landfills decreases [34]. Some controlled conditions are required in producing biopolymer where the microorganism needs certain nutrients and a suitable environment for its development [33].

Cellulose and alginate are the common biomaterials used in preparing hydrogel for the immobilization of microalgae [35,36]. CNF as the medium for immobilization of algae and cyanobacteria shows better outcomes as it is in nanostructure with a high aspect ratio. Moreover, it is fully transparent to penetrate light sources for cell growth [35]. At the same time, alginate has suitable biocompatibility, simple handling, and can use thin layer immobilization where it can provide constant light for capturing cell structure [35,36]. Further study has been conducted on the loading and releasing of protein in BNC hydrogel using a high-speed technique [37]. It is claimed that the adsorption process increases rapidly with the high-speed approach, but the release is slower compared to the adsorption technique. This occurs due to instability of the protein, thickness of fiber, size of pores, and density.

High-speed methods produce a higher thickness of fiber that partially close the pores and a higher density that cause more obstacle in transport. However, a robust cell is required during the 3D bioprinting process to ensure that those cells can endure physical and biological stresses such as shear stress and unstable pH [9]. A study on superabsorbent hydrogel was performed using sodium carboxymethyl cellulose (NaCMC) and cellulose in the presence of NaOH/urea aqueous with epichlorohydrin (ECH) as the crosslinker [38]. It is suitable for smart swelling and controllable delivery of protein.

Besides that, the loading and release of BSA from injectable polysaccharides-based hydrogel, prepared from cation octa(γ-chloroammoniumpropyl) silsesquioxane (OCAPS), chitosan, and oxidized hydroxypropyl cellulose show a successful analysis [39]. The release of protein is analyzed using an ultraviolet spectrophotometer. The formulation for encapsulation efficiency (EE) and drug loading efficiency (LE) is mentioned in Equations (1) and (2).

$$EE = \frac{m_1 - m_2}{m_1} \times 100\% \quad (1)$$

$$LE = \frac{m}{M} \times 100\% \quad (2)$$

m_1 is referring to the total mass of BSA added, m_2 is the mass of BSA in the solution, m is the total mass of BSA in the drug-loaded hydrogel, and M is the mass of hydrogel after drying. Their findings stated that the EE of hydrogel is approximately 100% and has an efficient encapsulation.

Other than that, there is a study on the photo-controlled release of BSA from photo-response hydrogel (PR-gel). It is developed from a combination of 4arm-PEG and azobenzene into CNF [40]. The release of BSA was conducted under the presence of ultraviolet light, which will cause some changes in the hydrogel structure. The suitable wavelength using UV radiation is 365 nm, while visible light is 400–500 nm. Leakage and diffusion of protein from the hydrogel structure and *trans*-to-*cis* photoisomerization are the possible causes of an increase in the release rate of protein.

The release of protein using BSA and lysozyme from sodium alginate-bamboo-bacterial cellulose hydrogel shows that the lysozyme release is faster during the early phase [41]. The release rate control by electrostatic adsorption gives suitable pH-dependent release. Meanwhile, BSA has a lower release rate because of stronger hydrophobic adsorption generated by lignin. The hydrogel used in this study has an excellent ability to support hydrophilic protein-based drugs, and it does not show any cell toxicity during the process.

Another study is on applying multi-layered spheres for protein-based drug release using alginate and cellulose nanocrystals (CNC) [42]. The first and third layers of the sphere consist of alginate, while the second layer combines CNC and bovine serum albumin (BSA). BSA was loaded into the multi-layered sphere by incubation in the gastric fluid, and the release was analyzed using a protein assay kit. It is explained that multi-layer spherical hydrogel can manage and make a sustained protein release into the gastric environment.

Bacterial nanocellulose (BNC) can be used in drug delivery to observe the loading and release behavior of protein albumin [43]. BNC for this study is being generated by *Gluconacetobacter xylinus* using bottom-up approaches by culturing the aerobic bacteria on the hydrogel. The high ability for protein absorption is analyzed on pure BNC compared to wood pulp cellulose [43]. The study concludes that BNC is the best choice of carrier for protein uptake. Despite that, some improvement needs to be made as the rate of protein uptake by hydrogel is faster than the rate of release. The loading rate of the drug depends on the time of loading and concentration of protein while released based on the amount of loaded protein. Releasing protein requires additional time compared to the loading process due to the concentration gradient, which causes a low diffusion rate. The application of BNC shows similar behavior as the other hydrogels with more environmental and user-friendly features.

Another study on protein enrichment uses superabsorbent polymer (SAPs) hydrogel for membrane technologies [44]. The protein enzyme (BSA) shows an increase in concentration through the forward osmosis process. The study explained that the protein is intact and has a low rate of denatured due to less membrane fouling, high performance of dewatering, and no issues related to reverse draw diffusion. Besides that, an increase in the contact area of the membrane, hydrogel mass, protein concentration, and number of setups in series would produce better enrichment factors. Briefly, a summary of protein immobilization for uptake and release in the hydrogel is simplified in Table 1.

Table 1. Review on the use of protein albumin with biomaterial-based hydrogel.

Type of Protein	Hydrogel	Release and Uptake	Refs.
Protein albumin	Bacterial nanocellulose, BNC	• Adsorption technique has slow uptake and faster release. • High-speed technique has faster uptake and slow release.	[37,43]

Table 1. *Cont.*

Type of Protein	Hydrogel	Release and Uptake	Refs.
BSA	Superabsorbent polymer hydrogel: Carboxymethyl cellulose, CMC, and cellulose (Crosslinker: NaOH/urea aqueous with epichlorohydrin (ECH))	• Release behavior is a biphasic release pattern that is gradually and controlled by the hydrogel. • Release time is controlled by the content of CMC.	[38]
BSA	Injectable polysaccharides-based hydrogel: cation octa (γ-chloroammoniumpropyl) silsesquioxane (OCAPS), chitosan, and oxidized hydroxypropyl cellulose	The encapsulation efficiency of hydrogel is approximately 100%.	[39]
BSA	Photo-response hydrogel (PR-gel): 4arm-PEG and azobenzene into CNF	Photo-controlled released BSA using UV and visible light • UV shows faster release. • Visible light shows slow release. • Increased released rate of protein: • Albumin outflow and diffusion from the hydrogel • Burst release can keep dosage in its effective range.	[40]
BSA	Multi-layered sphere using alginate and CNC	Sustained release of the protein into the gastric environment.	[42]
BSA	Sodium alginate-bamboo-bacterial cellulose hydrogel	Lower released rate due to stronger hydrophobic adsorption generated by lignin.	[41]
Lysozyme	Sodium alginate bamboo-bacterial cellulose hydrogel	Faster released at the early phase (conducted by electrostatic adsorption) and suitable pH-dependent released.	[41]

Furthermore, gel entrapment is one of the most practical active immobilization techniques for algae in which the substrates are made up of biomaterial such as polymers, proteins, and polysaccharides [45]. Polyacrylamide hydrogel has the potential to be the medium for the immobilization of microalgae [46]. The sizes of pores play important roles during immobilization, which increases the pores sizes causing the microalgae to move freely. As the pore sizes reduce, the microalgae can be kept in their best state for about 20 days. Polyethylenimine (PEI) can be used as an immobilization medium for microalgae due to its biocompatibility [47]. The additional crosslinker impacts the growth of microalgae and the rate of immobilization. Epichlorohydrin shows a better attachment to microalgae compared to diethylene glycol diglycidyl ether (DGDE). This explains suitable crosslinker is required for a suitable attachment of microalgae.

Several studies have been conducted for the immobilization of bacteria on biomaterial for various applications. Physical adsorptions were used to immobilize *Bacillus velezensis* strain on the microsphere of sodium alginate (SA)/polyvinyl alcohol (PVA)/nano zinc oxide in the treatment of slaughter wastewater [48]. The higher degradation rate of chemical oxygen demand and lower growth rate of the pathogen have resulted from the application of immobilized bacteria on the combination of biomaterial. Besides that, bioluminescence bacteria can be immobilized in the alginate beads to sense explosive material [49]. Even though the detection using bacteria is successful, the exposure of engineered microbes might impact marine life.

For the removal of polycyclic aromatic hydrocarbons (PAHs), soil washing can be carried out efficiently by immobilizing the bacteria in PVA-SA hydrogel beads [50]. The removal rate is higher and faster at a medium concentration of bacteria, and the hydrogel can be reused, which reduces the cost and waste generation. The bacteria in the immobilized hydrogel show better performance on adsorption and degradable activity than unattached bacteria. Immobilization of the *Clostridium intestinale* strain on the alginate hydrogel beads produces more biological hydrogen and reduces the formation of the end product [51].

Cell entrapment is a suitable method for the production of hydrogen as it is not destructed in the hydrogel and grows at a steady state. Alginate hydrogel and electrospun polystyrene nanofiber were used as the substrate for immobilizing the bacteria cell. It can provide suitable binding, cell protection, high cell activity, less restriction for cell growth, and lower toxic generation [52]. Immobilization using the layer by layer technique gives better stability to the structure, suitable thermal resistance, and lowered discharge potential of the immobilized cell. Figure 2 shows the layer by layer technique for immobilized bacteria on the hydrogel. Besides that, functional living ink (Flink) can be obtained by encapsulating the living bacteria cell in biocompatible hydrogels such as hyaluronic acid, k-carrageenan, and fumed silica [53].

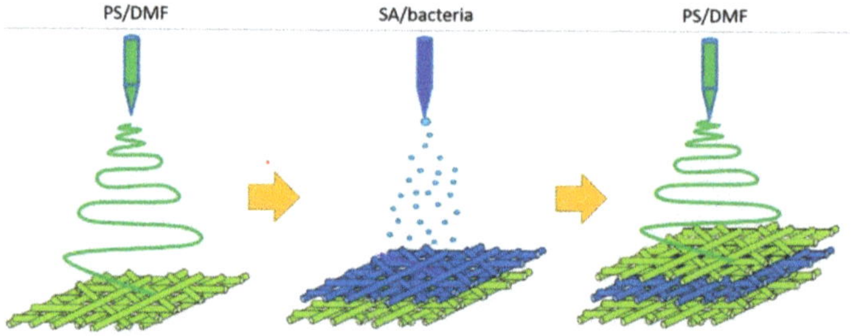

Figure 2. Preparation of alginate hydrogel for the immobilization of bacteria in the layer. Reproduced from [52] which is licensed under a Creative Commons Attribution-(CC BY 4.0) International License (http://creativecommons.org/licenses/by/4.0/, accessed on 17 April 2022).

4. 3D Bioprinting Application

Additive manufacturing has been widely used in various industries and has focused on medical applications for the past few years. Additive manufacturing in the medical field is widely known as 3D bioprinting. The significant difference between medical and non-medical fields application depends on the material used as printing ink. Three-dimensional bioprinting used biomaterial with living tissues or cells in the printing process where it required additional attention and care in handling the bioink. Applying additive manufacturing in the medical field has improved the development and technology of this industry. Tissue engineering and drug delivery are areas where 3D bioprinting is widely used. Figure 3 shows some fields and industries that started applying printing techniques to ease and faster the production process.

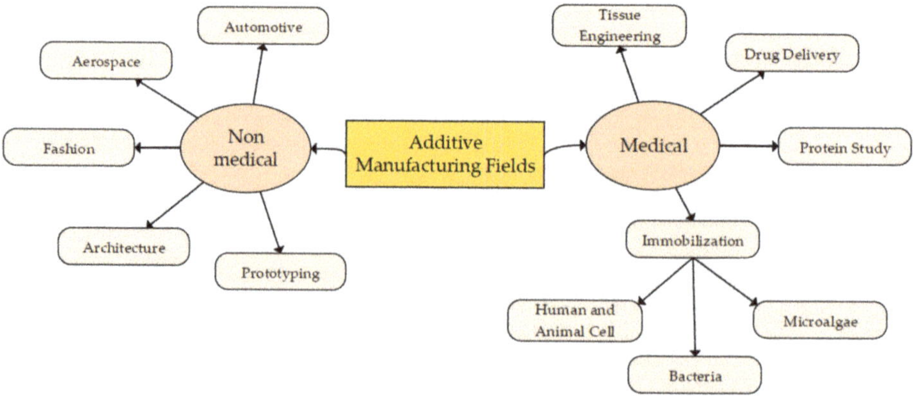

Figure 3. The fields involve the application of additive manufacturing for more efficient and advanced products.

4.1. Tissue Engineering

Three-dimensional bioprinting in tissue engineering has seen rapid development. Tissue engineering refers to a field in which engineering uses living things such as cells and tissues. Biomaterial will be the primary resource in constructing new tissues and repairing damaged tissues or organs. Tissue engineering has become one of the important fields in biomedical application due to its high potential in functionalizing biomaterials and exploring various interesting medical areas. Figure 4 shows the successful application of a 3D-printed hydrogel bone scaffold for replacing the tissue in rats and rabbits. Three-dimensional bioprinting in tissue engineering should have outstanding biocompatibility and be suitable for human use without causing any rejection or harm.

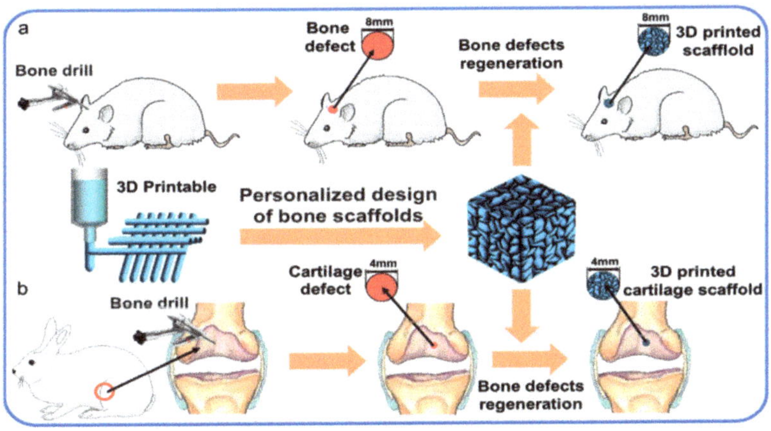

Figure 4. Printed bone scaffolds can replace the bone and cartilage defects in rat and rabbit, (**a**) regeneration of skull defect in rats; (**b**) articular cartilage regeneration in rabbits. Reproduced from [54] which is licensed under a Creative Commons Attribution-(CC BY 4.0) International License (http://creativecommons.org/licenses/by/4.0/, accessed on 17 April 2022).

Some studies have implemented 3D bioprinting in tissue engineering using other biomaterials. Inkjet bioprinting to print alginate hydrogel shows promising results and process efficiency [55]. They developed their inkjet printer using a bio clean bench to reduce contamination in printed material. The printer does not require any electrical energy during printing. It can produce a smaller printed material and understand the effect of beads size in 3D bioprinting.

Additionally, printing vascular tissue raised some concerns on selected criteria such as chemically modifying the biomaterial, controlling the size of pores, and involvement of growth factors [56]. It also stated that additional support is required during the printing of vascular tissue, such as for sacrificial ink, which can be removed upon printing. Recently, the conductive material in tissue engineering has been applied with 3D printing techniques [57]. The use of conductive material as ink can produce a more functionalized printed product that can reduce the rejection by the human body.

4.2. Drug Delivery

A drug delivery system commonly refers to the transport of drugs to the targeted area in a controlled manner. The specific untreated area in the body is being detected, and the drug is released based on body requirements. The loading techniques of the drug into the 3D-printed hydrogel scaffold for in vitro and in vivo study have been illustrated in Figure 5. For the pre-loading technique, the drug is introduced with the biomaterial before printing, while it is introduced once the hydrogel scaffold has been printed for direct loading [58].

Previously, many issues related to the uncontrolled release of the drug have been identified, and this situation can lead to side effects and further harm to the consumers.

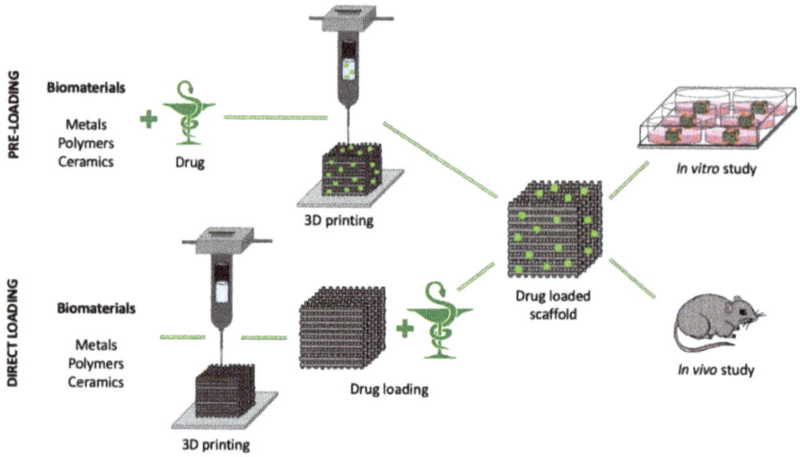

Figure 5. Loading of drug into the scaffold before and after printing for the in vitro and in vivo study. Reproduced from [58] which is licensed under a Creative Commons Attribution-(CC BY 4.0) International License (http://creativecommons.org/licenses/by/4.0/, accessed on 17 April 2022).

Recent studies by researchers have shown that applying biomaterial and 3D bioprinting has a high possibility of encountering problems related to the drug delivery process. Three-dimensional bioprinting has been widely used for this system as it enables the modification of the drugs and gives a faster production period. There are few mechanisms for a controlled released system of drugs, such as temporal-controlled, distribution-controlled, and erosion-controlled [59]. The drug delivery system is divided into two major groups, which are oral and transdermal delivery systems. The oral drug delivery system is administered into the body by passing through the oral system, while transdermal delivery is through the skin. Each controlled release and drug delivery system has its roles and benefit for the pharmaceutical field.

Further studies on the solid drug dosage form have continued to determine the effect of geometry and composition on the dissolution of the drug [11]. The composition will highly affect the dissolution of the drug compared to its surface area and geometry. Mini tablets generated from passive diffusion provided smoother surface structure and minor surface defect compared to hot-melt extrusion as it depends on the quality and density of the filament. Hot-melt extrusion can obtain higher drug loadings and weights but lower filament homogeneity. Moreover, this also led to complexity in getting the desired filament diameter.

4.3. Protein Study

Protein is one of the critical nutrients abundantly present in the human body. Every cell has protein for the growth of the cell [60]. Protein is a substance made up of amino acids, a group of organic molecules. There are a few types of protein, such as collagen, gelatin, silk, polysaccharides, milk, and wool protein [61]. The application of protein is quite complex due to its three-dimensional structure and strong molecular force.

Protein expresses some unique characteristics and properties to ensure that it has the ability to work with cellulose-based hydrogel. The solubility of the protein in water depends on its structure and pH. Higher acidity and alkalinity tend to increase the solubility of the protein. Protein can react with various chemical and mineral acids to form other products. It is being tested in multiple methods such as the Biuret test and Pauly test [62,63].

A study on the loading of protein was conducted in BSA solution for three days and then released using phosphate buffer solution (PBS) [38]. The release behavior of the protein shows a biphasic release pattern in which it starts with a gradual release then followed by a slow release. It is claimed that larger-sized hydrogel pores provide faster protein release and are highly capable as the polymeric carrier for protein transport.

4.4. Immobilization of Microalgae

Algae is an aquatic organism that will undergo photosynthesis to produce useful material. In general, a light source will aid algae in converting water and carbon dioxide into biomass. Commonly, algae are divided into two main groups, which are macroalgae and microalgae, and have complex structures [64,65]. Size and cellular structure are the characteristics used to classify the algae into the two main groups [66]. Macroalgae, larger-sized algae, have a multicellular structure, while microalgae are unicellular cells and smaller in size [67]. Naturally, algae can be obtained from ponds, oceans, and wastewater, but for commercial production, it can be cultivated in open or closed systems [66].

Recently, the production process used a minimum cost high yield method, and biomass production can increase up to 98% using open system cultivation [68,69]. The cultivation process of microalgae is faster where it can produce twice the actual amount within 24 h [70]. Besides that, the quality and amount of light supply, irradiance, photoperiod, and seasons of the area can also impact the growth rate [71].

Because of algae's characteristics, it has been used in various applications and fields such as the feedstock for biodiesel, source of carbon for fermentation, wastewater treatment, atmospheric carbon dioxide mitigation, and the pharmaceutical industry [66,72]. Accessible production and high yield of microalgae have gained the attention of researchers worldwide to apply algae in several potential sectors. Microalgae's advantage is fast adapting to an unfamiliar environment and can be produced in large quantities in less time, maintenance, and cost [69]. Lipids, proteins, and carbohydrates are the main components produced from microalgae biomass in huge amounts and are the primary producers of food sources for the other aquatic organisms [72,73].

Hydrogel as the medium for immobilization of microalgae has been proven in some studies based on its ability to grow. The soft structure with high water content can support the growth and cell activity for culturing microalgae. For immobilizing microalgae, two approaches can be applied, which are the encapsulation of the microalgae in the polymer matrix and creating substrate from natural sources before seeding the microalgae [46,74]. Nanoporous silica matrix is an example of a polymer matrix, while agar, alginate, and cellulose can be used as the source to fabricate substrates such as hydrogel.

Bold basal medium (BBM), a freshwater algae medium, is an important medium for growing algae. The ability to immobilize microalgae can be understood by inoculating the hydrogel or agar with the microalgae species and then further culturing with BBM [46]. Another method can be performed by preparing agar with BBM followed by inoculating microalgae on the top of the agar surface [75].

4.5. Immobilization of Bacteria

Bacteria, the most simple and abundant organism with more than 5000 species that have been identified, hold a critical function in supporting the ecosystem and the environment [76]. The classification of bacteria depends mainly on the metabolic and genetic characteristics as their structure does not differ much from each other.

A few factors affect the growth of bacteria, such as pH, temperature, nutrients, oxygen level, concentration, and species of bacteria [77]. Bacteria have been applied in many industries due to their potential and properties such as biological hydrogen, wastewater treatment, soil washing, biofilms, and biosensors [50,51,78–80]. Immobilization of bacteria on the surfaces depends on the characteristic of the material, the bond between surface and linker, physiological medium, and properties of bacteria [80]. Bacteria can be immobilized in two different techniques, which are through attachment to hard surfaces and hydrogel. Adsorption, crosslinking, cell entrapment, and cell encapsulation are the common techniques applied in the immobilization process.

In the immobilization of bacterial cells, 3D printing has the potential to be a new technique for encapsulating cells in less processing time. Multiphoton lithography, one of the laser printing methods, has been used to encapsulate bacteria cells within the BSA hydrogel [81]. Chemical crosslinking of the polypeptides requires gelatin from BSA and photo-sensing molecules at a suitable temperature. Bacteria will then be encapsulated in the thermal control BSA hydrogel and starts to grow within the 3D structure. Bacteria can grow rapidly in the culture medium without showing any potential for cell death because of exposure to extracellular factors. The micro-3D printing technique can be used as the most potent approach in rapidly producing an abundant amount of complex organisms with precise dimension and resolution. Despite this, it requires more costly equipment and is less suitable for a high production rate.

Multimaterial direct ink writing technique was used in fabricating the hydrogel digitally, and it gives unlimited choices of shapes and compositions. The bacteria can grow freely in the printed structure, form any desired shapes, and have a higher survival rate. The size of the hydrogel will increase up to 1.5 times larger than the actual size after immersion in the bacterial medium due to some deviation in the concentration of ions between hydrogel and culture medium.

4.6. Immobilization of Human and Animal Cells

Recently, 3D printing has changed its direction toward the medical field, especially for human applications. Many studies were performed involving human and animal cell tissues. The characteristics and properties of cell tissues call for extensive attention as it requires an appropriate working environment and handling procedure for cell survival. The growth of cell tissues need compatible surrounding with enough supply of nutrient, and hydrogel is one of the choices suitable to mimic the body tissue and provides all necessities. The application of hydrogel and 3D bioprinting will be thoroughly discussed in this section.

The hydrogel is made up of PVA, SA, and MXene without the presence of any chemical crosslinked [82]. MXene is the transition metal carbide capable of incubation and can improve the conductivity properties of the hydrogel. The conductivity of the hydrogel is analyzed by attaching the LED sensor to the hydrogel. It is found that the LED sensor increases in brightness once the hydrogel is compressed, and the response time is slower. This hydrogel can detect motion responses, such as at the knee or wrist, by converting the movement signal into an electrical signal. The sensor can be located directly on the human body for reading and analysis. These findings show that hydrogel has suitable conductivity and works well for sensors based on the stress exerted on its structure.

A study has been conducted using phenol-grafted polyglucuronic acid (PGU) as bioinks for 3D bioprinting [83]. The ability of the hydrogel to be used in cell culture was analyzed using mouse fibroblast and human hepatoma cells. PGU is one of the potential microbial polysaccharides for bioprinting ink as it requires less pre- and post-processing. The gelation of PGU-Ph solution required horseradish peroxidase (HRP)-catalyzed reaction and hydrogen peroxide. A faster gelation time is observed with the increases in the concentration of PGU-Ph and decreases in the concentration of hydrogen peroxide. Cell viability of mouse fibroblast and human hepatoma cells in PGU-Ph hydrogel shows high viability percentage of 95% and 94%, respectively, on day 2 and kept constant until day 11. Mouse fibroblast does not form any cell aggregates, while human hepatoma cells aggregation kept increasing during the culture process, which means that the cell kept growing in the hydrogel.

Recently, there have been a few studies on the involvement of 3D bioprinting in scar therapies. Besides that, printing hamster cells using a thermal inkjet is successful as it is represented by green fluorescent during the viability study [20]. Cell printing can be used as one of the methods of delivery to the targeted cell. It is stated that the cell is in suitable condition and shape during and after the printing process. The formation of transient pores can automatically fix the defect and produce a suitable transfection efficiency.

A model on human hypertrophic scars was constructed by producing bioink from decellularized extracellular matrix and hydrogel from alginate and gelatin [84]. Human hypertrophic scar cells show potential characteristics for drug testing and exhibit the same profile as scar at a specific gene and protein level. Preformed cellular aggregates (PCA) are the bioprinting technique used to form a scars model that can help in tissue morphogenesis. It is suitable for spontaneous multicellular processes and can reduce modeling time. The viability of the cell is considered high, which is more than 90%. Moreover, it does not express any changes in cell morphology. However, it increases the rate of proliferation and the ability to migrate. This study successfully explains the formation of scars by using hydrogel and a 3D bioprinting technique that can mimic the native scars and be applied in drug delivery.

Other than that, there is a study on reducing the potential of scar contraction using a dermal extracellular matrix (dECM) in a 3D-printed dermal analog [85]. Bioink dermal extracellular matrix powder (dECMp) is prepared from dECM with the same material composition and morphological characteristics. Crosslinked printed dermal analogs (PDA) show a great cytocompatibility (>75%) through the analysis with L929 fibroblast compared to the control medium. It can minimize the contraction of the wound and the formation of scars. The crosslinking used is glutaraldehyde (GA) solution, and its concentration will control the printed shape and strengthen the tensile properties. Hydrogel in scars therapies shows an excellent characteristic for application with drugs and lowers the contraction of scars.

Moreover, there is some study on microparticles and beads for better efficiency as the size is reduced and the surface area increases. Alginates microparticles can be fabricated using multichannel 3D printing for protein encapsulation. They are suitable for human mesenchymal stem cells once combined with a collagen scaffold [86]. The printing devices were built with more than one channel for ink deposition based on the phases, and it uses droplet-based techniques. Shear stress is the factor that forms microparticles with spherical shapes, while the size of microparticles is affected by the flow rate of mineral oil and alginate solution. The microparticles were used to study the release of BSA, and it shows the same pattern, which is the burst release for the first five hours and then continued with a fast release. Burst release should be controlled as it can concentrate protein in one place and lower the performance of protein encapsulation. Human mesenchymal stem cells show suitable viability after five days in collagen scaffolds with alginate microparticles, as more than 90% of cells can be analyzed from a cell staining kit. Furthermore, the microstructure of alginates does not affect the morphology and spread potential of cells.

Another study uses hydrogel granules from the extrusion of hyaluronic acid hydrogel through porous nylon fabric in assisting human-induced pluripotent stem cell (hiPSC)-derived neural networks [87]. Three-dimensionally printed tools able to reduce time and produce repetitive hydrogel granules for the extrusion process. Cell viability in granules hydrogel is better than in bulk hydrogel, and it can encourage cell attachment and extension. Granules have an excellent ability to match any tissue type due to boundary intersection. Moreover, the size of granules hydrogel will influence its performance as the smaller size provides more active space than larger granules. In vitro cell culture treatment of neurology disease and drug screening can be performed efficiently by applying the multiphase cell encapsulation of hydrogel.

5. Cellulose-Based as Biomaterial Ink

5.1. Preparation for Biomaterial Ink

Cellulose-based biomaterials express great potential in 3D bioprinting. CMC can be used as bioinks due to its excellent shear adjustment, alteration of viscosity, and attribute of shape [88]. For example, the application of direct ink writing as the printing technique for water-based ink of NaCMC with humidity control [89]. Suitable print head and nozzle humidity are crucial in printing for robustness and repeatability of the process. Moreover, it can avoid faster drying of the ink bridge and ensure a higher quality of printed surface as the humid surrounding might increase the water content.

Whereas nanocellulose has great potential to be printed due to its ability to pass through a micro-sized nozzle and release a precise shape. Nanocellulose is a natural fiber extracted from cellulose with a size less than 100 nm in diameter and has up to micrometers in length [90]. The preparation of nanocellulose is divided into three effective techniques, which are mechanical disintegration, chemical reaction, and biological reaction [91]. A brief explanation of the common techniques applied is listed in Figure 6. The type of nanocellulose is determined by the method of extraction. Mechanical disintegration results in nanofibrillated cellulose while extraction using chemicals produces CNC, and biological reaction is used to produce BNC [3].

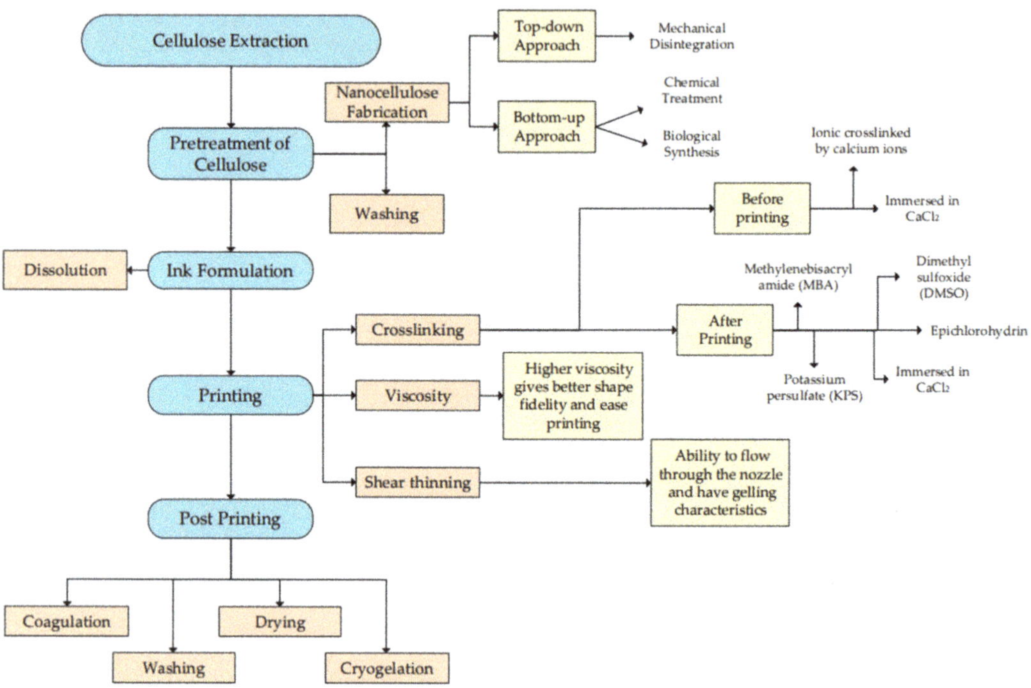

Figure 6. Preparation techniques in producing cellulose and nanocellulose by top-down and bottom-up approaches involve mechanical disintegration, chemical treatment, and biological synthesis.

Mechanical disintegration is used to break down cellulose pulp into smaller sizes and commonly uses fibril delamination to improve the mechanical properties of extracted nanocellulose. The standard mechanical treatment is ball milling, blending, ultrasonic, and electrospinning. The mechanical technique has its disadvantages where high energy is required for the process and can lead to a low amount and quality of the product [92]. High-pressure homogenization is one of the efficient processes. A tiny nozzle will deliver the cellulose slurry at high pressure and change into nanocellulose due to impact and shear force [92]. Because of the very small nozzle size, clogging might be the main problem, and it can be controlled by slightly reducing the size of cellulose prior to the process.

Other than that, microfluidization requires a pump to supply pressure and a chamber as a place for the shear process to occur [92]. Besides that, ultrasonication is a process that uses sound energy to cause agitation of particles. It applies an extreme oscillation using the hydrodynamic force of ultrasound [93]. Another technique under mechanical disintegration is grinding and cryocrushing. The grinding process has two grindstones, which are static and rotating grinding stones. The slurry will flow in between the two stones to form nanosized fiber [94]. Cryocrushing involves immersing cellulosic fiber in liquid nitrogen and turning it into small pieces by high pressure from mortar and pestle.

Besides that, the chemical treatment can also produce nanocellulose, and some possible techniques are acid hydrolysis, carboxylation, sulphonation, and ionic liquid. Extracting cellulose using ionic liquid is environmentally friendly as most of the solvent can be recovered. The main and commonly used technique for chemical treatment is acid hydrolysis [90]. The ordered region will undergo a hydrolyzation process using acid, while the disordered will remain without any treatment. It involves chemicals such as sodium hydroxide, sulfuric acid, and sodium bicarbonate [3]. The suspension is centrifuged and decanted with deionized water a few times until it obtains a neutral pH.

In the biological approach, enzymatic hydrolysis can be applied for pretreatment due to its potential to produce less chemical waste and energy consumption. Enzymatic hydrolysis uses the application of enzymes in digesting and transforming the structure of cellulose, and typically, this process takes a longer time [90]. Less adjustment on structure, excellent crystallinity, and thermal stability are expected when the cellulose undergoes a suitable pretreatment before the extraction [95].

5.2. Curing of 3D-Printed Cellulose Hydrogel

Hydrogels are a set of polymer chains bonded with a covalent or physical bond and can form a structure like mesh by applying a crosslinker [96]. Hydrogel is a gel-like structure, and it is water-resistant [97]. Some hydrogels are known as intelligent hydrogels as they can be temperature-sensitive, pH-sensitive, glucose-sensitive, stimuli-sensitive, protein-sensitive, and superabsorbent due to their unique characteristics [96,98]. The hydrogel can maintain its three-dimensional structure as it is a water-swollen polymer and contains a hydrophilic functional group in its structure. Due to the wet and soft structure of hydrogel, relatively low friction is required. Moreover, the hydrogels can self-heal based on their electrostatic attraction [97]. They will not dissolve in any solution and hold a high amount of water or any biological fluid [99].

There are two categories of hydrogel based on the type of crosslinker, which is physical hydrogel, where the formation is reversible and chemical hydrogel for permanent formation [100,101]. Physical hydrogel is produced from a non-covalent crosslinker, while chemical hydrogel is produced from covalent bonding. The presence of crosslinkers will affect the physical properties of the polymer and hydrogel. Table 2 describes the example of the crosslinking process that will affect hydrogel properties.

Table 2. Review on the effect and application of crosslinking toward hydrogel formation.

Crosslinker Material	Hydrogel Composition	Processing Technique	Benefit	Refs.
$CaCl_2$	CNF and alginates	Adding $CaCl_2$ to the bioink solution.	• Increase viscosity of ink for suitable shape fidelity. • Low tendency in shape deformation.	[22]
	CMC/SA/chitosan	Soak the film in the $CaCl_2$ for 2 min.	• Can improve tensile strength until certain limits.	[102]
	Alginic acid sodium salt and methylcellulose	Immersion in crosslinking solution for 10 min.		[36]

Table 2. *Cont.*

Crosslinker Material	Hydrogel Composition	Processing Technique	Benefit	Refs.
Ca^{2+}	CNF	Ionic crosslinking before printing.	• Improve viscosity of ink. • Stable printed material before photo-crosslinked. • Suitable shape fidelity of PEG-CNFs hydrogel.	[14]
	CNF, alginates, and colloidal lignin particle nanocomposites scaffold	Stored in Dulbecco's phosphate buffer solution (DPBS) for 7 days with Ca^{2+} and Mg^{2+} ions.	• Suitable shape stability. • High swelling ratio.	[103]
	PVA and SA	1st: Crosslinked PVA by freeze-thawing method 2nd: Crosslinked sodium alginates by immersed in $CaCl_2$.	• Faster hydrogel response. • Suitable mechanical properties.	[82]
ECH	CMC and cellulose	Addition of crosslinker into hydrogel solution at 30 °C and stir for 2 h. Kept for 12 h at 60 °C for the formation of the gel. Hydroxyl group of cellulose and CMC crosslinked by nucleophilic attack.	• Commonly used as crosslinking for carbohydrates. • Improve swelling ratio.	[38]
Disulfide bonds	Hyaluronic acid/carboxymethyl cellulose-based hydrogels (HA/CMC)	Oxidation reaction of dissolved oxygen in solution between the thiol groups a 37 °C.	• Better structure of hydrogel. • Improve swelling ratio.	[104]
Thiol–ene photoreaction of norbornene groups	PEG and CNF	Under visible light by exposing the blue light for 3 min at 460 nm and 25 mW cm^{-1}.	• Better mechanical properties of hydrogels. • Enhanced stability. • Stable for cell incubation in 2 weeks.	[14]
Dimethyl sulfoxide (DMSO) and N, N′-Methylenebisacrylamide (MBA)	Cotton cellulose	Immersion in the crosslinked solution after printing.	Provides better ability for reswelling and compression properties.	[105]
N,N′-methylenebisacrylamide (MBA) and potassium persulfate (KPS)	CNC−poly(acrylamide)	Stir hydrogel and crosslinker in a nitrogen bubble for deoxygenation in 10 min.	• Fix the topological network. • Suitable viscoelastic characteristic of hydrogel.	[106]

Natural hydrogel is gaining more attention for its application in the medical field [98]. There are a few types of natural hydrogel from alginate, cellulose, chitosan, and protein. The hydrogel of crosslinked alginate and gelatin is better due to its low stiffness, fast degradation reaction, excellent metabolic activity, and suitable adhesive properties compared to the unlinked alginate. Chitosan has gained recognition in drug delivery due to its smart delivery and properties that are high in biocompatibility and biodegradability. Moreover, it enables injection, direct formation of gel, and regeneration of tissue. For protein-based hydrogel, it expresses biological function with high compatibility. Suitable strength, shear-thinning, and the potential to respond to a stimulus are some other advantages of using protein-based hydrogel. Protein-sensitive hydrogel might be helpful in the advanced therapeutics field [96].

Hydrogel can be produced by using a few techniques such as photopolymerization and radiation with the application of physical or chemical crosslinking [74]. For example, CNC-based hydrogel can be prepared from various methods such as homogenization, cyclic freeze-thaw processing, free radical polymerization, and UV/ion-mediated crosslinking [107]. Forming hydrogel using ultraviolet light in polymerization techniques can control the behavior of reactions and produce suitable quality hydrogel in less time [46]. The summary of common methods available in hydrogel fabrication is summarized in Figure 7.

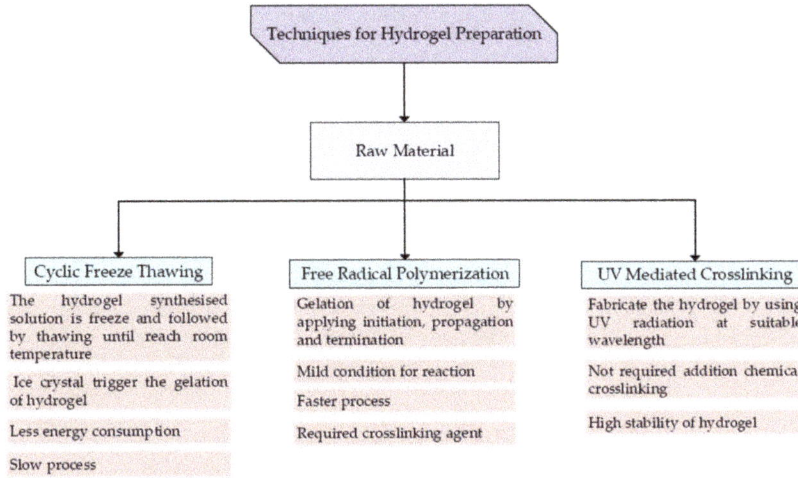

Figure 7. Few techniques for hydrogel formation.

Preparation of hydrogel from polyethylene glycol and cellulose nanofibers requires mixing of solution, sonication, and storage in a dark place [14]. Then, it will be added with calcium chloride before being placed in the mold for 3 min under blue light. Moreover, the homogenization process can be applied as the final process in extracting cellulose to form a cellulose-based hydrogel. A high-pressure cell homogenizer with the circulation of cooling water can reduce the potential of damage generated from excess heat released from the process [108].

A freeze-thaw method is another example of hydrogel production. An insoluble gel is produced in the early phase of the phases separation process and will generate ice crystals [101]. The ice crystal generated from the freezing process will trigger gelation of the solution, where it tends to be a crosslinker point [99]. A porous matrix is formed in a freeze-thaw method as the poor region of polymer will be surrounded by a rich gel region. The freezing process can be conducted at −18 °C while thawing at room temperature [99]. The cycle for freezing and thawing, temperature, and duration are the critical parameters in the hydrogel formation [101,109]. Gelling of CNC using the freeze-thaw method can be carried out in the water and polar organic solvent, which produces hydrogel and organogel, respectively [110]. Even though this method consumes more time, it has become one of the standard methods as it requires less energy and is environmentally friendly [101].

Another technique for hydrogel production is free radical polymerization, which has a faster rate of polymerization and can be used in an aqueous solution [101]. The crucial step for this method is the application of the initiator in polymerizing the monomer. Moreover, there are possibilities for fabricating hydrogel using radiation-induced crosslinking under mild conditions without the presence of toxins from the crosslinking agent. The production of CNC-poly(acrylamide) (PAM) hydrogel can apply this method [106]. Acrylamide was used in the free radical polymerization with the presence of some solid reagent, followed by deoxygenation using bubbles of nitrogen and stirring before adding to the mold.

Ultraviolet radiation can be used as one of the techniques for producing cellulose hydrogel. As an example, UV radiation in aqueous media was used for the formation of poly (N, N-dimethylacrylamide) (PDMA)/cellulose hydrogel [111]. The PDMA is attached to the CNC surface with the help of UV radiation in the water-soluble photoinitiator. This method does not need any extra chemical crosslinker. The interaction between PDMA and CNC polymer is a reversible physical interaction and has a suitable stability of the covalent bond. Other than that, the fabrication of CNC hydrogel uses the exposure of UV radiation under 365 nm (wavelength) [112]. For further analysis, the gel requires repeated purification to remove the excess material and freeze-drying using nitrogen for a better result.

5.3. Recent Trends of 3D Bioprinting Cellulose

All the published articles related to cellulose, 3D printing, and 3D bioprinting from Scopus were analyzed using VOSviewer for the bibliometric analysis and visualization research trends for the past 10 years. Based on Scopus, the analysis of the publication trends related to the keywords is increasing rapidly. Figure 8a shows the publication trend generated from Scopus. The number of documents only being specified to the articles type documents. The trends show that 3D printing of cellulose has a significant increase over the past 5 years compared to the 3D bioprinting of cellulose. The gap between these two keywords is huge due to the involvement of 3D bioprinting with cellulose is still new and less research. The co-occurrence analysis by VOS viewer is specified to 20 minimum number of occurrences of keywords. The generated cluster shows 3 clusters with 103 thresholds. Figure 8b shows the visualization of co-occurrences analysis by VOSviewer. Cellulose and 3D printing show the highest repetition represented by the size of the sphere, while 3D bioprinting is smaller. This can be related to the trend in Figure 8b.

(a)

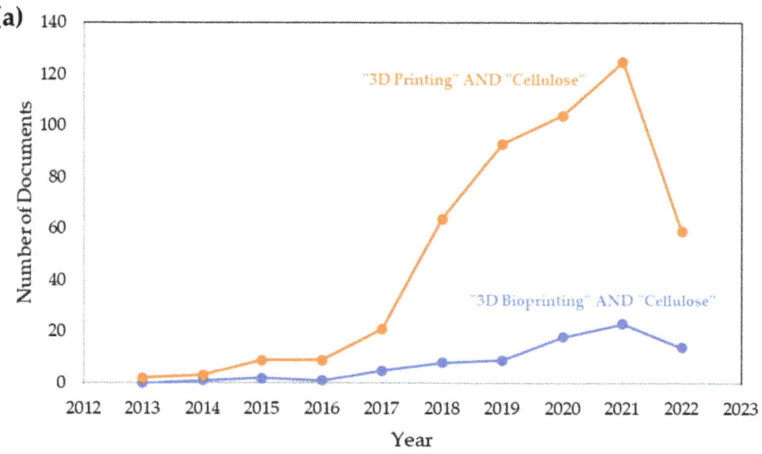

(b)

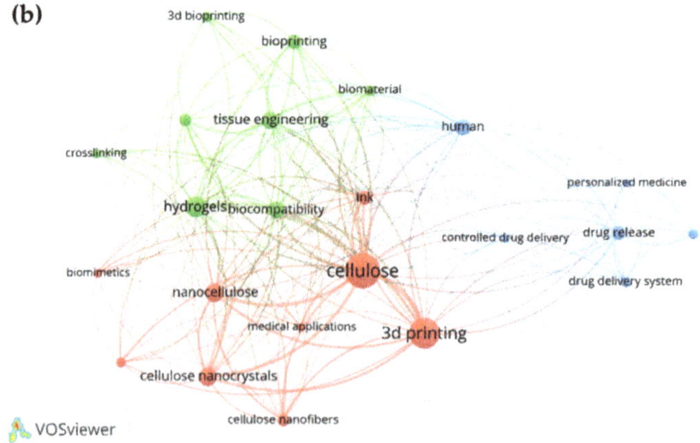

Figure 8. (**a**) Publication of article by year (Source: www.scopus.com, accessed on 17 April 2022) and (**b**) the co-occurrences analysis by all keywords.

Recently, there has been some implementation of 3D bioprinting with nanomaterials to study drug delivery systems. The filament used in fused deposition modeling is melted using hot-melt extrusion and then soaked in nanocellulosic solution to observe the loading process [113]. It is mentioned that the infill percentage only affects the mean weight of printed tablets. The weight decreases once the infill percentage becomes lower. A comparison of 50% and 100% infill percentages indicate that the printed material's diameter and height are unaffected. Furthermore, choices of polymer are important in determining the drug content and the percentage of drug loading. The drug loading can be more advanced when the drug in a solid form is produced from the transformation of nanocapsules liquid suspension.

Drug delivery of 5-Fluorouracil using 3D-printed cellulose macrofibrils/calcium carbonate has shown an efficient control of drug release [114]. The cellulose macrofibrils were printed using 3D printing techniques and mixed with calcium carbonate at different ratios. The adsorption of the drug by the 3D-printed sample can be controlled by the presence of $CaCO_3$. Cellulose macrofibrils can control the possibilities of rapid release. Besides that, laminating nanocellulose with calcium carbonate also shows an excellent adsorption and desorption process [13]. There is a significant difference in the percentage of adsorption and release of drugs for both nanocellulose and calcium carbonate. The adsorption of calcium carbonate is 75% higher in a shorter time.

For 3D protein printing, an early model printer has been developed using some components from the HP660C printer [115]. The printer is being modified and specified according to the requirement for protein printing, and software was used to control the location of printing. Ink cartridges used for protein printing are the original printer cartridges with multiple cleaning processes to avoid contamination. The study found that 25% of cells died after incubating for three days due to cell dehydration.

The application of 3D bioprinting produces a very neat and suitable structure for immobilization, which can affect the distribution of light and nutrients [36]. Cultivation of microalgae on bacterial cellulose is expected to merge the properties of cellulose with the photosynthetic criteria of microalgae [116]. The study found that microalgae have a huge potential for 3D bioprinting as they can be printed in layers with multiple geometries and sizes. The printed microalgae have stable physical properties as they can handle physical changes and high water content in the surrounding.

Despite the advantages, this technique might affect the physiological state of the cell. A study has been performed on fabricating hydrogel from alginate and methylcellulose to immobilize *Chlamydomonas reinhardtii* [117]. The immobilization of microalgae was observed by the increase in cell number due to the growth of cells in hydrogel under a photosynthetic environment. Moreover, the release of oxygen indicates a suitable photosynthesis process during immobilization and cell growth. Furthermore, the fabrication and crosslinking of the hydrogel are not affected by the existence of the microalgae. Methylcellulose will ease the printing process and is released after crosslinking. Releasing it will create micropores as the growth spot for algae.

Hydrogel produced from alginate and nanocellulose gives some insight into the advancement in hydrogel properties. A combination of alginate and CMC for wound dressing application can have better ink for printing due to its ability to form gelation and the suitable properties of viscosity and elasticity [118]. Moreover, the combination can lower the potential of syneresis and provide better mechanical and compression properties for resistance against deformation [119]. Fast gelation of the hydrogel can be performed with the presence of Ca^{2+} from the carboxyl group in alginates and CMC [120]. A mixture of alginates and CMC can provide better printing due to the higher viscosity of the ink solution. In tissue engineering, the combination of polyethylene glycol polymer with TEMPO-oxidized nanocellulose fiber found that the crosslinked material has excellent elastic recovery [14]. It can restore its original shape once being released from the extrusion. The composition of the ink causes variations in mechanical properties, swelling properties, and fractionation of hydrogel.

Other than that, 3D bioprinting in producing human chondrocytes has been studied using a combination of nanocellulose and alginate as the biomaterial [22]. The viability of the cell is compared between a printed and non-printed material. It is found that there is no change in the cell viability on the first day, but it starts to increase viability after seven days of the cell being cultured. Moreover, this combination of bioink is highly compatible with 3D bioprinting. Three-dimensional bioprinting can improve the tissue engineering fields by reducing the time consumed and providing a great product. The acceptance of printed material by the human body has increased with the use of hydrogel and high biocompatibility biomaterial in the printing process.

6. Conclusions

In conclusion, the fabrication of cellulose-based 3D-printed products has been extensively explored. In contrast, cellulose-based hydrogel shows great properties such as suitable tensile strength, elasticity, viscosity, and is lightweight. In the future, it is expected to be extensively applied in the medical and pharmaceutical industry as it has suitable criteria for biological application, and many recent studies are directed toward 3D bioprinting. Moreover, the direction of additive manufacturing starts to lead toward smart materials that respond to certain stimuli or parameters. Besides that, the application of hydrogel for the immobilization and growth of microalgae, bacteria, human and animal cells, and the properties to imitate the extracellular matrix makes it possible for any biological study.

Author Contributions: Conceptualization, M.S.S.; formal analysis, W.N.L.W.J.; writing—original draft preparation, W.N.L.W.J.; writing—review and editing, W.N.L.W.J., P.M.A., H.K. and M.S.S.; funding acquisition, M.S.S. All authors have read and agreed to the published version of the manuscript.

Funding: This work was supported by the Universiti Kebangsaan Malaysia (DIP-2020-006) and MRUN-RAKAN RU-2019-004/4.

Institutional Review Board Statement: Not applicable.

Informed Consent Statement: Not applicable.

Data Availability Statement: Not applicable.

Conflicts of Interest: The authors declare no conflict of interest.

References

1. Mukherjee, P.; Rani, A.; Saravanan, P. Polymeric materials for 3D bioprinting. In *3D Printing Technology in Nanomedicine*; Elsevier: Amsterdam, The Netherlands, 2019; pp. 63–81. [CrossRef]
2. Goel, A.; Meher, M.K.; Gulati, K.; Poluri, K.M. Fabrication of biopolymer-based organs and tissues using 3D bioprinting. In *3D Printing Technology in Nanomedicine*; Elsevier: Amsterdam, The Netherlands, 2019; pp. 43–62. [CrossRef]
3. Mehanny, S.; Abu-El Magd, E.E.; Ibrahim, M.; Farag, M.; Gil-San-Millan, R.; Navarro, J.; El Habbak, A.E.H.; El-Kashif, E. Extraction and Characterization of Nanocellulose from Three Types of Palm Residues. *J. Mater. Res. Technol.* **2021**, *10*, 526–537. [CrossRef]
4. Thayer, P.; Martinez, H.; Gatenholm, E. History and trends of 3D bioprinting. In *3D Bioprinting Principles and Protocols*; Springer: Berlin/Heidelberg, Germany, 2020; Volume 2140, pp. 3–18.
5. Kačarević, Ž.P.; Rider, P.M.; Alkildani, S.; Retnasingh, S.; Smeets, R.; Jung, O.; Ivanišević, Z.; Barbeck, M. An Introduction to 3D Bioprinting: Possibilities, Challenges and Future Aspects. *Materials* **2018**, *11*, 2199. [CrossRef] [PubMed]
6. Marija, Đ.; Obeid, S.; Mad, M.; Cviji, S.; Ibri, S. Paracetamol Extended Release FDM 3D Printlets: Evaluation of Formulation Variables on Printability and Drug Release. *Int. J. Pharm.* **2020**, *592*, 120053. [CrossRef]
7. Ioannidis, K.; Danalatos, R.I.; Tsaniras, S.C.; Kaplani, K.; Lokka, G.; Kanellou, A.; Papachristou, D.J.; Bokias, G.; Lygerou, Z.; Taraviras, S. A Custom Ultra-Low-Cost 3D Bioprinter Supports Cell Growth and Differentiation. *Front. Bioeng. Biotechnol.* **2020**, *8*, 1–13. [CrossRef] [PubMed]
8. Gao, C.; Lu, C.; Jian, Z.; Zhang, T.; Chen, Z.; Zhu, Q.; Tai, Z.; Liu, Y. 3D Bioprinting for Fabricating Artificial Skin Tissue. *Colloids Surf. B: Biointerfaces* **2021**, *208*, 112041. [CrossRef]
9. Murphy, S.V.; Atala, A. 3D Bioprinting of Tissues and Organs. *Nat. Biotechnol.* **2014**, *32*, 773–785. [CrossRef]
10. Wang, Q.; Sun, J.; Yao, Q.; Ji, C. 3D Printing with Cellulose Materials. *Cellulose* **2018**, *25*, 4275–4301. [CrossRef]

11. Ayyoubi, S.; Cerda, J.R.; Fernandez-Garcia, R.; Knief, P.; Lalatsa, A.; Healy, A.M.; Serrano, D.R. 3D Printed Spherical Mini-Tablets: Geometry versus Composition Effects in Controlling Dissolution from Personalised Solid Dosage Forms. *Int. J. Pharm.* **2021**, *597*, 120336. [CrossRef]
12. Markstedt, K.; Sundberg, J.; Gatenholm, P. 3D Bioprinting of Cellulose Structures from an Ionic Liquid. *3d Print. Addit. Manuf.* **2014**, *1*, 115–121. [CrossRef]
13. Mohan, D.; Khairullah, N.F.; How, Y.P.; Sajab, M.S. 3D Printed Laminated CaCO 3—Nanocellulose Films as Controlled-Release 5-Fluorouracil. *Polymers* **2020**, *12*, 986. [CrossRef]
14. Monfared, M.; Mawad, D.; Rnjak-Kovacina, J.; Stenzel, M.H. 3D Bioprinting of Dual-Crosslinked Nanocellulose Hydrogels for Tissue Engineering Applications. *J. Mater. Chem. B* **2021**, *9*, 6163–6175. [CrossRef] [PubMed]
15. Ibrahim, F.; Mohan, D.; Sajab, M.S.; Bakarudin, S.B.; Kaco, H. Evaluation of the Compatibility of Organosolv Lignin-Graphene Nanoplatelets with Photo-Curable Polyurethane in Stereolithography 3D Printing. *Polymers* **2019**, *11*, 1544. [CrossRef] [PubMed]
16. Anandakrishnan, N.; Ye, H.; Guo, Z.; Chen, Z.; Mentkowski, K.I.; Lang, J.K.; Rajabian, N.; Andreadis, S.T.; Ma, Z.; Spernyak, J.A.; et al. Fast Stereolithography Printing of Large-Scale Biocompatible Hydrogel Models. *Adv. Healthc. Mater.* **2021**, *10*, 1–12. [CrossRef] [PubMed]
17. Carnero, B.; Bao-Varela, C.; Gómez-Varela, A.I.; Álvarez, E.; Flores-Arias, M.T. Microfluidic Devices Manufacturing with a Stereolithographic Printer for Biological Applications. *Mater. Sci. Eng. C* **2021**, *129*, 1–14. [CrossRef] [PubMed]
18. Gong, Y.; Bi, Z.; Bian, X.; Chen, G.; Zhang, X. Study on Linear Bio-Structure Print Process Based on Alginate Bio-Ink in 3D Bio-Fabrication. *Bio-Des. Manuf.* **2020**, *3*, 109–121. [CrossRef]
19. Li, J.; Chen, M.; Fan, X.; Zhou, H. Recent Advances in Bioprinting Techniques: Approaches, Applications and Future Prospects. *J. Transl. Med.* **2016**, *14*, 1–15. [CrossRef]
20. Cui, X.; Dean, D.; Ruggeri, Z.M.; Boland, T. Cell Damage Evaluation of Thermal Inkjet Printed Chinese Hamster Ovary Cells. *Biotechnol. Bioeng.* **2010**, *106*, 963–969. [CrossRef]
21. Tan, C.T.; Liang, K.; Ngo, Z.H.; Dube, C.T.; Lim, C.Y. Application of 3d Bioprinting Technologies to the Management and Treatment of Diabetic Foot Ulcers. *Biomedicines* **2020**, *8*, 441. [CrossRef]
22. Markstedt, K.; Mantas, A.; Tournier, I.; Martínez Ávila, H.; Hägg, D.; Gatenholm, P. 3D Bioprinting Human Chondrocytes with Nanocellulose-Alginate Bioink for Cartilage Tissue Engineering Applications. *Biomacromolecules* **2015**, *16*, 1489–1496. [CrossRef]
23. Keriquel, V.; Oliveira, H.; Rémy, M.; Ziane, S.; Delmond, S.; Rousseau, B.; Rey, S.; Catros, S.; Amédée, J.; Guillemot, F.; et al. In Situ Printing of Mesenchymal Stromal Cells, by Laser-Assisted Bioprinting, for in Vivo Bone Regeneration Applications. *Sci. Rep.* **2017**, *7*, 1778. [CrossRef]
24. Devillard, R.; Pages, E.; Correa, M.M.; Keriquel, V.; Remy, M.; Kalisky, J.; Ali, M.; Guillotin, B.; Guillemot, F. Cell Patterning by Laser-Assisted Bioprinting. *Methods Cell Biol.* **2014**, *119*, 159–174. [CrossRef] [PubMed]
25. Guillotin, B.; Ali, M.; Ducom, A.; Catros, S.; Keriquel, V.; Souquet, A.; Remy, M.; Fricain, J.; Guillemot, F. Laser-assisted bioprinting for tissue engineering. In *Biofabrication*; Elsevier: Amsterdam, The Netherlands, 2013; pp. 95–118. [CrossRef]
26. Hölzl, K.; Lin, S.; Tytgat, L.; Van Vlierberghe, S.; Gu, L.; Ovsianikov, A. Bioink Properties before, during and after 3D Bioprinting. *Biofabrication* **2016**, *8*, 032002. [CrossRef] [PubMed]
27. Smallman, R.E.; Bishop, R.J. Chapter 13—Biomaterials. In *Modern Physical Metallurgy and Materials Engineering*, 6th ed.; Butterworth-Heinemann: Oxford, UK, 1999; pp. 394–405. [CrossRef]
28. Ng, I.C.; Pawijit, P.; Tan, J.; Yu, H. Anatomy and physiology for biomaterials research and development. In *Encyclopedia of Biomedical Engineering*; Elsevier: Amsterdam, The Netherlands, 2019; Volume 1, pp. 225–236. [CrossRef]
29. Kuhn, L.T. Biomaterials. In *Introduction to Biomedical Engineering*; Elsevier: Amsterdam, The Netherlands, 2005; pp. 255–312. [CrossRef]
30. Raghavendra, G.M.; Varaprasad, K.; Jayaramudu, T. Biomaterials: Design, development and biomedical applications. In *Nanotechnology Applications for Tissue Engineering*; Elsevier: Amsterdam, The Netherlands, 2015; pp. 21–44. [CrossRef]
31. Gopinathan, J.; Noh, I. Recent Trends in Bioinks for 3D Printing. *Biomater. Res.* **2018**, *22*, 11. [CrossRef] [PubMed]
32. Gungor-Ozkerim, P.S.; Inci, I.; Zhang, Y.S.; Khademhosseini, A.; Dokmeci, M.R. Bioinks for 3D Bioprinting: An Overview. *Biomater. Sci.* **2018**, *6*, 915–946. [CrossRef]
33. Mohan, S.; Oluwafemi, O.S.; Kalarikkal, N.; Thomas, S.; Songca, S.P. Biopolymers—Application in Nanoscience and Nanotechnology. *Recent Adv. Biopolym.* **2016**, *1*, 47–66. [CrossRef]
34. Maraveas, C. Production of Sustainable and Biodegradable Polymers from Agricultural Waste. *Polymers* **2020**, *12*, 1127. [CrossRef]
35. Jämsä, M.; Kosourov, S.; Rissanen, V.; Hakalahti, M.; Pere, J.; Ketoja, J.A.; Tammelin, T.; Allahverdiyeva, Y. Versatile Templates from Cellulose Nanofibrils for Photosynthetic Microbial Biofuel Production. *J. Mater. Chem. A* **2018**, *6*, 5825–5835. [CrossRef]
36. Krujatz, F.; Lode, A.; Brüggemeier, S.; Schütz, K.; Kramer, J.; Bley, T.; Gelinsky, M.; Weber, J. Green Bioprinting: Viability and Growth Analysis of Microalgae Immobilized in 3D-Plotted Hydrogels versus Suspension Cultures. *Eng. Life Sci.* **2015**, *15*, 1–28. [CrossRef]
37. Müller, A.; Wesarg, F.; Hessler, N.; Müller, F.A.; Kralisch, D.; Fischer, D. Loading of Bacterial Nanocellulose Hydrogels with Proteins Using a High-Speed Technique. *Carbohydr. Polym.* **2014**, *106*, 410–413. [CrossRef]
38. Chang, C.; Duan, B.; Cai, J.; Zhang, L. Superabsorbent Hydrogels Based on Cellulose for Smart Swelling and Controllable Delivery. *Eur. Polym. J.* **2010**, *46*, 92–100. [CrossRef]

39. Zhang, X.; Morits, M.; Jonkergouw, C.; Ora, A.; Valle-Delgado, J.J.; Farooq, M.; Ajdary, R.; Huan, S.; Linder, M.; Rojas, O.; et al. Three-Dimensional Printed Cell Culture Model Based on Spherical Colloidal Lignin Particles and Cellulose Nanofibril-Alginate Hydrogel. *Biomacromolecules* **2020**, *21*, 1875–1885. [CrossRef]
40. Dai, L.; Lu, J.; Kong, F.; Liu, K.; Wei, H.; Si, C. Reversible Photo-Controlled Release of Bovine Serum Albumin by Azobenzene-Containing Cellulose Nanofibrils-Based Hydrogel. *Adv. Compos. Hybrid Mater.* **2019**, *2*, 462–470. [CrossRef]
41. Ji, L.; Zhang, F.; Zhu, L.; Jiang, J. An In-Situ Fabrication of Bamboo Bacterial Cellulose/Sodium Alginate Nanocomposite Hydrogels as Carrier Materials for Controlled Protein Drug Delivery. *Int. J. Biol. Macromol.* **2021**, *170*, 459–468. [CrossRef] [PubMed]
42. Yoon, H.S.; Yang, K.; Kim, Y.M.; Nam, K.; Roh, Y.H. Cellulose Nanocrystals as Support Nanomaterials for Dual Droplet-Based Freeform 3D Printing. *Carbohydr. Polym.* **2021**, *272*, 118469. [CrossRef]
43. Müller, A.; Ni, Z.; Hessler, N.; Wesarg, F.; Müller, F.A.; Kralisch, D.; Fischer, D. The Biopolymer Bacterial Nanocellulose as Drug Delivery System: Investigation of Drug Loading and Release Using the Model Protein Albumin. *J. Pharm. Sci.* **2012**, *101*, 2271–2280. [CrossRef] [PubMed]
44. Gawande, N.; Mungray, A.A. Superabsorbent Polymer (SAP) Hydrogels for Protein Enrichment. *Sep. Purif. Technol.* **2015**, *150*, 86–94. [CrossRef]
45. Moreno-Garrido, I. Microalgae Immobilization: Current Techniques and Uses. *Bioresour. Technol.* **2008**, *99*, 3949–3964. [CrossRef] [PubMed]
46. González-Delgado, Á.D.; Barajas-Solano, A.F.; Peralta-Ruiz, Y.Y. Microalgae Immobilization Using Hydrogels for Environmental Applications: Study of Transient Photopolymerization. *Chem. Eng. Trans.* **2016**, *47*, 457–462. [CrossRef]
47. Vasilieva, S.; Shibzukhova, K.; Morozov, A.; Solovchenko, A.; Bessonov, I.; Kopitsyna, M.; Lukyanov, A.; Chekanov, K.; Lobakova, E. Immobilization of Microalgae on the Surface of New Cross-Linked Polyethylenimine-Based Sorbents. *J. Biotechnol.* **2018**, *281*, 31–38. [CrossRef]
48. Deng, J.; Jia, M.; Zeng, Y.Q.; Li, W.; He, J.; Ren, J.; Bai, J.; Zhang, L.; Li, J.; Yang, S. Enhanced Treatment of Organic Matter in Slaughter Wastewater through Live Bacillus Velezensis Strain Using Nano Zinc Oxide Microsphere. *Environ. Pollut.* **2022**, *292*, 1–7. [CrossRef]
49. Shemer, B.; Shpigel, E.; Hazan, C.; Kabessa, Y.; Agranat, A.J.; Belkin, S. Detection of Buried Explosives with Immobilized Bacterial Bioreporters. *Microb. Biotechnol.* **2020**, *14*, 251–261. [CrossRef] [PubMed]
50. Chen, W.; Zhang, H.; Zhang, M.; Shen, X.; Zhang, X.; Wu, F.; Hu, J.; Wang, B.; Wang, X. Removal of PAHs at High Concentrations in a Soil Washing Solution Containing TX-100 via Simultaneous Sorption and Biodegradation Processes by Immobilized Degrading Bacteria in PVA-SA Hydrogel Beads. *J. Hazard. Mater.* **2021**, *410*, 124533. [CrossRef]
51. Güngörmüşler, M.; Tamayol, A.; Levin, D.B. Hydrogen Production by Immobilized Cells of Clostridium Intestinale Strain URNW Using Alginate Beads. *Appl. Biochem. Biotechnol.* **2021**, *193*, 1558–1573. [CrossRef] [PubMed]
52. Grzywaczyk, A.; Zdarta, A.; Jankowska, K.; Biadasz, A.; Zdarta, J.; Jesionowski, T.; Kaczorek, E.; Smułek, W. New Biocomposite Electrospun Fiber/Alginate Hydrogel for Probiotic Bacteria Immobilization. *Materials* **2021**, *14*, 3861. [CrossRef] [PubMed]
53. Schaffner, M.; Rühs, P.A.; Coulter, F.; Kilcher, S.; Studart, A.R. 3D Printing of Bacteria into Functional Complex Materials. *Sci. Adv.* **2017**, *3*, 1–9. [CrossRef]
54. Xu, Y.; Zhang, F.; Zhai, W.; Cheng, S.; Li, J.; Wang, Y. Unraveling of Advances in 3D-Printed Polymer-Based Bone Scaffolds. *Polymers* **2022**, *14*, 566. [CrossRef] [PubMed]
55. Nakamura, M.; Nishiyama, Y.; Henmi, C. 3D Micro-fabrication by inkjet 3D biofabrication for 3D tissue engineering. In Proceedings of the 2008 International Symposium on Micro-NanoMechatronics and Human Science, Nagoya, Japan, 6–9 November 2008; pp. 451–456. [CrossRef]
56. Zhu, W.; Ma, X.; Gou, M.; Mei, D.; Zhang, K.; Chen, S. 3D Printing of Functional Biomaterials for Tissue Engineering. *Curr. Opin. Biotechnol.* **2016**, *40*, 103–112. [CrossRef]
57. Zhou, J.; Vijayavenkataraman, S. 3D-Printable Conductive Materials for Tissue Engineering and Biomedical Applications. *Bioprinting* **2021**, *24*, e00166. [CrossRef]
58. Limongi, T.; Susa, F.; Allione, M.; Fabrizio, E.D. Drug Delivery Applications of Three-Dimensional Printed (3DP) Mesoporous Scaffolds. *Pharmaceutics* **2020**, *12*, 851. [CrossRef]
59. Bhatt, P.; Trehan, S.; Inamdar, N.; Mourya, V.K.; Misra, A. Polymers in drug delivery: An update. In *Applications of Polymers in Drug Delivery*; Elsevier: Amsterdam, The Netherlands, 2021. [CrossRef]
60. Hermann, J.R. Protein and the Body. *Agric. Sci. Nat. Resour.* **2005**, *3163*, 1–4.
61. Schiffman, J.D.; Schauer, C.L. A Review: Electrospinning of Biopolymer Nanofibers and Their Applications. *Polym. Rev.* **2008**, *48*, 317–352. [CrossRef]
62. Aryal, S. Proteins—Definition, Properties, Structure, Classification, Functions. Available online: https://microbenotes.com/proteins-properties-structure-classification-and-functions (accessed on 31 December 2021).
63. Proteins—Physical & Chemical Properties | A-Level Biology Revision Notes. Available online: https://alevelbiology.co.uk/notes/proteins-physical-chemical-properties (accessed on 31 December 2021).
64. Beetul, K.; Gopeechund, A.; Kaullysing, D.; Mattan-Moorgawa, S.; Puchooa, D.; Bhagooli, R. Challenges and opportunities in the present era of marine algal applications. In *Algae-Organisms for Imminent Biotechnology*; Thajuddin, N., Dhanasekaran, D., Eds.; IntechOpen: London, UK, 2016; Volume 40. [CrossRef]

65. Barkia, I.; Saari, N.; Manning, S.R. Microalgae for High-Value Products towards Human Health and Nutrition. *Mar. Drugs* **2019**, *17*, 304. [CrossRef] [PubMed]
66. Khan, M.I.; Shin, J.H.; Kim, J.D. The Promising Future of Microalgae: Current Status, Challenges, and Optimization of a Sustainable and Renewable Industry for Biofuels, Feed, and Other Products. *Microb. Cell Factories* **2018**, *17*, 1–21. [CrossRef]
67. Satpati, G.G.; Pal, R. Microalgae—Biomass to Biodiesel: A Review. *J. Algal Biomass Util.* **2018**, *9*, 11–37.
68. Hossain, S.M.Z. Biochemical Conversion of Microalgae Biomass into Biofuel. *Chem. Eng. Technol.* **2019**, *42*, 2594–2607. [CrossRef]
69. Raja, R.; Shanmugam, H.; Ganesan, V.; Carvalho, I.S. Biomass from Microalgae: An Overview. *Oceanogr. Open Access* **2014**, *2*, 1–7. [CrossRef]
70. Poirier, D.R.; Michaux, E.; Sumiko Fulleringer, M. Design of a small scale algae cultivation system to produce biodiesel. In *Design Proposal*; McGill University: Montreal, QC, Canada, 2008.
71. Singh, S.P.; Singh, P. Effect of Temperature and Light on the Growth of Algae Species: A Review. *Renew. Sustain. Energy Rev.* **2015**, *50*, 431–444. [CrossRef]
72. Chew, K.W.; Yap, J.Y.; Show, P.L.; Suan, N.H.; Juan, J.C.; Ling, T.C.; Lee, D.J.; Chang, J.S. Microalgae Biorefinery: High Value Products Perspectives. *Bioresour. Technol.* **2017**, *229*, 1–43. [CrossRef]
73. Lauritano, C.; Andersen, J.H.; Hansen, E.; Albrigtsen, M.; Escalera, L.; Esposito, F.; Helland, K.; Hanssen, K.O.; Romano, G.; Ianora, A. Bioactivity Screening of Microalgae for Antioxidant, Anti-Inflammatory, Anticancer, Anti-Diabetes, and Antibacterial Activities. *Front. Mar. Sci.* **2016**, *3*, 68. [CrossRef]
74. Caliari, S.R.; Burdick, J.A. A Practical Guide to Hydrogels for Cell Culture. *Nat. Methods* **2016**, *13*, 405–414. [CrossRef]
75. Kumar, V.; Nanda, M.; Verma, M. Application of Agar Liquid-Gel Transition in Cultivation and Harvesting of Microalgae for Biodiesel Production. *Bioresour. Technol.* **2017**, *243*, 163–168. [CrossRef] [PubMed]
76. Maidak, B.L.; Larsen, N.; McCaughey, M.J.; Overbeek, R.; Olsen, G.J.; Fogel, K.; Blandy, J.; Woese, C.R. The ribosomal database project. *Nucleic Acids Res.* **1994**, *22*, 3485–3487. [CrossRef]
77. Maier, R.M. Bacterial growth. In *Environmental Microbiology*, 2nd ed.; Elsevier: Amsterdam, The Netherlands, 2009. [CrossRef]
78. Wang, T.; Wusigale; Kuttappan, D.; Amalaradjou, M.A.; Luo, Y.; Luo, Y. Polydopamine-Coated Chitosan Hydrogel Beads for Synthesis and Immobilization of Silver Nanoparticles to Simultaneously Enhance Antimicrobial Activity and Adsorption Kinetics. *Adv. Compos. Hybrid Mater.* **2021**, *4*, 696–706. [CrossRef]
79. Costerton, J.W.; Stewart, P.S.; Greenberg, E.P. Bacterial Biofilms: A Common Cause of Persistent Infections. *Science* **1999**, *284*, 1318–1322. [CrossRef] [PubMed]
80. Deliorman, M.; Wolfenden, M.L.; Suo, Z.; Beech, I.B.; Yang, X.; Avci, R. Immobilization and Trapping of Living Bacteria and Applications in Corrosion Studies. In *Understanding Biocorrosion: Fundamentals and Applications*; Woodhead Publishing Limited: Sawston, UK, 2014. [CrossRef]
81. Connell, J.L.; Ritschdorff, E.T.; Whiteley, M.; Shear, J.B. 3D Printing of Microscopic Bacterial Communities. *Appl. Phys. Sci.* **2013**, *110*, 18380–18385. [CrossRef] [PubMed]
82. Wang, T.; Wang, J.; Li, Z.; Yue, M.; Qing, X.; Zhang, P.; Liao, X.; Fan, Z.; Yang, S. PVA/SA/MXene Dual-Network Conductive Hydrogel for Wearable Sensor to Monitor Human Motions. *J. Appl. Polym. Sci.* **2022**, *139*, 51627. [CrossRef]
83. Sakai, S.; Kotani, T.; Harada, R.; Goto, R.; Morita, T.; Bouissil, S.; Dubessay, P.; Pierre, G.; Michaud, P.; Boutachfaiti, R.E.; et al. Development of Phenol-Grafted Polyglucuronic Acid and Its Application to Extrusion-Based Bioprinting Inks. *Carbohydr. Polym.* **2022**, *277*, 118820. [CrossRef]
84. Bin, Y.; Dongzhen, Z.; Xiaoli, C.; Jirigala, E.; Wei, S.; Zhao, L.; Tian, H.; Ping, Z.; Jianjun, L.; Yuzhen, W.; et al. Modeling Human Hypertrophic Scars with 3D Preformed Cellular Aggregates Bioprinting. *Bioact. Mater.* **2022**, *10*, 247–254. [CrossRef]
85. Chen, L.; Li, Z.; Zheng, Y.; Zhou, F.; Zhao, J.; Zhai, Q.; Zhang, Z.; Liu, T.; Chen, Y.; Qi, S. 3D-Printed Dermis-Specific Extracellular Matrix Mitigates Scar Contraction via Inducing Early Angiogenesis and Macrophage M2 Polarization. *Bioact. Mater.* **2022**, *10*, 236–246. [CrossRef]
86. Trinh, K.T.L.; Le, N.X.T.; Lee, N.Y. Microfluidic-Based Fabrication of Alginate Microparticles for Protein Delivery and Its Application in the in Vitro Chondrogenesis of Mesenchymal Stem Cells. *J. Drug Deliv. Sci. Technol.* **2021**, *66*, 102735. [CrossRef]
87. Hsu, C.-C.; George, J.H.; Waller, S.; Besnard, C.; Nagel, D.A.; Hill, E.J.; Coleman, M.D.; Korsunsky, A.M.; Cui, Z.; Ye, H. Increased Connectivity of HiPSC-Derived Neural Networks in Multiphase Granular Hydrogel Scaffolds. *Bioact. Mater.* **2022**, *9*, 358–372. [CrossRef] [PubMed]
88. Zennifer, A.; Senthilvelan, P.; Sethuraman, S.; Sundaramurthi, D. Key Advances of Carboxymethyl Cellulose in Tissue Engineering & 3D Bioprinting Applications. *Carbohydr. Polym.* **2021**, *256*, 117561. [CrossRef] [PubMed]
89. Estelle, K.T.; Gozen, B.A. Humidity-Controlled Direct Ink Writing for Micro-Additive Manufacturing with Water-Based Inks. *J. Manuf. Processes* **2021**, *69*, 583–592. [CrossRef]
90. Phanthong, P.; Reubroycharoen, P.; Hao, X.; Xu, G.; Abudula, A.; Guan, G. Nanocellulose: Extraction and Application. *Carbon Resour. Convers.* **2018**, *1*, 32–43. [CrossRef]
91. Shak, K.P.Y.; Pang, Y.L.; Mah, S.K. Nanocellulose: Recent Advances and Its Prospects in Environmental Remediation. *Beilstein J. Nanotechnol.* **2018**, *9*, 2479–2498. [CrossRef]

92. Abdul Khalil, H.P.S.; Davoudpour, Y.; Islam, M.N.; Mustapha, A.; Sudesh, K.; Dungani, R.; Jawaid, M. Production and Modification of Nanofibrillated Cellulose Using Various Mechanical Processes: A Review. *Carbohydr. Polym.* **2014**, *99*, 649–665. [CrossRef]
93. Cheng, Q.; Wang, S.; Rials, T.G. Composites: Part A Poly (Vinyl Alcohol) Nanocomposites Reinforced with Cellulose Fibrils Isolated by High Intensity Ultrasonication. *Compos. Part A* **2009**, *40*, 218–224. [CrossRef]
94. Siro, I.; Plackett, D. Microfibrillated Cellulose and New Nanocomposite Materials: A Review. *Cellulose* **2010**, *17*, 459–494. [CrossRef]
95. Moniruzzaman, M.; Ono, T. Separation and Characterization of Cellulose Fibers from Cypress Wood Treated with Ionic Liquid Prior to Laccase Treatment. *Bioresour. Technol.* **2013**, *127*, 132–137. [CrossRef]
96. Peppas, N.A.; Slaughter, B.V.; Kanzelberger, M.A. Hydrogels. *Polym. Sci. A Compr. Ref.* **2012**, *9*, 385–395. [CrossRef]
97. Sudhakar, C.K.; Upadhyay, N.; Jain, A.; Verma, A.; Charyulu, R.N.; Jain, S. Hydrogels-Promising Candidates for Tissue Engineering. In *Nanotechnology Applications for Tissue Engineering*; Elsevier: Amsterdam, The Netherlands, 2015. [CrossRef]
98. Korah, L.V.; Anilkumar, G.; Thomas, S. Hydrogels, DNA, and RNA Polypeptides for the Preparation of Biomaterials. In *Fundamental Biomaterials: Polymers*; Elsevier: Amsterdam, The Netherlands, 2018. [CrossRef]
99. Gonzalez, J.S.; Ludueña, L.N.; Ponce, A.; Alvarez, V.A. Poly (Vinyl Alcohol)/Cellulose Nanowhiskers Nanocomposite Hydrogels for Potential Wound Dressings. *Mater. Sci. Eng.* **2014**, *34*, 54–61. [CrossRef] [PubMed]
100. Mallakpour, S.; Tukhani, M.; Hussain, C.M. Recent Advancements in 3D Bioprinting Technology of Carboxymethyl Cellulose-Based Hydrogels: Utilization in Tissue Engineering. *Adv. Colloid Interface Sci.* **2021**, *292*, 102415. [CrossRef] [PubMed]
101. Kayra, N.; Aytekin, A.O. Synthesis of Cellulose-Based Hydrogels: Preparation, Formation, Mixture, and Modification. *Polym. Polym. Compos. A Ref. Ser.* **2019**, *14*, 407–434.
102. Lan, W.; He, L.; Liu, Y. Preparation and Properties of Sodium Carboxymethyl Cellulose/Sodium Alginate/Chitosan Composite Film. *Coatings* **2018**, *8*, 291. [CrossRef]
103. Zhang, X.; Meng, Y.; Shen, W.; Dou, J.; Liu, R.; Jin, Q.; Fang, S. PH-Responsive Injectable Polysaccharide Hydrogels with Self-Healing, Enhanced Mechanical Properties Based on POSS. *React. Funct. Polym.* **2020**, *70*, 282–287. [CrossRef]
104. Deng, S.; Li, X.; Yang, W.; He, K.; Ye, X. Injectable in Situ Cross-Linking Hyaluronic Acid/Carboxymethyl Cellulose Based Hydrogels for Drug Release. *J. Biomater. Sci. Polym. Ed.* **2018**, *29*, 1643–1655. [CrossRef]
105. Hu, X.; Yang, Z.; Kang, S.; Jiang, M.; Zhou, Z.; Gou, J.; Hui, D.; He, J. Cellulose Hydrogel Skeleton by Extrusion 3D Printing of Solution. *Nanotechnol. Rev.* **2020**, *9*, 345–353. [CrossRef]
106. Yang, J.; Han, C.; Zhang, X.; Xu, F.; Sun, R. Cellulose Nanocrystals Mechanical Reinforcement in Composite Hydrogels with Multiple Cross-Links: Correlations between Dissipation Properties and Deformation Mechanisms. *Macromolecules* **2014**, *47*, 4077–4086. [CrossRef]
107. Du, H.; Liu, W.; Zhang, M.; Si, C.; Zhang, X.; Li, B. Cellulose Nanocrystals and Cellulose Nanofibrils Based Hydrogels for Biomedical Applications. *Carbohydr. Polym.* **2019**, *209*, 130–144. [CrossRef]
108. Wang, H.; Zuo, M.; Ding, N.; Yan, G.; Zeng, X.; Tang, X.; Sun, Y.; Lei, T.; Lin, L. Preparation of Nanocellulose with High-Pressure Homogenization from Pretreated Biomass with Cooking with Active Oxygen and Solid Alkali. *ACS Sustain. Chem. Eng.* **2019**, *7*, 9378–9386. [CrossRef]
109. Butylina, S.; Geng, S.; Laatikainen, K.; Oksman, K. Cellulose Nanocomposite Hydrogels: From Formulation to Material Properties. *Fronteirs* **2020**, *8*, 655. [CrossRef] [PubMed]
110. Lewis, L.; Hatzikiriakos, S.G.; Hamad, W.Y.; Maclachlan, M.J. Freeze—Thaw Gelation of Cellulose Nanocrystals. *ACS Macro Lett.* **2019**, *8*, 48–491. [CrossRef]
111. Yang, J.; Han, C.; Xu, F.; Sun, R. Simple Approach to Reinforce Hydrogels with Cellulose Nanocrystals. *Nanoscale* **2014**, *6*, 5934–5943. [CrossRef] [PubMed]
112. Yang, J.; Zhang, X.; Xu, F. Design of Cellulose Nanocrystals Template-Assisted Composite Hydrogels: Insights from Static to Dynamic Alignment. *Macromolecules* **2015**, *48*, 1231–1239. [CrossRef]
113. Beck, R.C.R.; Chaves, P.S.; Goyanes, A.; Vukosavljevic, B.; Buanz, A.; Windbergs, M.; Basit, A.W.; Gaisford, S. 3D Printed Tablets Loaded with Polymeric Nanocapsules: An Innovative Approach to Produce Customized Drug Delivery Systems. *Int. J. Pharm.* **2017**, *528*, 268–279. [CrossRef]
114. Mohan, D.; Teong, Z.K.; Sajab, M.S.; Hidayatul, N.; Kamarudin, N.; Kaco, H. Intact Fibrillated 3D-Printed Cellulose Macrofibrils/CaCO 3 for Controlled Drug Delivery. *Polymers* **2021**, *13*, 1912. [CrossRef]
115. Wilson, W.C.; Boland, T. Cell and Organ Printing 1: Protein and Cell Printers. *Anat. Rec.* **2003**, *272*, 491–496. [CrossRef] [PubMed]
116. Balasubramanian, S.; Yu, K.; Meyer, A.S.; Karana, E.; Aubin-Tam, M.-E. Bioprinting of Regenerative Photosynthetic Living Materials. *Adv. Funct. Mater.* **2021**, *31*, 2170222. [CrossRef]
117. Lode, A.; Krujatz, F.; Brüggemeier, S.; Quade, M.; Schütz, K.; Knaack, S.; Weber, J.; Bley, T.; Gelinsky, M. Green Bioprinting: Fabrication of Photosynthetic Algae-Laden Hydrogel Scaffolds for Biotechnological and Medical Applications. *Eng. Life Sci.* **2015**, *15*, 177–183. [CrossRef]
118. Milojevic, M.; Harih, G.; Vihar, B.; Vajda, J.; Gradisnik, L.; Zidaric, T.; Kleinschek, K.S.; Maver, U.; Maver, T. Hybrid 3D Printing of Advanced Hydrogel-Based Wound Dressings with Tailorable Properties. *Pharmaceutics* **2021**, *13*, 564. [CrossRef]

119. Aarstad, O.; Heggset, E.B.; Pedersen, I.S.; Bjørnøy, S.H.; Syverud, K.; Strand, B.L. Mechanical Properties of Composite Hydrogels of Alginate and Cellulose Nanofibrils. *Polymers* **2017**, *9*, 378. [CrossRef] [PubMed]
120. Habib, A.; Sathish, V.; Mallik, S.; Khoda, B. 3D Printability of Alginate-Carboxymethyl Cellulose Hydrogel. *Materials* **2018**, *11*, 454. [CrossRef] [PubMed]

polymers

MDPI

Article

Relation between Droplet Size Distributions and Physical Stability for Zein Microfluidized Emulsions

Jenifer Santos [1,*], Luis Alfonso Trujillo-Cayado [1,*], Francisco Carrillo [1], María Luisa López-Castejón [1] and María Carmen Alfaro-Rodríguez [2]

[1] Departamento de Ingeniería Química, Escuela Politécnica Superior, Universidad de Sevilla, c/Virgen de África, 7, E41011 Sevilla, Spain; cfuente@us.es (F.C.); llcastejon@us.es (M.L.L.-C.)

[2] Departamento de Ingeniería Química, Facultad de Química, Universidad de Sevilla, c/Profesor García González, 1, E41012 Sevilla, Spain; alfaro@us.es

* Correspondence: jsantosgarcia@us.es (J.S.); ltrujillo@us.es (L.A.T.-C.); Tel.: +34-954-556447 (J.S.); +34-954-556441 (L.A.T.-C.)

Abstract: Zein, a subproduct of the food industry and a protein, possesses limited applications due to its high hydrophobic character. The objective of this research was to investigate the influence of homogenization pressure and cycles on the volumetric mean diameter ($D_{4,3}$), span values, and Turbiscan Stability Index (TSI) using the response surface methodology for microfluidized emulsions containing zein as a unique stabilizer. Results showed that homogenization pressure seems to be the most influential parameter to obtain enhanced physical stability and droplet size distributions, with the optimum being 20,000 psi. Interestingly, the optimum number of cycles for volumetric diameter, span value, and TSI is not the same. Although a decrease of $D_{4,3}$ with number of cycles is observed (optimum three cycles), this provokes an increase of span values (optimum one cycle) due to the recoalescence effect. Since physical stability is influenced by $D_{4,3}$ and span, the minimum for TSI is observed at the middle level of the cycles (2 cycles). This work highlights that not only volumetric diameter, but also span value must be taken into consideration in order to obtain stable zein emulsions. In addition, this study wants to extend the limited knowledge about zein-based emulsions processed with a Microfluidizer device.

Keywords: emulsion; droplet size distribution; microfluidization; Pickering emulsion; response surface methodology; Turbiscan Stability Index; zein

Citation: Santos, J.; Trujillo-Cayado, L.A.; Carrillo, F.; López-Castejón, M.L.; Alfaro-Rodríguez, M.C. Relation between Droplet Size Distributions and Physical Stability for Zein Microfluidized Emulsions. *Polymers* **2022**, *14*, 2195. https://doi.org/10.3390/polym14112195

Academic Editors: Antonio M. Borrero-López and Alexey Iordanskii

Received: 26 February 2022
Accepted: 13 May 2022
Published: 28 May 2022

1. Introduction

Emulsions are biphasic systems, where droplets are dispersed into a continuous medium. These systems are thermodynamically unstable. However, they can be kinetically stable. In emulsions, there are different destabilization processes, such as flocculation, creaming, sedimentation, coalescence, Ostwald ripening, and phase inversion. Droplet size distribution is an emulsion property that can determine its physical stability since it influences rheology, creaming rate, and microstructure. Interestingly, droplet size distribution can be quantified using different parameters, e.g., Sauter diameter, volumetric diameter, span, or uniformity. There are several studies that relate these parameters with physical stability and rheology [1–3]. Normally, smaller medium droplet sizes and narrower droplet size distributions provokes higher viscosities and enhanced physical stability.

Zein protein, a food by-product obtained from corn, has attracted much attention due to its multiple applications [4]. Among these, it is important to highlight the development of zein-based products as drug delivery or encapsulation systems [5,6]. However, zein has limited application for stabilizing Pickering emulsions due to its high hydrophobic character [7,8]. In addition, zein is only soluble in highly alkaline (pH > 11) solutions [9] and ethanol. Hence, Pickering emulsions stabilized by zein particles alone are scarcely efficient and they could cream easily. Several attempts have been made to solve this problem, e.g., by the incorporation of polysaccharides or combination with other proteins to modify the zein-based composite particles, aiming to enhance the physical stability and properties of zein-based Pickering emulsions [10,11]. Interestingly, zein shows a helical wheel quaternary structure [12]. Different treatments (physical, chemical, and enzymatic) are used to modify the structure and conformation of proteins, which can modify the physicochemical and functional properties [13].

Different mechanical devices have been used in order to develop emulsions with an appropriate droplet size distribution: rotor-stators, microfluidizers, ultrasonics, or high-pressure valve homogenizers [14]. The base of these devices is the use or an external energy to break large droplets into smaller ones during emulsification. Microfluidization is specialised for the formation of uniform emulsions or nanoemulsions. This device uses a high-pressure pump that forces the dispersed and continuous phases through an interaction chamber composed by small channels named microchannels. An impingement area along these microchannels produces fine droplets, normally smaller than one micron [15]. In addition, microfluidization can be used as a physical modification method for proteins and dietary fibers. In this way, ternary and quaternary structures of the proteins can be modified as a result of high shear forces, resulting in an improvement of functional properties of proteins [16]. Microfluidization has been used to enhance functional properties of various food products, such as hazelnut skin fiber [17], high methoxyl pectin [18], and whey protein [19]. The study of microfluidization or ultrasonication applied to the development of zein-based emulsions is very limited [20,21].

The response surface methodology (RSM) was first reported by Box and Wilson (1951) [21]. Nowadays, this methodology is used in multiple fields, such as microbiology [22], chemical engineering [23], or environmental sciences [24]. The base of RSM is to use some designed experiments to explore the relations between independent variables (X) and the response variable (Y) and obtain an optimum using linear or second-degree models. This statistical technique models and analyses problems in which a response of interest (Y) is influenced by other variables (X_1, X_2, ...).

The main objective of this work was to evaluate how the span parameter and volumetric diameter of zein-based emulsions can influence the physical stability of emulsions using the response surface methodology. The minimum of volumetric diameter and span parameter were determined and related to the physical stability measured by Turbiscan Stability Index (TSI). This study wants to extend the limited knowledge about concentrated emulsions stabilized only with zein and developed by microfluidization. It also contributes to the study of microfluidized food grade emulsions in connection with their droplet size distributions and physical stability, which is of paramount importance for their handling properties.

2. Materials and Methods

2.1. Materials

Zein protein was provided by Sigma Aldrich (St. Louis, MO, USA). Sunflower oil containing 40 wt.% of oleic acid was obtained from Coreysa company (Sevilla, Spain). All emulsions were prepared using deionized water.

2.2. Functional Properties Determination: Solubility and Zeta Potential

Suspensions of 1 g zein/100 mL were prepared by adding the protein to water adjusted to various pH values ranging from 1 to 13 with 0.5 M HCl or 0.5 M NaOH. Afterwards, zein solubility was quantified as described by Peterson [25]. For the determination of the Z-potential, a Zetasizer Nano ZS (Malvern Instruments, Malvern, UK) was used, using the Smoluchowski equation [26]. The measures were carried out in triplicate.

2.3. Microfluidization of Food Emulsions Formulated with Zein Protein

The formulation used for the preparation of emulsions was 0.5 wt.% of zein, 50 wt.% of sunflower oil and deionized water. Firstly, the continuous phase was prepared by dispersing zein protein into deionized water and adjusting the pH to 11.5 [9]. Then, a coarse emulsion (batches of 250 g) was prepared, at room temperature, using a Silverson L5M (Silverson, Chesham, UK) for 90 s at 8000 rpm. Finally, finer emulsions were homogenized using a Microfluidizer M110P (Microfluidics company, Westwood, MA, USA) at different processing parameters (Table 1). The microfluidizer device was used with a configuration of Y + Z and a refrigeration temperature of 20 °C.

Table 1. Experimental design, processing parameters, volumetric mean diameters ($D_{4,3}$), span and Turbiscan Stability Index (TSI) values for all emulsions studied.

Sample	X_1	X_2	Pressure	Cycles	$D_{4,3}$ (μm)	span	TSI
1	−1	−1	5000	1	9.65 ± 0.87	1.980	16.35
2	−1	0	5000	2	7.58 ± 0.55	1.925	13.51
3	−1	1	5000	3	6.57 ± 0.51	1.718	13.97
4	0	−1	15,000	1	1.72 ± 0.11	1.015	6.22
5	0	0	15,000	2	1.22 ± 0.07	1.051	6.02
6	0	0	15,000	2	1.25 ± 0.09	1.071	6.09
7	0	0	15,000	2	1.27 ± 0.08	1.06	6.05
8	0	1	15,000	3	0.97 ± 0.06	1.629	6.10
9	1	−1	25,000	1	2.59 ± 0.18	1.228	6.75
10	1	0	25,000	2	1.94 ± 0.16	1.455	6.58
11	1	1	25,000	3	1.71 ± 0.13	1.702	6.55

2.4. Design of Experiments

An experimental design and response surface methodology were used to analyze the relationship between the dependent variables (volumetric mean diameter, span, and Turbiscan Stability Index) and independent variables (number of cycles and homogenization pressure). The experimental design consisted of three levels and two factors, generating 32 experiments. This results in 9 experiments and 2 additional replicates of the central point (see Table 1). Every experiment was conducted by duplicate. All the data were analyzed with a one-way analysis of variance (ANOVA) at a 95% confidence level. All the experimental design and data analyses were performed using the Echip software (Experimentation by Design, Wilmington, DE, USA).

2.5. Laser Diffraction Measurements

In order to characterise droplet size distribution of the emulsions developed, a Malvern Mastersizer 2000 (Malvern, UK) was used. Furthermore, volumetric diameter ($D_{4,3}$) was used in order to quantify and compare the mean diameters of the emulsions developed. Finally, span parameter values were used to quantify the polydispersity of the droplets created.

$$D_{4,3} = \sum_{i=1}^{N} n_i d_i^4 / \sum_{i=1}^{N} n_i d_i^3 \quad (1)$$

$$\text{span} = \frac{D_{90} - D_{10}}{D_{50}} \quad (2)$$

where d_i is the droplet diameter, N is the total number of droplets, ni is the number of droplets having a diameter di, and d_{90}, d_{50}, d_{10} are the diameters at 90%, 50%, and 10% cumulative volume.

2.6. Multiple Light Scattering Technique

Turbiscan Stability Index (TSI) is a parameter that allows physical instability to be quantified and compared. It has been calculated following Equation (3):

$$\text{TSI} = \sum_{j} \left| \text{scan}_{\text{ref}}(h_j) - \text{scan}_i(h_j) \right| \quad (3)$$

where $scan_{ref}$ and $scan_i$ are the initial backscattering value and the backscattering value at a specific time, respectively, and h_j is a specific height in the measuring cell.

In order to obtain this parameter, backscattering measurements were carried out for the samples developed at different aging times using Turbiscan Lab Expert (Formulaction, France).

3. Results and Discussion

Figure 1 shows the solubility of zein in water and the Z-potential as a function of pH. On the one hand, zein is in a low solubility condition throughout the investigated pH range from 2.5 to 7. In addition, a significant increase for solubility was found at pH values higher than 8. At these pH values, zein side chains contain more negative net charge, influencing its structural stability, i.e., giving rise to higher solubility. This behavior could be explained by the electrostatic interactions produced by the side chains with ionizing properties as well as hydrogen bond formation with the solvent. On the other hand, the potential value Z follows a downward trend as the pH increases, starting the highest value coinciding with pH 2.5 (lowest value evaluated) and reaching its minimum at a pH of 11.5. The most significant zeta potential drop is between pH 8 and pH 9.5. From this, it can be deduced that the isoelectric point of the protein will be in this range. This value coincides with what was found by other authors in the literature who establish that the isoelectric point is between 5 and 9 [27]. However, increasing the pH to 11.5 results in an improved solubility. Hence, pH 11.5 is selected for the evaluation of emulsion development. High ionic strength decreased the solubility and emulsifying activity of zein suspensions [28].

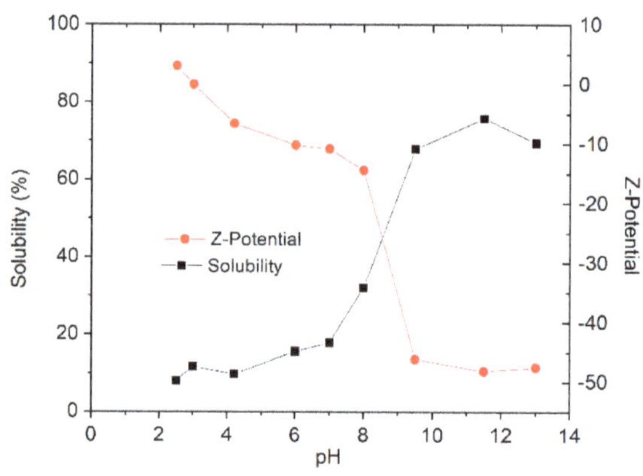

Figure 1. Zein solubility and zeta potential values as a function of pH.

Figure 2 shows the droplet size distribution for emulsions containing 50 wt.% sunflower oil and 0.5 wt.% of zein (ration 1:10) as a function of microfluidization parameters studied (pressure and cycles). First of all, the droplet size distribution of the pre-emulsion is bimodal, while microfluidized emulsions show a monomodal distribution. Secondly, the use of microfluidization provoked a clear reduction of droplet sizes, thus proving its importance. Furthermore, there is a decrease of droplet size when pressure increases from 5000 to 15,000 psi after just one cycle. However, an increase of droplet size is observed above 15,000 psi. This fact is a clear sign of recoalescence due to over-processing. This usually occurs using microfluidization in emulsions where the protein/surfactant does not entirely cover the oil-water interface [29]. Hence, this result suggests that there is a lack of protein in the interface. In order to obtain a deeper insight into these results, Figure 3 is illustrated.

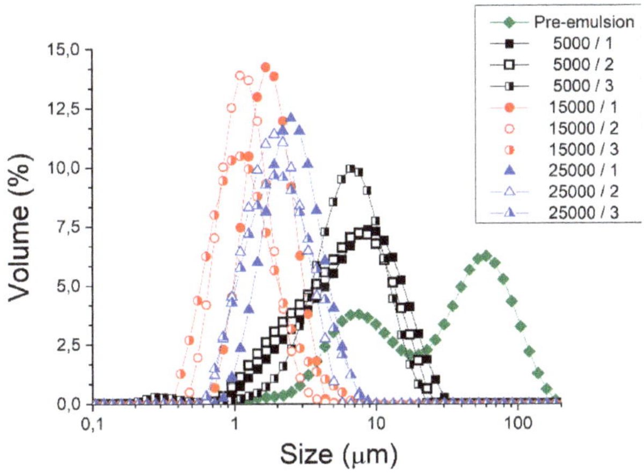

Figure 2. Droplet size distributions for emulsions developed formulated with zein as a function of homogenization pressure and passes in the microfluidizer.

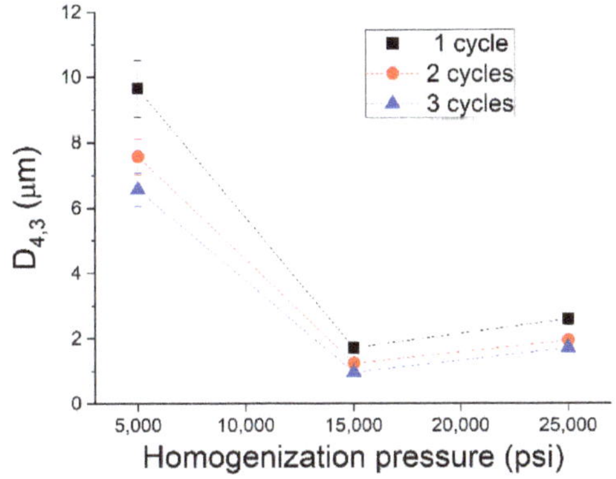

Figure 3. Volumetric mean diameters ($D_{4,3}$) for emulsions developed formulated with zein as a function of homogenization pressure and number of cycles in the microfluidizer.

Variation of the volumetric diameter ($D_{4,3}$) with homogenization pressure for different numbers of cycles is shown in Figure 3. The clear decrease in droplet size from 5000 to 15,000 psi is also observed here. In addition, a small increase in droplet size highlights the slight recoalescence from 15,000 to 25,000 psi. Whereas a big reduction of droplet size is observed from one to two cycles at 5000 psi. This is not noticed at 15,000 or 25,000 psi, and could be due to the lack of protein concentration to stabilize smaller droplets.

Results obtained for volumetric diameter and span values as a function of homogenization pressure (P) and cycles (C) are shown in Table 1. This table also illustrates the experiments design that has been carried out. These results have been modelled and optimized using the response surface methodology (RSM). On the one hand, the relation between volumetric diameter with microfluidized parameters is indicated in Figure 4 and in Equation (4). Equation (4) states that volumetric diameter fits a quadratic function of P and C with a determination coefficient (R^2) of 0.99. This coefficient suggests the great correlation between the experimental results and the model.

$$D_{4,3}(\mu m) = 1.22 - 2.92 \cdot P - 0.73 \cdot C + 0.55 \cdot P \cdot C + 3.57 \cdot P^2 \quad (4)$$

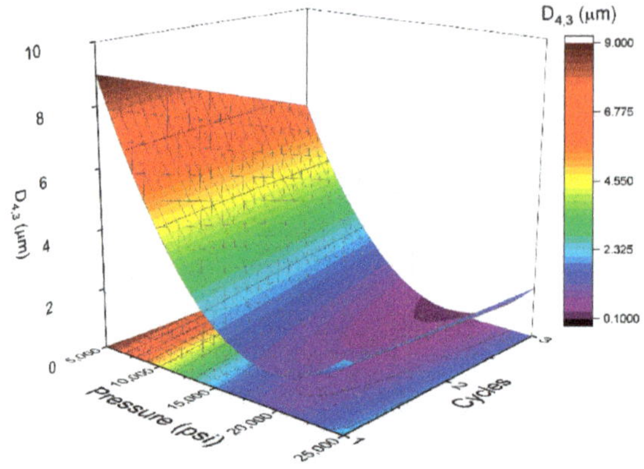

Figure 4. Response surface 3D plot of volumetric diameter as a function of pressure and cycles used in the microfluidizer.

Furthermore, *F* critical (F_{crit}), which determines the significance of the groups of variables, is higher than *F* lack of fit (F_{lof}) with $p = 0.05$, which tests how well the model fits the data. This fact is a clear indication of the suitability of the model. Volumetric diameter was sensitive to the homogenization pressure and number of cycles. Analyzing the coefficients of Equation (4), homogenization pressure seems to be more influential than cycles for volumetric diameter. These trends are clearly seen in Figure 4. At intermediate pressures, and especially at high pressures, the influence of the number of cycles on the diameters is not very significant. However, regardless of the number of cycles, volumetric diameters vary significantly with homogenization pressure. Taking this model into account, the minimum of volumetric diameter was at 20,000 psi and three cycles.

On the other hand, the relation between span values obtained with homogenization pressure and number of cycles is indicated in Figure 5 and in Equation (5). This correlation is also a quadratic one with more influence of homogenization pressure. In this case, the value of R^2 is 0.89, showing a good fit. This fit presents a minimum at 20,000 psi and one cycle. In addition, there is an increase of span values from one to three cycles at 15,000 and 25,000 psi (see Table 1). The latter could be due to the recoalescence effect abovementioned. The increase of number of cycles provoked a reduction of volumetric diameter but also an increase of span values. This fact has been reported for other emulsions containing proteins [30]. It is important to notice that physical stability of emulsions is not only influenced by mean diameters, but also by span values [31].

$$\text{span} = 1.126 - 0.211{\cdot}P + 0.137{\cdot}C + 0.184{\cdot}P{\cdot}C + 0.481{\cdot}P^2 \quad (5)$$

Figure 6 shows the variation of backscattering (BS) with height of the measuring cell as a function of aging time for the pre-emulsion. There is a big drop in the lower part of the measuring cell, that is related to a clarification process. Hence, the droplets are moving to the upper part of the vial. This is the definition of the creaming process. This process occurs in 1 hour and a half in this pre-emulsion. Therefore, the pre-emulsion showed a very poor stability due to its wide droplet size distribution centered at a big droplet size.

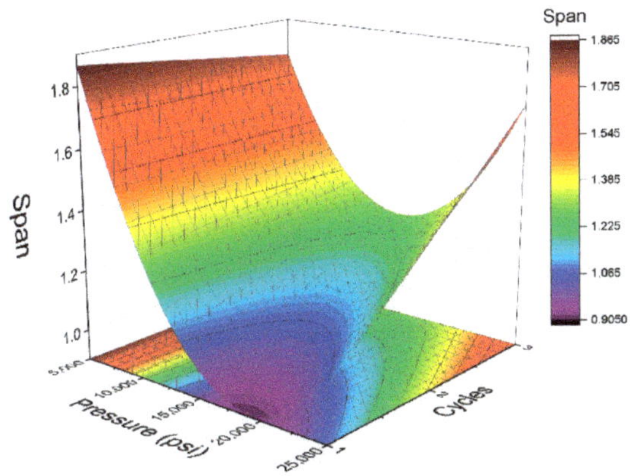

Figure 5. Response surface 3D plot of span values as a function of pressure and cycles used in the microfluidizer.

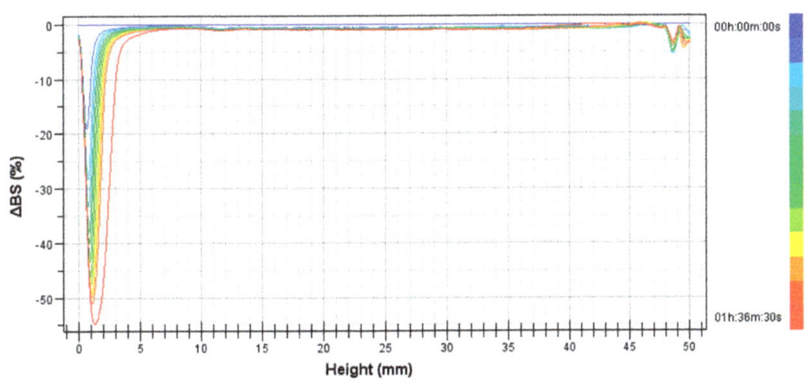

Figure 6. Variation of Backscattering (ΔBS) with height of the measuring cell as a function of aging time for the pre-emulsion.

Figure 7 illustrates the variation of backscattering (ΔBS) profile with height of the measuring cell as a function of aging time for the emulsion processed at 15,000 psi and two cycles. A decrease of ΔBS with storage time in the bottom and upper part of the vial is observed. These facts are a clear indication of a clarification process in the bottom part and a higher concentration of droplets in the upper part, i.e., a creaming process. Two big drops are observed, pointing to the destabilization process by creaming. Compared to the pre-emulsion (Figure 6), it seems that emulsions processed using the microfluidizer showed an enhanced physical stability.

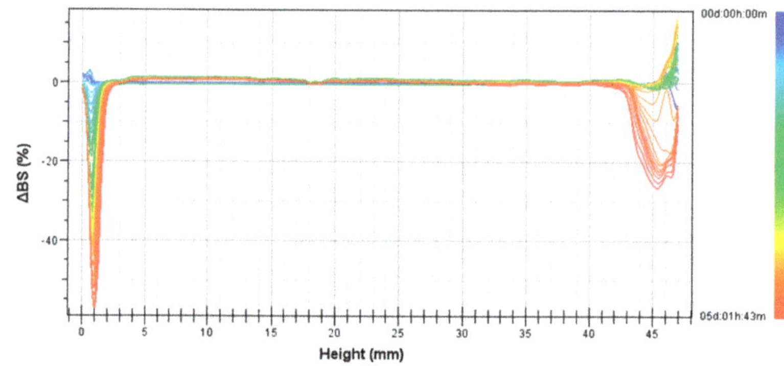

Figure 7. Variation of Backscattering (ΔBS) with height of the measuring cell as a function of aging time for the microfluidized emulsion processed at 15,000 psi and two cycles.

Figure 8 illustrates the relation between TSI values with homogenization pressure and number of cycles. In addition, Equation (6) states that TSI values follows a quadratic function with a determination coefficient (R^2) of 0.99. Furthermore, the homogenization pressure is also the most influential variable. The trend of the TSI with respect to homogenization pressure and the number of cycles observed in Figure 8 is similar to that shown in Figure 4 for volumetric diameter. Thus, while the TSI is heavily influenced by pressure regardless of the number of cycles, it does not vary significantly with the number of cycles at intermediate and high pressures. The minimum is observed at 20,000 psi and two cycles.

$$TSI = 5.89 - 3.99 \cdot P - 0.45 \cdot C + 0.55 \cdot P \cdot C + 4.88 \cdot P^2 \quad (6)$$

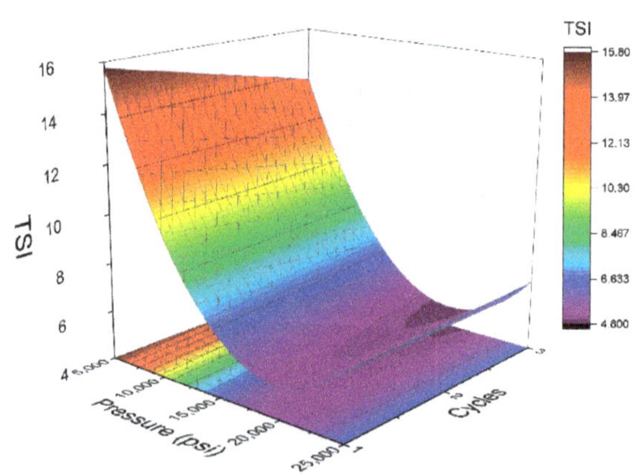

Figure 8. Response surface 3D plot of Turbiscan Stability Index (TSI) values as a function of pressure and cycles used in the microfluidizer.

Analyzing every minimum of the models, it is interesting to note that 20,000 psi is a key factor to obtain a good droplet size distribution and enhanced physical stability. However, while the minimum volumetric diameter is obtained at three cycles, the minimum span is obtained at one cycle and the minimum TSI value at two cycles. This fact is explained by the recoalescence that these emulsions present. There are more cycles in the microfluidizer and a lower volumetric diameter, but higher span values. Interestingly, physical stability is influenced by the two factors. In this way, the minimum for TSI is observed at the middle level of the cycles (two cycles).

4. Conclusions

Recently, the interest in the use of zein in the development of emulsions and bio-based delivery systems has been increasing. In this study, a solubility study has proven that pH 11.5 is suitable to prepare emulsions with zein. Optimization of the microfluidization parameters (homogenization pressure and cycles) in order to minimize the volumetric diameter and the span value of concentrated emulsions containing zein and sunflower oil was carried out. It has been proven that microfluidization provokes monomodal distributions regardless of homogenization pressure and number of cycles, improving the droplet size distributions (DSD) of the pre-emulsion. In addition, overprocessing has been observed for emulsions which have been summited above 15,000 psi, suggesting that zein does not entirely cover the interface. The response surface methodology has proven its importance to obtain not only a clear minimum of volumetric diameter, span value, and TSI, but also the key processing parameter (homogenization pressure) that influences these properties. Interestingly, the minimum of volumetric diameter, span, and TSI are not the same. This fact points out that volumetric diameter and span have to be taken into consideration in order to obtain stable zein emulsions. This study has revealed the impact of microfluidization on concentrated emulsions formulated only with zein as stabilizer, highlighting the importance of the selected pressure to develop these systems.

Author Contributions: Conceptualization, J.S. and M.L.L.-C.; methodology, M.C.A.-R.; software, F.C.; validation, L.A.T.-C. and F.C.; formal analysis, F.C.; investigation, J.S. and M.L.L.-C.; resources, L.A.T.-C.; data curation, J.S.; writing—original draft preparation, J.S.; writing—review and editing, L.A.T.-C.; visualization, L.A.T.-C.; supervision, M.C.A.-R.; project administration, M.C.A.-R.; funding acquisition, J.S. All authors have read and agreed to the published version of the manuscript.

Funding: This research received no external funding.

Institutional Review Board Statement: Not applicable.

Informed Consent Statement: Not applicable.

Data Availability Statement: Not applicable.

Acknowledgments: The financial support received (Perfeccionamiento Posdoctoral) from the University of Sevilla and from Junta de Andalucía (Contrato Doctores PAIDI 2020) are kindly acknowledged.

Conflicts of Interest: The authors declare no conflict of interest.

References

1. Chanamai, R.; McClements, D.J. Dependence of Creaming and Rheology of Monodisperse Oil-in-Water Emulsions on Droplet Size and Concentration. *Colloids Surf. A Physicochem. Eng. Asp.* **2000**, *172*, 79–86. [CrossRef]
2. Gallegos, C.; Valencia, C.; Partal, P.; Franco, J.M.; Maglio, O.; Abrahamsson, M.; Brito-de la Fuente, E. Droplet-Size Distribution and Stability of Commercial Injectable Lipid Emulsions Containing Fish Oil. *Am. J. Health-Syst. Pharm.* **2012**, *69*, 1332–1335. [CrossRef] [PubMed]
3. Martin-Piñero, M.J.; García, M.C.; Muñoz, J.; Alfaro-Rodriguez, M.-C. Influence of the Welan Gum Biopolymer Concentration on the Rheological Properties, Droplet Size Distribution and Physical Stability of Thyme Oil/W Emulsions. *Int. J. Biol. Macromol.* **2019**, *133*, 270–277. [CrossRef] [PubMed]
4. Giteru, S.G.; Ali, M.A.; Oey, I. Recent progress in understanding fundamental interactions and applications of zein. *Food Hydrocoll.* **2021**, *120*, 106948. [CrossRef]
5. Li, M.; Yu, M. Development of a nanoparticle delivery system based on zein/polysaccharide complexes. *J. Food Sci.* **2021**, *85*, 4108–4117. [CrossRef]
6. Champrasert, O.; Orfila, C.; Suwannaporn, P. Acrylamide mitigation using zein–polysaccharide complex particles. *Food Hydrocoll.* **2022**, *124*, 107317. [CrossRef]
7. Altan, A.; Aytac, Z.; Uyar, T. Carvacrol Loaded Electrospun Fibrous Films from Zein and Poly (Lactic Acid) for Active Food Packaging. *Food Hydrocoll.* **2018**, *81*, 48–59. [CrossRef]
8. Zou, Y.; van Baalen, C.; Yang, X.; Scholten, E. Tuning Hydrophobicity of Zein Nanoparticles to Control Rheological Behavior of Pickering Emulsions. *Food Hydrocoll.* **2018**, *80*, 130–140. [CrossRef]
9. Kasaai, M.R. Zein and Zein-Based Nano-Materials for Food and Nutrition Applications: A Review. *Trends Food Technol.* **2018**, *79*, 184–197. [CrossRef]
10. Li, M.; He, S. Utilization of Zein-Based Particles in Pickering Emulsions: A Review. *Food Rev. Int.* **2021**, 1–21. [CrossRef]

11. Santos, J.; Alcaide-González, M.A.; Trujillo-Cayado, L.A.; Carrillo, F.; Alfaro-Rodríguez, M.C. Development of Food-Grade Pickering Emulsions Stabilized by a Biological Macromolecule (Xanthan Gum) and Zein. *Int. J. Biol. Macromol.* **2020**, *153*, 747–754. [CrossRef] [PubMed]
12. Argos, P.; Pedersen, K.; Marks, M.D.; Larkins, B.A. A Structural Model for Maize Zein Proteins. *J. Biol. Chem.* **1982**, *257*, 9984–9990. [CrossRef]
13. Cabra, V.; Arreguin, R.; Vazquez-Duhalt, R.; Farres, A. Effect of Alkaline Deamidation on the Structure, Surface Hydrophobicity, and Emulsifying Properties of the Z19 α-Zein. *J. Agric. Food Chem.* **2007**, *55*, 439–445. [CrossRef] [PubMed]
14. Rayner, M.; Dejmek, P. *Engineering Aspects of Food Emulsification and Homogenization*; CRC Press: Boca Raton, FL, USA, 2015.
15. Ramos, O.L.; Pereira, R.N.; Cerqueria, M.A.; Teixeira, J.A.; Vicente, A.A. *Advances in Processing Technologies for Bio-Based Nanosystems in Food*; CRC Press: Boca Raton, FL, USA, 2019; Volume 1, ISBN 1351710052.
16. Iordache, M.; Jelen, P. High Pressure Microfluidization Treatment of Heat Denatured Whey Proteins for Improved Functionality. *Innov. Food Sci. Emerg. Technol.* **2003**, *4*, 367–376. [CrossRef]
17. Cikrikci, S.; Demirkesen, I.; Mert, B. Production of Hazelnut Skin Fibres and Utilisation in a Model Bakery Product. *Qual. Assur. Saf. Crop. Foods* **2016**, *8*, 195–206. [CrossRef]
18. Chen, J.; Liang, R.-H.; Liu, W.; Liu, C.-M.; Li, T.; Tu, Z.-C.; Wan, J. Degradation of High-Methoxyl Pectin by Dynamic High Pressure Microfluidization and Its Mechanism. *Food Hydrocoll.* **2012**, *28*, 121–129. [CrossRef]
19. Box, G.E.P.; Wilson, K.B. On the Experimental Attainment of Optimum Conditions. *J. R. Stat. Soc. Ser. B Methodol.* **1951**, *13*, 1–45. [CrossRef]
20. Santos, J.; Trujillo-Cayado, L.A.; Alcaide, M.D.Á.; Alfaro, M.D.C. Impact of Microfluidization on the Emulsifying Properties of Zein-Based Emulsions: Influence of Diutan Gum Concentration. *Materials* **2021**, *14*, 3695. [CrossRef]
21. Santos, J.; Trujillo-Cayado, L.A.; Carrello, H.; Cidade, M.T.; Alfaro, M.C. Optimization of sonication parameters to obtain food emulsions stabilized by zein: Formation of zein–diutan gum/zein–guar gum complexes. *J. Sci. Food Agric.* **2022**, *102*, 2127–2134. [CrossRef]
22. Shi, L.-E.; Ying, G.-Q.; Zhang, X.-Y.; Tang, Z.-X.; Chen, J.-S.; Xiong, W.-Y.; Liu, H.-Z. Medium Optimization for 5′-Phosphodiesterase Production from Penicillium Citrinum Using Response Surface Methodology. *Food Technol. Biotechnol.* **2007**, *45*, 126–133.
23. Margeta, D.; Faraguna, F.; Sertić Bionda, K.; Jukić, A. Modeling of Dibenzothiophene Oxidative Desulfurization Using Response Surface Methodology. *Pet. Sci. Technol.* **2018**, *36*, 68–74. [CrossRef]
24. Mochi, V.T.; Pacheco, J.R.; Cremasco, M.A. Response Surface Methodology Applied to Ozone Generation. *Ozone Sci. Eng.* **2010**, *32*, 372–378. [CrossRef]
25. Peterson, G.L. A Simplification of the Protein Assay Method of Lowry et al. Which Is More Generally Applicable. *Anal. Biochem.* **1977**, *83*, 346–356. [CrossRef]
26. Tan, W.F.; Koopal, L.K.; Weng, L.P.; van Riemsdijk, W.H.; Norde, W. Humic Acid Protein Complexation. *Geochim. Et Cosmochim. Acta* **2008**, *72*, 2090–2099. [CrossRef]
27. Shukla, R.; Cheryan, M. Zein: The Industrial Protein from Corn. *Ind. Crop. Prod.* **2001**, *13*, 171–192. [CrossRef]
28. Wu, Y.V. Emulsifying activity and emulsion stability of corn gluten meal. *J. Sci. Food Agric.* **2001**, *81*, 1223–1227. [CrossRef]
29. Jafari, S.M.; He, Y.; Bhandari, B. Production of Sub-Micron Emulsions by Ultrasound and Microfluidization Techniques. *J. Food Eng.* **2007**, *82*, 478–488. [CrossRef]
30. Jafari, S.M.; He, Y.; Bhandari, B. Optimization of Nano-Emulsions Production by Microfluidization. *Eur. Food Res. Technol.* **2007**, *225*, 733–741. [CrossRef]
31. Santos, J.; Trujillo-Cayado, L.A.; Calero, N.; Muñoz, J. Physical Characterization of Eco-Friendly O/W Emulsions Developed through a Strategy Based on Product Engineering Principles. *AIChE J.* **2014**, *60*, 2644–2653. [CrossRef]

polymers MDPI

Article

Relating Amorphous Structure to the Tear Strength of Polylactic Acid Films

Yutaka Kobayashi [1,*], Akira Ishigami [1,2] and Hiroshi Ito [1,2,*]

1 Research Center for GREEN Materials and Advanced Processing (GMAP), 4-3-16 Jonan, Yonezawa, Yamagata 992-8510, Japan; akira.ishigami@yz.yamagata-u.ac.jp
2 Graduate School of Organic Materials Science, Yamagata University, 4-3-16 Jonan, Yonezawa, Yamagata 992-8510, Japan
* Correspondence: kobayashi.y@yz.yamagata-u.ac.jp (Y.K.); ihiroshi@yz.yamagata-u.ac.jp (H.I.); Tel.: +81-(238)-263430 (Y.K.); +81-(238)-263081 (H.I.)

Abstract: Compared with polyolefins that are used as single-use plastics, polylactic acid (PLA) has a lower tear strength in films. The relationship between the tear strength and the higher-order structure of films was investigated using PLA films that absorbed moisture at 30 °C and 95% relative humidity (RH) or that had been annealed under reduced pressure conditions. Although the mobile amorphous (MAm) amount did not change under high humidity, the film became brittle due to enthalpy relaxation. The crystallization by annealing also caused embrittlement, and the MAm amount decreased to 10%. The displacement until tearing is lowered from 2.5 to 0.5 mm in both cases. However, in situ retardation measurements revealed that there was a significant difference in the fracture morphology of the torn tip. When crystallized, the molecular chains and crystals are oriented in the tensile direction of the film, and a fragmented structure is observed in the ligament. Embrittlement due to enthalpy relaxation caused a weak orientation perpendicular to the tensile direction of the film, and cracks occurs along with this orientation.

Keywords: biopolymers; tear strength; amorphous structure

Citation: Kobayashi, Y.; Ishigami, A.; Ito, H. Relating Amorphous Structure to the Tear Strength of Polylactic Acid Films. *Polymers* **2022**, *14*, 1965. https://doi.org/10.3390/polym14101965

Academic Editors: Antonio M. Borrero-López, Concepción Valencia-Barragán, Esperanza Cortés Triviño, Adrián Tenorio-Alfonso and Clara Delgado-Sánchez

Received: 29 April 2022
Accepted: 9 May 2022
Published: 11 May 2022

1. Introduction

Disposable packaging containers are often used for food distribution. In order to preserve food, packaging materials are required to be sterile and have gas permeability to prevent food spoilage. Petroleum-derived polymers such as polyolefins are inexpensive and are often used as disposable packaging materials. When used packaging is disposed of in the natural environment, these plastics take a long time to decompose into water and carbon dioxide. Additionally, when released into the ocean, it creates a great load on the ecosystem [1].

Polylactic acid (PLA) has been expected to replace polyolefins as a packing material because it is produced from biological raw materials [2]. PLA can be composted and decomposed easily in humid and warm conditions. However, even after one year in the ocean, PLA hardly shows any decomposition [3–5]. The marine environment has the following several characteristics that make it distinct from soil and compost: a low temperature, high salt content (conductivity), high pressure, and a low nutritional level (nitrate, etc.). In addition, there are fluctuations in different seasons or regions, and the water temperature also changes depending on the depth in the vertical direction. These characteristics also affect the activity of the microorganisms in the ocean.

On the other hand, commercial PLA consists of the following two enantiomers: L- and D-lactic acid, and the D-content is usually less than 5% for maintaining the crystallinity in a molded product. PLA needs to crystallize for thermal stability because food containers are frequently used in microwave ovens. When poly L-lactic acid (PLLA) and poly D-lactic acid (PDLA) are blended in equal amounts, stereocomplex (SC) crystals form [6–8]. A molded product with SC crystals has excellent heat resistance. However, the hydrolytic degradation of SC PLA is not faster than that of normal PLA, so utilizing PDLA does not solve the microplastic problem in the ocean at lower temperatures and with fewer microorganisms. When PLLA contains a few percent of D-form, the crystallization rate significantly reduces, and the hydrolysis rate increases. Generally, the presence of a trace amount of comonomer decreases the crystallinity of the random copolymer and changes its physical properties. In the case of high-density polyethylene (HDPE) and polypropylene (PP), their films are more flexible. According to the PE study [9,10], long-chain branches and high molecular weight components that prolong the relaxation time reduce the tear strength. Whether the crystallinity is the same in PE grades, PE with the smaller comonomer content exhibits a higher tear strength. As the D content in PLA increases, the tensile elongation increases, too. However, the elongation is much smaller than that of PE.

The reason for the brittleness of PLA is that the glass transition temperature (Tg) is approximately 60 °C, which is higher than room temperature. Even if L- and D-lactic acid are copolymerized or SC crystals form, Tg does not change. Thus, various strengthening methods have been studied for packaging containers. There are several methods that can be used to strengthen PLA [7,11–13], such as controlling the molecular structure of PLA by means of copolymerization with monomers other than lactic acid, blending plasticizers, or biodegradable polymers into PLA, and using natural materials as fillers [14]. Blending petroleum-derived polymers and fillers have even been attempted [15,16]. Furthermore, strengthening has been performed by utilizing polymer reactions such as cross-linking between different kinds of polymers, grafting reactions, and transesterification reactions. However, the degradability of this strengthened PLA in the ocean has scarcely been investigated.

Water is essential for the microbial and enzymatic degradation of PLA in the natural environment. During the process of hydrolysis, the higher-order structure of PLA is considered in a three-phase crystal, mobile amorphous (MAm), and rigid amorphous (RAm) model. The ratio of the three components changes during the process [17]. When hydrolyzed at 80 °C, which is a higher temperature than the Tg of PLA, it makes it difficult for the crystals to decompose, and the RAm structure promotes decomposition [18]. In our previous study, PLA films crystallized in the presence of water, even when the films that crystallized without water had the same crystallinity, and were susceptible to hydrolysis in a 70 °C and 95% RH environment [19]. Therefore, more detailed experiments are needed on higher-order structures and during hydrolysis.

As mentioned above, PLA has many problems, and we cannot say it is a marine-degradable polymer. However, in the development of food containers using marine-degradable polymers, which is our ultimate goal, we believe that it will be a starting point for finding a way to achieve both crystalline and degradable properties. Generally, a highly hydrophilic polymer is easily hydrolyzable. Therefore, increasing the toughness by absorbing water is the most appropriate method for improving the tearability of the packaging material. This study deals with the relationship between the higher-order structure of the PLA film, which is adjusted by water absorption and annealing, and the tear strength. In this study, we focused on the tear strength of packaging films such as plastic bags as an example of single-use plastics. The behavior of amorphous PLA in a constant temperature and humidity environment was quantitatively analyzed by the change in tear strength, which rapidly becomes brittle due to enthalpy relaxation. The deformation occurring at the tear tip was observed in situ using the latest instrument with an optical array. It is clarified that the mechanism of crack generation differs between the cases where PLA has crystallized and the case where PLA is embrittled by enthalpy relaxation.

2. Experimental Setup

For the PLA, Total Corbion (Rayong, Thailand) L175 with a melt mass flow rate (MFR) of 3 g/10 min at 190 °C was used. A compression-molding machine manufactured by Imoto Seisakusho (Kyoto, Japan) was used to prepare the film samples. The PLA pellets were pressurized at 6 MPa for 1 min after heating and melting at 190 °C for 4 min under reduced pressure; then, they were cooled at 20 °C for 2 min with a pressure of 6 MPa. The film thickness was about 100 μm. To change the higher-order structure, the PLA films were annealed at a predetermined temperature for 2 h under reduced pressure conditions in an AS ONE (Osaka, Japan) AW-250N vacuum oven. The constant temperature and humidity treatments were carried out at 30 °C and 95% relative humidity (RH) for a predetermined amount of time with a Tokyo Rika Kikai (Tokyo, Japan) KCL-2000W.

After adjusting the conditions of the film as stated above, a sample with the shape shown in Figure 1 was cut out from the film and used. A tear test was performed using a Sanko (Nagoya, Japan) ISL-T300 at a span of 14 mm and a tensile speed of 1 mm/min at room temperature. The measurement was repeated three times, and the average was taken as the reported value. A Photonic Lattice Inc. (Miyagi, Japan) WPA-micro and WPA-200 were used for the in-situ retardation measurements during the tear test to capture two-dimensional (2-D) mapping images of the retardation and the polarizing angles. The behavior of the tearing films was recorded by a video camera and then captured in a freeze-frame picture. After the tear test, the appearance of the torn films was taken using a microphotograph with a KEYENCE (Osaka, Japan) VHX-950F.

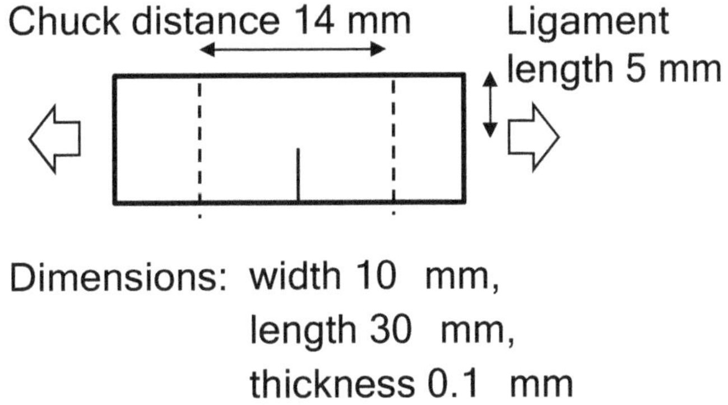

Figure 1. Illustration of the shape and dimensions of the film used for the tear test.

Thermal characterization was performed using differential scanning calorimetry (DSC) with a TA Instruments (New Castle, DE, USA) Q200. The same samples were used during the tear test. The change in specific heat (ΔCp) at the Tg was measured by increasing the temperature from 0 to 210 ° C at 10 ° C/min. The mobile amorphous (MAm) portion was quantified from the ratio of the change in the ΔCp at the Tg to the completely amorphous ΔCp 0.531 J/gK [20].

3. Results and Discussion

3.1. The Tear Strength of PLA after Conditioning in a Humid Chamber or an Oven

Polyolefins are non-polar and non-hygroscopic, and disposable plastic bags made from this material have a strong tear strength. On the other hand, PLA has high polarity and is easily affected by water. Thus, the film was conditioned at a high humidity of 95% RH and at 30 °C, and the changes in the tear strength were then measured.

Figure 2 shows the load and the displacement of the crosshead during the tear test of the PLA film conditioned in the humidity chamber. Since the film thickness was unified to 100 μm, the maximum load indicated the difference in yield strength between films. Compared to the film (0 h) immediately after molding, the film that had been adjusted under high humidity conditions for one hour showed increased maximum strength, but the amount of displacement until fracture did not change. After conditioning for 6 h or more, the film was torn immediately after reaching the maximum load.

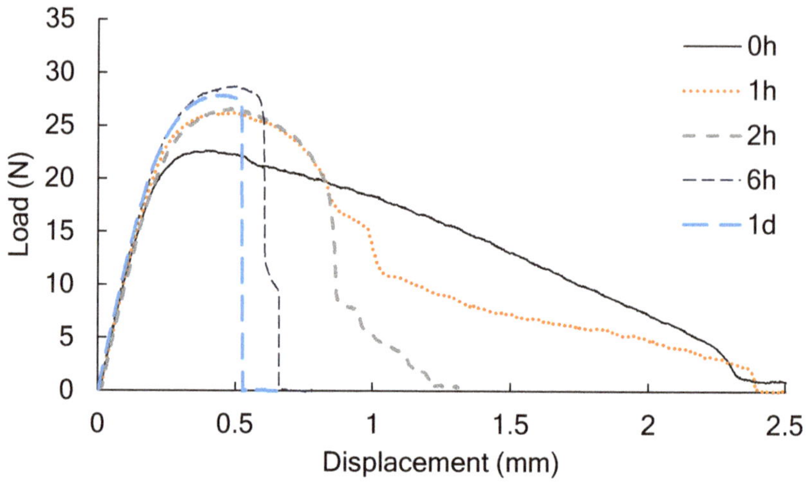

Figure 2. The load and the displacement of the crosshead of the PLA film conditioned in the humidity chamber for a predetermined amount of time from 0 h to 1 d in the tear test.

Figure 3 shows the load and the displacement in the tear test of the PLA film that was annealed at a predetermined temperature under reduced pressure. The reference sample (as molded) was prepared under the same conditions as the sample marked 0 h in Figure 1. After annealing at 80 °C for 2 h, the maximum load did not change, but the amount of displacement until tearing occurred decreased. When the annealing temperature was 90 °C or higher, the maximum load increased, and the amount of displacement decreased, and the tearability of the PLA film deteriorated.

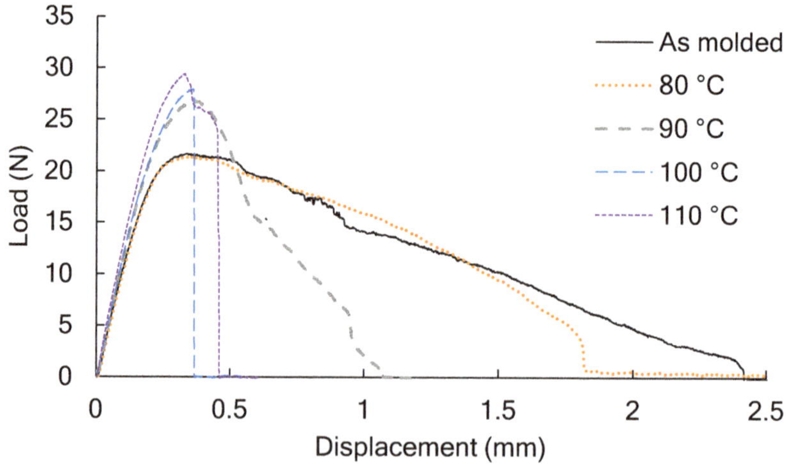

Figure 3. The load and the displacement of the PLA film annealed at a predetermined temperature from 80 to 110 °C under reduced pressure conditions in the tear test.

3.2. The Appearance of Films after the Tear Test

The produced amorphous PLA film was transparent, and no changes were observed in the transparency, even when moisture below the Tg was observed. On the other hand, when annealing was performed at the Tg or higher, the transparent film became opaque as the annealing temperature increased. Despite these differences in appearance, the elongation at the ligament was uniformly reduced in the tear test. Thus, the torn film was observed under a microscope to determine the differences in the fracture morphology.

Figure 4 shows traces of tearing in the film that was conditioned in the humidity chamber at 30 °C and at 95% RH. The ligament of the 0 h film was necked and stretched into a thin film, and it was then cut off. After conditioning in the humidity chamber at 30 °C and 95% RH for 1 h or more, black flaws occurred parallel to the tearing direction; that is, tearing occurred in the direction perpendicular to the tensile direction. This black flaw is considered to be a large crack. As the conditioning time increased, the thin-film portion formed by necking became smaller, and the 1 d film had a brittle fracture morphology.

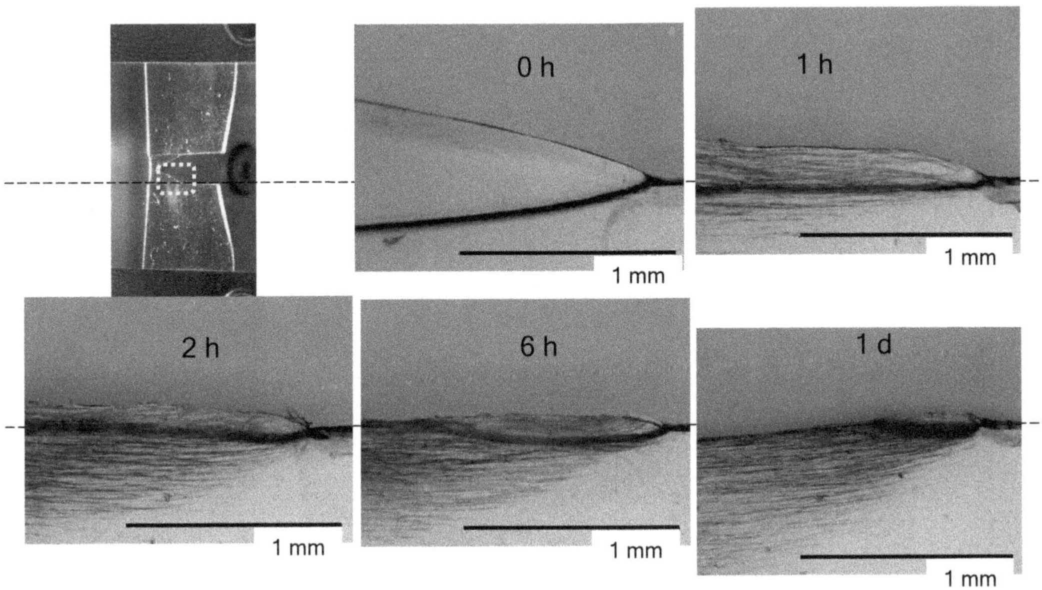

Figure 4. Photographs showing indications that there are tears in the film conditioned in a humidity chamber at 30 °C and 95% RH for a predetermined amount of time from 0 h to 1 d.

Figure 5 shows traces of tearing in the film annealed under reduced pressure conditions. After annealing at 80 °C for 2 h, no changes were observed in the traces of tearing compared to the mold film. Necking and stretching were observed in the thin film. When annealing was performed at temperatures of 90 °C or higher, no necking deformation was observed. In addition, the black flaws that were originally observed in Figure 4 did not appear.

3.3. Morphological Changes at the Tip of the Tear

The fracture morphology of the torn film changed significantly depending on the pretreatment conditions of the film. Because PLA is a semi-crystalline polymer, crystallization is proceeded by annealing. Thus, it is natural for brittle fractures to appear after annealing. In the case of the water-absorbing PLA, no changes were observed in the transparency in spite of the embrittlement. As such, it can be inferred that the crack growth mechanism is different in the amorphous and crystalline states. Thus, using 2-D retardation mapping, we analyzed the distribution of the orientation and the direction of the fast axis of the birefringence ellipsoid [21,22].

Figure 6 shows the distribution of the retardation at the tip of the tear and the direction of the high-speed axis of the birefringent ellipsoid, showing a displacement of approximately 0.4 mm. Figure 6a is a video capture image of the entire film. Necking occurred in the ligament, and it began to tear.

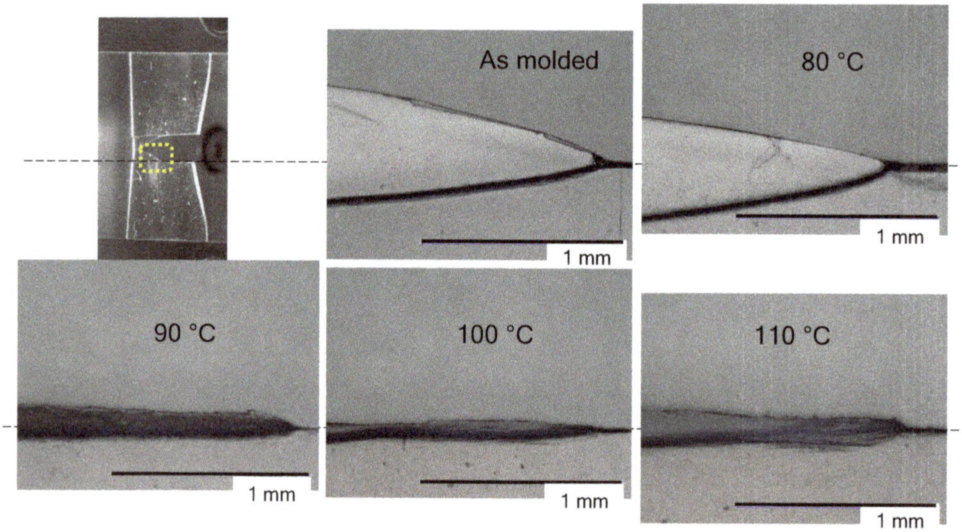

Figure 5. Photographs showing indications that there are tears in the film annealed at a predetermined temperature from 80 to 110 °C under reduced pressure.

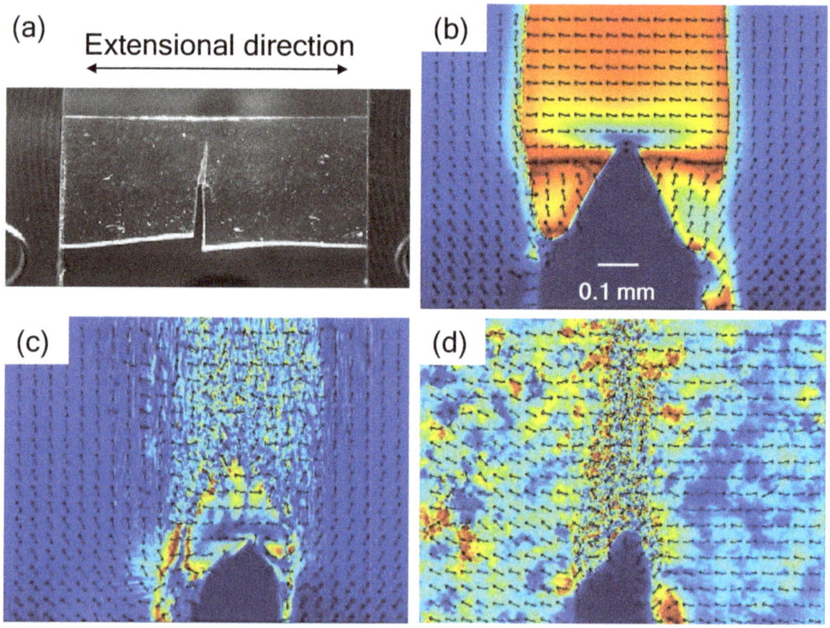

Figure 6. Images of the in-situ retardation measurements at the tip of the tear: (**a**) video capture, (**b**) 2-D mapping of the as-molded film, (**c**) 2-D mapping of the film-absorbed moisture after 6 h in a humidity chamber, (**d**) 2-D mapping of the film annealed at 100 °C for 2 h under reduced pressure conditions.

Figure 6b is the in-situ retardation mapping of the torn tip of the as-molded film during the tear test. The color change from ultramarine blue to dark red corresponds to the retardation changes taking place from 0 nm to 275 nm. When the film thickness is reduced by necking, the actual birefringence may be high due to the molecular orientation, but the observed retardation may be small due to a thinner thickness. Repeating retardation is also observed above 275 nm [22]. Although the film thickness was thin in the necking portion, the degree of retardation was large. That is, the stress was high during necking. The oriented direction coincided with the tensile direction. A local orientation perpendicular to the tensile direction was observed at the tip of the tear.

Figure 6c is the in-situ retardation mapping of the moisture that was absorbed by the film while in a humidity chamber for 6 h at 30 °C and at 95% RH. A striped pattern can be observed in the entire portion of the ligament that is perpendicular to the extension direction, and slight necking occurred at the tip of the tear. Furthermore, from the direction of the arrow indicating the high-speed axis, it can be seen that the oriented fragments are randomly spaced in the ligament.

Figure 6d shows the in-situ retardation mapping of the film that was annealed at 100 °C for 2 h under reduced pressure conditions. No uniform necking was observed in the ligament, and fragmentation occurred. As compared to (b) and (c), the orientation was parallel to the extensional direction at sites other than the ligament. From the comparison, it can be assumed that there is a relationship between the orientation perpendicular to the tension in (c) and the direction of the black flaws in Figure 4.

3.4. Changes in Heat Capacity at the Glass Transition Temperature

Generally, the crystallinity increases and the amorphous portion decreases as the semi-crystalline polymer becomes more embrittled. Thus, using DSC, the amount of MAm was calculated from the change in ΔCp at the Tg. Figure 7a shows the change in ΔCp near the Tg of the film-absorbed moisture at 30 °C and at 95% RH for a predetermined time. As shown in Figure 4, a brittle fracture was observed on the PLA film, but the ΔCp hardly changed, indicating that the amount of Mam was constant. On the other hand, as shown in Figure 7b, when annealing was performed under the reduced pressure conditions, the ΔCp decreased, indicating that the amount of MAm decreased. Thus, in the tear test, the higher-order structure of the film was different, even in the films showing the same brittle fracture.

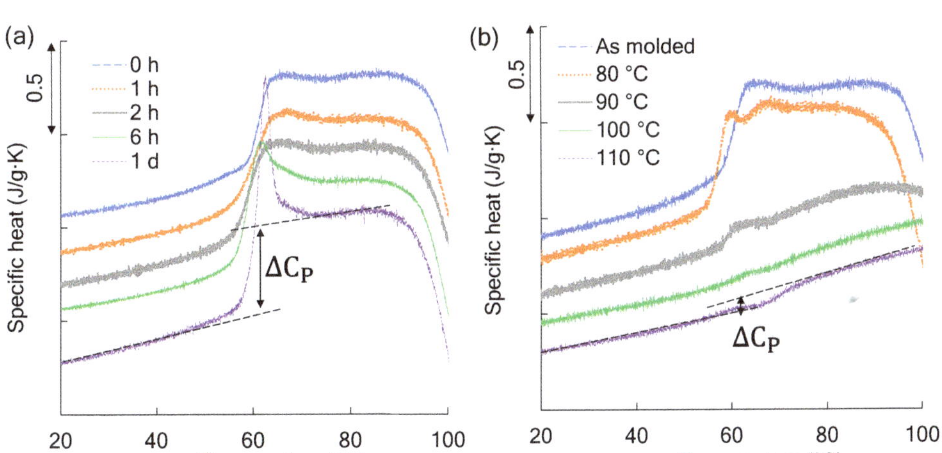

Figure 7. Changes in specific heat (ΔCp) near the glass transition temperature: (**a**) the film conditioned in a humidity chamber for a predetermined amount of time from 0 h to 1 d and (**b**) the film annealed at a predetermined temperature from 80 to 110 °C for 2 h under reduced pressure conditions.

3.5. Effect of Water Absorption on the Tear Test

When PLA is used in plastic bags, there is a problem in that the tear strength is lower than that of existing polyolefins. Therefore, it is necessary to understand the tear behavior of PLA film. As shown in Figures 2 and 3, the relationship between the displacement at the breaking point and the maximum load measured in the tear test is plotted in Figure 8. The as-molded film had a maximum load from 22 to 23 N and a displacement of 2 mm or more. When annealing was performed at the Tg temperature or higher under reduced pressure conditions, the maximum load increased to approximately 27 N, and the displacement decreased to approximately 0.5 mm. Via this annealing, the film changed from an amorphous state to a semi-crystalline state, with a crystallinity of 30% [19]. This increase in rigidity and the decrease in elongation associated with such crystallization are common.

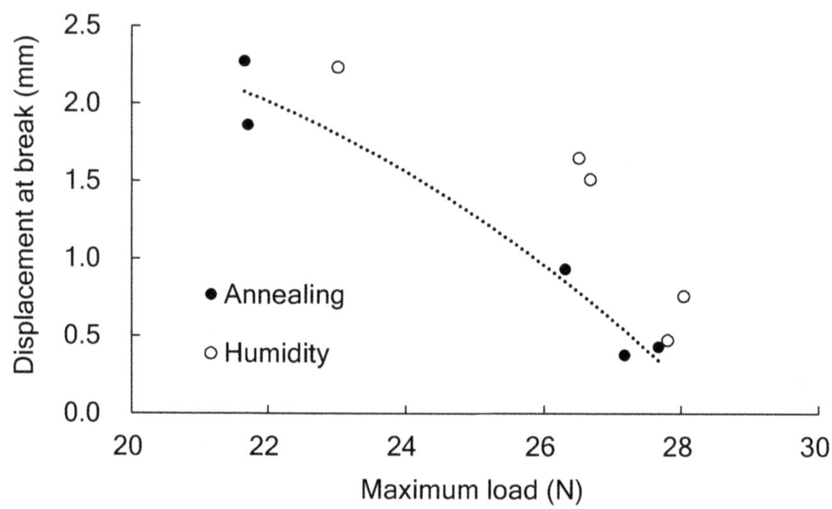

Figure 8. The relationship between the displacement at the breaking point and the maximum load measured in the tear test.

On the other hand, the maximum load and the displacement in the tear test changed in the same way as they did during crystallization by annealing when the PLA film was conditioned in a humidity chamber at 30 °C and 95% RH. Engineered general-purpose plastics such as polyamide 6 and polyethylene terephthalate (PET) lose their rigidity as they absorb more water because water molecules act as plasticizers [23,24]. However, PLA becomes more brittle under high humidity conditions, as shown in Figure 4. The reason for this is the enthalpy relaxation that is indicated in the ΔCp at the Tg shown in Figure 7a. Although PLA is embrittled by enthalpy relaxation [25,26], the effects of water on enthalpy relaxation are unclear. From this study, it was clarified that PLA embrittles within one hour under high humidity conditions. Polycarbonate (PC) is an example when considering the cause of the rapid progress of PLA enthalpy relaxation due to water. PC has high toughness, but the ductile-brittle transition occurs due to enthalpy relaxation. When the temperature of physical aging closely approaches Tg, the yield stress increases in a short time, and then the brittle fracture occurs [25]. Even if the temperature of physical aging is Tg or less, the molecular chains move at a higher temperature, and PC reaches the thermodynamically stable state in a short time. It is considered that the physical aging of PLA is promoted by increasing molecular motion due to water absorption even at a constant temperature.

3.6. Influence of Mobile Amorphous on the Fracture Generation

Figure 9 shows the relationship between the amount of MAm in the PLA film and the maximum load in the tear test. The enthalpy-relaxed film has a higher amount of

MAm than the film that was crystallized by annealing. Regardless of the differences in the amount of MAm, the film broke due to brittleness when the maximum load was about 27 N. Therefore, considering the practical use of PLA in plastic bags, no differences can be observed in the tearing characteristics regardless of the higher-order structure of PLA in the amorphous or semi-crystalline state. However, the behavior of fracture at the tip of the tear is very different between the two. In Figure 6d, the extensional direction and the orientation of the film coincide with each other. Generally, molecular chains or lamella crystals are arranged in the extensional direction. However, Figure 6c shows how the orientation is perpendicular to the extensional direction. In this case, the retardation that was generated in the film was very low, so the orientation of the amorphous molecular chains was small. Since this orientation coincides with the direction of the flaws and cracks generated around the ligament, it may indicate a pre-fracture stage such as the widening of the spaces between amorphous molecular chains.

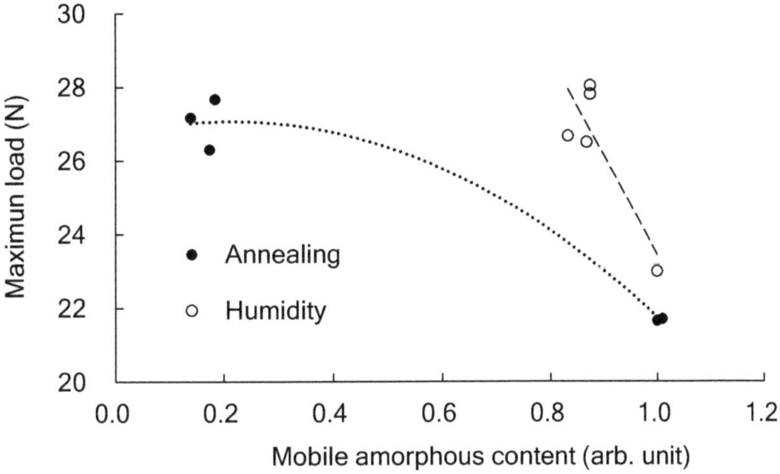

Figure 9. The relationship between the mobile amorphous amount (MAm) in the PLA film and the maximum load in the tear test.

4. Conclusions

The tear test was carried out at room temperature, which is lower than the Tg of PLA, meaning that the elongation in the break is smaller than that of polyolefins, which has a Tg that is lower than the test temperature. However, when using PLA as a packaging material, the essential problem is that the film ligament demonstrates little necking activity due to the embrittlement caused by crystallization or enthalpy relaxation. Therefore, in order to strengthen PLA, a polymer blend using a plasticizer that modifies the amorphous part and a biodegradable polymer with a low Tg that will strengthen the semi-crystalline part has been studied [7,11–13]. For polystyrene and ABS resin, which are synthetic polymers, techniques for modifying them using elastomers have been accumulated. However, a biodegradable elastomer with a high-impact modification performance has yet to be found. From this study, copolymers or a miscible polymer blend that slow down the enthalpy relaxation of PLA are also recommended as toughening techniques.

Author Contributions: Conceptualization, H.I.; methodology, Y.K.; formal analysis, Y.K.; writing—original draft preparation, Y.K.; writing—review and editing, A.I. and H.I.; supervision, H.I. All authors have read and agreed to the published version of the manuscript.

Funding: This work is based on results obtained from a project, JPNP18016, commissioned by the New Energy and Industrial Technology Development Organization (NEDO).

Institutional Review Board Statement: Not applicable.

Informed Consent Statement: Not applicable.

Data Availability Statement: The data presented in this study are available on request from the corresponding author.

Acknowledgments: The authors thank Y. Yamada for his experimental support.

Conflicts of Interest: The authors declare no conflict of interest.

References

1. Masry, M.; Rossignol, S.; Gardette, J.-L.; Therias, S.; Bussière, P.-O.; Wong-Wah-Chung, P. Characteristics, Fate, and Impact of Marine Plastic Debris Exposed to Sunlight: A Review. *Mar. Pollut. Bull.* **2021**, *171*, 112701. [CrossRef] [PubMed]
2. Freeland, B.; McCarthy, E.; Balakrishnan, R.; Fahy, S.; Boland, A.; Rochfort, K.D.; Dabros, M.; Marti, R.; Kelleher, S.M.; Gaughran, J. A Review of Polylactic Acid as a Replacement Material for Single-Use Laboratory Components. *Materials* **2022**, *15*, 2989. [CrossRef]
3. Napper, I.E.; Thompson, R.C. Environmental Deterioration of Biodegradable, Oxo-Biodegradable, Compostable, and Conventional Plastic Carrier Bags in the Sea, Soil, and Open-Air Over a 3-Year Period. *Environ. Sci. Technol.* **2019**, *53*, 4775–4783. [CrossRef] [PubMed]
4. Manfra, L.; Marengo, V.; Libralato, G.; Costantini, M.; De Falco, F.; Cocca, M. Biodegradable Polymers: A Real Opportunity to Solve Marine Plastic Pollution? *J. Hazard. Mater.* **2021**, *416*, 125763. [CrossRef] [PubMed]
5. Wang, G.-X.; Huang, D.; Ji, J.-H.; Völker, C.; Wurm, F.R. Seawater-Degradable Polymers—Fighting the Marine Plastic Pollution. *Adv. Sci.* **2021**, *8*, 2001121. [CrossRef]
6. Oyama, H.T.; Abe, S. Stereocomplex Poly(Lactic Acid) Alloys with Superb Heat Resistance and Toughness. *ACS Sustain. Chem. Eng.* **2015**, *3*, 3245–3252. [CrossRef]
7. Nofar, M.; Sacligil, D.; Carreau, P.J.; Kamal, M.R.; Heuzey, M.-C. Poly (Lactic Acid) Blends: Processing, Properties and Applications. *Int. J. Biol. Macromol.* **2019**, *125*, 307–360. [CrossRef]
8. Tsuji, H. Poly(Lactic Acid) Stereocomplexes: A Decade of Progress. *Adv. Drug Deliv. Rev.* **2016**, *107*, 97–135. [CrossRef]
9. Gupta, P.; Wilkes, G.L.; Sukhadia, A.M.; Krishnaswamy, R.K.; Lamborn, M.J.; Wharry, S.M.; Tso, C.C.; DesLauriers, P.J.; Mansfield, T.; Beyer, F.L. Does the Length of the Short Chain Branch Affect the Mechanical Properties of Linear Low Density Polyethylenes? An Investigation Based on Films of Copolymers of Ethylene/1-Butene, Ethylene/1-Hexene and Ethylene/1-Octene Synthesized by a Single Site Metallocene Catalyst. *Polymer* **2005**, *46*, 8819–8837. [CrossRef]
10. Guichon, O.; Séguéla, R.; David, L.; Vigier, G. Influence of the Molecular Architecture of Low-Density Polyethylene on the Texture and Mechanical Properties of Blown Films. *J. Polym. Sci. Part B Polym. Phys.* **2003**, *41*, 327–340. [CrossRef]
11. Zhao, X.; Hu, H.; Wang, X.; Yu, X.; Zhou, W.; Peng, S. Super Tough Poly(Lactic Acid) Blends: A Comprehensive Review. *RSC Adv.* **2020**, *10*, 13316–13368. [CrossRef]
12. Hamad, K.; Kaseem, M.; Ayyoob, M.; Joo, J.; Deri, F. Polylactic Acid Blends: The Future of Green, Light and Tough. *Prog. Polym. Sci.* **2018**, *85*, 83–127. [CrossRef]
13. Rasal, R.M.; Janorkar, A.V.; Hirt, D.E. Poly(Lactic Acid) Modifications. *Prog. Polym. Sci.* **2010**, *35*, 338–356. [CrossRef]
14. Asadollahzadeh, M.; Mahboubi, A.; Taherzadeh, M.J.; Åkesson, D.; Lennartsson, P.R. Application of Fungal Biomass for the Development of New Polylactic Acid-Based Biocomposites. *Polymers* **2022**, *14*, 1738. [CrossRef]
15. Gigante, V.; Bosi, L.; Parlanti, P.; Gemmi, M.; Aliotta, L.; Lazzeri, A. Analysis of the Damage Mechanism around the Crack Tip for Two Rubber-Toughened PLA-Based Blends. *Polymers* **2021**, *13*, 53. [CrossRef]
16. Aliotta, L.; Vannozzi, A.; Canesi, I.; Cinelli, P.; Coltelli, M.-B.; Lazzeri, A. Poly(Lactic Acid) (PLA)/Poly(Butylene Succinate-Co-Adipate) (PBSA) Compatibilized Binary Biobased Blends: Melt Fluidity, Morphological, Thermo-Mechanical and Micromechanical Analysis. *Polymers* **2021**, *13*, 218. [CrossRef]
17. Rocca-Smith, J.R.; Whyte, O.; Brachais, C.-H.; Champion, D.; Piasente, F.; Marcuzzo, E.; Sensidoni, A.; Debeaufort, F.; Karbowiak, T. Beyond Biodegradability of Poly(Lactic Acid): Physical and Chemical Stability in Humid Environments. *ACS Sustain. Chem. Eng.* **2017**, *5*, 2751–2762. [CrossRef]
18. Limsukon, W.; Auras, R.; Smith, T. Effects of the Three-Phase Crystallization Behavior on the Hydrolysis of Amorphous and Semicrystalline Poly(Lactic Acid)S. *ACS Appl. Polym. Mater.* **2021**, *3*, 5920–5931. [CrossRef]
19. Kobayashi, Y.; Ueda, T.; Ishigami, A.; Ito, H. Changes in Crystal Structure and Accelerated Hydrolytic Degradation of Polylactic Acid in High Humidity. *Polymers* **2021**, *13*, 4324. [CrossRef]
20. Henricks, J.; Boyum, M.; Zheng, W. Crystallization Kinetics and Structure Evolution of a Polylactic Acid during Melt and Cold Crystallization. *J. Therm. Anal. Calorim.* **2015**, *120*, 1765–1774. [CrossRef]
21. Uetani, K.; Koga, H.; Nogi, M. Estimation of the Intrinsic Birefringence of Cellulose Using Bacterial Cellulose Nanofiber Films. *ACS Macro Lett.* **2019**, *8*, 250–254. [CrossRef]
22. Sato, T.; Araki, T.; Sasaki, Y.; Tsuru, T.; Tadokoro, T.; Kawakami, S. Compact Ellipsometer Employing a Static Polarimeter Module with Arrayed Polarizer and Wave-Plate Elements. *Appl. Opt.* **2007**, *46*, 4963–4967. [CrossRef]
23. Jabarin, S.A.; Lofgren, E.A. Effects of Water Absorption on Physical Properties and Degree of Molecular Orientation of Poly (Ethylene Terephthalate). *Polym. Eng. Sci.* **1986**, *26*, 620–625. [CrossRef]

24. Rubin, J.; Andrews, R.D. Effect of Solvent Treatments on the Mechanical Properties of Nylon 6. *Polym. Eng. Sci.* **1968**, *8*, 302–309. [CrossRef]
25. Ohara, A.; Kodama, H. Correlation between Enthalpy Relaxation and Mechanical Response on Physical Aging of Polycarbonate in Relation to the Effect of Molecular Weight on Ductile-Brittle Transition. *Polymer* **2019**, *181*, 121720. [CrossRef]
26. Pan, P.; Zhu, B.; Inoue, Y. Enthalpy Relaxation and Embrittlement of Poly(l-Lactide) during Physical Aging. *Macromolecules* **2007**, *40*, 9664–9671. [CrossRef]

polymers

MDPI

Article

Amelioration of Obesity in Mice Fed a High-Fat Diet with Uronic Acid–Rich Polysaccharides Derived from *Tremella fuciformis*

Chun-Hui Chiu [1,2], Kai-Chu Chiu [3] and Li-Chan Yang [4,*]

1 Graduate Institute of Health Industry and Technology, Research Center for Chinese Herbal Medicine and Research Center for Food and Cosmetic Safety, College of Human Ecology, Chang Gung University of Science and Technology, Taoyuan City 333, Taiwan; chchiu@mail.cgust.edu.tw
2 Department of Traditional Chinese Medicine, Chang Gung Memorial Hospital, Keelung 204, Taiwan
3 Master Program for Pharmaceutical Manufacture, China Medical University, Taichung 406, Taiwan; pto2398gq141@gmail.com
4 Department of Pharmacy, School of Pharmacy, China Medical University, Taichung 406, Taiwan
* Correspondence: yang@mail.cmu.edu.tw; Tel.: +886-4-22053366; Fax: +886-4-22053764

Abstract: Obesity is rapidly becoming an emerging disease in developing countries due to the Westernization of societies and lifestyle changes. This study evaluated the ameliorative effect of acidic heteropolysaccharides derived from *Tremella fuciformis* (TFPS) on high-fat diet (HFD; 34.9% fat)-induced obesity in mice. The TFPS exhibited high uronic acid content and high viscosity in water. The structural characteristics of TFPS showed that average molecular weight was 679 kDa, and the monosaccharide composition was galactose, glucose, fructose, xylose, fucose, and mannose at a ratio of 1.0:6.5:10.0:18.5:30.5:67.5. In an in vivo study, HFD-induced obese C57BL/6 mice were orally given a TFPS treatment at 1 and 2 g/kg of body weight for 8 weeks. The TFPS treatment significantly reduced features of obesity in the mice, namely weight gain, feed efficiency, body fat percentage, and serum cholesterol level and increased fecal lipid content, compared with mice fed an HFD with water. In addition, TFPS exhibited the inhibition of cholesterol micelles in vitro in a concentration-dependent manner. In conclusion, the TFPS treatment ameliorated the diet-induced obesity in the mice, presumably reducing fat absorption in the intestine by interfering with viscous TFPS.

Keywords: high-fat diet; obesity; polysaccharide; *Tremella fuciformis*; uronic acid; viscosity

Citation: Chiu, C.-H.; Chiu, K.-C.; Yang, L.-C. Amelioration of Obesity in Mice Fed a High-Fat Diet with Uronic Acid–Rich Polysaccharides Derived from *Tremella fuciformis*. *Polymers* **2022**, *14*, 1514. https://doi.org/10.3390/polym14081514

Academic Editors: Antonio M. Borrero-López, Concepción Valencia-Barragán, Esperanza Cortés Triviño, Adrián Tenorio-Alfonso and Clara Delgado-Sánchez

Received: 1 March 2022
Accepted: 6 April 2022
Published: 8 April 2022

1. Introduction

Obesity is a condition in which the body is over deposited due to changes in physiological or biochemical functions, resulting in weight gain [1]. Obesity is rapidly becoming an emerging disease in developing countries due to the Westernization of societies and lifestyle changes [2,3]. Overweight and obesity are also world problems. According to the World Health Organization, approximately 1 billion people are overweight or obese [2]. Obesity, classified in terms of the body mass index and waist–hip ratio, has several associated comorbidities, namely diabetes mellitus, hypertension, degenerative osteoarthritis, and infertility [3]. Along with insufficient physical activity, the Westernization of diets has contributed to the global obesity epidemic. Fruits, vegetables, and whole grains are being replaced by readily accessible foods high in saturated fat, sugar, and refined carbohydrates [2]. An increased intake of dietary fiber appears to be effective in treating obesity [4]. Fiber-rich food is usually satisfying without being calorically dense [5]. Smith et al. [6] have reported that the supplementing a normal diet with gel-forming fibers increases satiation probably by slowing gastric emptying. Apart from a beneficial effect during caloric restriction, dietary fiber may improve some of the metabolic aberrations seen in obesity [5,6].

Tremella fuciformis, known as silver ear mushroom or white jelly fungus, is an edible basidiomycetous jelly mushroom that has been favored in Asia for centuries. The edible part of *T. fuciformis* is transparent to white with a jelly-like taste and is full of water soluble and indigestible polysaccharides. Polysaccharides are regarded as the major bioactive components of *T. fuciformis*, and studies have reported their bio-activities, such as humoral immune responses, antitumor, hypoglycernic, and hypolipidemic activities [7]. The structural characteristics of the polysaccharides derived from *T. fuciformis* contain glucuronoxylomannans, which consist of a linear (1 → 3)-linked-α-d-mannose backbone, mainly with β-d-xylose and β-d-glucuronic acid in the side chains [8,9]. A study [10] reported that these polysaccharides inhibited adipocyte differentiation through suppressing the expression of PPAR gamma, C/EBP alpha, and leptin; however, no study has investigated whether *T. fuciformis* polysaccharide supplements have anti-obesity effects in vivo. Therefore, this study examined the anti-obesity effects of the acidic heteropolysaccharides derived from *T. fuciformis* in a Western diet (high-fat diet) in mice.

2. Materials and Methods

2.1. Preparation of T. fuciformis Polysaccharides

The fruiting bodies of *T. fuciformis* were obtained from Well Youth Biotech Co., Ltd. (Taichung, Taiwan). Sawdust was used as the main cultivation substrate. The moisture content in the fresh fruiting bodies was 89–93%. To prepare the polysaccharides referred to by Yang et al., previously published [11], the fruiting bodies were cut into pieces and extracted by 10-fold of distilled water by weight at 80 °C for 4 h. Filtration was applied to remove debris, and 4-fold volume of 95% ethanol at 4 °C was added to the filtrate for 12 h to precipitate crude polysaccharides. A commercial total dietary fiber assay kit (K-TDFR, Megazyme, Ireland) was applied to the crude polysaccharides for removing the starch and proteins and obtained the indigestible *T. fuciformis* polysaccharide (TFPS) according to the AOAC 993.19 method.

2.2. Characterization of Indigestible TFPS

The TFPS were purified and fractionated using anion-exchange chromatography on a DEAE-650M column (Tosho, Tokyo, Japan). The DEAE column was eluted with 20 mM of Tris–HCl (pH 7.8), followed by a sodium chloride gradient (0–0.33 M). The carbohydrate elution profiles of the TFPS were analyzed according to the phenol-sulfuric acid method, with glucose as the standard, and measured at 490 nm [12]. In a chemical analysis, the protein content of the elution was measured using a modified Bradford method assay with bovine serum albumin as the standard [13], whereas the carbohydrate content of the elution was determined using the phenol-sulfuric acid method [12]. The content of uronic acid was determined using the m-hydroxydiphenyl method, with glucuronic acid as the standard [14]. In a nuclear magnetic resonance (NMR) analysis, the major fraction from the DEAE elution (5 mg) was dissolved in deuterium oxide (0.5 mL), and ^{1}H-NMR spectra were recorded on a Bruker DRX-600 spectrometer (Bruker BioSpin, Billerica, MA, USA) at 20 °C using 3-(trimethylsilyl)-propionic 2,2,3,3,-d4 acid sodium salt as an internal reference (δ 0.00 ppm). The relative molecular mass (Mr) was determined using high performance-size exclusion chromatography (HP-SEC), using a TSKgel PWHguard column with PWH (75 mm × 7.5 mm i.d.; TOSOH, Tokyo, Japan), TSKgel G4000PWXL, and G2500PWXL (300 mm × 7.8 mm i.d., TOSOH) columns connected in series. The columns were eluted with 0.3 N of a sodium nitrate solution containing 0.02% sodium azide at a flow rate of 0.8 mL/min at 65 °C. Peaks were detected using an interferometric refractometer (Wyatt Technology, CA, USA). The average Mr was estimated by comparison with the retention time of the pullulan standard P-82 kit (molecular weights of standards: P-800, 708 kDa; P-400, 375 kDa; P-200, 200 kDa; P-100, 107 kDa; P-50,47.1 kDa; P-20, 21.1 kDa; P-10, 9.6 kDa; and P-5, 5.9 kDa; Shodex, Kawasaki, Japan) and blue dextran 2000 (GE Healthcare, Uppsala, Sweden). The determination of molecular weights was also analyzed by dynamic light scattering (nanoPartica SZ-100V2, HORIBA, Kyoto, Japan). In a monosaccharide composition analysis, samples were hydrolyzed using a methanolysis method [15] for uronic acids analysis and trifluoroacetic acid (TFA) hydrolysis for neutral sugars. The resulting sample was analyzed using high-performance anion-exchange chromatography with pulsed amperometric detection (HPAEC-PAD). The peaks were detected using an 817 Bioscan PAD detector (Metrohm, Zofin-gen, Switzerland) and a CarboPac PA1 column (Dionex, Sunnyvale, CA, USA). The eluent consisted of 10 mM of sodium hydroxide and 1 mM of barium acetate (Sigma–Aldrich, MO, USA), pumped at a flow rate of 0.5 mL/min.

2.3. Determination of Viscosity and Inhibition on Cholesterol Micellization

Various concentrations of the TFPS were dissolved in distilled water (0–1.5%), and the viscosity was measured using a Sine-wave Vibro Viscometer SV-10 (A&D Co., Ltd., Tokyo, Japan) at a constant frequency of 30 Hz. The sample volume used in this experiment was 35 mL for each sample, measured at 25 °C, 37 °C, 50 °C, and 75 °C, respectively. As for the determination of inhibition on cholesterol micellization, a 15 mM phosphate buffered saline (PBS) (pH 7.4) containing 0.5 mM cholesterol, 4.8 mM lecithin, and 13.2 mM taurocholic acid was treated under ultrasonication at 37 °C for 24 h to prepare a cholesterol micellar solution [16]. In addition, 0.5 mL of TFPS solution in various concentrations (0–1.5%, *w*/*v*) was mixed with equal volume cholesterol micellar and incubated at 37 °C for 1 h. After incubation, the mixtures were centrifuged at 3000 rpm for 10 min and the supernatant was passed through a Millex-GP Syringe filter (0.22 μm, Millipore, Darmstadt, Germany), and the cholesterol in the filtrate was measured by a commercial cholesterol analysis kit (Fortress diagnostics, Antrim, UK). The inhibition of micellization was calculated by comparing the cholesterol level of the control group (TFPS 0 μg/mL) with the cholesterol level while the solution contained TFPS.

2.4. Animals and Experimental Design

The experimental animals received humane care, and the study protocols complied with the institutional guidelines of China Medical University for the use of laboratory animals (protocol No.: CMUIACUC-2017-286). The animals were housed in an air-conditioned room (21–24 °C) under 12 h of light (8:00 a.m.–8:00 p.m.) and were given free access to food pellets and water throughout the study. Eight-week-old male C57BL/6 mice were purchased from the National Laboratory Animal Center (Taipei, Taiwan). The mice were acclimated for 1 week and randomly divided into four groups ($n = 12$); normal diet (ND), high-fat diet (HFD) (control), HFD TFPS (low dose; HFD-TL), and HFD TFPS (high dose; HFD-TH). The ND mice were fed a maintenance diet (2.85 kcal/g, Altromin 1320, Altromin Spezialfutter GmbH & Co. KG, Im Seelenkamp, Germany), whereas the HFD mice were fed an HFD (D12492, Research Diet Inc., Brunswick, NJ, USA), which consisted of 34.9% fat and 5.24 kcal/g. The ND and HFD (control) mice were orally given distilled water (10 mL/kg body weight), and the HFD-TL and HFD-TH mice were given the TFPS at 1 and 2 g/kg of body weight, respectively, every day for 8 weeks using oral gavage. Body weight and food intake were measured twice a week, and unconsumed food was discarded. The formula for calculating calorie efficiency was as follows:

$$\text{kcal efficiency} = (\text{weight gain} \div \text{kcal intake}) \times 100\%$$

The feed efficiency was calculated as follows:

$$\text{Feed efficiency} = (\text{weight gain} \div \text{food intake}) \times 100\%$$

After 8 weeks of treatment, fresh feces were collected and stored at −20 °C until the fecal lipid content was analyzed. The mice were fasted overnight and sacrificed. Blood samples were collected and then centrifuged at 2000× *g* for 15 min for serum collection. Gonadal, retroperitoneal, and mesenteric fat were removed and weighed. The sum of the fat was the total body weight. The formula for calculating the body fat percentage was as follows:

$$\text{Body fat percentage} = (\text{total body fat} \div \text{body weight}) \times 100\%$$

2.5. Serum Biochemistry Analysis

Levels of serum triacylglycerol (TG) were determined using a commercial kit (Roche, Indianapolis, IN, USA) and measured using an Automatic Biochemical Analyzer (COBAS MIRA PLUS, Roche, Indianapolis, IN, USA). Serum total cholesterol (T-Cho), high-density lipoprotein cholesterol (HDL-C), and low-density lipoprotein cholesterol (LDL-C) were determined using the peroxidase-antiperoxidase method and a commercial kit (BXC0262, BXC0442, BXC0432; Fortress Diagnostics Limited, UK). Non-esterified fatty acid (NEFA) content in the serum was determined using an assay kit (FA115, Randox, Antrim, UK), as per the manufacturer's instructions.

2.6. Histology of the Gonadal Adipose Tissue

Gonadal adipose tissue was fixed with neutral 10% formaldehyde buffer and embedded in paraffin. Sections (5 μm) were stained with hematoxylin and eosin. The diameter of the adipose cells was scored and analyzed using Image-Pro (Plus 6, Media Cybernetics, Rockville, MD, USA).

2.7. Fecal Lipid Content Determination

After TFPS treatment for 8 weeks, all the mice were housed in individual metabolism cages containing a grid-floor and a facility for separate collection of feces for 3 days. Food consumption was also monitored on a daily basis when mice were in metabolism cages. The method for fecal lipid determination was as Kraus et al. [17] described before. Fecal samples of each animal were collected and weighted. Briefly, 1000 mg feces per mice were powderized using the tissue grinder, and then 5 mL of normal saline was added to 1000 mg of powderized feces in and vortexed. Next, 5 mL of Chloroform in methanol (2:1 by volume) was added to extract the lipids. The suspensions were centrifuged at $1000\times g$ for 10 min at room temperature and then the chloroform/methanol phase was removed carefully to a weighted glass tube for evaporation under a fume hood. Then the tubes were weighted to obtain the lipid mass per 1000 mg of feces. To obtain the mouse's average fecal lipid excretion per day, the lipid mass was multiplied by the total weight of feces in grams.

2.8. Statistical Analyses

The results were expressed as mean ± standard error. The statistical significance was evaluated using one-way analysis of variance (ANOVA), followed by Dunnett's post hoc test using SPSS/11.5 software. Values were considered statistically significant at $p < 0.05$.

3. Results

3.1. Preparation and Characterization of the TFPS

The yield of the crude water-soluble polysaccharides was 60.1% ± 1.2% on a dry basis of *T. fuciformis*. TFPS was 95.2% ± 2.3% to the crude water-soluble polysaccharides. The TFPS consisted of 98.3% carbohydrates and less than 0.5% protein. The uronic acid content was 26.7% to TFPS on a carbohydrate basis.

The TFPS were fractioned using the DEAE-650M column to obtain the major fraction of the contained acidic polysaccharides (98%), which were collected according to the total carbohydrate elution profiles (Figure 1). The uronic acid content was determined from the elution; the results revealed that the TFPS were acidic polysaccharides rich in uranic acid (26.7% to total sugars). The ^{1}H NMR (500 MHz) revealed the structural characteristics of the TFPS, as shown in Figure 2. A strong signal observed at 4.79 ppm corresponded to D_2O, whereas signals at δ 5.12 and 5.49 ppm were consistent with the presence of α-d-annopyranose. A signal at δ 4.37 ppm potentially corresponded to β-d-xylopyranose.

The signal for β-d-glucopyranuronic acid was not detected due to an overlap with the HOD peak [18]. Moreover, an obvious signal at δ 2.11 ppm corresponded to the CH_3 moiety of the acetyl group [18]. Signals at δ 1.09~1.17 ppm potentially corresponded to the H2 of α-d-fucose, a deoxysugar. According to the calibration curves derived from the pullulan standards, the results of the HP-SEC analysis revealed that the average molecular weight of the TFPS was 679 kDa and dynamic light scattering analysis showed that the molecular weight of the TFPS was 736 kDa. The HPAEC-PAD analysis of the monosaccharide composition revealed that the TFPS were composed of galactose, glucose, fructose, xylose, fucose, mannose, and glucuronic acid at a ratio of 0.5: 3.6:5.5:10.1:16.7:36.9:26.7 (Supplementary Figure S1).

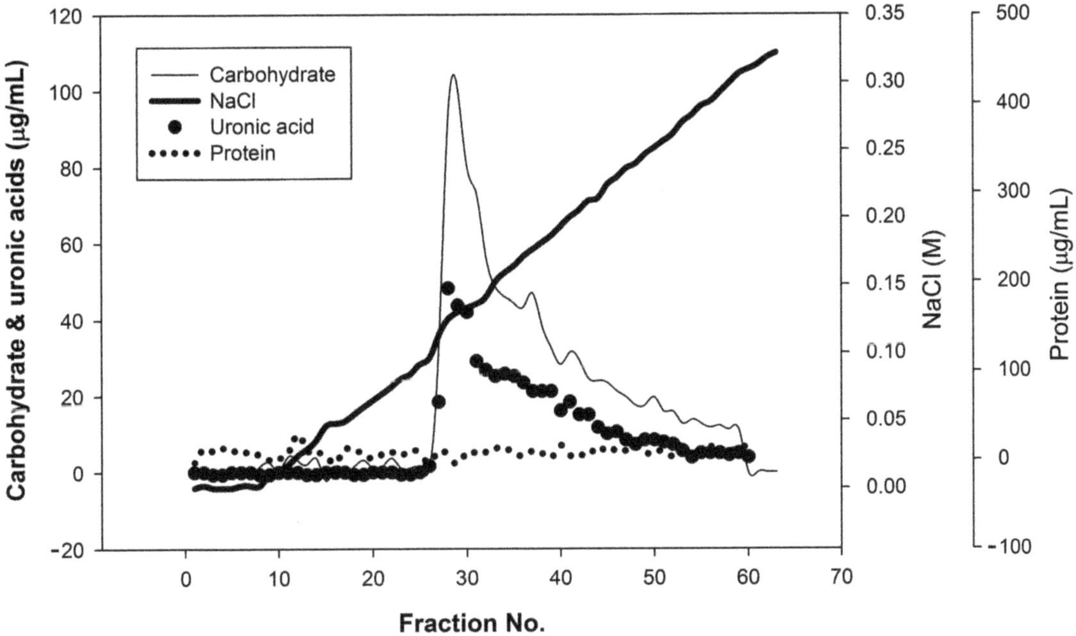

Figure 1. Anion-exchange chromatogram of water-soluble polysaccharides from *Tremella fuciformis* (TFPS) on a DEAE-650M column and eluted with 20 mM tris buffer (pH 7.8), followed by a sodium chloride gradient (0–0.32 M); the carbohydrate, uronic acid and protein contents were determined through a phenol–sulfuric acid method, *m*-hydroxydiphenyl method and Bradford method, respectively.

3.2. Polysaccharide Viscosity and Effect on Cholesterol Micellization

The relationship between the TFPS concentration and viscosity in an aqueous solution is presented in Figure 3A. The viscosity of the TFPS solution increased as the concentration increased; additionally, the viscosity decreased in equivalent concentration of TFPS solution when the temperature increased. The viscosity of a 1.0% solution was approximately 88.3, 78.5, and 54.3 mPa.S at 25, 50, and 75 °C, respectively. The TFPS aqueous solution at a concentration of 1.8% became gelatinous without fluidity at 25 °C. In addition, the cholesterol micellization was inhibited while TFPS was added to cholesterol micelles solution in a dose-dependent manner (Figure 3B). When TFPS concentration increased, the date of viscosity concurred with the inhibition of cholesterol micellar.

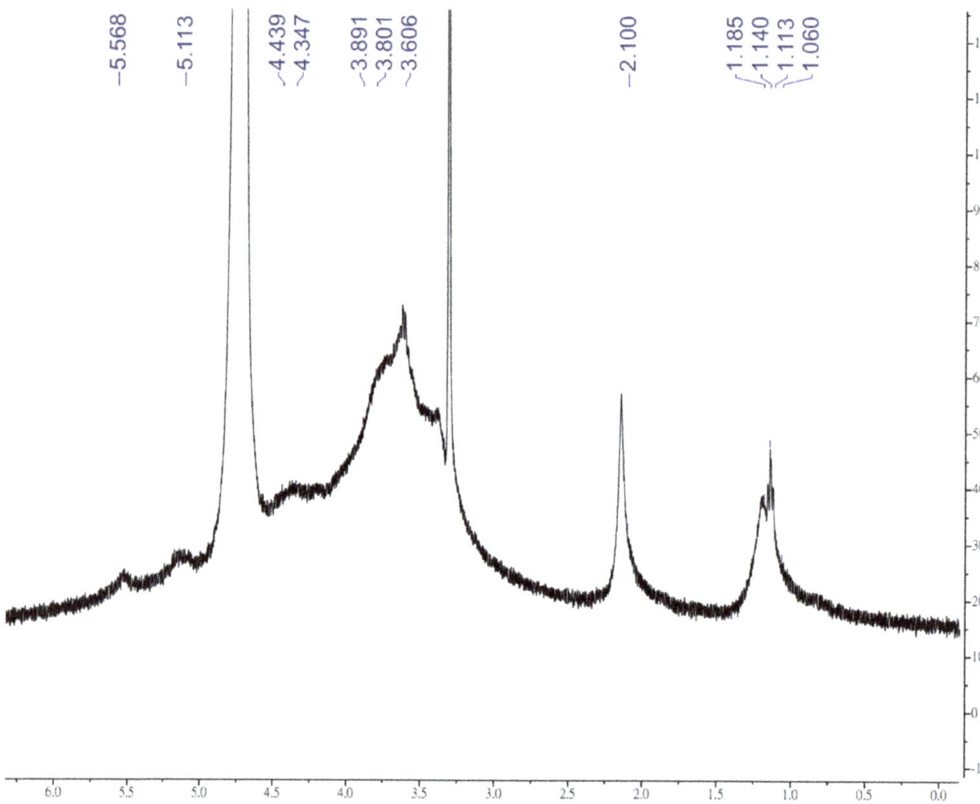

Figure 2. ^{1}H NMR spectra of polysaccharides from *Tremella fuciformis* (TFPS).

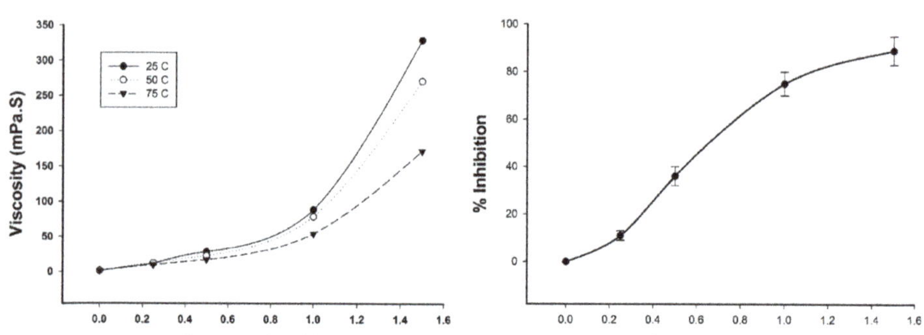

Figure 3. (**A**) Viscosity of aqueous solutions of polysaccharides from *Tremella fuciformis* (TFPS) in water at 25 °C, 37 °C, 50 °C, and 75 °C. (**B**) Effect of TFPS on the solubility of cholesterol micellar in vitro (n = 3).

3.3. Body Weight, Food Intake, and Tissue Weight

No differences were observed in the body weight of the mice among the groups at the start of the experiment. As shown in Figure 4, the body weight of the mice in the HFD groups was significantly higher than that of the mice in the ND group at day 7. Moreover, the body weight of the mice in the HFD-TL and HFD-TH groups was significantly lower than that of the mice in the control group from day 21 to the end of the experiment. No significant difference in body weight was observed between the HFD-TL and HFD-TH groups. Table 1 presents the food and calorie intake among the groups during the experiment. As shown, no significant differences were observed among the groups, even between the HFD and ND groups. Table 2 presents the weight gain and feed efficiency of the HFD and ND groups. The weight gain of the mice in the control group increased significantly compared with the mice in the ND group. The TFPS treatment significantly reduced the weight gain of the mice compared with the mice in the control group, which were also fed the HFD. No notable difference in total food intake was observed; however, the feed efficiency in the control group was 5.2-fold higher than that in the ND group. Moreover, the feed efficiency in the HFD-TL and HFD-TH groups was 0.71- and 0.65-fold higher than that in the control group, respectively.

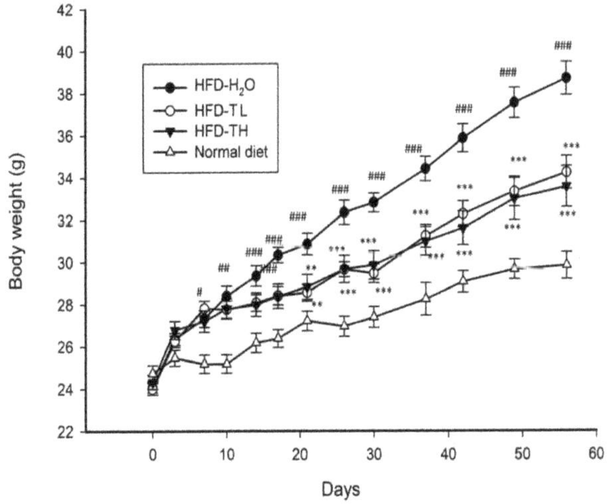

Figure 4. Body weight of mice in different groups. # $p < 0.05$, ## $p < 0.01$, and ### $p < 0.001$ while high-fat diet (HFD)-H_2O group compared with the normal diet-H_2O group. ** $p < 0.01$ and *** $p < 0.001$, while HFD-TFPS group compared with the HFD-H_2O group. HFD-TL and HFD-TH groups received oral administration of 1 and 2 g/kg TFPS, respectively.

At week 8, the weights of absolute adipose tissue were markedly greater in the HFD groups than in the ND group. Figure 5 shows that the weights of the gonadal, perirenal, and adipose tissues decreased significantly in both the TFPS-treated groups compared with the HFD-H_2O group. The TFPS treatment exhibited effects in reducing the fat weight and body weight percentages (Figure 5).

Table 1. Food and calorie intake profiles at Week 8.

Weeks	Food Intake (g/day)			
	Normal Diet	High-Fat-Diet	High-Fat-Diet	High-Fat-Diet
	H_2O	H_2O	TFPS 1 g/kg	TFPS 2 g/kg
Week 1	4.1 ± 0.5	2.2 ± 0.1	2.4 ± 0.1	2.4 ± 0.2
Week 2	4.1 ± 0.2	2.3 ± 0.1	2.3 ± 0.1	2.2 ± 0.0
Week 3	4.5 ± 0.6	2.3 ± 0.1	2.2 ± 0.1	2.2 ± 0.1
Week 4	4.1 ± 0.4	2.3 ± 0.2	2.2 ± 0.1	2.2 ± 0.3
Week 5	4.1 ± 0.2	2.5 ± 0.1	2.3 ± 0.1	2.4 ± 0.3
Week 6	4.8 ± 0.8	2.5 ± 0.1	2.4 ± 0.2	2.3 ± 0.2
Week 7	3.2 ± 0.2	2.5 ± 0.2	2.4 ± 0.1	2.3 ± 0.1
Week 8	4.5 ± 0.2	2.5 ± 0.0	2.4 ± 0.1	2.3 ± 0.2

Table 1. *Cont.*

Weeks	Food Intake (g/day)			
	Normal Diet	High-Fat-Diet	High-Fat-Diet	High-Fat-Diet
	H_2O	H_2O	TFPS 1 g/kg	TFPS 2 g/kg
	Kcal intake (kcal/day)			
Week 1	11.8 ± 0.8	11.7 ± 0.2	12.4 ± 0.3	12.4 ± 0.6
Week 2	11.8 ± 0.3	12.0 ± 0.3	12.2 ± 0.4	11.8 ± 0.1
Week 3	12.9 ± 1.0	12.0 ± 0.3	11.6 ± 0.2	11.4 ± 0.3
Week 4	11.7 ± 0.6	12.3 ± 0.6	11.7 ± 0.4	11.6 ± 0.8
Week 5	11.8 ± 0.3	12.9 ± 0.3	12.3 ± 0.3	12.7 ± 1.0
Week 6	13.7 ± 1.3	13.2 ± 0.2	12.5 ± 0.5	12.0 ± 0.7
Week 7	12.2 ± 0.2	13.0 ± 0.6	12.7 ± 0.3	12.0 ± 0.2
Week 8	13.1 ± 0.3	13.1 ± 0.1	12.8 ± 0.4	12.1 ± 0.5

All data are presented as mean ± standard error (n = 12).

Table 2. Effects of the TFPS on weight gain and feed efficiency in mice.

Treatments		Dose (g/kg)	Weight Gain (g)	Feed Efficiency (%)
Normal diet	H_2O	–	4.9 ± 0.4	2.1 ± 0.3
High-fat diet	H_2O	–	14.6 ± 0.6 ###	11.0 ± 0.7 ###
	TFPS	1.0	10.3 ± 0.4 **	7.8 ± 0.2 ***
	TFPS	2.0	9.2 ± 0.5 ***	7.1 ± 0.2 ***

All data are presented as mean ± standard error (n = 12). ### $p < 0.001$ high-fat diet (HFD)-H_2O group compared with the normal diet-H_2O group. ** $p < 0.01$ and *** $p < 0.001$, while HFD-TFPS groups compared with the HFD-H_2O group.

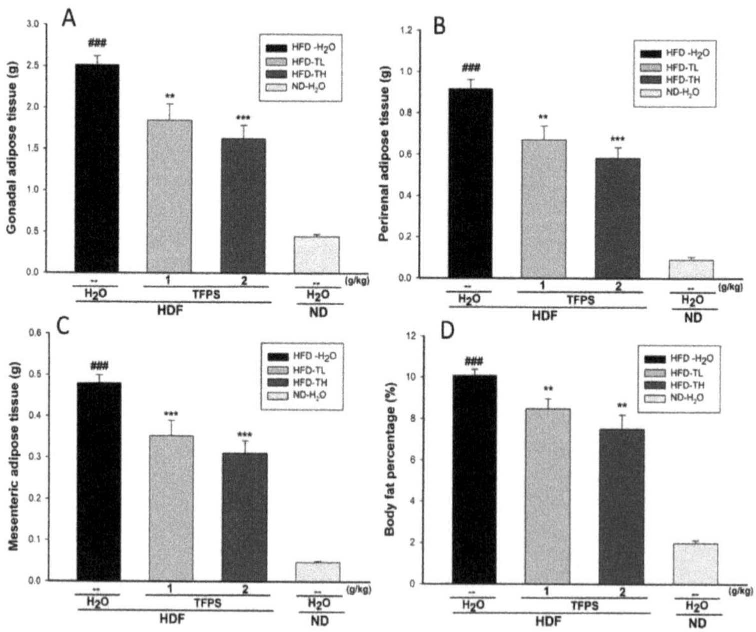

Figure 5. Effects of the TFPS treatment on the weights of the (**A**) gonadal adipose tissue, (**B**) perirenal adipose tissue, and (**C**) mesenteric adipose tissue as well as (**D**) body fat percentage. Mice were fed an HFD or ND, and the HFD mice were treated with water (*p.o.*) or TFPS (*p.o.*) for 8 weeks. All values are means ± standard error (n = 12). ### $p < 0.001$ while HFD-H_2O group compared with the ND-H_2O group; ** $p < 0.01$ and *** $p < 0.001$ while HFD-TFPS groups compared with the HFD-H_2O group.

3.4. Serum Lipid and Glucose Levels and Fecal Lipid Levels

No significant differences were found in the blood glucose levels among the groups at the end of the experiment. However, the blood sugar levels exhibited an increasing trend in the HFD groups compared with the ND group. Table 3 presents the serum lipid levels of the mice. The TG, T-Cho, LDL-C, and NEFA levels were significantly higher in the HFD-vehicle group compared with the ND group. The results revealed that the TFPS treatment reduced the T-Cho levels. Although the TFPS treatment did not significantly change the serum lipid levels, the HDL-C/LDL-C ratio was similar to that in the ND group. The HDL-C/LDL-C ratio was 1.75 and 1.26 in the ND and HFD-H_2O groups, respectively. The results revealed that the HFD reduced the ratio in these groups. Moreover, the HDL-C/LDL-C ratio in the HFD-TL and HFD-TH groups was 1.44 and 1.47, respectively, significantly higher than that in the HFD-vehicle group. The results showed that fecal lipid in ND group was 8.9 ± 0.5 mg/day/mice and in HFD-H_2O group was 11.2 ± 0.4 mg/day/mice. Fecal lipid content was 10.8 ± 0.6 and 9.1 ± 0.4 in the HFD-TL and HFD-TH group, respectively. A high dose of TFPS administration could significantly decrease the fecal lipid content compared to HFD-H_2O group.

Table 3. Serum lipid and glucose levels of mice in different groups.

Parameters	ND + H_2O	HFD + H_2O	HFD + TL	HFD + TH
TG (mg/dL)	92.4 ± 3.0	112.9 ± 4.2 ###	120.8 ± 5.0	122.1 ± 4.4
T-Cho (mg/dL)	61.1 ± 0.9	75.2 ± 2.0 ###	66.5 ± 3.0 *	64.8 ± 1.6 ***
LDL-C (mg/dL)	19.1 ± 1.6	30.9 ± 2.1 ###	26.6 ± 3.8	25.9 ± 3.4
HDL-C (mg/dL)	33.5 ± 3.5	38.9 ± 4.5	38.2 ± 5.8	38.2 ± 2.9
NEFA (mmol/L)	0.92 ± 0.23	1.40 ± 0.21 #	1.25 ± 0.20	1.22 ± 0.13
Glc (mg/dL)	143.9 ± 10.2	159.1 ± 15.5	154.5 ± 19.7	160.0 ± 18.1

$p < 0.001$ HFD-H_2O group compared with the ND-H_2O group. * $p < 0.05$ and *** $p < 0.001$ HFD-TFPS groups compared with the HFD-H_2O group. TFPS 1 g/kg and 2 g/kg was orally administered in the HFD-TL and HFD-TH groups, respectively.

3.5. Histology of Gonadal Adipose Tissue

As shown in Figure 6, the HFD induced adipocyte hypertrophy in the gonadal adipose tissue compared with the ND-H_2O group. The TFPS-treated groups exhibited less hypertrophy than the HFD-H_2O group. The results from the other mice were similar to those presented in Figure 6.

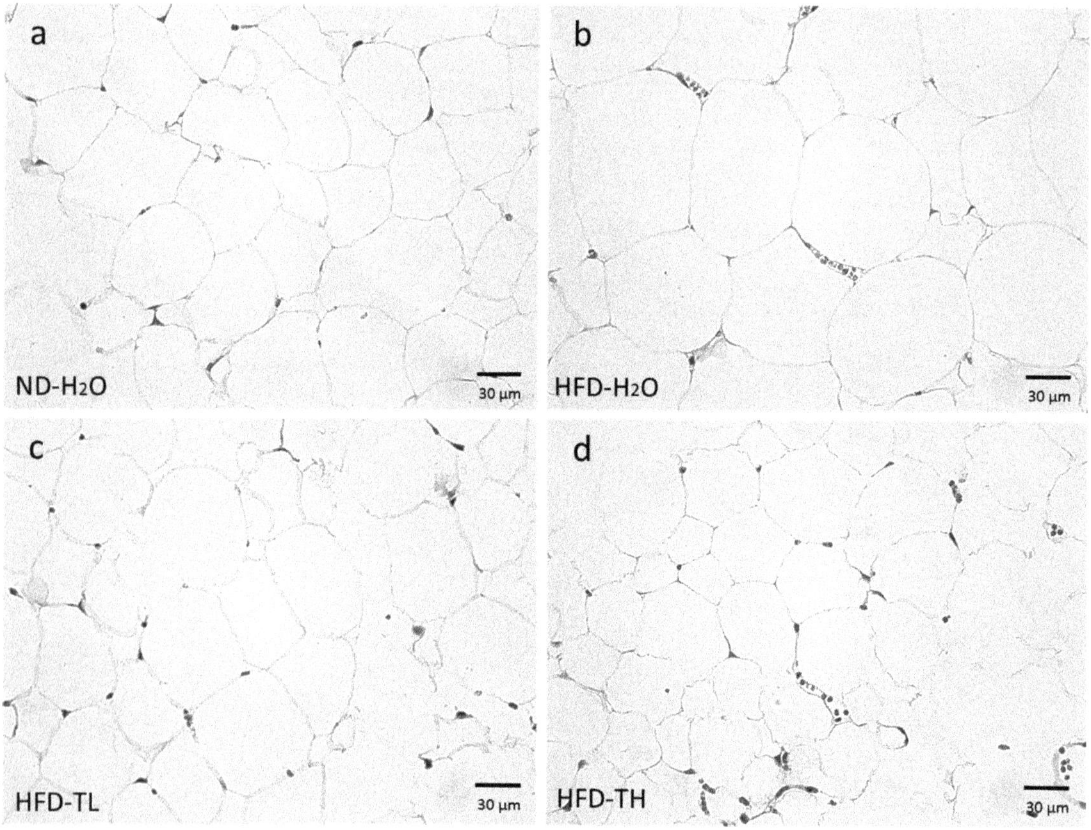

Figure 6. Histology of the gonadal adipose tissue of mice in the (**a**) ND-H2O, (**b**) HFD- H2O, (**c**) HFD-TL (TFPS 1 g/kg), and (**d**) HFD-TH (TFPS 2 g/kg) groups. Each image is representative of the nine mice (10 × 40).

4. Discussion

The species of *T. fuciformis* used in this study is called T8, which was selected from numerous species of *T. fuciformis* that have been collected in the mountains of Taiwan for several years. T8 has a larger fruiting body than other species and has natural fragrance smells like jasmine and lily floral. As one of the foremost medicinal and culinary mushrooms of China and Taiwan, *T. fuciformis* has the most unusual method of cultivation compared to other mushrooms, whereby it utilizes the nutrients from a common wood-decomposing ascomycete fungus, *Hypoxylon archeri*. The cultivation substrate used for T8 in this study was sawdust with rice bran as a supplementary nutrient, whereas cotton-seed hull is another common cultivation substrate used.

The structural characterizations of the TFPS showed that mannose followed by fucose were the major monosaccharide constituents. Moreover, the TFPS were rich in glucuronic acid. The average molecular weight was 679 kDa. Another study [19] that investigated the polysaccharides in different phases of *T. fuciformis* reported that the polysaccharides from the mycelial phase had 13.3% (*w/w*) uronic acid. Yui et al. [8] reported that the heteropolysaccharide isolated from *T. fuciformis* has a linear backbone of 1,3-linked mannose with a highly beta-D-glucuronic acidic residue. The result of glyosidic linkage showed that the backbone of TFPS was mannan to glucuronic acid and fucose side chains are attached, and responded to the linkage types that Yui et al., reported [8]. In the current study, the monosaccharide composition was comparable to other reports [8,19]. The portion of fucose was relatively high in the TFPS and was similar to the portion of mannose. However, other studies [8,9,20] have reported that xylose was the second highest monosaccharide detected in the polysaccharides from *T. fuciformis*. The possible reasons for the difference in the monosaccharide composition among TFPS and other polysaccharides isolated *T. fuciformis* may be related to species, cultivation methods, or cultivation substrates. High fucose content seems to be one of the structural characteristics of TFPS derived from T8.

The average molecular weight (Mr.) of the TFPS was 679 kDa, which is similar to those of *T. fuciformis* polysaccharides presented in other studies. The Mrs of the aqueous extracts of *T. fuciformis* polysaccharides reported by Zhang et al. [21] and Liu et al. [18] were 599 and 582 kDa, respectively.

High molecular weight polymers greatly increase the viscosity of liquids in which they are dissolved. The intrinsic viscosity of a polymer relates to Mr, as per the Mark–Houwink rule, and MW distribution also affects the viscosity [22]. It is generally agreed that there is a positive relationship between intrinsic viscosity and degree of polymerization [23]. An increase in viscosity depends on the nature of both a polymer and solvent. The high portion of glucuronic acids in the TFPS caused a high charge density and water molecules to bond strongly, relative to the strength of the water–water interaction [24]. The results showed that the TFPS had a high molecular weight and high uronic acid/carbohydrate ratio, which might have caused their high viscosity. Moreover, the high viscosity of the TFPS in D_2O potentially affected the ^1H NMR signals from 3.4 to 4.5 ppm to be low in resolution (Figure 2). However, the HNMR spectra of the TFPS in this study were similar to those reported by Liu et al. [18].

The most recommended method for maintaining a healthy weight is to exercise and consume a healthy diet. Alternative weight loss treatments include medicines, surgery, and dieting. However, these treatments have side effects. Another alternative treatment for obesity is ingesting large amounts of dietary fiber. Numerous studies [25–27] have reported that an increase in dietary fiber can control body weight. Alginate, a matrix polysaccharide of brown algae, is a potential treatment for obesity [25,28]. Moreover, alginate has been added to beverages and cereal bars to increase satiety, reduce calorie intake, reduce fat digestion, and increase weight loss [25,29]. Oats, a cereal rich in β-glucans, have also exhibited obesity prevention and health benefits in metabolic syndrome [30,31]. Moreover, studies have shown that diets rich in fiber improve glycemic control in type 2 diabetes [32], reduce LDL-C in hypercholesterolemia [33,34], and contribute positively to long-term weight management [6,27]. The mechanism is believed to be due to the viscosity generated by β-glucans or other water-soluble dietary fibers, which influence gastrointestinal mechanisms that mediate satiety [27,35]. One of the possible mechanisms indicated that soluble polysaccharides could interfere with the absorption of cholesterol in the intestines [27]. Moreover, Pau-Roblot et al. [36] examined the relationship between polysaccharides uronic acid sequences and lipid molecules and found that uronic acids, like galacturonic and glucuronic acids in polysaccharides, have the ability to entrap cholesterol. This might explain why the TFPS in the present study, which were rich in uronic acid, affected the body lipid accumulation in the HFD mice.

As reported [29,30,37], the polysaccharides derived from oats, plantago, and brown algae have the ability to reduce lipid uptake and hyperlipidemia in vivo. However, these polysaccharides do not have a similar backbone, and all of them have high viscosity. A study [6] reported that viscous fibers could be an adjunct to regular dietary treatment for obesity. Gel-forming fibers are particularly effective in reducing high LDL-cholesterol without changing the HDL-fraction [6]. Two structural characterizations of the TFPS, high uronic acid content and high viscosity, likely interfered with lipid absorption in the intestines of the mice on the HFD. In this study, TFPS exhibited the inhibition of cholesterol micelles in vitro and reduced the serum cholesterol level that was induced by the HFD in vivo. Both free or esterified cholesterol from diet should be emulsified by bile salts before absorption [38]. Various reports showed that polysaccharides with high viscosity and gel-forming property could influence the cholesterol absorption process. The possible mechanism may be that viscosity affects the diffusion of dietary mixed aggregates at the intestinal lumen, limiting cholesterol bio-accessibility. In addition, the interactions between polysaccharides and bile salts may reduce their emulsifying power towards cholesterol [39]. Heaton [40] has proposed that dietary fiber is a physiologic obstacle to energy intake by followed 3 mechanisms: (1) fiber replaces available calories and nutrients from the diet; (2) fiber increased satiety; and (3) fiber decreases the absorption efficiency of the small intestine. The third point is consistent with the assumption of this study that TFPS exhibits anti-obesity activity through interfering with lipid absorption in the intestine and results in increasing fecal lipid content. Moreover, the TFPS treatment increased the serum HDL-C/LDL-C ratio, which was similar to that in the normal group.

The results showed that the TFPS treatment reduced the weight gain in the HFD mice. Food and calorie intake in the two TFPS groups and HFD-vehicle group showed no significant difference. Moreover, the feed efficiencies of the two TFPS groups were significantly lower than that in the HFD-vehicle group. These results also revealed that the TFPS treatment did not affect appetite but influenced the lipid absorption in the HFD mice.

As mentioned, the daily calorie intake of the HFD-H_2O group was not different from that of the ND group. Nevertheless, the body fat percentage, weight gain, and feed efficiency between those groups varied. The results confirmed that the different diets influenced the body fat accumulation of the mice. The TFPS treatment inhibited the HFD-induced weight gain, body fat accumulation, and serum lipid in vivo.

5. Conclusions

T. fuciformis contains a high amount of polysaccharides, which are rich in uronic acids and exhibit high viscosity when dissolved in water. The polysaccharides isolated from T8 have the potential to treat HFD-induced obesity. Administration of the TFPS treatment reduced body weight gain, body fat percentage, and serum total cholesterol in the HFD mice. The TFPS treatment also reduced hypertrophy in adipose tissue. Thus, the results suggest that the health benefits of TFPS in HFD-induced obesity may be associated with its influence on gut lipid absorption.

Supplementary Materials: The following supporting information can be downloaded at: https://www.mdpi.com/article/10.3390/polym14081514/s1, Figure S1. Monosaccharide composition analysis by HPAEC-PAD. TFPS was hydrolysis by 2M TFA.

Author Contributions: Conceptualization, L.-C.Y. and C.-H.C.; Data curation, L.-C.Y. and C.-H.C.; Formal analysis, C.-H.C. and K.-C.C.; Methodology, L.-C.Y. and K.-C.C.; Resources, L.-C.Y. and C.-H.C.; Supervision, L.-C.Y.; Writing—original draft, L.-C.Y. and C.-H.C.; Writing—review & editing, L.-C.Y. All authors have read and agreed to the published version of the manuscript.

Funding: This research was funded by China Medical University Hospital DMR-106-229 and China Medical University CMU106-N-23.

Institutional Review Board Statement: The animal use protocol has been reviewed and approved by the Institutional Animal Care and Use Committee (IACUC) of China Medical University. The approved protocol number was CMUIACUC-2017-286.

Acknowledgments: Thanks to Hsiao-Chu Huang for their assistance in chemical analysis.

Conflicts of Interest: The authors declare no conflict of interest.

References

1. Esposito, K.; Chiodini, P.; Colao, A.; Lenzi, A.; Giugliano, D. Metabolic Syndrome and Risk of Cancer A systematic review and meta-analysis. *Diabetes Care* **2012**, *35*, 2402–2411. [CrossRef] [PubMed]
2. Keller, K.B.; Lemberg, L. Obesity and the metabolic syndrome. *Am. J. Crit. Care* **2003**, *12*, 167–170. [CrossRef] [PubMed]
3. Ogunbode, A.M.; Ladipo, M.; Ajayi, I.O.; Fatiregun, A.A. Obesity: An emerging disease. *Niger. J. Clin. Pract.* **2011**, *14*, 390–394. [CrossRef] [PubMed]

4. de Godoy, M.R.C.; Kerr, K.R.; Fahey, G.C. Alternative Dietary Fiber Sources in Companion Animal Nutrition. *Nutrients* **2013**, *5*, 3099–3117. [CrossRef] [PubMed]
5. Barry, K.A.; Middelbos, I.S.; Vester Boler, B.M.; Dowd, S.E.; Suchodolski, J.S.; Henrissat, B.; Coutinho, P.M.; White, B.A.; Fahey, G.C., Jr.; Swanson, K.S. Effects of dietary fiber on the feline gastrointestinal metagenome. *J. Proteome Res.* **2012**, *11*, 5924–5933. [CrossRef] [PubMed]
6. Smith, U. Dietary fibre, diabetes and obesity. *Int. J. Obes.* **1987**, *11* (Suppl. 1), 27–31. [CrossRef]
7. Wang, X.; Zhang, Z.; Zhao, M. Carboxymethylation of polysaccharides from Tremella fuciformis for antioxidant and moisture-preserving activities. *Int. J. Biol. Macromol.* **2015**, *72*, 526–530. [CrossRef]
8. Yui, T.; Ogawa, K.; Kakuta, M.; Misaki, A. Chain Conformation of a Glucurono-Xylo-Mannan Isolated from Fruit Body of Tremella-Fuciformis Berk. *J. Carbohyd. Chem.* **1995**, *14*, 255–263. [CrossRef]
9. De Baets, S.; Vandamme, E.J. Extracellular Tremella polysaccharides: Structure, properties and applications. *Biotechnol. Lett.* **2001**, *23*, 1361–1366. [CrossRef]
10. Jeong, H.J.; Yoon, S.J.; Pyun, Y.R. Polysaccharides from edible mushroom hinmogi (*Tremella fuciformis*) inhibit differentiation of 3T3-L1 adipocytes by reducing mRNA expression of PPAR gamma, C/EBP alpha, and leptin. *Food Sci. Biotechnol.* **2008**, *17*, 267–273.
11. Yang, L.C.; Lu, T.J.; Hsieh, C.C.; Lin, W.C. Characterization and immunomodulatory activity of polysaccharides derived from Dendrobium tosaense. *Carbohydr. Polym.* **2014**, *111*, 856–863. [CrossRef] [PubMed]
12. Dubois, M.; Gilles, K.; Hamilton, J.K.; Rebers, P.A.; Smith, F. A colorimetric method for the determination of sugars. *Nature* **1951**, *168*, 167. [CrossRef] [PubMed]
13. Bradford, M.M. A rapid and sensitive method for the quantitation of microgram quantities of protein utilizing the principle of protein-dye binding. *Anal. Biochem.* **1976**, *72*, 248–254. [CrossRef]
14. Blumenkrantz, N.; Asboe-Hansen, G. New method for quantitative determination of uronic acids. *Anal. Biochem.* **1973**, *54*, 484–489. [CrossRef]
15. Roberts, E.J.; Godshall, M.A.; Clarke, M.A.; Tsang, W.S.C.; Parrish, F.W. Methanolysis of Polysaccharides—A New Method. *Carbohydr. Res.* **1987**, *168*, 103–109. [CrossRef]
16. Nishi, S.; Saito, Y.; Souma, C.; Kato, J.; Koaze, H.; Hironaka, K.; Kojima, M. Suppression of serum cholesterol levels in mice by Adzuki bean polyphenols. *Food Sci. Technol. Res.* **2008**, *14*, 217–220. [CrossRef]
17. Kraus, D.; Yang, Q.; Kahn, B.B. Lipid Extraction from Mouse Feces. *Bio-Protocol* **2015**, *5*, e1375. [CrossRef]
18. Liu, J.; Meng, C.G.; Yan, Y.H.; Shan, Y.N.; Kan, J.; Jin, C.H. Structure, physical property and antioxidant activity of catechin grafted Tremella fuciformis polysaccharide. *Int. J. Biol. Macromol.* **2016**, *82*, 719–724. [CrossRef]
19. Zhu, H.Y.; Yuan, Y.; Liu, J.; Zheng, L.S.; Chen, L.G.; Ma, A.M. Comparing the sugar profiles and primary structures of alkali-extracted water-soluble polysaccharides in cell wall between the yeast and mycelial phases from Tremella fuciformis. *J. Microbiol.* **2016**, *54*, 381–386. [CrossRef]
20. Fraser, C.G.; Jennings, H.J.; Moyna, P. Structural analysis of an acidic polysaccharide from Tremella mesenterica NRRL Y-6158. *Can. J. Biochem.* **1973**, *51*, 219–224. [CrossRef]
21. Zhang, Z.; Wang, X.; Zhao, M.; Qi, H. Free-radical degradation by $Fe^{2+}/Vc/H_2O_2$ and antioxidant activity of polysaccharide from Tremella fuciformis. *Carbohydr. Polym.* **2014**, *112*, 578–582. [CrossRef]
22. Striegel, A.M. Viscometric Detection in Size-Exclusion Chromatography: Principles and Select Applications. *Chromatographia* **2016**, *79*, 945–960. [CrossRef]
23. Harland, W.G. Relation between Intrinsic Viscosity and Degree of Polymerization. *Nature* **1952**, *170*, 667. [CrossRef]
24. Collins, K.D. Charge density-dependent strength of hydration and biological structure. *Biophys. J.* **1997**, *72*, 65–76. [CrossRef]
25. Houghton, D.; Wilcox, M.D.; Chater, P.I.; Brownlee, I.A.; Seal, C.J.; Pearson, J.P. Biological activity of alginate and its effect on pancreatic lipase inhibition as a potential treatment for obesity. *Food Hydrocoll.* **2015**, *49*, 18–24. [CrossRef]
26. Lairon, D. Dietary fiber and control of body weight. *Nutr. Metab. Cardiovasc. Dis.* **2007**, *17*, 1–5. [CrossRef]
27. Pereira, M.A.; Ludwig, D.S. Dietary fiber and body-weight regulation. Observations and mechanisms. *Pediatr. Clin. N. Am.* **2001**, *48*, 969–980. [CrossRef]
28. Georg Jensen, M.; Kristensen, M.; Astrup, A. Effect of alginate supplementation on weight loss in obese subjects completing a 12-wk energy-restricted diet: A randomized controlled trial. *Am. J. Clin. Nutr.* **2012**, *96*, 5–13. [CrossRef]
29. Paxman, J.R.; Richardson, J.C.; Dettmar, P.W.; Corfe, B.M. Alginate reduces the increased uptake of cholesterol and glucose in overweight male subjects: A pilot study. *Nutr. Res.* **2008**, *28*, 501–505. [CrossRef]
30. Chang, H.C.; Huang, C.N.; Yeh, D.M.; Wang, S.J.; Peng, C.H.; Wang, C.J. Oat prevents obesity and abdominal fat distribution, and improves liver function in humans. *Plant Foods Hum. Nutr.* **2013**, *68*, 18–23. [CrossRef]
31. El Khoury, D.; Cuda, C.; Luhovyy, B.L.; Anderson, G.H. Beta glucan: Health benefits in obesity and metabolic syndrome. *J. Nutr. Metab.* **2012**, *2012*, 851362. [CrossRef] [PubMed]
32. Brennan, C.S. Dietary fibre, glycaemic response, and diabetes. *Mol. Nutr. Food Res.* **2005**, *49*, 560–570. [CrossRef] [PubMed]
33. Truswell, A.S. Meta-analysis of the cholesterol-lowering effects of dietary fiber. *Am. J. Clin. Nutr.* **1999**, *70*, 942–943. [CrossRef] [PubMed]
34. Anderson, J.W.; Davidson, M.H.; Blonde, L. Long-term cholesterol-lowering effects of psyllium as an adjunct to diet therapy in the treatment of hypercholesterolemia. *Am. J. Clin. Nutr.* **2000**, *71*, 1433–1438. [CrossRef]

35. Rebello, C.J.; Chu, Y.F.; Johnson, W.D.; Martin, C.K.; Han, H.; Bordenave, N.; Shi, Y.; O'Shea, M.; Greenway, F.L. The role of meal viscosity and oat beta-glucan characteristics in human appetite control: A randomized crossover trial. *Nutr. J.* **2014**, *13*, 49. [CrossRef]
36. Pau-Roblot, C.; Courtois, B.; Courtois, J. Interactions between polysaccharides uronic acid sequences and lipid molecules. *Comptes Rendus Chim.* **2010**, *13*, 443–448. [CrossRef]
37. Hu, J.L.; Nie, S.P.; Wu, Q.M.; Li, C.; Fu, Z.H.; Gong, J.; Cui, S.W.; Xie, M.Y. Polysaccharide from seeds of Plantago asiatica L. affects lipid metabolism and colon microbiota of mouse. *J. Agric. Food Chem.* **2014**, *62*, 229–234. [CrossRef]
38. Gunness, P.; Gidley, M.J. Mechanisms underlying the cholesterol-lowering properties of soluble dietary fibre polysaccharides. *Food Funct.* **2010**, *1*, 149–155. [CrossRef]
39. Silva, I.M.V.; Machado, F.; Moreno, M.J.; Nunes, C.; Coimbra, M.A.; Coreta-Gomes, F. Polysaccharide Structures and Their Hypocholesterolemic Potential. *Molecules* **2021**, *26*, 4559. [CrossRef]
40. Heaton, K.W. Food Fiber as an Obstacle to Energy-Intake. *Lancet* **1973**, *2*, 1418–1421. [CrossRef]

polymers

MDPI

Article

Sequencing Biodegradable and Potentially Biobased Polyesteramide of Sebacic Acid and 3-Amino-1-propanol by MALDI TOF-TOF Tandem Mass Spectrometry

Paola Rizzarelli [1,*], Stefania La Carta [2], Emanuele Francesco Mirabella [1], Marco Rapisarda [1] and Giuseppe Impallomeni [1]

[1] Istituto per i Polimeri, Compositi e Biomateriali, Consiglio Nazionale delle Ricerche, Via Paolo Gaifami 18, 95126 Catania, Italy; emanuelefrancesco.mirabella@cnr.it (E.F.M.); marco.rapisarda@ipcb.cnr.it (M.R.); giuseppe.impallomeni@cnr.it (G.I.)

[2] STMicroelectronics Srl, Stradale Primosole, 50, 95121 Catania, Italy; stefanialacarta@gmail.com

* Correspondence: paola.rizzarelli@cnr.it

Abstract: Biodegradable and potentially biobased polyesteramide oligomers (PEA-Pro), obtained from melt condensation of sebacic acid and 3-amino-1-propanol, were characterized by nuclear magnetic resonance (NMR), matrix assisted laser desorption/ionization-time of flight/time of flight-mass spectrometry/mass spectrometry (MALDI-TOF/TOF-MS/MS), thermogravimetric analysis (TGA), and pyrolysis-gas chromatography/mass spectrometry (Py-GC/MS). NMR analysis showed the presence of hydroxyl and amino terminal groups as well as carboxylic groups of the sebacate moiety. Hydroxyl and carboxyl termination had the same abundance, while the amine termination was 2.7-times less frequent. Information regarding the fragmentation pathways and ester/amide bond sequences was obtained by MALDI-TOF/TOF-MS/MS analysis performed on sodiated adducts of cyclic species and linear oligomers. Different end groups did not influence the observed fragmentation. Three fragmentation pathways were recognized. The β-hydrogen-transfer rearrangement, which leads to the selective scission of the –O–CH_2– bonds, was the main mechanism. Abundant product ions originating from –CH_2–CH_2– (β–γ) bond cleavage in the sebacate moiety and less abundant ions formed by –O–CO– cleavages were also detected. TGA showed a major weight loss (74%) at 381 °C and a second degradation step (22% weight loss) at 447 °C. Py-GC/MS performed in the temperature range of 350–400 °C displayed partial similarity between the degradation products and the main fragments detected in the MALDI-TOF/TOF-MS/MS experiments. Degradation products derived from amide bonds were related to the formation of CN groups, in agreement with the literature.

Keywords: polyesteramides; biodegradable polymers; biobased polymers; mass spectrometry; MALDI; tandem mass spectrometry; sequencing; characterization; Py-GC/MS

Citation: Rizzarelli, P.; La Carta, S.; Mirabella, E.F.; Rapisarda, M.; Impallomeni, G. Sequencing Biodegradable and Potentially Biobased Polyesteramide of Sebacic Acid and 3-Amino-1-propanol by MALDI TOF-TOF Tandem Mass Spectrometry. *Polymers* **2022**, *14*, 1500. https://doi.org/10.3390/polym14081500

Academic Editors: Antonio M. Borrero-López, Concepción Valencia-Barragán, Esperanza Cortés Triviño, Adrián Tenorio-Alfonso and Clara Delgado-Sánchez

Received: 15 March 2022
Accepted: 5 April 2022
Published: 7 April 2022

1. Introduction

Aliphatic polyesteramides represent a class of biodegradable and potentially biobased polymers. They combine the mechanical and thermal properties of polyamides with the high biodegradability of polyesters, leading to materials with better properties and processability compared to polyesters with similar structures [1].

The detailed characterization of synthetic polymers plays an important role in defining the chemical structure and its influence on material properties, as well as its utilization. During the last few years, matrix-assisted laser desorption/ionization time-of-flight mass spectrometry (MALDI-TOF MS) has become a routine analytical technique for the analysis of synthetic polymers, providing more and more detailed structural information with high sensitivity, thanks to improved technologies and analytical strategies [2,3]. It has been successfully used in both characterization and degradation studies, with some limitations in the accuracy of the measured molecular weight related to the polydisperse nature of polymers, especially when dealing with high mass range [4]. Despite MALDI being a soft ionization technique, it is possible to obtain fragment ion spectra through the so-called post-source decay (PSD) [5]. MALDI-PSD has been successfully applied for peptide sequencing and biopolymers analysis, but to a lesser extent for synthetic polymers, due to their greater complexity. The development of the MALDI-TOF/TOF (Matrix-assisted laser desorption ionization time-of-flight/time-of-flight) mass spectrometer, with the aim of simplifying and improving the experimental procedures and acquiring more reliable information, bypassed many limits of the PSD and collision-induced dissociation (CID) in a conventional MALDI instrument. In a MALDI-TOF/TOF mass spectrometer operating in tandem mass spectrometry mode (MALDI-TOF/TOF-MS/MS), the high-speed ions separated in the first TOF analyser are selected by a timed ion gate, collided with atoms or molecules, and accelerated for analysis in the second TOF mass analyser fitted with a two-stage reflector [6].

MS/MS is increasingly applied to analyse synthetic polymers since it can provide information on chain-end or in-chain substituents, discriminate isobaric and isomeric species, differentiate linear and cyclic polymers, and establish macromolecular connectivity, sequences and architectures [7,8]. In fact, in several cases, single-stage mass data may not be sufficient to unequivocally establishing the polymer structure. For confident structural assignments, fragmentation studies provide initial information to help understand tandem mass spectra. In fact, knowledge of the fragmentation mechanisms of polymer ions provides guidelines on how to obtain the desired information from the product ions detected in MS/MS spectra and to deduce the real macromolecular architecture [9]. Amongst the MS/MS tools, the collision-induced dissociation (CID) approach represents the preferred technique for the deep structural characterization of gas-phase ions.

MALDI-TOF/TOF-MS/MS has been successfully applied to the analysis of biodegradable, as well as polyesters and polyesteramides [10,11]. It provided structural information regarding the sequence of ester and amide bonds in synthetic polyesteramides [12,13] and the fragmentation pathways in biodegradable polyesters [14,15].

In this work, Nuclear magnetic resonance (NMR) analysis showed the nature of the terminal groups and allowed their quantification. MALDI-TOF MS revealed the presence of cyclic species and linear oligomers with the main end groups and additional ones which eluded NMR analysis. MALDI-TOF/TOF-MS/MS was employed to investigate the ester/amide bond sequences and the fragmentation mechanisms of a polyesteramide (PEA-Pro) from the melt condensation of 3-amino-1-propanol and sebacic acid that can be produced from a renewable source (castor oil). The MALDI-TOF/TOF-MS/MS analysis was performed on cyclic species and linear oligomers with different end groups, selecting as precursor ions the sodium adducts of the oligomers identified in the MALDI-TOF/TOF spectra.

Furthermore, pyrolysis-gas chromatography/MS (Py-GC/MS) was applied to confirm the results acquired by MALDI-TOF/TOF. Pyrolysis technique hyphenated to GC/MS is the method most frequently used for polymer characterization and for studying their thermal degradation processes [16]. Py-GC/MS provides valuable information about the molecular structure, and it is a particularly useful method for the analysis of complex blends of traditional and biodegradable macromolecules [17,18]. Moreover, we used thermogravimetric analysis (TGA) to check the degradation steps and establish the best experimental conditions in Py-GC/MS analyses.

2. Materials and Methods

2.1. Materials

Low molecular weight polyesteramide (PEA-Pro) was synthesized by melt polymerization starting from sebacic acid and 3-amino-1-propanol, using stoichiometric amounts of reagents as described in [19]. 2-(4-Hydroxyphenylazo)benzoic acid (HABA), reagents, and solvents were obtained from Sigma-Aldrich Chemical Co. (Milan, Italy) and used as received.

2.2. Nuclear Magnetic Resonance (NMR)

NMR analysis was conducted with a ^{UNITY}INOVA Agilent spectrometer (Santa Clara, CA, USA) operating at 500 MHz (^{1}H), using standard Agilent pulse sequences. Spectra were elaborated with the MestreNova software (Mestrelab Research, Santiago de Compostela, Spain). The sample was dissolved in deuterated dimethylsulfoxide (6d-DMSO) and spectra acquired at 50 °C and calibrated with the solvent residual ^{1}H signal set at 2500 ppm.

2.3. MALDI Sample Preparation

The polyesteramide sample was dissolved in 1,1,1,3,3,3-hexafluoro-2-propanol (HFIP) at a concentration of di 3 mg/mL. HABA (0.1 M in HFIP) was used as the matrix. Appropriate volumes of polymer solution and matrix solution were mixed to obtain 1:1, 1:2, 1:3 ratios (sample/matrix, v/v). An aliquot of 1 μL of each sample/matrix mixture was spotted onto the MALDI sample holder and slowly dried to allow matrix crystallization.

2.4. MALDI-TOF/MS Analysis

An Applied Biosystems 4800 MALDI TOF/TOF™ Analyser (Framingham, MA, USA) mass spectrometer was employed in this study to acquire MALDI and MS/MS spectra with and without using a collision gas (air or argon). This TOF/TOF instrument is equipped with a Nd:YAG laser with 355 nm wavelength, pulse of <500 ps, and 200 Hz repetition rate in both MS and MS/MS modes. The precursor ion selector of the analyser has a mass resolution of about 400. All measurements were performed in automatic mode. For MS/MS experiments, the potential difference between the source acceleration voltage and the collision cell was set at 2 kV. MALDI-MS e MALDI-TOF/TOF-MS/MS spectra were recorded in reflector positive ion mode. MS and MS/MS data were processed using Data Explorer 4.4 (Applied Biosystems, Framingham, MA, USA). Mass resolution (FWMH) was about 13,000 in MS mode and ranged from 1500 ($m/z < 500$) to 5000 ($500 < m/z < 1300$) in MS/MS mode.

2.5. Thermogravimetric Analysis (TGA)

TGA was performed with a TGA Q500 (TA Instruments, New Castle, DE, USA), using a platinum pan. Dynamic measurement was performed from 50 to 800 °C at 10 °C/min, under a nitrogen atmosphere (flow rate 60 mL/min). Sample weight was approximately 5 mg. The weight loss percent and its derivate (DTG) was recorded as a function of temperature.

2.6. Pyrolysis-Gas Chromatography/Mass Spectrometry Analysis (Py-GC/MS)

Py-GC/MS was performed on a small amount of sample (about 0.1 mg). It was placed, without pre-treatment, in the crucible immediately prior to analysis. Py-GC/MS was performed using a Multi-Shot Pyrolizer (EGA/PY-303D, Frontier Labs Ltd., Fukushima, Japan), connected to a GC system GC-2020 (Shimadzu Italia S.r.l., Milan, Italy), coupled with to a triple quadrupole mass spectrometer detector and with electronic ionization (70 eV) Mass Detector TQ8040 (Shimadzu Italia S.r.l., Milan, Italy).

The gas chromatography was equipped with Ultra Alloy® Metal Capillary Column (Frontier Labs, stationary phase 5% di-phenyl-methylpolysiloxane, with an inner diameter of 250 μm, a film thickness of 0.25 μm and a length of 30 m). Interfaces of Py-GC and GC/MS were kept at 300 °C and 250 °C, respectively. Several experiments were performed setting different combination steps of pyrolysis between 350 and 450 °C. Oven temperature was held at 50 °C for 1 min, increased to 100 °C at 30 °C/min, then held at 100 °C for 5 min, ramped from 100 °C to 300 °C at 10 °C/min and finally it was maintained at 300 °C for 10 min. The carrier gas was helium at a controlled flow of 1.78 mL/min. The split ratio was 1/50 of the total flux. Mass range was set from m/z 35 up to 500. Blanks were carried by placing the crucible empty in the furnace and performing pyrolysis in the conditions mentioned above. Structure assignment was carried out running the similarity search routine of the instrument.

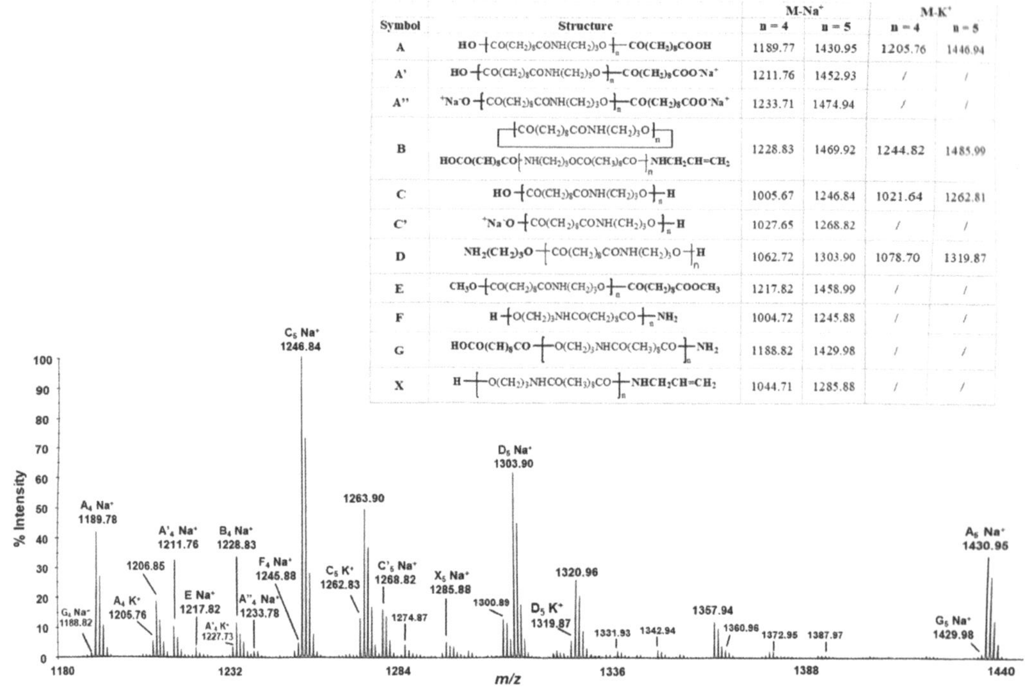

Symbol	Structure	M-Na+		M-K+	
		n = 4	n = 5	n = 4	n = 5
A	HO–[CO(CH2)8CONH(CH2)3O]n–CO(CH2)8COOH	1189.77	1430.95	1205.76	1446.94
A'	HO–[CO(CH2)8CONH(CH2)3O]n–CO(CH2)8COO−Na+	1211.76	1452.93	/	/
A''	+Na−O–[CO(CH2)8CONH(CH2)3O]n–CO(CH2)8COO−Na+	1233.71	1474.94	/	/
B	cyclo-[CO(CH2)8CONH(CH2)3O]n HOCO(CH)8CO[NH(CH2)3OCO(CH3)8CO]n NHCH2CH=CH2	1228.83	1469.92	1244.82	1485.99
C	HO–[CO(CH2)8CONH(CH2)3O]n–H	1005.67	1246.84	1021.64	1262.81
C'	+Na−O–[CO(CH2)8CONH(CH2)3O]n–H	1027.65	1268.82	/	/
D	NH2(CH2)3O–[CO(CH2)8CONH(CH2)3O]n–H	1062.72	1303.90	1078.70	1319.87
E	CH3O–[CO(CH2)8CONH(CH2)3O]n–CO(CH2)8COOCH3	1217.82	1458.99	/	/
F	H–[O(CH2)3NHCO(CH2)8CO]n–NH2	1004.72	1245.88	/	/
G	HOCO(CH)8CO–[O(CH2)3NHCO(CH3)8CO]n–NH2	1188.82	1429.98	/	/
X	H–[O(CH2)3NHCO(CH3)8CO]n–NHCH2CH=CH2	1044.71	1285.88	/	/

Figure 3. Enlarged portion of the MALDI/TOF-TOF mass spectrum in reflector mode of the PEA-Pro sample.

Fifteen MALDI ions in Figure 3 can be assigned unambiguously due to eight oligomers.

1. Sodiated and potassiated linear di-carboxyl terminated chains (species A Na^+, *m/z* 1189.77–1430.95; species A K^+, *m/z* 1205.76–1446.94) and the corresponding sodium salt (A′ Na^+, *m/z* 1211.76–1452.93) and disodium salt of the same species (A″ Na^+, *m/z* 1233.71–1474.94);
2. Sodiated and potassiated cyclic PEA-Pro chains (species B Na^+, *m/z* 1228.83; B K^+, *m/z* 1244.82).
3. Sodiated and potassiated linear oligomers terminated with carboxyl at one end and amino-alcohol groups at the other end (species C Na^+, *m/z* 1246.84; species C K^+, *m/z* 1262.81), and the corresponding sodium salt of the same species (C′ Na+, *m/z* 1268.82);
4. Sodiated and potassiated linear di-amino-alcohol terminated oligomers (species D Na^+ *m/z* 1303.90, species D K^+, *m/z* 1319.87);
5. Sodiated linear oligomers terminated with dimethyl ester of sebacic acid at both the ends (species E Na^+ *m/z* 1217.82);
6. Sodiated linear oligomers terminated with amino-alcohol at one end and $-NH_2$ group at the other end (species F Na^+ *m/z* 1245.88);
7. Sodiated linear oligomers terminated with sebacic acid at one end and $-NH_2$ group at the other end (species G Na^+ *m/z* 1429.98);
8. Sodiated linear oligomers terminated with amino alcohol at one end and olefin derived by amino alcohol at the other end (species X Na^+ *m/z* 1285.88).

The structures of these eight oligomers, identified in the inset of Figure 3 are predictable according to the method of synthesis and thermal degradation mechanisms (species G and F) [20]. Moreover, ions at *m/z* 1360.96 are reasonably due to end groups bearing ether bonds, i.e., $H[U]_4NH(CH_2)_3O(CH_2)_3NH_2$. Likewise, ions at *m/z* 1342.94 can be assigned to cyclic species with two ether bonds in the macromolecular chain. Noticeably, oligomers of unknown structure are also observed (*m*/*z* 1263.90, 1274.87, 1300.89, 1320.96, 1331.93, 1357.94, 1372.95, 1387.97) and their presence could be related to additional side reactions taking place in the polymerization process.

3.3. MALDI-TOF/TOF-MS/MS Analysis

Four sodiated oligomers were selected as precursor ions to perform MALDI-TOF/TOF-MS/MS analysis: three linear oligomers terminated, respectively, by two carboxylic groups (*m*/*z* 1189.8, species A, Figure 3), a carboxyl and an amino-alcohol group (*m*/*z* 1246.8, species C, Figure 3), and two amino-alcohol groups (*m*/*z* 1303.9, species D, Figure 3) and, finally, a cycle (*m*/*z* 1228.8, species B, Figure 3).

All the MALDI-TOF/TOF-MS/MS spectra acquired show a similar (most abundant) series of product ions. Then, the fragmentation pathway is not affected by different end groups.

As in previous works [21], the β-hydrogen transfer rearrangement appears to be the main fragmentation mechanism (Scheme 1). It is a typical mechanism of thermal degradation in polyesters [16] and involves the transfer of a hydrogen in β position to the carboxyl group of the diacid via a six-membered cyclic transition state. As a consequence, two ions are obtained, one bearing a carboxylic acid end group and the other terminated with an olefin end group (CH_2=$CHCH_2NHCO$–). The sodium affinity of the species formed in the dissociation reaction can result in a different abundance of the product ions in the MALDI-TOF/TOF-MS/MS spectrum.

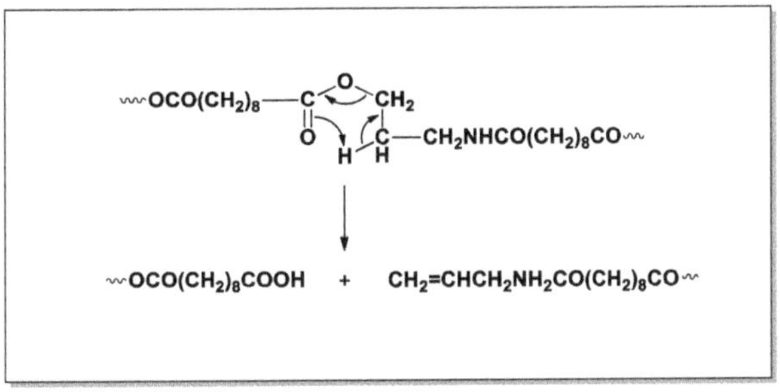

Scheme 1. β-hydrogen transfer mechanism.

Figure 4 displays the MALDI-TOF/TOF-MS/MS spectrum of the sodiated di-carboxyl-terminated oligomers at *m*/*z* 1189.8 of the PEA-Pro sample together with the *m*/*z* values of the two main fragmentation pathways. The main product ions at *m*/*z* 225.10, 466.27, 707.44 and 948.59 and their complementary ions at *m*/*z* 264.16, 505.33 746.49 and 987.66 are due to β-hydrogen-transfer rearrangement, Figure 4a, typical for polyesters in thermal degradation. The less abundant product ions at *m*/*z* 336.20, 577.38, and 818.56 and their complementary 633.45, 392.23 and 151.10 arise from –CH_2–CH_2– (β–γ) bond cleavage in the sebacate moiety, Figure 4b. This scission is supported by the formation of product ions bearing a α,β-unsaturated ester or amide end group (CH_2=$CHCOO$– or CH_2=$CHCONH$–). Furthermore, unsaturated and saturated fragments should both be formed, but the saturated fragments are significantly less abundant. In Figure 4 only the *m*/*z* values of the unsaturated fragments are reported; the *m*/*z* values of the complementary saturated ions can be easily obtained. The mass spectrum in Figure 4 also displays clusters of ions at an interval of 14 (CH_2) *m*/*z* units (e.g., *m*/*z* 336.17, 350.14, 364.41, 378.37, 392.41) that are typical of hydrocarbons and long-chain acids. All assigned structures are shown in Table S1.

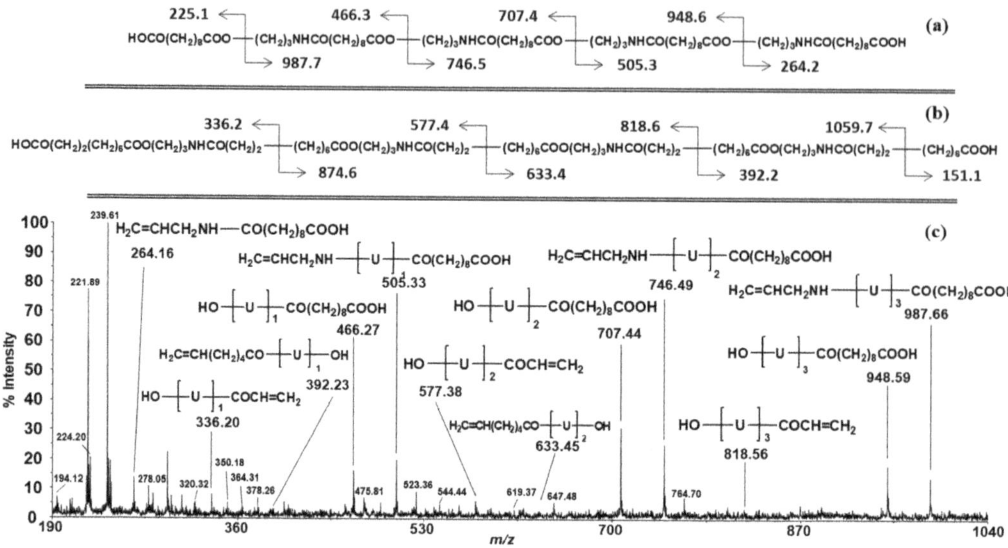

Figure 4. m/z values of the sodiated product ions originating from: (**a**) the β-hydrogen transfer and (**b**) the $-CH_2-CH_2-$ (β–γ) bond cleavages in the sebacate moiety for the sodiated precursor ions at m/z 1189.8. (**c**) MALDI-TOF/TOF-MS/MS spectrum of the sodiated dicarboxyl-terminated oligomers at m/z 1189.8 of the PEA-Pro sample (U = $-NH(CH_2)_3OCO(CH_2)_8CO-$ or $-O(CH_2)_3NHCO(CH_2)_8CO-$; collision gas = argon).

Similar results were obtained from the MALDI-TOF/TOF-MS/MS analysis of the other precursor ions analysed. The mass spectrum of the sodiated carboxyl and amino-alcohol terminated oligomers at m/z 1246.8 of the PEA-Pro sample is reported in Figure 5.

As seen before, the most intense signals are due to ions originated by the cleavage of $-O-CH_2-$ bonds, because of a β-hydrogen-transfer rearrangement. However, this time, the end groups of the chain are different, leading to two different series' of signals, Figure 5a, depending on the orientation of the 3-amino-1-propanol moiety (ions at m/z 225.10, 466.27, 707.44, 948.60, 1189.78 and their complementary at m/z 1044.75, 803.54, 562.38, 321.21, 80.03 and ions at m/z 264.16, 505.33, 746.49, 987.66, 1228.83 and their complementary at m/z 1005.68, 764.49, 523.33, 282.17). In the same way, the $-CH_2-CH_2-$ bond cleavage in the sebacate moiety gives rise to two different series' of ions, Figure 5b. α,β–unsaturated amide or ester, not discernible by MS, are originated as a function of the different orientation of the amino-alcohol (respectively, ions at m/z 95.0, 336.2, 577.3, 818.5, 1059.7 and their complementary at m/z 202.1, 449.3, 690.5, 931.6, 1172.8 and ions at m/z 151.1, 392.2, 633.4, 874.6, 1115.7 and their complementary at m/z 152.1, 393.2, 634.4, 875.6, 1116.7). Additionally, in this case, groups of ions differing by 14 m/z units are present (e.g., m/z 350.22, 364.24, 378.26 and 392.27), resulting from fragmentation of different $-CH_2-CH_2-$ bonds in hydrocarbon chain of the sebacic moiety. Ions at m/z 280.17, 521.36, and 762.51, Figure 5c, can be assigned to product ions whose end groups can be formed by $-O-CO-$ bond cleavages that lead to the formation of two ions: one bearing an aldehyde end group and the other with a $HOCH{=}CH(CH_2)_2NHCO-$/$HCO(CH_2)_3NHCO-$ end group. Product ions bearing end groups originating from $-CH_2-CH_2-$ ($\beta-\gamma$) bond cleavage, in the amino alcohol moiety, are also detected (e.g., m/z 251.12, 492.29 and 733.54). Several isobaric structures that differ by the position of a $-CH_2-$ group can be due to some of these ions. All the structures assigned are reported in Table S1.

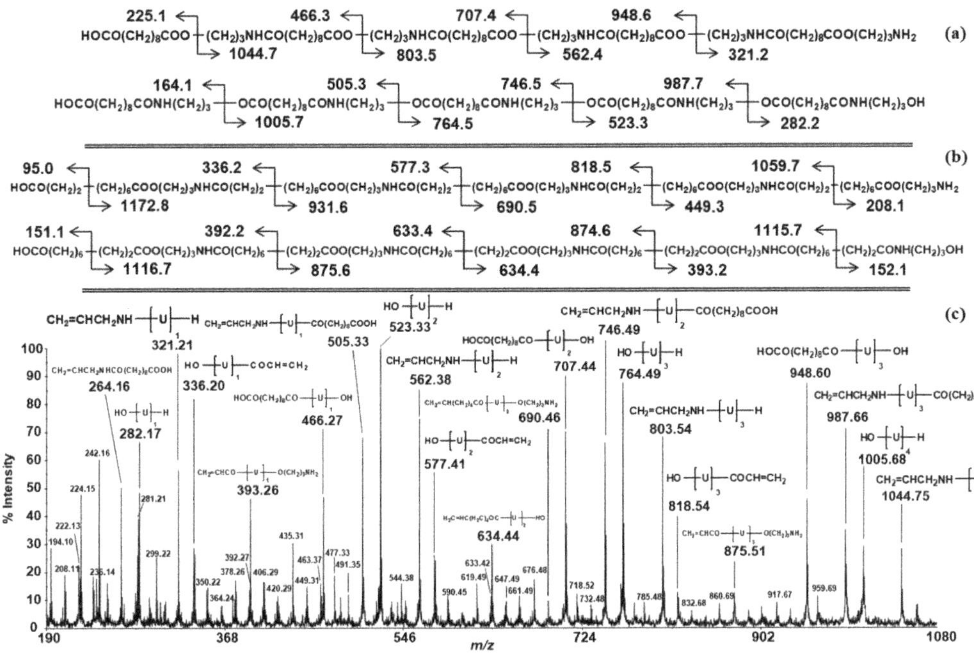

Figure 5. m/z values of the sodiated product ions originating from: (**a**) the β-hydrogen transfer and (**b**) the $-CH_2-CH_2-$ ($\beta-\gamma$) bond cleavages in the sebacate moiety for the sodiated precursor ions at m/z 1246.8. (**c**) MALDI-TOF/TOF-MS/MS spectrum of the sodiated carboxyl and amino-alcohol terminated oligomers at m/z 1246.8 of the PEA-Pro sample (U= $-NH(CH_2)_3OCO(CH_2)_8CO-$ or $-O(CH_2)_3NHCO(CH_2)_8CO-$; collision gas = argon).

Figure 6 displays the MALDI-TOF/TOF-MS/MS spectrum of the sodiated cyclic oligomers at m/z 1228.8 of the PEA-Pro sample. Cyclic oligomers may undergo ring opening at several backbone positions. However, since β-hydrogen-transfer is the main fragmentation mechanism, it is reasonable to assume that it is responsible for ring-opening reaction, resulting in linear chains terminated with carboxylic groups at one end and olefin group (CH_2=CH(CH_2)NHCO–) at the other end. Once again, different orientations of the 3-amino-1-propanol moiety led to two different series' of ions for each fragmentation mechanism. In Figure 6, we report the m/z values of the product ions originating from: (a) β-hydrogen transfer and (b) –CH_2–CH_2– (β–γ) bond cleavages in the sebacate moiety.

An enlarged portion of the same MALDI-TOF/TOF-MS/MS spectrum in the range m/z 190–330 is shown in Figure 7. Based on the structures of the identified ions, it is possible to exclude a regular alternate ester/amide sequence, i.e., infer that the orientation of 3-amino-1-propanol along the polymer chain is random. In fact, the ions at m/z 193, 239, 305 and 253 can be formed from ester/ester sequences (Figure 7a), and the structure of the ions at m/z 263, 319 and 303 can be originated by amide/amide sequences in the polymer chain (Figure 7b). Finally, the presence of the ions at m/z 224, 250, 264, 248 and 280 provides evidence for the existence of ester/amide sequences (Figure 7c). These ions arise from two bond cleavages of the polymer chains and the formation of some of them should involve high-energy processes.

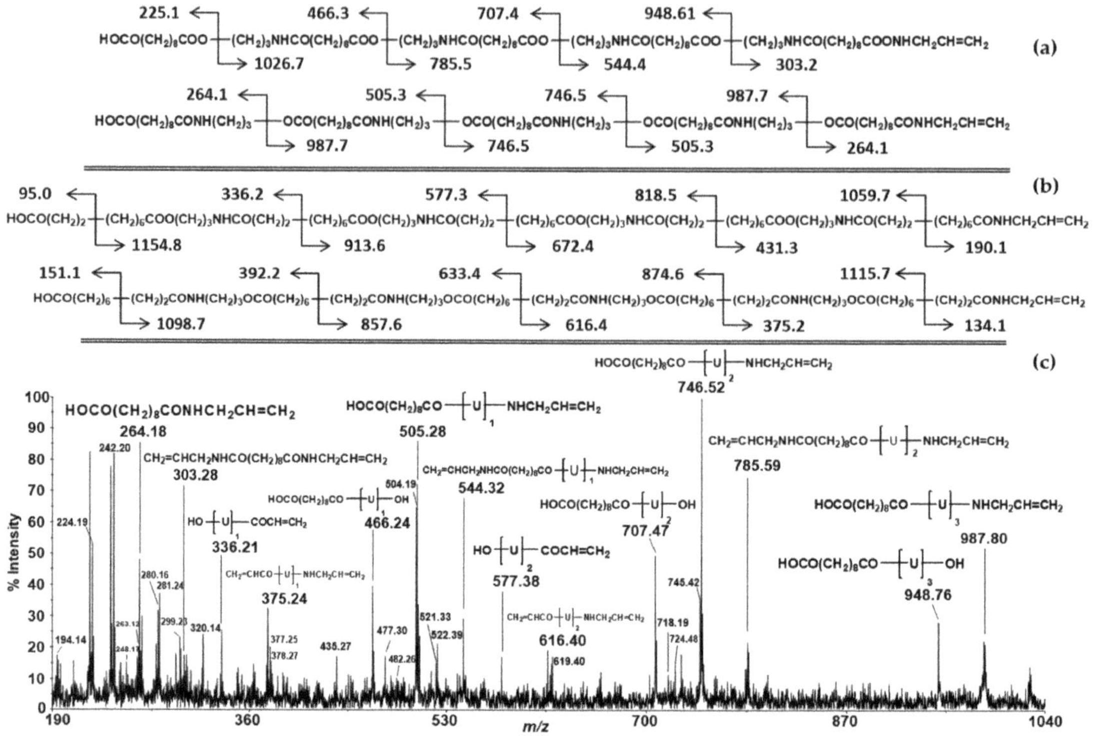

Figure 6. m/z values of the sodiated product ions originating from: (**a**) the β-hydrogen transfer and (**b**) the $-CH_2-CH_2-$ ($\beta-\gamma$) bond cleavages in the sebacate moiety for the sodiated precursor ions at m/z 1228.8. (**c**) MALDI-TOF/TOF-MS/MS spectrum of the sodiated cyclic oligomers at m/z 1228.8 of the PEA-Pro sample (U= $-NH(CH_2)_3OCO(CH_2)_8CO-$ or $-O(CH_2)_3NHCO(CH_2)_8CO-$; collision gas = argon).

In the same way, MALDI-TOF/TOF-MS/MS performed on the sodiated diamino alcohol terminated chains at m/z 1303.9 confirms that the ester and amide bonds are distributed randomly in the polymer chains. Figure 8 shows an enlarged portion of the MALDI-TOF/TOF-MS/MS spectrum in the range m/z 190–330. Additionally, in this case, it is possible identify sodiated product ions originated from: (a) ester/ester, (b) amide/amide and (c) ester/amide sequences.

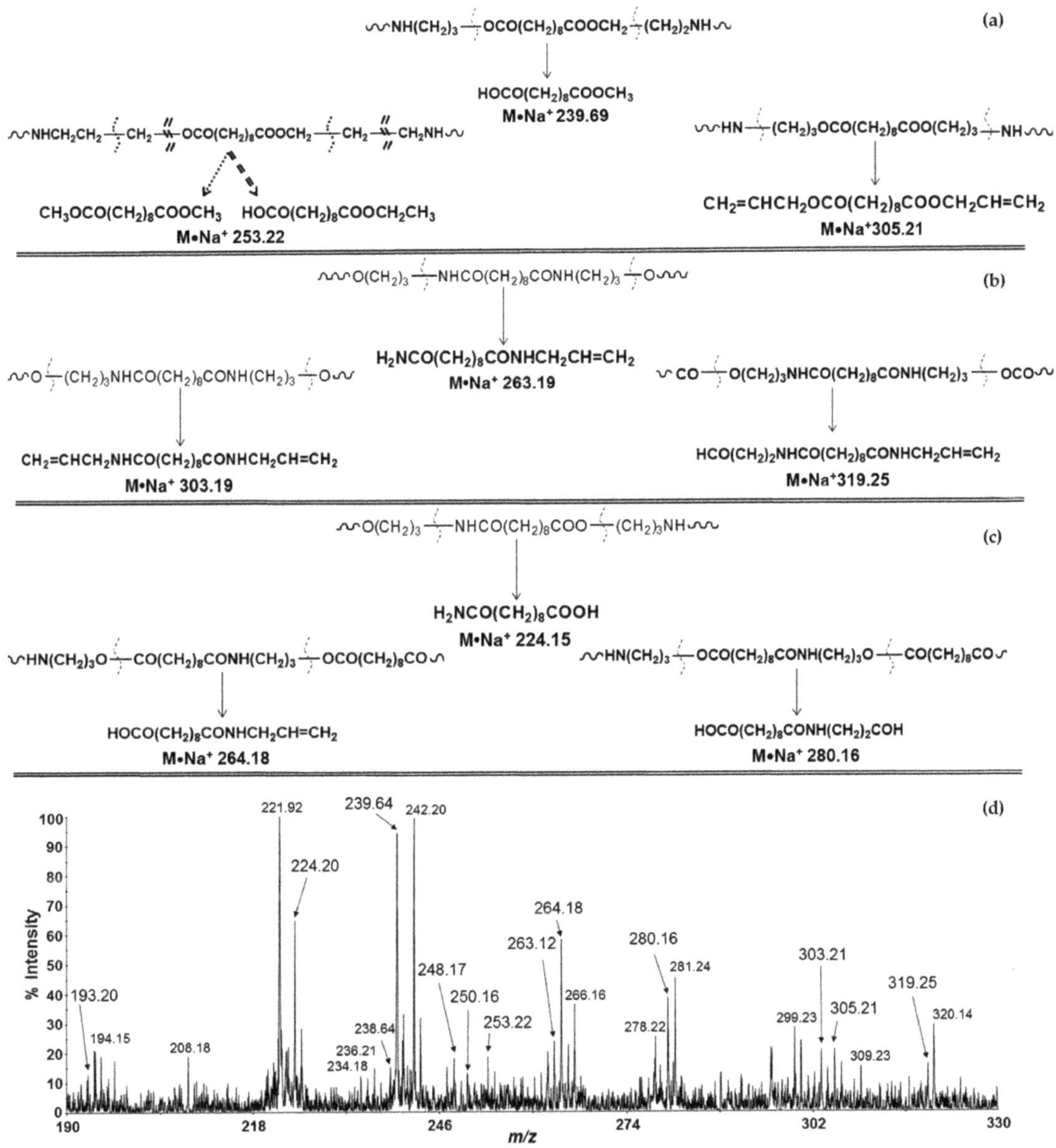

Figure 7. Bond cleavages leading to the diagnostic product ions from: (**a**) ester/ester, (**b**) amide/amide, and (**c**) ester/amide sequences. (**d**) Enlarged portion of MALDI-TOF/TOF-MS/MS spectrum of the sodiated cyclic oligomers at m/z 1228.8 of the PEA-Pro sample (collision gas = argon).

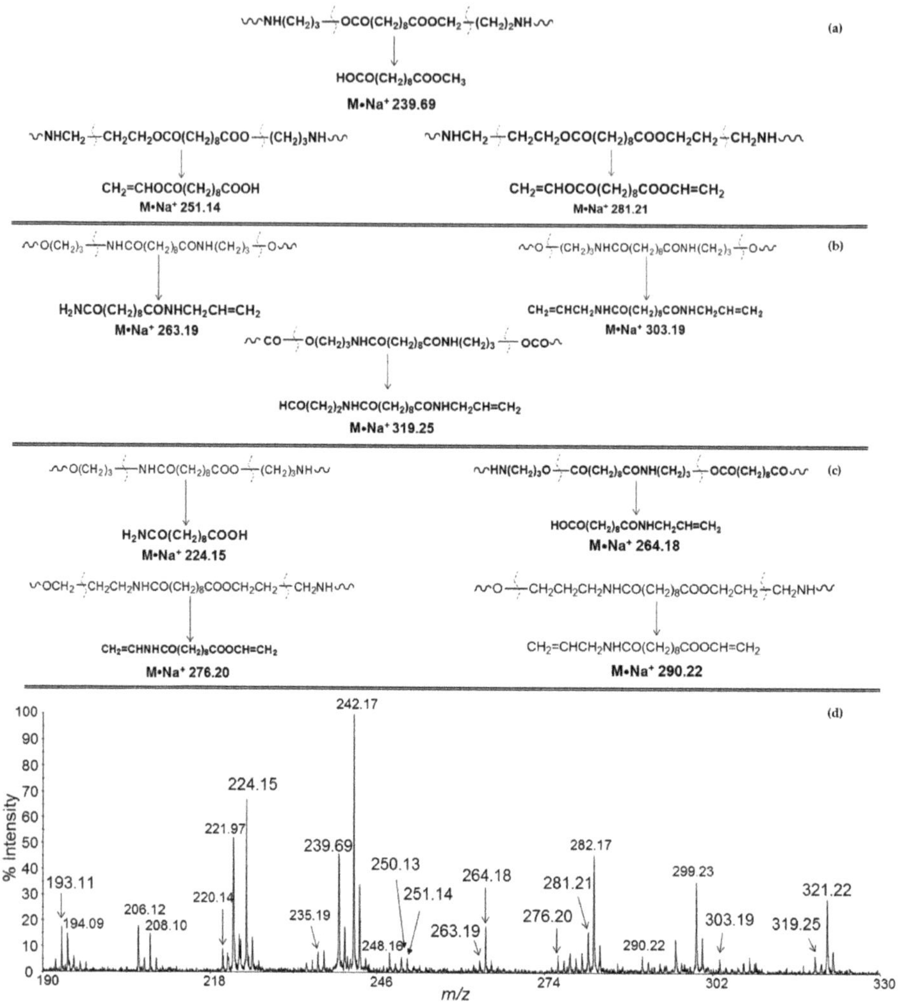

Figure 8. Bond cleavages leading to the diagnostic product ions from: (**a**) ester/ester, (**b**) amide/amide, and (**c**) ester/amide sequences. (**d**) Enlarged portion of MALDI-TOF/TOF-MS/MS spectrum of the sodiated dicarboxyl-terminated oligomers at m/z 1303.9 of the PEA-Pro sample (collision gas = argon).

3.4. Thermogravimetric Analysis (TGA)

TGA provides much unique information that can be usefully combined with those obtained by other techniques, such as Py-GC/MS [18,22]. The thermal stability of PEA-Pro was evaluated by TGA at a heating rate of 10 °C/min under nitrogen atmosphere. In Figure 9, the weight loss (TGA, green trace) and its derivative (DTG, blue trace) as a function of temperature are presented together. Weight loss starts at about 150 °C, probably due to oligomers present in the sample and possibly to adsorbed water. The synthesis was performed at 180 °C [19] and this can justify the formation of degradation products, in agreement with the presence of ions detected in the mass spectra (Figure 3, species G and F) and of unassigned ions. Thermal degradation was found to happen in two main stages with respective maximum decomposition rates at around 380 °C and 446 °C. The first step (380 °C) corresponds to a weight loss of 74% and could be caused by the degradation of polyester segments; the second step (447 °C), with a mass loss of 22%, is related to the degradation of the amide segments [23]. Information about the degradation steps in TGA were used to optimize experimental conditions in Py-GC/MS analyses.

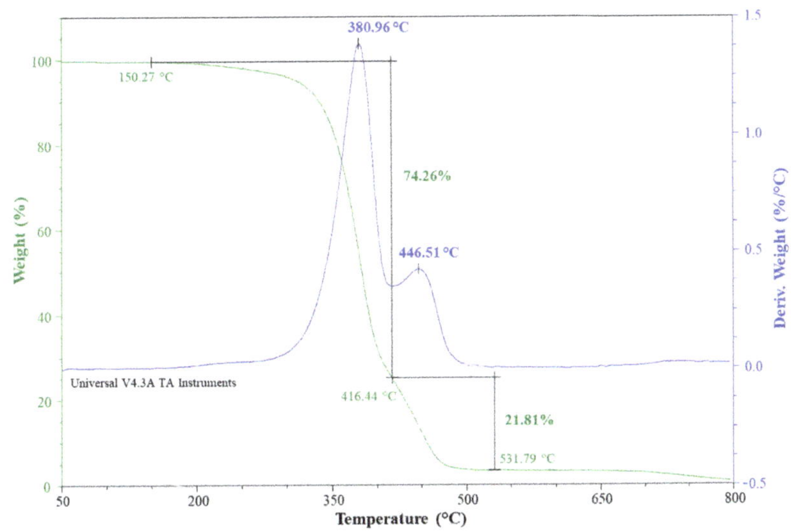

Figure 9. TGA (green trace) and DTG (blue trace) (heating rate 10 °C/min) for the PEA-Pro sample.

3.5. *Pyrolysis-Gas Chromatography/Mass Spectrometry Analysis (Py-GC/MS)*

The structure and thermal degradation of PEA-Pro was also investigated by pyrolysis coupled with GC/MS for the separation and detection/identification of the products of the pyrolysis. Two typical chromatograms of the pyrolysates of the PEA-Pro at 350 °C and 400 °C are shown in Figure S3a,b, respectively. At 450 °C, no signals were detected.

Possible compounds of the pyrolysis at 350 °C and 400 °C are listed in Table S2.

More degradation products were detected at 400 °C. Several vinyl-terminated compounds were identified, due to β-hydrogen scissions (occurring in β positions to both ester and amide bonds) and consequent rearrangements, and to dehydration processes. Additionally, carboxylic acids, alcohols, and as well as nitriles were assigned, in agreement with the literature [24]. At 350 °C, the main degradation product (RT = 22.049 min) originated from a β Hydrogen transfer mechanism or intramolecular cyclization accompanied by water loss. At higher degradation temperature (400 °C), the relative area of this signal (RT = 22.024 min) was lower, suggesting that the formation of those products is less favoured at higher temperatures. Overall, Py-GC/MS showed partial similarity between the degradation products and the main fragmentation mechanism detected by the MALDI tandem MS technique.

4. Conclusions

MALDI-TOF/TOF-MS/MS has proven to be a convenient technique for the structural characterization of a synthetic polyesteramide sample, providing detailed information on the fragmentation pathways and the ester/amide bond sequences in the polymer chains. The presence of different end groups does not influence the fragmentation of sodiated PEA-Pro oligomers, and similar series of product ions were observed in the MALDI-TOF/TOF-MS/MS spectra. Based on the structures of the product ions identified in the present work (Table S1), two main fragmentation cleavages have been proposed to occur most recurrently in PEA-Pro. In agreement with previous studies on similar polyesteramide samples, a β-hydrogen transfer rearrangement appears to be the most important fragmentation mechanism occurring in these MALDI-TOF/TOF-MS/MS experiments carried out with and without collision gas. The $-CH_2-CH_2-$ (β–γ) bond cleavage in the sebacate moiety is reasonably driven by the formation of product ions bearing α,β-unsaturated ester or amide end groups. For each cleavage, two product ions are expected to appear in the MS/MS spectrum. However, the cleavage of some of the bonds yields only one ion. This should suggest that considerable differences exist in the sodium affinity of the product ions. In fact, when fragmentation occurs, the sodium cation can remain on one or the other of the two fragments, but the sodium affinity of the fragments most likely affects the distribution of the sodium cation between the species formed. MALDI-TOF/TOF-MS/MS analysis provided structural information concerning bond sequences in the polymer chains. The several new product ions present in MALDI-TOF/TOF-MS/MS spectra recorded using argon as the collision gas, above all in the low-mass range, proved diagnostic and made it possible to ascertain the presence of random sequences of ester and amide bonds in the polyesteramide sample.

Furthermore, TGA of PEA-Pro highlighted two main degradation steps at 380 °C (weight loss 74%) and 446 °C (weight loss 22%). Additionally, structural characterization performed by Py-GC/MS provided a partial similarity between the degradation products and the main fragmentation mechanism detected by the MALDI technique.

NMR analysis showed the presence of hydroxyl and amino terminal groups as well carboxylic groups of the sebacate moiety. As a plus, NMR showed that the 3-amino-1-propanol termination may present a free hydroxyl or amine group but, from a quantitative point of view, that hydroxyl and carboxyl termination have the same abundance, while the amine termination is 2.7-times less frequent.

Supplementary Materials: The following supporting information can be downloaded at: https://www.mdpi.com/article/10.3390/polym14081500/s1, Figure S1: Partial g-COSY spectrum of PEA-Pro in 6d-DMSO at 50 °C showing the 4.2–1.0 ppm range. Figure S2: Partial g-COSY spectrum of PEA-Pro in 6d-DMSO at 50 °C showing the 8.0–2.6 ppm range. Figure S3: Total ion current (TIC) trace of pyrolysis products of PEA-Pro at (**a**) 350 °C and (**b**) 400 °C; Table S1: Structure assignments of product ions appearing in the MALDI-TOF/TOF-MS/MS spectra of the PEA-Pro sample (precursor ions at *m/z* 1189.8, 1228.8, 1246.8 and 1303.9). Table S2: Structure assignment of the degradation products derived from the pyrolysis of the PEA-pro at 350 and 400 °C.

Author Contributions: Conceptualization, P.R.; investigation, P.R., S.L.C., E.F.M., M.R. and G.I.; resources, P.R.; data curation, P.R., S.L.C., E.F.M., M.R. and G.I.; writing—original draft preparation, P.R. and S.L.C.; writing—review and editing, P.R. and G.I.; visualization, P.R., S.L.C., M.R. and G.I.; supervision, P.R.; project administration, P.R.; funding acquisition, P.R. All authors have read and agreed to the published version of the manuscript.

Funding: This research was funded by PO FESR 2014–2020. Action 1.1.5, project "New therapeutic strategies in ophthalmology: bacterial, viral and microbial infections-NUSTEO", CUP: G68I18000700007-application code 08CT2120090065. POR FESR Sicily 2020, Call 11/2017, project "Polymeric systems: innovative aspects and applications in the biomedical and agri-food fields–SPIN OFF of Polymers", CUP: G67B17000190009–application code 2014.IT.05.SFOP.014/3/10.4/9.2.10/0009, partially granted this research.

Institutional Review Board Statement: Not applicable.

Informed Consent Statement: Not applicable.

Data Availability Statement: The data presented in this study are available on request from the corresponding author.

Acknowledgments: Many thanks are due to Roberto Rapisardi for his continuous and skilful technical assistance.

Conflicts of Interest: The authors declare no conflict of interest. The funders had no role in the design of the study; in the collection, analyses, or interpretation of data; in the writing of the manuscript, or in the decision to publish the results.

References

1. Rodriguez, R.; Franco, M.; Puiggali, J. Biodegradable Poly(Ester Amide)s: Synthesis and Applications. In *Biodegradable Polymers: Processing, Degradation and Applications*; Felton, G.P., Ed.; Nova Science Publishers: Hauppauge, NY, USA, 2011; pp. 207–272, ISBN 978-1-61209-534-9. Available online: http://hdl.handle.net/2117/13036 (accessed on 11 March 2022).
2. Li, L. (Ed.) *MALDI Mass Spectrometry for Synthetic Polymer Analysis*; John Wiley & Sons: Hoboken, NJ, USA, 2010; Volume 175, ISBN 978-0-470-56722-7.
3. Yoo, H.-J.; Kim, D.-H.; Shin, D.; Oh, Y.; Lee, S.; Lee, J.Y.; Choi, Y.-J.; Lee, K.-S.; Kim, Y.; Cho, K. Recent developments in pre-treatment and analytical techniques for synthetic polymers by MALDI-TOF mass spectrometry. *Anal. Methods* **2020**, *12*, 5767–5800. [CrossRef] [PubMed]
4. Byrd, H.C.M.; McEwen, C.N. The Limitations of MALDI-TOF Mass Spectrometry in the Analysis of Wide Polydisperse Polymers. *Anal. Chem.* **2000**, *72*, 4568–4576. [CrossRef] [PubMed]
5. Nielen, M.W.F. Maldi time-of-flight mass spectrometry of synthetic polymers. *Mass Spectrom. Rev.* **1999**, *18*, 309–344. [CrossRef]
6. Medzihradszky, K.F.; Campbell, J.M.; Baldwin, M.A.; Falick, A.M.; Juhasz, P.; Vestal, M.L.; Burlingame, A.L. The Characteristics of Peptide Collision-Induced Dissociation Using a High-Performance MALDI-TOF/TOF Tandem Mass Spectrometer. *Anal. Chem.* **2000**, *72*, 552–558. [CrossRef] [PubMed]
7. Yol, A.M.; Dabney, D.E.; Wang, S.-F.; Laurent, B.A.; Foster, M.D.; Quirk, R.P.; Grayson, S.M.; Wesdemiotis, C. Differentiation of linear and cyclic polymer architectures by MALDI tandem mass spectrometry (MALDI-MS2). *J. Am. Soc. Mass Spectrom* **2013**, *24*, 74–82. [CrossRef] [PubMed]
8. Crotty, S.; Gerişlioğlu, S.; Endres, K.J.; Wesdemiotis, C.; Schubert, U.S. Polymer architectures via mass spectrometry and hyphenated techniques: A review. *Anal. Chim. Acta* **2016**, *932*, 1–21. [CrossRef] [PubMed]
9. Wesdemiotis, C.; Solak, N.; Polce, M.J.; Dabney, D.E.; Chaicharoen, K.; Katzenmeyer, B.C. Fragmentation pathways of polymer ions. *Mass Spectrom. Rev.* **2011**, *30*, 523–559. [CrossRef] [PubMed]
10. Rizzarelli, P.; Rapisarda, M.; Valenti, G. Mass spectrometry in bioresorbable polymer development, degradation and drug-release tracking. *Rapid Commun. Mass Spectrom.* **2020**, *34*, e8697. [CrossRef] [PubMed]
11. Rizzarelli, P.; Rapisarda, M. Tandem Mass Spectrometry in the Analysis of Biodegradable Polymers. In *Mass Spectrometry: Theory and Applications*; Nichols, W.O., Ed.; Nova Science Publishers, Inc.: Hauppauge, NY, USA, 2021; pp. 127–181, ISBN 978-1-53619-790-7.
12. Rizzarelli, P.; Puglisi, C.; Montaudo, G. Sequence determination in aliphatic poly(ester amide)s by matrix-assisted laser desorption/ionization time-of-flight and time-of-flight/time-of-flight tandem mass spectrometry. *Rapid Commun. Mass Spectrom.* **2005**, *19*, 2407–2418. [CrossRef] [PubMed]
13. Rizzarelli, P.; Puglisi, C. Structural characterization of synthetic poly(ester amide)from sebacic acid and 4-amino-1-butanol by matrix-assisted laser desorption ionization time-of-flight/time-of-flight tandem mass spectrometry. *Rapid Commun. Mass Spectrom.* **2008**, *22*, 739–754. [CrossRef] [PubMed]
14. Rizzarelli, P.; Puglisi, C.; Montaudo, G. Matrix-assisted laser desorption/ionization time-of-flight/time-of-flight tandem mass spectra of poly(butylene adipate). *Rapid Commun. Mass Spectrom.* **2006**, *20*, 1683–1694. [CrossRef] [PubMed]
15. Rizzarelli, P. Matrix-assisted laser desorption ionization time-of-flight/time-of-flight tandem mass spectra of biodegradable polybutylenesuccinate. *Rapid Commun. Mass Spectrom.* **2013**, *27*, 2213–2225. [CrossRef] [PubMed]
16. Rizzarelli, P.; Carroccio, S. Role of Mass Spectrometry in the Elucidation of Thermal Degradation Mechanisms in Polymeric Materials. In *Reactions and Mechanisms in Thermal Analysis of Advanced Materials*; Tiwari, A., Raj, B., Eds.; John Wiley & Sons: Hoboken, NJ, USA, 2015; pp. 221–258, ISBN 978-1-119-11771-1. [CrossRef]
17. Tsuge, S.; Ohtani, H. Structural characterization of polymeric materials by pyrolysis-GC/MS. *Polym. Degrad. Stab.* **1997**, *58*, 109–130. [CrossRef]
18. Rizzarelli, P.; Rapisarda, M.; Perna, S.; Mirabella, E.F.; La Carta, S.; Puglisi, C.; Valenti, G. Determination of polyethylene in biodegradable polymer blends and in compostable carrier bags by Py-GC/MS and TGA. *J. Anal. Appl. Pyrolysis* **2016**, *117*, 72–81. [CrossRef]
19. Rizzarelli, P.; Cirica, M.; Pastorelli, G.; Puglisi, C.; Valenti, G. Aliphatic poly(ester amide)s from sebacic acid and aminoalcohols of different chain lengths: Synthesis, characterization and soil burial degradation. *Polym. Degrad. Stab.* **2015**, *121*, 90–99. [CrossRef]
20. Levchik, S.V.; Weil, E.D.; Lewin, M. Thermal decomposition of aliphatic nylons. *Polym. Int.* **1999**, *48*, 532–557. [CrossRef]
21. Rizzarelli, P.; Carroccio, S. Modern mass spectrometry in the characterization and degradation of biodegradable polymers. *Anal. Chim. Acta* **2014**, *808*, 18–43. [CrossRef]
22. Rizzarelli, P.; Piredda, G.; La Carta, S.; Mirabella, E.F.; Valenti, G.; Bernet, R.; Impallomeni, G. Characterization and laser-induced degradation of a medical grade polylactide. *Polym. Degrad. Stab.* **2019**, *169*, 108991–109004. [CrossRef]
23. Papadopoulos, L.; Klonos, P.A.; Kluge, K.; Zamboulis, A.; Terzopoulou, Z.; Kourtidou, D.; Magaziotis, A.; Chrissafis, K.; Kyritsis, A.; Bikiaris, D.N.; et al. Unlocking the potential of furan-based poly(ester amide)s: An investigation of crystallization, molecular dynamics and degradation kinetics of novel poly(ester amide)s based on renewable poly(propylene furanoate). *Polym. Chem.* **2021**, *12*, 5518–5534. [CrossRef]
24. Tsuge, S.; Ohtani, H.; Watanabe, C. *Pyrolysis-GC/MS Data Book of Synthetic Polymers: Pyrograms, Thermograms and MS of Pyrolyzates*, 1st ed.; Elsevier: Oxford, UK, 2012; ISBN 9780444538932.

polymers

Review

Recent Strategies for the Immobilization of Therapeutic Enzymes

Chen-Yuan Zhu [1], Fei-Long Li [1], Ye-Wang Zhang [1,*], Rahul K. Gupta [2], Sanjay K. S. Patel [2,*] and Jung-Kul Lee [2,*]

[1] School of Pharmacy, Jiangsu University, Zhenjiang 212013, China; zhuchenyuann@126.com (C.-Y.Z.); feilongli2018@outlook.com (F.-L.L.)
[2] Department of Chemical Engineering, Konkuk University, Seoul 05029, Korea; guptarahul9m@gmail.com
* Correspondence: zhangyewang@ujs.edu.cn (Y.-W.Z.); sanjaykspatel@gmail.com (S.K.S.P.); jkrhee@konkuk.ac.kr (J.-K.L.); Tel.: +82-2-450-3505 (J.-K.L.)

Abstract: Therapeutic enzymes play important roles in modern medicine due to their high affinity and specificity. However, it is very expensive to use them in clinical medicine because of their low stability and bioavailability. To improve the stability and effectiveness of therapeutic enzymes, immobilization techniques have been employed to enhance the applications of therapeutic enzymes in the past few years. Reported immobilization techniques include entrapment, adsorption, and covalent attachment. In addition, protein engineering is often used to improve enzyme properties; however, all methods present certain advantages and limitations. For carrier-bound immobilization, the delivery and release of the immobilized enzyme depend on the properties of the carrier and enzyme. In this review, we summarize the advantages and challenges of the current strategies developed to deliver therapeutic enzymes and provide a future perspective on the immobilization technologies used for therapeutic enzyme delivery.

Keywords: adsorption; carrier; covalent attachment; entrapment; immobilization; protein engineering; therapeutic enzyme

Citation: Zhu, C.-Y.; Li, F.-L.; Zhang, Y.-W.; Gupta, R.K.; Patel, S.K.S.; Lee, J.-K. Recent Strategies for the Immobilization of Therapeutic Enzymes. *Polymers* **2022**, *14*, 1409. https://doi.org/10.3390/polym14071409

Academic Editors: Clara Delgado-Sánchez, Adrián Tenorio-Alfonso, Esperanza Cortés Triviño, Concepción Valencia-Barragán and Antonio M. Borrero-López

Received: 25 February 2022
Accepted: 23 March 2022
Published: 30 March 2022

1. Introduction

Immobilization processes have been effectively demonstrated to improve the properties of biocatalysts, including whole cells and enzymes, for useful biotechnological applications, such as environmental and medical sectors [1–3]. Enzymes have been used as therapeutic agents in the last century. Initially, proteolytic enzymes, such as pepsin were used for the treatment of gastrointestinal disorders, such as lysosomal storage disease. With the development of biotechnology, enzymes have been used to treat a wide range of diseases, including genetic diseases, cancer, infectious diseases, burn debridement, miscellany, and clotting [4].

Owing to their high activity, unique selectivity for target substrates, and catalytic capabilities, enzymes are more advantageous than traditional drugs [5]. Although there are a number of enzymes that could potentially be applied in clinical medicine, only a small amount can be used because of the body's capacity to develop defensive responses to extraneous proteins, which results in their rapid clearance [6]. As proteins, enzymes are costly and also easily degraded because of their large size and limited distribution within the host [7]. To protect therapeutic enzymes from degradation, scientists have attempted to modify or immobilize enzymes. Immobilization is an important strategy for stabilizing enzymes, especially for the delivery of therapeutic enzymes. In the present review, the various immobilization methods developed for therapeutic enzymes are discussed and summarized.

2. Immobilization of Therapeutic Enzymes

There are three main immobilization methods used for enzymes: adsorption, entrapment, and covalent attachment [8–11]. Rapidly developing strategies, such as molecular techniques, protein engineering, and nanotechnology have also been reported for the immobilization of enzymes with therapeutic applications [3,9,11–14].

2.1. Entrapment

The entrapment of enzymes is an important delivery strategy for therapeutic enzymes. In this method, enzymes can be entrapped in hydrogels, liposomes, red blood cells (RBCs), or deoxyribonucleic acid (DNA) cages and then transported to the target tissue.

2.1.1. Entrapment with Hydrogels

Chitosan and chitin are linear polysaccharides with *N*-acetyl-D-glucosamine and D-glucosamine units at different ratios. Owing to their exceptional ease in forming hydrogels and remarkable affinity for proteins, chitin- and chitosan-based materials have the advantages of good biocompatibility and biodegradability, non-toxicity, and physiological inertness [15–17]. Enzymes can be entrapped without requiring chemical activation and cross-linker or bifunctional agents.

Lipase has been reported to be immobilized inside chitosan hydrogel beads; moreover, the entrapment efficiency was approximately 51%, and the activity of immobilized lipase for freeze-dried beads was 1110 U/mL [18,19]. Chitosan hydrogels have low solubility and are highly pH-dependent, which could lead to undesired enzyme release in the stomach [20]. Therefore, graft copolymerization with acyl, alkyl, monomeric, and polymeric moieties of chitosan is required for applications in enzyme immobilization to allow the entrapped enzyme to be delivered more accurately; for this purpose, magnetite could be introduced for the matrix (Table 1) [21–23].

Carboxymethylcellulose (CMC) is also a good carrier for the preparation of hydrogels. Prepared CMC gels could be used as supports to carry superoxide dismutase (SOD) for wound healing, and the resistance of native SOD to hydrogen peroxide inactivation and the effects of applying SOD-CMC hydrogels to open wounds on the backs of rats were examined [24]. The results showed that SOD entrapped in the hydrogel could be used to shorten the wound healing time (19 and 18 days for SOD and SOD-CMC 50%, respectively) and was also more resistant to peroxide inactivation when compared with the native enzyme. CMC can be used to form network beads with alginate for protein entrapment, and the network beads were found to shrink in gastric conditions and swell in intestinal conditions. Therefore, the controlled delivery of an enzyme to a particular part of the gastrointestinal tract was demonstrated [25].

Hydrogels are three-dimensional matrices composed of a cross-linked polymer network and hydrophilic polymer chains that resemble soft human tissue because of the immobilization of large amounts of water [26]. For this reason, hydrogels are one of the best carriers for therapeutics because they have good biocompatibility and can be used for a number of biomedical applications. Degradable dextran hydrogels have excellent biological compatibility and can be used as a long-term delivery system with enzyme release patterns varying over wide time ranges: from days to months [27,28].

For example, L-asparaginase, which is used as a chemotherapeutic agent for the treatment of acute lymphoblastic leukemia, could cause side effects, such as allergic or hypersensitivity reactions. However, its immobilization with dextran hydrogels could improve its delivery and eliminate allergic reactions. In addition, to solve the inactivation of enzymes in organic solvents, dextran microspheres could be used, which could be formed in a poly(ethylene glycol) (PEG) water system [29,30].

Selecting the proper PEG molecular weight, concentration, and degree of substitution is essential for the entrapment and delivery of enzymes. Dextran can be incorporated with other biomaterials, such as alginate [31] or chitosan [19], for the entrapment of proteins or enzymes. PEG-b-polylactic acid was used for the encapsulation of β-galactosidase as polymersomes (~145 nm) and showed 72% residual activity. The resulting biocatalyst was found to be effective in the treatment of lysosomal storage disorders of the brain [32]. Alkaline phosphatase immobilized through composite bioceramic and hydrogel assembly exhibited enzyme loading up to 48%; thus, it could be a promising candidate for effective drug delivery, especially for bone reconstruction (Table 1) [33].

Alginate is a binary linear heteropolymer containing 1,4-linked β-D-mannuronic and α-L-guluronic acid residues. Barrias et al. [20] reported that entrapped glucocerebrosidase in alginate microspheres retained full activity and exhibited a broader pH-dependent activity profile. Moreover, the optimum pH was shifted toward the alkaline region by approximately 0.5 units, and the enzyme could also be released in a sustained manner. Glucose oxidase (GOX) has also been entrapped in alginate microspheres either chemically or physically [34], and nano-organized coatings were applied on the microspheres to stabilize the immobilized enzyme in a biological environment. Alginate composites can be prepared by incorporating other matrices, such as nano-hydroxyapatite for sustained drug release [23]. If an appropriate alginate, with an optimal pore size and degradation rate, is chosen for the entrapment of biological macromolecules, the release kinetics can be controlled [35].

In addition to these natural biomaterials, biodegradable organic polymers have been used for drug delivery for approximately 40 years. Poly(lactide-co-glycolide) (PLGA) and poly(caprolactone) are two notable biodegradable organic polymers [36–39]. Giovagnoli et al. [38,40] reported that SOD and catalase could be entrapped in PLGA microspheres and slowly released (Table 1). The immobilized enzymes significantly enhanced the viability and function of cultured neonatal pancreatic porcine cell clusters, suggesting that this technique could provide new strategies for introducing transplant-suitable pancreatic cell tissue for the treatment of type 1 diabetes mellitus.

When entrapped inside PLGA microspheres without denaturation, enzymes could be released in a sustained manner for 2 months. Giovagnoli et al. [40] also entrapped SOD and catalase in PLGA microspheres using a water/oil/water double emulsion method. Their results showed that immobilized SOD was better stabilized in PLGA microspheres and that hydrogels were useful for delivering SOD in its native form, making this a promising new depot system.

The chemical structures of these naturally degradable hydrogels are presented in Figure 1. All of these hydrogels exhibit common characteristics. They are natural polymers that can be dissolved in water to form hydrogels. Moreover, these hydrogels are typically biocompatible and can be used in the human body without leading to severe rejection reactions. Furthermore, their chemical and physical properties render them as good candidates for the immobilization and delivery of therapeutic enzymes [41].

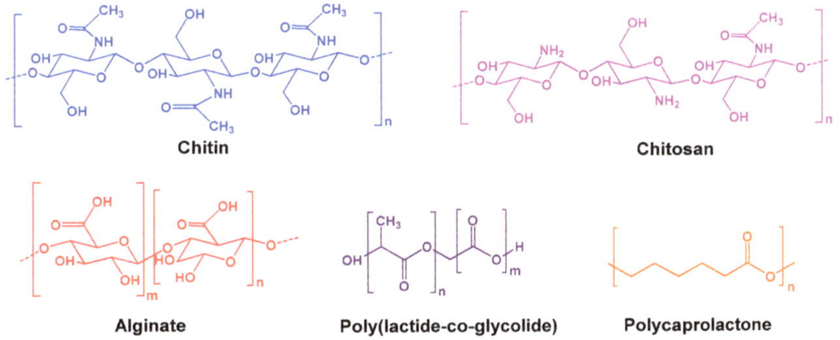

Figure 1. The chemical structures of certain biocompatible hydrogels used for therapeutic enzyme entrapment.

2.1.2. Liposomal Entrapment

Liposomes are colloidal particles that are formed when phospholipids are dispersed in an aqueous solution [42,43]. Depending on their composition, liposomes can be extremely stable, biocompatible, or biodegradable. Liposome-encapsulated drugs usually have a controlled release pattern, which could be used for the entrapment of enzymes (Figure 2a). The chemical properties, preparation, and applications of liposomes in drug delivery were previously reviewed [44]. In these studies, liposome encapsulation was used to protect β-galactosidase and other enzymes from degradation; it also lowered the risk of immunological reactions and enhanced the possibility of membrane fusion and the delivery of the enzymes into cells [18,45,46].

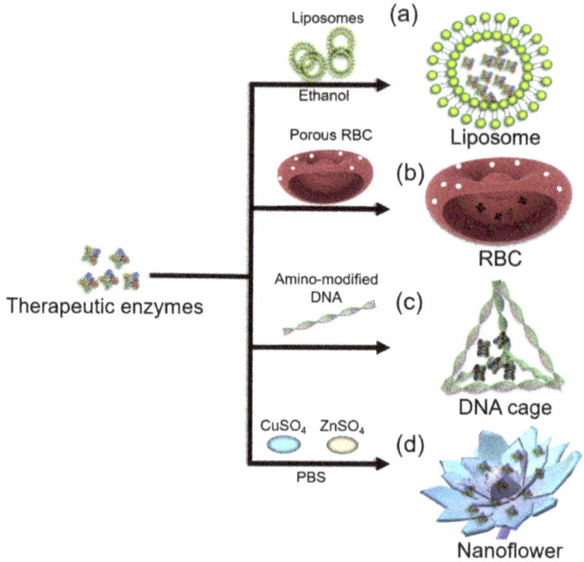

Figure 2. Illustration of entrapment approaches for therapeutic enzyme immobilization using (**a**) liposomes, (**b**) red blood cells (RBCs), (**c**) DNA cages, and (**d**) metal–protein hybrids (nanoflowers). Here, PBS denotes phosphate-buffered saline.

Alternating adsorption of layers of natural polymers, such as alginate, chitosan, and phospholipid vesicles, resulted in enhanced encapsulation efficiency of the protein [46]. The entrapment efficiency depends on the interactions between the enzymes and the phospholipid bilayer. These results indicated that liposomes coated with 10 layers of polyelectrolytes could encapsulate more than 80% of the loaded protein. The enzyme liposome preparation and catalytic efficiency as well as the kinetic properties and stability have also been previously reviewed [46,47].

The reverse-phase evaporation technique is widely used for the preparation of liposomes. When this strategy is used, all components that remain after the evaporation of the organic solvent form organogels. Subsequently, nanovesicles were formed via the dispersion of the organogels in pure water under shaking conditions, and liposomes were formed via the addition of the remaining aqueous solution that contained therapeutic enzymes to the organogel under high agitation using a rotary evaporator [48]. The entrapment process had a recondite impact on the enzymes—especially the catalytic efficiency.

2.1.3. RBCs

Compared with extraneous drug carriers, endogenous carriers have various ligands that can recognize targets that are expressed preferentially at the diseased site. Furthermore, they can serve as efficient and safe carriers for targeted drug delivery. Carriers, including RBCs, macrophages, stem cells, platelets, exosomes, albumin, transferrin, and apolipoproteins, were recently reviewed [49]. Among these carriers, RBCs are the most appealing because they are the most abundant cells in human blood, with a lifespan of 100–120 days (Figure 2b).

Many traditional methods have already been employed to entrap enzymes into RBCs. The most adopted techniques include drug-induced endocytosis, electroporation [50], and dialysis [51]. Only 0.2–0.3% of the added enzymes, such as β-glucosidase and β-galactosidase can be entrapped into RBCs using the hypotonic dilution method. In addition, some higher-molecular-weight substances, such as hemoglobin, ferritin, dextran, and colloidal gold, can enter RBCs using the drug-induced endocytosis method, which presents an entrapment rate of 20%.

Compared with the dialysis and endocytosis methods, electroporation is based on the decrease in the cell membrane barrier function that occurs upon the exposure of cells to a strong electric field pulse (electropermeabilization), which can significantly increase the permeability of the RBC membrane and lead to higher entrapment efficiency [52]. Using these methods to obtain large pores or perturbations on the cell membrane, several enzymes, including L-asparaginase, acetaldehyde dehydrogenase, and alcohol dehydrogenase, have been successfully loaded into RBCs. He et al. [53] reported that L-asparaginase can be entrapped with cell-penetrating peptides, resulting in a 4% enzyme loading efficiency with about 70% of activity being retained after encapsulation. Compared with the intravenous administration of non-encased β-glucuronidase, RBC-encased enzyme activity was significantly extended [54].

2.1.4. DNA Cage

DNA is an ideal material for the fabrication of rigid structures because its assembly is easy to execute and can be controlled by base-pairing [55]. DNA Td nanocages can be easily prepared and connected by programmable DNA linkers [56]. For the construction of DNA cages, strategies, such as modular self-assembly, the hierarchical assembly of branched DNA tiles, the one-pot approach, and DNA origami-based strategies were previously reviewed [57]. DNA cages can be functionalized with a multitude of chemical tags for target attachment, making them suitable as drug carriers [58]. Erben et al. [59] reported the encapsulation of a single protein molecule into a DNA cage with the protein conjugated to the end of an oligonucleotide through a surface amine.

This protein–DNA complex was combined with oligonucleotides s2–s4 to form a tetrahedron where the protein molecule was attached to one edge (Figure 2c). In addition, the position of the protein could be controlled by altering the sequence of the edges of the DNA cage [52]. Kim et al. [60] presented a method to encapsulate enzymes in a DNA cage that could transform its conformation depending on the pH, allowing reversible control of the enzyme accessibility to the surrounding environment.

If DNA cages can be modified with a proper tag, the delivery of encapsulated enzymes to their target is possible. For therapeutic enzymes, the use of DNA cages as immobilization carriers is alluring. Simultaneous loading of cytosine deaminase (as a prodrug enzyme) and doxorubicin (drug) in polymeric nanoparticles performed through tri-block co-polymers containing click nucleic acids can be effectively used for the co-delivery of chemotherapeutic agents for broad therapeutic applications (Table 1) [61].

2.1.5. Metal–Protein Hybrids

Proteins immobilized through the metal–protein hybrid system are known as nanoflowers owing to their flower-like structural morphology [8,9,62,63]. Metal–protein hybrid nanoflowers can be easily synthesized under the harsh conditions required for the modified nanomaterials to immobilize enzymes [64]. Self-assembly of enzyme and copper sulfate at room temperature for an incubation period of 3 days in phosphate buffer showed efficient immobilization of laccase with relative activity up to 650% compared with that of the free enzyme [62].

Epinephrine is excreted in the urine of patients with pheochromocytoma, a rare tumor of the adrenal gland. Laccase-based nanoflowers resulted in three-fold faster oxidation of epinephrine than the free enzyme and showed a diagnostic epinephrine detection range of 0.01–1.0 μg/mL. Thus, the detection of epinephrine using laccase-based nanoflowers could reach the diagnostic range of 0.01–1 mg/mL in 24 h [62,63]. Similarly, copper, zinc, calcium, magnesium, cobalt, and manganese based-enzyme immobilization systems have been reported [64,65]. Typically, most nanoflowers are synthesized using copper or calcium ions.

Although the formation of hybrid nanoflowers using copper ions is facile and does not involve high energy consumption, it is a lengthy process (approximately 3 days). Therefore, it is necessary to shorten the formation time of the hybrid nanoflowers. Zinc ions react with the phosphate radical faster than copper ions. Typically, the formation of biomolecule–inorganic hybrid nanoflowers involves three steps as follows. First, primary copper phosphate crystals are formed. During the growth step, large agglomerates of biomolecules and primary crystals are formed, and flower-like petals appear.

Finally, branched flower-like structures are formed via anisotropic growth (Figure 2d) [66]. Nanoflower hybrids synthesized using different enzymes exhibited significant improvements in enzyme properties, including efficiency, kinetics, substrate specificity, and stability [9,63]. The effective immobilization of enzymes as nanoflowers has been associated with: (i) large surface areas, (ii) stable structural morphology to reduce significant mass-transfer limitations, and (iii) well-organized confinement of enzyme molecules within the hybrid structure for enhanced co-operative influence [9,63,64]. Metal-based protein hybrid nanoflowers have been applied for efficient drug delivery, as biosensors, and for various medical approaches, including high-resolution images of cells and for monitoring traceable drug delivery systems [62].

Hao et al. [67] reported a mild biomimetic method for calcium hydrogen phosphate ($CaHPO_4$) nanoflower synthesis using uricase and horseradish peroxidase (HRP). The hybrid nanoflowers were potentially integrated by a hyaluronic acid dissolvable microneedle system (uricase and HRP-$CaHPO_4$@HA MN) to accomplish effective hyperuricemia treatment through transdermal drug delivery. The bio-friendly immobilization of SOD into vaterite calcium carbonate crystals through encapsulation showed a high loading efficiency of 93% [68].

The resulting hybrid exhibited controlled release of SOD at physiologically relevant ionic strength, completely retaining its biological activity, which demonstrates that it that can be applied for ophthalmology applications. Similarly, zeolite imidazole framework-8 (ZIF-8) has been used as a promising candidate for biomedical therapies. Bai et al. [69] reported that ZIF-8 was fabricated using the one-pot embedding GOX and HRP strategy (ZIF-8@GOX/HRP) for synergetic cancer therapy. Based on cascade catalysis, ZIF-8@GOX/HRP demonstrated an impressive tumor suppression rate (Table 1).

2.2. *Adsorption*

The adsorption of enzymes onto or into carriers can be accomplished physically or through ionic interactions, allowing for a weak interaction between the enzyme and carrier to be formed. Generally, the immobilization process is mild, and the enzyme properties can be maximally maintained. Thus, many therapeutic enzymes have been immobilized onto a variety of carriers.

2.2.1. Inorganic Carriers

Silica nanoparticles are good carriers for the immobilization of therapeutic enzymes because of their good biocompatibility [70] (Figure 3a). The size and shape of nanoparticles will affect the adsorption patterns, protein structure, and function. Many enzymes, such as lysozyme [70], lipase [71,72], nattokinase [73], asparaginase [74], SOD [75,76], β-galactosidase [77], glucocerebrosidase [78], binase [79], hyaluronidase (HAses) [80], GOX, and HRP [69], have been reported to be immobilized onto nanoparticles. Among these enzymes, the activity and stability of glucocerebrosidase are compromised in patients with Gaucher disease (metabolic storage disorder) because these patients do not properly degrade glucosylceramide (GlcCer), which accumulates and can lead to health problems, including bone lesions.

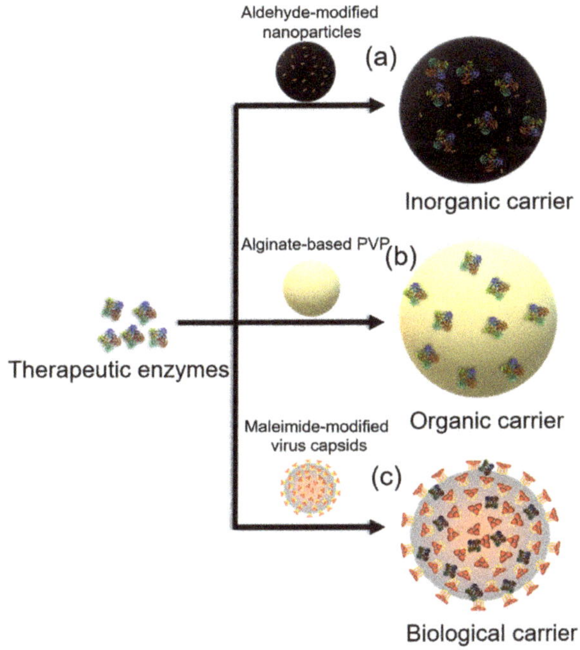

Figure 3. Illustration of adsorption approaches for therapeutic enzyme immobilization on (**a**) silica nanoparticles, (**b**) poly (N-vinyl-2 pyrrolidone) (PVP) microspheres, and (**c**) virus capsids.

Enzyme efficiency was significantly increased up to 230% at higher pH values (7–8) after immobilization using self-assembled monolayer nanoparticles (Table 1) [81]. The adsorption of proteins or enzymes onto or into carriers depends on the properties of the enzyme and carrier. The pore size, pH, and temperature are crucial factors for the adsorption of enzymes. Mesoporous silica nanoparticles have been widely explored because of their high stability under physiological conditions, controllable pore size, extensive surface area, and simplicity of synthesis [70]. Metal nanoparticles, such as gold nanoparticles [82] and silver nanoparticles [83] have been fabricated for the immobilization of β-galactosidase and other enzymes.

Through the surface modification of these nanoparticles, the interaction between enzymes and nanoparticles can be modulated, and their biocompatibility can be improved. Magnetic silica nanoparticles have also received considerable attention because the carried enzyme can be delivered to its target under an external magnetic field. To reduce the toxicity of the magnetic silica nanoparticles, surface modification is required. Patel et al. [84] synthesized a novel type of spherical and porous composite from reduced graphene oxide and magnetic materials to support the immobilization of enzymes. According to the results, this material has the advantages of higher immobilization efficiency and lower acute toxicity. In all the adsorption processes, structural changes may occur, and the enzyme loading will also depend on the surface area [81].

Ordered mesoporous silica nanoparticles have demonstrated effective bioactivity and drug delivery systems owing to their high biocompatibility, broad porosity, and structural morphology [80,85]. Chen et al. [80] reported the stepwise delivery of a biocatalyst and prodrug through multi-responsive nanoplatforms, such as HAase@SiO_2@prodrug for tumor therapeutics with high efficacy and negligible cytotoxicity of particles toward normal tissue. The synthesized hybrid HAase@SiO_2@prodrug contained nearly 22% and 48% of HAase and prodrug, respectively.

Similarly, halloysite nanotubes (HNTs) provide efficient support for the immobilization of SOD and protamine biomacromolecules via self-assembly and act as antioxidant materials [76]. HNTs accommodated 10 and 100 mg of SOD and protamine sulfate polyelectrolyte per gram of particles, respectively. The strong interactions (electrostatic, hydrophobic, and hydrogen bonding) between the HNTs and the biomacromolecules were found to be beneficial to stop enzyme leakage and the HNTs retained two-fold higher residual SOD activity after immobilization [76]. The immobilization method on HNTs with binase, the RNase from *Bacillus pumilus*, displayed a maximum enzyme loading of 85% and a two-fold enhancement in cytotoxicity toward tumor colon cells (Table 1) [79].

2.2.2. Organic Carriers

Organic carriers for the adsorption of therapeutic enzymes in drug delivery systems have also been widely studied. Natural polymers, such as alginate, pectin, chitosan, and carrageenan, have been used for the adsorption of the therapeutic enzymes [5,20,22,86]. Polyhydroxyalkanoate (PHA), an important biodegradable polymer, is synthesized by many Gram-positive and Gram-negative bacteria from at least 75 different genera (Figure 3b) [73,87].

When PHA was used as the carrier for the immobilization of nattokinase (which is a fibrinolytic enzyme involved in the hydrolysis of blood thrombi in vivo and the conversion of plasminogen to plasmin) the activity of nattokinase increased by 20% [73]. Moreover, the stability of nattokinase at 4 °C was excellent. Alginate-based biopolymers have also been developed for sustained drug release [23]. In addition to these natural polymers, biodegradable polyesters, such as polylactic acids, poly(caprolactone), and poly(lactic-co-glycolic acid) PLGA, can be synthesized for the adsorption of enzymes [32,40,61]. Compared with natural organic polymers, synthesized polymers exhibit regular and controlled lengths, molecular weights, and specific properties.

Choi et al. [88] reported a biocompatible system of immobilized urease-based nanomotors to treat bladder diseases that can penetrate deeply into the bladder wall and are stable for a long period. The urease-powered nanomotors exhibited 86% residual activity of the free enzyme and showed no cytotoxicity (1 mg/mL) in human bladder cells (RT4). Similarly, alginate was successfully applied for papain immobilization and used as a wound healer because of its accelerating therapeutic features. Papain remained very stable over a long period of 28 days and retained a residual activity of 80% [89]. In contrast, Jamwal et al. [90] reported the synthesis of a novel glucose-responsive system based on GOX and dextran particles that exhibited insulin-controlled release of 90% under artificial intestinal fluid conditions. However, this system can potentially be applied to subcutaneous insulin therapy (Table 1).

2.2.3. Biological Carriers

The natural ability of viruses to adhere to and enter specific cell types could be exploited for the delivery of therapeutic agents to cytosols. Thus, virosomes have also been used as ideal carriers for the transportation of enzymes (Figure 3c) [91]. However, a major limitation of using virosomes as delivery platforms is the potential for a large immune response from the host following exposure to the virosome surface antigen. Extracellular vesicles, virosomes, and liposomes are typical carriers for the delivery of enzymes. As mentioned above, the liposome-entrapped enzyme could have a disease-triggered drug release, and liposomes can remain stable for weeks [92].

Bacteria can be used to produce all of the components of extracellular vesicles for the production of complex therapeutics and carriers, which can greatly reduce the cost of production and purification [91]. Recently, a review on liposomes, biopolymers, and virus-like particles for the delivery of enzymes was also presented (Table 1) [92,93]. Compared with synthesized carriers, these biological carriers have more attractive properties, such as high biocompatibility, biodegradability, and adjustable structural characteristics, thus, allowing for a simple and low-cost production strategy.

2.3. Covalent Attachment

2.3.1. Conjugation onto Polymers

Polymers are usually used as carriers for the immobilization of therapeutic enzymes, and the immobilized enzyme can be delivered in vivo or used for wound healing. Poly (N-vinylpyrrolidone) (PVP) and poly(vinyl alcohol) (PVA) have been reported for the conjugation of proteins or enzymes, such as SOD [78]. Compared with PVP and PVA, PEG is the standard polymer for the conjugation of enzymes because it results in decreased enzymatic degradation of its payload and reduced immunogenicity. Initially, PEG was produced via the reaction of mono-methoxy-PEG (molecular weight (Mw) 5000) with cyanuric chloride to form PEG dichlorotriazine.

One of the two chlorine atoms of PEG dichlorotriazine would then be displaced by nucleophilic amino acid units, such as lysine, serine, tyrosine, cysteine, and histidine. The remaining chlorine atom was not electrophilic but could react to cause nonspecific crosslinking of the protein; thus, multiple PEG units would become attached to the protein (Figure 4a) [94]. There are two methods for the conjugation of proteins onto polymers: anionic and cationic routes. Furthermore, the prepared samples were also nearly monodispersed. In the past 20 years, several PEGylated enzymes have been approved by the Food and Drug Administration.

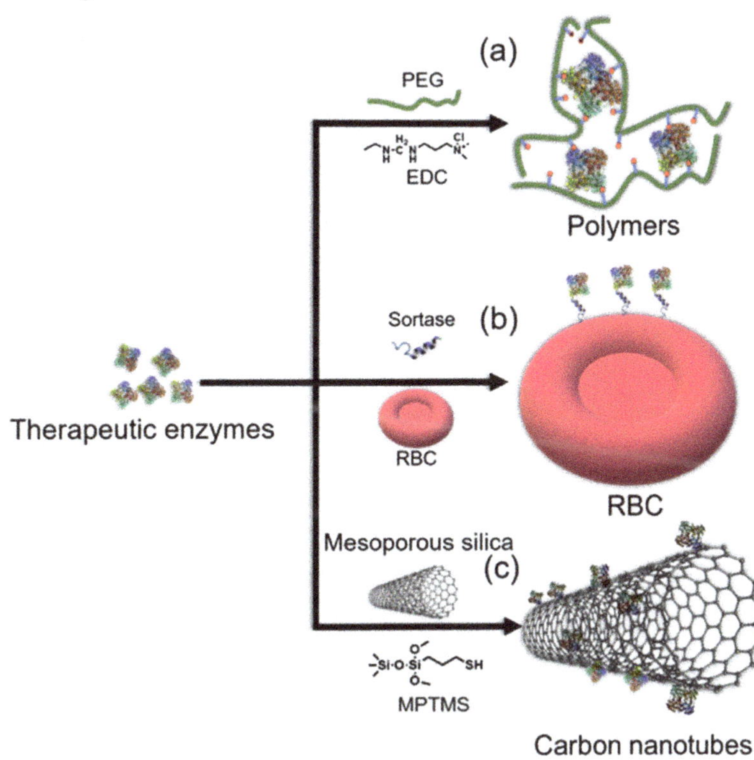

Figure 4. Illustration of covalent approaches for therapeutic enzyme immobilization via conjugation onto (**a**) polyethylene glycol (PEG), (**b**) red blood cells (RBC), and (**c**) mesoporous silica. Here, EDC and MPTMS denote 1-ethyl-3-(3-dimethyl aminopropyl)carbodiimide and mercaptopropyltrimethoxysilane, respectively.

With the development of polymer preparation, the conjugation of enzymes onto polymers will become simpler. In previous studies, PEGylation was used to significantly increase the stability of enzymes, decrease immunogenicity, and prolong circulation time.

However, the non-biodegradability of the polymers is a significant limitation of PEGylated enzymes, and this limitation also drove the development of new degradable polymers [52].

Magnetic polymer nanoparticles were synthesized by the polymerization of iron oxide (Fe_3O_4), 2-hydroxyethyl methacrylate, and glycidyl methacrylate to covalently immobilize L-asparaginase [95]. After immobilization, L-asparaginase retained 74.7% residual activity in artificial serum medium. It covered 50% activity after 10 h at 55 °C and retained 85% activity after eight cycles of reuse, which can be potentially applied for cancer treatment. The covalent immobilization of lysozyme on a polymeric agarose matrix provided beneficial potential as a bacteriolytic activity that can be effectively used against pathogens (Table 1) [96].

2.3.2. Covalent Immobilization onto Nanocarriers

Nanocarriers, especially nanoparticles [97,98] and carbon nanotubes (CNTs) [99], are promising supports for the immobilization of enzymes because they have a large surface area for drug delivery. Ansari and Husain [98] compiled a list of nanomaterials used for the immobilization of enzymes, and the potential for the immobilization of therapeutic enzymes has been widely explored. Inorganic and organic nanocarriers, such as silica and ferric oxide nanoparticles, chitosan, and alginate nanomaterials as well as CNTs with different sizes, have been prepared and modified for the immobilization of enzymes (Figure 4b) [99].

For example, Fe_3O_4@chitosan nanoparticles have been used as carriers for the immobilization of penicillin G acylase [22]. This is an important enzyme for the pharmaceutical industry; it is used to hydrolyze penicillin G to produce 6-aminopenicillanic acid and 7-amino deacetoxycephalosporanic acid and to synthesize β-lactam antibiotics. The immobilized penicillin G acylase presented higher activity, better reusability, and higher thermal stability compared with the free penicillin G acylase over broad pH and temperature ranges. Many enzymes, such as GOX [100] and HRP [101], have been immobilized on nanoparticles as biosensors with the advantages of high sensitivity, fast response, low cost, small size, and low weight, and have been used in clinical applications.

The shape and facet configuration influences the enzyme loading and release. Immobilized enzymes on nanoparticles showed a broader working pH and enhanced stability compared with the native enzyme [97]. The hybrid nano-biocatalyst synthesized by biomimetic silica (Si) nanoparticles, magnetic nanoparticles (MNP), polyethyleneimine (PEI), and HRP as BioSi@T_HRP_MNP_PEI exhibited high enzyme immobilization of 78% [102]. Remarkably, the resulting hybrid BioSi@T_HRP_MNP_PEI improved the thermal stability of the enzyme ~280 times, which can potentially be used in directed enzyme prodrug therapy (DEPT) and other related biotechnological applications (Table 1).

2.3.3. Albumin and RBCs as Carriers

Albumin and RBCs are natural carriers and can remain in the blood for between two to three weeks, which is far longer than any polymer [103,104]. As versatile carriers for therapeutic or diagnostic enzymes, enzymes being carried on albumin have been applied for the diagnosis and treatment of cancer, diabetes, and infectious diseases [103]. Enzymes can be engineered to bind, conjugate, or genetically fuse to albumin [104].

For example, human butyrylcholinesterase [105] and other molecular drugs, such as hormones, cytokines, growth factors, and peptides, have been reported to be fused to albumin with half-lives far longer than their unfused counterparts. The surface of RBCs can be modified to contain modified proteins on the plasma membrane, which can be labeled in a sortase-catalyzed reaction under native conditions without inflicting damage on the target membrane or cell (Figure 4c) (Table 1) [106]. However, the limitations of using RBCs as carriers include the macrophage-mediated destruction of enzymes-carrying RBCs from long-term circulation, the possibility of enzymes damaging the cells, membrane leakage, and the inability to control the release of enzymes.

Table 1. Descriptions of various therapeutic enzymes immobilized through different methods.

Techniques	Support	Enzyme	Descriptions	Reference
Encapsulation	Alginate	Glucocerebrosidase	The immobilized enzyme presented broader activity than the free enzyme at higher pH values. Localized delivery of the enzyme was observed; 40% of the enzyme was loaded within microspheres after 24 h.	[20]
	PEG-b-polylactic acid	β-galactosidase	Polymersomes showed 72% efficiency and restricted release at physiological pH (7.4). Enzymatic-based treatment to the brain in lysosomal storage disorders was demonstrated.	[32]
	Calcium carbonate	Alkaline Phosphatase	Enzyme loading of up to 48% was observed on particles. This approach could be a promising candidate for effective drug delivery, especially for bone reconstruction.	[33]
	Poly(D,L-lactide)	Catalase and superoxide dismutase	Compared with the control enzymes, the entrapped enzymes enhanced the in vitro viability and function of isolated neonatal pancreatic porcine cell clusters.	[38,40]
	Mouse RBCs	Human erythropoietin	Human RBCs presented higher encapsulation yield (22%) than mouse RBCs (14%) and an efficient cell recovery of 70%. The stability of the encapsulated enzyme was highly dependent on the experimental immobilization conditions. Carrier RBCs presented higher hypoosmotic resistance than regular RBCs, and they efficiently released the immobilized enzyme in suspension.	[51]
	PEG-CNA-PLGA	Cytosine deaminase	PEG-CNA-PLGA nanoparticles are efficiently used for the co-immobilization of doxorubicin (a chemotherapy drug) and cytosine deaminase (prodrug enzyme). This system could be used for the codelivery of chemotherapy agents.	[61]
	$CaHPO_4$ nanoflower	Uricase and HRP	The synthesized hybrid, uricase and HRP-$CaHPO_4$@HA MN, showed maximum encapsulation up to 71%. The resulting hybrid could be used for the treatment of hyperuricemia.	[67]
	Calcium carbonate crystals	Superoxide dismutase	A high loading efficiency of 93% was achieved on particles, and they exhibited an excellent controlled release of the enzyme at physiologically relevant ionic strength without changes in its biological activity. It could potentially be used for SOD delivery in ophthalmology.	[68]
	ZIF-8	GOX and HRP	ZIF-8@Gox/HRP preserved high residual activity up to 2.1 fold. The hybrids showed selective tumor cell growth inhibition.	[69]
	Magnetic hydrogel	L-asparaginase	The immobilization yield of the enzyme was as high as 90%. When the enzyme was immobilized within the polymeric shell, its activity remained unchanged after six months of storage at 4 °C. Moreover, the high biocompatibility of the support can be helpful for the delivery of the enzyme to tumor tissues.	[74]
Adsorption	Polyhydroxyalkanoate	Nattokinase	Enzyme activity increased by 20% after immobilization; moreover, the immobilized enzyme was stable up to 70 °C and did not lose its residual activity after 25 days of storage at 4 °C.	[73]
	HNT-PSP	Superoxide dismutase	HNT-PSP-SOD showed a two-fold higher residual activity without any enzyme leakage. This system can potentially be applied for inflammatory bowel disease therapy.	[76]
	HNTs	Binase	An enzyme loading of 85% was achieved on HNTs, and it exhibited a two-fold enhanced cytotoxicity toward tumor colon cells.	[79]
	Self-assembled monolayers	Glucocerebrosidase	Enzyme efficiency significantly increased up to 230% at higher pH (7–8) values after immobilization.	[81]

Table 1. *Cont.*

Techniques	Support	Enzyme	Descriptions	Reference
	Polydopamine	Urease	The biocompatible properties and long-lasting effectiveness of this system can be used for various bladder disease diagnoses.	[88]
	Alginate	Papain	The enzyme showed stability over 28 days with 80% residual activity. It improved the therapeutic features and thereby increased wound healing.	[89]
	Acryloyl cross linked dextrandialdehyde	Glucose oxidase	In vitro insulin controlled release of about 70% under artificial intestinal fluid conditions was demonstrated. It can be effectively used for therapeutic applications.	[90]
	Immuno-virosomes	Lysozyme	Approximately 75% of the enzyme cargo was delivered to the cytoplasm, avoiding the endocytic pathway.	[92]
Covalent	Poly(ethylene glycol)	Adenosine deaminase	The PEGylation process dramatically increased the stability of the enzyme, decreased the immunogenicity, and prolonged the circulation time.	[52]
	Fe_3O_4@chitosan	Penicillin G acylase	The immobilized enzyme presented higher activity, better reusability, and higher thermal stability than the free enzyme over wide pH and temperature ranges. The conversion yield of 72% was recorded for the synthesis of amoxicillin using the immobilized enzyme at 25 °C.	[22]
	Magnetic poly (2-hydroxyethyl methacrylate and glycidyl methacrylate)	L-asparaginase	Immobilized enzyme retained 50% activity after 10 h under thermophilic conditions (55 °C) and 85% residual activity preserved after 8 cycles of reuses. It could be a promising candidate for cancer treatment.	[95]
	Polymeric agarose matrix	Lysozyme	Almost complete immobilization achieved. It is a promising candidate for the synthesis of complementary material for medical applications in extracorporeal therapy.	[96]
	BioSi@THRP_MNP_PEI	HRP	The immobilization yield was achieved up to 78% and the stability improved ~280 times. It can potentially be used for DEPT and other related biotechnological applications.	[102]
	Mouse RBCs	L-asparaginase	The immobilization of enzymes on RBCs lowered the development of antibody titers by >1000-fold and extended the pharmacodynamic effects of the enzyme drug by approximately ten-fold when administered to mice.	[106]

3. Protein Engineering

Protein engineering can be used to modify the primary amino acid sequence of enzymes to improve enzyme structure, stability, and function. By combining protein engineering and immobilization, enzyme properties can be efficiently improved [14,107]. Protein engineering allows site-specific immobilization of enzymes at any desired residue. Smith et al. [108] developed a protein residue-explicit covalent immobilization technique for a stability enhancement system, which uses an amber codon substitution to enable rationally directed non-canonical amino acid incorporation for site-specific immobilization of enzymes.

This can be used to effectively improve stability during covalent immobilization. Later, the same group used this method to immobilize lysozymes with orientation control, allowing for more than a 50% improvement in activity compared with the randomly immobilized enzymes [109]. Raliski et al. [110] reported the utilization of unnatural amino acid technologies to introduce biorthogonal handles in a site-specific fashion for protein immobilization. When introducing genes into RBCs encoding surface proteins that can be covalently and site-specifically modified on the cell surface, the RBCs can be used as carriers for the immobilization of enzymes, including sortase for drug delivery [106].

Compared to traditional immobilization methods, protein engineering allows for orientation-control immobilization [111]. The immobilized enzyme often has high stability and recovery because the immobilization process is totally designable and controllable. The genetic modification of xenobiotic reductases (XenB) from *Pseudomonas putida* facilitated effective immobilization on the gold surface for potential human use in cancer prodrug (CB1954) therapy as a chemotherapy strategy via magnetic nanoparticle-based DEPT [112].

4. Conclusions

Compared with industrial enzymes, therapeutic enzymes are more specialized regarding their potential applications in humans. To safely use therapeutic enzymes, immobilization techniques must be reliable, non-toxic, and without side effects. Biocompatibility with the human body is the most important consideration when immobilization is performed. Thus, natural and artificial biopolymers are still the primary choice for the adsorption or entrapment of therapeutic enzymes.

Due to their high carrying capability as well as easy and controllable preparation methods, nanocarriers will likely become widely used in clinical and diagnostic applications. With the development of computational and molecular techniques, protein engineering will also fulfill important functions in the immobilization of enzymes. The combination of protein engineering and specific immobilization will make the targeted delivery of therapeutic enzymes much easier.

Author Contributions: Conceptualization, Y.-W.Z., S.K.S.P. and J.-K.L.; data curation, C.-Y.Z., F.-L.L. and R.K.G.; writing—original draft preparation, Y.-W.Z., C.-Y.Z., F.-L.L., R.K.G., S.K.S.P. and J.-K.L.; writing—review and editing, Y.-W.Z., S.K.S.P. and J.-K.L.; funding acquisition, Y.-W.Z., S.K.S.P. and J.-K.L. All authors have read and agreed to the published version of the manuscript.

Funding: This research was supported by Basic Science Research Program through the National Research Foundation of Korea (NRF) funded by the Ministry of Science, ICT & Future Planning (NRF-2019R1C1C11009766, NRF-2020R1A4A2002854, and 2021M3A9I5023254).

Institutional Review Board Statement: Not applicable.

Informed Consent Statement: Not applicable.

Data Availability Statement: Data sharing not applicable.

Acknowledgments: This work was supported by the KU Research Professor Program of Konkuk University.

Conflicts of Interest: The authors declare no conflict of interest.

References

1. Keçili, R.; Say, R.; Yavuz, H. Synthesis and characterization of pseudo-affinity ligand for penicillin acylase purification. *Int. J. Biol. Macromol.* **2006**, *39*, 250–255. [CrossRef] [PubMed]
2. Patel, S.K.S.; Jeon, M.S.; Gupta, R.K.; Jeon, Y.; Kalia, V.C.; Kim, S.C.; Cho, B.-K.; Kim, D.R.; Lee, J.-K. Hierarchical macro-porous particles for efficient whole-cell immobilization: Application in bioconversion of greenhouse gases to methanol. *ACS Appl. Mater. Interfaces* **2019**, *11*, 18968–18977. [CrossRef] [PubMed]
3. Subjakova, V.; Oravczova, V.; Hianik, T. Polymer nanoparticles and nanomotors modified by DNA/RNA aptamers and antibodies in targeted therapy of cancer. *Polymers* **2021**, *13*, 341. [CrossRef] [PubMed]
4. Robertson, J.G. Enzymes as a special class of therapeutic target: Clinical drugs and modes of action. *Curr. Opin. Struct. Biol.* **2007**, *17*, 674–679. [CrossRef] [PubMed]
5. Klein, M.D.; Langer, R. Immobilized enzymes in clinical medicine: An emerging approach to new drug therapies. *Trends Biotechnol.* **1986**, *4*, 179–186. [CrossRef]
6. Friboulet, A. From enzymes to new biocatalysts: Towards new therapeutic strategies. *Actual. Chim.* **2003**, *11–12*, 15–19.
7. Holcenberg, J.S. Enzymes as drugs. *Annu. Rev. Pharmacol. Toxicol.* **1977**, *17*, 97–116. [CrossRef]
8. Patel, S.K.S.; Otari, S.V.; Li, J.; Kim, D.R.; Kim, S.C.; Cho, B.-K.; Kalia, V.C.; Kang, Y.C.; Lee, J.-K. Synthesis of cross-linked protein-metal hybrid nanoflowers and its application in repeated batch decolorization of synthetic dyes. *J. Hazard. Mater.* **2018**, *347*, 442–450. [CrossRef]
9. Patel, S.K.S.; Choi, H.; Lee, J.-K. Multi-metal based inorganic–protein hybrid system for enzyme immobilization. *ACS Sustain. Chem. Eng.* **2019**, *7*, 13633–13638. [CrossRef]
10. Danial, E.N.; Alkhalaf, M.I. Co-immobilisation of superoxide dismutase and catalase using an in vitro encapsulation protocol. *J. King Saud Univ. Sci.* **2020**, *32*, 2489–2494. [CrossRef]
11. El-Newehy, M.H.; El-Hamshary, H.; Salem, W.M. Solution blowing spinning technology towards green development of urea sensor nanofibers immobilized with hydrazone probe. *Polymers* **2021**, *13*, 531. [CrossRef] [PubMed]

12. Gao, H.; Li, J.; Sivakumar, D.; Kim, T.-S.; Patel, S.K.S.; Kalia, V.C.; Kim, I.-W.; Zhang, Y.-W.; Lee, J.-K. NADH oxidase from Lactobacillus reuteri: A versatile enzyme for oxidized cofactor regeneration. *Int. J. Biol. Macromol.* **2019**, *123*, 629–636. [CrossRef] [PubMed]
13. Wahab, R.A.; Elias, N.; Abdullah, F.; Ghoshal, S.K. On the taught new tricks of enzymes immobilization: An all-inclusive overview. *React. Funct. Polym.* **2020**, *152*, 104613. [CrossRef]
14. Pagolu, R.; Singh, R.; Shanmugam, R.; Kondaveeti, S.; Patel, S.K.S.; Kalia, V.C.; Lee, J.-K. Site-directed lysine modification of xylanase for oriented immobilization onto silicon dioxide nanoparticles. *Bioresour. Technol.* **2021**, *331*, 125063.
15. Krajewska, B. Application of chitin- and chitosan-based materials for enzyme immobilizations: A review. *Enzyme Microb. Technol.* **2004**, *35*, 126–139. [CrossRef]
16. Li, H.; Dong, W.-F.; Zhou, J.-Y.; Xu, X.-M.; Li, F.-Q. Triggering effect of N-acetylglucosamine on retarded drug release from a lectin-anchored chitosan nanoparticles-in-microparticles system. *Int. J. Pharm.* **2013**, *449*, 37–43. [CrossRef]
17. Wang, F.; Chen, L.; Zhang, D.; Jiang, S.; Shi, K.; Huang, Y.; Li, R.; Xu, Q. Methazolamide-loaded solid lipid nanoparticles modified with low-molecular weight chitosan for the treatment of glaucoma: Vitro and vivo study. *J. Drug Target.* **2014**, *22*, 849–858. [CrossRef]
18. Betigeri, S.S.; Neau, S.H. Immobilization of lipase using hydrophilic polymers in the form of hydrogel beads. *Biomaterials* **2002**, *23*, 3627–3636. [CrossRef]
19. Wang, X.-Y.; Jiang, X.-P.; Li, Y.; Zeng, S.; Zhang, Y.-W. Preparation Fe_3O_4@chitosan magnetic particles for covalent immobilization of lipase from *Thermomyces lanuginosus*. *Int. J. Biol. Macromol.* **2015**, *75*, 44–50. [CrossRef]
20. Barrias, C.C.; Lamghari, M.; Granja, P.L.; Miranda, M.C.S.; Barbosa, M.A. Biological evaluation of calcium alginate microspheres as a vehicle for the localized delivery of a therapeutic enzyme. *Biomed. Mater. Res. A* **2005**, *74*, 545–552. [CrossRef]
21. Bhavsar, C.; Momin, M.; Gharat, S.; Omri, A. Functionalized and graft copolymers of chitosan and its pharmaceutical applications. *Expert Opin. Drug Deliv.* **2017**, *14*, 1189–1204. [CrossRef] [PubMed]
22. Ling, X.-M.; Wang, X.-Y.; Ma, P.; Yang, Y.; Qin, J.-M.; Zhang, X.-J.; Zhang, Y.-W. Covalent immobilization of penicillin G acylase onto Fe_3O_4@chitosan magnetic nanoparticles. *J. Microbiol. Biotechnol.* **2016**, *26*, 829–836. [CrossRef] [PubMed]
23. Saquib, M.; Nayak, A.; Singh, M.; Tabish, M.; Ansari, M.T.; Ara, T.J. Alginate-based bipolymeric-nanobioceramic composite matrices for sustained drug release. *Int. J. Biol. Macromol.* **2016**, *83*, 71–77.
24. Chiumiento, A.; Lamponi, S.; Barbucci, R.; Dominguez, A.; Perez, Y.; Villalonga, R. Immobilizing Cu,Zn-superoxide dismutase in hydrogels of carboxymethylcellulose improves its stability and wound healing properties. *Biochem. Mosc.* **2006**, *71*, 1324. [CrossRef]
25. Bannikova, A.; Rasumova, L.; Evteev, A.; Ivan, E.; Kasapis, S. Protein-loaded sodium alginate and carboxymethyl cellulose beads for controlled release under simulated gastrointestinal conditions. *Int. J. Food Sci. Technol.* **2017**, *52*, 2171–2179. [CrossRef]
26. Zelikin, A.N.; Ehrhardt, C.; Healy, A.M. Materials and methods for delivery of biological drugs. *Nat. Chem.* **2016**, *8*, 997–1007. [CrossRef]
27. Franssen, O.; Vandervennet, L.; Roders, P.; Hennink, W.E. Degradable dextran hydrogels: Controlled release of a model protein from cylinders and microspheres. *J. Control. Release* **1999**, *60*, 211–221. [CrossRef]
28. Alexandre, N.; Amorim, I.; Caseiro, A.R.; Pereira, T.; Alvites, R.; Rêma, A.; Gonçalves, A.; Valadares, G.; Costa, E.; Santos-Silva, A.; et al. Long term performance evaluation of small-diameter vascular grafts based on polyvinyl alcohol hydrogel and dextran and MSCs-based therapies using the ovine pre-clinical animal model. *Int. J. Pharm.* **2016**, *513*, 332–346. [CrossRef]
29. Crommelin, D.J.A.; Storm, G.; Jiskoot, W.; Stenekes, R.; Mastrobattista, E.; Hennink, W.E. Nanotechnological approaches for the delivery of macromolecules. *J. Control. Release* **2003**, *87*, 81–88. [CrossRef]
30. Nguyen, H.X.; O'Rear, E.A. Biphasic release of protein from polyethylene glycol and polyethylene glycol/modified dextran microspheres. *J. Biomed. Mater. Res. A* **2013**, *101*, 2699–2705. [CrossRef]
31. Jadhav, S.B.; Singhal, R.S. Pullulan-complexed α-amylase and glucosidase in alginate beads: Enhanced entrapment and stability. *Carbohydr. Polym.* **2014**, *105*, 49–56. [CrossRef] [PubMed]
32. Kelly, J.M.; Gross, A.L.; Martin, D.R.; Byrne, M.E. Polyethylene glycol-b-poly(lactic acid) polymersomes as vehicles for enzyme replacement therapy. *Nanomedicine* **2017**, *12*, 2591–2606. [CrossRef] [PubMed]
33. Abalymov, A.; Poelvoorde, L.V.; Atkin, V.; Skirtach, A.G.; Konrad, M.; Parakhonskiy, B. Alkaline phosphatase delivery system based on calcium carbonate carriers for acceleration of ossification. *ACS Appl. Bio Mater.* **2020**, *3*, 2986–2996. [CrossRef] [PubMed]
34. Zhu, H.; Srivastava, R.; Brown, J.Q.; McShane, M.J. Combined physical and chemical immobilization of glucose oxidase in alginate microspheres improves stability of encapsulation and activity. *Bioconjug. Chem.* **2005**, *16*, 1451–1458. [CrossRef]
35. Gombotz, W.R.; Wee, S. Protein release from alginate matrices. *Drug Deliv. Rev.* **2012**, *64*, 194–205. [CrossRef]
36. Zhang, H.-Y.; Firempong, C.K.; Wang, Y.-W.; Xu, W.-Q.; Wang, M.-M.; Cao, X.; Zhu, Y.; Tong, S.-S.; Yu, J.-N.; Xu, X.-M. Ergosterol-loaded poly(lactide-co-glycolide) nanoparticles with enhanced in vitro antitumor activity and oral bioavailability. *Acta Pharmacol. Sin.* **2017**, *37*, 834–844. [CrossRef]
37. Xiong, F.; Hu, K.; Yu, H.; Zhou, L.; Song, L.; Zhang, Y.; Shan, X.; Liu, J.; Gu, N. A functional iron oxide nanoparticles modified with PLA-PEG-DG as tumor-targeted MRI contrast agent. *Pharm. Res.* **2017**, *34*, 1683–1692. [CrossRef]
38. Giovagnoli, S.; Luca, G.; Casaburi, I.; Blasi, P.; Macchiarulo, G.; Ricci, M.; Calvitti, M.; Basta, G.; Calafiore, R.; Rossi, C. Long-term delivery of superoxide dismutase and catalase entrapped in poly(lactide-co-glycolide) microspheres: In vitro effects on isolated neonatal porcine pancreatic cell clusters. *J. Control. Release* **2005**, *107*, 65–77. [CrossRef]

39. Paik, D.-H.; Choi, S.-W. Entrapment of protein using electrosprayed poly(d,l-lactide-co-glycolide) microspheres with a porous structure for sustained release. *Macromol. Rapid Commun.* **2014**, *35*, 1033–1038. [CrossRef]
40. Giovagnoli, S.; Blasi, P.; Ricci, M.; Rossi, C. Biodegradable microspheres as carriers for native superoxide dismutase and catalase delivery. *AAPS PharmSciTech* **2004**, *5*, 1–9. [CrossRef]
41. Demers, N.; Agostinelli, E.; Averill-Bates, D.A.; Fortier, G. Immobilization of native and poly(ethylene glycol)-treated ('PEGylated') bovine serum amine oxidase into a biocompatible hydrogel. *Biotechnol. Appl. Biochem.* **2001**, *33*, 201–207. [CrossRef] [PubMed]
42. Yuan, L.; Geng, L.; Ge, L.; Yu, P.; Duan, X.; Chen, J.; Chang, Y. Effect of iron liposomes on anemia of inflammation. *Int. J. Pharm.* **2013**, *454*, 82–89. [CrossRef] [PubMed]
43. Yang, L.; Xin, J.; Zhang, Z.; Yan, H.; Wang, J.; Sun, E.; Hou, J.; Jia, X.; Lv, H. TPGS-modified liposomes for the delivery of ginsenoside compound K against non-small cell lung cancer: Formulation design and its evaluation in vitro and in vivo. *J. Pharm. Pharmacol.* **2016**, *68*, 1109–1118. [CrossRef] [PubMed]
44. Vemuri, S.; Rhodes, C.T. Preparation and characterization of liposomes as therapeutic delivery systems: A review. *Pharm. Acta Helv.* **1995**, *70*, 95–111. [CrossRef]
45. Chen, G.Y.; Li, D.H.; Jin, Y.; Zhang, W.Y.; Teng, L.R.; Bunt, C.; Wen, J.Y. Deformable liposomes by reverse-phase evaporation method for an enhanced skin delivery of (+)-catechin. *Drug Dev. Ind. Pharm.* **2014**, *40*, 260–265. [CrossRef] [PubMed]
46. Haidar, Z.S.; Hamdy, R.C.; Tabrizian, M. Protein release kinetics for core–shell hybrid nanoparticles based on the layer-by-layer assembly of alginate and chitosan on liposomes. *Biomaterials* **2008**, *29*, 1207–1215. [CrossRef]
47. Jahadi, M.; Khosravi-Darani, K. Liposomal encapsulation enzymes: From medical applications to kinetic characteristics. *Mini-Rev. Med. Chem.* **2017**, *17*, 366–370. [CrossRef]
48. Liang, J.F.; Li, Y.T.; Yang, V.C. Biomedical application of immobilized enzymes. *J. Pharm. Sci.* **2000**, *89*, 979–990. [CrossRef]
49. Parayath, N.; Amiji, M. Therapeutic targeting strategies using endogenous cells and proteins. *J. Control. Release* **2017**, *258*, 81–94. [CrossRef]
50. Lizano, C.; Sanz, S.; Luque, J.; Pinilla, M. In vitro study of alcohol dehydrogenase and acetaldehyde dehydrogenase encapsulated into human erythrocytes by an electroporation procedure. *Biochim. Biophys. Acta BBA Gen. Subj.* **1998**, *1425*, 328–336. [CrossRef]
51. Garín, M.-I.; López, R.-M.; Sanz, S.; Pinilla, M.; Luque, J. Erythrocytes as carriers for recombinant human erythropoietin. *Pharm. Res.* **1996**, *13*, 869–874. [CrossRef] [PubMed]
52. Lainka, E.; Hershfield, M.S.; Santisteban, I.; Bali, P.; Seibt, A.; Neubert, J.; Friedrich, W.; Niehues, T. Polyethylene gly-col-conjugated adenosine deaminase (ADA) therapy provides temporary immune reconstitution to a child with de-layed-onset ADA deficiency. *Clin. Diagn. Lab. Immonol.* **2005**, *12*, 861–866.
53. He, H.; Ye, J.; Wang, Y.; Liu, Q.; Chung, H.S.; Kwon, Y.M.; Shin, M.C.; Lee, K.; Yang, V.C. Cell-penetrating peptides meditated encapsulation of protein therapeutics into intact red blood cells and its application. *J. Control. Release* **2014**, *176*, 123–132. [CrossRef] [PubMed]
54. Magnani, M.; Pierigè, F.; Rossi, L. Erythrocytes as a novel delivery vehicle for biologics: From enzymes to nucleic acid-based therapeutics. *Ther. Deliv.* **2012**, *3*, 405–414. [CrossRef] [PubMed]
55. Seeman, N.C. DNA in a material world. *Nature* **2003**, *421*, 427–431. [CrossRef]
56. Goodman, R.P.; Schaap, I.A.T.; Tardin, C.F.; Erben, C.M.; Berry, R.M.; Schmidt, C.F.; Turberfield, A.J. Rapid chiral assembly of rigid DNA building blocks for molecular nanofabrication. *Science* **2005**, *310*, 1661–1665. [CrossRef]
57. Chandrasekaran, A.R.; Levchenko, O. DNA Nanocages. *Chem. Mater.* **2016**, *28*, 5569–5581. [CrossRef]
58. Saccà, B.; Niemeyer, C.M. Functionalization of DNA nanostructures with proteins. *Chem. Soc. Rev.* **2011**, *40*, 5910. [CrossRef]
59. Erben, C.M.; Goodman, R.P.; Turberfield, A.J. Single-molecule protein encapsulation in a rigid DNA cage. *Angew. Chem. Int. Ed.* **2006**, *45*, 7414–7417. [CrossRef]
60. Kim, S.H.; Kim, K.-R.; Ahn, D.-R.; Lee, J.E.; Yang, E.G.; Kim, S.Y. Reversible regulation of enzyme activity by pH-responsive encapsulation in DNA nanocages. *ACS Nano* **2017**, *11*, 9352–9359. [CrossRef]
61. Harguindey, A.; Roy, S.; Harris, A.W.; Fairbanks, B.D.; Goodwin, A.P.; Bowman, C.N.; Cha, J.N. Click nucleic acid mediated loading of prodrug activating enzymes in PEG−PLGA nanoparticles for combination chemotherapy. *Biomacromolecules* **2019**, *20*, 1683–1690. [CrossRef] [PubMed]
62. Ge, J.; Lei, J.; Zare, R.N. Protein–inorganic hybrid nanoflowers. *Nat. Nanotechnol.* **2012**, *7*, 428–432. [CrossRef] [PubMed]
63. Batule, B.S.; Park, K.S.; Kim, M.I.; Park, H.G. Ultrafast sonochemical synthesis of protein- inorganic nanoflowers. *Int. J. Nanomed.* **2015**, *10*, 137–142.
64. Cui, J.; Jia, S. Organic–inorganic hybrid nanoflowers: A novel host platform for immobilizing biomolecules. *Coord. Chem. Rev.* **2017**, *352*, 249–263. [CrossRef]
65. Shende, P.; Kasture, P.; Gaud, R.S. Nanoflowers: The future trend of nanotechnology for multi-applications. *Artif. Cells Nanomed. Biotechnol.* **2018**, *46*, 1–10. [CrossRef]
66. Li, Z.; Zhang, Y.; Su, Y.; Ouyang, P.; Ge, J.; Liu, Z. Spatial co-localization of multi-enzymes by inorganic nanocrystal–protein complexes. *Chem. Commun.* **2014**, *50*, 12465–12468. [CrossRef]
67. Hao, Y.; Li, H.; Cao, Y.P.; Chen, Y.W.; Lei, M.Y.; Zhang, T.Y.; Xiao, Y.; Chu, B.Y.; Qian, Z.Y. Uricase and horseradish peroxidase hybrid $CaHPO_4$ nanoflower integrated with transcutaneous patches for treatment of hyperuricemia. *J. Biomed. Nanotechnol.* **2019**, *15*, 951–965. [CrossRef]

68. Binevski, P.V.; Balabushevich, N.G.; Uvarova, V.I.; Vikulina, A.S.; Volodkin, D. Bio-friendly encapsulation of superoxide dismutase into vaterite $CaCO_3$ crystals. Enzyme activity, release mechanism, and perspectives for ophthalmology. *Colloids Surf. B Biointerfaces* **2019**, *181*, 437–449. [CrossRef]
69. Bai, J.; Peng, C.; Guo, L.; Zhou, M. Metal−organic framework-integrated enzymes as bioreactor for enhanced therapy against solid tumor via a cascade catalytic reaction. *ACS Biomater. Sci. Eng.* **2019**, *5*, 6207–6215. [CrossRef]
70. Vertegel, A.A.; Siegel, R.W.; Dordick, J. Silica nanoparticle size influences the structure and enzymatic activity of adsorbed lysozyme. *Langmuir* **2004**, *20*, 6800–6807. [CrossRef]
71. Liu, W.; Zhou, F.; Zhang, X.-Y.; Li, Y.; Wang, X.-Y.; Xu, X.-M.; Zhang, Y.-W. Preparation of Magnetic Fe_3O_4@SiO_2 nanoparticles for immobilization of lipase. *J. Nanosci. Nanotechnol.* **2014**, *14*, 3068–3072. [CrossRef] [PubMed]
72. Shi, Y.; Liu, W.; Tao, Q.-L.; Jiang, X.-P.; Liu, C.-H.; Zeng, S.; Zhang, Y.-W. Immobilization of lipase by adsorption onto magnetic nanoparticles in organic solvents. *J. Nanosci. Nanotechnol.* **2016**, *16*, 601–607. [CrossRef] [PubMed]
73. Deepak, V.; Pandian, S.B.R.K.; Kalishwaralal, K.; Gurunathan, S. Purification, immobilization, and characterization of nattokinase on PHB nanoparticles. *Bioresour. Technol.* **2009**, *100*, 6644–6646. [CrossRef] [PubMed]
74. Teodor, E.; Litescu, S.C.; Lazar, V.; Somoghi, R. Hydrogel-magnetic nanoparticles with immobilized L-asparaginase for biomedical applications. *J. Mater. Sci. Mater. Med.* **2009**, *20*, 1307–1314. [CrossRef] [PubMed]
75. Chen, Y.-P.; Chen, C.-T.; Hung, Y.; Chou, C.-M.; Liu, T.-P.; Liang, M.-R.; Chen, C.-T.; Mou, C.-Y. A new strategy for intracellular delivery of enzyme using mesoporous silica nanoparticles: Superoxide dismutase. *J. Am. Chem. Soc.* **2013**, *135*, 1516–1523. [CrossRef] [PubMed]
76. Katana, B.; Rouster, P.; Varga, G.; Muráth, S.; Glinel, K.; Jonas, A.M.; Szilagyi, I. Self-assembly of protamine biomacromolecule on halloysite nanotubes for immobilization of superoxide dismutase enzyme. *ACS Appl. Bio Mater.* **2020**, *3*, 522–530. [CrossRef]
77. Condori, J.; Acosta, W.; Ayala, J.; Katta, V.; Flory, A.; Martin, R.; Radin, J.; Cramer, C.L.; Radin, D.N. Enzyme replacement for GM1-gangliosidosis: Uptake, lysosomal activation, and cellular disease correction using a novel β-galactosidase: RTB lectin fusion. *Mol. Genet. Metab.* **2016**, *117*, 199–209. [CrossRef]
78. Ribeiro, C.C.; Barrias, C.C.; Barbosa, M.A. Calcium phosphate-alginate microspheres as enzyme delivery matrices. *Biomaterials* **2004**, *25*, 4363–4373. [CrossRef]
79. Khodzhaeva, V.; Makeeva, A.; Ulyanova, V.; Zelenikhin, P.; Evtugyn, V.; Hardt, M.; Rozhina, E.; Lvov, Y.; Fakhrullin, R.; Ilinskaya, O. Binase immobilized on halloysite nanotubes exerts enhanced cytotoxicity toward human colon adenocarcinoma cells. *Front. Pharmacol.* **2017**, *8*, 631. [CrossRef]
80. Chen, L.; Liu, Z.; Jin, R.; Yang, X.; Bai, Y.; Liu, S.; Chen, X. Stepwise co-delivery of an enzyme and prodrug based on a multi-responsive nanoplatform for accurate tumor therapy. *J. Mater. Chem. B* **2018**, *6*, 6262–6268. [CrossRef]
81. Barrias, C.C.; Martins, M.A.C.L.; Miranda, M.A.C.S.; Barbosa, M.A. Adsorption of a therapeutic enzyme to self-assembled monolayers: Effect of surface chemistry and solution pH on the amount and activity of adsorbed enzyme. *Biomaterials* **2005**, *26*, 2695–2704. [CrossRef] [PubMed]
82. Rana, S.; Bajaj, A.; Mout, R.; Rotello, V.M. Monolayer coated gold nanoparticles for delivery applications. *Adv. Drug Deliv. Rev.* **2012**, *64*, 200–216. [CrossRef] [PubMed]
83. Gunasekaran, T.; Nigusse, T.; Dhanaraju, M.D. Silver nanoparticles as real topical bullets for wound healing. *J. Am. Coll. Clin. Wound Spec.* **2011**, *3*, 82–96. [CrossRef] [PubMed]
84. Patel, S.K.S.; Choi, S.H.; Kang, Y.; Lee, J.-K. Eco-friendly composite of Fe_3O_4-reduced grapene oxide particles for efficient enzyme immobilization. *ACS Appl. Mater. Interfaces* **2017**, *9*, 2213–2222. [CrossRef]
85. Manzano, M.; Vallet-Regi, M. New developments in ordered mesoporous materials for drug delivery. *J. Mater. Chem.* **2010**, *20*, 5593–5604. [CrossRef]
86. Zahirinejad, S.; Hemmati, R.; Homaei, A.; Dinari, A.; Hosseinkhani, S.; Mohammadi, S.; Vianello, F. Nano-organic supports for enzyme immobilization: Scopes and perspectives. *Colloids Surf. B Biointerfaces* **2021**, *204*, 111774. [CrossRef]
87. Kalia, V.C.; Patel, S.K.S.; Shanmugam, R.; Lee, J.-K. Polyhydroxyalkanoates: Trends and advances toward biotechnological applications. *Bioresour. Technol.* **2021**, *326*, 124737. [CrossRef]
88. Choi, H.; Cho, S.H.; Hahn, S.K. Urease-powered polydopamine nanomotors for intravesical therapy of bladder diseases. *ACS Nano* **2020**, *14*, 6683–6692. [CrossRef]
89. Filho, R.N.F.M.; Vasconcelos, N.F.; Andrade, F.K.; Roas, M.D.F.; Vieira, R.S. Papain immobilized on alginate membrane for wound dressing application. *Colloids Surf. B Biointerfaces* **2020**, *194*, 111222. [CrossRef]
90. Jamwal, S.; Rama, B.; Ranote, S.; Dharela, R.; Chauhan, G.S. New glucose oxidase-immobilized stimuli-responsive dextran nanoparticles for insulin delivery. *Int. J. Biol. Macromol.* **2019**, *123*, 968–978. [CrossRef]
91. Cusi, M.G. Applications of influenza virosomes as a delivery system. *Hum. Vaccines* **2006**, *2*, 1–7. [CrossRef] [PubMed]
92. Kumar, M.; Mani, P.; Pratheesh, P.; Chandra, S.; Jeyakkodi, M.; Chattopadhyay, P.; Sarkar, D.P.; Sinha, S. Membrane fusion medi-ated targeted cytosolic drug delivery through scFv engineered sendai viral envelopes. *Curr. Mol. Med.* **2015**, *15*, 386–400. [CrossRef] [PubMed]
93. Koyani, R.; Pérez-Robles, J.; Cadena-Nava, R.D.; Vazquez-Duhalt, R. Biomaterial-based nanoreactors, an alternative for enzyme delivery. *Nanotechnol. Rev.* **2017**, *6*, 405–419. [CrossRef]
94. Alconcel, S.N.S.; Baas, A.S.; Maynard, H.D. FDA-approved poly(ethylene glycol)-protein conjugate drugs. *Polym. Chem.* **2011**, *2*, 1442–1448. [CrossRef]

95. Orhan, H.; Uygun, D.A. Immobilization of L-asparaginase on magnetic nanoparticles for cancer treatment. *Appl. Biochem. Biotechnol.* **2020**, *191*, 1432–1443. [CrossRef]
96. Levashov, P.A.; Matolygina, D.A.; Ovchinnikova, E.D.; Adamova, I.Y.; Dmitrieva, O.A.; Pokrovsky, N.S.; Eremeev, N.L. A novel method of covalent lysozyme immobilization for the development of materials for medical applications. *Rus. J. Bioorgan. Chem.* **2019**, *45*, 101–106. [CrossRef]
97. Patel, S.K.S.; Choi, S.H.; Kang, Y.; Lee, J.-K. Large-scale aerosol-assisted synthesis of biofriendly Fe_2O_3 yolk–shell particles: A promising support for enzyme immobilization. *Nanoscale* **2016**, *8*, 6728–6738. [CrossRef]
98. Ansari, S.A.; Husain, Q. Potential applications of enzymes immobilized on/in nano materials: A review. *Biotechnol. Adv.* **2012**, *30*, 512–523. [CrossRef]
99. Ali, S.S.; Morsy, R.; El-Zawawy, N.A.; Fareed, M.F.; Bedaiwy, M.Y. Synthesized zinc peroxide nanoparticles (ZnO_2-NPs): A novel antimicrobial, anti-elastase, anti-keratinase, and anti-inflammatory approach toward polymicrobial burn wounds. *Int. J. Nanomed.* **2017**, *12*, 6059–6073. [CrossRef]
100. Zhou, X.; Guo, S.; Gao, J.; Zhao, J.; Xue, S.; Xu, W. Glucose oxidase-initiated cascade catalysis for sensitive impedimetric aptasensor based on metal-organic frameworks functionalized with Pt nanoparticles and hemin/G-quadruplex as mimicking peroxidases. *Biosens. Bioelectron.* **2017**, *98*, 83–90. [CrossRef]
101. Fritzen-Garcia, M.B.; Monteiro, F.F.; Cristofolini, T.; Acuña, J.J.S.; Zanetti-Ramos, B.G.; Oliveira, I.R.W.Z.; Soldi, V.; Pasa, A.A.; Creczynski-Pasa, T.B. Characterization of horseradish peroxidase immobilized on PEGylated polyurethane nanoparticles and its application for dopamine detection. *Sens. Actuators B—Chem.* **2013**, *182*, 264–272. [CrossRef]
102. Correa, S.; Puertas, S.; Gutiérrez, L.; Asín, L.; de la Fuente, J.M.; Grazú, V.; Betancor, L. Design of stable magnetic hybrid nanoparticles of Si-entrapped HRP. *PLoS ONE* **2019**, *14*, e0214004. [CrossRef] [PubMed]
103. Elsadek, B.; Kratz, F. Impact of albumin on drug delivery—New applications on the horizon. *J. Control. Release* **2012**, *157*, 4–28. [CrossRef] [PubMed]
104. Sleep, D.; Cameron, J.; Evans, L.R. Albumin as a versatile platform for drug half-life extension. *Biochim. Biophys. Acta* **2013**, *1830*, 5526–5534. [CrossRef] [PubMed]
105. Cohen-Barak, O.; Wildeman, J.; van de Wetering, J.; Hettinga, J.; Schuilenga-Hut, P.; Gross, A.; Clark, S.; Bassan, M.; Gilgun-Sherki, Y.; Mendzelevski, B.; et al. Safety, pharmacokinetics, and pharmacodynamics of TV-1380, a novel mutated butyrylcholinesterase treatment for cocaine addiction, after single and multiple intramuscular injections in healthy subjects. *J. Clin. Pharmacol.* **2015**, *55*, 573–583. [CrossRef]
106. Shi, J.; Kundrat, L.; Pishesha, N.; Bilate, A.; Theile, C.; Maruyama, T.; Dougan, S.K.; Ploegh, H.L.; Lodish, H.F. Engineered red blood cells as carriers for systemic delivery of a wide array of functional probes. *Natl. Acad. Sci. USA* **2014**, *111*, 10131–10136. [CrossRef]
107. Singh, R.K.; Tiwari, M.K.; Singh, R.; Lee, J.-K. From protein engineering to immobilization: Promising strategies for the upgrade of industrial enzymes. *Int. J. Mol. Sci.* **2013**, *14*, 1232–1277. [CrossRef]
108. Smith, M.T.; Wu, J.C.; Varner, C.T.; Bundy, B.C. Enhanced protein stability through minimally invasive, direct, covalent, and site-specific immobilization. *Biotechnol. Prog.* **2013**, *29*, 247–254. [CrossRef]
109. Wu, J.C.Y.; Hutchings, C.H.; Lindsay, M.J.; Werner, C.J.; Bundy, B.C. Enhanced enzyme stability through site-directed covalent immobilization. *J. Biotechnol.* **2015**, *193*, 83–90. [CrossRef]
110. Raliski, B.K.; Howard, C.A.; Young, D.D. Site-specific protein immobilization using unnatural amino acids. *Bioconjug. Chem.* **2014**, *25*, 1916–1920. [CrossRef]
111. Redeker, E.S.; Ta, D.T.; Cortens, D.; Billen, B.; Guedens, W.; Adriaensens, P. Protein engineering for directed immobilization. *Bioconj. Chem.* **2013**, *24*, 1761–1777. [CrossRef] [PubMed]
112. Ball, P.; Halliwell, J.; Anderson, S.; Gwenin, V.; Gwenin, C. Evaluation of two xenobiotic reductases from *Pseudomonas putida* for their suitability for magnetic nanoparticle-directed enzyme prodrug therapy as a novel approach to cancer treatment. *Microbiologyopen* **2020**, *9*, e1110. [CrossRef] [PubMed]

polymers

MDPI

Review

Lignocellulosic Materials for the Production of Biofuels, Biochemicals and Biomaterials and Applications of Lignocellulose-Based Polyurethanes: A Review

Antonio M. Borrero-López *, Concepción Valencia and José M. Franco

Pro2TecS—Chemical Process and Product Technology Research Center, Departamento de Ingeniería Química, Escuela Técnica Superior de Ingeniería, Campus de "El Carmen", Universidad de Huelva, 21071 Huelva, Spain; barragan@uhu.es (C.V.); franco@uhu.es (J.M.F.)
* Correspondence: am.borrero@diq.uhu.es; Tel.: +34-959-217-633

Abstract: The present review is devoted to the description of the state-of-the-art techniques and procedures concerning treatments and modifications of lignocellulosic materials in order to use them as precursors for biomaterials, biochemicals and biofuels, with particular focus on lignin and lignin-based products. Four different main pretreatment types are outlined, i.e., thermal, mechanical, chemical and biological, with special emphasis on the biological action of fungi and bacteria. Therefore, by selecting a determined type of fungi or bacteria, some of the fractions may remain unaltered, while others may be decomposed. In this sense, the possibilities to obtain different final products are massive, depending on the type of microorganism and the biomass selected. Biofuels, biochemicals and biomaterials derived from lignocellulose are extensively described, covering those obtained from the lignocellulose as a whole, but also from the main biopolymers that comprise its structure, i.e., cellulose, hemicellulose and lignin. In addition, special attention has been paid to the formulation of bio-polyurethanes from lignocellulosic materials, focusing more specifically on their applications in the lubricant, adhesive and cushioning material fields. High-performance alternatives to petroleum-derived products have been reported, such as adhesives that substantially exceed the adhesion performance of those commercially available in different surfaces, lubricating greases with tribological behaviour superior to those in lithium and calcium soap and elastomers with excellent static and dynamic performance.

Keywords: lignocellulose; lignin; cellulose; biofuels; biomaterials; biochemicals; lubricating greases; adhesives; cushioning materials; rheology

Citation: Borrero-López, A.M.; Valencia, C.; Franco, J.M. Lignocellulosic Materials for the Production of Biofuels, Biochemicals and Biomaterials and Applications of Lignocellulose-Based Polyurethanes: A Review. *Polymers* **2022**, *14*, 881. https://doi.org/10.3390/polym14050881

Academic Editor: Alejandro Sosnik

Received: 26 January 2022
Accepted: 21 February 2022
Published: 23 February 2022

1. Lignocellulose as Renewable Raw Material: Sources and General Chemical Description

It is well known that with the increasing issues of climate change, waste management and unstoppable resource exhaustion, politics and research efforts need to be combined in the search for new materials and sources that can replace fossil fuels and non-renewable resources currently in use, which besides generally include hazardous/toxic manufacture protocols and problematic end-of-life. The anthropogenic imprint on global temperature has already been reported as an increase of 0.87 °C, and it is expected to be around 1.5 °C between 2030 and 2050, temperatures which may imply severe changes in the worldwide climate, increasing the probability of drought and heavy precipitation in determined regions, along with many increasing risks in fields such as health, food security, water supply, etc. [1]. In order to mitigate these increments and minimize the human impact, stronger policies need to be applied; in addition, worldwide research must provide the technology and resources necessary for the replacement of the contaminating sources by biomaterials and harmless product manufacturing [2,3]. It is at this point that lignocellulosic sources can play a fundamental role as a consequence of their natural origin, ubiquitous production all over the world, minimum carbon footprint and the interesting properties of their main components [3–6].

Lignocellulosic biomass comprises the skeleton of all the living plants on Earth from their roots, leaves and stalks to their fruits and flowers. Wood is generally divided into two main groups, hardwood and softwood. Whereas hardwood refers to wood coming from angiosperm trees, such as oak, eucalyptus and beech, softwood originates from gymnosperm trees, such as conifers. Both of them are mainly composed of the joined combination of three natural polymers, i.e., cellulose, hemicellulose and lignin, and are currently widely targeted as they comprise promising renewable materials for bioproduct performance and biofuels [6–11]. Their content range varies between 40 and 50% of cellulose, 15 and 30% of lignin and 25 and 30% of hemicellulose; nonetheless, these concentrations depend significantly on the type of biomass selected, part and age of the plant, growth conditions [12,13] and part of the cellulose wall [14] (see Table 1).

Table 1. Composition of representative lignocellulosic feedstocks [12].

Feedstocks	Carbohydrate Composition (% Dry Weight)		
	Cellulose	Hemicellulose	Lignin
Barley hull	24	36	19
Barley straw	36–43	24–33	6.3–9.8
Bamboo	49–50	18–20	23
Banana waste	13	15	14
Corn cob	32.3–45.6	39.8	6.7–13.9
Corn stover	35.1–39.5	20.7–24.6	11.0–19.1
Cotton	85–95	5–15	0
Cotton stalk	31	11	30
Coffee pulp	33.7–36.9	44.2–47.5	15.6–19.1
Douglas fir	35–48	20–22	15–21
Eucalyptus	45–51	11–18	29
Hardwood stems	40–55	24–40	18–25
Rice straw	29.2–34.7	23–25.9	17–19
Rice husk	28.7–35.6	11.96–29.3	15.4–20
Wheat straw	35–39	22–30	12–16
Wheat bran	10.5–14.8	35.5–39.2	8.3–12.5
Grasses	25–40	25–50	10–30
Newspaper	40–55	24–39	18–30
Sugarcane bagasse	25–45	28–32	15–25
Sugarcane tops	35	32	14
Pine	42–49	13–25	23–29
Poplar wood	45–51	25–28	10–21
Olive tree biomass	25.2	15.8	19.1
Jute fibres	45–53	18–21	21–26
Switchgrass	35–40	25–30	15–20
Winter rye	29–30	22–26	16.1
Oilseed rape	27.3	20.5	14.2
Softwood stem	45–50	24–40	18–25
Oat straw	31–35	20–26	10–15
Nut shells	25–30	22–28	30–40
Sorghum straw	32–35	24–27	15–21
Tamarind kernel powder	10–15	55–65	-
Water hyacinth	18.2–22.1	48.7–50.1	3.5–5.4
Miscanthus *	37	29	10
Oat husks **	23	35	25

* Adapted from [15]. ** Adapted from [16].

Cellulose comprises the most abundant biopolymer on Earth, as it is the most important skeletal component of plant cell walls (see Table 1). It is formed by the union of

D-glucose units via β-1,4 glycosidic linkages, forming a semicrystalline fibrous structure which can surpass polymerization degrees of 9000 units [17].

On the other hand, hemicellulose does not possess a defined structure, as it consists of a combination of several diverse monomers, i.e., xylose, arabinose, mannose, galactose, rhamnose, glucose, etc. [18], whose concentrations depend on the biomass. Unlike cellulose, it is composed of an amorphous and branched polymer in which monomer units are usually within the range of 500–3000 units [19,20]. The combination of the diverse units usually generates four main hemicellulose structures, i.e., xylan, xyloglucan, galactomannan and galactoglucomannan [21].

Lignin consists of a highly entangled biopolymer based on several phenyl propane aromatic units, known as monolignols, together with other aromatic and non-aromatic units. Basically, these monolignols are three, i.e., coniferyl alcohol, sinapyl alcohol and paracoumaryl alcohol, shown in Figure 1.

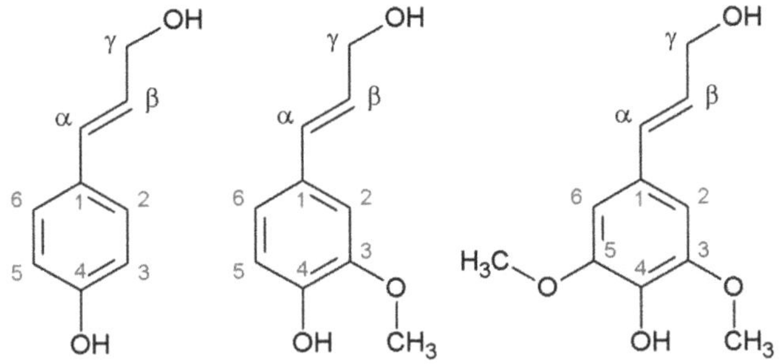

Figure 1. Main monolignol units. p-coumaryl alcohol, coniferyl alcohol and sinapyl alcohol, from left to right.

However, the possibilities of the lignin structure are immense, as just by the combination of those three units, and keeping in mind the different resonance structures that can be generated for further reaction (see Figure 2), a complex and wide range of bonding alternatives is available, making lignin structure extremely difficult to predict [22]. Nonetheless, thanks to powerful nuclear magnetic resonance (NMR) techniques, such as the two-dimensional NMR (2D-NMR) or ^{13}C NMR, some of the principal sequences have been elucidated, which are shown in Figure 3 [23–26]. Further information and extended work in 2D-NMR structural information of lignin can be seen elsewhere [27]. ^{13}C NMR detailed information for lignin structure can also be found in the literature [28,29].

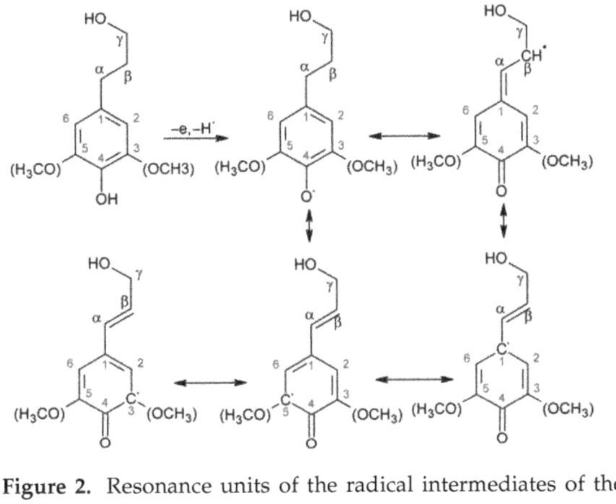

Figure 2. Resonance units of the radical intermediates of the diverse monolignol units during lignin synthesis.

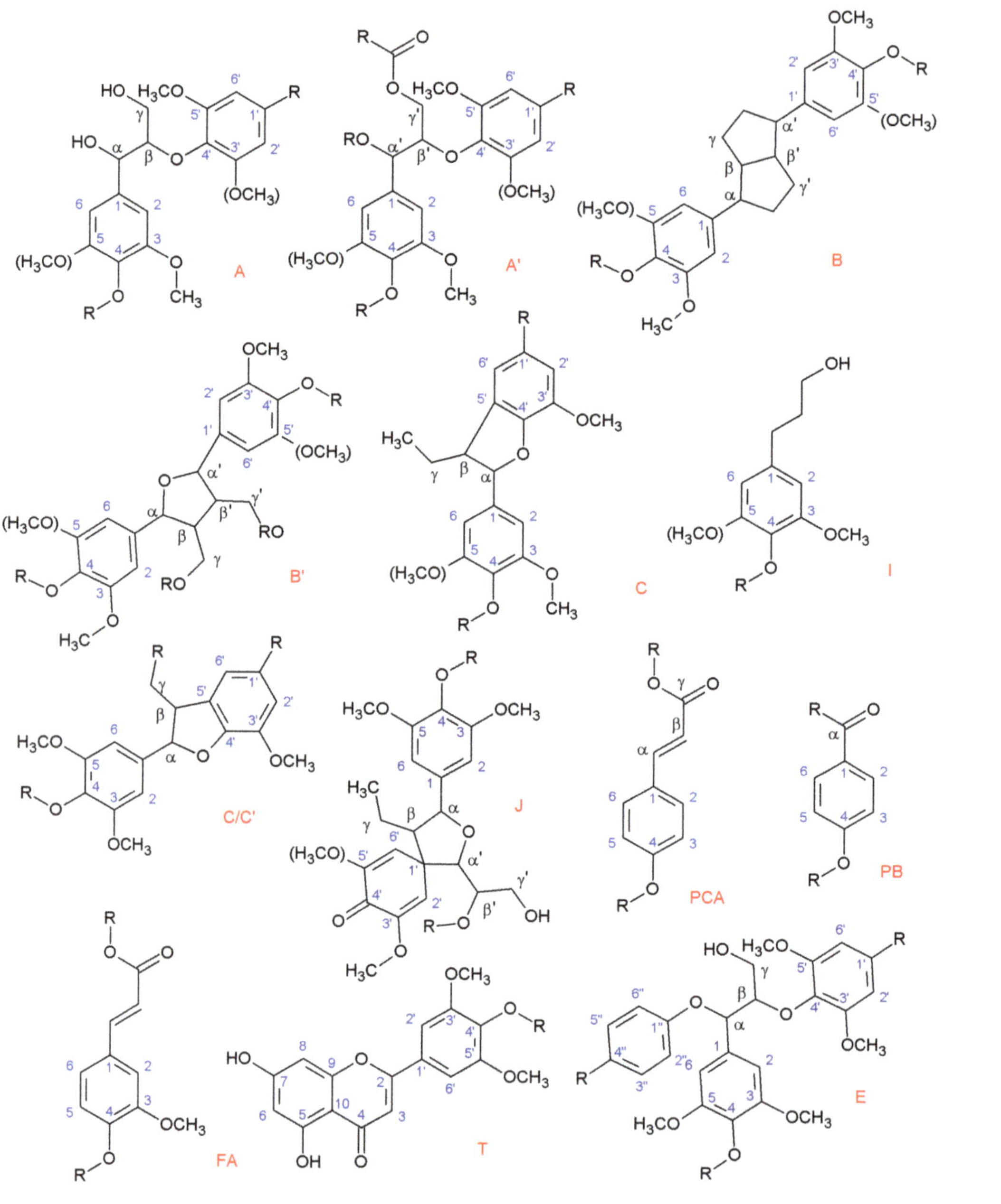

Figure 3. Main lignin structures identified by NMR. (R may indicate both aliphatic and aromatic chains.) (A) β-O-4 alkyl-aryl ethers; (A′) β-O-4 alkyl-aryl ethers with acylated γ′-OH with p-coumaric acid; (B) resinols; (B′) di-c-acylated mono-tetrahydrofuran structure formed by β–β′ coupling and subsequent a-O-a′ bonding (R, acetyl/p-coumaroyl); (C) phenylcoumarans; (I) p-hydroxycinnamyl alcohol end-groups; (C/C′) γ-acetylated phenylcoumaran (R, acetyl) (J) spirodienones (β-1′); (PCA) p-coumarates; (PB) p-hydroxybenzoate; (FA) ferulates; (T) tricin incorporation into the lignin polymer through a G-type β-O-4 linkage; (E) α,β-diaryl ethers (α-O-4/β-O-4).

2D-NMR technique also lets the different percentage of structural units be estimated [23,24,26,30,31]. Thus, the quantification of these interunit linkages has been widely reported, and as can be observed, β-O-4 is outlined as the most prevalent bond type (see Table 2) [32].

Table 2. Approximate percentages of linkages found in softwood and hardwood lignin [33].

Linkage Type	Approximate Percentage (%)	
	Softwood	Hardwood
β-O-4	45–50	60
5–5	18–25	5
β-5	9–12	6
4-O-5	4–8	7
β-1	7–10	7
β-β	3	3

All these three materials together produce a highly developed 3D network, which has made the separation of these biopolymers an extremely interesting topic, due to its inherent relationship with the papermaking industry. It has been the case that meticulous research has been carried out in order to improve this separation and hence decrease energy costs [34]. The most followed procedure, the Kraft process, considers both hemicellulose and lignin as low-value byproducts, both being generally used as low-value fuels to recover part of the energy consumed during the process.

Nonetheless, the full and/or alternative utilisation of the three components would lead to a much better economic performance, which is the basis of the biorefinery concept. An example of how a paper industry could lead to complete exploitation of the raw biomass can be found elsewhere [35].

2. Pretreatments of Lignocellulosic Biomass

Lignocellulosic biomass has naturally developed protection against enzymatic and pathogen activity through a great entangled network in which the three biopolymers are covalent- or hydrogen-bonded, conforming to a great resilient structure [36], which now plays against technological needs when the disruption of the plant cell wall is aimed at [37]. In order to ease the proper separation among the different lignocellulosic biopolymers, a wide range of pretreatments is available [38]. Furthermore, when specific purposes that require determined properties are aimed at, the application of pretreatments of a different nature becomes necessary. These pretreatments have been classified into four categories: physical, thermal, chemical and biological pretreatments [37,39]. Nevertheless, within a determined process, changes in parameter conditions may anyway lead to diverse structural changes [40]. A descriptive scheme summarizing the main content of this section has been included as Scheme 1.

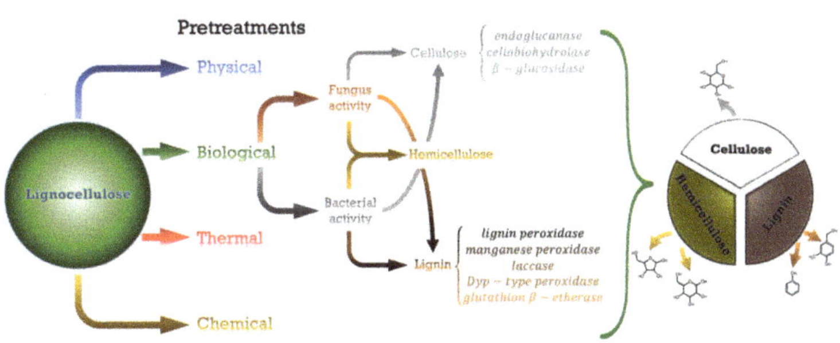

Scheme 1. Main pretreatments of lignocellulose and products obtained, with special attention to the enzymes involved in both fungi and bacterial biological processes of cellulose and lignin degradation.

2.1. Physical Pretreatments

The physical pretreatments include those processes which aim to disrupt the plant cell wall mechanically, reducing the particle size and exposing a higher surface for later purposes. Frequently, these are often preliminary stages after which other pretreatments may be applied. Diverse ball milling procedures are some of the most researched examples [4,37].

2.2. Thermal Pretreatments

Among the most used thermal pretreatments, examples such as steam explosion and hydrothermolysis (also known as autohydrolysis) can be found. By taking advantage of water at high temperatures, hemicellulose has been almost completely recovered [4], while lignin and cellulose have been mildly modified, making them more accessible for further treatments. However, the energy requirements in order to implement this technique industrially still remain too high [39].

2.3. Chemical Pretreatments

With the chemical pretreatments, the solubilisation of the biopolymers which comprise the lignocellulosic biomass into different solvents is sought. The vast majority of the lignocellulosic separation processes are chemically based, and can be classified into acidic, alkaline and oxidative pretreatments.

The three of them mainly affect both hemicellulose and lignin, leading to great solubilisation of these two biopolymers and making cellulose available to a greater extent for subsequent treatments [37]. For some authors, steam explosion and hydrothermolysis are considered acidic pretreatments as a consequence of the acidic characteristics of water at high temperatures [37]. Detailed information about the diverse chemical pretreatment methods can be found elsewhere [4,36–39].

2.4. Biological Pretreatments

In opposition to the other types of pretreatments, biological procedures have the advantages of being low-energy and low-chemical-consuming processes, eco-friendly and without the formation of inhibitors such as aldehydes, furfurals and phenolics [36,37,39]. Carried out through the action of either fungi or bacteria, the degradation capacity relies on the production of a great variety of enzymes, which are able to digest the three biopolymer types. The main parameters affecting biological pretreatments are the incubation temperature and time, moisture content, pH, aeration, inoculum concentration, particle size and type of microorganism. Moreover, the biomass type also plays a fundamental role [41]. Hereby, the correct scalability of the pretreatment to an industrial process lies in the correct selection of the above-mentioned parameters [39,41]. Although the varying biopolymer degradation by both fungi and bacteria will be discussed in the following sections separately, it is worth noticing that the synergistic activity of the different enzymes is responsible for the complete degradation carried out by the microorganisms [42].

2.4.1. Biodegradation by Fungus Activity

Fungi are able to secrete a wide range of enzymes with the capacity to degrade the three biopolymer types of plant cell walls. Within that range, enzymes can be divided into two types: hydrolytic, able to degrade both cellulose and hemicellulose, and oxidative, mainly responsible for lignin degradation [43]. Fungi can also be divided into three diverse groups, soft-, white- and brown-rot fungi, a classification that is based on the degradation mechanism pattern for each lignocellulose biopolymer [44,45]. Therefore, the white-rot fungi are known to successfully degrade the three lignocellulosic biopolymers. The brown-rot fungi attack is mainly centred on the holocellulose instead, whilst action in lignin is only limited. Finally, the soft-rot fungi demonstrate no effect on lignin [45].

Cellulose Biodegradation by Fungus Activity

Regarding cellulose biodegradation, those enzymes with the ability to digest cellulose are called cellulases. Three are the main cellulase groups generated by fungi, i.e., endoglucanases, cellobiohydrolases and β-glucosidases [44]. Endoglucanases are known to randomly attack the amorphous region, opening suitable locations for the subsequent hydrolisation of crystalline structures by cellobiohydrolases and β-glucosidases, which are known to act synergistically [44,45].

Hemicellulose Biodegradation by Fungus Activity

As a consequence of the greater heterogeneity of this biopolymer in comparison to cellulose, a higher number of enzymes is necessary in order to properly degrade the biopolymer. Each one of the typical monomers that compose hemicellulose, i.e., xylose, mannose, galactose, arabinose, etc., has its own range of enzymes suitable for its proper degradation and transformation [46]. Generally, at least two types of enzymes may be present for each monomer, a first one responsible for the transformation of the hemicellulose chain into oligosaccharides and a second one which acts for the ulterior degradation to the monomers and acetic acid [44].

Lignin Biodegradation by Fungus Activity

The most common oxidative enzymes produced by fungi are phenol oxidases, from which lignin and manganese peroxidases and laccases have been more deeply studied. The first ones are responsible for the degradation of the non-phenolic units, whereas the second ones are known to attack both phenolic and non-phenolic units. Laccases, however, only degrade phenolics and other electron-rich groups. The different routes for lignin degradation by using both laccases and peroxidases have been described elsewhere [47]. The lignin breakdown by those enzymes could lead to the production of important aromatic chemicals. For instance, a wide range of aromatic carboxylic acids and acyclic 2,4-hexadiene-1,6-dioic acids were found when spruce-based lignin was degraded by *Phanerochaete chrysosporium*. In this particular case, lignin degradation is occurring by C_α–C_β oxidative cleavage as suggested by the benzoic-acid derivative nature of the compounds obtained [48]. Detailed information about the scheme and further information about fungi-based lignocellulosic degradation can be found elsewhere [44,47,49].

2.4.2. Biodegradation by Bacterial Activity

By producing different types of enzymes, bacteria are also able to degrade lignocellulosic biomass in the same way that fungi do. However, as bacteria generally do not produce lignanases, the aromatic biopolymer constitutes a barrier for many of these bacteria. Such is the interest in the bacterial ability to degrade lignocellulose that some genetic modifications have been carried out in order to improve the degradation capacity [50] or target some specific degradation products [51,52]. The search for bacterial activity related to lignocellulosic degradation has traditionally been performed in the animal gastrointestinal system, but interesting microorganisms have also been found in landfill sites lately [53,54].

Cellulose Biodegradation by Bacterial Activity

Many specific bacteria have shown the ability to degrade cellulose, such as those coming from the genera *Sporocytophaga, Trichonympha, Cellulomonas Erwinia, Clostridium, Acetivibrio, Thermobifida, Mucilaginibacter, Bacteroides, Streptomyces, Cytophaga, Butyrivibrio, Fibrobacter, Pedobacter, Ruminococcus, Methanobrevibacter, Caldicel-lulosiruptor* and *Clostridium* [21,55–57], which come from both aerobic and anaerobic types of bacteria. Generally, likewise, fungi and aerobic bacteria possess the three types of enzymes acting synergistically [57], while the digestion by anaerobic procedures is based on the formation of complexes called cellulosomes (calcium- and-thiol-dependent multicomponent complexes) acting on bacteria's surface [58,59]. Cellulases have also been divided into families that share a distinctive catalytic core, thus exhibiting a similar reaction mechanism, i.e., either a single substitution with the inversion of the configuration or a double substitution leading to the maintenance of the β-arrangement at the anomeric carbon [60]. Cellulases possess very particular structures, where, along with the usual catalytic domain, many also include domains related to the substrate, cell or cellulosomes binding, the last one leading to the formation of these enzyme-based complexes [60]. These bindings may avoid the elimination of the enzyme from the substrate, conduct hydrolysis to specific domains or facilitate the recovery of the digestion products [60].

Furthermore, often bacteria and microorganisms do not possess the three types, but they act synergistically between them instead. The enzymatic cellulose degradation is affected by both the structural characteristics of the biopolymer (crystallinity, degree of polymerization, etc.) and their own acting enzymes [42,61]. The enzyme-related factors which affect cellulose degradation are enzyme origin, temperature, specific product inhibition, binding to the substrate, activity balance for synergism, specific activity and both enzyme processability and compatibility [42].

Hemicellulose Biodegradation by Bacterial Activity

Once more, species from both aerobic and anaerobic bacteria were identified as hemicellulose degraders, counting *Ochrobactrum*, *Bacillus*, *Paenibacillus*, *Acinetobacter*, *Thermomonospora*, *Clostridia*, *Streptomyces*, *Cellvibrio* and *Pseudomonas* between the hemicellulase-producer genera [59,62]. In the same way as fungi, complete hemicellulose degradation is accomplished by the synergistic operation of a vast range of enzymes because of the inherent variability of the hemicellulose biopolymer. Only for xylan, a wide range of enzymes has been reported as being mandatory for the degradation completion; these enzymes have been included in Table 3, together with a short explanation of their mode of action [62,63].

Table 3. Enzymes involved in the hydrolysis of complex heteroarabinoxylans.

Enzyme	Objective
Endo-β-1,4-xylanase	β-1,4-xylose linkage hydrolysation of the xylan backbone
Exo-xylanase	β-1,4-xylose linkage hydrolysation, releasing xylobiose
β-Xylosidase	Release short-chain xylooligosaccharides and xylose from xylobiose
α-L-Arabinofuranosidase	Terminal non-reducing α-arabinofuranose hydrolysation from arabinoxylans
α-Glucuronidase	Release glucuronic acid from glucuronoxylans
Acetylxylan esterase	Acetylester bond hydrolysation in acetyl xylans
Ferulic acid esterase	Feruloylester bond hydrolysation in xylans
ρ-Coumaric acid esterase	ρ -coumaryl ester bond hydrolysation in xylans

On the other hand, some xylanases are known not to provoke the breakdown of glycosidic linkages until a proper debranching has been performed. However, those debranching enzymes often require partial hydrolysis before a proper breakage can be obtained. Hence, these findings highlight the intricate complexity of hemicellulose degradation, which, apart from the great numbers of enzymes involved, also requires a careful equilibrium and synergistic operation between the different enzymes [59,63,64].

Lignin Biodegradation by Bacterial Activity

Although the ability to degrade lignin is exclusive to a few bacteria genera such as *Streptomyces*, *Rhodococcus*, *Nocardia* or several *Sphingomonas*, *Pseudomonas*, *Enterobacter* and *Actinomyces* species [21,51,65], it has often been reported to be comparable to that shown by well-known lignin-degrader fungi [65]. Furthermore, the difficulties found in the genetic modification together with the low enzymatic yields usually observed in fungi have propelled the interest in the lignin-degrader bacteria [66]. The vast majority of these bacteria have been found within the digestive system of termites and other insects, though some other important species have also been found in soil and decaying vegetation [65–67].

Despite the enzymology of bacterial lignin digestion not being as developed as deeply as with fungi, some studies have helped to elucidate some of the enzymes which play a significant role in the lignin degradation by bacteria. Thus, four different types of bacterial lignin-degrader enzymes have been found up to this date [66]. Multi-copper oxidase enzymes, also known as laccases, constitute the first one, which have been found in species from the *Streptomyces*, *Ochrobactrum*, *Pseudomonas*, *Paenibacillus* and *Amycolaptosis* genera. Laccases have been found in very different living organisms, in which they play different roles. Hence, they are related to morphogenesis, pathogen-host interaction, stress defence and lignin degradation in fungi, whereas they play an opposite role in plants, where they are responsible for growth by lignin biosynthesis. Functions such as morphogenesis, copper homeostasis and pigmentation have been found in bacteria, whereas they are related to the sclerotisation of the cuticle in insects [54].

Other species from the *Rhodococcus*, *Enterobacter*, *Saccharomonospora*, *Pseudomonas*, *Amycolatopsis* and *Thermobifida* genera have otherwise shown the production of Dyp-type peroxidases [67,68], which form the second type of bacterial lignin-degrader enzymes. Recently, a lignin-oxidising manganese superoxide dismutase enzyme was found in *Sphingobacterium* sp. T2, which establishes the third type of enzymes. The last group is formed by the glutathione-dependent β-etherase enzymes, which are known to act through the β-aryl ether linkage rupture [66]. The visual appearance of the active sites of the different lignin-degrader enzymes can be found elsewhere [66].

The main aromatic species found through the biological pretreatment with lignin-degrader bacteria have been summarised in Table 4, which also includes fungi-related data for this production.

Table 4. Aromatic products detected from lignin breakdown [51,65,66].

Compound		Fungal Lignin Degrader	Bacterial Lignin Degrader
Benzoic acid	4-hydroxy	*P. chrysosporium*	*A. aneurinilyticus*, A. sp. 75iv3
	4-hydroxy-3-methoxy	*P. chrysosporium*	*A. aneurinilyticus*
	4-hydroxy-3-methoxy-6-carboxy	*P. chrysosporium*	
	4-hydroxy-3-methoxy-5-carboxy		*P. putida*, *R. jostii* RHA1
	3,4-dimethoxy	*P. chrysosporium*	
	3,4-dimethoxy-2-carboxy	*P. chrysosporium*	
	2-hydroxy-3-methoxy		*A. aneurinilyticus*, *P. putida*
	2,3-dihydroxy		
	2,3,4-trihydroxy		*Bacillus sp.*
Benzaldehyde	4-hydroxy-3-methoxy		*S. paucimobilis*
	3,4,5-trimethoxy		*Bacillus sp.*
Cinnamic acid	4-hydroxy		*Bacillus sp.*
	4-hydroxy-3-methoxy		*Bacillus sp.*, *P. putida*, *R. jostii* RHA1
Biphenyl-5,5′-dicarboxylic acid, 2,2′-dihydroxy, 3,3′-dimethoxy	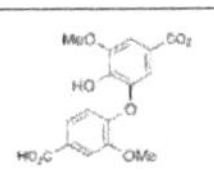	*P. chrysosporium*	
Diphenyl ether		*P. chrysosporium*	
Propiophenone-3′-hydroxy	4-hydroxy-3-methoxy		*S. paucimobilis. P. putida*, *R. jostii* RHA1
Acetophenone	4-hydroxy-3-methoxy		Soil metabolite
Phenol	2-methoxy		Soil metabolite
	2-methoxy-4-vinyl		Soil metabolite
Vanillin			*R. jostii* RHA1, *A.* sp. 75iv3

It is also worth mentioning that such products are not often directly obtained from the ligno-cellulosic chain; instead, oxidised polymeric intermediates are found. *Streptomyces viridosporus*,

Amycolatopsis sp. 75iv3 and *Thermobifida fusca* have been shown to produce a water-soluble intermediate described as acid-precipitable poly-phenolic polymeric lignin (APPL) [65,66]. However, the production of either the polymeric intermediates or the direct phenolic-based molecules is based on the cleavage of the interunit linkages that join lignin. Thus, as diverse bonds have been found to form lignin, diverse pathways for lignin breakdown have also been reported, where the β-ketoadipate pathway has been considered the most usual one [54,65,66]. Diverse enzymatic pathways regarding the main interunits that comprise lignin can be found elsewhere [65,69].

As a summary for the pretreatment section, the main pretreatment methods with remarkable characteristics and a description of their advantages/disadvantages have been included in Table 5.

Table 5. Comparison of the different pretreatment methods used for lignocellulosic degradation methods. Good or bad performance regarding cost, toxic byproduct formation and applicability are marked by ✓and ×, respectively [61].

Pretreatment	Cost	Toxic Byproduct	Applicable to a Wide Range of Biomass	Remarks
Acid pretreatment	✓	×	✓	Inhibitors limited by dilute acid use
Freezing	✓	✓	×	Freezing/thawing cycles
Milling	✓	✓	✓	Used for bioethanol and biogas production
Liquid hot water	✓	✓	×	High water and energy inputs
Organic solvent (Organosolv)	×	×	✓	Low boiling point of the solvent. Solvent recycling is required
Oxidation	×	✓	✓	High cost of ozone generation. Ozone handling is required
Steam explosion	×	×	✓	High cost of steam generation
Extrusion	✓	✓	✓	Hydrolysis efficiency is improved
Wet oxidation	×	✓	×	Less water use as no washing is required
CO_2 explosion	×	✓	✓	High cost for pressure maintenance
Microwave irradiation	×	✓	✓	More effective than conventional heating
Ultrasound	×	✓	✓	Low temperature and time required
Ammonium fibre expansion	×	✓	✓	Less effective for biomass with high lignin contents
Ionic liquid	×	✓	✓	Stability and reuse. Instability may cause contamination
Biological pretreatment	✓	✓	✓	Increases delignification. Able to reduce polymerisation
Hydrothermal liquefaction	×	✓	✓	Lignocellulosic materials are depolymerised into bio-oil, biogas, biochar and water-soluble compounds

3. Lignocellulosic Materials for the Production of Biofuels, Biochemicals and Biomaterials

As previously mentioned, the great availability of a wide range of biopolymers and renewable characteristics has propelled the research into biofuels, biochemicals and biomaterials based on lignocellulosic components [70]. As an overview of this section, the different procedures, products and fields of application of lignocellulose-based biofuels, biochemicals and biomaterials are outlined in Scheme 2.

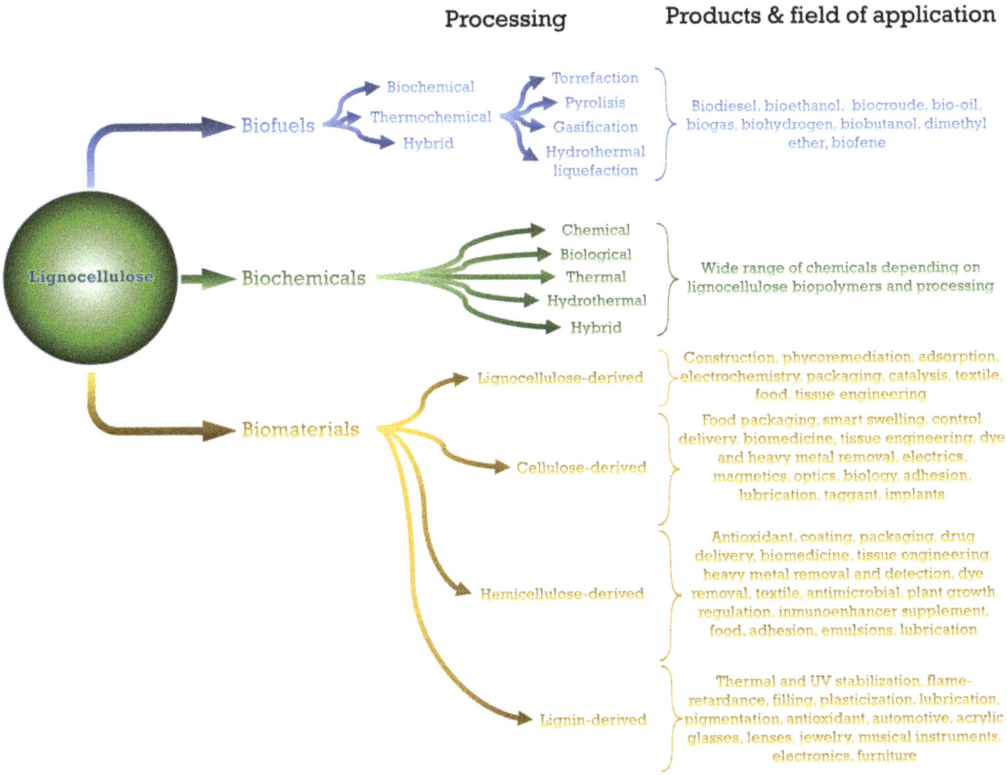

Scheme 2. Main procedures, products and fields of application of biofuels, biochemicals and biomaterials derived from lignocellulose.

3.1. Biofuels

Compared to petroleum-based fuels, biofuels possess advantages such as renewability, sustainability, availability, biodegradability, safety, neutral greenhouse effects and negligible SO_x and reduced NO_x gas emissions [71]. Lignocellulose represents the only sustainable, low-cost and scalable eco-friendly option for industrial fuel production [72]. Furthermore, it also represents a great opportunity for increasing the domestic energy production in those countries with large biomass supplies and/or land availability to produce energy crops [73]. The main drawback is found in lignin degradation, as it is considered the more energy-consuming step of the production process, due to the resilience of this biopolymer. Moreover, the cellulose efficiency, enzymatic and biomass costs and composition are other parallel parameters that critically affect the development of suitable technologies [72–74]. Furthermore, the obtaining of high-quality biofuels faces other problems, such as many of the most effective solvents for biomass pretreatment being simultaneously incompatible with enzymatic development, whereas those microbes with the highest yields in biofuel production do not often use the sugars present in hydrolysates as substrates [73]. Therefore, as mentioned before and widely discussed, massive attention has been paid to suitable pretreatments for advanced purposes.

Lately, effort has been focused on genome editing technologies as a powerful tool for understanding and developing an integrated system to produce fuels in fast and lower-energy-consuming processes [73]. Thus, some of the inhibitory compounds usually produced in natural plants, such as ferulic acid, can be substantially reduced by genomic editing; nonetheless, negative effects on crop yields, costs and environmental impacts can likewise take place. Another approach that has been substantially developed is the preferential growth of cellulose in detriment of hemicellulose and lignin, as most industrial microbes generally take advantage of hexoses instead of pentoses for biofuel production. Lignin content reduction and composition homogenisation have also been targeted, the latter allowing less complex product mixtures to be generated, thus higher-value molecules are obtained. On the other hand, other strategies have been based on the incorporation of unusual monomers, which potentiates chain elongation or the incorporation of interchangeable linkages, resulting, once more, in higher saccharification yields [73].

Nevertheless, the use of lignocellulosic biomass as biofuel competes with the use as food supply. Hence, a step further can be taken when residual lignocellulosic biomass is considered, as it can be transformed into what is called advanced biofuels, i.e., biofuels which significantly reduce greenhouse emissions simultaneously with the preservation of the common use of landfills, and therefore do not compete with food or feed commerce [70]. In this sense, deep research and many industrial projects which range from aerospace to common fuels, biogas, bioethanol or biodiesel have been accomplished or are being carried out currently [70,72]. In order to obtain those products, three different routes have been targeted, i.e., thermochemical, biochemical and hybrid conversion. An overview of these alternatives can be found elsewhere [70]. Therefore, the action of microorganisms is once again taking the lead as a renewable and environmentally friendly pathway for the conversion of biowaste into biofuels [75].

The thermomechanical route includes a variety of thermal treatments, from relatively low severity to strong processes where high temperatures are applied. Hence, torrefaction, pyrolysis, hydrothermal liquefaction and gasification are found among them.

The biochemical route relies on the enzymatic digestion of the different biopolymers that comprise lignin, which has been thoroughly described in previous sections, by either fungi or bacteria. Biomass digestion can lead to the production of small sugars, which can be directly used as fuel. However, a suitable separation process is generally needed as a consequence of the huge variety of products involved in lignocellulosic degradation. In addition, by following the biological route, certain compounds that cannot be obtained by chemical routes are produced, opening new areas for advanced biofuels [73].

Often, the products from biochemical routes are further converted by catalytic or thermomechanical processes into higher-added-value fuels. Thus, the hybrid route is accomplished.

The main biofuels that have been studied up to date are summarised in Table 6, including the energy that could be obtained from them. More specific details about raw materials, final characteristics and conversion routes can be seen in the indicated references.

Table 6. Main biofuels, the associated energy and corresponding processing routes reported in the literature.

Biofuel	Lower Heating Value	References
Biodiesel	32.6 MJ/L	[76]
Bioethanol	21.2 MJ/L	[76,77]
Biocrude	35.0 MJ/kg	[76]
Bio-oil	40 MJ/kg	[76,78]
Biogas	13–17 MJ/m^3	[71,79]
Biohydrogen	13 MJ/m^3	[71,79]
Biobutanol	27.8 MJ/L	[80]

3.2. Biochemicals

As a consequence of the great variety of monomeric units and linkages that comprise the three biopolymers, the range of biochemicals that can be obtained from them constitutes an even wider range. These biochemicals critically depend on the original biopolymer, i.e., lignin generally provides outstanding aromatic-based compounds, whereas sugars resulting from the hydrolysis of cellulose and hemicellulose may produce valuable six- and five-carbon-derived products [81]. In addition, the biochemicals obtained are dependent on the processing protocol; therefore, multiple biological and chemical processes, as well as combinations of both, have been studied in order to formulate different chemicals. Special attention is being paid to the bioengineering of these microorganisms; thus, more specific compounds can be targeted. More detailed information about this topic can be found elsewhere [82].

In the case of cellulose and hemicellulose, many different products can be obtained [82,83]. Werpy and Petersen [84] analysed more than 50 compounds, from which they found glycerol, acetic acid, levulinic acid, 3-hydroxybutyrolactone, glutamic acid, malic acid, itaconic acid, aspartic acid, oxalic acid, 3-hydroxy propionic acid, succinic acid, fumaric acid, 2,5-furan dicarboxylic acid, glucaric acid, sorbitol and xylitol/arabinitol to be among the more interesting ones. Most of them can also act as building blocks for the development of fine chemicals and derived compounds. Table 7 depicts the main processes to obtain those species, together with the main derived products that can be obtained from them.

Table 7. Main building blocks and their derivatives obtained from cellulose and hemicellulose [84].

Compound	Production	Derived Products
Succinic, fumaric and malic acid	Biofermentation	Tetrahydrofuran (THF), 1,4-butanediol, 2-pyrrolidone, o-butyrolactone, N-methyl-2-pyrrolidone (NMP)
2,5-Furan dicarboxylic acid	Chemical (oxidative dehydration of glucose) and biological	(2,5-Bis(aminomethyl)-tetrahydrofuran, 2,5-dihydroxymethyl-tetrahydrofuran, 2,5-dihydroxymethyl-furan
3-Hydroxy propionic acid	Biofermentation	1,3-Propanediol, acrylic acid, acrylamide
Aspartic acid	Chemical and biological pathways	2-Amino-1,4-butanediol, aspartic anhydride, 3-aminotetrahydrofuran, amino-γ-butyrolactone
Glucaric acid	Chemical (starch oxidation by nitric acid or bleach)	Glucaro-γ-lactone, polyhydroxypolyamides, glucarodilactone, glucaro-δ-lactone
Glutamic acid	Biofermentation	Glutaminol, glutaric acid, norvoline, 1,5-pentandiol, 5-amino-1-butanol
Itaconic acid	Chemical and biofermentation	3-Methylpyrrolidine, 3- & 4-methyl NMP, 3-methyl THF, 2-methyl-1,4-butanediol.
Levulinic acid	Chemical (acid decomposition of six-carbon sugars)	Diphenolic acid, 2-methyl-THF, b-acetylacrylic acid, 1,4-pentanediol
3-Hydroxybutyrolactone	Chemical (oxidative degradation of starch)	3-Hydroxytetrahydrofuran, 3-aminotetrahydrofuran, acrylate-lactone
Glycerol	Transesterification (via chemical or biological pathways)	Glyceric acid, 1,3-propanediol, propylene glycol
Sorbitol	Chemical (glucose hydrogenation)	Isosorbide, propylene glycol, ethylene glycol, 1,4-sorbitan
Xylitol/arabinitol	Chemical (hydrogenation of xylose and arabinose) and biological	Xylaric acid, propylene glycol, ethylene glycol, lactic acid

Other important cellulose- and hemicellulose-derived products are 5-hydroxymethyl furfural and furfural, which have been reported as being obtained by hydrothermal carbonisation of those biopolymers, respectively. Both derived furans can also be precursors of many other chemicals, biofuels and pharmaceutical and agrochemical products [81,83,85].

A detailed summary including the different chemical structures of both precursors and final products can be found elsewhere [83].

Regarding lignin, as mentioned above, a vast range of aromatic-based chemicals can also be obtained by its decomposition and transformation. However, the success of the suitable formation of chemicals from lignin relies on several main aspects, i.e., lignin fractionation from raw biomass, proper degradation, depolymerisation, transformation into high-value-added compounds and further separation. Depending on depolymerisation conditions, diverse products can be obtained, as shown in Table 8.

Table 8. Main procedures from lignin depolymerisation along with main products obtained.

Depolymerisation	Procedures	Products	Refs
Non-reductive depolymerisation	Thermal, hydrothermal, oxidative, acid and base catalysed, solvolytic	Vanillin, syringaldehyde, acetosyringone, guaiacylacetone, p-hydroxylated phenol acetovanillone, syringol, guaiacol, phenol, catechol, alkylcatechols, creosol, p-hydroxybenzaldehyde, vanillic, protocatechuic, syringic, homovanillic and p-hydroxybenzoic acid, aliphatic carboxylic acids (succinic, acetic and formic acid)	[81,86]
Reductive depolymerisation	Hydroprocessing, liquid phase reforming	Cresol, xylenol, phenol with long alkyl chains, p-substituted methoxyphenols,	[86]

The different processes do not only lead to the formation of different species but also the monomer yields depend on the procedure characteristics [86].

Similar to cellulose and hemicellulose, several of the compounds obtained can be considered end products, while many others can likewise act as building blocks for possible upgraded compounds [83,86].

3.3. Biomaterials

Due to the extensive variety of molecular species obtained from lignocellulosic degradation, the possibilities of derived biomaterials are massive. Werpy and Petersen [84] included in their work an exhaustive review of the biomass components, their primary degradation products, main intermediates and a brief description of derived bioproducts and uses. Hence, lignocellulose derivatives may play a significant role in areas such as general industry, transportation, textiles, packaging and other food vessels, environment, plastic replacers, stationery, house and leisure items, health and hygiene.

However, those biomaterials that make use of the interesting biopolymer network characteristics without further fractionation into derived products are not included among those uses; thus, they will be discussed separately. When considering lignocellulose, frequently research has taken advantage of lignin, cellulose and hemicellulose biopolymers separately; however, the whole biomass has also been considered.

3.3.1. Lignocellulose-Derived Biomaterials

There is an immense range of biosource applications currently, boosted by research possibilities, variety and exceptional properties. Some of them will be discussed further on in this section; nonetheless, a detailed revision of many of these applications is beyond the scope of this work.

As an example, for wheat and barley straw, advanced biomaterials for construction materials, phycoremediation of wastewater, cement properties enhancer and fibres in concrete were reported [87–92]. Straws and stalks from other sources were also focused on for board production, being potentially suitable for beaverboard, packing materials, one-use tableware or seeding devices [93,94]. Graphitised lignocellulose extracted from bamboo has also been used lately for electromagnetic wave absorbers [95].

One of the main applications that have been developed is the use of biomass as an adsorbent. In this sense, Rocha et al. [96] studied the adsorption of metal ions such as Cu(II), Zn(II), Hg(II) and Cd(II) on rice-straw-derived solutions. By the formation of biochar from rice straw, other metal ions such as Pb and Zn can also be adsorbed [97], as well as nitrogen and phosphorous [98]. Abdel-Aal et al. [99] reported the ability of rice straw in the treatment of wastewater including several commercial dyes instead. Likewise, banana peel- and palm-flower-waste-based derived products were also able to remove methylene blue and malachite green dyes from polluted solutions [100,101]. Other studies obtained proper adsorbents from residual products such as orange peel and sugarcane bagasse [102]. A detailed review of this matter has been recently published [103].

On the other hand, electrocatalytic activity was also studied on biomass-derived materials. Thus, Castro-Gutiérrez et al. [104] produced tannin-derived carbon materials, while Liu et al. [105] created soybean straw-based Fe-N co-doped porous carbons, both exhibiting excellent properties in electrochemical applications. Ma et al. [106] demonstrated that cornstalks and pomelos skins efficiently act as carbon sources for the construction of cathode catalysts for microbial fuel cells. Another biosource, watermelon, was used by Wu et al. [107] to create hydrogels and aerogels with electrochemical applications. Other biomass sources suitable for electrochemical applications are sawdust or grasses [108].

Composites containing residual lignocellulosic biosources have also been targeted. In those, lignocellulose acts as a reinforcing filler and avoids problems such as lack of flexibility or respiratory illnesses [109]. Bugatti et al. [110] showed how tomato peels could form proper composites with halloysite nanotubes for packaging applications. Ita-Nagy et al. [111] demonstrated that sugarcane bagasse fibres also properly reinforce composite structure. Pinhao and pecan nutshells were also used for reinforced composites preparation. The pinhao-nutshell-based composite exhibited lower water absorption capacity than the petcan-based one, based on the enhanced hydrophobic character of the pinhao-based composite [112]. Fibres from tropical maize and sweet sorghum bagasse were also studied as composite additives [113].

The use of biomass as catalysts or catalyst supports has also been deeply studied, as is the case for soybean and other biomass with high protein content [114].

In textile, bamboo fibres can provide comfortability, good dyeing and appealing characteristics. Hemp can also be utilised for textile application, as well as for making sacks and ropes, degumming, etc. [115].

In food, tomato peels have acted as an enhancer for colour and antioxidant properties for yoghurts [116], whereas tomato peel fibres have demonstrated the ability to produce a suitable network for edible gels, with enhanced stability and texture [117].

A singular case can be considered when lignocellulosic biomass acts as a hydrogel precursor, which has led to interesting applications being found in fields such as film formation, high-strength filaments, tissue engineering, among many others [118].

Nonetheless, the number of applications and studies is boosted when the different lignocellulosic biopolymers are considered separately. In the following sections, cellulose-, hemicellulose- and lignin-derived biomaterials have been examined.

3.3.2. Cellulose-Derived Biomaterials

The singular structure of cellulose, together with its possibilities of being modified by chemical reactions or converted into alkyl-derived or nanosystems, has made a great range of biomaterials available currently from this biopolymer.

The high number of hydroxyl groups present in cellulose has attracted research attention to the formation of cellulose-based hydrogels. Even though the cellulose ability to be dissolved in water is limited, the development of many suitable solvents has caused hydrogels with stunning properties to be obtained, with applications such as food packaging, smart swelling, controlled delivery and biomedical applications [119–121].

The well-known ability of some bacteria to produce cellulose (bacterial cellulose) has also been leveraged for the performance of hydrogels. In general, good tensile and compressive properties are shown, together with high water-absorption capacity, crystallinity and biocompatibility, which have caused bacterial-cellulose-based hydrogels to be focused on bio-applications such as dental and meniscus implants, or tissue engineering scaffolds [119].

Hydrogels from alkyl-cellulose-derived products have also been analysed, such as hydroxypropyl-, hydroxypropylmethyl-, carboxymethyl- or methyl-cellulose. The alkyl-derived chains introduce regions where physical crosslinking may be dampened; thus, chemical crosslinking has frequently been targeted, which has provided hydrogels with new characteristics, for instance, pH dependence in sorption capacity. These hydrogels have been shown to be valuable in water body elimination through the absorption of water in the stomach [122], dye elimination [123], food and drugs [119].

Nonetheless, the capacities of cellulose-based hydrogels can be further boosted by the combination of other synthetic or natural polymers. Hence, heavy metal elimination or food and tissue engineering applications have been targeted by the combination with chitosan, starch or alginate, respectively [119].

On the other hand, inorganic materials have also been added to the cellulose-based hydrogel structures, with applications in fields such as electricity, magnetics, optics and biology. A summarised insight on the potential applications of cellulose-based hydrogels can be found elsewhere [118].

However, not only hydrogels have been reported as being developed from cellulose. Aerogels, usually obtained by freeze-drying of hydrogels or supercritical drying with CO_2, have also been produced. In the same way as with hydrogels, cellulose, bacterial cellulose and many derived systems (from alkylated compounds to nanosystems) have been reported to produce aerogels, affecting both the synthesis process and final properties [124].

Aerogels possess characteristics such as very low density (up to 0.5 mg/mL), high specific surface area (up to 975 m^2/g) and highly porous structures (up to 99.9% porosity), while keeping good mechanical characteristics, which have propelled their use in applications such as shock absorbers, acoustic and thermal insulation, oil absorption, biomedical devices and implants, conductivity enhancers or carriers of metal nanoparticles and oxides [124–126].

On the other hand, by the combination with oily systems, oleogels have also been prepared. Once more, prepared through the use of cellulose and cellulose-derived materials, a vast range of products has been documented. In the food industry, it is ethyl cellulose which has attracted the most attention due to its appealing thickening properties, though pristine cellulose, methylcellulose and hydroxypropyl methylcellulose have also been studied [127]. Nonetheless, the rheology-modifier characteristics have been leveraged for its use in a wider range of applications such as binders, films, adhesives, lubricating greases and hot blends [128,129].

The case of lubricating greases remains especially appealing, as, regardless of the extensive work found in literature, the industry continues to employ almost entirely petroleum-based products and uses lithium- and other metal-based soaps as thickening agents. Hence, studies that explore the use of cellulose pulp as a thickener can be found [130–132], but also pristine cellulose [133,134] and cellulose derivatives [133–135]. A wide range of these systems has been demonstrated to impart suitable rheological and excellent tribological properties, along with appropriate mechanical stability comparable to lithium-based lubricating grease benchmarks.

As already introduced in this section, the formation of nanostructures from cellulose is also a very appealing approach for the development of potential systems with a broad range of applications. By the formation of nanofibres, excellent properties of water-based hydrogels have been shown, as only a very low concentration is needed in order to obtain good rheological properties. Furthermore, the formation of films and nanocomposites has also been extensively reported. Thus, applications such as reinforcing agent for paper, greaseproof paper, thermosetting resins, strengthened composites, obesity-precautionary thickener, suspension stabiliser, sanitary products, wound dressing, coatings, etc. can be found [136,137].

The formation of cellulose nanocrystals has also been extensively studied, with the majority of applications based on the formation of composites. Nonetheless, the suspensions containing these cellulose nanocrystals have shown nematic chirality, which boosts their applications in fields such as NMR spectroscopy and optical taggants [136,138].

In addition, not only mechanical and chemical processes have been documented to successfully produce cellulose nanostructures. Instead, some bacteria have also exhibited the possibility of directly obtaining cellulose nanostructures with the only presence of hydroxyl moieties as functional groups. The high yield for a biological process (up to 40%)

and unique structure have propelled its use in fields such as regenerative medicine, wound healing, implants, membranes, films and barrier layers [136].

3.3.3. Hemicellulose-Derived Biomaterials

Even though hemicellulose does not possess the significance and properties of cellulose and cellulose derivatives, there are also many studies that take advantage of hemicellulose structure to produce interesting biomaterials. Due to the fact that hemicellulose is not formed by a single type of biopolymer, the possibilities are again raised. The main target of hemicellulose-based biomaterials has been based on the production of antioxidant agents, hydrogels and films for uses such as coatings, packaging or biomedicine; nonetheless, some other less common uses have also been studied.

Hydrogels have been mainly centred on drug delivery [139–141], tissue engineering and environmental protection, and have demonstrated pH, ionic strength, media composition and organic-solvent-dependent behaviour. Thus, the removal of heavy metals such as Ni (II), Cu(II), Pb (II), Cd(II), Pd(II) and Zn(II) or sulfadimidine has been successfully achieved by these hydrogels [142–144]. On the contrary, good adhesion in tissues like liver has also been documented, in which they may play a good replacement role in detriment to more expensive and less available tissue and organ transplants [145]. In drug delivery, xylan- and galactomannan-based microcapsules have been shown to be excellent colon-specific carriers [140,146,147], while xyloglucan mucoadhesive and surface tailoring properties have led to the development of many specific studies [141]. Galactomannans have also been shown to be useful for drug delivery by the formation of an aerogel structure [148].

Hemicellulose-based films were also produced, whose significance is based on their outstanding properties against oxygen permeability, thus conforming to a great replacement for oxygen-sensitive food packaging. Xylan, arabinoxylan, glucomannan and galactoglucomannan, alone, modified or by combination with other biopolymers, are some of the biopolymers from hemicellulose which have shown suitable properties for film production [143,149,150].

The same oxygen permeability provides hemicellulose with interesting antioxidant and antimoisture properties, which make them good replacements as coatings for food packaging.

Between the less common uses, there is still a vast range to be found. Peng et al. [151] demonstrated that hemicellulose acts as a stabiliser for the formation of silver nanoparticles. Jiang et al. [152], instead, used hemicellulose for the synthesis of quantum dots to detect Ag(I) and L-Cysteyne in aqueous solutions. Farhat et al. [153] were able to produce hemicellulose from bleached hardwood pulp and switchgrass, which, crosslinked with zirconium, exhibited excellent adhesive properties. Xylan has been analysed by Ebringerova [154], who evaluated other potential applications such as textile printing, antimicrobial additive, plant growth regulator, inmunoenhancing supplement, additives and thickening agent in food. On the other hand, xyloglucan has also been reported as a texture enhancer, binder, dye absorption, emulsion stabilizer, syneresis control and food additive [155]. Some of these applications are shared by galactomannan-derived products, which have been shown to act as binders, texture modifiers, emulsifiers, lubricators or stabilisers, mainly in the food industry [156]. Galactoglucomannan has also been studied as a food additive because of its prebiotic properties [157].

3.3.4. Lignin-Derived Biomaterials

The unique characteristics of lignin have propelled both research and industry to focus on this biosource acting as an additive for a wide range of polymers, such as polypropylene, polystyrene, polyethylene, polyamide, PVC, poly(vinyl alcohol), among other bio and synthetic polymers [158]. Thus, thermal and UV stabiliser, flame-retardant, reinforcing filler, plasticiser, lubricant, colour-adding pigment and antioxidant are among the most significant properties that lignin incorporation can enhance or modify [159], producing interesting biomaterials that can be applied in fields such as thermoplastic, thermoset, bioplastic and rubber composites, aerogels, carbon fibres and foams [158].

For instance, lignin addition in polyethylene and polystyrene matrices up to 20% has been shown not to modify processability whilst significantly improving resistance against photodegradation [160]. In another study, Mishra et al. [161] demonstrated that lignin provides stronger resistance to UV in PVC films acting as UV absorber. Yang et al. [162] also showed that poly(methyl methacrylate) thermal and mechanical properties are enhanced by the lignin addition against UV light. Therefore, applications in fields such as automotive, acrylic glasses, vehicles and lenses have been targeted. On the other hand, different lignin fraction response to UV light as a function of the solvent utilised has also been studied, showing UV resistance to be dependent on the lignin extraction method [163,164]. In Sheng et al. [164], the phenolic hydroxyl group content of several residual lignins turned out to be crucial to their performance as antioxidants.

Thermal resistance is frequently another weak point in conventional polymers and composites; thus, lignin addition as enhancer of the heat resistance has been a deeply studied topic. Canetti et al. [165] demonstrated that lignin addition from 5 to 10% in polypropylene blends successfully improves the thermal resistance. Working on natural rubber modification, Gregorová et al. [166] showed that the lignin addition also increased the thermo-oxidative long-term resistance. On the contrary, Tavares et al. [167] reported that a simple 1% lignin addition reinforces the poly(butylene adipate-co-terephthalate) matrix. On the other hand, Lisperguer et al. [168] demonstrated that lignin addition to recycled polystyrene could provide similar characteristics to the native polymer. Such is the interest in the improvement of the thermal properties by lignin addition of blends, composites and copolymers that Sen et al. [169] wrote an extensive review where lignin modification by different processes and the thermal response of the products obtained were evaluated.

Flame prevention is another powerful characteristic that lignin can provide to a biomaterial. Thus, De Chirico et al. [170] worked with lignin and various derivatives, which were able to provide the polypropylene matrix with enhanced combustion time and char yields and reduced both heat liberation and mass loss rate while saving mechanical properties. The flame retardant properties of polylactic-acid-based biopolymers were likewise improved by the use of lignin nanoparticles functionalised with diethyl (2-(triethoxysilyl) ethyl) phosphonate [171]. Another renewable polymer, polybutylene succinate, was also successfully treated with nitrogen-and-phosphorous-doped lignin systems, improving both heat release rate and total heat release in around 30%, properties that could lead this biopolymer to be applied in wider fields [172]. Recently, some reviews concerning the flame retardant possibilities of lignin and its derivatives, as well as their future prospects, have been reported [173,174].

Lignin acting as reinforcing filler has also been extensively documented. Therefore, Ikeda et al. [175] showed that lignin potentially improves tensile stress and storage moduli of natural rubber, while reducing the dissipative loss. The tensile and flexural modulus of polylatic acid, poly(3-hydroxybutyrate) and thermoplastic elastomers after lignin addition were also raised [176]. Mechanical properties were also generally improved when lignin nanoparticle-poly (diallyldimethylammonium chloride) complexes were added to natural rubber [177]. In a study developed by Rozman et al. [178], coconut fibre and polypropylene were mixed using lignin as compatibiliser. The mixtures exhibited better flexural properties when lignin was included. On the other hand, Tanjung et al. [179] demonstrated that lignin addition to polypropylene/chitosan composites successfully increased tensile strength, elongation at break, Young's modulus and impact strength. Extensive reports have been documented regarding lignin acting as filler for bioplastics, thermoplastic and thermoset composites [158]. Regarding bioplastics, Yang et al. [180] revised the most recent literature in lignin-reinforced bioplastics made of cellulose, protein, starch, polylatic acid and polyhydroxybutyrate. Recently, an extensive review aiming to compile the latest advances in lignin-reinforcing properties in the rubber industry has also been published [181].

As mentioned above, lignin can likewise act as plasticiser, which has been achieved frequently through structural modifications. Therefore, lignosulfonate has been used as the usual plasticiser for concrete. Through other modifications like alkylation, the plasticising effect of lignin has also been achieved [169]. Working with PVC and different molecular weight lignin fractions, those with the lowest ones were reported to act as plasticisers by Yue et al. [182]. On the other hand, a novel process consisting of an alkali-O_2 oxidation technology for lignin revalorisation has recently been addressed for plasticisation, (LigniOx), which was included in the annual report of the top 20 innovative bio-based products form the European Commission [183]. Moreover, by further modifications, superplasticisers have been targeted, which could replicate the performance of well-known commercial naphthalene plasticisers [184].

The presence of hydroxyl groups in lignin has provided it with the possibility of establishing H-bonds, which has been leveraged for its use as an additive for lubricants. Hereby, lignin has been added to many different base oils, in which the H-bonds formed have proved to generally decrease both wear and friction. Mu et al. [185] developed fully bio-based lubricants by utilising lignin and ionic liquids, demonstrating outstanding tribological and anticorrosive properties on both aluminium and iron surfaces. Working with polyethylene glycol as the base oil in a more recent study, Mu et al. [186] compared lignins from different origins and extraction processes as additives, highlighting hydrogen bonding and molecular weight as crucial factors on thermal and lubricating properties. Up to 93.8% wear reduction was reported by the lignin incorporation. In another study, Hua et al. [187] created lignin-based green lubricants with excellent properties in diamond-like carbon-steel contact compared to commercial lubricants, showing lignin-based lubricants to be useful on surfaces with different characteristics. Cortés-Triviño et al. [188] tested a lignin-enriched residue coming from sugarcane bioethanol production to further valorise it by mixing with castor oil epoxidised at different degrees, showing the possibility to tune the rheological and lubricant properties thereof of the mentioned mixtures. The production of nanofibres from lignin by using the electrospinning technique has likewise permitted the use of these materials as oil structuring [189]. In addition, not only liquid lubricants but also semi-solid ones, i.e., lubricating greases, have been developed by using lignin [190,191].

Even though lignin is almost colourless in wood, its separation from cellulose and hemicellulose finally turns it into a dark brown colour. For this reason, Balasubramanian et al. [192] used lignin as brown pigment and tested it in leather, where, apart from dyeing the surface properly, it exhibited compatibilisation with common products for leather finishing. On the other hand, Araújo et al. [193] used lignin nanoparticles to encapsulate blue pigments, which increased both solubilisation and stabilisation, making them more suitable for industrial applications.

Within lignin roles in plants resides the antioxidant capacity as a consequence of its aromatic structure, which has likewise been used for the incorporation of antioxidant properties in biomaterials production, with applications in cosmetics, healthcare, agricultural products and pharmaceuticals [194,195]. Once more, the varied lignin characteristics depending on both the origin and extraction method have allowed adjustable antioxidant properties to be obtained [194,196,197]. For instance, Li et al. [198] used different solvents (ether, ethyl acetate, methanol, acetone and dioxane/water) in order to study the antioxidant properties of the different fractions obtained. The results showed that the higher the dissolving ability of the chemicals, the lower the antioxidant capacity of the resulting lignin fraction. Instead, Ma et al. [199] investigated the antioxidant activity as a function of pH in the lignin extraction method. High pHs led to low lignin content and low phenolic content, which exhibited low antioxidant activity. On the contrary, low pHs turned into lower molecular weight lignin fractions with high phenolic hydroxyl content, showing excellent antioxidant properties and highlighting the crucial effect of both molecular weight and phenolic hydroxyl group content in antioxidant properties.

Industrially, lignin has been used as a dispersant, binder or chelator, as a substitute of phenolics powder resins, in polyurethane foams and epoxy resins or as biodispersant, among others. Thus, some North American companies have implemented lignin addition in products such as automotive brake pads and moulds and oriented strand boards. The company Bioconsult Gesellschaft fuer Biotechnologie GmbH has developed a lignin-based system able to control the microbial growth in industrial wastewater ambits [200]. The company Nippon Paper Group has released diverse lignin-based biomaterials, SAN X®, VANILLEX®, PEARLLEX®, which can act as mentioned above according to their characteristics [201]. More concretely, the company TECNARO has been able to use lignin in thermoplastics with applications in fields such as jewellery, construction, musical instruments, electronics, furniture, etc. The Prisma Renewable Composites Company developed a system named evolUTIATM, able to create lignins with similar characteristics from different plant sources, which has let them use lignin in biomaterials such as plastics, elastomers and carbon fibres. Their first product is BioLANTM, which has been declared as a substitute for ABS with enhanced mechanical and UV-resistant properties.

4. Lignocellulose-Based Polyurethanes

Due to the ubiquitous production, environmentally friendly character and interesting properties of either lignocellulosic sources as a whole or sources separated into their main components, extensive research has been devoted to the formation of polyurethanes based on lignocellulosic materials [202]. For instance, the addition of wheat straw in polyethylene-glycol-based PU foams was shown to provide both better thermal resistance and compressive strength compared to those systems only based on polyethylene glycol. Moreover, biodegradability was likewise enhanced [203]. Wheat straw was once more used in combination with castor oil to produce suitable PU coatings for controlled-release fertiliser. Apart from the great controlled-release characteristics, the PUs exhibited good degradability and high density [204]. On the other hand, barley-straw-derived foams for insulation were also produced, which demonstrated water intakes up to 986% after 48 h, better than those obtained with synthetic PUs [205]. PU foams were also considered by Ertas et al. [206] but were obtained from *Eucalyptus camaldulensis* and *Pinus sylvestris* instead. Other biosources considered for the production of rigid foams were cotton stalk, pine bark and apricot stone [207,208]. Moreover, polyurethane films were successfully elaborated by employing lignocellulose materials such as wheat starch by the reaction with isophorone diisocyanate, which exhibit suitable mechanical properties to replace petrochemical substitutes [209]. Nonetheless, the range of available PU for different applications has been greatly expanded by the use of cellulose, hemicellulose and lignin separately.

4.1. Cellulose-Based Polyurethanes

Cellulose-based polyurethanes have been produced by either taking advantage of the biopolymer structure and hydroxyl functional groups or by the cleavage and modification of the structure in order to obtain intermediate products for PU formation. The last one is the case, for instance, for isosorbide and 2,3-butanediol. Both derived from glucose, Calvo-Correas et al. [210] demonstrated they can act as copolymers in PU synthesis for film formulations, which can compete with non-renewable ones. On the other hand, Wei et al. [211] elaborated a synthetic route to produce adipic acid from cellulose, from which Nylon 66 or PUs can be obtained. Another approach was the transformation of cellulose into what is called cellulose-based ionic liquids. These systems were transformed into PUs from which excellent membranes were produced, able to adequately separate CO_2/CH_4 from natural gas sidestreams [212].

Cellulose has been frequently employed as Pickering emulsifying agent, from which porous monoliths can be formed. However, poor mechanical properties have been generally observed. Nonetheless, by the chemical interaction with diisocyanates, proper robustness can be obtained. Moreover, the system provides suitable and fast absorption capacity and tunable wettability from hydrophilicity/oleophilicity to hydrophobicity/oleophilicity [213]. Cellulose has also been used as grafting copolymer, which has led to interesting shape-memory products for smart applications [214]. Cellulose from industrial furniture waste has also been tested to absorb dye within a PU foam matrix. For the three dyes utilised, Methylene blue, Procion yellow and Procion red, the kinetic studies suggested a pseudo-second-order absorption model. The maximum removal values were around 70, 90 and 80 wt.%, respectively [215].

Using cellulose fibres, the reinforcement of polyurethane composites has been targeted [216,217]. Hadjad et al. [218] incorporated up to 30 wt.% of cellulose fibres and studied the conductivity and capacitance alterations of the composites. Up to 10 wt.% concentration, the main electrical characteristics were unaffected. However, cellulose can also be used as a chain extender due to the high molecular weight usually reported; thus, Ikhwan et al. [219] developed cellulose fibres and polyethylene-glycol-based PUs, whose relative concentration demonstrated a strong influence on thermal properties. In order to provide PU composites with superior performance, nanofibres have been utilised. The use of nanofibres has provided a fast response (less than 1 min) in shape recovery, which propels the potential use of these nanocomposites for biomedical applications [220]. The improvement of properties by using cellulose nanofibers was also observed for flame retardant applications. Therefore, by assembling with anionic vermiculite, outstanding transparency, resistance to oxygen pass and record fire resistance characteristics have been reported [221]. On the other hand, aerogels with excellent flame retardant properties were also obtained by combination with hydroxyapatite [222].

Polyurethanes were also prepared by using bacterial cellulose. In Urbina et al. [223], biocompatible PU nanocomposites were once more reinforced by bacterial cellulose incorporation, displaying outstanding shape memory and mechanical results. Nonetheless, they could not be compared to the use of nanofibers, as 93% recovery took place in around 3 min.

The use of cellulose nanocrystals was also focused on for the reinforcement of PU nanocomposites. An enhancement on both Young modulus and stress at break was usually recorded [224,225].

4.2. Hemicellulose-Based Polyurethanes

In the same way as has been happening with its glucose-based homonym, research has been devoted to the production of bio-PUs by following two pathways; using the complete hemicellulose structure and derived constituents or dividing it into smaller units, which can be further used as polyols. Hence, following the last approach, xylitol or furfural have been obtained and applied for PU formation by liquefaction or oxypropilation [226]. Along with xylitol, sorbitol was also produced by Robinson et al. [227]. On the other hand, Samavi and Rakshit [228] utilised hemicellulose liquor to produce epoxidised microbial oil, which could be further used as polyol in PU production. A summary of the main processes which can lead to hemicellulose division into polyols has been explained elsewhere [229].

Taking advantage of the hemicellulose and derivatives structure instead, Cheng et al. [230] reported the use of xylan to produce PUs with enhanced thermal stability. On the other hand, arabinoxylan was used to achieve PU films with application in the packaging field [231], or coatings by crosslinking with glutaraldehyde, which exhibited comparable properties to polyvinyl alcohol, becoming a sustainable substitute [232]. Simultaneously, studies using arabinogalactan have helped polyurethane scaffolds to improve cell attachment yields [233,234]. On the contrary, another hemicellulose-constituent biopolymer, galactomannan, has been shown to provide PUs with excellent characteristics for drug release in specific body parts [235]. Nonetheless, the most used hemicellulose-derived biopolymer is glucomannan. Improving mechanical performance, the use of glucomannan in composites and nanocomposites has been thoroughly studied. Its combination with waterborne PUs has also provided excellent mechanical performance and thermal properties due to the strong H-bonding between the PU and glucomannan [236]. A summary of glucomannan-based products with their potential applications and techniques used for characterisation is included herein below as Table 9.

Table 9. Glucomannan-based materials, characterisation performed and future application prospects in various fields [236].

Components	Potential Applications
Chitosan/Konjac glucomannan (KGM)	Membrane with superior dehydration
KGM/Chitosan	Food industry, biomaterial matrix, biomedical material
KGM/Ethyl cellulose	Films for food packaging
Glucomannan–Chitosan–Nisin	Active packaging material
KGM/Gellan gum	Food packaging material
KGM/Poly(acrylic acid)	Specific drug delivery
KGM/Polyacrylamide/Sodium xanthate	Hydrogels for drug delivery
KGM/poly(methacrylic acid)	Specific drug delivery
KGM/Polyvinyl alcohol	Pervaporation dehydration, food package film
KGM/Xanthan gum	Gels for delivery systems, specific drug delivery
KGM/Alginate/Chitosan	Controlled release
KGM/Carboxymethyl cellulose	Emulsion stabiliser
KGM/Curdian	Food films and coatings
KGM/Poly(aspartic acid)	Carrier for drug delivery
KGM/Cellulose	Separation
KGM/Whey protein	Edible food films
KGM/Sodium alginate	Food films
KGM/Gelatin	Specific drug delivery
KGM/Starch	Edible food films & coatings
KGM/Poly(diallydimethylammonium chloride)	Antibacterial in biomedicine
KGM/xanthan gum	Drug delivery
KGM-graft-Polyacrylamide-co-sodium xanthate	Flocculant
KGM	Coating

Furthermore, in a study elaborated by Shao et al. [237], no division but the whole hemicellulose extracted from corncob was used to perform PU films. These were compared to films that originated from cellulose, lignin and diverse mixes between the three biopolymers, obtaining for the hemicellulose-based PU films the highest glass transition temperature.

4.3. Lignin-Based Polyurethanes

Similar to the previous biopolymers, both the lignin as a whole structure or derivatives and small compounds obtained from it have successfully produced PUs. Thus, the production of ferulic acid and cresol-based monomers from lignin yielded PUs with superior thermal characteristics [238]. In the same way, vanillin-derived systems were also used as polyols for PU formation [239].

Extensive research has been carried out regarding the use of lignin in polyurethane formation due to the availability of hydroxyl groups, which makes it an interesting biopolyol replacer. Moreover, both aromatic and aliphatic hydroxyl groups have been demonstrated to react with diisocyanates [240]; however, some of these groups may be hindered due to steric encumbrance based on the network ordering and self-association [241]. Generally, lignin has been used alone or accompanied by other synthetic or natural polyols, while often it is first modified to enhance overall hydroxyl group value and aliphatic hydroxyl value or improve access to them. In this sense, demethylation stands as a very useful technique, through which lignin causes methyl groups to be replaced by hydrogen, finally obtaining higher hydroxyl values. Another technique extensively used is hydroxyalkylation, which introduces primary and secondary alcohols to the lignin network by combination with different compounds [241]. Another significant process that is intended to enhance hydroxyl value is phenolation. Since diisocyanates also possess a great affinity to amine groups, lignin amination and nitration have likewise been focused on [241–243]. Nonetheless, the high molecular weight and structure stand for a too-high viscosity and difficulties with reactivity; thus, depolymerisation is often targeted as well.

In any case, lignin can provide exceptional fire resistance, crosslink density, ultraviolet stability, biodegradability, insulation, compression, antioxidant, thermal and reinforcing properties to PUs [200,238,241,244,245], as well as high availability and low cost, the reason why foams constitute one of the main formulations for lignin-based PUs, where it can act either as a polyol or as a filler. The mechanical performance is usually improved as a consequence of the more entangled generated systems [245]; however, lignin is also known to increase stiffness, and therefore brittleness, due to the aromatic and low flexible structural configuration, which has been usually reported for concentrations larger than 30 wt.% [241]. As a consequence, flexible chains like castor oil, butanediol and polypropylene oxide have been proved valuable for improving this characteristic. Another approach followed has been to depolymerise native lignin to reduce its molecular weight [244]. It is also worth mentioning that different lignin-extraction procedures, origin, biomass source, etc. can likewise lead to lignin with different properties, which extends to derived PUs [240,246,247]. Therefore, new pathways are always being studied, such as aldehyde-assisted fractionation, which provides lignin with enhanced solubility, functionality and purity compared to the well-extended Kraft process [248]. Despite the mentioned disadvantages, many authors have proved that lignin-based foams act similarly or even better than those petroleum-based ones [238,241]. However, the use of lignin in PUs is not limited to foams. Elastomers, adhesives, coatings and more special products are also frequently targeted [241,245]. Regarding coating applications, Griffini et al. [249] utilised non-modified lignin to produce suitable films with enhanced formation ability, adhesion on diverse substrates, tensile properties and hydrophobic behaviour. Nonetheless, other properties like antioxidation, gas impermeability and UV resistance have also been reported for lignin-PU-based coatings [250]. Thus, Xie et al. [251] developed PUs by using different diisocyanates and Ag nanoparticles, which showed excellent thermal and mechanical properties with very high lignin content (more than 40%), while excellent antibacterial properties were provided by the metal addition. On the other hand, Fuqiang and Xiangjiao [252] developed UV-cured lignin PUs with improved hydrophobicity. In a more complex combination, Hu et al. [253] utilised the combined benefits of carbon nanotubes, Fe_3O_4 nanoparticles and lignin to develop PUs with excellent performance as electromagnetic shields. In general, weathering performance was also improved by the lignin inclusion into the polymeric matrix [241].

To a lesser extent, lignin has also been used to produce composites that can act as tent fabrics [254], potentiometric chemical sensors [255], chromatographic and chemoselective membrane applications [256] and stereolithography 3D printing ink.

4.4. *Lignocellulose-Based Polyurethane Adhesives*

Since PU arrival, great effort has been made for the production of high-performance PU-based adhesives. As can be observed within the landscape of adhesives established by Burchardt [257], PUs plays a very significant role throughout the whole mechanical/adhesion response spectra. Nonetheless, recent advances in political and environmental restrictions are endangering the use of commercial PU adhesives since they are usually formed by the combination of non-renewable polyols and harmful diisocyanates. As already mentioned above, it is in the polyol substitution where biomass can play a significant role; however, other techniques aiming to reach 100% bio-based adhesives are also to be discussed in this section.

Once more, the crosslinking ability of lignocellulose biopolymers, low cost and renewable character have been leveraged in order to produce bio-based PU adhesives. Therefore, liquefied products have demonstrated a great ability to develop a wide range of products [258]. Thus, the liquefaction of two plant species, China fir (*Cunninghamia lanceolata*) and Taiwan acacia (*Acacia confusa*) was carried out to subsequently produce PU adhesives by combining up to three types of poliisocyanates, i.e., Desmodur L (based on toluene diisocyanate and trimethylol propane), Desmodur N (based on hexamethylene diisocyanate) and poly-4,4′-diphenylmethane diisocyanate. The study allowed one to conclude that liquefied Taiwan acacia and Desmodur L exhibit the best adhesion performance [259]. Different sources, such as the Japanese cherry blossom or Sakura (*Prunus cerasus*) [260], kenaf (*Hibiscus cannabinus* L.) [261] or Sugi (*Criptmeria Japonica*) [262] were likewise tested after liquefaction to produce polyurethane adhesives with poly-4,4′-diphenylmethane diisocyanate. Nonetheless, the fractionation of lignocellulose into its main components is much preferred for PU adhesive production and is discussed in the next sections.

4.4.1. Cellulose-Based Polyurethane Adhesives

Cellulose and its derivatives have also been employed to produce bioadhesives. Thus, cellulose acetate in combination with CO was used by Tenorio-Alfonso et al. in several studies [263–265], from which a wide range of concentrations, NCO:OH ratios and diverse diisocyanates were tested, always exhibiting excellent properties, comparable to commercial adhesives. Similar to lignocellulose, liquefaction was also used to formulate adhesives from cellulose pulp; thus, sugar-beet-derived pulp adhesives with up to 87% biosource exhibited good adhesion ability, improved by increasing diisocyanate/biosource ratio. The stability of sugar beet pulp adhesives was better than another biomass also tested [266].

4.4.2. Hemicellulose-Based Polyurethane Adhesives

As previously mentioned, hemicellulose is formed by a combination of different monomers, such as xylose, arabinose, etc. Balcioglu et al. [267] utilised xylose for PU-adhesive formulations, obtaining adequate biocompatible adhesives with strength up to 415 kPa, at 15% xylose concentration. A patent has been registered where xylan inclusion in adhesives reported strengthening of the bonding [268]. Although any hemicellulose-derived polyol can actually be used as a precursor for PU-based adhesives, no such studies have been performed to our knowledge.

4.4.3. Lignin-Based Polyurethane Adhesives

Lignin has also been widely studied over the last few decades for PU adhesion, as demonstrated by the vast number of publications and patents centred on this topic [241,242,256,269,270]. The use of lignin allows one to overcome typical PU-adhesive drawbacks, such as delamination resistance, cohesive failure and gap-filling properties. Moreover, it generally enhances the thermal behaviour of the adhesive [241].

To mention some of the greatest advances reported, lignin in combination with vanillin-derived diisocyanates was able to withstand more than 9 MPa break at failure onto glass surface, while the results for other surfaces (wood, aluminium and steel) were much lower (around 1 MPa). Moreover, the use of diverse lignin sources demonstrated adhesive properties to be tailored [271]. Griffini et al. [249] also exhibited outstanding adhesive properties of lignin-based PUs; however, the maximum strength was found in wood specimens (more than 9 MPa), while glass still exhibited very good results (7.6 MPa), as well as aluminium and steel, showing around 1 MPa. Very good results in wood-wood surface, but also in metal-metal and metal-textile contact were also reported for lignin-and-castor-oil-based adhesives, where the lignin precedence, either from barley or wheat straw, played a crucial role on the adhesive performance. In this work, biological modifications were also targeted, allowing the tailoring of the adhesion values by the careful selection of both the lignin precedence and the *Streptomyces* strain [270]. As hydroxyl groups are the main bonding point to form proper urethane bonds with diisocyanates, other studies have been focused on primarily increasing the hydroxyl group index of lignin by hydroxypropylation or demethylation in order to create a stronger adhesive network [256,272,273]. In this sense, improved tensile strength of more than 200% reference values was obtained by hydroxypropylation [272]. On the other hand, biological modifications with laccase were leveraged by Ibrahim et al. [274], who tested both uninoculated and inoculated lignin with other natural sources such as chitosan and soy protein as well as with polyethyleneimine to produce suitable adhesives. These exhibited close values to petroleum-based ones and withstood hard thermal treatments with no significant performance loss. Regarding systematic studies aiming to elucidate main parameters for adhesion evaluation, Lima-García et al. [275] tested different kinds of lignins, Kraft and organosolv, resulting in particle size, lignin source and dispersion processing being key parameters for adequate incorporation of lignin into the adhesive, which exhibited more than 50% better mechanical properties than non-lignin-based ones. In another study, it was demonstrated that both isocyanate and lignin source together with solvent were critical for adhesive strength performance [269]. Looking for synergistic effects, the combination of epoxy-modified lignin with PU emulsion also demonstrated significant enhancement of the adhesive properties of lignin-based adhesives [276].

4.5. Lignocellulose-Based Polyurethane Lubricating Greases

As already mentioned in previous sections, the future prospects of the lubricating grease industry rely on the discovery of suitable thickening agents and base oils that can replace the metal-based soaps and mineral oils usually utilised for those purposes, respectively (see Table 10).

Table 10. Most-used base oils and thickeners for lubricating greases.

Base Oils	Thickeners
Mineral oil	Sodium soap
Synthetic oil	Calcium soap
Diester	Lithium soap
Silicone liquid	Aluminium soap
Phosphate ester	Lithium complex
Fluorinated silicone	Calcium complex
Chlorinated silicone	Aluminium complex
Polyglycol	Bentonite
Castor oil	Silicon oxide
	Carbon/graphite
	Polyurea
	Polyethylene
	Indanthrene dye
	Phthalocyanine dye

In history, the use of natural materials as lubricating greases was widely carried out since the first mention of the topic, dating back to Ancient Egypt. Nonetheless, the discovery of metal-based soaps and mineral and synthetic oils crucially changed the course of lubricating grease production (see Table 11) [277].

Table 11. Lubricating grease evolution through history [277,278].

Date	Event
About 1400 BC	The use of animal fat and limestone for lubrication of axes of Hittite chariots was started. In the same period, lubricants based on olive oil and limestone were used in Ancient Egypt
1845	A lubricant consisting of mineral oil, animal fat and limestone was invented in the United States
1853	The first sodium lubricant based on beef fat appeared in the United Kingdom
1912	The production of calcium lubricants with the dispersion medium based on mineral oil was started in Japan
1938	Lithium lubricants were developed. They quickly conquered the world and were used as multipurpose lubricants
1954	Invention of complex aluminium lubricants for operation at high temperatures
1955	Invention of urea greases in the United States
1960–Present	Resurgence of vegetable-oil-based lubricating greases

In this sense, once more, natural resources can play a decisive role, providing suitable replacers for both main lubricating grease components, aiming to fulfil current industrial demands and enhancing biodegradation properties of lubricants based on non-renewable resources. Via polyurethane formation, structural networks able to withstand and thicken oils can be obtained, making them suitable for lubricating purposes. Some of the most used natural sources for PUs formation will be discussed in the following sections regarding their application as lubricating greases.

Both barley and wheat straws were considered for lubricating grease production by biological modification with *Streptomyces*, demonstrating excellent rheological and tribological properties. The straw origin once more was crucial, as the *Streptomyces* activity was extremely dependent on the biosource, producing completely different effects on the rheological behaviour. Friction coefficients of around 0.09 (around 0.02 lower than commercially available lithium-based lubricating grease formulations) were obtained [279]. Likewise, some of the main constituents, i.e., cellulose and lignin, have also been reported to be used as polyurethane-type thickening agents.

4.5.1. Cellulose-Based Polyurethane Lubricating Greases

Several studies have taken advantage of the polyol structure of cellulose and its derivatives for the production of lubricating greases. Thus, PU formulations proposed as bio-based lubricating greases were obtained by functionalisation of methylcellulose with hexamethylene diisocyanate (HDI) and further mixing with castor oil. The different concentrations and methylcellulose/HDI ratios led to tunable characteristics, covering a wide range of lubricating grease consistencies [280,281]. Other studies have tested α-cellulose, 2-hydroxyethyl cellulose, methyl 2-hydroxyethyl cellulose and cellulose acetate propionate as precursors for PU-based lubricating grease formulations. The results let one conclude that the balance and size of the polar and non-polar groups turned out to be crucial in order to predict the rheological behaviour of the bio-based PUs [282]. Nonetheless, less purified cellulose-based systems as industrial cellulose pulps were also evaluated as an oil thickening agent, providing a more economical and industrial approach. Therefore, both *Eucalyptus globulus* and *Pinus radiata* cellulose pulps were assessed, the first one with two different purity grades, obtained by semi-mechanical cooking and commercial Kraft process. Cellulose pulps obtained from the soda pulping process of previously biologically treated wheat and barley straws with different *Streptomyces* strains were also targeted as thickening agents via PU formation, in this case by following a greener one-step approach. The modifications produced by the enzymatic activity of these strains were successfully correlated to the rheological properties of the lubricating greases [283]. All of them exhibited gel-like rheological characteristics with well-developed plateau regions within the frequency range studied, and microstructures very similar to those found in commercial lithium lubricating greases [284]. On the other hand, the tribological properties of many of these systems were also evaluated, as well as the mechanical stability, showing that most of them had suitable values for being considered potential replacers of traditional petroleum-based lubricating greases [283–285].

4.5.2. Lignin-Based Polyurethane Lubricating Greases

Lignin, as shown in previous sections, can easily act as a lubricant additive in composites and other materials, in both liquid and solid lubricants. Nonetheless, in this section, those studies related to the production of lignin-based PUs with lubricating properties are summarised. Thus, Ma et al. [286] developed lignin-based PUs, with lignin acting as an emulsifier for [BMIm]PF6 ionic liquid lubricant. The tribological results demonstrated very low friction coefficient and wear scars, as well as excellent thermal stability. Several studies concerning lignin-based PU formulations acting as gelling agents in castor oil media with promising applications as lubricants have also been reported. Thus, in a preliminary study, several diisocyanates were tested, showing HDI having the best properties to act as a crosslinker, resulting in rheological characteristics similar to those observed for commercial lithium-soap-based lubricating greases [287]. In a greener approach, a one-step method for the preparation of lignin-thickened lubricating greases, free of solvents or catalysts, was carefully addressed (evaluating temperature, agitation speed, etc.), it being possible to obtain proper formulations not only with HDI but also with toluene diisocyanate (TDI), at room temperature and relatively low agitation speed [288]. The thickening potential of a modified lignin fraction derived from the enzymatic activity of *Streptomyces* strains in wheat straw was also evaluated. In this sense, the−OH group concentration increase due to the enzymatic digestion of wheat straw led to higher viscoelastic functions of the obtained oleogels [289]. Alternatively, the direct use of the laccase enzyme to modify lignin was also aimed at, producing significant differences between commercial- and residual lignin-based lubricating greases [290]. Furthermore, several residual lignin fractions from bioethanol or Kraft pulping processes of different species were valorised, exhibiting good properties as lubricant thickeners but dependent on the lignin origin, composition and extraction method [291–293]. Some of these formulations were tested for biodegradability and ecotoxicity, showing generally that there is no harm or danger to their disposal [294]. On the other hand, lignin-based lubricating greases formed with a polyurea-polyurethane thickener have also been patented lately [295]. Moreover, other patented lubricants also include lignin as a dispersing agent and polyurethane as a polymeric binder in their formulations [296].

4.6. *Lignocellulose-Based Polyurethane Elastomers*

The substitution or at least minimisation of the use of petroleum-based sources by the inclusion of bio-based plant-derived components in PU elastomers is considered in this section. Different types of natural vegetable sources have been reviewed and discussed hereunder.

Lignocellulose-based PU elastomers have been widely studied due to their low cost, high availability, biodegradability and excellent performance. Once more, the properties of the related elastomers have shown great dependency on origin, composition, growth conditions, etc. Nonetheless, the lack of industrial advance in the topic relies on incompatibilities related to differences in hydrophilicity, compromising suitable dispersion [297].

There are several lignocellulosic sources that have been analysed in literature for this purpose. For instance, sawdust has been tested for PU elastomeric formulations. Thus, acoustic panels with soundproof properties have been obtained from the combination of some components, sawdust and PU elastomer binder being two of them [298]. A close material, wood flour, was also used for composite formation with elastomeric properties, which was prepared in filaments by 3D-printing technology. The tensile strength exhibited a minimum when wood flour concentration varied between 0 and 40%, while elongation at break was seriously compromised by increasing concentration [299]. In another study, Mengeloğlu and Çavuş [300] tested different lignocellulose sources as fillers, such as teak wood flour and rice husk, which were previously milled and screened before being used. The results showed that a 15% filler concentration potentially increased mechanical properties, while a 30% concentration caused a dramatic drop in the already mentioned behaviour. In general terms, the performance of teak-based systems was slightly better than that of rice husk. By using rice straw, the liquefied products from it have also led to the formation of polyurethane elastomers [301]. Following this tendency, straw liquefied products were also used by Zhao et al. [302] to perform PU elastomers where special attention to the obtaining procedure and its relation to mechanical performance were paid. In the following sections, the use of cellulose, hemicellulose and lignin in PU elastomeric formulations is discussed in more detail.

4.6.1. Cellulose-Based Polyurethane Elastomers

Cellulose and cellulose-derived materials have been known to act as proper fillers and endow materials with thorough properties, due to the fibrous structure and hydrogen bonding possibilities. Regarding PU elastomers, the same can be applied; nonetheless, the use of micro and nanocellulose has been more lately targeted. For instance, microfibrillated cellulose has been leveraged as casting material for PU formation, thus acting as tough elastomers for composite reinforcing [303]. The addition of up to 5 wt.% of microcrystalline cellulose to a PU based on MDI, butanediol and polytetramethylene glycol turned out a better performance of more than 650% tensile strength compared to the reference system [304]. Pei et al. [305] demonstrated that by incorporating cellulose nanocrystals in an MDI-poly(tetramethylene glycol)-based PU only at 1 wt.% concentration, stress at break could be 8-fold increased. In another work, the use of cellulose nanowhiskers prepared from microcrystalline cellulose allowed the formulation of shape-memory PU elastomers, triggered by water [306]. The addition of cellulose nanocrystals also demonstrated the improvement of the rheological properties of PU elastomers, with an increase of several orders of magnitude of the still unreacted suspensions, due to strong interactions between the nanocrystals and the polyol [307]. In another study, Saralegi et al. [308] developed from elastomeric to rigid bio-based PUs by increasing cellulose nanocrystal concentration. Interestingly, Lee et al. [309] incorporated cellulose nanofibres to a PU-urea matrix and reported a decrease in the elastic modulus by increasing cellulose concentration up to 1% and a further increase at 2%. Nonetheless, the mechanical properties were anyway enhanced. In castor oil/PEG-based PUs, the inclusion of cellulose nanocrystals obtained from *Eucalyptus globulus* was likewise positive for mechanical properties enhancement, reporting shifts from 5 to 12 MPa and from 1 to 5 MPa of tensile strength and Young modulus respectively, in comparison with the cellulose-free system [310].

4.6.2. Hemicellulose-Based Polyurethane Elastomers

Only a few studies proposed either hemicellulose derivatives or the monomers that comprise its structure as suitable polyols for the preparation of PU elastomers. Glucomannan was used in different concentrations to tune castor-oil-based PU properties from elastomeric to very rigid systems within the range of 5–90 wt.% concentration [311,312]. In a related work, films with elastomeric properties were elaborated by modifying glucomannan molecular weight [313]. It is also well known that very interesting furan derivatives can be obtained from hemicellulose, like furfural. Such compounds have also been used as sources for polyurethane formation, some of them serving as elastomers [83].

4.6.3. Lignin-Based Polyurethane Elastomers

Lignin and lignin derivatives have also been occasionally used as polyols and fillers for elastomers production [314]. One of the most important derived chemicals that can be obtained from lignin is vanillin. Apart from the already mentioned uses in previous sections, PU elastomers have likewise been generated. Vanillin inclusion propelled mechanical properties, causing Young modulus and strain at break to increase up to around 130 and 150% respectively [315]. Another lignin derivative that has been used as a copolymer in PU elastomers is lignin-derived polycarboxylic acid. With content of only 2.5%, the Young modulus and tensile stress at different strains (100 and 300%) were augmented 384, 135 and 90% respectively [316].

In the development of lignin-based PU elastomers, frequently lignin acted as polyol, becoming a structural part of the formulation, while sometimes it is only cast or mixed, acting as filler. As examples of the first case, improved thermal and mechanical properties were reported by the use of unmodified lignin as polyol along with poly (propylene glycol) and TDI as a crosslinker, yielding PUs with outstanding tensile performance (176 MPa, 33 MPa and 1394% of Young modulus, tensile strength and elongation at break respectively) by the addition of up to 40% lignin. In addition, the better performance of low-molecular-weight lignin fractions was reported [317]. The use of acetylated lignin and polyethylene glycol (PEG) as copolymers allowed elongations at break higher than 2000% to be obtained [318]. Likewise, PEG with oxidised lignin was used by Zhang et al. [319], resulting once more in great tensile performance. On the other hand, Liu et al. [320] took advantage of enzymatic procedures to alter lignin structure, which decreased molecular weight, thus imparting stronger mechanical properties to the network, i.e., up to 60 MPa tensile strength. Elastomers with excellent cushioning properties which could well act as coatings or shoe soles were likewise obtained by the inclusion of lignin in castor-oil-based PUs, showing an outstanding 88-fold enhancement at the stress at break in compressive performance, along with an improvement of the strain at failure from 50 to 93%, compared to the lignin-free system. The tensile properties were also enhanced by 17- and 7-fold regarding the Young modulus and stress at break, respectively. The dynamic properties of these elastomers were also tested, showing excellent results even for high loads and long-time essays [245]. Using lignin as filler, Ciobanu et al. [321] performed PU elastomers based on poly(ethyleneadipate), ethylene glycol, and MDI where lignin was cast using DMF as solvent at concentrations from 4 to 23%. The results indicated, once more, improved performance on mechanical properties with only a 10 wt.% lignin concentration, where the tensile strength and elongation at break were augmented up to 370 and 160% respectively. Acting as filler, lignin is usually responsible for enhancing mechanical properties, such as Young modulus or compressive performance [173,241,322]. In general, the higher the lignin concentration, the higher both the storage and Young moduli up to a certain concentration depending on the formulation, since lignin acts as a rigid component [241]. In this sense, storage modulus has been raised 6-fold by the addition of up to 60% lignin to polypropylene-glycol-based PUs [323]. In another study, Ortíz-Serna et al. [247] evaluated both the thermal and mechanical properties of Kraft and organosolv lignin acting as fillers for PU elastomers based on castor oil and butanediol, exhibiting great differences between these products. Culebras et al. [324] tested Alcell organosolv and hydroxypropyl-modified Kraft hardwood lignin for the same purpose, being the first one able to achieve excellent mechanical properties, with Young modulus of 80 GPa, comparable to commercial elastomers. Other authors tested other different lignins, such as wheat straw soda-derived, Indulin AT softwood kraft, Protobind 1000 soda, corn stover or sulphite hardwood lignins [241,325]. Frigerio et al. [326] found particle size, polar surface and agglomeration tendency to be key points in lignin performance as filler. Non-modified lignin was demonstrated to carry out the functions of both crosslinker and reinforcing agent in silicone-based PUs, matching some of the properties of commercially available silicone PUs [327]. Some of these elastomers can act as sealants, materials in surfboards, car interiors, household items and sporting indoor courts [328].

Finally, the three main bio-polyurethane types studied, i.e., adhesives, lubricating greases and elastomers, have been included in Figure 4, with the background showing biomaterial testing or application.

Figure 4. Main bio-polyurethanes studied, i.e., adhesives, lubricating greases and elastomers, from left to right, with testing or application as background.

5. Conclusions

Due to emerging environmental concern during the last few decades, lignocellulose is playing a fundamental role in research. Either by its use as a whole, fractionated into its main biopolymers or by the derived production of small molecules, great focus is being paid due to its environmentally friendly alternative and ubiquitous production. On the other hand, the residual character of some lignocellulose-derived fractions boosts possibilities for their valorisation, especially lignin. The present review highlights the crucial importance of the study of the lignocellulose extraction, separation and decomposition, and provides an insight into these modifications, mainly based on thermal, mechanical, chemical and biological pretreatments. Special attention is paid to biological processes whether by the action of fungi or bacteria, due to the greener and more environmentally friendly characteristics. By selecting a determined type of fungi or bacteria, some of the fractions may remain unaltered, while others may be decomposed. Therefore, the possibility to obtain different final products becomes massive, depending also on the biomass selected.

Furthermore, this review focusses its efforts on the use of lignocellulose, fractionated or not, in the production of biofuels, biochemicals and biomaterials. Thus, a wide range of biofuels, from the well-known biodiesel, biogas and bioethanol to other emerging species like dimethyl ether, biohydrogen, biofene®, etc., have been produced. Moreover, the last technologies regarding genome-editing approaches and biological procedures are presented. Considering biochemicals, an extremely wide range of products has been analysed and obtained in literature, and here we collect and separate them into those obtained from cellulose, hemicellulose and lignin. Biomaterials from lignocellulose have also been targeted massively lately, and many different products, like films, coatings, membranes, hydro-, oleo- and aero-gels, emulsifiers, fillers, flocculants, etc. have been produced, with many different applications, in fields such as transportation, textile, packaging, environmental remediation, plastic replacers, house and leisure items, health and hygiene, etc. Among the wide range of biomaterials, the emerging and outstanding characteristics of polyurethanes have attracted the attention of this review, especially in fields that are not usually deeply considered, such as adhesion, grease lubrication and elastomeric cushioning. Therefore, we have reported high-performance alternatives to traditional petroleum-derived products in those areas, such as adhesives that substantially exceed the adhesion performance of commercially available formulations in different surfaces, lubricating greases with tribological behaviour superior to that of lithium and calcium soap-based lubricants, and elastomers with excellent static and dynamic performance.

Overall, this review provides an extensive overview of the current state-of-the-art of the pretreatments for lignocellulosic materials, and the development of lignocellulose-derived biofuels, biochemical and biomaterials, with special attention to PU-based formulations for adhesion, lubricating greases and elastomeric cushioning fields.

Author Contributions: Conceptualisation A.M.B.-L.; investigation A.M.B.-L., C.V. and J.M.F.; writing—original draft preparation, A.M.B.-L.; writing—review and editing A.M.B.-L., C.V. and J.M.F.; supervision and project administration C.V. and J.M.F.; funding acquisition J C.V. and J.M.F. All authors have read and agreed to the published version of the manuscript.

Funding: This work is part of a research project (RTI2018-096080-B-C21) funded by MCIN/AEI/10.13039/501100011033 and by "ERDF A way of making Europe". A.M.B.-L. research activity was funded by Ministerio de Educación, Cultura y Deporte, grant number FPU16/03697. The financial support is gratefully acknowledged.

Institutional Review Board Statement: Not applicable.

Informed Consent Statement: Not applicable.

Data Availability Statement: Not applicable.

Conflicts of Interest: The authors declare no conflict of interest.

References

1. Peace, A.H.; Carslaw, K.S.; Lee, L.A.; Regayre, L.A.; Booth, B.B.B.; Johnson, J.S.; Bernie, D.J. Effect of aerosol radiative forcing uncertainty on projected exceedance year of a 1.5 °C global temperature rise. *Environ. Res. Lett.* **2020**, *15*, 0940a6. [CrossRef]
2. Cheng, C.-L.; Lo, Y.-C.; Lee, K.-S.; Lee, D.-J.; Lin, C.-Y.; Chang, J.-S. Biohydrogen production from lignocellulosic feedstock. *Bioresour. Technol.* **2011**, *102*, 8514–8523. [CrossRef]

3. Himmel, M.E.; Ding, S.-Y.; Johnson, D.K.; Adney, W.S.; Nimlos, M.R.; Brady, J.W.; Foust, T.D. Biomass Recalcitrance: Engineering Plants and Enzymes for Biofuels Production. *Science* **2007**, *315*, 804–807. [CrossRef] [PubMed]
4. Mosier, N.; Wyman, C.; Dale, B.; Elander, R.; Lee, Y.Y.; Holtzapple, M.; Ladisch, M. Features of promising technologies for pretreatment of lignocellulosic biomass. *Bioresour. Technol.* **2005**, *96*, 673–686. [CrossRef]
5. Bussemaker, M.; Trokanas, N.; Koo, L.; Cecelja, F. Ontology Modelling for Lignocellulosic Biomass: Composition and Conversion. *Comput. Aided Chem. Eng.* **2018**, *43*, 1565–1570. [CrossRef]
6. Huber, G.W.; Iborra, S.; Corma, A. Synthesis of Transportation Fuels from Biomass: Chemistry, Catalysts, and Engineering. *Chem. Rev.* **2006**, *106*, 4044–4098. [CrossRef] [PubMed]
7. Hendriks, A.T.W.M.; Zeeman, G. Pretreatments to enhance the digestibility of lignocellulosic biomass. *Bioresour. Technol.* **2009**, *100*, 10–18. [CrossRef]
8. Li, C.; Zhao, X.; Wang, A.; Huber, G.; Zhang, T. Catalytic Transformation of Lignin for the Production of Chemicals and Fuels. *Chem. Rev.* **2015**, *115*, 11559–11624. [CrossRef]
9. Binder, J.B.; Raines, R.T. Simple Chemical Transformation of Lignocellulosic Biomass into Furans for Fuels and Chemicals. *J. Am. Chem. Soc.* **2009**, *131*, 1979–1985. [CrossRef]
10. Domínguez-Robles, J.; Espinosa, E.; Savy, D.; Rosal, A.; Rodriguez, A. Biorefinery Process Combining Specel® Process and Selective Lignin Precipitation using Mineral Acids. *BioResources* **2016**, *11*, 7061–7077. [CrossRef]
11. Fitzpatrick, M.; Champagne, P.; Cunningham, M.F.; Whitney, R.A. A biorefinery processing perspective: Treatment of lignocellulosic materials for the production of value-added products. *Bioresour. Technol.* **2010**, *101*, 8915–8922. [CrossRef]
12. Menon, V.; Rao, M. Trends in bioconversion of lignocellulose: Biofuels, platform chemicals & biorefinery concept. *Prog. Energy Combust. Sci.* **2012**, *38*, 522–550. [CrossRef]
13. Yang, S.-T. Bioprocessing—From Biotechnology to Biorefinery. In *Bioprocessing for Value-Added Products from Renewable Resources*; Elsevier: Amsterdam, The Netherlands, 2007; pp. 1–24.
14. Fan, L.T.; Lee, Y.-H.; Gharpuray, M.M. The nature of lignocellulosics and their pretreatments for enzymatic hydrolysis. In *Bioprocesses and Applied Enzymology*; Springer Science and Business Media LLC: Berlin/Heidelberg, Germany, 1982; pp. 157–187.
15. Oginni, O.; Singh, K.; Zondlo, J.W. Pyrolysis of dedicated bioenergy crops grown on reclaimed mine land in West Virginia. *J. Anal. Appl. Pyrolysis* **2017**, *123*, 319–329. [CrossRef]
16. Schmitz, E.; Karlsson, E.N.; Adlercreutz, P. Warming weather changes the chemical composition of oat hulls. *Plant Biol.* **2020**, *22*, 1086–1091. [CrossRef]
17. Klemm, D.; Heublein, B.; Fink, H.-P.; Bohn, A. Cellulose: Fascinating Biopolymer and Sustainable Raw Material. *Angew. Chem. Int. Ed.* **2005**, *44*, 3358–3393. [CrossRef]
18. Alternative Fuels from Biomass Sources. Available online: https://www.e-education.psu.edu/egee439/node/664 (accessed on 20 September 2021).
19. Scheller, H.V.; Ulvskov, P. Hemicelluloses. *Annu. Rev. Plant Biol.* **2010**, *61*, 263–289. [CrossRef]
20. Gibson, L.J. The hierarchical structure and mechanics of plant materials. *J. R. Soc. Interface* **2012**, *9*, 2749–2766. [CrossRef]
21. De Souza, W.R. Microbial Degradation of Lignocellulosic Biomass. In *Sustainable Degradation of Lignocellulosic Biomass—Techniques, Applications and Commercialization*; Chandel, A., Ed.; IntechOpen: London, UK, 2013; pp. 207–248.
22. Amen-Chen, C.; Pakdel, H.; Roy, C. Production of monomeric phenols by thermochemical conversion of biomass: A review. *Bioresour. Technol.* **2001**, *79*, 277–299. [CrossRef]
23. Martínez Ángel, T.; Rencoret, J.; Marques, G.; Gutiérrez, A.; Ibarra, D.; Jimenez-Barbero, J.; del Río, J.C. Monolignol acylation and lignin structure in some nonwoody plants: A 2D NMR study. *Phytochemistry* **2008**, *69*, 2831–2843. [CrossRef]
24. Sun, S.-L.; Wen, J.-L.; Ma, M.-G.; Li, M.-F.; Sun, R.-C. Revealing the Structural Inhomogeneity of Lignins from Sweet Sorghum Stem by Successive Alkali Extractions. *J. Agric. Food Chem.* **2013**, *61*, 4226–4235. [CrossRef]
25. Heikkinen, S.; Toikka, M.M.; Karhunen, P.T.; Kilpeläinen, I.A. Quantitative 2D HSQC (Q-HSQC) via Suppression of J-Dependence of Polarization Transfer in NMR Spectroscopy: Application to Wood Lignin. *J. Am. Chem. Soc.* **2003**, *125*, 4362–4367. [CrossRef] [PubMed]
26. Santos, J.I.; Fillat, U.; Martín-Sampedro, R.; Ballesteros, I.; Manzanares, P.; Ballesteros, M.; Eugenio, M.E.; Ibarra, D. Lignin-enriched Fermentation Residues from Bioethanol Production of Fast-growing Poplar and Forage Sorghum. *Bioresources* **2015**, *10*, 5215–5232. [CrossRef]
27. Heitner, C.; Dimmel, D.; Schmidt, J. *Lignin and Lignans: Advances in Chemistry*; CRC Press: Boca Raton, FL, USA, 2010.
28. Kringstad, B.K.P.; Mörck, R. ^{13}C-NMR Spectra of Kraft Lignins. *Holzforschung* **1983**, *37*, 237–244. [CrossRef]
29. Xia, Z.; Akim, A.L.G.; Argyropoulos, D.S. Quantitative ^{13}C NMR Analysis of Lignins with Internal Standards. *J. Agric. Food Chem.* **2001**, *49*, 3573–3578. [CrossRef] [PubMed]
30. Santos, J.I.; Martin-Sampedro, R.; Fillat, Ú.; Oliva, J.M.; Negro, M.J.; Ballesteros, M.; Eugenio, M.E.; Ibarra, D. Evaluating Lignin-Rich Residues from Biochemical Ethanol Production of Wheat Straw and Olive Tree Pruning by FTIR and 2D-NMR. *Int. J. Polym. Sci.* **2015**, *2015*, 314891. [CrossRef]
31. Zhang, L.; Gellerstedt, G. Quantitative 2D HSQC NMR determination of polymer structures by selecting suitable internal standard references. *Magn. Reson. Chem.* **2007**, *45*, 37–45. [CrossRef]
32. Zakzeski, J.; Bruijnincx, P.C.A.; Jongerius, A.L.; Weckhuysen, B.M. The Catalytic Valorization of Lignin for the Production of Renewable Chemicals. *Chem. Rev.* **2010**, *110*, 3552–3599. [CrossRef]

33. Patil, N.D.; Tanguy, N.R.; Yan, N. Lignin Interunit Linkages and Model Compounds. In *Lignin in Polymer Composites*; Elsevier BV: Amsterdam, The Netherlands, 2016; pp. 27–47.
34. Lawrence, A.; Thollander, P.; Karlsson, M. Drivers, Barriers, and Success Factors for Improving Energy Management in the Pulp and Paper Industry. *Sustainability* **2018**, *10*, 1851. [CrossRef]
35. Alonso, D.M.; Hakim, S.H.; Zhou, S.; Won, W.; Hosseinaei, O.; Tao, J.; Garcia-Negron, V.; Motagamwala, A.H.; Mellmer, M.A.; Huang, K.; et al. Increasing the revenue from lignocellulosic biomass: Maximizing feedstock utilization. *Sci. Adv.* **2017**, *3*, e1603301. [CrossRef]
36. Baruah, J.; Nath, B.K.; Sharma, R.; Kumar, S.; Deka, R.C.; Baruah, D.C.; Kalita, E. Recent Trends in the Pretreatment of Lignocellulosic Biomass for Value-Added Products. *Front. Energy Res.* **2018**, *6*, 141. [CrossRef]
37. Sousa, L.D.C.; Chundawat, S.P.; Balan, V.; Dale, B.E. 'Cradle-to-grave' assessment of existing lignocellulose pretreatment technologies. *Curr. Opin. Biotechnol.* **2009**, *20*, 339–347. [CrossRef]
38. McMillan, J.D. Pretreatment of Lignocellulosic Biomass. In *ACS Symposium Series*; American Chemical Society (ACS): Washington, DC, USA, 1994; pp. 292–324.
39. Yang, B.; Wyman, C.E. Pretreatment: The key to unlocking low-cost cellulosic ethanol. *Biofuels Bioprod. Biorefin.* **2008**, *2*, 26–40. [CrossRef]
40. El Hage, R.; Brosse, N.; Sannigrahi, P.; Ragauskas, A. Effects of process severity on the chemical structure of Miscanthus ethanol organosolv lignin. *Polym. Degrad. Stab.* **2010**, *95*, 997–1003. [CrossRef]
41. Sindhu, R.; Binod, P.; Pandey, A. Biological pretreatment of lignocellulosic biomass—An overview. *Bioresour. Technol.* **2016**, *199*, 76–82. [CrossRef]
42. Yang, B.; Dai, Z.; Ding, S.-Y.; Wyman, C.E. Enzymatic hydrolysis of cellulosic biomass. *Biofuels* **2011**, *2*, 421–449. [CrossRef]
43. Andlar, M.; Rezić, T.; Marđetko, N.; Kracher, D.; Ludwig, R.; Šantek, B. Lignocellulose degradation: An overview of fungi and fungal enzymes involved in lignocellulose degradation. *Eng. Life Sci.* **2018**, *18*, 768–778. [CrossRef]
44. Sánchez, C. Lignocellulosic residues: Biodegradation and bioconversion by fungi. *Biotechnol. Adv.* **2009**, *27*, 185–194. [CrossRef]
45. Coughlan, M.P. Cellulose Degradation by Fungi. In *Microbial Enzymes and Biotechnology*; Springer Science and Business Media LLC: Berlin/Heidelberg, Germany, 1990; pp. 1–36.
46. Dekker, R.F.H. Biodegradation of the Hemicelluloses. In *Biosynthesis and Biodegradation of Wood Components*; Higuchi, T., Ed.; Academic Press: Cambridge, MA, USA, 1985; pp. 505–533. [CrossRef]
47. Martínez, A.T.; Speranza, M.; Ruiz-Dueñas, F.J.; Ferreira, P.; Camarero, S.; Guillén, F.; Martinez, M.J.; Gutiérrez Suárez, A.; del Río Andrade, J.C. Biodegradation of lignocellulosics: Microbial, chemical, and enzymatic aspects of the fungal attack of lignin. *Int. Microbiol.* **2005**, *8*, 195–204.
48. Chen, C.-L.; Chang, H.-M.; Kirk, T.K. Carboxylic Acids Produced Through Oxidative Cleavage of Aromatic Rings During Degradation of Lignin in Spruce Wood by Phanerochaete Chrysosporium. *J. Wood Chem. Technol.* **1983**, *3*, 35–57. [CrossRef]
49. Shi, Z.; Han, C.; Zhang, X.; Tian, L.; Wang, L. Novel Synergistic Mechanism for Lignocellulose Degradation by a Thermophilic Filamentous Fungus and a Thermophilic Actinobacterium Based on Functional Proteomics. *Front. Microbiol.* **2020**, *11*, 539438. [CrossRef]
50. Thornbury, M.; Sicheri, J.; Slaine, P.; Getz, L.J.; Finlayson-Trick, E.; Cook, J.; Guinard, C.; Boudreau, N.; Jakeman, D.; Rohde, J.; et al. Characterization of novel lignocellulose-degrading enzymes from the porcupine microbiome using synthetic metagenomics. *PLoS ONE* **2019**, *14*, e0209221. [CrossRef]
51. Sainsbury, P.D.; Hardiman, E.M.; Ahmad, M.; Otani, H.; Seghezzi, N.; Eltis, L.D.; Bugg, T.D.H. Breaking Down Lignin to High-Value Chemicals: The Conversion of Lignocellulose to Vanillin in a Gene Deletion Mutant of *Rhodococcus jostii* RHA1. *ACS Chem. Biol.* **2013**, *8*, 2151–2156. [CrossRef]
52. Strachan, C.R.; Singh, R.; VanInsberghe, D.; Ievdokymenko, K.; Budwill, K.; Mohn, W.W.; Eltis, L.; Hallam, S.J. Metagenomic scaffolds enable combinatorial lignin transformation. *Proc. Natl. Acad. Sci. USA* **2014**, *111*, 10143–10148. [CrossRef]
53. Ransom-Jones, E.; McCarthy, A.J.; Haldenby, S.; Doonan, J.; McDonald, J.E. Lignocellulose-Degrading Microbial Communities in Landfill Sites Represent a Repository of Unexplored Biomass-Degrading Diversity. *mSphere* **2017**, *2*, e00300-17. [CrossRef]
54. Granja, R.; Persinoti, G.F.; Squina, F.M.; Bugg, T.D.H. Functional genomic analysis of bacterial lignin degraders: Diversity in mechanisms of lignin oxidation and metabolism. *Appl. Microbiol. Biotechnol.* **2020**, *104*, 3305–3320. [CrossRef]
55. López-Mondéjar, R.; Zühlke, D.; Becher, D.; Riedel, K.; Baldrian, P. Cellulose and hemicellulose decomposition by forest soil bacteria proceeds by the action of structurally variable enzymatic systems. *Sci. Rep.* **2016**, *6*, 25279. [CrossRef] [PubMed]
56. Gupta, P.; Samant, K.; Sahu, A. Isolation of Cellulose-Degrading Bacteria and Determination of Their Cellulolytic Potential. *Int. J. Microbiol.* **2012**, *2012*, 578925. [CrossRef] [PubMed]
57. Himmel, M.E.; Adney, W.S.; Baker, J.O.; Nieves, R.A.; Thomas, S.R. Cellulases: Structure, Function, and Applications. In *Handbook on Bioethanol*; Routledge: London, UK, 2018; pp. 143–161.
58. Coughlan, M.P. Mechanisms of cellulose degradation by fungi and bacteria. *Anim. Feed Sci. Technol.* **1991**, *32*, 77–100. [CrossRef]
59. Shallom, D.; Shoham, Y. Microbial hemicellulases. *Curr. Opin. Microbiol.* **2003**, *6*, 219–228. [CrossRef]
60. Béguin, P.; Aubert, J.-P. The biological degradation of cellulose. *FEMS Microbiol. Rev.* **1994**, *13*, 25–58. [CrossRef]
61. Cheah, W.Y.; Sankaran, R.; Show, P.L.; Ibrahim, T.N.B.T.; Chew, K.W.; Culaba, A.; Chang, J.-S. Pretreatment methods for lignocellulosic biofuels production: Current advances, challenges and future prospects. *Biofuel Res. J.* **2020**, *7*, 1115–1127. [CrossRef]

62. Schäfer, A.; Konrad, R.; Kuhnigk, T.; Kampfer, P.; Hertel, H.; König, H. Hemicellulose-degrading bacteria and yeasts from the termite gut. *J. Appl. Bacteriol.* **1996**, *80*, 471–478. [CrossRef]
63. Saha, B.C.; Bothas, R.J. Enzymology of xylan degradation. In *Biopolymers: Utilizing Natures Advanced Materials*; American Chemical Society: Washington, DC, USA, 1999. [CrossRef]
64. Brigham, J.S.; Adney, W.S.; Himmel, M.E. Hemicellulases: Diversity and Applications. In *Handbook on Bioethanol*; CRC Press: Boca Raton, FL, USA, 2018; pp. 119–141. [CrossRef]
65. Bugg, T.D.H.; Ahmad, M.; Hardiman, E.M.; Rahmanpour, R. Pathways for degradation of lignin in bacteria and fungi. *Nat. Prod. Rep.* **2011**, *28*, 1883–1896. [CrossRef]
66. Bugg, T.D.; Williamson, J.J.; Rashid, G.M. Bacterial enzymes for lignin depolymerisation: New biocatalysts for generation of renewable chemicals from biomass. *Curr. Opin. Chem. Biol.* **2020**, *55*, 26–33. [CrossRef]
67. Ahmad, M.; Roberts, J.N.; Hardiman, E.M.; Singh, R.; Eltis, L.D.; Bugg, T.D.H. Identification of DypB from *Rhodococcus jostii* RHA1 as a Lignin Peroxidase. *Biochemistry* **2011**, *50*, 5096–5107. [CrossRef]
68. Rahmanpour, R.; Rea, D.; Jamshidi, S.; Fülöp, V.; Bugg, T.D. Structure of *Thermobifida fusca* DyP-type peroxidase and activity towards Kraft lignin and lignin model compounds. *Arch. Biochem. Biophys.* **2016**, *594*, 54–60. [CrossRef]
69. Del Cerro, C.; Erickson, E.; Dong, T.; Wong, A.R.; Eder, E.K.; Purvine, S.O.; Mitchell, H.D.; Weitz, K.K.; Markillie, L.M.; Burnet, M.C.; et al. Intracellular pathways for lignin catabolism in white-rot fungi. *Proc. Natl. Acad. Sci. USA* **2021**, *118*, e2017381118. [CrossRef]
70. Balan, V.; Chiaramonti, D.; Kumar, S. Review of US and EU initiatives toward development, demonstration, and commercialization of lignocellulosic biofuels. *Biofuels Bioprod. Biorefin.* **2013**, *7*, 732–759. [CrossRef]
71. Velvizhi, G.; Goswami, C.; Shetti, N.P.; Ahmad, E.; Pant, K.K.; Aminabhavi, T.M. Valorisation of lignocellulosic biomass to value-added products: Paving the pathway towards low-carbon footprint. *Fuel* **2021**, *313*, 122678. [CrossRef]
72. Wi, S.G.; Cho, E.J.; Lee, D.-S.; Lee, S.J.; Lee, Y.J.; Bae, H.-J. Lignocellulose conversion for biofuel: A new pretreatment greatly improves downstream biocatalytic hydrolysis of various lignocellulosic materials. *Biotechnol. Biofuels* **2015**, *8*, 228. [CrossRef]
73. Baral, N.R.; Sundstrom, E.R.; Das, L.; Gladden, J.; Eudes, A.; Mortimer, J.C.; Singer, S.W.; Mukhopadhyay, A.; Scown, C.D. Approaches for More Efficient Biological Conversion of Lignocellulosic Feedstocks to Biofuels and Bioproducts. *ACS Sustain. Chem. Eng.* **2019**, *7*, 9062–9079. [CrossRef]
74. Baral, N.R.; Kavvada, O.; Mendez-Perez, D.; Mukhopadhyay, A.; Lee, T.S.; Simmons, B.A.; Scown, C.D. Techno-economic analysis and life-cycle greenhouse gas mitigation cost of five routes to bio-jet fuel blendstocks. *Energy Environ. Sci.* **2019**, *12*, 807–824. [CrossRef]
75. Bhujbal, S.K.; Ghosh, P.; Vijay, V.K.; Rathour, R.; Kumar, M.; Singh, L.; Kapley, A. Biotechnological potential of rumen microbiota for sustainable bioconversion of lignocellulosic waste to biofuels and value-added products. *Sci. Total Environ.* **2022**, *814*, 152773. [CrossRef] [PubMed]
76. Demirbas, A. Competitive liquid biofuels from biomass. *Appl. Energy* **2011**, *88*, 17–28. [CrossRef]
77. Danso, B.; Ali, S.S.; Xie, R.; Sun, J. Valorisation of wheat straw and bioethanol production by a novel xylanase- and cellulase-producing Streptomyces strain isolated from the wood-feeding termite, Microcerotermes species. *Fuel* **2021**, *310*, 122333. [CrossRef]
78. Isahak, W.N.R.W.; Hisham, M.W.; Yarmo, M.A.; Hin, T.-Y.Y. A review on bio-oil production from biomass by using pyrolysis method. *Renew. Sustain. Energy Rev.* **2012**, *16*, 5910–5923. [CrossRef]
79. Prasad, S.; Dhanya, M.S.; Gupta, N.; Kumar, A. Biofuels from biomass: A sustainable alternative to energy and environment, *Biochem. Cell. Arch.* **2012**, *12*, 255–260.
80. Gottumukkala, L.D.; Mathew, A.K.; Abraham, A.; Sukumaran, R.K. Biobutanol Production: Microbes, Feedstock, and Strategies. In *Biofuels: Alternative Feedstocks and Conversion Processes for the Production of Liquid and Gaseous Biofuels*; Elsevier BV: Amsterdam, The Netherlands, 2019; pp. 355–377.
81. Borrero-López, A.M.; Masson, E.; Celzard, A.; Fierro, V. Modelling the reactions of cellulose, hemicellulose and lignin submitted to hydrothermal treatment. *Ind. Crop. Prod.* **2018**, *124*, 919–930. [CrossRef]
82. Lee, S.Y.; Kim, H.U.; Chae, T.U.; Cho, J.S.; Kim, J.W.; Shin, J.H.; Kim, D.I.; Ko, Y.-S.; Jang, W.D.; Jang, Y.-S. A comprehensive metabolic map for production of bio-based chemicals. *Nat. Catal.* **2019**, *2*, 18–33. [CrossRef]
83. Isikgor, F.H.; Becer, C.R. Lignocellulosic biomass: A sustainable platform for the production of bio-based chemicals and polymers. *Polym. Chem.* **2015**, *6*, 4497–4559. [CrossRef]
84. Werpy, T.; Holladay, J.; White, J. Top Value Added Chemicals from Biomass: I. In *Results of Screening for Potential Candidates from Sugars and Synthesis Gas*; U.S. Department of Energy Office of Scientific and Technical Information: Oak Ridge, TN, USA, 2004.
85. Mamman, A.S.; Lee, J.-M.; Kim, Y.-C.; Hwang, I.T.; Park, N.-J.; Hwang, Y.K.; Chang, J.-S.; Hwang, J.-S. Furfural: Hemicellulose/xylosederived biochemical. *Biofuels Bioprod. Biorefin.* **2008**, *2*, 438–454. [CrossRef]
86. Schutyser, W.; Renders, T.; Van Den Bosch, S.; Koelewijn, S.-F.; Beckham, G.T.; Sels, B.F. Chemicals from lignin: An interplay of lignocellulose fractionation, depolymerisation, and upgrading. *Chem. Soc. Rev.* **2018**, *47*, 852–908. [CrossRef]
87. Sharma, J.; Kumar, S.; Kumar, V.; Malyan, S.K.; Mathimani, T.; Bishnoi, N.R.; Pugazhendhi, A. Upgrading of microalgal consortia with CO_2 from fermentation of wheat straw for the phycoremediation of domestic wastewater. *Bioresour. Technol.* **2020**, *305*, 123063. [CrossRef]

88. Ahmad, M.R.; Chen, B.; Duan, H. Improvement effect of pyrolyzed agro-food biochar on the properties of magnesium phosphate cement. *Sci. Total Environ.* **2020**, *718*, 137422. [CrossRef]
89. Chen, Y.-A.; Yang, H.; Ouyang, D.; Liu, T.; Liu, D.; Zhao, X. Construction of electron transfer chains with methylene blue and ferric ions for direct conversion of lignocellulosic biomass to electricity in a wide pH range. *Appl. Catal. B Environ.* **2020**, *265*, 118578. [CrossRef]
90. Ahmadi, R.; Souri, B.; Ebrahimi, M. Evaluation of wheat straw to insulate fired clay hollow bricks as a construction material. *J. Clean. Prod.* **2020**, *254*, 120043. [CrossRef]
91. Martos, S.; Mattana, S.; Ribas, A.; Albanell, E.; Domene, X. Biochar application as a win-win strategy to mitigate soil nitrate pollution without compromising crop yields: A case study in a Mediterranean calcareous soil. *J. Soils Sediments* **2020**, *20*, 220–233. [CrossRef]
92. Luhar, S.; Cheng, T.-W.; Luhar, I. Incorporation of natural waste from agricultural and aquacultural farming as supplementary materials with green concrete: A review. *Compos. Part B Eng.* **2019**, *175*, 107076. [CrossRef]
93. Chen, H. Lignocellulose biorefinery product engineering. In *Lignocellulose Biorefinery Engineering*, 1st ed.; Woodhead Publishing Limited: Cambridge, UK, 2015; pp. 125–165.
94. Basta, A.; El-Sayed, E.S.A.; El-Saied, H. Lignocellulosic Materials in Building Elements. Part IV—Economical Manufacture and Improvement of Properties of Light-Weight Agro-Panels. *Int. J. Polym. Mater. Polym. Biomater.* **2004**, *53*, 709–723. [CrossRef]
95. Lou, Z.; Wang, Q.; Kara, U.I.; Mamtani, R.S.; Zhou, X.; Bian, H.; Yang, Z.; Li, Y.; Lv, H.; Adera, S.; et al. Biomass-Derived Carbon Heterostructures Enable Environmentally Adaptive Wideband Electromagnetic Wave Absorbers. *Nano-Micro Lett.* **2021**, *14*, 11. [CrossRef]
96. Rocha, C.G.; Zaia, D.A.M.; Alfaya, R.V.D.S.; Alfaya, A.A.D.S. Use of rice straw as biosorbent for removal of Cu(II), Zn(II), Cd(II) and Hg(II) ions in industrial effluents. *J. Hazard. Mater.* **2009**, *166*, 383–388. [CrossRef] [PubMed]
97. Zhao, M.; Dai, Y.; Zhang, M.; Feng, C.; Qin, B.; Zhang, W.; Zhao, N.; Li, Y.; Ni, Z.; Xu, Z.; et al. Mechanisms of Pb and/or Zn adsorption by different biochars: Biochar characteristics, stability, and binding energies. *Sci. Total Environ.* **2020**, *717*, 136894. [CrossRef] [PubMed]
98. Dai, Y.; Wang, W.; Lu, L.; Yan, L.; Yu, D. Utilization of biochar for the removal of nitrogen and phosphorus. *J. Clean. Prod.* **2020**, *257*, 120573. [CrossRef]
99. Abdelaal, S.; Gad, Y.; Dessouki, A. Use of rice straw and radiation-modified maize starch/acrylonitrile in the treatment of wastewater. *J. Hazard. Mater.* **2006**, *129*, 204–215. [CrossRef] [PubMed]
100. Nethaji, S.; Sivasamy, A.; Thennarasu, G.; Saravanan, S. Adsorption of Malachite Green dye onto activated carbon derived from Borassus aethiopum flower biomass. *J. Hazard. Mater.* **2010**, *181*, 271–280. [CrossRef]
101. Liu, R.-L.; Liu, Y.; Zhou, X.-Y.; Zhang, Z.-Q.; Zhang, J.; Dang, F.-Q. Biomass-derived highly porous functional carbon fabricated by using a free-standing template for efficient removal of methylene blue. *Bioresour. Technol.* **2014**, *154*, 138–147. [CrossRef]
102. Licona-Aguilar, Á.I.; Lois-Correa, J.A.; Torres, A.; Domínguez-Crespo, M.A.; Dorantes-Rosales, H.J.; Garcia-Zaleta, D.S. Sugarcane Bagasse-, Orange Peel-Derived Adsorbent Materials: Thermal and Morphological Studies. *J. Nanosci. Nanotechnol.* **2020**, *20*, 4563–4573. [CrossRef]
103. Aragaw, T.A.; Bogale, F.M. Biomass-Based Adsorbents for Removal of Dyes from Wastewater: A Review. *Front. Environ. Sci.* **2021**, *9*, 764958. [CrossRef]
104. Castro-Gutiérrez, J.; Díez, N.; Sevilla, M.; Izquierdo, M.; Celzard, A.; Fierro, V. Model carbon materials derived from tannin to assess the importance of pore connectivity in supercapacitors. *Renew. Sustain. Energy Rev.* **2021**, *151*, 111600. [CrossRef]
105. Liu, Y.; Su, M.; Li, D.; Li, S.; Li, X.; Zhao, J.; Liu, F. Soybean straw biomass-derived Fe-N co-doped porous carbon as an efficient electrocatalyst for oxygen reduction in both alkaline and acidic media. *RSC Adv.* **2020**, *12*, 6763–6771. [CrossRef]
106. Ma, M.; Dai, Y.; Zou, J.-L.; Wang, L.; Pan, K.; Fu, H.-G. Synthesis of Iron Oxide/Partly Graphitized Carbon Composites as a High-Efficiency and Low-Cost Cathode Catalyst for Microbial Fuel Cells. *ACS Appl. Mater. Interfaces* **2014**, *6*, 13438–13447. [CrossRef]
107. Wu, X.-L.; Wen, T.; Guo, H.-L.; Yang, S.; Wang, X.; Xu, A.-W. Biomass-Derived Sponge-like Carbonaceous Hydrogels and Aerogels for Supercapacitors. *ACS Nano* **2013**, *7*, 3589–3597. [CrossRef]
108. Gupta, G.K.; De, S.; Franco, A.; Balu, A.M.; Luque, R. Sustainable Biomaterials: Current Trends, Challenges and Applications. *Molecules* **2015**, *21*, 48. [CrossRef]
109. Iqbal, H.M.N.; Kyazze, G.; Keshavarz, T. Advances in the Valorization of Lignocellulosic Materials by Biotechnology: An Overview. *BioResources* **2013**, *8*, 3157–3176. [CrossRef]
110. Bugatti, V.; Brachi, P.; Viscusi, G.; Gorrasi, G. Valorization of Tomato Processing Residues Through the Production of Active Bio-Composites for Packaging Applications. *Front. Mater.* **2019**, *6*, 34. [CrossRef]
111. Ita-Nagy, D.; Vázquez-Rowe, I.; Kahhat, R.; Quispe, I.; Chinga-Carrasco, G.; Clauser, N.M.; Area, M.C. Life cycle assessment of bagasse fiber reinforced biocomposites. *Sci. Total Environ.* **2020**, *720*, 137586. [CrossRef]
112. Engel, J.B.; Mac Ginity, M.; Luchese, C.L.; Tessaro, I.C.; Spada, J.C. Reuse of Different Agroindustrial Wastes: Pinhão and Pecan Nutshells Incorporated into Biocomposites Using Thermocompression. *J. Polym. Environ.* **2020**, *28*, 1431–1440. [CrossRef]
113. Chen, B.; Luo, Z.; Chen, H.; Chen, C.; Cai, D.; Qin, P.; Cao, H.; Tan, T. Wood Plastic Composites from the Waste Lignocellulosic Biomass Fibers of Bio-Fuels Processes: A Comparative Study on Mechanical Properties and Weathering Effects. *Waste Biomass Valorization* **2018**, *11*, 1701–1710. [CrossRef]

114. Chen, W.-F.; Iyer, S.; Iyer, S.; Sasaki, K.; Wang, C.-H.; Zhu, Y.; Muckerman, J.T.; Fujita, E. Biomass-derived electrocatalytic composites for hydrogen evolution. *Energy Environ. Sci.* **2013**, *6*, 1818–1826. [CrossRef]
115. Chen, H. Integrated industrial lignocellulose biorefinery chains. In *Lignocellulose Biorefinery Engineering*; Elsevier BV: Amsterdam, The Netherlands, 2015; pp. 219–245.
116. Horuz, T.İ.; Belibağli, K.B. Encapsulation of tomato peel extract into nanofibers and its application in model food. *J. Food Process. Preserv.* **2019**, *43*, e14090. [CrossRef]
117. Niu, Y.; Xia, Q.; Gu, M.; Yu, L. Interpenetrating network gels composed of gelatin and soluble dietary fibers from tomato peels. *Food Hydrocoll.* **2019**, *89*, 95–99. [CrossRef]
118. Ajdary, R.; Tardy, B.L.; Mattos, B.D.; Bai, L.; Rojas, O.J. Plant Nanomaterials and Inspiration from Nature: Water Interactions and Hierarchically Structured Hydrogels. *Adv. Mater.* **2021**, *33*, e2001085. [CrossRef]
119. Chang, C.; Zhang, L. Cellulose-based hydrogels: Present status and application prospects. *Carbohydr. Polym.* **2011**, *84*, 40–53. [CrossRef]
120. Zhao, G.; Lyu, X.; Lee, J.; Cui, X.; Chen, W.-N. Biodegradable and transparent cellulose film prepared eco-friendly from durian rind for packaging application. *Food Packag. Shelf Life* **2019**, *21*, 100345. [CrossRef]
121. Kabir, S.M.F.; Sikdar, P.P.; Haque, B.; Bhuiyan, M.A.R.; Ali, A.; Islam, M.N. Cellulose-based hydrogel materials: Chemistry, properties and their prospective applications. *Prog. Biomater.* **2018**, *7*, 153–174. [CrossRef] [PubMed]
122. Sannino, A.; Madaghiele, M.; Lionetto, M.G.; Schettino, T.; Maffezzoli, A. A cellulose-based hydrogel as a potential bulking agent for hypocaloric diets: Anin vitro biocompatibility study on rat intestine. *J. Appl. Polym. Sci.* **2006**, *102*, 1524–1530. [CrossRef]
123. Yan, L.; Shuai, Q.; Gong, X.; Gu, Q.; Yu, H. Synthesis of Microporous Cationic Hydrogel of Hydroxypropyl Cellulose (HPC) and its Application on Anionic Dye Removal. *CLEAN—Soil Air Water* **2009**, *37*, 392–398. [CrossRef]
124. Long, L.-Y.; Weng, Y.-X.; Wang, Y.-Z. Cellulose Aerogels: Synthesis, Applications, and Prospects. *Polymers* **2018**, *10*, 623. [CrossRef]
125. Tripathi, A.; Parsons, G.N.; Rojas, O.J.; Khan, S.A. Featherlight, Mechanically Robust Cellulose Ester Aerogels for Environmental Remediation. *ACS Omega* **2017**, *2*, 4297–4305. [CrossRef]
126. Tripathi, A.; Parsons, G.N.; Khan, S.A.; Rojas, O.J. Synthesis of organic aerogels with tailorable morphology and strength by controlled solvent swelling following Hansen solubility. *Sci. Rep.* **2018**, *8*, 2106. [CrossRef]
127. Pușcaș, A.; Mureșan, V.; Socaciu, C.; Muste, S. Oleogels in Food: A Review of Current and Potential Applications. *Foods* **2020**, *9*, 70. [CrossRef]
128. Cortés-Triviño, E.; Valencia, C.; Delgado, M.A.; Franco, J.M. Thermo-rheological and tribological properties of novel bio-lubricating greases thickened with epoxidized lignocellulosic materials. *J. Ind. Eng. Chem.* **2019**, *80*, 626–632. [CrossRef]
129. Gravelle, A.J.; Marangoni, A.G.; Davidovich-Pinhas, M. Ethylcellulose Oleogels. In *Edible Oleogels*; Elsevier BV: Amsterdam, The Netherlands, 2018; pp. 331–362.
130. Nuñez, N.; Martín-Alfonso, J.E.; Eugenio, M.E.; Valencia, C.; Díaz, M.J.; Franco, J.M. Preparation and Characterization of Gel-like Dispersions Based on Cellulosic Pulps and Castor Oil for Lubricant Applications. *Ind. Eng. Chem. Res.* **2011**, *50*, 5618–5627. [CrossRef]
131. Núñez, N.; Martín-Alfonso, J.; Valencia, C.; Sánchez, M.; Franco, J.M. Rheology of new green lubricating grease formulations containing cellulose pulp and its methylated derivative as thickener agents. *Ind. Crops Prod.* **2012**, *37*, 500–507. [CrossRef]
132. Cortés-Triviño, E.; Valencia, C.; Delgado, M.A.; Franco, J.M. Rheology of epoxidized cellulose pulp gel-like dispersions in castor oil: Influence of epoxidation degree and the epoxide chemical structure. *Carbohydr. Polym.* **2018**, *199*, 563–571. [CrossRef] [PubMed]
133. Sánchez, R.; Fiedler, M.; Kuhn, E.; Franco, J.M. Tribological characterization of green lubricating greases formulated with castor oil and different biogenic thickener agents: A comparative experimental study. *Ind. Lubr. Tribol.* **2011**, *63*, 446–452. [CrossRef]
134. Sánchez, R.; Franco, J.; Delgado, M.; Valencia, C.; Gallegos, C. Rheological and mechanical properties of oleogels based on castor oil and cellulosic derivatives potentially applicable as bio-lubricating greases: Influence of cellulosic derivatives concentration ratio. *J. Ind. Eng. Chem.* **2011**, *17*, 705–711. [CrossRef]
135. Alfonso, J.E.M.; Núñez, N.; Valencia, C.; Franco, J.; Díaz, M. Formulation of new biodegradable lubricating greases using ethylated cellulose pulp as thickener agent. *J. Ind. Eng. Chem.* **2011**, *17*, 818–823. [CrossRef]
136. Klemm, D.; Kramer, F.; Moritz, S.; Lindström, T.; Ankerfors, M.; Gray, D.; Dorris, A. Nanocelluloses: A New Family of Nature-Based Materials. *Angew. Chem. Int. Ed.* **2011**, *50*, 5438–5466. [CrossRef]
137. Mittal, N.; Ansari, F.; Gowda, V.K.; Brouzet, C.; Chen, P.; Larsson, P.T.; Roth, S.V.; Lundell, F.; Wågberg, L.; Kotov, N.A.; et al. Multiscale Control of Nanocellulose Assembly: Transferring Remarkable Nanoscale Fibril Mechanics to Macroscale Fibers. *ACS Nano* **2018**, *12*, 6378–6388. [CrossRef]
138. Peng, B.L.; Dhar, N.; Liu, H.L.; Tam, K.C. Chemistry and applications of nanocrystalline cellulose and its derivatives: A nanotechnology perspective. *Can. J. Chem. Eng.* **2011**, *89*, 1191–1206. [CrossRef]
139. Farhat, W.; Venditti, R.; Mignard, N.; Taha, M.; Becquart, F.; Ayoub, A. Polysaccharides and lignin based hydrogels with potential pharmaceutical use as a drug delivery system produced by a reactive extrusion process. *Int. J. Biol. Macromol.* **2017**, *104*, 564–575. [CrossRef] [PubMed]
140. Da Silva, A.E.; Rodrigues, H.; Gomes, M.C.S.; Eleamen, E.; Nagashima, T.; Egito, E.S.T. Xylan, a Promising Hemicellulose for Pharmaceutical Use. In *Products and Applications of Biopolymers*; IntechOpen: London, UK, 2012. [CrossRef]

141. Pardeshi, C.V.; Kulkarni, A.D.; Belgamwar, V.S.; Surana, S.J. Xyloglucan for drug delivery applications. In *Fundamental Biomaterials: Polymers*; Woodhead Publishing: Sawston, UK, 2018; pp. 143–169. [CrossRef]
142. Wan, J.; Liu, L.; Ayub, K.S.; Zhang, W.; Shen, G.; Hu, S.; Qian, X. Characterization and adsorption performance of biochars derived from three key biomass constituents. *Fuel* **2020**, *269*, 117142. [CrossRef]
143. Peng, P.; She, D. Isolation, structural characterization, and potential applications of hemicelluloses from bamboo: A review. *Carbohydr. Polym.* **2014**, *112*, 701–720. [CrossRef] [PubMed]
144. Ayoub, A.; Venditti, R.A.; Pawlak, J.J.; Salam, A.; Hubbe, M.A. Novel Hemicellulose–Chitosan Biosorbent for Water Desalination and Heavy Metal Removal. *ACS Sustain. Chem. Eng.* **2013**, *1*, 1102–1109. [CrossRef]
145. Seo, S.-J.; Park, I.-K.; Yoo, M.-K.; Shirakawa, M.; Akaike, T.; Cho, C.-S. Xyloglucan as a synthetic extracellular matrix for hepatocyte attachment. *J. Biomater. Sci. Polym. Ed.* **2004**, *15*, 1375–1387. [CrossRef]
146. Silveira, J.L.M.; Bresolin, T.M.B. Pharmaceutical use of galactomannans. *Quím. Nova* **2011**, *34*, 292–299. [CrossRef]
147. Landin, M.; Echezarreta, M.M. Galactomannans: Old and new pharmaceutical materials. In *Polysaccharides: Development, Properties and Applications*; Nova Science: Hauppauge, NY, USA, 2010.
148. Rossi, B.; Ponzini, E.; Merlini, L.; Grandori, R.; Galante, Y.M. Characterization of aerogels from chemo-enzymatically oxidized galactomannans as novel polymeric biomaterials. *Eur. Polym. J.* **2017**, *93*, 347–357. [CrossRef]
149. Hansen, N.M.L.; Plackett, D. Sustainable Films and Coatings from Hemicelluloses: A Review. *Biomacromolecules* **2008**, *9*, 1493–1505. [CrossRef]
150. Prakobna, K.; Kisonen, V.; Xu, C.; Berglund, L.A. Strong reinforcing effects from galactoglucomannan hemicellulose on mechanical behavior of wet cellulose nanofiber gels. *J. Mater. Sci.* **2015**, *50*, 7413–7423. [CrossRef]
151. Peng, H.; Yang, A.; Xiong, J. Green, microwave-assisted synthesis of silver nanoparticles using bamboo hemicelluloses and glucose in an aqueous medium. *Carbohydr. Polym.* **2013**, *91*, 348–355. [CrossRef] [PubMed]
152. Jiang, X.; Huang, J.; Chen, T.; Zhao, Q.; Xu, F.; Zhang, X. Synthesis of hemicellulose/deep eutectic solvent based carbon quantum dots for ultrasensitive detection of Ag+ and L-cysteine with "off-on" pattern. *Int. J. Biol. Macromol.* **2020**, *153*, 412–420. [CrossRef] [PubMed]
153. Farhat, W.; Venditti, R.; Quick, A.; Taha, M.; Mignard, N.; Becquart, F.; Ayoub, A. Hemicellulose extraction and characterization for applications in paper coatings and adhesives. *Ind. Crop. Prod.* **2017**, *107*, 370–377. [CrossRef]
154. Ebringerova, A. The potential of xylans as biomaterial resources. In *Polysaccharide Building Blocks: A Sustainable Approach to the Development of Renewable Biomaterials*; Wiley: Hoboken, NJ, USA, 2006. [CrossRef]
155. Martínez-Ibarra, D.M.; López-Cervantes, J.; Machado, D.I.S.; Sanches-Silva, A. Chitosan and Xyloglucan-Based Hydrogels: An Overview of Synthetic and Functional Utility. In *Chitin-Chitosan—Myriad Functionalities in Science and Technology*; IntechOpen: London, UK, 2018. [CrossRef]
156. Prajapati, V.D.; Jani, G.K.; Moradiya, N.G.; Randeria, N.P.; Nagar, B.J.; Naikwadi, N.N.; Variya, B.C. Galactomannan: A versatile biodegradable seed polysaccharide. *Int. J. Biol. Macromol.* **2013**, *60*, 83–92. [CrossRef]
157. Polari, L.; Ojansivu, P.; Mäkelä, S.; Eckerman, C.; Holmbom, B.; Salminen, S. Galactoglucomannan Extracted from Spruce (*Picea abies*) as a Carbohydrate Source for Probiotic Bacteria. *J. Agric. Food Chem.* **2012**, *60*, 11037–11043. [CrossRef]
158. Faruk, O.; Sain, M. *Lignin in Polymer Composites*; Elsevier: Amsterdam, The Netherlands, 2016. [CrossRef]
159. Gordobil, O. New Products from Lignin. Ph.D. Thesis, Universidad del País Vasco, San Sebastián, Spain, 2018.
160. Pucciariello, R.; Villani, V.; Bonini, C.; D'Auria, M.; Vetere, T. Physical properties of straw lignin-based polymer blends. *Polymer* **2004**, *45*, 4159–4169. [CrossRef]
161. Mishra, S.B.; Mishra, A.; Kaushik, N.; Khan, M.A. Study of performance properties of lignin-based polyblends with polyvinyl chloride. *J. Mater. Process. Technol.* **2007**, *183*, 273–276. [CrossRef]
162. Yang, W.; Rallini, M.; Wang, D.-Y.; Gao, D.; Dominici, F.; Torre, L.; Kenny, J.M.; Puglia, D. Role of lignin nanoparticles in UV resistance, thermal and mechanical performance of PMMA nanocomposites prepared by a combined free-radical graft polymerization/masterbatch procedure. *Compos. Part A Appl. Sci. Manuf.* **2018**, *107*, 61–69. [CrossRef]
163. Park, S.Y.; Kim, J.-Y.; Youn, H.J.; Choi, J.W. Utilization of lignin fractions in UV resistant lignin-PLA biocomposites via lignin-lactide grafting. *Int. J. Biol. Macromol.* **2019**, *138*, 1029–1034. [CrossRef]
164. Sheng, Y.; Ma, Z.; Wang, X.; Han, Y. Ethanol organosolv lignin from different agricultural residues: Toward basic structural units and antioxidant activity. *Food Chem.* **2021**, *376*, 131895. [CrossRef]
165. Canetti, M.; Bertini, F.; De Chirico, A.; Audisio, G. Thermal degradation behaviour of isotactic polypropylene blended with lignin. *Polym. Degrad. Stab.* **2006**, *91*, 494–498. [CrossRef]
166. Gregorová, A.; Košíková, B.; Moravčík, R. Stabilization effect of lignin in natural rubber. *Polym. Degrad. Stab.* **2006**, *91*, 229–233. [CrossRef]
167. Tavares, L.B.; Rosa, D.D.S. Stabilization effect of kraft lignin into PBAT: Thermal analyses approach. *Matéria* **2019**, *24*, 12405. [CrossRef]
168. Lisperguer, J.; Nuñez, C.; Perez-Guerrero, P. Structure and Thermal Properties of Maleated Lignin-Recycled Polystyrene Composites. *J. Chil. Chem. Soc.* **2013**, *58*, 1937–1940. [CrossRef]
169. Sen, S.; Patil, S.; Argyropoulos, D. Thermal properties of lignin in copolymers, blends, and composites: A review. *Green Chem.* **2015**, *17*, 4862–4887. [CrossRef]

170. De Chirico, A.; Armanini, M.; Chini, P.; Cioccolo, G.; Provasoli, F.; Audisio, G. Flame retardants for polypropylene based on lignin. *Polym. Degrad. Stab.* **2003**, *79*, 139–145. [CrossRef]
171. Chollet, B.; Lopez-Cuesta, J.-M.; Laoutid, F.; Ferry, L. Lignin Nanoparticles as A Promising Way for Enhancing Lignin Flame Retardant Effect in Polylactide. *Materials* **2019**, *12*, 2132. [CrossRef]
172. Chen, S.; Lin, S.; Hu, Y.; Ma, M.; Shi, Y.; Liu, J.; Zhu, F.; Wang, X. A lignin-based flame retardant for improving fire behavior and biodegradation performance of polybutylene succinate. *Polym. Adv. Technol.* **2018**, *29*, 3142–3150. [CrossRef]
173. Yang, H.; Yu, B.; Xu, X.; Bourbigot, S.; Wang, H.; Song, P. Lignin-derived bio-based flame retardants toward high-performance sustainable polymeric materials. *Green Chem.* **2020**, *22*, 2129–2161. [CrossRef]
174. Mandlekar, N.; Cayla, A.; Rault, F.; Giraud, S.; Salaün, F.; Malucelli, G.; Guan, J.-P. An Overview on the Use of Lignin and Its Derivatives in Fire Retardant Polymer Systems. In *Lignin—Trends and Applications*; Poletto, M., Ed.; IntechOpen: London, UK, 2018.
175. Ikeda, Y.; Phakkeeree, T.; Junkong, P.; Yokohama, H.; Phinyocheep, P.; Kitano, R.; Kato, A. Reinforcing biofiller "Lignin" for high performance green natural rubber nanocomposites. *RSC Adv.* **2017**, *7*, 5222–5231. [CrossRef]
176. Al Mamun, A.; Nikousaleh, M.A.; Feldmann, M.; Rüppel, A.; Sauer, V.; Kleinhans, S.; Heim, H.-P. Lignin Reinforcement in Bioplastic Composites. In *Lignin in Polymer Composites*; Elsevier BV: Amsterdam, The Netherlands, 2016; pp. 153–165.
177. Jiang, C.; He, H.; Jiang, H.; Ma, L.; Jia, D.M. Nano-lignin filled natural rubber composites: Preparation and characterization. *Express Polym. Lett.* **2013**, *7*, 480–493. [CrossRef]
178. Rozman, H.; Tan, K.; Kumar, R.; Abubakar, A.; Ishak, Z.M.; Ismail, H. The effect of lignin as a compatibilizer on the physical properties of coconut fiber–polypropylene composites. *Eur. Polym. J.* **2000**, *36*, 1483–1494. [CrossRef]
179. Tanjung, F.A.; Husseinsyah, S.; Hussin, K. Chitosan-filled polypropylene composites: The effect of filler loading and organosolv lignin on mechanical, morphological and thermal properties. *Fibers Polym.* **2014**, *15*, 800–808. [CrossRef]
180. Yang, J.; Ching, Y.C.; Chuah, C.H. Applications of Lignocellulosic Fibers and Lignin in Bioplastics: A Review. *Polymers* **2019**, *11*, 751. [CrossRef]
181. Aini, N.A.M.; Othman, N.; Hussin, M.H.; Sahakaro, K.; Hayeemasae, N. Lignin as Alternative Reinforcing Filler in the Rubber Industry: A Review. *Front. Mater.* **2020**, *6*, 329. [CrossRef]
182. Yue, X.; Chen, F.; Zhou, X.; He, G. Preparation and Characterization of Poly (vinyl chloride) Polyblends with Fractionated Lignin. *Int. J. Polym. Mater. Polym. Biomater.* **2012**, *61*, 214–228. [CrossRef]
183. European Commission, Directorate-General for Research and Innovation. *Study on Support to R & I Policy in the Area of Bio-Based Products and Services on the Top 20 Innovative Bio-Based Products*; Publications Office: Luxembourg, 2019. [CrossRef]
184. Yu, G.; Li, B.; Wang, H.; Liu, C.; Mu, X. Preparation of Concrete Superplasticizer by Oxidation-Sulfomethylation of Sodium Lignosulfonate. *Bioresources* **2012**, *8*, 1055–1063. [CrossRef]
185. Mu, L.; Shi, Y.; Guo, X.; Ji, T.; Chen, L.; Yuan, R.; Brisbin, L.; Wang, H.; Zhu, J. Non-corrosive green lubricants: Strengthened lignin–[choline][amino acid] ionic liquids interaction via reciprocal hydrogen bonding. *RSC Adv.* **2015**, *5*, 66067–66072. [CrossRef]
186. Mu, L.; Wu, J.; Matsakas, L.; Chen, M.; Rova, U.; Christakopoulos, P.; Zhu, J.; Shi, Y. Two important factors of selecting lignin as efficient lubricating additives in poly (ethylene glycol): Hydrogen bond and molecular weight. *Int. J. Biol. Macromol.* **2019**, *129*, 564–570. [CrossRef]
187. Hua, J.; Shi, Y. Non-corrosive Green Lubricant with Dissolved Lignin in Ionic Liquids Behave as Ideal Lubricants for Steel-DLC Applications. *Front. Chem.* **2019**, *7*, 857. [CrossRef]
188. Cortés-Triviño, E.; Valencia, C.; Franco, J.M. Thickening Castor Oil with a Lignin-Enriched Fraction from Sugarcane Bagasse Waste via Epoxidation: A Rheological and Hydrodynamic Approach. *ACS Sustain. Chem. Eng.* **2021**, *9*, 10503–10512. [CrossRef]
189. Borrego, M.; Martín-Alfonso, J.E.; Sánchez, M.C.; Valencia, C.; Franco, J.M. Electrospun lignin-PVP nanofibers and their ability for structuring oil. *Int. J. Biol. Macromol.* **2021**, *180*, 212–221. [CrossRef]
190. Cortés-Triviño, E.; Valencia, C.; Delgado, M.A.; Franco, J.M. Modification of Alkali Lignin with Poly(Ethylene Glycol) Diglycidyl Ether to Be Used as a Thickener in Bio-Lubricant Formulations. *Polymers* **2018**, *10*, 670. [CrossRef]
191. Rubio-Valle, J.; Sánchez, M.; Valencia, C.; Martín-Alfonso, J.; Franco, J. Electrohydrodynamic Processing of PVP-Doped Kraft Lignin Micro- and Nano-Structures and Application of Electrospun Nanofiber Templates to Produce Oleogels. *Polymers* **2021**, *13*, 2206. [CrossRef]
192. Balasubramanian, P.; Ramalingam, S.; Javid, M.A.; Rao, J.R. Lignin Based Colorant: Modified Black Liquor for Leather Surface Coating Application. *J. Am. Leather Chem. Assoc.* **2018**, *113*, 311–317.
193. Araújo, P.; Costa, A.; Fernandes, I.; Mateus, N.; de Freitas, V.; Sarmento, B.; Oliveira, J. Stabilization of bluish pyranoanthocyanin pigments in aqueous systems using lignin nanoparticles. *Dye. Pigment.* **2019**, *166*, 367–374. [CrossRef]
194. Ugartondo, V.; Mitjans, M.; Vinardell, M.P. Comparative antioxidant and cytotoxic effects of lignins from different sources. *Bioresour. Technol.* **2008**, *99*, 6683–6687. [CrossRef]
195. Espinoza-Acosta, J.L.; Torres-Chávez, P.I.; Ramírez-Wong, B.; López-Saiz, C.-M.; Montaño-Leyva, B. Antioxidant, antimicrobial, and antimutagenic properties of technical lignins and their applications. *BioResources* **2016**, *11*, 5452–5481. [CrossRef]
196. Mahmood, Z.; Yameen, M.; Jahangeer, M.; Riaz, M.; Ghaffar, A.; Javid, I. Lignin as Natural Antioxidant Capacity. In *Lignin—Trends and Applications*; IntechOpen: London, UK, 2018. [CrossRef]
197. Jiang, B.; Zhang, Y.; Gu, L.; Wu, W.; Zhao, H.; Jin, Y. Structural elucidation and antioxidant activity of lignin isolated from rice straw and alkali-oxygen black liquor. *Int. J. Biol. Macromol.* **2018**, *116*, 513–519. [CrossRef]

253. Hu, W.; Zhang, J.; Liu, B.; Zhang, C.; Zhao, Q.; Sun, Z.; Cao, H.; Zhu, G. Synergism between lignin, functionalized carbon nanotubes and Fe_3O_4 nanoparticles for electromagnetic shielding effectiveness of tough lignin-based polyurethane. *Compos. Commun.* **2021**, *24*, 100616. [CrossRef]
254. Zhang, Y.; Li, T.-T.; Lou, C.-W.; Lin, J.-H. Facile method for tent fabrics with eco-friendly/durable properties using waterborne polyurethane/lignin: Preparation and evaluation. *J. Ind. Text.* **2020**, 1528083720931884. [CrossRef]
255. Gonçalves, S.S.L.; Rudnitskaya, A.; Sales, A.J.; Costa, L.M.C.; Evtuguin, D.V. Nanocomposite Polymeric Materials Based on Eucalyptus Lignoboost® Kraft Lignin for Liquid Sensing Applications. *Materials* **2020**, *13*, 1637. [CrossRef]
256. Karunarathna, M.S.; Smith, R.C. Valorization of Lignin as a Sustainable Component of Structural Materials and Composites: Advances from 2011 to 2019. *Sustainability* **2020**, *12*, 734. [CrossRef]
257. Burchardt, B. Advances in polyurethane structural adhesives. In *Advances in Structural Adhesive Bonding*; Elsevier: Amsterdam, The Netherlands, 2010; pp. 35–65.
258. Jiang, W.; Kumar, A.; Adamopoulos, S. Liquefaction of lignocellulosic materials and its applications in wood adhesives—A review. *Ind. Crop. Prod.* **2018**, *124*, 325–342. [CrossRef]
259. Lee, W.-J.; Lin, M.-S. Preparation and application of polyurethane adhesives made from polyhydric alcohol liquefied Taiwan acacia and China fir. *J. Appl. Polym. Sci.* **2008**, *109*, 23–31. [CrossRef]
260. Mori, R. Inorganic–organic hybrid biodegradable polyurethane resin derived from liquefied Sakura wood. *Wood Sci. Technol.* **2015**, *49*, 507–516. [CrossRef]
261. Juhaida, M.; Paridah, M.; Hilmi, M.M.; Sarani, Z.; Jalaluddin, H.; Zaki, A.M. Liquefaction of kenaf (*Hibiscus cannabinus* L.) core for wood laminating adhesive. *Bioresour. Technol.* **2010**, *101*, 1355–1360. [CrossRef]
262. Tohmura, S.-I.; Li, G.-Y.; Qin, T.-F. Preparation and characterization of wood polyalcohol-based isocyanate adhesives. *J. Appl. Polym. Sci.* **2005**, *98*, 791–795. [CrossRef]
263. Tenorio-Alfonso, A.; Sánchez, M.C.; Franco, J.M. Impact of moisture curing conditions on the chemical structure and rheological and ultimate adhesion properties of polyurethane adhesives based on castor oil and cellulose acetate. *Prog. Org. Coat.* **2021**, *161*, 106547. [CrossRef]
264. Tenorio-Alfonso, A.; Pizarro, M.; Sánchez, M.; Franco, J. Assessing the rheological properties and adhesion performance on different substrates of a novel green polyurethane based on castor oil and cellulose acetate: A comparison with commercial adhesives. *Int. J. Adhes. Adhes.* **2018**, *82*, 21–26. [CrossRef]
265. Tenorio-Alfonso, A.; Sánchez, M.C.; Franco, J.M. Synthesis and mechanical properties of bio-sourced polyurethane adhesives obtained from castor oil and MDI-modified cellulose acetate: Influence of cellulose acetate modification. *Int. J. Adhes. Adhes.* **2019**, *95*, 102404. [CrossRef]
266. Jasiūnas, L.; Peck, G.; Bridžiuvienė, D.; Miknius, L. Mechanical, thermal properties and stability of high renewable content liquefied residual biomass derived bio-polyurethane wood adhesives. *Int. J. Adhes. Adhes.* **2020**, *101*, 102618. [CrossRef]
267. Balcioglu, S.; Parlakpinar, H.; Vardi, N.; Denkbas, E.B.; Karaaslan, M.G.; Gulgen, S.; Taslidere, E.; Koytepe, S.; Ates, B. Design of Xylose-Based Semisynthetic Polyurethane Tissue Adhesives with Enhanced Bioactivity Properties. *ACS Appl. Mater. Interfaces* **2016**, *8*, 4456–4466. [CrossRef]
268. Rackham, R.A. Use of Xylan to Improve Bond Strength. European Patent EP 1265969B1, 12 October 2005.
269. Hsu, O.H.H.; Glasser, W.G. Polyurethane adhesives and coatings from modified lignin. *J. Appl. Polym. Sci. Appl. Polym. Symp.* **1975**, *378*, 297–307.
270. Borrero-López, A.M.; Valencia, C.; Domínguez, G.; Eugenio, M.E.; Franco, J.M. Rheology and adhesion performance of adhesives formulated with lignins from agricultural waste straws subjected to solid-state fermentation. *Ind. Crop. Prod.* **2021**, *171*, 113876. [CrossRef]
271. De Haro, J.C.; Allegretti, C.; Smit, A.T.; Turri, S.; D'Arrigo, P.; Griffini, G. Biobased Polyurethane Coatings with High Biomass Content: Tailored Properties by Lignin Selection. *ACS Sustain. Chem. Eng.* **2019**, *7*, 11700–11711. [CrossRef]
272. Chen, Y.; Zhang, H.; Zhu, Z.; Fu, S. High-value utilization of hydroxymethylated lignin in polyurethane adhesives. *Int. J. Biol. Macromol.* **2020**, *152*, 775–785. [CrossRef]
273. Gouveia, J.R.; Antonino, L.D.; Garcia, G.E.S.; Tavares, L.B.; Santos, A.N.B.; dos Santos, D.J. Kraft lignin-containing polyurethane adhesives: The role of hydroxypropylation on thermomechanical properties. *J. Adhes.* **2020**, *97*, 1423–1439. [CrossRef]
274. Ibrahim, V.; Mamo, G.; Gustafsson, P.-J.; Hatti-Kaul, R. Production and properties of adhesives formulated from laccase modified Kraft lignin. *Ind. Crop. Prod.* **2013**, *45*, 343–348. [CrossRef]
275. García, J.L.; Pans, G.; Phanopoulos, C. Use of lignin in polyurethane-based structural wood adhesives. *J. Adhes.* **2018**, *94*, 814–828. [CrossRef]
276. Wang, S.; Yu, Y.; Di, M. Green Modification of Corn Stalk Lignin and Preparation of Environmentally Friendly Lignin-Based Wood Adhesive. *Polymers* **2018**, *10*, 631. [CrossRef]
277. Lyadov, A.S.; Maksimova, Y.M.; Shakhmatova, A.S.; Kirillov, V.V.; Parenago, O.P. Urea (Polyurea) Greases. *Russ. J. Appl. Chem.* **2018**, *91*, 885–894. [CrossRef]
278. Chandra Sharma, U.; Singh, N. Biogreases for Environment Friendly Lubrication. In *Environonmental Science & Engineering*; Sustainable Development; Studium Press: New Delhi, India, 2019; Volume 1.

279. Borrero-López, A.M.; Blánquez, A.; Valencia, C.; Hernández, M.; Arias, M.E.; Franco, J.M. Influence of solid-state fermentation with Streptomyces on the ability of wheat and barley straws to thicken castor oil for lubricating purposes. *Ind. Crops Prod.* **2019**, *140*, 111625. [CrossRef]
280. Gallego, R.; Arteaga, J.F.; Valencia, C.; Franco, J.M. Chemical modification of methyl cellulose with HMDI to modulate the thickening properties in castor oil. *Cellulose* **2013**, *20*, 495–507. [CrossRef]
281. Gallego, R.; Arteaga, J.; Valencia, C.; Franco, J. Rheology and thermal degradation of isocyanate-functionalized methyl cellulose-based oleogels. *Carbohydr. Polym.* **2013**, *98*, 152–160. [CrossRef] [PubMed]
282. Gallego, R.; Arteaga, J.; Valencia, C.; Franco, J. Thickening properties of several NCO-functionalized cellulose derivatives in castor oil. *Chem. Eng. Sci.* **2015**, *134*, 260–268. [CrossRef]
283. Borrero-López, A.M.; Valencia, C.; Blánquez, A.; Hernández, M.; Eugenio, M.E.; Franco, J.M. Cellulose Pulp- and Castor Oil-Based Polyurethanes for Lubricating Applications: Influence of *Streptomyces* Action on Barley and Wheat Straws. *Polymers* **2020**, *12*, 2822. [CrossRef]
284. Gallego, R.; Arteaga, J.F.; Valencia, C.; Díaz, M.J.; Franco, J.M. Gel-Like Dispersions of HMDI-Cross-Linked Lignocellulosic Materials in Castor Oil: Toward Completely Renewable Lubricating Grease Formulations. *ACS Sustain. Chem. Eng.* **2015**, *3*, 2130–2141. [CrossRef]
285. Gallego, R.; Cidade, M.T.; Sánchez, R.; Valencia, C.; Franco, J. Tribological behaviour of novel chemically modified biopolymer-thickened lubricating greases investigated in a steel–steel rotating ball-on-three plates tribology cell. *Tribol. Int.* **2016**, *94*, 652–660. [CrossRef]
286. Ma, Y.; Li, Z.; Wang, H.; Li, H. Synthesis and optimization of polyurethane microcapsules containing [BMIm]PF6 ionic liquid lubricant. *J. Colloid Interface Sci.* **2019**, *534*, 469–479. [CrossRef]
287. Borrero-López, A.M.; Valencia, C.; Franco, J.M. Rheology of lignin-based chemical oleogels prepared using diisocyanate crosslinkers: Effect of the diisocyanate and curing kinetics. *Eur. Polym. J.* **2017**, *89*, 311–323. [CrossRef]
288. Borrero-López, A.M.; Valencia, C.; Franco, J.M. Green and facile procedure for the preparation of liquid and gel-like polyurethanes based on castor oil and lignin: Effect of processing conditions on the rheological properties. *J. Clean. Prod.* **2020**, *277*, 123367. [CrossRef]
289. Borrero-López, A.M.; Blánquez, A.; Valencia, C.; Hernández, M.; Arias, M.E.; Eugenio, M.E.; Fillat, Ú.; Franco, J.M. Valorization of Soda Lignin from Wheat Straw Solid-State Fermentation: Production of Oleogels. *ACS Sustain. Chem. Eng.* **2018**, *6*, 5198–5205. [CrossRef]
290. Domínguez, G.; Blánquez, A.; Borrero-López, A.M.; Valencia, C.; Eugenio, M.E.; Arias, M.E.; Rodríguez, J.; Hernández, M. Eco-Friendly Oleogels from Functionalized Kraft Lignin with Laccase SilA from *Streptomyces ipomoeae*: An Opportunity to Replace Commercial Lubricants. *ACS Sustain. Chem. Eng.* **2021**, *9*, 4611–4616. [CrossRef]
291. Borrero-López, A.M.; Martín-Sampedro, R.; Ibarra, D.; Valencia, C.; Eugenio, M.E.; Franco, J.M. Evaluation of lignin-enriched side-streams from different biomass conversion processes as thickeners in bio-lubricant formulations. *Int. J. Biol. Macromol.* **2020**, *162*, 1398–1413. [CrossRef]
292. Borrero-López, A.M.; Valencia, C.; Ibarra, D.; Ballesteros, I.; Franco, J.M. Lignin-enriched residues from bioethanol production: Chemical characterization, isocyanate functionalization and oil structuring properties. *Int. J. Biol. Macromol.* **2021**, *195*, 412–423. [CrossRef]
293. Borrero-López, A.M.; Santiago-Medina, F.J.; Valencia, C.; Eugenio, M.E.; Martin-Sampedro, R.; Franco, J.M. Valorization of Kraft Lignin as Thickener in Castor Oil for Lubricant Applications. *J. Renew. Mater.* **2018**, *6*, 347–361. [CrossRef]
294. Fajardo, C.; Blánquez, A.; Domínguez, G.; Borrero-López, A.; Valencia, C.; Hernández, M.; Arias, M.; Rodríguez, J. Assessment of Sustainability of Bio Treated Lignocellulose-Based Oleogels. *Polymers* **2021**, *13*, 267. [CrossRef]
295. Litters, T.; Hahn, F.; Goerz, T.; Erkel, H.J. Process for the Preparation of Polyurea-Thickened Lignin Derivative-Based Lubricating Greases, Such Lubricant Greases and Use Thereof. U.S. Patent US10604721B2, 9 March 2016.
296. Aureliano Perez, J.R.; Vaught, K.A.; Hansen, G.P. Low Volatile Organic Content Lubricant. U.S. Patent US7524797B1, 28 April 2009.
297. Pasquini, D. Fully Green Elastomer Composites. In *Advances in Elastomers II*; Springer: Berlin/Heidelberg, Germany, 2013. [CrossRef]
298. Ducharme, R. Multi-composite acoustic panel. *J. Acoust. Soc. Am.* **2009**, *125*, 4104. [CrossRef]
299. Bi, H.; Ren, Z.; Guo, R.; Xu, M.; Song, Y. Fabrication of flexible wood flour/thermoplastic polyurethane elastomer composites using fused deposition molding. *Ind. Crop. Prod.* **2018**, *122*, 76–84. [CrossRef]
300. Mengeloğlu, F.; Çavuş, V. Preparation of thermoplastic polyurethane-based biocomposites through injection molding: Effect of the filler type and content. *BioResources* **2020**, *15*, 5749–5763. [CrossRef]
301. Guodong, T.; Yu, Z.; Jiming, X.; Jianying, Z.; Yiwei, W. Rice Straw Liquefying Method and Method for Synthesizing Polyurethane Elastomer from Rice Straw Liquefying Product. China Patent CN103965485A, 6 August 2014.
302. Zhao, S.L.; Xue, Z.H.; Bin Li, Y.; Huang, J.T. Study on Straw Liquefied Product Synthetizing Polyurethane Elastomer. *Adv. Mater. Res.* **2013**, *821–822*, 977–980. [CrossRef]
303. Yan, Q.; Zhao, S.; Kang, H.; Zhang, S. Thiol-assisted bioinspired deposition of polyurethane onto cellulose as robust elastomer for reinforcing soy protein-based composites. *J. Appl. Polym. Sci.* **2020**, *137*, 49176. [CrossRef]
304. Wu, Q.; Henriksson, M.; Liu, X.; Berglund, L. A High Strength Nanocomposite Based on Microcrystalline Cellulose and Polyurethane. *Biomacromolecules* **2007**, *8*, 3687–3692. [CrossRef] [PubMed]

305. Pei, A.; Malho, J.-M.; Ruokolainen, J.; Zhou, Q.; Berglund, L.A. Strong Nanocomposite Reinforcement Effects in Polyurethane Elastomer with Low Volume Fraction of Cellulose Nanocrystals. *Macromolecules* **2011**, *44*, 4422–4427. [CrossRef]
306. Zhu, Y.; Hu, J.; Luo, H.; Young, R.J.; Deng, L.; Zhang, S.; Fan, Y.; Ye, G. Rapidly switchable water-sensitive shape-memory cellulose/elastomer nano-composites. *Soft Matter* **2012**, *8*, 2509–2517. [CrossRef]
307. Aranguren, M.I.; Marcovich, N.E.; Salgueiro, W.; Somoza, A. Effect of the nano-cellulose content on the properties of reinforced polyurethanes. A study using mechanical tests and positron anihilation spectroscopy. *Polym. Test.* **2013**, *32*, 115–122. [CrossRef]
308. Saralegi, A.; Rueda, L.; Martin, L.; Arbelaiz, A.; Eceiza, A.; Corcuera, M. From elastomeric to rigid polyurethane/cellulose nanocrystal bionanocomposites. *Compos. Sci. Technol.* **2013**, *88*, 39–47. [CrossRef]
309. Lee, M.; Heo, M.H.; Lee, H.-H.; Kim, Y.-W.; Shin, J. Tunable softening and toughening of individualized cellulose nanofibers-polyurethane urea elastomer composites. *Carbohydr. Polym.* **2017**, *159*, 125–135. [CrossRef]
310. Gao, Z.; Peng, J.; Zhong, T.; Sun, J.; Wang, X.; Yue, C. Biocompatible elastomer of waterborne polyurethane based on castor oil and polyethylene glycol with cellulose nanocrystals. *Carbohydr. Polym.* **2012**, *87*, 2068–2075. [CrossRef]
311. Lu, Y.; Zhang, L. Morphology and mechanical properties of semi-interpenetrating polymer networks from polyurethane and benzyl konjac glucomannan. *Polymer* **2002**, *43*, 3979–3986. [CrossRef]
312. Gao, S.; Zhang, L.; Cao, J. Synthesis and characterization of poly(ester urethane)/nitrokonjac glucomannan semi-interpenetrating polymer networks. *J. Appl. Polym. Sci.* **2003**, *90*, 2224–2228. [CrossRef]
313. Lu, Y.; Zhang, L.; Zhang, X.; Zhou, Y. Effects of secondary structure on miscibility and properties of semi-IPN from polyurethane and benzyl konjac glucomannan. *Polymer* **2003**, *44*, 6689–6696. [CrossRef]
314. Li, H.; Liang, Y.; Li, P.; He, C. Conversion of biomass lignin to high-value polyurethane: A review. *J. Bioresour. Bioprod.* **2020**, *5*, 163–179. [CrossRef]
315. Gang, H.; Lee, D.; Choi, K.-Y.; Kim, H.-N.; Ryu, H.; Lee, D.-S.; Kim, B.-G. Development of High Performance Polyurethane Elastomers Using Vanillin-Based Green Polyol Chain Extender Originating from Lignocellulosic Biomass. *ACS Sustain. Chem. Eng.* **2017**, *5*, 4582–4588. [CrossRef]
316. Wang, R.; Zhou, B.; Zhu, Y.; Wang, Z. Effects of Lignin-Derived Polycarboxylic Acids on the Properties of Waterborne Polyurethane Elastomers. *Int. J. Polym. Sci.* **2018**, *2018*, 7989367. [CrossRef]
317. Li, H.; Sun, J.-T.; Wang, C.; Liu, S.; Yuan, D.; Zhou, X.; Tan, J.; Stubbs, L.; He, C. High Modulus, Strength, and Toughness Polyurethane Elastomer Based on Unmodified Lignin. *ACS Sustain. Chem. Eng.* **2017**, *5*, 7942–7949. [CrossRef]
318. Jeong, H.; Park, J.; Kim, S.; Lee, J.; Ahn, N.; Roh, H.-G. Preparation and characterization of thermoplastic polyurethanes using partially acetylated kraft lignin. *Fibers Polym.* **2013**, *14*, 1082–1093. [CrossRef]
319. Zhang, Y.; Yan, R.; Ngo, T.-D.; Zhao, Q.; Duan, J.; Du, X.; Wang, Y.; Liu, B.; Sun, Z.; Hu, W.; et al. Ozone oxidized lignin-based polyurethane with improved properties. *Eur. Polym. J.* **2019**, *117*, 114–122. [CrossRef]
320. Liu, W.; Fang, C.; Wang, S.; Huang, J.; Qiu, X. High-Performance Lignin-Containing Polyurethane Elastomers with Dynamic Covalent Polymer Networks. *Macromolecules* **2019**, *52*, 6474–6484. [CrossRef]
321. Ciobanu, C.; Ungureanu, M.; Ignat, L.; Popa, V. Properties of lignin–polyurethane films prepared by casting method. *Ind. Crop. Prod.* **2004**, *20*, 231–241. [CrossRef]
322. Kai, D.; Tan, M.J.; Chee, P.L.; Chua, Y.K.; Yap, Y.L.; Loh, X.J. Towards lignin-based functional materials in a sustainable world. *Green Chem.* **2016**, *18*, 1175–1200. [CrossRef]
323. Lang, J.M.; Shrestha, U.M.; Dadmun, M. The Effect of Plant Source on the Properties of Lignin-Based Polyurethanes. *Front. Energy Res.* **2018**, *6*, 4. [CrossRef]
324. Culebras, M.; Beaucamp, A.; Wang, Y.; Clauss, M.M.; Frank, E.; Collins, M.N. Biobased Structurally Compatible Polymer Blends Based on Lignin and Thermoplastic Elastomer Polyurethane as Carbon Fiber Precursors. *ACS Sustain. Chem. Eng.* **2018**, *6*, 8816–8825. [CrossRef]
325. Li, H.; Sivasankarapillai, G.; McDonald, A.G. Lignin valorization by forming toughened thermally stimulated shape memory copolymeric elastomers: Evaluation of different fractionated industrial lignins. *J. Appl. Polym. Sci.* **2015**, *132*, 41389. [CrossRef]
326. Frigerio, P.; Zoia, L.; Orlandi, M.; Hanel, T.; Castellani, L. Application of Sulphur-Free Lignins as a Filler for Elastomers: Effect of Hexamethylenetetramine Treatment. *BioResources* **2013**, *9*, 1387–1400. [CrossRef]
327. Zhang, J.; Chen, Y.; Sewell, P.; Brook, M.A. Utilization of softwood lignin as both crosslinker and reinforcing agent in silicone elastomers. *Green Chem.* **2015**, *17*, 1811–1819. [CrossRef]
328. Engelmann, G.; Ganster, J. Lignin Reinforcement in Thermosets Composites. In *Lignin in Polymer Composites*; Elsevier BV: Amsterdam, The Netherlands, 2016; pp. 119–151.

polymers

MDPI

Article

Effect of Antibacterial Peptide Microsphere Coating on the Microbial and Physicochemical Characteristics of *Tricholoma matsutake* during Cold Storage

Hongli Li, Yan Feng, Peng Zhang, Mingwei Yuan * and Minglong Yuan *

National and Local Joint Engineering Research Center for Green Preparation Technology of Biobased Materials, Yunnan Minzu University, Kunming 650500, China; hongli@vip.163.com (H.L.); fy971129@163.com (Y.F.); newworldopen@163.com (P.Z.)

* Correspondence: 041808@ymu.edu.cn (M.Y.); yml@ynni.edu.cn (M.Y.); Tel.: +86-189-8718-8989 (Minglong Yuan)

Abstract: The effect of novel antimicrobial peptides (AMPs) and antimicrobial peptide microspheres (AMS) on the physicochemical and microbial quality of *Tricholoma matsutake* wild edible mushrooms was investigated. In the experiments, 1.0 g/L, 0.5 g/L of AMS, and 1.0 g/L AMPs were used as preservatives. Mushrooms coated with 1.0 g/L and 0.5 g/L of AMS as a preservative had better physicochemical and sensory qualities than did mushrooms coated with 1.0 g/L of AMPs. In the experiment, 1.0 g/L of blank microspheres without cathelicidin-BF-30 (PLGA-1.0) and distilled water was used as the control. Samples with these two treatments had minimal changes in texture, weight loss, total bacteria count, and sensory attributes. Research results suggests that the use of AMS can maintain the quality of *Tricholoma matsutake* wild edible mushrooms and could extend the postharvest life to 20 d.

Keywords: antimicrobial peptide (microspheres); *Tricholoma matsutake*; physicochemical quality; microbial quality

Citation: Li, H.; Feng, Y.; Zhang, P.; Yuan, M.; Yuan, M. Effect of Antibacterial Peptide Microsphere Coating on the Microbial and Physicochemical Characteristics of *Tricholoma matsutake* during Cold Storage. *Polymers* **2022**, *14*, 208. https://doi.org/10.3390/polym14010208

Academic Editors: Clara Delgado-Sánchez, Concepción Valencia-Barragán, Esperanza Cortés Triviño, Adrián Tenorio-Alfonso and Antonio M. Borrero-López

Received: 1 December 2021
Accepted: 27 December 2021
Published: 5 January 2022

1. Introduction

Mushrooms have been widely used since ancient times as food and for medicinal or other functional purposes. In recent years, there has been an increase in the worldwide consumption of fresh wild mushrooms (e.g., *Lactarius deliciosus Boletus edulis Cantharellus* spp. *Hygrophorus* spp. and *Tricholoma matsutake* spp.) because of their delicate flavor, trace minerals, and texture [1]. *Tricholoma matsutake* is a fungus that belongs to the subgenus *Tricholoma*, and it is widely distributed in Asian countries, such as China, Korea, and Japan [2,3]. A large number of bioactive substances extracted from the fruiting bodies of *T. matsutake*, such as volatile compounds [4], polysaccharides [5], and polysaccharide–protein complex fractions [6,7], have been found to have immunomodulating and antioxidant properties [8,9]. *Tricholoma matsutake* is one of more precious and relatively expensive species throughout the world, exhibiting a characteristic and delicate flavor as well as several biological activities, such as sterol-lowering, anti-oxidant, immunomodulating, and anti-tumor effects in humans.

However, mushrooms perish rapidly and have a limited shelf-life of only 1 to 3 days at room temperature [10]. The high respiration rate, lack of physical protection to avoid water loss, and changes from microbial attack are often associated with a rapid decrease in mushroom quality [11]. An extended shelf-life is a key factor for making any food commodities more profitable and commercially available for longer periods of time at the best possible quality. The producer will benefit from a longer shelf-life by being able to sell the product in markets over greater distances [12]. There is a general trend in mushroom preservation research toward the development of preservation techniques that are practical and operable in order to reduce damaging to food products. These damages include mushroom loss of quality, contributing to their deterioration through browning, cap opening, stipe elongation, cap diameter increase, weight loss, and texture changes [13].

According to several authors, a coating could delay mushroom spoilage with minimum changes to the physiochemical and sensory quality of the mushrooms [14–16]. Edible coatings can maintain the quality of fruits and vegetables by functioning as solute, vapor, and gas barriers. These coatings could help decrease moisture loss and slow respiration by reducing fruit oxygen uptake from the environment [17]. Tiwari outlined the various types of animal and plant microorganisms isolated from antimicrobial applications in the food industry [18]. Edible coatings have been widely used to extend the shelf-life of various agricultural products [19]. Moreover, the product can be protected from microbial and mechanical damage, maintain a beautiful appearance, and prevent the escape of volatile substances [20]. Edible coatings derived from biodegradable and biomass can better meet the needs of customers and environmental package [21]. Coating preservation technology has focused on surface coating technology and the different types of coatings that are used to protect food from the effects of environmental factors, such as light, oxygen, and microorganisms [22]. Some summaries have been published on the effects of keeping perishables fresh by using coating technology, along with the shortcomings, future development trends, and applications. Research has found that the bacteriostatic action of antimicrobial peptides is achieved by destroying bacterial cell membranes and playing a role in antibacterial mechanisms, which is apparently not easy in the creation of medicines [23].

In recent years, antimicrobial peptides (AMPs) have been used as potential new antibiotics to solve the problem of resistance, acting as a kind of biological epidemic prevention system against exogenous pathogens of natural antibacterial peptides, which are small molecules composed of 12–100 amino acids [24,25]. Once in a target microbial membrane, the peptide kills target cells through diverse mechanisms [26]. Thus, antimicrobial peptides are widely applied in the fields of medicine, agriculture, aquaculture, and food industry because of their small molecular weight [27–29]. Although antibacterial peptides may have unprecedented application prospects for edible fungus such as precious wild fungus—as an antistaling agent in particular—antibacterial peptides have not yet been used in the preservation of wild fungus.

Nevertheless, due to its short half-life, AMPs can only be stable for 2.5 h in rat plasma [30]. Embedding it into a polymer microsphere material can protect it from the effects of the external environment and improve its half-life. Polylactic acid polymer microsphere materials play an important role in the field of drug-controlled release. In recent years, the structure has become more complex, and new materials have shown excellent performance. There is increasing interest in using PLA for food packaging applications. PLA is a biodegradable polymer that can be produced from annually renewable resources, such as sugar beet or corn starch [11]. Additionally, combinations of natural polymer materials with synthetic polymer materials integrate the advantages of both, with polypeptides, proteins, nucleic acids, and other new drugs employing controlled release applications.

In this work, the antimicrobial peptide microspheres and antimicrobial peptides as preservatives were carried out for the preservation experiments of Tricholoma matsutake. The purpose is to test the fresh-keeping effect of antimicrobial peptides and whether the form of microspheres can maintain the fresh-keeping effect for a longer time. The weight loss and firmness measurement were used to characterize the macroscopic properties of mushrooms. Changes were detected in the internal components of mushrooms, including their chemical properties and enzymes.

2. Materials and Methods

2.1. Materials

Tricholoma matsutake, as wild edible mushrooms, were harvested from a local grove in Kunming (Kunming, China). The mushrooms were transported to the laboratory immediately after harvest. Antibacterial peptides included cathelicidin-BF-30 (Sequence:

KFFRKLKKSVKKRAKEFFKKPRVIGVSIPF, W = 3638.57 Da, peptide purity = 95.74%), which was purchased from China Peptides Co., Ltd. (ShangHai, China). Cathelicidin-BF-30 was found to exert broad antimicrobial activity against bacteria and to exhibit excellent inhibitory activity [31]. Wang [32] reported that cathelicidin-BF-30 could be an excellent therapeutic agent for acne vulgaris and that cathelicidin-BF-30 is harmless to the human body. Li [33] also showed that AMS had a proliferation effect on the cells and that the microspheres were not toxic.

2.2. The Preparation of the Antimicrobial Peptide Microspheres

The method for making the antimicrobial peptide microspheres was an emulsified curing method. First, 0.5 g poly(lactic-co-glycolic acid) (PLGA; 75:25) was added into 5 mL dichloromethane (DCM), and then, 50 mg of Tween80 was also added in an ice bath at a rate of 0.5 mL/min with rapid dispersion of 1 min. To obtain the colostrum, the material quickly was transferred into the solution with 20 mL polyvinyl alcohol (PVA, 5% g/mL) and continued rapid dispersion for 4 min. The emulsion with a 5% isopropyl alcohol solution (400 mL) was mixed at low speed for 3 h. At 7000 RPM and 4 °C, the mixture was centrifuged for 10 min; afterwards, the microspheres were collected and washed three times with pure water, and then, they were employed in rapid lyophilization at 55 °C.

2.3. Scanning Electron Microscopy Measurements

Scanning electron microscopy (Hitachi, S-3000 N, Tokyo, Japan) was used to analyze the microsphere morphology. A small amount of solution was collected before centrifugation and diluted with pure water, and the particle size distribution was measured with a laser particle size analyzer. The average value of each sample was measured 3 times.

2.4. Pretreatment and Preservation of Mushrooms

Each sample was sorted according to the shape and maturity of the mushrooms. Damaged or rotten mushrooms were removed. Then, the *Tricholoma matsutake* mushrooms were carefully cleaned by hand to remove mud and pine needles from the surface. The mushrooms were randomly divided into four different treatments. They were dipped into different preservatives (1.0 g/L and 0.5 g/L AMS and 1.0 g/L AMPs for 2 min at 20 °C) with a dip in distilled water used as a control. A fan generating a low-speed air flow was used to dry the *Tricholoma matsutake* mushrooms. The *Tricholoma matsutake* mushrooms were placed on a crisper (115 × 115 × 45) that was constructed from low-density polyethylene (LDPE) film, which was punched with four holes (Φ = 7 mm) and stored at 4 ± 1 °C for 16 days. The quality of the mushrooms was determined initially and after 4, 8, 12, 16, and 20 days. Three replicates from each group were randomly selected and sampled as described below.

2.5. Weight Loss

Weight loss was determined by the weight of the mushroom before and after storage. It was expressed as the percentage of weight loss with respect to the initial weight. Weight loss was determined gravimetrically.

2.6. Firmness Measurement

The firmness of the mushrooms was measured with a texture analyzer (TA-XT, Stable Micro System Ltd., London, UK) using a 2 mm-diameter cylindrical probe. Samples were penetrated to a depth of 5 mm. The speed of the probe was 2.00 mm s^{-1} during both the pretest and penetration. From the force versus time curves, the firmness was defined as the maximum force (Newton N). There were, on average, 10 mushrooms in each package.

2.7. Analysis of Chemical Properties

The total sugars in the mushrooms were determined according to the methods described by Miller [34] and Dubois et al. [35]. The determination of total ascorbic acid was performed as described by Hanson et al. [36] based on coupling 2,4-dinitrophenylhydrazine (DNPH) with the ketonic groups of dehydroascorbic acid through the oxidation of ascorbic acid by 2,6-dichlorophenolindophenol (DCIP), providing a yellow/orange color in acidic conditions. Mushroom tissues (10 g) were blended with 80 mL of 5% meta-phosphoric acid in a homogenizer and centrifuged. After centrifuging, 2 mL of the supernatant were poured into a 20 mL test tube containing 0.1 mL of 0.2% 2,6-DCIP sodium salt in water, 2 mL of 2% thiourea in 5% meta-phosphoric acid, and 1 mL of 4% 2,4-DNPH in 9 N sulfuric acid. The mixtures were kept in a water bath at 37 °C for 3 h and then placed in an ice bath for 10 min. Then, 5 mL of 85% sulfuric acid was added, and the mixtures were kept at room temperature for 30 min before being read at 520 nm.

2.8. Enzyme Assays

To analyze the enzyme activities, mushroom tissues (5.0 g) were removed using a sharp knife and then homogenized with 20 mL of 0.5 M sodium phosphate buffer (pH 6.5) containing 20 g/L of polyvinylpyrrolidone to restrict oxidation in the samples. After centrifugation for 20 min at 8000 rpm and 4 °C, the supernatant was collected and used as a crude enzyme extract for the polyphenol oxidase (PPO) and peroxidase (POD) assays.

The PPO activity was assessed using the oxidation of p-phenylendiamine by catechol as a substrate [37]. PPO was measured immediately after the extraction to avoid degradation of the enzymes. The absorbance was measured at 420 nm with a UV-vis spectrophotometer (T90, Beijing Purkinje General Instrument Co. Ltd., Beijing, China). One unit of PPO activity was defined as the amount of enzyme that caused a 0.001 increase in absorbance per minute in 1 mL in the reaction mixture.

The POD activity was measured based on the spectrophotometric measurement by using a modified method based on those described by Gao [38]. The reaction mixture for the determination of POD activity consisted of 50 mM sodium phosphate buffer (pH 6.0), 5 mM guaiacol, 5 mM H_2O_2, and 50 μL of tissue extract. One unit of POD activity was defined as the amount of enzyme that caused a change of 0.01 per minute at an absorbance at 470 nm under the specified conditions. The specific PPO and POD activities were expressed as Umg^{-1} protein. Measurements were replicated three times. The PPO and POD activities of the samples were calculated using the following formula:

$$\mathrm{E} = \frac{\Delta OD \times V_T}{0.01 \times V_1 \times t \times F_W}$$

V_T: The total volume of the sample solution (mL).
V_1: Test sample volume (mL).
F_W : Weight of the sample (g).

2.9. Microbiological Analysis

Methods for the determination of microorganisms were reported by Simón and Gonzalez-Fandos [39]. All samples were analyzed for the mesophilic and psychrophilic bacteria counts. Twenty-five grams of mushrooms was removed aseptically from the package, weighed, and homogenized in a sterile stomacher bag for 2 min, with 225 mL of 0.1% peptone water. Further decimal dilutions were made with the same diluent. Aerobic counts were determined on plate count agar. The plates were incubated at 37 °C for 2 days for mesophilic bacteria and at 4 °C for 7 days for psychrophilic bacteria.

2.10. Statistical Analysis

All measurements were replicated three times, and the mean values ± standard deviations were reported for each case. An analysis of variance (ANOVA) was performed using a SPSS statistical computer software package (SPSS version 13.0). The significance of differences between mean values was assessed using the Duncan's multiple range test at a significance level of $p < 0.05$.

3. Results and Discussion

3.1. The Morphology and Particle Size of Antimicrobial Peptide Microspheres

The experimental microsphere production rate was 78.6%, which was close to 80%, and the yield can be considered relatively high. Figure 1 shows the SEM images of freeze-dried microspheres; the microsphere surface was smooth, there were fewer adhesions, and the particle size distribution was not uniform, but this is typical of a complex emulsion/solvent extraction of microspheres. Only through the membrane emulsification method could the microsphere particle size be made uniform.

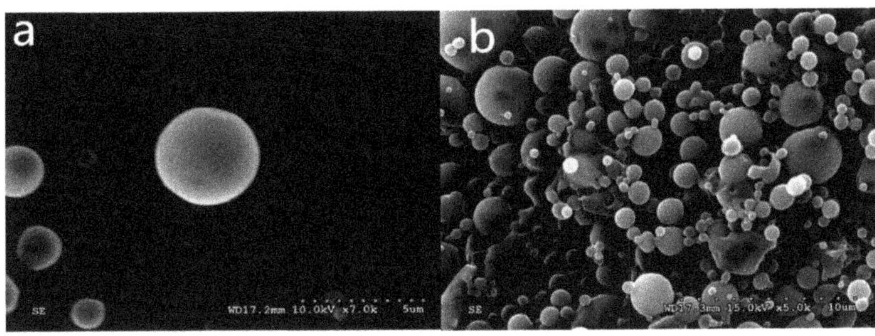

Figure 1. The antimicrobial peptide-PLGA microspheres SEM results; (**a**,**b**) correspond to different amplification lyophilized microspheres, multiples of (**a**) ×7000, (**b**) ×5000.

The particle size distribution range was 0.5–3 μm, and the polydispersity was 0.239 (Figure 2). Most of the particle size distribution was in the range of 800–1400 nm. A monodisperse system can make the dose more reliable and more accurate. The carrier of a particle size less than 1 micron can be avoided by capillary blockage and filtration, which can be effectively absorbed by the cells [40]. Such a particle size distribution is suitable for drug delivery systems [41].

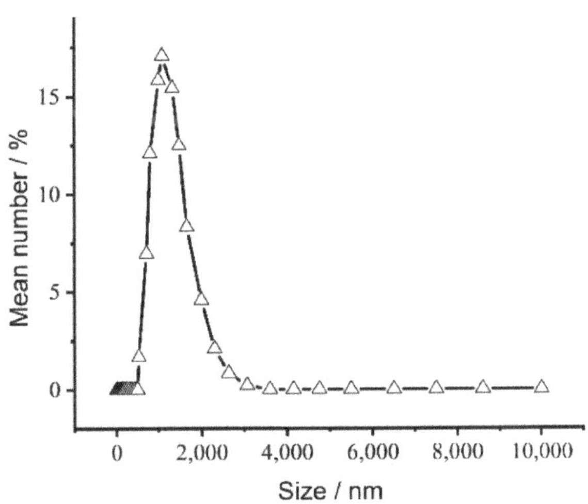

Figure 2. Particle size distribution of the freeze-dried microspheres.

3.2. Weight Loss

All samples lost weight over the 20-day storage period (Figure 3). This loss increased as the storage period progressed for all treatments. After 20 days of storage, the mushrooms that were coated with AMS had 2.72% and 2.93% weight loss, respectively, compared with the 3.34% and 3.25% weight loss values observed in the control and antimicrobial peptide groups. The weight losses were all less than 3.5% during storage. However, after 4 d of storage, the weight loss was relatively higher for the control samples than for those with an antimicrobial peptide microsphere coating (0.5 g/L and 1 g/L). Mushroom weight loss is mainly caused by water transpiration and CO_2 loss during respiration. The thin skin of shiitake mushrooms makes them susceptible to rapid water loss, resulting in shriveling and deterioration [42]. Mushrooms can endure surface shrinkage. Additionally, there can be fresh mushroom loss due to transpiration, which decreases the weight and quality and affects the normal physiological processes, as occurs with increased enzyme activity and respiration intensity with decreased resistance to disease.

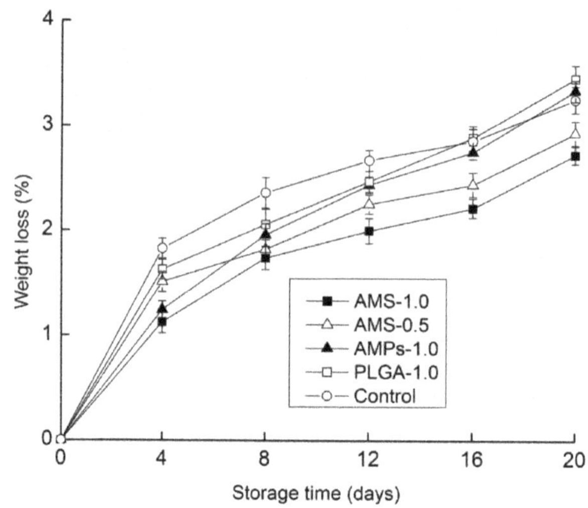

Figure 3. Effect of different antimicrobial coating on weight loss changes of *Tricholoma matsutake* stored at 4 ± 1 °C for 20 days. Each data point is the mean of three replicate samples.

3.3. Firmness Measurement

Change in firmness is considered one of the chief problems in the postharvest deterioration of mushrooms [43]. Mushrooms suffer a rapid loss of firmness during senescence, which strongly contributes to its short postharvest life and susceptibility to fungal contamination. The texture of *Tricholoma matsutake* is often the first of many quality attributes that the consumer judges; it is extremely important in overall product acceptance. Mushrooms suffer a rapid loss of firmness during senescence, which strongly contributes to its short postharvest life and susceptibility to fungal contamination. From a quality perspective, the texture of *Tricholoma matsutake* is an important parameter that the consumer judges. At harvest, mushrooms are firm, crisp, and tender. However, they soften during postharvest deterioration [44]. During the process of storage, the preservative effect on *Tricholoma matsutake* hardness is shown in Figure 4. At the start of storage, the firmness increases slightly, which is likely from mushroom water loss. Subsequently, the hardness begins to decline, and softening can occur because of the degradation of cell walls in postharvest mushrooms in response to bacterial enzymes and the increased activity of endogenous autolysins [45]. Microorganisms, such as pseudomonas, degrade mushrooms by breaking down the intracellular matrix and reducing the central vacuole, resulting in partially collapsed cells and a loss of turgor. This type of bacterial-induced softening was observed in control samples, but it was inhibited by antimicrobial peptide coating treatments. The maintenance of firmness in the mushrooms that were treated with antimicrobial peptide coatings could be due to their higher antifungal activity or to the covering of the cuticle and lenticels, which reduced infection, respiration, and other senescence processes during storage. The loss of moisture during storage was responsible for firmness loss. With respect to firmness on the eighth day of storage, the control group and PLA-1.0 group had obvious reduced firmness compared with the other three groups. In 20 days, the hardness of the mushrooms with AMS-1.0 as a preservative decreased more slowly than did that of the other four groups; therefore, AMS-1.0 can be very effective in slowing the process of mushrooms becoming soft.

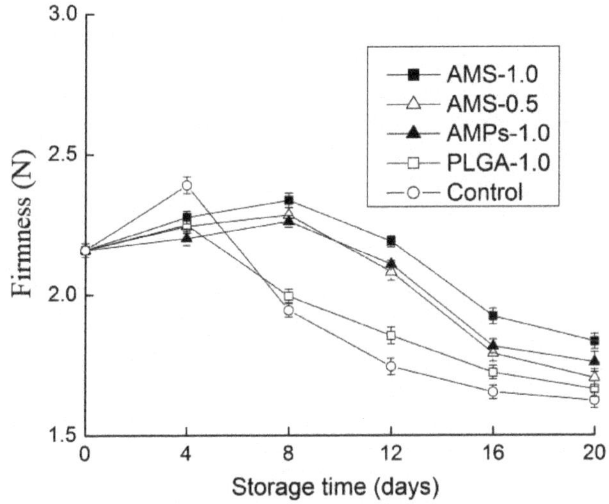

Figure 4. Effect of different antimicrobial coating on texture changes of *Tricholoma matsutake* stored at 4 ± 1 °C for 20 days. Each data point is the mean of three replicate samples.

3.4. Analysis of Chemical Properties

As one of the nutritional indexes, total sugar can reflect the process of storage and quality changes in *Tricholoma matsutake*. Total soluble sugars are not only the main photosynthates in higher plants, but they are also the main form of carbohydrate metabolism and temporary storage. Five freshness modes of *Tricholoma matsutake* storage in the process of changing the total sugar content are analyzed. With early storage, the total sugar content increased (Figure 5). The increase in total sugar content in the observed samples may be due to the high respiration rate and maturation of mushrooms during storage, and it is consistent with the existing research results [15]. The total sugar levels in the antimicrobial peptide microsphere-coated (1.0 g/L) samples increased at a significantly different rate ($p < 0.05$) compared with the other three groups over the first 12 d period, resulting in a 109.8% higher than initial concentration at the termination of the experiment. The highest total sugar content was 16.54 mg/g. Total sugar began to decline in the control group and PLGA-1.0 group on the eighth day, whereas the total sugar content in the remaining three groups decreased. During storage, the continuous respiration of Tricholoma matsutake consumed microbial nutrients, resulting in the decrease in total sugar content. In the period from day 12 to day 20, the total sugar contents of the control group and PLGA-1.0 group were significantly lower than those of the other three groups.

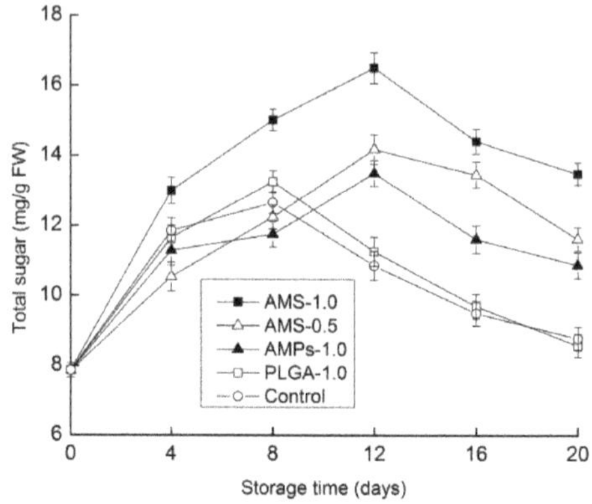

Figure 5. Effect of different antimicrobial coatings on total sugar changes of *Tricholoma matsutake* stored at 4 ± 1 °C for 20 days. Each data point is the mean of three replicate samples.

Figure 6 shows the ascorbic acid level changes in *Tricholoma matsutake* after 20 days of storage. Even for *Tricholoma matsutake* mushrooms that were treated with both AMPs and AMS, the ascorbic acid levels decreased. The starting value for ascorbic acid was 42.9 mg/Kg, and it significantly decreased by the fourth day. On the eighth day, the control and PLGA-1.0 groups had somewhat lower ascorbic acid levels, although these levels were not significantly lower than those of the other three groups. These findings were in agreement with a study by Ayranci and Tunc [46]. Storage temperature was the main factor affecting the fluctuation of ascorbic acid content, because the temperature directly affects the enzyme activity. Keeping the appropriate low temperature in the storage environment of Tricholoma matsutake was conducive to the preservation of ascorbic acid.

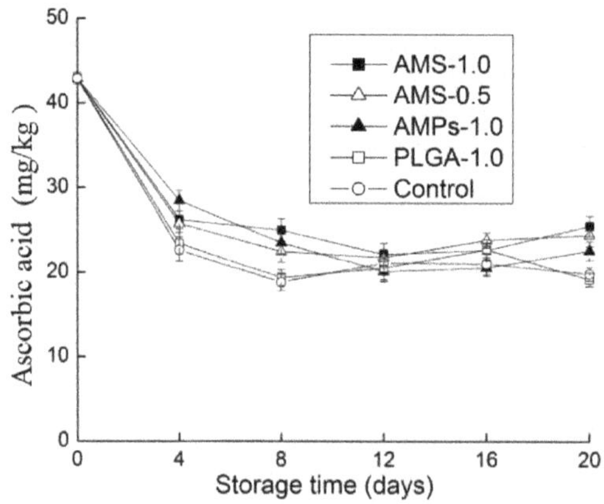

Figure 6. Effect of different antimicrobial coatings on ascorbic acid changes of *Tricholoma matsutake* stored at 4 ± 1 °C for 20 days. Each data point is the mean of three replicate samples.

3.5. Enzyme Assays

It is generally accepted that browning is due to the oxidation of phenolics and is caused by PPO and POD, resulting in the formation of brown-colored substances. This may be due to the oxidation of edible fungus polyphenols in the body into quinone substances, which form through a series of reactions, i.e., the Browning reactions. Figure 7a shows the influence of different preservative agents on the *Tricholoma matsutake* PPO activity. With an increase in storage time, the preservation of *Tricholoma matsutake* PPO activity had a unimodal curve change in this way. The different preservative agents for *Tricholoma matsutake* had different effects on the PPO activity. The PPO activity of both the control group and PLGA-1.0 group in the first eight days increased more rapidly, peaking within 12 days, whereas the other three groups peaked on the 16th day. The hydrogen POD and PPO activity had a similar trend as the storage time increased. Within 12 days, the POD activity of the four groups peaked (Figure 7b). When the concentration of AMS was 1.0 g/L for *Tricholoma matsutake*, the PPO activity and POD activity were significantly lower than those observed in the other three groups. It was responsible for the oxidation of phenolic compounds and the formation of brown-colored melanins, preventing the formation of brown patches and improving the appearance and color of the mushrooms [47]. At the same time, the AMS could improve the effect of the enzyme. These results suggested that the AMS coating treatment was more effective in retarding mushroom sensory deterioration.

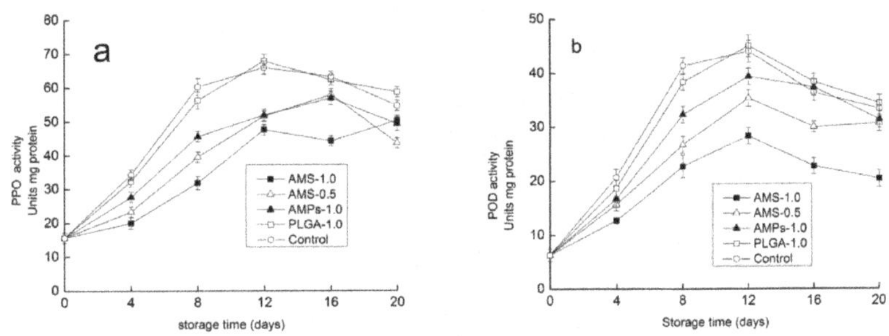

Figure 7. Effect of different antimicrobial coatings on PPO activity (**a**) and POD activity (**b**) changes of *Tricholoma matsutake* stored at 4 ± 1 °C for 20 days. Each data point is the mean of three replicate samples.

3.6. Microbiological Analysis

The changes in the total aerobic bacteria counts for *Tricholoma matsutake* during storage are shown in Figure 8. In all groups, the total levels of mesophilic and psychrophilic bacteria counts steadily increased throughout the storage period. Mesophilic and psychrophilic bacteria predominated during storage in all of the analyzed samples. At harvest, the mushrooms had 1.51 and 3.94 $\log_{10}$cfug^{-1} for psychrophilic and mesophilic bacteria, respectively. These levels were higher than those reported by Cliffe-Byrnes and O'Beirne [48] for water-rinsed white button mushrooms and by Jiang et al. [49] for shiitake mushrooms. This was probably because *Tricholoma matsutake* mushrooms were harvested from a local grove and mud, and many pine needles had adhered to the stems and caps of the wild mushrooms. Following 20 days of cold storage, the microbial counts in the antimicrobial peptide and antimicrobial peptide microsphere treatment groups were increased throughout the storage period. However, their growth was slower that of the controls. AMS coating was more effective in reducing the microbial counts than other treatments. For all groups, the psychrophilic bacteria counts increased by less than two orders throughout the storage period. Mushrooms from the control group and PLGA-1.0 group had tiny brown spots on day 12 that developed into dark zones, which are characteristic of pseudomonas spoilage, by day 16. The mushrooms were highly decayed at this point, and the end of their shelf life was due to microbial spoilage. There were no differences between the AMS and the AMPs in their response to psychrophilic bacteria, but with a high concentration of AMS solution, the antimicrobial peptide preservation effect was much better for mesophilic bacteria. However, overall, the coating and control groups showed obvious differences in the number of psychrophilic bacteria. Thus, antimicrobial peptides might have stronger effects on psychrophilic bacteria.

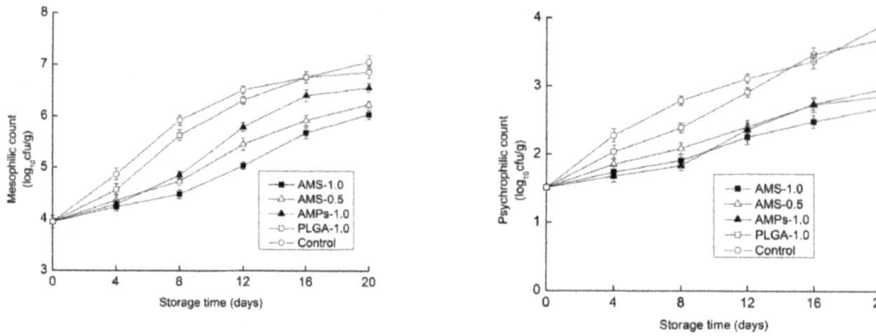

Figure 8. Effect of different antimicrobial coatings on mesophilic and psychrophilic count changes of *Tricholoma matsutake* stored at 4 ± 1 °C for 20 days.

3.7. Sensorial Analysis

The average values for the sensory attributes are listed in Table 1. All of the selected sensory attributes gradually decreased as the storage period advanced. This supported the validity of the chosen descriptors as indicators of mushroom deterioration.

Table 1. Effect of different treatments on the color of *Tricholoma matsutake* mushrooms stored at 4 ± 1 °C for 20 days.

Treatments	Appearance	Spoilage	Odor
0 day			
	9.93 ± 0.54	9.76 ± 0.47	10.0
4 days			
AMS-1.0	9.24 ± 0.47 ^{a}	9.42 ± 0.16 ^{a}	9.83 ± 0.19 ^{a}
AMS-0.5	9.32 ± 0.62 ^{a}	9.45 ± 0.53 ^{a}	9.82 ± 0.25 ^{a}
AMPS-1.0	9.20 ± 0.25 ^{a}	9.51 ± 0.67 ^{a}	9.89 ± 0.22 ^{a}
PLGA-1.0	9.12 ± 0.16 ^{a}	9.46 ± 0.25 ^{a}	9.85 ± 0.31 ^{a}
Control	9.07 ± 0.17 ^{a}	9.43 ± 0.55 ^{a}	9.74 ± 0.72 ^{a}
8 days			
AMS-1.0	8.63 ± 0.83 ^{a}	8.12 ± 0.82 ^{a}	8.79 ± 0.56 ^{a}
AMS-0.5	8.51 ± 0.74 ^{a}	8.07 ± 0.14 ^{a}	8.83 ± 0.49 ^{a}
AMPS-1.0	8.74 ± 0.42 ^{a}	8.19 ± 0.26 ^{a}	8.57 ± 0.31 ^{a}
PLGA-1.0	7.72 ± 0.37 ^{b}	7.45 ± 0.63 ^{b}	8.31 ± 0.31 ^{b}
Control	7.46 ± 0.26 ^{b}	7.24 ± 0.37 ^{b}	8.25 ± 0.89 ^{a}
12 days			
AMS-1.0	8.04 ± 0.33 ^{a}	7.24 ± 0.53 ^{a}	8.05 ± 0.63 ^{a}
AMS-0.5	7.87 ± 0.59 ^{a}	7.16 ± 0.33 ^{a}	8.13 ± 0.46 ^{a}
AMPS-1.0	7.93 ± 0.35 ^{a}	7.09 ± 0.74 ^{a}	7.24 ± 0.53 ^{a}
PLGA-1.0	6.52 ± 0.14 ^{b}	6.33 ± 0.44 ^{b}	6.57 ± 0.26 ^{b}
Control	6.05 ± 0.63 ^{b}	6.04 ± 0.92 ^{b}	6.43 ± 0.48 ^{b}
16 days			
AMS-1.0	7.33 ± 0.52 ^{a}	6.88 ± 0.74 ^{a}	7.67 ± 0.27 ^{a}
AMS-0.5	7.04 ± 0.71 ^{ab}	6.53 ± 0.81 ^{ab}	7.25 ± 0.48 ^{ab}
AMPS-1.0	6.82 ± 0.35 ^{b}	6.24 ± 0.42 ^{b}	6.53 ± 0.31 ^{b}
PLGA-1.0	5.02 ± 0.37 ^{c}	5.13 ± 0.05 ^{c}	4.93 ± 0.11 ^{c}
Control	4.87 ± 0.83 ^{c}	4.82 ± 0.22 ^{c}	4.86 ± 0.61 ^{c}
20 days			
AMS-1.0	6.81 ± 0.56 ^{a}	6.32 ± 0.63 ^{a}	6.83 ± 0.44 ^{a}
AMS-0.5	6.13 ± 0.68 ^{ab}	6.14 ± 0.58 ^{a}	6.47 ± 0.49 ^{a}
AMPS-1.0	5.67 ± 0.14 ^{b}	5.75 ± 0.15 ^{a}	5.52 ± 0.58 ^{b}
PLGA-1.0	3.24 ± 0.72 ^{c}	3.57 ± 0.63 ^{c}	3.38 ± 059 ^{c}
Control	3.13 ± 0.32 ^{c}	3.66 ± 0.44 ^{b}	3.14 ± 0.27 ^{c}

$^{a–c}$ Values followed by different letters in the same column are significantly different ($p < 0.05$), where a is the lowest value. Data are presented as mean ± standard deviation ($n = 3$).

There was no significant ($p > 0.05$) difference among the groups on day 0 of storage. The application of AMS did not influence the sensorial analysis of the samples compared with the PLGA-1.0 group and control group. A similar observation with no significant difference between the samples and controls was observed on day 4 of storage. The off-odor of mushrooms coated with AMS-1.0 and AMS-0.5 was significantly ($p < 0.05$) less than those coated with AMPs-1.0 after 16 days of storage. The growth of microorganisms may accelerate the decay of the mushroom. This phenomenon and the trend of microbial quantity was the same.

The appearance of mushrooms coated with antimicrobial peptides was significantly ($p < 0.05$) better than the control group and PLGA-1.0 group after 8 days of storage. Mushrooms in the control group and PLGA-1.0 group became unacceptable on day 16. The general acceptability of AMP-treated mushrooms fell below the limit of marketability at day 16. However, a better trend was observed for AMS-0.5 and AMS-1.0-treated mushrooms. AMS-0.5 and AMS-1.0-treated mushrooms were still acceptable in a marketable condition and received a score of 6 at the end of the storage time. The results suggested that the AMS-0.5 and AMS-1.0 treatments were more effective in retarding the deterioration of the general acceptability of mushrooms compared with the other test samples.

Fresh Tricholoma matsutake can be stored in the refrigerator freezer for 3–5 days. The experimental results showed that after fresh-keeping treatment, it was still acceptable for 8 days. To some extent, the purpose of prolonging the shelf life was achieved.

4. Conclusions

In this study, novel AMPs and AMS coating materials were applied for the preservation of *Tricholoma matsutake* mushrooms in storage at 4 ± 1 °C. The AMPs and AMS coatings had beneficial effects on the physicochemical and physiological quality of the mushrooms compared to the control treatment. The AMS coating had a significant effect on the texture, PPO activity, POD activity, total sugar, and microbiological analysis of *Tricholoma matsutake* wild edible mushrooms, although its effect on weight loss and ascorbic acid was limited. The result suggested that AMS coatings maintain the firmness of mushrooms and improve the postharvest quality during cold storage. These findings also suggested that AMS are promising as an edible coating for use in commercial postharvest applications to prolong the storage life of button mushrooms. Therefore, the AMS coating may provide an attractive alternative to improve the preservation qualities of *Tricholoma matsutake* during extended storage. The existing research results show that the long-term preservation of wild fungi in the form of microspheres has application prospects in order to improve the shelf-life of Tricholoma matsutake and other expensive mushroom varieties.

Author Contributions: This paper was accomplished based on the collaborative work of the authors. H.L. performed the experiments, analyzed the data, interpreted the experimental results, and wrote the paper. Y.F. performed the experiments and samples performance test. P.Z. performed samples performance test and data validation. M.Y. (Mingwei Yuan) and M.Y. (Minglong Yuan) supervised the entire research progress and contributed to the experimental design. All authors have read and agreed to the published version of the manuscript.

Funding: This work was supported by the National Natural Science Foundation of China (52163013, 81760644) and the Yunnan Science and Technology Project: Major Science and Technology Specialties of Biological Medicine (2018ZF008) and Yunnan Ten Thousand Talents Program-Special Program for Top Young Talents (YNWR-QNBJ-2019-094).

Institutional Review Board Statement: Not applicable.

Informed Consent Statement: Not applicable.

Data Availability Statement: The data presented in this study are available on request from the corresponding author.

Conflicts of Interest: The authors declare no conflict of interest.

References

1. Venturini, M.E.; Reyes, J.E.; Rivera, C.S.; Oria, R.; Blanco, D. Microbiological quality and safety of fresh cultivated and wild mushrooms commercialized in Spain. *Food Microbiol.* **2011**, *28*, 1492–1498. [CrossRef]
2. Bergius, N.; Danell, E. The Swedish matsutake (*Tricholoma nauseosum*, syn *T. matsutake*): Distribution, abundance and ecology. Scandinav. *J. For. Res.* **2000**, *15*, 318–325. [CrossRef]
3. Guerin-Laguette, A.; Vaario, L.M.; Gill, W.M.; Lapeyrie, F.; Matsushita, N.; Suzuki, K. Rapid in vitro ectomycorrhizal infection on Pinus densiflora roots by Tricholoma matsutake. *Mycoscience* **2000**, *41*, 389–393. [CrossRef]
4. Cho, I.H.; Choi, H.K.; Kim, Y.S. Different in the volatile composition of pine-mushrooms (*Tricholoma matsutake Sing*) according to their grades. *J. Agric. Food Chem.* **2006**, *54*, 4820–4825. [CrossRef] [PubMed]
5. Ding, X.; Tang, J.; Cao, M.; Guo, C.X.; Zhang, X.; Zhong, J.; Zhang, J.; Sun, Q.; Feng, S.; Yang, Z.R.; et al. Structure elucidation and antioxidant activity of a novel polysaccharide isolated from Tricholoma matsutake. *Int. J. Biol. Macromol.* **2010**, *47*, 271–275. [CrossRef] [PubMed]
6. Mau, J.L.; Lin, H.C.; Song, S.F. Antioxidant properties of several specialty mushrooms. *Food Res. Int.* **2002**, *35*, 519–526. [CrossRef]
7. Hoshi, H.; Yagi, Y.; Iijima, H.; Matsunaga, K.; Ishihara, Y.; Yasuhara, T. Isolation and characterization of a novel immunomodulatory a-glucan–protein complex from the mycelium of Tricholoma matsutake in basidiomycetes. *J. Agric. Food Chem.* **2005**, *53*, 8948–8956. [CrossRef]
8. Liu, P. Anticancer and mechanism research of glucoprotein MTSGS1 from Tricholoma matsutake hyphal body. *Med. Biotech.* **2001**, *8*, 284–287.
9. Ebina, T.; Kubota, T.; Ogamo, N.; Matsunaga, K. Antitumor effect of a peptide–glucan preparation extracted from a mycelium of Tricholoma matsutake (S. Ito and Imai) Sing. *Biotherapy* **2002**, *16*, 255–259.
10. Fernandes, Â.; Antonio, A.L.; Barreira, J.C.; Botelho, M.L.; Oliveira, M.B.P.; Martins, A.; Ferreira, I.C.F.R. Effects of gamma irradiation on the chemical composition and antioxidant activity of *Lactarius deliciosus* L. wild edible mushroom. *Food Bioprocess Technol.* **2013**, *6*, 2895–2903. [CrossRef]
11. Han, L.L.; Qin, Y.Y.; Liu, D.; Chen, H.Y.; Li, H.L.; Yuan, M.L. Evaluation of biodegradable film packaging to improve the shelf-life of Boletus edulis wild edible mushrooms. *Innov. Food Sci. Emerg.* **2015**, *29*, 288–294. [CrossRef]
12. Akram, K.; Kwon, J.H. Food Irradiation for Mushrooms: A Review. *J. Korean Soc. Appl. Biol. Chem.* **2010**, *53*, 257–265. [CrossRef]
13. Fernandes, Â.; Antonio, A.L.; Barreira, J.; Oliveira, M.B.P.P.; Martins, A.; Ferreira, I.C.F.R. Effects of gamma irradiation on physical parameters of Lactarius deliciosus wild edible mushrooms. *Postharvest Biol. Technol.* **2012**, *74*, 79–84. [CrossRef]
14. Jiang, T.J.; Feng, L.F.; Li, J.R. Changes in microbial and postharvest quality of shiitake mushroom (*Lentinus edodes*) treated with chitosan–glucose complex coating under cold storage. *Food Chem.* **2012**, *131*, 780–786. [CrossRef]
15. Jafri, M.; Jha, A.; Bunkar, D.S.; Ram, R.C. Quality retention of oyster mushrooms (*Pleurotus florida*) by a combination of chemical treatments and modified atmosphere packaging. *Postharvest Biol. Technol.* **2013**, *76*, 112–118. [CrossRef]
16. Chien, P.J.; Sheu, F.; Lin, H.R. Coating citrus (*Murcott tangor*) fruit with low molecular weight chitosan increases postharvest quality and shelf life. *Food Chem.* **2007**, *100*, 1160–1164. [CrossRef]
17. Ojagh, S.M.; Rezaei, M.; Razavi, S.H.; Hosseini, S.M.H. Effect of chitosan coatings enriched with cinnamon oil on the quality of refrigerated rainbow trout. *Food Chem.* **2010**, *120*, 193–198. [CrossRef]
18. Jiang, T.; Feng, L.; Zheng, X.; Li, J. Physicochemical responses and microbial characteristics of shiitake mushroom (*Lentinus edodes*) to gum arabic coating enriched with natamycin during storage. *Food Chem.* **2013**, *138*, 1992–1997. [CrossRef] [PubMed]
19. Baldwin, E.A.; Nisperos, M.O.; Chen, X.; Hagenmaier, R.D. Improving storage life of cut apples and potato with ediblecoating. *Postharvest Biol. Technol.* **1996**, *9*, 151–163. [CrossRef]
20. Hadar, A.; Rina, G.; Ron, P.; Elena, P. Development of polysaccharides-based edible coatings for citrus fruits: A layer-by-layer approach. *Food Chem.* **2015**, *166*, 465–472.
21. Dhall, R.K. Advances in edible coatings for fresh fruits and vegetables: A review. *Crit. Rev. Food Sci.* **2013**, *53*, 435–450. [CrossRef]
22. Chawengkijwanich, C.; Kopermsub, P. Coating technology for food preservation. *Prog. Food Preserv.* **2012**, *16*, 111–127.
23. Zasloff, M. Antimicrobial peptides of multicellular organisms. *Nature* **2002**, *415*, 389–395. [CrossRef] [PubMed]
24. Dubin, A.; Mak, P.; Dubin, G.; Rzychon, M.; Stec-Niemczyk, J.; Wladyka, B.; Maziarka, K.; Chmiel, D. New generation of peptide antibiotics. *Acta Biochim. Pol.* **2005**, *52*, 633–638. [CrossRef]
25. Yang, D.; Chertov, O.; Oppenheim, J.J. The role of mammalian antimicrobial peptides and proteins in awakening of innate host defenses and adaptive immunity. *Cell. Mol. Life Sci.* **2001**, *58*, 978–989. [CrossRef] [PubMed]
26. Boman, H.G. Peptide antibiotics and their role in innate immunity. *Annu. Rev. Immunol.* **1995**, *13*, 61–92. [CrossRef] [PubMed]
27. Jung, D.; Powers, J.P.; Straus, S.K.; Hancock, R.E. Lipid-specific binding of the calcium-dependent antibiotic daptomycin leads to changes in lipid polymorphism of model membranes. *Chem. Phys. Lipids* **2008**, *154*, 120–128. [CrossRef] [PubMed]
28. Sallum, U.W.; Chen, T.T. Inducible resistance of fish bacterial pathogens to the antimicrobial peptide cecropin B. *Antimicrob. Agents Chemother.* **2008**, *52*, 3006–3012. [CrossRef] [PubMed]
29. Sudagidan, M.; Yemenicioglu, A. Effects of nisin and lysozyme on growth inhibition and biofilm formation capacity of Staphylococcus aureus strains isolated from raw milk and cheese samples. *J. Food Prot.* **2012**, *75*, 1627–1633. [CrossRef]
30. Wang, Y.P.; Hong, J.; Liu, X.H.; Yang, H.L.; Liu, R.; Wu, J.; Wang, A.L.; Lin, D.H.; Lai, R. Snake Cathelicidin from Bungarus fasciatus Is a Potent Peptide Antibiotics. *PLoS ONE* **2008**, *3*, e3217. [CrossRef]

31. Tian, Y.W.; Wang, H.; Li, B.; Ke, M.Y.; Wang, J.; Dou, J.; Zhou, G.L. The cathelicidin-BF Lys^{16} mutant Cbf-K_{16} selectively inhibits non-small cell lung cancer proliferation in vitro. *Oncol. Rep.* **2013**, *30*, 2502–2510. [CrossRef]
32. Wang, Y.P.; Zhang, Z.Y.; Chen, L.L.; Guang, H.J.; Li, Z.; Yang, H.L.; Li, J.X.; You, D.W.; Yu, H.N.; Lai, R. Cathelicidin-BF, a Snake Cathelicidin-Derived Antimicrobial Peptide, Could Be an Excellent Therapeutic Agent for Acne Vulgaris. *PLoS ONE* **2011**, *6*, e22120. [CrossRef]
33. Li, L.L.; Wang, Q.F.; Li, H.L.; Yuan, M.W.; Yuan, M.L. Preparation, Characterization, In Vitro Release and Degradation of Cathelicidin-BF-30-PLGA Microspheres. *PLoS ONE* **2014**, *9*, e100809. [CrossRef]
34. Mille, G.L. Use of dinitrosalicylic acid reagent for determination of reducing sugar. *Anal. Biochem.* **1959**, *31*, 426–428.
35. Dubois, M.; Gilles, K.A.; Hamilton, J.K. Colorimetric method for determination of sugar and related substances. *Anal. Biochem.* **1956**, *28*, 350–356. [CrossRef]
36. Hanson, P.M.; Yang, R.Y.; Wu, J.; Chen, J.T.; Ledesma, D.; Tsou, C.S.C.; Lee, T.C. Variation for antioxidant activity and antioxidants in tomato. *J. Am. Soc. Hortic. Sci.* **2004**, *129*, 704–711. [CrossRef]
37. Mohapatra, D.; Frias, J.M.; Oliveira, F.A.R.; Bira, Z.M.; Kerry, J. Development and validation of a model to predict enzymatic activity during storage of cultivated mushrooms (*Agaricus bisporus* spp.). *J. Food Eng.* **2008**, *86*, 39–48. [CrossRef]
38. Gao, M.; Feng, L.; Jiang, T. Browning inhibition and quality preservation of button mushroom (*Agaricus bisporus*) by essential oils fumigation treatment. *Food Chem.* **2014**, *149*, 107–113. [CrossRef] [PubMed]
39. Simón, A.; Gonzalez-Fandos, E. Influence of modified atmosphere packaging and storage temperature on the sensory and microbiological quality of fresh peeled white asparagus. *Food Control* **2011**, *22*, 369–374. [CrossRef]
40. Brigger, I.; Dubernet, C.; Couvreur, P. Nanoparticles in cancer therapy and diagnosis. *Adv. Drug Deliv. Rev.* **2002**, *54*, 631–651. [CrossRef]
41. Zhang, L.; Jeong, Y.I.; Zheng, S.; Suh, H.; Kang, D.H.; Kim, I. Fabrication of Microspheres via Solvent Volatization Induced Aggregation of Self-Assembled Nanomicellar Structures and Their Use as a pH-Dependent Drug Release System. *Langmuir* **2012**, *29*, 65–74. [CrossRef]
42. Simón, A.; González-Fandos, E.; Vázquez, M. Effect of washing with citric acid and packaging in modified atmosphere on the sensory and microbiological quality of sliced mushrooms (*Agaricus bisporus* L.). *Food Control* **2010**, *21*, 851–856. [CrossRef]
43. Khan, Z.U.; Aisikaer, G.; Khan, R.U.; Bu, J.; Jiang, Z.; Ni, Z.; Ying, T. Effects of composite chemical pretreatment on maintaining quality in button mushrooms (*Agaricus bisporus*) during postharvest storage. *Postharvest Biol. Technol.* **2014**, *95*, 36–41. [CrossRef]
44. Oliveira, F.; Sousa-Gallagher, M.J.; Mahajan, P.V.; Teixeira, J.A. Development of shelf-life kinetic model for modified atmosphere packaging of fresh sliced mushrooms. *J. Food Eng.* **2012**, *111*, 466–473. [CrossRef]
45. Zivanovic, S.; Buescher, R.W.; Kim, K.S. Textural changes in mushroom (*Agaricus bisporus*) associated tissue ultrastructure and composition. *J. Food Sci.* **2000**, *65*, 1404–1408. [CrossRef]
46. Ayranci, E.; Tunc, S. A method for the measurement of the oxygen permeability and the development of edible films to reduce the rate of oxidative reactions in fresh foods. *Food Chem.* **2003**, *80*, 423–431. [CrossRef]
47. Srivastava, A.; Singh, A.; Balaji Raja, R.; Arunachalam, D.K. Shelf-life extension of fresh Mushrooms (*Agaricus bisporus*) by application of Tomato paste. *Int. J. Eng. Sci. Technol.* **2010**, *2*, 783–786.
48. Cliffe-Byrnes, V.; O'Beirne, D. Effects of washing treatment on microbial and sensory quality of modified atmosphere (MA) packaged fresh sliced mushroom (*Agaricus bisporus*). *Postharvest Biol. Technol.* **2008**, *48*, 283–294. [CrossRef]
49. Jiang, T.J. Effect of alginate coating on physicochemical and sensory qualities of button mushrooms (*Agaricus bisporus*) under a high oxygen modified atmosphere. *Postharvest Biol. Technol.* **2013**, *76*, 91–97. [CrossRef]

polymers

MDPI

Article

Preparation and Characterization of an Injectable and Photo-Responsive Chitosan Methacrylate/Graphene Oxide Hydrogel: Potential Applications in Bone Tissue Adhesion and Repair

Daniela N. Céspedes-Valenzuela [1,†], Santiago Sánchez-Rentería [1,†], Javier Cifuentes [1], Mónica Gantiva-Diaz [1,2], Julian A. Serna [1], Luis H. Reyes [3], Carlos Ostos [4], Christian Cifuentes-De la Portilla [2], Carolina Muñoz-Camargo [1,*] and Juan C. Cruz [1,*]

[1] Grupo de Investigación en Nanobiomateriales, Ingeniería Celular y Bioimpresión (GINIB), Department of Biomedical Engineering, Universidad de los Andes, Bogota 111711, Colombia; dn.cespedes@uniandes.edu.co (D.N.C.-V.); s.sanchezr2@uniandes.edu.co (S.S.-R.); jf.cifuentes10@uniandes.edu.co (J.C.); mr.gantiva@uniandes.edu.co (M.G.-D.); ja.serna10@uniandes.edu.co (J.A.S.)
[2] Grupo de Investigación en Biomecánica (IBIOMECH), Department of Biomedical Engineering, Universidad de los Andes, Bogota 111711, Colombia; cc.cifuentes@uniandes.edu.co
[3] Department of Chemical and Food Engineering, School of Engineering, Universidad de Los Andes, Carrera 1 No. 18A-12, Bogota 111711, Colombia; lh.reyes@uniandes.edu.co
[4] Grupo CATALAD, Instituto de Química, Universidad de Antioquia, Medellin 050010, Colombia; carlos.ostos@udea.edu.co
* Correspondence: c.munoz2016@uniandes.edu.co (C.M.-C.); jc.cruz@uniandes.edu.co (J.C.C.); Tel.: +57-13-394-949 (ext. 1789) (J.C.C.)
† These authors contributed equally to this work.

Citation: Céspedes-Valenzuela, D.N.; Sánchez-Rentería, S.; Cifuentes, J.; Gantiva-Diaz, M.; Serna, J.A.; Reyes, L.H.; Ostos, C.; Cifuentes-De la Portilla, C.; Muñoz-Camargo, C.; Cruz, J.C. Preparation and Characterization of an Injectable and Photo-Responsive Chitosan Methacrylate/Graphene Oxide Hydrogel: Potential Applications in Bone Tissue Adhesion and Repair. *Polymers* **2022**, *14*, 126. https://doi.org/10.3390/polym14010126

Academic Editors: Antonio M. Borrero-López, Concepción Valencia-Barragán, Esperanza Cortés Triviño, Adrián Tenorio-Alfonso and Clara Delgado-Sánchez

Received: 30 November 2021
Accepted: 16 December 2021
Published: 30 December 2021

Abstract: As life expectancy continues to increase, the inevitable weakening and rupture of bone tissue have grown as concerns in the medical community, thus leading to the need for adhesive materials suitable for bone repair applications. However, current commercially available adhesives face certain drawbacks that prevent proper tissue repair, such as low biocompatibility, poor adhesion to wet surfaces, and the need for high polymerization temperatures. This work aims to develop an injectable and photo-responsive chitosan methacrylate/graphene oxide (ChiMA/GO) adhesive nanocomposite hydrogel of high biocompatibility that is easy to apply by simple extrusion and that offers the possibility for in situ polymer and physiological temperatures. The nanocomposite was thoroughly characterized spectroscopically, microscopically, rheologically, thermally, and through mechanical, textural, and biological assays to fully evaluate its correct synthesis and functionalization and its performance under physiological conditions that mimic those observed in vivo. In addition, a finite element analysis (FEA) simulation was used to evaluate its performance in femur fractures. Results suggest the material's potential as a bioadhesive, as it can polymerize at room temperature, shows superior stability in physiological media, and is capable of withstanding loads from body weight and movement. Moreover, the material showed remarkable biocompatibility as evidenced by low hemolytic and intermediate platelet aggregation tendencies, and high cytocompatibility when in contact with osteoblasts. The comprehensive studies presented here strongly suggest that the developed hydrogels are promising alternatives to conventional bone adhesives that might be further tested in vivo in the near future.

Keywords: bioadhesive; chitosan methacrylate; graphene oxide; bone repair; photocrosslinking

1. Introduction

Comminuted fractures are complex bone lesions that involve the break or splinter of a bone into multiple fragments, usually after being subjected to loads far superior to those it can resist [1]. These injuries require a surgical treatment to allow the bone to properly heal and regain its ability to sustain strain, which is generally based on the reduction of the fractured surfaces, coupling of processed fragments, and stabilization by fixation implants or adhesive materials [2–6]. The main disadvantage of internal fixation devices (e.g., screws, plates, and wires) is the need for a second procedure to remove non-resorbable materials after the initial intervention, which usually results in additional complications such as refractures, infections, and other wound healing disorders [7–9]. In contrast, adhesive materials, which bond fragments together through surface adhesion and internal cohesion, are intended to be replaced gradually by re-growing bone, thus preventing the need for additional interventions. Moreover, these adhesives enable uniform distribution of physical forces over the bonded area, avoiding physical stress and tissue damage during the fixation and repair processes [10–12]. Over the past two decades, engineered biomaterials have been tailored to yield desirable biocompatibility and reduced foreign body responses, therefore emerging as promising next-generation adhesives for biomedical applications [11,12].

Misnomered as an adhesive, polymethyl methacrylate (PMMA) or bone cement is considered the gold standard for implant fixation in orthopedic surgery. Because it lacks intrinsic adhesive properties, bone cement relies on the close mechanical interlock between irregular surfaces to transfer forces from bone-to-bone or bone-to-prosthesis [13–15]. However, PMMA also shows low adhesion on wet surfaces, which makes it largely unsuitable for living bone tissue applications. Additionally, it induces persistent inflammation, which is undesirable for bone regeneration, and demands high polymerization temperatures (66–80 °C), a condition that might lead to tissue necrosis [13–16]. Attempts for overcoming these limitations have focused mainly on adding excipients such as powdered bioactive compounds to enhance biocompatibility and consequently accelerate the fixation process. This occurs at the expense of sacrificing partially its mechanical and handling properties [16–19]. A next generation of clinically usable bone adhesives must therefore comply with a few requirements, including (1) being easily sterilizable and biocompatible, (2) being capable of properly stabilizing the bone fragments for at least two weeks, (3) avoiding major interference with the bone healing processes, (4) being easy to apply, and (5) degrading without generating toxic metabolites [11,12,20].

Hydrogels might be an alternative to tackle the challenges presented above. These polymers are hydrophilic, three-dimensional networks capable of absorbing, retaining, and releasing water under different physical and chemical stimuli. Either synthetic or biobased, hydrogels have emerged as powerful platforms for tissue engineering and drug delivery [21,22]. Depending on the type of monomers, pendant chains, and degree of crosslinking, hydrogels can undergo reversible sol–gel phase transitions (and consequently changes in their properties) in response to specific stimuli such as temperature, pH, and light [21,23]. For example, several authors have reported tuning the mechanical properties of hydrogels through photo-crosslinking for 3D bioprinting and in situ deposition applications [24,25]. Additionally, hydrogels have proven suitable for wound healing as they are non-irritant, permeable to metabolites, and provide a moist and cell-friendly environment [26,27]. This healing potential has been combined with the remarkable adhesiveness of various biobased polymers to develop therapeutic wound dressings and bioadhesives for skin, intestine, and bone applications [26–33].

One of the most employed natural polymers for therapeutic hydrogels is chitosan, a biomaterial that can be obtained from the partial deacetylation of naturally abundant chitin [34–37]. Along with a strong inherent adhesiveness and biocompatibility, chitosan exhibits desirable hemostatic, antimicrobial, and analgesic properties, making it an attractive material for numerous applications in nanomedicine [38,39] and regenerative medicine [40]. Moreover, chitosan can be further engineered for specific functionalities by conjugating functional molecules to the free acetamide and hydroxyl groups linked to the glucopyranose rings in their structure, which are susceptible to nucleophilic attacks [37]. For instance, the conjugation of methacrylic acid (or methacrylation) has been used to enhance the adhesive properties of chitosan for dental applications [41]. Complementarily, graphene oxide (GO), a two-dimensional nanomaterial obtained from the exfoliation-oxidation of graphite, has been incorporated into chitosan hydrogels and scaffolds to form nanocomposites that further improve mechanical properties [42–44]. GO is biocompatible, biodegradable, antibacterial, physiologically stable, and sensitive to temperature and pH changes to elicit drug release, but several reports have also discussed its promising potential as reinforcement to enhance mechanical response and colloidal stability, mimic tissue microenvironments, and promote bone regeneration [45–48]. Moreover, the abundance of oxygen-rich functional groups on its surface makes it suitable for relatively simple chemical functionalization schemes and protein adsorption [42–44,49,50]. Nevertheless, recent reports have also shown that GO can bioaccumulate and induce inflammatory and cytotoxic responses, and thus only low doses are recommended for biological applications.

This work was dedicated to embedding GO in a covalently photocrosslinked network of chitosan methacrylate (ChiMA) to develop a biobased adhesive nanocomposite hydrogel that promotes bonding of fractured bone tissue under normal physiological conditions. Extensive physicochemical and biological characterization tests were carried out on the synthesized material to evaluate comprehensively its potential as a bioadhesive, including mechanical response, swelling and degradation, microscopic imaging, methacrylation level, and biocompatibility. Moreover, we conducted in silico analyses of the mechanical performance to gain a more detailed mechanistic understanding of the bioadhesive response under different load and compression conditions. The material put forward here holds much promise as a platform that can be tailored for matching the mechanical properties of different tissues through light-directed polymerization, and that is suitable for in situ deposition through extrusion-based methods.

2. Materials and Methods

2.1. Materials

High-density chitosan, methacrylic acid (MA), N-[3-(dimethylamino)-propyl]-N′-ethyl carbodiimide hydrochloride (EDC), N-hydroxysulfosuccinimide (NHS), dimethylformamide (DMF), riboflavin (RF), phosphate-buffered saline (PBS), 3-(4,5-Dimethyl-2-thiazolyl)-2,5-diphenyl-2H-tetrazolium bromide (MTT), and triton were purchased from Sigma-Aldrich (St. Louis, MO, USA). High-glucose Dulbecco's modified Eagle medium (DMEM) was purchased from Gibco (Amarillo, TX, USA). Hydrogen peroxide (H_2O_2), sodium hydroxide (NaOH), glacial acetic acid, sulfuric acid, phosphoric acid, potassium permanganate (K_2MnO_4), hydrochloric acid (HCl), and Tris-HCl were purchased from PanReac AppliChem (Chicago, IL, USA). A commercial lactate dehydrogenase (LDH) cytotoxicity assay kit was purchased from Novus Biologicals (Littleton, CO, USA). Fetal bovine serum (FBS) was purchased from BioWest (Riverside, MO, USA), and graphite flakes were purchased from Graphene Supermarket (Ronkonkoma, NY, USA). The *Cercopithecus aethiops* kidney (Vero) cell line (ATCC® CCL-81) was obtained from the American Type Culture Collection (Manassas, VA, USA). The Normal Human Osteoblast (NHOst) cell line (CC-2538) and Osteoblast Growth and Differentiation Basal Medium (OBM) were purchased from Lonza Bioscience (Basel, Switzerland). Polyester fiber and bovine femur samples were purchased from local shops.

2.2. Chitosan Methacrylation

Methacryloyl groups from MA were covalently bound to the pendant primary amines of chitosan monomers through EDC/NHS-mediated activation, following a modification of the protocol put forward by Shen et al. (Figure 1) [51]. High-density chitosan was solubilized in acetic acid 0.17 M at 3.5 mg/mL for 10 min under magnetic stirring. Furthermore, MA, EDC, and NHS were mixed in 3 mL of DMF 99.8% (*v*/*v*) at 1:1, 1:2, and 1:4 molar ratios with respect to the estimated free-amine groups in chitosan and activated for 15 min at 37 °C before their addition to the chitosan solution. These mixtures were then left to react for 24 h at 60 °C under continuous magnetic stirring at 600 RPM. The synthesized

ChiMA formulations were dialyzed for 48 h against acetic acid 0.17 M and lyophilized for 48 h before sterilization with ethylene oxide.

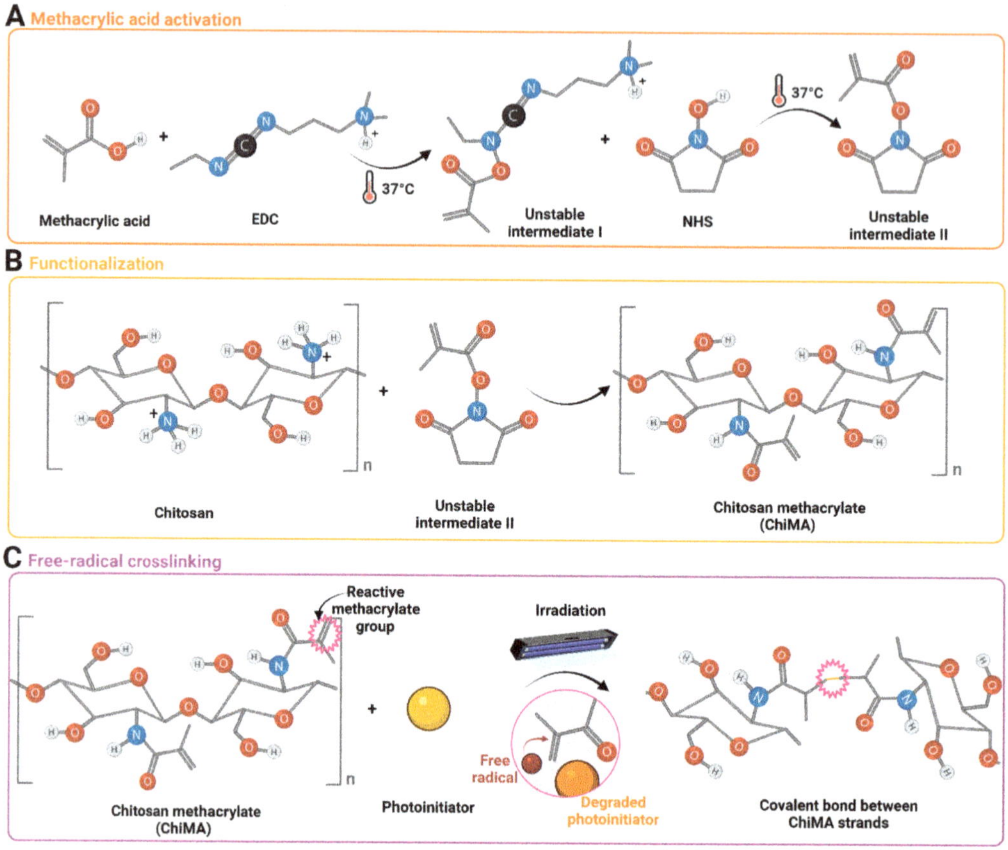

Figure 1. (**A**) Activation of methacrylic acid mediated by EDC and NHS. (**B**) Chemical conjugation of methacryloyl groups to the free amines of glucosamine units in chitosan monomers. (**C**) Methacryloyl group destabilization and crosslinking induced by the light-directed degradation of a photoinitiator (riboflavin).

The conjugation of methacryloyl groups to chitosan was evaluated by Fourier-transform infrared spectroscopy (FTIR) using an A250 FTIR (Bruker, Germany) instrument in the spectral range of 4000–400 cm^{-1}. Chemical modifications were also assessed by X-ray Photoelectron Spectroscopy (XPS) using a NAP-XPS spectrometer (SPECS Surface Nano Analysis GmbH) equipped with a PHOIBOS 150-1D-DLD analyzer and a monochromatic Al-Kα source (1486.7 eV, 13 kV, 100 W). The survey spectra were obtained after five cycles using a pass energy of 100 eV and an energy step of 1 eV. For high-resolution spectra, the setup was fixed at 20 cycles, 20 eV pass, and 0.1 eV energy steps. An electron flood gun at 3 eV and 20 μA was employed for charge compensation. The C1s at 284.6 eV were selected for binding energy calibration. The fitting was performed via a Shirley-type background and optimized in the least-square method under well-established conditions of 30 GL ratio (Gaussian–Lorentzian peak shape), fullwidth half-maximum (FWHM), and peak position from the literature. A semi-quantification approach for the methacrylate functional groups was conducted from the decomposition C1s sub-peak components associated with the CO– and COO– species and N1s sub-peak component associated with the –O–N–C species. The atomic ratio composition was performed based on corrected peak areas for each spectrum and the pristine chitosan sample as a reference.

2.3. Synthesis of Graphene Oxide

GO was synthesized by graphite's coupled exfoliation/oxidation, following Tour's method (Figure 2) [52]. Briefly, a mixture of 90 mL of sulfuric acid 98% (*v*/*v*) and 10 mL of phosphoric acid 85% (*v*/*v*) was slowly added to 0.75 g of graphite flakes and 4.5 g of potassium permanganate and left to react under constant magnetic stirring at 50 °C. After 12 h, 150 mL of type I water ice cubes and 3 mL of hydrogen peroxide 30% (*w*/*w*) were added to the viscous solution, thus inducing a color change from dark purple to gold yellow. The obtained GO was sonicated for 5 min in an ultrasonic bath at an amplitude of 38% and a frequency of 40 kHz and washed successively aided by filtration with polyester fiber. Each washing cycle involved centrifugation at 4000 RPM for 4 h followed by resuspension in a washing solution containing 50 mL of HCl 30% (*v*/*v*), 50 mL of ethanol 96% (*v*/*v*), and 50 mL of type I water. The final pellet was resuspended in type I water and lyophilized for 24 h.

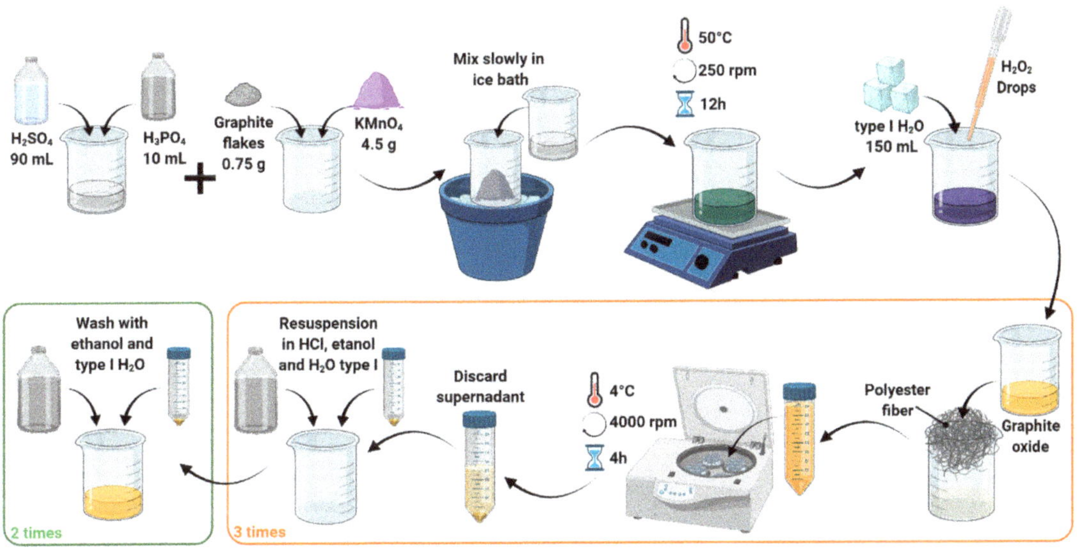

Figure 2. Synthesis of graphene oxide by the coupled exfoliation/oxidation of graphite, as described by Marcano's modification of the Tour's method.

The proper synthesis of GO was confirmed through FTIR, Raman spectroscopy, thermogravimetric analysis (TGA), and transmission electron microscopy (TEM). FTIR spectra were recorded with an A250 FTIR (Bruker, Germany) instrument in the spectral range of 4000–400 cm^{-1}, while Raman spectra were recorded with an XploRA Confocal Raman Microscope (Horiba Scientific, Japan). TGA profiles were collected with a TG analyzer (TA Instruments, New Castle, DE, USA) using a temperature ramp of 25–600 °C at a heating rate of 10 °C/min under a nitrogen atmosphere, and TEM imaging was conducted at 15,000× with a Tecnai F30 Microscope (Fei Company, Hillsboro, OR, USA).

2.4. Preparation of the Bioadhesive

Lyophilized ChiMA samples were resuspended at 40 mg/mL and 60 mg/mL in acetic acid 0.02 M. In parallel, a working solution of DMEM supplemented with 10% (*v*/*v*) FBS, 0.1% (*w*/*v*) RF, and Tris-HCl 0.1 M was prepared and adjusted to a pH of 8.5 by dripping NaOH 5M slowly. Resuspended ChiMA was mixed at a 1:1 volume ratio with the working solution, thus accounting for the GO-free formulations (ChiMA2% and ChiMA3%). For a GO-laden hydrogel, the nanomaterial was dispersed at 1 mg/mL in the working solution before forming the mixture (ChiMA3%GO) [53]. The overall process is shown in Figure 3.

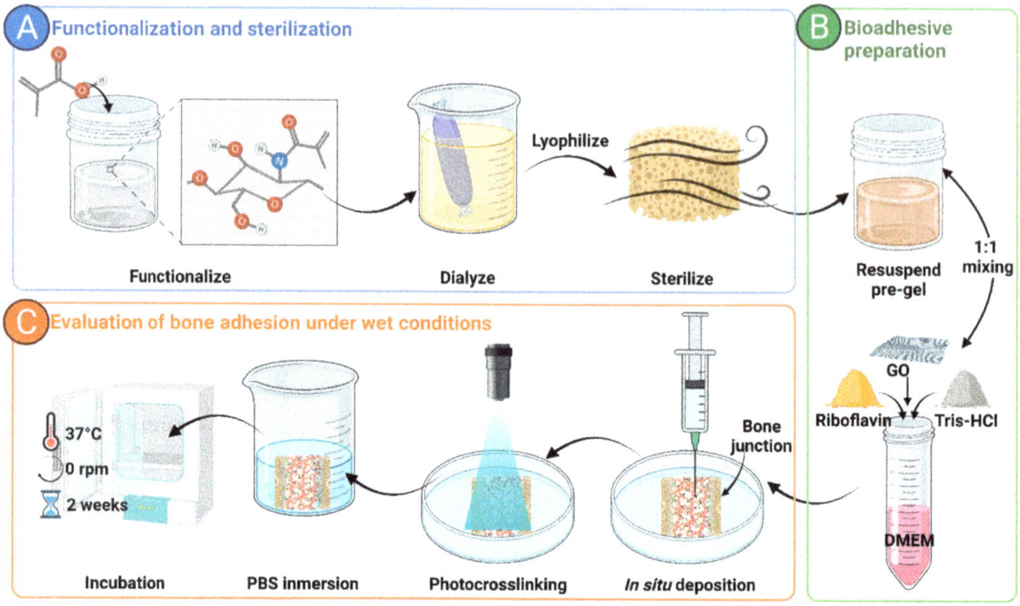

Figure 3. General methodology employed for the development of the bioadhesive. (**A**) Functionalization and sterilization of chitosan methacrylate (ChiMA). (**B**) Preparation of the bioadhesive with graphene oxide (GO) and riboflavin. (**C**) Evaluation of the adhesive potential under wet conditions.

Hydrogels were loaded into 3 mL syringes and manually ejected through a 21-gauge needle to assess their potential to form continuous filaments during extrusion. After deposition, samples were irradiated for 10 min at 62 mW/cm^2 with blue light (420–460 nm) to induce covalent crosslinking of methacryloyl groups in the main backbone of ChiMA [24,53].

2.5. Rheological Evaluation

Experiments to evaluate the rheological behavior of bioadhesive samples were performed in a Discovery Series Hybrid Rheometer-1 (TA Instruments, New Castle, DE, USA) using a parallel plate geometry with a 20 mm gap. Changes in both the storage (G') and loss (G'') moduli of photocrosslinked samples were assessed through flow, frequency, time, and temperature sweeps. The flow sweep experiments were conducted from 0.01 to 100 Hz at 1% strain and room temperature. Frequency sweep experiments were performed between 0.01 and 100 rad/s at room temperature. Last, coupled time/temperature sweep experiments were carried under oscillatory mode, with a constant strain of 1% and 10 rad/s with a temperature ramp of 20 °C/min between 15 °C and 37 °C. Shear-thinning properties of the hydrogels were estimated by fitting the viscosity (η) vs. shear rate (γ) data to the power law model (Equation (1)):

$$\eta = K\gamma^{n-1} \quad (1)$$

2.6. GO Dispersion

The nanomaterial's self-fluorescence at 405 nm was imaged through confocal microscopy to evaluate GO dispersion within the hydrogel. ChiMA3%GO samples were imaged at 20× with an FV1000 Confocal Microscope (Olympus, Tokyo, Japan), and particle count and area were then analyzed aided by ImageJ® software. The spatial distribution of

GO sheets within the nanocomposite hydrogel was determined by a Z-stack reconstruction of images captured at different depth positions [53].

2.7. Morphological Analysis of Hydrogels

The polymeric microstructure of the hydrogels was imaged before and after photocrosslinking via scanning electron microscopy (SEM). ChiMA3%GO samples were imaged at 500× and 1500× magnification with a JSM 6490-LV microscope (JEOL, Tokyo, Japan) under vacuum conditions and a 20 kV accelerating voltage. Pore size distribution was determined with the aid of ImageJ® software.

2.8. Qualitative Adhesion Test

The adhesion potential of the hydrogels was qualitatively assessed by bonding bone samples in a buffer simulating physiological conditions. Bovine femur fragments were washed by being immersed in PBS 1 × with media changes every 2 h. After 6 h, fragments were dried, cut into cube-shaped specimens, and stored at 4 °C until further use. Then 500 mL of the hydrogels were deposited and photocrosslinked between two bone specimens. The bonded hydrogel–bone complex was then immersed in PBS 1×, incubated at 37 °C, and checked daily until bone samples were fully detached.

2.9. Mechanical Adhesion

The mechanical adhesion of the hydrogels was evaluated through the tensile testing of adhesive butt-joint specimens. Bovine femur fragments were washed by immersion in PBS 1X with media changes every 2 h. After 6 h, fragments were dried and cut into dog bone-shaped specimens with a Dremel 4000 moto tool (Bosch Power Tools, Stuttgart, Germany). These specimens were then shopped in half at the center of their longitudinal axis and stored at 4 °C until further use. The viable bonding area of the specimens was determined from microscopy images. Then 500 µL of the hydrogels were deposited and photocrosslinked between the two half-dogbones to assemble the butt-joint specimen. Tensile testing was conducted in a planar uniaxial machine (Bose, Electroforce) with a 500 N load cell at a constant displacement rate of 0.1 mm/s [54].

2.10. Texture Analysis

The firmness of the hydrogels was evaluated with a TA.HDplusC texture analyzer (Stable Micro Systems, Godalming, UK) before and after sample irradiation. Hydrogels were molded into 20 mm diameter and 25 mm height cylinder-shaped constructs. This test measured compression force at a 1.0 mm/s speed and 15 mm penetration length using a 10 mm cylindrical probe [49].

2.11. Swelling and Degradation

Deposited samples of the hydrogels were weighted, immersed in DMEM supplemented with 10% (v/v) FBS, and incubated at 37 °C. For swelling, sample weight was monitored every 12 h until no further increase in weight was observed. For degradation, sample weight was monitored every 12 h from maximum weight until the hydrogel was fully dissolved in the medium. The percentage of swelling and degradation was estimated following Equation (2):

$$Swelling\ or\ degradation\ (\%) = \frac{W_f - W_0}{W_0} \times 100\% \quad (2)$$

2.12. Hemolysis and Platelet Aggregation

To assess the hemolytic activity of hydrogels, blood samples from a healthy donor were collected in EDTA tubes and centrifuged at 1800 RPM for 5 min to separate the plasma and replace it with PBS. This procedure was repeated until a purified erythrocyte precipitate was obtained. Then 100 mL of a 10% (v/v) PBS-diluted erythrocyte solution was mixed with 100 mL of each extract and incubated for 1 h at 37 °C. The samples were then centrifuged at 3000 RPM for 5 min, and 100 mL from the supernatant was seeded by triplicate in a 96-well microplate and read at 450 nm in a Multiskan FC microplate reader (Thermo Fisher Scientific, Waltham, MA, USA). Triton 100-X and PBS 1X were used as positive and negative controls, respectively. The percentage of hemolytic activity was estimated following Equation (3):

$$Platelet\ aggregation(\%) = \frac{Abs_s - Abs_{(-)}}{Abs_{(+)} - Abs_{(-)}} \times 100\% \quad (3)$$

For platelet aggregation, blood samples from a healthy donor were collected in sodium citrate tubes and centrifuged at 1000 RPM for 15 min to retrieve the platelet-rich plasma (PRP). Then 100 mL aliquots of each hydrogel were seeded by triplicate with 100 mL of PRP in a 96-well microplate and left under agitation for 1 h. The unaggregated platelets in the supernatant (100 mL) were removed and seeded in a separate microplate with 10 mL of Triton 100-X and left under agitation for 15 min. This microplate was then centrifuged at 1000 RPM for 15 min, and 50 mL of each well was transferred to another microplate. Finally, 50 mL of LDH reagent was pipetted into each of the microplate wells, and the absorbance was then read at 493 nm in a Multiskan FC microplate reader (Thermo Fisher Scientific, Waltham, MA, USA). Platelet aggregation was estimated by comparing absorbances to an LDH calibration curve.

2.13. Cytotoxicity

The cytotoxicity of the hydrogels was determined by an indirect contact assay with MTT. First, hydrogel samples were mixed with FBS-free OBM at 25% (*v*/*v*) and incubated for 4 h at 37 °C. Extracts for the assay were obtained by collecting the supernatant of the centrifuged mixtures and preparing dilutions at 25%, 12.5%, 6.25%, 3.13%, and 1.56% (*v*/*v*). Parallelly, 100 µL of NHOst cells was seeded in a 96-well culture plate at a density of 1×10^6 cells/mL and incubated at 37 °C and 5% CO_2 for 24 h. Culture media were then removed and replaced with FBS-free OBM. Hydrogel extracts were added by triplicate to cell-seeded wells, and the culture plate was then incubated at 37 °C and 5% CO_2 for another 48 h. Then 10 µL of MTT (5 mg/mL) was added to each well, and cells were incubated at 37 °C and 5% CO_2 for 2 h. The culture medium was removed, and 100 µL of DMSO was added to dissolve the formed formazan crystals. Absorbance was read at 595 nm in a Multiskan FC microplate reader (Thermo Fisher Scientific, Waltham, MA, USA) [55]. The MTT assay was also performed on Vero cell cultures.

2.14. Finite Element Analyses of Bioadhesive-Repaired Bones

Finite element analyses were performed in the software Abaqus/CAE® (Dassault Systemes, Vélizy-Villacoublay, France) to simulate the stress distribution experienced by bioadhesive-repaired femurs during movement. Injured bones were modeled by generating transverse and oblique ruptures on a reconstruction of the distal epiphysis of the femur (obtained from the Embodi 3D repository) in the 3D Slicer® software (The Slicer Community, Boston, MA, USA). To assure mesh convergence, 127,118 and 156,861 C3D4 tetrahedral domain elements were built in the software Ansys® ICEM CFD (Ansys, Canonsburg, PA, USA) for transverse and oblique ruptures, respectively. As an initial resting state, a vertical load of 571 N was distributed over 24 nodes in the medial end of the femur to simulate the patient's body weight. Additionally, 16 nodes were distributed over the distal end of the femur to represent articular constraints.

The built models were subjected to two kinds of movements: internal rotation of the knee, and momentum from the hip (Figure 4). To simulate knee rotation, a rotational load of 5 N was distributed over 20 nodes in the internal and external flanks of the femur. Similarly, a load of 5 N was exerted diagonally (y+, z−) over the same nodes to simulate hip momentum. The distributions of minimum principal, maximum principal, and von Mises stresses were evaluated for each model. All the simulations were performed with parameters either recovered from the mechanical experiments conducted here or reported for commercial bone cements. Supplementary Table S1 summarizes the mechanical properties of bone tissue, the hydrogel nanocomposite bioadhesive, and bone cement [56,57].

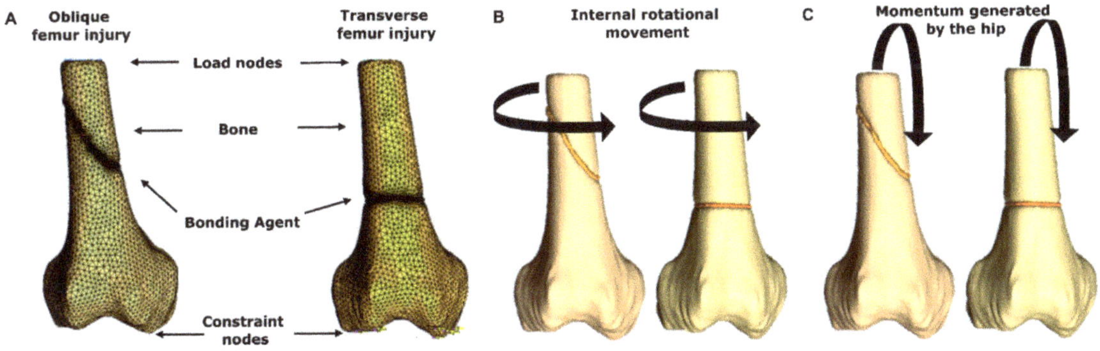

Figure 4. Geometries employed for the FEA simulations. (**A**) Components of the geometries along with the rendered mesh. (**B**) Load direction during internal rotation of the knee. (**C**) Load direction during hip momentum.

2.15. Statistical Analysis

All experimental data were collected in triplicate, for which mean and standard deviation were calculated. Statistical analysis was performed using two-way ANOVA and Tukey's test for pairwise comparison in GraphPad Prism (GraphPad Software Inc, San Diego, CA, USA). $p < 0.05$ was considered statistically significant.

3. Results and Discussion

3.1. Chemical Functionalization of Chitosan

The chemical modification of chitosan was confirmed by changes in the FTIR spectrum of ChiMA (Supplementary Figure S1), as the presence of peaks at 1680 cm^{-1} and 670 cm^{-1} corresponded to the C=O stretching and C=C bending vibrations present in the amide-bound methacryloyl groups [58,59]. Peaks at 1580 cm^{-1}, 1410 cm^{-1}, and 2950 cm^{-1} represented N–H bending, O–H bending, and C–H stretching vibrations, attributed to the glucosamine and N-acetylglucosamine monomers of chitosan [60,61].

XPS survey spectra for the Chitosan sample and ChiMA samples for 1:1, 1:2, and 1:4 molar ratios are shown in Supplementary Figure S2A. The major peaks could be associated with C1s, N1s, and O1s signals, and there was no evidence of chitosan degradation caused either by the XPS setup conditions or the chemical modification process. After this, high-resolution spectra for C, N, and O elements were recorded on the pristine sample and the modified ones under the same setup conditions. The decomposition peaks and corresponding fitting curves for each element and the samples are shown in Supplementary Figure S2B. For the case of C1s, the signal was decomposed into three sub-peak components. The first at lower energy values could be assigned to C–C, C=C, and C–N bonds, the next to C–O– and C–O–C bonds, and the last, at higher energy values to –O–C–O– bonds. For the case of N1s, the signal was also decomposed into three sub-peak components. The first at lower energy values could be assigned to –C–N– bonds, the next to –O–N–C– bonds, and the last to ionic NH+ species, in a good agreement with functional groups existing for chitosan and ChiMA samples. Finally, the O1s signal could be resolved into two sub-peak components attributed to C=O at low binding energy values, and C–O–C or other chemisorbed –C–O– species at higher ones. Results confirmed the successful conjugation of methacryloyl groups in the polymeric backbone of chitosan with the highest yield obtained for the 1:2 ChiMA sample. Moreover, for the elemental analysis quantification, the Relative Sensitivity Factors (RSF) were used to scale the measured peak areas, and the N1s peak area of the 75% deacetylated chitosan sample was selected for a stoichiometric calculation after considering that adventitious carbon is comparative for all samples. The calculated elemental composition of the chitosan and ChiMA samples is summarized in Table 1. Notably, the carbon contents in chitosan seemed to be higher than expected due to the presence of acetyl groups after the required dissolution of chitosan in 1% v/v acetic acid before conducting the conjugation reaction. The ChiMA 1:1 sample showed a small unidentified sub-peak component at lower binding energy values for C1s and N1s, which can be attributed to surface charge compensation artifacts during the XPS run, which was neglected for the calculations. The increments in the sub-peak components of carbonyl and amide groups relative to unmodified chitosan are summarized in Table 2. These findings are consistent with previous reports in the literature and support the notion that the 1:2 ratio provides the highest methacrylation level [41].

Table 1. Surface chemical analysis and elemental composition (%) of the chitosan and ChiMA 1:1, 1:2, and 1:4 samples. Details of C1s, N1s, and O1s decomposed peaks and related binding energies (BE) and FWHM values.

Sample	Peak	BE (eV)	FWHM	Corrected Area RSF	Ratio (%)
ChiMA	N1s	398.83 399.55 400.98	1.08 1.04 1.84	842	6.10
	C1s	284.57 286.07 287.78	1.16 1.43 1.31	8594	62.28
	O1s	531.58 532.44	1.36 1.4	4835	31.62
ChiMA 1:1	N1s	398.78 399.42 401.09	1.25 1.34 2.19	698	6.61
	C1s	284.56 285.97 287.57	1.28 1.41 1.29	5884	55.71
	O1s	530.93 532.16	1.89 1.51	3980	37.68
ChiMA 1:2	N1s	398.91 399.53 401.45	1.46 1.63 1.43	728	7.59
	C1s	284.67 286.21 287.82	1.42 1.42 1.49	5390	56.17
	O1s	531.62 532.53	1.85 1.41	3478	36.24
ChiMA 1:4	N1s	398.69 399.41 401.22	1.43 1.36 1.69	864	7.36
	C1s	284.54 286.10 287.68	1.43 1.41 1.33	6827	58.16
	O1s	531.35 532.39	1.78 1.41	4048	34.48

ChiMA: chitosan methacrylate.

Table 2. Semi-quantification of sub-peak component variation (%) for carbonyl and amide groups in ChiMA samples with respect to the pristine chitosan.

	ChiMA 1:1	ChiMA 1:2	ChiMA 1:4
Carbonyl sub-peak (D%)	24	30	16
Amide sub-peak (D%)	17	21	12

ChiMA: chitosan methacrylate.

3.2. Synthesis of Graphene Oxide

The correct synthesis and oxidation level of GO were confirmed by FTIR, Raman spectroscopy, TGA, and TEM analyses, as shown in Figure 5. The FTIR spectra (Figure 5A) proved the correct oxidation of graphite, as GO exhibits multiple peaks related to oxidative functional groups [62,63]. Peaks at 3392 cm^{-1} and 1226 cm^{-1} corresponded to O–H and C–OH stretching, while peaks at 1740 cm^{-1} and 1050 cm^{-1} represented C=O stretching and C–O bending, respectively. The peak at 1620 cm^{-1} could be attributed to C=C aromatic stretching. The absence of peaks near 1250 cm^{-1} confirmed that the material was not partially reduced into reduced graphene oxide (rGO) [50,62–64].

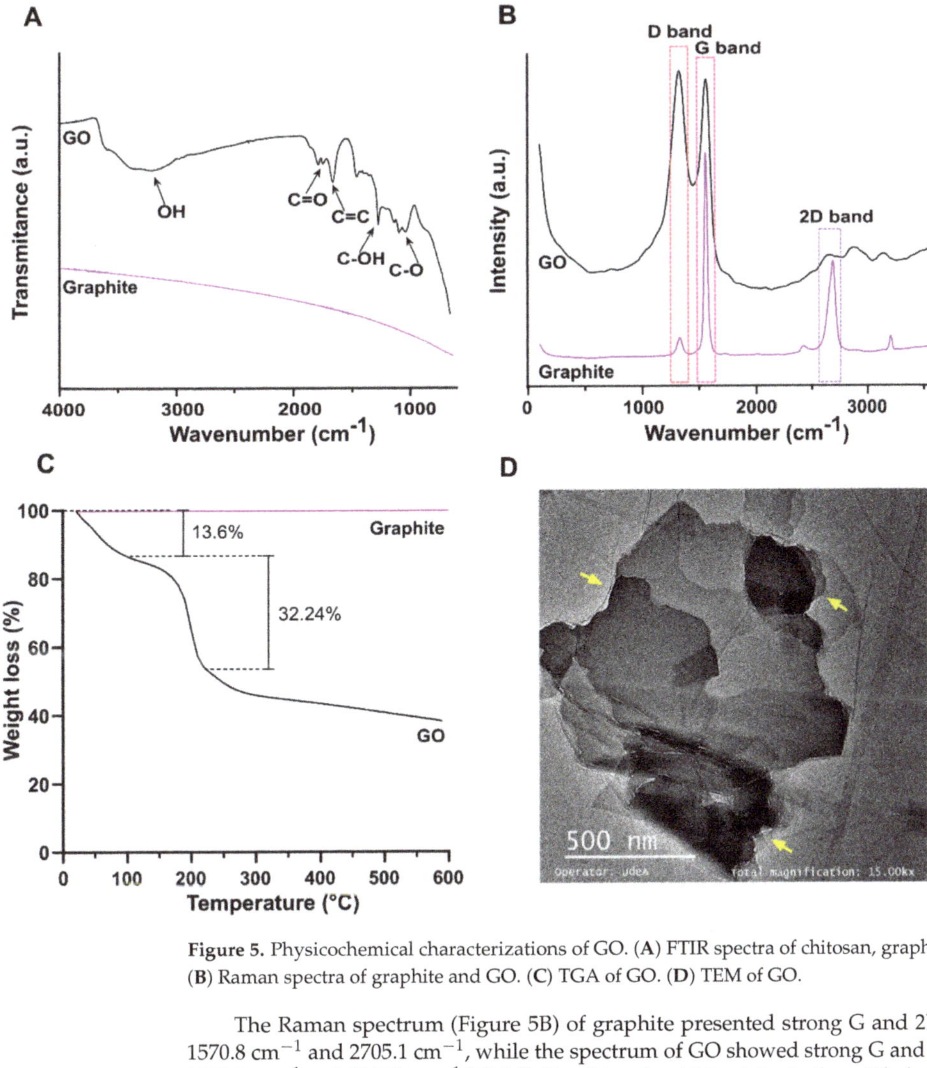

Figure 5. Physicochemical characterizations of GO. (**A**) FTIR spectra of chitosan, graphite, and GO. (**B**) Raman spectra of graphite and GO. (**C**) TGA of GO. (**D**) TEM of GO.

The Raman spectrum (Figure 5B) of graphite presented strong G and 2D bands at 1570.8 cm^{-1} and 2705.1 cm^{-1}, while the spectrum of GO showed strong G and D bands at 1588.9 cm^{-1} and 1345.8 cm^{-1} [65,66]. The G band, which relates to the sp2 hybridization in carbon, most likely shifted to 1588.9 due to the oxidation of graphite [65]. Similarly, the noticeable growth of the D band, which results from vacancies or dislocations that disrupt sp2 carbon layers, suggests the transition of some carbon atoms in graphene to and sp3 hybridization to accept functional groups [67]. The I(D)/I(G) ratio for GO corresponded to

1.03, which was most likely due to a high oxidation level [64]. Furthermore, the weakening of the 2D band can be explained by breaking the stacking order along the c-axis in graphite due to the oxidation reaction, thus confirming the high level of oxidation in the synthesized GO [64]. Similarly, the I(2D)/I(G) ratio for GO corresponded to 0.4, demonstrating the graphene-like structure of the synthesized material [68].

The TGA thermogram of GO (Figure 5C) presented three noticeable weight losses. A first weight loss of 13.6% from room temperature to 100 °C could be attributed to the evaporation of bound water. A second weight loss of 32.24% between 100 and 200° could be attributed to removal of oxygen-rich species such as C–O, C–OH, and C=O, further confirming the high level of oxidation of the synthesized GO [69,70]. The TEM micrograph of GO nanosheets (Figure 5D) demonstrated the characteristic flake-like structure of the material [71]. Different levels of opacity in the image revealed non-uniform stacking of multiple GO layers. Regions of higher transparency indicated thinner films, while darker regions indicated greater stacking [72]. Moreover, the disordered and unwrinkled structure of the GO sheets could be attributed to the high abundance of oxygen-rich functional groups on the surface [73].

3.3. Rheological Evaluation

Three potential formulations for the bioadhesive were successfully prepared: ChiMA2%, ChiMA3%, and ChiMA3%GO. These hydrogels presented an apparent homogeneous consistency and integrity, suggesting the materials' potential to endure mild alkaline pH adjustments instead of regular chitosan, which deprotonates and aggregates into insoluble clusters [51]. Consequently, all formulations allowed extrusion through a 21-gauge blunt needle, although only ChiMA3% and ChiMA3%GO could form continuous filaments, while ChiMA2% dripped ununiformly, forming drops due to low viscosity and, consequently, lack of consistency (Figure 6A).

Figure 6B–E present the changes in the storage modulus (G′) and loss modulus (G″) of the hydrogels when subjected to different stimuli. The flow, temperature, and frequency experiments were performed on irradiated samples, as exposure to blue light (Figure 6C) significantly improved both moduli, favoring mechanical stability and a solid-like behavior [51,60]. The flow sweep (Figure 6B) shows that the hydrogels exhibited lower viscosity at higher shear rates, characteristic of pseudoplastic fluids. This shear-thinning behavior could be further confirmed by the power-law fittings, shown in Supplementary Figure S3, yielding values of $n < 1$ for all formulations [60,74]. Moreover, temperature increments between 15 °C and 37 °C (Figure 6D) caused a slight decrease in both G′ and G″, implying that less energy was required for deformation and thus suggesting the weakening of crosslinking bonds [75,76]. However, according to an ANOVA one-way test, these temperature-induced changes in the moduli were non-statistically significant (p-value > 0.9999). Furthermore, the frequency sweep test (Figure 6E) showed that both moduli were frequency-dependent, as evidenced by the decrease of G′ and the increase of G″ with increasing angular frequency. Hydrogels maintained a predominant solid-like behavior until G′ crosses G″ (gel point), after which they acquired a predominant liquid-like behavior [77].

In general, all hydrogels presented a higher G′ and a lower G″, confirming the material's potential to store deformation energy with small amounts of dissipation from internal friction [60]. The ChiMA2% formulation yielded low G′ and G″, suggesting that ChiMA's concentration was insufficient to form a mechanically stable polymer network. In contrast, ChiMA3% presented desirable values for both moduli, exhibiting a sustained response to the studied stimuli. ChiMA3%GO yielded the highest G′, confirming that the addition of GO to the hydrogels significantly improves their mechanical strength, however potentially compromising their swelling capacity [78].

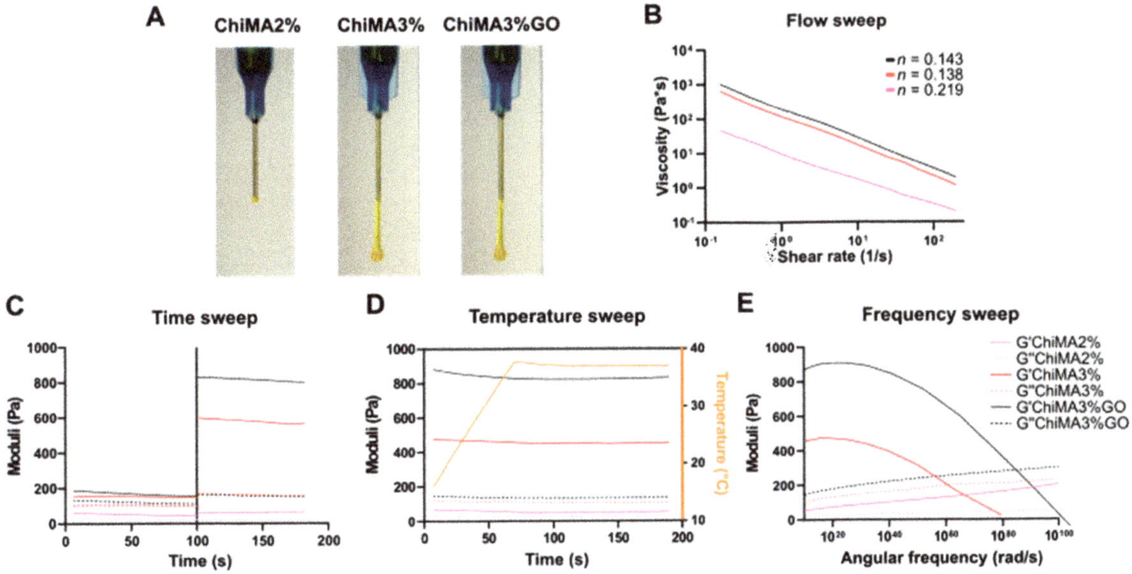

Figure 6. (**A**) Manual injection of the hydrogels through a 21-gauge needle. (**B**) Flow sweep of photocrosslinked samples. (**C**) Time sweeps before and after blue-light photocrosslinking. (**D**) Temperature sweep of photocrosslinked samples between 5 °C and 37 °C. (**E**) Frequency sweep of photocrosslinked samples.

3.4. GO Dispersion and Morphological Structure of the Hydrogels

Dispersion of GO in the ChiMA hydrogel matrix was observed in Z-stack reconstructions from confocal microscopy images. As evidenced in Figure 7A,B, the nanomaterial's dispersion was limited by its own aggregation, as the particle area distribution exhibited a right tail centered at 0.386 μm^2, but with agglomerates as large as 566.805 μm^2. To address this issue, we sought to promote the adsorption of hydrophilic serum proteins from culture media to the surface of GO nanosheets, as proposed by Rueda-Gensini et al. [53]. Thus, GO was mixed with FBS-supplemented DMEM before its addition to the ChiMA pre-gel, yielding the homogeneous dispersion presented in Figure 7C,D. In this case, the particle area distribution exhibited a right tail centered at 0.024 μm^2 with agglomerates as large as 44.640 μm^2, which corroborates a much higher dispersion of GO in the ChiMA hydrogels [53].

Similarly, the morphology of the polymeric matrix in the ChiMA3%GO hydrogel was observed before and after photocrosslinking via SEM, as shown in Supplementary Figure S4. Under the effect of blue light, riboflavin degrades and generates free radicals that destabilize the alkene bonds in ChiMA, thus inducing covalent crosslinking between adjacent strands and causing the microstructure to collapse into smaller pores [74]. Differences in size were visually perceptible and statistically significant, as pore diameter varied between 2.9 μm and 28.2 μm, with a mean of 13.4 μm for non-irradiated samples. In comparison, irradiated samples varied from 0.4 μm to 12.7 μm with an average of 3.3 μm. However, crosslinked samples also exhibited an uneven microstructure, which could be attributed to the formation of imine bonds between GO and the free amine radicals in unmethacrylated units of N-acetyl glucosamine [51,79,80].

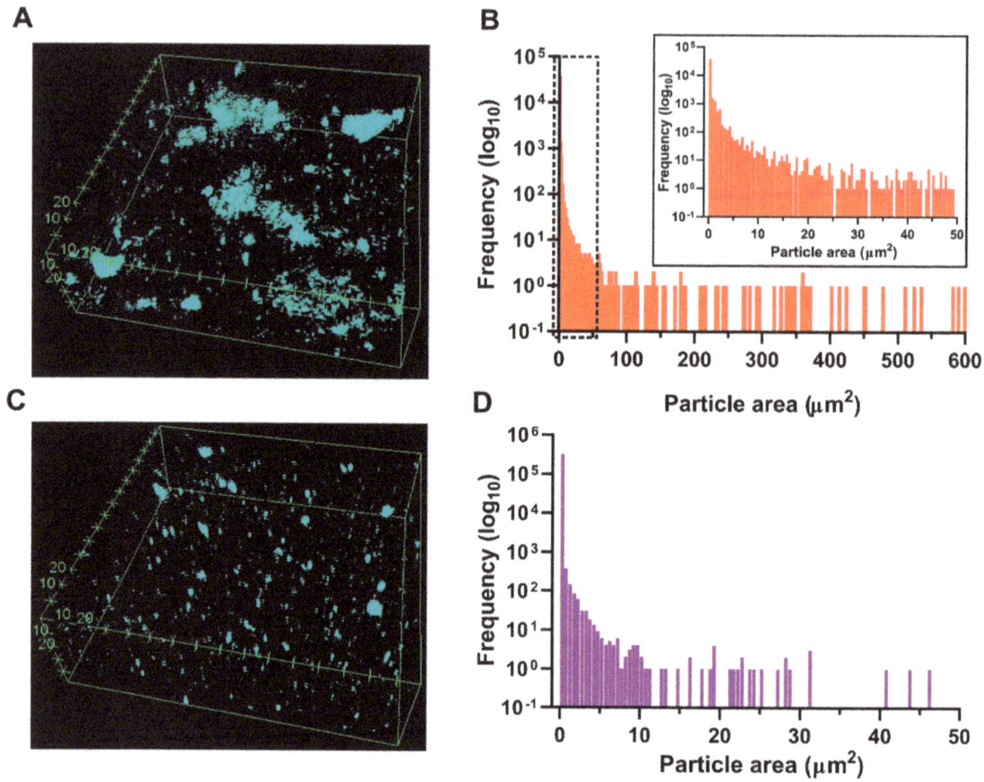

Figure 7. Dispersion of GO and morphological structure of the nanocomposite hydrogel. (**A**) Dispersion of serum-free GO in a ChiMA matrix. (**B**) Distribution of serum-free GO particle area, which shows proper tail distribution centered at 0.386 mm^2. (**C**) Dispersion of serum-doped GO in a ChiMA matrix. (**D**) Distribution of serum-doped GO particle area, which shows proper tail distribution centered at 0.024 mm^2.

3.5. Mechanical and Adhesion Evaluation

The results for the qualitative and quantitative evaluation of the hydrogels' adhesion are presented in Figure 8. Initially, adhesion was qualitatively assessed by bonding two bone fragments in a buffer simulating physiological conditions. The ChiMA3% hydrogel maintained adhesion under wet conditions for only ten days, while the ChiMA3%GO hydrogel successfully maintained bonding for over two weeks. Moreover, this hydrogel progressively darkened for days, thus reflecting GO's reduction in the warm environment (Figure 8A). Further texture analysis was performed with the hydrogels encompassing hardness, compressibility, cohesiveness, and adhesiveness measurements. Hardness, determined as the maximum peak force during the first compression cycle, was studied to estimate the required force to produce deformation of the gels [81]. As evidenced in Figure 8B, ChiMA3%GO samples yielded significantly higher values ($p < 0.01$) than the ChiMA3% samples, confirming GO's contribution to the structural stability of the hydrogels. Moreover, hardness was significantly improved by the crosslinking stimuli, mainly by GO's reduction during incubation. This shift from low to high hardness is a desirable behavior, as it indicates that unstimulated samples will be easily spreadable, while after stimulation, the hydrogel will be retained on the surface [82]. The hardness of ChiMA3% was null after two-weeks of incubation because of the sample degradation. Compressibility (Figure 8C) was estimated as the work required to deform the material during the first compression [81]. This property was significantly enhanced by irradiation and incubation for the ChiMA3%GO samples, favoring the material's potential to endure greater compression loads. Once again, the shift from low to high values increased spreadability and resistance before and after irradiation, respectively [81]. The differences between the GO-free samples were non-statistically significant.

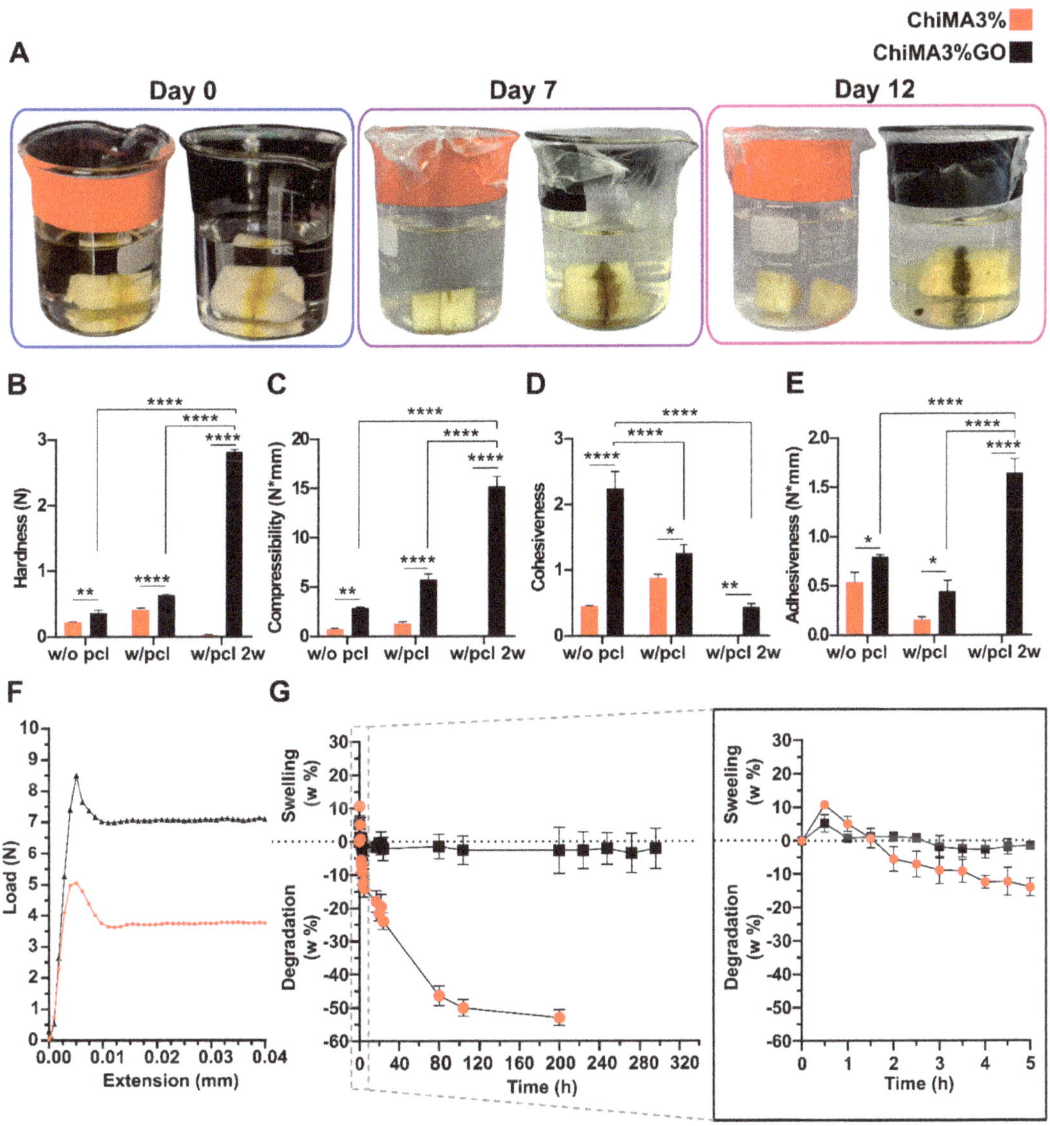

Figure 8. Mechanical and adhesive evaluation of the ChiMA3% and ChiMA3%GO hydrogels. (**A**) Qualitative adhesion of bone specimens at 0, 7, and 12 days. (**B**) Hardness, (**C**) compressibility, (**D**) cohesiveness, and (**E**) adhesiveness of the hydrogels before and after photocrosslinking and 2-weeks of incubation at 37 °C (two-way ANOVA * $p \leq 0.05$ ** $p \leq 0.01$ *** $p \leq 0.001$ **** $p \leq 0.0001$). (**F**) Tensile strength of butt-joint specimens. (**G**) Swelling and degradation of the hydrogels.

Cohesiveness reflects the reconstruction ability of gels after application and was determined as the ratio of the area under the force–time curve on the second compression cycle to that produced on the first one (Figure 8D) [81]. The addition of GO into the hydrogel significantly improved the material's cohesiveness, thus favoring its manipulation. However, cohesiveness was reduced during both crosslinking stages, limiting the structural recovery of the samples [83].

Adhesiveness represents the work required to overcome the attractive forces between the surfaces of the hydrogel and the probe and was estimated as the negative force area for the first compression cycle (Figure 8E) [81]. The addition of GO significantly increases the hydrogel's stickiness, thus improving its chances of remaining attached to the surface of bones [84]. Although this property was partially reduced after crosslinking, it was largely recovered after incubation. This final improvement of adhesiveness suggests that the hydrogel's application could potentially shorten the treatment length and improve patient compliance [85].

Finally, a mechanical tensile test was made to evaluate the adhesive's performance over butt-joined specimens. As evidenced in Figure 8F, ChiMA3% can withstand 5 N loads. At the same time, ChiMA3%GO can endure over 8.5 N. These values are well within the range reported for other bioadhesives and further confirm that the addition of GO significantly improves ChiMA's capacity to bear tensile loads.

3.6. Swelling and Degradation

Figure 8G shows the swelling and degradation profiles of both hydrogels under the simulated physiological conditions. Although limited, both hydrogels achieved their maximum swelling capacity within the first hour, yielding maximum differences of 10.73% for ChiMA3% and 5.06% for ChiMA3%GO. Similarly, results showed that over 50% of the ChiMA3% hydrogel was degraded within two weeks, while the ChiMA3%GO hydrogel seems unaffected by exposure to the medium. These results support the notion that the addition of GO into the ChiMA matrix enhances its mechanical stability, although at the expense of its swelling capacity. However, although this limited swelling restricts the potential of loading functional molecules within the microstructure's pores, immobilization or adsorption on the GO's surface could be alternatively exploited for controlled drug delivery in future applications.

3.7. Hemolysis and Platelet Aggregation

Both hydrogels presented a non-hemolytic behavior, as the hemolysis percentage approached 1% for ChiMA3% and 2% for ChiMA3%GO (Figure 9A). Despite GO's reported dose-dependent hemolytic activity, its incorporation into a biocompatible polymer matrix such as chitosan might have helped reduce such a tendency [86,87]. Similarly, platelet aggregation was estimated to be 28.89% for ChiMA3% and 20.48% for ChiMA3%GO, implying that they can both be classified as intermediate aggregates (Figure 9B). Aggregation is favorable for bone regeneration, as platelets promote migration and proliferation of osteogenic cells, increase blood vessel formation, and induce inflammatory reactions [88,89]. Although higher aggregation was expected, recent reports have suggested that methacryloyl modifications may inhibit platelet aggregation by overshadowing the polymeric backbone of chitosan [90,91].

3.8. Cytotoxicity

Results from the MTT assays revealed that the hydrogels exhibited promising cytocompatibility (Figure 9C,D). None of the evaluated concentrations yielded cell viability levels below 50% after 24 h and 30% after 72 h for the NHOst cells. Moreover, viability increased as sample concentration decreased, reaching over 80% for 6.3% (v/v) concentrations and below. This sharp shift in viability suggests that the material induces cytotoxicity in a dose-dependent manner. In addition, microplate in vitro assays may broadly predict material–cell culture interactions but fail to represent clearance phenomena, and thus some hydrogel components such as riboflavin may accumulate in a non-representative population of cells [92]. Importantly, viability appeared to be enhanced in samples with GO, suggesting that its addition may impede the diffusion and accumulation of riboflavin in the culture medium. This can happen both by ChiMA3%GO's smaller porosity and ChiMA3%'s unrestricted degradation. Similar results were obtained for the Vero cell cultures (Supplementary Figure S6).

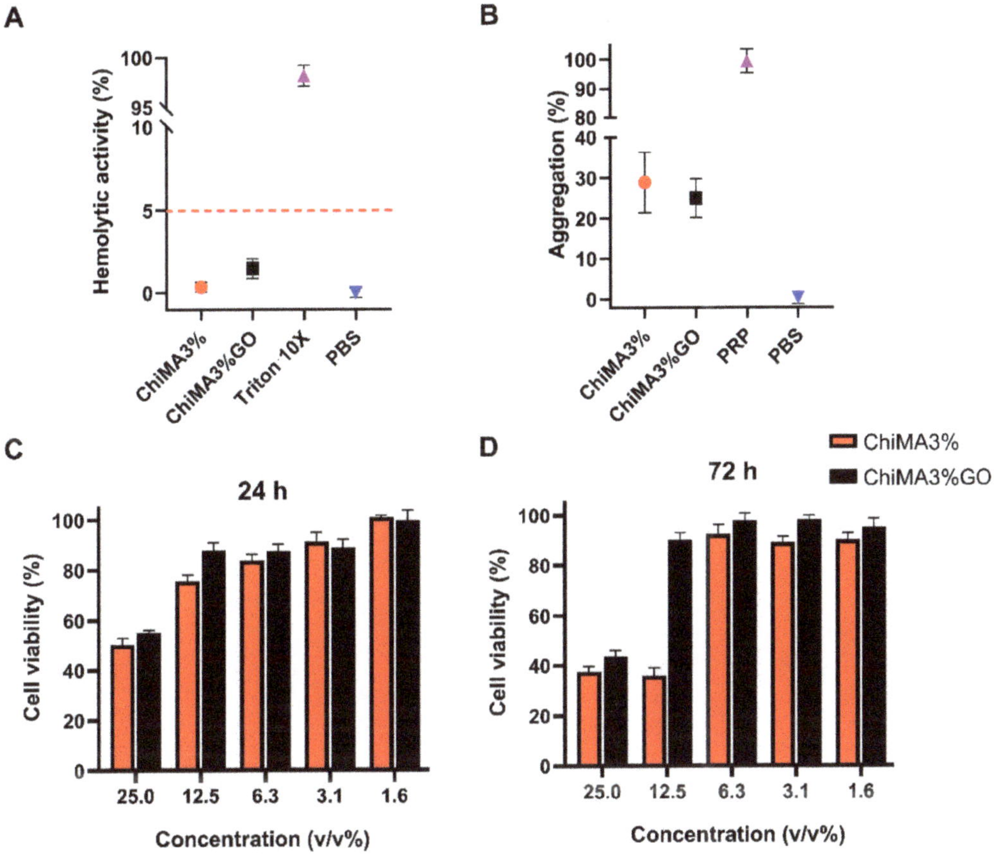

Figure 9. Biocompatibility evaluation of the ChiMA3% and ChiMA3%GO hydrogels. (**A**) Hemolytic activity and (**B**) platelet aggregation of the adhesive hydrogels. Cell viability of NHOst cells when exposed to the hydrogels for (**C**) 24 h and (**D**) 72 h.

3.9. Simulations

Simulation results for the knee rotation and hip momentum models are shown in Figures 10 and 11. The transverse lesion (Figure 10A) showed increased values for the minimum principal (9.88%) and maximum principal (1.18%) stresses and a decrease in the von Mises stress (−5.28%). In contrast, the oblique lesion in knee rotation (Figure 10B) yielded increased values for the minimum principal (1389.76%), maximum principal (17.98%), and von Mises (554.55%) stresses with respect to the commercial bone cement. Additionally, the transverse lesion in hip momentum (Figure 11A) yielded a decrease in the minimum principal (−5.67%), maximum principal (−5.34%), and the von Mises (−0.79%) stresses. On the contrary, the oblique lesion (Figure 11B) exhibited increased values for the minimum principal (467.76%), maximum principal (58.68%), and von Mises (279.35%) stresses.

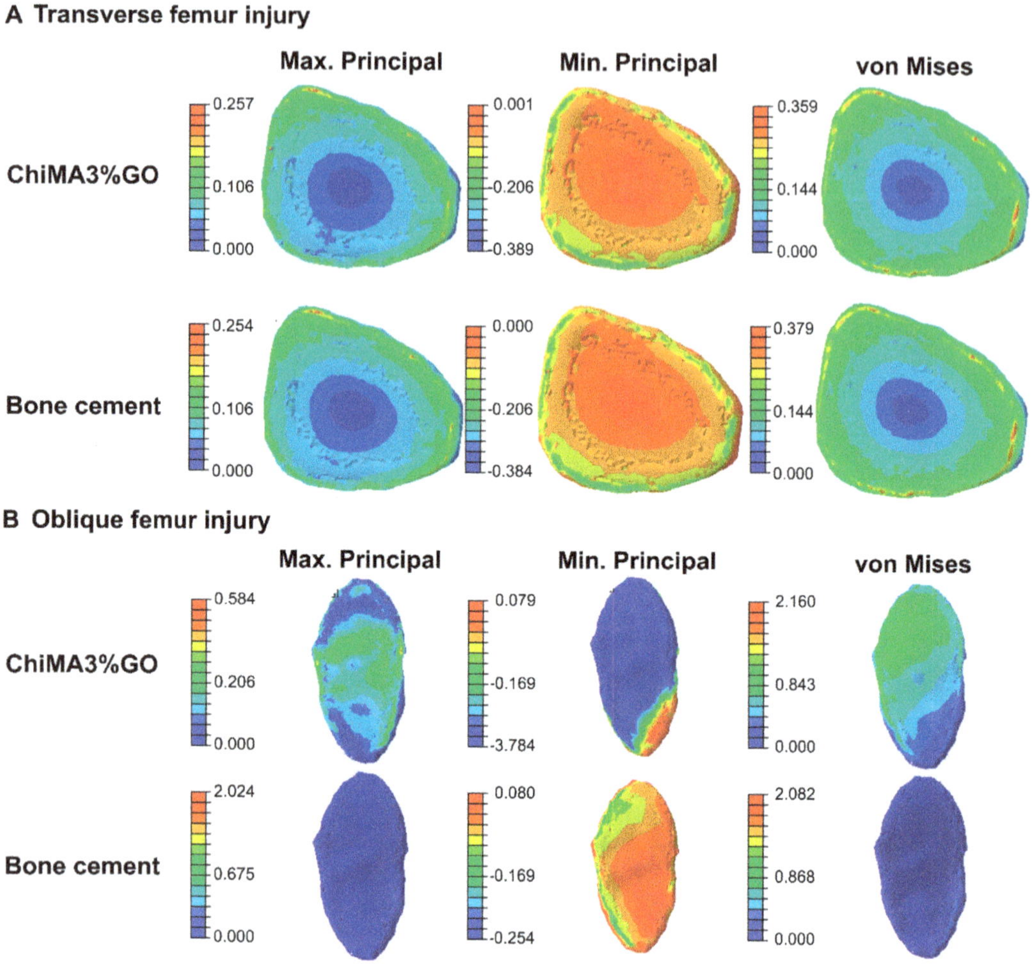

Figure 10. Distribution of tensile and compressive forces over ChiMA3%GO- and bone cement-repaired femurs under the effect of internal knee rotation. Maximum principal, minimum principal, and von Mises stresses are shown for (**A**) the transverse lesion and (**B**) the oblique lesion.

The obtained results revealed that the femur distribution of tension and compression forces was similar for both treatments. However, during internal rotation, the presence of the bioadhesive led to a more uniform distribution when compared with the bone cement, thereby avoiding the generation of stress concentrators. The absence of such areas prevents ruptures in the bonding material, which favors the adhesion's endurance. Moreover, by generating a more uniform distribution of loads, the bioadhesive can enhance the healing process by enabling the continuous transmission of forces and mechanical signals between the cells present in the bonded bone segments [93]. Young's moduli differences explain changes in the distribution of tension and compression forces between the two treatments. The bioadhesive's relative improved elasticity can be beneficial for bone repair, as it prevents stress shielding. This phenomenon arises from tissue adaptation to reduced loads and leads to bone density loss [94]. By putting these results together, it is possible to confirm the bioadhesive's suitability for bone tissue repair applications.

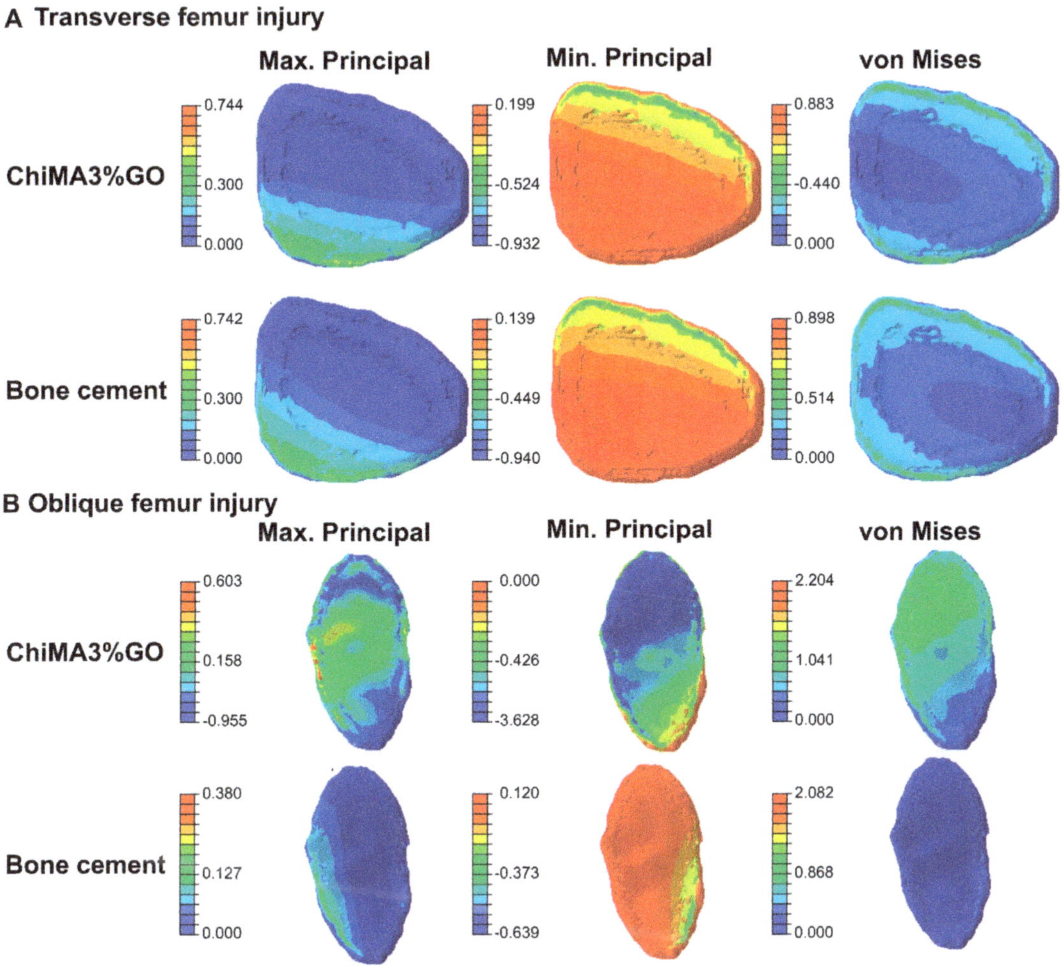

Figure 11. Distribution of tensile and compressive forces over ChiMA3%GO- and bone cement-repaired femurs under the effect of hip momentum. Maximum principal, minimum principal, and von Mises stresses are shown for (**A**) the transverse lesion and (**B**) the oblique lesion.

4. Conclusions

Here, we put forward a ChiMA/GO photoresponsive and adhesive nanocomposite hydrogel to overcome major issues of conventional bone adhesives, i.e., low biocompatibility, poor performance under physiological conditions, difficult applicability, and limited possibilities for tuning mechanical properties in situ. Addition of GO to ChiMA hydrogels led to improved mechanical integrity and resistance, which were crucial for withstanding bone-to-bone adhesion in a simulated physiological medium for over two weeks. XPS and FTIR analyses along with SEM images allowed us to correlate photo-crosslinking by short blue light exposure at room temperature, with key performance attributes such as pseudoplastic response, superior textural properties, and tensile resistance of about 8.5 N. Cytocompatibility assays corroborated cell viability levels above 80% for concentrations below 6% (v/v) of the hydrogel. The material showed intermediate platelet aggregation, which is advantageous for enhanced tissue regeneration. In addition, the hydrogels exhibited hemolytic tendencies well below 5%. In addition to these major attractive attributes, we demonstrated that the materials could be spread easily on surfaces and tended to remain attached to them. Finally, in silico studies showed the superior performance of the bioadhesives in femur fractures when compared to a commercially available cement for both oblique and transverse lesions. This was attributed to the hydrogel's ability to distribute the applied loading more uniformly. Even though the developed formulations incorporated relatively well exfoliated GO, future work should focus on maximizing the dispersion, as it has been demonstrated to be a key parameter to assure a much higher material stability. This is a critical attribute, as it directly impacts load distribution and resistance to wet conditions, and the possibility for more uniform bone repair processes enabled by osteoinduction/osteoconduction induced by GO distributed homogeneously at the bone-to-bone interfaces.

We are confident that the results put forward here pave the way for developing new functional nanocomposite hydrogels for applications in orthopedics. Next steps will be dedicated to exploring the treatment of bone lesions in vivo. This will be a major milestone to assure moving from the bench to the bedside in the near future.

Supplementary Materials: The following are available online at https://www.mdpi.com/article/10.3390/polym14010126/s1, Table S1. Material properties of bone, bioadhesive, and bone cement, Figure S1: FTIR of graphite, graphene oxide, chitosan methacrylate (ChiMA), and chitosan methacrylate mixed with graphene oxide (ChiMAGO), Figure S2: XPS spectra of ChiMA samples to assess the degree of methacrylation, Figure S3: Flow sweep power-law model fitting, Figure S4: SEM imaging of the hydrogel structure. Figure S5: Platelet aggregation calibration curve, Figure S6: Cell viability of Vero cells when exposed to the hydrogels for (A) 24 h and (B) 72 h, Figure S7: Distribution of tensile and compressive forces over ChiMA3%-, ChiMA3%GO-, and bone cement-repaired femurs under the effect of internal knee rotation, Figure S8: Distribution of tensile and compressive forces over ChiMA3%, ChiMA3%GO, and bone cement-repaired femurs under the effect of hip momentum.

Author Contributions: Conceptualization, J.C.C. and C.M.-C.; methodology, D.N.C.-V., S.S.-R., J.C., J.A.S. and M.G.-D.; software, M.G.-D.; validation, D.N.C.-V. and S.S.-R.; formal analysis, D.N.C.-V., S.S.-R. and C.O.; investigation, D.N.C.-V., S.S.-R. and M.G.-D.; resources, J.C.C., C.M.-C., L.H.R. and C.C.-D.l.P.; data curation, D.N.C.-V., S.S.-R. and J.C.; writing—original draft preparation, D.N.C.-V. and S.S.-R.; writing—review and editing, D.N.C.-V., S.S.-R., J.C., M.G.-D., J.A.S., C.O., C.C.-D.l.P., C.M.-C., L.H.R. and J.C.C.; visualization, D.N.C.-V.; supervision, J.C.C., C.M.-C. and C.C.-D.l.P.; project administration, J.C.C.; funding acquisition, J.C.C., L.H.R. and C.M.-C. All authors have read and agreed to the published version of the manuscript.

Funding: This research was funded by Minciencias Grant 689-2018. Additional funding was provided by the 2019 Fundación Santa Fe de Bogotá-Uniandes grant "Production of recombinant antimicrobial peptides to modify materials of biomedical interest".

Institutional Review Board Statement: Not applicable.

Informed Consent Statement: Not applicable.

Acknowledgments: The authors would like to thank the Department of Biomedical Engineering at Universidad de Los Andes and the XPS Laboratory at Universidad de Antioquia for providing access to laboratory infrastructure.

Conflicts of Interest: The authors declare no conflict of interest.

References

1. Filardi, V. Healing of tibial comminuted fractures by the meaning of an innovative intramedullary nail. *J. Orthop.* **2019**, *16*, 145–150. [CrossRef] [PubMed]
2. Thakeb, M.F.; Mahran, M.A.; El-Motassem, H.M. Bone transport for the management of severely comminuted fractures without bone loss. *Strateg. Trauma Limb. Reconstr.* **2016**, *11*, 19–24.
3. Massoud, E.I.E. Repair of comminuted fracture of the lower patellar pole. *Ulus Travma Acil Cerrahi Derg* **2017**, *23*, 150–155. [PubMed]
4. Zwingmann, J.; Neumann, M.V.; Hammer, T.O.; Reising, K.; Südkamp, N.P. Comminuted Fracture of Elbow—Ostheosynthesis vs. Total Joint Replacement. *Acta Chir. Orthop. Traumatol. Cech.* **2016**, *83*, 231–237.

5. Kakazu, R.; Archdeacon, M.T. Surgical Management of Patellar Fractures. *Orthop. Clin. N. Am.* **2016**, *47*, 77–83. [CrossRef] [PubMed]
6. Farrar, M.T. Bone adhesives for trauma surgery: A review of challenges and developments. *Int. J. Adhes. Adhes.* **2012**, *33*, 89–97. [CrossRef]
7. Onche, I.I.; Osagie, O.E.; INuhu, S. Removal of orthopaedic implants: Indications, outcome and economic implications. *J. West Afr. Coll. Surg.* **2011**, *1*, 101–112.
8. Haseeb, M.; Butt, M.F.; Altaf, T.; Muzaffar, K.; Gupta, A.; Jallu, A. Indications of implant removal: A study of 83 cases. *Int. J. Health Sci.* **2017**, *11*, 1–7.
9. Iwata, T.; Nozawa, S.; Maeda, M.; Akiyama, H. New Technique for Removal of Screws with Damaged Heads. *Orthopedics* **2017**, *40*, e911–e914. [CrossRef] [PubMed]
10. Endres, K.; Marx, R.; Tinschert, J.; Wirtz, D.C.; Stoll, C.; Riediger, D.; Smeets, R. A new adhesive technique for internal fixation in midfacial surgery. *Biomed. Eng. Online* **2008**, *7*, 16. [CrossRef]
11. Sánchez-Fernández, M.J.; Hammoudeh, H.; Félix Lanao, R.P.; van Erk, M.; van Hest, J.C.M.; Leeuwenburgh, S.C.G. Bone-adhesive materials: Clinical requirements, mechanisms of action, and future perspective. *Adv. Mater. Interfaces* **2019**, *6*, 1802021. [CrossRef]
12. Böker, K.O.; Richter, K.; Jäckle, K.; Taheri, S.; Grunwald, I.; Borcherding, K.; von Byern, J.; Hartwig, A.; Wildemann, B.; Schilling, A.F.; et al. Current State of Bone Adhesives-Necessities and Hurdles. *Materials* **2019**, *12*, 3975. [CrossRef] [PubMed]
13. Vaishya, R.; Chauhan, M.; Vaish, A. Bone cement. *J. Clin. Orthop. Trauma* **2013**, *4*, 157–163. [CrossRef] [PubMed]
14. Frazer, R.Q.; Byron, R.T.; Osborne, P.B.; West, K.P. PMMA: An essential material in medicine and dentistry. *J. Long. Term. Eff. Med. Implants* **2005**, *15*, 629–639. [CrossRef] [PubMed]
15. Arora, M.; Chan, E.K.; Gupta, S.; Diwan, A.D. Polymethylmethacrylate bone cements and additives: A review of the literature. *World J. Orthop.* **2013**, *4*, 67–74. [CrossRef]
16. Tsukeoka, T.; Suzuki, M.; Ohtsuki, C.; Sugino, A.; Tsuneizumi, Y.; Miyagi, J.; Kuramoto, K.; Moriya, H. Mechanical and histological evaluation of a PMMA-based bone cement modified with gamma-methacryloxypropyltrimethoxysilane and calcium acetate. *Biomaterials* **2006**, *27*, 3897–3903. [CrossRef] [PubMed]
17. Guerra, N.B.; Hernandez, M.L.; Santos, R.G. Acrylic bone cement modified with hydroxiapatyte/vinyl acetate: Mechanical, thermoanalytical characterization and in vitro bioactivity. *Polímeros* **2010**, *20*, 98–106. [CrossRef]
18. Montaño, C.J.; Campos, T.P.R.; Lemos, B.R.S.; Yoshida, M.I.; Almeida, N.G.S.; Aguilar, M.T.P.; Lima, C.V. Effects of hydroxyapatite on PMMA-HAp cement for biomedical applications. *Biomed. Mater. Eng.* **2020**, *31*, 191–201. [CrossRef] [PubMed]
19. de Souza Leão, R.; de Moraes, S.L.D.; de Luna Gomes, J.M.; Lemos, C.A.A.; da Silva Casado, B.G.; do Egito Vasconcelos, B.C.; Pellizzer, E.P. Influence of addition of zirconia on PMMA: A systematic review. *Mater. Sci. Eng. C Mater. Biol. Appl.* **2020**, *106*, 110292. [CrossRef]
20. Holmes, D. Closing the gap. *Nature* **2017**, *550*, S194–S195. [CrossRef]
21. Majee, S.B. *Emerging Concepts in Analysis and Applications of Hydrogels*; Books on Demand; IntechOpen: London, UK, 2016.
22. Hanafy, N.A.N.; Leporatti, S.; El-Kemary, M.A. Mucoadhesive hydrogel nanoparticles as smart biomedical drug delivery system. *Appl. Sci.* **2019**, *9*, 825. [CrossRef]
23. Norouzi, M.; Nazari, B.; Miller, D.W. Injectable hydrogel-based drug delivery systems for local cancer therapy. *Drug Discov. Today* **2016**, *21*, 1835–1849. [CrossRef] [PubMed]
24. Serna, J.A.; Florez, S.L.; Talero, V.A.; Briceño, J.C.; Muñoz-Camargo, C.; Cruz, J.C. Formulation and Characterization of a SIS-Based Photocrosslinkable Bioink. *Polymers* **2019**, *11*, 569. [CrossRef] [PubMed]
25. Kim, E.; Kim, M.H.; Song, J.H.; Kang, C.; Park, W.H. Dual crosslinked alginate hydrogels by riboflavin as photoinitiator. *Int. J. Biol. Macromol.* **2020**, *154*, 989–998. [CrossRef] [PubMed]
26. Francesko, A.; Petkova, P.; Tzanov, T. Hydrogel Dressings for Advanced Wound Management. *Curr. Med. Chem.* **2018**, *25*, 5782–5797. [CrossRef]
27. Chen, T.; Chen, Y.; Rehman, H.U.; Chen, Z.; Yang, Z.; Wang, M.; Li, H.; Liu, H. Ultratough, Self-Healing, and Tissue-Adhesive Hydrogel for Wound Dressing. *ACS Appl. Mater. Interfaces* **2018**, *10*, 33523–33531. [CrossRef]
28. Qu, J.; Zhao, X.; Liang, Y.; Zhang, T.; Ma, P.X.; Guo, B. Antibacterial adhesive injectable hydrogels with rapid self-healing, extensibility and compressibility as wound dressing for joints skin wound healing. *Biomaterials* **2018**, *183*, 185–199. [CrossRef]
29. Chen, J.; Wang, D.; Wang, L.H.; Liu, W.; Chiu, A.; Shariati, K.; Liu, Q.; Wang, X.; Zhong, Z.; Webb, J.; et al. An Adhesive Hydrogel with Load-Sharing Effect as Tissue Bandages for Drug and Cell Delivery. *Adv. Mater.* **2020**, *32*, e2001628. [CrossRef] [PubMed]
30. Hasani-Sadrabadi, M.M.; Sarrion, P.; Pouraghaei, S.; Chau, Y.; Ansari, S.; Li, S.; Aghaloo, T.; Moshaverinia, A. An engineered cell-laden adhesive hydrogel promotes craniofacial bone tissue regeneration in rats. *Sci. Transl. Med.* **2020**, *12*, 534. [CrossRef]
31. Qiao, Y.; Liu, X.; Zhou, X.; Zhang, H.; Zhang, W.; Xiao, W.; Pan, G.; Cui, W.; Santos, H.A.; Shi, Q. Gelatin Templated Polypeptide Co-Cross-Linked Hydrogel for Bone Regeneration. *Adv. Health Mater.* **2020**, *9*, e1901239. [CrossRef]
32. Ding, J.; He, R.; Zhou, G.; Tang, C.; Yin, C. Multilayered mucoadhesive hydrogel films based on thiolated hyaluronic acid and polyvinylalcohol for insulin delivery. *Acta Biomater.* **2012**, *8*, 3643–3651. [CrossRef]
33. Tran, P.H.L.; Tran, T.T.D. Mucoadhesive Formulation Designs for Oral Controlled Drug Release at the Colon. *Curr. Pharm. Des.* **2021**, *27*, 540–547. [CrossRef] [PubMed]
34. Muxika, A.; Etxabide, A.; Uranga, J.; Guerrero, P. Chitosan as a bioactive polymer: Processing, properties and applications. *Int. J. Biol. Macromol.* **2017**, *105*, 1358–1368. [CrossRef] [PubMed]

35. Qasim, S.B.; Zafar, M.S.; Najeeb, S.; Khurshid, Z.; Shah, A.H.; Husain, S.; Rehman, I.U. Electrospinning of Chitosan-Based Solutions for Tissue Engineering and Regenerative Medicine. *Int. J. Mol. Sci.* **2018**, *19*, 407. [CrossRef]
36. Husain, S.; Al-Samadani, K.H.; Najeeb, S.; Zafar, M.S.; Khurshid, Z.; Zohaib, S.; Qasim, S.B. Chitosan Biomaterials for Current and Potential Dental Applications. *Materials* **2017**, *10*, 602. [CrossRef]
37. Patrulea, V.; Ostafe, V.; Borchard, G.; Jordan, O. Chitosan as a starting material for wound healing applications. *Eur. J. Pharm. Biopharm.* **2015**, *97*, 417–426. [CrossRef]
38. Wang, W.; Meng, Q.; Li, Q.; Liu, J.; Zhou, M.; Jin, Z.; Zhao, K. Chitosan Derivatives and Their Application in Biomedicine. *Int. J. Mol. Sci.* **2020**, *21*, 487. [CrossRef]
39. Ibrahim, H.M.; El-Zairy, E.M.R. Chitosan as a biomaterial—Structure, properties, and electrospun nanofibers. *Concepts Compd. Altern. Antibact.* **2015**, *1*, 81–101.
40. Gonzalez-Melo, C.; Garcia-Brand, A.J.; Quezada, V.; Reyes, L.H.; Muñoz-Camargo, C.; Cruz, J.C. Highly Efficient Synthesis of Type B Gelatin and Low Molecular Weight Chitosan Nanoparticles: Potential Applications as Bioactive Molecule Carriers and Cell-Penetrating Agents. *Polymers* **2021**, *13*, 4078. [CrossRef] [PubMed]
41. Diolosà, M.; Donati, I.; Turco, G.; Cadenaro, M.; Di Lenarda, R.; Breschi, L.; Paoletti, S. Use of methacrylate-modified chitosan to increase the durability of dentine bonding systems. *Biomacromolecules* **2014**, *15*, 4606–4613. [CrossRef] [PubMed]
42. Zapata, M.E.V.; Tovar, C.D.G.; Hernandez, J.H.M. The Role of Chitosan and Graphene Oxide in Bioactive and Antibacterial Properties of Acrylic Bone Cements. *Biomolecules* **2020**, *10*, 1616. [CrossRef]
43. Shamekhi, M.A.; Mirzadeh, H.; Mahdavi, H.; Rabiee, A.; Mohebbi-Kalhori, D.; Baghaban, E.M. Graphene oxide containing chitosan scaffolds for cartilage tissue engineering. *Int. J. Biol. Macromol.* **2019**, *127*, 396–405. [CrossRef]
44. Prakash, J.; Prema, D.; Venkataprasanna, K.S.; Balagangadharan, K.; Selvamurugan, N.; Venkatasubbu, G.D. Nanocomposite chitosan film containing graphene oxide/hydroxyapatite/gold for bone tissue engineering. *Int. J. Biol. Macromol.* **2020**, *154*, 62–71. [CrossRef]
45. Rokaya, D.; Srimaneepong, V.; Qin, J.; Siraleartmukul, K.; Siriwongrungson, V. Graphene Oxide/Silver Nanoparticle Coating Produced by Electrophoretic Deposition Improved the Mechanical and Tribological Properties of NiTi Alloy for Biomedical Applications. *J. Nanosci. Nanotechnol.* **2019**, *19*, 3804–3810. [CrossRef]
46. Rokaya, D.; Srimaneepong, V.; Thunyakitpisal, P.; Qin, J.; Rosa, V.; Sapkota, J. Potential Applications of Graphene-Based Nanomaterials in Biomedical, Dental, and Implant Applications. *Adv. Dent. Implantol. Nanomater. Allied Technol. Appl.* **2021**, *1*, 77–105.
47. Srimaneepong, V.; Rokaya, D.; Thunyakitpisal, P.; Qin, J.; Saengkiettiyut, K. Corrosion Resistance of Graphene oxide/Silver Coatings on Ni-Ti alloy and Expression of IL-6 and IL-8 in Human Oral Fibroblasts. *Sci. Rep.* **2020**, *10*, 3247. [CrossRef] [PubMed]
48. Pipattanachat, S.; Qin, J.; Rokaya, D.; Thanyasrisung, P.; Srimaneepong, V. Biofilm inhibition and bactericidal activity of NiTi alloy coated with graphene oxide/silver nanoparticles via electrophoretic deposition. *Sci Rep.* **2021**, *11*, 14008. [CrossRef]
49. Patarroyo, J.L.; Fonseca, E.; Cifuentes, J.; Salcedo, F.; Cruz, J.C.; Reyes, L.H. Gelatin-Graphene Oxide Nanocomposite Hydrogels for *Kluyveromyces lactis* Encapsulation: Potential Applications in Probiotics and Bioreactor Packings. *Biomolecules* **2021**, *11*, 922. [CrossRef] [PubMed]
50. Raslan, A.; Saenz, D.B.L.; Ciriza, J.; Pedraz, J.L. Graphene oxide and reduced graphene oxide-based scaffolds in regenerative medicine. *Int. J. Pharm.* **2020**, *580*, 119226. [CrossRef]
51. Shen, Y.; Tang, H.; Huang, X.; Hang, R.; Zhang, X.; Wang, Y.; Yao, X. DLP printing photocurable chitosan to build bio-constructs for tissue engineering. *Carbohydr. Polym.* **2020**, *235*, 115970. [CrossRef] [PubMed]
52. Marcano, D.C.; Kosynkin, D.V.; Berlin, J.M.; Sinitskii, A.; Sun, Z.; Slesarev, A.; Alemany, L.B.; Lu, W.; Tour, J.M. Improved Synthesis of Graphene Oxide. *ACS Nano* **2010**, *4*, 4806–4814. [CrossRef]
53. Rueda-Gensini, L.; Serna, J.A.; Cifuentes, J.; Cruz, J.C.; Muñoz-Camargo, C. Graphene Oxide-Embedded Extracellular Matrix-Derived Hydrogel as a Multiresponsive Platform for 3D Bioprinting Applications. *Int. J. Bioprinting* **2021**, *7*, 353. [CrossRef]
54. Cedano, F.J.; Pinzón, L.M.; Narváez, D.M.; Castro, C.I.; Moreno-Serrano, C.L.; Tabima, D.M.; Salcedo, F.; Briceno, J.C.; Casas-Rodriguez, J.P. Evaluation of a water-resistant and biocompatible adhesive with potential use in bone fractures. *J. Adhes. Sci. Technol.* **2016**, *31*, 1480–1495. [CrossRef]
55. Lopez-Barbosa, N.; Suárez-Arnedo, A.; Cifuentes, J.; Gonzalez Barrios, A.F.; Silvera Batista, C.A.; Osma, J.F.; Muñoz-Camargo, C.; Cruz, J.C. Magnetite OmpA Nanobioconjugates as Cell-Penetrating Vehicles with Endosomal Escape Abilities. *ACS Biomater. Sci. Eng.* **2019**, *6*, 415–424. [CrossRef]
56. Li, S.; Sun, G.X.; Chang, S.M.; Yang, C.S.; Li, Y.; Niu, W.; Zhang, C. Simulated postoperative weight-bearing after fixation of a severe osteoporotic intertrochanteric fracture. *Int. J. Clin. Exp. Med.* **2017**, *10*, 8438–8448.
57. Harper, E.J.; Bonfield, W. Tensile characteristics of ten commercial acrylic bone cements. *J. Biomed. Mater. Res.* **2002**, *53*, 605–616. [CrossRef]
58. Sharma, K.; Kaith, B.S.; Kumar, V.; Kalia, S.; Kapur, B.K.; Swart, H.C. Corrigendum to: A comparative study of the effect of Ni9+ and Au8+ ion beams on the properties of poly(methacrylic acid) grafted gum ghatti films. *Radiat. Phys. Chem.* **2014**, *99*, 97. [CrossRef]
59. Krylova, V.; Dukštienė, N. The structure of PA-Se-S-Cd composite materials probed with FTIR spectroscopy. *Appl. Surf. Sci.* **2019**, *470*, 462–471. [CrossRef]

60. Coussot, P. Introduction to rheology and fluid mechanics. In *Mudflow Rheology and Dynamics*; Taylor & Francis, Routledge: Abington, UK, 2017; pp. 25–44.
61. Banerjee, C.; Ghosh, S.; Sen, G.; Mishra, S.; Shukla, P.; Bandopadhyay, R. Study of algal biomass harvesting using cationic guar gum from the natural plant source as flocculant. *Carbohydr. Polym.* **2013**, *92*, 675–681. [CrossRef]
62. Yoo, H.J.; Mahapatra, S.S.; Cho, J.W. High-Speed Actuation and Mechanical Properties of Graphene-Incorporated Shape Memory Polyurethane Nanofibers. *J. Phys. Chem. C* **2014**, *118*, 10408–10415. [CrossRef]
63. Strankowski, M.; Włodarczyk, D.; Piszczyk, L.; Strankowska, J. Polyurethane Nanocomposites Containing Reduced Graphene Oxide FTIR Raman, and XRD Studies. *J. Spectrosc.* **2016**, *2016*, 1–6. [CrossRef]
64. Krishnamoorthy, K.; Veerapandian, M.; Yun, K.; Kim, S.-J. The chemical and structural analysis of graphene oxide with different degrees of oxidation. *Carbon* **2013**, *53*, 38–49. [CrossRef]
65. Krishnamoorthy, K.; Veerapandian, M.; Mohan, R.; Kim, S.-J. Investigation of Raman and photoluminescence studies of reduced graphene oxide sheets. *Appl. Phys. A* **2011**, *106*, 501–506. [CrossRef]
66. Eda, G.; Chhowalla, M. Chemically Derived Graphene Oxide: Towards Large-Area Thin-Film Electronics and Optoelectronics. *Adv. Mater.* **2010**, *22*, 2392–2415. [CrossRef]
67. Muzyka, R.; Drewniak, S.; Pustelny, T.; Chrubasik, M.; Gryglewicz, G. Characterization of Graphite Oxide and Reduced Graphene Oxide Obtained from Different Graphite Precursors and Oxidized by Different Methods Using Raman Spectroscopy. *Materials* **2018**, *11*, 1050. [CrossRef] [PubMed]
68. Arul, R.; Osterbeek, R.N.; Robertson, J.; Xu, G.; Jin, J.; Simpson, M.C. The mechanism of direct laser writing of graphene features into graphene oxide films involves photoreduction and thermally assisted structural rearrangement. *Carbon* **2016**, *99*, 423–431. [CrossRef]
69. Feng, H.; Cheng, R.; Zhao, X.; Duan, X.; Li, J. A low-temperature method to produce highly reduced graphene oxide. *Nat. Commun.* **2013**, *4*, 1–8. [CrossRef] [PubMed]
70. Guo, H.L.; Wang, X.F.; Qian, Q.Y.; Wang, F.B.; Xia, X.H. Synthesis of graphene-based nanosheets via chemical reduction of exfoliated graphite oxide. *Carbon* **2007**, *45*, 1558–1565.
71. Song, M.; Yu, L.; Wu, Y. Simple Synthesis and Enhanced Performance of Graphene Oxide-Gold Composites. *J. Nanomater.* **2012**, *2012*, 37. [CrossRef]
72. Stobinski, L.; Lesiak, B.; Malolepszy, A.; Mazurkiewicz, M.; Mierzwa, B.; Zemek, J.; Bieloshapka, I. Graphene oxide and reduced graphene oxide studied by the XRD TEM and electron spectroscopy methods. *J. Electron. Spectrosc. Relat. Phenom.* **2014**, *195*, 145–154. [CrossRef]
73. Aziz, M.; Halim, F.S.A.; Jaafar, J. Preparation and Characterization of Graphene Membrane Electrode Assembly. *J. Technol.* **2014**, 146–152. [CrossRef]
74. Paxton, N.; Smolan, W.; Böck, T.; Melchels, F.; Groll, J.; Jungst, T. Proposal to assess printability of bioinks for extrusion-based bioprinting and evaluation of rheological properties governing bioprintability. *Biofabrication* **2017**, *9*, 044107. [CrossRef]
75. Sandolo, C.; Matricardi, P.; Alhaique, F.; Coviello, T. Effect of temperature and cross-linking density on rheology of chemical cross-linked guar gum at the gel point. *Food Hydrocoll.* **2009**, *23*, 210–220. [CrossRef]
76. Zhang, X.; Xu, L.; Wei, S.; Zhai, M.; Li, J. Stimuli responsive deswelling of radiation synthesized collagen hydrogel in simulated physiological environment. *J. Biomed. Mater. Res. A* **2012**, *101*, 2191–2201. [CrossRef]
77. Sánchez-Cid, P.; Jiménez-Rosado, M.; Alonso-González, M.; Romero, A.; Perez-Puyana, V. Applied Rheology as Tool for the Assessment of Chitosan Hydrogels for Regenerative Medicine. *Polymers* **2021**, *13*, 2189. [CrossRef] [PubMed]
78. Jafarigol, E.; Salehi, M.B.; Mortaheb, H.R. Preparation and assessment of electro-conductive poly(acrylamide-co-acrylic acid) carboxymethyl cellulose/reduced graphene oxide hydrogel with high viscoelasticity. *Chem. Eng. Res. Des.* **2020**, *162*, 74–84. [CrossRef]
79. Sahraei, R.; Ghaemy, M. Synthesis of modified gum tragacanth/graphene oxide composite hydrogel for heavy metal ions removal and preparation of silver nanocomposite for antibacterial activity. *Carbohydr. Polym.* **2017**, *157*, 823–833. [CrossRef]
80. Tohamy, H.-A.S.; El-Sakhawy, M.; Kamel, S. Carboxymethyl Cellulose-Grafted Graphene Oxide/Polyethylene Glycol for Efficient Ni(II) Adsorption. *J. Polym. Environ.* **2020**, *29*, 859–870. [CrossRef]
81. Tuğcu-Demiröz, F. Development of in situ poloxamer-chitosan hydrogels for vaginal drug delivery of benzydamine hydrochloride: Textural mucoadhesive and in vitro release properties. *Marmara Pharm. J.* **2017**, *21*, 762–770. [CrossRef]
82. Kakkar, P.; Madhan, B. Fabrication of keratin-silica hydrogel for biomedical applications. *Mater. Sci. Eng.* **2016**, *66*, 178–184. [CrossRef]
83. Karavana, S.Y.H.; Güneri, P.; Ertan, G. Benzydamine hydrochloride buccal bioadhesive gels designed for oral ulcers: Preparation rheological, textural, mucoadhesive and release properties. *Pharm. Dev. Technol.* **2009**, *14*, 623–631. [CrossRef]
84. Da Silva, J.B.; Ferreira, S.; Reis, A.; Cook, M.; Bruschi, M. Assessing Mucoadhesion in Polymer Gels: The Effect of Method Type and Instrument Variables. *Polymers* **2018**, *10*, 254. [CrossRef]
85. Sezer, A.D.; Cevher, E.; Hatipoğlu, F.; Oğurtan, Z.; Baş, A.L.; Akbuğa, J. Preparation of Fucoidan-Chitosan Hydrogel and Its Application as Burn Healing Accelerator on Rabbits. *Biol. Pharm. Bull.* **2008**, *31*, 2326–2333. [CrossRef]
86. Figueroa, T.; Aguayo, C.; Fernandez, K. Design and Characterization of Chitosan-Graphene Oxide Nanocomposites for the Delivery of Proanthocyanidins. *Int. J. Nanomed.* **2020**, *15*, 1229–1238. [CrossRef]

87. Vallabani, N.V.; Mittal, S.; Shukla, R.K.; Pandey, A.K.; Dhakate, S.R.; Pasricha, R.; Dhawan, A. Toxicity of Graphene in Normal Human Lung Cells (BEAS-2B). *J. Biomed. Nanotechnol.* **2011**, *7*, 106–107. [CrossRef] [PubMed]
88. Sharif, P.S.; Abdollahi, M. The Role of Platelets in Bone Remodeling. *Inflamm. Allergy Drug Targets* **2010**, *9*, 393–399. [CrossRef] [PubMed]
89. Nurden, A.T. Platelets and wound healing. *Front. Biosci.* **2008**, *13*, 3525. [CrossRef] [PubMed]
90. Chou, T.-C.; Fu, E.; Wu, C.-J.; Yeh, J.-H. Chitosan enhances platelet adhesion and aggregation. *Biochem. Biophys. Res. Commun.* **2003**, *302*, 480–483. [CrossRef]
91. Blinc, A.; Božič, M.; Vengust, R.; Stegnar, M. Methyl-methacrylate bone cement surface does not promote platelet aggregation or plasma coagulation in vitro. *Thromb. Res.* **2004**, *114*, 179–184. [CrossRef] [PubMed]
92. Flieger, M.; Bandouchova, H.; Cerny, J.; Chudíčková, M.; Kolarik, M.; Kovacova, V.; Martínková, N.; Novák, P.; Šebesta, O.; Stodůlková, E. Vitamin B2 as a virulence factor in Pseudogymnoascus destructans skin infection. *Sci. Rep.* **2016**, *6*, 33200. [CrossRef]
93. Delgado-Calle, J.; Riancho, J.A. Mecanobiología celular y molecular del tejido óseo. *Rev. Osteoporos. Metab. Miner.* **2013**, *5*, 51–56. [CrossRef]
94. Yan, L.; Lim, J.L.; Lee, J.W.; Tia, C.S.H.; O'Neill, G.K.; Chong, D.Y.R. Finite element analysis of bone and implant stresses for customized 3D-printed orthopaedic implants in fracture fixation. *Med. Biol. Eng. Comput.* **2020**, *58*, 921–931. [CrossRef] [PubMed]

polymers

MDPI

Article

Populus alba L., an Autochthonous Species of Spain: A Source for Cellulose Nanofibers by Chemical Pretreatment

David Ibarra [1], Raquel Martín-Sampedro [1], Bernd Wicklein [2], Antonio M. Borrero-López [3], Concepción Valencia [3], Ana Valdehíta [4], José M. Navas [4] and María E. Eugenio [1,*]

1 Forest Research Center (INIA, CSIC), Ctra. de la Coruña Km 7.5, 28040 Madrid, Spain; ibarra.david@inia.es (D.I.); raquel.martin@inia.es (R.M.-S.)
2 Materials Science Institute of Madrid (ICMM), Consejo Superior de Investigaciones Científicas (CSIC), Sor Juana Inés de la Cruz 3, 28049 Madrid, Spain; bernd@icmm.csic.es
3 Pro2TecS—Chemical Process and Product Technology Research Centre, Departamento de Ingeniería Química, ETSI, Campus de "El Carmen", Universidad de Huelva, 21071 Huelva, Spain; am.borrero@diq.uhu.es (A.M.B.-L.); barragan@diq.uhu.es (C.V.)
4 Environment and Agronomy Department (INIA, CSIC), Ctra. de la Coruña Km 7.5, 28040 Madrid, Spain; ana.valdehita@inia.es (A.V.); jmnavas@inia.es (J.M.N.)
* Correspondence: mariaeugenia@inia.es; Tel.: +34-913473948

Abstract: In order to identify new sustainable sources for producing cellulose nanofibers (CNFs), fast-growing poplar (*Populus alba* L.) wood was evaluated herein. For that purpose, bleached poplar kraft pulp was produced and submitted to TEMPO (2,2,6,6-tetramethylpiperidine-1-oxyl radical) mediated oxidation (TEMPO-ox) chemical pretreatment followed by microfluidization. The resulting CNFs were thoroughly characterized, including a rheological study at different pH values. Poplar CNFs showed properties comparable to eucalypt CNFs (reference material for CNFs production), showing high carboxylate content (1048 $\pm$ 128 μmol g^{-1}), fibrillation yield (87.3% $\pm$ 8.1%), optical transmittance (83% at 700 nm) and thermal stability (up to more than 200 °C). Regarding the rheological study, whereas pH from 4 to 10 did not produce significant changes in rheological behavior, a reduction of pH down to 1 led to an order-of-magnitude increase on the viscoelastic functions. Therefore, poplar CNF shows potential in the pH-sensitive hydrogels application field. Finally, the possible ecotoxicity of poplar CNF was assessed. The decrease in cell viability was very low so that only concentrations causing a 10% cytotoxicity could be calculated for the assay detecting alterations in cell metabolism (10 μg mL^{-1}) and plasma membrane integrity (60 μg mL^{-1}).

Keywords: cellulose nanofibers; chemical pretreatment; cytotoxicity; nanocellulose; *Populus alba* L.; rheology

Citation: Ibarra, D.; Martín-Sampedro, R.; Wicklein, B.; Borrero-López, A.M.; Valencia, C.; Valdehíta, A.; Navas, J.M.; Eugenio, M.E. *Populus alba* L., an Autochthonous Species of Spain: A Source for Cellulose Nanofibers by Chemical Pretreatment. *Polymers* **2022**, *14*, 68. https://doi.org/10.3390/polym14010068

Academic Editor: Jem-Kun Chen

Received: 4 November 2021
Accepted: 22 December 2021
Published: 25 December 2021

1. Introduction

The search for new and abundant sources of lignocellulosic biomass to supply biorefineries, for bioproducts and biofuels production, is currently on the rise in order to eliminate our dependence on fossil fuels [1]. In this context, the Short Rotation Coppice system (SRC) is one of the most effective options for producing lignocellulosic biomass. One of the advantages of the SRC system, from a logistic point of view, is its tendency to be localized, in space and time [2]. Among the raw materials suitable for obtaining biomass through SRC are species within the genus *Populus* [3]. Moreover, these species show a hardness and plasticity suitable for growing in marginal areas affected by salinity, high temperatures and dry climates, such as Mediterranean conditions [3]. These materials have contents of 42–49% in cellulose, higher than in other hardwood such as eucalypt [4], together with 16–23% in hemicellulose and 21–29% in lignin [5,6]. The high sugar content of different species and hybrids of genus *Populus* has promoted their use as a source of biomass in lignocellulosic biorefineries. For example, these raw materials have been used to obtain fermentable sugars [5,6], from which bioethanol [7] and xylitol are produced [8].

Moreover, different bioproducts such as activated carbon for water defluorination [9] or adhesives [10] have also been obtained from its ligninolytic fraction. However, these materials have scarcely been studied to produce cellulose-based products, such as nanocellulose.

Nanocellulose is currently in demand due not only to its specific properties, i.e., renewably sourced, biodegradable, and biocompatible, but also to its excellent characteristics [11]. In particular, nanocellulose and more specifically cellulose nanofibers (CNFs) are characterized by properties such as transparency, low density, high mechanical resistance, thermal and surface stability, and versatility of chemical modification [12]. These characteristics make it an ideal material to be used in several applications such as paper; construction materials; packaging; paints; medical applications; and flexible, printed electronics among others [13–16].

CNFs form long, flexible fiber networks, which are composed of microfibrils with a nano-size diameter that depends on the method used to obtain them [17]. CNFs can be obtained by mechanical microfibrillation processes (using refiner, grinder, homogenizer, microfluidizer, etc.) from cellulose pulp, although with high energy consumption. Chemical, biological and mechanical pretreatments have been studied to reduce the energy consumption of the microfibrillation process in order to allow the implementation of the CNFs production at an industrial scale [14]. Among the chemical pretreatments, TEMPO (2,2,6,6-tetramethylpiperidine-1-oxyl radical) mediated oxidation (TEMPO-ox) has been selected as one of the most effective in reducing the energy consumption in the further mechanical microfibrillation treatment. Under alkaline conditions, the system TEMPO/NaBr/NaClO can oxidize the C6-primary hydroxyl groups found on the surface of cellulose microfibers to carboxyl groups [18]. To carry out this reaction, only NaClO and NaOH are consumed [19]. This reaction negatively charges the fibril surface creating a repulsive effect between adjacent fibrils. Consequently, the efficiency of the separation of the individual nanofibers is increased in the subsequent mechanical microfibrillation treatment, reducing the energy consumed in the whole process [20].

Not only the technology used but also the starting raw material has a marked influence on the characteristic of the CNFs obtained [21]. Therefore, the search for new raw materials as alternatives to traditional woody species (e.g., eucalypt, birch, pine, spruce, etc.) used for CNFs production is a research line of interest to open even more the range of applications of the CNF [14]. As it was mentioned before, to our knowledge, only a few studies about CNFs production from different species and hybrids of genus *Populus* have been reported and none of them have used poplar biomass obtained by SRC or TEMPO-ox process as pre-treatment. For example, Qi et al. [22] have used poplar residues (branches) taken from the suburbs of the city of TaiAn (China). Then, nanocellulose was produced from chemically purified cellulose via combined mechanical treatments of grinding and high-pressure homogenization. The nanocellulose obtained by these authors was applied in the preparation of nanopaper. Zhao et al. [23] studied the thermostability of nanocellulose obtained from poplar wood supplied by a pilot plant located in Jiangsu, China. These authors produced nanocellulose from purified cellulose of poplar wood via mechanical treatment with high pressure-grinding. Wu et al. [24] have studied the nanocellulose production from poplar (*Populus tomentosa*) catkin fiber via sonication. Finally, Ibarra et al. [25] produced microfibrillated cellulose from fast-growing poplar using a physical pretreatment (PFI refining) before the microfibrillation stage, showing similar properties to that obtained from a commercial industrial eucalypt pulp. Then, films from this microfibrillated cellulose were produced with high mechanical properties and low wettability.

Considering all the above, the objective of this work is to prove the use of *Populus alba* L. "PO-10-10-20", an autochthonous genotype of Spain currently tested in SRC plantations for biomass production, as a source for CNFs production. To this end, TEMPO-ox reaction as pre-treatment and subsequent defibrillation in a high-pressure microfluidizer was used. Resulting CNFs were thoroughly characterized (carboxylate content, fibrillation yield, optical transmittance and thermal stability), including X-ray diffraction and Fourier transform infrared spectroscopy (FTIR). Moreover, an extensive rheological study modifying the pH of the CNFs was also carried out. Finally, the possible ecotoxicity of poplar CNFs was also assessed.

2. Materials and Methods

2.1. Raw Materials and Chemicals

The autochthonous genotype of *P. alba* L., "PO-10-10-20", was supplied by the Silviculture and Forest Management Department of INIA, CSIC (Madrid, Spain). The short rotation planting protocol in Mediterranean conditions described in Sixto et al. [26] was used for planting this genotype. The plantation was cut after three years of growth (first rotation) and the resulting material was used to produce the CNF.

CNF from eucalypt was produced in previous work [21] and was used as a reference hydrogel.

Sodium hypochlorite solution (NaClO), TEMPO, sodium bromide (NaBr), hydrochloric acid (HCl), sodium hidroxyde (NaOH), sodium sulfide (Na_2S), were purchased from Sigma–Aldrich (Steinheim, Germany) and magnesium sulphate ($MgSO_4$), hydrogen peroxide (H_2O_2) and diethylenetriaminepentaacetic acid (DTPA) from Merck Life Science S.L (Madrid, Spain).

For cell culture, in the cytoxicity analysis, L-Glutamine (200 mM), penicillin and streptomycin (P/S) (10,000 U mL^{-1} each), trypsin-ethylene diamine tetraacetic acid (EDTA) (200 mg/mL), and Eagle's Minimum Essential Medium (EMEM) cell culture and non-essential aminoacids, (NEAA) were purchased from Lonza (Barcelona, Spain). Phenol-red free serum-free Minimum Essential Medium (MEM) was supplied by Gibco (Life Technologies, Madrid, Spain). Among the reactants used for determining cytotoxicity, AlamarBlue reagent and 5-carboxyfluorescein diacetate, acetoxy methyl ester (CFDA-AM) and neutral red were purchased from Invitrogen (Madrid, Spain).

2.2. Bleached Pulp Production

P. alba L. "PO-10-10-20" chips were subjected to a kraft pulping process using a 15-litre batch cylindrical reactor which incorporates a computer-controlled electric jacket heater. To ensure proper agitation, the reactor was equipped with a rotating shaft. Pulping conditions were as follows [27]: 17% active alkali, 20% sulphate (related to sulphur content), 170 °C, 120 min, and a liquid/solid ratio of 7:1. The resulting material from the pulping was subjected to a wet disintegration process for 30,000 revolutions and then screened through a 0.16 mm mesh.

After pulping, a standard industrial totally chlorine-free (TCF) bleaching sequence consisting of OOQPoP was carried out, where O was a delignification stage with oxygen (held twice), Q was a chelating stage, and Po and P were hydrogen peroxide bleaching stages with or without oxygen, respectively. Conditions of both oxygen delignification stages were: 6 MPa of oxygen pressure, 60 min, 98 °C, 0.5% $MgSO_4$ o.d.p. (oven-dry pulp), 1.5% NaOH o.d.p., and 20% of consistency. The oxygen steps were carried out using the same reactor described above for the pulping process. The chelating step was performed using 0.3% DTPA at 85 °C for 60 min (pH 5–6). Finally, bleaching conditions in the first hydrogen peroxide step was: 2% NaOH, 0.5% $Na_2Si_2O_3$, 0.1% $MgSO_4$ and 3% H_2O_2 for 140 min at 105 °C pressurized with oxygen at 6 kg cm^{-2} while the second step was carried out using the same chemical reagents but during 180 min at 98 °C without pressure [28]. Pulps samples were washed after each stage using distilled water and stored at 4 °C.

2.3. CNF Production

The bleached pulp obtained from *P. alba* L. "PO-10-10-20", was first subjected to a demineralization process. For this purpose, 45 g of bleached pulp was soaked in a diluted solution of HCl (4500 mL) at pH $\sim$2 for 20 min at room temperature with mechanical stirring. Subsequently, the pulp was washed with distilled water repeatedly by filtration. For CNF production, the already demineralized pulp was subjected to an oxidation process with TEMPO as described by [18]. Thus, the pulp was dispersed in a solution containing 0.016 g/g o.d.p. of TEMPO and 0.1 g/g o.d.p. of NaBr to which a solution of 10 mmol/g o.d.p. NaClO was added maintaining the pH of the total dispersion at 10 by adding 0.5 M NaOH, until no decrease in pH was observed. After that, the dispersion containing the oxidized fibers was filtered and washed up to pH 7, adjusted to a 2% consistency and stored at 4 °C. To obtain the final product (CNF), the resulting dispersion at 2% was subjected to a mechanical process, passing three times by a high-pressure microfluidizer (Microfluidizer M-110EH, Microfluidics Corp., Westwood, MA, USA) using interaction chambers of 100 and 200 μm. Obtained CNF (referenced as poplar CNF) was stored at 4 °C.

2.4. Pulp and CNF Characterization

2.4.1. Chemical Composition

The procedure provided by The National Renewable Energies Laboratory [29] was used to determine the chemical composition of poplar pulp and CNF. Thus, the analysis of carbohydrates and lignin was carried out by means of acid hydrolysis. From the liquid fraction resulting from the hydrolysis, carbohydrates were quantified by liquid chromatography (HPLC) on an Agilent Technologies 1260 HPLC (Agilent, Waldbronn, Germany) using an Agilent Hi-PlexPb column operated at 70 °C with ultrapure water as a pumped mobile phase at a rate of 0.6 mL min^{-1}. This HPLC was equipped with a refractive index detector (Agilent, Waldbronn, Germany). On the other hand, the residual solid from hydrolysis is denominated insoluble acid lignin (Klason lignin).

2.4.2. Fibrillation Yield

The determination of the fibrillation yield of poplar CNF was carried out by centrifugation. For this purpose, a 0.1% CNF dispersion was centrifuged for 20 min at 4500 rpm in order to separate the nanofibrillated material (present in the supernatant fraction) from the partially or non-fibrillated one (sediment) as described by Besbes et al. [30]. The fibrillation yield (nanofibrillated part) was calculated by Equation (1):

$$\text{Yield (\%)} = 100\ (1 - (\text{Wp}/\text{Ws}) \quad (1)$$

where Wp is the weight of the dry sediment after centrifugation (g o.d.p.) and Wp is the weight of the CNF sample before centrifugation (g o.d.p.). Mean values and standard deviations were calculated from triplicates.

2.4.3. Optical Transmittance

To determine the optical transmittance of the sample, a 0.1% (*w*/*v*) dispersion of poplar CNF was prepared in water. The transmittance was measured using a Lambda 365 UV—vis spectrometer (PerkinElmer, Waltham, MA, USA) from 285 to 750 nm.

2.4.4. Polymerization Degree

The polymerization degree (DPv) of the pulp and the poplar CNF, was calculated using the Mark–Houwink Equation (2):

$$\mu = \text{Kp} \times \text{DPv}^{\text{a}} \quad (2)$$

where μ is the viscosity measured in the samples according to ISO 5351-2 using a cupri-etilendiamine (CED) solution and where a and Kp are constant (0.80 mL g^{-1} and 1.70, respectively) [31]. Mean values and standard deviations were calculated from triplicates.

2.4.5. Carboxylate Content

The conductometric titration method described by Fras et al. [32] with some modifications was used for the determination of the carboxylate content of the poplar CNF. By means of an ion-exchange treatment, the sodium carboxylate groups present in the sample were the first converted into free carboxyl. For this purpose, 0.10 g o.d.p. of the sample was dispersed in 185 mL of a 1 mM NaCl solution by stirring under a nitrogen atmosphere for 30 min at 25 °C. By adding 0.1 M HCl the pH of the solution was adjusted to <3. Subsequently, the titration was performed with the addition of 0.05 M NaOH at 30 s intervals. A Crison GLP 31 conductivity meter (Crison, L'Hospitalet de Llobregat, Spain) was used to measure conductivity. Equation (3) was used to determine the carboxylate content of the sample:

$$\text{Carboxylate content (mmol L}^{-1}) = \text{c (V2}-\text{V1)/w} \tag{3}$$

where c is the concentration of NaOH (M), V1 and V2 are the volume of NaOH (mL) consumed in the first and second point of intersection respectively, w is the dry weight of the sample (g o.d.p.). Mean values and standard deviations were calculated from triplicates.

2.4.6. X-ray Diffraction (XRD)

The crystallinity of the poplar pulp and CNF was analyzed by X-ray powder diffraction. The equipment used was Bruker D8 Advance (Bruker, Billerica, MA, USA) which was equipped with a Cu anode (radiation CuKα) and a Ni filter. The preparation of the sample for such an analysis requires the preparation of a film by casting. Equation (4) was used to calculate the crystallinity index (CI) by the Segal method:

$$CI = [(I_{002}-I_{AM})/I_{002}]100 \tag{4}$$

where I_{AM} is the intensity at approximately 2θ = 18.5° attributed to the amorphous phase and I_{002} is the intensity of the (002) reflection.

Using the reflection (002) according to the Scherrer equation, the crystallite size t was calculated:

$$t = k \times \lambda/(\beta \times \cos\theta) \tag{5}$$

where λ is the radiation wavelength, k = 0.9 is the correction factor, θ is the diffraction angle and β is the full width at half maximum.

2.4.7. Cytotoxicity Analysis

The stock suspension of CNFs (poplar CNF and eucalypt CNF used as reference) were prepared following the NANOGENOTOX dispersion [33]. CNFs cytotoxicity was studied in vitro by means of the topminnow fish hepatoma cell line (PLHC-1) obtained from the American Type Culture Collection (Manassas, VA, USA). PLHC-1 was maintained in EMEM (with NEAA and Na Pyruvate) supplemented with 1% L-glutamine, 1% P/S and 5% FBS at 30 °C with 5% CO_2.

PLHC-1 cells were seeded in flat-bottomed 96-well plates (Greiner Bio-One GmbH, Germany) at a density of 5×10^4 cells mL^{-1}. Following 24 h, cells were exposed to a concentration range of 0 to 256 μg mL^{-1} of CNFs (dilution factor 2) for 72 h. Although exposure concentrations may not reflect environmentally realistic scenarios, the concentration range chosen allowed us to construct the logistic curve from which inhibitory concentration (IC, concentration causing a given percentage of cell growth inhibition) values were calculated. Control wells were treated with medium or medium plus BSA. Sodium dodecyl sulphate (SDS, 65.84 to 500 μM, dilution factor 1.5) served as a positive cytotoxicity control. At least three independent experiments were carried out with a minimum of three replicates of each CNFs concentration in each plate. Cell viability was measured through three different cytotoxicity assays applied on the same set of cells according to a modified version [34,35] of a protocol described by Dayeh et al. [36]. The three cytotoxicity assays were the AlamarBlue assay, informing about alterations in the cellular metabolism; the CFDA-AM assay, which detects disruption of the plasma membrane integrity; and the Neutral Red assay which reflects lysosomal misfunctioning. The experimental results were compared to their corresponding control values using one-way repeated measures analysis of variance (ANOVA; $p < 0.05$) followed by a post hoc Dunnett's test.

2.4.8. Thermogravimetric Analysis (TGA)

The thermogram of the dried poplar CNF was recorded from room temperature to 600 °C by a steady rate of 10 °C min^{-1} under a nitrogen atmosphere in a Q-50 (TA Instrument Waters, New Castle, DE, USA) apparatus.

2.4.9. Fourier-Transform Infrared Spectroscopy (FTIR)

The FTIR spectra of poplar pulp and dried CNF were performed by using a JASCO FT/IR 460 Plus spectrometer (Jasco Inc., Tokyo, Japan), equipped with an accessory single reflection diamond. The spectrum was obtained within the range of 600 to 4000 cm^{-1} wavenumber at a resolution of 1 cm^{-1} and 400 scans.

2.4.10. Zeta Potential and Particle Size

Prior to these tests, a modification of poplar CNF was carried out by adding either 1M hydrochloric acid or 1M sodium hydroxide solutions in the case lower or higher pHs were targeted, respectively. Then, both zeta potential and particle size were measured using a Zetasizer Nano series (Nano ZS, Malvern, UK) with CNF concentrations of 0.05% wt. The measurements were repeated at least three times, showing mean values and standard deviations.

2.4.11. Rheological Characterization

The rheological properties of the pH-modified poplar CNFs were measured by small-amplitude oscillatory shear (SAOS) and viscous flow tests. For that, frequencies from 0.03 to 100 rad·s^{-1} and increasing stepped shear rate ramps from 0.03 to 100 s^{-1} were respectively applied in both the Rheoscope (Thermo Haake, Vreden, Germany) and the ARES (Rheometric Scientific, Leaderhead, UK) rheometers. Rough plate–plate geometries of 20 and 25 mm diameter, respectively, and 1 mm gap were set as testing conditions. All rheological measurements were performed twice, showing mean values and standard deviations.

3. Results and Discussion

3.1. Poplar CNF Composition

In this work, cellulose nanofibers (CNFs) were produced from *P. alba* L., "PO-10-10-20", an autochthonous genotype of the Mediterranean area. This material was planted following the Short Rotation Coppice (SRC) system. The chemical composition of the samples obtained at the third year of growth is shown in Table 1. *P. alba* L. "PO-10-10-20" raw material showed a similar glucan content than other materials from genus *Populus* previously studied [4,5,37]. Thus, Sannigrahi et al. [4] found a cellulose content of about 42–49% for different species and hybrids of poplar. In the same way, Ibarra et al. [5] referenced the glucan content in different poplar clones between 41.4 and 46.3%; and Martín-Davison et al. [37] between 35.5 and 40.4% for different poplar hybrids. With respect to hemicelluloses content, these authors have described a content of 16-23% [4,5,37]. Similar hemicelluloses content, 22.7%, was observed in the case of *P. alba* L. "PO-10-10-20". The presence of around 25% of hemicelluloses has been reported to be beneficial for CNFs production, improving the efficiency of nanofibrillation [38]. The reason behind this finding is that hemicelluloses, mainly xylans, prevent microfibril aggregations by stabilizing the fiber suspension, which improves further fibrillation. Moreover, hemicelluloses contribute to the swelling of the fibers and this is one of the factors favoring the fibrillation process [20].

On the other hand, *P. alba* L. "PO-10-10-20" showed lower Klason lignin content (18.7%) than those previously shown for several hybrids and poplar clones (21–29%) [4,5,37]. In principle, this lower lignin content could also be beneficial for the use of this poplar material in the production of CNFs, because when the lignin content of the starting material is higher than 20%, the fibrillation yield decreases [39]. However, a residual amount of lignin is necessary to ensure good fibrillation as the antioxidant capacity of such lignin prevents the re-bonding of previously broken covalent bonds [40,41].

Table 1. Chemical composition of *P. alba* L. "PO-10-10-20" and its corresponding bleached pulp and CNF.

	Extractives (% o.d.p.)	Klason Lignin (% o.d.p.)	Soluble Lignin (% o.d.p.)	Glucans (% o.d.p.)	Xylans (% o.d.p.)
P. alba L. "PO-10-10-20"	6.3 ± 0.5	18.7 ± 0.1	4.0 ± 0.2	39.7 ± 0.4	22.7 ± 0.2
Bleached pulp	-	0.8 ± 0.6	0.4 ± 0.3	74.0 ± 2.1	24.0 ± 1.9
CNF	-	0.4 ± 0.3	0.2 ± 0.1	76.2 ± 1.7	16.0 ± 1.9

Mean values and standard deviations were calculated from triplicates.

Kraft cooking and TCF bleaching processes were applied to obtain a material with the optimum quantities of lignin and hemicelluloses which favor subsequent fibrillation. After these processes, a decrease in lignin content (resulting in a 0.8% of Klason lignin), an increase in carbohydrates (reaching values of up to 74% in glucans and 24% in xylans), and a total elimination of extractives were observed in the resultant bleached pulp (Table 1). These values are very similar to those obtained by other authors using other kinds of biomass (including eucalypt as reference material) and who subsequently obtained CNFs with high yields [21,42–44].

Finally, as already demonstrated by other authors, during the TEMPO-ox reaction used as pre-treatment to obtain CNF, a delignification occurs (due to TEMPO itself and NaClO) together with a decrease of xylans [45,46]. This effect was observed herein for the resulting CNFs (Table 1), reaching lignin and xylans values of 0.4% and 16%, respectively. Consequently, the glucan content was risen up to 76.2%.

3.2. Poplar CNF Characterization

The degree of polymerization was measured for the pulp and CNF samples. A large decrease in this parameter (86.5%) was observed for the CNF sample, reaching a DPv value of 280 ± 81 vs. 2076 ± 51 found in the starting bleached pulp. Many authors have observed this strong decrease in the polymerization degree due to either physical or chemical pretreatments prior to microfluidization [18,47]. For example, for eucalypt, Fillat et al. [21] found a reduction of 87.53% in DPv using TEMPO-ox reaction as a pretreatment. However, Ibarra et al. [25] showed a reduction of 39% in DPv using physical pretreatment (PFI mill) prior to the microfluidization step. Normally, the decrease observed in this parameter is more pronounced in the case of TEMPO-ox reaction than in the physical pretreatments. This is because in the physical pretreatments case the reduction of DPv occurs due to shear forces generated during refining and microfluidization, while in the case of chemical treatment there is also cellulose degradation caused by TEMPO-ox reaction [21].

Regarding nanofibrillation yield, poplar CNF showed a value of 87.3% ± 8.1%, indicating that almost all of the fibers in the starting material have been transformed to nano-sized. The nanofibrillation yield values for CNFs obtained by chemical TEMPO-ox pretreatment tend to be higher (80–100%) than those obtained when only physical processes are used (10–60%) [21,25,44,47,48]. This is due to the fact that sodium carboxylate groups are formed during the TEMPO-ox reaction. Then, the negative charges generated on the fiber's surface induce repulsive forces between them, favoring the subsequent microfluidization process [20]. Therefore, the carboxylate content of CNF samples is an important factor which exerts influence on the nanofibrillation yield [21]. In the case of poplar CNF, a carboxylate content of 1048 ± 128 μmol g^{-1} was found. This value is similar to those reported by other authors for CNF samples obtained from olive tree pruning (1038 μmol g^{-1}), eucalypt (1043 μmol g^{-1}) and elm (1178 μmol g^{-1}), using also a TEMPO-ox pretreatment [21,44]. However, when only physical processes were used to produce CNFs, the carboxylate content values were lower than 100 μmol g^{-1} [44].

Nevertheless, the carboxylate content is not the only parameter that influences the nanofibrillation yield. Thus, among the authors who used TEMPO-ox pretreatment, Fillat et al. [21] showed values close to 96% in nanofibrillation yield for CNF obtained from eucalypt, while Jiménez-López et al. [44] obtained a nanofibrillation yield of 100% with the same raw material and pretreatment. Therefore, the difference in the nanofibrillation values found by these authors could be due to a higher number of passes in the microfluidizer [44]. On the other hand, Fillat et al. [21] found nanofibrillation yield values of 48% for CNF obtained from a bleached olive tree pruning pulp using the same process (including TEMPO-ox pretreatment and microfluidization). Therefore, not only the total applied process (including both pretreatment and the fibrillation process and their conditions) but also the starting material determines the nanofibrillation yield values of the resulting CNFs.

Optical transmittance is related to the light scattering of fibers with dimensions in the visible light range. Thus, a transmittance of 100% corresponds to highly fibrillated nanofibers [20,30]. From the wavelength scanning of the poplar CNF sample, lower transmittance was obtained at 400 nm compared with 700 nm. This effect could be due to the presence of bundles in the poplar CNF sample, which can absorb at 400 nm, but not at 700 nm [45]. Poplar CNF showed its highest transmittance at 700 nm (83%). This value is lower than that found by Fillat et al. [21] for eucalypt CNF (96%) and higher than that of olive tree pruning CNF (67%). It is known that the transmittance is directly related to the nanofibrillation yield; the more nano-size fibers are in a CNF sample, the higher is the transmittance. Accordingly, nanofibrillation yield values for poplar, eucalypt and olive tree pruning CNF samples were 87.3%, 97% and 48%, respectively [21].

On the other hand, Figure 1 shows the X-ray diffraction patterns of the poplar pulp and CNF samples where the typical reflections corresponding to I_β cellulose allomorph can be observed. Thus, (101) and (002) reflections of the crystalline planes were observed at ~15.2 and ~22.7 2θ degrees and the ((10$\bar{1}$)) reflection was indicated as a shoulder at ~16.5 2θ degrees. Moreover, the diffractograms did not show the three reflections at 12.2, 20.8 and 21.4θ degrees, corresponding to (1$\bar{1}$0), (110) and (200), respectively of cellulose II, suggesting that TEMPO-ox pretreatment followed by microfluidization did not induce the transformation of cellulose I into cellulose II. The crystallinity index and the crystallite size determined from these X-ray patterns were 93.8 ± 0.2% and 9.4 ± 0.4 nm for bleached pulp and 74.4 ± 1.6% and 7.2 ± 0.1 nm for CNF sample, respectively. In general, there is the notion that the TEMPO-ox pretreatment prior to microfluidization step can partially damage the crystalline cellulose structure [21], explaining the reduction of the crystallinity index from bleached pulp to CNF sample. Moreover, it is well known that the microfluidization step can affect the crystallinity of the resultant cellulose nanofibers due to the rupture of the highly ordered crystalline structure or to the hornification phenomenon of the cellulose nanofibers during this process [48,49]. The reduction of the crystallite size from bleached pulp to CNF during TEMPO-ox pretreatment and subsequent microfluidization has already been reported for bleached eucalypt and olive tree pruning pulps [21]. This effect is commonly attributed to the degradation of the amorphous phase during oxidation [50]. These crystallinity index and the crystallite size values described for poplar CNF are only slightly lower than those reported by Jiménez-López et al. [44] for TEMPO-ox CNF obtained from eucalypt (77 % and 8.2 nm) and elm (84 % and 9.4 nm). These values are also in the range of those reported by other authors for TEMPO-ox CNF obtained from different raw materials: 64–74% and 2.3–9.4 nm [17,20,44,51].

A thermogravimetric analysis (TGA) of the cellulose nanofibers from poplar was also performed (Figure 2). It can be observed that apart from the initial moisture loss, the poplar CNF sample exhibited outstanding thermal stability up to more than 200 °C, as previously reported by other authors with different CNFs [21,23]. The main weight losses take place around 250–300 °C with two incipient peaks centered at 242 and 295 °C. The first one is mainly due to dehydration of the cellulose chain and decarboxylation of sodium carboxylate groups, while the second one is mainly a consequence of the rupture of glycosidic linkages [52]. This last peak is shown at a lower temperature than that of cellulose pulp (typically between 300 °C and 350 °C), probably due to the lower diameter of nanofibers compared to cellulose fibers, which promote a faster degradation due to higher surface area exposed to heat [53,54]. From the different events, several characteristic parameters have been outlined, i.e., the temperature at which the degradation begins

(T_{onset}), the temperature at maximum derivative weight loss (T_{max}) and the temperature at which the event is considered finished (T_{final}). The weight loss during the events along with the remaining undegraded weight was also reported, all of them shown in Table 2. Similar T_{onset} and T_{max} were found for TEMPO-ox cellulose nanofibers from elm and eucalypt by Jiménez-López et al. [44]. However, these authors found higher T_{final}. In agreement with other authors, the char residue obtained for TEMPO-ox poplar CNFs (28%) is higher than that typically found in non-chemical pretreated cellulose nanofibers (18–26%). This fact is due to the presence of carboxylate groups, which leads to the formation of carbon dioxide during decarboxylation that reacts with water (produced during dehydration of cellulose) forming sodium carbonate [44,55].

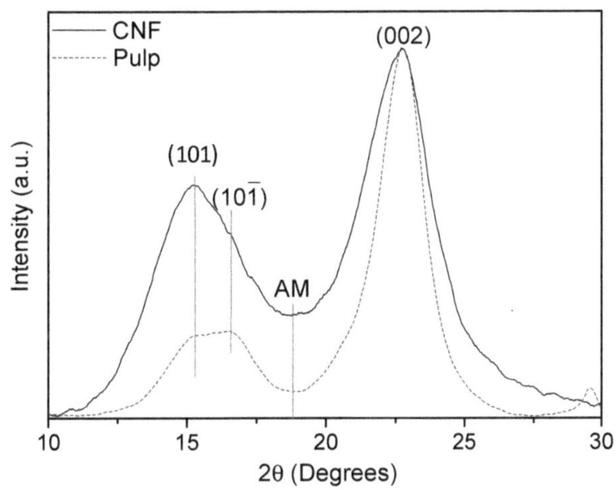

Figure 1. X-ray diffraction pattern of the poplar bleached pulp and CNF.

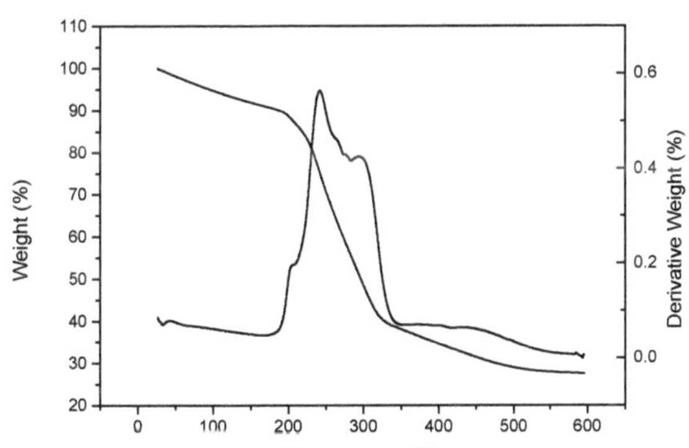

Figure 2. TGA of poplar CNF.

Table 2. Thermogravimetric analysis results of poplar CNF.

	T_{onset} (°C)	T_{max} (°C)	T_{final} (°C)	ΔW (%)	Residue (%)
Poplar CNF	218/270	242/295	256/320	32/40	28

Poplar pulp and dried CNF samples have also been analyzed by FTIR (Figure 3). The spectra show the typical bands that conform to the cellulose structure. The wide band centered around 3329 cm^{-1} is related to the hydroxyl groups, whereas the small band at 2895 cm^{-1} is due to asymmetric and symmetric C-H stretching vibrations [56]. The band detected at 1645 cm^{-1} in poplar pulp is attributed to the O-H stretching vibration of absorbed water. The three bands at 1163, 1101, and 1026 cm^{-1} observed in both CNF and pulp samples spectra can also be assigned to cellulose. Firstly, the band at 1163 cm^{-1} is related to pyranosyl rings among C-O asymmetric bands, the band at 1101 cm^{-1} is attributed to the C-OH skeletal vibration, while the band at 1026 cm^{-1} is related to C-O stretching vibration [57]. Moreover, the glycosidic bond vibration was observed at 893 cm^{-1} in both spectra, and the bands at 1361–1370 and 710 cm^{-1} also observed in both spectra indicate the presence of I_β cellulose allomorph [58], in agreement with X-ray diffraction analysis.

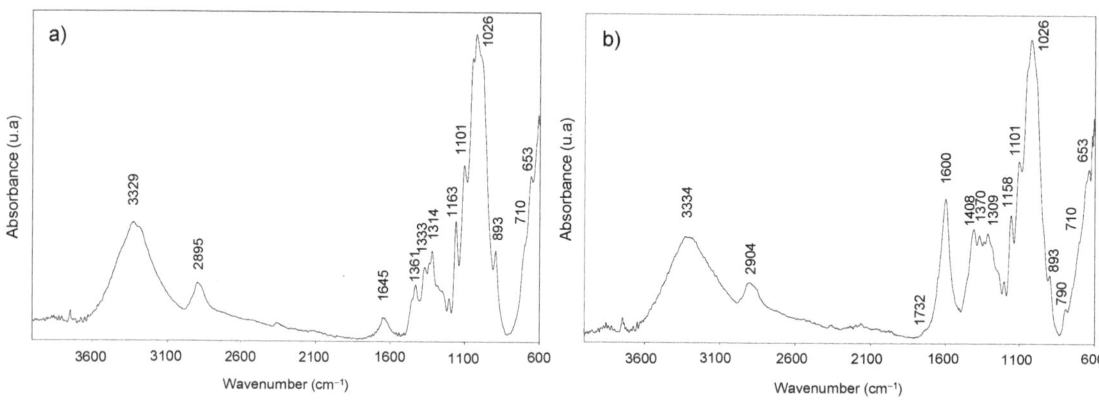

Figure 3. FTIR spectra of poplar pulp (**a**) and CNF (**b**).

As the poplar CNF was obtained by the TEMPO-oxidation pretreatment, the stretching band from sodium carboxylate groups was prominently shown at 1600 cm^{-1} [21], supporting the oxidation of C6 primary hydroxyl groups of cellulose during the TEMPO-oxidation reaction. When these COONa groups are transformed into free carboxyl groups (COOH), this band would shift to 1732 cm^{-1}, as it is observed in the CNF spectrum. Moreover, the band at 1408 cm^{-1} also indicates the presence of COONa, since it is associated with the C-O symmetric stretching of dissociated carboxyl groups [59].

3.3. Cytotoxicity Assays of Poplar CNF

Although CNFs are isolated from cellulosic sources, biomasses with no or low toxicity, either the nanoscale or chemical modifications of nanofibers might influence their toxicity [60]. Then, the cytotoxicity assays of CNF samples obtained from *P. alba* L. "PO-10-10-20" and eucalypt (as reference CNF) were carried out. In general, the two CNF samples caused limited cytotoxicity to PLHC-1 cells so that an IC50 could not be calculated for any of the CNF samples with any of the cytotoxicity assays (Figure 4). In the case of the eucalypt CNF, no decrease of cell viability was detected with any of the tests so that even the calculation of an IC10 was not possible. Poplar CNF provoked a concentration-dependent decrease in cell viability allowing the calculation of IC10 values with AlamarBlue (IC10 = 10 $\mu g\ mL^{-1}$) and CFDA-AM (IC10 = 60 $\mu g\ mL^{-1}$) assays. No effect was observed at the lysosomal level by means of the Neutral Red assay. Such results suggest a very low toxicity of the tested CNF samples although additional in vivo experiments would be necessary to confirm the actual ecotoxicity of these substances. Of particular interest is the slight difference observed in the reduction of cell viability caused by both CNF samples (eucalypt and poplar). So far, considering the lack of significant differences observed in the properties of these cellulose nanofibers, it is difficult to hypothesize about the mechanisms underlying the slightly higher toxicity of poplar CNF. The investigation about the origin of this increase of cytotoxicity was out of the scope of the present study but it merits further efforts in the future.

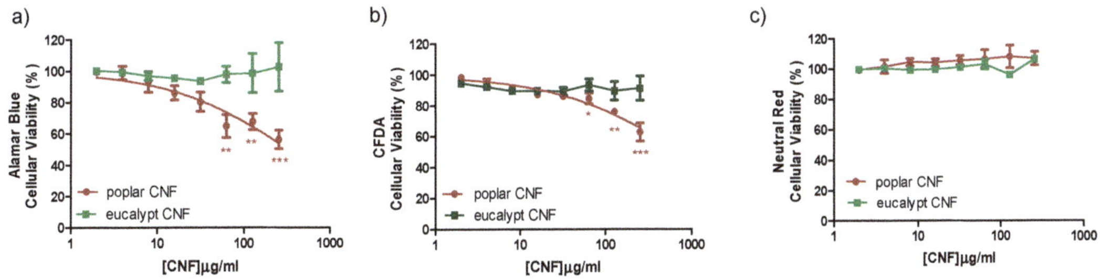

Figure 4. Effect of CNF samples on PLHC cell viability. Cells were exposed to increasing concentrations of CNFs for 72 h. Cytotoxicity of CNFs was assessed by means of the AlamarBlue assay (**a**), CFDA-AM assay (**b**), NRU assay (**c**). Points represent the mean and standard error of the mean (SEM) of at least three independent experiments. Statistically significant differences with respect to the vehicle control (one-way rmANOVA, Dunnett's Post-hoc test) are indicated as followed: * $p < 0.05$, ** $p < 0.01$ and *** $p < 0.001$.

Different studies have discussed the toxicity of different CNF samples, and in general none of them have shown evidence of the toxic effect on human health or the environment. Then, Vartiainen et al. [61] assayed the health and environmental safety of CNF isolated from birch via mechanical grinding. For that, mouse and human cells and luminescent bacteria were used as tested models, concluding that the tested CNF samples appeared hazardous to neither human health nor the ecological environment. Lopes et al. [62] assessed the in vitro biological responses elicited by wood-derived CNF samples (isolated from bleached sulfite softwood dissolving pulp by enzymatic hydrolysis combined with mechanical shearing and high-pressure homogenizationand), using human dermal fibroblasts, lung MRC-5 cells and THP-1 macrophage cells. Furthermore, they studied if the presence of surface-charged groups (i.e., carboxymethyl groups) on CNF samples could induce distinct biological responses. The assay data presented the absence of cytotoxic effects associated with the exposure to unmodified and carboxymethylated-modified CNFs. Finally, Alexandrescu et al. [63] produced nanofibrillated materials from *Eucalyptus* and *Pinus radiata* pulp fibers using a homogenizer without pretreatment and with TEMPO-ox pretreatment. Cytotoxicity tests were applied on the different CNF samples obtained, which showed that the nanofibrils do not exert acute toxic phenomena on the tested fibroblast cells (3T3 cells).

3.4. CNF Zeta Potential, Particle Size and Rheological Properties of Poplar CNF

Both zeta potential and particle size analysis of the different pH-modified poplar CNF samples were included in Figure 5. By slightly increasing or decreasing the pH from the neutral sample, higher values of zeta potential were obtained compared to the very low values obtained at neutral pH (−60 mV). Therefore, slightly poorer stability was exhibited, but still stable enough since the values were lower than −40 mV, which is the limit for colloidal stabilization [64]. Nonetheless, when the pH was reduced to 1, poplar CNFs gelled and formed clots even at the low 0.05 wt % solution concentration, exhibiting values of zeta potential close to 0 (isoelectric point), results which were in agreement with previous studies within the field [65,66].

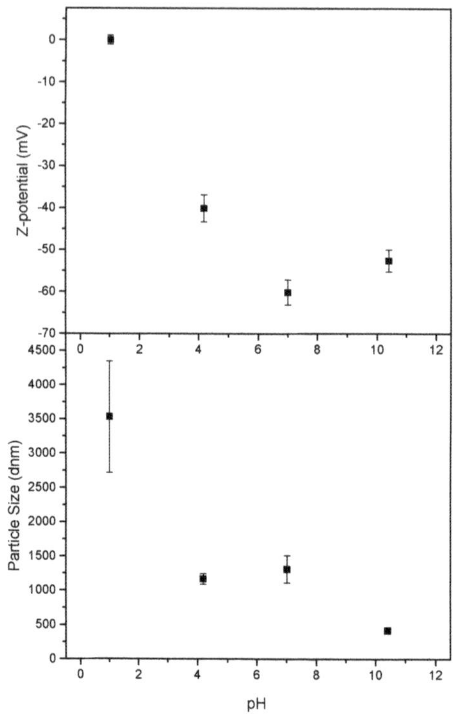

Figure 5. Zeta potential (**top**) and particle size (**bottom**) of nanocellulose dispersions as a function of pH.

However, the particle size did not follow the same trend. At the highest pH the particle size was significantly reduced, while only a small size decrease was observed at pH 4.17. It should be taken into account that the particle size determined here (by DLS assays) is not the real size of the nanofibers, but the hydrodynamic radius of an equivalent sphere that diffuses at the same rate as the nanofibers. Furthermore, this hydrodynamic radius will depend on the extent of particle-particle interactions in the aqueous suspension, which strongly depends on the particles' concentration and their surface chemistry and resulting hydration shells [67]. Therefore, these particle sizes values give us only an estimation of the hydrodynamic size of the CNF samples but they do give us an indication of nanofibers aggregation caused by changes in pH. Thus, the particle size measured for the sample at pH 1 demonstrated the agglomeration seen with the naked eye, as a much larger size was reported. The general aggregation trend observed when reducing the pH is explained as a consequence of the cellulose carboxyl groups protonation, leading to a reduction in surface charge and hence diminishing electrostatic repulsions between fibers, also increasing the possibility of crosslinking through hydrogen bonding [49,66,68]. This is important to keep in mind if the application of the obtained poplar CNF requires significant pH changes.

On the other hand, the rheological properties of the so-performed CNFs are also significantly influenced by the effect of pH, as can be clearly shown in Figure 6. Thus, the lower the pH, the higher the SAOS functions observed as previously reported Hujaya et al. [68] (Figure 6a). Nonetheless, this effect was not linear, as very close viscoelastic moduli values were obtained between pH 4.17, 7 and 10, unlike pH 1, which was able to raise viscoelastic functions in more than an order of magnitude (see Figure 6a). Regardless of the pH, the poplar CNFs presented a well-developed plateau zone, with predominant elastic characteristics due to the high separation of around an order of magnitude between storage (G′) and loss (G″) moduli, as already shown by other CNF-based hydrogels [68,69].

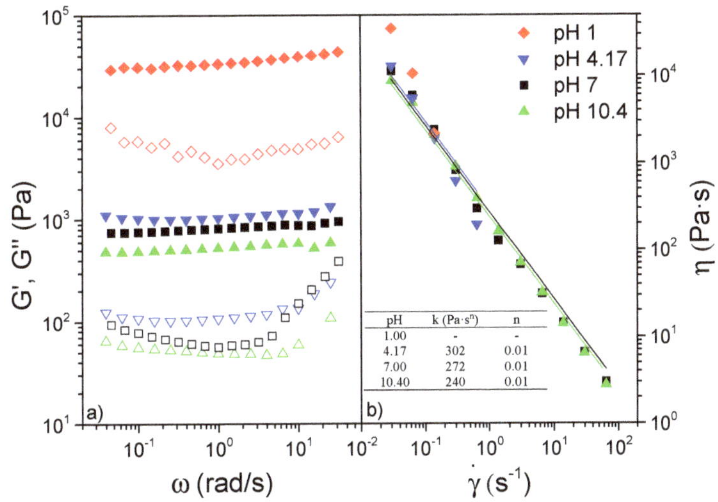

Figure 6. (**a**) Storage (filled symbols) and loss (empty symbols) functions and (**b**) viscosity of poplar CNF according to pH. Inset: values of K and n as a function of pH.

The subtle differences observed for viscoelastic functions between pHs 4.17, 7 and 10.4 were even lower when viscous flow measurements were considered (see Figure 6b). The shear-thinning behavior observed fitted properly the power-law model included in Equation (5):

$$\eta = K \times \dot{\gamma}^{n-1} \tag{6}$$

where K and n are the consistency and flow indices, respectively. Once more, the stiffer samples generated when decreasing pH yielded higher values of the consistency index, whereas the flow index values close to 0 (see inset Figure 6b) demonstrate poplar CNFs to possess an extremely pronounced shear-thinning behavior [69,70]. Nonetheless, the extreme toughness exhibited by the pH 1 CNF did not allow a proper fit to be obtained as a consequence of the expulsion of the sample during the test.

4. Conclusions

This study has demonstrated the feasibility of obtaining cellulose nanofibers (CNFs) from fast-growing poplar, *P. alba* L., genotype "PO-10-10-20", using chemical TEMPO-mediated oxidation followed by microfluidization. The resulting CNFs showed high carboxylate content, fibrillation yield, optical transmittance and crystallinity, properties very similar to those obtained with a reference raw material, eucalypt. Moreover, the rheological study of the poplar CNF at different pH values determined its potential in pH-sensitive hydrogel applications. Whereas pH from 4 to 10 did not produce significant changes in rheological behavior, a reduction of pH down to 1 led to an order-of-magnitude increase on the viscoelastic functions. Finally, a very low toxicity of the tested CNF samples was observed from in vitro assays.

Author Contributions: Conceptualization, D.I. and M.E.E.; methodology, D.I., R.M.-S., B.W., A.M.B.-L., A.V., J.M.N., C.V. and M.E.E.; software, D.I., R.M.-S., B.W., A.V., A.M.B.-L. and M.E.E.; validation, M.E.E., R.M.-S., B.W., J.M.N., C.V. and D.I.; formal analysis, D.I., R.M.-S., B.W., A.M.B.-L., A.V., J.M.N., C.V. and M.E.E.; investigation, D.I., R.M.-S., B.W. and M.E.E.; resources, D.I., R.M.-S., B.W., J.M.N., C.V. and M.E.E.; data curation, D.I., R.M.-S., B.W., A.M.B.-L., A.V. and M.E.E.; writing—original draft preparation, D.I., R.M.-S., B.W., A.M.B.-L., A.V., J.M.N., C.V. and M.E.E.; writing—review and editing, D.I., R.M.-S., B.W., A.M.B.-L., A.V., J.M.N., C.V. and M.E.E.; visualization, D.I., R.M.-S. and M.E.E.; supervision, M.E.E. and D.I.; project administration, M.E.E. and D.I.; funding acquisition, M.E.E. and D.I. All authors have read and agreed to the published version of the manuscript.

Funding: This research was funded by Comunidad de Madrid and MCIU/AEI/FEDER, EU via Projects SUSTEC-CM S2018/EMT-4348 and RTI2018-096080-B-C22, respectively.

Institutional Review Board Statement: Not applicable.

Informed Consent Statement: Not applicable.

Data Availability Statement: Not applicable.

Acknowledgments: Silviculture and Forest Management Department (Forest Research Center, INIA, CSIC) is acknowledged for fast-growing poplar materials.

Conflicts of Interest: The authors declare no conflict of interest.

References

1. Arevalo-Gallegos, A.; Ahmad, Z.; Asgher, M.; Parra-Saldivar, R.; Iqbal, H. Lignocellulose: A sustainable material to produce value-added products with a zero waste approach. A review. *Int. J. Biol. Macromol.* **2017**, *99*, 308–318. [CrossRef] [PubMed]
2. Sixto, H.; Cañellas, I.; van Arendonk, J.; Ciria, P.; Camps, F.; Sánchez, M.; Sánchez-González, M. Growth potential of different species and genotypes for biomass production in short rotation in Mediterranean environments. *For. Ecol. Manag.* **2015**, *354*, 291–299. [CrossRef]
3. Oliveira, N.; Pérez-Cruzado, C.; Cañellas, I.; Rodríguez-Soalleiro, R.; Sixto, H. Poplar Short Rotation Coppice Plantations under Mediterranean Conditions: The Case of Spain. *Forests* **2020**, *11*, 1352. [CrossRef]
4. Sannigrahi, P.; Ragauskas, A.J. Poplar as a feedstock for biofuels: A review of compositional characteristics. *Biofuels Bioprod. Biorefining* **2010**, *4*, 209–226. [CrossRef]
5. Ibarra, D.; Eugenio, M.E.; Cañellas, I.; Sixto, H.; Martín-Sampedro, R. Potential of different poplar clones for sugar production. *Wood Sci. Technol.* **2017**, *51*, 669–684. [CrossRef]
6. Jiménez-López, L.; Martín-Sampedro, R.; Eugenio, M.E.; Santos, J.I.; Sixto, H.; Cañellas, I.; Ibarra, D. Co-production of soluble sugars and lignin from short rotation white poplar and black locust crops. *Wood Sci. Technol.* **2020**, *54*, 1617–1643. [CrossRef]
7. Zhu, J.; Chen, L.; Gleisner, R.; Zhu, J.Y. Co-production of bioethanol and furfural from poplar wood via low temperature ($\leq$90 °C) acid hydrotropic fractionation (AHF). *Fuel* **2019**, *254*, 115572. [CrossRef]
8. Dalli, S.S.; da Silva, S.S.; Uprety, B.K.; Rakshit, S.K. Enhanced Production of Xylitol from Poplar Wood Hydrolysates Through a Sustainable Process Using Immobilized New Strain Candida tropicalis UFMG BX 12-a. *Appl. Biochem. Biotechnol.* **2017**, *182*, 1053–1064. [CrossRef]
9. Bonyadi, Z.; Kumar, P.S.; Foroutan, R.; Kafaei, R.; Arfaeinia, H.; Farjadfard, S.; Ramavandi, B. Ultrasonic-assisted synthesis of Populus alba activated carbon for water defluorination: Application for real wastewater. *Korean J. Chem. Eng.* **2019**, *36*, 1595–1603. [CrossRef]
10. Wang, S.; Shuai, L.; Saha, B.; Vlachos, D.G.; Epps, T.H. From Tree to Tape: Direct Synthesis of Pressure Sensitive Adhesives from Depolymerized Raw Lignocellulosic Biomass. *ACS Cent. Sci.* **2018**, *4*, 701–708. [CrossRef]
11. Li, T.; Chen, C.; Brozena, A.H.; Zhu, J.Y.; Xu, L.; Driemeier, C.; Dai, J.; Rojas, O.J.; Isogai, A.; Wågberg, L.; et al. Developing fibrillated cellulose as a sustainable technological material. *Nature* **2021**, *590*, 47–56. [CrossRef]
12. Thakur, V.; Guleria, A.; Kumar, S.; Sharma, S.; Singh, K. Recent advances in nanocellulose processing, functionalization and applications: A review. *Mater. Adv.* **2021**, *2*, 1872–1895. [CrossRef]
13. Trache, D.; Tarchoun, A.F.; Derradji, M.; Hamidon, T.S.; Masruchin, N.; Brosse, N.; Hussin, M.H. Nanocellulose: From Fundamentals to Advanced Applications. *Front. Chem.* **2020**, *8*, 392. [CrossRef] [PubMed]
14. Khiari, R.; Rol, F.; Brochier Salon, M.-C.; Bras, J.; Belgacem, M.N. Efficiency of Cellulose Carbonates to Produce Cellulose Nanofibers. *ACS Sustain. Chem. Eng.* **2019**, *7*, 8155–8167. [CrossRef]
15. Rivadeneyra, A.; Marín-Sánchez, A.; Wicklein, B.; Salmerón, J.F.; Castillo, E.; Bobinger, M.; Salinas-Castillo, A. Cellulose nanofibers as substrate for flexible and biodegradable moisture sensors. *Compos. Sci. Technol.* **2021**, *208*, 108738. [CrossRef]
16. Wicklein, B.; Diem, A.M.; Knöller, A.; Cavalcante, M.S.; Bergström, L.; Bill, J.; Burghard, Z. Dual-Fiber Approach toward Flexible Multifunctional Hybrid Materials. *Adv. Funct. Mater.* **2018**, *28*, 1704274. [CrossRef]
17. Jiang, F.; Kondo, T.; Hsieh, Y.-L. Rice Straw Cellulose Nanofibrils via Aqueous Counter Collision and Differential Centrifugation and Their Self-Assembled Structures. *ACS Sustain. Chem. Eng.* **2016**, *4*, 1697–1706. [CrossRef]
18. Shinoda, R.; Saito, T.; Okita, Y.; Isogai, A. Relationship between Length and Degree of Polymerization of TEMPO-Oxidized Cellulose Nanofibrils. *Biomacromolecules* **2012**, *13*, 842–849. [CrossRef]
19. Isogai, A. Wood nanocelluloses: Fundamentals and applications as new bio-based nanomaterials. *J. Wood Sci.* **2013**, *59*, 449–459. [CrossRef]

20. Saito, T.; Nishiyama, Y.; Putaux, J.-L.; Vignon, M.; Isogai, A. Homogeneous Suspensions of Individualized Microfibrils from TEMPO-Catalyzed Oxidation of Native Cellulose. *Biomacromolecules* **2006**, *7*, 1687–1691. [CrossRef]
21. Fillat, Ú.; Wicklein, B.; Martín-Sampedro, R.; Ibarra, D.; Ruiz-Hitzky, E.; Valencia, C.; Sarrión, A.; Castro, E.; Eugenio, M.E. Assessing cellulose nanofiber production from olive tree pruning residue. *Carbohydr. Polym.* **2018**, *179*, 252–261. [CrossRef]
22. Qi, Y.; Zhang, H.; Xu, D.; He, Z.; Pan, X.; Gui, S.; Dai, X.; Fan, J.; Dong, X.; Li, Y. Screening of Nanocellulose from Different Biomass Resources and Its Integration for Hydrophobic Transparent Nanopaper. *Molecules* **2020**, *25*, 227. [CrossRef]
23. Zhao, G.; Du, J.; Chen, W.; Pan, M.; Chen, D. Preparation and thermostability of cellulose nanocrystals and nanofibrils from two sources of biomass: Rice straw and poplar wood. *Cellulose* **2019**, *26*, 8625–8643. [CrossRef]
24. Wu, Y.; Sun, M.; Wu, X.; Shi, T.; Chen, H.; Wang, H. Preparation of Nanocellulose Aerogel from the Poplar (*Populus tomentosa*) Catkin Fiber. *Forests* **2019**, *10*, 749. [CrossRef]
25. Ibarra, D.; Martín-Sampedro, R.; Wicklein, B.; Fillat, Ú.; Eugenio, M.E. Production of Microfibrillated Cellulose from Fast-Growing Poplar and Olive Tree Pruning by Physical Pretreatment. *Appl. Sci.* **2021**, *11*, 6445. [CrossRef]
26. Sixto, H.; Hernández, M.J.; Ciria, P.; Carrasco, J.E.; Cañellas, I. *Manual de Cultivo de Populus spp. Para la Producción de Biomasa con Fines Energéticos*; Monografias INIA: Madrid, Spain, 2010.
27. Kang, K.Y.; Jo, B.M.; Oh, J.S.; Mansfield, S.D. The effects of biopulping on chemical and energy consumptionduring Kraft pulping of hybrid poplar. *Wood Fiber Sci.* **2003**, *35*, 594–600.
28. Ibarra, D.; Camarero, S.; Romero, J.; Martínez, M.J.; Martínez, A.T. Integrating laccase–mediator treatment into an industrial-type sequence for totally chlorine-free bleaching of eucalypt kraft pulp. *J. Chem. Technol. Biotechnol.* **2006**, *81*, 1159–1165. [CrossRef]
29. *NREL P-510-42618-Determination of Structural Carbohydrates and Lignin in Biomass*; Laboratory Analytical Procedure; National Renewable Energy Laboratory: Golden, CO, USA, 2011.
30. Besbes, I.; Alila, S.; Boufi, S. Nanofibrillated cellulose from TEMPO-oxidized eucalyptus fibres: Effect of the carboxyl content. *Carbohydr. Polym.* **2011**, *84*, 975–983. [CrossRef]
31. Kasaai, M.R. Comparison of various solvents for determination of intrinsic viscosity and viscometric constants for cellulose. *J. Appl. Polym. Sci.* **2002**, *86*, 2189–2193. [CrossRef]
32. Fras, L.; Stana-Kleinschek, K.; Ribitsch, V.; Sfiligoj-Smole, M.; Kreze, T. Quantitative Determination Of Carboxyl Groups In Cellulose Polymers Utilizing Their Ion Exchange Capacity And Using A Complexometric Titration. *Mater. Res. Innov.* **2004**, *8*, 145–146. [CrossRef]
33. Jensen, K.A.; Kembouche, Y.; Christiansen, E.; Jacobsen, N.; Wallin, H.; Guiot, C.; Spalla, O. Final Protocol for Producing Suitable Manufactured Nanomaterial Exposure Media—Report The generic NANOGENOTOX Dispersion Protocol—Standard Operation Procedure (SOP) and Background Documentation. July 2011. Available online: https://www.anses.fr/en/system/files/nanogenotox_deliverable_5.pdf (accessed on 1 November 2021).
34. Lammel, T.; Boisseaux, P.; Fernández-Cruz, M.-L.; Navas, J.M. Internalization and cytotoxicity of graphene oxide and carboxyl graphene nanoplatelets in the human hepatocellular carcinoma cell line Hep G2. *Part. Fibre Toxicol.* **2013**, *10*, 27. [CrossRef] [PubMed]
35. Lammel, T.; Navas, J.M. Graphene nanoplatelets spontaneously translocate into the cytosol and physically interact with cellular organelles in the fish cell line PLHC-1. *Aquat. Toxicol.* **2014**, *150*, 55–65. [CrossRef]
36. Dayeh, V.R.; Bols, N.C.; Tanneberger, K.; Schirmer, K.; Lee, L.E.J. The Use of Fish-Derived Cell Lines for Investigation of Environmental Contaminants: An Update Following OECD's Fish Toxicity Testing Framework No. 171. *Curr. Protoc. Toxicol.* **2013**, *56*, 1–5. [CrossRef] [PubMed]
37. Martín-Davison, J.S.; Ballesteros, M.; Manzanares, P.; Sepúlveda, X.P.-B.; Vergara-Fernández, A. Effects of Temperature on Steam Explosion Pretreatment of Poplar Hybrids with Different Lignin Contents in Bioethanol Production. *Int. J. Green Energy* **2015**, *12*, 832–842. [CrossRef]
38. Chaker, A.; Alila, S.; Mutjé, P.; Vilar, M.R.; Boufi, S. Key role of the hemicellulose content and the cell morphology on the nanofibrillation effectiveness of cellulose pulps. *Cellulose* **2013**, *20*, 2863–2875. [CrossRef]
39. Rojo, E.; Peresin, M.S.; Sampson, W.W.; Hoeger, I.C.; Vartiainen, J.; Laine, J.; Rojas, O.J. Comprehensive elucidation of the effect of residual lignin on the physical, barrier, mechanical and surface properties of nanocellulose films. *Green Chem.* **2015**, *17*, 1853–1866. [CrossRef]
40. Fall, A.B.; Burman, A.; Wågberg, L. Cellulosic nanofibrils from eucalyptus, acacia and pine fibers. *Nord. Pulp Pap. Res. J.* **2014**, *29*, 176–184. [CrossRef]
41. Solala, I.; Iglesias, M.C.; Peresin, M.S. On the potential of lignin-containing cellulose nanofibrils (LCNFs): A review on properties and applications. *Cellulose* **2020**, *27*, 1853–1877. [CrossRef]
42. Sánchez-Gutiérrez, M.; Espinosa, E.; Bascón-Villegas, I.; Pérez-Rodríguez, F.; Carrasco, E.; Rodríguez, A. Production of Cellulose Nanofibers from Olive Tree Harvest—A Residue with Wide Applications. *Agronomy* **2020**, *10*, 696. [CrossRef]
43. Rajan, K.; Djioleu, A.; Kandhola, G.; Labbé, N.; Sakon, J.; Carrier, D.J.; Kim, J.-W. Investigating the effects of hemicellulose pre-extraction on the production and characterization of loblolly pine nanocellulose. *Cellulose* **2020**, *27*, 3693–3706. [CrossRef]
44. Jiménez-López, L.; Eugenio, M.E.; Ibarra, D.; Darder, M.; Martín, J.A.; Martín-Sampedro, R. Cellulose nanofibers from a dutch elm disease-resistant ulmus minor clone. *Polymers* **2020**, *12*, 2450. [CrossRef]
45. Balea, A.; Merayo, N.; Fuente, E.; Delgado-Aguilar, M.; Mutje, P.; Blanco, A.; Negro, C. Valorization of corn stalk by the production of cellulose nanofibers to improve recycled paper properties. *BioResources* **2016**, *11*, 3416–3431. [CrossRef]

46. Okita, Y.; Saito, T.; Isogai, A. TEMPO-mediated oxidation of softwood thermomechanical pulp. *Holzforschung* **2009**, *63*, 529–535. [CrossRef]
47. Espinosa, E.; Arrebola, R.I.; Bascón-Villegas, I.; Sánchez-Gutiérrez, M.; Domínguez-Robles, J.; Rodríguez, A. Industrial application of orange tree nanocellulose as papermaking reinforcement agent. *Cellulose* **2020**, *27*, 10781–10797. [CrossRef]
48. Delgado-Aguilar, M.; González Tovar, I.; Tarrés, Q.; Alcalá, M.; Pèlach, M.À.; Mutjé, P. Approaching a Low-Cost Production of Cellulose Nanofibers for Papermaking Applications. *BioResources* **2015**, *10*, 5345–5355. [CrossRef]
49. Jiang, F.; Hsieh, Y.-L. Chemically and mechanically isolated nanocellulose and their self-assembled structures. *Carbohydr. Polym.* **2013**, *95*, 32–40. [CrossRef]
50. Liu, C.; Li, B.; Du, H.; Lv, D.; Zhang, Y.; Yu, G.; Mu, X.; Peng, H. Properties of nanocellulose isolated from corncob residue using sulfuric acid, formic acid, oxidative and mechanical methods. *Carbohydr. Polym.* **2016**, *151*, 716–724. [CrossRef] [PubMed]
51. Montanari, S.; Roumani, M.; Heux, L.; Vignon, M.R. Topochemistry of Carboxylated Cellulose Nanocrystals Resulting from TEMPO-Mediated Oxidation. *Macromolecules* **2005**, *38*, 1665–1671. [CrossRef]
52. Tanaka, A.; Seppänen, V.; Houni, J.; Sneck, A.; Pirkonen, P. BIOREFINERY. Nanocellulose characterization with mechanical fractionation. *Nord. Pulp Pap. Res. J.* **2012**, *27*, 689–694. [CrossRef]
53. Rantuch, P.; Chrebet, T. Thermal decomposition of cellulose insulation. *Cellul. Chem. Technol.* **2014**, *48*, 461–467.
54. Gaspar, D.; Fernandes, S.N.; de Oliveira, A.G.; Fernandes, J.G.; Grey, P.; Pontes, R.V.; Pereira, L.; Martins, R.; Godinho, M.H.; Fortunato, E. Nanocrystalline cellulose applied simultaneously as the gate dielectric and the substrate in flexible field effect transistors. *Nanotechnology* **2014**, *25*, 94008. [CrossRef]
55. Lichtenstein, K.; Lavoine, N. Toward a deeper understanding of the thermal degradation mechanism of nanocellulose. *Polym. Degrad. Stab.* **2017**, *146*, 53–60. [CrossRef]
56. Tsuboi, M. Infrared spectrum and crystal structure of cellulose. *J. Polym. Sci.* **1957**, *25*, 159–171. [CrossRef]
57. Indarti, E.; Marwan; Rohaizu, R.; Wanrosli, W.D. Silylation of TEMPO oxidized nanocellulose from oil palm empty fruit bunch by 3-aminopropyltriethoxysilane. *Int. J. Biol. Macromol.* **2019**, *135*, 106–112. [CrossRef]
58. Sugiyama, J.; Persson, J.; Chanzy, H. Combined infrared and electron diffraction study of the polymorphism of native celluloses. *Macromolecules* **1991**, *24*, 2461–2466. [CrossRef]
59. Fujisawa, S.; Okita, Y.; Fukuzumi, H.; Saito, T.; Isogai, A. Preparation and characterization of TEMPO-oxidized cellulose nanofibril films with free carboxyl groups. *Carbohydr. Polym.* **2011**, *84*, 579–583. [CrossRef]
60. Xue, Y.; Mou, Z.; Xiao, H. Nanocellulose as a sustainable biomass material: Structure, properties, present status and future prospects in biomedical applications. *Nanoscale* **2017**, *9*, 14758–14781. [CrossRef] [PubMed]
61. Vartiainen, J.; Pöhler, T.; Sirola, K.; Pylkkänen, L.; Alenius, H.; Hokkinen, J.; Tapper, U.; Lahtinen, P.; Kapanen, A.; Putkisto, K.; et al. Health and environmental safety aspects of friction grinding and spray drying of microfibrillated cellulose. *Cellulose* **2011**, *18*, 775–786. [CrossRef]
62. Lopes, V.R.; Sanchez-Martinez, C.; Strømme, M.; Ferraz, N. In vitro biological responses to nanofibrillated cellulose by human dermal, lung and immune cells: Surface chemistry aspect. *Part. Fibre Toxicol.* **2017**, *14*, 1. [CrossRef] [PubMed]
63. Alexandrescu, L.; Syverud, K.; Gatti, A.; Chinga-Carrasco, G. Cytotoxicity tests of cellulose nanofibril-based structures. *Cellulose* **2013**, *20*, 1765–1775. [CrossRef]
64. Carneiro-Da-Cunha, M.G.; Cerqueira, M.A.; Souza, B.W.S.; Teixeira, J.A.; Vicente, A.A. Influence of concentration, ionic strength and pH on zeta potential and mean hydrodynamic diameter of edible polysaccharide solutions envisaged for multinanolayered films production. *Carbohydr. Polym.* **2011**, *85*, 522–528. [CrossRef]
65. Valo, H.; Arola, S.; Laaksonen, P.; Torkkeli, M.; Peltonen, L.; Linder, M.B.; Serimaa, R.; Kuga, S.; Hirvonen, J.; Laaksonen, T. Drug release from nanoparticles embedded in four different nanofibrillar cellulose aerogels. *Eur. J. Pharm. Sci.* **2013**, *50*, 69–77. [CrossRef] [PubMed]
66. Fall, A.B.; Lindström, S.B.; Sundman, O.; Ödberg, L.; Wågberg, L. Colloidal stability of aqueous nanofibrillated cellulose dispersions. *Langmuir* **2011**, *27*, 11332–11338. [CrossRef] [PubMed]
67. Qua, E.H.; Hornsby, P.R.; Sharma, H.S.S.; Lyons, G. Preparation and characterisation of cellulose nanofibres. *J. Mater. Sci.* **2011**, *46*, 6029–6045. [CrossRef]
68. Hujaya, S.D.; Lorite, G.S.; Vainio, S.J.; Liimatainen, H. Polyion complex hydrogels from chemically modified cellulose nanofibrils: Structure-function relationship and potential for controlled and pH-responsive release of doxorubicin. *Acta Biomater.* **2018**, *75*, 346–357. [CrossRef]
69. Aaen, R.; Simon, S.; Wernersson Brodin, F.; Syverud, K. The potential of TEMPO-oxidized cellulose nanofibrils as rheology modifiers in food systems. *Cellulose* **2019**, *26*, 5483–5496. [CrossRef]
70. Costa, V.L.D.; Costa, A.P.; Simões, R.M.S. Nanofibrillated cellulose rheology: Effects of morphology, ethanol/acetone addition, and high NaCl concentration. *BioResources* **2019**, *14*, 7636–7654. [CrossRef]

polymers MDPI

Review

A Comprehensive Review on Natural Fibers: Technological and Socio-Economical Aspects

Azizatul Karimah [1], Muhammad Rasyidur Ridho [1], Sasa Sofyan Munawar [1], Ismadi [1], Yusup Amin [1], Ratih Damayanti [2], Muhammad Adly Rahandi Lubis [1], Asri Peni Wulandari [3], Nurindah [4], Apri Heri Iswanto [5,6,*], Ahmad Fudholi [7,8], Mochamad Asrofi [9,10], Euis Saedah [11], Nasmi Herlina Sari [12], Bayu Rizky Pratama [13], Widya Fatriasari [1,*], Deded Sarip Nawawi [14,*], Sanjay Mavinkere Rangappa [15,*] and Suchart Siengchin [15]

1. Research Center for Biomaterials, National Research and Innovation Agency (BRIN), Jl Raya Bogor KM 46, Cibinong 16911, Indonesia; karimahazizatul@gmail.com (A.K.); rasyidmuhammad0505@gmail.com (M.R.R.); sasa001@brin.go.id (S.S.M.); isma011@brin.go.id (I.); yusu007@brin.go.id (Y.A.); muha142@brin.go.id (M.A.R.L.)
2. Forest Products Research and Development Center, Ministry of Environment and Forestry, Bogor 16610, Indonesia; ratih_turmuzi@yahoo.com
3. Department of Biology, Faculty of Mathematics and Science, University of Padjajaran, Jatinangor 45363, Indonesia; asri.peni@unpad.ac.id
4. Indonesian Sweetener and Fiber Crops Research Institute (ISFCRI), Ministry of Agriculture, Malang 65152, Indonesia; nurarindatta@gmail.com
5. Department of Forest Product, Faculty of Forestry, Universitas Sumatera Utara, Medan 20155, Indonesia
6. JATI-Sumatran Forestry Analysis Study Center, Jl. Tridarma Ujung No. 1, Kampus USU, Medan 20155, Indonesia
7. Solar Energy Research Institute, Universiti Kebangsaan Malaysia, Bangi 43600, Malaysia; a.fudholi@ukm.edu.my
8. Research Centre for Electrical Power and Mechatronics, National Research and Innovation Agency (BRIN), Kawasan LIPI Cisitu, Bandung 40135, Indonesia
9. Department of Mechanical Engineering, Kampus Tegalboto, University of Jember, Jember 68121, Indonesia; asrofi.net@gmail.com
10. Center for Development of Advanced Science and Technology (CDAST), Advanced Materials Research Group, Kampus Tegalboto, University of Jember, Jember 68121, Indonesia
11. Indonesia Natural Fiber Council (DSI), Gedung Smesco/SME Tower Lt. G (APINDO UMKM Hub), Jl Gatot Subroto Kav. 94 Pancoran, Jakarta Selatan 12780, Indonesia; euis@inafiber.id
12. Department of Mechanical Engineering, Faculty of Engineering, University of Mataram, Mataram 001016, Indonesia; n.herlinasari@unram.ac.id
13. The Graduate School, Kasetsart University, Chatuchak, Bangkok 10903, Thailand; bayu_rizkypratama@yahoo.com
14. Department of Forest Products, Faculty of Forestry and Environment, IPB University, Bogor 16680, Indonesia
15. Natural Composites Research Group Lab, Department of Materials and Production Engineering, The Sirindhorn International Thai-German Graduate School of Engineering (TGGS), King Mongkut's University of Technology North Bangkok, Bangkok 10800, Thailand; suchart.s.pe@tggs-bangkok.org

* Correspondence: apri@usu.ac.id (A.H.I.); widya.fatriasari@lipi.go.id or widy003@brin.go.id (W.F.); dsnawawi@apps.ipb.ac.id (D.S.N.); mcemrs@gmail.com (S.M.R.)

Citation: Karimah, A.; Ridho, M.R.; Munawar, S.S.; Ismadi; Amin, Y.; Damayanti, R.; Lubis, M.A.R.; Wulandari, A.P.; Nurindah; Iswanto, A.H.; et al. A Comprehensive Review on Natural Fibers: Technological and Socio-Economical Aspects. *Polymers* **2021**, *13*, 4280. https://doi.org/10.3390/polym13244280

Academic Editors: Clara Delgado-Sánchez, Concepción Valencia-Barragán, Esperanza Cortés Triviño, Adrián Tenorio-Alfonso and Antonio M. Borrero-López

Received: 5 November 2021
Accepted: 30 November 2021
Published: 7 December 2021

Abstract: Asian countries have abundant resources of natural fibers, but unfortunately, they have not been optimally utilized. The facts showed that from 2014 to 2020, there was a shortfall in meeting national demand of over USD 2.75 million per year. Therefore, in order to develop the utilization and improve the economic potential as well as the sustainability of natural fibers, a comprehensive review is required. The study aimed to demonstrate the availability, technological processing, and socio-economical aspects of natural fibers. Although many studies have been conducted on this material, it is necessary to revisit their potential from those perspectives to maximize their use. The renewability and biodegradability of natural fiber are part of the fascinating properties that lead to their prospective use in automotive, aerospace industries, structural and building constructions, bio packaging, textiles, biomedical applications, and military vehicles. To increase the range of applications, relevant technologies in conjunction with social approaches are very important. Hence, in the future, the utilization can be expanded in many fields by considering the basic characteristics and appropriate technologies of the natural fibers. Selecting the most prospective natural fiber for creating national products can be assisted by providing an integrated management system from a digitalized information on potential and related technological approaches. To make it happens, collaborations between stakeholders from the national R&D agency, the government as policy maker, and academic institutions to develop national bioproducts based on domestic innovation in order to move the circular economy forward are essential.

Keywords: natural fibers; socio-economic assessment; technological aspects; sustainability; renewable resources

1. Introduction

Scientists, researchers, and practitioners around the world have recently been working to maximize the potential of natural fibers to create the most sustainable, biodegradable, and high-quality natural fiber products [1–3]. Natural fibers, which are renewable and ecologically acceptable sources of raw materials for producing environmentally friendly products, have played a significant part in human civilization [1]. Natural fibers have many advantages over synthetic fibers, including lower density, lighter weight, lower cost, biodegradability, minimal health hazards during processing, abundant availability and ease of availability, low investment at low cost for production, lower energy requirements, and lower CO_2 emissions, indicating that they have great potential as a substitute for glass, carbon, or other synthetic fibers. Natural fibers are desirable bio-sourced materials as an alternative to non-sustainable glass and carbon fiber reinforced composites owing to their availability and technical viability.

From the physical and mechanical properties point of view, natural fiber has relatively high tensile strength and Young's modulus, good thermal, good acoustic insulation characteristics, and high electrical resistant [1–8]. Furthermore, chemical properties of natural fibers, such as high cellulose content, have a strong relationship with tensile properties, crystallinity, and density [1,7]. Notwithstanding, natural fibers have some drawbacks that need to be enhanced, such as low impact strength, non-uniformity in quality and price, poor moisture resistance, low durability, low compatibility, low adhesion efficiency, moisture absorption, and poor wettability [9–14]. Therefore, to achieve adequate uses and overcome some natural fiber limitations such as biocompatibility and hydrophilic properties, appropriate technologies should be applied for instance by surface modifications and chemical treatment methods [3].

Natural fibers have been successfully applied to a wide range of applications, for instance, furniture, automotive, electronic industries, and building construction. According to Ahmed et al., the wear resistance of *Areva javanica* fiber brake pads is 16% higher than acrylic fiber-based brake pad; hence, the *A. javanica* fiber can be used as a possible substitute for synthetic acrylic fiber-based disc brake pads [15]. One example in the automotive sector is the utilization of hybrid kenaf-glass reinforced epoxy composite for passenger car bumper beams [16]. Chandramohan and Bharanichandar [17] also develop natural fiber reinforced polymer composites for automobile accessories and conclude that one of the best materials is the hybrid of sisal and rosella. Plastic/wood fiber composites are used in a variety of applications, including decks, docks, window frames, and molded panel components [18]. Furthermore, corn husk fiber/polyester composites have also been used as sound absorbers to replace glass fiber composites [19].

According to all the development technologies and the availability of natural fibers mentioned above, the utilization of natural fibers can improve economic growth and the well-being of citizens. Industries that use natural fibers as a raw material provide a major source of revenue. Various studies have already been conducted on the use of natural fibers, namely, as reinforced composites in biomedical applications [20–22], automotive devices [23], aerospace [24], and textile resources [25]. This study provides more information about the potential of Indonesian natural fibers from technological and socio-economical aspects.

2. Overview on Natural Fibers

The most common classification for natural fibers is from botanical forms. Natural fibers can be classified into five types [26]: Other forms include (1) bast fibers (for example abaca, sisal, pineapple), (2) leaf fibers (such as ramie, flax, kenaf), (3) seed fibers (coir, cotton, and kapok), (4) grass and reeds (wheat, corn, and rice), and (5) wood and roots. A more comprehensive list of fiber classifications can be found in Figure 1. A variety of fibers are produced by several plants. For example, jute, flax, hemp, and kenaf have both bast and core fibers, whereas agave, coconut, and oil palm have both fruit and stem fibers. Both stem and hull fibers can be found in cereal grains [27].

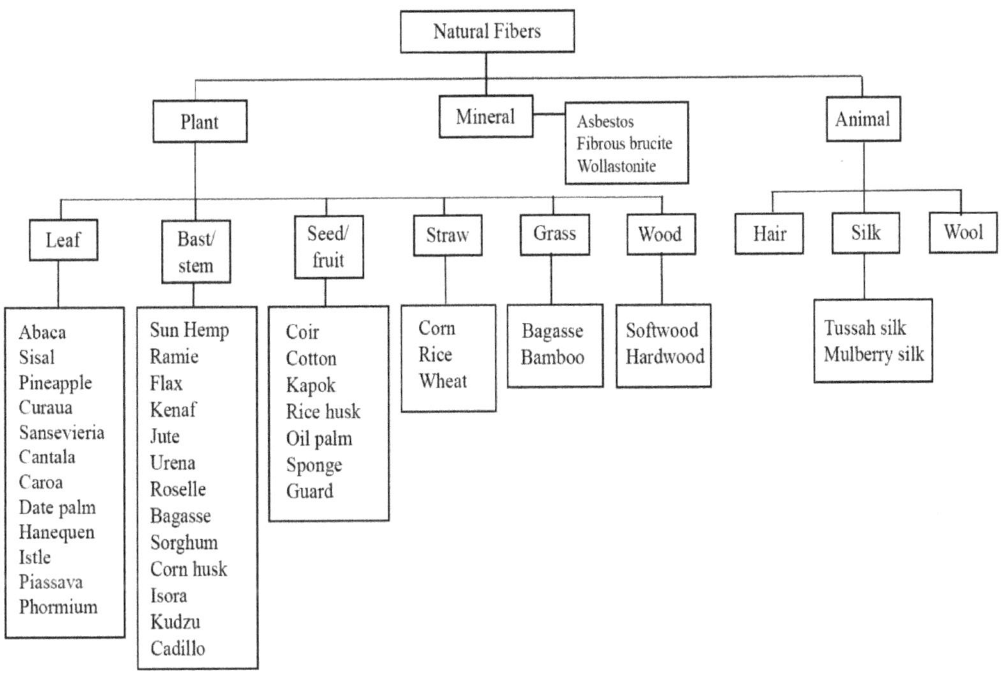

Figure 1. Schematic representation of fiber classification, reprint with permission from ref. [26]. Copyright © 2021 Woodhead Publishing Limited.

The musa plants (*Musa acuminata*) are native to the South-East Asia and belong to the Musaceae family [6,28,29]. This plant produces biomasses that are categorized as useful materials with high fiber materials, such as bunches, pseudo-stems, leaves, and stalk [30]. Banana is widely available in tropical countries such as Malaysia and South India [31]. It is the fourth most important crop in developing countries [32]; meanwhile, tropical and subtropical countries also have sufficient natural resources [6,28,29]. However, approximately 88.84% of the waste with a high fiber content was discarded [33]. The tree becomes waste after one season of fruit harvesting and cutting it allows for the growth of new plants [34]. After the banana trees have been cut down, they are dried and processed to extract the fiber [34,35]. High low elongation at break, light weight, good fire resistance, strong moisture absorption, low density, high tensile strength, and modulus are some of the advantages of banana fiber [36].

Abaca (*Musa textilis*) fiber is classified as leaf fiber in some classifications, but it is classified as a stem fiber in others, especially those derived from pseudo stem [37]. This plant grows to a height of 4.5 to 7.5 m and has a stem diameter of 12.5 to 20 cm. The stem was surrounded by a tangle of large, piled-up fronds. Fibers with a width and length of 12.5 to 30 cm and 1.5 to 2.5 m, respectively, are found in the leaf fronds [38]. The abaca fiber has high tensile strength and is resistant to banding and sea water. The position of the frond on the stem influences its fiber quality, with the outer part being stronger than the inner part. Additionally, abaca fiber has excellent flotation properties [38]. Abaca is widely planted in Talaud, North Sulawesi, and from this location, five local superior varieties were released in October 2019 [39].

Pineapple leaf fibers (*Ananas comosus* L. Merr.) are one of the waste materials available in South East Asia, India, and South America that has not been fully explored [40]. In addition, after Philippines and Thailand, Indonesia contributes 23% of pineapple production. Almost all of Indonesia's regions have ideal tropical climates for growing pineapple plants. On the other hand, the pineapple plantation needs more development by applying superior varieties and appropriate cultivation techniques [41]. These plants grow in moist and dry climates, as well as tropical and subtropical regions [42]. However, these plants tend to grow at temperatures between 18–45 °C and at altitude below 800 mdpl, due to high altitude and extreme temperatures influencing the size and quality of pineapple plants [40]. With a fiber yield of 1.55% to 2.5% [43], pineapple leaf fibers could become a new source of raw material for industries [40], such as polymer composites reinforcement [42], and could also be used to replace synthetic fibers [40,43]. Additionally, they have a softer texture than other vegetable fibers [43], as well as a high strength and smooth surface [42].

Bamboo is a type of lignocellulose material from the grasses (Graminae) family that has a wide range application as a potential fiber source. Bamboo fiber is a type of natural fiber that aligns in longitudinal directions, according to Zakikhani et al. [44] and Wang and Chen [45]. After India and China, Indonesia is ranked the third place in bamboo production [46]. Bamboo has the highest productivity [47], is easy to grow, and harvests quickly when compared to other non-wood forest products [48,49]. Indonesia has 160 bamboo species, 88 of which are endemic [50]. Betung bamboo (*Dendrocalamus asper* (Schult.f)) is one of the most common species in Indonesia [51]. Betung bamboo has better fiber morphology and physical–chemical properties compared to the other species, followed by the yellow bamboo (*Bambusa vulgaris* Scharader ex Wendland), andong bamboo (*Gigantochloa pseudoarundinaceae* (Steud) Widjaja), tali bamboo (*Gigantochloa apus* (Schutz)), and black bamboo (*Gigantochloa atroviolacea* Widjaja) [52]. In the textile industry, this plant is used in two ways: to produce natural (bast) fiber by physical and chemical treatment and to spin regenerated (pulp) fiber through retting bamboo plant into pulp [53].

Cotton fiber (*Gossypium* sp.) is a type of fruit fiber that is used as a primary raw material for textile, health, and beauty products [54]. The cotton production was about 2.558 tons, with a plantation area of 11.287 Ha [55]. Even though the domestic cotton demand is increasing, the supply is not keeping up. Low cotton production and less farmer interest in planting are to blame [56]. Cotton is classified into three classes based on the length and smoothness of the fibers: long, medium, and short cotton fibers with lengths of 1.5, 0.5 to 1.3, and 0.3 to 1 inch, respectively [20]. Training on cultivation techniques and the use of superior seeds are needed to increase cotton productivity [57].

Ramie (*Boehmeria nivea* S. Gaud) is a kind of compatible fiber, making it is simple to be combined with a variety of other fibers [25]. Ramie is a fast-grown and branchless plant that can reach a height of 1 to 2 m. The stem-extracted bast fibers are the strongest and the longest natural bast fibers [58]. The productivity of ramie fiber is determined by the stem's height, diameter, skin thickness, and fiber yield (fiber content per stem). Since the fiber production can be carried out every 2 months, harvesting is conducted 5 to 6 times per year in tropical areas like Indonesia. China grass crude fiber contains around 2–4% of fresh stalk, 1–3% of degummed fibers, and 1–2% of hemp top [59,60]. Furthermore, ramie production is about 100 thousand tons per year, which is higher than abaca fiber production, at about 70 thousand tons per year [61]. According to Soeroto [62], these plants could grow in Indonesia's middle to highland areas, with the highest productivity in the highlands (700 mdpl) from 2.5 to 3.0 tones/ha/year [63]. Ramie productivity per hectare is much higher than cotton [64] and its fiber quality is higher than cotton with the color and luster comparable to natural silk. Furthermore, ramie absorbs 12% of water while cotton absorbs just 8% [65].

Sisal (*Agave sisalana* L.) is a good fiber-producing plant that can be grown on dry land and is resistant to soil with a high salt content [66]. Because of its ease of cultivation and fast renewal times, sisal fiber accounts for half of all textile production [67]. Indonesia produces 500 tons of sisal fiber per year, which is obtained from the plant's leaves [61]. According to Mukherjee and Satyanarayana [67], sisal plants can produce 200 to 250 leaves per year, with each leaf containing 1000 to 1200 fiber bundles. Furthermore, each bundle contains 87.25% water, 4% fiber, 0.75% cuticle, and 8% dry matter. Sisal fibers are characterized by their hardness, strength, and yellowish white color. Each leaf of 1000 fiber bundles contains 4% fiber [68], which has not been used to its full potential [37].

Coconut (*Cocos nucifera*) is a plantation plant that produces fiber from its fruit for use in furniture, crafts, and probably polymer composite reinforcements [69]. Coir fiber can be obtained from coconut, and it has the thickest, the most resistant, and the lowest decomposition rate of all-natural fibers. It is ideal for rope production due to its high strength [70]. Fruit cultivation and pruning produce a significant amount of coir fiber [71,72]. Furthermore, coconut cultivation produces coir and pith, which account for around 35% of the total weight of the crop [73].

Sansevieria is a genus of ornamental plants in the Agavaceae family that grows from lowlands to highlands in the tropics and sub tropics [74]. This plant is xerofit with thick leaves due to high moisture content [75] with a spherical, half-shaped leaf style round, stiff as a blade, short curved, and sunken fleshy. The leaves have smooth and corrugated margins, and the tips are tapered, pointy, and blunt [76]. In addition, fibers from Sansevieria were extracted using the retting method [14]. It may be used as textile raw materials, an absorbent of pollutants, and cancer cell inhibitor [76–80]. It is also used to treat diseases like stomach pain, earaches, diarrhea, hemorrhoids, fungi, scabies infections [81], as well as for anti-inflammatories, analgesics, antipyretics, antioxidants, and antimicrobial activity [82], and as raw material for handicrafts [83].

Jute fibers (*Corchorus capsularis* and *C. olitorius*) are off-white to brown in color and range in length from 1 to 4 m, which is obtained from the bast or skin of the plant. Jute fibers with a large amount of cellulose, high tensile strength, and low extensibility could be grown in 4 to 6 months. They have better fabric breathability, are free of narcotics or odor, have strong insulating and anti-static properties, low thermal conductivity, and mild moisture recovery. Jute fibers are appealing because they are biodegradable, recyclable and environmentally friendly [84]. According to Suliyanthini [38], these fibers have very low creep, are brittle, and have a coarse nature that limits the fineness of the yarn. Packaging, sack material, tapestry coatings, electrical insulation, an industrial fiber for carpet coatings, electrical, rigging, tarpaulin, roofing materials, automotive manufacture, and straps are some of these fiber applications [38,85]. Because of broken hair that may cause food contamination, jute fiber is not appropriate for certain forms of food [38]. Jute plants have short, tall, straight stems with leaves at the top of the tree, and the fibers are derived from them. The jute tree grows to a height of 1.5 to 4.8 m and has a stem diameter of 1.25 to 2.0 cm [38]. Furthermore, the retting method may be used to draw these fibers [86].

Kenaf fiber comes from the *Hibiscus cannabinus* L. plant stem, which has been grown since 1979/1980 as part of the ISKARA (intensification of community sack community) program [87]. These plants are adaptable and can be grown on a variety of surfaces, including peat [88] and flooded soil [88,89]. Depending on the variety and growing climate, kenaf productivity can range from 2.0 to 4.0 tons of dry fiber/ha [90]. It is an annual plant with a stem diameter of 1.25 cm and a height of 2.5 to 3.75 m [38]. According to previous records, India and Pakistan are the world's top kenaf producers. Tropical and subtropical climates with high humidity, heavy rain, and no strong winds are ideal for these plants.

They thrive in loose, well-draining soil and are planted similarly to jute. They can be harvested 4 to 5 months after they begin to bloom [38,91].

Bombix mori caterpillar cocoons are used to make silk fiber [35,92]. Due to its high tensile strength, strong degree of resilience, elasticity, flexibility, biodegradation, and great biocompatibility, it is considered a possible biomaterial that supports cell attachment and proliferation [92–98]. Silk fibers are strong, smooth and crease resistant, with a high capacity to absorb water. Furthermore, these fibers are used in the manufacture of women's clothes, socks, ties, and tissue engineering [35]. They are considered as the most desirable and coveted fibers because of their relative rarity, unique luster, softness, and drape [99]. Wool is the most essential animal fiber, and it is obtained from sheep, in either a staple or short form. It contains keratin protein, lanolin (an external lipid), and a small amount of internal wool lipid (about 1.5%) [100,101]. Clothing, sweaters, blankets, rugs, weaving, and knitting all use wool as a raw material [35].

Collagen is a connective tissue extracellular matrix that is derived from the skin and bones of animals and comprises 30% protein [102]. It is commonly used in biomedicine, medicinal food, food and drug growth, and cosmetics [103]. It has been used as a homeostatic agent, bone tissue regeneration, membrane oxygenator, contraception (barrier method), implant, and drug delivery system in biomedicine. In the cosmetics industry, collagen is used as an emulsifier and foaming agent in the food industry, while in the field of cosmetics, it becomes an active ingredient used to avoid the incidence of premature aging (anti-aging) [103]. Chemical processes and the combination of both enzymatic and chemical processes [104] have been used to isolate collagen (acid-soluble collagen and pepsin-soluble collagen) [105]. To make certain products, collagen fibers in animal skin are processed into leather through a tanning process [103]. Footwear, clothing, gloves, leather goods, heavy leather, and upholstery are all made of leather [35].

Corn is the most common crop found in every region of many Asian countries. The potential of corn plants in providing natural fiber is very high, such as the stems, leaves, and skins of corn. Cornhusk fiber contains cellulose, hemicellulose, and lignin of 46.15%, 33.79%, and 8.92%, respectively. The tensile strength value of corn husk fiber is 169.49 MPa, which is higher than the tensile strength of glass fiber, which is 1.7–3.5 MPa [7]. Modification of corn husk fiber using sodium hydroxide (NaOH) solution with a concentration of 0.5–8% is known to reduce the hydrophilic properties, and increase the crystallinity, tensile strength, and thermal resistance of the fibers [7]. The addition of corn husk fiber to polymer composites can increase the tensile strength, bending strength, and toughness properties of the polyester composite [19,106]. Although immersed in water and exposed to ultraviolet (UV) light, the mechanical properties of the corn husk fiber composite were still quite high compared to the "pandan wangi" fiber-reinforced composite. In several previous studies, this corn husk fiber composite was found suitable to be used as a substitute for wood, soundproofing panels, and building materials [19,106]. Figure 2 shows some natural fiber resources.

Figure 2. Some natural fiber resources (Pic courtesy authors collection).

3. Technological Perspective of Natural Fibers Processing

Temperature, humidity, height, growing site, local climatic conditions, season, and harvesting are all factors influencing the quality and structure of natural fibers [39,107–109]. Handling method, storage period and condition, and harvested plant portion can all affect fiber qualities [107–109]. Some of the elements should be closely monitored to acquire the best fiber characteristics.

Natural fibers include kapok, ramie, pineapple, sansevieria, kenaf, abaca, sisal, and coconut fiber, as well as bamboo [110]. Bamboo has a greater ultimate strength [111] than other fiber bioresources like jowar and sisal, which may be tested in single unit fiber or fiber-bundle tests. Practically, the last test is preferable because it is easier to administer and yields faster findings [110]. The mechanical properties of fibers were also influenced by their microstructure and chemical composition (cellulose, hemicellulose, and lignin) [112], and the fiber-cross-sectional area became the key variable controlling the fiber strength [110]. Aside from that, the extraction method and chemical treatment have an impact on fiber tensile strength. Deka et al. [113] observed that alkali soaking increased the tensile strength of *Parthenium hysterophorus* fiber.

According to De Farias et al. [114], cellulose content has a significant impact on tensile strength and Young's modulus. The lignin content, on the other hand, had an inverse effect on those strength. Hemicellulose, pectin, and wax, like cellulose, play a role in specific Young's modulus. The moisture gain of fiber is related to the quantity of hemicellulose and lignin [114]. Microfibril angle (MFA) has a negative and positive relationship with the pectin and lignin content, as well as hemicellulose, respectively, while failure strain value was affected by hemicellulose, lignin, and pectin content, respectively. Cellulose and pectin have a positive effect on density, while wax has a negative effect. Based on this knowledge, it is crucial to investigate the chemical composition of fiber as well as its mechanical and physical qualities.

Glass fiber and natural fibers are extensively distributed, with glass fibers being non-renewable and non-recyclable and natural fibers being the opposite. Natural fibers do not abrade the machine and are not harmful to the lungs when inhaled. According to their disposal viewpoint, glass fiber is a non-biodegradable material, whereas natural fibers are the opposite. Natural fibers have lower tensile strength than synthetic fibers, but they have several advantages, such as not being fractured during processing, equivalent stiffness, and specific strength to glass fibers [115]. They also have less Young's Modulus and density [116] as well as less energy, density, and cost consumption [117] than synthetic fiber.

To match the use of natural fiber, it was necessary to understand the physical-mechanical properties. Natural fibers have porous qualities, which might make it is difficult to estimate a realistic density. Glass fiber has a higher density of 2.4 g/cm^3 than natural fiber, which has a density of 1.2–1.6 g/cm^3. As a result, it can be used to make light-weight composites [58]. The increase in porosity has a proportionate relationship with the lumen size and density of fiber [110]. Fiber bundle diameter tends to rise as density decreases. The specific toughness of bananas, hemp, pineapples, and jute fibers is high. Natural fibers are suitable as reinforcing components in composites because of their unique specific stiffness and tensile strength. Fiber bundle diameter tends to rise as density decreases. The specific toughness of bananas, hemp, pineapples, and jute fibers is high. Natural fibers are suitable as reinforcing components in composites because of their unique specific stiffness and tensile strength. Some of natural fiber, such as jute and sisal fiber, have the potential to replace glass and carbon fibers [118] in composites that demand a high strength-to-weight ratio and weight reduction in that application [119] due to their ease of availability and low cost.

Ramie, sansevieria, pineapple, sisal, and kenaf had low strain (2–6%) but high stress, while *Cocos nucifera* husk fiber had high stress (24%) but low strain [110]. Jute fibers displayed a similar pattern on the stress-strain curve of pineapple fiber [86]. Ramie bast fiber has cellulose content (69–97%) and a low spiral angle (7–12%), as well as a high molecular weight (69–97%), resulting in good mechanical properties [120–124]. Sisal fiber also possesses excellent porosity, tensile strength, bulk, folding strength, and absorbency [90]. Bast fibers were found to have low stiffness, but great strength and elongation, as well as elastic recovery. These fibers are widely available, inexpensive, and function well [125]. The mechanical strength of pseudo-stem banana fiber was also discovered. The flexural and tensile strength of glass fibers were enhanced when sisal or jute fiber were mixed in [31]. The addition of hay fiber, milkweed fiber, kusha grass, and sisal fiber boosted the tensile strength of polypropylene composite [126].

The principal constituents of lignocellulose, which included the cell walls, are cellulose, hemicellulose, and lignin. Aside from that, ash, silica, pectin, waxes, and water-soluble compounds [127–129] and oil [128,129] can be found in natural fibers. Plant growth conditions, harvesting period, geographical considerations, and fiber extraction technology all affect the chemical component within the same plant species [127]. To assure the quality of the manufacturing process, the impact of various plant material properties must be analyzed [130]. The fiber characteristic is determined by the angle of the microfibrils and their placement within the cell wall [131]. Figure 3 depicts the position of chemical components in the cell wall (a) and layer position in the secondary cell wall of a plant. Cellulose uses lignin and pectin as glue to join with hemicellulose. The cellulose content of cell wall increases from primary layer (S1) to secondary layer (S2), while the lignin content decreases. Hemicellulose content is found in equal amounts in each layer. The S2 layer is primarily responsible for the physical and mechanical strength of fibers. It has lower microfibrillar angle, higher cellulose content, and contribute to improve fiber strength properties [132,133].

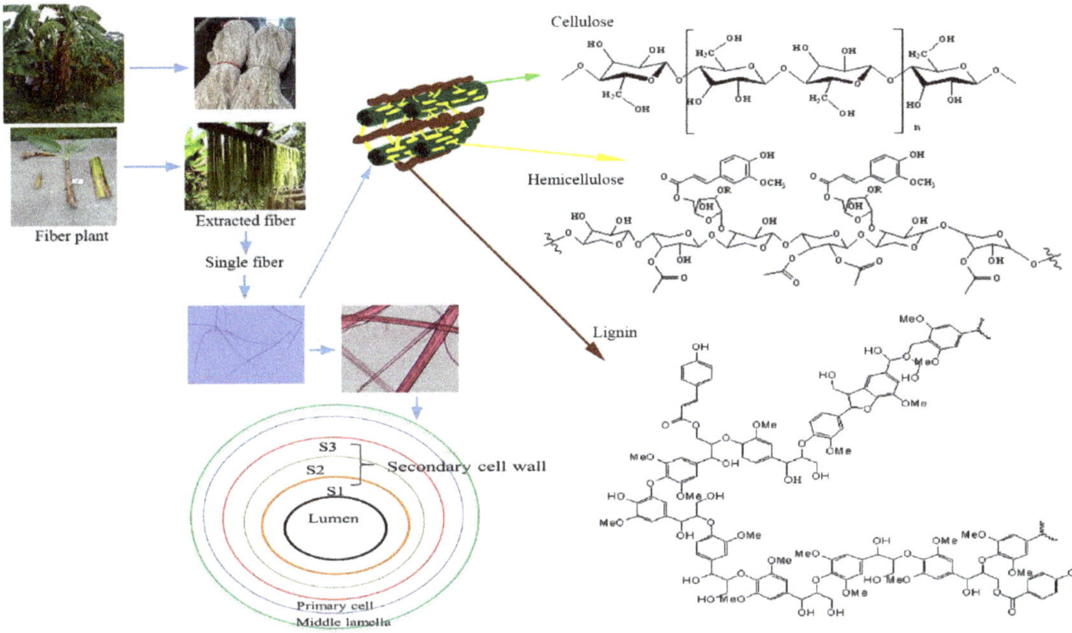

Figure 3. The hierarchical cell wall structure of lignocellulosic biomass (adapted from [134], Copyright © 2021 Elsevier Inc., License Number: 5193031271486) Cell wall layer position of plants (modified from [135], Copyright © 2021 Elsevier Ltd., License Number: 5193040344374).

Harvesting method, plant age, sample position in plant, environmental growth condition, and extraction fiber methodology all play role in this diversity, as previously indicated. Cotton linter, cotton, ramie, *Mamordica charantia*, flax, Henequen, and palmyrah are examples of natural fiber with a higher cellulose content than others. Because fibers with low lignin and high cellulose content have high tensile strength, numerous factors might influence this number, and therefore the relationship is not necessarily linear. The crystalline domains of cellulose have a substantial impact on the tensile strength, with more cellulose crystallinity resulting in higher fiber strength. The position of lignin on biomass affects tensile strength because it is wedged between cellulose and hemicellulose.

Some methods have been established in preparation of natural fiber which are summarized in Table 1. Dew retting and water retting process are two common techniques to separate the plant fibers that require about 14 to 28 days to degrade the waxes, pectin, hemicellulose, and lignin [131].

Table 1. Preparation method for producing the extracted natural bast fibers.

Introducing Methods	Advantages	Disadvantages	References
Dew retting	• Relative ecofriendly or less energy process by using bacteria and moisture in the plant for separating individual fibers from the plant • Common in areas with heavy night dew and warm days, as well as areas with water shortages.	• Excessive retting brings difficulties in separating individual fibers or tend to weak the fiber strength • Required long processing time (2–3 weeks) depending on climatic condition • Obtaining dark fiber and poor quality	[15,136]

Table 1. *Cont.*

Introducing Methods	Advantages	Disadvantages	References
Water retting	• Produce fiber with high cellulose content, which gives the fiber a higher tensile strength • Produce fiber with lower density which is suitable for low weight composite applications • Duration of process for 7–14 days	• Need surface treatment as initial step to increase the surface roughness • High cost • High water treatment maintenance	[17,137–143]
Mechanical extraction	• Produce a significant amount of acceptable quality of fibers • Short time process duration	• Damage fiber cell wall structures, resulting in dislocations, kink bands or node that have a negative impact on tensile mechanical properties and may even compromise composite performance • High cost	[15,134,141]
Chemical treatment	• Produce fiber with high cellulose content, higher tensile strength, thermal stability, and crystallinity index • The surface roughness of the fiber relatively good (based on SEM analysis) • Enhance the physicochemical properties of the fibers	• Some chemical treatment waste can pollute the environment	[17,137–142]

Figure 3 depicts the structures of cellulose, lignin, and hemicellulose [134]. The cellulose content of a plant is determined by its age and species. Cellulose has hexose sugar with greater thermal stability than hemicellulose's branched structure. It is made up of a linear chain of glucose units linked together by (β (1→4) bonds with a high degree of polymerization (DP). Cellulosic plant fibers have high moisture absorption capacity and poor dimensional stability when exposed to water [144]. The hydrophobicity and hydrophilicity of fibers, as well as their interaction with the matrix, may alter fiber-matrix adhesion with natural fibers as reinforcement [145]. Different cell wall polymers of lignocellulosic materials influence the degradability and properties of natural fibers [146]. Furthermore, cellulose affects natural fiber strength, but lignin prevents UV breakdown and char production [1].

Hemicellulose, which has a low DP, is the third most abundant cell wall constituent of lignocellulosic biomass after cellulose and lignin. As a result, this biopolymer dissolves more quickly in the liquid fraction during biomass pretreatment. Pentose is the most prevalent hemicellulose in non-wood plants, hence a larger concentration effects fiber fibrillation, which raises the bonding potential of pulp sheets [147]. High hemicellulose content improves fiber flexibility during paper sheet usage, allowing it to swell and expand to a large surface area. The higher the crystallinity of cellulose, the less hemicellulose present. As a result, a low hemicellulose content promotes cellulose in the amorphous zone [148]. Thermal, biological, and moisture degradation, as well as absorption, are all caused by hemicelluloses in natural fibers [146]. Hemicellulose material has a favorable relationship with moisture sorption and biodegradation. In addition, hollow natural fibers have a lower/lighter bulk density and contain more water. The degree of crystallinity, orientation, swelling behavior, tensile strength, and porosity of fibers are all affected by their moisture content. As a result, the higher the moisture absorption, the higher the risk of microbial attack [149].

Excess amorphous material such as lignin, pectin, hemicellulose, wax, and cellulose are removed during the alkali process by adding NaOH. Reactive dye fixation is improved by the presence of –OH and –COOH groups in natural fibers [150]. The alkali treatment (adding NaOH) removes excess amorphous content such as cellulose, lignin, hemicellulose, pectin, and wax from natural fibers (*Pennisetum orientale* grass), whereas the alkali treatment removes excess amorphous content such as lignin, pectin, hemicellulose, wax, and cellulose.

As a result, the thermal stability and density of the NaOH-treated fibers were higher than those of the untreated fibers and the HCl-treated fibers [151].

Lignin is a biological substance that helps plants maintain their structural integrity [152]. It is also the second most abundant biopolymer with an aromatic molecular structure after cellulose and forms an ester connection with hemicellulose. Lignin molecules include three active functional groups: namely, coniferyl alcohol (G), p-coumaryl alcohol (H), and synapyl alcohol (S). The most frequent connection in lignin is the aryl ether linkage (β-O-4), which accounts for nearly half of all links. This connection is more easily cleaved during lignin conversion and depolymerization. In natural fibers, lignin is also implicated in UV degradation and the formation of char [146]. Coir has a larger microfibrillar angle, as well as a reduced proportion of hemicellulose and cellulose, which impacts plant qualities like strength, durability, damping, wear, weather resistance, and high elongation at break [58].

Since lignin is an unwanted component in raw materials in pulping, it is typically removed during the pulping process in the pulp and paper industry. A complete delignification process could produce pulp with desirable Kappa numbers. Furthermore, delignification of samples with higher lignin content requires a significant quantity of chemical energy. Delignification of samples with decreased lignin content is possible under lower chemical charges and temperatures [147]. Binder-less fiber board with a high lignin content could be used as an adhesive source. It could also be extracted for use in high-value products including adhesives, biosurfactants, antibacterial agents, fine chemicals, lignosulfonate, and so on [1,153–157].

Cellulose content is one of the most important components in assessing the mechanical and physical properties of natural fibers; it is one of three main components (cellulose, hemicellulose, and lignin). The DP was reduced because of excessive chemical treatments such as pulping and bleaching. The number of glucose molecules in one cellulose chain is measured in DP. Depending on the cellulose source, the length of the cellulose polymer chain varies greatly. Plants with a DP of more than 10,000, for example, are vascular cellulose plants. The amount of DP released by a plant is determined by the process used to isolate and treat it. Pure cellulose has a DP of over 10,000 in most cases. Microcrystalline cellulose is another example, which is a high-level crystalline cellulose after going through hydrolyzed acid. Microcrystalline cellulose has a DP value of between 300 and 600 [158,159]. The DP value of microcrystalline cellulose ranges from 300 to 600. This is due to the strong chemical treatment that results in the breakage of the short cellulose chains, and it also influences the crystallinity, mechanical properties, and morphology of cellulose [160,161].

The crystallinity index, or degree of crystallinity, is used to determine the physical and mechanical properties of natural fibers. X-ray diffraction (XRD) at the crystalline peak at 2θ (22.6°) for diffraction intensity I_{200} (crystalline region) and 18° for diffraction intensity I_{am} (amorphous region) can be used to measure the degree of crystallinity of cellulose. The peak height method can be used to calculate the crystallinity [162,163].

4. Social and Economic Aspects on Utilization of Natural Fibers

Since prehistoric times, natural fibers have played a vital role in human society as a sustainable and ecologically beneficial source of raw materials that are easily degraded into environmentally friendly items and have the ability to absorb enough moisture. Natural fibers have a variety of fascinating properties, including low density, light weight, low cost, biodegradability, abundant accessibility, minimal health hazards during processing, relatively good basic strength and modulus, good thermal and acoustic insulation characteristics, physical properties, and ease of availability [125,131].

Natural fibers have been favored over synthetic fibers because of their superior qualities [125,131]. Natural fiber has been used as a raw material in a variety of industries, including aerospace, automotive, marine, building and construction, sports and leisure items, electronic appliances, military vehicles, biomedical purposes [10,11,15,16,21,22,164–166] as shown in Figure 4: Natural fiber applications are also increasing in textiles, packaging, printed goods, filters, automobiles, furniture, particleboard, insulation board, and other materials [167–170]. Woven-kenaf aramid and pineapple leaves were used in military vehicles, especially for ballistic purposes [166], and hard armor plate [166,171], respectively. In biomedical applications, natural fibers are in fiber-reinforced composites (FRC), such as various clinical fields [172] as described in Table 2. Hemp and sisal have been reported for utilization as cementitious construction and fancy materials in the construction field [173,174]. In the biomedical field, the most promising natural fiber candidate is undoubtedly cellulose, in the form of nanofibers. Nanocellulose has a variety of biomedical applications, including drug delivery, vascular grafts, skin tissue regeneration, antimicrobial membranes, medical implants, biosensors and diagnostics, and scaffolds [175]. Several methods have been developed to improve the compatibility of natural fibers and polymer matrices in order to enhance the physical and mechanical properties of targeted bioproducts. However, the acceptability of natural fiber and biocomposite materials by the human body is a critical requirement that must be addressed [176].

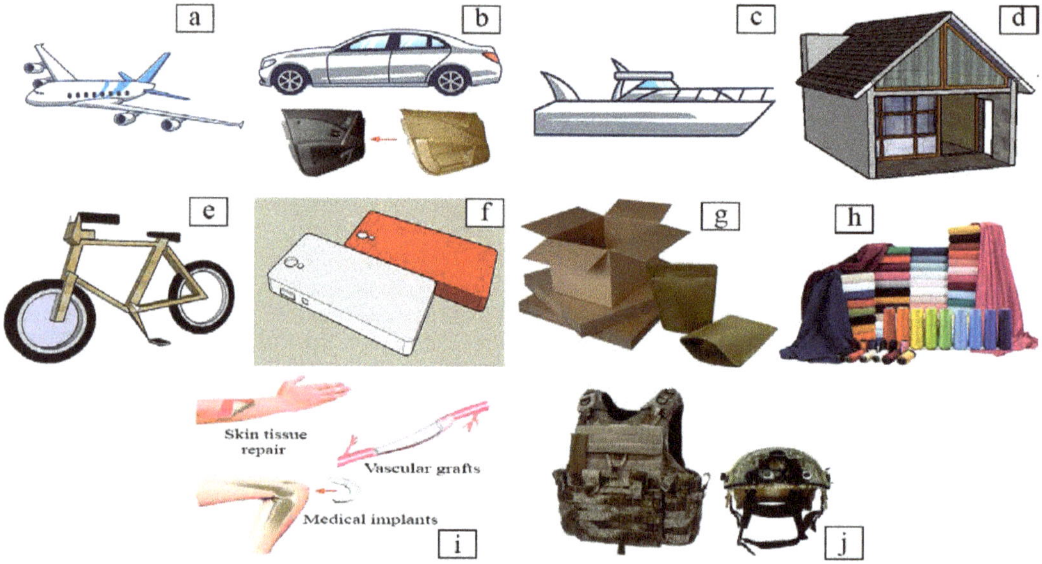

Figure 4. Potential application of natural fibers in many sectors such as (**a**) aerospace, (**b**) automotive, (**c**) marine such as boat hulls, (**d**) building and construction such as insolation board, (**e**) sport and leisure goods, (**f**) electronic appliances such as handphone casing, (**g**) paper and packaging, (**h**) textile, (**i**) biomedical, and (**j**) military fields.

Table 2. Application of natural fiber composites in biomedical field.

Specific Area Application of Fiber Composite	Source of Fiber	References
• Blood bag • Drug/gene delivery scaffold	Pineapple, rambutan and banana skin	[177]
• Ancient medicine • Modern functional food	Flax and flaxseed oil	[178]
• Wound dressings	Flax	[179]
• Drug delivery	Cotton	[179]
• Wound healing	*Bombyx mori* silk	[180]

Table 2. *Cont.*

Specific Area Application of Fiber Composite	Source of Fiber	References
• Tissue engineering • Drug delivery • Wound dressing • Medical implants • Cardiovascular implants • Scaffolds for tissue engineering	Pineapple leaf	[181,182]
• Prosthodontics • Orthopedics • Cosmetic orthodontics	-	[176]
• Dental application	-	[20–22]

Natural fibers are a type of biomaterial used for reinforcement of polymer-based composites. Some agricultural plants, including ramie, sisal, and pineapple leaf fiber [107–110,183] and hybrid fibers of Egyptian and Qatari palm trees [176] and woven cotton fabrics [184] have reported used as bioresources of FRC. The manufacture of natural fiber composite materials or eco-friendly composites has become a popular topic as people become more aware of environmental sustainability. To minimize material weight, natural fibers may be a suitable option for replacing synthetic materials. Natural fiber reinforced polymer and resin composites have been widely used in a variety of industries, including automotive and aviation interior components, as well as military vehicles [166,185–187]. Because of their high specific qualities at a lower cost than synthetic fibers, they are appealing for several applications.

Miller [167] mentions the usage of hemp fiber in textile manufacture. The mechanical properties of the bio-based textile composites studied in this review are like those of some traditional materials. The use of pineapple leaf fiber as a reinforcement in the fabrication of yam starch films with packing potential was defined by Mahardika et al. [168]. Asrofi et al. [188] created a bioplastic made of tapioca starch and sugarcane stem fiber for reinforcement. The interior components of an automobile are composed of hemp fiber/polypropylene composites [169], while kenaf and wheat straw were used as vehicle spall-liners and quarter trim panel storage [189]. Natural fiber mats, aluminum sheets, and epoxy resins provide excellent electromagnetic interference prevention while keeping high mechanical qualities in hybrid composites [170]. Good specific properties, low cost, low density, good formability and processability, good mechanical properties, and a plentiful and sustainable source of raw materials are all the benefits of using natural fibers over synthetic fibers. Natural fibers, on the other hand, have a high moisture sensitivity [190,191]. The development of natural fiber composites in a variety of applications has paved new avenues in both academia and industry for the future applications of sustainable natural fibers.

As previously stated, several of the shortcomings of natural fibers should be addressed during the optimization of natural fiber applications. When used as a composite, the hydrophilic nature of natural fibers makes it difficult to adhere to a hydrophobic matrix, resulting in poor mechanical characteristics and processability [112]. Surface treatment methods applied include chemical and enzymatic treatments, corona treatment, and coupling agent addition [119,120,192–197]. Furthermore, the handling of the interfacial region before processing with thermoplastics at a temperature up to 200 °C, the interfacial treatment (surface treatment resins, additives, and coating) must be reinforced to address the low degradation temperature of natural fibers [107,198].

The Indonesian government has taken steps to encourage the use of natural fibers, such as appointing an institution to focus on the development of natural fibers and establishing a multi-stakeholder research community, namely, the Indonesian Ramie Consortium (KORI), to study specific natural fibers, primarily ramie. Ramie is a type of natural fiber that has become a national priority in Indonesia for widespread use. Figure 5 depicts the strategy for manufacturing of ramie development in Indonesia.

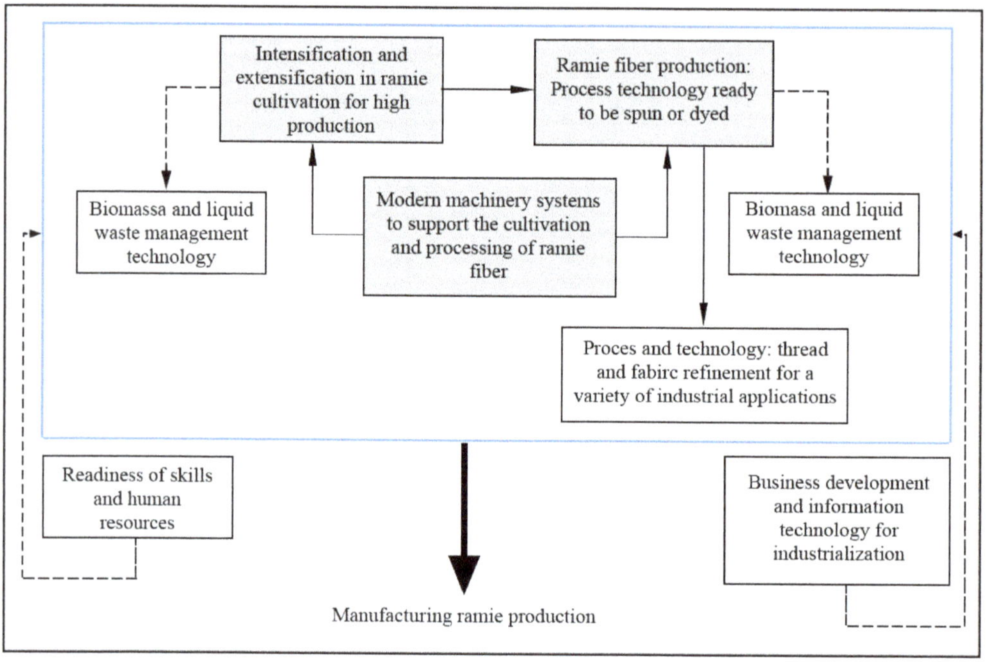

Figure 5. Manufacturing integration strategy in the ramie production system in Indonesia.

Manufacturing integration strategies in ramie processing systems to support large-scale production are to be developed with an emphasis on three main sub-systems: cultivation, fiber processing technology, and machining. The ramie-based industry will be able to support the functional value of the fiber or fabric of ramie for functional enhancement of the products. Strategies in business concepts and supply of human resources with competence in all aspects of processing systems will be able to support the realization of the manufacturing of ramie production in Indonesia. This research strategy is currently supported by the Indonesian government in the National Research Program, for the period 2020–2024.

Additionally, the Indonesia Natural Fiber Council (DSI) was founded in Indonesia to assist scientists, policymakers, and other stakeholders in the development of bioproducts generated from natural fibers. DSI proposed a road map for the Indonesian fiber sector from 2020 to 2024, with abaca, kenaf, bamboo, pineapple, sisal, cotton, and ramie as types of promising fiber to be further developed [199]. Furthermore, biduri (*Calotropis gigantea*) is a natural fiber that has the potential to be developed in Indonesia as a thermal and acoustic insulation material and filler material [200] and for winter jacket [201]. Biduri fiber production is predicted to be around 3.6 tons per hectare per year [200]. Some bioproducts, such as biopellets, food, textiles, biocomposites, and ecofriendly shoes, have been launched into the Indonesian market as a result of continued efforts. In addition, several small local businesses extract fiber from fresh pineapple leaves using basic techniques such as retting followed by decortication for clotes, handycraft, and other items. However, the process output is still low, with 2.5 kg of air-dry pineapple fiber produced from 100 kg of fresh leaf fiber and 97.5% of decorticator waste that has yet to be used (visualized in Figure 6).

Banana stems are treated in a similar way to make banana fibers in this local enterprise. Until now, cotton has been the main fiber source in the Indonesian textile industry, but the qualities of local cotton have not met the requirements, so nearly all of it is imported, while the other fibers have been used to their full potential. Considering the potency and challenge, continual efforts to disseminate information about the various uses of natural fibers in the community are required.

Figure 6. The process production of extracted fiber from fresh leaf pineapple, (**a**) fresh leaf fiber, (**b**) decorticator process, (**c**) wet extracted fiber, (**d**) decorticator waste, (**e**) dried fiber that ready for spinning.

Natural fibers play an important role in improving the quality of human life. However, waste can be generated during the product life cycle and during the processing of natural fibers into bioproducts. To achieve the most efficient utilization of resources, waste management should be conducted continuously by recycling and/or upcycling of waste, aside from innovation in the design of bioproducts. Shanmugam et al. [202] recognized recycling and the use of bio-based constituents as essential issues in adopting a circular economy (CE). CE adheres to the principles of reduce, reuse, recycle, and replace. CE is beneficial to the environment, economy, and society when used in FRC manufacturing. Given the numerous sustainability challenges confronting our societies, transitioning to a circular economy and closing resource loops through recycling is a viable solution [203]. Figure 7 proposes a CE concept based on natural fibers that is more considered than a linear economy concept for future resource conservation and environmental balance. The CE approach is gaining traction and has been proposed in some fields, such as carbon fiber manufacturing [204], agricultural sector [205] and biomass biorefinery [206], for gradually reducing energy consumption during the manufacturing process. Biomaterials in the CE present numerous challenges for the industry in terms of developing new network and commercial opportunities while remaining focused on consumer demands [206].

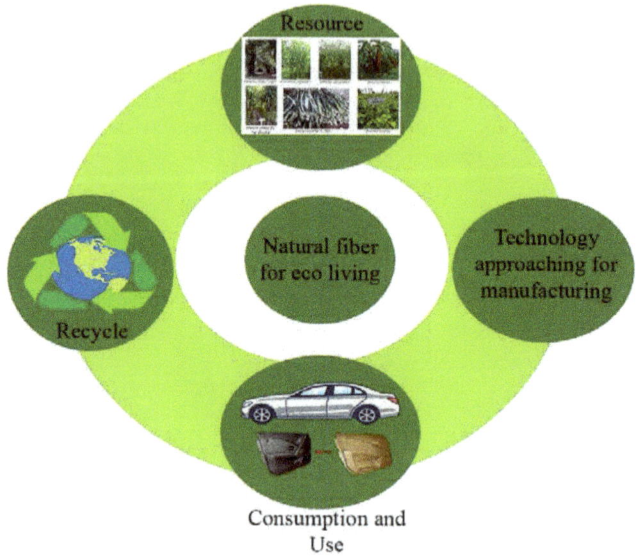

Figure 7. Circular economy concept to conserve the natural fiber for future eco living.

The development of an information system for Indonesian natural fibers, as well as collaboration with a variety of stakeholders such as research and development institutions, industries, policymakers (local and national), and universities, are ongoing efforts to bring Indonesian local industry independence. National innovation products made from natural fibers can be created by local industry in the future and sold at least locally, with Indonesians consuming them.

Natural fiber as lignocellulosic biomass has an economic chance to meet industrial needs, depending on the processing level that has been made to make its derivative products, including its market to accomplish. According to Ruamsook and Evelyn [207], there are four levels in which biomass can be processed and turned into value-added goods before being sold to potential demand markets (Figure 8). Farmers become the first important people actors to create their biomass as the major components of industrial needs, as indicated in this picture. Commodities such as corn, wheat, cotton, and hay, as well as other crop farms such as paddy, are the possible resources of rubber and polymer markets. Many industrial polymers and plastics are still made from non-renewable oil and gas resources today. This would cause a supply shock when non-renewable resources are depleted, causing the processed product to bubble to an unacceptably high price [207]. As a result, the growing interest in bio-based polymer and plastic products derived from renewable sources creates a market opportunity for biomass in exchange for enhanced environmental support in reducing climate change pressure.

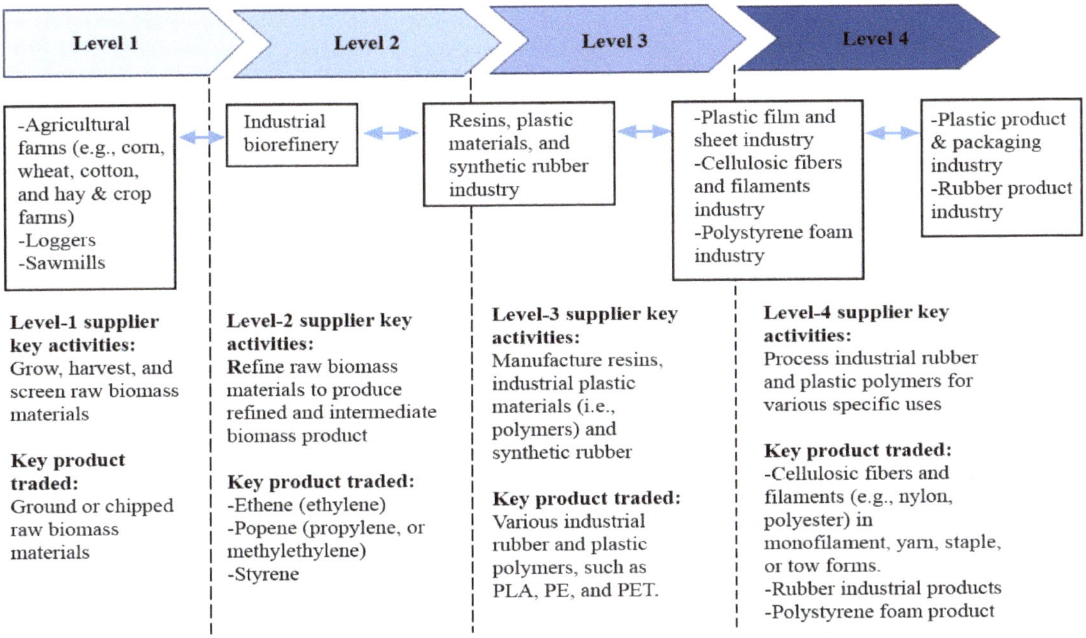

Figure 8. Simplified example of multi-level rubber and polymer markets for biomass (adapted from [207]).

Farmers will benefit from the potential use of paddy waste as an alternative source of packaging because they have been heavily reliant on agriculture without any additional revenue, as most farmers are still subsistence and have a low-middle income. On the other hand, this will contribute to reducing future climate change issues by allowing farmers to benefit from better climate conditions through sustainable agriculture. Nonetheless, the government must promote this innovation to increase economic potential and to provide an instrument for industrial businesses to improve their knowledge of the sustainable industrial environment. It may not be easy, but once the government steps in to regulate the industrial ecosystem by paying more attention to reducing plastic waste and implementing sustainable bioplastic for bio-packaging for both large industrial ecosystems and small-medium enterprises, it could have structural potential. One of the studies that uses paddy-waste as bioplastic is the usage of rice straw cellulose (*Oryza sativa*) as bioplastic by the pulping process and phase inversion method [208].

Building an ecosystem of sustainable industry, particularly for consumer behavior, is to use more sustainable packaging or bio-packaging that can be created from natural fiber. As it is known that the production of food packaging made from plastic as well as styrofoam is about 14.000 tons per year, it has affected the use of plastic packaging for food [203]. However, this material is not environmentally friendly and could cause a significant impact both on the user and the environment after its usage, with a long-term impact on climate change. Thus, the government could develop bio packages as a particular potential both for reducing environmental issues and improving the economic opportunity of farmers from the paddy waste produced. Basically, this novel innovation could be started by a small-scale industry where a group of farmers could start to process the paddy waste materials where generally they would not be sold except for burning.

On the other hand, the development of bio-packages made of paddy straw benefits not only farmers in terms of economic earnings, but also consumers, as they have paid more for the environmentally friendly food package to have both future-health preservation as well as the original flavor and scent of stored food from this bio-package material, compared to a conventional plastic food package, which has more influence on their food, particularly when the foods are still at high temperatures. With these kinds of benefits, coupled with intense regulatory aspects, consumers would indeed be willing to pay more to get this type of food package if the government could guarantee that it would not harm them and be rigid in implementing environmental policy in general. If this is accomplished, the introduction of novel food packages made of paddy straw will be imminent, benefiting farmers who are the primary source of this material. Many countries have taken steps to encourage the use of natural fibers, such as appointing an institution to focus on the development of natural fibers and establishing a multi-stakeholder research community.

5. Future Prospects

Many different types of natural fibers are potential raw materials for bioproducts, but it is necessary to select the most locally viable fiber before attempting to employ it in an industry. In several fields, using local potential natural fiber for industrial purposes helps to lessen reliance on imported products. Some efforts will be made to speed up the exploitation of local raw materials for local industry, thereby assisting in the establishment of local industry, mostly on a small and medium scale. The wide diversity of natural fiber characteristics remains a difficulty in manufacturing consistent quality of bioproducts. Therefore, understanding the features is essential when processing natural fiber effectively. Processing of natural fiber with environmentally friendly technology and appropriate procedures should be used in the future as environmental concerns grow for the preservation of sustainable nature. Kenaf and ramie, bamboo, bananas, and pineapple have all been used as commercial bioproducts for a variety of industries, including automotive, building materials, handicrafts, and textiles. Up to now, there has been no exposed intermediate industry in natural fiber; therefore, the biocomponent industry has been initiated in recent years and is able to create environmentally friendly footwear products. This footwear can be built with biocomponents from natural bioresources such as natural fiber. In summary, viability, suitable technology, and social issues in the development of natural fibers are significant components that, when controlled together, can stimulate the use of bioproducts made from natural fibers. It is critical that respective ministries should be involved in natural fiber development and prioritize the agenda to ensure a strong supply chain and sustainability of natural fiber production. The Ministry of Agriculture should be in charge of providing sufficient land for planting as well as the necessary technology for a successful harvest. While the Ministry of Industry can provide the manufacturing technology and create the product diversification design.

6. Conclusions

Natural fibers with enticing properties such as lower density, lighter weight, biodegradability, good specific strength and modulus, good thermal insulation, good acoustic insulation, and high electrical resistance can be used for various applications. Furniture, automotive (car bumper beam, disc brake pads), electronic industries (automobile components), and building construction (molded panel components, window frames) are only a few of the applications for natural fibers. Biomedical application refers to the potential use of natural fibers to cover tissue engineering, biomedical implants, and drug delivery systems that must be biocompatible with the human body. Although natural fibers such as bananas, abaca, pineapples, bamboo, cotton, ramie, sisal, coconut, sansevieria, jute, and kenaf are plentiful and readily available, industrial applications require access to the most readily available. Ramie is being used as a model for developing a national priority bioproduct based on textile innovation, which is facilitated through government support. This way, it will coordinate the efforts of all stakeholders, including industry, research and development agencies, and farmers, in order to maximize benefits. A circular economy that is sustainable can be achieved by developing a biocomponent industry for bioproducts. The establishment of an information system for natural fibers, as well as collaboration with a variety of stakeholders such as research and development institutions, companies, policymakers (local and national), and universities, are ongoing efforts to increase the use of natural fibers in a sustainable circular economy.

Author Contributions: Conceptualization, W.F., S.M.R. and A.F.; methodology, W.F., A.K., S.M.R. and M.A.R.L.; investigation, D.S.N., A.K., W.F., A.H.I. and S.M.R., resources, W.F., R.D., D.S.N.; writing—original draft preparation, A.K., W.F., M.A.R.L., A.H.I., I., R.D., A.P.W., A.F., M.A., N.H.S., B.R.P., N., S.M.R., Y.A. and E.S.; writing—review and editing, W.F., S.M.R., A.H.I., M.A.R.L., R.D. and S.S.M.; validation, M.A.R.L., A.H.I., W.F., S.M.R. and A.F.; visualization, R.D., Y.A., S.S.M., I., A.K., W.F. and M.R.R.; supervision, W.F., A.H.I., S.S., S.M.R. and D.S.N.; project administration, W.F. and D.S.N.; funding acquisition, W.F., R.D., D.S.N. and A.H.I. All authors have read and agreed to the published version of the manuscript.

Funding: This study was funded by Program Research of National Research Priorities (PRN) flagship for University FY 2021 with contract number 7457/IT3.L1/PT.01.03/M/B/2021. It was also supported by Productive Innovative Research Activities (RISPRO) Mandatory Themed Education Fund Management Institution (LPDP), National Research Priorities (PRN) FY 2020/2021, Ministry of Finance in collaboration with Ministry of Research and Technology/National Research and Innovation Agency (Kemenristek-BRIN), Republic of Indonesia, under grant No. 273/E1/PRN/2020 and B-4528/IPH/KS.02.04/VII/2020.

Institutional Review Board Statement: Not applicable.

Informed Consent Statement: Not applicable.

Data Availability Statement: The data presented in this study are available on request from the corresponding author.

Acknowledgments: Authors appreciate for the facilities, scientific and technical assistance provided by the Advanced Characterization Laboratories Cibinong—Integrated Laboratory of Bioproducts, Indonesian Institute of Sciences via E-Layanan Sains Lembaga Ilmu Pengetahuan Indonesia.

Conflicts of Interest: The authors declare no conflict of interest.

Abbreviations

CE	circular economy
DP	degree of polymerization
DSI	Indonesia natural fiber council
FRC	fiber-reinforced composites
G	coniferyl alcohol
H	synapyl alcohol
I_{am}	Intensity of amorphous region
ISKARA	intensification of community sack community
KORI	Indonesian Ramie Consortium
MFA	microfibril angle
NaOH	sodium hydroxide
PE	polyethylene
PET	polyethylene terephthalate
PLA	polylactic acid
S	p-coumaryl alcohol
S1	primary layer
S2	secondary layer
S1, S2, S3	secondary cell wall 1, 2, 3
SEM	scanning electron microscopy
UV	ultraviolet
XRD	X-ray diffraction

References

1. Karimah, A.; Ridho, M.R.; Munawar, S.S.; Adi, D.S.; Ismadi; Damayanti, R.; Subiyanto, B.; Fatriasari, W.; Fudholi, A. A review on natural fibers for development of eco-friendly bio-composite: Characteristics, and utilizations. *J. Mater. Res. Technol.* **2021**, *13*, 2442–2458. [CrossRef]
2. Madhu, P.; Sanjay, M.R.; Senthamaraikannan, P.; Pradeep, S.; Siengchin, S.; Jawaid, M.; Kathiresan, M. Effect of various chemical treatments of *Prosopis juliflora* fibers as composite reinforcement: Physicochemical, thermal, mechanical, and morphological properties. *J. Nat. Fibers* **2020**, *17*, 833–844. [CrossRef]
3. Vinod, A.; Sanjay, M.R.; Suchart, S.; Jyotishkumar, P. Renewable and sustainable biobased materials: An assessment on biofibers, biofilms, biopolymers and biocomposites. *J. Clean. Prod.* **2020**, *258*, 120978. [CrossRef]
4. Sari, N.H.; Fajrin, J.; Suteja; Fudholi, A. Characterisation of swellability and compressive and impact strength properties of corn husk fibre composites. *Compos. Commun.* **2020**, *18*, 49–54. [CrossRef]
5. Zhang, Z.; Zhang, J.; Li, S.; Liu, J.; Dong, M.; Li, Y.; Lu, N.; Lei, S.; Tang, J.; Fan, J.; et al. Effect of graphene liquid crystal on dielectric properties of polydimethylsiloxane nanocomposites. *Compos. Part B Eng.* **2019**, *176*, 107338. [CrossRef]
6. Li, K.; Fu, S.; Zhan, H.; Zhan, Y.; Lucia, L. Analysis of the chemical composition and morphological structure of banana pseudo-stem. *Bioresources* **2010**, *5*, 10. [CrossRef]
7. Herlina Sari, N.; Wardana, I.N.G.; Irawan, Y.S.; Siswanto, E. Characterization of the chemical, physical, and mechanical properties of NaOH-treated natural cellulosic fibers from corn husks. *J. Nat. Fibers* **2018**, *15*, 545–558. [CrossRef]
8. Madhu, P.; Sanjay, M.R.; Pradeep, S.; Subrahmanya Bhat, K.; Yogesha, B.; Siengchin, S. Characterization of cellulosic fibre from *Phoenix pusilla* leaves as potential reinforcement for polymeric composites. *J. Mater. Res. Technol.* **2019**, *8*, 2597–2604. [CrossRef]
9. Jawaid, M.; Abdul Khalil, H.P.S. Cellulosic/synthetic fibre reinforced polymer hybrid composites: A review. *Carbohydr. Polym.* **2011**, *86*, 1–18. [CrossRef]
10. Sanjay, M.R.; Siengchin, S. Natural fibers as perspective materials. *KMUTNB Int. J. Appl. Scie. Tech.* **2018**, *11*, 233.
11. Sanjay, M.R.; Arpitha, G.R.; Naik, L.L.; Gopalakrishna, K.; Yogesha, B. Applications of natural fibers and its composites: An overview. *Nat. Res.* **2016**, *7*, 108. [CrossRef]
12. Bavan, D.S.; Kumar, G.C.M. Potential use of natural fiber composite materials in India. *J. Reinf. Plast. Compos.* **2010**, *29*, 3600–3613. [CrossRef]
13. Araújo, J.R.; Waldman, W.R.; De Paoli, M.A. Thermal properties of high density polyethylene composites with natural fibres: Coupling agent effect. *Polym. Degrad. Stab.* **2008**, *93*, 1770–1775. [CrossRef]
14. Kant, R.; Alagh, P. Extraction of fiber from *Sansevieria trifasciata* plant and its properties. *IJSR* **2015**, *7*, 2547–2549.
15. Ahmed, M.J.; Balaji, M.A.S.; Saravanakumar, S.S.; Sanjay, M.R.; Senthamaraikannan, P. Characterization of *Areva javanica* fiber—A possible replacement for synthetic acrylic fiber in the disc brake pad. *J. Ind. Text.* **2018**, *49*, 294–317. [CrossRef]
16. Davoodi, M.M.; Sapuan, S.M.; Ahmad, D.; Ali, A.; Khalina, A.; Jonoobi, M. Mechanical properties of hybrid kenaf/glass reinforced epoxy composite for passenger car bumper beam. *Mater. Des.* **2010**, *31*, 4927–4932. [CrossRef]
17. Chandramohan, D.; Bharanichandar, J. Natural fiber reinforced polymer composites for automobile accessories. *Am. J. Environ. Sci.* **2013**, *9*, 494–504. [CrossRef]
18. John, M.J.; Thomas, S. Biofibres and Biocomposites. *Carbohydr. Polym.* **2008**, *71*, 343–364. [CrossRef]
19. Sari, N.H.; Wardana, I.N.G.; Irawan, Y.S.; Siswanto, E. Corn husk fiber-polyester composites as sound absorber: Nonacoustical and acoustical properties. *Adv. Acoust. Vib.* **2017**, *2017*, 4319389. [CrossRef]
20. Murdiyanto, D. Potensi serat alam tanaman Indonesia sebagai bahan fiber reinforced composite kedokteran gigi. *J. Mater. Kedokt. Gigi* **2017**, *6*, 14–22. [CrossRef]
21. Scribante, A.; Vallittu, P.K.; Özcan, M. Fiber-reinforced composites for dental applications. *BioMed Res. Int.* **2018**, *2018*, 4734986. [CrossRef]
22. Adekomaya, O.; Majozi, T. 23-Industrial and biomedical applications of fiber reinforced composites. In *Fiber Reinforced Composites*; Joseph, K., Oksman, K., George, G., Wilson, R., Appukuttan, S., Eds.; Woodhead Publishing: Sawston, UK, 2021; pp. 753–783.
23. Schlöesser, T.P. Natural Fiber Reinforced Automotive Parts. In *Natural Fibers, Plastics and Composites*; Wallenberger, F.T., Weston, N.E., Eds.; Springer: Boston, MA, USA, 2004; pp. 275–285.
24. Mansor, M.R.; Nurfaizey, A.H.; Tamaldin, N.; Nordin, M.N.A. 11-Natural fiber polymer composites: Utilization in aerospace engineering. In *Biomass, Biopolymer-Based Materials, and Bioenergy*; Verma, D., Fortunati, E., Jain, S., Zhang, X., Eds.; Woodhead Publishing: Kidlington, UK, 2019; pp. 203–224.
25. Novarini, E.; Sukardan, M.D. Potensi serat rami (*Boehmeria nivea* S. Gaud) sebagai bahan baku industri tekstil dan produk tekstil dan tekstil teknik. *Arena Tekst.* **2015**, *30*, 113–122. [CrossRef]
26. Rowell, R.M. *The Use of Biomass to Produce Bio-Based Composites and Building Materials*; Woodhead Publishing Limited: Oxford, UK, 2014.
27. Rowell, R.M. *Natural Fibers: Types and Properties*; Woodhead Publishing Limited: Boca Raton, FL, USA, 2008.
28. Aziz, N.A.A.; Ho, L.-H.; Azahari, B.; Bhat, R.; Cheng, L.H.; Ibrahim, M.N.M. Chemical and functional properties of the native banana (*Musa acuminata × balbisiana Colla cv.* Awak) pseudo-stem and pseudo-stem tender core flours. *Food Chem.* **2011**, *128*, 748–753. [CrossRef]

29. Sango, T.; Cheumani Yona, A.M.; Duchatel, L.; Marin, A.; Kor Ndikontar, M.; Joly, N.; Lefebvre, J.-M. Step–wise multi–scale deconstruction of banana pseudo–stem (*Musa acuminata*) biomass and morpho–mechanical characterization of extracted long fibres for sustainable applications. *Ind. Crop. Prod.* **2018**, *122*, 657–668. [CrossRef]
30. Zuluaga, R.; Putaux, J.-L.; Restrepo, A.; Mondragon, I.; Gañán, P. Cellulose microfibrils from banana farming residues: Isolation and characterization. *Cellulose* **2007**, *14*, 585–592. [CrossRef]
31. Sapuan, S.M.; Leenie, A.; Harimi, M.; Beng, Y.K. Mechanical properties of woven banana fibre reinforced epoxy composites. *Mater. Des.* **2006**, *27*, 689–693. [CrossRef]
32. Arias, P.; Dankers, C.; Liu, P.; Pilkauskas, P. *The World Banana Economy 1985–2002*; Food and Agriculture Organization of United Nations: Rome, Italy, 2003.
33. Zuluaga, R.; Rios, A.; Mauricio, A.; Casas, A.; Ramırez, M.; Kerguel, H.; Ganán, P. Aprovechamiento de losresiduos fibrosos de la Agroindustria bananera. In Proceedings of the VII Jornadas de Investigación, Medellín, Colombia, 5–8 May 2003; Universidad Pontifucia Bolivariana: Medellín, Colombia, 2003; pp. 267–276.
34. Ramesh, M. *Hemp, Jute, Banana, Kenaf, Ramie, Sisal Fibers*; Elsevier Ltd.: Amsterdam, The Netherlands, 2018.
35. Suparno, O. Upaya potensi dan masa depan serat alam indonesia sebagai bahan baku aneka industri. *J. Teknol. Ind. Pertan.* **2020**, *30*, 221–227.
36. Gupta, U.S.; Dhamarikar, M.; Dharkar, A.; Tiwari, S.; Namdeo, R. Study on the effects of fibre volume percentage on banana-reinforced epoxy composite by finite element method. *Adv. Compos. Hybrid Mater.* **2020**, *3*, 530–540. [CrossRef]
37. Nurnasari, E.; Nurindah. Karakteristik kimia serat buah, serat batang, dan serat daun. *Bul. Tanam. Tembakau Serat Miny. Ind.* **2017**, *9*, 64–72. [CrossRef]
38. Suliyanthini, D. *Ilmu Tekstil*; Rajawali Perss: Jakarta, Indonesia, 2016.
39. Nebangka, M.; Sumayku, B.R.A.; Pongoh, J. Potensi pengembangan pisang abaka (*Musa textilis* Nee) di Pulau Karakelang. *COCOS* **2020**, *1*, 1–11.
40. Jawaid, M.; Asim, M.; Paridah, T.; Nasir, M. *Processing, Properties and Applications*; Springer Nature Pte. Ltd.: Singapore, 2020.
41. Hadiyati, S.; Indriyani, N.L.P. *Petunjuk Teknis Budidaya Nanas*; Balai Penelitian Buah Tropika: Solok, Indonesia, 2008.
42. Setyawan, P.D.; Sari, N.H.; Putra, D.G.P. Pengaruh orientasi dan fraksi volume serat daun nanas (*Ananas comosus*) terhadap kekuatan putus komposit polyester tak jenuh (UP). *J. Din. Tek. Mesin* **2012**, *2*, 31–32.
43. Pandit, P.; Pandey, R.; Singha, K.; Shrivastava, S.; Gupta, V.; Jose, S. Pineapple Leaf Fibre: Cultivation and Production. In *Pineapple Leaf Fibers, Green Energy and Technology*; Mohammad, J., Mohammad, A., Paridah, M.T., Mohammed, N., Eds.; Springer Nature Pte. Ltd.: Singapore, 2020.
44. Zakikhani, P.; Zahari, R.; Sultan, M.T.H.; Majid, D.L. Extraction and Preparation of Bamboo Fibre-Reinforced Composites. *Mater. Des.* **2014**, *63*, 820–828. [CrossRef]
45. Wang, G.; Chen, F. Development of bamboo fiber-based composites. In *Advanced High Strength Natural Fibre Composites in Construction*; Elsevier Ltd.: Amsterdam, The Netherlands, 2016.
46. Lobovikov, M.; Paudel, S.; Piazza, M.; Ren, H.; Wu, J. *World Bamboo Resources: A Thematic Study Prepared in the Framework of the Global Forest Resources Assessment*; Non Wood Forest Products; Food and Agricultural Organization of the United Nations: Rome, Italy, 2007.
47. Kant, P. *Should Bamboos and Palms Be Included in Cdm Forestry Projects?* IGREC Working Paper: New Delhi, India, 2010.
48. Scurlock, J.M.O.; Dayton, D.C.; Hames, B. Bamboo: An overlooked biomass resource? *Biomass Bioenergy* **2000**, *19*, 229–244. [CrossRef]
49. Gratani, L.; Crescente, M.F.; Fabrini, L.G.; Digiulio, E. Growth pattern and photosynthetic activity of different bamboo species growing in the botanical garden of Rome. *Flora-Morphol. Distrib. Funct. Ecol. Plants* **2008**, *203*, 77–84. [CrossRef]
50. Damayanti, R.; Jasni; Sulastiningsih, I.M.; Djarwanto; Suprapti, S.; Pari, G.; Basri, E.; Komarayati, S.; Abdurahman. *Atlas Bambu Indonesia 1*; IPB Press: Bogor, Indonesia, 2019.
51. Dransfield, S.; Widjaja, E.A. *Plant Resources of South-East Asia No.7 Bamboos*; Backhuys Publishers: Leiden, The Netherlands, 1995.
52. Fatriasari, W.; Hermiati, E. Analisis morfologi serat dan sifat fisis-kimia pada enam jenis bambu sebagai bahan baku pulp dan kertas. *J. Ilmu Dan Teknol. Has. Hutan* **2008**, *67–72*, 2.
53. Yueping, W.; Ge, W.; Haitao, C.; Genlin, T.; Zheng, L.; Feng, X.Q.; Xiangqi, Z.; Xiaojun, H.; Xushan, G. Structures of bamboo fibre for textiles. *Text. Res. J.* **2010**, *80*, 334–343. [CrossRef]
54. Subyakto. Teknologi pengendalian berbasis ekologi dalam mendukung pengembangan kapas. *J. Litbang Pertan.* **2011**, *30*, 81–86.
55. Ditjenbun. *Komoditas Kapas di Indonesia 2011–2013*; Direktorat Jenderal Perkebunan, Kementerian Pertanian RI: Jakarta, Indonesia, 2013.
56. Sumartini, S.; Sulistyowati, E.; Mulyani, S.; Abdurrakhman. Skrining galur kapas (*Gossypium hirsutum* L.) toleran terhadap kekeringan PEG-6000 pada fase kecambaH. *J. Littri* **2013**, *19*, 139–146. [CrossRef]
57. Ditjenbun. *Statistik Perkebunan Indonesia Komoditas Kapas*; Direktorat Jenderal Perkebunan, Kementerian Pertanian RI: Jakarta, Indonesia, 2015.
58. Thyavihalli Girijappa, Y.G.; Mavinkere Rangappa, S.; Parameswaranpillai, J.; Siengchin, S. Natural fibers as sustainable and renewable resource for development of eco-friendly composites: A comprehensive review. *Front. Mater.* **2019**, *6*, 226. [CrossRef]
59. Berger, J. *Fibre Crops: Their Cultivation and Manuring*; Centre d'Etude de I'Azote: Zurich, Switzerland, 1969.

60. Suratman, W.; Murdoko; Darwis, S.N. Tinjauan kemungkinan pengembangan rami di Indonesia. In Proceedings of the Seminar Nasional Rami, Malang, Indonesia, 1993.
61. Eichhorn, S.J.; Baillie, C.A.; Zafeiropoulos, N.; Mwaikambo, L.Y.; Ansell, M.P.; Dufresne, A.; Entwistle, K.M.; Herrera-Franco, P.J.; Escamilla, G.C.; Groom, L.; et al. Review: Current international research into cellulosic fibres and composites. *J. Mater. Sci.* **2001**, *36*, 2107–2131. [CrossRef]
62. Soeroto, H. *Cultur Technik Boehmeria nivea L. Gaud.*; Balai Besar Penyelidikan Pertanian, 1956.
63. Setyo-Budi, U.; Hartati, R.S.; Purwanti, R.D. *Biologi Tanaman Rami (Boehmeria nivea L. Gaud)*; Monograf Balittas Rami: Malang, Indonesia, 2005.
64. Sumantri, R.H.L. Haramay (Ramie), Penanaman, Pemeliharaan dan Kegunaan. 1984. Available online: http://scholar.unand.ac.id/37471/4/DAFTAR%20PUSTAKA.pdf (accessed on 4 November 2021).
65. Brink, M.; Escobin, R. *Plant Resources of South-East Asia*; Backhuys Publisher: Leiden, The Netherlands, 2003.
66. Santoso, B. Peluang pengembangan Agave sebagai sumber serat alam. *Perspektif* **2009**, *8*, 84–95.
67. Mukherjee, P.S.; Satyanarayana, K.G. Structure and properties of some vegetable fibres. *J. Mater. Sci.* **1986**, *21*, 51–56. [CrossRef]
68. Sanjaygowda, M.; Rangappa, S.M.; Juwaid, M.; Shivana, P.; Basavegowda, Y.; Saba, N. *Potential of Natural/Synthetic Hybrid Composites for Aerospace Applications*; Woodhead Publishing Series in Composites Science and Engineering; Woodhead Publishing: Duxford, UK, 2018.
69. Amin, M.; Samsudi, R. Pemanfaatan limbah serat sabut kelapa sebagai bahan pembuat helm pengendara kendaraan roda dua. *Pros. Semin. Nas. Int.* **2010**, *3*, 314–318.
70. Sapuan, S.M.; Ismail, H.; Zainudin, E.S. Natural fiber reinforced vinyl ester and vinyl polymer composite. In *Development, Characterization and Applications*; Woodhead Publishing: Duxford, UK, 2018.
71. van Dam, J.E.G.; van den Oever, M.J.A.; Teunissen, W.; Keijsers, E.R.P.; Peralta, A.G. Process for production of high density/high performance binderless boards from whole coconut husk: Part 1: Lignin as intrinsic thermosetting binder resin. *Ind. Crop. Prod.* **2004**, *19*, 207–216. [CrossRef]
72. Elseify, L.A.; Midani, M.; Shihata, L.A.; El-mously, H. Review on cellulosic fibers extracted from date palms (*Phoenix dactylifera*, L.) and their applications. *Cellulose* **2019**, *26*, 2209–2232. [CrossRef]
73. Thampan, P.K. *Handbook on Coconut Palm*; Oxford and IBH Publishing Co.: New Delhi, India, 1991.
74. Heyne, K. *Tumbuhan Berguna Indonesia, I*; Koperasi Karyawan Departemen Kehutanan: Jakarta, Indonesia, 1987.
75. Puwanto, A.W. *Sansevieria Trifasciata Flora Cantik Penyerap Racun*; Kanisius: Yogyakarta, Indonesia, 2006.
76. Rosanti, D. Pengaruh pupuk majemuk dalam merangsang pertumbuhan tunas pada stek daun Sansevieria trifasciata. *J. Sainmatika* **2011**, *8*. [CrossRef]
77. Trubus, T. *Sansevieria*; PT Trubus Swadaya, 2008; Available online: http://digilib.unimus.ac.id/files//disk1/156/jtptunimus-gdl-alfianrizk-7783-5-daftarp-a.pdf (accessed on 4 November 2021).
78. Huda, Z.M. Ampuhnya Si Penyedot Polutan. Available online: http://www.gogreenschool.com (accessed on 4 November 2021).
79. Agromedia, R. *Tip Jitu Merawat Tanaman Hias Popular*; PT Agromedia Pustaka: Jakarta, Indonesia, 2010.
80. Franz, J.B. Tangkis Renggutan Gas Polutan. Available online: http://agriculturesupercamp.wordpress.com (accessed on 4 November 2021).
81. Hakim, L.; Widyorini, R.; Nugroho, W.D.; Prayitno, T.A. Anatomical, chemical, and mechanical properties of fibrovascular bundles of Salacca (Snake Fruit) frond. *BioResources* **2019**, *14*, 7943–7957.
82. Philip, D.; Kaleena, P.K.; Valivittan, K.; Kumar, G. Phytochemical screening and amtimicrobial activity of (*Sansevieria roxburghiana* schult). *J. Middle-East J. Sci. Res.* **2011**, *10*, 512–518.
83. Rikara, D. Menjilati Polusi Dengan Lidah Mertua. Available online: http://id.wordpress.com/tag/tanaman-hias/ (accessed on 4 November 2021).
84. Chand, N.; Fahim, M. Jute reinforced polymer composites. In *Tribology of Natural Fiber Polymer Composites*; Elsevier: Amsterdam, The Netherlands, 2008; pp. 108–128.
85. Bismarck, A.; Mishra, S.; Lampke, T. *Plant Fibers as Reinforcement for Green Composites: Natural Fibers, Biopolymers, and Biocomposite*; CRC Press Taylor and Francis Group: Boca Raton, FL, USA, 2005.
86. Ramamoorthy, S.K.; Skrifvars, M.; Persson, A. A review of natural fibers used in biocomposites: Plant, animal, and regenerated cellulose fibers. *Polym. Rev.* **2015**, *55*, 107–162. [CrossRef]
87. Sudjindro. Arah pengembangan kenaf di Indonesia menyongsong bangkitnya serat alam dunia 2009. *War. Penelit. Pengemb. Tanam. Ind.* **2008**, *14*, 20–22.
88. Sudjindro; Marjani; Heliyanto, B.; Sunardi, D. Galur harapan kenaf adaptif di lahan Bonorowo Kabupaten Lamongan. *J. Penelit. Tanam. Ind.* **2001**, *7*, 31–34. [CrossRef]
89. Sudjindro; Marjani; Heliyanto, B.; Purwati, R.D. Uji Daya Hasil Galur-Galur Kenaf (*Hibiscus cannabinus* L.). In Proceedings of the Simposium V Peripi, 8–9 September 1998.
90. Sudjindro; Marjani. Pemuliaan tanaman kenaf (*Hibiscus cannabinus* L.). In *Monograf Kenaf*; 2009.
91. Sudjindro. Prospek serat alam untuk bahan baku kertas uang. *Perspektif* **2011**, *10*, 92–104.
92. Luong, T.-H.; Dang, T.-N.; Ngoc, O.P.; Thi, D.-T.; Nguyen, T.-H.; Toi, V.V.; Duong, H.T.; Son, H. Investigation of the silk fiber extraction process from the vietnam natural bombyx mori silkworm cocoon. In Proceedings of the 5th International Conference on Biomedical Engineering in Vietnam, Ho Chi Minh City, Vietnam, 16–18 June 2014; pp. 325–328.

93. Kearns, V.; MacIntosh, A.C.; Crawford, A.; Hatton, P.V. Silk-based biomaterials for tissue engineering. *Top. Tissue Eng.* **2008**, *4*, 1–19.
94. Altman, G.H.; Diaz, F.; Jakuba, C.; Calabro, T.; Horan, R.L.; Chen, J.; Lu, H.; Richmon, J.; Kaplan, D.L. Silk-based biomaterials. *Biomaterials* **2003**, *24*, 401–416. [CrossRef]
95. Yang, C.; Wang, B. Biodegradation of silk biomaterials. *Int. J. Mol. Sci.* **2009**, *10*, 1514–1524.
96. Wang, Y.; Kim, H.-J.; Vunjak-Novakovic, G.; Kaplan, D.L. Stem cell-based tissue engineering with silk biomaterials. *Biomaterials* **2006**, *27*, 6064–6082. [CrossRef]
97. Hakimi, O.; Vollart, F.V.; Carr, A.J. Evaluation of silk as a scaffold for musculoskeletal regeneration—The path from the laboratory to clinical trials. In *Comprehensive Biotechnology*; Elsevier: Amsterdam, The Netherlands, 2011.
98. Mandal, B.B.; Kundu, S.C. Cell proliferation and migration in silk fibroin 3D scaffolds. *Biomaterials* **2009**, *30*, 2956–2965. [CrossRef] [PubMed]
99. Babu, K.M. Silk: Processing, Properties and Application. In *The Textile Institute Book Series*; Elsevier: Amsterdam, The Netherlands, 2019.
100. Coderch, L.; de la Maza, A.; Soriano, C.; Erra, P.; Parra, J.L. Chromatographic characterization of internal polar lipids from wool. *J. Am. Oil Chem. Soc.* **1995**, *72*, 715–720. [CrossRef]
101. Schaefer, H.; Redelmeier, T.E. Skin Barrier: Principles in percutaneous penetration. *Arch. Dermatol.* **1997**, *133*, 924. [CrossRef]
102. Gelse, K.; Poschlb, E.; Aigner, T. Collagens-structure, function, and biosynthesis. *Adv. Drug Deliv. Rev.* **2003**, *55*, 1531–1546. [CrossRef]
103. Suparno, O.; Prasetyo, N.B. Isolation of collagen from chicken feet with hydroextraction method and its physico-chemical characterisation. In Proceedings of the ICDALC 2018, Bogor, Indonesia, 20–21 September 2018; pp. 1–11.
104. Karim, A.A.; Bhat, R. Fish gelatin: Properties, challenges, and prospects as an alternative to mammalian gelatins. *Food Hydrocoll.* **2009**, *23*, 563–576. [CrossRef]
105. Aberoumand, A. Comparative study between different methods of collagen extraction from fish and its properties. *World Appl. Sci. J.* **2012**, *16*, 316–319.
106. Sari, N.H.; Pruncu, C.I.; Sapuan, S.M.; Ilyas, R.A.; Catur, A.D.; Suteja, S.; Sutaryono, Y.A.; Pullen, G. The effect of water immersion and fibre content on properties of corn husk fibres reinforced thermoset polyester composite. *Polym. Test.* **2020**, *91*, 106751. [CrossRef]
107. Vinod, A.; Vijay, R.; Singaravelu, D.L. ThermoMechanical characterization of *Calotropis gigantea* stem powder-filled jute fiber-reinforced epoxy composites. *J. Nat. Fibers* **2018**, *15*, 648–657. [CrossRef]
108. Jothibasu, S.; Mohanamurugan, S.; Vijay, R.; Lenin Singaravelu, D.; Vinod, A.; Sanjay, M.R. Investigation on the mechanical behavior of areca sheath fibers/jute fibers/glass fabrics reinforced hybrid composite for light weight applications. *J. Ind. Text.* **2018**, *49*, 1036–1060. [CrossRef]
109. Vijay, R.; Lenin Singaravelu, D.; Vinod, A.; Sanjay, M.R.; Siengchin, S.; Jawaid, M.; Khan, A.; Parameswaranpillai, J. Characterization of raw and alkali treated new natural cellulosic fibers from *Tridax procumbens*. *Int. J. Biol. Macromol.* **2019**, *125*, 99–108. [CrossRef]
110. Munawar, S.S.; Umemura, K.; Kawai, S. Manufacture of oriented board using mild steam treatment of plant fiber bundles. *J. Wood Sci.* **2008**, *54*, 369–376. [CrossRef]
111. Chen, H.; Miao, M.; Ding, X. Influence of moisture absorption on the interfacial strength of bamboo/vinyl ester composites. *Compos. Part A Appl. Sci. Manuf.* **2009**, *40*, 2013–2019. [CrossRef]
112. Okubo, K.; Fujii, T.; Yamamoto, Y. Development of bamboo-based polymer composites and their mechanical properties. *Compos. Part A Appl. Sci. Manuf.* **2004**, *35*, 377–383. [CrossRef]
113. Deka, H.; Misra, M.; Mohanty, A. Renewable resource based "all green composites" from kenaf biofiber and poly(furfuryl alcohol) bioresin. *Ind. Crop. Prod.* **2013**, *41*, 94–101. [CrossRef]
114. de Farias, J.G.; Cavalcante, R.C.; Canabarro, B.R.; Viana, H.M.; Scholz, S.; Simão, R.A. Surface lignin removal on coir fibers by plasma treatment for improved adhesion in thermoplastic starch composites. *Carbohydr. Polym.* **2017**, *165*, 429–436. [CrossRef]
115. Yousif, B.F.; Shalwan, A.; Chin, C.W.; Ming, K.C. Flexural properties of treated and untreated kenaf/epoxy composites. *Mater. Des.* **2012**, *40*, 378–385. [CrossRef]
116. Yan, L.; Chouw, N.; Huang, L.; Kasal, B. Effect of alkali treatment on microstructure and mechanical properties of coir fibres, coir fibre reinforced-polymer composites and reinforced-cementitious composites. *Constr. Build. Mater.* **2016**, *112*, 168–182. [CrossRef]
117. Li, X.; Tabil, L.; Panigrahi, S. Chemical treatments of natural fiber for use in natural fiber-reinforced composites: A review. *J. Polym. Environ.* **2007**, *15*, 25–33. [CrossRef]
118. Hajiha, H.; Sain, M.; Mei, L.H. Modification and Characterization of Hemp and Sisal Fibers. *J. Nat. Fibers* **2014**, *11*, 144–168. [CrossRef]
119. George, M.; Mussone, P.G.; Bressler, D.C. Surface and thermal characterization of natural fibres treated with enzymes. *Ind. Crop. Prod.* **2014**, *53*, 365–373. [CrossRef]
120. Poyyamozhi, V.S.; Kadirvel, R. The value of banana stalk as a feed for goats. *Anim. Feed Sci. Technol.* **1986**, *15*, 95–100. [CrossRef]
121. Viswanathan, K.; Kadirvel, R.; Chandrasekaran, D. Nutritive value of banana stalk (*Musa cavendishi*) as a feed for sheep. *Anim. Feed Sci. Technol.* **1989**, *22*, 327–332. [CrossRef]

122. Chiena, E. Estudio del ensilado del raquis de banana (*Musa acuminata* Colla, subgrupo 'Cavendish') para la alimentaci óndel ganado caprino en las Islas Canarias. *Rev. Fac. Agron.* **1999**, *16*, 291–305.
123. Pothan, L.A.; Thomas, S. Polarity parameters and dynamic mechanical behaviour of chemically modified banana fiber reinforced polyester composites. *Compos. Sci. Technol.* **2003**, *63*, 1231–1240. [CrossRef]
124. Pothan, L.A.; Thomas, S.; Groeninckx, G. The role of fibre/matrix interactions on the dynamic mechanical properties of chemically modified banana fibre/polyester composites. *Compos. Part A Appl. Sci. Manuf.* **2006**, *37*, 1260–1269. [CrossRef]
125. Vinod, A.; Vijay, R.; Singaravelu, D.L.; Sanjay, M.R.; Siengchin, S.; Moure, M.M. Characterization of untreated and alkali treated natural fibers extracted from the stem of Catharanthus roseus. *Mater. Res. Express* **2019**, *6*, 085406. [CrossRef]
126. Idicula, M.; Boudenne, A.; Umadevi, L.; Ibos, L.; Candau, Y.; Thomas, S. Thermophysical properties of natural fibre reinforced polyester composites. *Compos. Sci. Technol.* **2006**, *66*, 2719–2725. [CrossRef]
127. Venkateshwaran, N.; ElayaPerumal, A.; Alavudeen, A.; Thiruchitrambalam, M. Mechanical and water absorption behaviour of banana/sisal reinforced hybrid composites. *Mater. Des.* **2011**, *32*, 4017–4021. [CrossRef]
128. Quintana, G.; Velásquez, J.; Betancourt, S.; Gañán, P. Binderless fiberboard from steam exploded banana bunch. *Ind. Crop. Prod.* **2009**, *29*, 60–66. [CrossRef]
129. Mohapatra, D.; Mishra, S.; Sutar, N. Banana and its by-product utilisation: An overview. *J. Sci. Ind. Res.* **2010**, *69*, 323–329.
130. Bello, K.; Sarojini, B.K.; Narayana, B.; Rao, A.; Byrappa, K. A study on adsorption behavior of newly synthesized banana pseudo-stem derived superabsorbent hydrogels for cationic and anionic dye removal from effluents. *Carbohydr. Polym.* **2018**, *181*, 605–615. [CrossRef] [PubMed]
131. Sanjay, M.R.; Siengchin, S.; Parameswaranpillai, J.; Jawaid, M.; Pruncu, C.I.; Khan, A. A comprehensive review of techniques for natural fibers as reinforcement in composites: Preparation, processing and characterization. *Carbohydr. Polym.* **2019**, *207*, 108–121. [CrossRef]
132. Jordan, W.; Chester, P. Improving the properties of banana fiber reinforced polymeric composites by treating the fibers. *Procedia Eng.* **2017**, *200*, 283–289. [CrossRef]
133. Ray, D.; Nayak, L.; Ammayappan, L.; Shambhu, V.; Nag, D. Energy conservation drives for efficient extraction and utilization of banana fibre. *Int. J. Emerg. Technol. Adv. Eng.* **2013**, *3*, 296–310.
134. Bertella, S.; Luterbacher, J.S. Lignin functionalization for the production of novel materials. *Trends Chem.* **2020**, *2*, 440–453. [CrossRef]
135. Kabir, M.M.; Wang, H.; Lau, K.T.; Cardona, F. Chemical treatments on plant-based natural fibre reinforced polymer composites: An overview. *Compos. Part B Eng.* **2012**, *43*, 2883–2892. [CrossRef]
136. Sathishkumar, T.P.; Navaneethakrishnan, P.; Shankar, S.; Rajasekar, R. Characterization of new cellulose sansevieria ehrenbergii fibers for polymer composites. *Compos. Interfaces* **2013**, *20*, 575–593. [CrossRef]
137. Arthanarieswaran, V.P.; Kumaravel, A.; Saravanakumar, S.S. Physico-chemical properties of alkali-treated *Acacia leucophloea* fibers. *Int. J. Polym. Anal. Charact.* **2015**, *20*, 704–713. [CrossRef]
138. Indran, S.; Raj, R.E.; Daniel, B.S.S.; Saravanakumar, S.S. Cellulose powder treatment on *Cissus quadrangularis* stem fiber-reinforcement in unsaturated polyester matrix composites. *J. Reinf. Plast. Compos.* **2015**, *35*, 212–227. [CrossRef]
139. Jayaramudu, J.; Maity, A.; Sadiku, E.R.; Guduri, B.R.; Varada Rajulu, A.; Ramana, C.V.V.; Li, R. Structure and properties of new natural cellulose fabrics from *Cordia dichotoma*. *Carbohydr. Polym.* **2011**, *86*, 1623–1629. [CrossRef]
140. Kommula, V.P.; Reddy, K.O.; Shukla, M.; Marwala, T.; Reddy, E.V.S.; Rajulu, A.V. Extraction, modification, and characterization of natural ligno-cellulosic fiber strands from napier grass. *Int. J. Polym. Anal. Charact.* **2016**, *21*, 18–28. [CrossRef]
141. Saravanakumar, S.S.; Kumaravel, A.; Nagarajan, T.; Moorthy, I.G. Effect of chemical treatments on physicochemical properties of *Prosopis juliflora* fibers. *Int. J. Polym. Anal. Charact.* **2014**, *19*, 383–390. [CrossRef]
142. Rashid, B.; Leman, Z.; Jawaid, M.; Ghazali, M.J.; Ishak, M.R. Physicochemical and thermal properties of lignocellulosic fiber from sugar palm fibers: Effect of treatment. *Cellulose* **2016**, *23*, 2905–2916. [CrossRef]
143. Rangappa, S.; Siengchin, S.; Parameswaranpillai, J.; Jawaid, M.; Ozbakkaloglu, T. Lignocellulosic fiber reinforced composites: Progress, performance, properties, applications, and future perspectives. *Polym. Compos.* **2021**, 1–47. Available online: https://onlinelibrary.wiley.com/doi/full/10.1002/pc.26413 (accessed on 4 November 2021). [CrossRef]
144. Silva, G.; Kim, S.; Aguilar, R.; Nakamatsu, J. Natural fibers as reinforcement additives for geopolymers—A review of potential eco-friendly applications to the construction industry. *Sustain. Mater. Technol.* **2020**, *23*, e00132. [CrossRef]
145. Yan, L.; Kasal, B.; Huang, L. A review of recent research on the use of cellulosic fibres, their fibre fabric reinforced cementitious, geo-polymer and polymer composites in civil engineering. *Compos. Part B Eng.* **2016**, *92*, 94–132. [CrossRef]
146. Azwa, Z.N.; Yousif, B.F.; Manalo, A.C.; Karunasena, W. A review on the degradability of polymeric composites based on natural fibres. *Mater. Des.* **2013**, *47*, 424–442. [CrossRef]
147. Ferdous, T.; Quaiyyum, M.A.; Bashar, S.; Jahan, M.S. Anatomical, morphological and chemical characteristics of kaun straw (*Seetaria-Italika*). *Nord. Pulp Pap. Res. J.* **2020**, *35*, 288–298. [CrossRef]
148. Wan, J.; Wang, Y.; Xiao, Q. Effects of hemicellulose removal on cellulose fiber structure and recycling characteristics of eucalyptus pulp. *Bioresour. Technol.* **2010**, *101*, 4577–4583. [CrossRef]
149. Joseph, S.; Sreekala, M.S.; Oommen, Z.; Koshy, P.; Thomas, S. A comparison of the mechanical properties of phenol formaldehyde composites reinforced with banana fibres and glass fibres. *Compos. Sci. Technol.* **2002**, *62*, 1857–1868. [CrossRef]
150. Cesarino, I.R.G.; Bronzato, F.; Leao, A. *Pineapple Leaf Fibers*; Springer: Berlin, Germany, 2020.

151. Vijay, R.; Vinod, A.; Lenin Singaravelu, D.; Sanjay, M.R.; Siengchin, S. Characterization of chemical treated and untreated natural fibers from *Pennisetum orientale* grass—A potential reinforcement for lightweight polymeric applications. *Int. J. Lightweight Mater. Manuf.* **2021**, *4*, 43–49. [CrossRef]
152. Mohanty, A.K.; Misra, M.; Drzal, L.T. *Natural Fibers, Biopolymer and Biocomposites*; CRC Press: London, UK, 2005.
153. Fatriasari, W.; Hamzah, F.N.; Pratomo, B.I.; Fajriutami, T.; Ermawar, R.A.; Falah, F.; Laksana, R.P.B.; Ghozali, M.; Iswanto, A.H.; Hermiati, E.; et al. Optimizing the synthesis of lignin derivatives from *Acacia mangium* to improve the enzymatic hydrolysis of kraft pulp sorghum bagasse. *Int. J. Renew. Energy Dev.* **2020**, *9*, 227–235. [CrossRef]
154. Solihat, N.; Sari, F.; Falah, F.; Ismayati, M.; Lubis, M.; Fatriasari, W.; Santoso, E.; Syafii, W. Lignin as an active biomaterial: A review. *J. Sylva Lestari* **2021**, *9*, 1–22. [CrossRef]
155. Fatriasari, W.; Nurhamzah, F.; Raniya, R.; Laksana, R.; Anita, S.H.; Iswanto, A.H.; Hermiati, E. Enzymatic hydrolysis performance of biomass by the addition of a lignin based biosurfactant. *J. Korean Wood Sci. Technol.* **2020**, *48*, 651–665.
156. Falah, F.; Lubis, M.; Triastuti; Fatriasari, W.; Sari, F. Utilization of lignin from the waste of bioethanol production as a mortar additive. *J. Sylva Lestari* **2020**, *8*, 326–339. [CrossRef]
157. Solihat, N.; Raniya, R.; Fajriutami, T.; Iswanto, A.; Fatriasari, W.; Fudholi, A. Design and performance of amphiphilic lignin derivatives in enzymatic hydrolysis of sweet sorghum bagasse for bioethanol production. *BioResources* **2021**, *16*, 5875–5889. [CrossRef]
158. Zhang, J.; Song, H.; Lin, L.; Zhuang, J.; Pang, C.; Liu, S. Microfibrillated cellulose from bamboo pulp and its properties. *Biomass Bioenergy* **2012**, *39*, 78–83. [CrossRef]
159. Shlieout, G.; Arnold, K.; Müller, G. Powder and mechanical properties of microcrystalline cellulose with different degrees of polymerization. *AAPS PharmSciTech* **2002**, *3*, 45–54. [CrossRef] [PubMed]
160. Jonoobi, M.; Oksman Niska, K.; Harun, J.; Misra, M. Chemical composition, crystallinity, and thermal degradation of bleached and unbleached kenaf bast (*Hibiscus cannabinus*) pulp and nanofibers. *BioResources* **2009**, *4*, 626–639.
161. Granstrom, M. *Cellulose Derivatives: Synthesis, Properties and Applications*; University of Helsinki: Helsinki, Finland, 2009.
162. Asrofi, M.; Abral, H.; Kasim, A.; Pratoto, A.; Mahardika, M.; Park, J.-W.; Kim, H.-J. Isolation of Nanocellulose from Water Hyacinth Fiber (WHF) Produced via Digester-Sonication and Its Characterization. *Fibers Polym.* **2018**, *19*, 1618–1625. [CrossRef]
163. Park, S.; Baker, J.O.; Himmel, M.E.; Parilla, P.A.; Johnson, D.K. Cellulose crystallinity index: Measurement techniques and their impact on interpreting cellulase performance. *Biotechnol. Biofuels* **2010**, *3*, 1–10. [CrossRef]
164. Puglia, D.; Biagiotti, J.; Kenny, J.M. A Review on Natural Fibre-Based Composites—Part II. *J. Nat. Fibers* **2005**, *1*, 23–65. [CrossRef]
165. Nurazzi, N.M.; Asyraf, M.; Khalina, A.; Abdullah, N.; Aisyah, H.A.; Rafiqah, S.A.; Sabaruddin, F.A.; Kamarudin, S.H.; Norrrahim, M.; Ilyas, R.A.; et al. A Review on Natural Fiber Reinforced Polymer Composite for Bullet Proof and Ballistic Applications. *Polymers* **2021**, *13*, 646. [CrossRef]
166. Yahaya, R.; Sapuan, S.M.; Jawaid, M.; Leman, Z.; Zainudin, E.S. Investigating ballistic impact properties of woven kenaf-aramid hybrid composites. *Fibers Polym.* **2016**, *17*, 275–281. [CrossRef]
167. Miller, S.A. Natural fiber textile reinforced bio-based composites: Mechanical properties, creep, and environmental impacts. *J. Clean. Prod.* **2018**, *198*, 612–623. [CrossRef]
168. Mahardika, M.; Abral, H.; Kasim, A.; Arief, S.; Hafizulhaq, F.; Asrofi, M. Properties of cellulose nanofiber/bengkoang starch bionanocomposites: Effect of fiber loading. *LWT* **2019**, *116*, 108554. [CrossRef]
169. Al-Oqla, F.M.; Sapuan, S.M.; Ishak, M.R.; Nuraini, A.A. A decision-making model for selecting the most appropriate natural fiber—Polypropylene-based composites for automotive applications. *J. Compos. Mater.* **2015**, *50*, 543–556. [CrossRef]
170. Xia, C.; Yu, J.; Shi, S.Q.; Qiu, Y.; Cai, L.; Wu, H.F.; Ren, H.; Nie, X.; Zhang, H. Natural fiber and aluminum sheet hybrid composites for high electromagnetic interference shielding performance. *Compos. Part B Eng.* **2017**, *114*, 121–127. [CrossRef]
171. Luz, F.S.D.; Garcia Filho, F.D.C.; Oliveira, M.S.; Nascimento, L.F.C.; Monteiro, S.N. Composites with Natural Fibers and Conventional Materials Applied in a Hard Armor: A Comparison. *Polymers* **2020**, *12*, 1920. [CrossRef]
172. Murali Mohan, S.; Mahesh Gowda, E.; Shashidhar, M.P. Clinical evaluation of the fiber post and direct composite resin restoration for fixed single crowns on endodontically treated teeth. *Med. J. Armed Forces India* **2015**, *71*, 259–264. [CrossRef]
173. Anidha, S.; Latha, N.; Muthukkumar, M. Effect of polyaramid reinforced with sisal epoxy composites: Tensile, impact, flexural and morphological properties. *J. Mater. Res. Technol.* **2020**, *9*, 7947–7954. [CrossRef]
174. Zhou, X.; Saini, H.; Kastiukas, G. Engineering properties of treated natural hemp fiber-reinforced concrete. *Front. Built Environ.* **2017**, *3*, 33. [CrossRef]
175. Luzi, F.; Puglia, D.; Torre, L. *Natural Fiber Biodegradable Composites and Nanocomposites: A Biomedical Application*; Elsevier Ltd.: Kidlington, UK, 2019.
176. Namvar, F.; Jawaid, M.; Md Tahir, P.; Mohamad, R.; Azizi, S.; Khodavandi, A.; Rahman, H.S.; Nayeri, M.D. Potential use of plant fibres and their composites for biomedical applications. *BioResources* **2014**, *9*, 5688–5706. [CrossRef]
177. Ahad, N.; Rozali, F.; Hanif, N.; Rosli, N. Oils and water absorption behavior of natural fibers filled TPU composites for biomedical applications. *J. Eng. Res. Educ.* **2018**, *10*, 16–21.
178. Goyal, A.; Sharma, V.; Upadhyay, N.; Gill, S.; Sihag, M. Flax and flaxseed oil: An ancient medicine & modern functional food. *J. Food Sci. Technol.* **2014**, *51*, 1633–1653. [CrossRef] [PubMed]
179. Kulma, A.; Skórkowska-Telichowska, K.; Kostyn, K.; Szatkowski, M.; Skała, J.; Drulis-Kawa, Z.; Preisner, M.; Żuk, M.; Szperlik, J.; Wang, Y.F.; et al. New flax producing bioplastic fibers for medical purposes. *Ind. Crop. Prod.* **2015**, *68*, 80–89. [CrossRef]

180. Song, J.; Chen, Z.; Liu, Z.; Yi, Y.; Tsigkou, O.; Li, J.; Li, Y. Controllable release of vascular endothelial growth factor (VEGF) by wheel spinning alginate/silk fibroin fibers for wound healing. *Mater. Des.* **2021**, *212*, 110231. [CrossRef]
181. Cherian, B.M.; Leão, A.L.; de Souza, S.F.; Thomas, S.; Pothan, L.A.; Kottaisamy, M. Isolation of nanocellulose from pineapple leaf fibres by steam explosion. *Carbohydr. Polym.* **2010**, *81*, 720–725. [CrossRef]
182. Giri, J.; Adhikari, R.; Campus, T. A brief review on extraction of nanocellulose and its application. *Nepal J. Online* **2013**, *9*, 81–87. [CrossRef]
183. Alharbi, M.A.H.; Hirai, S.; Tuan, H.A.; Akioka, S.; Shoji, W. Effects of chemical composition, mild alkaline pretreatment and particle size on mechanical, thermal, and structural properties of binderless lignocellulosic biopolymers prepared by hot-pressing raw microfibrillated *Phoenix dactylifera* and *Cocos nucifera* fibers and leaves. *Polym. Test.* **2020**, *84*, 106384. [CrossRef]
184. Daunton, C.; Kothari, S. A history of materials and practices for wound management. *Wound Manag.* **2012**, *20*, 174–186.
185. Asim, M.; Saba, N.; Jawaid, M.; Nasir, M. 12-Potential of natural fiber/biomass filler-reinforced polymer composites in aerospace applications. In *Sustainable Composites for Aerospace Applications*; Jawaid, M., Thariq, M., Eds.; Woodhead Publishing: Kidlington, UK, 2018; pp. 253–268.
186. Potluri, R. Natural Fiber-Based Hybrid Bio-composites: Processing, Characterization, and Applications. In *Green Composites: Processing, Characterisation and Applications for Textiles*; Muthu, S.S., Ed.; Springer: Singapore, 2019; pp. 1–46.
187. Arockiam, N.J.; Jawaid, M.; Saba, N. 6-Sustainable bio composites for aircraft components. In *Sustainable Composites for Aerospace Applications*; Elsevier: Selangor, Malaysia, 2018; pp. 109–123.
188. Asrofi, M.; Sapuan, S.M.; Ilyas, R.A.; Ramesh, M. Characteristic of composite bioplastics from tapioca starch and sugarcane bagasse fiber: Effect of time duration of ultrasonication (Bath-Type). *Mater. Today Proc.* **2021**, *46*, 1626–1630. [CrossRef]
189. Baltazar-Y-Jimenez, A.; Sain, M. *Natural Fibres for Automotive Applications*; Woodhead Publishing Limited: New Delhi, India, 2012.
190. Moudood, A.; Rahman, A.; Khanlou, H.M.; Hall, W.; Öchsner, A.; Francucci, G. Environmental effects on the durability and the mechanical performance of flax fiber/bio-epoxy composites. *Compos. Part B Eng.* **2019**, *171*, 284–293. [CrossRef]
191. Fuentes, C.A.; Ting, K.W.; Dupont-Gillain, C.; Steensma, M.; Talma, A.G.; Zuijderduin, R.; Van Vuure, A.W. Effect of humidity during manufacturing on the interfacial strength of non-pre-dried flax fibre/unsaturated polyester composites. *Compos. Part A Appl. Sci. Manuf.* **2016**, *84*, 209–215. [CrossRef]
192. George, M.; Mussone, P.G.; Alemaskin, K.; Chae, M.; Wolodko, J.; Bressler, D.C. Enzymatically treated natural fibres as reinforcing agents for biocomposite material: Mechanical, thermal, and moisture absorption characterization. *J. Mater. Sci.* **2016**, *51*, 2677–2686. [CrossRef]
193. Mohammed, L.; Ansari, M.N.M.; Pua, G.; Jawaid, M.; Islam, M.S. A Review on natural fiber reinforced polymer composite and its applications. *Int. J. Polym. Sci.* **2015**, *2015*, 243947. [CrossRef]
194. Rohit, K.; Dixit, S. A Review—Future aspect of natural fiber reinforced composite. *Polym. Renew. Resour.* **2016**, *7*, 43–59. [CrossRef]
195. Mahesha, G.T.; Satish, S.B.; Vijaya, K.M.; Bhat, K.S. Preparation of unidirectional *Grewia serrulata* fiber-reinforced polyester composites and evaluation of tensile and flexural properties. *J. Nat. Fibers* **2016**, *13*, 547–554. [CrossRef]
196. Safri, S.N.A.; Sultan, M.T.H.; Jawaid, M.; Jayakrishna, K. Impact behaviour of hybrid composites for structural applications: A review. *Compos. Part B Eng.* **2018**, *133*, 112–121. [CrossRef]
197. Yang, X.; Wang, K.; Tian, G.; Liu, X.E.; Yang, S. Evaluation of chemical treatments to tensile properties of cellulosic bamboo fibers. *Eur. J. Wood Wood Prod.* **2018**, *76*, 1303–1310. [CrossRef]
198. Shih, Y.-F. Mechanical and thermal properties of waste water bamboo husk fiber reinforced epoxy composites. *Mater. Sci. Eng. A* **2007**, *445*, 289–295. [CrossRef]
199. Kemenperin. Kemenperin: Kemenperin Akselerasi Penggunaan Serat Alam Jadi Bahan Baku Industri. Available online: https://kemenperin.go.id/artikel/22183/ (accessed on 4 November 2021).
200. Sukardan, D.M.; Natawijaya, D.; Prettyanti, P.; Cahyadi; Novarini, E. Characterization of the fiber from biduri (*Calotropis gigantea*) and the identification of it's utilization possibility as a textile fiber. *Arena Tekst.* **2016**, *31*, 51–62.
201. Sana, A.W.; Noerati, N.; Sugiyana, D.; Sukardan, D.M. The application of biduri natural fiber (*Calotropis gigantea*) as an insulative filler material in winter jacket. *Arena Tekst.* **2020**, *35*, 1–12.
202. Shanmugam, V.; Mensah, R.A.; Försth, M.; Sas, G.; Restás, Á.; Addy, C.; Xu, Q.; Jiang, L.; Neisiany, R.E.; Singha, S.; et al. Circular economy in biocomposite development: State-of-the-art, challenges and emerging trends. *Compos. Part C Open Access* **2021**, *5*, 100138. [CrossRef]
203. Martina, R.A.; Oskam, I.F. Practical guidelines for designing recycling, collaborative, and scalable business models: A case study of reusing textile fibers into biocomposite products. *J. Clean. Prod.* **2021**, *318*, 128542. [CrossRef]
204. Khayyam, H.; Naebe, M.; Milani, A.S.; Fakhrhoseini, S.M.; Date, A.; Shabani, B.; Atkiss, S.; Ramakrishna, S.; Fox, B.; Jazar, R.N. Improving energy efficiency of carbon fiber manufacturing through waste heat recovery: A circular economy approach with machine learning. *Energy* **2021**, *225*, 120113. [CrossRef]
205. Velasco-Muñoz, J.F.; Mendoza, J.M.F.; Aznar-Sánchez, J.A.; Gallego-Schmid, A. Circular economy implementation in the agricultural sector: Definition, strategies and indicators. *Resour. Conserv. Recycl.* **2021**, *170*, 105618. [CrossRef]
206. Awasthi, M.K.; Sarsaiya, S.; Patel, A.; Juneja, A.; Singh, R.P.; Yan, B.; Awasthi, S.K.; Jain, A.; Liu, T.; Duan, Y.; et al. Refining biomass residues for sustainable energy and bio-products: An assessment of technology, its importance, and strategic applications in circular bio-economy. *Renew. Sustain. Energy Rev.* **2020**, *127*, 109876. [CrossRef]

207. Ruamsook, K.; Thomchick, E. Market opportunity for lignocellulosic biomass. *NewBio* **2014**, 1–108. Available online: https://farm-energy.extension.org/wp-content/uploads/2019/04/Biomass-Market-Opportunity_Final-2014_0.pdf (accessed on 4 November 2021).
208. Pratiwi, R.; Rahayu, D.; Barliana, M.I. Pemanfaatan selulosa dari limbah jerami padi (*Oryza sativa*) sebagai bahan bioplastik. *Indones. J. Pharm. Sci. Technol.* **2016**, 3, 83–91. [CrossRef]

polymers

MDPI

Article

Preparation, Characterisation and Antibacterial Activity of Carvacrol Encapsulated in Gellan Gum Hydrogel

Adila Mohamad Jaafar [1,2,*], Norafida Hasnu [1], Zulkarnain Zainal [1], Mas Jaffri Masarudin [3], Mohd Mokrish Md. Ajat [4], Min Min Aung [2,5] and Marwah Rayung [5]

1 Department of Chemistry, Faculty of Science, Universiti Putra Malaysia, Serdang 43400, Malaysia; norafidahasnu@gmail.com (N.H.); zulkar@upm.edu.my (Z.Z.)
2 Unit of Chemistry, Centre of Foundation Studies for Agriculture Science, Universiti Putra Malaysia, Serdang 43400, Malaysia; minmin_aung@upm.edu.my
3 Department of Cell and Molecular Biology, Faculty of Biotechnology and Biomolecular Science, Universiti Putra Malaysia, Serdang 43400, Malaysia; masjaffri@upm.edu.my
4 Department of Veterinary Pre Clinical Science, Faculty of Veterinary Medicine, Universiti Putra Malaysia, Serdang 43400, Malaysia; mokhrish@upm.edu.my
5 Institute of Forestry and Forest Products, Universiti Putra Malaysia, Serdang 43400, Malaysia; marwahrayung@yahoo.com
* Correspondence: adilamj@upm.edu.my

Citation: Jaafar, A.M.; Hasnu, N.; Zainal, Z.; Masarudin, M.J.; Md. Ajat, M.M.; Aung, M.M.; Rayung, M. Preparation, Characterisation and Antibacterial Activity of Carvacrol Encapsulated in Gellan Gum Hydrogel. *Polymers* **2021**, *13*, 4153. https://doi.org/10.3390/polym13234153

Academic Editors: Luminita Marin, Antonio M. Borrero-López, Concepción Valencia-Barragán, Esperanza Cortés Triviño, Adrián Tenorio-Alfonso and Clara Delgado-Sánchez

Received: 21 September 2021
Accepted: 5 November 2021
Published: 27 November 2021

Abstract: Recently, the antibacterial properties of Carvacrol (Carv) have been significantly reported. However, due to the unstable properties of Carv under various environment conditions, research approaches tailored towards its widespread and efficient use in various antimicrobial applications are scarce. Here, we discuss progress towards overcoming this challenge by utilising the encapsulation of Carv in gellan gum hydrogels to form thin films (GG-Carv) containing different concentrations of Carv (0.01–0.32 M). FTIR spectrum of GG-Carv revealed that both functional groups from GG and Carv existed. The carbon, hydrogen and nitrogen elemental analysis further supported the encapsulation of Carv with the changes in the element percentage of GG-Carv. Both swelling and degradation percentage increased with time and the decreasing patterns were observed as the concentration of Carv increased. In an antibacterial study, GG-Carv exhibited significant antibacterial activity against *E. coli* with the clear inhibition zone of 200 mm and the detection of bacterial growth showed enhancement with continuous decline throughout the study as compared to free-standing Carv.

Keywords: encapsulated carvacrol; polymer; gellan gum hydrogel; antibacterial activity; *E. coli*

1. Introduction

The essential oils from plant materials are potentially useful as a reservoir for antimicrobial compounds. They are well known for having wide applications in folk medicine, fragrance industries, preservation and food flavouring. As recognised in recent reports, essential oils containing compounds such as carvacrol, eugenol and thymol (phenolic compounds) demonstrate high antibacterial potential [1]. The antimicrobial properties of the plant volatile oils and their constituents have been explored and they extend use of these plants in potential applications such as medical procedures, cosmetics, food and pharmaceutical industries [2]. The most compelling finding is that essential oils are agents for alternative approaches towards controlling the spread of pathogenic organisms as well as in the mitigation of the development of antibiotic resistance [3].

In this context, we highlight carvacrol (Carv), which is found in the aromatic leaves and flowering plant of both thyme (*Thymus vulgaris*) and oregano (*Origanum vulgare*). Interestingly, Carv has shown effective antibacterial activity and has been proven to be a potential agent in the treatment of infections, safe for human and animal health [4]. The researchers worldwide have investigated the wide spectrum of the antibacterial activity of Carv against various types of microorganisms such as *C. albicans* [5], *P. fluorescens* [6], *L. plantarum*,

S. cerevisiae, B. cinerea [7], *S. aureus* [8], *Clostridium perfringens* [9], *Salmonella enterica* [10], *L. monocytogenes* and *E. coli* [11]. This monoterpenic phenol has also emerged for its multiple of biological properties such as anti-microbial [12], effective adjuvant for preterm labour [13], bio-film inhibitor [14], anti-inflammatory [15], antioxidant [16], antitumor [17], analgesic [18], anti-parasitic [19], anti-mutagenic [20], anti-cancer [21], anti-viral [22] insecticidal [23] and fish dietary additives [24].

The host, hydrogel, is a three-dimensional, hydrophilic, polymeric network that is capable of imbibing a large amount of water. In order to maintain the three-dimensional structure, the polymer chains of hydrogels are usually cross-linked either chemically or physically [25]. It is highly permeable to various drug compounds, able to withstand acidic environments and has high swelling properties, which can release entrapped molecules through their web-like surfaces [26]. The component of hydrogel, gellan gum, is a microbial polysaccharide that is derived from *Sphingomonas elodea*, previously known as *Pseudomonas elodea*. Significantly, gellan gum is nontoxic, biocompatible and biodegradable, and the resulting hydrogels are transparent and stable [27]. To date, this biopolymer-based hydrogel has been gaining substantial attention as a potential carrier in controlled release studies [28].

Based on the foregoing, it is believed that the encapsulation technology provides stability and protection to enhance the effectiveness due to the fact that Carv is unstable in harsh environment conditions. Carv is volatile, easily evaporates and prone to degradation during the processing, owing to direct exposure to heat, pressure, light or oxygen [29]. The goal of this study is to characterise Carv encapsulation efficiency in a hydrogel polymer system—gellan gum (GG)—as an antibacterial agent. In essence, the embedding of active hydrophobic compounds (chemical immobilisation or physical encapsulation) requires the use of emulsifying techniques to produce the hydrophilic films via aqueous dispersions to compensate for its lack of solubility. Previous studies by Yadava [30] and Vargas [31] encapsulating lovastatin and thyme essential oil, respectively, report similar processes. The Carv is encapsulated in biodegradable gellan gum hydrogel as an alternative way to extend its shelf life and control the release manner, thereby maximising the usage of the compound. Thus, in this study, the research was carried out to prepare the Carv encapsulated in gellan gum hydrogel in the form of a thin film and further characterise it for antibacterial application.

The effectiveness of Carv against *E. coli* is well established [32]. This antibacterial activity of Carv has been attributed to its considerable effects on the structural and functional properties of cytoplasmic membrane [7]. The mechanism of action is believed to be associated with the damage to the cell membrane. Thus, due to the hydrophobic nature of carvacrol, it interacts with the lipid bilayer of the cytoplasmatic membrane and aligns itself between the fatty acid chains. This causes the expansion and destabilisation of the membrane structure, increasing the fluidity and permeability, which finally inhibit the cell growth [33]. A high degree of swelling resulted in a higher water content. This affects the increase in fluidity and permeability and thereby has high expectation to inhibit cell growth [34].

2. Materials and Methods

2.1. Materials

The chemicals used in this study were glycerine (1,2,3-Propanetriol) (>96%), gelzan (gellan gum) (>96%), carvacrol (2-Methyl-5-(1-methylethyl)-phenol) (>98%) and calcium chloride ($CaCl_2$) (>96%). All chemicals were obtained from Sigma-Aldrich (St. Louis, MO, USA). Sodium dihydrogen orthophosphate (NaH_2PO_4) ($\geq$98%) was purchased from BDH Chemicals Ltd. Poole, UK, sodium hydrogen carbonate ($NaHCO_3$) ($\geq$99.8%) was provided by Fisher Brand (Leicestershire, UK), sodium chloride (NaCl) ($\geq$99%) was obtained from AnalaR (Poole, UK) and potassium chloride (KCl) ($\geq$99.5%) was purchased from HmbG Chemicals (Darmstadt, Germany). All chemicals were used directly without any purification. The strain of *Escherichia coli* (*E. coli* Dh5alpha), the plasmid of which carries gene resistance against ampicillin, was used in the antibacterial activity study.

2.2. Preparation of Carvacrol Gellan Gum Thin Films (GG-Carv)

GG-Carv was synthesised via in situ drug loading, in which the Carv was first put to disperse in deionised water (18 MΩ cm) to a specific concentration and mixed with the dissolved 1 g of gellan gum. This mixture was then mixed with $CaCl_2$ to establish the physical crosslinking. The solution was stirred at 500 rpm using a hotplate set at a temperature of 80 °C and mixed for 2 h to ensure homogeneity, and then 5 mL of glycerine was added as plasticiser. The resulting solution was poured into the petri dish and left in the oven for 48 h at 35 °C and then stored in a desiccator at room temperature for further characterisation.

In this study, GG-Carv was encapsulated with Carv at various concentrations; 0.01, 0.02, 0.04, 0.08, 0.16 and 0.32 M and are hereafter referred as GG-Carv 01, GG-Carv 02, GG-Carv 04, GG-Carv 08, GG-Carv 16 and GG-Carv 32, respectively.

2.3. Characterisations

Fourier transform infrared (FTIR) spectra of the samples were recorded in the range of 4000–400 cm^{-1} on a Perkin-Elmer 1752X Spectrophotometer (Llantrisant, UK) with the KBr disc method, in conjunction with the elemental analysis, LECO CHNS-932 Analyser (St. Joseph, MI, USA). The surface morphology and the cross-section of sample were observed by using Variable Pressure Scanning Electron Microscopy (VPSEM). This was used for the examination of the samples that are not compatible with high vacuum conditions. The samples were prepared and freeze-dried at −80 °C for 2 days to expel the moisture. Then, the samples were fixed on the aluminium stubs and gold coated before imaging. The model used in this study was LEO 1455 VPSEM (VPSEM, Kensington, UK).

2.4. The Study of Swelling Percentage

In this study, Pseudo Extra Cellular Fluid (PECF) buffer solution was used and for the preparation of the buffer, an amount of 0.68 g of NaCl, 0.22 g of KCl, 2.5 g of $NaHCO_3$ and 0.35 g of NaH_2PO_4 were dissolved in 100 mL of deionised water. The solution was stirred until fully dissolved and the resulting solution was then adjusted to pH 5.5 using 10% nitric acid.

Water uptake of GG-Carv with the dimensions of 2 cm × 2 cm was measured by weighing the dried films (W_d) prior to immersion in 20 mL of PECF buffer solution with pH 5.5 at room temperature. The subsequent weight was recorded every 24 h. The films were removed after 72 h, wiped gently with a tissue to expel surface solution and then weighed (W_w). The percentage of water uptake was then determined from the equilibrium swelling ratio:

$$Swelling\ Percentage\ (\%) = \frac{W_w - W_d}{W_d} \times 100 \quad (1)$$

where W_w is the weight of wet samples and W_d is the weight of dry samples.

2.5. The Study of Degradation Percentage

Degradation of GG-Carv was measured by weighing its initial weight of 1.0 g (W_i) and leaving it on a Petri dish at room temperature. The degradation test was performed at room temperature considering to the potential of dermal applications and the common storage temperature. The subsequent weight was recorded every day until a constant weight (W_f) pattern was observed. The percentage of degradation was then determined from the equilibrium degradation ratio:

$$Degradation\ Percentage\ (\%) = \frac{W_f - W_i}{W_i} \times 100 \quad (2)$$

where W_f is the final weight of samples and W_i is the initial weight of dry samples.

2.6. The Study of Antibacterial Activity

The antibacterial efficiency of Carv was evaluated using two methods, which are disc diffusion test and detection of bacterial growth by optical density.

2.6.1. Disc Diffusion Test

This study was evaluated by following the disc diffusion test method. The GG-Carv with a diameter of 6 mm (similar to the disc size) was placed onto the surface of the Luria Bertani (LB) agar plate incorporated with ampicillin (Amp) and seeded with 100 μL of *E. coli* culture. The plates were incubated at 37 °C for 18 h and the noticeable bacterial inhibition zone around the polymeric thin film was then observed and measured. The standard disc with impregnated antibiotics, ampicillin and kanamycin (Kana), were also assayed as the negative and positive control, respectively.

2.6.2. Detection of Bacterial Growth by Optical Density (OD)

The Antibacterial Effect on Cell Viability

Here, 10 mL of LB broth incorporated with ampicillin was seeded with 100 μL of *E. coli* and incubated for 16 h prior to the addition of GG-Carv 08 with dimensions of 10 × 10 mm and 0.08 M Carv. The *E. coli* culture without any further addition was used as the negative control. The OD of each tube was then measured for every subsequent hour at a wavelength of 600 nm against the standard medium.

The Antibacterial Effect on Cell Growth

Here, 10 mL of LB broth with the addition of ampicillin was seeded with 100 μL of *E. coli* culture. The GG-Carv 08 and 0.08 M Carv was directly placed into the bacterial culture. The OD of each tube was then measured every subsequent hour. The *E. coli* culture without any further addition was used as the negative control.

3. Results

3.1. Fourier Transform Infrared Spectroscopy (FTIR) Analysis

Shifting or disappearances of the frequency of functional characteristic peaks indicate the interaction between polymer and drug [35]. The hydrophilicity, which was reported to be the main factor in a hydrogel's ability to swell, is influenced by the presence of hydroxyl, carboxyl, sulphonic, amidic and primary amidic functional groups [36]. Thus, the IR peaks of pure GG and GG with Carv are compared.

Chemical structures of the samples were characterised by FTIR (Figure 1). In general, Carv (Figure 1a) showed characteristic peaks at 3360.88 cm^{-1} (phenolic–OH group), 2958.46 cm^{-1} (C–H stretching), 1583 and 1511 cm^{-1} (C–C ring stretching), 1421 cm^{-1} (O–H bending), 1359 cm^{-1} (isopropyl group), 1242 cm^{-1} (C–O stretching) and 864 cm^{-1} (aromatic ring). Meanwhile, the peaks of pure GG (Figure 1b) can be seen at 3273 cm^{-1} (O–H stretching), 2933 cm^{-1} (C–H stretching), 1625 cm^{-1} (C=C stretching), 1427 cm^{-1} (C–H bending), 1033 cm^{-1} (C–O stretching) and 919 cm^{-1} (C–H bending).

From the results obtained (Figure 1c–g), all GG-Carv samples showed peaks in the ranges of 3266–3290 cm^{-1} (O–H stretching) and 2887–2933 cm^{-1} (C–H stretching), which belong to both gellan gum hydrogel and Carv. Furthermore, the peaks at 1637–1642 cm^{-1} and 1408–1415 cm^{-1} (C–C ring stretching), 1033–1036 cm^{-1} (C–O stretching) and 850–918 (aromatic ring), which belong to Carv exist in all GG-Carv samples, reflect the existence of Carv in the gellan gum hydrogel polymer.

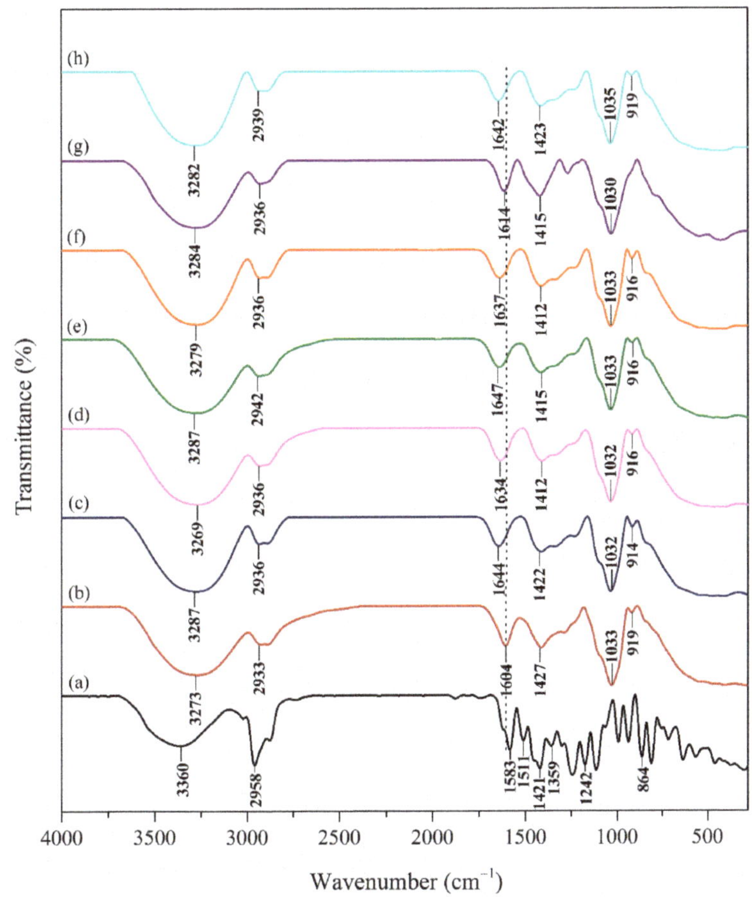

Figure 1. FTIR spectra of (**a**) carvacrol, (**b**) pure GG, (**c**) GG-Carv 01, (**d**) GG-Carv 02, (**e**) GG-Carv 04, (**f**) GG-Carv 08, (**g**) GG-Carv 16 and (**h**) GG-Carv 32.

3.2. Elemental Analysis

Hypothetically, gellan gum contains the carbon and hydrogen; therefore, to determine the carbon content of gellan gum containing Carv, the resulting carbon percentage will be subtracted from the carbon content of free-standing gellan gum. Table 1 shows the weight percentage of carbon, C and hydrogen, H for pure GG and encapsulated GG with various concentrations of Carv. From Table 1, it can be observed that in GG-Carv, the content of C showed an increasing pattern as the concentration of Carv increased. This inclining amount resulted due to the encapsulation of Carv anion and caused the content of C to increase. Similarly, the H content in GG-Carv exhibited the increasing pattern as the concentration of Carv increased.

Table 1. Elemental analysis of weight percentage of carbon, C and hydrogen, H for pure GG and encapsulated GG-Carv with various concentrations of Carv.

Material	Weight Percentage (%)	
	C	H
Pure GG	20.33	8.97
GG-Carv 01	22.52	9.08
GG-Carv 02	23.76	9.29
GG-Carv 04	26.77	9.52
GG-Carv 08	27.42	9.90
GG-Carv 16	28.45	10.21
GG-Carv 32	29.62	10.40

3.3. *Variable Pressure Scanning Electron Microscopy (VPSEM) Analysis*

VPSEM micrographs were used to study the surface and cross sectional area of GG-Carv. The observation was made at 1000 times magnification. This powerful technique is widely used to capture the characteristic "network" structure in hydrogels [37], encapsulation potential [38], supporting information for release study [39] and crosslinking density [40].

3.3.1. Surface Morphology

A clear network structure can be observed on the surface morphology of pure GG (Figure 2a). Meanwhile, GG-Carv (Figure 2b–g) exhibited a round-shaped structure scattered evenly, which is possibly due to the Carv binding to the surface of gellan gum hydrogel. The appearances of these structures were more abundant as the concentration of Carv increased with an average diameter of 5 to10 μm.

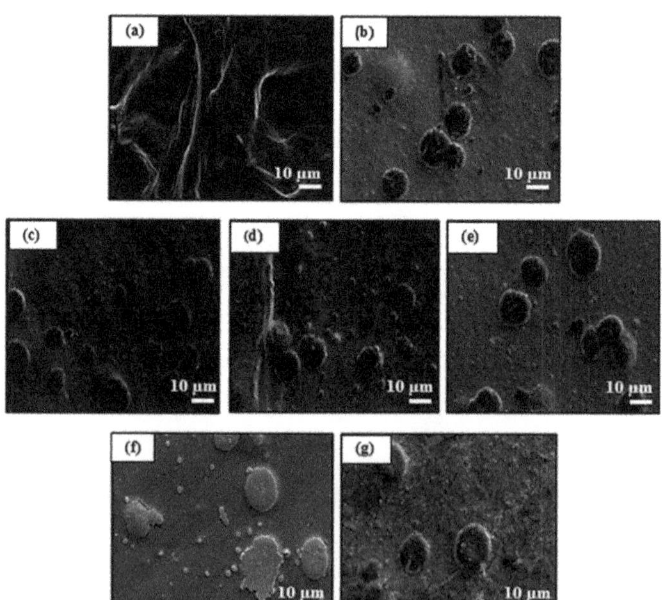

Figure 2. VPSEM surface micrograph at 1000× magnification of (**a**) pure GG film, (**b**) GG-Carv 01, (**c**) GG-Carv 02, (**d**) GG-Carv 04, (**e**) GG-Carv 08, (**f**) GG-Carv 16 and (**g**) GG-Carv 32.

3.3.2. Cross Sectional Morphology

Unpacked layers' structure can be observed in the cross-sectional morphology of pure GG film (Figure 3a). Meanwhile, GG-Carv (Figure 3b–g) displayed very compact layers as the concentration of Carv increased. This can be explained due to congestion of Carv molecules residing in between the layers.

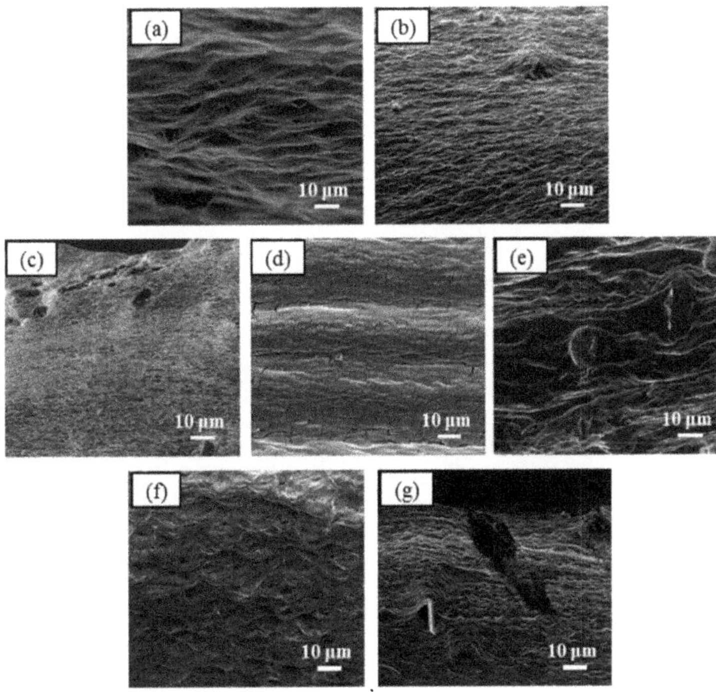

Figure 3. VPSEM cross sectional micrograph at 1000× magnification of (**a**) pure GG film, (**b**) GG-Carv 01, (**c**) GG-Carv 02, (**d**) GG-Carv 04, (**e**) GG-Carv 08, (**f**) GG-Carv 16 and (**g**) GG-Carv 32.

3.4. Swelling Percentage

Hydrogels are three-dimensional, hydrophilic, polymeric networks capable of imbibing large amounts of water or biological fluids [41]. They also provide the effect of becoming a semi-permeable membrane. When hydrogels are treated in an open system and placed in excess solutions, swelling may occur [42]. Similarly, gellan gum has the tendency to incur high swelling behaviours, but this trend will reduce as more Carv is added into the system for encapsulation. In order to accommodate more of the compound, the swelling rate of gellan gum reduces.

The result showed that the swelling percentage (Figure 4) increased with time. Table 2 shows the swelling percentage of gellan gum with Carv loading concentrations. As higher concentrations of Carv were used, the absorption of the solutions was less observed. This was reflected in GG-Carv 32, where the highest concentration of Carv caused the lowest swelling percentage due to the formation of more rigid structure of gellan gum hydrogel. Carv is known as a hydrophobic phenolic compound [43]. Thus, the resistance effect towards the solutions which account for the hydrophobicity of Carv also resulted in decreased swelling. Consequently, the higher the concentration of Carv, the higher the expected water resistance of the film.

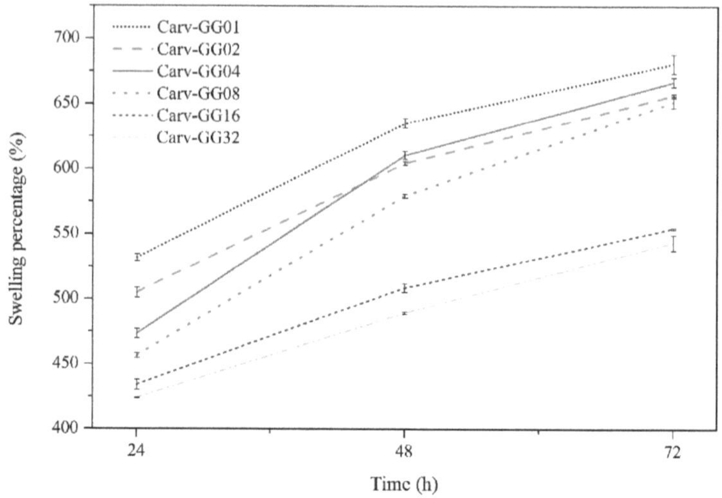

Figure 4. The swelling percentages of GG-Carv 01, GG-Carv 02, GG-Carv 04, GG-Carv 08, GG-Carv 16 and GG-Carv 32.

Table 2. The effect of Carv loading on swelling percentage of GG.

Materials	Carv (M)	Swelling Percentage (%)
GG-Carv 01	0.01	680
GG-Carv 02	0.02	670
GG-Carv 04	0.04	650
GG-Carv 08	0.08	645
GG-Carv 16	0.16	550
GG-Carv 32	0.32	540

3.5. Degradation Percentage

Most of the degradation study of gellan gum was usually achieved in vivo through the action of enzymes and in vitro [44] in accordance with their application in tissue engineering. However, to understand the degradation behaviour of polymers aimed to be used on the skin, it is important to predict and ultimately be tuned in to their condition at common room temperature for humans. One study reported that a modified gellan gum reacted with methacrylic anhydride and calcium added has an optimised thiol/ene ratio and mixed-mode crosslinking mechanism, which yielded to stiffer hydrogels related to an increase in fibroblast proliferation [45]. Modification of the fabrication of the gellan gum material will truly affect the degradation and the swelling properties in finding the appropriate application.

The percentage of degradation (Figure 5) was found to increase with time. However, it was inversely proportional to the concentration. As the concentration of Carv increased, the degradation percentage decreased. Similar to the swelling results, this might be explained by the formation of a more rigid structure of gellan gum hydrogel occurring in GG-Carv at higher concentrations. Hence, this stability resulted in more durable GG-Carv against the environmental conditions as the concentration increased.

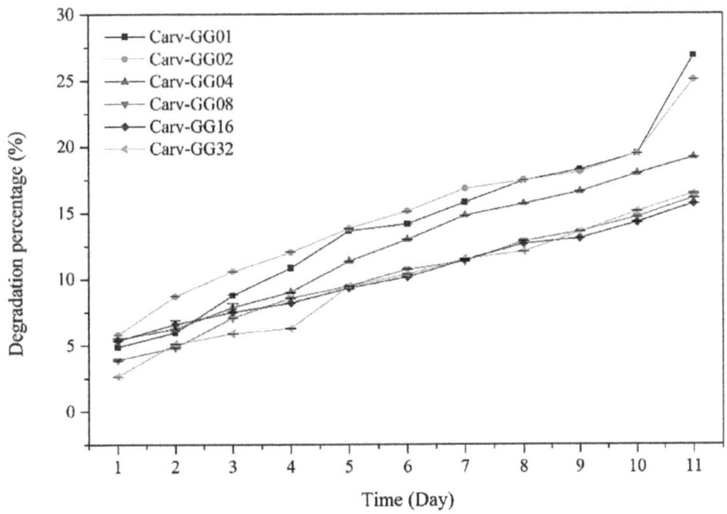

Figure 5. The time-based degradation of GG-Carv at different Carv loadings.

3.6. *The Antibacterial Activity*

We chose to evaluate the antibacterial activities of GG-Carv 08 through qualitative and quantitative studies against the same type of bacteria, *Escherichia coli* (*E. coli* Dh5alpha) strain, the plasmid of which carries gene resistance against ampicillin. A qualitative study was conducted by following the disc diffusion method. Meanwhile, the quantitative study was examined through the detection of bacterial growth by optical density (OD) in two conditions, the antibacterial effect against cell viability and cell growth.

3.6.1. Disc Diffusion Test

The purpose of the qualitative studies conducted by disc diffusion test was to examine the efficiency of gellan gum hydrogel to release the encapsulated Carv and inhibit bacteria. The inhibitory effects of GG-Carv 08 against *E. coli* were then compared with the standard antibiotics, with kanamycin as a positive control and ampicillin as a negative control. After 18 h of incubation at 37 °C, results showed that pure GG demonstrated no sign of an inhibition zone (Figure 6d), reflecting that GG does not possess antibacterial properties on its own. The same observation was found on the negative control, ampicillin, which also exhibited no sign of an inhibition zone, which proved that the *E. coli* were resistant to ampicillin, and it behaves accordingly (Figure 6c).

In contrast, GG-Carv 08 displayed a significant antibacterial activity with a clear inhibition zone of 200 mm (Figure 6a), which in this case was larger than the inhibition zone of the standard antibiotic, kanamycin, with only 130 mm (Figure 6b). This finding directly highlights the potential of Carv, a natural compound, to be utilised as an antibacterial agent and as an alternative approach to conventional antibiotics. This result also indicates that gellan gum hydrogel works as effective carrier and is capable of releasing carvacrol in a sustained manner within 18 h.

The effectiveness of Carv against *E. coli* is well established [32]. This antibacterial activity of Carv has been attributed to its considerable effects on the structural and functional properties of cytoplasmic membrane [7]. The mechanism of action is believed to be associated with damage to the cell membrane. Thus, due to the hydrophobic nature of carvacrol, it interacts with the lipid bilayer of the cytoplasmatic membrane and aligns itself between the fatty acid chains. This causes the expansion and destabilisation of the membrane structure, increasing the fluidity and permeability, which finally inhibit the cell growth [33]. A high degree of swelling resulted in a higher water content. This affects the increase in fluidity and permeability thereby has a high expectation to inhibit cell growth [34].

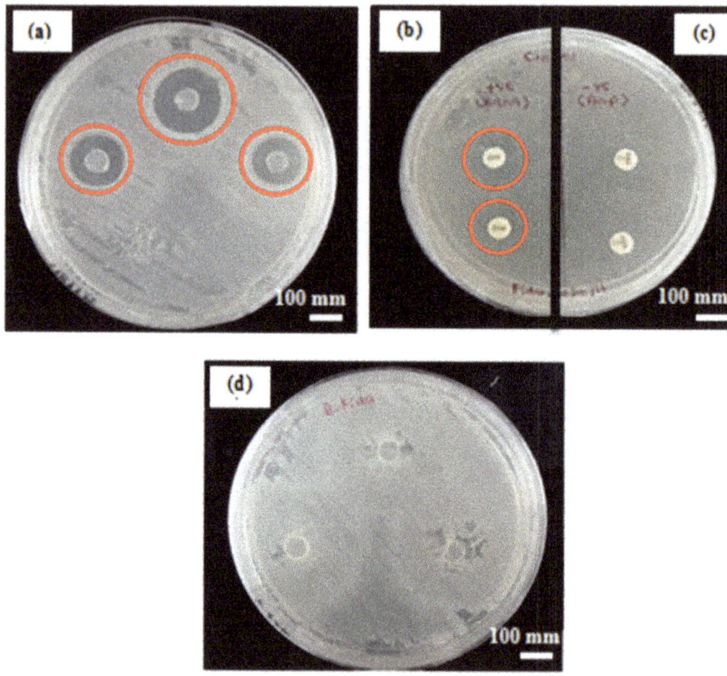

Figure 6. The appearance inhibition zone after 18 h of incubation for (**a**) GG-Carv 08, (**b**) positive control (Kanamycin), (**c**) negative control (Ampicillin) and (**d**) pure GG.

3.6.2. Detection of Bacterial Growth by Optical Density (OD)

In order to further investigate the antibacterial activity of GG-Carv 08, the quantitative test was performed through the detection of bacterial growth by optical density (OD) in two conditions, the antibacterial effect against cell viability and cell growth.

The Antibacterial Effects on Cell Viability

This study was performed with the intention to compare the efficiency of the encapsulated GG-Carv with free-standing Carv against bacteria that are in a state of very low metabolic activity and limited available nutrients. Based on the results shown in Figure 6, the growth curve of *E. coli* followed the normal phase of bacterial growth (Figure 7a), which in this case was used as the negative control. Meanwhile, the *E. coli* culture with the addition of free-standing Carv (Figure 7c) exhibited a decrease in bacterial cells but started to remain constant at the 21st hour. This can be explained by the environment conditions that might affect the efficiency of Carv and become deteriorated. As reported previously, Carv is a bioactive compound that easily evaporates and is prone to degradation, owing to its properties when exposed heat, pressure, light or oxygen [29].

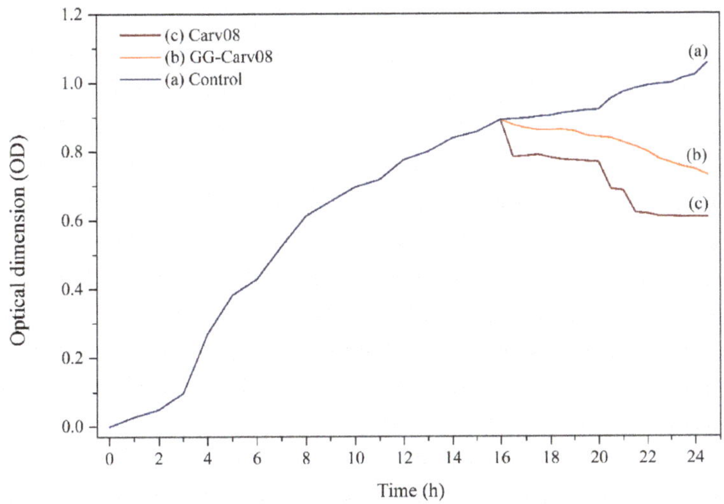

Figure 7. (**a**) The growth profile of *E. coli* culture without further addition and further addition of (**b**) GG-Carv 08 and (**c**) Carv 0.08 M, after 16 h incubation of *E. coli* culture.

Meanwhile, the addition of GG-Carv 08 to the *E. coli* culture (Figure 7b) demonstrated a gradual and continuous decrease in bacteria cells, but in a slow and sustained manner. This trend was expected to continue for extended hours until the maximum release of Carv could be achieved. Above all, this result reflects that the main function of the host, gellan gum hydrogel, had successfully controlled the release of Carv. In further applications, the prolonged antibacterial effect can be expected from GG-Carv to treat infectious bacteria.

The Antibacterial Effects on Cell Growth

This study demonstrated the antibacterial effects on cell growth. The growth phase of bacteria is defined as an increase in bacterial number in a population rather than in the size of individual cells. Thus, with the addition of GG-Carv 08 and Carv, the inhibition of the growth and the dead cells are considered to have disrupted or broken membranes due to the antibacterial activity of carvacrol against bacteria. This mechanism of action is believed to be associated with the damage caused to the cell membrane [46]. Thus, this study was performed with the intention to compare the efficiency of the encapsulated GG-Carv with the free-standing Carv against bacteria that are in a state of very high metabolic activity and high nutrients available.

According to the results, the growth curve of *E. coli* followed the normal phase of bacterial growth throughout the 10 h (Figure 8a), which in this case were used as the negative control. In contrast, the trends of *E. coli* culture with further addition of GG-Carv and free-standing Carv were similar with previous antibacterial effects on cell viability, respectively, both showing a reduction. The *E. coli* cultures with immersed GG-Carv showed the significant and continued decline of growth but still in controlled manner (Figure 8b). Meanwhile, the *E. coli* cultures with free-standing Carv exhibited a great decline, but started to remain constant at the 9th hour (Figure 8c). The results obtained in this study prove that encapsulated Carv performed better than the free-standing Carv. Hence, it has been proven that the encapsulation technology enhanced the release of Carv and the usage of this compound could be maximised in further applications.

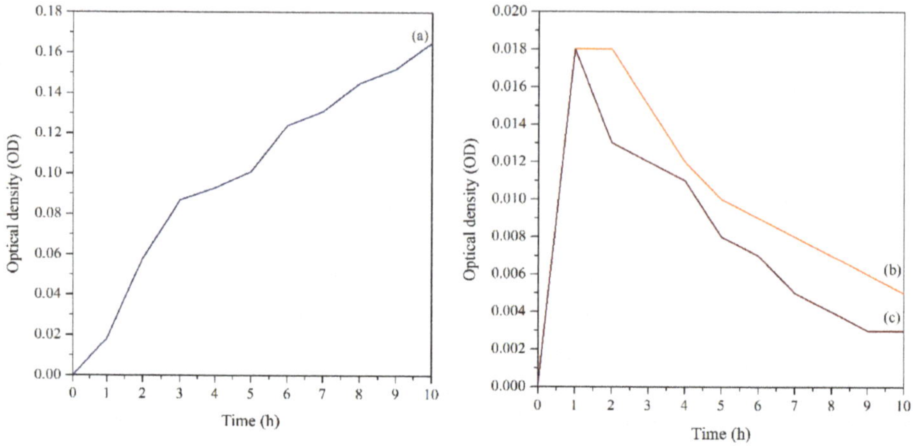

Figure 8. (**a**) The growth profile of *E. coli* culture without further addition and further addition of (**b**) GG-Carv 08 and (**c**) Carv 0.08 M, after 16 h incubation of *E. coli* culture.

4. Conclusions

The preparation of carvacrol encapsulated in gellan gum hydrogel in the form of a thin film (GG-Carv) was successfully achieved. This was confirmed by the FTIR spectrum of GG-Carv, which showed the combination of both functional groups from the hydrogel and Carv. The elemental analysis further supported the encapsulation with the observed changes to the element percentage. Similarly, the swelling percentage and percentage of degradation increased with the time in normal environmental conditions, but showed a decreasing trend with concentrations of Carv being loaded into the system. GG-Carv showed significant antibacterial activity against *E. coli*, with a clear inhibition zone of 200 mm. Subsequently, the detection of bacterial growth through optical density showed enhancement with continuous decline throughout the study as compared to free-standing Carv. Incorporating Carv together with gellan promotes the bioavailability and bio-delivery. Hence, the gellan gum hydrogel is proven to be an effective carrier of carvacrol for antibacterial applications. This study has generated fundamental knowledge of gellan gum hydrogel Carv films which could be used and further explored in appropriate bacterial conditions with enhanced properties for the development of antibacterial applications. Subsequently, this would pave the way towards potential applications as wound-healing patches as well as in the development of more sterile food packaging systems.

Author Contributions: Conceptualisation, A.M.J.; methodology, A.M.J. and M.J.M.; investigation, A.M.J. and N.H.; writing—original draft preparation, A.M.J. and N.H.; writing—review and editing, M.M.M.A., M.M.A. and M.R.; supervision, Z.Z. and M.J.M.; funding acquisition, A.M.J. All authors have read and agreed to the published version of the manuscript.

Funding: This research was funded by the Fundamental Research Grant Scheme (FRGS 5524557) and Universiti Putra Malaysia (IPS grant 9458100).

Institutional Review Board Statement: Not applicable.

Informed Consent Statement: Not applicable.

Data Availability Statement: All data are available in the main text.

Acknowledgments: The author would like to thank the staff of the Faculty of Science, UPM, for the technical support.

Conflicts of Interest: The authors declare no conflict of interest.

References

1. Meng, J.H.; Zhao, S.H.; Doyle, M.P.; Joseph, S.W. Antibiotic Resistance of *E. coli* O157: H7 and O157: NM Isolated from Animals, Food and Humans. *J. Food Prot.* **1998**, *61*, 1511–1514. [CrossRef] [PubMed]
2. Dorman, H.J.D.; Deans, S.G. Antimicrobial Agents from Plants: Antibacterial Activity of Plant Volatile Oils. *J. Appl. Microbiol.* **2000**, *88*, 308–316. [CrossRef] [PubMed]
3. Nostro, A.; Papalia, T. Antimicrobial Activity of Carvacrol: Current Progress and Future Prospectives. *Recent Pat. Antiinfect. Drug Discov.* **2012**, *7*, 28–35. [CrossRef]
4. Nostro, A.; Roccaro, A.S.; Bisignano, G.; Marino, A.; Blanco, A.R. Effect of Oregano, Carvacrol and Thymol on *S. aureus* and *S. epidermdis* Biofilms. *J. Med. Microbiol.* **2007**, *56*, 519–523. [CrossRef]
5. Chami, F.; Chami, N.; Bennis, S.; Bouchikhi, T.; Remmal, A. Oregano and Clove Essential Oils Induce Surface Alteration of *Saccharomyces cerevisiae*. *Phytother. Res.* **2005**, *19*, 405–408. [CrossRef] [PubMed]
6. Di Pasqua, R.; Betts, G.; Hoskins, N.; Edwards, M.; Ercolini, D.; Mauriello, G. Membrane Toxicity of Antimicrobial Compounds from Essential Oils. *J. Agric. Food Chem.* **2007**, *55*, 4863–4870. [CrossRef]
7. Arfa, B.A.; Combes, S.; Preziosi-Belloy, L.; Gontard, N.; Chalier, P. Antimicrobial Activity of Carvacrol Related to its Chemical Structure. *Lett. Appl. Microbiol.* **2006**, *43*, 149–154. [CrossRef] [PubMed]
8. Oussalah, M.; Caillet, S.; Saucier, L.; Lacroix, M. Inhibitory Effects of Selected Plant Essential Oils on the Growth of Four Pathogenic Bacteria: *E. coli* O157:H7, *Salmonella Typhimurium*, *Staphylococcus aureus* and *Listeria monocytogenes*. *Food Control* **2007**, *18*, 414–420. [CrossRef]
9. Juneja, V.K.; Friedman, M. Carvacrol, Cinnamaldehyde, Oregano oil, and Thymol Inhibit *Clostridium perfringens* Spore Germination and Outgrowth in Ground Turkey During Chilling. *J. Food Prot.* **2007**, *70*, 218–222. [CrossRef]
10. Ravishankar, S.; Zhu, L.; Reyna-Granados, J.; Law, B.; Joens, L.; Friedman, M. Carvacrol and Cinnamaldehyde Inactivate Antibiotic-Resistant *Salmonella enterica* in Buffer and on Celery and Oysters. *J. Food Prot.* **2010**, *73*, 234–240. [CrossRef]
11. Perez-Conessa, D.; Cao, J.; Chen, L.; Mclandsborough, L.; Weiss, J. Inactivation of *Listeria monocytogenes* and *Escherichia coli* O157:H7 Biofilms by Micelle Encapsulated Eugenol and Carvacrol. *J. Food Prot.* **2011**, *74*, 55–62. [CrossRef]
12. Shelef, L.A. Antimicrobial Effects of Spices. *J. Food Saf.* **1983**, *6*, 29–44. [CrossRef]
13. Muñoz-Pérez, V.M.; Ortiz, M.I.; Gerardo-Muñoz, L.S.; Cariño-Cortés, R.; Salas-Casas, A. Tocolytic Effect of the Monoterpenic Phenol Isomer, Carvacrol, on the Pregnant Rat Uterus. *Chin. J. Physiol.* **2020**, *63*, 204–210. [CrossRef] [PubMed]
14. Huma, J.; Firoz, A.A.; Iqbal, A. Prospects of Essential Oils in Controlling Pathogenic Biofilm. In *New Look to Phytomedicine*; Khan, M.S.A., Ahmad, I., Chattopadhyay, D., Eds.; Academic Press: Cambridge, MA, USA, 2019; Chapter 9; pp. 203–236.
15. Aligiannis, N.; Kalpoutzakis, E.; Mitaki, S.; Chinou, I.B. Composition and Antimicrobial Activity of the Essential oil of Two *Origanum* Species. *J. Agric. Food Chem.* **2001**, *49*, 4168–4170. [CrossRef] [PubMed]
16. Hajhashemi, V.; Ghannadi, A.; Pezeshkian, S.K. Antinociceptive and Anti-Inflammatory Effects of *Satureja hortensis* L. extracts and Essential Oil. *J. Ethnopharmacol.* **2002**, *7*, 82–83. [CrossRef]
17. Prieto, J.M.; Jacopini, P.; Cioni, P.; Chericoni, S. In vitro Activity of the Essential Oils of *Origanum vulgare*, *Satureja montana* and their main Constituents in Peroxynitrite-Induced Oxidative Processes. *Food Chem.* **2007**, *104*, 889–895. [CrossRef]
18. Karkabounas, S.; Kostoula, O.K.; Daskalou, T.; Veltsistas, P.; Karamousis, M.; Zelovitis, I. Anticarcinogenic and Antiplatelet Effects of Carvacrol. *Exp. Oncol.* **2006**, *28*, 121–125.
19. Wagner, H.; Wierer, M.; Bauer, R. In vitro Inhibition of Prostaglandin Biosynthesis by Essential Oils and Phenolic Compounds. *Planta Med.* **1986**, *52*, 184–187. [CrossRef]
20. Force, M.; Sparks, W.S.; Ronzio, R.A. Inhibition of Enteric Parasites by Emulsified Oil of Oregano in vivo. *Phytother. Res.* **2000**, *14*, 213–214. [CrossRef]
21. Aydin, S.; Basaran, A.A.; Basaran, N. The Effects of Thyme Volatiles on the Induction of DNA Damage by the Heterocyclic Amine IQ and Mitomycin C. *Mutat. Res.* **2005**, *581*, 43–53. [CrossRef]
22. Ravizza, R.; Gariboldi, M.B.; Molteni, R.; Monti, E. Linalool, a Plant-derived Monoterpene Alcohol, Reverses Doxorubicin Resistance in Human Breast Adenocarcinoma Cells. *Oncol. Rep.* **2008**, *20*, 625–630. [CrossRef] [PubMed]
23. Karpouhtsis, I.; Pardali, E.; Feggou, E.; Kokkini, S.; Scouras, Z.G.; Mavragani-Tsipidou, P. Insecticidal and Genotoxic Activities of Oregano Essential Oils. *J. Agric. Food Chem.* **1998**, *46*, 1111–1115. [CrossRef]
24. Park, B.S.; Choi, W.S.; Kim, J.H.; Kim, K.H.; Lee, S.E. Monoterpenes from Thyme (*Thymus vulgaris*) as Potential Mosquito Repellents. *J. Am. Mosq. Control. Assoc.* **2005**, *21*, 80–83. [CrossRef]
25. Park, M.R.; Chun, C.; Ahn, S.W.; Ki, M.H.; Cho, C.S.; Song, S.C. Sustained Delivery of Human Growth Hormone Using a Polyelectrolyte Complex-loaded Thermosensitive Polyphosphazene Hydrogel. *J. Control Release* **2010**, *147*, 359–367. [CrossRef] [PubMed]
26. Han, J.H.; Krochta, J.M.; Kurth, M.J.; Hsieh, Y.L. Lactitol-based Poly (Ether Polyol) Hydrogels for Controlled Release Chemical and Drug Delivery Systems. *J. Agric. Food Chem.* **2000**, *48*, 5278–5282. [CrossRef]
27. Kang, D.; Zhang, H.B.; Nitta, Y.; Fang, Y.P.; Nishinari, K. *Polysaccharides: Bioactivity and Biotechnology*; Springer: Cham, Switzerland, 2015; pp. 1627–1682.
28. Van Tomme, S.R.; Storm, G.; Hennink, W.E. In situ Gelling Hydrogels for Pharmaceutical and Biomedical Applications. *Int. J. Pharm.* **2008**, *355*, 1–18. [CrossRef]

29. Charlier, P.; Arfa, A.B.; Belloy, L.P.; Gontard, N. Carvacrol Losses from Soy Protein Coated Papers as a Function of Drying Conditions. *J. Appl. Polym. Sci.* **2007**, *106*, 611–620. [CrossRef]
30. Yadava, S.K.; Naik, J.B.; Patil, J.S.; Vinod, J.; Mokale, R.S. Enhanced Solubility and Bioavailability of Vovastatin using Stabilized form of Self-Emulsifying Drug Delivery System. *Colloids Surfaces A Physicochem. Eng. Asp.* **2015**, *481*, 63–71. [CrossRef]
31. Perdones, Á.; Chiralt, A.; Vargas, M. Properties of Film-Forming Dispersions and Films based on Chitosan Containing Basil or Thyme Essential Oil. *Food Hydrocolloids* **2016**, *57*, 271–279. [CrossRef]
32. Helander, I.M.; Alakomi, H.L.; Latva-Kala, K.; Mattila-Sandholm, T.; Pol, I.; Smid, E.J. Characterization of the Action of Selected Essential Oil Components on Gram-Negative Bacteria. *J. Agric. Food Chem.* **1998**, *46*, 3590–3595. [CrossRef]
33. Ultee, A.; Bennik, M.H.; Moezelaar, R. The Phenolic Hydroxyl Group of Carvacrol is Essential for Action Against the Food-Borne Pathogen Bacillus cereus. *Appl. Environ. Microbiol.* **2002**, *68*, 1561–1568. [CrossRef]
34. Raza, S.; Matuła, K.; Karoń, S.; Paczesny, J. Resistance and Adaptation of Bacteria to Non-Antibiotic Antibacterial Agents: Physical Stressors, Nanoparticles, and Bacteriophages. *Antibiotics* **2021**, *10*, 435. [CrossRef]
35. Prakash, S.J.; Santhiagu, A.; Jasemine, S. Preparation, Characterization and in vitro Evaluation of Novel Gellan Gum-Ralxifene HCl Nanoparticles. *J. Pharma.* **2014**, *2*, 63–71.
36. Ganji, F.; Abdekhodei, M.J. Chitosan-g-PLGA Copolymer as a Thermosensitive Membrane. *Carbohydr. Polym.* **2010**, *80*, 740–746. [CrossRef]
37. Figueroa-Pizano, M.D.; Vélaz, I.; Peñas, F.J.; Zavala-Rivera, P.; Rosas-Durazo, A.J.; Maldonado-Arce, A.D.; Martínez-Barbosa, M.E. Effect of Freeze-Thawing Conditions for Preparation of Chitosan-Poly (Vinyl Alcohol) Hydrogels and Drug Release Studies. *Carbohydrate Polym.* **2018**, *195*, 476–485. [CrossRef] [PubMed]
38. Alesa Gyles, D.; Pereira Júnior, A.D.; Diniz Castro, L.; Santa Brigida, A.; Nobre Lamarão, M.L.; Ramos Barbosa, W.L.; Carréra Silva Júnior, J.O.; Ribeiro-Costa, R.M. Polyacrylamide-Metilcellulose Hydrogels Containing *Aloe barbadensis* Extract as Dressing for Treatment of Chronic Cutaneous Skin Lesions. *Polymers* **2020**, *12*, 690. [CrossRef] [PubMed]
39. Zheng, Y.; Liang, Y.; Zhang, D.; Sun, X.; Liang, L.; Li, J.; Liu, Y.N. Gelatin-Based Hydrogels Blended with Gellan as an Injectable Wound Dressing. *ACS Omega* **2018**, *3*, 4766–4775. [CrossRef]
40. Wei, Z.; Chen, Y.; Wijaya, W.; Cheng, Y.; Xiao, J.; Huang, Q. Hydrogels Assembled from Ovotransferrin Fibrils and Xanthan Gum as Dihydromyricetin Delivery Vehicles. *Food Funct.* **2020**, *11*, 1478–1488. [CrossRef]
41. Peppas, N.A.; Bures, P.; Leobandung, W.; Ichikawa, H. Hydrogels in Pharmaceutical Formulations. *Eur. J. Pharm. Biopharm.* **2000**, *50*, 27–46. [CrossRef]
42. Huang, C.; Leavitt, T.; Bayer, L.R.; Orgill, D.P. Effect of Negative Pressure Wound Therapy on Wound Healing. *Curr. Probl. Surg.* **2014**, *51*, 301–331. [CrossRef]
43. Veldhuizen, E.J.A.; Tjeerdsma-Van Bokhoven, J.L.M.; Zweijtzer, C.; Burt, S.A.; Haagsman, H.P. Structural Requirements for the Antimicrobial Activity of Carvacrol. *J. Agric. Food Chem.* **2006**, *54*, 1874–1879. [CrossRef]
44. Oliveira, J.T.; Santos, T.C.; Martins, L.; Picciochi, R.; Marques, A.P.; Castro, A.G. Gellan Gum Injectable Hydrogels for Cartilage Tissue Engineering Applications: In Vitro Studies and Preliminary In Vivo Evaluation. *Tissue Eng. Part A* **2010**, *16*, 343–353. [CrossRef] [PubMed]
45. Xu, Z.; Li, Z.; Jiang, S.; Bratlie, K.M. Chemically Modified Gellan Gum Hydrogels with Tunable Properties for Use as Tissue Engineering Scaffolds. *ACS Omega* **2018**, *3*, 6998–7007. [CrossRef] [PubMed]
46. Sikkema, J.; de Bont, J.A.M.; Poolman, B. Mechanisms of Membrane Toxicity of Hydrocarbons. *Microbiol. Rev.* **1995**, *59*, 201–222. [CrossRef] [PubMed]

polymers

MDPI

Article

An Insight into a Sustainable Removal of Bisphenol A from Aqueous Solution by Novel Palm Kernel Shell Magnetically Induced Biochar: Synthesis, Characterization, Kinetic, and Thermodynamic Studies

Kamil Kayode Katibi [1,2], Khairul Faezah Yunos [3,*], Hasfalina Che Man [2], Ahmad Zaharin Aris [4,5], Mohd Zuhair Mohd Nor [3] and Rabaah Syahidah Azis [6,7]

1 Department of Agricultural and Biological Engineering, Faculty of Engineering and Technology, Kwara State University, Malete 23431, Nigeria; kamil.katibi@kwasu.edu.ng
2 Department of Biological and Agricultural Engineering, Faculty of Engineering, University Putra Malaysia, Serdang 43400, Selangor, Malaysia; hasfalina@upm.edu.my
3 Department of Food and Process Engineering, Faculty of Engineering, University Putra Malaysia, Serdang 43400, Selangor, Malaysia; zuhair@upm.edu.my
4 Department of Environment, Faculty of Forestry and Environment, University Putra Malaysia, Serdang 43400, Selangor, Malaysia; zaharin@upm.edu.my
5 Material Processing and Technology Laboratory (MPTL), Institute of Advance Technology (ITMA), University Putra Malaysia, Serdang 43400, Selangor, Malaysia
6 Department of Physics, Faculty of Science, University Putra Malaysia, Serdang 43400, Selangor, Malaysia; rabaah@upm.edu.my
7 Materials Synthesis and Characterization Laboratory (MSCL), Institute of Advanced Technology (ITMA), University Putra Malaysia, Serdang 43400, Selangor, Malaysia
* Correspondence: kfaezah@upm.edu.my; Tel.: +60-1-8231-4746

Citation: Katibi, K.K.; Yunos, K.F.; Man, H.C.; Aris, A.Z.; Mohd Nor, M.Z.; Azis, R.S. An Insight into a Sustainable Removal of Bisphenol A from Aqueous Solution by Novel Palm Kernel Shell Magnetically Induced Biochar: Synthesis, Characterization, Kinetic, and Thermodynamic Studies. *Polymers* **2021**, *13*, 3781. https://doi.org/10.3390/polym13213781

Academic Editor: Antonio M. Borrero-López

Received: 28 September 2021
Accepted: 24 October 2021
Published: 31 October 2021

Abstract: Recently Bisphenol A (BPA) is one of the persistent trace hazardous estrogenic contaminants in the environment, that can trigger a severe threat to humans and environment even at minuscule concentrations. Thus, this work focused on the synthesis of neat and magnetic biochar (BC) as a sustainable and inexpensive adsorbent to remove BPA from aqueous environment. Novel magnetic biochar was efficiently synthesized by utilizing palm kernel shell, using ferric chloride and ferrous chloride as magnetic medium via chemical co-precipitation technique. In this experimental study, the influence of operating factors comprising contact time (20–240 min), pH (3.0–12.0), adsorbent dose (0.2–0.8 g), and starting concentrations of BPA (8.0–150 ppm) were studied in removing BPA during batch adsorption system using neat biochar and magnetic biochar. It was observed that the magnetically loaded BC demonstrates superior maximum removal efficiency of BPA with 94.2%, over the neat biochar. The functional groups (FTIR), Zeta potential, vibrating sample magnetometer (VSM), surface and textural properties (BET), surface morphology, and mineral constituents (FESEM/EDX), and chemical composition (XRD) of the adsorbents were examined. The experimental results demonstrated that the sorption isotherm and kinetics were suitably described by pseudo-second-order model and Freundlich model, respectively. By studying the adsorption mechanism, it was concluded that π-π electron acceptor–donor interaction (EAD), hydrophobic interaction, and hydrogen bond were the principal drives for the adsorption of BPA onto the neat BC and magnetic BC.

Keywords: neat biochar; palm kernel shell; magnetic biochar; adsorption mechanism

1. Introduction

Recently, the safety and quality of drinking water has been problematic owing to rapid urban development induced by human activities. The upsurge of some micro-contaminants identified as endocrine disrupting compounds (EDCs) has elicited increasing concerns over the supply of safe and clean drinking water [1–3]. Particularly, Bisphenol

A (BPA; 2,2-bis(4-hydroxyphenyl)propane) is an intermediate and highly essential raw material highly utilized in the manufacture of several polycarbonate plastics and epoxy resins (plastic component), food cans, polyester fibers, and thermal paper as well as other materials in industry, thus it frequently emerges in numerous products for daily use, such as electronic equipment, water-pipes, toys, or paper [4,5]. Globally, BPA is classified among the most ever-present and most extensively produced synthetic chemical compounds in manufacturing today, with more than three million tons generated annually [6]. Markedly, the manufacture and sale of plastics comprising BPA has been prohibited by the Brazilian government since 2012, owing to its environmental universality, elevated industrial output, and toxicological consequence, and it has been classified as an important contaminant in water purification in many regions and nations [7]. Furthermore, exposure to BPA is through tableware and polycarbonate bottles, including those employed for new-born formula milk, and by means of epoxy resin coatings inside beverage and food containers, dental fissure sealants, adhesives, epoxy-based surface coatings, canned goods, household dust, and printing inks [5,6] and massively released into the water environment [8]. BPA in plastic containers can hydrolyse and pollute groundwater via landfill leachate [9]. Risen plastic industries trigger profuse discharge of BPA into the environment and it is detected in drinking and surface water [10]. More unfortunately, BPA has undesirable ecological impacts that are associated with dysfunction of the hormone system in animals and humans, even at very minuscule concentration. Specifically, an estrogenic action of BPA can provoke cancer [11–13], neurological challenges, diabetes, tumors, obesity, immune effects, heart disease in humans, damaged reproductive function, disturbance of the normal hormone functions, and undesired physiological state in animals and humans, biomagnification, and bioaccumulation through food chain or food web in human beings [14,15]. Besides, the dysfunction and damages caused by BPA are long-term [16]. Therefore, releasing water-containing BPA into waterways without adequate treatment poses a damaging impact on the environment (humans and aquatic) [17]. Owing to the frequent extensive utilization of BPA-based products by humans with their associated negative impacts on public health and ecosystem, hence, the removal of BPA became unavoidable. Extensive efforts have been employed towards extenuating the undesirable effects and environmental hazards by reducing the contaminants' concentration using various treatment technologies, including precipitation, coagulation, activated sludge, biological filter, and constructed wetlands, with inadequate effect on BPA removal [18–21]. Some relatively advanced treatment techniques, such as ozonation and advanced oxidation [22,23], Fenton oxidation [24], membrane technology [25,26], photocatalysis degradation [27], enzymatic degradation [28,29], and adsorption technique utilizing activated carbon [30–32] exhibit a good BPA removal from wastewater and drinking water. Yet, the above treatment approaches still have significant drawbacks, such as high capital and maintenance cost, complex treatment procedures with increased operating expenses, costly capital expenses (CAPEX), generation of toxic by-products which hamper their application in developing nations [33]. Noticeably, among the aforementioned water-treatment approaches, adsorption technique is found more suitable, environmentally friendly, relatively economical, robust, and simple, and it could efficiently be employed in large-scale applications devoid of generating by-products in the environment [34,35]. Adsorption method has attracted wide attention and has been extensively considered as a promising technology for removing BPA pollutant during the past two decades, due to its ease of configuration and application, high efficiency, insensitivity to poisonous compounds, low cost, and comparatively small footprint as compared to other water-treatment methods [31].

Several adsorbents are available, however, carbon materials have been reported to perform remarkably owing to their several benefits, such as enormous surface area, superior stability, and outstanding removal efficiency [36]. Activated carbons generated from biomass wastes are often used as adsorbent during the adsorption process for the removal of emerging pollutants [37,38], since it can efficiently and rapidly remove BPA as a result of its abundant surface area as well as adsorptive capacity; but, its preparation is highly expensive. Thus, various studies have considered reliable resources, low-cost, and environment-friendly materials as alternative adsorbents [38,39]. Thus far, traditional carbonaceous materials, such as BC, activated carbon, graphene oxide together with its derivates, and carbon nanotubes, have been studied as promising adsorbent materials for hydrophobic organic pollutants, including pharmaceutical compounds and EDCs [40]. Furthermore, biochar as an adsorbent has porous structure comparable with activated carbon, which is the most generally utilized and effective sorbent for the removal of various contaminants from water globally [41].

Biochar, a major pyrogenic by-product obtained from the complete or partial pyrolysis of naturally plenteous biomass under oxygen-deficient conditions [42], has several macropores, hence it could attain a superior adsorption capacity [43]. More importantly, BC exhibits superior unique properties, such as augmented surface functional groups, porous structure, abundant specific surface area, eco-friendly, low-cost, ample inherent mineral components, high cation exchange capacity, and, notably, is efficient in the removal of various hydrophobic and hydrophilic organic pollutants owing to its high aromaticity and hydrophobicity [40,44,45]. These properties have made BC to be the best and most promising precursor over other adsorbent materials for various contaminants removal. Synchronically, after pyrolysis, BC can be reapplied to sorb contaminants in water; thus, it has abundant utilization value [46]. However, the minuscule particle size of BC in addition to its lower density makes its regeneration, separation, and recovery more problematic after adsorption, and this could undermine its recycling capacity and industrial applications. In order to subdue these difficulties, few studies have considered synthesis of magnetically recyclable biochar (magnetic biochar) via the implantation of iron oxide (Fe_3O_4 and Fe_2O_3) [47]. For instance, Lu et al. [48] studied the removal of BPA using N-doped ulva prolifera (marine macroalgae). Over 90% of BPA was eliminated and the sorption capacity of 9.19 mg/g was attained within 4 hrs. In another application, Heo et al. [49] synthesized $CuZnFe_2O_4$ composite biochar using bamboo to enhance the removal of BPA and sulfamethoxazole (SMX) from aqueous solution. The adsorption capacity of 263.2 mg/g was recorded when CZF–biochar adsorbent was applied to remove BPA. It was observed that the sorption capacity was improved for SMX and BPA when CZF nanomaterials were impregnated on the surface of biochar. Furthermore, Wang et al. [16] use grapefruit peel biochar to remove BPA under variable adsorbent dosage, pH, and contact time. The authors reported that pH has a significant influence on the adsorption of BPA and that almost 100% BPA removal was achieved at a pH (6).

In Malaysia, the agro-industrial sector generates a substantial volume of biomass solid wastes, predominantly from palm oil mills which produce the huge expanse of biomass, including palm kernel shells (PKS), empty fruit bunches, oil palm fibers, and palm oil mill effluents [50]. Malaysia recorded the largest export of 19.9 million tonnes (mt) from oil palm biomass residues in 2017 [51]. With the expansion of palm oil production in Malaysia, the volume of residue generated has correspondingly risen. Approximately 50–70 tonnes of biomass residues could be produced from a hectare of oil palm plantation [52]. Thus, the palm oil industry is presently generating around 50 mt of dry oil palm residues annually and attain 100 mt per year by 2020 [53]. Notably, PKS has the maximum commercial consumption value as compared to other oil palm biomass [54]. Besides, the specific characteristics of PKS-BC (palm kernel shell biochar), particularly its porous structure, enhancement of functional groups, its huge specific surface area, and improved mineral component, make it suitable for use as adsorbent material [42]. Despite this huge potential, PKS is considered as biomass waste, generating superfluous waste of resources and environmental nuisance [55].

There is still a lack of reports on the adsorption of BPA using PKS magnetic BC, as well as its adsorption mechanism which is yet to be explored and needs to be investigated. In view of this highlight, PKS biomass waste that is available in huge quantities in Malaysia was applied as a precursor for the synthesis of magnetic BC as a sustainable and inexpensive adsorbent for the removal of BPA from aqueous solutions via adsorption procedure.

Magnetic BC was synthesized using the chemical co-precipitation method. A batch adsorption test was employed to evaluate the maximum adsorption capacity of BC produced from PKS. The effects of essential factors, in particular contact time, adsorbent dosage, ionic strength, initial BPA concentration, and pH, on the adsorption of BPA by NBC (neat biochar) and MBC (magnetic biochar) were studied. The adsorption mechanism of BPA by synthesized BC was examined together with the analysis of the NBC and MBC, to boost the study of the adsorption of BPA by NBC and MBC. The adsorption isotherm was stimulated via Freundlich and Langmuir models. The adsorption kinetics and reusability of the novel synthesized PKS-BC were also investigated.

2. Materials and Methods

2.1. Materials and Chemicals

All chemicals: Bisphenol A (99% purity) reagent, ferric chloride hexahydrate ($FeCl_36H_2O$), ferrous sulfate heptahydrate ($FeSO_47H_2O$), and sodium hydroxide (NaOH) (AR) utilized in this study were of analytical grade and purchased from Sigma-Aldrich (West Chester, PA, USA) and applied without additional purification. Distilled water (DW) was utilized for the preparation of all the aqueous solutions during the entire experiment. The BPA solution utilized in the experiment was carefully prepared prior to each experiment to avoid possible microbial degradation. The chemical and physical properties of BPA are presented in Table 1. The PKS biochar was obtained from AMR Environmental Sdn Bhd, located at Johor, Bahru, Malaysia.

Table 1. Properties of Bisphenol A (BPA).

Compound Name	Lipid-Water Partition Coefficient (Log K_{ow})	Molecular Mass (g/mol.)	Chemical Structure	Molecular Formula	pKa
Bisphenol A	3.32	228	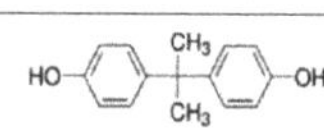	$C_{15}H_{16}O_2$	10.1

2.2. Synthesis of Magnetic Biochar

The impregnation of biochar with magnetite using co-precipitation of ferric and ferrous salts on carbonaceous materials method was according to the procedure employed in previous studies [16]. Initially, the PKS-BC sample was grounded and modified using high energy ball milling (HEBM) for 3 hrs to achieve a nano-sized biochar. The obtained ball-milled powdered BC was passed through 50–63 µm sieves and washed with distilled water and ethanol three times. Afterwards, the PKS-BC was oven-dried overnight at 110 °C. The obtained dried BC was collected in a bottle container, sealed, and labelled as NBC and kept in a desiccator for further use. In the next phase, 8.5 g of the dried PKS-BC was soaked in a 100 mL of $FeCl_3$ of 0.25 $mol{\cdot}L^{-1}$ and 0.125 $mol{\cdot}L^{-1}$ of $FeSO_4$ solution and uniformly mixed via the magnetic stirrer. The mixtures were subjected to heat at 60 °C with constant stirring to ensure that Fe^{3+} and Fe^{2+} could saturate into BC and later allowed to cool to 40 °C. Subsequently, 100 mL of 1 $mol{\cdot}L^{-1}$ of sodium hydroxide (NaOH) solution was dropwisely added to the solution and stirred until the pH attained the range of (10–11), and the colour of the solution transform from brownish green to black. Fe_3O_4 precipitation was formed on the surface of carbon materials under this alkaline condition. Upon the completion of the reaction, the NaOH-impregnated biochar solution was adequately stirred for 60 min to achieve better dispersion and homogenous mixture. The resultant mixture was separated by an external magnet and rinsed with distilled water until the solution pH closed to 7 and then oven-dried at 70 °C for 12 h. The final modified BC sample was grounded and collected into a bottle container, sealed, and labelled as MBC and kept in a desiccator thereafter.

2.3. Characterization of Neat and Magnetic Biochar

The magnetic properties and magnetization curve of MBC was determined using a vibrating sample magnetometer (Lakeshore 7404, Westerville, OH, USA) with an applied field between $-10{,}000$ and 10,000 Oe at room temperature, from American Quantum.

Surface functional groups of the BC samples (before and after modification) were confirmed using Fourier transform infrared (FTIR) spectroscopy (Perkin Elmer, 1650 Spectrometer) within the scanned range of 400 to 4000 cm^{-1} using the attenuated total reflection method.

The textural properties such as surface area, pore volume, and pore size distribution of the samples were determined by the standard N_2 physisorption procedures using Micrometrics analyzer (Tristar II Plus model) and determined in line with the Brunauere Emmentte Teller (BET) technique with degassed temperature of 350 °C.

X-ray diffraction (XRD) was used to analyse the phase and chemical composition of the prepared BC adsorbents at Cu Kα radiation (2θ spectrum = 20–80°; phase = 0.05° 2θ; time per step = 0.2 s) via X-ray powder diffractometer (Rigaku MiniFlex 600). The compositions and surface morphologies of the samples were detected by field emission scanning electron microscopy (FESEM, Zeiss ULTRA 55) and energy dispersive spectroscopy (EDS, Bruker/Quanta 200), Westerville, OH, USA).

Also, the pH of BC was assessed viz: BC was blended with distilled water at (1:10) mass ratio, agitated magnetically for 30 min, and subsequently kept for 60 min. Afterwards, the pH of BC was determined by a pH5S Spear pH tester (T531009086, Shanghai, China).

2.4. Analysis of Surface Chemistry of Biochar (pH_{pzc})

The surface chemistry of neat and magnetically-modified biochar was performed experimentally using the pH at the point of zero charge according to the procedure of [56] with slight modifications. Ten samples of varying pH (2.0–11.0) were prepared using a 0.01 M aqueous solution of sodium acetate ($C_2H_3NaO_2$) as the base electrolyte. The pH of the solution was adjusted using either 0.1 M of NaOH or 0.1 M HCl aqueous solution. Then, 0.1 g of magnetic and neat samples of BC were carefully added to each 20 mL of the prepared solutions in a conical flask placed in a swing agitator at a speed of 120 rpm at 298 K room temperature and stirred for 48 h. Afterwards, the resultant supernatant was then decanted, and its pH was determined. The value of pH_{pzc} was computed from a plot of pH of the initial solution against pH of the supernatant [57,58]. Thus, the zeta potentials of neat biochar (NBC) and magnetic biochar (MBC) were examined via zeta potential analyzer (Zetasizer Nano Plus 3, Zeta/nano particle analyzer, Malvin, Austin, TX, USA).

2.5. Adsorption Experiment

Definite amount of NBC and MBC adsorbents (50.0 mg) and 125.0 mL of BPA solution were weighed and added to a set of 250 mL; Erlenmeyer conical flasks sealed externally using aluminium foil to avoid probable photodegradation. The mixture was then placed on an orbital shaker (HY-8, Shanghai, China) and agitated mechanically at 160 rpm at room temperature for 24 h to achieve equilibrium condition. Upon the completion of the agitation process and at pre-specified times, the solution was rapidly removed and filtered using 0.4-µm membrane and the absorbance of the supernatants were analyzed spectrophotometrically at a wavelength of 276 nm using a double beam UV-visible spectrophotometer (Shimadzu UV-1800, Kyoto, Japan) with a pair of 10 mm matched quartz cells, and the concentrations were converted into the established BPA calibration (standard) curve [59,60].

In the single-variable experiment, the effects of adsorbents (BC) doses (0.2–0.8 g; BPA = 20 ppm), pH range were adjusted by 0.01 mol/L HCl and 0.01 mol/L NaOH solution (3.0–12.0; BPA = 20 ppm), contact time (20–240 min; BPA = 20 ppm), ionic strength (Na^+ concentration: 0, 0.2, 0.4, 0.6, 0.8, and 1.0 mol/L, BPA = 20 ppm); initial BPA concentrations (8–150 ppm), and reaction temperature (298.15, 318.15, and 338.15 K), on the adsorption of

BPA by NBC and MBC were studied. The amount of BPA adsorbed on BC adsorbent was determined using the following Equation (1):

$$q_t = \frac{(C_S - C_t)}{M} \times V \tag{1}$$

where, C_s and C_t are the starting and residual concentrations of BPA in mg/L, M is the mass of the adsorbent (g); and V represent the solution volume (L).

For the adsorption thermodynamic tests, 0.5 g of the adsorbent and 100 mL of BPA solution were added to 250 mL conical flask to study the adsorption thermodynamic of MBC for BPA. The mixtures were stirred in a temperature-regulated incubator shaker (Excella E24 Incubator shaker series, USA) at a speed of 150 rpm, pH (6) for 60 min adsorption time, and the adsorption temperature were varied between (298.15 K–338.15 K).

2.6. Statistical Analysis

For the adsorption experiment, each experimental samples and group were carried out in triplicate, and the average value of the data was considered. The removal efficiency of BPA ($Re\%$) and quantity of BPA adsorbed (q_e) onto BC adsorbent were determined.

The adsorption isotherms, kinetics (20–240 min), and thermodynamics (298.15, 318.15, and 338.15 K) were studied. Each test was carried out in triplicate, and the average results were considered.

2.7. Reusability and Regeneration of MBC

The reusability studies on the prospect of desorbing of BPA compounds from MBC are indispensable based on environmental and economical perspective, in addition to industrial applicability [61]. With the purpose of achieving regeneration of MBC, an efficient and green eluent was considered. The reusability of MBC at the starting BPA concentration of 20 ppm was assessed by rinsing BPA-ladened MBC with 0.1-M NaOH and distilled water and subjected to ultrasonication, and then followed by oven-drying at 60 °C overnight after magnetic field separation [49]. The MBC was repetitively used five times, and the values of q_e was noted accordingly.

2.8. Adsorption Isotherm

Adsorption isotherm was employed to compute the quantity of adsorbed BPA on BC based on BPA concentration at a steady temperature. The application is based on standardizing the amount of BPA adsorbed by the mass of BC adsorbent and this confirms comparison with Isotherm models.

2.9. Langmuir Adsorption Isotherm

The Langmuir model relies on the assumption that uptake of BPA takes place on a homogenous surface via monolayer adsorption with no interaction between adsorbed materials. It also presumes that all sorption sites are 'correspondingly active', and the surface is robustly homogeneous [62]. Principally, the larger values of R^2 signify the significance of the adsorption model for contaminants removal in water. The value of R_L between 0 and 1 implies that the isotherm is advantageous. The isotherm is unsuitable if $R_L > 1$, linear if $R_L = 1$, irreversible if $R_L = 0$, and favorable if R_L lies between 0 to 1. Arithmetically, the Langmuir isotherm model is denoted in Equation (2):

$$\frac{1}{q_e} = \frac{1}{K_L q_{max.}} \cdot \frac{1}{C_e} + \frac{1}{q_{max.}} \tag{2}$$

where C_e represent BPA concentration at equilibrium (mg/L); $q_{max.}$(mg/g) is the maximum single-layer adsorption capacity of the BC adsorbent, q_e (mg/g) is the quantity of BPA adsorbed; K_L (mg/g) represent the Langmuir constant closely related to adsorption capacity for overall monolayer coverage; the intercept ($1/K_L$), slope ($1/q_{max.}$), and q_{max}. could be evaluated from a plot of $1/q_e$ versus $1/C_e$ for BPA adsorption onto NBC and MBC.

2.10. Freundlich Adsorption Isotherm

Freundlich isotherm model was employed to study the uneven distribution on the surface of the absorbent, which is heterogeneousness in the adsorption process [63]. The model indicates that the adsorption energy proportionally declines on the endpoint of the adsorption centres of an adsorbent [64]. The degree of correlation between adsorption and solution concentration relies on the adsorption intensity, n. The adsorption conditions can be chemical (n < 1), favourable physical process (n > 1) and favourable linear (n = 1). The K_F and n are Freundlich constants and were found from the graph as the intersection and slope, respectively. Scientifically, the Freundlich isotherm model can be described in Equation (3). The graph of ln q_e against ln C_e provide a linear plot with an intercept logK_F and slope 1/n, from which R^2, K_F and n, can be computed, respectively [65].

$$\log q_e = \log K_f + \frac{1}{n} \log C_e \tag{3}$$

3. Results

3.1. Results and Discussion

Characterization and Analysis of Synthesized Biochar Adsorbents

The surface chemistry and morphological characterization of BC sorbents are excellent indicators of adsorption propensity. Hence, VSM, FTIR, FESEM, EDX, and BET surface analyses, together with the point of zero charge of the adsorbents (NBC and MBC) were also evaluated.

3.2. Morphological Analysis of Synthesized Biochar

FESEM analysis was conducted to examine the surface morphology of the BC samples. The presence of abundant pores were observed on the surface of the BC, which was substantiated by the FESEM micrographs (Figure 1a–c). Images of NBC, MBC (before and after adsorption) are described in Figure 1b,c. The FESEM analysis of the MBC (prior to adsorption and after adsorption) was performed to observe the morphological structure along with the particle size and size distribution of MBC, as presented in (Figure 1b–e), respectively. From Figure 1a–c, the images showed the particles obtainable are spherical in shape with a single uniform aggregate and small non-uniform agglomerates. In Figure 1a, larger particles with a porous and rougher surface were noticed for NBC. Also, the particles possess more cleavages with few connected to bulks. Furthermore, it can be observed from the FESEM micrographs (Figure 1a,b) of NBC and MBC, that the surface of NBC is relatively coarse and rough since the spherical shaped particles with particle size more than 28.32 nm were agglomerated and attached to each other and the pore form is not fully developed. Contrastingly, the surface of MBC is comparatively smooth, with substantial porous structure than NBC, and the surface is filled with numerous nano-iron oxide particles. The dispersal of Fe_3O_4 nanoparticles on the surface of BC is relatively uniform. The surface of MBC developed shinier and smoother surface after adsorbing BPA (Figure 1c). The average particle size of MBC prior to adsorption as depicted in Figure 1d was computed to be 28.32 nm. Conversely, after adsorption of BPA, the average particle size rose to 46.65 nm as showed in Figure 1e, confirming that the adsorption process was taking place.

3.3. Elemental Analysis of Synthesized PKS Biochar

For this procedure, 1.0 g of each of the samples (NBC and MBC) was utilized for the EDX analysis. This test studied different component elements existing in each of the BC samples. The EDX test performed reveal the elemental compositions and distribution both in the NBC and MBC. Figure 2 presents the elemental analysis of the NBC and MBC. As revealed in Figure 2, the existence of peaks matches with the carbon (C), oxygen (O), iron (Fe), and silicon (Si) elements. The intensity of C peak is greater than O, Si, Al, and Fe, demonstrating the elevated content of C in NBC (Figure 2a). Also, three peaks displayed at

0.2, 0.5, and 6.40 keV to substantiate the binding energies of Fe_3O_4 nanoparticles [66]. The elemental composition of novel MBC was presented in Table 2 with the mass ratio of C, O, and Fe was 32.52, 27.23, and 38.14%, respectively.

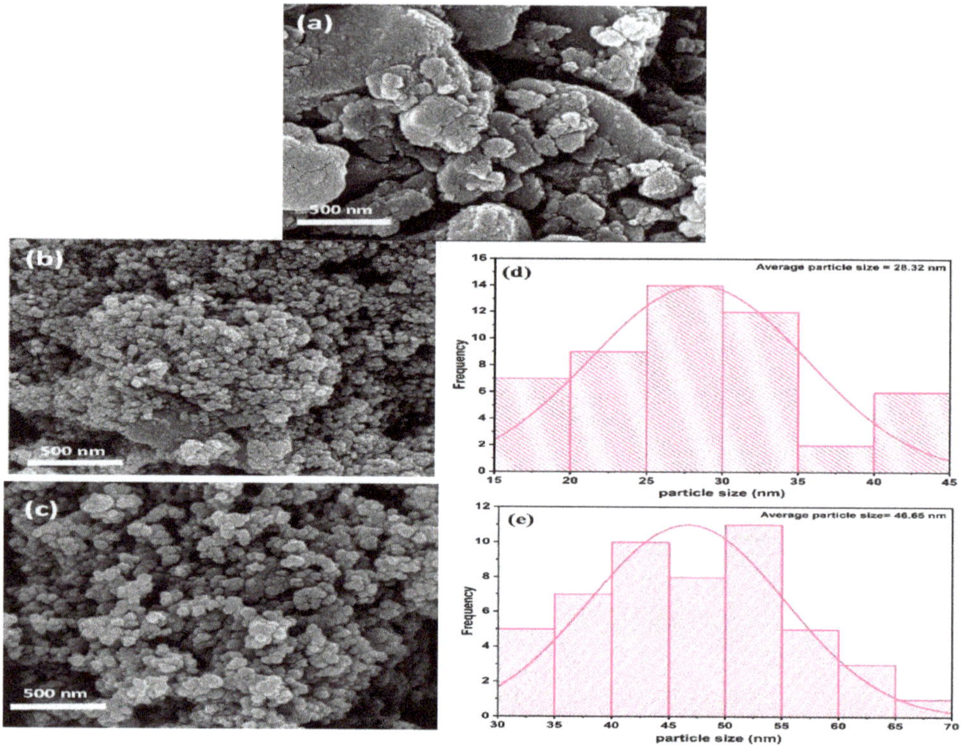

Figure 1. FESEM (field emission scanning electron microscopy) images of NBC (**a**), MBC before adsorption (**b**), magnetic biochar after adsorption (**c**), average particle size distribution before (**d**), and after adsorption (**e**).

Table 2. Elemental constituents of magnetic biochar.

Elements	C	O	Si	Fe	Total
Mass ratio (%)	32.52	27.23	2.11	38.14	100.00
Atom ratio (%)	52.40	32.93	1.45	13.22	100.00

The elements of NBC include C, O, Si, Al, and Fe, while the element observed in MBC were C, O, Si, and Fe. Noticeably, from Figure 2b,c, the Al element entirely disappeared after the magnetic modification, and also C, and Si weight components were significantly reduced. This indicates that chemical reactions occur during the modification. Conversely, Fe and O weight composition were substantially increased in modified MBC as compared to NBC. Particularly, the oxidation process considerably influences the obtainability of C active sites, in addition to the structural reformation of the adsorbent [67]. This indicates that the oxidation activity of the carbonaceous material justified the reduction in the weight composition of the trace elements, and also the formation of more acidic oxygen-containing functional groups on the surface [68]. These processes play a significant role in the development of more active sites for effectual adsorption process. The mass fraction of Fe in MBC was 38.14%. The high composition of Fe_3O_4 further proved the success of the magnetic modification of BC, which agreed with the FTIR and XRD analysis.

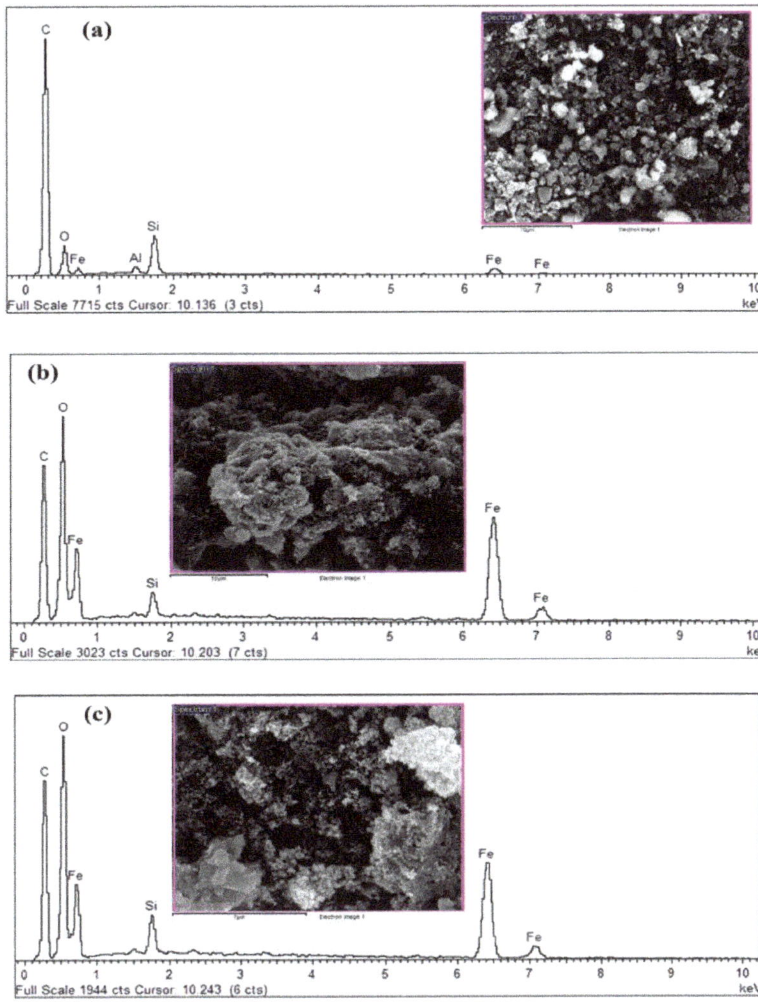

Figure 2. EDX spectra of the as-synthesized NBC (**a**), and MBC prior to adsorption (**b**), and after adsorption (**c**).

3.4. BET Surface Area Analysis of Biochar

The nitrogen adsorption isotherm of NBC and MBC is presented in Figure 3. Essentially, the physical factors of BC samples that can influence the adsorptive removal of organic contaminants comprises of effective surface area, total pore diameter and pore volume. Surface area properties of the NBC and MBC were assessed via BET analysis. The BET results illustrate the adsorbate–adsorbent relationships of that adsorbed molecules which are gathered around on the surface of MBC [69]. The results of effective BET surface area, pore diameter and pores volume for both NBC and MBC are presented in Table 3. Noticeably from Table 3, the NBC exhibited a superior surface area of 536.54 m^2/g, while MBC had a surface area of 362.07 m^2/g. The decline in the surface area of MBC was possible since the MBC was comprised of a moderate surface area of Fe_3O_4 and an elevated surface area carbonaceous, and superfluous iron (Fe) nanoparticles loading could capture and clog some of the active sorption sites and pores of the NBC [70,71]. This reduction in surface area of MBC composites as compared to NBC has also been reported in previous studies [72,73]. Similarly, the pore size and pore volume of MBC are higher than that of NBC, which is better beneficial for the sorption of BPA.

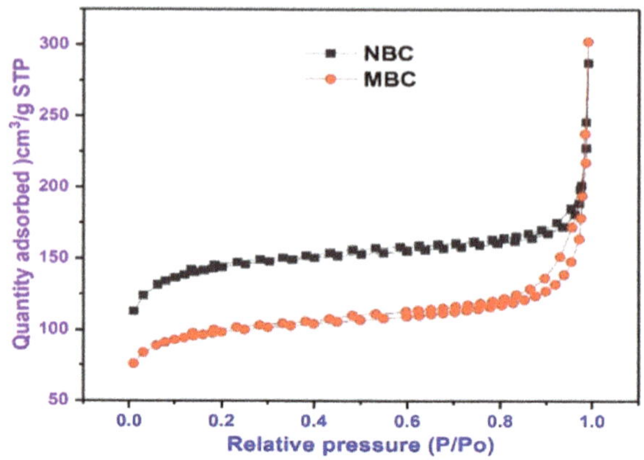

Figure 3. Nitrogen adsorption–desorption isotherm.

Table 3. Physico-chemical properties of neat biochar and magnetic biochar.

Materials	Pore Diameter (Å)	Pore Volume (cm^3/g)	BET Specific Surface Area (m^2/g)	pH_{pzc}
NBC	15.516	0.416260	536.5398	4.829
MBC	24.427	0.442203	362.0673	5.612

Similarly, the total pore volume of NBC and MBC found were, respectively, 0.416260 cm^3/g, and 0.442203 cm^3/g, while MBC modified by Fe_3O_4 exhibits a better porosity, and presents a substantial capacity of pollutant adsorption, while after modification with Fe_3O_4, the existence of Fe_3O_4 between layers of MBC upsurges the heterogeneity of the adsorbent, thereby resulting in a superior porosity [74]. The impregnation of magnetite on BC has no significant influence on the pore-volume, surface area, and mean aperture. Iron-amendment can either decrease or increase the surface area of an adsorbent and this relies on the initial surface area value of the adsorbent and the proportions of the Fe_3O_4 particles [75]. The infrequent blocking of surface micropores from magnetite groups may also be responsible for the insignificant reduction in surface area. Thus, the adsorption capacity of modified biochar material is improved. This finding was further corroborated by the FESEM images and EDS spectra. Figure 3 displays the N_2 adsorption–desorption isotherms of N_2 at 77 K of NBC and MBC. The findings revealed that at moderately high relative pressures, the adsorption isotherm rose relatively than at relatively low pressures. This implies that the major adsorption takes place at moderately increased pressures and suggests that the material is highly porous with a narrow size distribution.

3.5. Magnetic Properties of As-Synthesized Magnetic Biochar

The vibrating sample magnetometer (VSM) procedure was used to determine the hysteresis loop at room temperature. Most importantly it is employed to quantify the magnetic properties of a materials with respect to magnetic field, time, and temperature. Besides, VSM analysis provides information about whether the magnetization is perpendicular or parallel to the plane described by the substrate. The hysteresis loop of the synthesized MBC was obtained by plotting the magnetization (emu/g) against the magnetic field (Oe). Figure 4 reveals the magnetic hysteresis loops used to analyze the magnetic properties of MBC. As illustrated in the loop observed in Figure 4, the magnetization sharply increased with the decrease of the mass ration of BC. The curve signified distinctive super-paramagnetic properties. The synthesized MBC exhibited magnetization M values of 6.4882 emu/g which revealed that the magnetization of MBC can be altered by the mass ration of BC: magnetite (Fe_3O_4) nanoparticles. The super-paramagnetic properties was triggered by the alteration of BC with magnetite (Fe_3O_4) nano-particles, which could ensure the MBC can be readily recovered from the suspended solution through external magnetic field, which makes the replicated use of the MBC in the actual wastewater purification system viable [72]. Thus, novel MBC was easily detached using an external magnetic field, as illustrated by the inset. The result of magnetization obtained in this study is consistent with previous studies reported from [76–78].

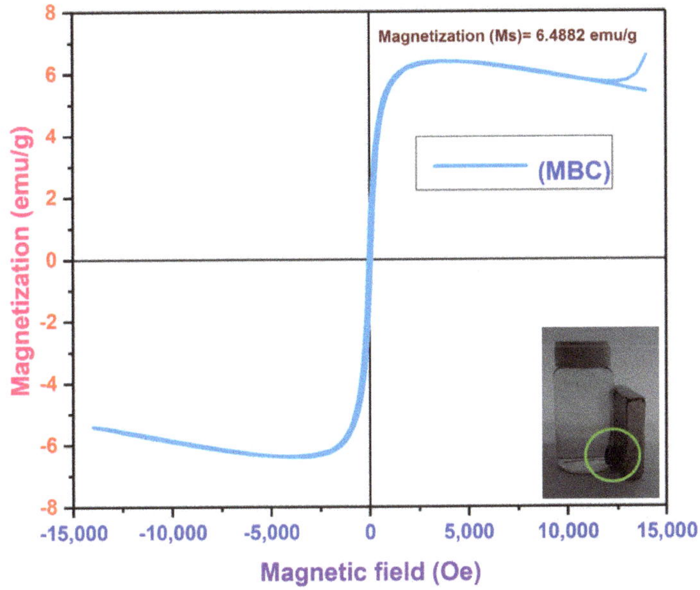

Figure 4. Magnetization curve for MBC.

3.6. XRD Analysis of Synthesized Biochar

X-ray diffraction analysis of a material describe the size and the nature of the planes of the synthesized material. The crystalline structures of NBC and MBC were analyzed using XRD as indicated in Figure 5. XRD is an efficient technique to verify the presence of Fe_3O_4 in the synthesized MBC [79]. Figure 5 reveals the XRD patterns of both NBC and MBC. The diffraction spectra of the synthesized BC samples showed the existence of magnetite (Fe_3O_4). The diffraction peaks at 2θ of 30.044°, 36.66°, 36.57°, 42.45°, 57.26°, 61.54°, and 62.76° are indexed to the (200), (016), (220), (232), and (040) hkL planes, respectively, which correspond satisfactorily with the database of Fe_3O_4 standard card Inorganic Crystal Structure Database (ICSD No. 98-007-7864) with a space group of P12/c1 and lattice parameter (a = b = c) of 28.644 Å and confirms the signature peaks of a hexagonal unit cell Fe_3O_4, respectively. No impurity peak is observed in the XRD pattern, which indicates that the Fe_3O_4 particles are highly crystalline hexagonal spinel structure. As observed from the spectra, all the diffraction peaks are designated to the magnetic hexagonal structure. There are no other peaks associated with another material detected from the XRD result, which confirmed that the nano-magnetite is pure magnetite (Fe_3O_4). The XRD spectra of Fe_3O_4 in this study is analogous to other studies from [80,81], and both Fe_3O_4 and BC patterns were overlapped in a XRD spectra of MBC, demonstrating successful synthesis of MBC composite.

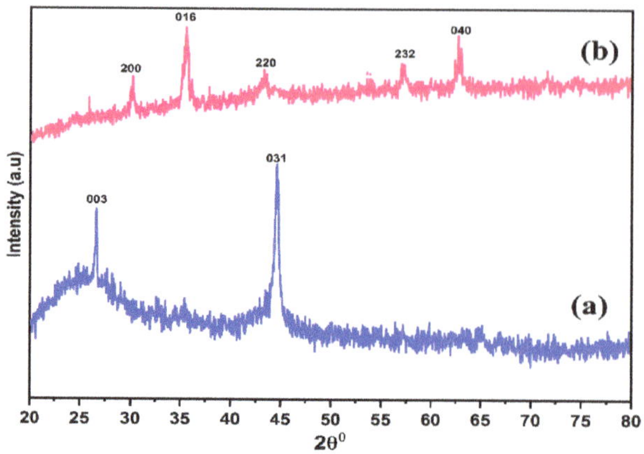

Figure 5. X-ray diffraction (XRD) patterns of NBC (**a**) and MBC (**b**).

3.7. Analysis of Functional Group

The surface functional groups are the central chemical variable of BC material that influence the BPA adsorption. FTIR spectroscopy was employed to substantiate the modification process and acquire the information on the existence of different functional groups on the surface of the material [82]. The spectrum results and various bands in the spectra signifying vibration of functional groups were illustrated in Figure 6 for NBC and MBC, respectively. During the $Fe^{3+}/Fe^{2+}/NaOH$ process, the surface of the char aids the nucleation of iron oxide precipitation. As indicated in Figure 6, the existence of functional groups of –COOH and –OH group are responsible for the binding of the iron oxide and iron hydroxide particles in the solution, and then bonded to the char [83]. This could be symbolized as char-O-Fe_xO_y (Fe_3O_4/Fe_2O_3). In this context, the bold 'O' was initially a hydroxyl group on the surface and signify a chemical bond between iron oxide particle and the char phases. This bonding may conceivably be combined with some mechanical interlocking between H-bonding (such as between Fe–OH at the metal oxide surface with C–OH of the char surfaces) and phases in addition to some columbic interfaces. These interactions could strongly bind magnetic iron oxide to the BC materials [84].

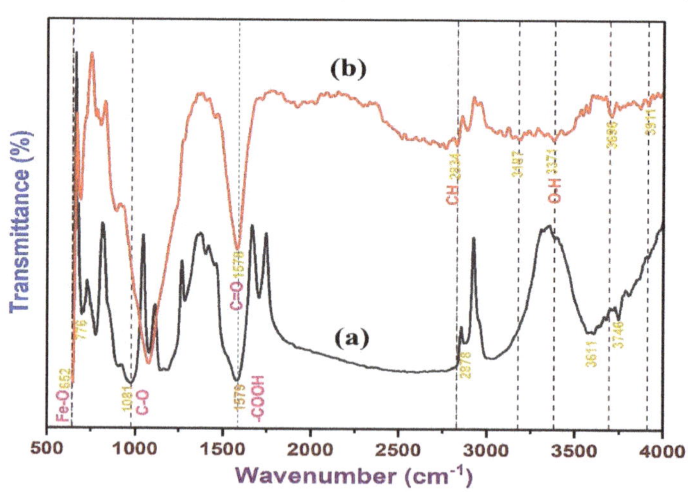

Figure 6. Fourier transform infrared spectra of NBC (**a**) and MBC (**b**).

Also, as indicated in Figure 6 of the FTIR spectra, there are limited peaks that surfaced for both NBC and MBC. The peaks at 2834, and 2878 cm^{-1} were distinctive peaks of the C-H bond [82]. Similarly, the peak at 1081 cm^{-1} was the C-O stretching vibration in the composition of carbohydrate, polysaccharide, or aromatic ether. The peaks close to 1578 and 1579 cm^{-1} were the stretching vibration peaks of C=C, C-H, and of C=O on the aromatic ring [85]. The results go along with the BC characteristics that not only contain porous structure but also abundant active adsorption sites for BPA removal [86]. Besides, a noticeable peak at 652 cm^{-1} was observed for MBC, which correspond to the typical stretching peak of Fe-O [87]. This indicated that the Fe_3O_4 nanoparticles have been efficaciously loaded on the surface of BC as revealed by the FESEM results. The spectra results revealed that magnetic alteration improved the varieties and number of functional groups of the BC, which may influence the adsorption of BPA.

3.8. Determination of Electrokinetic Charge (pH_{pzc}) of Biochar

pH_{pzc} is the pH at which the net surface charge on the surface of an adsorbent is zero. It is a critical variable in examining the efficacy of the adsorption systems. When the pH is greater than the pH_{pzc}, the adsorbent surface acquires negative charge, hence repelling or attracting organic contaminants, in line with their cationic or anionic functional groups [88]. The point of zero charges was investigated to describe the surface chemistry of MBC and NBC. The values of point of zero charge (pH_{pzc}) of neat and MBC in relation to the solution pH were computed and displayed in Figure 7. According to Figure 7, the pH_{pzc} at the point where the change in pH (initial pH-final pH) equal to zero for the MBC and NBC were 5.61 and 4.81, respectively. It is observed that the pH_{pzc} of MBC (5.61) is higher than that of NBC (4.81). This increase may be due to the introduction of iron oxides (Fe_3O_4) on the surface of NBC [89]. Hence, higher pH_{pzc} is more beneficial for the adsorption of BPA anions, which are often existed in acidic solution [90]. The zeta potential of all samples steadily reduced with rising pH indicating more net negative surface charge at elevated pH. Similarly, the zeta potential of both NBC and MBC was both positive and negative within the whole pH range studied (2–11) as can be seen from Figure 7. Evidently, both MBC and NBC are positively charged when the pH is lower than 6.0. Once the pH is increase, the surface charge of the two biochar materials swiftly transformed from positive to negative. As pH consistently increase, the surface charges of both NBC and MBC risen and, noticeably, the zeta potential of NBC and MBC were −28.56 mV and −22.47 mV, which shows that both biochar samples were negatively charged (see Figure 7). Though, the surface of NBC is slightly more negatively charged, and the electrostatic attraction between the negatively charged NBC and HBPA improves the adsorption. This finding is in good agreement with previous studies [16,91].

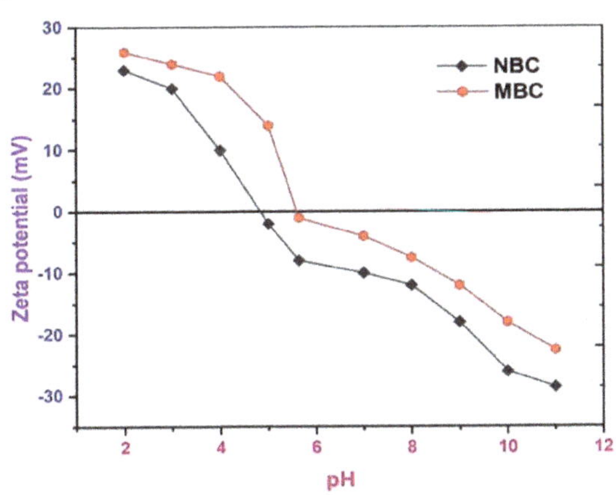

Figure 7. Zeta potential of NBC and MBC at different pH.

3.9. Influence of Working Conditions on the Adsorption of BPA

The operating variables for instance contact time, solution pH, adsorbent dosage, ionic strength, BPA concentration, and other variables can influence the surface properties of the adsorbent surface and its BPA binding ability [88,92]. Hence, a batch adsorption test was carried out on the as-synthesized BC adsorbents to study the influence of these operating variables on the adsorption capacity and removal efficiency for BPA uptake.

3.10. Effect of pH on BPA Adsorption

The pH is one of the main parameters that control the removal of compounds present in aqueous environment using solid adsorbents. The solution pH is among the most essential variables that determine the elimination of various compounds existing in aqueous environment utilizing solid adsorbent materials and the optimization of adsorption process. The effect of pH on adsorption was reliant on the target contaminants and nature of absorbent [42]. It influences not only the speciation of the adsorbate, but also the level of ionization and adsorbent surface charge [90,93,94]. Accordingly, most of the investigations involved in contaminants adsorption onto BC strongly considered the effect of solution pH. To investigate the influence of different solution pH upon adsorption of BPA on the surface of NBC and MBC, the tests were performed in the pH range of 3.0–12.0, BPA concentration 20 ppm, adsorbent dose 0.5 g, and temperature 20 °C, and the results were illustrated in Figure 8a. As indicated in the figure, the sorption of BPA by NBC and MBC is clearly pH reliant. The highest adsorption capacity of NBC to BPA take place at pH (3.0), which is attributable to the development of electron receiver-giver interaction (ERG) between BPA and NBC, together with a robust hydrogen bond [95,96]. When pH rose from 6.0 to 7.0, the adsorption capacity of NBC to BPA slightly increased further, since BPA starts to moderately dissociate, and BPA in the solution is no more in molecular form, nevertheless few $HBPA^-$ still occurs. The electrostatic attraction between NBC and $HBPA^-$ enhances the adsorption process due to the positively charged surface of NBC. Further surge in pH led to a subsequent reduction in the adsorption capacity of NBC to BPA, owing to weakness in the interaction of the hydrogen bonding and π-π electron receiver–giver(ERG) between BPA and NBC when the solution pH is higher than the BPA acid dissociation constant [97]. Conversely, the adsorption capacity of BPA by MBC rose initially and after that it declined with the upsurge of pH greater than 6.0, which implied that the adsorption of BPA on BC was largely dependent on the solution pH of the system. Thus, the considerable rise in BPA adsorption on MBC between pH 3.0–6.0 was possibly attributed to the electrostatic interaction between the positively charged $HBPA^+$ species and negatively charged surface of MBC in the solution. The surface of MBC may develop positively charged at low pH because of the protonation reaction ($H^+_{(aq)} + -ROH_{(surf)} \leftrightarrow -ROH^{2+}_{(surf)}$) on the surface of MBC [98]. When the solution pH is increased, a negatively charged surface of MBC with plentiful active biding sites emerge owing to deprotonation reaction ($OH^-_{(aq)} + -ROH_{(surf)} \leftrightarrow H_2O + {}^-RO^-_{(surf)}$) on the surface of MBC. Hence, the enhanced BPA adsorption by MBC at pH 3.0–6.0 was not only controlled by the electrostatic interaction mechanism [99], but possibly caused by the surface reduction/complexation of BPA species onto MBC, and the decline of BPA adsorption at pH >6.0 following the repulsion interaction between negatively charged MBC and dissociated bisphenolate anions (BPA^{2-} and $HBPA^-$) species. The maximum adsorption capacity take place at pH of 6.0.

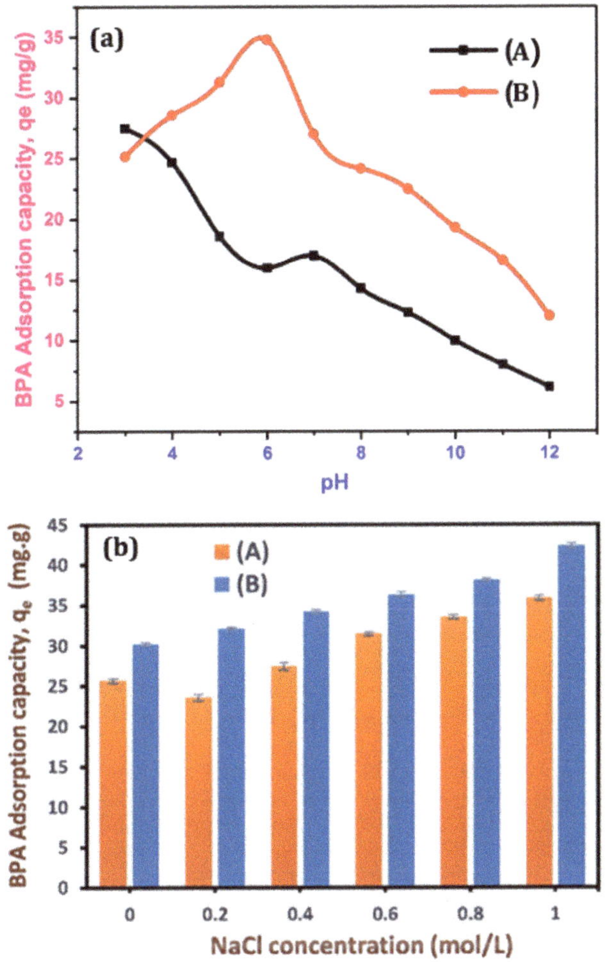

Figure 8. Effect of solution pH (**a**), and ionic strength (**b**) on the adsorption of BPA by NBC and MBC. (BPA concentration: 20 ppm; contact time: 60 min; biochar dose: 0.5 g; and temperature: 293.15 K).

3.11. Effect of Ionic Strength on BPA Adsorption

Generally, the water body comprises such a complex system that salts and organic contaminants frequently coexist in wastewater which may influence the removal of the contaminants. Hence, the profile of ionic strength influence on BPA adsorption by the as-synthesized BC was also investigated using 0–1.0 mol/L NaCl, BPA concentration 20 ppm, and the experimental findings are presented in Figure 8b. Initially, it could be noticed in Figure 8b, the increase in NaCl concentration led to a decline in the adsorption capacity of NBC to BPA and later a rise. Noticeably, the surge in high ionic strength (Na^+ concentration) considerably increased the adsorption capacity of BPA to MBC (Figure 8b). The BPA adsorption capacity of BC increased when the Na^+ concentration rose from 0 to 1.0 mol/L. Analogously, Zhou et al. [100], found that the upsurge in ionic strength with various ionic species ($CaCl_2$ or NaCl) yielded increase in the adsorption capacity of BPA when peat was applied as an adsorbent. Similarly, erstwhile study has also reported that the increased ionic strength could improve the adsorption of organic contaminants into carbonaceous adsorbents, such as BC, owing to the screening effect of the surface charge generated by the addition of salt [101]. The BPA adsorption capacity by BC adsorbents rose steadily with the increase in the NaCl concentration. The plausible explanations for this surge could be due to the penetration of ions into the diffusion dual layer around NBC and MBC surfaces and lessen the repulsion between the adsorbents, thereby stimulating the squeezing-out effect (i.e., nanoparticle aggregation), which caused a decline in the adsorption capacity of BPA [102], enhanced activity coefficient of hydrophobic organic contaminants, which leads to a salting-out effect (reduction in solubility), and, therefore, was favorable to the adsorption of BPA [103]. Thus, the decreased adsorption capacity when the NaCl concentration is at 0.2 mol./L might be caused by the competition between BPA and the low concentration of salt solution for the accessible active sites of adsorbents, corroborating that the squeezing-out effect was stronger than the salting-out effect. Conversely, as the NaCl concentration increases, the adsorption capacity enhanced evidently, suggesting that the salting-out effect improved unceasingly [104].

3.12. Effects of Biochar Dose on the Removal of BPA

Adsorbent dosage (AD) is an important parameter in an adsorption process. It controls the adsorbate–adsorbent equilibrium of the adsorption procedure. The removal efficiency based on the adsorbent dosage was studied at other test conditions, BPA concentration 20 ppm, varying dosages (0.2–0.8 g), pH 6.0, contact time (60 min), and adsorption temperature 20 °C. The influence of BC dose on the adsorption of BPA by NBC and MBC is illustrated in Figure 9. The removal rate of BPA enhanced with the rise in BC dose. When BC dosage was 0.2 g, the removal efficiencies of BPA by NBC and MBC were 48.45% and 60.0%, respectively. Similarly, when the BC dose rose to 0.5 g, the removal efficiency substantially rose to 85.97% and 94.2%, respectively. The increased removal rate is because of the augmented BC, which also expands the operational specific surface area of adsorption, together with increase in the active pore (binding) site of adsorption [105]. This outstanding adsorption performance was attributable to the presence of plentiful active sites and various interactions between BC adsorbent and BPA, specifically ion exchange and complexation electrostatic interaction. Afterward, further increase in BC dosage has no influence on the removal rate. This is because surplus adsorbent overlaps the effective active sites on BC and competes for limited solutes [42].

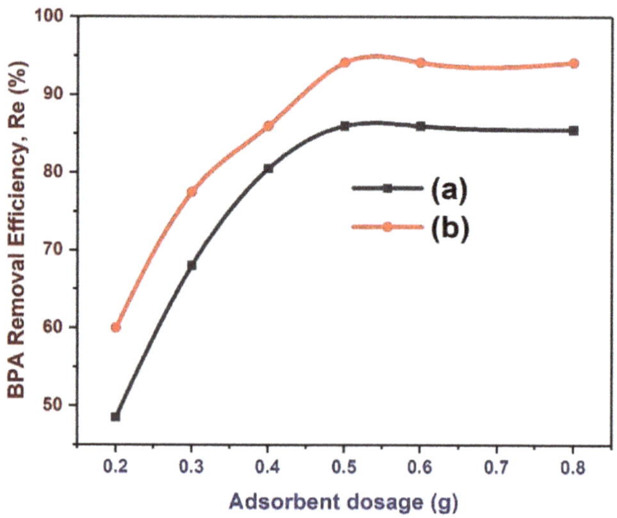

Figure 9. Effects of BC (a: NBC, b: MBC) dose on the removal of BPA. BPA removal efficiency by BC under varying doses: 0.2–0.8 g; BPA concentration: 20 ppm; temperature: 293.15 K; pH: 6.0; and contact time: 60 min.

3.13. *Effects of Initial Concentration and Contact Time on the Removal of BPA*

The influence of initial BPA concentration and contact time on the adsorption of BPA from aqueous solution was investigated. The experiments were performed in the presence of a constant dose of adsorbent (0.5 g) of BC at room temperature, at pH (6.0) with various initial concentrations of BPA in the stock solution varied between 8.0 to 150.0 ppm, at different time intervals up to 240 min. The result of effect of initial concentration on BPA removal is presented in Figure 10. The initial concentration offers the stimulating force required to overcome the mass transfer wall between the adsorbate and the adsorbent media [106]. Thus, a higher initial concentration may enhance the efficacy of the adsorption process. Evidently, BPA removal is lower at a small concentration due to smaller amount of adsorbates in the solution to dominate active sites on the adsorbent and the amount of BPA adsorbed rises with the upsurge in BPA concentration. Thus, increase in initial BPA concentration, led to gradual rise in the removal efficiency of BPA, since the increased BPA concentration can improve the adsorption drive between the solute and the adsorbent [107,108]. The rise in the concentration of BPA conforms with an increase in BPA removal as shown in Figure 9. When equilibrium is attained, the adsorbent turns out to be saturated. This result is in line with previous study reported by Wang and Zhang [16].

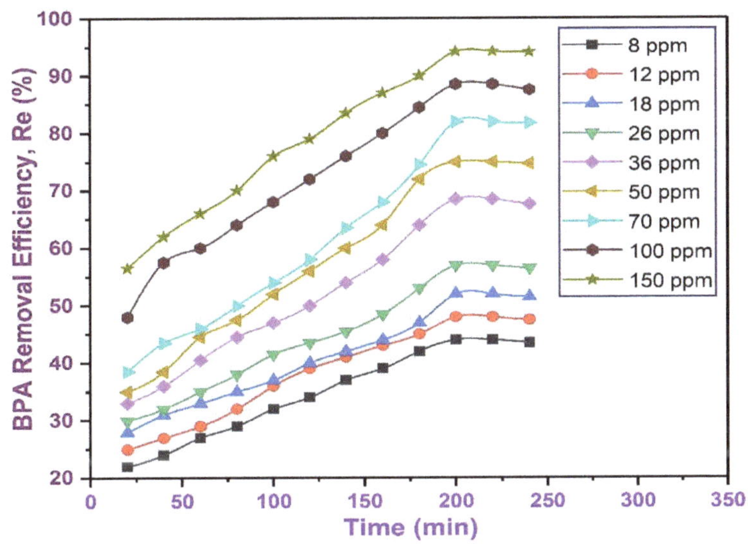

Figure 10. Effects of initial BPA concentrations on the removal efficiency under varying initial concentrations (8–150 ppm) at constant pH (6.0), dosage (0.5 g), and varying contact times (20–240 min).

The test conditions evaluating the influence of contact time on the adsorption of BPA at the test conditions, BPA concentration 20 ppm, adsorbent dosage 0.5, and pH 6.0. Similarly, the result of the effects of contact time on the adsorption capacity and the removal efficiency are illustrated in Figure 11. Noticeably, more than 93% of BPA became adsorbed in about 60 min. For contact time beyond 60 min, the per cent removal of BPA remains stable, because the active sorption sites have been saturated on the adsorbent surface.

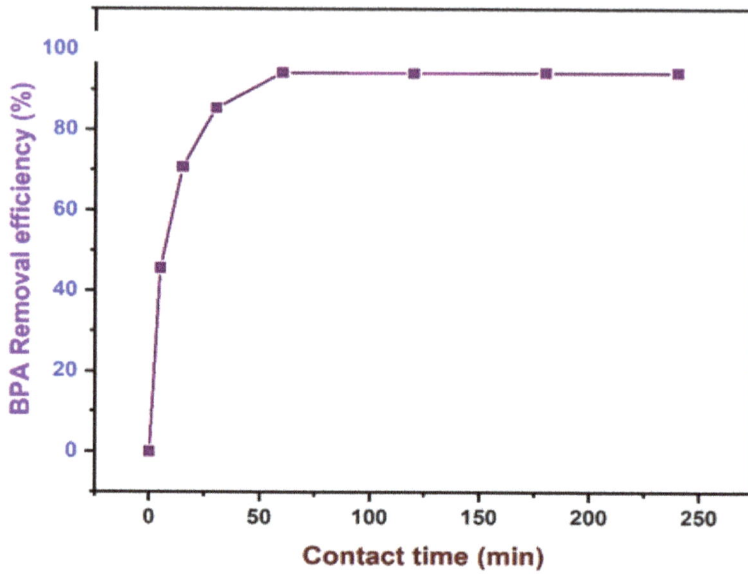

Figure 11. Effects of contact time (20–240 min) on the per cent removal of BPA, at 0.5 g dosage, BPA concentration 20 ppm, temperature (293.15 K), and pH (6.0).

3.14. BPA Sorption Isotherm Study

Sorption isotherm models are indispensable for recognizing the mechanisms of the adsorption process based on fundamental characteristics and numerical derivations. Also, the isothermal models are used to study the interrelatedness between the adsorbate and the adsorbent. The Freundlich and Langmuir isotherm models are normally employed and were fitted to the adsorption experimental data. The Freundlich isotherm is appropriate to both multilayer and monolayer adsorption and assumes that the adsorbates are adsorbed onto the heterogeneous surface of an adsorbent [109]. The Langmuir isotherm assumes monolayer adsorption on a uniform surface with a limited amount of adsorption sites [110].

The Langmuir isotherm model presumes that every molecule has a stable adsorption enthalpy and activation energy and signifies homogeneous adsorption. Correspondingly, the Freundlich isotherm model is experimental and regards the surface to be heterogeneous [111]. Two regular isotherm models were applied to fit the BPA adsorption isotherms on the as-synthesised BC (Figures 12 and 13). The models were employed to better investigate the BPA adsorption mechanism and performance. The fitting data are illustrated in Figures 12 and 13, and the applicable fitting parameters of these models were summarized in Table 3. The BPA adsorption experimental data were fitted to the Freundlich and Langmuir isotherm models with Origin 8.0.

The fitting of the experimental data into the isotherm models illustrates the adsorption process by the correlation coefficient R^2 and constants. To accomplish this, log q_e is plotted against log C_e based on the linear form of the Freundlich model (Figure 12) [112]. For the Langmuir model linear expression, $1/q_e$ versus $1/C_e$ is plotted (Figure 13) [113,114]. The computed q_{max} values of the Langmuir model from the graph of $1/q_e$ against $1/C_e$ was lower than the q_e values from the experiments (see Table 4), indicating that the experimental adsorption data fit this model. However, the R^2 value of Langmuir is lower than that of the Freundlich model.

The variables revealed that the Freundlich model (R^2 = 0.88964–0.9195) offers a superior fit to the data than the Langmuir model (R^2 = 0.7535–0.85608), as showed by the linear regression values. Additionally, the heterogeneity coefficient (1/n) of the two BC falls above 1 (1/n >1), which implied satisfactory physical adsorption process [115]. This trend was also corroborated by several studies [116,117] using BC synthesized from eucalyptus forest residues, sugarcane bagasse, castor meal, water hyacinth, green pericarp of coconut, and MBC to remove various contaminants. Hence, BC generated from various biomasses at different temperatures were observed to have various adsorption mechanisms and adsorption abilities [118]. In this work, BPA adsorption onto NBC and MBC was excellently fit using the Freundlich model in the determination of the efficacy of NBC and MBC for the removal of BPA, which demonstrates that the adsorption take place as a heterogeneous surface multilayer. A synopsis of the isotherm variables is presented in Table 4.

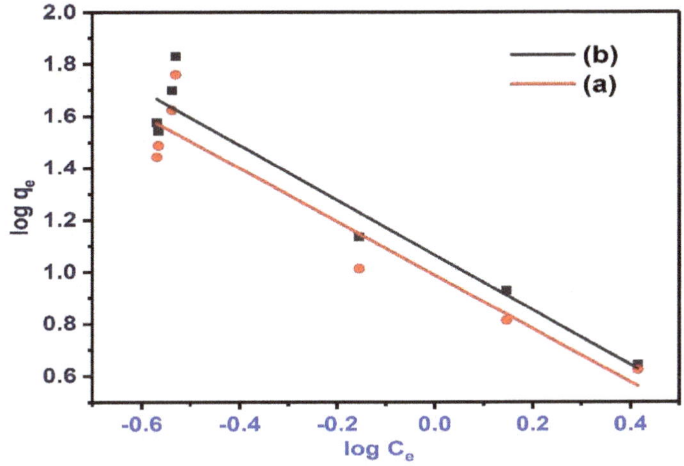

Figure 12. Freundlich isotherm for BPA adsorption onto NBC (a) and MBC (b) at pH 6.0, dose 0.5 g, contact time 240 min, and initial BPA concentration 50 ppm.

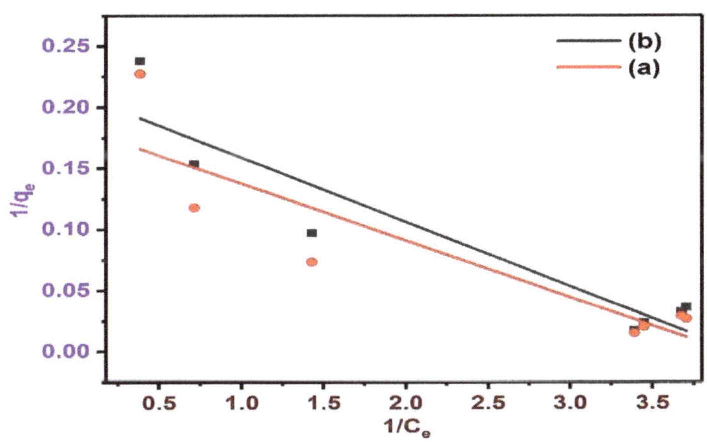

Figure 13. Langmuir isotherm on BPA adsorption onto NBC (a) and MBC (b) at 0.5 g dose, pH 6.0, contact time 240 min, and initial BPA concentration 50 ppm.

Table 4. Adsorption Isotherms variables.

Adsorbents	Freundlich Model			Langmuir Model			
	1/n	K_F(L·mg^{-1})	R^{2*}	q_{max}(mg/g)	K_L(L·mg^{-1})	R_L	R^{2**}
NBC	1.02974	9.718964	0.88964	5.438033607	3.95207	0.00509	0.7535
MBC	1.05679	11.62947	0.9195	4.72701489	4.01652	0.005	0.85608

Note: 1/n, K_F, and R^{2*} were computed from a plot of logq_e versus logC_e (Equation (5)), while K_L, qmax, and R^{2**} were calculated by plotting $1/q_e$ against $1/C_e$ (Equation (4)), respectively, using the data from Figures 10 and 11.

3.15. BPA Adsorption Kinetic Studies

Kinetic study offers essential information on the mechanism of adsorption and influencing mechanism of adsorption process as either chemical reaction or mass transfer to attain optimal working conditions for industrial scale [119]. With a view to examine the kinetic mechanism triggering BPA adsorption, experimental data were fitted to the pseudo-first order (PFO) [120] and pseudo-second order (PSO) [121] linearized models using Equations (4) and (5), respectively:

$$ln(q_e - q_t) = ln(lnq_e - K_1\ t) \tag{4}$$

$$\frac{1}{q_t} = \frac{1}{K_2 q_e{}^2} \times \frac{1}{t} + \frac{1}{q_e} \tag{5}$$

where q_t and q_e are the amounts of BPA adsorbed at time t and equilibrium, respectively (mg/g), t is the contact time (min), and K_1 and K_2 are the rate constants of the PFO and PSO kinetic models, respectively (1/min). In batch process, sorption kinetics is described by different models based on adsorption equilibrium including the PSO (pseudo-first order) and PSO (pseudo-second order) kinetic models. In this context, it is used to model the kinetics of the adsorption of BPA onto BC.

Figure 14a,b illustrates the graphs of ($ln(q_e - q_t)$ against time) and (time/q_t against time) for the pseudo-first order and pseudo-second order kinetic models, respectively. The computed variables and the experimental data of the two models are presented in Table 5. Noticeably the coefficient of correlation value (R^2 = 0.9937) for the PSO model is of better linearity and higher in comparison with the PFO (R^2 =0.8515) for BPA adsorption, which was also noticed for a similar compound (sulfamethoxazole) adsorption by functionalized BC [122]. Similarly, the computed q_e (mg/g) for the PSO model is close to the experimental q_e value (see Table 5). Hence, PSO model satisfactorily fits better the experimental results than the PFO kinetic model. The PSO kinetic model suggests that the chemisorption can be regarded as a rate-controlling phase during the adsorption procedure [3,123]. Similarly, the chemisorption takes place via electron-exchange or sharing between BPA and BC [124]. The elevated rate constants of the PSO model for BPA can be attributed to rapid interaction with the active adsorption sites of MBC. This finding agrees with previous studies on the BPA adsorption on the following adsorbents: algal BC [48], magnetic $CuZnFe_2O_4$–BC composite [125], modified organo-montmorillonites [126].

Table 5. Kinetic study result for BPA adsorption.

Kinetic Models	Parameters	Value	Linear Regression
PSO $\frac{1}{q_t} = \frac{1}{K_2 q_e{}^2} \times \frac{1}{t} + \frac{1}{q_e}$ where K_2 is rate constant (g·mg^{-1}·min^{-1})	K_2 (g·mg^{-1}·min^{-1}) Comp. q_e (mg/g) Exp. q_e (mg/g) R^2	0.005289 47.15266 37.75 0.99376	$y = 1.556 + 0.0583x$
PFO $\ln(q_e - qt) = \ln(q_e) - K_1 t$	K_1 (min^{-1}) Comp. q_e (mg/g) Exp. q_e (mg/g) R^2	3.50×10^{-6} 12.003837 37.75 0.8515	$y = 0.00383 + 8.4x$

NB: PSO is pseudo-second order; PFO is pseudo-first order.

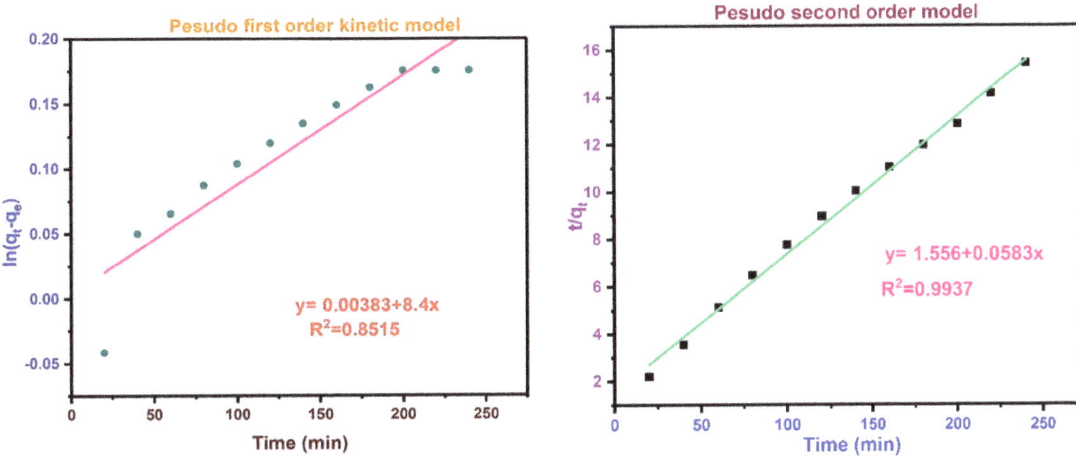

(a) (b)

Figure 14. (**a**) Pseudo-first order model for adsorption of BPA by MBC, and (**b**) pseudo-second order model for adsorption of BPA by MBC.

3.16. Adsorption Thermodynamics

Thermodynamic studies are employed to decipher any reaction in a better approach and reveals whether the adsorption process is an intended process or a spontaneous process [127]. It also indicates the influence of temperature on the adsorption process. Generally, exothermal and endothermal sorption processes are the two common processes. If the sorption declines with increasing temperature, it implies the exothermal sorption process whereas if the sorption increases with rising temperature, it signifies that the sorption is an endothermal process. In this study, the temperature influence on BPA adsorption onto MBC was studied by performing adsorption tests at varying temperatures, viz., 298.15, 318.15, and 338.15 K, at optimum pH (6.0) and adsorbent dosage 0.5 g, respectively. The thermodynamic variables such as enthalpy change ($\Delta H°$), standard Gibb's free energy change ($\Delta G°$), and change in entropy ($\Delta S°$), were computed and summarized in Table 6 using the following equations:

$$\Delta G_O = -RT \ln K_L \tag{6}$$

$$K_L = \frac{q_e}{C_e} \tag{7}$$

$$\Delta S^{\circ} = \Delta H^{\circ} - \Delta G^{\circ}/T \tag{8}$$

where R is the gas constant (8.314 J/mol^{-1}K^{-1}), q_e is the adsorption capacity (mg/g), C_e is the equilibrium concentration (mg/L) and T is the actual temperature (K). The y-intercept and the slope of the linear fit following a plot of ln K_L versus 1/T were employed to calculate the values of $\Delta H°$ and $\Delta S°$ via the Van't Hoff plot (Figure 15).

Table 6. Computed thermodynamic variables for BPA adsorption on MBC at varying temperature.

Adsorbate	Temperature (K)	lnK_L	$\Delta G°$ (kJ·mol^{-1})	$\Delta H°$ (kJ·mol^{-1})	$\Delta S°$ (J·mol·K^{-1})
BPA	298.15	0.526093278	−1.304090065	51.22768066	176.1349999
	318.15	1.803593997	−4.770684859		
	338.15	2.971634746	−8.354391819		

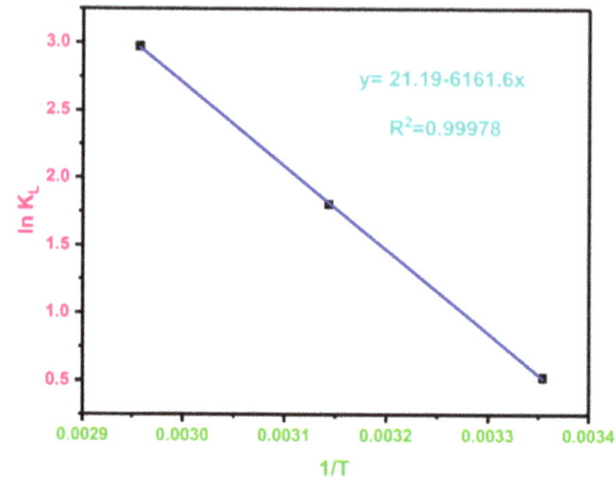

Figure 15. Thermodynamic plot of lnK_L versus 1/T for adsorption of BPA on MBC.

As the sorption temperature rises, the values of ΔG° turn out to be more negative, which implies that BPA is better effectively adsorbed to MBC at elevated temperatures. The negative ΔG° indicate that all adsorption processes were spontaneous [128]. The positive ΔH° (51.23 kJ·mol^{-1}) indicated that the interaction between BPA and MBC was an endothermic process which possibly attributed to the competitive dissolution of BPA in aqueous solution [129]. The positive ΔS° and negative ΔG0 values suggested the spontaneity of the adsorption process. The higher positive ΔS° (176.13 J·mol^{-1}·K^{-1}) as indicated in Table 5, further demonstrate increasing randomness at the adsorbate-adsorbent interface as a result of free water molecules [130]. Thus, the determined positive ΔS° and negative ΔG° values at experimented temperatures substantiated the BPA adsorption on MBC and indicated the spontaneity of the sorption process, besides the entropy effect should be the primary leading force for the adsorption of BPA on MBC [131].

3.17. Investigation of Biochar Regeneration and Reusability

Recyclability of the adsorbents is strongly essential to preserve process costs down and for the industrial-scale application. Five phases of adsorption–desorption tests were undertaken to examine the reusability of the as-synthesized MBC in accordance with previously adopted experimental procedure [125]. The MBC was repetitively applied five times, and the q_e values were recorded, as illustrated in Figure 16. The result indicated that q_e was slightly reduced during the adsorption–desorption experiment. Upon the completion of five cycles of reusability test, the removal efficiency of BPA only reduced by 12.85% as compared with the first cycle. The slight reduction in adsorption capacity during multiple adsorption–desorption tests, was due to the partial desorption of BPA in MBC surface, competitive available sorption sites, loss of solid in solution, and the elution of iron oxide nanoparticles and energetic substances on MBC surface [49,132]. This reveals that the MBC could be an effectual, economical, and environmental benign adsorbent with superior re-usability, which can be practically applied in BPA removal procedures. The recyclability test reveals that MBC can be applied repeatedly in wastewater purification as an efficient adsorbent.

Table 7 presents the comparison of specific surface area characteristics, adsorption capacity of BPA, and the magnetic intensity of the synthesized MBC material in the current study in comparison with various studies.

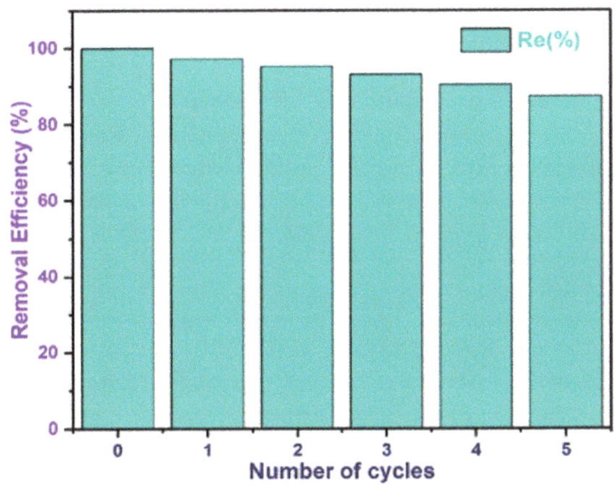

Figure 16. Regeneration of magnetic biochar for BPA removal. Experimental conditions: m = 0.5 g; [BPA] = 20 ppm; and temperature (293.15 K).

Table 7. Comparison of specific surface area, magnetic strength, and adsorption capacity of BPA on different adsorbents from previous literature with current study.

Adsorbents	Surface Area (m^2/g)	Magnetic Strength (emu/g)	Regeneration (%)	Adsorption Capacity (mg/g)	Reference
ulva prolifera (marine macroalgae)	25.43	ND	ND	84.19	[48]
pomelo peel	889.8	ND	ND	26.25	[133]
magnetic composite sepiolite	NA	14.1	NA	36.30	[134]
bamboo	61.5	37.6 and 32.6	7.6 and 8.2	263.2	[125]
sewage sludge wheat straw	65.03	ND		196.91	[135]
dried pineapple	84.89	12.83	34.93	101.16	[47]
corn straw	313.88	14.5	ND	46.90	[111]
local reed biomass	154.79	ND	ND	9.92	[136]
grapefruit peel	20.732	30.60	20	229.19	[16]
magnetic biochar palm kernel shell	362.0673	6.4882	12.85	37.64	Current study

NA = Not Available; ND = Not determined.

Though few studies reported higher adsorption capacity than the current study despite their low surface area as indicated in Table 7, this is because previous studies have described the performance of MBC to be significantly influenced by the nature of auxiliary and raw materials, starting contaminant concentration, pyrolysis temperature, competitive anions, nature of modifier, reaction temperature, sorption time, and various synthesis methods [86,136,137]. As highlighted in Table 7, the maximum adsorption capacity recorded in the current study is comparable and in line with previous studies using different BC adsorbents. Hence, it can be inferred that the MBC synthesized from palm kernel shell is a promising, efficient, and essential precursor (adsorbent) for the removal of BPA from aqueous environment. The sorption capacity recorded from the current study using magnetic PKS-BC could be stems from highly developed pore structure, superfluous surface area, smaller pore size, considerable surface functional group, high sorptive capacity, and super paramagnetism. It is noteworthy that studies on the adsorption of BPA onto MBC synthesized in-house from palm kernel shell biomass is still very limited. Besides, the adsorbent demonstrated superior regeneration efficacy, and the resultant MBC could be reclaimed numerous times.

3.18. Controlling Mechanism for BPA Removal

The schematic diagram of mechanisms of adsorption of BPA (adsorbate) onto as-synthesized NBC and MBC is depicted in Figure 17. The average pore size of the as-synthesized BC falls below 50 nm, indicating mesoporous. Also, BPA molecules can penetrate the pores of the as-synthesized BC via pore filling. As stated earlier, the adsorption capacity of NBC and MBC to BPA is largely influenced by pH. In acidic environments, the phenolic hydroxyl groups in the chemical structure of BPA protonate, and thereby producing electrostatic repulsion with the positively charged surfaces of both NBC and MBC. Also, since BC adsorbent comprises carbon, silicon, iron, and oxygen groups, these elements C, Si, Fe, and O groups are vastly electronegative because of the sufficiency of available lone pair of electrons, which exhibits binding capabilities toward the BPA as indicated in Figure 2. The carboxyl, hydroxyl, carbonyl, and amine functional groups (as identified in Figure 5) facilitate the affinity of BPA molecules and their adsorption on the surface of BC. Such sorption mechanism is an electron giver–receiver type according to the unbalanced electrons' supply between BPA compound and the BC functional groups. The hydroxyl functional group ($^{-}$OH) on the BC's surface also develops potent hydrogen bonds with the C-H and $^{-}$OH on the molecular composition of BPA. Also, the C=O and $^{-}$COOH acidic functional groups of the as-synthesized BC can serves as electron receivers, producing π-π electron giver–receiver interactions (EGR) with BPA. In alkaline environments, the surface of MBC becomes negatively charged, which retains the electrostatic repulsion with the dissociated BPA^{2-} and $HBPA^{-}$, whereas NBC acts oppositely. Similarly, the π-π electron giver-receiver interaction (EGR), as well as hydrogen bonds between BPA and as-synthesized BC, would be severely weakened, thereby making it hard for BC to adsorb BPA. Since, BPA exhibits robust hydrophobicity and can be intermixed with the hydrophobic site on NBC and MBC surfaces. Hence, a hydrophobic interaction is correspondingly a central influential force for the adsorption of BPA onto NBC and MBC.

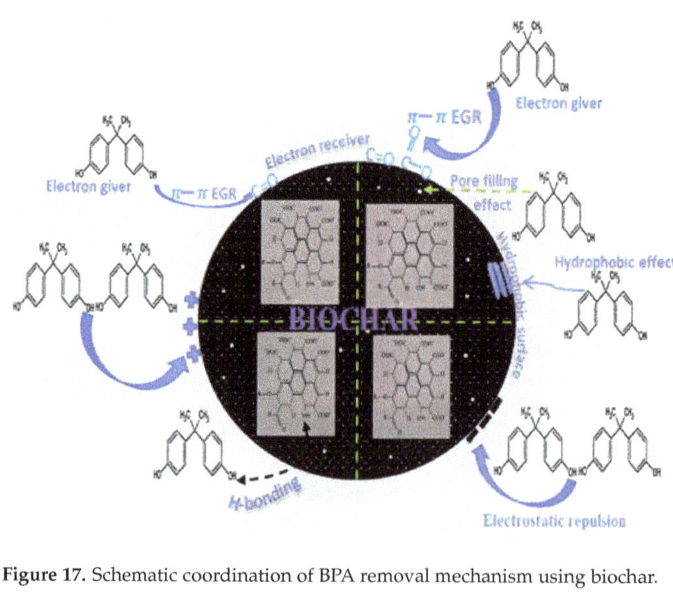

Figure 17. Schematic coordination of BPA removal mechanism using biochar.

4. Conclusions

In this work, a novel magnetic palm kernel shell biochar was efficiently synthesized via magnetic modification and applied to remove BPA from aqueous solution. The results of VSM, XRD, FTIR, BET, FESEM/EDX, and Zeta potential revealed that the Fe_3O_4 nanoparticles was effectually engrafted on the surface of biochar. The experimental findings demonstrated the adsorption isotherm could be better fitted by Freundlich model, whereas the adsorption kinetic data was controlled by pseudo-second order model. The as-synthesized BC also demonstrated a good magnetic strength for facile recovery, superior recyclability, high selectivity, and eco-friendly. It was noticed that BPA removal is greatly influenced by adsorbent dose, pH, and contact time. The BPA adsorption does increases with reduction of pH, with a maximum adsorption at pH 6 for MBC. Thermodynamic study revealed that BPA adsorption on MBC is endothermic process and spontaneous due to $+\Delta H$ and $-\Delta G$, respectively. The principal mechanisms for BPA adsorption on the prepared BC comprised of electrostatic interactions (π-π electron acceptor– donor interactions), hydrophobic interaction, and H-bonding. Therefore, it can be deduced from this study that novel MBC is efficient and practical for the removal of BPA from aqueous solution with benefit of being sustainable, and abundantly available.

Author Contributions: K.K.K. designed the project, performed the study, and wrote the manuscript; K.F.Y. supervised the study, and helped acquire funding; R.S.A. contributed to nanoparticles synthesis and characterization; thanks M.Z.M.N., for the software support; A.Z.A. and H.C.M. were principal investigators and helped acquire funding; and A.Z.A. contributed to chemicals. All authors have read and agreed to the published version of the manuscript.

Funding: This work was funded by Universiti Putra Malaysia under FRGS with grant number 5540204.

Institutional Review Board Statement: Not applicable.

Informed Consent Statement: Not applicable.

Data Availability Statement: The data presented in this study are available on request from the corresponding author.

Acknowledgments: The Universiti Putra Malaysia is hereby appreciated. The sponsorship received from the Tertiary Education Trust Fund (TETFund) through the Kwara State University, Malete, Nigeria, is also hereby thankfully acknowledged.

Conflicts of Interest: The authors declare no conflict of interest.

References

1. Ahsan, M.A.; Islam, M.T.; Hernandez, C.; Kim, H.; Lin, Y.; Curry, M.L.; Gardea-Torresdey, J.; Noveron, J.C. Adsorptive Removal of Sulfamethoxazole and Bisphenol A from Contaminated Water using Functionalized Carbonaceous Material Derived from Tea Leaves. *J. Environ. Chem. Eng.* **2018**, *6*, 4215–4225. [CrossRef]
2. Vandenberg, L.N.; Luthi, D.; Quinerly, D.A. Plastic bodies in a plastic world: Multi-disciplinary approaches to study endocrine disrupting chemicals: We dedicate this manuscript in memory of Dr. Theo Colborn: Mentor, colleague and friend. *J. Clean. Prod.* **2017**, *140*, 373–385. [CrossRef]
3. Rathnayake, S.I.; Xi, Y.; Frost, R.L.; Ayoko, G.A. Environmental applications of inorganic-organic clays for recalcitrant organic pollutants removal: Bisphenol A. *J. Colloid Interface Sci.* **2016**, *470*, 183–195. [CrossRef] [PubMed]
4. Michałowicz, J. Bisphenol A–Sources, toxicity and biotransformation. *Environ. Toxicol. Pharmacol.* **2014**, *37*, 738–758. [CrossRef]
5. Hoekstra, E.J.; Simoneau, C. Release of Bisphenol A from Polycarbonate-A Review. *Crit. Rev. Food Sci. Nutr.* **2013**, *53*, 386–402. [CrossRef]
6. Mileva, G.; Baker, S.L.; Konkle, A.T.M.; Bielajew, C. Bisphenol-A: Epigenetic reprogramming and effects on reproduction and behavior. *Int. J. Environ. Res. Public Health* **2014**, *11*, 7537–7561. [CrossRef]
7. vom Saal, F.S.; Nagel, S.C.; Coe, B.L.; Angle, B.M.; Taylor, J.A. The estrogenic endocrine disrupting chemical bisphenol A (BPA) and obesity. *Mol. Cell. Endocrinol.* **2012**, *354*, 74–84. [CrossRef]
8. Su-Hua, W.; Bing-zhi, D.; Yu, H. Adsorption of bisphenol A by polysulphone membrane. *Desalination* **2010**, *253*, 22–29. [CrossRef]
9. Wintgens, T.; Gallenkemper, M.; Melin, T. Removal of endocrine disrupting compounds with membrane processes in wastewater treatment and reuse. *Water Sci. Technol.* **2004**, *50*, 1–8. [CrossRef]
10. Rajasärkkä, J.; Pernica, M.; Kuta, J.; Lašňák, J.; Šimek, Z.; Bláha, L. Drinking water contaminants from epoxy resin-coated pipes: A field study. *Water Res.* **2016**, *103*, 133–140. [CrossRef]
11. Yüksel, S.; Kabay, N.; Yüksel, M. Removal of bisphenol A (BPA) from water by various nanofiltration (NF) and reverse osmosis (RO) membranes. *J. Hazard. Mater.* **2013**, *263*, 307–310. [CrossRef]

12. Wee, S.Y.; Aris, A.Z.; Yusoff, F.M.; Praveena, S.M. Occurrence and risk assessment of multiclass endocrine disrupting compounds in an urban tropical river and a proposed risk management and monitoring framework. *Sci. Total Environ.* **2019**, *671*, 431–442. [CrossRef]
13. Nazifa, T.H.; Kristanti, R.A.; Ike, M.; Kuroda, M.; Hadibarata, T. Occurrence and distribution of estrogenic chemicals in river waters of Malaysia. *Toxicol. Environ. Health Sci.* **2020**, *12*, 65–74. [CrossRef]
14. Shekhar, S.; Sood, S.; Showkat, S.; Lite, C.; Chandrasekhar, A.; Vairamani, M.; Barathi, S.; Santosh, W. Detection of phenolic endocrine disrupting chemicals (EDCs) from maternal blood plasma and amniotic fluid in Indian population. *Gen. Comp. Endocrinol.* **2017**, *241*, 100–107. [CrossRef]
15. Colin, A.; Bach, C.; Rosin, C.; Munoz, J.F.; Dauchy, X. Is drinking water a major route of human exposure to alkylphenol and bisphenol contaminants in France? *Arch. Environ. Contam. Toxicol.* **2014**, *66*, 86–99. [CrossRef]
16. Wang, J.; Zhang, M. Adsorption characteristics and mechanism of bisphenol a by magnetic biochar. *Int. J. Environ. Res. Public Health* **2020**, *17*, 1075. [CrossRef]
17. Gadupudi, C.K.; Rice, L.; Xiao, L.; Kantamaneni, K. Endocrine Disrupting Compounds Removal Methods from Wastewater in the United Kingdom: A Review. *Science* **2019**, *1*, 15. [CrossRef]
18. Rodriguez-Narvaez, O.M.; Peralta-Hernandez, J.M.; Goonetilleke, A.; Bandala, E.R. Treatment technologies for emerging contaminants in water: A review. *Chem. Eng. J.* **2017**, *323*, 361–380. [CrossRef]
19. Stackelberg, P.E.; Gibs, J.; Furlong, E.T.; Meyer, M.T.; Zaugg, S.D.; Lippincott, R.L. Efficiency of conventional drinking-water-treatment processes in removal of pharmaceuticals and other organic compounds. *Sci. Total Environ.* **2007**, *377*, 255–272. [CrossRef]
20. Zielińska, M.; Bułkowska, K.; Cydzik-Kwiatkowska, A.; Bernat, K.; Wojnowska-Baryła, I. Removal of bisphenol A (BPA) from biologically treated wastewater by microfiltration and nanofiltration. *Int. J. Environ. Sci. Technol.* **2016**, *13*, 2239–2248. [CrossRef]
21. Gunnarsson, L.; Adolfsson-Erici, M.; Björlenius, B.; Rutgersson, C.; Förlin, L.; Larsson, D.G.J. Comparison of six different sewage treatment processes-Reduction of estrogenic substances and effects on gene expression in exposed male fish. *Sci. Total Environ.* **2009**, *407*, 5235–5242. [CrossRef]
22. Hu, Z.; Wen, X.; Si, X. Pre-ultrafiltration or pre-ozonation for EDCs removal in a combined ultrafiltration and ozonation process. *J. Chem. Technol. Biotechnol.* **2016**, *91*, 2929–2934. [CrossRef]
23. Fan, X.; Tao, Y.; Wang, L.; Zhang, X.; Lei, Y.; Wang, Z.; Noguchi, H. Performance of an integrated process combining ozonation with ceramic membrane ultra-filtration for advanced treatment of drinking water. *Desalination* **2014**, *335*, 47–54. [CrossRef]
24. Li, S.; Zhang, G.; Wang, P.; Zheng, H.; Zheng, Y. Microwave-enhanced Mn-Fenton process for the removal of BPA in water. *Chem. Eng. J.* **2016**, *294*, 371–379. [CrossRef]
25. Zahari, A.M.; Shuo, C.W.; Sathishkumar, P.; Yusoff, A.R.M.; Gu, F.L.; Buang, N.A.; Lau, W.J.; Gohari, R.J.; Yusop, Z. A reusable electrospun PVDF-PVP-MnO_2 nanocomposite membrane for bisphenol A removal from drinking water. *J. Environ. Chem. Eng.* **2018**, *6*, 5801–5811. [CrossRef]
26. Katibi, K.K.; Yunos, K.F.; Man, H.C.; Aris, A.Z.; Zuhair, M.; Syahidah, R. Recent Advances in the Rejection of Endocrine-Disrupting Compounds from Water Using Membrane and Membrane Bioreactor Technologies: A Review. *Polymers* **2021**, *13*, 329. [CrossRef] [PubMed]
27. Garg, A.; Singhania, T.; Singh, A.; Sharma, S.; Rani, S.; Neogy, A.; Yadav, S.R.; Sangal, V.K.; Garg, N. Photocatalytic Degradation of Bisphenol-A using N, Co Codoped TiO_2 Catalyst under Solar Light. *Sci. Rep.* **2019**, *9*, 1–13. [CrossRef] [PubMed]
28. Becker, D.; Rodriguez-mozaz, S.; Insa, S.; Schoevaart, R.; Barcelo, D.; de Cazes, M.; Belleville, M.P.; Marcano, J.S.; Misovic, A.; Oehlmann, J.; et al. Removal of endocrine disrupting chemicals in wastewater by enzymatic treatment with fungal laccases. *Org. Process. Res.Dev.* **2017**, *21*, 480–491. [CrossRef]
29. Hou, J.; Dong, G.; Ye, Y.; Chen, V. Enzymatic degradation of bisphenol-A with immobilized laccase on TiO_2 sol-gel coated PVDF membrane. *J. Memb. Sci.* **2014**, *469*, 19–30. [CrossRef]
30. Bastami, T.R.; Entezari, M.H. Activated carbon from carrot dross combined with magnetite nanoparticles for the efficient removal of p-nitrophenol from aqueous solution. *Chem. Eng. J.* **2012**, *210*, 510–519. [CrossRef]
31. Grassi, M.; Kaykioglu, G.; Belgiorno, V.; Lofrano, G. Emerging Compounds Removal from Wastewater. *Green Chem. Sustain.* **2012**, 15–38. [CrossRef]
32. Saber, S.E.M.; Jamil, S.N.A.M.; Abdullah, L.C.; Choong, T.S.Y.; Ting, T.M. Insights into thep-nitrophenol adsorption by amidoxime-modified poly(acrylonitrile-co-acrylic acid): Characterization, kinetics, isotherm, thermodynamic, regeneration and mechanism study. *RSC Adv.* **2021**, *11*, 8150–8162. [CrossRef]
33. Katibi, K.K.; Yunos, K.F.; Man, H.C.; Aris, A.Z.; Zuhair, M.; Nor, M.; Azis, R.S.; Umar, A.M. Contemporary Techniques for Remediating Endocrine-Disrupting Compounds in Various Water Sources: Advances in Treatment Methods and Their Limitations. *Polymers* **2021**, *13*, 3229. [CrossRef]
34. Haddad, M.; Oie, C.; Vo Duy, S.; Sauvé, S.; Barbeau, B. Adsorption of micropollutants present in surface waters onto polymeric resins: Impact of resin type and water matrix on performance. *Sci. Total Environ.* **2019**, *660*, 1449–1458. [CrossRef]
35. Park, C.M.; Han, J.; Chu, K.H.; Al-Hamadani, Y.A.J.; Her, N.; Heo, J.; Yoon, Y. Influence of solution pH, ionic strength, and humic acid on cadmium adsorption onto activated biochar: Experiment and modeling. *J. Ind. Eng. Chem.* **2017**, *48*, 186–193. [CrossRef]
36. Hanigan, D.; Zhang, J.; Herckes, P.; Krasner, S.W.; Chen, C.; Westerhoff, P. Adsorption of N-nitrosodimethylamine precursors by powdered and granular activated carbon. *Environ. Sci. Technol.* **2012**, *46*, 12630–12639. [CrossRef]

37. Wirasnita, R.; Hadibarata, T.; Yusoff, A.R.M.; Yusop, Z. Removal of bisphenol a from aqueous solution by activated carbon derived from oil palm empty fruit bunch. *Water. Air. Soil Pollut.* **2014**, *225*, 2148. [CrossRef]
38. Chang, K.L.; Hsieh, J.F.; Ou, B.M.; Chang, M.H.; Hseih, W.Y.; Lin, J.H.; Huang, P.J.; Wong, K.F.; Chen, S.T. Adsorption Studies on the Removal of an Endocrine-Disrupting Compound (Bisphenol A) using Activated Carbon from Rice Straw Agricultural Waste. *Sep. Sci. Technol.* **2012**, *47*, 1514–1521. [CrossRef]
39. Bolong, N.; Ismail, A.F.; Salim, M.R.; Matsuura, T. A review of the effects of emerging contaminants in wastewater and options for their removal. *Desalination* **2009**, *239*, 229–246. [CrossRef]
40. Ahmed, M.B.; Zhou, J.L.; Ngo, H.H.; Guo, W. Adsorptive removal of antibiotics from water and wastewater: Progress and challenges. *Sci. Total Environ.* **2015**, *532*, 112–126. [CrossRef]
41. Chen, W.; Parette, R.; Zou, J.; Cannon, F.S.; Dempsey, B.A. Arsenic removal by iron-modified activated carbon. *Water Res.* **2007**, *41*, 1851–1858. [CrossRef] [PubMed]
42. Tan, X.; Liu, Y.; Zeng, G.; Wang, X.; Hu, X.; Gu, Y.; Yang, Z. Application of biochar for the removal of pollutants from aqueous solutions. *Chemosphere* **2015**, *125*, 70–85. [CrossRef]
43. Huggins, T.M.; Haeger, A.; Biffinger, J.C.; Ren, Z.J. Granular biochar compared with activated carbon for wastewater treatment and resource recovery. *Water Res.* **2016**, *94*, 225–232. [CrossRef]
44. Ahmed, M.B.; Zhou, J.L.; Ngo, H.H.; Guo, W. Insight into biochar properties and its cost analysis. *Biomass Bioenergy* **2016**, *84*, 76–86. [CrossRef]
45. Mohan, D.; Sarswat, A.; Ok, Y.S.; Pittman, C.U. Organic and inorganic contaminants removal from water with biochar, a renewable, low cost and sustainable adsorbent-A critical review. *Bioresour. Technol.* **2014**, *160*, 191–202. [CrossRef]
46. Wang, S.; Gao, B.; Zimmerman, A.R.; Li, Y.; Ma, L.; Harris, W.G.; Migliaccio, K.W. Removal of arsenic by magnetic biochar prepared from pinewood and natural hematite. *Bioresour. Technol.* **2015**, *175*, 391–395. [CrossRef]
47. Liao, T.; Li, T.; Su, X.; Yu, X.; Song, H.; Zhu, Y.; Zhang, Y. La(OH)3-modified magnetic pineapple biochar as novel adsorbents for efficient phosphate removal. *Bioresour. Technol.* **2018**, *263*, 207–213. [CrossRef]
48. Lu, J.; Zhang, C.; Wu, J.; Luo, Y. Adsorptive Removal of Bisphenol A Using N-Doped Biochar Made of Ulva prolifera. *Water. Air Soil Pollut.* **2017**, *228*, 1–9. [CrossRef]
49. Heo, J.; Yoon, Y.; Lee, G.; Kim, Y.; Han, J.; Park, C.M. Enhanced adsorption of bisphenol A and sulfamethoxazole by a novel magnetic CuZnFe2O4–biochar composite. *Bioresour. Technol.* **2019**, *281*, 179–187. [CrossRef] [PubMed]
50. Yahayu, M.; Abas, F.Z.; Zulkifli, S.E.; Ani, F.N. Utilization of Oil Palm Fiber and Palm Kernel Shell in Various Applications. *Sustain. Technol. Manag. Agric. Wastes* **2018**, 45–56. [CrossRef]
51. Loh, S.K. The potential of the Malaysian oil palm biomass as a renewable energy source. *Energy Convers. Manag.* **2017**, *141*, 285–298. [CrossRef]
52. Umar, M.S.; Jennings, P.; Urmee, T. Generating renewable energy from oil palm biomass in Malaysia: The Feed-in Tariff policy framework. *Biomass Bioenergy* **2014**, *62*, 37–46. [CrossRef]
53. AI, M. *National Biomass Strategy 2020: New Wealth Creation for Malaysia's Palm Oil Industry*; Agensi Inovasi Malaysia: Kuala Lumpur, Malaysia, 2011.
54. Kong, S.H.; Loh, S.K.; Bachmann, R.T.; Zainal, H.; Cheong, K.Y. Palm kernel shell biochar production, characteristics and carbon sequestration potential. *J. Oil Palm Res.* **2019**, *31*, 508–520. [CrossRef]
55. Zahraee, S.M.; Golroudbary, S.R.; Shiwakoti, N.; Kraslawski, A.; Stasinopoulos, P. An investigation of the environmental sustainability of palm biomass supply chains via dynamic simulation modeling: A case of Malaysia. *J. Clean. Prod.* **2019**, *237*, 117740. [CrossRef]
56. Oliveira, L.C.A.; Rios, R.V.R.A.; Garg, V.K.; Lago, R.M. Clay–iron oxide magnetic composites for the adsorption of contaminants in water. *Appl. Clay Sci.* **2003**, *22*, 169–177. [CrossRef]
57. Han, Z.; Sani, B.; Akkanen, J.; Abel, S.; Nybom, I.; Karapanagioti, H.K.; Werner, D. A critical evaluation of magnetic activated carbon's potential for the remediation of sediment impacted by polycyclic aromatic hydrocarbons. *J. Hazard. Mater.* **2015**, *286*, 41–47. [CrossRef]
58. Mohan, D.; Sarswat, A.; Singh, V.K.; Alexandre-Franco, M.; Pittman, C.U. Development of magnetic activated carbon from almond shells for trinitrophenol removal from water. *Chem. Eng. J.* **2011**, *172*, 1111–1125. [CrossRef]
59. Petrova, B.; Budinova, T.; Tsyntsarski, B.; Kochkodan, V. Removal of aromatic hydrocarbons from water by activated carbon from apricot stones. *Chem. Eng. J.* **2010**, *165*, 258–264. [CrossRef]
60. Xie, J.; Wang, Z.; Wu, D.-Y.; Li, C.-J. [Adsorption of phenol chemicals by surfactant-modified zeolites]. *Huan Jing Ke Xue Huanjing Kexue* **2012**, *33*, 4361–4366.
61. Cheng, M.; Jiang, J.; Wang, J.; Fan, J. Highly Salt Resistant Polymer Supported Ionic Liquid Adsorbent for Ultrahigh Capacity Removal of p-Nitrophenol from Water. *ACS Sustain. Chem. Eng.* **2019**, *7*, 8195–8205. [CrossRef]
62. Abdelnaeim, M.Y.; El Sherif, I.Y.; Attia, A.A.; Fathy, N.A.; El-Shahat, M.F. Impact of chemical activation on the adsorption performance of common reed towards Cu(II) and Cd(II). *Int. J. Miner. Process.* **2016**, *157*, 80–88. [CrossRef]
63. Vassileva, P.S.; Radoykova, T.H.; Detcheva, A.K.; Avramova, I.A.; Aleksieva, K.I.; Nenkova, S.K.; Valchev, I.V.; Mehandjiev, D.R. Adsorption of Ag^+ ions on hydrolyzed lignocellulosic materials based on willow, paulownia, wheat straw and maize stalks. *Int. J. Environ. Sci. Technol.* **2016**, *13*, 1319–1328. [CrossRef]

64. Rangabhashiyam, S.; Anu, N.; Giri Nandagopal, M.S.; Selvaraju, N. Relevance of isotherm models in biosorption of pollutants by agricultural byproducts. *J. Environ. Chem. Eng.* **2014**, *2*, 398–414. [CrossRef]
65. Muhammad, M.; Choong, T.S.Y.; Chuah, T.G.; Yunus, R.; Yap, Y.H.T. Adsorption of β-carotene onto mesoporous carbon coated monolith in isopropyl alcohol and n-hexane solution: Equilibrium and thermodynamic study. *Chem. Eng. J.* **2010**, *164*, 178–182. [CrossRef]
66. Yew, Y.P.; Shameli, K.; Miyake, M.; Bt Ahmad Khairudin, N.B.; Bt Mohamad, S.E.; Hara, H.; Bt Mad Nordin, M.F.; Lee, K.X. An eco-friendly means of biosynthesis of superparamagnetic magnetite nanoparticles via marine polymer. *IEEE Trans. Nanotechnol.* **2017**, *16*, 1047–1052. [CrossRef]
67. Liu, C.; Tang, Z.; Chen, Y.; Su, S.; Jiang, W. Characterization of mesoporous activated carbons prepared by pyrolysis of sewage sludge with pyrolusite. *Bioresour. Technol.* **2010**, *101*, 1097–1101. [CrossRef] [PubMed]
68. Ghani, Z.A.; Yusoff, M.S.; Zaman, N.Q.; Zamri, M.F.M.A.; Andas, J. Optimization of preparation conditions for activated carbon from banana pseudo-stem using response surface methodology on removal of color and COD from landfill leachate. *Waste Manag.* **2017**, *62*, 177–187. [CrossRef]
69. Sulaiman, S.; Azis, R.S.; Ismail, I.; Man, H.C.; Rosdi, N. Rapid Adsorption of Magnetite Nanoparticles from Recycled Mill Scale Waste as Potential Adsorbent for Removal of Cu(II) Ions. *Solid State Phenom.* **2021**, *317*, 270–275. [CrossRef]
70. Dong, H.; Deng, J.; Xie, Y.; Zhang, C.; Jiang, Z.; Cheng, Y.; Hou, K.; Zeng, G. Stabilization of nanoscale zero-valent iron (nZVI) with modified biochar for Cr(VI) removal from aqueous solution. *J. Hazard. Mater.* **2017**, *332*, 79–86. [CrossRef] [PubMed]
71. Dong, H.; Zhang, C.; Hou, K.; Cheng, Y.; Deng, J.; Jiang, Z.; Tang, L.; Zeng, G. Removal of trichloroethylene by biochar supported nanoscale zero-valent iron in aqueous solution. *Sep. Purif. Technol.* **2017**, *188*, 188–196. [CrossRef]
72. Zhou, X.; Liu, Y.; Zhou, J.; Guo, J.; Ren, J.; Zhou, F. Efficient removal of lead from aqueous solution by urea-functionalized magnetic biochar: Preparation, characterization and mechanism study. *J. Taiwan Inst. Chem. Eng.* **2018**, *91*, 457–467. [CrossRef]
73. Reguyal, F.; Sarmah, A.K.; Gao, W. Synthesis of magnetic biochar from pine sawdust via oxidative hydrolysis of $FeCl_2$ for the removal sulfamethoxazole from aqueous solution. *J. Hazard. Mater.* **2017**, *321*, 868–878. [CrossRef]
74. Hamedi, A.; Trotta, F.; Zarandi, M.B.; Zanetti, M.; Caldera, F.; Anceschi, A.; Nateghi, M.R. In situ synthesis of MIL-100(Fe) at the surface of fe3o4@ac as highly efficient dye adsorbing nanocomposite. *Int. J. Mol. Sci.* **2019**, *20*, 5612. [CrossRef]
75. Kalderis, D.; Kayan, B.; Akay, S.; Kulaksiz, E.; Gözmen, B. Adsorption of 2,4-dichlorophenol on paper sludge/wheat husk biochar: Process optimization and comparison with biochars prepared from wood chips, sewage sludge and HOG fuel/demolition waste. *J. Environ. Chem. Eng.* **2017**, *5*, 2222–2231. [CrossRef]
76. Zhang, X.; Lv, L.; Qin, Y.; Xu, M.; Jia, X.; Chen, Z. Removal of aqueous Cr(VI) by a magnetic biochar derived from Melia azedarach wood. *Bioresour. Technol.* **2018**, *256*, 1–10. [CrossRef]
77. Rong, X.; Xie, M.; Kong, L.; Natarajan, V.; Ma, L.; Zhan, J. The magnetic biochar derived from banana peels as a persulfate activator for organic contaminants degradation. *Chem. Eng. J.* **2019**, *372*, 294–303. [CrossRef]
78. Zhao, T.; Ma, X.; Cai, H.; Ma, Z.; Liang, H. study on the adsorption of $CuFe_2O_4$-loaded corncob biochar for Pb(II). *Molecules* **2020**, *25*, 3456. [CrossRef]
79. Fe, O.; Lin, Q. RSC Advances Removal of Cu (II) from aqueous solution using. *RSC Adv.* **2017**, *7*, 53135–53144. [CrossRef]
80. Wang, L.; Gan, K.; Lu, D.; Zhang, J. Hydrophilic Fe3O4@C for High-Capacity Adsorption of 2,4-Dichlorophenol. *Eur. J. Inorg. Chem.* **2016**, *2016*, 890–896. [CrossRef]
81. Jun, B.M.; Kim, Y.; Han, J.; Yoon, Y.; Kim, J.; Park, C.M. Preparation of activated biochar-supported magnetite composite for adsorption of polychlorinated phenols from aqueous solutions. *Water* **2019**, *11*, 1899. [CrossRef]
82. Saleh, S.; Kamarudin, K.B.; Ghani, W.A.W.A.K.; Kheang, L.S. Removal of Organic Contaminant from Aqueous Solution Using Magnetic Biochar. *Procedia Eng.* **2016**, *148*, 228–235. [CrossRef]
83. Mohan, D.; Kumar, H.; Sarswat, A.; Alexandre-Franco, M.; Pittman, C.U. Cadmium and lead remediation using magnetic oak wood and oak bark fast pyrolysis bio-chars. *Chem. Eng. J.* **2014**, *236*, 513–528. [CrossRef]
84. Shan, D.; Deng, S.; Zhao, T.; Wang, B.; Wang, Y.; Huang, J.; Yu, G.; Winglee, J.; Wiesner, M.R. Preparation of ultrafine magnetic biochar and activated carbon for pharmaceutical adsorption and subsequent degradation by ball milling. *J. Hazard. Mater.* **2016**, *305*, 156–163. [CrossRef] [PubMed]
85. Su, J.; Lyu, T.; Yi, H.; Bi, L.; Pan, G. Superior arsenate adsorption and comprehensive investigation of adsorption mechanism on novel Mn-doped $La_2O_2CO_3$ composites. *Chem. Eng. J.* **2020**, *391*, 123623. [CrossRef]
86. Li, X.; Wang, C.; Zhang, J.; Liu, J.; Liu, B.; Chen, G. Preparation and application of magnetic biochar in water treatment: A critical review. *Sci. Total Environ.* **2020**, *711*, 134847. [CrossRef]
87. Fan, H.; Ma, X.; Zhou, S.; Huang, J.; Liu, Y.; Liu, Y. Highly efficient removal of heavy metal ions by carboxymethyl cellulose-immobilized Fe_3O_4 nanoparticles prepared via high-gravity technology. *Carbohydr. Polym.* **2019**, *213*, 39–49. [CrossRef]
88. Kim, E.; Jung, C.; Han, J.; Her, N.; Park, C.M.; Jang, M.; Son, A.; Yoon, Y. Sorptive removal of selected emerging contaminants using biochar in aqueous solution. *J. Ind. Eng. Chem.* **2016**, *36*, 364–371. [CrossRef]
89. Zhang, X.; Qian, J.; Pan, B. Fabrication of Novel Magnetic Nanoparticles of Multifunctionality for Water Decontamination. *Environ. Sci. Technol.* **2016**, *50*, 881–889. [CrossRef]
90. Li, M.; Liu, Q.; Guo, L.; Zhang, Y.; Lou, Z.; Wang, Y.; Qian, G. Cu(II) removal from aqueous solution by Spartina alterniflora derived biochar. *Bioresour. Technol.* **2013**, *141*, 83–88. [CrossRef]

91. Fang, Q.; Chen, B.; Lin, Y.; Guan, Y. Aromatic and hydrophobic surfaces of wood-derived biochar enhance perchlorate adsorption via hydrogen bonding to oxygen-containing organic groups. *Environ. Sci. Technol.* **2014**, *48*, 279–288. [CrossRef]
92. Gupta, V.K.; Nayak, A. Cadmium removal and recovery from aqueous solutions by novel adsorbents prepared from orange peel and Fe_2O_3 nanoparticles. *Chem. Eng. J.* **2012**, *180*, 81–90. [CrossRef]
93. Zhang, Z.B.; Cao, X.H.; Liang, P.; Liu, Y.H. Adsorption of uranium from aqueous solution using biochar produced by hydrothermal carbonization. *J. Radioanal. Nucl. Chem.* **2013**, *295*, 1201–1208. [CrossRef]
94. Regmi, P.; Garcia Moscoso, J.L.; Kumar, S.; Cao, X.; Mao, J.; Schafran, G. Removal of copper and cadmium from aqueous solution using switchgrass biochar produced via hydrothermal carbonization process. *J. Environ. Manag.* **2012**, *109*, 61–69. [CrossRef]
95. Ahmed, M.B.; Zhou, J.L.; Ngo, H.H.; Guo, W.; Johir, M.A.H.; Sornalingam, K.; Belhaj, D.; Kallel, M. Nano-Fe0 immobilized onto functionalized biochar gaining excellent stability during sorption and reduction of chloramphenicol via transforming to reusable magnetic composite. *Chem. Eng. J.* **2017**, *322*, 571–581. [CrossRef]
96. Ahmed, M.B.; Zhou, J.L.; Ngo, H.H.; Guo, W.; Johir, M.A.H.; Belhaj, D. Competitive sorption affinity of sulfonamides and chloramphenicol antibiotics toward functionalized biochar for water and wastewater treatment. *Bioresour. Technol.* **2017**, *238*, 306–312. [CrossRef] [PubMed]
97. Ahmed, M.B.; Zhou, J.L.; Ngo, H.H.; Johir, M.A.H.; Sornalingam, K. Sorptive removal of phenolic endocrine disruptors by functionalized biochar: Competitive interaction mechanism, removal efficacy and application in wastewater. *Chem. Eng. J.* **2018**, *335*, 801–811. [CrossRef]
98. Duan, S.; Liu, X.; Wang, Y.; Shao, D.; Alharbi, N.S.; Alsaedi, A.; Li, J. Highly efficient entrapment of U(VI) by using porous magnetic $Ni_{0.6}Fe_{2.4}O_4$ micro-particles as the adsorbent. *J. Taiwan Inst. Chem. Eng.* **2016**, *65*, 367–377. [CrossRef]
99. Sun, Y.; Zhang, R.; Ding, C.; Wang, X.; Cheng, W.; Chen, C.; Wang, X. Adsorption of U(VI) on sericite in the presence of Bacillus subtilis: A combined batch, EXAFS and modeling techniques. *Geochim. Cosmochim. Acta* **2016**, *180*, 51–65. [CrossRef]
100. Zhou, Y.; Lu, P.; Lu, J. Application of natural biosorbent and modified peat for bisphenol a removal from aqueous solutions. *Carbohydr. Polym.* **2012**, *88*, 502–508. [CrossRef]
101. Fontecha-Cámara, M.A.; López-Ramón, M.V.; Álvarez-Merino, M.A.; Moreno-Castilla, C. Effect of surface chemistry, solution pH, and ionic strength on the removal of herbicides diuron and amitrole from water by an activated carbon fiber. *Langmuir* **2007**, *23*, 1242–1247. [CrossRef]
102. Zhou, Z.; Liu, Y.-G.; Liu, S.-B.; Liu, H.-Y.; Zeng, G.-M.; Tan, X.-F.; Yang, C.-P.; Ding, Y.; Yan, Z.-L.; Cai, X.-X. Sorption performance and mechanisms of arsenic(V) removal by magnetic gelatin-modified biochar. *Chem. Eng. J.* **2017**, *314*, 223–231. [CrossRef]
103. Li, M.-F.; Liu, Y.-G.; Zeng, G.-M.; Liu, S.-B.; Hu, X.-J.; Shu, D.; Jiang, L.-H.; Tan, X.-F.; Cai, X.-X.; Yan, Z.-L. Tetracycline absorbed onto nitrilotriacetic acid-functionalized magnetic graphene oxide: Influencing factors and uptake mechanism. *J. Colloid Interface Sci.* **2017**, *485*, 269–279. [CrossRef]
104. Liu, S.; Li, M.; Liu, Y.; Liu, N.; Tan, X.; Jiang, L. Supported activated magnetic biochar: Adsorption behavior and mechanism. *J. Taiwan Inst. Chem. Eng.* **2019**, *102*, 330–339. [CrossRef]
105. Garg, V.K.; Gupta, R.; Yadav, A.B.; Kumar, R. Dye removal from aqueous solution by adsorption on treated sawdust. *Bioresour. Technol.* **2003**, *89*, 121–124. [CrossRef]
106. Tuzen, M.; Sari, A. Biosorption of selenium from aqueous solution by green algae (*Cladophora hutchinsiae*) biomass: Equilibrium, thermodynamic and kinetic studies. *Chem. Eng. J.* **2010**, *158*, 200–206. [CrossRef]
107. Lee, C.G.; Park, J.A.; Choi, J.W.; Ko, S.O.; Lee, S.H. Removal and Recovery of Cr(VI) from Industrial Plating Wastewater Using Fibrous Anion Exchanger. *Water. Air Soil Pollut.* **2016**, *227*, 287. [CrossRef]
108. Idrees, M.; Batool, S.; Kalsoom, T.; Yasmeen, S.; Kalsoom, A.; Raina, S.; Zhuang, Q.; Kong, J. Animal manure-derived biochars produced via fast pyrolysis for the removal of divalent copper from aqueous media. *J. Environ. Manag.* **2018**, *213*, 109–118. [CrossRef] [PubMed]
109. Haghdoost, G.; Aghaie, H.; Monajjemi, M. Investigation of Langmuir and Freundlich Adsorption Isotherm of Co^{2+} Ion by Micro Powder of Cedar Leaf. *Orient. J. Chem.* **2017**, *33*, 1569–1574. [CrossRef]
110. Boparai, H.K.; Joseph, M.; Carroll, D.M.O. Kinetics and thermodynamics of cadmium ion removal by adsorption onto nano zerovalent iron particles. *J. Hazard. Mater.* **2011**, *186*, 458–465. [CrossRef] [PubMed]
111. Khan, Z.H.; Gao, M.; Qiu, W.; Islam, M.S.; Song, Z. Mechanisms for cadmium adsorption by magnetic biochar composites in an aqueous solution. *Chemosphere* **2020**, *246*, 125701. [CrossRef] [PubMed]
112. Sounthararajah, D.P.; Loganathan, P.; Kandasamy, J.; Vigneswaran, S. Effects of humic acid and suspended solids on the removal of heavy metals from water by adsorption onto granular activated carbon. *Int. J. Environ. Res. Public Health* **2015**, *12*, 10475–10489. [CrossRef]
113. Silas, K.; Ghani, W.A.W.A.K.; Choong, T.S.Y.; Rashid, U. Breakthrough studies of Co_3O_4 supported activated carbon monolith for simultaneous SO2/NOx removal from flue gas. *Fuel Process. Technol.* **2018**, *180*, 155–165. [CrossRef]
114. Edet, U.A.; Ifelebuegu, A.O. Kinetics, Isotherms, and Thermodynamic Modeling of the Adsorption of Phosphates from Model Wastewater Using Recycled Brick Waste. *Processes* **2020**, *8*, 665. [CrossRef]
115. Radnia, H. Isotherm and Kinetics of Fe(II) Adsorption onto Chitosan in a Batch Process. *Iran. J. Energy Environ.* **2011**, *2*, 250–257. [CrossRef]
116. Mubarak, N.M.; Alicia, R.F.; Abdullah, E.C.; Sahu, J.N.; Haslija, A.B.A.; Tan, J. Statistical optimization and kinetic studies on removal of Zn2+ using functionalized carbon nanotubes and magnetic biochar. *J. Environ. Chem. Eng.* **2013**, *1*, 486–495. [CrossRef]

117. Doumer, M.E.; Rigol, A.; Vidal, M.; Mangrich, A.S. Removal of Cd, Cu, Pb, and Zn from aqueous solutions by biochars. *Environ. Sci. Pollut. Res.* **2016**, *23*, 2684–2692. [CrossRef]
118. Van Hien, N.; Valsami-Jones, E.; Vinh, N.C.; Phu, T.T.; Tam, N.T.T.; Lynch, I. Effectiveness of different biochar in aqueous zinc removal: Correlation with physicochemical characteristics. *Bioresour. Technol. Reports* **2020**, *11*, 100466. [CrossRef]
119. Alyüz, B.; Veli, S. Kinetics and equilibrium studies for the removal of nickel and zinc from aqueous solutions by ion exchange resins. *J. Hazard. Mater.* **2009**, *167*, 482–488. [CrossRef] [PubMed]
120. Doğan, M.; Alkan, M.; Demirbaş, Ö.; Özdemir, Y.; Özmetin, C. Adsorption kinetics of maxilon blue GRL onto sepiolite from aqueous solutions. *Chem. Eng. J.* **2006**, *124*, 89–101. [CrossRef]
121. Ho, Y.S. Review of second-order models for adsorption systems. *J. Hazard. Mater.* **2006**, *136*, 681–689. [CrossRef]
122. Ahmed, M.B.; Zhou, J.L.; Ngo, H.H.; Guo, W.; Johir, M.A.H.; Sornalingam, K. Single and competitive sorption properties and mechanism of functionalized biochar for removing sulfonamide antibiotics from water. *Chem. Eng. J.* **2017**, *311*, 348–358. [CrossRef]
123. Park, Y.; Sun, Z.; Ayoko, G.A.; Frost, R.L. Bisphenol A sorption by organo-montmorillonite: Implications for the removal of organic contaminants from water. *Chemosphere* **2014**, *107*, 249–256. [CrossRef] [PubMed]
124. Zhou, N.; Chen, H.; Xi, J.; Yao, D.; Zhou, Z.; Tian, Y.; Lu, X. Biochars with excellent Pb(II) adsorption property produced from fresh and dehydrated banana peels via hydrothermal carbonization. *Bioresour. Technol.* **2017**, *232*, 204–210. [CrossRef]
125. Yang, Q.; Gao, M.; Luo, Z.; Yang, S. Enhanced removal of bisphenol A from aqueous solution by organo-montmorillonites modified with novel Gemini pyridinium surfactants containing long alkyl chain. *Chem. Eng. J.* **2016**, *285*, 27–38. [CrossRef]
126. Saeed, A.A.H.; Harun, N.Y.; Sufian, S.; Bilad, M.R.; Zakaria, Z.Y.; Jagaba, A.H.; Ghaleb, A.A.S.; Mohammed, H.G. Pristine and Magnetic Kenaf Fiber Biochar for Cd^{2+} Adsorption from Aqueous Solution. *Int. J. Environ. Res. Public Health* **2021**, *18*, 7949. [CrossRef] [PubMed]
127. Srivastava, V.; Sharma, Y.C.; Sillanpää, M. Response surface methodological approach for the optimization of adsorption process in the removal of Cr(VI) ions by $Cu_2(OH)_2CO_3$ nanoparticles. *Appl. Surf. Sci.* **2015**, *326*, 257–270. [CrossRef]
128. Chen, H.; Chen, Z.; Zhao, G.; Zhang, Z.; Xu, C.; Liu, Y.; Chen, J.; Zhuang, L.; Haya, T.; Wang, X. Enhanced adsorption of U(VI) and 241Am(III) from wastewater using Ca/Al layered double hydroxide@carbon nanotube composites. *J. Hazard. Mater.* **2018**, *347*, 67–77. [CrossRef] [PubMed]
129. Cazetta, A.L.; Pezoti, O.; Bedin, K.C.; Silva, T.L.; Paesano Junior, A.; Asefa, T.; Almeida, V.C. Magnetic Activated Carbon Derived from Biomass Waste by Concurrent Synthesis: Efficient Adsorbent for Toxic Dyes. *ACS Sustain. Chem. Eng.* **2016**, *4*, 1058–1068. [CrossRef]
130. Lingamdinne, L.P.; Choi, Y.L.; Kim, I.S.; Yang, J.K.; Koduru, J.R.; Chang, Y.Y. Preparation and characterization of porous reduced graphene oxide based inverse spinel nickel ferrite nanocomposite for adsorption removal of radionuclides. *J. Hazard. Mater.* **2017**, *326*, 145–156. [CrossRef]
131. Li, M.; Liu, H.; Chen, T.; Dong, C.; Sun, Y. Synthesis of magnetic biochar composites for enhanced uranium(VI) adsorption. *Sci. Total Environ.* **2019**, *651*, 1020–1028. [CrossRef]
132. Nguyen, V.H.; Van, H.T.; Nguyen, V.Q.; Dam, X.V.; Hoang, L.P.; Ha, L.T.; Ha, L.T. Magnetic Fe_3O_4 Nanoparticle Biochar Derived from Pomelo Peel for Reactive Red 21 Adsorption from Aqueous Solution. *J. Chem.* **2020**, *2020*. [CrossRef]
133. Yang, X.J.; Zhan, J.Y.; Yu, J.; Liu, G.; Huang, R.X. Adsorption features and kinetics of bisphenol A onto magnetic composite organic sepiolite. *Chin. J. Environ. Eng.* **2016**, *10*, 3597–3602.
134. Zhao, Y.; Zhang, R.; Liu, H.; Li, M.; Chen, T.; Chen, D.; Zou, X.; Frost, R.L. Green preparation of magnetic biochar for the effective accumulation of Pb(II): Performance and mechanism. *Chem. Eng. J.* **2019**, *375*, 122011. [CrossRef]
135. Song, X.; Zhang, Y.; Cao, N.; Sun, D.; Zhang, Z.; Wang, Y.; Wen, Y.; Yang, Y.; Lyu, T. Sustainable chromium (VI) removal from contaminated groundwater using nano-magnetite-modified biochar via rapid microwave synthesis. *Molecules* **2021**, *26*, 103. [CrossRef]
136. Yi, Y.; Huang, Z.; Lu, B.; Xian, J.; Tsang, E.P.; Cheng, W.; Fang, J.; Fang, Z. Magnetic biochar for environmental remediation: A review. *Bioresour. Technol.* **2020**, *298*, 122468. [CrossRef]
137. Wang, J.; Liao, Z.; Ifthikar, J.; Shi, L.; Chen, Z.; Chen, Z. One-step preparation and application of magnetic sludge-derived biochar on acid orange 7 removal via both adsorption and persulfate based oxidation. *RSC Adv.* **2017**, *7*, 18696–18706. [CrossRef]

polymers

MDPI

Article

Cytocompatibility of Polymethyl Methacrylate Honeycomb-like Pattern on Perfluorinated Polymer

Klaudia Hurtuková [1], Veronika Juřicová [1], Klára Fajstavrová [1], Dominik Fajstavr [1], Nikola Slepičková Kasálková [1], Silvie Rimpelová [2], Václav Švorčík [1] and Petr Slepička [1,*]

[1] Department of Solid State Engineering, University of Chemistry and Technology Prague, Technická 3, 166 28 Prague, Czech Republic; klaudia.hurtukova@vscht.cz (K.H.); veronika.juricova@vscht.cz (V.J.); klara.fajstavrova@vscht.cz (K.F.); dominik.fajstavr@vscht.cz (D.F.); nikola.kasalkova@vscht.cz (N.S.K.); vaclav.svorcik@vscht.cz (V.Š.)

[2] Department of Biochemistry and Microbiology, University of Chemistry and Technology Prague, Technická 3, 166 28 Prague, Czech Republic; silvie.rimpelova@vscht.cz

* Correspondence: petr.slepicka@vscht.cz; Tel.: +420-220-445-162

Citation: Hurtuková, K.; Juřicová, V.; Fajstavrová, K.; Fajstavr, D.; Slepičková Kasálková, N.; Rimpelová, S.; Švorčík, V.; Slepička, P. Cytocompatibility of Polymethyl Methacrylate Honeycomb-like Pattern on Perfluorinated Polymer. *Polymers* **2021**, *13*, 3663. https://doi.org/10.3390/polym13213663

Academic Editors: Antonio M. Borrero-López, Concepción Valen-cia-Barragán, Esperanza Cortés Triviño, Adrián Tenorio-Alfonso and Clara Delgado-Sánchez

Received: 30 September 2021
Accepted: 22 October 2021
Published: 24 October 2021

Publisher's Note: MDPI stays neutral with regard to jurisdictional claims in published maps and institutional affiliations.

Abstract: In this study, we present a simple approach for developing a biocompatible polymer scaffold with a honeycomb-like micropattern. We aimed to combine a plasma treatment of fluorinated ethylene propylene (FEP) substrate with an improved phase separation technique. The plasma exposure served for modification of the polymer surface properties, such as roughness, surface chemistry, and wettability. The treated FEP substrate was applied for the growth of a honeycomb-like pattern from a solution of polymethyl methacrylate (PMMA). The properties of the pattern were strongly dependent on the conditions of plasma exposure of the FEP substrate. The physico-chemical properties of the prepared pattern, such as changes in wettability, aging, morphology, and surface chemistry, were determined. Further, we have examined the cellular response of human osteoblasts (U-2 OS) on the modified substrates. The micropattern prepared with a selected combination of surface activation and amount of PMMA for honeycomb construction showed a positive effect on U-2 OS cell adhesion and proliferation. Samples with higher PMMA content (3 and 4 g) formed more periodic hexagonal structures on the surface compared to its lower amount (1 and 2 g), which led to a significant increase in the pattern cytocompatibility compared to pristine or plasma-treated FEP.

Keywords: polymer; plasma modification; PMMA; material morphology; honeycomb-like pattern; cell growth; cytocompatibility; FEP; cell adhesion

1. Introduction

Honeycombs are one of the most studied naturally occurring shapes/structures, which is evidenced by many new reported studies [1]. The impressive hexagonal shape of a honeycomb structure has fascinated many scientists for several decades, especially the research groups of Francois [2,3], Pirk, Zhang [1,4–7], or Bui et al. [8–10]. There were a lot of theories that competed for periodic structures. Two of them have got considerable attention in this field. The first theory states that a bee's six legs ideally fit onto a hexagon. On the contrary, the second theory supports a mathematical explanation that their geometry enables the coverage of the largest possible area using a minimum of building material [11,12]. However, the latest new theory confirms that hexagonal structures are formed due to highly elastic wax, elevated temperature, and surface tension, transforming circular shapes into hexagonal [12–14]. Based on this study, honeycomb-like patterns (HCP) have been identified as unique surfaces due to thousands of interconnected and oriented pores with high mechanical strength [15,16]. Based on the pore size and porosity, three-dimensional (3D) scaffolds resembling the in vivo shape and conditions can be created. This has great potential in biomedical applications, especially in tissue engineering. Porous scaffolds are essential for tissue nutrition, cell proliferation, and the formation of new viable tissues [15,17,18].

Moreover, the porous scaffolds have also protective and storage functions, the latter of which serves for depositing of adhesion molecules and growth factors. Another possible way of utilization is the formation of specific systems for drug delivery [19]. Otherwise, the HCP structures have found a great use in macro-, micro-, and nano-applications in electronics [20], electro-optics [21,22], bioanalytics [23], or in nanofabrication [24].

In several studies, the processes of HCP preparation by various methods including lithography or soft lithography [25–27], template techniques [27], emulsions, and plasma etching [28] have been described. However, one of the fundamental techniques invented by Francois et al. has become the so-called breath figure (BF) method. This method has proven to be cost-effective, uncomplicated, flexible, fast, and does not require specific skills to create a micro-/nano-pattern of honeycombs [2,3]. The principle consists of applying a polymer solution in a non-polar solvent to a substrate located in an environment with a relative humidity higher than 50%. A humid atmosphere can be ensured in two ways: static or dynamic. The static method requires atmosphere saturation with water vapor in a closed vessel, while in the dynamic method, the environment is created by the flow of moist air onto the substrate surface [29–31]. At the beginning of the process, the solvent is endothermically evaporated from the prepared solution. As the surface temperature decreases, water vapor begins to condense in the form of heterogeneous droplets on the surface of the substrate. After equilibration between the substrate and the environment, the condensed droplets begin to evaporate, thus, spontaneously leaving the hexagonal HCP structures on the substrate surface. The change of key units such as soluble substance, solvent, and humid air creates a wide range of highly ordered porous films with different pore sizes [28,32]. While the BF method requires a humid atmosphere for HCP formation, the newer method of improved phase separation (IPS) invented by Bui et al. [8] requires typical air conditions. The IPS technique uses two steps to prepare thin films on the surface of economically undemanding polymeric materials [8,33,34]. The method is based on a tertiary system: solvent (chloroform)–polymer–"no" solvent (methanol), in the correct volume ratio. To achieve the preparation of HCP structures is necessary to immerse the substrate in the prepared solution and then, spontaneously evaporate it in the air [16,35–37]. The advantage of this method, except for the humid environment and other active substances, is the exact ratio of solvents [8].

The development of new methods for creating scaffolds that are structurally and functionally similar to the extracellular matrix is one of the main challenges of tissue engineering. One possibility is to use polymeric materials, mainly due to their simplicity, flexibility, and low cost compared to other materials, and to prove their usefulness as extracellular matrix (ECM) replacements [38]. These properties and many others such as low density, environmental friendliness, and good physical and mechanical properties have been found in polymethyl methacrylate (PMMA) and its derivatives [35], which have been approved as bioinert and biocompatible polymers used in biomedical applications [36,37,39]. Nanocomposites based on PMMA exhibit excellent results in a body environment and have a protective role against corrosion and, thus, increase the durability of biomedical implants [40]. PMMA is commonly used in dentistry in dental prostheses [38,41] as well as bone cement [42–44]. Moreover, PMMA has been also used in the field of drug delivery or for the production of contact and implanted lenses, blood pumps and reservoirs, intravenous sets, and blood dialysis membranes [40,45].

In this study, we have focused on the preparation of HCP structures with the addition of PMMA in different amounts/ratios. Surface morphology, chemical composition, wettability, and cytocompatibility of the formed material were investigated for the particular HCP-coated foils. Samples with higher PMMA content formed more periodic hexagonal structures on the FEP surface, as well as cytocompatibility testing of cell culture was more efficient.

2. Materials and Methods

2.1. Materials and Chemicals

In this study, we used a PMMA polymer foil with a thickness of 50 μm and a density of 1.19 g·cm^{-3}, which was purchased from Goodfellow Ltd. (Cambridge, Huntingdon, UK). As solvents, we used methanol (MeOH, HPLC purity, M_r = 32.04 g·mol^{-1}, the density of 0.791 g·cm^{-3}; purchased from Penta, Prague, The Czech Republic), and chloroform ($CHCl_3$, stabilized with ~1% ethanol A.G., M_r = 119.38 g·mol^{-1}, the density of 1.48 g·cm^{-3}; purchased from Penta, Prague, The Czech Republic). As a substrate, we used fluorinated ethylene propylene polymer (FEP, the thickness of 50 μm, the density of 2.15 g·cm^{-3}; purchased from Goodfellow Ltd., Cambridge, Huntingdon, UK).

2.2. Pattern Preparation and Modification

The polymer film was treated by Ar^{+} plasma using the SCD 050 sputtering device from BAL-TEC (Baltec AG, Balzers, Liechtenstein). The purity of gas in the chamber was 99.997%, and the pressure was 8 Pa. The samples were placed onto a circular holder (an anode) with a diameter of 10 cm and a distance of 5 cm from the cathode. They were modified at 8 W for 240 s.

In the next step, HCP-like structures were formed using the IPS method via immersion of plasma-modified polymeric substrates in the prepared 100 mL solution for 10 s. The solution was a mixture of two solvents, namely chloroform and methanol, at a volume ratio of 85:15. Then, the amount of 1, 2, 3, and 4 g of PMMA foils was added to the mixture while stirring for 2–3 h at room temperature (22 °C), which gave a homogeneous solution. Then, the HCP-like structures were removed and left to air dry in Petri dishes. After complete evaporation of the solvents, the samples were prepared for further modification and subjected to examination by various analytical methods (see Section 2.3).

2.3. Analytical Methods

The wettability of all samples was studied via goniometric measurements of the contact angles, with a drop of distilled water applied to the surface of each sample (6 positions). The contact angles of the samples were determined using a See System goniometer (Advex Instruments, Brno, Czech Republic) at room temperature. A drop of water (8 μL) was applied onto each sample with a Transferpette® automatic pipette (Brand, Wertheim, Germany).

Atomic force microscopy (AFM) was used to study the surface morphology of the FEP substrate with HCP microstructures. The movement of the tip as it passed over the sample was recorded, and a point-by-point image of the surface was compiled. Sample surface morphology was examined by atomic force microscope (AFM) Dimension ICON (Bruker Corp., Billerica, MA, USA); the ScanAsyst mode in the air was used for determination. A silicon tip on a nitride lever SCANASYST-AIR with the spring constant of 0.4 N·m^{-1} was used. NanoScope Analysis software was applied for data processing. The mean roughness values (R_a) represent the average of the deviations from the center plane of the sample.

Scanning electron microscopy (SEM) and energy-dispersive X-ray spectroscopy (EDS) were used for detailed analysis of the morphology and chemical characterization of the FEP substrate with HCP microstructures. We used a scanning electron microscope LYRA3 GMU (Tescan, Brno, The Czech Republic) with the accelerating voltage of 10 kV for the electrons that bombarded the samples and an F-MaxN analyzer and SDD (Silicon Drift Detector) (Oxford Instruments, Abingdon, UK) with an applied accelerating voltage of 10 kV for EDS. Platinum (target, purity 99.999%, Safina, Vestec, The Czech Republic) was sputtered onto the samples before analysis using a Quorum Q300T sputtering device (Quorum Technologies, Laughton, East Sussex, England) at a current of 30 mA for 400 s.

2.4. Cell Culture

VAMPIRO U-2 OS cells supplied by Innoprot (Derio, Spain) were used to evaluate the material cytocompatibility. The VAMPIRO U-2 OS are cells derived from human osteosarcoma developed through stable transfection of U-2 OS cell line with tFP602 under puromycin resistance gene. The VAMPIRO U-2 OS cells exhibit red fluorescence since they express the red fluorescent protein gene sequence as free cytoplasmatic protein. The cells were cultured in high-glucose Dulbecco's modified Eagle medium (DMEM; Sigma-Aldrich, St. Louis, MO, USA) containing a stable L-glutamine dipeptide. The medium was supplemented with 10% fetal bovine serum (FBS; certified, Thermo Fisher Scientific, Waltham, MA, USA). The cells were cultivated in a sterile cell culture incubator at 37 °C, atmosphere with 5% CO_2 and 95% humidity. The cells were regularly passaged when they were at the exponential phase of growth. First, the cells were washed with 5 mL of sterile phosphate-buffered saline (PBS, pH 7.4). After its removal, the cells were detached from the underlying cell culture dish (VWR International, Radnor, PA, USA) with 1 mL of trypsin-EDTA (ethylenediaminetetraacetic acid).

2.5. Cell Seeding and Staining

To evaluate the material cytocompatibility, the VAMPIRO U-2 OS cells were used from passages 4–8 (counted from defrosting after the cell delivery). The prepared material samples were subjected to sterilization by immersion into 70% (volume percent) ethanol in water for 30 min. Then, the samples were inserted into cell culture plates with 12 wells (the diameter of 2.14 cm; VWR, Radnor, PA, USA), washed twice with sterile PBS, and weighted down by hollow PMMA cylinders (Zenit, Prague, The Czech Republic). Then, 1 mL of DMEM containing 15,000 cells per one squared centimeter was added to each well onto the evaluated samples in four biological replicates. The endpoints of cultivation were 24 and 96 h, after which the medium was removed, the cells were washed with prewarmed PBS, and fixed with a 4% methanol-free formaldehyde solution in PBS as described in [46]. The fixation occurred for 20 min. at 22 °C in the dark, then, the fixative was removed, and the cells were washed twice with cold PBS. This was followed by staining of the cell structures. Cell nuclei were stained with 0.5 $\mu g \cdot mL^{-1}$ of 4′,6-diamidino-2-phenylindole (DAPI; Sigma-Aldrich, St. Louis, MO, USA) and cell cytoskeleton was stained with 2 $\mu g \cdot mL^{-1}$ of phalloidin conjugated to Atto 488 (Sigma-Aldrich, St. Louis, MO, USA) in PBS for 15 min. in the dark at 22 °C. Then, the staining solution was removed, and the cells were washed with cold PBS. The standard deviation was calculated from four biological replicates each including twenty regions of interest, on which the number of grown cells was counted.

2.6. Fluorescence Microscopy

The fixed and stained VAMPIRO U-2 OS cell samples were subjected to wide-field fluorescence microscopy using an inverse fluorescence microscope Olympus IX-81 (Olympus, Tokyo, Japan) operated with xCellence software. The fluorescently labeled cells were examined at 100× (10× objective, NA of 0.30), 200× (20× objective, NA of 0.45), and 400× (40× objective, NA of 0.60) magnification. Cell cytoplasm, cytoskeleton, and nuclei were imaged by a triple quad filter DAPI/FITC/TRITC (Olympus, Tokyo, Japan). The images were captured by an EM-CCD camera (Hamamatsu, Honshu, Japan). Then, the fluorescence images were background-corrected, deconvolved (50%), and all three fluorescence channels (DAPI, FITC, TRITC) were merged.

3. Results

3.1. Surface Morphology Analysis Using SEM and AFM Method

In this study, we have aimed to prepare HCP structures with incorporated PMMA (different content of 1, 2, 3, and 4 g) on plasma-modified FEP matrices. Plasma treatment affects polymer structure resulting in free radical formation. This phenomenon can increase the movement of molecular particles (the ability of a polymer to diffuse) and stimulate the etching rate of the surface layer. The duration of Ar^+ particle deposition (plasma) significantly affects the polymer topography. Short exposure of a polymer to the plasma treatment creates nanometric structures while changing the polymer surface into a porous form for a longer time [47]. Based on previous research [16,25], the plasma treatment parameters for creating the suitable surface of HCP structures were selected: duration 240 s at the power of 8 W. The changes on the prepared HCP structures' surface morphology were analyzed by two different methods, AFM and SEM. Figure 1 represents the scans of the prepared samples acquired by the SEM analysis. The analyzed area is divided into two regions for samples with an amount of PMMA higher than 1 g. The first area shows the major heterogeneous site (herein called "major"), which confirms the occurrence of inhomogeneous hexagonal pores. While the second area (herein called "minor") detected 2 mm from the sample edges of the formed regular hexagonal microstructures. The phenomenon of the split surface is probably caused by different evaporation rates of the solvent mixture on the sample surface, and even though that the difference in patterns was not significant, it was still present on all examined samples, therefore, we wanted to point this out also in the description of Figure 1.

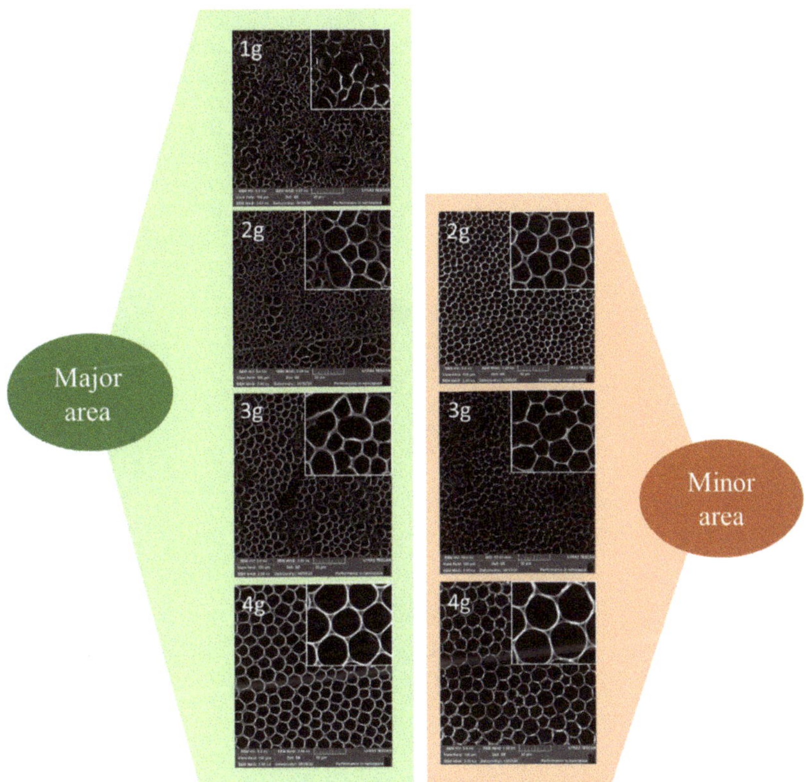

Figure 1. Scanning electron microscopy images of honeycomb-like pattern major and minor areas formed on plasma-modified fluorinated ethylene propylene containing different amounts (1–4 g) of incorporated polymethyl methacrylate. The scanned areas were 100 × 100 μm^2.

From the SEM images, it can be concluded that a sample containing 1 g of PMMA was the least favorable substrate for forming hexagonal structures. The remaining concentrations of PMMA confirmed the formation of HCP structures. As the amount of PMMA increased, the pore size also increased as well as the pattern homogeneity, and other smaller cavities were layered below each other to different levels. A slight difference in pore sizes was observed when comparing the major and minor areas of HCP on the sample surface. The minor area of the HCP was detected up to 2 mm from the outer diameter. Even the difference in pattern properties in the outer and inner areas was not significant for samples prepared with 3 and 4 g of PMMA, we wanted to introduce this fact in this article for all studied patterns. If we focus specifically on images containing 4 g of PMMA, we conclude that the major area provided pores of up to 4 μm in diameter and the minor area contained also larger pores of up to 10 μm in diameter. In the minor areas, when compared to the major areas, there was a more precise distribution of cavities of almost the same size. These trends were maintained also for samples with a polymer layer containing 2–4 g of PMMA. However, no considerable variation in pore formation was observed for samples formed from a solution containing 3 and 4 g of PMMA.

Minor areas of the samples with a polymer layer of 1–4 g PMMA were also studied by the AFM analysis as shown in Figure 2. From the results obtained with AFM and SEM, it can be concluded that the pore sizes on the substrates with the same concentration of PMMA remained similar for both studied methods. Therefore, both dimensions and shape of the pattern could be considered as verified. Furthermore, it should be added that the analyzed surface roughness and the height of the pores formed did not change dramatically with the increasing amount of PMMA. The only significant difference among the samples was on the substrate containing the highest content of PMMA (4 g). On the AFM images, we can observe the formation of pores with a slightly smaller diameter in the major area, compared to the edges of the sample (minor area), however, the differences were not significant for samples prepared with 3 and 4 g of PMMA. Even though that the AFM analysis of the sample's surface was repeated several times. We would like to point out that the SEM scans were acquired on larger sample areas than AFM scans, therefore, the measurement can be considered as more "accurate".

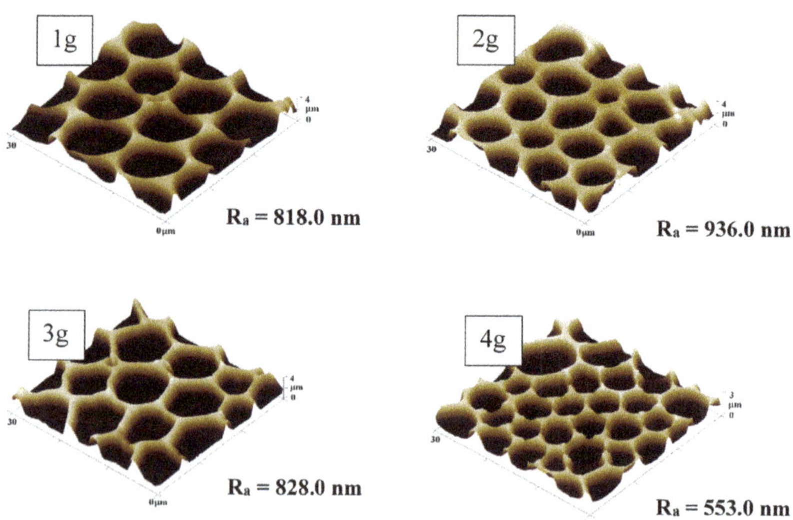

Figure 2. The surface morphology of honeycomb-like pattern minor areas on plasma-modified fluorinated ethylene propylene with different amounts (1–4 g) of incorporated polymethyl methacrylate. The scanned areas were 30 × 30 μm^2. R_a represents the average roughness in nm.

3.2. Surface Chemical Analysis

Qualitative and quantitative data on the elemental composition of the prepared substrates were determined by the EDS method and are summarized in Table 1 and Figure 3. The samples were analyzed before any modification (pristine FEP), after plasma deposition at 8 W for 240 s (plasma FEP), and after preparation of HCP structures from PMMA solution of various content 1–4 g (HCP + 1 g PMMA). The chemical distribution of major and minor areas of the prepared samples was studied by the EDS technique and described in Section 3.1.

Table 1. The atomic concentrations of fluorine, oxygen, and carbon in at. % were determined using the energy-dispersive X-ray spectroscopy for unmodified pristine fluorinated ethylene propylene (FEP), plasma-treated FEP, and with a honeycomb-like pattern (HCP) major and minor structure with incorporated polymethyl methacrylate (PMMA) of various content (1–4 g).

Sample	C [%]	O [%]	F [%]
Pristine FEP	45.5	0.0	54.5
Plasma-treated FEP	41.9	1.8	56.3
HCP + 1 g PMMA (major)	58.3	2.8	38.9
HCP + 2 g PMMA (major)	74.0	3.6	22.4
HCP + 3 g PMMA (major)	74.3	4.1	21.6
HCP + 4 g PMMA (major)	86.4	7.2	6.4
HCP + 2 g PMMA (minor)	93.6	6.3	0.1
HCP + 3 g PMMA (minor)	91.0	6.6	2.4
HCP + 4 g PMMA (minor)	92.9	7.0	0.1

From Table 1, it is evident that no oxygen element was detected on the pristine FEP compared to other modified samples. The explanation is the characteristic of the FEP polymer chain containing only C and F elements [48]. One of the reasons for the increased oxygen concentration is plasma treatment, which causes that the ionized Ar ions to hit the polymer surface. This process induces new functional groups that bind oxygen from the surrounding atmosphere after the sample is removed from the chamber or during the process if a slight amount of oxygen is present in the atmosphere in the chamber [49,50]. Another reason is the presence of PMMA honeycombs, which contain a large amount of oxygen in their structure [51]. The subsequent phase separation increases the carbon content and, conversely, decreases the fluorine concentration, confirming a polymeric PMMA layer formation. The presence of PMMA may limit the content of the F element, as shown in Table 1. The highest tested amount of PMMA (4 g) formed surface with minimum fluorine content, which was only 6.4%. It is interesting to note that if we observed a minor area of the sample (minor), the F element almost disappeared, i.e., the substrate was fully covered with a PMMA polymer layer. For a better interpretation of the surface's chemical composition, the scans from EDS of the sample with 2 g PMMA are attached in Figure 3.

3.3. *Surface Wettability*

Wettability, i.e., the contact angle values, is a crucial property for material characterization. It plays an essential role in applications of technological (cleaning, gluing, dyeing) [52,53] or medical processes (cell or bacterial adhesion) [54]. The value of a contact angle has a significant impact on the material surface and characterizes the hydrophilicity/hydrophobicity of a polymer [55].

The graph shown in Figure 4 provides data on the surface wettability of all evaluated samples during an aging period of 34 days. The aging study was performed on the unmodified FEP substrate (pristine FEP), plasma-modified FEP substrate at 8 W for 240 s (plasma FEP), and the substrate after plasma modification applied with a layer of PMMA at various amounts (1–4 g). The value of the contact angle of the unmodified FEP sample was 98.9°. As it is obvious from Figure 4, the samples with HCP had contact angles in the region of slight hydrophobicity ranging from 90–110° with fluctuation during the aging period. However, the slightly hydrophobic character remained unchanged and stable during the aging period. The most apparent change in contact angle values was detected for the plasma-modified FEP substrate. On the first day, the plasma FEP had a contact angle of 50.2°, and on the third day, it increased rapidly to 96.8°. Contact angle did not change significantly throughout the measurement and was around the values of the pristine FEP. This could be caused by the redirection of polar groups from the material's surface to its volume due to the preservation of samples in the ambient atmosphere. On the other hand, an atmosphere with relatively low humidity and temperature can slow down the material aging process. All FEP samples covered with PMMA maintained a hydrophobic surface with contact angles within 10° of the value of unmodified FEP (98.9°). The influencing of contact angle by various patterns including honeycomb structure was reported in [56], where also superhydrophibic effect was observed.

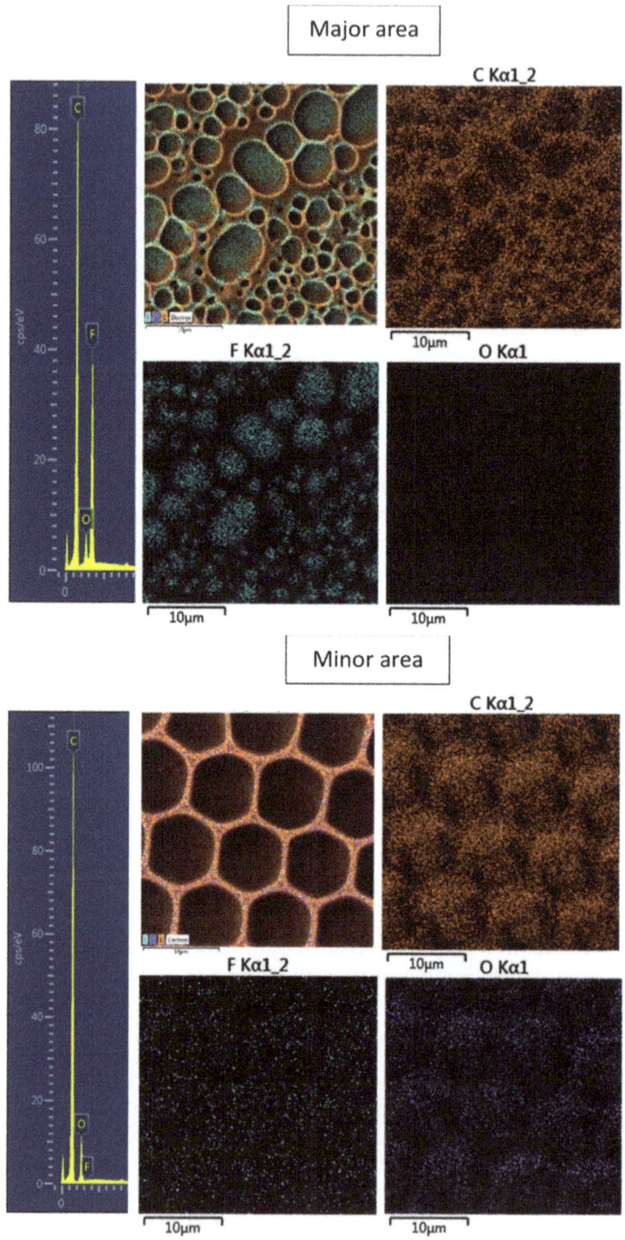

Figure 3. A spectrum from energy-dispersive X-ray spectroscopy (EDS) analysis and elemental maps with layered EDS images for the honeycomb-like pattern major and minor structures on plasma-treated fluorinated ethylene propylene.

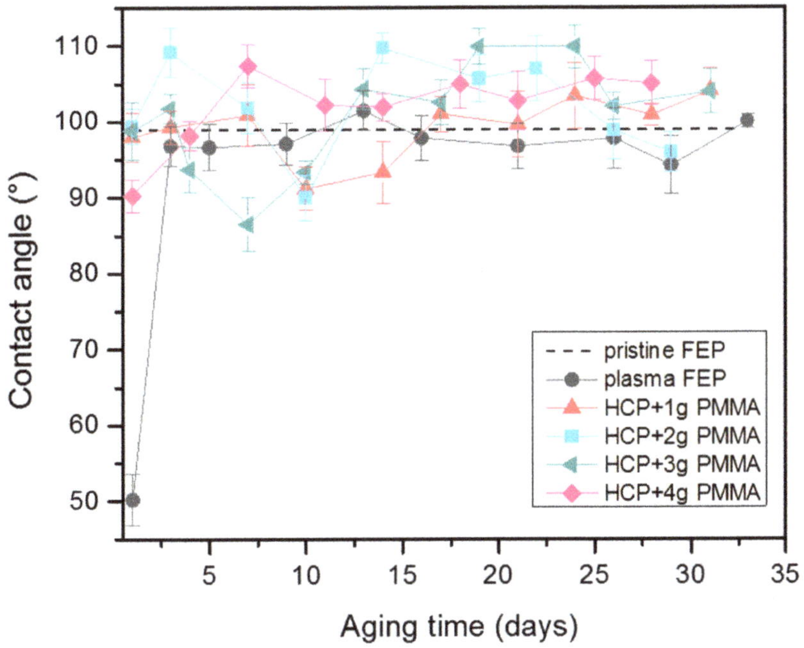

Figure 4. Water contact angles measurements over 34 days for pristine fluorinated ethylene propylene (FEP), plasma-treated FEP (plasma FEP), and honeycomb-like pattern (HCP) structure with incorporated polymethyl methacrylate (PMMA) of various content (1–4 g).

3.4. Cytocompatibility

Human cells derived from osteosarcoma stably transfected with a red fluorescent protein tFP602 (VAMPIRO U-2 OS cells) were chosen as a model cell line for cytocompatibility measurement of the prepared materials. Samples of pristine FEP, plasma-modified FEP, and FEP with a PMMA layer were subjected to this study. Tissue culture polystyrene (TCPS) was chosen as a control since this material is routinely used for tissue culture. From the plot in Figure 5, it is evident that after 24 h post-seeding, there was no statistically significant difference in the number of cells adhered on modified or unmodified samples compared to the control sample of TCPS. It can be caused by the cell adaptation process to the samples after their seeding [57]. After additional cultivation, 96 h post-seeding, cell proliferation was obvious on each evaluated sample. The lowest number of cells was detected on the unmodified FEP sample, which is consistent with our previous results, e.g., [58–60]. The cell number was increased on plasma-treated FEP 96 h after seeding when compared to pristine FEP. This was caused by the plasma exposure, which creates new oxygen-containing groups (carbonyl and carboxyl mainly) in combination with significant wettability change, enabling the design of excellent scaffolds with increased biocompatibility [60]. However, the best cytocompatibility was observed on FEP samples with HCP structures and PMMA layers, since a markedly pronounced increase in the cell number was detected. The number of cells of these samples was equal to that of the control TCPS sample. The improved VAMPIRO U-2 OS cell proliferation on the modified samples might be caused by the porous character of the HCP-modified sample surface or the presence of PMMA. The presence of 3D HCP structure mimics the extracellular matrix necessary as scaffolds for proper cell growth and proliferation [61], the effect also observed for laser induced periodic surface structures [24,62]. On the other hand, the PMMA material exhibits biocompatible properties, such as surface chemistry in combination with morphology properties [36,37,39,40] that could increase cell proliferation on the substrate surface compared to pristine and plasma-modified FEP.

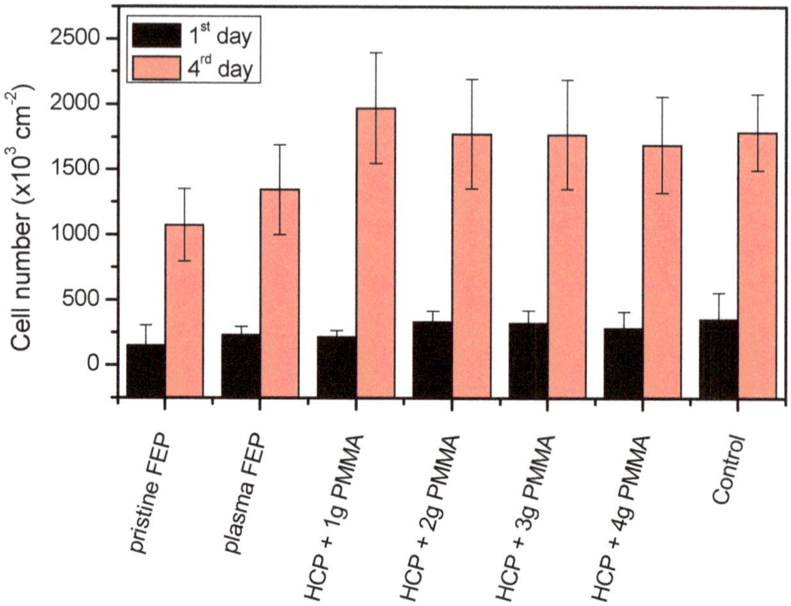

Figure 5. The number of adhered (24 h) and proliferated (96 h) VAMPIRO U-2 OS cells on a pristine and plasma-treated (plasma) fluorinated ethylene propylene (FEP) film and plasma-treated FEP films subsequently coated with honeycomb-like patterns formed from different polymethyl methacrylate (PMMA) solutions (1–4 g) and tissue culture polystyrene (TCPS, control).

Another evaluation of the cellular response, morphology, was studied on samples with cultured U-2 OS cells for 1 and 4 days, shown in Figure 6. Images of the cell samples were captured by fluorescence microscopy. After the first day, round cells were detected with irregular spreading on pristine FEP samples and with a layer of 3 g and 4 g of PMMA, which did not correspond to the shape of osteoblasts. In comparison, the control sample showed a cell morphology similar to the shape of osteoblasts. Samples with plasma modification and 1 g and 2 g PMMA had a cell shape similar to the control, but their distribution was not regular. After day 4, all prepared samples were completely covered with cells, except for the pristine FEP sample. This phenomenon could be due to the smooth surface of the unmodified FEP. The shape of the cells after day 4 can no longer be recognized in all sample types.

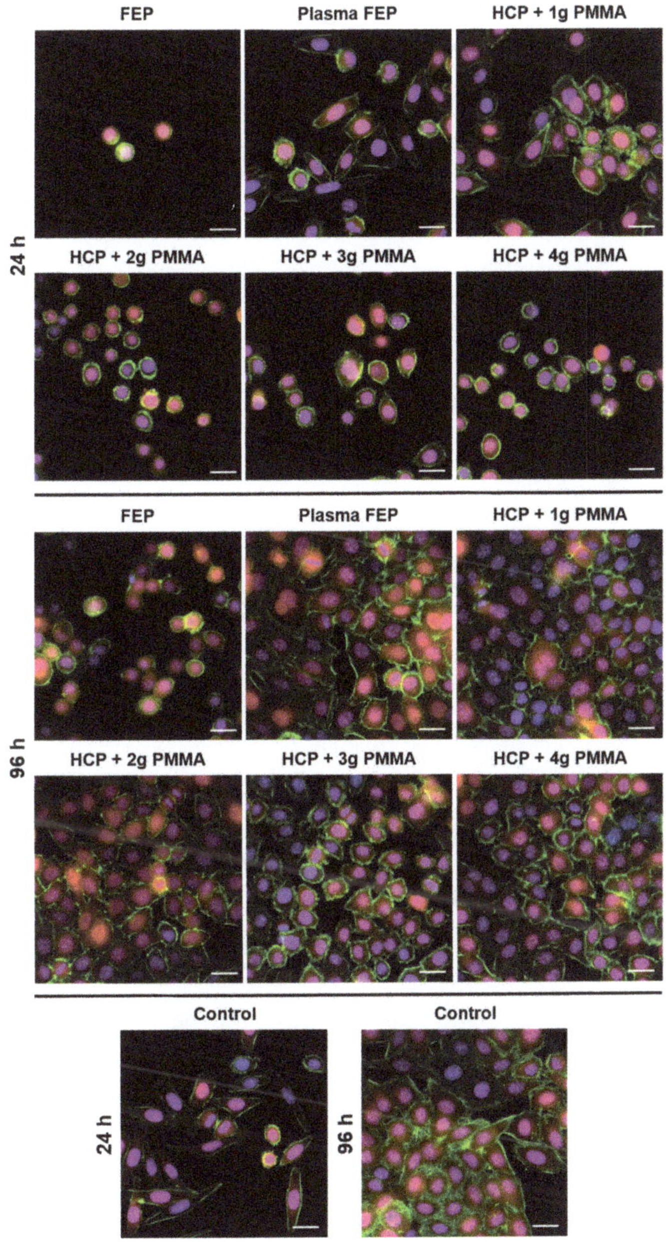

Figure 6. Fluorescence microscopy images of VAMPIRO U-2 OS cells (human cells derived from osteosarcoma, stably transfected with tFP602) growing on fluorinated ethylene propylene (FEP) treated by 8W plasma for 240 s and subsequently coated with a honeycomb-like pattern formed from polymethyl methacrylate (PMMA) solution containing its different amount (1, 2, 3, and 4 g). Left and right columns-cells 24 and 96 h post-seeding, respectively. As controls, tissue culture polystyrene (TCPS), pristine FEP, and plasma-treated (8 W, 240 s) FEP matrices were used. In green—cell cytoskeleton, in blue—cell nuclei, in red—cell cytoplasm. Scale bar represents 20 μm.

4. Conclusions

In this study, we prepared an HCP from PMMA on plasma-treated perfluorinated polymer FEP. The plasma treatment was confirmed to play a crucial role in the pattern formation since the surface wettability and chemistry were significantly altered and, thus, the process of IPS was successfully applied. The pattern aging revealed the stabilization of contact angle values in a range of 90–110° for all types of studied PMMA micropatterns. The surface morphology of the HCP was confirmed by both AFM and SEM, by which we revealed the formation of two areas of PMMA micropatterns on the FEP surface. The micropatterns slightly varied in their surface properties, which was verified by SEM surface mapping. Pristine unmodified FEP did not support the adhesion and proliferation of human osteoblasts (VAMPIRO U-2 OS cells) due to low surface wettability, roughness, and the absence of oxygen. Further, we have confirmed that the plasma exposure of FEP itself slightly improved the cytocompatibility of this substrate. The subsequent process of HCP formation from PMMA led to improvement of adhesion and growth of VAMPIRO U-2 OS cells, and the cells were able to grow in clusters according to the shape of the pattern even after 96 h of cultivation. This study also opens the wide possibility to prepare surfaces with the possibility of drug release due to an extreme increase in effective surface area for particular HCP structures.

Author Contributions: Conceptualization, P.S. and K.F.; methodology, S.R. and N.S.K.; validation, S.R. and K.H.; formal analysis, D.F. and V.Š.; investigation, S.R., V.J., D.F. and P.S.; data curation, V.Š. and N.S.K.; writing—original draft preparation, K.H. and V.J.; writing—review and editing, S.R. and P.S.; supervision, P.S. and K.F.; funding acquisition, P.S. All authors have read and agreed to the published version of the manuscript.

Funding: This research was funded by the GACR under project 19-02495S.

Institutional Review Board Statement: Not applicable.

Informed Consent Statement: Not applicable.

Data Availability Statement: Not applicable.

Acknowledgments: The authors would like to kindly acknowledge the *Polymers* journal and *Polymers* editorial office for providing us the APC waiver.

Conflicts of Interest: The authors declare no conflict of interest.

Abbreviations

3D	Three dimensional
AFM	Atomic force microscopy
BF	Breath figure
EDS	Energy-dispersive X-ray spectroscopy
FEP	Fluorinated ethylene propylene
HCP	Honeycomb-like pattern
HPLC	High-performance liquid chromatography
IPS	Improved phase separation
PMMA	Polymethyl methacrylate
SEM	Scanning electron microscopy
TCPS	Tissue culture polystyrene
U-2 OS	Human cells from osteosarcoma

References

1. Karihaloo, B.L.; Zhang, K.; Wang, J. Honeybee combs: How the circular cells transform into rounded hexagons. *J. R. Soc. Interface* **2013**, *10*, 20130299. [CrossRef]
2. Bekkar, F.; Bettahar, F.; Meghabar, R.; Hamadouche, M.; Moreno-Benitez, I.; Vilas-Vilela, J.L.; Ruiz-Rubio, L. Study of the capacity of poly(*N*-vinylcarbazole) derivatives to form honeycomb-like patterns. *J. Appl. Polym. Sci.* **2021**, *138*, 50975. [CrossRef]

3. Bui, V.T.; Ko, S.H. Large-scale fabrication of commercially available, nonpolar linear polymer film with a highly ordered honeycomb pattern. *ACS Appl. Mater. Interfaces* **2015**, *7*, 10541–10547. [CrossRef]
4. Svečnjak, L.; Chesson, L.A.; Gallina, A.; Maia, M.; Martinello, M.; Mutinelli, F.; Muz, N.M.; Nunes, F.M.; Saucy, F.; Tipple, B.J.; et al. Standard methods for *Apis mellifera* beeswax research. *J. Apic. Res.* **2019**, *58*, 1–108. [CrossRef]
5. Zhang, A.; Bai, H.; Li, L. Breath figure: A nature-inspired preparation method for ordered porous films. *Chem. Rev.* **2015**, *115*, 9801–9868. [CrossRef]
6. Zhang, S.; Ren, J.; Chen, S.; Luo, Y.; Bai, X.; Ye, L.; Yang, F.; Cao, Y. Large area electrochromic displays with ultrafast response speed and high contrast using solution-processable and patternable honeycomb-like polyaniline nanostructures. *J. Electroanal. Chem.* **2020**, *870*, 114248. [CrossRef]
7. Hepburn, H.R.; Muerrle, T.; Radloff, S.E. The cell bases of honeybee combs. *Apidologie* **2007**, *38*, 268–271. [CrossRef]
8. Bui, V.T.; Thi Thuy, L.; Chinh Tran, Q.; Nguyen, V.T.; Dao, V.D.; Sig Choi, J.; Choi, H. Ordered honeycomb biocompatible polymer films via a one-step solution immersion phase separation used as a scaffold for cell cultures. *Chem. Eng. J.* **2017**, *320*, 561–569. [CrossRef]
9. Dao, V.D.; Bui, V.T.; Choi, H.S. Pt-coated cylindrical micropatterned honeycomb Petri dishes as an efficient TCO-free counter electrode in liquid junction photovoltaic devices. *J. Power Sources* **2018**, *376*, 41–45. [CrossRef]
10. Bui, V.T.; Thi Thuy, L.; Choi, J.S.; Choi, H.S. Ordered cylindrical micropatterned Petri dishes used as scaffolds for cell growth. *J. Colloid Interface Sci.* **2018**, *513*, 161–169. [CrossRef]
11. Hales, T.C. The honeycomb conjecture. *Discret. Comput. Geom.* **2001**, *25*, 1–22. [CrossRef]
12. Hales, T.C. Dělové koule a včelí plásty. *Pokroky Mat. Fyziky Astron.* **2001**, *46*, 101–118.
13. Nazzi, F. The hexagonal shape of the honeycomb cells depends on the construction behavior of bees. *Sci. Rep.* **2016**, *6*, 28341. [CrossRef]
14. Pirk, C.W.W.; Hepburn, H.R.; Radloff, S.E.; Tautz, J. Honeybee combs: Construction through a liquid equilibrium process? *Naturwissenschaften* **2004**, *91*, 350–353. [CrossRef]
15. Sari, M.; Hening, P.; Chotiman, S.E.; Ana, D.I.; Yusuf, Y. Bioceramic hydroxyapatite-based scaffold with a porous structure using honeycomb as a natural polymeric Porogen for bone tissue engineering. *Biomater. Res.* **2021**, *25*, 2. [CrossRef] [PubMed]
16. Hurtuková, K.; Fajstavrová, K.; Rimpelová, S.; Vokatá, B.; Fajstavr, D.; Slepičková Kasálková, N.; Siegel, J.; Švorčík, V.; Slepička, P. Antibacterial properties of a honeycomb-like pattern with cellulose acetate and silver nanoparticles. *Materials* **2021**, *14*, 4051. [CrossRef]
17. Loh, Q.L.; Choong, B.; Choong, C. Three-dimensional scaffolds for tissue engineering applications: Role of porosity and pore size. *Tissue Eng. B Rev.* **2013**, *19*, 485–502. [CrossRef]
18. Tu, C.; Cai, Q.; Yang, J.; Wan, Y.; Bei, J.; Wang, S. The fabrication and characterization of poly(lactic acid) scaffolds for tissue engineering by improved solid–liquid phase separation. *Polym. Adv. Technol.* **2003**, *14*, 565–573. [CrossRef]
19. Calejo, M.T.; Ilmarinen, T.; Skottman, H.; Kellomäki, M. Breath figures in tissue engineering and drug delivery: State-of-the-art and future perspectives. *Acta Biomater.* **2017**, *66*, 44–66. [CrossRef]
20. Huang, K.; Pan, Q.; Yang, F.; Ni, S.; Wei, X.; He, D. Controllable synthesis of hexagonal WO_3 nanostructures and their application in lithium batteries. *J. Phys. D Appl. Phys.* **2008**, *41*, 155417. [CrossRef]
21. Ungár, T.; Gubicza, J.; Ribárik, G.; Borbély, A. Crystallite size distribution and dislocation structure determined by diffraction profile analysis: Principles and practical application to cubic and hexagonal crystals. *J. Appl. Crystallogr.* **2004**, *34*, 298–310. [CrossRef]
22. Niezgoda, S.R.; Kanjarla, A.K.; Beyerlein, I.J.; Tomé, C.N. Stochastic modeling of twin nucleation in polycrystals: An application in hexagonal close-packed metals. *Int. J. Plast.* **2014**, *56*, 119–138. [CrossRef]
23. Farjadian, F.; Azadi, S.; Mohammadi-Samani, S.; Ashrafi, H.; Azadi, A. A novel approach to the application of hexagonal mesoporous silica in solid-phase extraction of drugs. *Heliyon* **2018**, *4*, e00930. [CrossRef] [PubMed]
24. Slepicka, P.; Siegel, J.; Lyutakov, O.; Slepickova Kasalkova, N.; Kolska, Z.; Bacakova, L.; Svorcik, V. Polymer nanostructures for bioapplications induced by laser treatment. *Biotechnol. Adv.* **2018**, *36*, 839–855. [CrossRef] [PubMed]
25. Slepička, P.; Neználová, K.; Fajstavr, D.; Slepičková Kasálková, N.; Švorčík, V. Honeycomb-like pattern formation on perfluoroethylenepropylene enhanced by plasma treatment. *Plasma Process. Polym.* **2019**, *16*, 1900063. [CrossRef]
26. Karthaus, O.; Maruyama, N.; Cieren, X.; Shimomura, M.; Hasegawa, H.; Hashimoto, T. Water-assisted formation of micrometer-size honeycomb patterns of polymers. *Langmuir* **2000**, *16*, 6071–6076. [CrossRef]
27. Ke, B.B.; Van, L.S.; Zhang, W.X.; Xu, Z.K. Controlled synthesis of linear and comb-like glycopolymers for preparation of honeycomb-patterned films. *Polymer* **2010**, *51*, 2168–2176. [CrossRef]
28. Munoz-Bonilla, A.; Fernández-García, M.; Rodríguez-Hernández, J. Towards hierarchically ordered functional porous polymeric surfaces prepared by the breath figures approach. *Prog. Polym. Sci.* **2014**, *39*, 510–554. [CrossRef]
29. Liu, Q.; Wu, Y.; Li, Z. Facile preparation of super-hydrophobic fabrics composed of fibres with microporous or microspherical coatings using the static breath figure method. *Prog. Org. Coat.* **2020**, *149*, 105938. [CrossRef]
30. Fajstavrová, K.; Rimpelová, S.; Fajstavr, D.; Švorčík, V.; Slepička, P. PLLA honeycomb-like pattern on fluorinated ethylene propylene as a substrate for fibroblast growth. *Polymers* **2020**, *12*, 2436. [CrossRef]
31. Bui, V.T.; Ko, S.; Choi, H.S. A surfactant-free bio-compatible film with a highly ordered honeycomb pattern fabricated via an improved phase separation method. *Chem. Commun.* **2014**, *50*, 3817. [CrossRef]

32. Dou, Y.; Jin, M.; Zhou, G.; Shui, L. Breath figure method for construction of honeycomb films. *Membranes* **2015**, *5*, 399–424. [CrossRef] [PubMed]
33. Bui, V.T.; Dao, V.D.; Choi, H.S. Transferable thin films with sponge-like porous structure via improved phase separation. *Polymer* **2018**, *101*, 184–191. [CrossRef]
34. Qi, S.; Moffat, J.G.; Yang, Z. Early stage phase separation in pharmaceutical solid dispersion thin films under high humidity: Improved spatial understanding using probe-based thermal and spectroscopic nanocharacterization methods. *Mol. Pharm.* **2013**, *10*, 918–930. [CrossRef] [PubMed]
35. Shirzad, M.; Matbouei, A.; Fathi, A.; Rabiee, S.M. Experimental and numerical investigation of polymethyl methacrylate scaffolds for bone tissue engineering. *Proc. Inst. Mech. Eng. Part L J. Mater. Des. Appl.* **2020**, *234*, 586–594. [CrossRef]
36. Tan, H.Y.; Loke, W.K.; Nguyen, N.T. A reliable method for bonding polydimethylsiloxane (PDMS) to polymethylmethacrylate (PMMA) and its application in micropumps. *Sens. Actuators B* **2010**, *151*, 133–139. [CrossRef]
37. Zafar, M.S. Prosthodontic applications of polymethyl methacrylate (PMMA): An update. *Polymers* **2020**, *12*, 2299. [CrossRef] [PubMed]
38. Matbouei, A.; Fathi, A.; Rabiee, S.M.; Shirzad, M. Layered manufacturing of a three-dimensional polymethyl methacrylate (PMMA) scaffold used for bone regeneration. *Mater. Technol.* **2018**, *34*, 167–177. [CrossRef]
39. Ali, U.; Loke, W.K.; Bt. Abd Karim, K.J.; Buang, N.A. A Review of the properties and applications of poly(methyl methacrylate) (PMMA). *Polym. Rev.* **2015**, *55*, 678–705. [CrossRef]
40. Harb, S.V.; Bassous, N.J.; de Souza, T.A.C.; Trentin, A.; Pulcinelli, S.H.; Santilli, C.V.; Webster, T.J.; Lobo, A.O.; Hammer, P. Hydroxyapatite and β-TCP modified PMMA-TiO_2 and PMMA-ZrO_2 coatings for bioactive corrosion protection of Ti6Al4V implants. *Mater. Sci. Eng. C* **2020**, *116*, 111149. [CrossRef]
41. Gautam, R.; Singh, R.D.; Sharma, V.P.; Siddhartha, R.; Chand, P.; Kumar, R. Biocompatibility of polymethylmethacrylate resins used in dentistry. *J. Biomed. Mater. Res.* **2012**, *100B*, 1444–1450. [CrossRef]
42. Webb, J.C.J.; Spencer, R.F. The role of polymethylmethacrylate bone cement in modern orthopaedic surgery. *J. Bone Jt. Surg.* **2007**, *89-B*, 851–857. [CrossRef]
43. Arora, M.; Chan, E.K.; Gupta, S.; Diwan, A.D. Polymethylmethacrylate bone cements and additives: A review of the literature. *World J. Orthop.* **2013**, *4*, 67–74. [CrossRef] [PubMed]
44. Vallo, C.I.; Montemartini, P.E.; Lopez, J.M.P.; Cuadrado, T.R. Polymethylmethacrylate-based bone cement modified with hydroxyapatite. *J. Biomed. Mater. Res.* **1999**, *48*, 150–158. [CrossRef]
45. Samad, H.A.; Jaafar, M.; Othman, R.; Kawashita, M.; Razak, N.H.A. New bioactive glass-ceramic: Synthesis and application in PMMA bone cement composites. *Biomed. Mater. Eng.* **2011**, *21*, 247–258. [CrossRef]
46. Juřík, P.; Slepička, P.; Mistrík, J.; Janíček, P.; Rimpelová, S.; Kolská, Z.; Švorčík, V. Oriented gold ripple-like structures on poly-l-lactic acid. *Appl. Surf. Sci.* **2014**, *321*, 503–510. [CrossRef]
47. Hassan, A.; Abd El Aal, S.A.; Shehata, M.M.; El-Saftawy, A.A. Plasma etching and modification of polyethylene for improved surface structure, wettability and optical behavior. *Surf. Rev. Lett.* **2018**, *26*, 1850220. [CrossRef]
48. Ebnesajjad, S. Introduction to fluoropolymers. In *Applied Plastics Engineering Handbook*; Elsevier: Amsterdam, The Netherlands, 2017; pp. 55–71. [CrossRef]
49. Chu, P.K.; Chen, J.Y.; Wang, L.P.; Huang, N. Plasma-surface modification of biomaterials. *Mater. Sci. Eng. R Rep.* **2002**, *36*, 143–206. [CrossRef]
50. Slepička, P.; Trostová, S.; Slepičková Kasálková, N.; Kolská, Z.; Sajdl, P.; Švorčík, V. Surface modification of biopolymers by argon plasma and thermal treatment. *Plasma Process. Polym.* **2011**, *9*, 197–206. [CrossRef]
51. Li, R.; Wu, G.; Hao, Y.; Peng, J.; Zhai, M. Radiation degradation or modification of poly(tetrafluoroethylene) and natural polymers. In *Radiation Technology for Advanced Materials*; Elsevier: Amsterdam, The Netherlands, 2019; pp. 141–182. [CrossRef]
52. Furstner, R.; Barthlott, W.; Neinhuis, C.; Walzel, P. Wetting and self-cleaning properties of artificial superhydrophobic surfaces. *Langmuir* **2005**, *21*, 956–961. [CrossRef]
53. Wang, X.; Wang, F.; Yu, Z.; Zhang, Y.; Qi, C.; Du, L. Surface free energy and dynamic wettability of wood simultaneously treated with acidic dye and flame retardant. *J. Wood Sci.* **2017**, *63*, 271–280. [CrossRef]
54. Junkar, I. Interaction of cells and platelets with biomaterial surfaces treated with gaseous plasma. *Adv. Biomembr. Lipid Self-Assem.* **2016**, *36*, 25–59. [CrossRef]
55. Polini, A.; Yang, F. Chapter Hydrophilicity: Physicochemical characterization of nanofiber composites. In *Nanofiber Composites for Biomedical Applications*, 1st ed.; Elsevier Science & Technology: Amsterdam, The Netherlands, 2017; pp. 97–115. [CrossRef]
56. Sun, J.; Li, Y.; Liu, G.; Chu, F.; Chen, C.; Zhang, Y.; Tian, H.; Song, Y. Patterning a Superhydrophobic Area on a Facile Fabricated Superhydrophilic Layer Based on an Inkjet-Printed Water-Soluble Polymer Template. *Langmuir* **2020**, *36*, 9952–9959. [CrossRef]
57. Beijer, N.R.M.; Nauryzgaliyeva, Z.M.; Arteaga, E.M.; Pieuchot, L.; Anselme, K.; van de Peppel, J.; Vasilevich, A.S.; Groen, N.; Roumans, N.; Hebels, D.G.A.J.; et al. Dynamic adaptation of mesenchymal stem cell physiology upon exposure to surface micropatterns. *Sci. Rep.* **2019**, *9*, 9099. [CrossRef]
58. Fajstavrová, K.; Rimpelová, S.; Fajstavr, D.; Švorčík, V.; Slepička, P. Cell Behavior of Primary Fibroblasts and Osteoblasts on Plasma-Treated Fluorinated Polymer Coated with Honeycomb Polystyrene. *Materials* **2021**, *14*, 889. [CrossRef]
59. Neznalová, K.; Fajstavr, D.; Rimpelová, S.; Slepičková Kasálková, N.; Kolská, Z.; Švorčík, V.; Slepička, P. Honeycomb-patterned poly(L-lactic) acid on plasma-activated FEP as cell culture scaffold. *Polym. Deg. Stab.* **2020**, *181*, 109370. [CrossRef]

60. Slepička, P.; Peterková, L.; Rimpelová, S.; Pinker, A.; Slepičková Kasálková, N.; Kolská, Z.; Ruml, T.; Švorčík, V. Plasma activated perfluoroethylenepropylene for cytocompatibility enhancement. *Polym. Degrad. Stab.* **2016**, *130*, 277–287. [CrossRef]
61. Frantz, C.; Stewart, K.M.; Weaver, V.M. The extracellular matrix at a glance. *J. Cell Sci.* **2010**, *123*, 4195–4200. [CrossRef]
62. Slepicka, P.; Slepickova Kasalkova, N.; Siegel, J.; Kolska, Z.; Bacakova, L.; Svorcik, V. Nano-structured and functionalized surfaces for cytocompatibility improvement and bactericidal action. *Biotechnol. Adv.* **2015**, *33*, 1120–1129. [CrossRef]

polymers

MDPI

Article

Development of Biodegradable Bio-Based Composite for Bone Tissue Engineering: Synthesis, Characterization and In Vitro Biocompatible Evaluation

Muhammad Umar Aslam Khan [1,2,3,*], Saiful Izwan Abd Razak [1,2,4,*], Mohamed Nainar Mohamed Ansari [5,*], Razauden Mohamed Zulkifli [2], Nurliyana Ahmad Zawawi [2] and Muhammad Arshad [3]

[1] BioInspired Device and Tissue Engineering Research Group, School of Biomedical Engineering and Health Sciences, Faculty of Engineering, Universiti Teknologi Malaysia, Skudai 81300, Johor, Malaysia
[2] Department of Biosciences, Faculty of Science, Universiti Teknologi Malaysia, Skudai 81300, Johor, Malaysia; razauden@utm.my (R.M.Z.); nurliyana@utm.my (N.A.Z.)
[3] National Center for Physics, Nanosciences and Technology Department, Quaid-i-Azam University Islamabad Campus, Islamabad 44000, Pakistan; Arshad_pr2002@yahoo.com
[4] Centre of Advanced Composite Materials, Faculty of Engineering, Universiti Teknologi Malaysia, Skudai 81300, Johor, Malaysia
[5] Institute of Power Engineering, Universiti Tenaga Nasional, Kajang 43000, Selangor, Malaysia
* Correspondence: umar007khan@gmail.com (M.U.A.K.); saifulizwan@utm.my (S.I.A.R.); ansari@uniten.edu.my (M.N.M.A.)

Citation: Khan, M.U.A.; Razak, S.I.A.; Ansari, M.N.M.; Zulkifli, R.M.; Ahmad Zawawi, N.; Arshad, M. Development of Biodegradable Bio-Based Composite for Bone Tissue Engineering: Synthesis, Characterization and In Vitro Biocompatible Evaluation. *Polymers* **2021**, *13*, 3611. https://doi.org/10.3390/polym13213611

Academic Editors: Antonio M. Borrero-López, Concepción Valencia-Barragán, Esperanza Cortés Triviño, Adrián Tenorio-Alfonso and Clara Delgado-Sánchez

Received: 18 September 2021
Accepted: 15 October 2021
Published: 20 October 2021

Abstract: Several significant advancements in the field of bone regenerative medicine have been made in recent years. However, therapeutic options, such as bone grafts, have several drawbacks. There is a need to develop an adequate bone substitute. As a result, significant bone defects/injuries pose a severe challenge for orthopaedic and reconstructive bone tissue. We synthesized polymeric composite material from arabinoxylan (ARX), β-glucan (BG), nano-hydroxyapatite (nHAp), graphene oxide (GO), acrylic acid (AAc) through free radical polymerization and porous scaffold fabricated using the freeze-drying technique. These fabricated porous scaffolds were then coated with chitosan solution to enhance their biological activities. The complex structure of BG, nHAp, GO was studied through various characterization and biological assays. The structural, morphological, wetting and mechanical analyses were determined using FT-IR, XRD, XPS, SEM/EXD, water contact angle and UTM. The swelling (aqueous and PBS media) and degradation (PBS media) observed their behavior in contact with body fluid. The biological activities were conducted against mouse pre-osteoblast cell lines. The result found that BGH3 has desirable morphological, structural with optimum swelling, degradation, and mechanical behavior. It was also found to be cytocompatible against MC3T3-E1 cell lines. The obtained results confirmed that the fabricated polymeric scaffolds would be a potential bone substitute to regenerate defective bone with different loading bearing applications for bone tissue engineering.

Keywords: biopolymer; biomaterials; biodegradations; polysaccharide; bone tissue engineering

1. Introduction

Bone is a connective tissue supporting and protecting the rest of the body systems and organs by storing bone marrow, minerals and ions. Bone constantly remodels to adapt to the mechanical environments and heal the minor lesions [1]. However, certain bone defects of critical dimensions may cause delayed and harming bone function restoration. Bone-related impairments and diseases are approximately half of all chronic diseases in people 50 years above, a big challenge in medical science [2]. Over the last century, a wide range of activities, strategies, and surgical materials were used to study to regenerate defective bone tissue. Bone grafts are used in bone tissue application to restore function and aesthetics that were damaged due to bone defects. To mimic the bone tissue to be regenerated, the bone substitute's physicochemical and biomechanical properties should be recognized, including architecture, pore size distribution, biocompatibility, degradability, osteoconduction osteoinductivity and osteogenesis. Biomaterials are categorized based on their source (allogeneic, alloplastic, autogenous and xenogeneic, etc.), stoichiometry composition (ceramics and bioglass and polymers, etc.), or interaction with biological systems [3,4]. Sizeable segmental bone defects cannot be repaired by themselves, even though bones can heal and/or regenerate. The "gold standard" for repairing bone defects is still autologous bone grafting. Autologous bone grafting is affected by several issues (i.e., secondary damage, high donor site morbidity, minimal shape, autogenous bone insufficiency, etc.). Because of these flaws, it is not widely used in the clinical system [5].

A biodegradable scaffold is placed as a temporary implant into the impaired sites of defective bones to support and activate bone regeneration which is slowly degraded and replaced by the new bone in bone tissue engineering. Tissue engineering scaffolds were developed and tested using bioactive ceramics and polymers with similar chemical cosmetics of a natural bone, supporting osteogenesis and bonds with the host tissue [6,7]. Bioceramics, despite their beneficial biochemical activity, are brittle and biodegrade slowly, limiting their medical applications. In contrast, biopolymers offer several advantages over ceramics. Their mechanical and biodegradation features can be optimized significantly for some applications. These are especially well-suited to implant in the body and known for easy term formation into desired orientations and shapes [8,9]. Little mechanical hardness and shape retention breakdown are the two most severe issues with polymer scaffolds.

For use as scaffolds, many types of natural and synthetics polymers were investigated. The scaffolds used in tissue repairing are bone, tendon, cartilage, skeletal muscle, skin, ligament, vascular tissues, and neural tissues. Natural polymers, which are comprised of proteins (such as collagen, silk and fibrin, etc.) or polysaccharides (including chitosan, alginate, starch, and arabinoxylan, etc.), have restricted mechanical characteristics, immunogenic prospect, and demand [10–12]. Poly(glycolic acid) (PGA), poly(L-lactic acid) (PLLA), and their copolymers, such as poly(DL-lactic-co-glycolic acid), are some of the most common synthetic polymers (PLGA). However, these synthetic polymers prevent cellular growth in a 3D structure due to poor cell adhesion and hydrophobic surfaces [13]. They are also deficient in specific functional groups, which could be used to alter the surface further. Previous synthetic biopolymers implanted in vivo discharged acidic degradation products and triggered a long-term immune response harmful to the host tissues. Furthermore, during polymer erosion, some specific bulk hydrolyzing PLGA copolymers were observed to considerably diminish the osteogenesis of the regenerating bone [14].

Polysaccharides are the most common type of carbohydrate in foods. These are polymeric carbohydrates made up of monosaccharide units linked by glycosidic linkages that form long chains. Chitosan is a polysaccharide cationic biopolymer that belongs to the polysaccharide class of biodegradable biopolymers. It is non-toxic, biodegradable, and non-antigenic. Unlike synthetic polymers, chitosan has a hydrophobic nature that encourages cell adhesion, differentiation, and proliferation while causing only a minor inflammatory response when implanted [15]. Chitosan scaffolds are osteoconductive, meaning they can promote bone regeneration in vitro and in vivo. Despite its widespread recognition as a biocompatible compound, chitosan is mechanically fragile and unsound, and swelling causes it to lose its predetermined form for transplantation [16]. A natural polymer-based system has yet to be developed with a sufficiently porous scaffold with appropriate mechanical and biological properties comparable to or better than chitosan [17]. β-Glucans (beta-glucans) is a naturally occurring biopolymer composed of β-D-glucose polysaccharides. It is primarily found in bacteria, cereals, and fungi cell walls, and its physicochemical properties vary a lot depending on the source [18]. β-Glucans usually have a linear backbone with 1–3-glycosidic bonds. However, these differ in viscosity, solubility, molecular mass, branching structure, and gelation properties, resulting in many physiological effects [19].

Here we report a biodegradable porous composite scaffold, its synthesis and properties. Briefly, polymeric composites were synthesized using the free radical method from arabinoxylan (ARX), β-glucans (BG), nano-hydroxyapatite (nHAp), graphene oxide (GO), acrylic acid (AAc) and freeze-drying technique was used to fabricate porous scaffolds. These fabricated scaffolds were coated with chitosan to enhance their biological activities. X-ray diffraction (XRD), Fourier-transform infrared spectroscopy (FT-IR), scanning electron microscope (SEM), water contact angle, and universal testing machine (UTM) were used to characterize the structural, morphological, wetting analyses, and mechanical testing of the fabricated scaffolds, respectively. The swelling (aqueous and PBS media) and degradation analysis (PBS media) were conducted to investigate their behavior. The biological activities were determined against mouse pre-osteoblast (MC3T3-E1) cell lines.

2. Materials and Methods

2.1. Materials

B-glucan is a natural polymer extracted from barley flour and supplied by DSP Gokyo Food and Chemical Co., Ltd., Osaka, Japan. Acrylic acid (monomer), N, N′-methylene bisacrylamide (crosslinker) and graphene oxide (CAS# 763713-1G) were supplied by Sigma Aldrich Selangor, Malaysia. Hydroxyapatite nanoparticles (<100 nm particle size, >95%), phosphate buffer saline (PBS) solution and ethanol were supplied by Sigma-Aldrich, Selangor, Malaysia. All chemicals and reagents were used as received.

The osteoblast cells were purchased from the American Type Culture Collection (ATCC), Manassas, VA, USA. ThermoFisher Scientific (Waltham, MA, USA) supplied Alpha-MEM (α-MEM). L-glutamine penicillin/streptomycin and fetal bovine serum (FBS) were provided by Hyclone Laboratories Inc. (Logan, UT, USA).

2.2. Methods

2.2.1. Extraction and Purification of β-Glucan

β-glucan was extracted using a standard method with a slight modification [20]. Briefly, 100 g fine barley flour was dispersed into 500 mL deionized water and pH = 10 of the slurry was maintained by Na_2CO_3 (20%, *v*/*w*). For 30 min the suspension was stirred at 45 °C and centrifuged at 18,000× *g* for 20 min (4 °C and pH 4.5) to isolate protein contents. The precipitated proteins were removed, and equal amounts of absolute ethanol were added to the suspension with slow stirring to precipitate β-G. Then, the suspension was centrifuged (4000× *g*) for 15 min to get β-G precipitates and kept at 4 °C overnight. These were freeze-dried to have fine powder of β-G and packed in an airtight glass jar.

2.2.2. Extraction and Purification of Arabinoxylan

The psyllium (Plantago ovata) seed husk extracted arabinoxylan with slight modification by a well-reported method by Saghir et al. [21]. Briefly, the dust and stones were removed from psyllium seed husk (500 g) and put into deionized water (3 L) overnight. The swelled psyllium seed husk was then blended with NaOH soln. (2.5%) and stirring for 5 min. The insoluble seed husk was separated using a muslin cloth to get the pure gel. Then, it was coagulated by concentrated acetic acid. The resultant gel was neutralized using deionized water and freeze-dried to obtain a well-dried powder of arabinoxylan (ARX).

2.2.3. Polymeric Composite Synthesis

The polymeric composite materials were synthesized by free radical polymerization. Briefly, biopolymeric powder (ARX = 1 g and BG = 1 g) was dispersed in deionized water (25 mL) to have homogenized suspension. Different amounts of nHAp (1.4, 1.5 and 1.6) were dispersed in deionized water (10 mL) to have homogenized suspension by sonication. These all suspensions were transferred into a round bottom flask (two-necked) and stirred for 45 min. Then, GO (0.3 mg) was added to the reaction media and stirred for 30 min at 65 °C. The amount of GO was optimized in our previous studies. After, 65 °C, the monomer (AAc = 0.50 mL) and crosslinker (N,N-MBA = 0.05% of AAc) were added into well homogenized mixture. The free-radical polymerization reaction was initiated by adding potassium persulphate (0.05 g) and heated for 3 h at the same temperature under an inert environment with constant stirring. The reaction was stopped heating, and nitrogen gas was removed. The reaction media was cooled and vacuum filtered, and the residue was washed 3 times with deionized water to ensure the unreacted species were from the product. The synthesized polymeric composite was dried in the oven at 50 °C overnight and packed for further use, and the actual yield was found to be approximately 63.82%. The composition of materials is mentioned in Table 1. The codes were assigned after a different amount of nHAp. Each sample (2 g) dispersed into deionized water (5 mL) to make a homogenized slurry and shifted into aluminium moulds (h = 2 cm and d = 1 cm). These moulds were frozen at −40 °C for 24 h and lyophilized by freeze-drying methodology to have porous and crack-less well-dried scaffolds.

Table 1. Chemical composition of biomaterial to fabricate scaffolds.

Sample Name	ARX (g)	BG (g)	AAc (g)	GO (mg)	nHAp (g)
BGC1	1	1	0.05	0.03	1.4
BGC2	1	1	0.05	0.03	1.5
BGC3	1	1	0.05	0.03	1.6

3. Characterizations

3.1. Fourier-Transform Infrared Spectroscopy (FTIR)

The structural properties and chemical interactions were determined of fabricated scaffolds using FTIR (Shimadzu FTIR-8100A, Tokyo, Japan). The analysis was conducted using ATR mode, and 4000–400 cm^{-1} was scanning range. The average scan was 150 with a resolution of 4.0 cm^{-1}.

3.2. X-ray Diffraction (XRD)

The diffraction pattern of fabricated scaffolds was determined by X-ray diffractometry. The analysis was conducted using XRD (Bruker AXS D8 (Kontich, Belgium)) using CuKα radiation (λ = 1.540 Å) using the voltage at 30 kV with a current of 30 mA. The XRD spectrum was obtained with a fixed range of 10 to 60°. The scan speed was 1.5 min^{-1} and 0.03 o step size.

3.3. Scanning Electron Microscope (SEM)/Energy Dispersive X-ray (EDX)

The morphological analysis of fabricated scaffolds was conducted by SEM (JEOL-JSM 6480, Peabody, MA, USA). The SEM was coupled with EDS to analyze elemental composition. Before analysis, the fabricated scaffolds were gold-sputtered.

3.4. X-ray Photoelectron Spectroscopy (XPS)

The XPS analysis was performed by a Scienta-Omicron system coupled with a micro-focused monochromatic Al K-alpha X-ray source. The detailed survey scan analysis was performed by operating X-ray at 15 KeV with 700-micron spot size. The constant analyzer energy (CAE) was kept at 100 eV. The charging effects were avoided by applying a combined low energy/ion flood source. The Matrix software was used to perform data acquisition, and Igor pro was applied to analyze the XPS fitting procedure. The shrilly background was corrected, and Gaussian–Lorentzian line shape was used to study the Curve fitting of detailed spectra. The C1s were used as reference data by fixing binding energy at 284.8 eV.

3.5. Mechanical Testing

The mechanical testing was performed using a universal testing machine (UTM, Testometrics, Rochdale, UK) to obtain strain–stress curves. The strain–each fabricated cylindrical scaffold determined stress curve with dimensions 1.5 cm (width) and 1.7 cm (height) at 0.5 mm/min speed.

3.6. Wetting

The hydrophilicity/hydrophobicity was determined of fabricated scaffolds through wetting analysis. The contact angle meter (XCA-50) (VCA-Optima, AST Inc., Tacoma, WA, USA) determines the wetting behavior of fabricated scaffolds. The wetting was investigated at a different time interval to analyze the effect of time.

3.7. Swelling and Degradation Analysis

The swelling of well-dried fabricated scaffolds was determined in aqueous and PBS media at 37 °C. The fabricated samples of each scaffold were weighted (50 mg) and placed in corresponding media to determine swelling behavior. The samples were taken out after a specific interval, and the excess surface liquid was removed carefully by tissue paper to record actual weight. The scaffold sample was placed into media until equilibrium, and the swelling percentage was determined by Equation (1).

$$Swelling\ (\%) = \frac{W_s - W_d}{W_d} \times 100 \quad (1)$$

The degradation studies of fabricated scaffolds were observed in PBS media (pH 7.4 at 37 °C) for 30 days. The fabricated scaffold was weighed (50 mg) carefully and placed in PBS media. The scaffold samples were taken out after a specific time interval. The scaffolds were rinsed with deionized water and oven-dried weight. The percentage degradation was determined by Equation (2).

$$Degradation\ (\%) = \frac{W_i - W_t}{W_t} \times 100 \quad (2)$$

whereas: W_t = scaffold weight at "t" time, "W_i" initial scaffold weight.

3.8. In Vitro Biological Studies

3.8.1. Cell Morphology

The cell viability of the polymeric composite was studied against MC3T3-E1 cell lines in triplicates. The 24-well plate was coated by a fine layer of polymeric scaffold and sterilized using a UV light for 1 h. The cell lines were maintained in culture media (1% penicillin/streptomycin, 1% (2 mM) L-glutamine, 10% FBS and α-MEM. The cell lines (5000 cell/cm^2) were seeded in a 24-well plate triplicate by cell culture media. The 24-well plates were incubated for 24, 48 and 72 h under in vitro conditions, i.e., 37 °C, 5% CO_2 and 90% humidity. After removing cell culture media, the grown cells were fixed by absolute ethanol for 5 min at room temperature. The cell was gold-sputtered and SEM (JEOL-JSM 6480, Peabody, MA, USA) to capture the photographs at different intervals of time.

3.8.2. Cell Viability and Optical Density

The MC3T3-E1 cell lines were used to determine the scaffolds' cell viability and optical density by taking gelatin (0.1%) as +ive control, and 1% dimethyl sulfoxide (DMSO) was taken −ive control. These well plates were incubated at different time intervals (24, 48, and 72 h) under in vitro conditions (5% CO_2 at 37 °C with 90% humidity), as mentioned in our previous work [22]. The culture well plates were again incubated with neutral red (40 µg/mL) media for 2 h, Repetto et al. [23]. The excess NR stain was removed by washing cells with PBS solution for 20 min. These cells were destained by dye-staining solution (50% deionized water, 49% absolute ethanol and 1% glacial acetic acid) for 10 min. An absorbance microplate reader (ELx-800) (BioTek, Winooski, VT, USA) was used at 570 nm to record optical density. A fluorescence microscope (Nikon ECLIPS TS100) with a 488 nm excitation filter was used to study cell morphology. The background interference of the microscopy was fixed by Vital dye (fluorescein diacetate (FDA)), created by coated scaffolds. Equation (3) was used to calculate cell viability percentage.

$$Cell\ viability = \frac{OD_s}{OD_c} \times 100 \quad (3)$$

where OD_s = sample optical density and OD_c = control optical density.

3.9. Statistical Analysis

The statistical analysis was conducted by software (IBM SPSS Statistics 21) of the obtained data and presented in mean, standard error (S.E.) and have shown as Y-error bars in figures. ($n = 3$), and $p < 0.05$.

4. Results and Discussion

4.1. FTIR Analysis

The structural analysis of the porous scaffolds was analyzed by FTIR spectral profile as shown in Figure 1a. The vibration peaks at 571, 608 cm^{-1} and 1089 cm^{-1} present the triply degenerated P−O stretching and O−P−O bending (first two) of HAp. The characteristic bands from 560 to 600 cm^{-1} and 1000–1100 cm^{-1} exhibit moiety of calcium phosphate due to nHAp [22]. It was further noted that increasing amount of nHAp cause the appearance of 1146 and 1062 cm^{-1}, which are belonged to stretching vibrations of –C–O–C– and –C–O–, respectively, which is a clear indication that increasing nHAp amount decrease polymerization contents due to increased ceramic contents and decreased monomeric contents [24]. However, the –OH group confirmed HAp in the scaffold at 630 cm^{-1}. The band from 1750 to 1600 cm^{-1} is attributed to the C–O stretching of GO, NN-MBA and free hydroxyl (–OH) [25]. The peak at 2928 cm^{-1} is typical of saturated aliphatic C–H stretching vibrations, and broadband at 3100–3500 cm^{-1} is attributed to inter and intra-hydrogen bonding due to biopolymers (ARX, BG) [6]. The broadband also confirms the ceramic and other contents were linked via hydrogen bonding. These peaks and functional groups confirm the available interaction (covalent bonding, hydrogen and van der Waal forces, etc.) synthesis of the polymeric composite material via free-radical polymerization. Hence, the development of our desired composite scaffolds was demonstrated by the presence of these vibration peaks and bands.

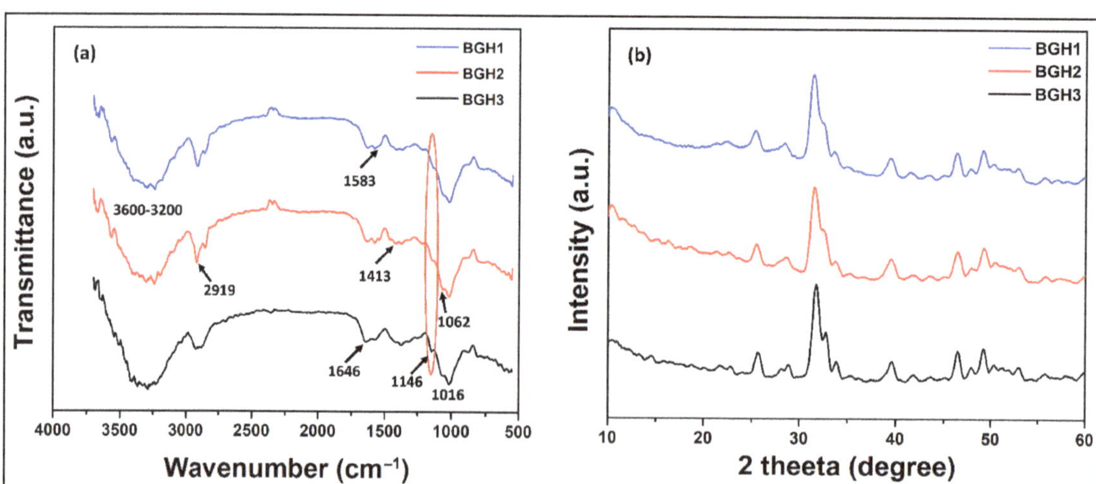

Figure 1. (a) FTIR spectra and (b) XRD pattern of porous scaffolds samples.

4.2. XRD Analysis

XRD analysis was conducted in the current research to observe the crystalline and phase analysis of inorganic content in polymeric scaffolds, as shown in Figure 1b. XRD pattern of polymeric scaffolds was compared by standard JCPDS (PDF-4-932) data for HAp. The peaks at 25.89, 31.86, 32.59, 39.77, 47.57 and 49.33 o have corresponding plans such as (201), (217), (300), (310), (222) and (213), and these are the characteristic peaks of hydroxyapatite. A sharp peak was observed for HAp with good crystallinity. The characteristic peak of HAp with the highest intensity was found at a 2θ value of 31.86° with corresponding planes (211) and indicated a highly crystalline form of HAp [6]. The XRD pattern of the polymeric matrix has peaked for nHAp, AAc, GO, BG and ARX. The polymeric and ceramic contents have retained their crystalline phase in polymeric scaffolds, confirming the substantial change in their crystal structure. It also indicates the uniform and fine distribution of HAp in the polymeric matrix. It is analogous to the natural bone in which HAp is homogeneously dispersed into a collagen matrix [26]. The week and distinguish peaks at 21.78, 23.81 and 26.34 represented characteristic peaks of the polymeric matrix that establish HAp incorporation in polymeric scaffolds. The dispersion of HAp in polymeric solution and more significant diffraction of HAp nanoparticles have improved the crystalline phase and intensity of polymeric scaffolds. However, characteristic peaks of the polymeric matrix were observed in FTIR spectra. The entire composite scaffold exhibited an enhanced sharpness and intensities of characteristic HAp peaks. The increase in crystal sizes of HAp particles when introduced into polymeric scaffold could be likened to these reasons [27]. The decrease in crystalline behavior of BG-co-AAc was observed in composite polymeric materials with poor peaks for BG, GO and AAc. It can be due to the formation of covalent bonding, hydrogen bonding and other weak interaction during free-radical polymerization. These were developed in the synthesis of polymeric composite materials that imprints HAp into the polymeric matrix.

4.3. XPS Analysis

The XPS spectra collected from the as-synthesized BGH-3 sample is shown in Figure 2. The survey scan collected from BGH-3 shown in Figure 2a identified the carbon, oxygen, calcium, and phosphorus as we expected. The high-resolution scans of C1s, O1s, Ca2p and P2p, are shown in Figure 2b–e. The relative atomic percentages of all elements present on the surface of the sample and the binding energies position of all deconvoluted peaks are given in Table 2. The deconvoluted peaks of C1s spectra confirm sp2 bonded carbon (graphene) and functional groups [28]. Ca2p and P2p peaks correspond to calcium phosphate, and these peak assignments are also reported in the literature [29]. The first peak of O1s confirms the oxide of Ca, and other peaks are related to OH species [29]. Hence, the available different binding energy level confirms the different interaction during the synthesis of polymeric composite material via free radical polymerization.

Table 2. The peak position and relative atomic% of the scaffold samples.

Elements	Peaks	Deconvoluted Peak Positions (eV)	Spin-Orbit Splitting	Bonding	Relative Atomic%
Carbon (C1s)	peak 1	284.6 ± 0.05	–	C=C (sp2 carbon)	62.90
	peak 2	285.4 ± 0.05	–	C–O	
	peak 3	287 ± 0.05	–	C≥O	
	peak 4	288.5 ± 0.05	–	O–C=O	
Oxygen (O1s)	peak 1	531.1 ± 0.05	–	O–P–O	
	peak 2	532.51 ± 0.05	–	–OH	25.89
Calcium (Ca2p)	peak 1	347.39 ± 0.05	$2P_{3/2}$	CaO	
	peak 2	350.97 ± 0.05	$2P_{1/2}$	CaCO3	6.09
Phosphorous (P2p)	peak 1	133 ± 0.05	$2P_{3/2}$	O–P–O	5.12
	peak 2	133.89 ± 0.05	$2P_{1/2}$	P–O	

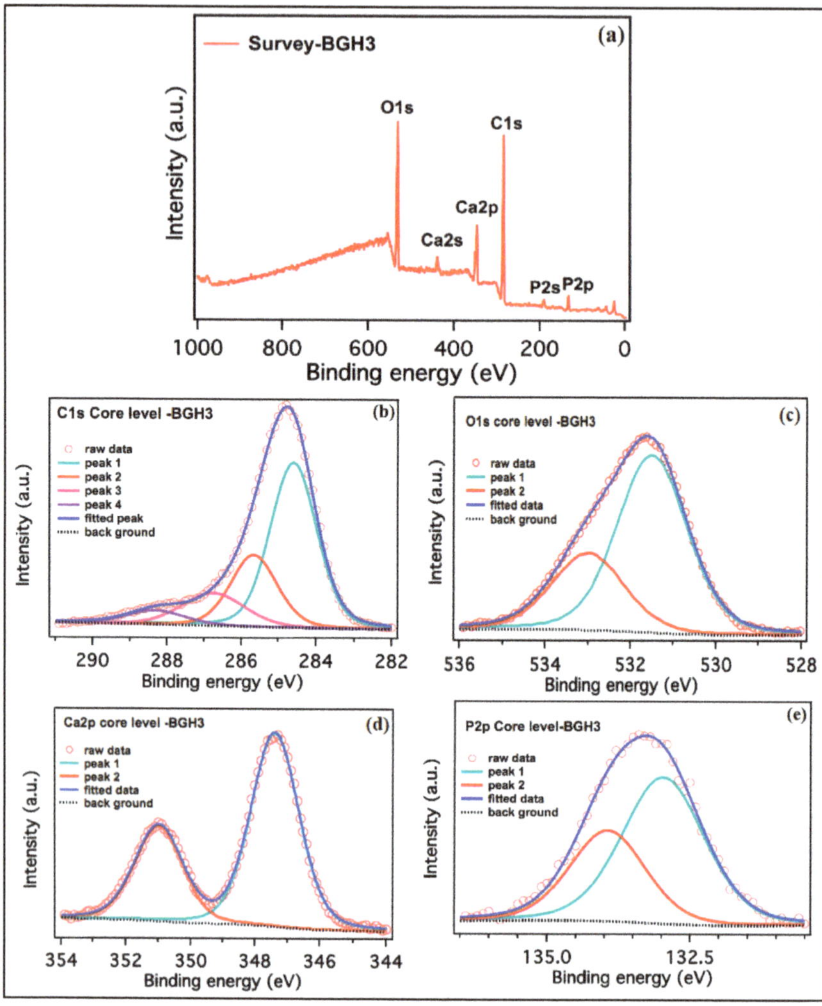

Figure 2. XPS spectra of porous scaffolds. (**a**) BGH-3 survey, (**b**) carbon, (**c**) oxygen, (**d**) calcium and (**e**) phosphorous high-resolution scans, respectively.

4.4. Morphological Analysis

The surface and porous morphologies are significant characters of any biomaterial for biomedical applications, especially bone tissue engineering. The rough surface and interconnected porosity are essential to proliferate, differentiate, and adhere to bone tissue support. These samples exhibit rough surface morphology, and it is also observed that increasing HAp amount caused interconnected porosity with regulated pore size, as shown in Figure 3. It is assumed as the polymer concentration increased as ceramic amount decreased, the viscosity of the solution increased, preventing it from flowing out, causing some pores to partially close.

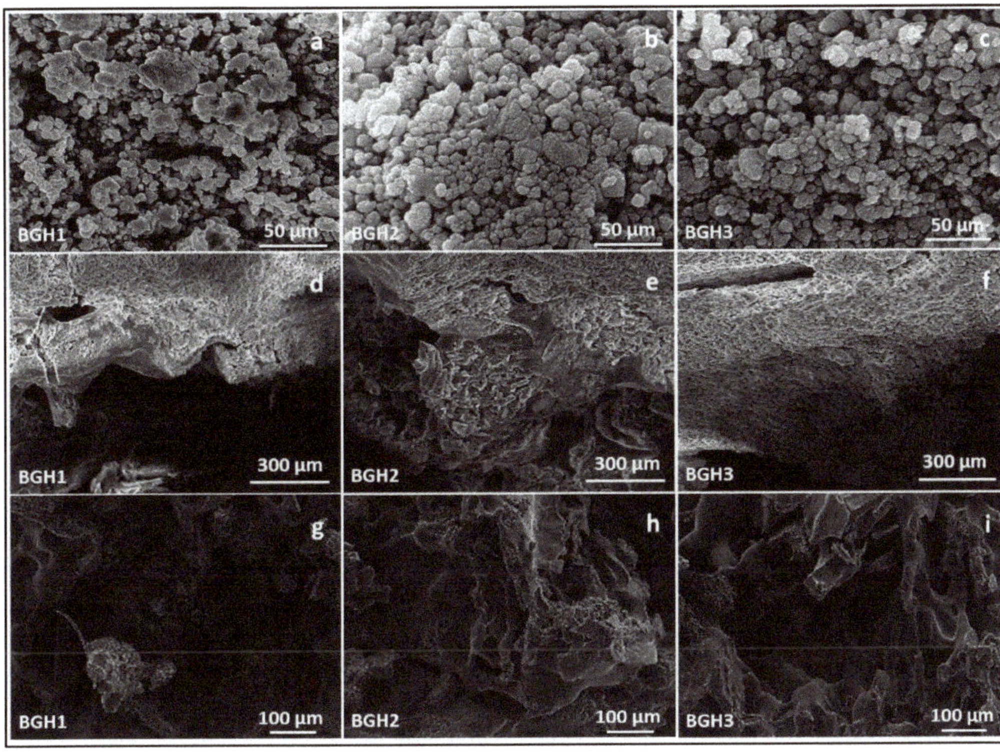

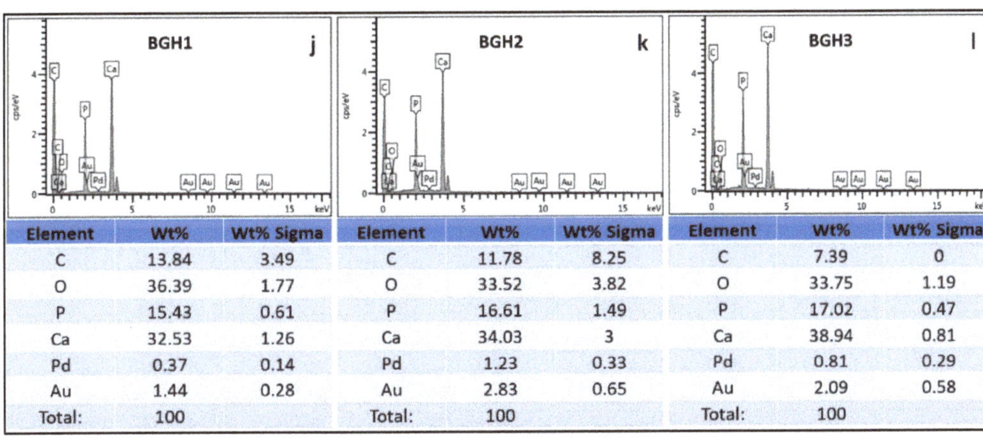

Element	Wt%	Wt% Sigma
C	13.84	3.49
O	36.39	1.77
P	15.43	0.61
Ca	32.53	1.26
Pd	0.37	0.14
Au	1.44	0.28
Total:	100	

Element	Wt%	Wt% Sigma
C	11.78	8.25
O	33.52	3.82
P	16.61	1.49
Ca	34.03	3
Pd	1.23	0.33
Au	2.83	0.65
Total:	100	

Element	Wt%	Wt% Sigma
C	7.39	0
O	33.75	1.19
P	17.02	0.47
Ca	38.94	0.81
Pd	0.81	0.29
Au	2.09	0.58
Total:	100	

Figure 3. SEM morphologies of composite materials (**a**–**c**), scaffolds' surface morphology at different scales (300 (**d**–**f**) and 100 (**g**–**i**) µm) and elemental analysis (**j**–**l**) by EDX.

The presence of adequate porosity in the scaffold is essential to regulate critical processes such as nutrient delivery to cells, metabolite dispersion, pH stability, and cell signalling. In the early stages, the size of the pores is a significant factor for evaluating the closeness of cells. The later stages of tissue growth allow for three-dimensional cell-cell communication and space for cells to organize themselves in three dimensions. Because cell seeding in the scaffold's centre and feeding the scaffolds' inner surfaces are limited, smaller pores have their constraints. On the other hand, larger pores affect the scaffold's stability and offer physical support for the cell seeding. Pores of sufficient size allow cells to migrate or adhere to a material's surface [30]. Table 1 shows that the porosity of all scaffolds was greater than 55%. It indicates that they have substantial porosity and are effective for cell adherence and proliferation. Researchers found that increasing the porosity of macroporous biphasic calcium phosphate ceramic samples improved bone growth, implying the importance of high porous scaffolds for better cell growth in the scaffold [31]. The porosity of polymeric scaffolds BGH1, BGH2 and BGH3 are 54.69, 71.24, and 82.96%, respectively. The different porosity (%) of the fabricated scaffolds confirm that increasing HAp amount have effect over porosity (%) that also affect the different interaction (covalent, hydrogen bonding and van der Waal forces, etc.) during free radical polymerization. Pore size is a critical factor to consider when designing a tissue engineering scaffold. In terms of attachment, matrix deposition, and differentiation, it has an impact on cellular activity. Within the field of bone tissue engineering, there appears to be widespread agreement that the optimal pore size for bone tissue scaffolds is between 100 and 500 micrometres for better cell adhesion and proliferation, as well as vascular ingrowth [32]. The EDX analysis confirms the increase in HAp intensity as HAp amount increases, and no foreign component is found that causes any toxicity.

4.5. Mechanical and Porosity

Porous scaffold must mimic the native extracellular matrix (ECM) with desirable mechanical behavior and porosity, an essential requirement for bone tissue regeneration. The mechanical properties of the scaffolds should be similar to those of native bone tissue. The mechanical strength can be determined by the impact resistance of polymeric scaffolds to retain structural integrity during implantation [33]. Multiple aspects of a biomaterial, including topography, particle and crystal size, chemical nature, and pore size distribution, can significantly affect the cellular behavior for biological efficiency. Almost all of these features are linked to the compositions of ceramic and polymeric components during synthesizing the composite materials [34].

The polymeric scaffolds' machinal properties were conducted via stress-strain curves, as shown in Figure 4a. Since GO and HAp are well-known materials to enhance the mechanical and structural integrity of materials. The amount of GO was optimized, and different nHAp amounts were used to enhance the mechanical behavior of fabricated scaffolds [35]. Hence, the increasing amount of nHAp has increased compression strength (minimum = 6.19 and maximum = 13.76 MPa) and Young's modulus (maximum = 113.75 and minimum = 35.73 MPa). It was noted that increased nHAp amount caused increased mechanical strength, but a decreased porosity was observed. The detailed properties of mechanical and porosity are given in Table 1. Since nHAp is a ceramic material and has several available active sites. The increasing nHAp caused increased mechanical strength due to crosslinking behavior that caused the close packing of polymeric scaffolds that reduced the porosity [15]. The relationship between porosity and Young's modulus of the polymeric scaffolds have shown in Figure 4b. The variable nHAp amount has changed the chemical structure by affecting interface and matrix grains. Optimized amount of GO and increased HAp amount enhanced mechanical properties due to grain size boundary. The nHAp provides a vast surface area that interacts with polymeric matrix and other materials to regulate the porosity of polymeric scaffolds [36]. The porosity, mechanical strength and Young's modulus has been summarized in Table 3.

Table 3. Summaries the mechanical and porosity of the polymeric scaffolds.

Samples	Compression Strength (MPa)	Young's Modulus (MPa)	Porosity (%)
BGH1	06.19 ± 1.3	35.73 ± 1.2	82.69 ± 1.1
BGH2	10.48 ± 1.1	56.07 ± 1.1	71.24 ± 1.2
BGH3	13.76 ± 1.1	113.75 ± 1.2	54.69 ± 1.2

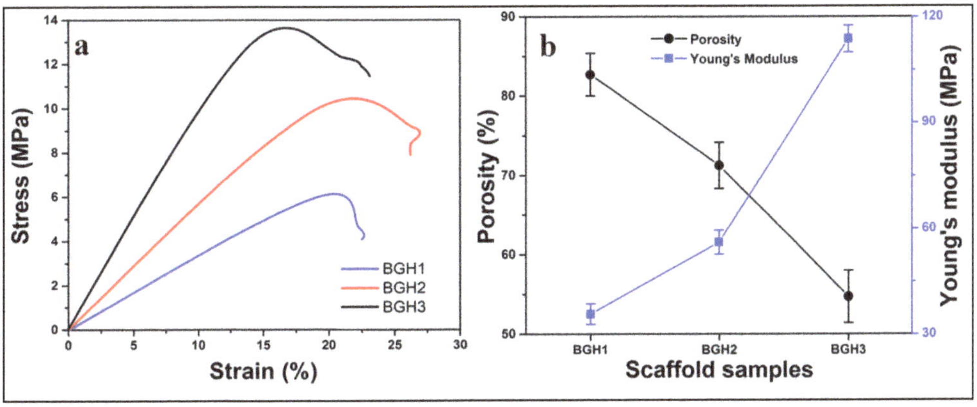

Figure 4. (**a**) Mechanical analysis via strain–stress curves, (**b**) relationship between porosity and Young's modulus relationship.

4.6. Wetting Analysis

Wetting of material surfaces is a crucial feature for tissue engineering, and it has an impact on cell-material interactions. Wetting of polymeric material is a vital property to achieve a suitable biological response. Wetting measurements are an essential part of the scientific analysis of biomaterial features. Contact angle measurements are one of the most common methods for determining the wetting behavior of polymeric biomaterial surfaces [37]. The hydrophilic surface is better for osteoblast growth and mineral deposition than the hydrophobic surface. The bioactivity and cytocompatibility of the scaffold are both critical factors and can be evaluated by the wetting behavior of scaffolds [38]. Figure 5 presents the wetting behavior concerning the time and quantity of nHAp. It was observed that an increasing amount of nHAp shifts hydrophobic behavior from hydrophilic. It is also worth mentioning that increasing water contact time shift hydrophobicity behavior towards hydrophilicity, as shown in Figure 6. Since these scaffolds possessed different wetting behavior due to different times. Hence, the scaffolds are hydrophilic, which have a water contact angle of less than 90°. This analysis indicates that the hydrophobicity can be modified by introducing hydrophilic materials and increasing water contact time with the polymeric scaffolds. The wetting behavior of the fabricated scaffolds has been summarized in Table 3.

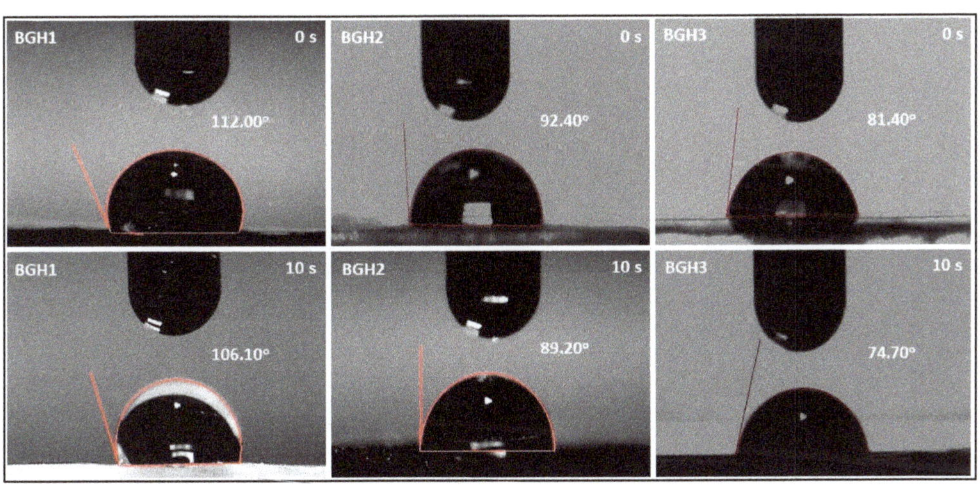

Figure 5. Wetting behavior of polymeric scaffolds via water contact angle.

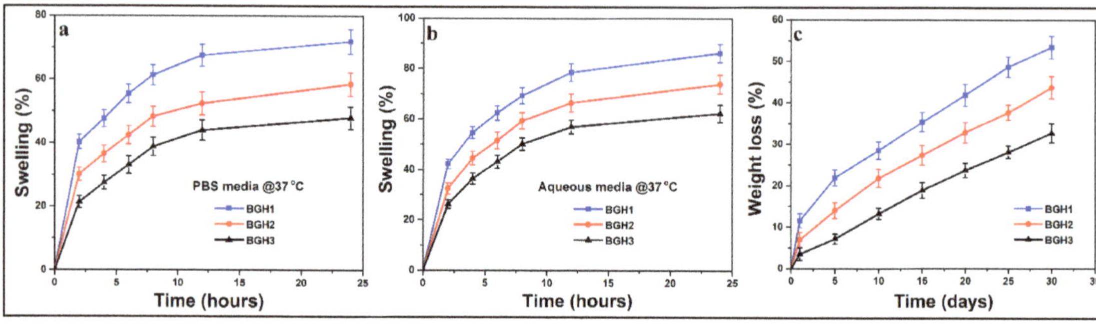

Figure 6. (**a**,**b**) swelling behavior of polymeric scaffold in different media (aqueous and PBS) and (**c**) in vitro biodegradation of polymeric scaffolds in PBS media.

4.7. Swelling and Degradation Analysis

The ability of biomaterials to swell is a crucial feature that facilitates the interaction of the biological system with biomaterials. It also regulates the sustained release of therapeutic agents. Swelling is a characteristic of polymeric materials that occurs when a solvent penetrates the void space of the polymeric chain network. The swelling causes the polymeric materials to expand. In some cases, external triggers (such as pH, ionic strength, temperature, etc.) can influence the swelling behavior of polymeric materials [39]. The polymeric content of the biomaterial is higher. It allows more fluids to exchange with the environment. Due to the large surface areas of the polymeric composite, it responds faster to exhibit swelling capacity than conventional polymeric composites [40].

Scaffold swelling behavior and structural integrity are critical for tissue engineering clinical use. In biological fluids, mostly biopolymers swell quickly. Initial swelling is desirable in the in vitro culture studies because the resulting increase in pore size enhances 3D cell adherence and proliferation. On the other hand, persistent swelling would result in a loss of mechanical integrity that develops compressive stress in the surrounding tissue [41]. The implantation site's pH value and media condition significantly impact the swelling behavior of a scaffold. Because of the different HAp nanoparticles amounts, the swelling behavior of these fabricated scaffolds varies. The scaffolds swelled more in deionized water and less in PBS solution, as shown in Figure 6a. The data show that increasing nHAp in the scaffolds reduced swelling in deionized water and PBS media at 37 °C, as presented in Figure 6b. Because there was less nHAp and more polymeric content in BGH-3, it swelled the least (deionized water = 62.30%, PBS media = 47.87%). Due to the highest content of nHAp, BGH-1 had the maximum swelling (deionized water = 86.20%, PBS media = 71.90%). The interesting feature where HAp nanoparticles behave like a physical crosslinker is responsible for different swelling characteristics of scaffolds. Increasing the quantity of HAp nanoparticles in a polymeric material may cause additional crosslinking, reducing the material's elasticity behavior [42].

The essential characteristics of polymeric composites for biomedical applications are degradability and biocompatibility. Polymeric composites are generally biocompatible and low in toxicity. As a result, they do not cause any unwanted biological responses at the molecular, cellular, or organ level. They contain more water, and the hydrophilic functional groups in the polymeric network are more numerous [43]. Because the polymeric substrates are arranged via various weak interactions, most polymeric composite-based drug delivery systems prepared by self-assembly are degradable. The degradation of scaffolds in PBS solution at 37 °C was evaluated, and degradation (%) was estimated by weight loss, as shown in Figure 6c. As the time spent immersed in the PBS solution increased, so did the weight loss. Because these are physical crosslinkers, increasing the amount of HAp nanoparticles has the reverse impact on degradation. The scaffold sample BGH1 showed the most degradation, while BGH3 showed the least. The crosslinking ability of the HAp nanoparticles is responsible for the different degradation properties of hybrid composite scaffolds [44]. It would be simple to conclude that increased HAp amount causes the scaffolds to crosslink more, resulting in a more compact structure that is less easily eroded. Increased crosslinking also changes the scaffold's hydrophobicity to a more hydrophobicity (Figure 6c). The swelling and biodegradation behavior of the fabricated scaffolds has been summarized in Table 4.

Table 4. Summaries the swelling, biodegradation and wetting analysis.

Samples	Swelling (Aqueous) (%)	Swelling (PBS) (%)	Biodegradation (%)	Wetting 0 s (degree)	Wetting 10 s (degree)
BGH1	86.20 ± 1.3	71.90 ± 1.2	32.72 ± 1.1	112	106.10
BGH2	73.87 ± 1.1	58.40 ± 1.1	43.69 ± 1.2	92.40	89.20
BGH3	62.30 ± 1.1	47.87 ± 1.2	53.38 ± 1.2	81.40	74.70

4.8. In Vitro Studies

4.8.1. Cell Morphology

The cell adherence and morphology of the polymeric scaffold were evaluated against MC3T3-E1 cell lines, as shown in Figure 7. These polymeric scaffolds are multifunctional due to several available functional groups (–OH, –OPO_3, –CH_3 and –COOH, etc.) of GO, HAp, and the polymeric matrix. This multifunctionality encourages cell adherence that supports cell viability and proliferation due to hydrogen bonding [6]. It was observed that increasing HAp enhances active sites that are helpful for integrin bonding with the surface of polymeric scaffolds and improve cell adherence, differentiation with a proper cylindrical shape. The increased hydrophilicity with desirable multifunctionality helped material-DNA communication, and increased nHAp encouraged osteogenesis to heal host bone [6]. The rough and porous morphology of polymeric scaffolds has a microstructural surface that develops a feasible microenvironment that helps cell adherence and cell spreading in a cylindrical shape, as shown in Figure 8. Hence all polymeric scaffolds have proper cylindrical and stretched morphology, and mature cell adherence and morphology were observed after 72 h for BGH3.

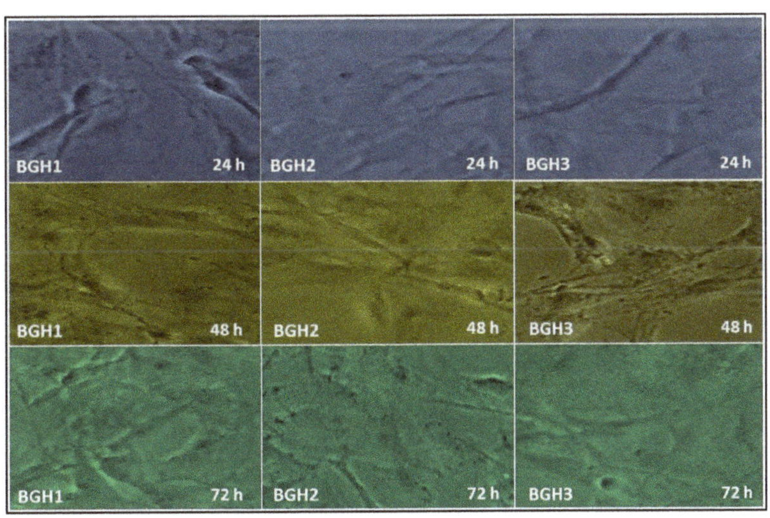

Figure 7. In vitro cell morphology of MC3T3-E1 against polymeric scaffolds (BGH1, BGH2 and BGH3) after different time intervals (24, 48, and 72 h).

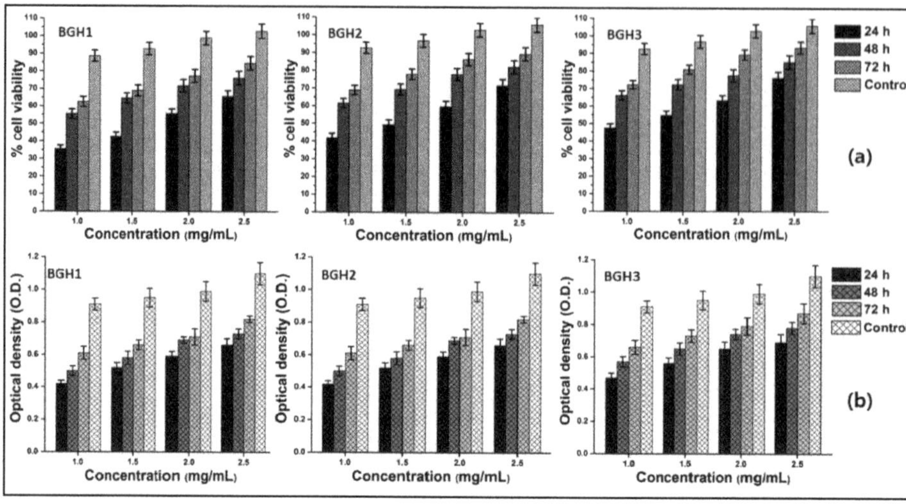

Figure 8. In vitro biocompatibility behavior of polymeric scaffolds against MC3T3-E1 at different concentrations (1, 1.5, 2 and 2.5 mg/mL) after different time intervals (24, 48 and 72 h). (**a**) cell viability and (**b**) optical density.

4.8.2. Cell Viability and Optical Density

The biological activities of polymeric were performed against MC3T3-E1 cell lines by in vitro analyses to determine cell viability (Figure 8a) assay and optical density (Figure 8b). The advancements in biomolecular science techniques have made it possible to collect osteoblast-precursor cells from various sources. The most common sources are bone marrow, umbilical blood, facial bones, subcutaneous fat, muscular tissue, osseous tissue, orthodontic pulp, and periosteum [45]. To study the concentration effect, the cell viability and optical density were performed using different concentrations (1, 1.5, 2 and 2.5 mg/mL). These polymeric scaffolds were incubated against MC3T3-E1 cell lines under in vitro analysis at different time intervals (24, 48 and 72 h). The cell viability, adherence and proliferation were increased with increasing HAp amount and maximum cell viability and proliferation after 72 h with mature cell adherence. The cylindrical cell shape was observed with proper adherence, and mature adherence was found after 72 h for BGH3. The increasing HAp amount with fixed GO amount increases the active sites that encourage cell adherence, which leads to cell viability and optical density, causing cell proliferation. Since the active sites (HAp) and oxygen-based (GO and BG) have several functional groups that improve microstructural behavior and micro-environmental for cell adherence that facilitate cell viability and proliferation as HAp is well known for osteogenesis [16]. The combined effect of BG, GO and HAp support cytocompatibility due to enhanced physicochemical properties. Hence, the polymeric scaffolds were biocompatible with improved cell viability and proliferation with mature cell adherence.

5. Conclusions

Excellent translational biomaterials demonstrate a ton of potential performance in the lab but have limitations in commercialization due to cast and other research aspects. Here we reported the synthesis of polymeric composite materials via the free radical method to fabricate polymeric scaffolds using the freeze-drying technique. The results demonstrate that the fabricated scaffolds have porous morphology, adjustable wetting, swelling, degradation and mechanical testing. In vitro bioactivities were performed against MC3T3-E1 at different times and concentrations. It was found that increasing ceramic (nHAp) amount enhances physicochemical and biomechanical characteristics. BGH3 was the best polymeric scaffold due to high mechanical (Young's modulus and compression) with appropriate porosity and surface morphology due to maximum HAp amount with fixed GO quantity. The increasing HAp amount enhances the cytocompatibility and biocompatibility of cell adherence, leading to cell viability and proliferation due to the highly interrelated porous scaffold structure. It was found from the results that these polymeric scaffolds have desirable structural, morphological and biocompatible properties with different load-bearing mechanical behaviors. These polymeric scaffolds would be promising biomaterials to regenerate fractured bone for bone tissue engineering applications.

Author Contributions: Conceptualization, M.U.A.K., S.I.A.R.; data curation, M.U.A.K., R.M.Z. and N.A.Z.; formal analysis, M.U.A.K., M.N.M.A. and M.A.; funding acquisition, M.U.A.K. and S.I.A.R.; investigation, M.N.M.A., S.I.A.R., R.M.Z. and N.A.Z.; methodology, M.U.A.K., S.I.A.R., R.M.Z. and N.A.Z.; project administration, S.I.A.R., R.M.Z. and N.A.Z.; resources, S.I.A.R., R.M.Z. and N.A.Z.; software, M.U.A.K.; supervision, S.I.A.R., R.M.Z. and N.A.Z.; validation, N.A.Z., S.I.A.R. and M.N.M.A.; visualization, M.U.A.K.; writing—original draft, M.U.A.K.; writing—review and editing, M.U.A.K. and S.I.A.R. All authors have read and agreed to the published version of the manuscript.

Funding: The Universiti Teknologi Malaysia for research grant number 02M44 and Legasi Megajaya Sdn Bhd (M) to provide the facilities. The innovative Research and Management Centre (iRMC), Universiti Tenaga Nasional, Malaysia to provide the publication funding (J510050002).

Institutional Review Board Statement: Not applicable.

Informed Consent Statement: Not applicable.

Data Availability Statement: Data available within the article.

Acknowledgments: The authors would like to thank the Innovative Research and Management Centre (iRMC), Universiti Tenaga Nasional, Malaysia, for providing research publication BOLD grant (J510050002), the Universiti Teknologi Malaysia for research grant number 02M44 and Legasi Megajaya Sdn Bhd (M) for providing the facilities.

Conflicts of Interest: All authors declare no conflict of interest.

References

1. Oryan, A.; Monazzah, S.; Bigham-Sadegh, A. Bone injury and fracture healing biology. *Biomed. Environ. Sci.* **2015**, *28*, 57–71.
2. Winkler, T.; Sass, F.; Duda, G.; Schmidt-Bleek, K. A review of biomaterials in bone defect healing, remaining shortcomings and future opportunities for bone tissue engineering: The unsolved challenge. *Bone Jt. Res.* **2018**, *7*, 232–243. [CrossRef] [PubMed]
3. Araújo, L.K.; Antunes, G.S.; Melo, M.M.; Castro-Silva, I.I. Brazilian dentists' perceptions of using bone grafts: An inland survey. *Acta Odontol. Latinoam.* **2020**, *33*, 165–173. [PubMed]
4. Pan, S.; Liu, B.; Min, Y.; Sun, J.Y.; Zhai, B.; Guo, X.H. Pre-Registration Assessment of Bone-Filling Products. *Orthop. Surg.* **2019**, *11*, 720–724. [CrossRef] [PubMed]
5. Dimitriou, R.; Mataliotakis, G.I.; Calori, G.M.; Giannoudis, P.V. The role of barrier membranes for guided bone regeneration and restoration of large bone defects: Current experimental and clinical evidence. *BMC Med.* **2012**, *10*, 81. [CrossRef]
6. Aslam Khan, M.U.; Haider, A.; Abd Razak, S.I.; Abdul Kadir, M.R.; Haider, S.; Shah, S.A.; Hasan, A.; Khan, R.; Khan, S.U.D.; Shakir, I. Arabinoxylan/graphene-oxide/nHAp-NPs/PVA bionano composite scaffolds for fractured bone healing. *J. Tissue Eng. Regen. Med.* **2021**, *15*, 322–335. [CrossRef]
7. Khan, M.U.A.; Al-Thebaiti, M.A.; Hashmi, M.U.; Aftab, S.; Abd Razak, S.I.; Abu Hassan, S.; Abdul Kadir, M.R.; Amin, R. Synthesis of silver-coated bioactive nanocomposite scaffolds based on grafted beta-glucan/hydroxyapatite via freeze-drying method: Anti-microbial and biocompatibility evaluation for bone tissue engineering. *Materials* **2020**, *13*, 971. [CrossRef]
8. Aslam Khan, M.U.; Mehboob, H.; Abd Razak, S.I.; Yahya, M.Y.; Mohd Yusof, A.H.; Ramlee, M.H.; Sahaya Anand, T.J.; Hassan, R.; Aziz, A.; Amin, R. Development of polymeric nanocomposite (xyloglucan-co-methacrylic acid/hydroxyapatite/sio2) scaffold for bone tissue engineering applications—In-Vitro antibacterial, cytotoxicity and cell culture evaluation. *Polymers* **2020**, *12*, 1238. [CrossRef]
9. Khan, M.U.A.; Yaqoob, Z.; Ansari, M.N.M.; Razak, S.I.A.; Raza, M.A.; Sajjad, A.; Haider, S.; Busra, F.M. Chitosan/Poly Vinyl Alcohol/Graphene Oxide Based pH-Responsive Composite Hydrogel Films: Drug Release, Anti-Microbial and Cell Viability Studies. *Polymers* **2021**, *13*, 3124. [CrossRef]
10. De Souza, F.F.P.; Cavalcante, F.L.; Castro-Silva, I.I.; da Silva, A.L.C.; Souza Filho, M.D.S.M.D. Poultry by-products as source of collagen, nanokeratin and bioapatite for biomedical use. *Rev. Ciência Agronômica* **2021**, *52*. [CrossRef]
11. Aslam Khan, M.U.; Abd Razak, S.I.; Al Arjan, W.S.; Nazir, S.; Sahaya Anand, T.J.; Mehboob, H.; Amin, R. Recent Advances in Biopolymeric Composite Materials for Tissue Engineering and Regenerative Medicines: A Review. *Molecules* **2021**, *26*, 619. [CrossRef] [PubMed]

12. Al-Arjan, W.S.; Aslam Khan, M.U.; Nazir, S.; Abd Razak, S.I.; Abdul Kadir, M.R. Development of Arabinoxylan-Reinforced Apple Pectin/Graphene Oxide/Nano-Hydroxyapatite Based Nanocomposite Scaffolds with Controlled Release of Drug for Bone Tissue Engineering: In-Vitro Evaluation of Biocompatibility and Cytotoxicity against MC3T3-E1. *Coatings* **2020**, *10*, 1120. [CrossRef]
13. Chen, G.; Ushida, T.; Tateishi, T. Poly (DL-lactic-co-glycolic acid) sponge hybridized with collagen microsponges and deposited apatite particulates. *J. Biomed. Mater. Res. Off. J. Soc. Biomater. Jpn. Soc. Biomater. Aust. Soc. Biomater. Korean Soc. Biomater.* **2001**, *57*, 8–14. [CrossRef]
14. Jin, S.; Xia, X.; Huang, J.; Yuan, C.; Zuo, Y.; Li, Y.; Li, J. Recent advances in PLGA-based biomaterials for bone tissue regeneration. *Acta Biomater.* **2021**, *127*, 56–79. [CrossRef]
15. Aslam Khan, M.U.; Al-Arjan, W.S.; Binkadem, M.S.; Mehboob, H.; Haider, A.; Raza, M.A.; Abd Razak, S.I.; Hasan, A.; Amin, R. Development of Biopolymeric Hybrid Scaffold-Based on AAc/GO/nHAp/TiO2 Nanocomposite for Bone Tissue Engineering: In-Vitro Analysis. *Nanomaterials* **2021**, *11*, 1319. [CrossRef]
16. Khan, M.U.A.; Abd Razak, S.I.; Mehboob, H.; Abdul Kadir, M.R.; Anand, T.J.S.; Inam, F.; Shah, S.A.; Abdel-Haliem, M.E.; Amin, R. Synthesis and characterization of silver-coated polymeric scaffolds for bone tissue engineering: Antibacterial and in vitro evaluation of cytotoxicity and biocompatibility. *ACS Omega* **2021**, *6*, 4335–4346. [CrossRef] [PubMed]
17. Lavanya, K.; Chandran, S.V.; Balagangadharan, K.; Selvamurugan, N. Temperature-and pH-responsive chitosan-based injectable hydrogels for bone tissue engineering. *Mater. Sci. Eng. C* **2020**, *111*, 110862. [CrossRef] [PubMed]
18. Bacic, A.; Fincher, G.B.; Stone, B.A. *Chemistry, Biochemistry, and Biology of 1–3 Beta Glucans and Related Polysaccharides*; Academic Press: Cambridge, MA, USA, 2009.
19. Oves, M.; Rauf, M.A.; Ansari, M.O.; Warsi, M.K.; Hussain, A.; Ismail, I.I. Polysaccharide-based nanocomposites for gene delivery and tissue engineering. In *Polysaccharide-Based Nanocomposites for Gene Delivery and Tissue Engineering*; Elsevier: Amsterdam, The Netherlands, 2021; pp. 103–129.
20. Wood, P.J.; Weisz, J.; Fedec, P.; Burrows, V.D. Large-scale preparation and properties of oat fractions enriched in (1-3)(1-4)-β-D-glucan. *Cereal Chem.* **1989**, *66*, 97–103.
21. Saghir, S.; Iqbal, M.S.; Hussain, M.A.; Koschella, A.; Heinze, T. Structure characterization and carboxymethylation of arabinoxylan isolated from Ispaghula (*Plantago ovata*) seed husk. *Carbohydr. Polym.* **2008**, *74*, 309–317. [CrossRef]
22. Khan, M.U.A.; Haider, S.; Shah, S.A.; Abd Razak, S.I.; Hassan, S.A.; Kadir, M.R.A.; Haider, A. Arabinoxylan-co-AA/HAp/TiO_2 nanocomposite scaffold a potential material for bone tissue engineering: An in vitro study. *Int. J. Biol. Macromol.* **2020**, *151*, 584–594. [CrossRef]
23. Repetto, G.; Del Peso, A.; Zurita, J.L. Neutral red uptake assay for the estimation of cell viability/cytotoxicity. *Nat. Protoc.* **2008**, *3*, 1125–1131. [CrossRef] [PubMed]
24. Zahedi, P.; Fallah-Darrehchi, M. Electrospun egg albumin-PVA nanofibers containing tetracycline hydrochloride: Morphological, drug release, antibacterial, thermal and mechanical properties. *Fibers Polym.* **2015**, *16*, 2184–2192.
25. Khan, M.U.A.; Raza, M.A.; Mehboob, H.; Kadir, M.R.A.; Abd Razak, S.I.; Shah, S.A.; Iqbal, M.Z.; Amin, R. Development and in vitro evaluation of κ-carrageenan based polymeric hybrid nanocomposite scaffolds for bone tissue engineering. *RSC Adv.* **2020**, *10*, 40529–40542. [CrossRef]
26. Zare-Harofteh, A.; Saber-Samandari, S.; Saber-Samandari, S. The effective role of akermanite on the apatite-forming ability of gelatin scaffold as a bone graft substitute. *Ceram. Int.* **2016**, *42*, 17781–17791. [CrossRef]
27. Saber-Samandari, S.; Nezafati, N.; Saber-Samandari, S. The effective role of hydroxyapatite based composites in anticancer drug delivery systems. *Crit. Rev. Ther. Drug Carr. Syst.* **2016**, *33*, 41–75. [CrossRef]
28. Daniyal, W.M.E.M.M.; Fen, Y.W.; Saleviter, S.; Chanlek, N.; Nakajima, H.; Abdullah, J.; Yusof, N.A. X-ray Photoelectron Spectroscopy Analysis of Chitosan–Graphene Oxide-Based Composite Thin Films for Potential Optical Sensing Applications. *Polymers* **2021**, *13*, 478. [CrossRef]
29. Uskoković, V. X-ray photoelectron and ion scattering spectroscopic surface analyses of amorphous and crystalline calcium phosphate nanoparticles with different chemical histories. *Phys. Chem. Chem. Phys.* **2020**, *22*, 5531–5547. [CrossRef]
30. Bakeri, G.; Ismail, A.F.; Shariaty-Niassar, M.; Matsuura, T. Effect of polymer concentration on the structure and performance of polyetherimide hollow fiber membranes. *J. Membr. Sci.* **2010**, *363*, 103–111. [CrossRef]
31. Zeng, H.; Pathak, J.L.; Shi, Y.; Ran, J.; Liang, L.; Yan, Q.; Wu, T.; Fan, Q.; Li, M.; Bai, Y. Indirect selective laser sintering-printed microporous biphasic calcium phosphate scaffold promotes endogenous bone regeneration via activation of ERK1/2 signaling. *Biofabrication* **2020**, *12*, 025032. [CrossRef] [PubMed]
32. Nájera-Romero, G.V.; Yar, M.; Rehman, I.U. Heparinized chitosan/hydroxyapatite scaffolds stimulate angiogenesis. *Funct. Compos. Mater.* **2020**, *1*, 9. [CrossRef]
33. Tran, T.; Hamid, Z.; Cheong, K. A Review of Mechanical Properties of Scaffold in Tissue Engineering: Aloe Vera Composites. In *Journal of Physics: Conference Series*; IOP Publishing: London, UK, 2018; p. 012080.
34. de Oliveira Lomelino, R.; Castro-Silva, I.I.; Linhares, A.B.R.; Alves, G.G.; de Albuquerque Santos, S.R.; Gameiro, V.S.; Rossi, A.M.; Granjeiro, J.M. The association of human primary bone cells with biphasic calcium phosphate (βTCP/HA 70: 30) granules increases bone repair. *J. Mater. Sci. Mater. Med.* **2012**, *23*, 781–788. [CrossRef] [PubMed]
35. Nie, J.; Zhou, J.; Huang, X.; Wang, L.; Liu, G.; Cheng, J. Effect of TiO_2 doping on densification and mechanical properties of hydroxyapatite by microwave sintering. *Ceram. Int.* **2019**, *45*, 13647–13655. [CrossRef]

36. Janovák, L.; Deák, Á.; Tallósy, S.P.; Sebők, D.; Csapó, E.; Bohinc, K.; Abram, A.; Pálinkó, I.; Dékány, I. Hydroxyapatite-enhanced structural, photocatalytic and antibacterial properties of photoreactive TiO_2/HAp/polyacrylate hybrid thin films. *Surf. Coat. Technol.* **2017**, *326*, 316–326. [CrossRef]
37. Agrawal, G.; Negi, Y.S.; Pradhan, S.; Dash, M.; Samal, S. Wettability and contact angle of polymeric biomaterials. In *Characterization of Polymeric Biomaterials*; Elsevier: Amsterdam, The Netherlands, 2017; pp. 57–81.
38. Liang, H.; Yang, Y.; Xie, D.; Li, L.; Mao, N.; Wang, C.; Tian, Z.; Jiang, Q.; Shen, L. Trabecular-like Ti-6Al-4V scaffolds for orthopedic: Fabrication by selective laser melting and in vitro biocompatibility. *J. Mater. Sci. Technol.* **2019**, *35*, 1284–1297. [CrossRef]
39. Guarrotxena, N.; Quijada-Garrido, I. Optical and swelling stimuli-response of functional hybrid nanogels: Feasible route to achieve tunable smart core@ shell plasmonic@ polymer nanomaterials. *Chem. Mater.* **2016**, *28*, 1402–1412. [CrossRef]
40. Spencer, D.; Shodeinde, A.; Beckman, D.; Luu, B.; Hodges, H.; Peppas, N. Biodegradable cationic nanogels with tunable size, swelling and pKa for drug delivery. *Int. J. Pharm.* **2020**, *588*, 119691. [CrossRef] [PubMed]
41. Alagoz, A.S.; Rodriguez-Cabello, J.C.; Hasirci, V. PHBV wet-spun scaffold coated with ELR-REDV improves vascularization for bone tissue engineering. *Biomed. Mater.* **2018**, *13*, 055010. [CrossRef]
42. Mirtaghavi, A.; Baldwin, A.; Tanideh, N.; Zarei, M.; Muthuraj, R.; Cao, Y.; Zhao, G.; Geng, J.; Jin, H.; Luo, J. Crosslinked porous three-dimensional cellulose nanofibers-gelatine biocomposite scaffolds for tissue regeneration. *Int. J. Biol. Macromol.* **2020**, *164*, 1949–1959.
43. Kankala, R.K.; Wang, S.-B.; Chen, A.-Z.; Zhang, Y.S. Self-Assembled Nanogels: From Particles to Scaffolds and Membranes. In *Handbook of Nanomaterials for Cancer Theranostics*; Elsevier: Amsterdam, The Netherlands, 2018; pp. 33–62.
44. Li, J.; Hu, Y.; Liu, W.; Weng, X.; Dong, X.; Zhang, X.; Zhou, W. High flux and hydrophilic fibrous ultrafiltration membranes based on electrospun titanium dioxide nanoparticles/polyethylene oxide/poly (vinylidene fluoride) composite scaffolds. *J. Nanosci. Nanotechnol.* **2017**, *17*, 9042–9049. [CrossRef]
45. Castro-Silva, I.I.; Zambuzzi, W.F.; de Oliveira Castro, L.; Granjeiro, J.M. Periosteal-derived cells for bone bioengineering: A promising candidate. *Clin. Oral Implant. Res.* **2012**, *23*, 1238–1242. [CrossRef]

Citric acid is a natural chemical substance that could be found in various fruits and vegetables, especially in citrus fruits. Lemons and limes are among citrus fruits that contain a higher concentration of citric acid [6]. Thanks to its adhesivity and green nature, citric acid could be a good candidate as the main binding agent for wood and wood-based products. Umemura et al. [7] reported that the good adhesivity and bonding properties are a result of the ester linkages formed as the carboxyl groups of citric acid reacts with the hydroxyl groups of the wood constituent. In fact, citric acid has been used as a main binding agent for several types of wood composites such as particleboard, fiberboard, plywood, and laminated veneer lumber [8]. Citric-acid-bonded particleboard is among the most extensively studied wood composites. In recent years, due to the persistent environmental issues, application of non-formaldehyde-based or green binders from renewable resources has become an irresistible general trend. Citric acid could serve ideally to address this problem.

In the production of particleboard, many additives could be added to the citric acid in order to improve the performance of the resultant boards. For example, Umemura et al. [9] added sucrose to citric acid to improve the bonding performance of the particleboard. On the other hand, teak particleboard was fabricated using citric acid and sucrose, and improvement in dimensional stability was reported [10]. Apart from sucrose, starch could also be added to citric acid to enhance the performance of particleboard. Widyorini et al. [11] fabricated petung bamboo (*Dendrocalamus* sp.) particleboard bonded with different ratios of citric acid and starch. It was found that an addition of 12.5 wt% starch could improve the dimensional stability and bending strength of the particleboard. The authors attributed the improvement to the strong hydrogen bond interactions between citric acid and starch. The authors also reported that the amount of amylose in starch is a deciding factor to the dimensional stability of particleboard. Starch with higher amylose tends to result in particleboard with better dimensional stability as the linear chains of amylose are able to form strong linkages that are responsible for better water resistance.

A variety of wood and non-wood materials have been used in the production of citric-acid-bonded particleboard such as bamboo [12], teak [10], sweet sorghum bagasse [13], Nipa fronds [14], alang-alang (*Imperata cylindrica*) [15], new giant reed (*Arundo Donax* L.) [16], sugarcane bagasse [17], and rubberwood [18]. However, to the best of the authors' knowledge, studies on the citric-acid-bonded particleboard made from oil palm biomasses such as oil palm trunk (OPT), oil palm frond (OPF), and empty fruit bunch (EFB) have yet to be reported. The effects of starch addition on the performance of citric acid as a binder for these oil palm biomasses is also unknown. Therefore, the objectives of this study were to investigate the performance of three oil palm biomasses bonded with citric and tapioca starch at different ratios.

2. Materials and Methods

2.1. Preparation of Materials

Three types of oil palm biomasses, namely, oil palm trunk (OPT), oil palm frond (OPF), and empty fruit bunch (EFB) were used in this study. Both OPT and OPF were collected from the felled oil palm trees in an oil palm field at Universiti Putra Malaysia. A portable wood chainsaw (STIHL M170, Waiblingen, Stuttgart, Germany) was used to reduce the size of the OPT and OPF before ground into particle sizes. On the other hand, fibers of EFB were supplied by a local company located at Dengkil, Selangor. The distribution of the particles' width and length are shown in Figure 1. Most of the OPT and OPF particles had a width ranging between 0.5 to 1.0 mm, while the majority of the EFB particles had a width less than 0.1 mm. In terms of length, the EFB particles used in this study were longer than that of OPT and OPF particles. All of the oil palm biomasses were dried in an oven at 60 °C for 7 days to achieve a 3% moisture content before particleboard fabrication. Citric acid in powder form was purchased from Evergreen Engineering & Resources, Semenyih, Selangor, Malaysia. Tapioca starch was purchased from a grocery store, Serdang, Selangor, Malaysia. Citric acid and tapioca starch were dissolved into distilled water to achieve a

60% resin solid content. The citric acid and tapioca starch mixture ratios were set at 100:0, 87.5:12.5, and 75:25 (w/w).

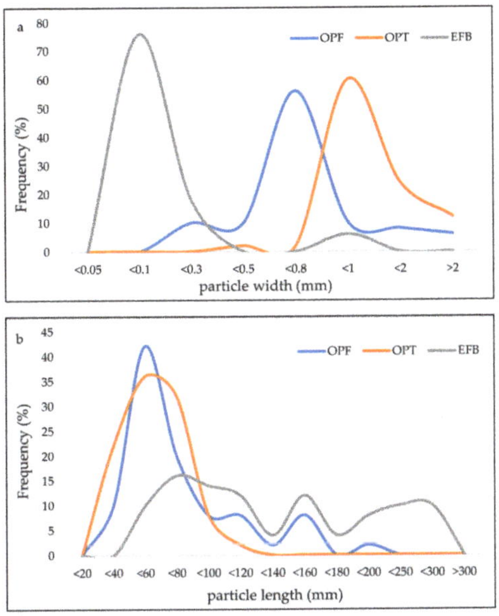

Figure 1. Distribution of width (**a**) and length (**b**) of oil palm frond (OPF), oil palm trunk (OPT), and empty fruit bunch (EFB) particles used in this study.

2.2. Fabrication of Particleboard

Single-layer particleboard was produced using citric acid and tapioca starch at different ratios (100:0, 87.5:12.5, and 75:25). Thirty percent resin loading based on the oven-dried weight of oil palm biomass particles were sprayed on the particles. The mixture was then mixed at room temperature for 5 min using a mixer machine (custom made). The mixing process was carried out carefully to ensure the mixture of citric acid and tapioca starch were spread evenly with the oil palm biomass particles. The mixed particle was then oven-dried in an oven at 80 °C for 12 h prior to hot pressing. The oven-dried particles were then hand-formed into a mat using a 340 mm × 340 mm wood mold. The mat was first pre-pressed to compact and then hot-pressed at 180 °C for 10 min with a 4 MPa pressure using a 100 Ton hydraulic laboratory hot press (Carver CMG 100H-15, Ontario, NY, USA). The pressing parameters were chosen according to the study by Lee et al. [8,19], where a pressing temperature of more than 180 °C for 10 min was preferential for citric-acid-bonded particleboard. A pair of 12-mm thick steel bar was used to control the thickness of the board during the hot-pressing process. Particleboards with final dimensions of 340 mm × 340 mm × 12 mm (length × width × thickness) with a target density of 650 kg/m^3 were produced. Three boards produced from a urea formaldehyde (UF) resin (65% solid content) was used as a control. Ten wt% resin loading was used to produce the control particleboard. One wt% ammonium chloride was added to the UF resin as a hardener. After hot pressing, the boards were conditioned in a conditioning room setting at a temperature of 20 ± 5 °C and a relative humidity of 65 ± 5% until a constant mass was attained. The conditioned particleboards were then trimmed and cut according to the dimension requirements stated in JIS A 5908:2003 [20] for physical and mechanical properties' evaluation. Figure 2 shows the cutting pattern and sample sampling for the tests. For every particleboard, 2 replicates of samples for each test were obtained. Three replications of particleboard were produced for every citric acid/tapioca starch ratio and UF (control) resin in order to obtain 5 replicates for every physical and mechanical test.

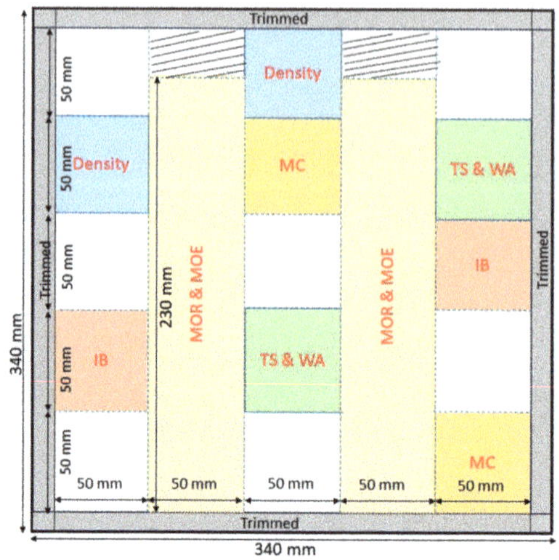

Figure 2. Cutting pattern and sample sampling for evaluation of density, moisture content (MC), thickness swelling (TS), water absorption (WA), internal bond (IB), modulus of rupture (MOR), and modulus of elasticity (MOE).

2.3. Properties Evaluation

2.3.1. Thermogravimetric Analysis (TGA) of Citric Acid/Tapioca Starch Binder

TGA of citric acid and topical starch mixture in powdered form was performed under a nitrogen atmosphere using the instrument TGA Q500 V20.13 Build 39 (TA Instruments, Pittsburgh, PA, USA) between 30 and 600 °C (10 °C min^{-1}) in the Laboratory of Biocomposite Technology, Institute of Tropical Forestry and Forest Products, Universiti Putra Malaysia, Serdang, Selangor.

2.3.2. Chemical Composition and Bulk Density of Oil Palm Biomasses

The determination of chemical content such as lignin, cellulose, hemicellulose, and extractives of the oil palm biomasses were conducted in accordance to the Technical Association of Pulp and Paper Industries (TAPPI) standard. The determination of lignin of the samples was conducted in accordance to TAPPI standard T222 os-74 [21], while the cellulose and hemicellulose was determined according to the procedure specified in TAPPI standard T203 os-74 [22] and Wise et al. [23], respectively. The bulk density of the particles of the oil palm biomasses were determined by using a cylinder method according to the study by Lee et al. [24]. A known weight of oil palm biomasses particles was put into a graduated cylinder half-filled with water. The amount of increment in water volume was recorded, and the bulk density was obtained by dividing the mass of the particles over volume increased.

2.3.3. Mechanical Properties of Particleboard

Test specimens were trimmed and prepared according to JIS A 5908:2003 [20] for properties' evaluation. Mechanical properties including the modulus of rupture (MOR), the modulus of elasticity (MOE), and the internal bond (IB) of the particleboard were evaluated in accordance with JIS A 5908:2003 [20]. The bending properties (MOR and MOE) were evaluated by using a three-point test on samples with dimensions of 230 mm × 50 mm × 12 mm (length × width × thickness). Meanwhile, the IB test was conducted on samples with dimensions 50 mm × 50 mm × 12 mm using a Universal Testing Machine (UTM, Instron-3366, Norwood, MA, USA). Five replicates were tested for each test.

2.3.4. Physical Properties of Particleboard

The physical properties of the particleboard including water absorption (WA) and thickness swelling (TS) produced were evaluated in accordance with JIS A 5908:2003 [20]. The samples having a width and length of 50 mm × 50 mm were immersed in water, and the weight and thickness of the immersed were weighed and measured at time intervals of 2 and 24 h. The changes in thickness and weight before and after immersion were expressed as a percentage (%). Cyclic aging treatment was performed according to Kusumah et al. [25]. The immersed samples (after 24 h) were subjected to cyclic aging treatment in the following sequence:

i. Drying at 105 °C for 10 h
ii. Immersion in water at 70 °C for 10 h
iii. Drying at 105 °C for 10 h
iv. Immersion in boiling water for 4 h
v. Drying at 105 °C for 10 h

The weight and thickness changes of the samples after every stage of the treatment were recorded. Five replicates were used for every test.

2.3.5. Statistical Analysis

The data collected were analyzed using Statistical Analysis System (SAS) software (SAS 9.4 solutions, Armonk, NY, United States). Analysis of variance (ANOVA) at a 95% confidence level ($p \leq 0.05$) was performed. Tukey's honest significant difference (HSD) test was used to further determine the significant level of average values for each treatment.

3. Results and Discussion

3.1. Chemical Composition and Bulk Density

The chemical composition and bulk density of the oil palm biomass particles used in this study are shown in Table 1. OPT had the highest lignin content of 26.47% compared with OPF (19.93%) and EFB (21.17%). Meanwhile, EFB had higher cellulose than that of OPT and OPF, while OPF had the highest content of hemicellulose (35.32%). The extractive content for these three oil palm biomasses ranged from 1.54 to 4.80%. The values reported are in agreement with the values reported in Abdul Khalil et al. [26]. They had a high cellulose content, making these oil palm biomasses very suitable in fabricating polymer composites [27].

Table 1. Chemical composition and bulk density of oil palm frond, oil palm trunk, and empty fruit bunch particles.

Biomasses	Oil Palm Frond	Oil Palm Trunk	Empty Fruit Bunch
Lignin (%)	19.93 ± 0.26	26.47 ± 0.46	21.17 ± 0.88
Holocellulose (%)	75.26 ± 0.21	71.99 ± 0.48	75.49 ± 0.67
Cellulose (%)	39.95 ± 0.23	42.47 ± 0.19	45.30 ± 0.41
Hemicellulose (%)	35.32 ± 0.41	29.52 ± 0.41	30.19 ± 0.38
Extractive (%)	4.80 ± 0.07	1.54 ± 0.01	3.34 ± 0.12
Moisture content (%)	7.41 ± 0.01	5.82 ± 0.01	6.70 ± 0.01
Bulk density (kg/m^3)	560.80 ± 15.73	460.00 ± 11.91	670.20 ± 22.40

Note: values after "±" are standard deviations.

As for bulk density, EFB had the highest bulk density of 670.20 kg/m^3, while the bulk densities for OPF and OPT were 560.80 kg/m^3 and 460.00 kg/m^3, respectively. The findings were comparable to the study by Abdul Khalil et al. [26] who reported that the bulk density of OPF, OPT, and EFB ranged between 600–1200 kg/m^3, 500–1100 kg/m^3, and 700–1550 kg/m^3, respectively. However, the bulk density of oil palm biomasses reported in this study was higher than other agricultural plants such as sweet sorghum bagasse (125 kg/m^3) [25], kenaf core (118 kg/m^3) [28], and bamboo (200 kg/m^3) [29].

3.2. Thermogravimetric Analysis (TGA)

Figure 3 displays the thermogravimetric (TG) and derivative thermogravimetric (DTG) curves of binder mixed at different ratios of citric acid and tapioca starch. It can be seen from Figure 3 that pure citric acid started to degrade at around 150 °C and peaked at 224 °C (Figure 3b). The observation is very close to the study by Silva et al. [30] who reported that the decomposition of citric acid occurred between a temperature range of 150 to 220 °C, and the DTG peak was at 210 °C. The authors stated that 95% of the transformation of citric acid occurred within this temperature range. Nevertheless, the addition of starch reduced the initial degradation temperature of the mixture of citric acid and tapioca starch, as seen in Figure 3a. From the DTG curves shown in Figure 3b, one can see that the decomposition temperature was reduced to 210 °C when 12.5% and 25% tapioca starch was added. The observation indicates a lower thermal stability after the addition of starch [31]. A study by Ali et al. [32] revealed that the tapioca starch started to degrade at about 62 °C under a nitrogen atmosphere and continued to lose weight until 500 °C. Therefore, it might be one of the reasons that causes a reduction in the thermal stability of citric acid after the addition of starch.

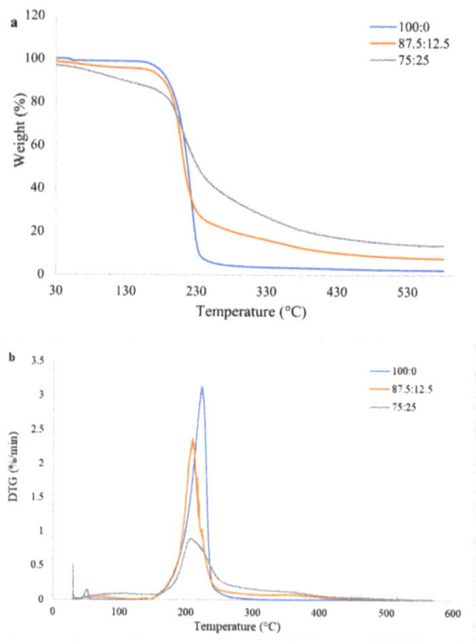

Figure 3. (**a**) Thermogravimetric (TG) and (**b**) derivative thermogravimetric (DTG) curves of binder mixed at different ratios of citric acid and tapioca starch.

The particleboard was hot-pressed at 180 °C in this study. At this temperature, pure citric acid experienced a weight loss of 5.9%, while the mixture of citric acid and tapioca starch at 87.5:12.5 and 75:25 recorded weight losses of 9.0% and 15.8%, respectively. The event implies a lower thermal stability of the mixture at a lower temperature. However, the mixture of citric acid and tapioca starch showed a higher resistance at a higher temperature, and their final residue was much higher than that of pure citric acid. Overall, the mixture of citric acid and tapioca starch at a ratio of 75:25 had the highest T_{80} (temperature for 80% weight loss), which was 388.3 °C, followed by 87.5:12.5 (281.4 °C) and pure citric acid (228.0 °C). The final residues were 13.7%, 7.6%, and 2.2%, respectively. This might be due to the crosslinking of citric acid with starch through covalent bonding. As a higher degree of crosslinking with citric acid was achieved at higher starch loadings, the final residue also increased correspondingly [31]. Additionally, a higher residue could be a result of carbonized material and the ash of starch at a high temperature [32].

3.3. Mechanical Properties of the Particleboard

Table 2 demonstrates the mechanical properties of the particleboard produced in this study.

Table 2. Modulus of rupture (MOR), modulus of elasticity (MOE), and internal bonding (IB) of oil palm biomass particleboard bonded with different binders.

Binder	MOR (N/mm^2)	MOE (N/mm^2)	IB (N/mm^2)
	Oil palm frond		
UF	6.89 ± 1.57 ^{d,e,f}	972.75 ± 51.5 ^{e}	0.22 ± 0.06 ^{d,e}
100CA:0 starch	7.32 ± 0.65 ^{c,d,e}	2299.49 ± 250.38 ^{a}	0.52 ± 0.13 ^{a}
87.5CA:12.5 starch	9.28 ± 0.63 ^{a,b}	2471.98 ± 143.67 ^{a}	0.46 ± 0.16 ^{ab}
75 CA:25 starch	10.24 ± 0.53^{a}	2329.84 ± 114.77^{a}	0.34 ± 0.07^{bcd}
	Oil palm trunk		
UF	6.44 ± 0.45 ^{e,f,g}	578.12 ± 60.48 ^{f}	0.25 ± 0.02 ^{c,d,e}
100CA:0 starch	8.75 ± 0.93 ^{a,b,c}	1711.96 ± 150.82 ^{b,c}	0.44 ± 0.02 ^{a,b}
87.5CA:12.5 starch	8.24 ± 0.75 ^{b,c,d}	1873.79 ± 62.28 ^{b}	0.40 ± 0.03 ^{a,b}
75 CA:25 starch	5.64 ± 0.52 ^{f,g,h}	1500.05 ± 65.13 ^{c,d}	0.37 ± 0.04 ^{b,c}
	Empty fruit bunch		
UF	4.49 ± 0.33 ^{h}	854.04 ± 63.81 ^{e}	0.12 ± 0.01 ^{e}
100CA:0 starch	5.08 ± 0.35 ^{g,h}	1287.28 ± 93.51 ^{d}	0.35 ± 0.02 ^{b,c,d}
87.5CA:12.5 starch	7.15 ± 0.92 ^{c,d,e,f}	1285.17 ± 50.57 ^{d}	0.25 ± 0.03 ^{c,d,e}
75 CA:25 starch	4.49 ± 0.40 ^{h}	772.51 ± 64.69 ^{e,f}	0.23 ± 0.03 ^{c,d,e}

Note: values after "±" are standard deviations. Within the same column, mean values followed by the different letters $^{a–h}$ were significantly different at $p \leq 0.05$.

UF-bonded OPF particleboard had an MOR and an MOE of 6.89 N/mm^2 and 972.75 N/mm^2, respectively. The MOR and MOE values increased as citric acid was used as a binder. It is interesting to note that the MOR of the OPF particleboard increased when higher starch was added to the citric acid. As for the MOE, the addition of 12.5% starch increased the MOE of the particleboard, but it started to decrease when 25% starch was added. On the other hand, OPT particleboard experienced the same trend as the OPF particleboard in terms of MOE. However, OPT particleboard showed a decreasing MOR trend as the starch addition increased. The MOR of OPT particleboard bonded with pure citric acid and 87.5CA:12.5 starch did not differ significantly. UF-bonded EFB particleboard had the lowest MOR value of 4.49 N/mm^2. However, EFB bonded with citric acid exhibited a higher MOR, particularly those bonded with 87.5CA:12.5 starch, which has a statistically higher MOR value than UF-bonded EFB particleboard. The MOE values of the citric-acid-bonded EFB particleboard also displayed a decreasing trend with increasing starch addition, but no significant different was detected between particleboard bonded with pure citric acid and 87.5CA:12.5 starch.

Citric-acid-bonded particleboard had a higher IB strength than that of UF-bonded particleboard. A very consistent trend was noticeable for the IB strength of citric-acid-bonded particleboard as the IB values decreased with increasing starch addition. The maximum IB strength of pure citric-acid-bonded OPF, OPT, and EFB particleboard were 0.52 N/mm^2, 0.44 N/mm^2, and 0.35 N/mm^2, respectively. However, with the addition of 12.5% starch, the IB strength decreased around 11.5%, 10%, and 28.6% for OPF, OPT, and EFB particleboard, respectively. The IB strength of the particleboard experienced further decrement when 25% starch was added. All the particleboard bonded with citric acid achieved the minimum requirement of IB strength (0.15 N/mm^2) for type 8 particleboard as specified in JIS A 5908. Overall, with the exception of some outliers, the addition of 12.5% starch was beneficial to the bending strength of the particleboard. Further starch addition (25%) could cause a disastrous effect to the mechanical strength of the fabricated particleboard.

Among the oil palm biomasses, particleboard made from OPF particles seemed to have the highest mechanical properties. This may be due to fact that OPF possesses thick fiber walls. Better bonding could be achieved as the lumen to cell wall ratio is high in OPF, which means OPF particles could be easily compressed closely together [33]. Both OPF and OPT particleboard have significantly better mechanical properties than that of EFB particleboard. One of the probable reasons leading to this observation might be the bulk density of the material itself. Boruszewski et al. [34] reported in their study that the particleboard manufactured from poplar with a lower bulk density achieved a more compacted structure. As a result, minimum voids existed in the boards and correspondingly lead to better mechanical strength. As shown in Table 1, EFB had the highest bulk density followed by OPF and OPT. Lower bulk density inevitably leads to a higher compaction ratio, and the mechanical properties of the boards also increased proportionately with an increasing compaction ratio [24]. In addition, the EFB was reported to have traces of oil on the fiber surfaces [35]. These oil traces caused the EFB to have lower wettability, which prevents the adhesive to be spread evenly on the surfaces of the particleboard. Consequently, it is very difficult for the EFB particleboard to achieve an effective bonding, and subsequently this leads to poor mechanical strength.

3.4. Physical Properties

Thickness swelling (TS) and water absorption (WA) of the particleboard fabricated in this study are listed in Table 3.

Table 3. Thickness swelling (TS) and water absorption (WA) of oil palm biomasses particleboard bonded with different binders after 2 h and 24 h immersion in water.

Binder	TS_{2h} (%)	TS_{24h} (%)	WA_{2h} (%)	WA_{24h} (%)
Oil palm frond				
UF	48.07 ± 1.15^{d}	71.22 ± 2.51^{e}	109.60 ± 5.31^{h}	140.01 ± 2.85^{g}
100CA:0 starch	6.73 ± 0.63^{a}	12.85 ± 1.00^{b}	$49.46 \pm 2.40^{a,b}$	$63.69 \pm 1.24^{a,b}$
87.5CA:12.5 starch	12.27 ± 0.47^{b}	20.84 ± 1.11^{c}	$48.48 \pm 2.81^{a,b}$	$62.60 \pm 2.87^{a,b}$
75 CA:25 starch	18.77 ± 1.10^{c}	22.01 ± 1.99^{c}	63.91 ± 2.15^{d}	81.46 ± 5.00^{d}
Oil palm trunk				
UF	44.19 ± 3.08^{d}	56.96 ± 4.12^{d}	102.10 ± 1.43^{g}	128.01 ± 4.03^{f}
100CA:0 starch	6.65 ± 0.51^{a}	12.11 ± 0.70^{b}	46.97 ± 1.91^{a}	57.25 ± 1.15^{a}
87.5CA:12.5 starch	6.95 ± 0.40^{a}	7.28 ± 1.00^{a}	$59.56 \pm 1.99^{c,d}$	73.01 ± 1.55^{c}
75 CA:25 starch	7.38 ± 0.84^{a}	$10.60 \pm 0.60^{a,b}$	$55.00 \pm 0.84^{b,c}$	$68.01 \pm 1.37^{b,c}$
Empty fruit bunch				
UF	90.86 ± 5.11^{e}	144.02 ± 4.55^{f}	176.64 ± 6.58^{i}	185.55 ± 4.71^{h}
100CA:0 starch	16.53 ± 0.44^{c}	20.11 ± 0.94^{c}	93.29 ± 3.17^{f}	96.68 ± 3.07^{e}
87.5CA:12.5 starch	20.04 ± 1.34^{c}	24.63 ± 0.78^{c}	81.79 ± 2.89^{e}	87.24 ± 3.63^{d}
75 CA:25 starch	19.96 ± 1.07^{c}	23.01 ± 1.27^{c}	$87.26 \pm 2.45^{e,f}$	123.02 ± 6.09^{f}

Note: values after "±" are standard deviations. Within the same column, mean values followed by the different letters, $^{a-h}$ were significantly different at $p \leq 0.05$.

After 2 h immersion in water, the TS_{2h} of the UF-bonded OPF, OPT, and EFB particleboards were 48.07%, 44.19%, and 90.86%, respectively. Meanwhile, TS_{24h} values were 71.22%, 59.96%, and 144.02%, respectively. The observation is unsurprising as UF resin is known to be very instable at high relative humidity owing to its susceptible aminomethylene linkage [36]. However, the TS values of the particleboard dropped significantly when citric acid was used as a binder. The TS_{24h} values of OPF, OPT, and EFB particleboards bonded with pure citric acid (100:0) were approximately one-sixth, one-fifth, and one-seventh of their respective UF counterparts. Both OPF and EFB particleboard bonded with pure citric acid had significantly lower TS_{24h} values than the particleboard bonded with starch addition. The findings were in agreement with Widyorini et al. [11] who reported that the TS of bamboo particleboard increased along with increasing starch content. The authors attributed the increment of the TS to the solubility of starch in water. However, OPT particleboard bonded with 87.5CA:12.5 starch exhibited the lowest TS_{24h} value of 7.28%.

Similarly, particleboard bonded with citric acid had a significantly lower WA_{2h} (46.97% to 93.29%) and WA_{24h} (57.25% to 123.02%) than UF-bonded particleboard (102.10% to 176.64% and 128.01% to 185.55%, respectively). The findings indicate that the citric acid inhibited water absorption by the particleboard during the water-immersion test [25]. Among oil palm biomasses, EFB particleboard had higher WA values compared to that of the OPF and the OPT particleboards. The OPT particleboard had the lowest TS and WA values. Hashim et al. [37] stated that the particleboard made from non-trunk particles typically had higher TS or poor dimensional stability. As discussed previously, the surfaces of EFB are hydrophobic due to the existence of oil. This hydrophobic characteristic leads to poor resin spreading and prevents good particle–particle bonding [38]. Therefore, EFB particleboard has a very low dimensional stability.

The thickness and weight changes of oil palm biomass particleboards during cyclic aging treatment are shown in Figures 4–6.

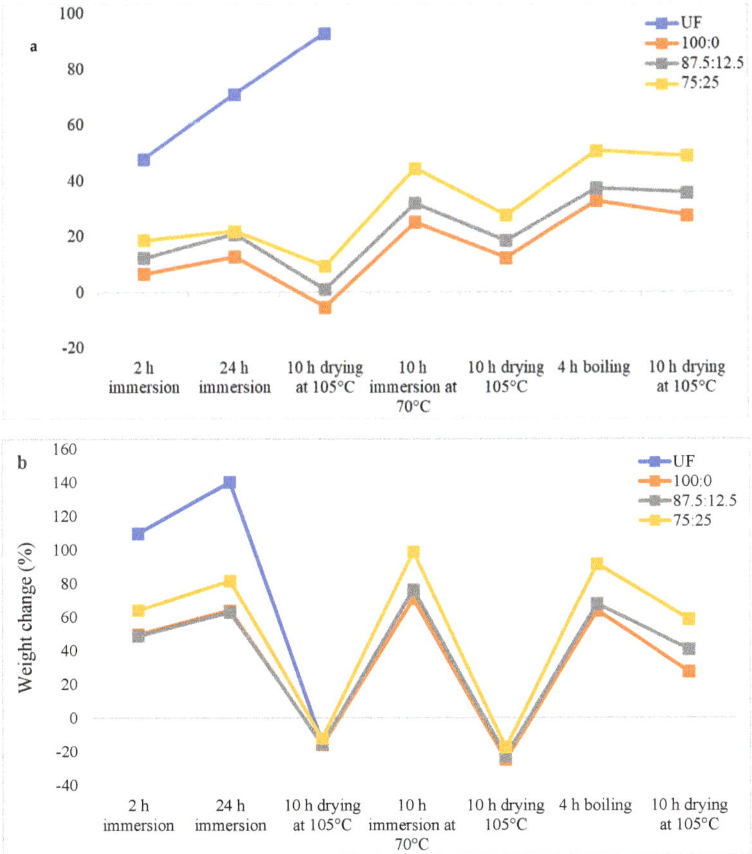

Figure 4. Thickness change (**a**) and weight change (**b**) of oil palm frond particleboard during cyclic-aging treatment.

As shown in Figure 4, UF-bonded OPF particleboard was disintegrated after 24 h of immersion in water followed by drying at 105 °C for 10 h, which indicates very poor dimensional stability. In comparison, OPF particleboard bonded with citric acid displayed a much better dimensional stability compared to that of the UF-bonded particleboard. After boiling in hot water for 4 h, the thickness changes of the particleboard bonded with pure citric acid (100:0) was 32.82%. Meanwhile, particleboard bonded with 87.5CA:12.5 starch and 75CA:25 starch recorded thickness-change values of 37.18% and 50.77%, respectively. The trend of weight change (Figure 4b) during and after the cyclic-aging treatment mirrored that of the thickness change as the particleboard bonded with pure citric acid exhibited the lowest weight-change value.

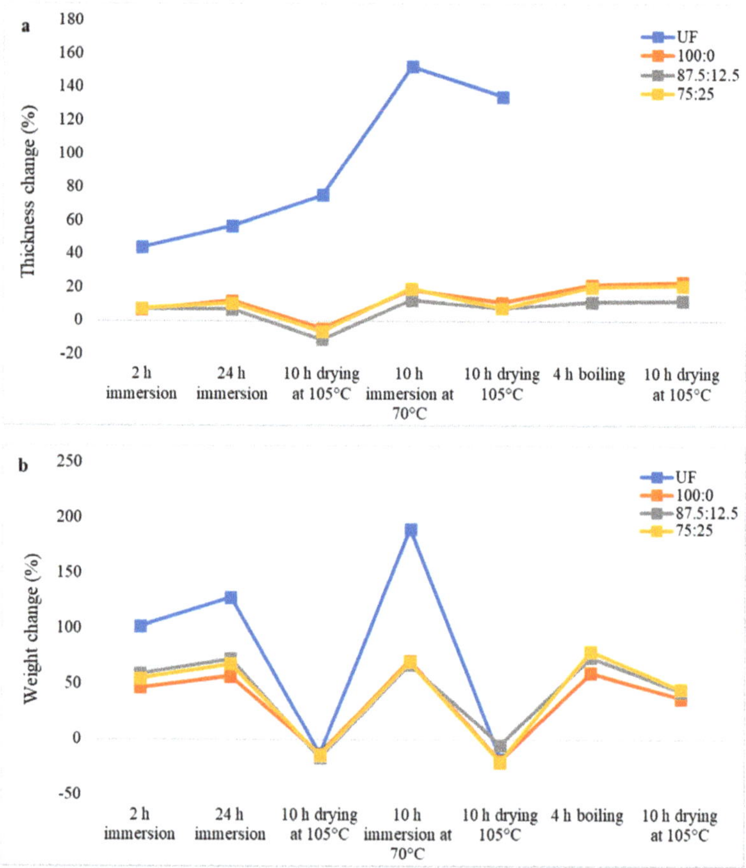

Figure 5. Thickness change (**a**) and weight change (**b**) of oil palm trunk particleboard during cyclic-aging treatment.

UF-bonded OPT particleboards are the most dimensionally stable ones compared to that of the OPF and EFB particleboards. UF-bonded OPT particleboard experienced disintegration after the second drying stage at 105 °C, right after immersion in warm water at 70 °C for 10 h. Dissimilar to OPF particleboard, OPT particleboard bonded with 87.5CA:12.5 starch displayed the lowest thickness change of 11.69% after being boiled in hot water for 4 h (Figure 5a). The thickness change values of citric-acid-bonded OPT particleboard was more than two-fold lower than its OPF counterparts. Nevertheless, OPT particleboard bonded with pure citric acid exhibited the lowest weight change after 4 h of boiling in hot water (Figure 5b).

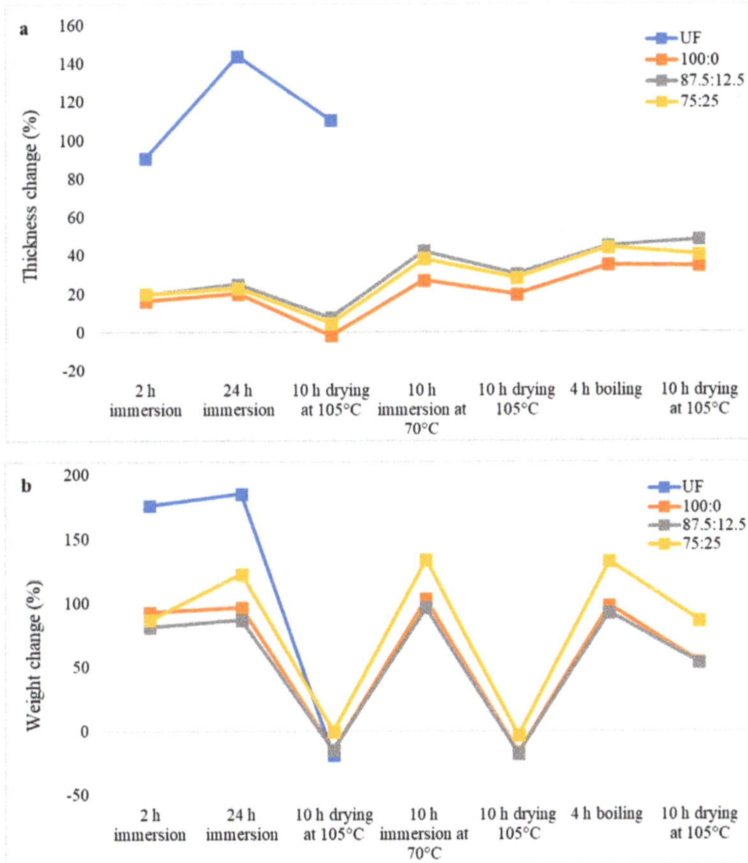

Figure 6. Thickness change (**a**) and weight change (**b**) of empty fruit bunch particleboard during cyclic-aging treatment.

For EFB particleboard, UF-bonded particleboard experienced disintegration after the first drying stage at 105 °C, which was similar to UF-bonded OPF particleboard. The thickness change value of EFB particleboard bonded with pure citric acid was 35.08% after being boiled in hot water for 4 h (Figure 6a). Meanwhile, the weight values of EFB particleboard bonded with citric acid showed a very drastic change during each stage of the treatment (Figure 6b).

Generally, in most of the cases, the thickness and weight change values increased along with increasing starch content. OPT particleboard was the only exception, where the particleboard bonded with 87.5CA:12.5 starch displayed the lowest thickness change. It can be concluded that the addition of starch contributed to the increment of water-soluble compounds in the particleboards. As a result, the water resistance of the adhesive reduced as the particleboards were subjected to repeated immersion cycles. The water penetrated into the particleboard and weakened the bond of the particleboard [20]. Apart from that, the reduction in thermal stability after the addition of starch might also contribute to this observation. The particleboard was hot-pressed at 180 °C in this study. Based on the TGA curves shown in Figure 3, at this temperature, pure citric acid experienced a weight loss of 5.9%, while the mixture of citric acid and tapioca starch at 87.5:12.5 and 75:25 recorded weight losses of 9.0% and 15.8%, respectively. Therefore, the citric acid compounds that contribute to the adhesiveness became lesser in the particleboard after the addition of starch, especially when 25% of starch was added.

With the exception of UF-bonded particleboard, all the particleboard bonded with citric acid and the mixture of citric acid and tapioca starch maintained their original shape after cyclic-aging treatment (Table 4). A darker color was observed for the citric-acid-bonded particleboard after the cyclic-aging treatment. On the other hand, UF-bonded particleboard either displayed a huge extent of swelling or complete disintegration after the cyclic-aging treatment.

Table 4. Visual change of oil palm biomass particleboard before and after being subjected to cyclic-aging treatment.

Binder	Before	After
	Oil palm frond	
UF		 
100CA:0 starch		
87.5CA:12.5 starch		
75CA:25 starch		
	Oil palm trunk	
UF		 
100CA:0 starch	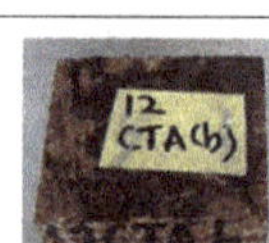	

Table 4. *Cont.*

Binder	Before	After
87.5CA:12.5 starch		
75CA:25 starch		
	Empty fruit bunch	
UF		 
100CA:0 starch		
87.5CA:12.5 starch		
75CA:25 starch		

4. Conclusions

The effects of the addition of starch on the properties of oil palm biomass particleboard bonded with citric acid were investigated in this study. Generally, the addition of starch reduced the thermal stability of citric acid when subjected to lower temperatures. However, the mixture of citric acid and tapioca starch showed higher resistance at higher temperatures as indicated by the higher amount of residues compared to that of pure citric acid. In terms of the physical and mechanical properties of the particleboard produced, an addition of 12.5% starch might be beneficial. All of the citric-acid-bonded particleboard showed significantly better performance than that of UF-bonded particleboard. Some of the particleboard displayed better bending strength when bonded with a mixture of citric acid and tapioca starch at ratio of 87.5:12.5. Although some particleboards showed a decreasing trend after an addition of 12.5% starch, the decrement was, however, insignificant compared to those bonded with pure citric acid. The addition of 25% starch was unfavorable as it resulted in an adverse effect on the mechanical properties of the particleboard. On the other hand, a similar observation was obtained for physical properties of the particleboard. The addition of 12.5% starch slightly increased the TS and WA of the particleboard (although some showed a slight decrement) due to the solubility characteristic of starch in water. The increment was, however, acceptable. When subjected to the cyclic-aging treatment, citric-acid-bonded particleboard displayed superior dimensional stability as they maintained their original shape after the treatment. On the contrary, all the UF-bonded particleboard disintegrated half way through the treatment. Among oil palm biomasses, OPT and OPF exhibited better performance than that of EFB in terms of mechanical strength and dimensional stability. In conclusion, citric acid is a promising binder for particleboard as well as other wood composites. The addition of 12.5% starch could be beneficial to the mechanical properties, but addition beyond that loading is not recommended.

Author Contributions: Conceptualization, S.H.L. and P.B.; methodology, S.H.L.; software, W.C.L.; validation, S.H.L., W.C.L. and S.S.O.A.-E.; formal analysis, S.H.L.; investigation, R.Z.; resources, P.B.; data curation, R.Z.; writing—original draft preparation, R.Z. and S.H.L.; writing—review and editing, S.H.L., W.C.L., S.S.O.A.-E. and Z.I.; visualization, S.S.; supervision, P.B. and S.S.; project administration, P.B.; funding acquisition, P.B. and S.H.L. All authors have read and agreed to the published version of the manuscript.

Funding: This research was funded by the Fundamental Research Grant Scheme (FRGS 2019-1), Reference code: FRGS/1/2019/WAB07/UPM/02/3 by the Ministry of Higher Education, Malaysia and the Higher Institution Center of Excellence (HICoE).

Institutional Review Board Statement: Not applicable.

Informed Consent Statement: Not applicable.

Data Availability Statement: Not applicable.

Acknowledgments: The authors would like to acknowledge the Library of Universiti Putra Malaysia for providing access to the literature.

Conflicts of Interest: The authors declare no conflict of interest.

References

1. Malaysian Palm Oil Board (MPOB). Oil Palm Planted Area 2020. Available online: https://bepi.mpob.gov.my/index.php/en/area/area-2020/oil-palm-planted-area-as-at-dec-2020 (accessed on 18 August 2021).
2. Mohammad Padzil, F.N.M.; Lee, S.H.; Ainun, Z.M.A.; Lee, C.H.; Abdullah, L.C. Potential of oil palm empty fruit bunch resources in nanocellulose hydrogel production for versatile applications: A review. *Materials* **2020**, *13*, 1245. [CrossRef] [PubMed]
3. Khalil, H.S.A.; Alwani, M.S.; Omar, A.K.M. Chemical composition, anatomy, lignin distribution, and cell wall structure of Malaysian plant waste fibers. *BioResources* **2006**, *1*, 220–232. [CrossRef]
4. Griffin, W.; Michalek, J.; Matthews, H.; Hassan, M. Availability of biomass residues for co-firing in Peninsular Malaysia: Implications for cost and GHG emissions in the electricity sector. *Energies* **2014**, *7*, 804–823. [CrossRef]
5. Lathrop, E.C.; Naffziger, T.R. Evaluation of fibrous agricultural residue for structural building board products-III. A process for the manufacture of high-grade products from wheat straw. *Tappi* **1994**, *32*, 319–330.
6. Penniston, K.L.; Nakada, S.Y.; Holmes, R.P.; Assimos, D.G. Quantitative assessment of citric acid in lemon juice, lime juice, and commercially-available fruit juice products. *J. Endourol.* **2008**, *22*, 567–570. [CrossRef] [PubMed]
7. Umemura, K.; Ueda, T.; Munawar, S.S.; Kawai, S. Application of citric acid as natural adhesive for wood. *J. Appl. Polym. Sci.* **2012**, *123*, 1991–1996. [CrossRef]
8. Lee, S.H.; Tahir, P.M.; Lum, W.C.; Tan, L.P.; Bawon, P.; Park, B.D.; Al Edrus, S.S.O.; Abdullah, U.H. A review on citric acid as green modifying agent and binder for wood. *Polymers* **2020**, *12*, 1692. [CrossRef] [PubMed]
9. Umemura, K.; Sugihara, O.; Kawai, S. Investigation of a new natural adhesive composed of citric acid and sucrose for particleboard. *J. Wood Sci.* **2013**, *59*, 203–208. [CrossRef]
10. Widyorini, R.; Nugraha, P.; Rahman, M.; Prayitno, T. Bonding ability of a new adhesive composed of citric acid-sucrose for particleboard. *BioResources* **2016**, *11*, 4526–4535. [CrossRef]
11. Widyorini, R.; Umemura, K.; Kusumaningtyas, A.R.; Prayitno, T.A. Effect of starch addition on properties of citric acid-bonded particleboard made from bamboo. *BioResources* **2017**, *12*, 8068–8077.
12. Widyorini, R.; Umemura, K.; Isnan, R.; Putra, D.R.; Awaludin, A.; Prayitno, T.A. Manufacture and properties of citric acid-bonded particleboard made from bamboo materials. *Eur. J. Wood Wood Prod.* **2016**, *74*, 57–65. [CrossRef]
13. Kusumah, S.S.; Umemura, K.; Guswenrivo, I.; Yoshimura, T.; Kanayama, K. Utilization of sweet sorghum bagasse and citric acid for manufacturing of particleboard II: Influences of pressing temperature and time on particleboard properties. *J. Wood Sci.* **2017**, *63*, 161–172. [CrossRef]

14. Santoso, M.; Widyorini, R.; Prayitno, T.A.; Sulistyo, J. Bonding performance of maltodextrin and citric acid for particleboard made from nipa fronds. *J. Korean Wood Sci. Technol.* **2017**, *45*, 432–443.
15. Syamani, F.A.; Kusumah, S.S.; Astari, L.; Prasetiyo, K.W.; Wibowo, E.S.; Subyakto. Effect of pre-drying time and citric acid content on *Imperata cylindrica* particleboards properties. *IOP Conf. Ser. Earth Environ. Sci.* **2018**, *209*, 01203. [CrossRef]
16. Ferrandez-Garcia, M.T.; Ferrandez-Garcia, C.E.; Garcia-Ortuño, T.; Ferrandez-Garcia, A.; Ferrandez-Villena, M. Experimental evaluation of a new giant reed (*Arundo Donax* L.) composite using citric acid as a natural binder. *Agronomy* **2019**, *9*, 882. [CrossRef]
17. Liao, R.; Xu, J.; Umemura, K. Low density sugarcane bagasse particleboard bonded with citric acid and sucrose: Effect of board density and additive content. *BioResources* **2016**, *11*, 2174–2185. [CrossRef]
18. Zhou, H.X.; Hua, L.S.; Tahir, P.M.; Ashaari, Z.; Al-Edrus, S.S.O.; Ibrahim, N.A.; Abdullah, L.C.; Mohamad, S.F. Physico-Mechanical and Biological Durability of Citric Acid-Bonded Rubberwood Particleboard. *Polymers* **2021**, *13*, 98.
19. Lee, S.H.; Lum, W.C.; Zaidon, A.; Maminski, M. Microstructural, mechanical and physical properties of post heat-treated melamine-fortified urea formaldehyde-bonded particleboard. *Eur. J. Wood Wood Prod.* **2015**, *73*, 607–616. [CrossRef]
20. JIS A 5908. *Particleboards. Japanese Industrial Standards (JIS)*; Japanese Standards Association: Tokyo, Japan, 2003.
21. TAPPI Standard T222 os-74. *Acid-Insoluble Lignin in Wood and Pulp*; Technical Association of the Pulp and Paper Industry (TAPPI): Atlanta, Georgia, 1974.
22. TAPPI Standard T203 os-74. *Alpha-, Beta-, Gamma-Cellulose in Pulp*; Technical Association of the Pulp and Paper Industry (TAPPI): Atlanta, Georgia, 1974.
23. Wise, L.E.; Murphy, M.; D'Addieco, A.A. Chlorite holocellulose, its fractionation and bearing on summative wood analysis and on studies on the hemicelluloses. *Paper Trade J* **1946**, *122*, 35–43.
24. Lee, S.H.; Ashaari, Z.; Chen, L.W.; San, H.P.; Peng, T.L.; Chow, M.J.; Chai, E.W.; Chin, K.L. Properties of particleboard with oil palm trunk as core layer in comparison to three-layer rubberwood particleboard. *J. Oil Palm Res.* **2015**, *27*, 67–74.
25. Kusumah, S.S.; Umemura, K.; Yoshioka, K.; Miyafuji, H.; Kanayama, K. Utilization of sweet sorghum bagasse and citric acid for manufacturing of particleboard I: Effects of pre-drying treatment and citric acid content on the board properties. *Ind. Crop. Prod.* **2016**, *84*, 34–42. [CrossRef]
26. Abdul Khalil, H.P.S.; Jawaid, M.; Hassan, A.; Paridah, M.T.; Zaidon, A. Oil palm biomass fibres and recent advancement in oil palm biomass fibres based hybrid biocomposites. In *Composites and Their Applications*, 1st ed.; Hu, N., Ed.; InTechOpen: London, UK, 2012.
27. John, M.J.; Francis, B.; Varughese, K.T.; Thomas, S. Effect of chemical modification on properties of hybrid fiber biocomposites. *Compos.-A Appl. Sci. Manuf.* **2008**, *39*, 352–363. [CrossRef]
28. Grigoriu, A.; Passialis, C.; Voulgaridis, E. Experimental particleboards from kenaf plantations grown in Greece. *Holz. Roh. Werkst.* **2000**, *58*, 309–314. [CrossRef]
29. Papadopoulos, A.N.; Hill, C.A.S.; Gkaraveli, A.; Ntalos, G.A.; Karastergiou, S.P. Bamboo chips (Bambusa vulgaris) as an alternative lignocellulosic raw material for particleboard manufacture. *Holz. Roh. Werkst.* **2004**, *62*, 36–39. [CrossRef]
30. Silva, N.T.; Nascimento, N.F.; Cividanes, L.S.; Bertran, C.A.; Thim, G.P. Kinetics of cordierite crystallization from diphasic gels. *J. Solgel. Sci. Technol.* **2008**, *47*, 140–147. [CrossRef]
31. Chen, W.C.; Judah, S.N.M.S.M.; Ghazali, S.K.; Munthoub, D.I.; Alias, H.; Mohamad, Z.; Abd Majid, R. The Effects of Citric Acid on Thermal and Mechanical Properties of Crosslinked Starch Film. *Chem. Eng. Trans.* **2021**, *83*, 199–204.
32. Ali, I.; Ahmad, M.; Ganat, T. Development of a new formulation for enhancing the rheological and filtration characteristics of low-solids WBMs. *J. Pet. Sci. Eng.* **2021**, *205*, 108921. [CrossRef]
33. Hashim, R.; Nadhari, W.N.A.W.; Sulaiman, O.; Kawamura, F.; Hiziroglu, S.; Sato, M.; Sugimoto, T.; Seng, T.G.; Tanaka, R. Characterization of raw materials and manufactured binderless particleboard from oil palm biomass. *Mater. Des.* **2011**, *32*, 246–254. [CrossRef]
34. Boruszewski, P.; Borysiuk, P.; Mamiński, M.; Czechowska, J. Mat compression measurements during low-density particleboard manufacturing. *BioResources* **2016**, *11*, 6909–6919. [CrossRef]
35. Norul Izani, M.A.; Paridah, M.T.; Astimar, A.A.; Mohd, N.; Anwar, U.M.K. Mechanical and dimensional stability properties of medium-density fibreboard produced from treated oil palm empty fruit bunch. *J. Appl. Sci.* **2012**, *12*, 561–567. [CrossRef]
36. Ghani, A.; Bawon, P.; Ashaari, Z.; Wahab, M.W.; Hua, L.S.; Chen, L.W. Addition of propylamine as formaldehyde scavenger for urea formaldehyde-bonded particleboard. *Wood Res.* **2017**, *62*, 329–334.
37. Hashim, R.; Wan Nadhari, W.N.A.; Sulaiman, O.; Sato, M.; Hiziroglu, S.; Kawamura, F.; Sugimoto, T.; Seng, T.G.; Tanaka, R. Properties of binderless particleboard panels manufactured from oil palm biomass. *BioResources* **2012**, *7*, 1352–1365.
38. Osman, N.F.; Bawon, P.; Lee, S.H.; Zaki, P.H.; Al-Edrus, S.S.O.; Halip, J.A.; Atkhar, M.S.M. Characterization of Particleboard Made from Oil Heat-Treated Rubberwood Particles at Different Mixing Ratios. *BioResources* **2020**, *15*, 6795–6810. [CrossRef]

polymers

MDPI

Article

Influence of Formate Concentration on the Rheology and Thermal Degradation of Xanthan Gum

María José Martín-Alfonso *, Javier Mauricio Loaiza, Clara Delgado-Sánchez and Francisco José Martínez-Boza *

Centro de Investigación en Tecnología de Procesos y Productos Químicos (Pro2TecS), ETSI, Universidad de Huelva, 21071 Huelva, Spain; javiermauricio.loiza@diq.uhu.es (J.M.L.); clara.delgado@diq.uhu.es (C.D.-S.)
* Correspondence: mariajose.martin@diq.uhu.es (M.J.M.-A.); martinez@diq.uhu.es (F.J.M.-B.); Tel.: +34-959-219-993 (F.J.M.-B.)

Abstract: Xanthan gum solutions have gained increasing interest for their use as environmentally friendly chemicals in the oil industry. Xanthan is compatible with most concentrate brines used for controlling formation damage and fluid loss. Particularly, formate brines reinforce the ordered structure of the biopolymer in solution, gel strength, and the specific gravity of the resulting fluid. In this paper, we studied the effect of thermal aging on the rheological behavior of xanthan solutions as a function of the concentration in potassium formate. Ionic strength below a threshold concentration does not prevent the degradation of the structure of xanthan after being submitted to aging at 165 °C. Aged solutions show an important loss of strength in their mechanical properties, lower pH, and higher content in furfural and hydroxymethylfurfural. Highly concentrated formate brines are necessary to maintain the strength of the rheological properties after exposure to high-temperature environments.

Keywords: xanthan gum; potassium formate; rheological behavior; high-temperature aging

Citation: Martín-Alfonso, M.J.; Loaiza, J.M.; Delgado-Sánchez, C.; Martínez-Boza, F.J. Influence of Formate Concentration on the Rheology and Thermal Degradation of Xanthan Gum. *Polymers* **2021**, *13*, 3378. https://doi.org/10.3390/polym13193378

Academic Editor: Chin-San Wu

Received: 25 August 2021
Accepted: 26 September 2021
Published: 30 September 2021

1. Introduction

Biopolymers have been proven to be an interesting alternative for the development of sustainable and environmentally friendly fluids in many industrial activities. Polysaccharides, such as xanthan gum (XG), in brine solutions are commonly used in oilfields as rheology controllers. Its pseudoplastic characteristics and high resistance to shear degradation [1,2] cover the desired requirement of high viscosity at a low shear rate, which is necessary for maintaining drill cutting in suspension when the flow is stopped, avoiding settling to the bottom hole. At the same time, these characteristics cover the desired requirement of low viscosity at a high shear rate, which facilitates pumping circulation [3–7]. Particularly, formate salts have been used successfully since the late 1990s as completion fluids due to their high solubility in water and the ease with which one can prepare formulations with the required density to compensate the formation pressure [8,9].

The rheological versatility of XG in solution is due to the development of a time-dependent structure with a well-known order-disorder transition that leads to non-Newtonian behavior, which depends on concentration, temperature, pH, ionic strength, pyruvate and acetate content, etc. [10,11]. The rheological behavior of XG has been characterized as weak gel-like in linear oscillatory shear [12–15] and pseudoplastic in viscous flow [16]. This behavior turns it into a Newtonian liquid at higher temperatures with the disruption of the ordered structure above the conformational thermal transition [17].

The presence of salts in XG solutions not only improves the strength of the rheological behavior but also enhances the thermal resistance to degradations at high temperatures [8]. High ionic strength protects the helical structure, favoring the ordered state and increasing the order-disorder transition temperature, and consequently preventing XG chains from chemical attack [18], mainly in highly concentrated formate brines [9]. Despite the screening protection of the structure, degradation always takes place with thermal aging [7,19–21]. Degradation of the ordered xanthan heavily influences its rheological behavior, turning its gel-like properties to liquid, and consequently, there is the challenge of extending the performance window for engineering oilfield applications [22–26].

To go further into the role that formate salts play in the stabilization of xanthan solutions at high temperatures, in previous works, the rheological behavior of XG as a function of the XG concentration in concentrate brines has been studied, concluding that potassium formate enlarges the range of temperatures at which the solution behaves as a weak gel [9], enhancing its resistance to thermal degradation and improving the conservation of the pseudoplasticity [20].

In this paper, we focus on the effect that ionic strength has due to potassium formate on the rheological behavior of XG solutions, exploring the change in pH and the products of degradation, before and after being submitted to thermal aging. It was hypothesized that the screening effect and association promoted due to the high concentration in potassium formate would protect the ordered structure of XG in solution against the chemical attack at high temperature, avoiding the degradation of the non-Newtonian behavior.

2. Materials and Methods

2.1. Sample Preparation

A native xanthan solution, 2.0 wt%, was prepared by adding xanthan powder (Mw $\sim\cdot 10^6$ g/mol; Guinama S.L.,Valencia, Spain) without purification to distilled water; 0.01 wt% sodium azide (Sigma-Aldrich Co., Steinheim am Albuch, Germany) was added to the solution as a preservative. The sample was maintained at rest to fully hydrate for 24 h, after which the solution was stirred at room temperature in a Silverson mixer (Silverson Machines Ltd, Chesham, UK) (600 and 2000 rpm for 15 min, respectively). This stock solution was stored at room temperature.

Solutions of 100 mL of XG in potassium formate were made by taking 25 g of the native xanthan stock solution at 2.0 wt%, dissolving the corresponding quantities of reagent grade potassium formate (Sigma-Aldrich Co., Steinheim am Albuch, Germany), and adding distilled water to complete the volume of 100 mL. A magnetic stirrer (C-MAG HS 7, IKA-Werke GmbH, Staufen, Germany), at 500 rpm, was used to homogenize the solution.

The densities of the solutions were measured at 30 °C, taking the clear liquid after separating the xanthan gel by centrifugation at 6000 rpm (MAGNUS 22, Ortoalresa, Daganzo, Spain), using an Anton Paar DMA 5000 densimeter.

2.2. Rheological Measurements

The rheological characterization was carried out using a controlled stress rheometer Physica MCR-301 (Anton Paar, Seiersberg, Austria), equipped with coaxial cylinder geometry CC27 (26.266 mm inner diameter, 28.92 mm outer diameter, and 40.032 mm length), that was used for all measurements, at the temperature of 30 °C, which was controlled by a Peltier system C-PTD200.

The linear viscoelasticity region was determined by performing stress sweeps on each sample at a frequency of 1 rad/s, at the temperature of 30 °C. Frequency sweeps, from 0.1 to 100 rad/s, were carried out at shear stress inside the linear viscoelasticity region. At least two replicates of each test were undertaken. The standard deviation among replicates was within ±5%.

Steady-state viscosity curves were collected at 30 °C, increasing the shear rate (upward curve) and then decreasing the shear rate, with the data being collected over a period of 60 s at each shear rate and using the continuous shear rate mode option of the rheometer. The two sets of data were very similar (within 5% of each other). As such, only the downward flow curves are presented here.

2.3. Thermal Aging and Degradation

Samples for aging were placed in high-pressure containers (100 mL) made of 316L stainless steel and were pressurized with 50 bar of nitrogen at room temperature. The samples were statically aged at 165 °C for 16 h in a convection oven (Heraterm, Thermofisher, Waltham, MA, USA). Pressure and temperature were recorded during the whole aging period with the help of a pressure transmitter Sensotec DMP 331, a thermocouple type K, and a DAC device inet-555 (Omega Engineering, Norwalk, CT, USA). After aging, samples were cooled to room temperature and stored at 4 °C for the measurement of the rheology, density, and furfural/ hydroxymethylfurfural (F+HMF) content.

F+HMF was quantified before and after aging by ultraviolet-visible (UV-vis) spectroscopy, using a GENESYS 10 UV-Vis Spectrophotometer (Thermo Fisher Scientific, Waltham, MA, USA) equipped with quartz cuvettes of 1 cm light path. Samples were centrifuged (MAGNUS 22, Ortoalresa, Daganzo, Spain) to remove precipitates, and the supernatant was diluted and analyzed. A sweep for wavelengths from 200 to 800 nm was performed (results not shown) and the maximum peak was detected in the range 272–282 nm. A wavelength of 277 nm was selected for testing in accordance with the work by Heggset et al. [26]. The calibration curve for F+HMF was constructed from solutions of Sigma-Aldrich standard (Lot 185914/Lot W501808) containing a 1–20 ppm concentration of F+HMF fitted to the equation:

$$A = 0.149[\mathrm{F} + \mathrm{HMF}] + 0.066 \quad \left(R^2 = 0.999\right) \tag{1}$$

where A is the absorbance at 277 nm, which ranged from 0.122 to 1.037, and [F+HMF] is the furfural and hydroxymethylfurfural concentration in ppm. The samples were diluted 1/50 for spectrophotometric measurement. Each experimental value was the average of five results. Deviations from the respective means were all less than 1%.

3. Results and Discussion

3.1. Viscoelastic Behavior

Figure 1 shows both the storage and loss moduli versus frequency for fresh xanthan solutions in distilled water and in 15.00 M of potassium formate (HCOOK), at the temperature of 30 °C, well below the conformational thermal transition of XG in solution [15]. As can be observed, for both solutions, the values of the storage modulus are higher than those of the loss modulus at the same frequency. In addition, both storage and loss moduli show a weak dependence on frequency. This is more apparent for the solution in formate brine, which shows a lower slope in the evolution of moduli with frequency at low and intermediate frequencies. These observations demonstrate that these solutions behave as weak gels. Gel-like behavior is typical for XG solutions, at moderate concentrations, where the intermolecular associations control the response to linear deformations, as has been previously reported [12]. The presence of a high concentration of salts compacts the ordered structure of XG in solution, enhancing the molecular interactions and the association by hydrogen bonding [27]. This effect increases the strength of the viscoelastic response of the structure, leading to both the increase in the values of the storage and loss moduli and the decrease of the loss tangent (tan $\delta = G''/G'$) [28]. Similar behavior has been described for concentrate XG solutions in chloride and formate brine [9] and in sodium and calcium chloride [21].

Figure 2 shows the evolution of the viscoelastic moduli as a function of the potassium formate concentration for fresh XG solutions. As can be observed in Figure 2, the values of both the storage (Figure 2A) and loss modulus (Figure 2B) increase when the molar concentration of formate increases in the solution. At the same time, the loss tangent decreases with both the frequency and concentration, up to a minimum value of around 20 rad/s. After that, tan δ values increase suddenly (Figure 2C).

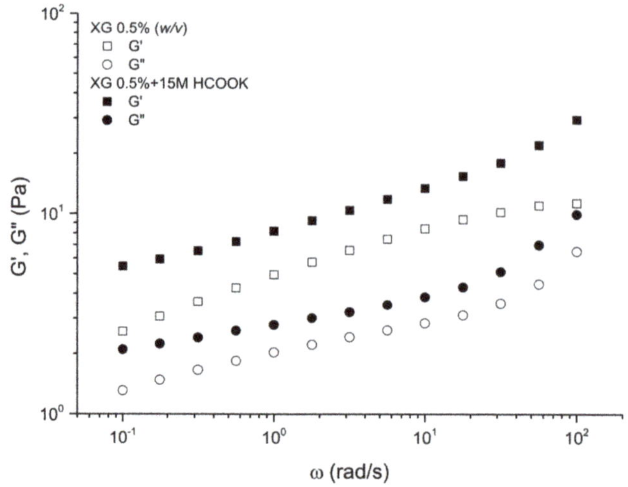

Figure 1. Storage and loss moduli versus the frequency for fresh XG solutions at 0.5% (*w/v*).

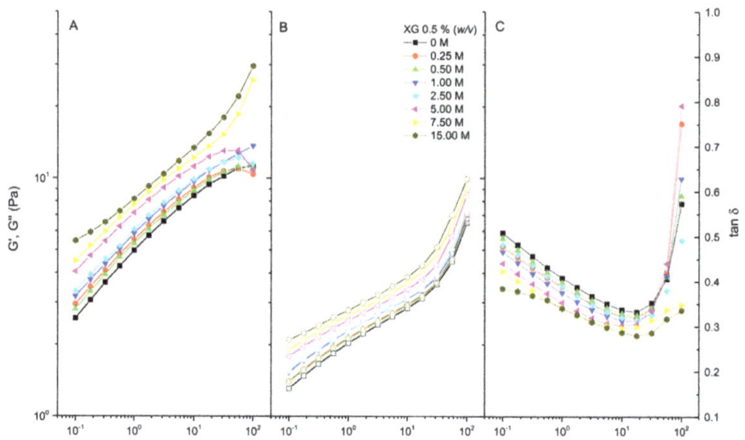

Figure 2. Evolution of the viscoelastic properties with the potassium formate concentration for fresh XG solutions at 0.5% (*w/v*) in HCOOK. (**A**) Storage modulus. (**B**) Loss modulus. (**C**) Loss tangent.

To quantitatively evaluate the effect of potassium formate on the viscoelasticity of these solutions, a plateau modulus ($G_N{}^0$) is defined as the value of the storage modulus at the frequency of the minimum value of the loss tangent. Figure 3A,B display the values of both the plateau modulus and loss tangent at the frequency where the minimum in the loss tangent appears, as a function of the formate concentration, respectively. As can be seen in Figure 3A, $G_N{}^0$ increases nearly linear (R^2 = 0.94) with the formate concentration from 9.4 Pa for the native solution to 15.4 Pa for the concentrated solution, respectively. Moreover, a slight decrease of the loss tangent is observed when increasing the formate concentration, from a value of 0.33 to 0.28 for native and concentrate solutions, respectively. Wyatt et al. [17,27] studied the effect of ionic strength on the rheology of XG solutions over a wide range of XG concentrations, observing that for moderate XG concentrations (above 2000 ppm), ionic strength plays different roles depending on the ionic concentration scale.

In contrast, for this XG solution studied at a moderate concentration (0.5% *w/v*), 5000 ppm), a clear increase in rheology was observed with the concentration of formate over the whole range of ionic strength explored. A screening effect and Colombian repulsion would explain the changes observed in rheology with respect to native XG solutions in the low region of salt concentration [29]. Nevertheless, for higher ionic strengths (above, 1.00 M), the screening effect would reinforce interactions due to hydrogen bonding that would be responsible for the observed slight increase in the viscoelasticity and viscous properties, in agreement with the hypothesis of Wyatt et al. [27] of strengthening the hydrogen bonding by screening of the charge in concentrate brines.

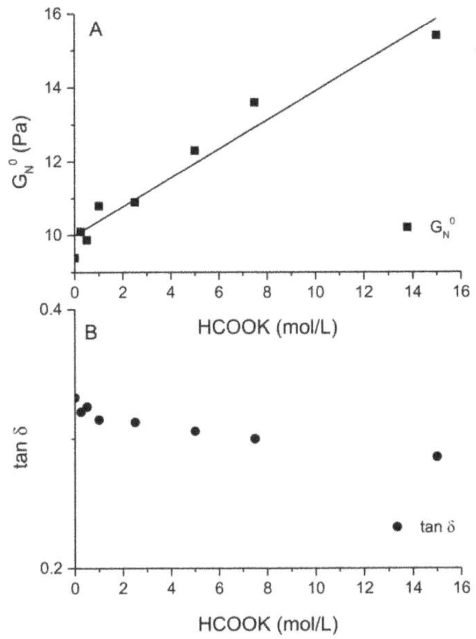

Figure 3. Evolution of the plateau modulus (**A**) and the minimum of tan δ (**B**) with potassium formate concentrations for fresh XG solutions at 0.5% (*w/v*).

3.2. Flow Behavior

At room temperature, below the conformational transition, XG in solution adopts a partial self-associated structure that is very sensitive to shear, showing a shear-thinning behavior due to the orientation of the polymer chains in the direction of flow as it has been previously reported [12,15,30]. Depending on the concentration and the experimental shear-rate windows, XG may show a tendency to reach a zero-shear limiting viscosity in the low shear-rate region; a power-law drop in viscosity at intermediate shear-rates that, in some cases, is followed by a trend towards an infinite shear-rate-limiting viscosity in the high shear-rate region. This shear-thinning behavior has been modelled by using different approaches, such as the power-law model [31], the Hershel-Bulkley model [15], the Carreau model [2], and the Cross model [9,11,20].

In this case, the parameters that characterize the shear-thinning behavior have been obtained from the Cross model Equation (2) [32] (viscosity η, shear rate $\dot{\gamma}$, limiting viscosities at low and high shear-rate η_o and η_∞, respectively, the flow index m, and the consistency index k. Values are shown in Table 1:

$$\eta = \eta_\infty + \frac{\eta_o - \eta_\infty}{1 + (k \cdot \dot{\gamma})^m} \tag{2}$$

Table 1. Cross model parameters for fresh samples XG 0.5% (*w/v*) as a function of the formate concentration.

		Fresh Xanthan 0.5% (*w/v*)		
$KHCO_2$ (mol/L)	η_0 (Pa·s)	η_∞ (Pa·s)	k (s)	m
0	145.98	6.02×10^{-3}	52.3	0.881
0.25	186.06	7.39×10^{-3}	61.5	0.894
0.5	221.43	7.47×10^{-3}	61.9	0.890
1	191.68	7.82×10^{-3}	63.1	0.891
2.5	216.60	9.34×10^{-3}	68.9	0.899
5	266.08	1.14×10^{-2}	73.7	0.907
7.5	328.67	1.40×10^{-2}	94.6	0.893
15	428.43	2.63×10^{-2}	129.7	0.904

The main effects of formate ionic strength on the flow behavior of these solutions are a slight increase in the values of the limiting viscosities (η_o and η_∞) and the consistency index, k, more significant for higher concentrations of potassium formate. Nevertheless, little effect was observed on the values of the flow index, which remains almost constant, around 0.90, with ionic strength.

These observations indicate that for this XG solution, the ionic strength due to potassium formate salts would reinforce the response of the solution in steady-state viscous flow; however, this does not significantly influence the degree of sensitivity to shear. This behavior is compatible with the hypothesis of Wyatt et al. [27] in the sense that the effect of the screening of charge above a certain ionic strength influences interactions that remain independent of the shear rate that flows the solution.

3.3. *Effect of Aging on the Rheological Behavior*

Despite XG in solution having suitable mechanical properties under shear and being considered as a good thickener in many oilfield operations, its resistance to high-temperature environments is a major drawback that limits high-pressure/high-temperature (HP/HT) applicability. The evaluation of the effect that thermal aging exerts on the rheology and microstructure of XG in solution would be a good strategy for determining both the time and applicability window in thermal aggressive environments.

The thermal resistance of XG solutions is enhanced by both the screening protection of charge due to the ionic strength of cations and the antioxidant capacity of the anion [33]. The thermal resistance of XG would improve if the ordered structure remains as the main conformation of the biopolymer in solution at high temperatures [19,21]. In this sense, in previous work, it was proved that brines not only protect XG in solution against thermal disruption, enlarging the temperature interval where the thermal transition takes place, but also that the weak-gel response dominates the rheological properties [9]. However, the original rheological behavior, at temperatures below the thermal transition, was not fully recovered after submitting the solution to high temperatures, even in the presence of concentrate brines. Furthermore, formate salt was the most suitable salt compared with chlorides for protecting the structure of XG in solution against thermal degradation, at concentrations above XG 0.5 wt% [8,20]. These solutions retained the minimum degree of pseudoplasticity to be suitable for engineering applications in aggressive temperature environments, such as drilling and completion operations in oilfields. Recently, these findings have also been corroborated by Wu et al. [21], studying fluids based on XG with side-chain modifications in sodium and calcium concentrate brines at high temperatures.

Here, XG 0.5% (*w/v*) in potassium formate solutions at concentrations below 1.00 M completely loses its pseudoplastic characteristics after being submitted to static aging for 16 h at 165 °C, resulting in an aged solution that behaves as a low viscosity Newtonian liquid. The residual viscosity of the aged solutions is around 6 mPa·s (see Table 2), slightly above the viscosity of potassium formate, which ranges from 1–8 mPa·s for 0.50 to 15.00 M, respectively. At higher concentrations of salt, as can be seen in Figure 4, the pseudoplasticity is partially recovered at 30 °C. This result suggests that some degree of the ordered structure remains after aging. The flow curves after aging were fitted to the Cross model, with the resulting parameters shown in Table 2.

Table 2. Cross model parameters for aged samples XG 0.5% (*w/v*) as a function of the formate concentration.

HCOOK (mol/L)	η_0 (Pa·s)	η_∞ (Pa·s)	k (s)	m	η_{0Ag}/η_{0Fr} (%)	k_{Ag}/k_{Fr} (%)	m_{Ag}/m_{Fr} (%)
0	6.5×10^{-3}		-	-			
0.25	6.4×10^{-3}		-	-			
0.50	6.7×10^{-3}		-	-			
1.00	7.8×10^{-3}		-	-			
2.50	2.200	1.00×10^{-3}	1.70	0.707	1.00	2.40	78.6
5.00	41.41	2.64×10^{-3}	19.3	0.817	15.2	26.1	90.1
7.50	120.60	7.50×10^{-3}	52.3	0.847	36.7	55.3	94.8
15.00	192.00	2.39×10^{-2}	77.1	0.874	44.8	59.4	96.7

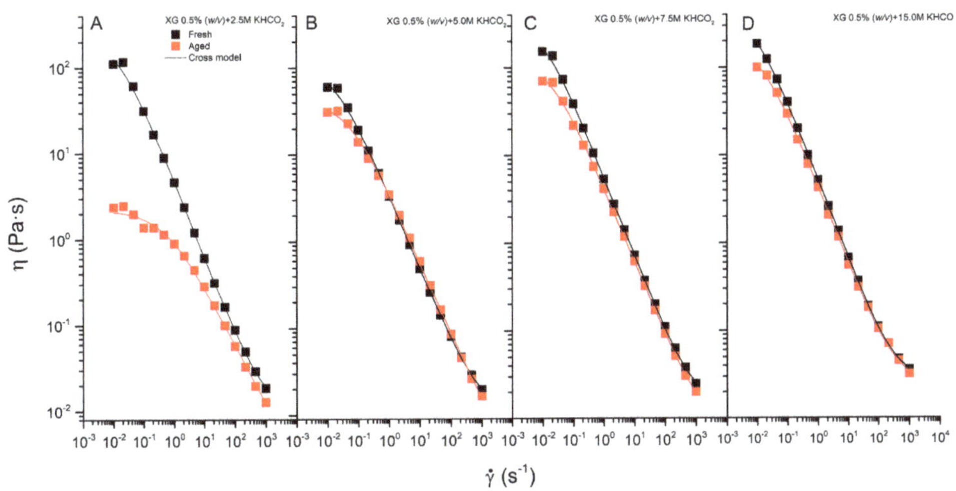

Figure 4. Effect of aging on the flow behavior for XG 0.5% (*w/v*) in potassium formate 2.50 M (**A**), 5.00 M (**B**), 7.50 M (**C**), and 15.00 M (**D**).

In general, the higher the concentrate in formate salt, the more resistant it is to degradation; however, some degree of degradation is always observed. To some extent, thermal aging modifies all parameters that characterize the pseudoplastic behavior of XG solutions, as can be deduced by comparing the values of the parameter in Table 1 (before aging) and Table 2 (after aging).

The effect of thermal degradation on the XG structure also changes the viscoelastic characteristic of the solution. In agreement with the previous results of viscous flow tests, XG solutions below the concentration of 1.00 M in potassium formate completely lose their viscoelastic properties, turning into a viscous liquid solution. This result is due to the high degradation of the polymeric structure. XG solution at the concentration of 2.50 M partially retains its original viscoelastic properties, as can be seen in Figure 5A, but after aging, a clear decrease in the values of moduli is noticed. In addition, at low frequencies, the solution shows an apparent tendency to the terminal region of the mechanical spectrum, with values of the viscous modulus higher than those of the storage modulus at the same frequency. In this region, the slope in log scale of the storage and loss moduli is 1.26 and 0.77, relatively close to 2 and 1, respectively. This would indicate that some degree of interaction remains among the degraded biopolymer chains in solution with longer relaxation times. At higher frequencies, a crossover point, at which the loss tangent becomes 1, appears followed by a flattening of the slope of the viscoelastic moduli versus frequency. This fact indicates that part of the weak-gel structure of xanthan remains in solution, becoming a more viscoelastic liquid [12].

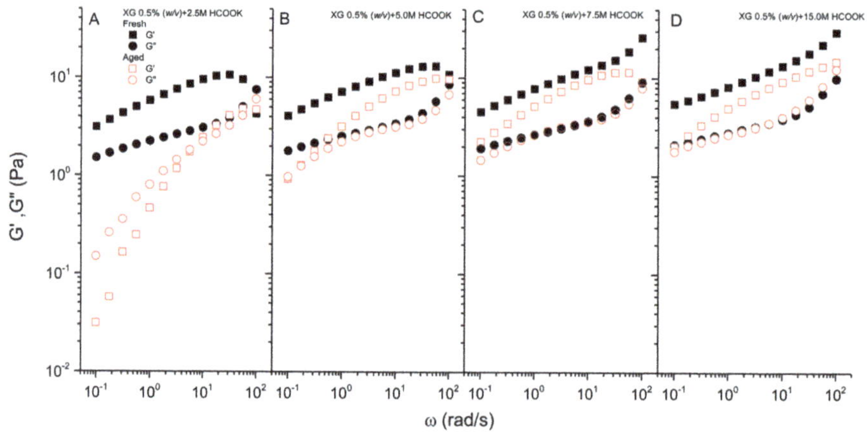

Figure 5. Effect of aging on the viscoelastic behavior for XG 0.5% (*w/v*) in potassium formate 2.50 M (**A**), 5.00 M (**B**), 7.50 M (**C**), and 15.00 M (**D**).

The increase in the formate concentration leads to an increase in the values of both moduli, enhancing the viscoelastic properties of the solution. This fact is more evident for the storage modulus, as can be seen in Figure 5B–D, which resists the effects of thermal aging very well. The results indicate less degradation with the increase in the concentration of salt. Consequently, the formate salt would protect against thermal degradation of the ordered structure, which would be responsible for the elastic properties of the solution. Moreover, the crossover point shifts to lower frequencies, extending the frequency windows where the solution displays weak-gel properties. Nevertheless, as is the case of steady-state flow, some degree of degradation is also observed even for concentrate brines. It is worth noting that the percentage of recovery in the values of G_N^0 ($100 \cdot G_N^0{}_{Aged} / G_N^0{}_{Fresh}$) after aging is 68%, 77%, and 68% for concentrations of 5.00, 7.50, and 15.00 M, respectively.

It is also worth highlighting that in agreement with previous results [15,28], positive deviations of the Cox-Merz rule (steady shear viscosity values lower than those of the complex viscosity at the same shear rate and frequency) have been observed for both fresh and aged XG solutions (i.e., for fresh XG in brine 7.50 M, at 10 s^{-1} or rad/s, η = 0.73 Pa·s, η^* = 1.27 Pa·s; for aged XG in brine 7.50 M, at 10 s^{-1} or rad/s, η = 0.62 Pa·s, η^* = 1.04 Pa·s). This suggests that the structural degradation due to thermal aging of XG does not change the thermorheological nature in solution [9].

3.4. Structure-Properties Relationships

The mechanisms of structural degradation for XG in solution include free-radical oxidation, acid-basic catalyzed hydrolysis, and enzymatic degradation [34–36]. It has been stated that the degradation of the structure would start with an attack on the side chains and would continue through the main chain with different effects on the rheological behavior [18]. Side-chain degradation would mainly affect the acetyl group, with there being less influence on the reduction of the elastic properties of the solution since the main ordered structure remains [21]. Further attacks to the main chain would drastically change the molecular weight and the rheological properties.

In this study, the changes in pH and the products of degradation were determined to address the changes in XG structure. The pH drop after high-temperature aging has been associated with both free-radical oxidation [35,37] and acid-basic hydrolysis [36,38]. The concentration of F+HMF in the aged solution would be a measure of the degradation of the main chain. At high temperatures, the degradation of hexoses and pentoses by acid-hydrolysis converts them into F+HMF, respectively. These compounds can be determined using UV-vis spectroscopy in polysaccharide solutions after aging. This has been related to chain degradation even in the presence of formate salt [26].

At 165 °C, the disordered conformation would be the main structure for XG in potassium formate solution at concentrations below 5.00 M [39]. For these solutions, XG in a disordered conformation would be very vulnerable to hydrolysis of the main chain [18,19], resulting in the loss of the rheological properties after 16 h of exposure at 165 °C in a low-oxygen environment. For these solutions, the ionic strength is not sufficient to raise the order-disorder transition temperature above the aging temperature and prevent XG chains from chemical attack [21].

In this regard, the evolution of pH and the increase in F+HMF content of the solutions after degradation are displayed in Figure 6A,B, respectively. As can be seen in Figure 6A, aging leads to a decrease in pH inversely proportional to the formate content. Native XG solution shows acidic pH and, consequently, the highest degree of degradation due to acid-hydrolysis would be expected. In addition, the highest content in F+HMF would also indicate the degradation of the main chain of XG, and consequently, the loss of the rheological properties. The concentration in potassium formate increases the pH of both fresh (R^2 = 0.99) and aged (R^2 = 0.96) solutions linearly up to 9.9 and 9.3 for the saturated solution, respectively. Degradation due to thermal aging acidifies the resulting solution; however, the pH drop is flattened when the concentration of potassium formate increases [26]. This observation would be explained assuming the high protection capacity of the ordered structure by formate salt [37] and subsequently the delay in degradation processes for concentrations higher than 5.00 M, where the ordered state remains at aging temperature. Since fresh samples were pressurized with N_2 at 50 bar, free radical oxidation would be minimized and, therefore, degradation by hydrolysis seems to be the most probable mechanism, showing a different extent depending on the formate concentration.

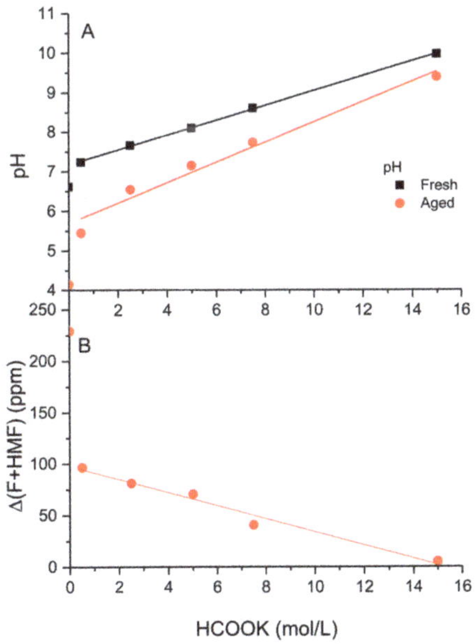

Figure 6. Degradation measured as: (**A**) Evolution of pH. (**B**) Furfural and hydroxymethyl/furfural content.

However, the content in F+HMF decreases suddenly with the presence of formate 0.50 M (from 235 to 100 ppm), and evolutes inversely and proportionally to the formate content with a slope of −6.9 M (R^2 = 0.97), being 5 ppm for 15.00 M. A high degree of resistance to thermal degradation with important maintenance of both pseudoplasticity and elastic properties (Figures 4 and 5, respectively) is observed for aged solutions in HCOOK above 7.50 M. These solutions show a (F+HMF) content below 50 ppm. Therefore, XG in potassium formate above 7.50 M (equivalent to 48.2 wt% in HCOOK and density at 30 °C of 1.310 g/mL) shows rheological behavior suitable for use in oilfield applications. It is worth noting that the density of the solution does not change after aging at 165 °C for 16 h, being very similar to that of the fresh solution for all concentrations of formate tested.

4. Conclusions

In linear viscoelasticity, XG at 0.5% (*w/v*) behaves as a weak gel, with values of the storage modulus being higher than those of the loss modulus at the same frequency. The gel strength increases with the concentration of formate in solution. The main effects on the flow behavior of the formate ionic strength are a slight increase in the values of the limiting viscosities, (η_o and η_∞) and the consistency index, k.

In general, the higher the concentrate in formate salt, the more resistant it is to degradation after aging. Aged XG solutions in formate below 1.00 M completely lose both their elasticity and their pseudoplasticity, resulting in a low-viscosity Newtonian liquid. XG in formate solutions below 5.00 M would develop a disordered structure that is very vulnerable to the hydrolysis attack of the main chain, resulting in a loss of rheological properties after 16 h at 165 °C.

At higher formate concentrations, XG retains its ordered structure responsible for the high rheological properties. This structure would be protected by high ionic strength against the thermal degradation of aging at 165 °C; however, some degree of degradation always takes place, lowering all the parameters that characterize the pseudoplastic behavior and the values of the storage modulus. Nevertheless, solutions above 7.50 M in formate retain the rheological properties enough to be suitable for use in oilfield applications.

Author Contributions: Conceptualization, M.J.M.-A., J.M.L., and C.D.-S.; validation, M.J.M.-A., J.M.L., and C.D.-S.; formal analysis, M.J.M.-A. and F.J.M.-B.; investigation, M.J.M.-A., J.M.L., C.D.-S.; resources, F.J.M.-B.; writing—original draft preparation, M.J.M.-A., J.M.L. and C.D.-S.; writing—review and editing, M.J.M.-A. and F.J.M.-B.; visualization, F.J.M.-B.; supervision, F.J.M.-B.; project administration, F.J.M.-B.; funding acquisition, F.J.M.-B. All authors have read and agreed to the published version of the manuscript.

Funding: This research was funded by the EU-FEDER Program, grant numbers P18-RT-4684 and CTQ-2017-89792-R.

Institutional Review Board Statement: Not applicable.

Informed Consent Statement: Not applicable.

Data Availability Statement: The raw/processed data required to reproduce these findings cannot be shared at this time due to technical or time limitations.

Conflicts of Interest: The authors declare no conflict of interest.

References

1. Wei, B.; Romero-Zerón, L.; Rodrigue, D. Mechanical properties and flow behavior of polymers for enhanced oil recovery. *J. Macromol. Sci. Part B Phys.* **2014**, *53*, 625–644. [CrossRef]
2. Jang, H.Y.; Zhang, K.; Chon, B.H.; Choi, H.J. Enhanced oil recovery performance and viscosity characteristics of polysaccharide xanthan gum solution. *J. Ind. Eng. Chem.* **2015**, *21*, 741–745. [CrossRef]
3. Asafa, K.A.; Shah, S.N. Rheology and flow characteristics of xanthan in calcium chloride brine. In Proceedings of the SPE/ICoTA Coiled Tubing and Well Intervention Conference and Exhibition, The Woodlands, TX, USA, March 2014. [CrossRef]
4. Hermoso, J.; Martínez-Boza, F.; Gallegos, C. Influence of viscosity modifier nature and concentration on the viscous flow behavior of oil-based drilling fluids at high pressure. *Appl. Clay Sci.* **2014**, *87*, 14–21. [CrossRef]

5. Hermoso, J.; Martínez-Boza, F.; Gallegos, C. Combined effect of pressure and temperature on the viscous behaviour of all-oil drilling fluids. *Oil Gas Sci. Technol.* **2014**, *69*, 1283–1296. [CrossRef]
6. Hermoso, J.; Martínez-Boza, F.; Gallegos, C. Influence of aqueous phase volume fraction, organoclay concentration and pressure on invert-emulsion oil muds rheology. *J. Ind. and Eng. Chem.* **2015**, *22*, 341–349. [CrossRef]
7. Boul, P.J.; Abdulquddos, S.; Thaemlitz, C.J. High performance brine viscosifiers for high temperatures. In Proceedings of the SPE Middle East Oil & Gas Show and Conference, Manama, Kingdom of Bahrain, March 2017. [CrossRef]
8. Howard, S.; Kaminski, L.; Downs, J. Xanthan stability in formate brines-Formulating non-damaging fluids for high temperature applications. In Proceedings of the SPE European Formation Damage Conference and Exhibition, Budapest, Hungary, June 2015. [CrossRef]
9. Reinoso, D.; Martin-Alfonso, M.J.; Luckham, P.F.; Martinez-Boza, F.J. Rheological characterisation of xanthan gum in brine solutions at high temperature. *Carbohydr. Polym.* **2019**, *203*, 103–109. [CrossRef] [PubMed]
10. Garcia-Ochoa, F.; Santos, V.E.; Casas, J.A.; Gómez, E. Xanthan gum: Production, recovery, and properties. *Biotechnol. Adv.* **2000**, *18*, 549–579. [CrossRef]
11. Fitzpatrick, P.; Meadows, J.; Ratcliffe, I.; Williams, P.A. Control of the properties of xanthan/glucomannan mixed gels by varying xanthan fine structure. *Carbohydr. Polym.* **2013**, *92*, 1018–1025. [CrossRef] [PubMed]
12. Rochefort, W.E.; Middleman, S. Rheology of xanthan gum: Salt, temperature, and strain effects in oscillatory and steady shear experiments. *J. Rheol.* **1987**, *31*, 337–369. [CrossRef]
13. Whitcomb, P.J.; Macosko, C.W. Rheology of xanthan gum. *J. Rheol.* **1978**, *22*, 493–505. [CrossRef]
14. Lee, H.C.; Brant, D.A. Rheology of concentrated isotropic and anisotropic xanthan solutions. 1. A rodlike low molecular weight sample. *Macromolecules* **2002**, *35*, 2212–2222. [CrossRef]
15. Choppe, E.; Puaud, F.; Nicolai, T.; Benyahia, L. Rheology of xanthan solutions as a function of temperature, concentration and ionic strength. *Carbohydr. Polym.* **2010**, *82*, 1228–1235. [CrossRef]
16. Marcotte, M.; Hoshahili, A.R.T.; Ramaswamy, H.S. Rheological properties of selected hydrocolloids as a function of concentration and temperature. *Food Res. Int.* **2001**, *34*, 695–703. [CrossRef]
17. Wyatt, N.B.; Liberatore, M.W. Rheology and viscosity scaling of the polyelectrolyte xanthan gum. *J. Appl. Polym. Sci.* **2009**, *114*, 4076–4084. [CrossRef]
18. Lambert, F.; Rinaudo, M. On the thermal stability of xanthan gum. *Polymer* **1985**, *26*, 1549–1553. [CrossRef]
19. Xie, W.; Lecourtier, J. Xanthan behaviour in water-based drilling fluids. *Polym. Degrad. Stab.* **1992**, *38*, 155–164. [CrossRef]
20. Reinoso, D.; Martín-Alfonso, M.J.; Luckham, P.F.; Martinez-Boza, F.J. Flow behavior and thermal resistance of xanthan gum in formate brine. *J. Petrol. Sci. Eng.* **2020**, *188*, 106881. [CrossRef]
21. Wu, M.; Shi, Z.; Ming, Y.; Wang, C.; Qiu, X.; Li, G.; Ma, T. Thermostable and rheological properties of natural and genetically engineered xanthan gums in different solutions at high temperature. *Int. J. Biol. Macromol.* **2021**, *182*, 1208–1217. [CrossRef] [PubMed]
22. Hermoso, J.; Jofore, B.D.; Martínez-Boza, F.J.; Gallegos, C. High pressure mixing rheology of drilling fluids. *Ind. Eng. Chem. Res.* **2012**, *51*, 14399–14407. [CrossRef]
23. Hermoso, J.; Martínez-Boza, F.J.; Gallegos, C. Modeling Pressure-Viscosity Behavior of Oil-Based Drilling Fluids. *Oil Gas Sci. Technol.* **2017**, *72*, 18. [CrossRef]
24. Bradshaw, R.J.; Hodge, R.M.; Wolf, N.O.; Knox, D.A.; Hudson, C.E.; Evans, E. Formate-based reservoir drilling fluid resolves high-temperature challenges in the Natuna Sea. In Proceedings of the International Symposium and Exhibition on Formation Damage Control, Lafayette, LA, USA, February 2006. [CrossRef]
25. Hermoso, J.; Martínez-Boza, F.J.; Gallegos, C. Organoclay influence on high pressure-high temperature volumetric properties of oil-based drilling fluids. *J. Pet. Sci. Eng.* **2017**, *151*, 13–23. [CrossRef]
26. Heggset, E.B.; Chinga-Carrasco, G.; Syverud, K. Temperature stability of nanocellulose dispersions. *Carbohydr. Polym.* **2017**, *157*, 114–121. [CrossRef] [PubMed]
27. Wyatt, N.B.; Gunther, C.M.; Liberatore, M.W. Increasing viscosity in entangled polyelectrolyte solutions by the addition of salt. *Polymer* **2011**, *52*, 2437–2444. [CrossRef]
28. Pelletier, E.; Viebke, C.; Meadows, J.; Williams, P.A. A rheological study of the order–disorder conformational transition of xanthan gum. *Biopolymers* **2001**, *59*, 339–346. [CrossRef]
29. Dobrynin, A.V.; Rubinstein, V. Theory of polyelectrolytes in solutions and at surfaces. *Prog. Polym. Sci.* **2005**, *30*, 1049–1118. [CrossRef]
30. Ross-Murphy, S.B. Structure-property relationships in food biopolymer gels and solutions. *J. Rheol.* **1995**, *39*, 1451–1463. [CrossRef]
31. Song, K.W.; Kim, Y.S.; Chang, G.S. Rheology of concentrated xanthan gum solutions: Steady shear flow behavior. *Fibers Polym.* **2006**, *7*, 129–138. [CrossRef]
32. Cross, M.M. Rheology of non-Newtonian fluids: A new flow equation for pseudoplastic systems. *J. Colloid Sci.* **1965**, *20*, 417–434. [CrossRef]
33. Clarke-Sturman, A.J.; Pedley, J.B.; Sturla, P.L. Influence of anions on the properties of microbial polysaccharides in solution. *Int. J. Biol. Macromol.* **1986**, *8*, 355–360. [CrossRef]
34. Ash, S.G.; Clarke-Sturman, A.J.; Calvert, R.; Nisbet, T.M. Chemical stability of biopolymer solutions. In Proceedings of the SPE Annual Technical Conference and Exhibition, San Francisco, CA, USA, October 1983. [CrossRef]

35. Wellington, S.L. Biopolymer solution viscosity stabilization-Polymer degradation and antioxidant use. *SPE J.* **1983**, *23*, 901–912. [CrossRef]
36. Seright, R.S.; Henrici, B.J. Xanthan stability at elevated temperatures. *SPE Reservoir Eng.* **1990**, *5*, 52–60. [CrossRef]
37. Downs, J.D. High-temperature stabilization of xanthan in drilling fluids by the use of formate salts. In *Physical Chemistry of Colloids and Interfaces in Oil Production*; Toulhoat, H., Lecourtier, J., Eds.; Editions TECHNIP: Paris, France, 1992; pp. 197–202. ISBN 2-7108-0618-5.
38. Callet, F.; Milas, M.; Rinaudo, M. On the role of thermal treatments on the properties of xanthan solutions. *Carbohydr. Polym.* **1989**, *11*, 127–137. [CrossRef]
39. Howard, S. *Formate Manual, Section B5, Compatibility with Additives*, 2nd ed.; Cabot Corporation: Boston, MA, USA, 2009.

polymers

MDPI

Article

Bioethanol Production and Alkali Pulp Processes as Sources of Anionic Lignin Surfactants

Rodrigo Álvarez-Barajas, Antonio A. Cuadri *, Francisco J. Navarro, Francisco J. Martínez-Boza and Pedro Partal

Pro2TecS-Chemical Process and Product Technology Research Centre, Department of Chemical Engineering, ETSI, Campus de "El Carmen", Universidad de Huelva, 21071 Huelva, Spain; rodrigo.alvarez@diq.uhu.es (R.Á.-B.); frando@diq.uhu.es (F.J.N.); martinez@diq.uhu.es (F.J.M.-B.); partal@uhu.es (P.P.)
* Correspondence: antonio.cuadri@diq.uhu.es; Tel.: +34-959-219-882

Abstract: Lignin is an abundant biopolymer with potential value-added applications that depend on biomass source and fractioning method. This work explores the use as emulsifiers of three native lignin-rich product coming from industrial bioethanol production and alkali or Kraft pulping. In addition to their distinctive characteristics, the different molecular organization induced by emulsification pH is expected to interact in various ways at the water-oil interface of the emulsion droplets. Initially, model oil-in-water (O/W) emulsions of a silicone oil will be studied as a function of lignin source, disperse phase concentration and emulsification pH. Once stablished the effect of such variables, emulsion formulations of three potential bitumen rejuvenators (waste vegetable cooking oil, recycled lubricating oil and a 160/220 penetration range soft bitumen). Droplet size distribution, Z-potential and viscous tests conducted on model emulsions have shown that emulsification pH strongly affects stabilization ability of the lignins tested. Regarding bitumen rejuvenators, lignin emulsification capability will be affected by surfactant source, pH and, additionally, by the dispersed phase characteristics. Lower Z-potential values shown by KL at pH 9 and 11 seem to facilitate emulsification of the less polar disperse phases formed by RLUB and bitumen. In any case, lower particle size and higher yield stress values were found for both bioethanol-derived lignins emulsifying RVO and RLUB at pH 13, which are expected to exhibit a longer stability.

Keywords: lignin; emulsion; rheology; microstructure; bitumen; product design

Citation: Álvarez-Barajas, R.; Cuadri, A.A.; Navarro, F.J.; Martínez-Boza, F.J.; Partal, P. Bioethanol Production and Alkali Pulp Processes as Sources of Anionic Lignin Surfactants. *Polymers* **2021**, *13*, 2703. https://doi.org/10.3390/polym13162703

Academic Editor: Jalel Labidi

Received: 22 July 2021
Accepted: 9 August 2021
Published: 13 August 2021

1. Introduction

Lignin is an abundant renewable polymer that, in many cases, ends up burned to obtain energy as, for instant, lignin contained in black liquor of Kraft process side stream [1,2]. However, potential value-added applications have been described elsewhere, such as antioxidants, antimicrobials or potential building block for chemical modifications, which arise from the high amount of chemicals sites present in the lignin structure [2]. Thus, the complex structure of polymeric lignin has been described elsewhere showing a wide variety of functional groups that depend on its biomass source and fractioning method [3,4]. Particularly, the latter affecting both lignin structural characteristics and industrial potential applications (e.g., surface located carboxylic and phenolic groups, contribute to the biosorption properties of lignin) [3–6].

Regarding biomass source, hardwood contains approximately 17 y 26% of lignin and is mostly composed of guaiacyl and syringyl units and traces of p-hydroxyphenyl, whereas softwood with around 25 and 31% of lignin mostly contains guaiacyl units, with a small amount of p-hydroxyphenyl units [6,7].

On the other hand, among fractioning methods, lignin derived from alkali or Kraft process is the most extended pulping technology and generates a high amount of residual lignin [8], which is characterized by ionizable groups as phenolic hydroxyl and sulfonic acid added to the lignin structure [9]. Such additional groups result from treatment of lignocellulosic raw material with NaOH and Na_2S, in an aqueous solution at high temperature [10,11]. As an alternative process, lignin-rich residues may be generated from the bioethanol production process, after biomass saccharification and fermentation [12]. In contrast with Kraft lignin, residual fraction after bioethanol production is composed mostly by lignin but a certain contamination of carbohydrates may be also present, related to non-hydrolysed cellulose during the saccharification [12].

As a result, given its availability as an industrial residue and the biodegradable character of lignin, this polymer is a promising functional polymer that can be involved in the development of novel sustainable and high added-value products. Among them, this work will explore the use of lignin as emulsifier in the manufacture of bituminous emulsions or, more interestingly, of emulsified asphalt rejuvenators. These products, commonly used in road applications, may represent a significant reduction of emissions (e.g., fumes and CO_2) and energy consumption during asphalt recycling operations, both widely demanded by pavement industry. Thus, the use of reclaimed or aged asphalt pavement (RAP) increases the cost-effectiveness and the sustainability of new asphalt mixes [13]. However, RAP rejuvenation is previously necessary due to the oxidative aging of the binder. Hence, soft virgin bitumen or, more significantly, the use of bio-rejuvenators and rejuvenators derived from waste recycling have been proposed due to their performances and sustainability, since they are recycling products that present a safer usage and, likely, more cost-efficient [13–15].

Lignin emulsification ability arises from its structure composed of a hydrophobic backbone and hydrophilic chains, which can allow the interfacial tension between aqueous and oil phases to be reduced. Therefore, lignin can stabilize emulsions by its adsorption in the oil-water interface, causing electrostatic and steric repulsion between oil droplets [16]. To that end, native lignin must be dissolved in water at high pH, being as macromolecules or in a polymeric form with negative net charge, i.e., as anionic surfactants, which are expected to form the so-called anionic bituminous emulsions.

Cationic emulsions are commonly used in road applications, since their positive droplet surface charge can be more easily destabilized by the most common "acidic" type aggregates (negatively charged), e.g., those with high silica content. Consequently, lignin needs to be modified by amination, turning it into modified cationic modified lignin with protonable groups soluble in acidic media [17]. However, the use of native lignins as anionic emulsifiers of rejuvenators may represent a promising, cheaper and more sustainable alternative for road recycling that involves high rates of reclaimed asphalt (RAP). Under such conditions, surface of RAP aggregates is mostly coated by aged bitumen, which needs to be rejuvenated by emulsified additives. The non-polar character of the aged bitumen may allow either the use of cationic or anionic emulsions [18].

With this aim, this work explores the use of three native lignins as anionic emulsifiers, which are an abundant by-product of industrial processes such as bioethanol production and alkali or kraft pulping. Their particular characteristics derived from their fractioning method, along with their different molecular organization induced by the emulsification pH, are expected to interact in diverse ways with the water-oil interface of the emulsion. Initially model O/W emulsions of silicone oil, with a selected viscosity similar to that of the bitumen during its emulsification, will be studied as a function of lignin source, disperse phase concentration and emulsification pH. Once stablished the effect of such variables, emulsion formulations of three potential bitumen rejuvenators (waste vegetable cooking oil, recycled lubricating oil and a 160/220 penetration range soft bitumen) will be optimized.

2. Experimental

2.1. Materials

Three commercial lignins derived from Kraft pulping and bioethanol production processes were used in this study. Kraft lignin (KL) was supplied by Sigma Aldrich (Huelva, Spain), with low sulfonate content (4% sulphur) and an average molecular weight, Mw~10,000. Bioethanol derived lignin-reach materials were provided by DONG Energy

(Denmark), with two purification degrees. This process uses a hydrothermal pre-treatment to open the cellulosic fibres followed by hydrolysis and fermentation to produce cellulosic ethanol. Depending on processing steps a regular (EthL-1) and a more purified (EthL-2) lignin are produced. As a result, EthL-1 contains 59% lignin, 20 wt.% carbohydrates and 7% proteins, with an average Mw~6800, whereas the most purified, EthL-2, contains 68 wt.% lignin, 19% carbohydrates and 9% proteins, with an average Mw~4600 [19].

Lignin emulsification ability was assessed using oils with different compositions such as silicone oil (FS100 from Esquim S.A., Barcelona, Spain), referred to as SIL, waste vegetable cooking oil (Biolia, Huelva, Spain), referred to as RVO, and a recycled mineral lubricating oil (ECO-350 from Sertego S.L., Madrid, Spain), composed of 66.4% paraffinic, 27.9% naphthenic and 5.7% aromatic compounds, referred to as RLUB. All of them exhibited low viscosity (below 100 mPa·s) at room temperature, which allowed us to easily reproduce, at ambient conditions, bitumen viscosity during its emulsification at high temperature, which is recommended to be lower than 200 mPa·s [20]. Additionally, a bitumen with a penetration grade of 160-220 (Cepsa S.A., Madrid, Spain) was emulsified.

2.2. Methods

Emulsification was performed with a homogenizer IKA T25 (Germany) coupled to a S25N–25F SK dispersion tool, at 20,000 rpm for 4 min. All studied oils were emulsified at room temperature, whereas bitumen was blended with the aqueous phase (water and surfactant) to prepare a premixture at 90 °C, prior to emulsification stage. Oil-in-water emulsions with oil concentrations (O%) between 40 and 70 wt.% were studied, while emulsifier concentration was fixed at 0.5 wt.%. Emulsification pH ranged from 9 to 13.

Rheological characterization was carried out by means of steady viscous flow tests at 30 °C. Measurements were conducted in a controlled-stress Haake RS150 (Germany), using a serrated plate-plate geometry PP35 (35 mm diameter, 1 mm gap) to avoid wall-slip phenomena. Steady shear tests were performed in a controlled-strain mode, ranging shear rate from 0.01 and 100 s^{-1}. An equilibration time of 120 s guaranteed steady state conditions at every selected shear rate.

Droplet surface charge was studied by Zeta potential tests, conducted in a Malvern Panalytical Zetasizer (UK). Three measurements were performed for each sample and the average value was reported. Emulsion droplet size distribution (DSD) was determined by laser diffraction in a Malvern Panalytical Mastersizer 2000 (UK). Sauter ($D_{3,2}$) and De-Brouckere ($D_{4,3}$) mean diameters were calculated as follows [21]:

$$D_{3,2} = \frac{\sum_i n_i d_i^3}{\sum_i n_i d_i^2} \quad (1)$$

$$D_{4,3} = \frac{\sum_i n_i d_i^4}{\sum_i n_i d_i^3} \quad (2)$$

where n_i is the number of droplets with diameter d_i.

3. Results and Discussion

3.1. Model Silicone Emulsions: Effect of Disperse Phase Concentration

A first step in the manufacture of any O/W emulsion consists of surfactant dissolution in an aqueous medium that forms the continuous phase of the emulsion. Regarding native lignins, it is well known their solubility in water under alkaline conditions [22]. Thus, all commercial lignin-based emulsifiers were dissolved at pH 13 for this part of the study, which assesses the effect of oil concentration on emulsion microstructure and viscosity.

Given that all emulsions were formulated with the same surfactant content (0.5 wt.% S), the effective concentration of lignin-based emulsifier in the continuous phase ranged from 0.83 wt.% S (in the emulsion with 40 wt.% O) up to 1.67 wt.% S, for the most concentrated emulsion with 70 wt.% O. As may be seen in Figure 1A, a higher surfactant concentration leads to an increase in the viscosity of the emulsion aqueous phase, more apparent for the most oil-concentrated emulsions. Interestingly, lignin products derived from bioethanol process give rise to more viscous aqueous phases than Kraft lignin.

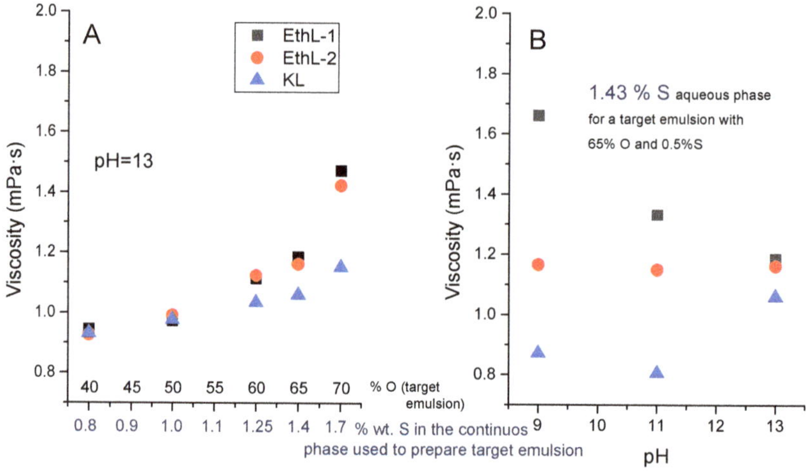

Figure 1. Viscosity of the emulsion continuous phase as a function of: (**A**) selected oil concentration of emulsion prepared at pH = 13; (**B**) emulsification pH for emulsions containing 65% O.

In this regard, Estigneev [23] stated the ratio of aliphatic/aromatic hydroxyl groups is the main factor determining the solubility of lignins in alkaline aqueous solutions and, more precisely, the amount of OH groups per aromatic ring. Likewise, Santos et al. [12] pointed out that lignin derived from bioethanol is characterized by a lower phenolic hydroxyl content than Kraft lignin. A higher amount of polar aromatic hydroxyl groups would limit the lignin agglomeration [22,24], and would be related to the lower viscosity exhibited by Kraft lignin solutions at high concentration (Figure 1A).

All lignin-based surfactants were able to emulsify silicone oil between 40 and 70 wt.% O at pH 13. Most emulsions showed a wide bimodal droplet size distribution (DSD), with a first peak of smaller height located at ca. 4 μm and a second main peak at around 20 μm (Figure 2).

However, as oil concentration increases the height of first peak decreases in emulsions stabilized by bioethanol-based lignins (Figure 2A) and, eventually, disappears for the most concentrated system stabilized by Kraft lignin (Figure 2B). Furthermore, second peak slightly shifts towards smaller diameters.

As oil concentration increases, the higher number of droplets along with the narrower DSD are expected to rise droplet-droplet interactions, leading to more viscous systems and with an apparent non-Newtonian character, as confirmed by Figure 3.

When the viscous response is plotted in the form of shear stress-versus-shear rate in log–log scale, a plateau region at intermediate shear rates is observed, which can be related to so-called apparent yield stress. This parameter will be used to study the effect of formulation on emulsion microstructure and stability. The Herschel-Bulkley model describes observed non-Newtonian flow behaviour fairly well:

$$\sigma = \sigma_o + k\,\dot{\gamma}^n \tag{3}$$

being σ_o the apparent yield stress, k the consistency index and n the flow index. Table 1 gathers the values of Herschel-Bulkley parameters calculated for silicone emulsions prepared with different oil concentrations.

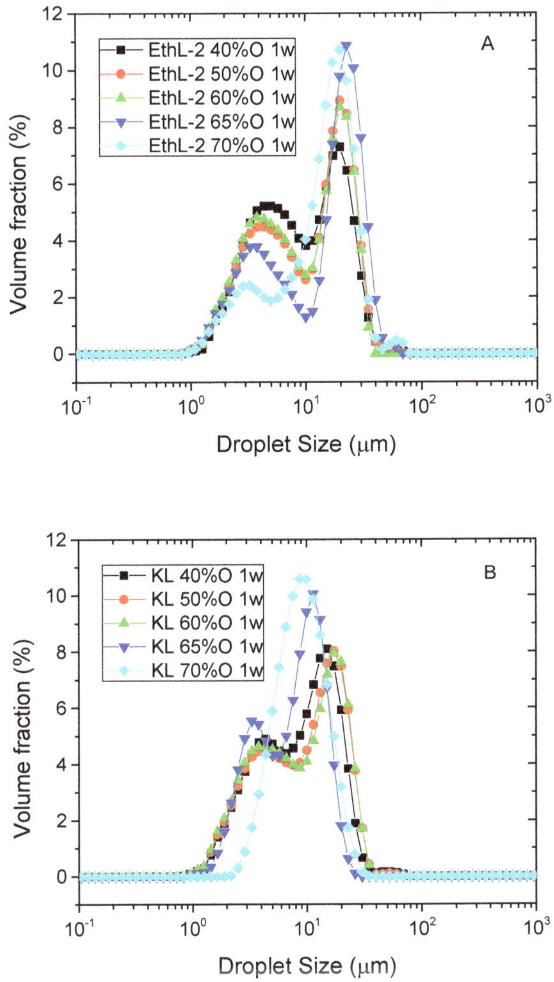

Figure 2. Effect of silicone oil concentration on droplet size distribution of emulsions stabilized by 0.5 wt.% bioethanol (**A**) or Kraft lignin (**B**) at pH = 13, after one week storage.

As may be seen, regardless of the oil concentration or lignin emulsifier considered, all n values are below 1, which evidences the shear-thinning behaviour of these systems. Such a viscous behaviour (characterized by a decrease of emulsion viscosity with shear rate) has been related to droplet deformation, shear-induced deflocculation and/or the non-Newtonian behaviour of the dispersed phase [25,26].

Given the range of shear rates studied and the Newtonian behaviour exhibited by the continuous phases shown in Figure 1, the observed shear thinning behaviour should be mainly attributed to a shear-induced deflocculation of the emulsion [27,28]. As expected, an increase in oil concentration leads to more viscous emulsions and with higher yields stress values (Figures 3 and 4B). However, if emulsions with the same concentration of disperse phase are compared, emulsions stabilized with bioethanol-derived emulsifiers show higher yield stress values than emulsions stabilized by KL, despite kraft lignin leads to smaller Sauter ($D_{3,2}$) (up to 65% O) and De-Brouckere ($D_{4,3}$) mean diameters (Figure 4B).

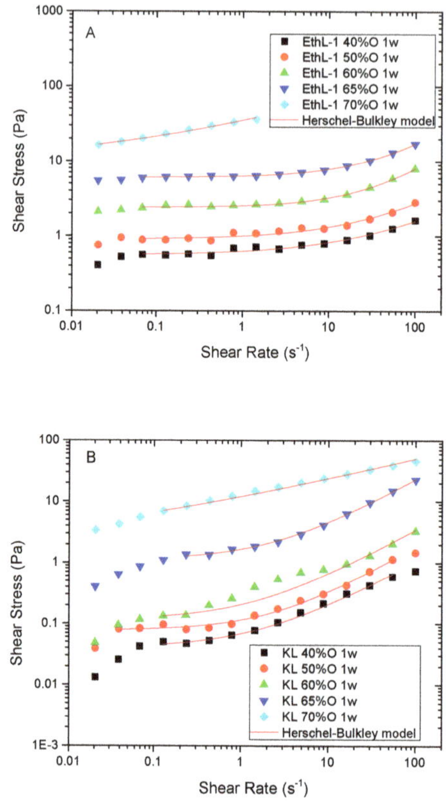

Figure 3. Effect of silicone oil concentration on viscous behaviour of emulsions stabilized by 0.5 wt.% bioethanol (**A**) or Kraft lignin (**B**) at pH = 13, after one week storage.

Table 1. Effect of oil concentration and lignin source on silicone emulsion viscous behaviour fitted to Herschel-Bulkley model. All emulsions were formulated at pH = 13 and measured after one week storage time.

Viscous Model Parameters	**EthL-1 70% O**	**EthL-1 65% O**	**EthL-1 60% O**	**EthL-1 50% O**	**EthL-1 40% O**
σ_o	10.05	6.02	2.4	0.90	0.56
K	25.16	0.30	0.13	0.10	0.07
n	0.35	0.79	0.82	0.64	0.58
Viscous Model Parameters	**EthL-2 70% O**	**EthL-2 65% O**	**EthL-2 60% O**	**EthL-2 50% O**	**EthL-2 40% O**
σ_o	4.65	4.30	2.70	0.77	0.5
K	14.9	0.33	0.17	0.04	0.04
n	0.31	0.74	0.80	0.81	0.75
Viscous Model Parameters	**KL 70% O**	**KL 65% O**	**KL 60% O**	**KL 50% O**	**KL 40% O**
σ_o	2.25	1.11	0.115	0.071	0.04
K	9.75	0.55	0.09	0.04	0.03
n	0.35	0.80	0.79	0.84	0.77

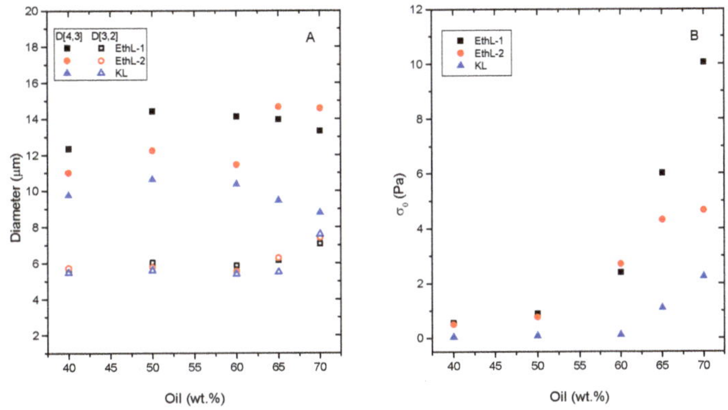

Figure 4. Effect of silicone oil concentration on emulsions parameters: (**A**) Sauter, $D_{3,2}$ and volumetric, $D_{4,3}$, mean diameters; and (**B**) yield stress values determined from Herschel-Bulkley model. All emulsions were emulsified at pH 13, stabilized by 0.5 wt.% lignin and stored for 1 week.

Initially, such smaller particle sizes should increase emulsion viscosity, due to the increase in specific surface area undergone by its disperse phase [27]. Nevertheless, droplet-droplet flocculation due to entanglements among emulsifier biomolecules seems to control rheological response of these systems. As a result, although KL shows a better emulsification ability at pH 13 (i.e., leading to smaller droplet size), bioethanol-derived lignins are able to form a more developed microstructure of flocculated droplets, which collapses a higher stress values (yield stress) (Figure 4A). Such a flocculated microstructure would be extended forming a three-dimensional network, which would contribute to emulsion stability, as it will discussed later. In this regard, it is worth noting that yield stress results are in good agreement with viscosities measured in Figure 1, suggesting that the microstructure formed by lignins in the continuous phase of the emulsion would play a significant role in its stability and rheology.

In any case, all emulsions formulated at pH 13 exhibit similar Z-potential values around −40 mV (Table 2), confirming the negative charge of the lignin stabilizing droplet interface and the anionic character of the emulsions formed.

Table 2. Effect of formulation on emulsion storage stability, Z potential values (ζ) and Herschel-Bulkley model parameters. Measurements were performed after 7 days storage.

Oil	Surfactant	Oil wt.%	pH	Minimum Storage Stability (Days)	σ_0 (Pa)	k (Pa s^n)	n (−)	ζ (mV)
SIL	KL	65	9	30	0.09	0.07	0.89	-36.7
SIL	KL	65	11	30	1.11	0.36	0.88	-56.8
SIL	KL	65	13	30	1.11	0.55	0.80	-45.8
SIL	EthL-1	65	13	30	6.02	0.29	0.79	-42.2
SIL	EthL-2	65	13	30	4.30	0.32	0.74	-46.4
VGO	KL	65	9	30	1.82	0.33	0.78	-12.7
VGO	KL	65	11	30	1.66	0.15	0.93	-22.2
VGO	KL	65	13	30	0.65	0.25	0.73	-39.7
VGO	EthL-1	65	13	30	0.72	0.22	0.80	-46.0
VGO	EthL-2	65	13	30	0.86	0.33	0.74	-47.3
RLUB	KL	65	9	30	2.03	4.99	0.48	-6.8
RLUB	KL	65	11	30	4.06	3.92	0.52	-36.7
RLUB	KL	65	13	7	-	-	-	-42.2
RLUB	EthL-1	65	13	30	5.11	0.23	0.86	-36.5
RLUB	EthL-2	65	13	30	2.62	0.56	0.69	-42.7
Bitumen	KL	65	9	7	-	-	-	-29.8
Bitumen	KL	65	11	30	0.30	0.14	0.82	-28.1

3.2. Model Silicone Emulsions: Effect of pH

The effect of pH on the viscosity of the emulsion continuous phase, for a selected oil concentration of 65 wt.%, is shown in Figure 1B. Aqueous phases of both lignin-based surfactants derived from bioethanol process showed higher viscosity than Kraft lignin solution at pH between 9 and 11, being more apparent the difference as pH is lower. However, only EthL-1 solution underwent a significant increase in viscosity as pH became less alkaline, whilst viscosity of KL and EthL-2 solutions hardly changed with pH. Observed viscosity changes may be related to different agglomeration behaviours. At high pH values, carboxyl as well as aromatic hydroxyl groups are deprotonated in Kraft lignin, and resultant repulsive Coulomb forces would lead to a better solubility and therefore, to a lower polymer aggregation and viscosity [22].

On the other hand, lining agglomeration is also due to intermolecular hydrogen bonds (formed by aliphatic and aromatic hydroxyl groups) and π–π interactions [22]. Aliphatic hydroxyl groups are able to form stronger hydrogen bonds than aromatic hydroxyls, thus agglomeration behaviour is expected to be more important in bioethanol-derived lignin, characterized by a lower phenolic hydroxyl content than KL [24]. In addition, Maitz et al. [22] pointed out that the π–π stacking effects increase with increasing number of methoxyl groups. When comparing EthL-1 to the more purified EthL-2 [19], this washed sample is expected to be partially demethylated and with a higher amount of polar aromatic hydroxyl groups which, in contrast to methoxyl groups, make more difficult π–π stacking and limit the agglomeration [22,24]. Therefore, its viscosity would be less affected by a pH decrease due to a less extended aggregation, as suggested by Figure 1B.

As a result, both bioethanol lignins would be more aggregated than KL, particularly at lower pH, as may be deduced from Figure 1. This fact seems to affect their functional properties as emulsifiers. Thus, in contrast with results obtained at pH 13, bioethanol-derived lignins were not capable of stabilizing emulsions with 65 wt.% oil at pH 9 and 11, at least, for one week.

Conversely, KL emulsifier stabilized silicone oil emulsions in the whole pH range tested (Figure 5). Showing monomodal DSD at pH 9 and 11 (Figure 5A) and lower mean particle sizes as medium alkalinity is raised. Specially, $D_{3,2}$ significantly drops at pH 13 (Figure 5B).

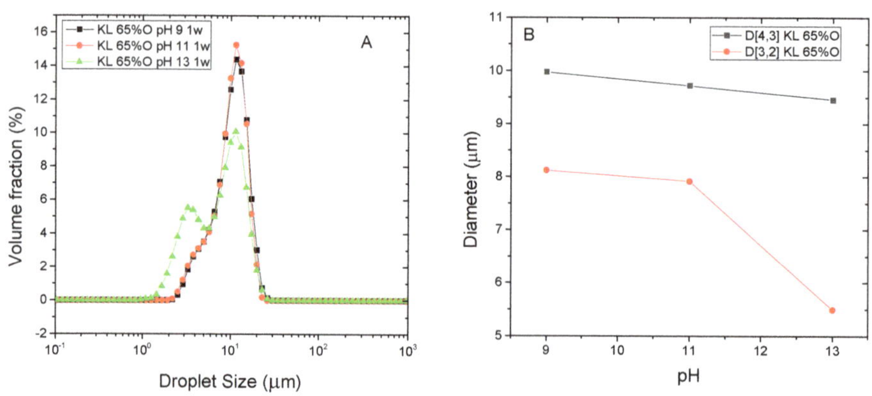

Figure 5. Droplet size distribution (**A**) and characteristic mean diameters (**B**) of silicone oil emulsions stabilized by Kraft lignin as function of pH. All emulsions were formulated with 65 wt.% O, stabilized by 0.5 wt.% lignin and measured after 1 week storage.

As may be seen in Table 2, all KL emulsions present negative values of Z-potential, higher at pH 13. At this pH, the highest zeta potential of KL stabilized emulsions would result from lignin deprotonated carboxylic, phenolic hydroxyl and acid sulfonic groups [9]. Under conditions less alkaline, in addition to carboxylic groups, sulfonic groups would contribute to the slightly lower zeta potential recorded at pH < 11 (Table 2), which would remain ionized at pH 11 and 9 due to their low pKa ≤ 2 [22,29–31]. This fact, in contrast with bioethanol-derived lignins, would allow stable emulsions to be obtained in the whole pH range tested, although emulsion rheology is affected at lower pH. All pH values studied led to shear-thinning behaviours that can be fitted to the Herschel-Bulkley model (Figure 6A), i.e., showing an apparent yield stress and n values below 1. However, emulsion bulk viscosity (and yield stress) at pH 9 significantly decreases if compared with more alkaline pH values (Figure 6B).

3.3. Emulsification of Bitumen Rejuvenators

Previous results obtained with O/W silicone model emulsions have pointed out all lignin-based surfactants present good functional properties as emulsifiers at pH 13 in a wide range of oil concentrations. On the contrary, only Kraft lignin extends its emulsification ability towards the lower pHs values of 11 a 9. Subsequently, considering these results, lignins were tested as emulsifiers of bitumen rejuvenators such as a waste vegetable oil (RVO), a recycled lubricant (RLUB) or a soft bitumen (160/220 penetration range).

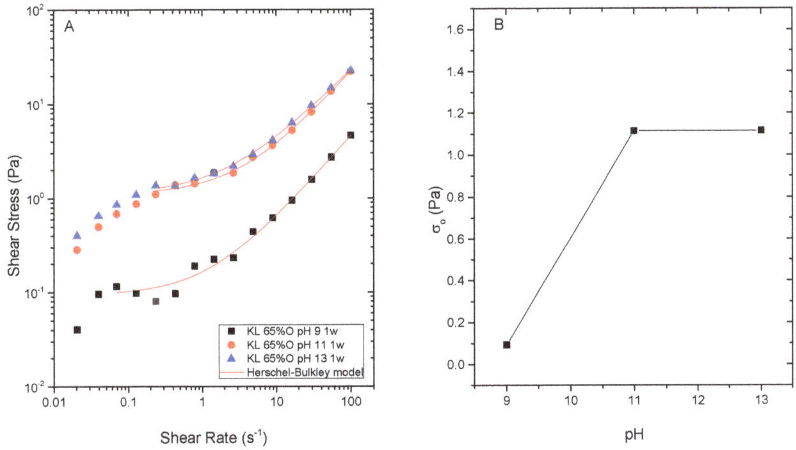

Figure 6. Effect of emulsification pH on silicone emulsions flow behaviour (**A**) and yield stress values (**B**) determined from Herschel-Bulkley model All emulsions were formulated with 65 wt.% O, stabilized by 0.5 wt.% Kraft lignin and measured after 1 week storage.

As seen with silicone oil, bioethanol-derived lignins exhibited a limited ability of emulsion stabilization, being only able to emulsify waste oils RVO and RLUB only at pH 13 (Figure 7A).

Furthermore, all bitumen emulsions stabilized by these lignins became unstable no matter pH considered. However, Kraft lignin was able to emulsify RVO and RLBU emulsions at pH values between 9 and 13, but the stability of RLBU emulsion was poorer at pH 13 (less than 30 days storage, as seen in Table 2). Likewise, stable bitumen emulsions were obtained only at pH 9 and 11 (Figure 7B, Table 2). All emulsions showed a wide DSD (Figure 7), where main peaks of RVO systems are located at lower diameters (at ca. 5 μm), while bitumen emulsions formulated showed the highest particle sizes. All emulsion presented negative values of Z-potential (negatively charged anionic emulsions) that decreased as pH was less alkaline (Table 2). Initially, a less charged lignin seems to favour emulsification of the less polar rejuvenators, bitumen and RLUB.

As found for silicone systems, rejuvenator emulsions exhibited a shear thinning behaviour with a trend to reach an apparent yield stress at intermediate shear rates in the log-log curve of shear stress, which has been fitted to the Herschel-Bulkley model (Figure 8).

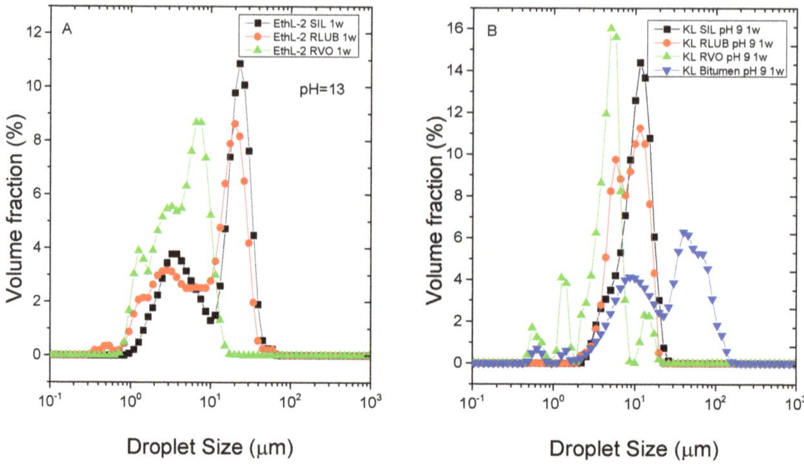

Figure 7. Droplet size distribution of bitumen rejuvenator emulsions stabilized by 0.5 wt.% bioethanol-derived (**A**) and Kraft (**B**) lignins.

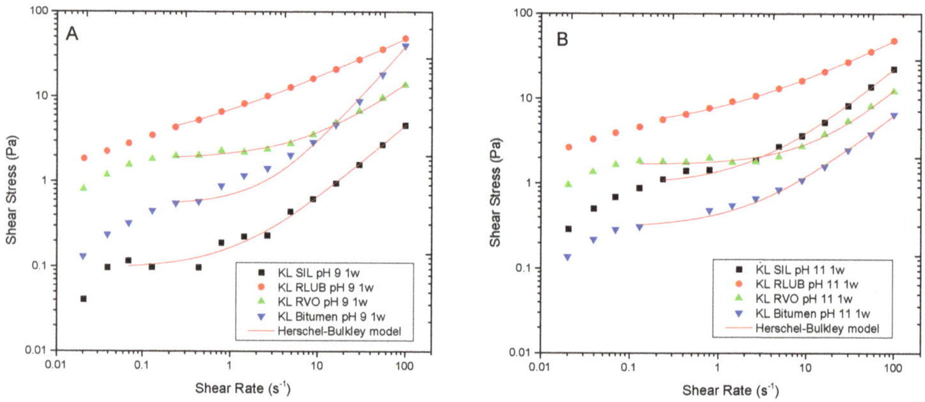

Figure 8. Effect of emulsification pH on the flow behaviour of bitumen emulsions stabilized by Kraft lignin after 1 week storage at pH 9 (**A**) and 11 (**B**).

As may be seen, in contrast with recycled oils RVO and RLUB, pH mainly affects viscosity of bitumen emulsions, being emulsions formulated at pH 9 more viscous than that formulated at pH 11. In any case, RVO and RLUB lead to emulsions with higher viscosity than bitumen. This result is in good agreement with $D_{3,2}$ value displayed in Figure 9A, where RVO and RLUB present the lower values of this mean diameter.

On the other hand, viscous and DSD combined results may be used for selecting best surfactants for every rejuvenator studied (Figure 9). Thus, although most emulsions exhibited a storage stability of at least 30 days, some emulsions formulated with KL and RLUB at pH 13 or with bitumen at pH 9, showed a shorter stability (Table 2). Both systems are characterized by large particle sizes along with low yield stress values (Figure 9), both factors would not prevent droplet motion, and an eventual phase separation. Conversely, lower particle size and higher yield stress values were found for both bioethanol-derived lignins emulsifying RVO and RLUB at pH 13, which are expected to exhibit enhanced stability. Figure 10 seems to confirm this assumption, since these systems hardly change their particle size with time. Regarding bitumen emulsions, best results are achieved with

KL at pH 11, although lignin emulsions are characterized by high $D_{3,2}$ and low σ_o values. Among all system studied, this emulsion undergoes the largest increase in particle size along storage time (Figure 10).

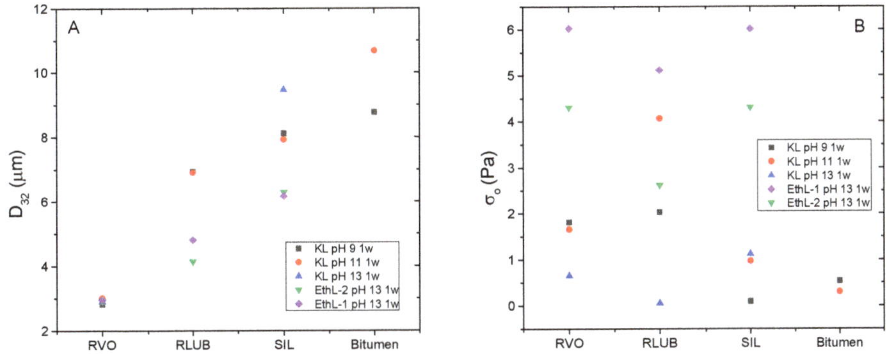

Figure 9. Sauter mean diameters (**A**) and yield stress values (**B**) of emulsions stabilized by bioethanol-derived and Kraft lignins formulated at different pH and stored for one week.

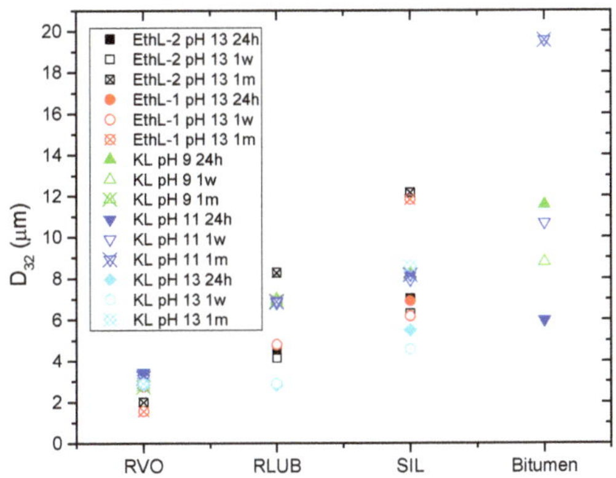

Figure 10. Evolution of emulsion droplet size with storage time as a function of the selected dispersed phase.

4. Concluding Remarks

Three native lignins have been assessed as emulsifiers of oil-in-water emulsion of three potential bitumen rejuvenators (waste vegetable cooking oil, recycled lubricating oil and 160/220 penetration range soft bitumen). Selected lignins are abundant by-products of industrial processes such as bioethanol production and alkali or Kraft pulping. Previously, lignin emulsification ability has been studied in model silicone emulsions as a function of disperse phase concentration and emulsification pH.

All lignin-based surfactants were able to emulsify silicone oil between 40 and 70 wt.% O at pH 13, giving rise to anionic emulsions with Z-potential values around −40 mV. Most resultant emulsions showed a wide bimodal droplet size distribution and a non-Newtonian viscous character, with shear thinning behaviour and an apparent yield stress. Thus, the observed viscous behaviour has been fitted to the Herschel-Bulkley model. Droplet-droplet flocculation due to entanglements among emulsifier biomolecules seems to control rheological response of these systems. Such an extended flocculated microstructure would contribute to emulsion stability.

Differences among lignins appear when a less alkaline pH is used during emulsification. Bioethanol-derived lignins are not able to stabilize emulsions with 65 wt.% oil at pH 9 and 11, for at least one week. Conversely, KL emulsifier stabilized silicone oil emulsions in the whole pH range tested. In this regard, different agglomeration behaviours have been observed depending on the fractioning method, where bioethanol-derived lignins would tend to aggregate more than Kraft lignin, particularly at lower pH. The expected higher phenolic hydroxyl content in KL, along with the presence of other deprotonated carboxylic and sulfonic groups, would prevent lignin aggregation and extend surfactant functional properties over a wider pH range, from 9 to 13.

Similarly, lignin emulsification ability of bitumen rejuvenators depends on surfactant source and pH, but now characteristics of the dispersed phase must be also taken into consideration. Bioethanol-derived lignins exhibited a limited ability of emulsion stabilization, being only capable of emulsifying waste oils RVO and RLUB at pH 13. On the contrary Kraft lignin was capable of emulsifying RVO and RLBU emulsions at pH values between 9 and 13. Likewise, stable bitumen emulsions were obtained only at pH 9 and 11. In this regard, the lower Z-potential values shown by KL at pH 9 and 11 seem to facilitate emulsification of the less polar disperse phases formed by RLUB and bitumen.

In any case, lower particle size and higher yield stress values were found for both bioethanol-derived lignins emulsifying RVO and RLUB at pH 13, which are expected to exhibit a longer stability.

Finally, although this work focuses its application on the important road paving industry, other possible applications could add value to this waste biopolymer as anionic surfactant for cleaning, food, personal care, cosmetic, paint, biomedical, chemical, agricultural and pharmaceutical industry, among others [29]. In this regard, it is worth noting that 100 million tons of lignin are produced annually (including bioethanol, Kraft, sulphite, soda, organosolv processes, among others), but less than 2% is used in commercial products with higher added value [6,29].

Author Contributions: Formal analysis, R.Á.-B. and A.A.C.; investigation, R.Á.-B. and F.J.M.-B.; methodology, A.A.C.; project administration, F.J.N., F.J.M.-B. and P.P.; supervision, F.J.N.; writing—review and editing, P.P. All authors have read and agreed to the published version of the manuscript.

Funding: This work is part of GreenAsphalt project (ref. 802C1800001), co-funded by FEDER European Programme (80%) and Junta de Andalucía (Consejería de Economía, Conocimiento, Empresas y Unversidades/Agencia-IDEA), and has been also co-funded by EPIT2019, Estrategia de Política de Investigación y Transferencia (Universidad de Huelva), Project UHU2019049.

Institutional Review Board Statement: Not applicable.

Informed Consent Statement: Not applicable.

Data Availability Statement: The raw/processed data required to reproduce these findings cannot be shared at this time due to technical or time limitations.

Conflicts of Interest: The authors declare no conflict of interest.

References

1. Bolker, I.H. *Natural and Synthetic Polymers: An Introduction*; Marcel Dekker, Inc.: New York, NY, USA, 1974.
2. Du, X.; Li, J.; Lindström, M. Modification of industrial softwood kraft lignin using Mannich reaction with and without phenolation pretreatment. *Ind. Crop. Prod.* **2014**, *52*, 729–735. [CrossRef]
3. Pilarska, A.A.; Wolna-Maruwka, A.; Niewiadomska, A.; Pilarski, K.; Adamski, M.; Grzyb, A.; Grządziel, J.; Gałązka, A. Silica/Lignin Carrier as a Factor Increasing the Process Performance and Genetic Diversity of Microbial Communities in Laboratory-Scale Anaerobic Digesters. *Energies* **2021**, *14*, 4429. [CrossRef]
4. Pilarska, A.A.; Wolna-Maruwka, A.; Pilarski, K. Kraft Lignin Grafted with Polyvinylpyrrolidone as a Novel Microbial Carrier in Biogas Production. *Energies* **2018**, *11*, 3246. [CrossRef]
5. Tejado, A.; Peña, C.; Labidi, J.; Echeverria, J.M.; Mondragon, I. Physico-chemical characterization of lignins from different sources for use in phenol–formaldehyde resin synthesis. *Bioresour. Technol.* **2007**, *98*, 1655–1663. [CrossRef]

6. Lobato-Peralta, D.R.; Duque-Brito, E.; Villafán-Vidales, H.I.; Longoria, A.; Sebastian, P.; Cuentas-Gallegos, A.K.; Arancibia-Bulnes, C.A.; Okoye, P.U. A review on trends in lignin extraction and valorization of lignocellulosic biomass for energy applications. *J. Clean. Prod.* **2021**, *293*, 126123. [CrossRef]
7. Rydholm, S.A. *Pulping Processes*; John Wiley & Sons Inc.: Hoboken, NJ, USA, 1965.
8. Carvajal, J.C.; Gómez, Á.; Cardona, C.A. Comparison of lignin extraction processes: Economic and environmental assessment. *Bioresour. Technol.* **2016**, *214*, 468–476. [CrossRef]
9. Klapiszewski, Ł.; Nowacka, M.; Milczarek, G.; Jesionowski, T. Physicochemical and electrokinetic properties of silica/lignin biocomposites. *Carbohydr. Polym.* **2013**, *94*, 345–355. [CrossRef]
10. Belgacem, M.N.; Gandini, A. *Monomers, Polymers and Composites from Renewable Resources*; Elsevier: Amsterdam, The Netherlands, 2008.
11. Collins, M.N.; Nechifor, M.; Tanasă, F.; Zănoagă, M.; McLoughlin, A.; Stróżyk, M.A.; Culebras, M.; Teacă, C.-A. Valorization of lignin in polymer and composite systems for advanced engineering applications—A review. *Int. J. Biol. Macromol.* **2019**, *131*, 828–849. [CrossRef]
12. Santos, J.I.; Fillat, U.; Martin-Sampedro, R.; Eugenio, M.E.; Negro, M.J.; Ballesteros, I.; Rodríguez, A.; Ibarra, D. Evaluation of lignins from side-streams generated in an olive tree pruning-based biorefinery: Bioethanol production and alkaline pulping. *Int. J. Biol. Macromol.* **2017**, *105*, 238–251. [CrossRef]
13. Cuciniello, G.; Mallegni, N.; Cappello, M.; Filippi, S.; Leandri, P.; Polacco, G.; Losa, M. Classification and selection of exhausted oils for rejuvenating bituminous blends. *Constr. Build. Mater.* **2021**, *278*, 122387. [CrossRef]
14. El-Shorbagy, A.M.; El-Badawy, S.M.; Gabr, A.R. Investigation of waste oils as rejuvenators of aged bitumen for sustainable pavement. *Constr. Build. Mater.* **2019**, *220*, 228–237. [CrossRef]
15. Zargar, M.; Ahmadinia, E.; Asli, H.; Karim, M.R. Investigation of the possibility of using waste cooking oil as a rejuvenating agent for aged bitumen. *J. Hazard. Mater.* **2012**, *233–234*, 254–258. [CrossRef]
16. Bai, L.; Greca, L.G.G.; Xiang, W.; Lehtonen, J.; Huan, S.; Nugroho, R.W.N.; Tardy, B.L.; Rojas, O.J. Adsorption and Assembly of Cellulosic and Lignin Colloids at Oil/Water Interfaces. *Langmuir* **2018**, *35*, 571–588. [CrossRef] [PubMed]
17. Yuliestyan, A.; García-Morales, M.; Moreno, E.; Carrera, V.; Partal, P. Assessment of modified lignin cationic emulsifier for bitumen emulsions used in road paving. *Mater. Des.* **2017**, *131*, 242–251. [CrossRef]
18. Yuliestyan, A.; Gabet, T.; Marsac, P.; García-Morales, M.; Partal, P. Sustainable asphalt mixes manufactured with reclaimed asphalt and modified-lignin-stabilized bitumen emulsions. *Constr. Build. Mater.* **2018**, *173*, 662–671. [CrossRef]
19. Landa, P.A.; Gosselink, R.J.A. Lignin-Based Bio-Asphalt. Patent WO 2019/092278 A1, 16 May 2019.
20. Read, J.; Whiteoak, D.; Hunter, R.N. *The Shell Bitumen Handbook*, 5th ed.; Thomas Telford Publishing: London, UK, 2003.
21. Piacentini, E. Droplet Size. In *Encyclopedia of Membranes*; Springer Science and Business Media LLC: Berlin/Heidelberg, Germany, 2016; pp. 591–592.
22. Maitz, S.; Schlemmer, W.; Hobisch, M.A.; Hobisch, J.; Kienberger, M. Preparation and Characterization of a Water-Soluble Kraft Lignin. *Adv. Sustain. Syst.* **2020**, *4*, 2000052. [CrossRef]
23. Evstigneev, E.I. Factors affecting lignin solubility. *Russ. J. Appl. Chem.* **2011**, *84*, 1040–1045. [CrossRef]
24. Zhao, W.; Xiao, L.-P.; Song, G.; Sun, R.-C.; He, L.; Singh, S.; Simmons, B.A.; Cheng, G. From lignin subunits to aggregates: Insights into lignin solubilization. *Green Chem.* **2017**, *19*, 3272–3281. [CrossRef]
25. Pal, R. Rheology of polymer-thickened emulsions. *J. Rheol.* **1992**, *36*, 1245–1259. [CrossRef]
26. Pal, R. Effect of droplet size on the rheology of emulsions. *AIChE J.* **1996**, *42*, 3181–3190. [CrossRef]
27. Partal, P.; Guerrero, A.; Berjano, M.; Gallegos, C. Influence of concentration and temperature on the flow behavior of oil-in-water emulsions stabilized by sucrose palmitate. *J. Am. Oil Chem. Soc.* **1997**, *74*, 1203–1212. [CrossRef]
28. Partal, P.; Martínez, I.; Muñoz, J.; Gallegos, C. Influence of thermal treatment on the flow of starch-based food emulsions. *Eur. Food Res. Technol.* **2003**, *217*, 17–22. [CrossRef]
29. Czaikoski, A.; Gomes, A.; Kaufmann, K.C.; Liszbinski, R.B.; de Jesus, M.B.; da Cunha, R.L. Lignin derivatives stabilizing oil-in-water emulsions: Technological aspects, interfacial rheology and cytotoxicity. *Ind. Crop. Prod.* **2020**, *154*, 112762. [CrossRef]
30. Konduri, M.K.R.; Fatehi, P. Production of Water-Soluble Hardwood Kraft Lignin via Sulfomethylation Using Formaldehyde and Sodium Sulfite. *ACS Sustain. Chem. Eng.* **2015**, *3*, 1172–1182. [CrossRef]
31. Yan, M.; Yang, D.; Deng, Y.; Chen, P.; Zhou, H.; Qiu, X. Influence of pH on the behavior of lignosulfonate macromolecules in aqueous solution. *Colloids Surf. A Physicochem. Eng. Asp.* **2010**, *371*, 50–58. [CrossRef]

polymers

MDPI

Article

Microwave Enabled Physically Cross Linked Sodium Alginate and Pectin Film and Their Application in Combination with Modified Chitosan-Curcumin Nanoparticles. A Novel Strategy for 2nd Degree Burns Wound Healing in Animals

Hafiz Muhammad Basit [1,2], Muhammad Ali [1,2], Mian Mufarih Shah [3], Shefaat Ullah Shah [1,2], Abdul Wahab [4], Hassan A. Albarqi [5], Abdulsalam A. Alqahtani [5], Ismail A. Walbi [6] and Nauman Rahim Khan [2,4,*]

1 Department of Pharmaceutics, Faculty of Pharmacy, Gomal University, DIKhan 29050, Pakistan; basitkhan053@gmail.com (H.M.B.); alipharm33@gmail.com (M.A.); shefaatbu@gmail.com (S.U.S.)
2 Gomal Centre for Skin/Regenerative Medicine and Drug Delivery Research, Faculty of Pharmacy, Gomal University, DIKhan 29050, Pakistan
3 Department of Medicine MTI, Hayatabad Medical Complex, Peshawar 25000, Pakistan; mianmufarih458@gmail.com
4 Department of Pharmacy, Kohat University of Science and Technology, Kohat 26000, Pakistan; wahabscholar@yahoo.com
5 Department of Pharmaceutics, College of Pharmacy, Najran University, Najran 55461, Saudi Arabia; haalbarqi@nu.edu.sa (H.A.A.); aamari@nu.edu.sa (A.A.A.)
6 Department of Clinical Pharmacy, College of Pharmacy, Najran University, Najran 55461, Saudi Arabia; iawalbi@nu.edu.sa
* Correspondence: naumanpharma@gmail.com; Tel.: +92-34-5983-4257

Citation: Basit, H.M.; Ali, M.; Shah, M.M.; Shah, S.U.; Wahab, A.; Albarqi, H.A.; Alqahtani, A.A.; Walbi, I.A.; Khan, N.R. Microwave Enabled Physically Cross Linked Sodium Alginate and Pectin Film and Their Application in Combination with Modified Chitosan-Curcumin Nanoparticles. A Novel Strategy for 2nd Degree Burns Wound Healing in Animals. *Polymers* **2021**, *13*, 2716. https://doi.org/10.3390/polym13162716

Academic Editors: Antonio M. Borrero-López, Concepción Valencia-Barragán, Esperanza Cortés Triviño, Adrián Tenorio-Alfonso and Clara Delgado-Sánchez

Received: 26 July 2021
Accepted: 10 August 2021
Published: 13 August 2021

Publisher's Note: MDPI stays neutral with regard to jurisdictional claims in published maps and institutional affiliations.

Abstract: This study reports microwave assisted physically cross-linked sodium alginate and pectin film and their testing in combination with modified chitosan-curcumin nanoparticles for skin tissue regeneration following 2nd degree burn wound. Film was formulated by solution casting method and physically cross-linked using microwave irradiation at frequency of 2450 MHz, power 750 Watt for different time intervals for optimization. The optimized formulation was analyzed for various physiochemical attributes. Afterwards, the optimized film and optimized modified chitosan-curcumin nanoparticles were tested in combination for skin regeneration potential following burn wound in vivo and skin samples extracted and tested for different attributes. The results indicated that the optimized film formulation (5 min microwave treatment) physicochemical attributes significantly enhanced addressing the properties required of a wound healing platform. The vibrational analysis indicated that the optimized film experienced significant rigidification of hydrophilic domains while the hydrophobic domains underwent significant fluidization which also resulted in significant increase in the transition temperatures and system enthalpies of both polymer moieties with microwave treatment. The combined film and nanoparticles application significantly increased protein content in the wounds which were evident from higher absorbance ratios of amide-I and amide-II (2.15 ± 0.001), significantly higher melting transition temperature and enthalpy ($\Delta T = 167.2 \pm 15.4$ °C, $\Delta H = 510.7 \pm 20.1$ J/g) and higher tensile strength (14.65 ± 0.8 MPa) with significantly enhanced percent re-epithelization (99.9934 ± 2.56) in comparison to other treatments. The combined application of film and nanoparticles may prove to be a new novel treatment strategy for 2nd degree burn wound healing.

Keywords: burn wound healing; polymeric film; microwave; nanoparticles-film combined application

1. Introduction

As the largest human organ, skin is often prone to easy damage thus compromising the barrier properties and enabling exogenous pathogens to infiltrate the wound and complicating the wound as well as delaying the wound healing process [1]. Burn wound treatment is still a challenge for biomedical scientists since they also leave behind scars which drastically compromise the psychological wellbeing of the patient. As per WHO statistics, a total of 11 million burn cases happen globally each year, of which 180,000 cases are fatal [2]. The skin tissue engineering is a prospective source of advanced therapy which is extensively studied these days to combat acute and chronic skin injuries [3]. It involves development of platforms with ability to either replace the entire tissue and/or help in creating suitable cell environment which help in restoring the normal function of the damaged tissue [4]. Some of the epidermal [5], dermal [6] and epidermal/dermal composite [7] skin substitute had already been marketed, but their high cost is the major hindrance for patients from a developing country like Pakistan. Thus, it is always imperative to look for an alternative treatment protocol to treat burn injuries.

Polymeric films have long been utilized as a wound healing platform for skin tissue regeneration applications [8] and various polymers has thus been used for this purpose [9–13]. Films are not only easy to prepare, but also their physicochemical properties can easily be modified to address the properties required of an ideal wound healing platform [14–17].

Sodium alginate and pectin are two natural polysaccharide polymers and due to their inherent biocompatibility, biodegradability, nontoxicity and non-immunogenicity, serve as an ideal candidates for fabricating wound healing platforms [18].They have long been employed in fabricating controlled drug delivery systems [19,20], for cellular encapsulation [21], as an immune stimulatory agent [22], in cosmetics [23], as scaffold in tissue engineering [24], wound dressing material [25], fabrication of bandages [26], as a disintegrating agent in tablets [27], bone regeneration scaffolds [24], as a dental impression material [28] and soft tissue engineering [29], for enhanced absorption of wound exudates [30], and as a protective barrier against bacterial infection [31]. For synergistic wound healing effect, alginate and pectin had also been employed as a blend with good physicochemical properties [32], were found to be noncytotoxic [33] and promote skin regeneration [34]. The simple mixing of polymers is often not enough to address the demerits associated with lone use of either polymer, thus cross linking using chemical cross linkers is often employed [32,35].

Physical cross linking using microwave represents a feasible cross-linking technique which is devoid of demerits associated with use of chemical cross linkers [36]. Microwave, an electromagnetic wave with frequencies in the range of 300 MHz to 300 GHz has numerous biomedical [37–43] and pharmaceutical applications [44–51], as well as applications in food processing [52]. Keeping in view the ability of microwave to affect the macromolecular rearrangement via hydrogen bonding, electrostatic interaction and/or hydrophobic interactions, it is envisaged that the physically cross-linked sodium alginate and pectin blend films using microwave may enhance their physicochemical properties thus making them best suited for wound healing applications. Thus, the objective of this study was to investigate the microwave enabled physically cross-linked polymer blend films alone and in combination with modified-chitosan curcumin nanoparticles to facilitate skin tissue regeneration following 2nd degree burn wound in animals. The modified-chitosan curcumin nanoparticles portion of our project is published earlier [53].

2. Materials and Methods

2.1. Chemicals

Sodium alginate (Sigma Aldrich, St. Louis, MO, USA) and pectin (Sigma Aldrich, St. Louis, MO, USA) were used for film preparation. Monobasic potassium phosphate (Sigma Aldrich, St. Louis, MO, USA), disodium hydrogen orthophosphate (Merck, Darmstadt, Germany), hydrochloric acid (Merck, Darmstadt, Germany), Tween80 and Glycerol were graciously supplied by Biolabs, Islamabad, Pakistan. All chemicals were used without any further purification.

2.2. Film Formulation

The film formulations were prepared using solution casting method. Briefly, an accurately weighed 2 g each of sodium alginate and pectin were dispersed separately in 93.9 g distilled water under continuous magnetic stirring. Following complete dissolution, 0.1% tween 80 and 2% glycerol were added and mixed thoroughly. Both the polymer solutions were mixed in another beaker in ratio of 1:1 and again stirred thoroughly till complete homogenization. Accurately weighed 50 g bubble free polymer mixture solution was poured into a glass petri dishes (diameter 8.5 cm, area 56.75 cm^2) and placed in an oven (DO-100NG, SH Scientific, Daejeon, Korea) at 40 ± 2 °C till complete drying. When required, the polymer mixture solution in a petri dish was subjected to microwave treatment at frequency of 2450 MHz at power of 750 Watt with varying treatment time for the purpose of optimization followed by incubation in oven for drying. The dried films were peeled off from the petri dishes, cut into specific dimensions (25 mm diameter) and stored in desiccator till subjection to physicochemical characterization tests described below. The formulations are given in Table 1.

Table 1. Different formulations of sodium alginate pectin blend film that were microwave treated.

Formulation	Sodium Alginate (*w*/*w*)	Pectin (*w*/*w*)	Glycerol (2%, *w*/*w*)	Tween-80 (0.1%*w*/*w*)	Distilled Water (*w*/*w*)	Microwave Treatment Time (Min)
Untreated	2 g	2 g	2 g	0.1 g	93.9 g	0
1 min	2 g	2 g	2 g	0.1 g	93.9 g	1
3 min	2 g	2 g	2 g	0.1 g	93.9 g	3
5 min	2 g	2 g	2 g	0.1 g	93.9 g	5

2.3. Moisture Adsorption

The moisture adsorption test was performed using an already reported method [54]. Briefly, the dried film samples with known dimensions (25 mm) were placed in a desiccator prefilled with calcium sulphate and incubated for 48 h (at RH of 0%), then their dried weight (*W*) was determined. Subsequently, the same samples were then transferred into another desiccator prefilled with saturated solution of potassium sulphate and incubated again for the next 48 h (at RH of 97%) enabling them to maximally absorb moisture. The samples were removed and weighed again individually to determine their moist weight (*W*0). The percent moisture adsorption was then calculated using the following relation [13].

$$MA\ (\%) = \frac{W0 - W}{W} \times 100 \quad (1)$$

The experiment was repeated at least three times for each film and results averaged.

2.4. Water Vapor Transmission (WVTR) and Vapor Permeability Rate (WVP)

The *WVTR* and *WVP* of the films were determined using the ASTM method (ASTM E96-93, 1993) with some modifications. Briefly, the film samples were cut into dimensions of 33 mm and tightly placed onto the neck of glass bottle prefilled with saturated solution of potassium sulphate for providing virtual relative humidity of 97%. The bottles were placed in a desiccator containing anhydrous calcium sulphate to maintain relative humidity of 0%. The desiccator containing the bottle was then transferred into an oven maintained at 25 ± 1 °C. Each bottle was individually weighed at time interval of 1 h for total of 12 h and the water vapor transmission rate across the film was calculated from the weight loss of the bottle. The average weight change as a function of time was plotted and slope determined from the linear regression of the curve. The experiment was repeated six times and results averaged. The *WVTR* and *WVP* were estimated from following equations [13].

$$WVTR = \frac{Slope}{Film\ area} \tag{2}$$

where slope is the slope of the graph of calculated from the weight loss vs. time curve and film area was 0.000903 m^2.

$$WVP = \frac{WVTR \times T}{\Delta P} \tag{3}$$

where T is the mean film thickness (mm); ΔP is the partial water vapor pressure difference (mmHg) across the two sides of the film specimen (the partial vapor pressure of water at 25 °C = 23.73 mmHg).

2.5. Erosion and Water Uptake

An accurately weighed (initial weight *Wi*) and known dimension (25 mm) of the film samples were placed in a petri-dish containing 20 mL phosphate buffer pH 7.4 simulating the pH of the wound bed conditions and incubated in an oven (BOV 50V, Biobase, Jinan, China) at 37 °C ± 0.2 °C for a duration of 30 min. At specific time intervals, film samples were removed from buffer solution and blot dried by gently sliding them on the surface of the petri dish to remove adhered moisture and weighed (*Wt*). Afterwards, the same film samples were placed in an oven at 40 ± 0.2 °C for at least 5 days. The dried films were then transferred to a desiccator maintained at 25 °C, enabling the films to achieve a constant weight and weighed again (*Wd*). The percent erosion (*E*) and percent water uptake (*WU*) were calculated using the following equation [13].

$$E\ (\%) = \frac{Wi - Wt}{Wi} \times 100 \tag{4}$$

$$WU\ (\%) = \frac{Wt - Wd}{Wd} \times 100 \tag{5}$$

The test was performed for each sample in triplicate and results averaged.

2.6. Tensile Strength

The mechanical strength of films was determined under room conditions of temperature and relative humidity using Universal testing machine (Testometrics, Rochdale, UK). The film samples were cut into strips, 7.5 cm in length and 3.5 cm in width, and fixed between the jaws of the machine. The initial grip separation and cross-head speed were set to 50 mm and 5 mm/min, respectively. Sample was pulled with 50 N loads. The maximum force to break the film was recorded. The experiment was repeated three times and results averaged.

2.7. Vibrational Spectroscopic Analysis

The characteristics peaks of the dried polymeric films were recorded by an ATR-FTIR spectrophotometer (UATR TWO, Perkin Elmer, Buckinghamshire, UK). Each film and/or powdered polymer was placed on the surface of the diamond crystal and clamped to ensure close contact and high sensitivity. All the samples were scanned over wave number range of 400 to 4000 cm^{-1} with an acquisition time of 2 min. Each sample was analyzed three times and results averaged.

2.8. Thermal Analysis

The changes in the transition temperature of the polymeric film were recorded via differential scanning calorimetry (PerkinElmer Thermal Analysis, Waltham, MA, USA). An accurately weighed 4 to 6 mg of the polymeric film was sealed in standard aluminum pan and heated from 0 to 300 °C under continuous flow of nitrogen gas at rate of 40 mL/min.

The characteristic peak temperature and enthalpy of the system were recorded. Each sample was analyzed three times and results averaged.

2.9. Morphology

Surface morphology of the films was analyzed using ultra-high resolution field emission scanning electron microscope (UHR-FESEM, MERLINN/344999-9001-030, Zeiss, Germany). The films were mounted onto stubs (12 mm diameter) using double-sided adhesive carbon tape. The films were then sputter coated with gold for 5 min (QUORUM Sputter Coater Q150R S, Quorum, UK) before placing in the chamber of the microscope and images acquired using SmartTiff software. Images were taken at magnification power of 100×, 500×, 1000×, 2000× and 3000×.

2.10. In Vivo Wound Healing

Healthy male Sprague Dawley rats (age of 3 months, weighing 220–250 g) were used for in vivo wound healing studies. For this purpose, rats were acclimatized for 7 days with free access to food and water under temperature conditions of 25 ± 2 °C and relative humidity at 65 ± 5%. Prior to wound infliction, all animals were kept on fasting for 24 h and were then randomly divided into four groups: untreated group, film, nanoparticles and film-nanoparticles combined group with a total of 8 animals per group. All animals were anesthetized by intra-muscular injection of ketamin (100 mg/kg) and xylazine (10 mg/kg) mixture followed by shaving of the back hairs with sharp blade, cleansed by ethanol swab and area marked followed by infliction of 2nd degree burn wound. An already reported method was used for 2nd degree burn wound infliction [55]. Briefly, contact partial thickness wound was inflicted in each animal on shaved dorsal region using deionized water preheated to 65 ± 5 °C through adhering a plastic tube (internal diameter 14.4 mm, external diameter 16.5 mm) on shaved region using UHU gum followed by addition of 6 mL of preheated water into it and kept for 1 min in 9 repetitive cycles. Following wound infliction, the formulations were applied onto the wounded area with the aid of standard gauze and 3 M adhesive tape. The untreated animal group received only standard gauze application. The treatment group (nanoparticles, film, film-nanoparticles combined) received either 120 mg nanoparticles application alone, or optimized film application alone or nanoparticles application followed by covering the wounded area with optimized film. The treatments were applied on daily basis and dressing changed accordingly till complete healing of the wound. When required, animals were sacrificed by cervical dislocation on day 14 and skin containing the wound surgically excised, washed with normal saline and stored at −20 °C till further use. All animal experimentation was conducted in accordance with the institutional ethics regulations adapting the international guidelines (OECD Environment, Health and Safety). The study protocol was approved by the institutional ethical review board vide reference No. 503/QEC/GU of the Gomal University Pakistan.

2.11. Wound Morphology

The surface morphology of the wound was recorded using a digital camera (Canon D5100, Tokyo, Japan) from a fixed distance and angle in the absence of formulation and/or gauze at intervals of day 0, day 2, day 7, day 10, day 14 and day 18. The size of the wound was determined by analyzing the photographs using Image J software (version.1.53h) and the percent re-epithelization determined using the following equation.

$$Reepithelization\ (\%) = \frac{(Wound\ size\ at\ t = 0)\ -\ (Wound\ size\ at\ t)}{Wound\ size\ at\ t = 0} \times 100 \quad (6)$$

where the wound size was an average measurement from the longest and shortest dimensions of the wounded area.

2.12. Skin Histology

The newly regenerated skin tissue containing the wound was subjected to histological examination. For this purpose, the stored skin samples were thawed at room temperature for 3 h followed by fixation in 10% formalin aqueous solution for 3 days under ambient conditions and then preceded by trimming of skin samples, washed with normal saline and dehydrated in ethanol. The dehydrated skin samples were cleansed using xylene and embedded in paraffin wax. Sections of 5 μm thickness were prepared using a microtome (HM-340E, Microm Inc., Boise, ID, USA) and subjected to H&E (hematoxylin and eosin) and Masson's trichrome separately. The slides were viewed, and respective sections photographed using an inverted microscope fitted with a camera (HDCCE—X5N).

2.13. Vibrational and Thermal Analysis of Skin

The vibrational spectra of dermal layer of the untreated as well as treated animal groups skin samples were recorded at a resolution of 16 cm^{-1} over a wave number range of 400 to 4000 cm^{-1} with an acquisition time of 2 min using ATR-FTIR (UATR TWO, Perkin Elmer, Buckinghamshire, UK). The corresponding ATR-FTIR spectra were compared to estimate extent of collagen deposition. For this purpose, the corresponding absorbance of amide-I and amide-II were recorded which originates from the protein contents of the skin. The ratio of absorbance of treated group to control group was used as a novel technique to estimate the extent of collagen and/or protein deposition. Each sample was analyzed three time and results averaged.

The skin samples containing wound were also subjected to thermal analysis using DSC, (Perkin Elmer, Thermal Analysis, Boston, MA, USA) for the purpose of estimating changes induced in the lipidic and proteinous domains of skin with various treatments applied in comparison to control skin samples. Briefly, an accurately weighed 3 mg of the full thickness skin containing wound was carefully trimmed and sealed in standard aluminum pan followed by subjection to thermal analysis in temperature range of 30 to 180 °C at a heating rate of 10 °C/min under continuous pulse of nitrogen gas at flow rate of 40 mL/min. The melting temperature (ΔT) and enthalpy (ΔH) corresponding to lipidic and proteinous domains were recorded. Each sample was analyzed at least three times and results averaged.

2.14. Tensile Strength of Skin Samples

The excised skin samples were trimmed into strips of 5 cm length and 2.5 cm width and subjected to tensile strength measurement (Testometric M-500, Rochdale, UK). The strips were affixed between the upper and lower jaw of the tensiometer and uniaxially pulled at test speed of 5 mm/s and withdrawal speed of 10 mm/s respectively using load of 30 kg. The maximum force required to break the skin samples and break point was recorded. Each skin sample was examined three times and results averaged.

2.15. Statistical Analysis

All values are expressed as a mean of three readings with respective standard deviation. Student's *t*-test and/or analysis of variance (ANOVA) followed by post-hoc analysis was used with the level of significance set at $p < 0.05$.

3. Results

3.1. Moisture Adsorption

The efficient moisture adsorption ability of film helps remove wound exudates from the wound bed which not only helps in wound debridement but also keeps the wound moist which is detrimental for fast skin regeneration [56]. Furthermore, a wet microenvironment also ensures fewer aberrations and the least scar formation [57,58]. The moisture adsorption results of untreated and microwave-treated sodium alginate and pectin blended films are shown in Figure 1. The moisture adsorption capacity of film is directly related to microwave treatment time, where significantly higher moisture (Student's *t*-test, $p < 0.05$)

adsorption was observed with blend treated for 5 min in comparison to untreated as well as 1 min and 3 min treated film, respectively.

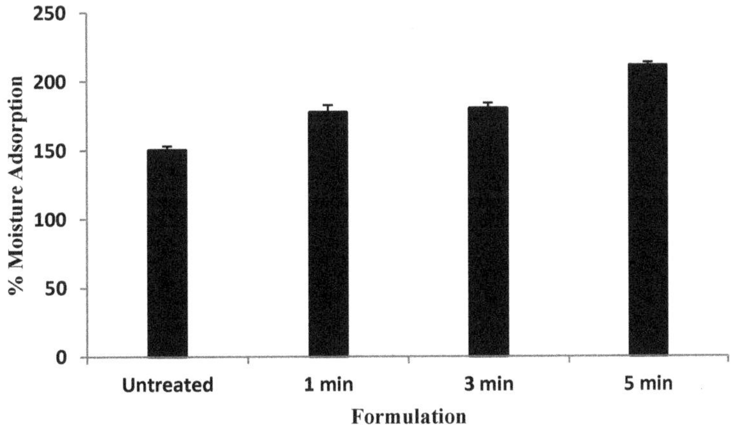

Figure 1. Percent moisture adsorption of various blend films.

3.2. Water Vapor Transmission Rate (WVTR) and Water Vapor Permeability (WVP)

The water vapor transmission and water vapor permeability potential are important physical characteristics of a polymeric film where reduced water loss from the wound bed into the external environment is of pivotal importance in facilitating the wound healing process [59]. The water vapor transmission and water vapor permeability results are shown in Table 2. It is evident from the results that the microwave treatment induced a significant increase (Student's *t*-test, $p < 0.05$) in the *WVTR* and *WVP* in comparison to untreated blend film.

Table 2. *WVTR* and *WVP* results of different formulations.

Formulation	Film Thickness (mm)	WVTR (g/m^2h)	WVP ($g\ mm/h{\cdot}m^2$)
Untreated blend	0.22 ± 0.03	180 ± 1.8	0.152 ± 0.06
1 min treated blend	0.18 ± 0.02	240 ± 0.6	0.164 ± 0.02
3 min treated blend	0.19 ± 0.03	256 ± 0.6	0.177 ± 0.03
5 min treated blend	0.20 ± 0.03	276 ± 0.5	0.191 ± 0.02

3.3. Erosion and Water Uptake

The percent erosion and water uptake govern the fate of wound healing platform upon application to moist wounds. The percent erosion and water uptake results of untreated and microwave-treated film formulations are given in Figure 2. The untreated blend film eroded at a significantly faster rate (Student's *t*-test, $p < 0.05$, Figure 2a) within 20 min of the immersion in phosphate buffer saline where it rapidly swelled and eroded because polymer hydrophilicity where chain relaxation proceeded quickly resulted in the loss of its structural integrity, and ultimately dissolution in the media [60], in comparison to microwave-treated blend film formulations where a significant reduction (Student's *t*-test, $p < 0.05$, Figure 2a) in the ability of blend erosion was observed. The water uptake results of the untreated and microwave-treated blend films are given in Figure 2b. A slight, though insignificant (ANOVA, $p > 0.05$), rise in the water uptake capacity of the microwave-treated blend film for 5 min was observed in comparison to untreated as well as 1 min and 3 min treated film formulations.

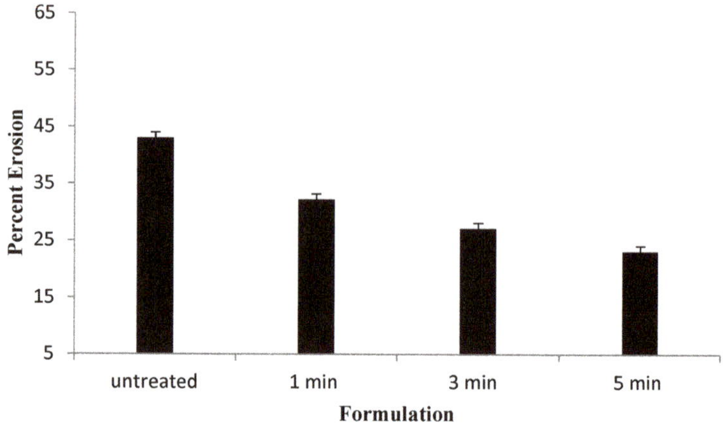

(a)

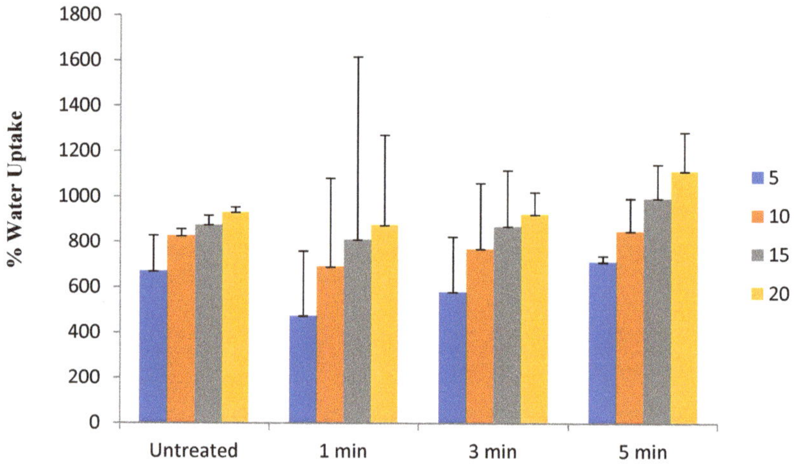

(b)

Figure 2. (**a**) Percent erosion; (**b**) water uptake of blend films.

3.4. Tensile Strength

The tensile strength results of the various film formulations are shown in Table 3. Microwave treatment resulted a gradual increase in the tensile strength in comparison to untreated formulation where a significant rise (Student's t-test, $p < 0.05$) was observed for formulation with 5 min microwave treatment.

Table 3. Mechanical properties of various film formulations.

Formulation	Tensile Strength (Mpa)	Elongation Break (%)	Elastic Modulus (Mpa)
Untreated blend	2.34 ± 0.24	40.02 ± 5.52	9.39 ± 12.50
1 min treated blend	5.21 ± 0.41	43.34 ± 2.98	60.28 ± 15.55
3 min treated blend	7.88 ± 0.04	50.01 ± 4.71	79.99 ± 25.98
5 min treated blend	9.94 ± 0.08	66.66 ± 3.29	86.78 ± 22.98

3.5. Vibrational Analysis

The ATR-FTIR spectrum of untreated and blend films with different microwave time treatments are shown in Figure 3. The treatment of blend film with microwave for different time intervals resulted in significant changes in the corresponding wave numbers of hydrophilic as well as hydrophobic domains of the formulation. The hydrophilic OH/NH domain (3290–3305 cm^{-1}) and C=O moiety (1604–1610 cm^{-1}) underwent significant rigidification upon microwave treatment where 5 min treatment (Figure 4d) resulted in significant reduction (Student's *t*-test, $p < 0.05$) in the corresponding wave number. Contrary to this, the hydrophobic asymmetric CH band (2920–2930 cm^{-1}) undergoes significant fluidization with 5 min microwave treatment of the blend film where a significant rise (Student's *t*-test, $p < 0.05$) in corresponding wave number from 2921.2 ± 1.15 cm^{-1} to 2931.2 ± 0.8 cm^{-1} was observed.

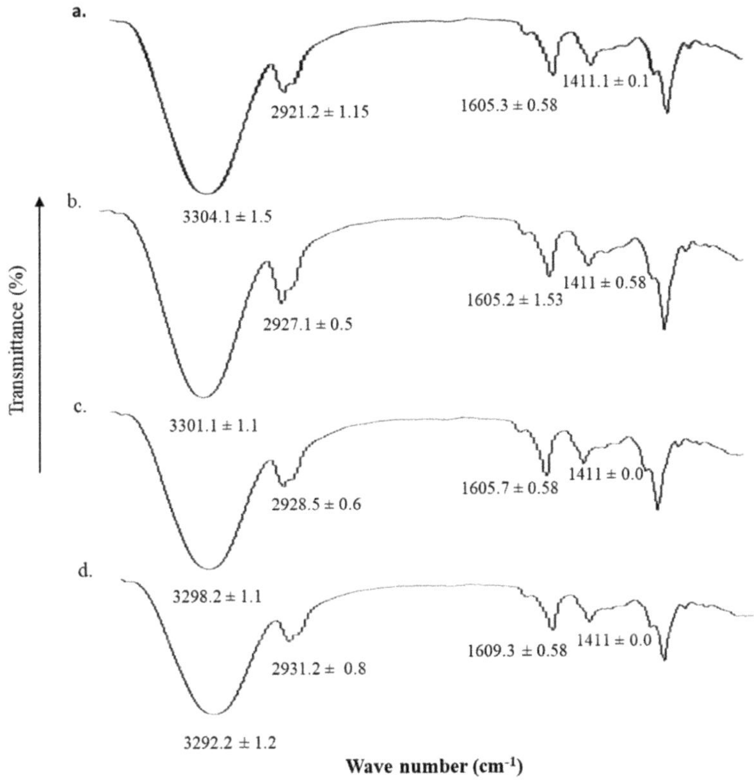

Figure 3. ATR-FTIR spectrum of (**a**) untreated, (**b**) 1 min, (**c**) 3 min, and (**d**) 5 min microwave treated pectin and sodium alginate blend film.

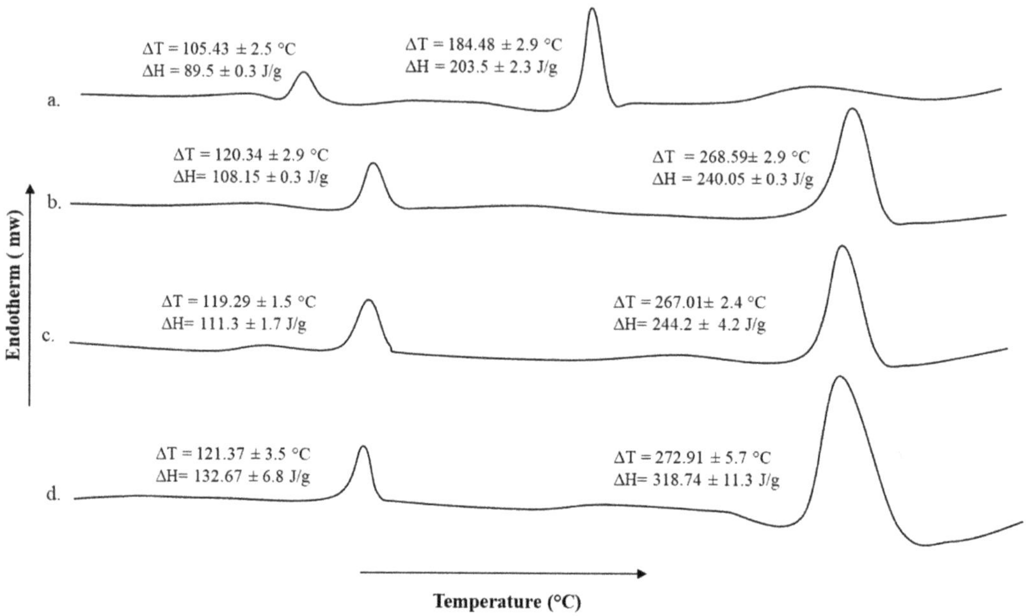

Figure 4. DSC thermogram of (**a**) untreated, (**b**) 1 min, (**c**) 3 min, and (**d**) 5 min microwave treated pectin and sodium alginate blend film.

3.6. Thermal Analysis

The DSC thermogram of untreated sodium alginate-pectin blend and microwave treated blends for different time intervals are shown in Figure 4. The pectin moiety showed transition melting temperature of 105.43 ± 2.5 °C while the transition melting temperature of 184.48 ± 2.9 °C is ascribed to the sodium alginate moiety. The transition temperature and energy required to induce those transitions underwent significant rise for both pectin as well as sodium alginate moieties upon microwave treatments (Figure 4b–d). All microwave treatments significantly increased the transition temperatures of both polymer moieties as well as enthalpies in comparison to untreated blend formulation but a significant increase (Student's *t*-test, $p < 0.05$, Figure 4d) in the enthalpies among various microwave treatments applied was observed with 5 min treatment, though the transition temperatures remain insignificant (Student's *t*-test, $p > 0.05$, Figure 4d).

3.7. Scanning Electron Microscopy

The surface morphological characteristics of untreated and microwave-treated blend films are shown in Figure 5. The untreated blend (Figure 5a) showed nonhomogeneous texture which could be due to migration of formulation constituents (tween 80 and/or glycerol) to the surface upon drying. The microwave treatment (Figure 5b–d) showed variable surface morphological results, where 1 min and 3 min microwave treatment showed surface adhesion of formulation constituents though to lesser extent than untreated one; formation of cracks (Figure 5c) depicting 3 min microwave treatment though facilitated polymer homogeneity but further drying process may have led to improper laying down of polymer fibers resulting in formation of spaces in the matrix. The 5 min microwave treatment (Figure 5d) resulted in more homogenous and uniform matrix deposition in a specific manner without any surface artifacts where even the further drying process did not render any destructive effect neither on the texture of the film nor on the formulation ingredients.

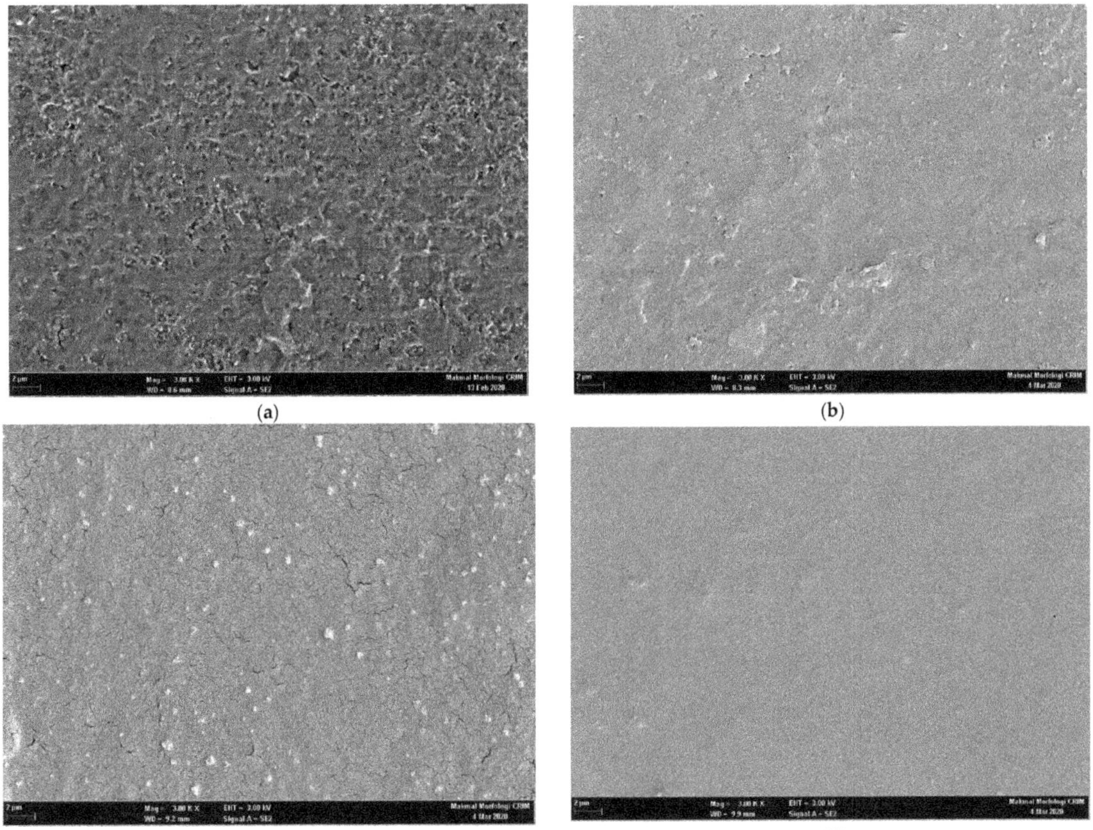

(**a**) (**b**)

(**c**) (**d**)

Figure 5. Scanning electron micrographs of (**a**) untreated, (**b**) 1 min, (**c**) 3 min, and (**d**) 5 min microwave treated pectin and sodium alginate blend film.

3.8. Wound Morphology

The wound healing progress following 2nd degree burn wound infliction of various animal groups is shown in Figure 6. The untreated animal group depicted slow percent reepithelization and wound closure where the complete wound healing did not affect till day 18 as observed with test groups. Among test groups, the chitosan-curcumin nanoparticles alone as well as in combination showed significantly faster wound closure and percent reepithelization in comparison to untreated and only film treated animal group (ANOVA, $p < 0.05$, Figure 6) where complete wound closure as well as scare minimization was observed. Complete reepithelization took place within 18 days post treatment with nanoparticles and film combined application (Figure 7), while the untreated animals group showed complete wound closure over a time span of 28 days.

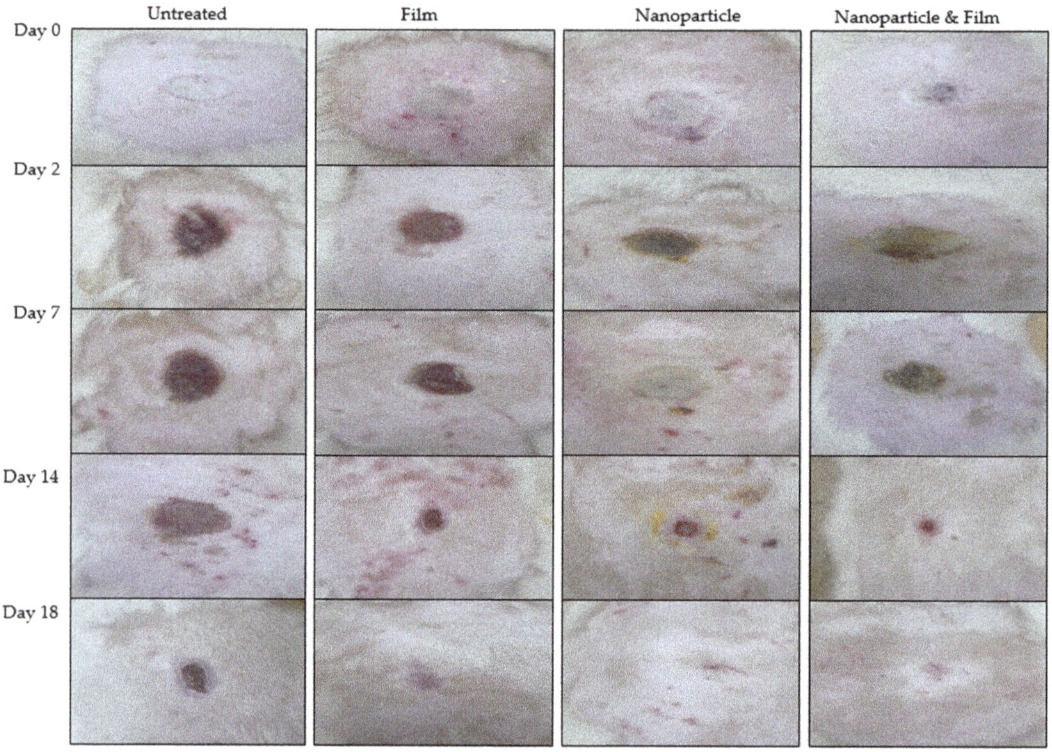

Figure 6. Wound images of rats with no applied scaffold (control) and treated by film, nanoparticles and combination of both on different days.

3.9. Vibrational Spectroscopic Analysis (ATR-FTIR) of Skin Samples

The ATR-FTIR spectra of dermis layer of untreated and various test treated animal skin samples are given in Figure 8. Vibrational spectroscopic method of analysis of skin samples was used as a novel method of estimation of extent of proteins deposition as a function of applied treatments. For the purpose, OH/NH, amide-I and amide-II bands wave numbers and corresponding absorbance ratios with respect of untreated skin samples were recorded and calculated where the amide-I and amide-II as well as OH/NH bands typically originate from the protein moieties of the skin [61]. The changes in the wave numbers and absorbance ratios were assigned to increase and/or decrease in the rigidification of respective domains depicting higher population of corresponding protein moieties in the skin. The absorbance ratios are given in Table 4. The wave numbers of OH/NH domains (3222–3294 cm^{-1}) in the untreated dermal layer significantly decreased (Students' t-test, $p < 0.05$) with film, nanoparticles alone as well as nanoparticle-film combine application suggesting rigidification of hydrophilic OH/NH moieties of the dermal layer depicting formation of more compact skin structure. The absorbance ratio of OH/NH of test samples to untreated samples increased depicting more rigidification of hydrophilic domains of the dermal moieties (Table 4). Similarly, the wave number of amide-I (C=O stretching, 1627–1638 cm^{-1}, Figure 8) moiety underwent significant decrease (Student's t-test, $p < 0.05$) with various treatments applied while the amide-II moiety wave number (1549 to 1553 cm^{-1}) did not show any significant changes (Students' t-test, $p > 0.05$).

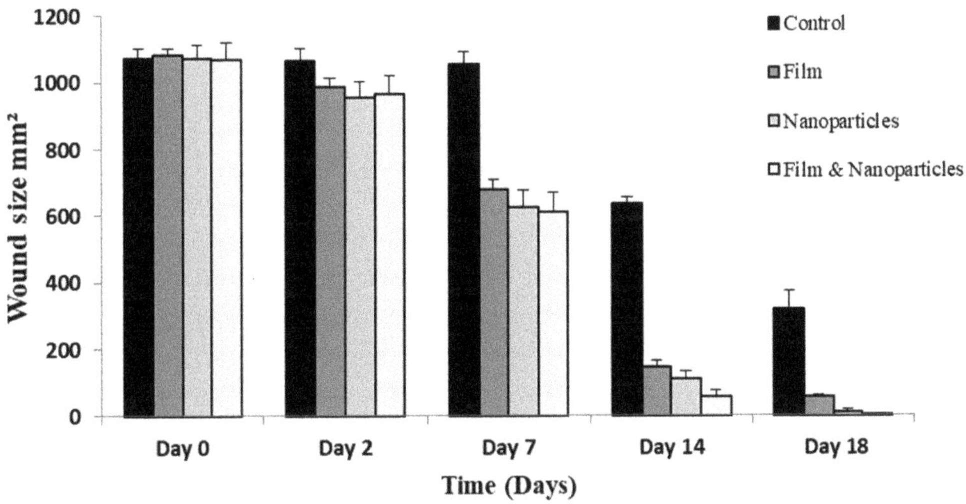

(a)

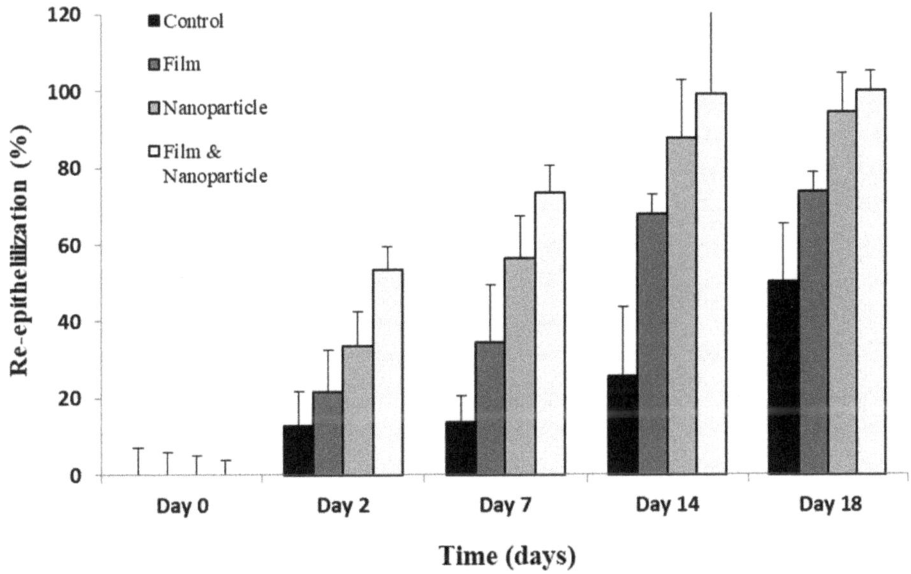

(b)

Figure 7. Profiles of (**a**) wound size and (**b**) state of re-epithelialization of rats.

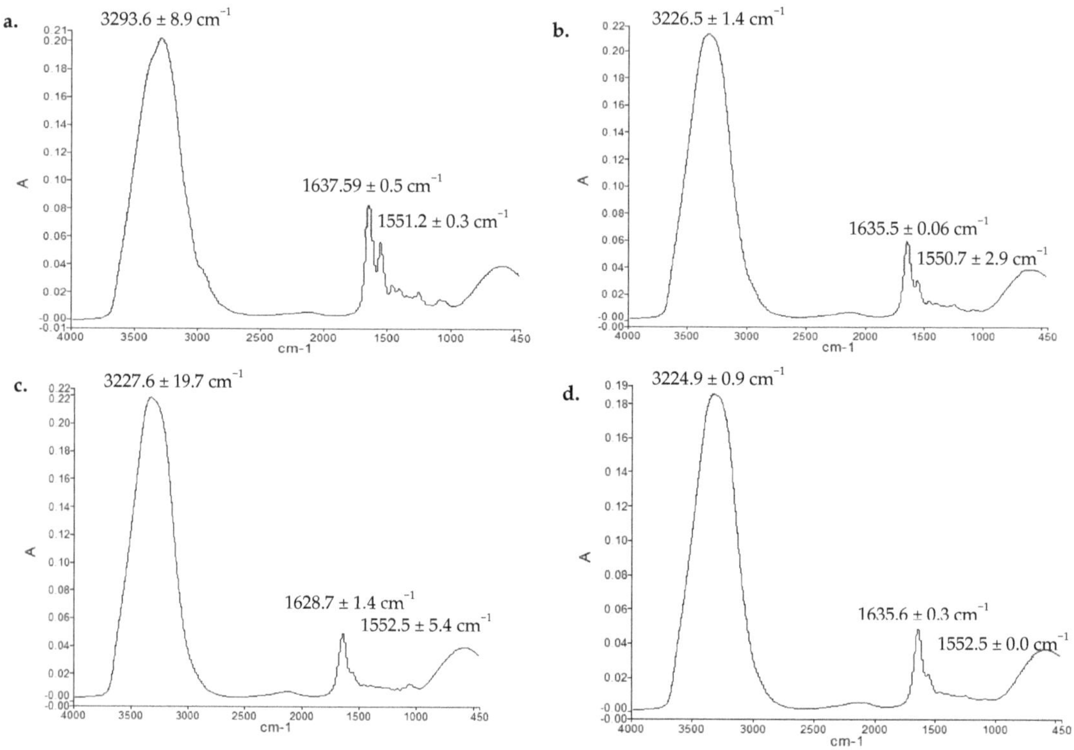

Figure 8. ATR-FTIR spectra of dermis of (**a**) control skin, skin treated by (**b**) film, (**c**) nanoparticles, (**d**) film and nanoparticle group.

Table 4. Absorbance ratios of various regions.

Samples	Absorbance Ratios of OH/NH of Test to Untreated Dermal Moieties	Absorbance Ratios of Amide-I (C=O Stretching) to Amide-II (C-N Bending) of Various Test Dermal Moieties
Film	0.91 ± 0.005	1.01 ± 0.05
Nanoparticle	1.05 ± 0.005	1.8 ± 0.01
Nanoparticle-film combined	1.12 ± 0.015	2.15 ± 0.001

3.10. Thermal and Mechanical Strength Analysis of Skin Samples

The DSC thermograms of test animal skin samples and untreated extracted from animals on 14 days of wounding are shown in Figure 9, while the tensile strength and percent elongation break are given in Table 5.

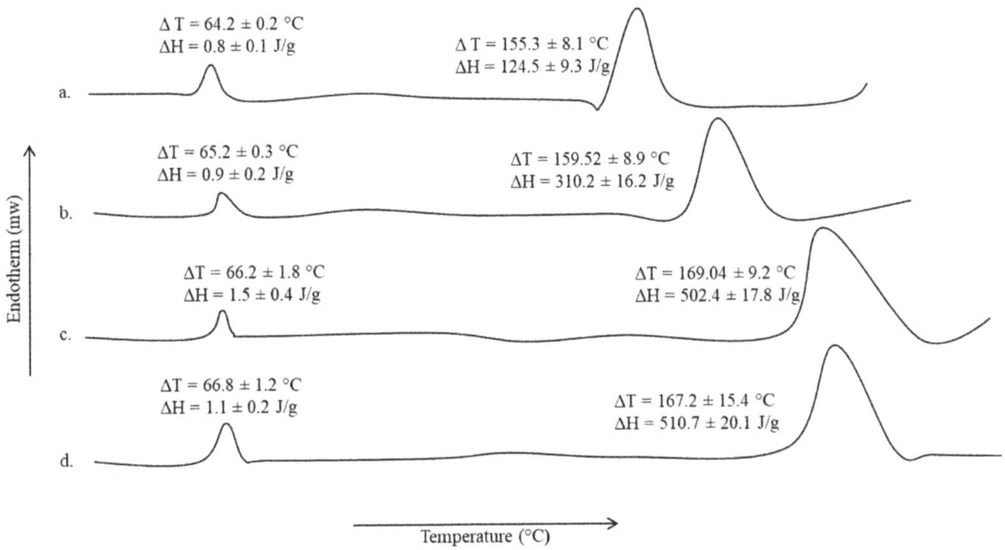

Figure 9. Thermograms of (**a**)untreated, (**b**) film treated, (**c**) nanoparticles treated, and (**d**) film-nanoparticles combined treated skin samples.

Table 5. Mechanical properties of skin sample.

Animal Groups	Tensile Strength (MPa)	Elongation Break (%)	Elastic Modulus (MPa)
Untreated	8.61 ± 0.7	12.09 ± 0.7	1.82 ± 2.1
Film	10.98 ± 0.6	14.89 ± 0.5	3.32 ± 2.9
Nanoparticles	13.78 ± 0.9	17.14 ± 1.5	4.10 ± 1.9
Nanoparticles-film combined	14.65 ± 0.8	17.99 ± 0.7	5.78 ± 1.7

A slight increase in the transition temperature and corresponding enthalpy, though insignificant (Student's *t*-test, $p > 0.05$) of lipidic domains (ΔT = 64.02 ± 0.2 °C, ΔH = 0.8 ± 0.1 J/g, Figure 9) of the skin samples was observed with all treatments applied in comparison to untreated animal skin sample. However, the proteinous domain (ΔT = 155.3 ± 8.1 °C, ΔH = 124.5 ± 9.3 J/g, Figure 9) underwent a significant rise not only in the corresponding transition temperature but also in the energy required to induce transition. The animal group treated with nanoparticles alone as well as in combination with film showed a significant rise in transition temperature and enthalpies (Student's *t*-test, $p < 0.05$) in comparison to untreated and only film treated groups. The nanoparticle-film combined application hence translated into a significant rise in the tensile strength as well as percent elongation break (Student's *t*-test, $p < 0.05$, Table 5) in comparison to untreated as well as other experimental animal groups.

3.11. Skin Histology

The hematoxylin and eosin (H&E) staining micrographs are shown in Figure 10 while Masson trichrome is shown in Figure 11. The H&E staining was performed on day 14 skin samples with an aim to elucidate the accumulation of inflammatory cells, regeneration of epidermis layer, blood vessels proliferation and tissue granulation formation. Control group showed poor epithelial regeneration along with sub-epithelial layer degeneration, and connective tissue fibers along with dermal glands were displaced in entire connective tissues (Figure 10a). Blend film treated skin exhibited partial epithelial regeneration followed by epidermal layer that regenerated over the granulation tissue and mild inflammatory infiltrate was noted (Figure 10b). Nanoparticles treated group with and without blend film application have connective tissue fibers that were properly arranged with many newly grown connective tissue, and entire dermal tissue in fiber bundles is visible. Connective tissue fibers were well developed and showed compactness in their structure without any displacement and sign of degeneration of tissue fibers (Figure 10d).

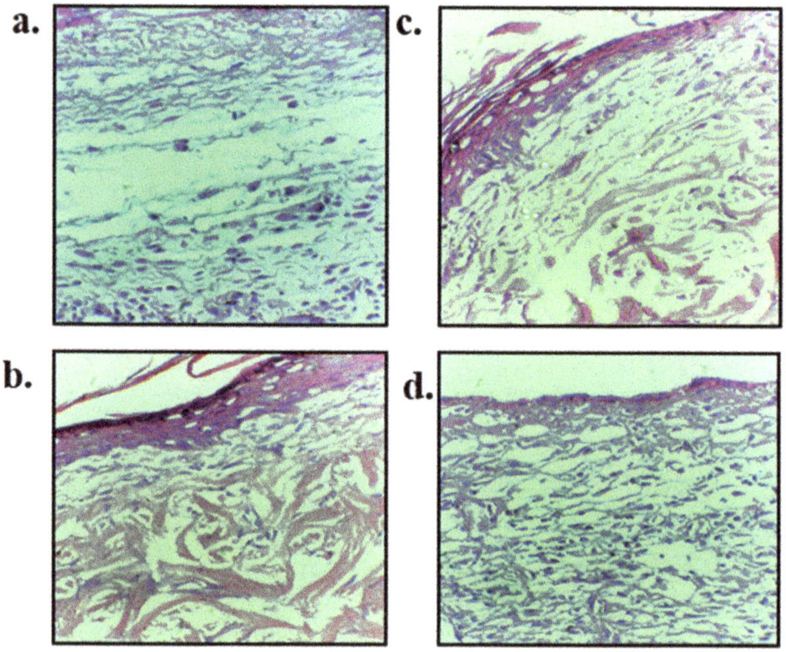

Figure 10. Hematoxylin and Eosin staining of skin samples extracted on day 14 of wounding on (**a**) untreated, (**b**) blend film, (**c**) nanoparticles and (**d**) nanoparticles-blend film combined treated skin samples.

The emphasis of using Masson trichrome on burn wound tissue samples of untreated and various treated groups was to check the epidermal and dermal connective tissue fibers, as it gives green color to the collagen fibers which is very helpful in determining the extent of collagen production as well as its alignment at the healing area on 14 days post wounding (Figure 11). A significant difference in the extent and alignment of collagen fibers deposition following burn wound treated with nanoparticles (Figure 11c) and combine nanoparticles and blend film group (Figure 11d) was observed in comparison to untreated (Figure 11a) and film treated groups (Figure 11b), which demonstrated good synergistic wound healing characteristics of chitosan and curcumin. The untreated group showed void space in the dermis layer with irregular arrangement and disintegration of collagen fibers. Dermis and sub-epidermal regions also showed too much distortion followed by the epidermal region where epithelial integrity was distorted (Figure 11a).

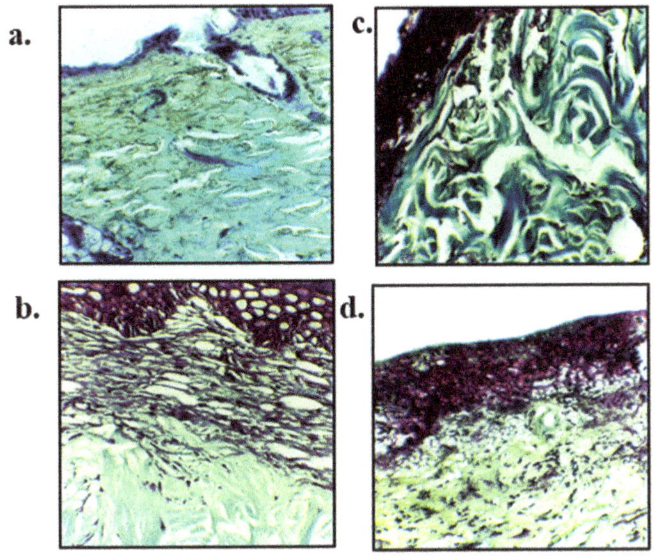

Figure 11. Masson trichrome staining of skin samples extracted on day 14 of wounding on (**a**) untreated, (**b**) blend film, (**c**) nanoparticles and (**d**) nanoparticles-blend film combined treated skin samples.

4. Discussion

This project investigated the film and nanoparticles combined application as a novel strategy to treat burn wounds. As shown in Figure 1, the moisture adsorption ability of the blended films was directly related to microwave treatment time. Microwaves are electromagnetic waves which interact with polar moieties of the polymer in a volumetric manner [62]. Both polymers, i.e., sodium alginate and pectin, contain a number of polar functional groups viz OH/NH_2, amide-I and amide-II, which may interact with each other upon microwave treatment preferably through hydrogen bonds [63]. Exposition of surface hydrophilic functional groups enables attraction and/or interaction with more water molecules [64]. The blend film also contained additional hydrophilic ingredients namely tween 80 and glycerol. The increase in the hydrophilicity of the blend film with microwave treatment might be due to arrangement of all hydrophilic components of the formulation towards the surface of the dried film matrix which enabled further increase in the attraction ability of water molecules towards the film surface. Furthermore, the polymer interlinking helps lay down the polymer fibers in a uniform fashion during drying process and results in formation of voids between polymer chains to form a controlled pore size throughout the polymer matrix [65]. Smaller pore size in the matrix translates into maximal retention and/or adsorption of moisture at the wound bed [66]. The pore size of the film is also considered important to minimize fibroblast/keratin in-growth into the film, thereby reducing secondary damage upon dressing changing [67]. The smaller pore size also enables reduction in the infiltration of opportunistic bacteria into wound bed, which translates into minimization of chances of secondary bacterial infections which may otherwise complicate the wound [68]. Furthermore, high surface adsorption capacity of film formulation also helps clear surface bacteria from the wound. The polymeric film enables bacterial adsorption possibly through biospecific/selective interaction (carbohydrate–protein, protein–protein) and nonspecific interactions (electrostatic or hydrophobic) [69] and hence film with high adsorption capacity may thus help in removing surface pathogenic bacteria from the wound bed [70]. The microwave based physically cross-linked sodium alginate and pectin blend film at frequency of 2450 MHz for 5 min is thus envisaged to promote skin tissue regeneration following 2nd degree burn by increasing the exudates removal from the wound bed due to controlled pore formation in the polymer matrix and hence prevent bacterial infiltration as well.

The reduced water loss from the wound bed into the external environment is of pivotal importance. Microwave treatment resulted in a significant increase in the *WVTR* and *WVP* (Table 2, Student's *t*-test, $p < 0.05$). The *WVTR* and *WVP* are directly related to the pore size formed in the polymer matrix upon drying [71], while reduced *WVTR* and *WVP* are considered essential to prevent the wound from drying out. At the same time, appropriate pore size is also essential to facilitate gaseous exchange (oxygen and carbon dioxide) between the wound bed and environment [13]. A poor exchange of gases between the wound bed and external environment may accumulate CO_2 leading to acidification of the wound media, which on one hand may directly inhibit the cell proliferation during angiogenesis and on the other hand may provide a favorable environment for the infiltration and growth of anaerobic bacteria [72]. In aerobic metabolism of glucose, generation of ATP's are major energy molecules which drive the majority of the cellular processes during the wound healing process [73].Since healing tissue requires an increased energy demand [74], appropriate pore size of the polymer film, enough to facilitate gaseous exchange between wound bed and external environment, is thus envisaged to promote skin tissue regeneration.

The microwave treatment significantly delayed the erosion ability of the film (Student's *t*-test, $p < 0.05$, Figure 2a), which could be attributed to crosslinking that restricts chain relaxation due to formation of egg box junctions among pectin and/or sodium alginate moieties, a time-dependent stabilization of the elicitor-active conformation which increases its biological activity [75]. Furthermore, reduction in the percent erosion with 5 min microwave treatment depicts formation of additional inter and intra polymer chains and/or polymer and other formulation moieties thus deterring the rapid entry of solvent molecules into the polymer matrix preventing rapid breakage of matrix structure [76]. Similarly, the water uptake capacity of the film determines the physicochemical properties of film like degradation, swelling, mechanical integrity, adhesivity, drug stability, drug release profile and biological response [77]. Reduced water uptake capacity determines the resistance of film to degradation in the wound physiological environment [78]. The water uptake ability of microwave-treated blend films (Figure 2b) demonstrated a slight increase in the water uptake capacity where the reason behind was not investigated but it is believed that microwave interaction with polar moieties of the polymers and other formulation ingredients may have resulted in the availability of more hydrophilic functional groups (OH/NH_2, ester, amide) from the sodium alginate, pectin and/or tween 80 and glycerol resulting in a slight increase in the hydrophilic interaction with the available water. Although, high water uptake and slow erosion are deemed favorable to withstand shear and stress conditions during application as well as removal of exudate and debridement of the wound bed which translate into faster skin regeneration [13].

The tensile strength of the film formulation determines its fate following application as well as its ability to withstand shear stress during packaging and transportation; hence blend use is often advocated in place of lone polymer, which has higher mechanical strength [79,80]. Polymeric films that are intended to be used for wound healing applications should have appropriate mechanical properties that can support cellular activities such as proliferation, migration, and angiogenesis, as well as protect structures found in native skin such as blood vessels, lymphatic systems and nerve bundles [81]. For this purpose, the film should have mechanical properties similar to those found in native tissue. In this regard, the values of the tensile strengths (5 to 40 MPa), the Young's modulus (4.5 to 25 MPa) and the percent elongation-to-break (35% to 120%) are considered appropriate for wound dressings [82] and sufficient to provide enough mechanical support for angiogenesis and tissue remodeling process during wound healing which also help in preventing stress shielding [83]. Microwave interacts with hydrophilic domains of the polymers (OH/NH_2, amide-I, amide-II) [84], which are responsible for development of extra inter and intra polymer linkages preferably through H-bonding and/or electrostatic interactions [85]. Formation of additional linkages translates into stiffness of the polymer matrix thus formed and enables deposition of polymer fibers in a specific geometric manner during drying process [86]. The decrease in the spaces between the polymer fibers upon treatment results in more resistance to breakage and hence translates into higher mechanical stability [87]. This was the reason that 5 min microwave treatment significantly enhanced the mechanical stability of the blend film in comparison to untreated as well as other treatments.

The changes in hydrophilic and hydrophobic domains of the blend film (Figure 3) can be attributed to the interaction between the polymer chains and/or other formulation moieties namely tween 80 and/or glycerol which is envisaged to increase the mechanical strength as well as elasticity of the blends with microwave treatment in comparison to untreated. The interaction between the polar moieties of sodium alginate and pectin and/or polymers and tween 80/glycerol upon microwave treatment could have resulted in the formation of new bonds between them which resulted in the shrinkage of hydrophilic domains (OH/NH_2, C=O, Figure 3b–d), which could aptly explain the blend film resistance to erosion and an insignificant rise in water uptake capacity due to reduction in the inter polymer chain spaces restricting rapid penetration of immersion medium molecules into polymer matrix. On the other hand, significant fluidization of hydrophobic moiety of the blend was observed which can aptly be attributed to the role of glycerol being a plasticizer in the blend film which resulted in a significant rise in the percent elongation as well as elastic modulus of the blend treated with microwave for 5 min [88]. The microwave activated polar functional groups and an increase in polar interactions between the polymers and/or other formulation entities translated into formation of new inter and intra molecular forces which might have enabled deposition of polymer chains in a layer-by-layer pattern to form a specific three-dimensional structure with controlled inter and intra polymer spaces [89].

Similarly, in thermal analysis (Figure 4) significant changes in the transition temperatures as well as enthalpies of the polymer moieties was observed with 5 min microwave treatment (Figure 4d). Since both pectin and sodium alginate have a number of hydrophilic functional groups with the ability to form inter and intra polymer H-bonding, this results in a rise in energy required to induce transition due to rise in propensity and extent of interaction between the various functional groups under microwave treatment [90]. Both pectin and sodium alginate are polysaccharide polymers and contain a number of OH and COOH functional groups which have the ability to form strong inter and intramolecular H-bonding [91]. Microwave interacts with polymers through polar functional groups in the polymer structure and elicit formation of additional linkages through H-bonding and/or electrostatic interaction [92]. The ability of microwave to develop a physically cross-linked structure resulted in a significant rise in the transition temperature as well as enthalpies of the blend which is envisaged to improve the physicochemical properties of the blend film.

The 5 min microwave treatment (Figure 5d) translated into a more homogenous and uniform matrix deposition in a specific manner without any surface artifacts where even the further drying process did not render any destructive effect, neither on the texture of the film nor on the formulation ingredients, suggesting 5 min microwave treatment of pectin and sodium alginate blend ensures proper amalgamation of all formulation ingredients which translates into smooth surface texture formation.

The modified chitosan-curcumin nanoparticles alone as well as in combination showed significantly faster wound closure and percent reepithelization in comparison to untreated and only film treated animal group (ANOVA, $p < 0.05$, Figures 6 and 7). Curcumin is a natural wound healing compound with remarkable anti-inflammatory and collagen deposition properties [93]. Curcumin is reported to reduce expression of pro-inflammatory cytokines namely tumor necrosis factor alpha (TNF-α) and interleukin-1 (IL-1) [94]. Furthermore, curcumin has the inherent ability to recruit M2-like macrophages in the white adipose tissues which translate into enhanced production of anti-inflammatory cytokines which are detrimental for the inflammatory response [95] and also inhibit nuclear factor (NF-kB), thus reducing inflammation at the wound site [96]. Curcumin is also reported to recruit wound site fibroblasts in vitro and in vivo by inciting proliferation and migration which is partially mediated by Dkk-1, where curcumin has been speculated to precisely control the Dkk-1 level in a temporospatial manner to regulate the downstream Wnt signaling pathway which is necessary to facilitate fibroblasts proliferation and migration at wound site [97]. The wounds are also associated with formation of reactive oxygen species (ROS) particularly hydrogen peroxide (H_2O_2) and superoxide (O_2^-) which are associated with prolonging the inflammatory phase of the wound healing due to damage caused to healing cells [98]. Curcumin acts as a potent antioxidant by scavenging ROS and suppression of transcription factors related to oxidation, which is conferred by its electron donating functional groups (i.e., the phenolic OH groups) [99].

Chitosan is a promising polymer with anti-bacterial [100], anti-inflammatory [101] and hemostatic properties [102]. The antimicrobial properties of the chitosan are due to its positive surface charge due to protonation of NH_2 functional groups which interact with negatively charged microbial cell membrane and/or cell wall and thus incite lysis [103]. The extent of positive surface charge on chitosan is directly related to its degree of deacetylation [104]. The chitosan variant used had a degree of deacetylation 91.41 $\pm$ 8.2, thus increased propensity of positive surface charge was imparted to nanoparticles. The chitosan-curcumin nanoparticles are thus envisaged to exert synergistic antimicrobial activity. Furthermore, chitosan has also been reported to facilitate wound healing by promoting growth of macrophages and fibroblasts where secretion of cytokines such as transforming growth factor-β (TGF-β), PDGF and IL-1 which translate into migration of macrophages into wound site, promoting fibroblasts proliferation and enhancing collagen secretion [105]. This could be the reason that chitosan-curcumin nanoparticles hasten percent reepithelization process and wound closure in comparison to untreated and film animal groups. The percent re-epithelization was significantly higher when nanoparticles and film were used in combination (Figure 7). The nanoparticle and film combined application facilitated rapid wound healing by preventing rapid water loss from the wound bed and facilitated easy gaseous exchange between the wound bed and external environment which translated into rapid wound closure and higher percent of reepithelization.

The vibrational spectroscopic analysis of skin samples and absorbance ratios of various absorption bands excised at specific time interval (Figure 8, Table 4) underwent significant changes in the propensity of skin proteinous domains. The spectral bands in vibrational analysis of samples are molecule-specific and are directly associated with the vibrations of particular chemical bonds and/or functional groups when exposed to IR radiations [106], while the absorbance values are directly related to the propensity and/or concentration of those particular molecules where higher absorbance values indicate higher intensity of that specific region [107]. The higher absorbance ratios observed with nanoparticles-film combined application clearly indicate that more proteins (predominantly collagen) are being deposited at the wound site as compared to other treatments applied [108].

As shown in Figure 9, nanoparticles application alone as well as in combination with film translated into a significant rise in the transition temperature as well as enthalpies of the proteinous domains of the skin samples as well as a significant rise in the mechanical strength of the skin (Table 5). Curcumin is found to increase collagen synthesis, cellular proliferation and hence increase collagen deposition for rapid wound healing [98].Chitosan, with high degree of deacetylation, promotes wound healing and increases fibroblast activity and uniform collagen deposition translating into higher tensile strength [109]. The increased transition temperature and corresponding enthalpy observed with nanoparticles treated group and nanoparticles-film combined treated group is thus the synergistic activity of curcumin as well as chitosan to promote skin tissue regeneration, while the film provided an additional support to keep the wound moist which is detrimental for rapid wound closure. The nanoparticle-film combined application hence translated into a significant rise in the tensile strength as well as percent elongation break (Student's t-test, $p < 0.05$, Table 5) in comparison to untreated as well as other experimental animal groups, which is envisaged to be due to synergistic action of curcumin and nanoparticles in ensuring rapid formation of collagen fibers and their uniform deposition at wound site.

The hematoxylin and eosin (H&E) staining micrographs (Figure 10) showed application of nanoparticles alone as well as in combination with film resulted in formation of properly arranged connective tissue fibers. Chitosan has been reported to possess antimicrobial [110], antioxidant [111] and immunomodulatory activity [112]. The moist environment of the wound post skin injury provides a favorable environment for the microbes to infiltrate the wound and hence induce secondary bacterial infection [113]. The antimicrobial activity of chitosan is directly linked to its degree of deacetylation as chitosan variant having a higher number of free NH_2 groups reported to possess higher antimicrobial activity due to their enhanced ability to interact electrostatically with the negatively charged microbial cell membrane [114]. Similarly, the antioxidant activity of the chitosan is also directly linked to its molecular weight and degree of deacetylation where low molecular weight and higher degree of deacetylation are reported to exert higher free radical scavenging ability [115]. Chitosan possesses the ability to inhibit proinflammatory cytokines and promote tissue granulation through fibroblast recruitment [116]. Friedman et al. investigated the P. *acnes* induced cytokines and chemokines and found a significant reduction in levels of IL-12p40 demonstrating complete reduction of IL-12 and IL-6 [117]. On the other hand, curcumin is a natural compound which possesses remarkable antioxidant, antimicrobial, and anti-inflammatory properties [118]. The anti-inflammatory activity of curcumin was reported to be through peroxisome proliferator-activated receptor gamma (PPAR-γ) [119]. Furthermore, curcumin also exerts its anti-inflammatory activity through T cells, B cells, neutrophils, natural killer cells, dendritic cells, and macrophages [120]. The synergistic anti-inflammatory activity of chitosan and curcumin is thus envisaged to promote rapid wound closure and formation of denser and more well-developed connective fibers and higher fibroblasts accumulation at wound site independent of film application (Figure 11d).

Similarly, a synergistic activity of chitosan and curcumin was observed when nanoparticles were applied in combination with film in the alignment at the healing area 14 days post wounding (Figure 11). Chitosan possesses the ability to regulate granulation tissue formation and promote angiogenesis, thus assuring correct deposition of collagen fibers which translate into correct repair of injured dermal tissue [121]; while curcumin also promotes collagen deposition by enhancing the fibroblast proliferation, vascular density and modulating the granulation tissue formation which facilitate extracellular matrix production at the wound site [122]. It also induces transforming growth factor-β and thus stimulates angiogenesis and accumulation of extracellular matrix continues through the remodeling phase of wound healing [123]. The synergistic collagen deposition ability of chitosan and curcumin translated into more compact collagen fiber deposition in case of nanoparticles alone as well as when applied in combination with blend film in comparison to untreated and film treated animal group.

5. Conclusions

This study explored the combinational use of modified chitosan-curcumin nanoparticles and microwave enabled physically cross-linked polymer blend film for skin regeneration potential following 2nd degree burn wound in animals. The microwave treatment significantly improved the physicochemical attributes of polymer blend film addressing the properties required of a wound healing platform. The cross linking between the two polymers was initiated by rigidification of hydrophilic and fluidization of hydrophobic domains which resulted in a significant change in the corresponding transition temperature as well as enthalpies of the polymer moieties. The combinational application of modified chitosan-curcumin nanoparticles hastens the skin regeneration process and translates into a significant increase in the extent of protein deposition as well as mechanical strength. The combined use of microwave enabled physically cross-linked sodium alginate and pectin film and modified chitosan-curcumin nanoparticles may open new horizons in treatment strategies of burn wounds.

Author Contributions: Conceptualization, N.R.K.; data curation, N.R.K.; formal analysis, H.M.B. and N.R.K.; funding acquisition, H.M.B., S.U.S. and N.R.K.; investigation, H.M.B., H.A.A. and N.R.K.; methodology, M.M.S., S.U.S. and N.R.K.; project administration, S.U.S. and N.R.K.; resources, H.A.A., A.A.A., I.A.W. and N.R.K.; software, H.M.B., M.A., A.W. and N.R.K.; supervision, S.U.S. and N.R.K.; validation, H.M.B., M.M.S. and N.R.K.; visualization, S.U.S. and N.R.K.; writing—original draft, H.M.B., M.A. and N.R.K.; writing—review and editing, H.M.B., S.U.S. and N.R.K. All authors have read and agreed to the published version of the manuscript.

Funding: This research was funded by Higher Education Commission of Pakistan, Grant Number 7493.

Institutional Review Board Statement: The study was conducted according to the international guidelines (OECD Environment, Health and Safety) and approved by the institutional ethical review board of Gomal University Pakistan (vide letter no. 503/QEC/GU, dated. 29 March 2019).

Informed Consent Statement: Not applicable.

Acknowledgments: The authors acknowledge the Faculty of Pharmacy Gomal University for facilities support and Higher Education Commission of Pakistan for financial support.

Conflicts of Interest: The authors declare no conflict of interest.

References

1. Jeschke, M.G.; van Baar, M.E.; Choudhry, M.A.; Chung, K.K.; Gibran, N.S.; Logsetty, S. Burn injury. *Nat. Rev. Dis. Primers* **2020**, *6*, 1–25. [CrossRef]
2. World Health Organization. *BASIC EMERGENCY CARE: Approach to the Acutely Ill and Injured*; World Health Organization: Geneva, Switzerland, 2018.
3. Boyce, S.T.; Lalley, A.L. Tissue engineering of skin and regenerative medicine for wound care. *Burn. Trauma* **2018**, 6. [CrossRef]
4. Chen, F.-M.; Liu, X. Advancing biomaterials of human origin for tissue engineering. *Prog. Polym. Sci.* **2016**, *53*, 86–168. [CrossRef]
5. Tottoli, E.M.; Dorati, R.; Genta, I.; Chiesa, E.; Pisani, S.; Conti, B. Skin wound healing process and new emerging technologies for skin wound care and regeneration. *Pharmaceutics* **2020**, *12*, 735. [CrossRef]
6. Ahadian, S.; Civitarese, R.; Bannerman, D.; Mohammadi, M.H.; Lu, R.; Wang, E.; Davenport-Huyer, L.; Lai, B.; Zhang, B.; Zhao, Y. Organ-on-a-chip platforms: A convergence of advanced materials, cells, and microscale technologies. *Adv. Healthc. Mater.* **2018**, *7*, 1700506. [CrossRef] [PubMed]
7. Castano, O.; Pérez-Amodio, S.; Navarro-Requena, C.; Mateos-Timoneda, M.Á.; Engel, E. Instructive microenvironments in skin wound healing: Biomaterials as signal releasing platforms. *Adv. Drug Deliv. Rev.* **2018**, *129*, 95–117. [CrossRef]
8. Tabriz, A.G.; Douroumis, D.; Boateng, J. 3D printed scaffolds for wound healing and tissue regeneration. *Ther. Dress. Wound Health Appl.* **2020**, *1*, 385–398.
9. Kulig, D.; Zimoch-Korzycka, A.; Jarmoluk, A.; Marycz, K. Study on alginate–chitosan complex formed with different polymers ratio. *Polymers* **2016**, *8*, 167. [CrossRef] [PubMed]
10. Valle, K.Z.M.; Saucedo Acuña, R.A.; Ríos Arana, J.V.; Lobo, N.; Rodriguez, C.; Cuevas-Gonzalez, J.C.; Tovar-Carrillo, K.L. Natural Film Based on Pectin and Allantoin for Wound Healing: Obtaining, Characterization, and Rat Model. *BioMed Res. Int.* **2020**, *2020*, 1–7. [CrossRef]
11. Torres, F.G.; Commeaux, S.; Troncoso, O.P. Starch-based biomaterials for wound-dressing applications. *Starch-Stärke* **2013**, *65*, 543–551. [CrossRef]
12. Escárcega-Galaz, A.A.; Sánchez-Machado, D.I.; López-Cervantes, J.; Sanches-Silva, A.; Madera-Santana, T.J.; Paseiro-Losada, P. Mechanical, structural and physical aspects of chitosan-based films as antimicrobial dressings. *Int. J. Biol. Macromol.* **2018**, *116*, 472–481. [CrossRef] [PubMed]
13. Ali, M.; Khan, N.R.; Basit, H.M.; Mahmood, S. Physico-chemical based mechanistic insight into surfactant modulated sodium Carboxymethylcellulose film for skin tissue regeneration applications. *J. Polym. Res.* **2020**, *27*, 1–11. [CrossRef]
14. Maver, T.; Maver, U.; Mostegel, F.; Griesser, T.; Spirk, S.; Smrke, D.M.; Stana-Kleinschek, K. Cellulose based thin films as a platform for drug release studies to mimick wound dressing materials. *Cellulose* **2015**, *22*, 749–761. [CrossRef]
15. Hubner, P.; Donati, N.; de Menezes Quines, L.K.; Tessaro, I.C.; Marcilio, N.R. Gelatin-based films containing clinoptilolite-Ag for application as wound dressing. *Mater. Sci. Eng. C* **2020**, *107*, 110215. [CrossRef] [PubMed]
16. Michalska-Sionkowska, M.; Walczak, M.; Sionkowska, A. Antimicrobial activity of collagen material with thymol addition for potential application as wound dressing. *Polym. Test.* **2017**, *63*, 360–366. [CrossRef]
17. Akkaya, N.E.; Ergun, C.; Saygun, A.; Yesilcubuk, N.; Akel-Sadoglu, N.; Kavakli, I.H.; Turkmen, H.S.; Catalgil-Giz, H. New biocompatible antibacterial wound dressing candidates; agar-locust bean gum and agar-salep films. *Int. J. Biol. Macromol.* **2020**, *155*, 430–438. [CrossRef]
18. Zhang, H.; Cheng, J.; Ao, Q. Preparation of Alginate-Based Biomaterials and Their Applications in Biomedicine. *Mar. Drugs* **2021**, *19*, 264. [CrossRef]

19. Soni, M.L.; Kumar, M.; Namdeo, K. Sodium alginate microspheres for extending drug release: Formulation and in vitro evaluation. *Int. J. Drug Deliv.* **2010**, *2*, 64–68. [CrossRef]
20. Sachan, N.K.; Pushkar, S.; Jha, A.; Bhattcharya, A. Sodium alginate: The wonder polymer for controlled drug delivery. *J. Pharm. Res.* **2009**, *2*, 1191–1199.
21. Luna, S.; Gomes, M.E.; Mano, J.; Reis, R. Development of a novel cell encapsulation system based on natural origin polymers for tissue engineering applications. *J. Bioact. Compat. Polym.* **2010**, *25*, 341–359. [CrossRef]
22. Dettmar, P.W.; Strugala, V.; Richardson, J.C. The key role alginates play in health. *Food Hydrocoll.* **2011**, *25*, 263–266. [CrossRef]
23. Shalaka, D.; Naik, S.; Amruta, A.; Parimal, K. Vitamin E loaded pectin alginate microspheres for cosmetic application. *J. Pharm. Res.* **2009**, *2*, 1098–1102.
24. Xia, Y.; Mei, F.; Duan, Y.; Gao, Y.; Xiong, Z.; Zhang, T.; Zhang, H. Bone tissue engineering using bone marrow stromal cells and an injectable sodium alginate/gelatin scaffold. *J. Biomed. Mater. Res. Part A* **2012**, *100*, 1044–1050. [CrossRef]
25. Devi, M.P.; Sekar, M.; Chamundeswari, M.; Moorthy, A.; Krithiga, G.; Murugan, N.S.; Sastry, T. A novel wound dressing material—fibrin–chitosan–sodium alginate composite sheet. *Bull. Mater. Sci.* **2012**, *35*, 1157–1163. [CrossRef]
26. Mohandas, A.; Sudheesh Kumar, P.; Raja, B.; Lakshmanan, V.-K.; Jayakumar, R. Exploration of alginate hydrogel/nano zinc oxide composite bandages for infected wounds. *Int. J. Nanomed.* **2015**, *10* (Suppl. S1), 53–66. [CrossRef] [PubMed]
27. Badgujar, B.; Mundada, A. The technologies used for developing orally disintegrating tablets: A review. *Acta Pharm.* **2011**, *61*, 117–139. [CrossRef]
28. Hamrun, N.; Rachman, S.A. Measuring sodium alginate content of brown algae species padina sp. as the basic matter for making dental impression material (irreversible hydrocolloid impression material). *J. Dentomaxillofac. Sci.* **2016**, *1*, 129–133. [CrossRef]
29. Wang, Y.; Wang, X.; Shi, J.; Zhu, R.; Zhang, J.; Zhang, Z.; Ma, D.; Hou, Y.; Lin, F.; Yang, J. A biomimetic silk fibroin/sodium alginate composite scaffold for soft tissue engineering. *Sci. Rep.* **2016**, *6*, 1–13. [CrossRef] [PubMed]
30. Rezvani Ghomi, E.; Khalili, S.; Nouri Khorasani, S.; Esmaeely Neisiany, R.; Ramakrishna, S. Wound dressings: Current advances and future directions. *J. Appl. Polym. Sci.* **2019**, *136*, 47738. [CrossRef]
31. Munarin, F.; Tanzi, M.C.; Petrini, P. Advances in biomedical applications of pectin gels. *Int. J. Biol. Macromol.* **2012**, *51*, 681–689. [CrossRef] [PubMed]
32. Shahzad, A.; Khan, A.; Afzal, Z.; Umer, M.F.; Khan, J.; Khan, G.M. Formulation development and characterization of cefazolin nanoparticles-loaded cross-linked films of sodium alginate and pectin as wound dressings. *Int. J. Biol. Macromol.* **2019**, *124*, 255–269. [CrossRef]
33. Rezvanian, M.; Amin, M.C.I.M.; Ng, S.-F. Development and physicochemical characterization of alginate composite film loaded with simvastatin as a potential wound dressing. *Carbohydr. Polym.* **2016**, *137*, 295–304. [CrossRef]
34. Gohil, R.M. Synergistic blends of natural polymers, pectin and sodium alginate. *J. Appl. Polym. Sci.* **2011**, *120*, 2324–2336. [CrossRef]
35. Sun, X.; Cameron, R.G.; Bai, J. Effect of spray-drying temperature on physicochemical, antioxidant and antimicrobial properties of pectin/sodium alginate microencapsulated carvacrol. *Food Hydrocoll.* **2020**, *100*, 105420. [CrossRef]
36. Sonker, A.K.; Verma, V. Influence of crosslinking methods toward poly (vinyl alcohol) properties: Microwave irradiation and conventional heating. *J. Appl. Polym. Sci.* **2018**, *135*, 46125. [CrossRef]
37. Hofmann, M.; Fischer, G.; Weigel, R.; Kissinger, D. Microwave-based noninvasive concentration measurements for biomedical applications. *IEEE Trans. Microw. Theory Tech.* **2013**, *61*, 2195–2204. [CrossRef]
38. Chih-Ho Hong, H.; Lupin, M.; O'Shaughnessy, K.F. Clinical evaluation of a microwave device for treating axillary hyperhidrosis. *Dermatol. Surg.* **2012**, *38*, 728–735. [CrossRef] [PubMed]
39. Eskandari, S.; Azimzadeh, A.; Bahar, M.; Naraghi, Z.S.; Javadi, A.; Khamesipour, A.; Mohamadi, A.M. Efficacy of microwave and infrared radiation in the treatment of the skin lesions caused by leishmania major in an animal model. *Iran. J. Public Health* **2012**, *41*, 80. [PubMed]
40. Vogl, T.J.; Naguib, N.N.; Lehnert, T.; Nour-Eldin, N.-E.A. Radiofrequency, microwave and laser ablation of pulmonary neoplasms: Clinical studies and technical considerations. *Eur. J. Radiol.* **2011**, *77*, 346–357. [CrossRef] [PubMed]
41. Hetta, O.M.; Shebrya, N.H.; Amin, S.K. Ultrasound-guided microwave ablation of hepatocellular carcinoma: Initial institutional experience. *Egypt. J. Radiol. Nucl. Med.* **2011**, *42*, 343–349. [CrossRef]
42. Castle, S.M.; Salas, N.; Leveillee, R.J. Initial experience using microwave ablation therapy for renal tumor treatment: 18-month follow-up. *Urology* **2011**, *77*, 792–797. [CrossRef] [PubMed]
43. Yu, M.-A.; Liang, P.; Yu, X.-L.; Cheng, Z.-G.; Han, Z.-Y.; Liu, F.-Y.; Yu, J. Sonography-guided percutaneous microwave ablation of intrahepatic primary cholangiocarcinoma. *Eur. J. Radiol.* **2011**, *80*, 548–552. [CrossRef]
44. Wong, T. Use of microwave in processing of drug delivery systems. *Curr. Drug Deliv.* **2008**, *5*, 77–84. [CrossRef] [PubMed]
45. Goksu, E.I.; Sumnu, G.; Esin, A. Effect of microwave on fluidized bed drying of macaroni beads. *J. Food Eng.* **2005**, *66*, 463–468. [CrossRef]
46. Romano, V.R.; Marra, F.; Tammaro, U. Modelling of microwave heating of foodstuff: Study on the influence of sample dimensions with a FEM approach. *J. Food Eng.* **2005**, *71*, 233–241. [CrossRef]
47. Verboven, P.; Datta, A.K.; Anh, N.T.; Scheerlinck, N.; Nicolaï, B.M. Computation of airflow effects on heat and mass transfer in a microwave oven. *J. Food Eng.* **2003**, *59*, 181–190. [CrossRef]

48. Meyer, D.E.; Shin, B.; Kong, G.; Dewhirst, M.; Chilkoti, A. Drug targeting using thermally responsive polymers and local hyperthermia. *J. Control. Release* **2001**, *74*, 213–224. [CrossRef]
49. Wong, T.; Wahab, S.; Anthony, Y. Drug release responses of zinc ion crosslinked poly (methyl vinyl ether-co-maleic acid) matrix towards microwave. *Int. J. Pharm.* **2008**, *357*, 154–163. [CrossRef] [PubMed]
50. Anuar, N.K.; Wong, T.W.; Taib, M.N. Microwave modified non-crosslinked pectin films with modulated drug release. *Pharm. Dev. Technol.* **2012**, *17*, 110–117. [CrossRef]
51. Vandelli, M.A.; Romagnoli, M.; Monti, A.; Gozzi, M.; Guerra, P.; Rivasi, F.; Forni, F. Microwave-treated gelatin microspheres as drug delivery system. *J. Control. Release* **2004**, *96*, 67–84. [CrossRef]
52. Shaheen, M.S.; El-Massry, K.F.; El-Ghorab, A.H.; Anjum, F.M. Microwave applications in thermal food processing. *Dev. Appl. Microw. Heat.* **2012**, 3–16. [CrossRef]
53. Basit, H.M.; Mohd Amin, M.C.I.; Ng, S.-F.; Katas, H.; Shah, S.U.; Khan, N.R. Formulation and Evaluation of Microwave-Modified Chitosan-Curcumin Nanoparticles—A Promising Nanomaterials Platform for Skin Tissue Regeneration Applications Following Burn Wounds. *Polymers* **2020**, *12*, 2608. [CrossRef] [PubMed]
54. Ghanbarzadeh, B.; Almasi, H. Physical properties of edible emulsified films based on carboxymethyl cellulose and oleic acid. *Int. J. Biol. Macromol.* **2011**, *48*, 44–49. [CrossRef] [PubMed]
55. Ramli, N.A.; Wong, T.W. Sodium carboxymethylcellulose scaffolds and their physicochemical effects on partial thickness wound healing. *Int. J. Pharm.* **2011**, *403*, 73–82. [CrossRef]
56. Kim, H.S.; Sun, X.; Lee, J.-H.; Kim, H.-W.; Fu, X.; Leong, K.W. Advanced drug delivery systems and artificial skin grafts for skin wound healing. *Adv. Drug Deliv. Rev.* **2019**, *146*, 209–239. [CrossRef]
57. Junker, J.P.; Kamel, R.A.; Caterson, E.; Eriksson, E. Clinical impact upon wound healing and inflammation in moist, wet, and dry environments. *Adv. Wound Care* **2013**, *2*, 348–356. [CrossRef]
58. Junker, J.P.; Caterson, E.; Eriksson, E. The microenvironment of wound healing. *J. Craniofac. Surg.* **2013**, *24*, 12–16. [CrossRef]
59. Altiok, D.; Altiok, E.; Tihminlioglu, F. Physical, antibacterial and antioxidant properties of chitosan films incorporated with thyme oil for potential wound healing applications. *J. Mater. Sci. Mater. Med.* **2010**, *21*, 2227–2236. [CrossRef]
60. Peles, Z.; Zilberman, M. Novel soy protein wound dressings with controlled antibiotic release: Mechanical and physical properties. *Acta Biomater.* **2012**, *8*, 209–217. [CrossRef]
61. Belbachir, K.; Noreen, R.; Gouspillou, G.; Petibois, C. Collagen types analysis and differentiation by FTIR spectroscopy. *Anal. Bioanal. Chem.* **2009**, *395*, 829–837. [CrossRef]
62. Wang, Z.; Zhao, Z.; Khan, N.R.; Hua, Z.; Huo, J.; Li, Y. Microwave assisted chitosan-polyethylene glycol hydrogel membrane synthesis of curcumin for open incision wound healing. *Die Pharm.-An Int. J. Pharm. Sci.* **2020**, *75*, 118–123.
63. Gonçalves, V.S.; Gurikov, P.; Poejo, J.; Matias, A.A.; Heinrich, S.; Duarte, C.M.; Smirnova, I. Alginate-based hybrid aerogel microparticles for mucosal drug delivery. *Eur. J. Pharm. Biopharm.* **2016**, *107*, 160–170. [CrossRef]
64. Monroe, J.; Barry, M.; DeStefano, A.; Aydogan Gokturk, P.; Jiao, S.; Robinson-Brown, D.; Webber, T.; Crumlin, E.J.; Han, S.; Shell, M.S. Water Structure and Properties at Hydrophilic and Hydrophobic Surfaces. *Annu. Rev. Chem. Biomol. Eng.* **2020**, *11*, 523–557. [CrossRef] [PubMed]
65. Pinto, G.D.; Rodrigues, J.; Bernardo, L. A Review on Thermoplastic or Thermosetting Polymeric Matrices Used in Polymeric Composites Manufactured with Banana Fibers from the Pseudostem. *Appl. Sci.* **2020**, *10*, 3023. [CrossRef]
66. Namviriyachote, N.; Lipipun, V.; Akkhawattanangkul, Y.; Charoonrut, P.; Ritthidej, G.C. Development of polyurethane foam dressing containing silver and asiaticoside for healing of dermal wound. *Asian J. Pharm. Sci.* **2019**, *14*, 63–77. [CrossRef] [PubMed]
67. Levina, E.M.; Kharitonova, M.A.; Rovensky, Y.A.; Vasiliev, J.M. Cytoskeletal control of fibroblast length: Experiments with linear strips of substrate. *J. Cell Sci.* **2001**, *114*, 4335–4341. [CrossRef]
68. Hiro, M.E.; Pierpont, Y.N.; Ko, F.; Wright, T.E.; Robson, M.C.; Payne, W.G. Comparative evaluation of silver-containing antimicrobial dressings on in vitro and in vivo processes of wound healing. *Eplasty* **2012**, *12*, 409–419.
69. Habash, M.; Reid, G. Microbial biofilms: Their development and significance for medical device—Related infections. *J. Clin. Pharmacol.* **1999**, *39*, 887–898. [CrossRef] [PubMed]
70. Wong, T.W.; Ramli, N.A. Carboxymethylcellulose film for bacterial wound infection control and healing. *Carbohydr. Polym.* **2014**, *112*, 367–375. [CrossRef] [PubMed]
71. Barbosa, G.P.; Debone, H.S.; Severino, P.; Souto, E.B.; da Silva, C.F. Design and characterization of chitosan/zeolite composite films—Effect of zeolite type and zeolite dose on the film properties. *Mater. Sci. Eng. C* **2016**, *60*, 246–254. [CrossRef]
72. Bishop, A. Role of oxygen in wound healing. *J. Wound Care* **2008**, *17*, 399–402. [CrossRef]
73. Ichioka, S.; Ando, T.; Shibata, M.; Sekiya, N.; Nakatsuka, T. Oxygen consumption of keloids and hypertrophic scars. *Ann. Plast. Surg.* **2008**, *60*, 194–197. [CrossRef] [PubMed]
74. Guo, S.A.; DiPietro, L.A. Factors affecting wound healing. *J. Dent. Res.* **2010**, *89*, 219–229. [CrossRef] [PubMed]
75. Cabrera, J.C.; Boland, A.; Messiaen, J.; Cambier, P.; Van Cutsem, P. Egg box conformation of oligogalacturonides: The time-dependent stabilization of the elicitor-active conformation increases its biological activity. *Glycobiology* **2008**, *18*, 473–482. [CrossRef]
76. Mashingaidze, F. Design of an Intravaginal Composite Polymeric System for the Reduction and Prevention of STI and HIV Transmission. Ph.D. Thesis, Department of Pharmacy and Pharmacology, University of the Witwatersrand, Johannesburg, South Africa, 2014.

77. Farzanian, K.; Ghahremaninezhad, A. On the effect of chemical composition on the desorption of superabsorbent hydrogels in contact with a porous cementitious material. *Gels* **2018**, *4*, 70. [CrossRef]
78. Rezvanian, M.; Ahmad, N.; Amin, M.C.I.M.; Ng, S.-F. Optimization, characterization, and in vitro assessment of alginate-pectin ionic cross-linked hydrogel film for wound dressing applications. *Int. J. Biol. Macromol.* **2017**, *97*, 131–140. [CrossRef] [PubMed]
79. Mishra, R.; Banthia, A.; Majeed, A. Pectin based formulations for biomedical applications: A review. *Asian J. Pharm. Clin. Res.* **2012**, *5*, 1–7.
80. Croisier, F.; Jérôme, C. Chitosan-based biomaterials for tissue engineering. *Eur. Polym. J.* **2013**, *49*, 780–792. [CrossRef]
81. Zaman, H.U.; Islam, J.; Khan, M.A.; Khan, R.A. Physico-mechanical properties of wound dressing material and its biomedical application. *J. Mech. Behav. Biomed. Mater.* **2011**, *4*, 1369–1375. [CrossRef]
82. Zahouani, H.; Pailler-Mattei, C.; Sohm, B.; Vargiolu, R.; Cenizo, V.; Debret, R. Characterization of the mechanical properties of a dermal equivalent compared with human skin in vivo by indentation and static friction tests. *Ski. Res. Technol.* **2009**, *15*, 68–76. [CrossRef] [PubMed]
83. Lancerotto, L.; Orgill, D.P. Mechanoregulation of angiogenesis in wound healing. *Adv. Wound Care* **2014**, *3*, 626–634. [CrossRef]
84. Wang, Y.; Zhang, X.; Qiu, D.; Li, Y.; Yao, L.; Duan, J. Ultrasonic assisted microwave synthesis of poly (Chitosan-co-gelatin)/polyvinyl pyrrolidone IPN hydrogel. *Ultrason. Sonochem.* **2018**, *40*, 714–719. [CrossRef]
85. Ravishankar, K.; Dhamodharan, R. Advances in chitosan-based hydrogels: Evolution from covalently crosslinked systems to ionotropically crosslinked superabsorbents. *React. Funct. Polym.* **2020**, *149*, 104517. [CrossRef]
86. Miao, C.; Hamad, W.Y. Cellulose reinforced polymer composites and nanocomposites: A critical review. *Cellulose* **2013**, *20*, 2221–2262. [CrossRef]
87. Kamoun, E.A.; Chen, X.; Eldin, M.S.M.; Kenawy, E.-R.S. Crosslinked poly (vinyl alcohol) hydrogels for wound dressing applications: A review of remarkably blended polymers. *Arab. J. Chem.* **2015**, *8*, 1–14. [CrossRef]
88. Rodríguez, M.; Osés, J.; Ziani, K.; Mate, J.I. Combined effect of plasticizers and surfactants on the physical properties of starch based edible films. *Food Res. Int.* **2006**, *39*, 840–846. [CrossRef]
89. Ballesteros, L.F.; Cerqueira, M.A.; Teixeira, J.A.; Mussatto, S.I. Production and physicochemical properties of carboxymethyl cellulose films enriched with spent coffee grounds polysaccharides. *Int. J. Biol. Macromol.* **2018**, *106*, 647–655. [CrossRef] [PubMed]
90. Sin, L.T.; Bee, S.T.; Wah, T.Y.; Chee, T.M.; Kadhum, A.A.H.; Rahmat, A.R. Microwave effects on montmorillonite reinforced polyvinyl alcohol-starch nanocomposit. *J. Vinyl Addit. Technol.* **2016**, *23*, E142–E151. [CrossRef]
91. Swamy, T.M.; Ramaraj, B.; Lee, J.H. Sodium alginate and its blends with starch: Thermal and morphological properties. *J. Appl. Polym. Sci.* **2008**, *109*, 4075–4081.
92. Sundaram, J.; Durance, T.D.; Wang, R. Porous scaffold of gelatin–starch with nanohydroxyapatite composite processed via novel microwave vacuum drying. *Acta Biomater.* **2008**, *4*, 932–942. [CrossRef]
93. Wang, Y.; Ying, T.; Li, J.; Xu, Y.; Wang, R.; Ke, Q.; Shen, S.G.; Xu, H.; Lin, K. Hierarchical micro/nanofibrous scaffolds incorporated with curcumin and zinc ion eutectic metal organic frameworks for enhanced diabetic wound healing via anti-oxidant and anti-inflammatory activities. *Chem. Eng. J.* **2020**, *402*, 126273. [CrossRef]
94. Wu, N.-C.; Wang, J.-J. Curcumin Attenuates Liver Warm Ischemia and Reperfusion–Induced Combined Restrictive and Obstructive Lung Disease by Reducing Matrix Metalloprotease 9 Activity. In *Transplantation Proceedings*; Elsevier: Amsterdam, The Netherlands, 2014; pp. 1135–1138.
95. Song, Z.; Revelo, X.; Shao, W.; Tian, L.; Zeng, K.; Lei, H.; Sun, H.S.; Woo, M.; Winer, D.; Jin, T. Dietary curcumin intervention targets mouse white adipose tissue inflammation and brown adipose tissue UCP1 expression. *Obesity* **2018**, *26*, 547–558. [CrossRef]
96. Merrell, J.G.; McLaughlin, S.W.; Tie, L.; Laurencin, C.T.; Chen, A.F.; Nair, L.S. Curcumin loaded poly (ε-caprolactone) nanofibers: Diabetic wound dressing with antioxidant and anti-inflammatory properties. *Clin. Exp. Pharmacol. Physiol.* **2009**, *36*, 1149. [CrossRef]
97. Dai, X.; Liu, J.; Zheng, H.; Wichmann, J.; Hopfner, U.; Sudhop, S.; Prein, C.; Shen, Y.; Machens, H.-G.; Schilling, A.F. Nano-formulated curcumin accelerates acute wound healing through Dkk-1-mediated fibroblast mobilization and MCP-1-mediated anti-inflammation. *NPG Asia Mater.* **2017**, *9*, e368. [CrossRef]
98. Panchatcharam, M.; Miriyala, S.; Gayathri, V.S.; Suguna, L. Curcumin improves wound healing by modulating collagen and decreasing reactive oxygen species. *Mol. Cell. Biochem.* **2006**, *290*, 87–96. [CrossRef] [PubMed]
99. Tapia, E.; Sánchez-Lozada, L.; García-Niño, W.; García, E.; Cerecedo, A.; García-Arroyo, F.; Osorio, H.; Arellano, A.; Cristóbal-García, M.; Loredo, M. Curcumin prevents maleate-induced nephrotoxicity: Relation to hemodynamic alterations, oxidative stress, mitochondrial oxygen consumption and activity of respiratory complex I. *Free. Radic. Res.* **2014**, *48*, 1342–1354. [CrossRef]
100. Laskar, K.; Faisal, S.M.; Rauf, A.; Ahmed, A.; Owais, M. Undec-10-enoic acid functionalized chitosan based novel nano-conjugate: An enhanced anti-bacterial/biofilm and anti-cancer potential. *Carbohydr. Polym.* **2017**, *166*, 14–23. [CrossRef]
101. Adnan, S.; Ranjha, N.M.; Hanif, M.; Asghar, S. O-Carboxymethylated chitosan; A promising tool with in-vivo anti-inflammatory and analgesic properties in albino rats. *Int. J. Biol. Macromol.* **2020**, *156*, 531–536. [CrossRef] [PubMed]
102. Hu, Z.; Lu, S.; Cheng, Y.; Kong, S.; Li, S.; Li, C.; Yang, L. Investigation of the effects of molecular parameters on the hemostatic properties of chitosan. *Molecules* **2018**, *23*, 3147. [CrossRef]
103. Xing, K.; Zhu, X.; Peng, X.; Qin, S. Chitosan antimicrobial and eliciting properties for pest control in agriculture: A review. *Agron. Sustain. Dev.* **2015**, *35*, 569–588. [CrossRef]

104. Wang, Q.Z.; Chen, X.G.; Liu, N.; Wang, S.X.; Liu, C.S.; Meng, X.H.; Liu, C.G. Protonation constants of chitosan with different molecular weight and degree of deacetylation. *Carbohydr. Polym.* **2006**, *65*, 194–201. [CrossRef]
105. Howling, G.I.; Dettmar, P.W.; Goddard, P.A.; Hampson, F.C.; Dornish, M.; Wood, E.J. The effect of chitin and chitosan on the proliferation of human skin fibroblasts and keratinocytes in vitro. *Biomaterials* **2001**, *22*, 2959–2966. [CrossRef]
106. Movasaghi, Z.; Rehman, S.; ur Rehman, D.I. Fourier transform infrared (FTIR) spectroscopy of biological tissues. *Appl. Spectrosc. Rev.* **2008**, *43*, 134–179. [CrossRef]
107. Palombo, F.; Cremers, S.G.; Weinberg, P.D.; Kazarian, S.G. Application of Fourier transform infrared spectroscopic imaging to the study of effects of age and dietary L-arginine on aortic lesion composition in cholesterol-fed rabbits. *J. R. Soc. Interface* **2009**, *6*, 669–680. [CrossRef] [PubMed]
108. Cheheltani, R.; McGoverin, C.M.; Rao, J.; Vorp, D.A.; Kiani, M.F.; Pleshko, N. Fourier transform infrared spectroscopy to quantify collagen and elastin in an in vitro model of extracellular matrix degradation in aorta. *Analyst* **2014**, *139*, 3039–3047. [CrossRef] [PubMed]
109. Minagawa, T.; Okamura, Y.; Shigemasa, Y.; Minami, S.; Okamoto, Y. Effects of molecular weight and deacetylation degree of chitin/chitosan on wound healing. *Carbohydr. Polym.* **2007**, *67*, 640–644. [CrossRef]
110. Sahariah, P.; Masson, M. Antimicrobial chitosan and chitosan derivatives: A review of the structure–activity relationship. *Biomacromolecules* **2017**, *18*, 3846–3868. [CrossRef]
111. Divya, K.; Smitha, V.; Jisha, M. Antifungal, antioxidant and cytotoxic activities of chitosan nanoparticles and its use as an edible coating on vegetables. *Int. J. Biol. Macromol.* **2018**, *114*, 572–577. [CrossRef]
112. Fong, D.; Hoemann, C.D. Chitosan Immunomodulatory Properties: Perspectives on the Impact of Structural Properties and Dosage. *Future Sci.* **2017**. [CrossRef]
113. Bessa, L.J.; Fazii, P.; Di Giulio, M.; Cellini, L. Bacterial isolates from infected wounds and their antibiotic susceptibility pattern: Some remarks about wound infection. *Int. Wound J.* **2015**, *12*, 47–52. [CrossRef]
114. Younes, I.; Rinaudo, M. Chitin and chitosan preparation from marine sources. Structure, properties and applications. *Mar. Drugs* **2015**, *13*, 1133–1174. [CrossRef] [PubMed]
115. Samar, M.M.; El-Kalyoubi, M.; Khalaf, M.; Abd El-Razik, M. Physicochemical, functional, antioxidant and antibacterial properties of chitosan extracted from shrimp wastes by microwave technique. *Ann. Agric. Sci.* **2013**, *58*, 33–41. [CrossRef]
116. Ueno, H.; Mori, T.; Fujinaga, T. Topical formulations and wound healing applications of chitosan. *Adv. Drug Deliv. Rev.* **2001**, *52*, 105–115. [CrossRef]
117. Friedman, A.J.; Phan, J.; Schairer, D.O.; Champer, J.; Qin, M.; Pirouz, A.; Blecher-Paz, K.; Oren, A.; Liu, P.T.; Modlin, R.L. Antimicrobial and anti-inflammatory activity of chitosan–alginate nanoparticles: A targeted therapy for cutaneous pathogens. *J. Investig. Dermatol.* **2013**, *133*, 1231–1239. [CrossRef] [PubMed]
118. Hussain, Z.; Thu, H.E.; Amjad, M.W.; Hussain, F.; Ahmed, T.A.; Khan, S. Exploring recent developments to improve antioxidant, anti-inflammatory and antimicrobial efficacy of curcumin: A review of new trends and future perspectives. *Mater. Sci. Eng. C* **2017**, *77*, 1316–1326. [CrossRef] [PubMed]
119. Siddiqui, A.M.; Cui, X.; Wu, R.; Dong, W.; Zhou, M.; Hu, M.; Simms, H.H.; Wang, P. The anti-inflammatory effect of curcumin in an experimental model of sepsis is mediated by up-regulation of peroxisome proliferator-activated receptor-γ. *Crit. Care Med.* **2006**, *34*, 1874–1882. [CrossRef] [PubMed]
120. Jagetia, G.C.; Aggarwal, B.B. "Spicing up" of the immune system by curcumin. *J. Clin. Immunol.* **2007**, *27*, 19–35. [CrossRef] [PubMed]
121. Shi, C.; Zhu, Y.; Ran, X.; Wang, M.; Su, Y.; Cheng, T. Therapeutic potential of chitosan and its derivatives in regenerative medicine. *J. Surg. Res.* **2006**, *133*, 185–192. [CrossRef]
122. Mehrabani, D.; Farjam, M.; Geramizadeh, B.; Tanideh, N.; Amini, M.; Panjehshahin, M.R. The healing effect of curcumin on burn wounds in rat. *World J. Plast. Surg.* **2015**, *4*, 29. [PubMed]
123. Thangapazham, R.L.; Sharma, A.; Maheshwari, R.K. Beneficial role of curcumin in skin diseases. *Mol. Targets Ther. Uses Curcumin Health Dis.* **2007**, 343–357.

polymers

MDPI

Article

Study of Mechanical Properties of PHBHV/Miscanthus Green Composites Using Combined Experimental and Micromechanical Approaches

Thibault Lemaire [1,*], Erica Gea Rodi [1,2], Valérie Langlois [2], Estelle Renard [2] and Vittorio Sansalone [1]

[1] MSME, UMR 8208, Univ. Paris Est Creteil, Univ. Gustave Eiffel, CNRS, 94010 Creteil, France; egea.rodi@gmail.com (E.G.R.); vittorio.sansalone@u-pec.fr (V.S.)
[2] ICMPE, CNRS UMR 7182, Univ. Paris Est Creteil, 94320 Thiais, France; langlois@u-pec.fr (V.L.); renard@icmpe.cnrs.fr (E.R.)
* Correspondence: lemaire@u-pec.fr

Citation: Lemaire, T.; Rodi, E.G.; Langlois, V.; Renard, E.; Sansalone, V. Study of Mechanical Properties of PHBHV/Miscanthus Green Composites Using Combined Experimental and Micromechanical Approaches. *Polymers* **2021**, *13*, 2650. https://doi.org/10.3390/polym13162650

Academic Editors: Concepción Valencia-Barragán, Esperanza Cortés Triviño, Adrián Tenorio-Alfonso, Clara Delgado-Sánchez and Antonio M. Borrero-López

Received: 30 April 2021
Accepted: 5 August 2021
Published: 10 August 2021

Publisher's Note: MDPI stays neutral with regard to jurisdictional claims in published maps and institutional affiliations.

Abstract: In recent years the interest in the realization of green wood plastic composites (GWPC) materials has increased due to the necessity of reducing the proliferation of synthetic plastics. In this work, we study a specific class of GWPCs from its synthesis to the characterization of its mechanical properties. These properties are related to the underlying microstructure using both experimental and modeling approaches. Different contents of Miscanthus giganteus fibers, at 5, 10, 20, 30 weight percent's, were thus combined to a microbial matrix, namely poly (3-hydroxybutyrate)-co-poly(3-hydroxyvalerate) (PHBHV). The samples were manufactured by extrusion and injection molding processing. The obtained samples were then characterized by cyclic-tensile tests, pycnometer testing, differential scanning calorimetry, Fourier transform infrared spectroscopy, X-ray diffraction, and microscopy. The possible effect of the fabrication process on the fibers size is also checked. In parallel, the measured properties of the biocomposite were also estimated using a Mori–Tanaka approach to derive the effective behavior of the composite. As expected, the addition of reinforcement to the polymer matrix results in composites with higher Young moduli on the one hand, and lower failure strains and tensile strengths on the other hand (tensile modulus was increased by 100% and tensile strength decreased by 23% when reinforced with 30 wt % of Miscanthus fibers).

Keywords: biocomposite; bio-sourced products; mechanical properties; micro-mechanics; mechanical testing

1. Introduction

In recent years, the market for wood plastic composites (WPCs) has grown exponentially thanks to the possibility to assure durability to the manufactured products without the use of toxic chemical agents or harmful fibers [1]. The different applications for these materials, such as in building and construction field or automotive, make this market highly fragmented, but at the same time very attractive for many research groups and industries. Today, most of these composites are constituted by matrices derived from oil such as polyethylene (PE) or polypropylene (PP), in which a small amount of process and property modifiers are added though extrusion or injection processes to enhance outdoor applications [2]. Even though natural cellulosic fibers have been successfully used with petroleum-derived polymers, the environmental benefits of natural fiber composites can be enhanced considerably if biodegradable polymers are used [3]. These biocomposites can be easily disposed of or composted at the end of their life without harming the environment, which is not possible with synthetic fiber-based polymer composites. In fact, the addition of natural fillers increases the renewable content in the final product without changing the biodegradability of the continuous phase and also allows the use of agroforestry and fruit wastes, marking an important transition of the economy from a linear to circular model [4,5].

Poly(3-hydroxyalkanoates) (PHAs) are a class of natural biodegradable polyesters accumulated by many bacteria as carbon and energy supply when an essential nutrient is limited [6,7]. Using various substrates, a wide variety of PHAs can be synthesized, differing notably in the length of their side chains [6,8]. Two types of PHAs can be distinguished: (i) short-chain-length PHAs, or scl-PHAs, possessing alkyl side chains with up to two carbon atoms, as the widely used poly(3-hydroxybutyrate-co-3-hydroxyvalerate) PHBHV, for example, is considered hereafter; (ii) medium-chain-length PHAs, or mcl-PHAs, with at least three carbon atoms in their side chains. PHAs have been suggested as green substitutes for conventional plastics, due to their synthesis from renewable resources and their biodegradation by enzymatic action [9]. Owing to their biocompatibility and biodegradability, PHAs proved to be good candidates for biomedical applications, including in the design of devices, biodegradable drug carriers, and tissue engineering (TE) scaffolds [10–13]. They have also been combined with natural cellulosic fibers such as hemp jute flax carnauba fibers, miscanthus, bamboo [14–17], pineapple fibers [18], recycled wood fiber [19], and cellulose nanowhisker [20,21] to prepare biocomposites.

Among all the existing PHAs, poly(3-hydroxybutyrate) (PHB) is certainly one of the most important. It possesses a melting point close to that of polypropylene, better oxygen barrier property, and similar mechanical properties [9]. However, its brittleness and narrow processing temperature window limit its application. To overcome the inferior properties of PHB, a variety of copolymers were synthesized by bioconversion, such as the poly(3-hydroxybutyrate-co-3-hydroxyvalerate) (PHBHV) with 12% of valerate units was chosen to realize composite materials. Among the different vegetable fibers that can be used, we chose Miscanthus (*Miscanthus giganteus*). This is a perennial crop, highly productive with a very efficient nitrogen-recycling system and a very interesting energy balance due to the absence of nitrogen [22] and recently used in composites [23–26]. A well-known phenomenon in these materials is the incompatibility between fibers and matrix caused by the hydrophobicity of the polymer and the hydrophilic nature of the fibers. This effect has serious consequences on the mechanical seal of the final material which is why compatibilizer agents, additives, or surface treatments that improve the cohesion at the matrix/fiber interface are often necessary [27–33]. Thus, we recently showed the possibility to improve the adhesion between poly(caprolactone) and vegetal fibers using photoactivated grafting of a bounding agent [34,35]. In this update paper, by working with unfunctionalized materials, we intend to present a complete study of PHBHV/Miscanthus composites from their synthesis to their testing and in silico characterization. More specifically, this paper is not an increment in the enhancement of the fiber compatibility with the matrix, but a necessary work providing a starting reference that combines an extensive experimental characterization of PHBHV/Miscanthus composites and its associated modelling tool.

Considering different content of Miscanthus fibers, the purpose of this study is to propose a rigorous method to produce these composites and then determine a reference protocol to depict their physical and mechanical properties. This characterization is performed using a twofold approach since, in addition to classical mechanical and physical testing, a micromechanical approach taking into account the geometry, volume fraction, and organization of the fibers in the matrix, is carried out to estimate the effective mechanical properties of the composite.

At this stage, the in silico methods are validated using non-functionalized media. This is necessary before introducing interfacial terms in the model to represent enhanced behaviors of functionalized composites. Then, in a further step, this modelling approach will be useful to derive a model-driven optimization of the composite for enhanced synthesis such as the one proposed by Rodi et al. [17]. After having briefly presented the composites processing, the different experimental and numerical characterization tools are introduced in the materials and methods section. These characterizations consist in scanning electron microscopy, mechanical testing, fibers size comparison, pycnometer testing, differential scanning calorimetry, Fourier transform infrared spectroscopy, and X-ray diffraction. Then the results are presented and finally discussed.

2. Materials and Methods

2.1. Materials

Poly(3-hydroxybutyrate-co-3-hydroxyvalerate) ($PHB_{88}HV_{12}$), containing 12% of valerate, was purchased from Goodfellow, Lille, France, in a pelletized form. *Miscanthus giganteus* (MIS) fibers were provided by Miscanplus, Digny, France. MIS fibers came from a 2014 spring crop roughly chopped and subsequently milled until fiber length ranged between 1 and 5 mm.

2.2. Composite Processing

Thermogravimetric analysis revealed that the MIS has ambient moisture of 5%. For this reason, prior to processing, both PHBHV and Miscanthus were dried in a conventional oven at 80 °C for 5 h in order to remove any moisture. In order to prevent the absorption of moisture, products were stored in a desiccator containing K_2HPO_4 prior to processing. To investigate the effect of the fiber content on composite mechanical properties, the following nominal values were investigated: 0, 5, 10, 20, and 30 wt %. The nominal mass content corresponds to the mass content of MIS at the beginning of the composite processing. As shown later in this paper (see Section 3.4), the fiber mass content of the final product may be slightly lower. Notwithstanding this observation, the classical use of nominal contents is adopted hereafter to present the results.

According to the nominal mass content of each phase, PHBHV and MIS were mixed together in a lab-scale twin-screw extruder (Minilab Thermo Scientific Haake, Waltham, MA, USA). The experiments were performed at 160 °C (T_E) with a screw speed of 60 rpm (n). The retention time for the pure matrix was 1 min; this time was increased to 2 min in order to fully disperse the fibers into the matrix. After recirculation, the extruded molten material was transferred by means of a preheated piston-cylinder assembly and was shot in the micro-injection unit (MiniJet Thermo Scientific Haake, Waltham, MA, USA) at the injection pressure (P_I) for 30 s. A maintenance pressure (P_M), lower than that used during the phase of injection, was applied for another 30 s. The collector and the mold temperatures were set at 165 °C (T_I) and 45 °C (T_m), respectively. Parameters during the phase of injection of the material were adjusted according to the increase in the polymer melt viscosity with the fiber content. Some of the final optimized parameters used for the entire process are resumed in Table 1.

Table 1. Some of the extrusion and injection molding parameters. T_E extrusion temperature; n rotational speed; T_I injection temperature; T_m mold temperature.

T_E [°C]	n [tr.min^{-1}]	T_I [°C]	T_m [°C]
160	60	165	45

2.3. Materials Characterization

2.3.1. Scanning Electron Microscope (SEM)

SEM observations were performed using a Merlin Carl Zeiss, Marly le Roi, France, scanning electron microscope. Prior to observation, the cross sections of specimens at different fiber content were sputter-coated with a thin layer of palladium in a Cressington (Watford, UK) 208 HR sputter-coater. Images were recorded with an acceleration voltage of 10 keV and at different magnifications.

2.3.2. Mechanical Properties

The mechanical properties of the composites were evaluated using an Instron 5965 Universal Testing Machine, Norwood, MA, USA, equipped with a cell load of 100 N. All specimens presented standard dimensions according to ASTM638 (West Conshohocken, PA, USA). Two types of tests were made. On the one hand, a simple traction test was set up at a rate of 5 mm/min in order to evaluate the mechanical behavior of the composites, typically its failure strain and tensile strength. Five samples were tested for each fiber content value to obtain standard deviation values. These tests were performed 2 days after the day of realization of the biocomposites. On the other hand, a cyclic traction test was set up to evaluate the Young modulus. Ten specimens for each fiber content value were tested in this setup. Prior to testing, specimens were stored for 8 days at 23 °C. The cyclic traction test was set up with increasing values of the maximum load applied by the testing machine from one cycle to another. The initial maximum state stress was set to 3 N and the final one to 10 N with an increment of 1 N from one cycle to another (i.e., 8 cycles overall). The lower and upper limits were related to the sensitivity of the experimental device threshold and to the yield stress of the PHBHV, respectively. We, moreover, checked a posteriori that the cyclic loading curves remain in the elastic domain. All cycles were made at a constant speed of 0.05 N/s.

2.3.3. Fiber-Size Distribution

After processing, specimens with different fiber content were solubilized three times in dichloromethane during 30 min. After filtration, the collected Miscanthus fibers were observed using a 3B Scientific Physics, Bartenheim, France, microscope at a magnification of 4×. A series of 10 observations was made per specimen.

2.3.4. Density Measurements

Average density of pure matrix and composites were evaluated using a helium AccuPyc 1330 Micromeritics, Norcross, GA, USA, pycnometer on around 40 mg of mass taken from the central section of the specimens used for tensile tests. The density calculated with this method was compared with that calculated using the ratio between the mass and the volume of the specimens.

2.3.5. Thermal Analyses

Differential scanning calorimetry experiments were performed on a PerkinElmer Diamond DSC Apparatus, Haguenau, France. Sample of around 10 mg sealed in aluminum pans were initially heated from −60 °C to 200 °C at 20 °C/min, cooled down rapidly, and then reheated in the same conditions used in the first heating run. Melting point (T_M) and melting enthalpy (ΔH_M) were determined during the first heating. The degree of crystallization (Xc) was then calculated using the equation

$$Xc\ (\%) = 100 \times \Delta H_M/(\Delta H_0 \times W) \quad (1)$$

where ΔH_0 corresponds to the melting enthalpy of a 100% crystalline PHBHV (146 J/g) and W is the polymer fraction present in the composite.

2.3.6. Fourier Transform Infrared Spectroscopy (FTIR)

Different biocomposites were prepared, varying the content of fibers (5 and 20 wt %) and the length of fibers (1000 and 45 µm). The specimens of PHBHV/MIS composites were then solubilized in dichloromethane in order to separate fibers from matrix. The collected fibers were extracted 3 times in 100 mL of dichloromethane at 54 °C, stirring for 30 min at 200 rpm. Fibers were then dried before analysis. Infrared spectra of the extracted fibers were recorded using a TENSOR27 Brucker (Champs sur Marne, France) apparatus equipped with an attenuated internal reflection accessory using a diamond crystal (Digi Tech DLATGS Detector, 32 scans, 4 cm^{-1}) in the range 500–4000 cm^{-1}. These spectra were then compared with that of raw Miscanthus fibers to assess the natural grafting of PHBHV on the MIS through the ratio R calculated as follows:

$$R = \frac{I_{1726cm^{-1}}}{I_{1604cm^{-1}}} \quad (2)$$

where I_{1726} corresponds to the intensity of carbonyl group of PHBHV and I_{1604} corresponds to the intensity of the esters present in the lignin structure.

FTIR is also an useful analysis tool to evaluate the crystallinity of the PHBHV after processing. For instance, the band at 1726 cm^{-1} is representative of the C=O stretch present in the highly crystalline structure of the matrix, while the small shoulder at 1740 cm^{-1} represents the same stretch in the amorphous region. Typically, the absorption bands at 1720, 1276, 1225, and 980 cm^{-1} are representatives of the crystalline regions, whereas the bands at 1740, 1452, and 1176 cm^{-1} are representatives of the amorphous ones [36,37]. In particular, the band around 1378 cm^{-1} corresponds to the symmetrical wagging of the CH3 groups and that at 1452 cm^{-1} to the asymmetric deformation of methylene groups. These bands are considered as insensitive to crystallinity and they can be good candidates to evaluate the crystallinity degree. Hereafter, the band at 1452 cm^{-1} is thus considered in addition to the one at 1225 cm^{-1}; the latter corresponding to the C-O-C stretching which is representative of the crystalline regions. Thus, we calculate the crystallinity index C_I

$$CI = \frac{I_{1225cm^{-1}}}{I_{1452cm^{-1}}} \tag{3}$$

where I_{1225} is assigned to the C-O-C stretching mode of the crystalline parts, and I_{1452} corresponds to the asymmetric deformation of the methylene groups (insensitive to crystallinity). This index provides qualitative information about all changes that may occur in the crystalline structure of the matrix.

2.3.7. X-ray Diffraction

Structural characterizations of Miscanthus fibers and PHBHV/MIS composites were determined by X-ray diffraction (XRD) using a D8 advance Bruker (Champs sur Marne, France) diffractometer operating at 40 kV and 40 mA with a CuK α radiation. The whole area investigated was in the range $2\theta \approx 5$–$40°$ at a scanning rate of 0.2°/min.

2.4. *Modelling*

Among the several methods predicting the elastic properties of fiber-reinforced composites, the rule of mixtures (ROM) is probably the quickest and easiest one. Using the elastic moduli E_F and E_M of the fiber and matrix phases, and the volume fraction of the fibers ϕ_F, the effective Young modulus of the composite reads:

$$E_C = \phi_F E_F + (1 - \phi_F) E_M \tag{4}$$

Much more sophisticated models were also developed to evaluate the effective elastic behavior of the composite using homogenization approaches. Homogenization theories can estimate the effective elastic tensor $\mathbb{C}_{hom}$ of a multiphase material based on information about its microstructural organization. Among others, continuum micromechanics [38,39] proved to be quite useful when dealing with composite materials of matrix-inclusion type [40,41]. Continuum micromechanics use the solution of the matrix-inclusion problem provided by Eshelby in the fifties [42] to estimate the effective elastic tensor $\mathbb{C}_{hom}$ of a multiphase material as [43]:

$$\mathbb{C}_{hom} = \sum_r \phi_r \mathbb{C}_r : \mathbb{A}_r \tag{5}$$

where ϕ_r, $\mathbb{C}_r$, and $\mathbb{A}_r$ are the volume fraction, (4th-order) elastic tensor and (4th-order) localization tensor of phase r, and the sum runs over all the constituent phases. The localization tensor $\mathbb{A}_r$ accounts for the nature and geometrical organization of the phase r within the effective matrix and its expression depends, in general, on the volume fraction and elastic tensor of all the phases. Different estimates of $\mathbb{C}_{hom}$ can be obtained by suitable choices of the effective matrix. As long as one actual phase can be identified as a "matrix" phase, the relevant estimate of $\mathbb{C}_{hom}$ is provided by the Mori–Tanaka model. The idea behind this approach is sketched in Figure 1.

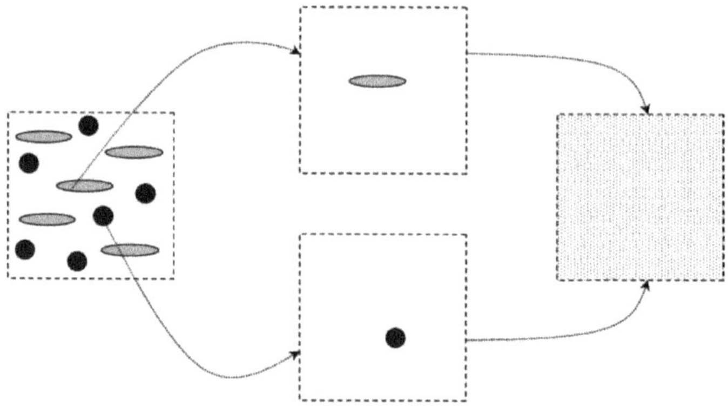

Figure 1. Schematic of the Mori–Tanaka model. On the left: actual, heterogeneous material; on the right: effective, homogeneous material. The intermediate step represents the homogenization procedure where individual inclusions are considered as embedded in the matrix phase and contribute to the overall elasticity of the homogenized material.

Our case is well described by the Mori–Tanaka model, since MIS fibers (inclusion phase) are disconnected with one another and fully embedded in the PHBHV (matrix phase). Thus, the information required by the model concerns the elastic tensors of the MIS and PHBHV, and the volume fraction and geometrical organization of the MIS fibers. We assumed both MIS and PHBHV to be elastic isotropic materials (see Table 2). The volume fraction of the MIS fibers ϕ_r was computed based on the measured mass fraction using the procedure outlined in the Appendix B. Eventually, we assumed the MIS fibers to be either cylinder shaped and aligned with the sample main axis (that is to say the extrusion direction) or spherical particles. Note that these two hypotheses lead to homogenized materials which are transversely isotropic and isotropic, respectively.

Table 2. Technical data of Young Modulus and Poisson coefficient for *Miscanthus giganteus* and PHBHV. [(a)] Adapted from Kaack et al. [44] [(b)] Experimental value. [(c)] Assumed.

Constituents	Young Modulus E [GPa]	Poisson Coefficient [−]
Miscanthus *giganteus*	4.5 [(a)]	0.3 [(c)]
PHBHV	1.0 [(b)]	0.3 [(c)]

3. Results

3.1. Mechanical Properties of Biocomposites

The tensile modulus, the tensile stress and the ultimate strain of the PHBHV/MIS composites were evaluated using classical tensile tests (see Table 3). For rather low nominal contents of fibers, typically 5, 10, and 20 wt %, the properties change rather slowly. It is necessary to reach 30 wt % of fibers mass content to observe a significant effect of the reinforcement in the matrix. This gradual increase in the tensile modulus is clearly visible comparing the initial slopes of the curves obtained from tensile tests (Figure 2).

Table 3. Results of tensile tests on PHBHV/MIS composites considering different fiber content (0, 5, 10, 20, 30 wt %). The standard deviation values are also presented.

Samples	Tensile Modulus [MPa]	Tensile Strength [MPa]	Ultimate Strain [%]
PHBHV	889 ± 41	22.0 ± 0.48	9.9 ± 1.1
$PHBHV_{95}MIS_5$	1074 ± 44	17.0 ± 1.54	4.2 ± 0.7
$PHBHV_{90}MIS_{10}$	1238 ± 74	16.8 ± 1.67	4.4 ± 0.9
$PHBHV_{80}MIS_{20}$	1267 ± 90	15.8 ± 0.77	3.9 ± 0.2
$PHBHV_{70}MIS_{30}$	1891 ± 172	16.9 ± 1.03	3.3 ± 0.7

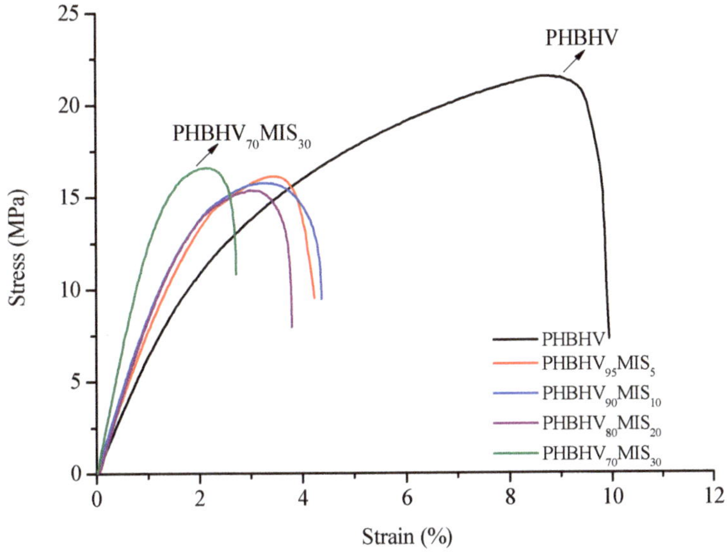

Figure 2. Strain-stress curves for PHBHV/MIS composites (the weight percents contents are indicated).

In parallel, a series of cyclic loading–unloading tensile tests was performed. First, a pure matrix sample was tested to quantify its damage limit. Through these tests, it was possible to identify the applied force beyond which damage appears in the matrix, which turned out to be about 10 N. This value was used to set up the upper limit of the eight-cycle loading–unloading tensile test performed on all the composite samples (see Section 2.3.2), in order to prevent rupture of the samples. For each test, the slope of each unloading phase was computed and used to obtain the tensile modulus of the composite. This last was definitively corrected using the procedure showed in Appendix B and the final values are showed in Table 4.

Table 4. Results of cyclic loading–unloading test on PHBHV/MIS composites at different fiber content (0, 5, 10, 20, 30 wt %). The standard deviation values are also presented.

Samples	Slope of Unloading Phase [N/mm]	Tensile Modulus [MPa]
PHBHV	115 ± 5.0	1012 ± 48
$PHBHV_{95}MIS_5$	134 ± 5.3	1185 ± 44
$PHBHV_{90}MIS_{10}$	159 ± 7.0	1408 ± 63
$PHBHV_{80}MIS_{20}$	196 ± 15.2	1736 ± 137
$PHBHV_{70}MIS_{30}$	228 ± 22.3	2032 ± 191

The slope, and thus the tensile modulus, increased with increasing fiber content, but also the standard deviations of these values increased. Furthermore, the values obtained with a simple tensile test are lower from those obtained with the cyclic procedure (Figure 3). Another important point of these tests is the difference between the slopes of the curve during the loading and unloading phases, the latter being more important (Figure 4).

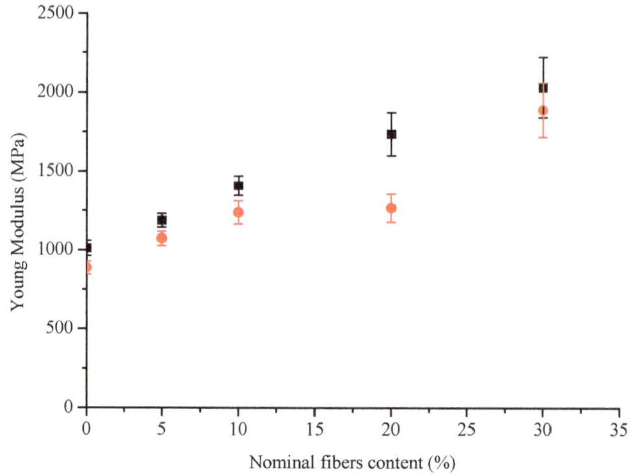

Figure 3. Young Modulus vs Nominal fibers content calculated with the two methods of loading–unloading tests (■) and traction tests (•).

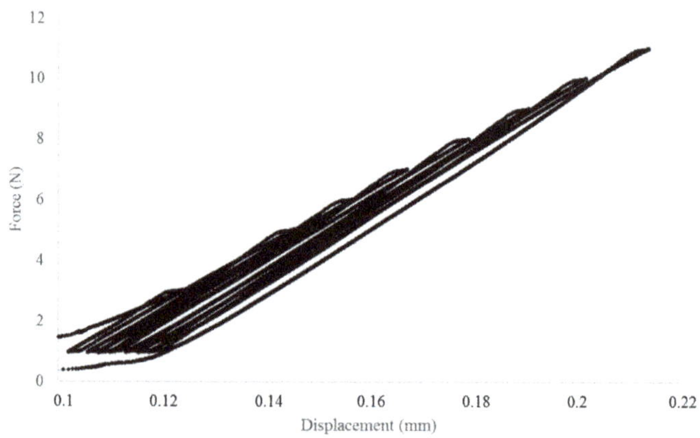

Figure 4. Loading–unloading cycle for a $PHBHV_{95}MIS_5$ specimen.

3.2. Scanning Electron Microscopy (SEM)

The cross sections of composites at different content of fibers were characterized by SEM at a magnification of 100; the results are shown in Figure 5. At lower fiber contents, typically at 5 and 10 wt %, the fibers are isolated in the matrix and perfectly identifiable. At higher fiber contents, fibers tend to form aggregates in all the section. In all samples, fibers pull out from the matrix.

3.3. Fiber-Size Distribution

Average values of width (D) and length (L) of the fibers and the L/D ratio are reported in Table 5. The fibers aspect ratio decreases when fiber content increases. This indicates that the composite synthesis based on an injection molding processing does damage the fibers for higher content. Indeed, due to the narrow pathway in the injector, it is reasonable to assume that high fiber contents correspond to stronger deterioration of the fibers depending on the injection conditions (see for instance Moritzer et al. [45] or Dupuis et al. [46]). In particular, it could be interesting in a future work to check the effect of annealing after composite processing.

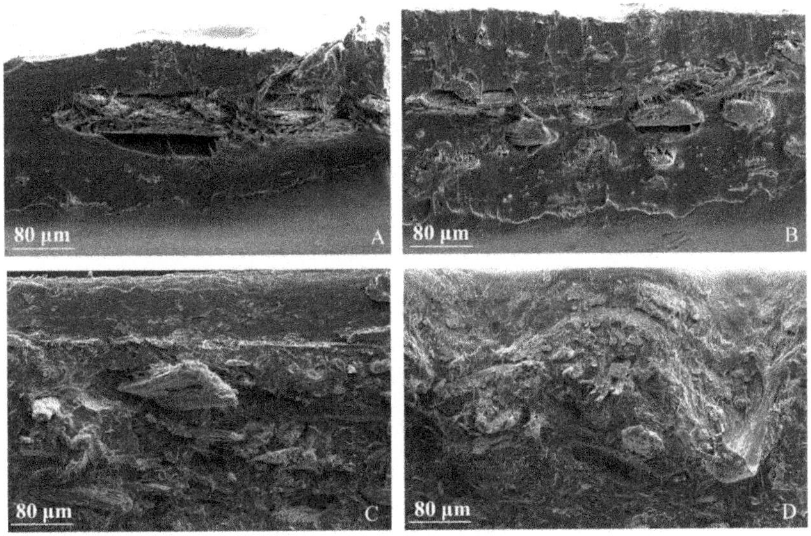

Figure 5. SEM images of the cross section of PHBHV/MIS composites: (**A**) $PHBHV_{95}MIS_5$; (**B**) $PHBHV_{90}MIS_{10}$; (**C**) $PHBHV_{80}MIS_{20}$; (**D**) $PHBHV_{70}MIS_{30}$.

Table 5. Evaluation of fiber-size distribution. The standard deviation values are also presented.

Samples	Width, D (mm)	Length, L (mm)	L/D
$PHBHV_{95}MIS_5$	0.27 ± 0.04	1.72 ± 0.33	6.4
$PHBHV_{90}MIS_{10}$	0.20 ± 0.06	1.11 ± 0.44	5.6
$PHBHV_{80}MIS_{20}$	0.16 ± 0.01	0.82 ± 0.10	5.1
$PHBHV_{70}MIS_{30}$	0.18 ± 0.08	0.80 ± 0.09	4.6

3.4. *Density of Miscanthus and Composites*

In Table 6, the fiber and composites densities are presented, introducing: ν_{MIS_n}, the nominal mass fraction of Miscanthus; ν_{MIS_m}, the measured mass fraction; $<\rho_{MIS}>$, the average density of Miscanthus; $<\rho_{C\_calc}>$, the calculated average density of composites; and $<\rho_{C\_exp}>$, the experimental average density of composites.

Table 6. Density values for MIS fibers and composites materials calculated with weight values [(a)] and with a Helium pycnometer [(b)]. The standard deviation values are also presented.

Samples	ν_{MIS_n}	ν_{MIS_m}	$<\rho_{MIS}>$ [g/cm^3]	$<\rho_{C\_calc}>$ [g/cm^3]	$<\rho_{C\_exp}>$ [g/cm^3]
PHBHV	0	0	-	1.25 [(a)]	1.083 [(b)]
$PHBHV_{95}MIS_5$	5	4.4 ± 0.1	0.75 ± 0.03	1.214 ± 0.003 [(a)]	1.233 [(b)]
$PHBHV_{90}MIS_{10}$	10	7.9 ± 1.6	0.76 ± 0.05	1.187 ± 0.002 [(a)]	1.176 [(b)]
$PHBHV_{80}MIS_{20}$	20	17.5 ± 2.1	0.96 ± 0.01	1.186 ± 0.005 [(a)]	1.184 [(b)]
$PHBHV_{70}MIS_{30}$	30	27.1 ± 0.2	0.93 ± 0.03	1.143 ± 0.012 [(a)]	1.240 [(b)]

First, using the procedure presented in Appendix B to calculate the volumetric fraction of the MIS fibers, the fiber mass M_F is obtained and then their density ρ_F is determined thanks to Equation (A4). The fiber mass measurements clearly show that discrepancies between the nominal and actual MIS mass contents do exist, the actual value being slightly lower. This trend may be due to a clogging effect at the injection point. Actually, during the injection process, the mixture is cooled quickly from $T_I = 165°$ in the collector to $T_m = 45°$ in the mold, leading to a progressive clogging at the injection point which increasingly prevents fibers from entering the mold. According to this hypothesis, the bulk of the mold should be richer in fibers than the space near the mold walls, the latter being essentially filled by pure matrix. This phenomenon is clearly visible in the SEM images of Figure 5A,B where there are no fibers in the boundary of the sample. The density of MIS roughly ranged from 0.7 to 1 g/cm^3 after the injection molding procedure. Note that the MIS fibers extracted from specimens at 5 and 10 wt % from specimens at 20 and 30 wt %, respectively, have similar densities. Secondly, the density of the biocomposites was evaluated through the pycnometer and compared with that obtained from a weighing procedure. Results obtained with the two methods are coherent for 5, 10, and 20 wt %, whereas they differ for the pure PHBHV and the 30 wt % case. We attribute this to a potential misuse of the pycnometer.

3.5. Evaluation of the Crystal Structure of PHBHV Matrix

FTIR and XRD investigations were conducted to evaluate the internal crystalline structure of PHBHV and different biocomposites with various contents of fibers (from 5 to 20 wt %) and different lengths of fibers (1000 and 45 μm).

On the one hand, the use of FTIR analysis can provide an indication on the crystalline behavior (see Section 2.3.6). As shown in Table 7, the crystallinity index C_I decreased from 1.07 for the neat matrix to 1 for a composite with 20 wt % of long fibers or of 5 wt % of short fibers. Thus, fibers length and content slightly affect the crystalline behavior of the composite. This may, thus, cause differences when measuring the mechanical properties of the different samples.

Table 7. Crystallinity parameters of PHBHV and its composites determined by FTIR-ATR analysis.

Sample	MIS Length [μm]	C_I
PHBHV	-	1.07
$PHBHV_{95}MIS_5$	1000	1.06
$PHBHV_{95}MIS_5$	45	1.00
$PHBHV_{80}MIS_{20}$	1000	1.00

On the other hand, XRD analysis was conducted on the neat matrix and composites with 20% of long fibers (Figure 6). PHBHV has a semi-crystalline nature with characteristics peaks at 2theta around 13°, 17°, 21°, 22°, 25°, and 27°, corresponding to planes (020), (110), (101), (111), (121), (040), respectively, in the orthorhombic crystalline lattice. The addition of MIS does not alter the basic crystal structure of PHBHV since the reflections are located at the same angle. Moreover, the evaluation of Bravais parameters showed that the lattice volume did not change.

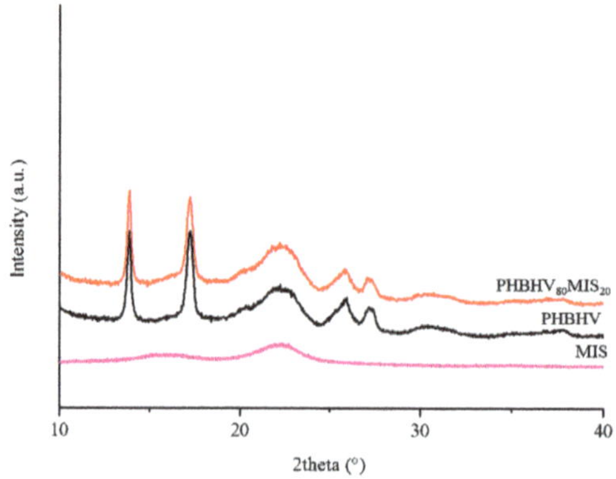

Figure 6. XRD diffractograms of Miscanthus (magenta), PHBHV neat matrix (black), and $PHBHV_{80}MIS_{20}$ composites (red).

3.6. FTIR Analysis to Assess Grafting Efficiency

In parallel, FTIR analysis can also be used to evaluate the quantity of PHBHV chains grafted onto MIS surface. That is why FTIR-ATR spectroscopy analyses were carried out on the fibers extracted from biocomposites $PHBHV_{95}MIS_5$. In Figure 7, for instance, we present the spectra of raw Miscanthus fibers and of the fibers extracted from biocomposites $PHBHV_{95}MIS_5$.

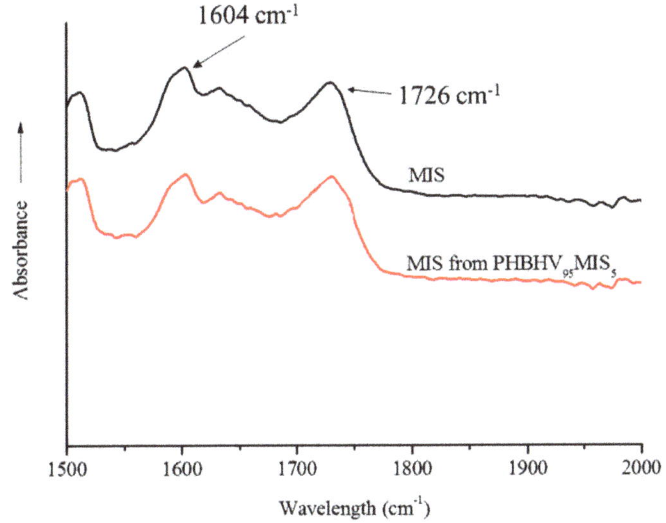

Figure 7. FTIR-ATR spectra of MIS (black curve) and MIS extracted from a composite $PHBHV_{95}MIS_5$ (red curve).

The ratio R between the peak at 1726 cm^{-1}, corresponding to the carbonyl group of the PHBHV, and the peak at 1604 cm^{-1}, corresponding to the esters of lignin, is then evaluated and reported in Table 8. Since this ratio for the composite is the same as the one obtained for the raw Miscanthus, no PHBHV was grafted onto the MIS surface during the process. This suggests a direct prospect of this study by improving the interfacial fibers/matrix properties.

Table 8. R values obtained by FTIR-ATR analysis as function of DCP content for raw fibers of 1000 μm and for fibers extracted from biocomposites $PHBHV_{95}MIS_5$. The standard deviation values are also presented.

Sample	R
MIS	1.3 ± 0.01
PHBHV95MIS5	1.3 ± 0.01

3.7. Results of Numerical Simulation

A Mori–Tanaka model was used to estimate the overall homogenized elastic modulus of PHBHV/MIS composites in the direction of application of the stress (horizontal direction E3 in Figure A1). In a first case, we assumed that the fibers had a cylindrical shape and were aligned along the axis of stress, leading to a transversely isotropic effective behavior. In a second case, we considered that fibers had a spherical shape, thus giving an isotropic system. Both models require knowledge of the volumetric fractions of fibers in the biocomposites and the mechanical properties of the constituents. In both models, both matrix and fibers are assumed to be isotropic. Then, their elastic behavior is fully described by their Young modulus and Poisson ratio.

The experimental values calculated with the loading–unloading cycles lie between the two simulations, which constitute the upper and lower limits for the elastic modulus E_3. The model with cylindrical fibers aligned in the direction of the stress seems to get closer to the experimental values than the model with spherical fibers. Moreover, results obtained from the cylindrical fiber model are perfectly overlapped with those obtained using the rule of mixtures (ROM). All these results are shown in Figure 8.

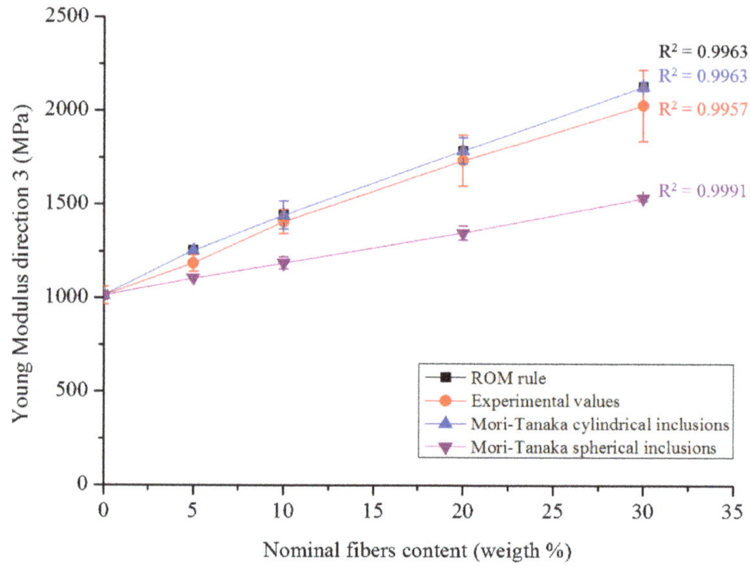

Figure 8. Results of numerical simulation: the values obtained with ROM rule are superposed with those obtained with the Mori–Tanaka with cylindrical inclusions.

4. Discussion

4.1. The Mechanical Behavior of PHBHV/MIS Composites

A familiar behavior in composite materials demonstrates an increase in Young's modulus and a decrease in failure strain and stress when increasing the fiber mass content. This general trend was showed in many previous works [44,47–50] and it seems to be independent from the nature of the polymer matrix. Biocomposites realized with other bio-based matrix, such as PLA, showed an increase by 20% when reinforced with 20 wt % of Miscanthus fibers and a slight decrease in Young Modulus for percentage over 20 wt %; the latter probably due to the formation of aggregates [26,51]. Comparing the tensile modulus and strength of the polypropylene (PP)-based wood fiber composites with PHBHV–wood fiber composites, the first ones present the higher order of tensile properties [52]. As reported in the literature, the modulus and the strength of a PP composite reinforced with 30 wt % of wood fibers are 3.33 GPa and 27.1 MPa, respectively [53]. In the present work, all the phenomena previously described were highlighted by simple tensile tests (see Table 3) and by loading–unloading tests (see Table 4). For our samples, the tensile modulus was increased by 100% and tensile strength decreased by 23% when reinforced with 30 wt % of Miscanthus fibers as compared to neat PHBHV. These results are in line with the classical observation that stiffer particles normally increase Young's modulus, but reduces tensile strength of the composite, often due to bad particle–matrix interface properties [34]. Although the tensile modulus was calculated with the two methods, the values obtained with tensile tests are lower from those obtained with the cyclic procedure (Figure 3). Another important point of these tests is the difference between the slopes of the curve during the loading and unloading phases, the latter being more important (Figure 4). These phenomena may be explained by the fact that the matrix is subjected to crystallization during time. As shown through the FTIR analysis, the crystalline behavior of the composite slightly depends on the fibers content and morphology. Moreover, PHBHV is a semi-crystalline polymer, and changes in the microstructure of the polymer are possible at temperatures higher than the glass transition one. In our case, the tests were carried out at a temperature (23 °C) that is slightly higher than the glass transition temperature of the neat matrix (5 °C), as evaluated by Lorenzini et al. [54]. A possible mechanism occurring during the traction tests consists of different deformation steps. In particular, in the first stage of deformation, changes occur exclusively in amorphous zones that stretch. Successively, the crystalline areas start to slide parallel to the traction axis pulled from the stretched amorphous zones and, at the end, the highly stretched polymer chains align with the axis.

Since the polymer chains may vary during the loading, slopes in the loading and unloading phases can be different from each other. The crystallization phenomenon occurring during time was checked a posteriori using both mechanical and thermal tests. as showed in Appendix C.

4.2. *Internal Morphology and Density of the Biocomposites*

The morphology of the biocomposites can be perfectly identified by SEM images. Moreover, optical microscopy can provide important information on the effect of the processing on the fibers. Generally, when a compatibilizing agent is not used, fibers pull out from the matrix because of the poor adhesion between the two constituents. In the case of the biocomposites manufactured by extrusion and injection molding, a multi-layer effect along the cross section was also observed. This phenomenon was likely caused by the flow of melted material in the mold during the phase of injection molding and has a clear effect on the dispersion of the fibers in the matrix. These latter are mainly located in the upper part of the cross section while pure matrix is visible in the lower part (see Figure 5A,B). This effect can be explained as the result of low interfacial adhesion between the fibers and the matrix.

The fabrication procedure, from milling to injection molding, strongly impacts the length of the fibers, preserving their width (see Table 5). At low contents of reinforcement in the matrix, the fibers appear intact and long, while at high levels they are reduced to a finer powder. Keeping the rotational speed and time of mixing constant during the extrusion process, the mechanical torque of the extruder machine was observed to increase with the fiber content. As expected, after processing, there is a reduction in the aspect ratio due to the breakup of the fibers during processing. Moreover, the density of the fibers, and consequently that of the composites, is influenced by the degree of compression imposed during the step of injection.

4.3. *Numerical Simulation*

For the numerical simulation, a Mori–Tanaka model was adopted considering either cylindrical or spherical inclusions (see Section 2.4) and then implemented in a personal code (see Sansalone et al. [55] for details). The hypothesis of cylindrical fibers seems to be confirmed from MEB observations, while the hypothesis of a transversal isotropic organization seems to be plausible and suggested by the method adopted to realize the biocomposites. In particular, the injection molding procedure requires the injection of the extruded material into a mold along the vertical direction. Fibers and polymer chains are forced to enter into the mold and to basically align with the flow direction. Accordingly, both the Young modulus calculated using the ROM (see Equation (4)) and the E_3 Young modulus provided by the Mori–Tanaka model with cylindrical inclusions correspond to the fiber direction. Although the transversely isotropic model with cylindrical inclusions is closer to the experimental results than the one with spherical inclusions (see Figure 8), it remains a relatively rough model, showing that the reality is different and more complex than assumed. The fibers are likely oriented in several directions, resulting in an anisotropic material. Moreover, the method used for manufacturing the materials causes a reduction in fiber size with effects on their shape, the latter not being perfectly cylindrical. As a first prospect, we could introduce damage effects in the model to take into account microcracking that occurs at the fiber-matrix interface. Another avenue of research to tackle the real geometry of the samples would consist of performing a direct finite element (FE) analysis using constructed geometries mimicking the composite, or grid built from 3D micro-CT images of the composite. Finally, if aiming at describing functionalized composites, a peculiar modelling of the force transmission trough the matrix/fiber interface would be necessary.

5. Conclusions and Perspectives

Biocomposites from Miscanthus giganteus fibers and PHBHV were fabricated using extrusion followed by injection molding. Tensile properties were evaluated using loading–unloading and simple traction tests. Young modulus increased slowly for low fiber contents, typically 5, 10, 20 wt %. This protocol could then easily be used for other type of green composites with other matrix and/or natural fibers. It is necessary to attain 30 wt % of fibers to observe a significant difference in the Young modulus. Although composites appear to be more rigid when compared to the pure matrix due to the presence of the reinforcement, they exhibit a decrease in the tensile strength. This effect can be justified by the lack of adhesion between the fibers and the matrix, which causes a loss of mechanical seal under tensile stress. An evidence of this lack of adhesion is that the fibers pull out from the composites instead of being totally immersed in the matrix, as shown by SEM images. Moreover, a difference in the Young modulus values calculated with the two methods (tensile tests and loading–unloading tests) was observed, indicating that PHBHV crystallize during time. This phenomenon was demonstrated by mechanical and thermal tests conducted a posteriori on the biocomposite with 5 wt % of raw fibers at different times.

The mechanical behavior of these materials was modeled by a two-phase Mori–Tanaka model where fibers were assumed either of cylindrical shape and oriented along the stress axis or spherical. The first model provides a better approximation of the experimental values of the Young Modulus, although, in reality, fibers are not perfectly cylindrical and oriented in one direction. Indeed, fiber size [55,56], shape [56], and orientation [55] may strongly affect the effective elastic properties of the composite and can explain the gap between the model and the experimental behavior. A better understanding of fibers orientation and shape via micro-CT images may be useful to implement a more accurate model. Moreover, the realization of PHBHV-based composites with different fibers size could be an interesting perspective in order to further improve the mechanical properties of these biocomposites. Lastly, the adhesion between fibers and matrix can be improved by the chemical modification of vegetal fibers or by a simple reactive extrusion [34].

Author Contributions: Study design: V.L., E.R. and T.L. Study conduct: E.G.R., E.R., V.L., V.S. and T.L. Results production: E.G.R., E.R., V.L., V.S. and T.L. Results interpretation: E.G.R., E.R., T.L., V.S. and V.L. Drafting manuscript: E.G.R., V.L. and T.L. Approving final version of manuscript: E.G.R., E.R., V.L., V.S. and T.L. All authors have read and agreed to the published version of the manuscript.

Funding: This work has been funded by LABEX MMCD (Multi-Scale Modelling & Experimentation of Materials for Sustainable Construction) and has benefited from a French government grant managed by ANR within the frame of the national program investments for the Future ANR-11-LABX-022-01. This work was partially funded by CNRS/IRP Coss&Vita between Fédération Francilienne de Mécanique (F2M, CNRS FR2609) and M&MoCS.

Institutional Review Board Statement: Not applicable.

Informed Consent Statement: Not applicable.

Data Availability Statement: All the data can be obtained according to the LABEX MMCD policy.

Conflicts of Interest: The authors declare no conflict of interest.

Appendix A. Identification of the Young Modulus of the Matrix

The Young modulus of the matrix was identified by inverse analysis which was then used as input data. Our raw input data are the slopes of the force/displacement curves of pure PHBHV samples in the unloading phase of the loading-unloading tests, see Table 4. The average value of these slopes, i.e., 115 N/mm, was assumed to be the stiffness of the PHBHV sample, K_s.

A 2D Comsol model was then developed to model the part of the actual sample between the clamps of the testing machine, see Figure A1. The effective thickness of the model was set to that of the actual sample, i.e., H = 0.93 mm, the material was defined as linearly elastic and homogeneous, and simple traction conditions were applied on the left and right boundaries. A parametric analysis was then performed by letting the Young modulus of the material, E_m, vary (while keeping fixed the Poisson ratio to 0.3), and the stiffness of the sample was computed as the ratio between the total force applied on the boundaries and the calculated elongation of the sample. The value of E_m best matching the experimental value was found to be about 1 GPa.

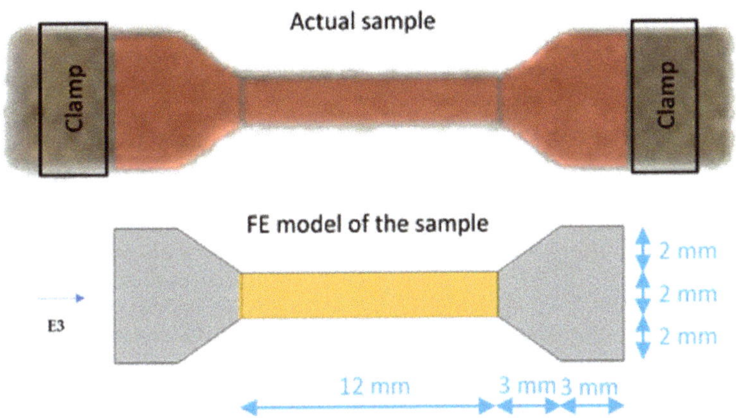

Figure A1. Actual sample (on the top) and 2D FE model of the sample (on the bottom; the "active" part is colored in yellow). The horizontal E3 direction is also presented.

Since the numerical models developed afterward focus only on the "active" part of the sample, i.e., the central rectangular region sized L × B = 12 mm × 2 mm, a shape factor Λ was calculated to transform the experimental values measured on the whole sample into values corresponding to its active part only. This shape factor is the ratio between the stiffness of the active sample, i.e.,

$$K_a = E_m \times (B \times H)/L \quad \text{(A1)}$$

and the stiffness of the sample K_s (numerically computed). It turned out that

$$\Lambda = K_a/K_s = 1.37. \quad \text{(A2)}$$

Assuming that this shape factor does not change when considering a composite sample, it allows computing the tensile modulus of composite material, E_c, based on the experimental value of the stiffness of the composite sample, K_c, namely:

$$E_c = \Lambda \times Kc \times L/(B \times H). \quad \text{(A3)}$$

Appendix B. Estimation of Volumetric Fraction of Fibers in the Specimens

The Mori–Tanaka method requires the knowledge of the volumetric fraction of the fibers in each composite specimen. This parameter $\phi_F = V_F/V_C$ is defined as the ratio between the volume of the fibers V_F and the total volume of the composite $V_C = V_F + V_M$, the volume of the matrix phase being V_M. Note that we can measure the composite volume V_C from the sample geometry. In order to obtain ϕ_F, we adopted an invasive procedure in order to separate the matrix from the fibers. First of all, every specimen manufactured by extrusion and injection molding was weighed and then solubilized under pressure and high temperature in 100 mL of dichloromethane with a rotation speed of 200 rpm for 30 min. After stirring, the solution was filtered and the solvent was evaporated in a static manner. Fibers were washed twice with the same procedure in order to eliminate the matrix attached. The matrix and the fibers were finally collected and their mass values M_M and M_F were obtained. Then, knowing the matrix density ρ_M from the commercial provider, the fiber density ρ_F, which may vary with processing due to the various degree of compression during the injection molding procedure, can finally be expressed as:

$$\rho_F = \frac{M_F}{V_C - \frac{M_M}{\rho_M}} \quad \text{(A4)}$$

Finally, the fiber volume $V_F = M_F/\rho_F$ can be deduced and thus the fiber volume content ϕ_F.

Appendix C. Effect of Time on the Mechanical Properties of PHBHV-Based Composites

In order to evaluate the change in mechanical properties due to the time, traction tests were carried out on specimens of $PHBHV_{95}MIS_5$ after 8 days and then after 8 months. Results of these tests are showed in Figure A2. An increase in modulus and a decrease in the elongation at break are visible after 8 months, confirming that the crystallization phenomenon during time observed from other researchers is valid also for our composites. In a second time, we used differential scanning calorimetry (DSC) as an additional analysis to evaluate the crystalline behavior of the biocomposites; the results obtained are showed in Figure A3. A shift in the melt temperature from 157 °C to 162 °C and an increase in the crystallinity degree from 28% to 34% are visible after 8 months. This result is a further confirmation of the crystallization phenomenon.

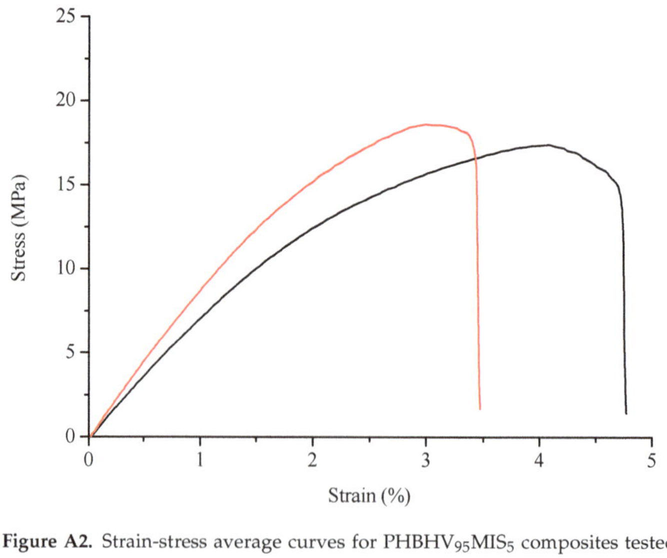

Figure A2. Strain-stress average curves for $PHBHV_{95}MIS_5$ composites tested after 8 days (black curve) and after 8 months (red curve).

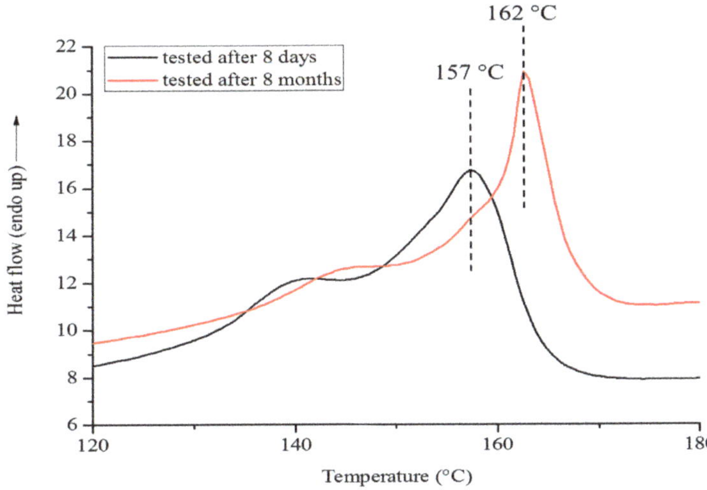

Figure A3. DSC first heating thermograms of $PHBHV_{95}MIS_5$ biocomposites tested after 8 days (black curve) and after 8 months (red curve).

References

1. Ashori, A. Wood plastic composites as promising green composites for automotive industries. *Bioresour. Technol.* **2008**, *99*, 4661–4667. [CrossRef]
2. Tasdemir, M.; Biltekin, H.; Caneba, G.T. Preparation and Characterization of LDPE and PP-Wood fiber composites. *J. Appl. Chem. Sci.* **2009**, *112*, 3095–3112. [CrossRef]
3. Gurunathan, T.; Mohanty, S.; Nayak, S.K. A review of the recent developments in biocomposites based on natural fibres and their application perspectives. *Compos. Part A Appl. Sci. Manuf.* **2015**, *77*, 1–25. [CrossRef]
4. Gowman, A.C.; Picard, M.C.; Lim, L.-T.; Misra, M.; Mohanty, A.K. Fruit Waste Valorization for Biodegradable Biocomposite Applications: A Review. *Bioresources* **2019**, *14*, 10047–10092. [CrossRef]
5. Torres, F.G.; Rodriguez, S.; Saavedra, A.C. Green Composite Materials from Biopolymers Reinforced with Agroforestry Waste. *J. Polym. Environ.* **2019**, *27*, 2651–2673. [CrossRef]
6. Müller, H.-M.; Seebach, D. Poly(hydroxyalkanoates): A Fifth Class of Physiologically Important Organic Biopolymers? *Angew. Chem. Int. Ed. Engl.* **1993**, *32*, 477–502. [CrossRef]
7. Lao, H.-K.; Renard, E.; Linossier, I.; Langlois, V.; Vallée-Rehel, K. Modification of Poly(3-hydroxybutyrate-co-3-hydroxyvalerate) Film by Chemical Graft Copolymerization. *Biomacromolecules* **2007**, *8*, 416–423. [CrossRef] [PubMed]
8. Brandl, H.; Gross, R.A.; Lenz, R.W.; Fuller, R.C. Plastics from bacteria and for bacteria: Poly(β-hydroxyalkanoates) as natural, biocompatible, and biodegradable polyesters. *Adv. Biochem. Eng. Biotechnol.* **1990**, *41*, 77–93. [PubMed]
9. Chen, G.-Q.; Patel, M.K. Plastics Derived from Biological Sources: Present and Future: A Technical and Environmental Review. *Chem. Rev.* **2012**, *112*, 2082–2099. [CrossRef] [PubMed]
10. Vergnol, G.; Sow, H.; Renard, E.; Haroun, F.; Langlois, V. Multilayer approach for tuning the drug delivery from poly(3-hydroxyalkanaoate)s coatings. *React. Funct. Polym.* **2012**, *72*, 260–267. [CrossRef]
11. Modjinou, T.; Lemechko, P.; Babinot, J.; Versace, D.-L.; Langlois, V.; Renard, E. Poly(3-hydroxyalkanoate) sulfonate: From nanoparticles toward water soluble polyesters. *Eur. Polym. J.* **2015**, *68*, 471–479. [CrossRef]
12. Vergnol, G.; Renard, E.; Haroun, F.; Guerin, P.; Seron, A.; Bureau, C.; Loirand, G.; Langlois, V. Electrografting of a biodegradable layer as a primer adhesion coating onto a metallic stent: In Vitro and in Vivo evaluations. *J. Mater. Sci. Mater. Med.* **2013**, *24*, 2729–2739. [CrossRef] [PubMed]
13. Babinot, J.; Guigner, J.-M.; Renard, E.; Langlois, V. Poly(3-hydroxyalkanoate)-derived amphiphilic graft copolymers for the design of polymersomes. *Chem. Commun.* **2012**, *48*, 5364–5366. [CrossRef]
14. Dufresne, A.; Dupeyre, D.; Paillet, M. Lignocellulosic flour-reinforced poly(hydroxybutyrate-co-valerate) composites. *J. Appl. Polym. Sci.* **2003**, *87*, 1302–1315. [CrossRef]
15. Christian, S.J.; Billington, S.L. Mechanical response of PHB-and cellulose acetate natural fiber-reinforced composites for construction applications. *Compos. Part B Eng.* **2011**, *42*, 1920–1928. [CrossRef]
16. Melo, J.D.D.; Carvalho, L.F.M.; Medeiros, A.M.; Souto, C.R.O.; Paskocimas, C.A. A biodegradable composite material based on polyhydroxybutyrate (PHB) and carnauba fibers. *Compos. Part B Eng.* **2012**, *43*, 2827–2835. [CrossRef]

17. Rodi, E.G.; Langlois, V.; Renard, E.; Sansalone, V.; Lemaire, T. Biocomposites Based on Poly(3-Hydroxybutyrate-co-3-Hydroxyvalerate) (PHBHV) and Miscanthus giganteus Fibers with Improved Fiber/Matrix Interface. *Polymers* **2018**, *10*, 509. [CrossRef] [PubMed]
18. Luo, S.; Netravali, A.N. Mechanical and thermal properties of environment-friendly "green" composites made from pineapple leaf fibers and poly(hydroxybutyrate-co-valerate) resin. *Polym. Compos.* **1999**, *20*, 367–378. [CrossRef]
19. Javadi, A.; Srithep, Y.; Lee, J.; Pilla, S.; Clemons, C.; Gong, S.; Turng, L.-S. Processing and characterization of solid and microcellular PHBV/PBAT blend and its RWF/nanoclay composites. *Compos. Part A Appl. Sci. Manuf.* **2010**, *41*, 982–990. [CrossRef]
20. Ten, E.; Bahr, D.F.; Li, B.; Jiang, L.; Wolcott, M.P. Effects of Cellulose Nanowhiskers on Mechanical, Dielectric, and Rheological Properties of Poly(3-hydroxybutyrate-co-3-hydroxyvalerate)/Cellulose Nanowhisker Composites. *Ind. Eng. Chem. Res.* **2012**, *51*, 2941–2951. [CrossRef]
21. Ten, E.; Jiang, L.; Wolcott, M.P. Preparation and properties of aligned poly(3-hydroxybutyrate-co-3-hydroxyvalerate)/cellulose nanowhiskers composites. *Carbohydr. Polym.* **2013**, *92*, 206–213. [CrossRef] [PubMed]
22. Fischer, G.; Prieler, S.; van Velthuizen, H. Biomass potentials of miscanthus, willow and poplar: Results and policy implications for Eastern Europe, Northern and Central Asia. *Biomass Bioenergy* **2005**, *28*, 119–132. [CrossRef]
23. Eschenhagen, A.; Raj, M.; Rodrigo, N.; Zamora, A.; Labonne, L.; Evon, P.; Welemane, H. Investigation of Miscanthus and Sunflower Stalk Fiber-Reinforced Composites for Insulation Applications. *Adv. Civ. Eng.* **2019**, *2019*, 9328087. [CrossRef]
24. Dias, P.P.; Jayasinghe, L.B.; Waldmann, D. Investigation of Mycelium-Miscanthus composites as building insulation material. *Results Mater.* **2021**, *10*, 100189. [CrossRef]
25. Bourmaud, A.; Pimbert, S. Investigations on mechanical properties of poly(propylene) and poly(lactic acid) reinforced by miscanthus fibers. *Compos. Part A* **2008**, *39*, 1444–1454. [CrossRef]
26. Gamon, G.; Evon, P.; Rigal, L. Twin-screw extrusion impact on natural fibre morphology and material properties in poly(lactic acid) based biocomposites. *Ind. Crop. Prod.* **2013**, *46*, 173–185. [CrossRef]
27. Selke, S.E.; Wichman, I. Wood fiber/polyolefin composites. *Compos. Part A Appl. Sci. Manuf.* **2004**, *35*, 321–326. [CrossRef]
28. Mokhothu, T.H.; John, M.J. Bio-based coatings for reducing water sorption in natural fibre reinforced composites. *Sci. Rep.* **2017**, *7*, 13335. [CrossRef] [PubMed]
29. Zhao, C.; Li, J.; He, B.; Zhao, L. Fabrication of hydrophobic biocomposite by combining cellulosic fibers with polyhydroxyalkanoate. *Cellulose* **2017**, *24*, 2265–2274. [CrossRef]
30. Al-Maharma, A.Y.; Al-Huniti, N. Critical Review of the Parameters Affecting the Effectiveness of Moisture Absorption Treatments Used for Natural Composites. *J. Compos. Sci.* **2019**, *3*, 27. [CrossRef]
31. Fazeli, M.; Florez, J.P.; Simão, R.A. Improvement in adhesion of cellulose fibers to the thermoplastic starch matrix by plasma treatment modification. *Compos. Part B Eng.* **2019**, *163*, 207–216. [CrossRef]
32. Teraube, O.; Agopian, J.-C.; Petit, E.; Metz, F.; Batisse, N.; Charlet, K.; Dubois, M. Surface modification of sized vegetal fibers through direct fluorination for eco-composites. *J. Fluor. Chem.* **2020**, *238*, 109618. [CrossRef]
33. Sánchez, M.L.; Patiño, W.; Cárdenas, J. Physical-mechanical properties of bamboo fibers-reinforced biocomposites: Influence of surface treatment of fibers. *J. Build. Eng.* **2020**, *28*, 101058. [CrossRef]
34. Rodi, E.G.; Mangeon, C.; Dessauw, E.; Sansalone, V.; Lemaire, T.; Renard, E.; Langlois, V. Functionalization of Miscanthus by Photoactivated Thiol–Ene Addition to Improve Interfacial Adhesion with Polycaprolactone. *ACS Sustain. Chem. Eng.* **2016**, *4*, 5475–5482. [CrossRef]
35. Mangeon, C.; Rodi, E.G.; Dessauw, E.; Samain, X.; Sansalone, V.; Lemaire, T.; Renard, E.; Langlois, V. Effet de la modification chimique des fibres végétales sur les propriétés des biocomposites biosourcés. *Revue des Composites et des Matériaux Avancés* **2017**, *27*, 11–29. [CrossRef]
36. Wei, L.; McDonald, A.G.; Stark, N.M. Grafting of bacterial polyhydroxybutyrate (PHB) onto cellulose via in situ reactive extrusion with dicumyl peroxide. *Biomacromolecules* **2015**, *16*, 1040–1049. [CrossRef]
37. Singh, S.; Mohanty, A.K.; Sugie, T.; Takai, Y.; Hamada, H. Renewable resource based biocomposites from natural fiber and polyhydroxybutyrate-co-valerate (PHBV) bioplastic. *Compos. Part A Appl. Sci. Manuf.* **2008**, *39*, 875–886. [CrossRef]
38. Nasser, S.N.; Hori, M. *Micromechanis: Overall Properties of Heterogeneous Materials*, 2nd ed.; Applied Mathematics and Mechanics: North-Holland, The Netherlands, 1999.
39. Qu, J.; Cherkaoui, M. *Fundamentals of Micromechanics of Solids*; Wiley & Sons: Hoboken, NJ, USA, 2006.
40. Tan, H.; Huang, Y.; Liu, C.; Geubelle, P.H. The Mori–Tanaka method for composite materials with nonlinear interface debonding. *Int. J. Plast.* **2005**, *21*, 1890–1918. [CrossRef]
41. Benveniste, Y. A new approach to the application of Mori-Tanaka's theory in composite materials. *Mech. Mater.* **1987**, *6*, 147–157. [CrossRef]
42. Eshelby, J.D. The Determination of the Elastic Field of an Ellipsoidal Inclusion, and Related Problems. In Proceedings of the Royal Society of London. Series, A. Mathematical and Physical Sciences; Royal Society: London, UK, 1957; Volume 241, p. 376.
43. Sansalone, V.; Naili, S.; Desceliers, C. A stochastic homogenization approach to estimate bone elastic properties. *Comptes Rendus Mécanique* **2014**, *342*, 326–333. [CrossRef]
44. Kaack, K.; Schwarz, K.U.; Brander, P.E. Variation in morphology, anatomy and chemistry of stems of Miscanthus genotypes differing in mechanical properties. *Ind. Crop. Prod.* **2003**, *17*, 131–142. [CrossRef]
45. Moritzer, E.; Heiderich, G.; Hirsch, A. Fiber Length Reduction during Injection Molding. *AIP Conf. Proc.* **2019**, *2055*, 070001.

46. Dupuis, A.; Pesce, J.J.; Ferreira, P.; Régnier, G. Fiber Orientation and Concentration in an Injection-Molded Ethylene-Propylene Copolymer Reinforced by Hemp. *Polymers* **2020**, *12*, 2771.
47. Johnson, M.; Tucker, N.; Barnes, S.; Kirwan, K. Improvement of the impact performance of a starch based biopolymer via the incorporation of Miscanthus giganteus fibres. *Ind. Crop. Prod.* **2005**, *22*, 175–186. [CrossRef]
48. Singh, S.; Mohanty, A.K. Wood fiber reinforced bacterial bioplastic composites: Fabrication and performance evaluation. *Compos. Sci. Technol.* **2007**, *67*, 1753–1763. [CrossRef]
49. Ahankari, S.S.; Mohanti, A.K.; Misra, M. Mechanical behaviour of agro-residue reinforced poly(3-hydroxybutyrate-co-3-hydroxyvalerate), (PHBV) green composites: A comparison with traditional polypropylene composites. *Compos. Sci. Technol.* **2011**, *71*, 653–657. [CrossRef]
50. Gunning, M.A.; Geever, L.M.; Killion, J.A.; Lyons, J.G.; Higginbotham, C.L. Mechanical and biodegradation performance of short natural fibre polyhydroxybutyrate composites. *Polym. Test.* **2013**, *32*, 1603–1611. [CrossRef]
51. Ragoubi, M.; George, B.; Molina, S.; Bienaimé, D.; Merlin, A.; Hiver, J.-M.; Dahoun, A. Effect of corona discharge treatment on mechanical and thermal properties of composites based on miscanthus fibres and polylactic acid or polypropylene matrix. *Compos. Part A Appl. Sci. Manuf.* **2012**, *43*, 675–685. [CrossRef]
52. Bledzki, A.K.; Jaszkiewicz, A. Mechanical performance of biocomposites based on PLA and PHBV reinforced with natural fibres—A comparative study to PP. *Compos. Sci. Technol.* **2010**, *70*, 1687–1696. [CrossRef]
53. Stark, N.M.; Berger, M. Effect of species and particle size on properties of wood–flour-filled polypropylene composites, functional fillers for thermoplastics, thermosets and elastomers. In Proceedings of the Functional Fillers for Thermoplastics and Thermosets, San Diego, CA, USA, 8–10 December 1997.
54. Lorenzini, C.; Renard, E.; Bensemhoun, J.; Babinot, J.; Versace, D.-L.; Langlois, V. High glass transition temperature bio-based copolyesters from poly(3-hydroxybutyrate-co-3-hydroxyvalerate) and isosorbide. *React. Funct. Polym.* **2013**, *73*, 1656–1661. [CrossRef]
55. Sansalone, V.; Trovalusci, P.; Cleri, F. Multiscale modeling of materials by a multifield approach: Microscopic stress and strain distribution in fiber–matrix composites. *Acta Mater.* **2016**, *54*, 3485–3492. [CrossRef]
56. Sansalone, V.; Naili, S.; Lemaire, T. Nanostructure and effective elastic properties of bone fibril. *Bioinspired Biomim. Nanobiomater.* **2012**, *1*, 154–165. [CrossRef]

polymers

MDPI

Article

Comparative Study of the Properties of Wood Flour and Wood Pellets Manufactured from Secondary Processing Mill Residues

Geeta Pokhrel [1,*], Yousoo Han [1,2] and Douglas J. Gardner [1,2,*]

[1] School of Forest Resources, 5755 Nutting Hall, University of Maine, Orono, ME 04469, USA; yousoo.han@maine.edu
[2] Advanced Structures and Composites Center, University of Maine, 35 Flagstaff Road, Orono, ME 04469, USA
* Correspondence: geeta.pokhrel@maine.edu (G.P.); douglasg@maine.edu (D.J.G.)

Citation: Pokhrel, G.; Han, Y.; Gardner, D.J. Comparative Study of the Properties of Wood Flour and Wood Pellets Manufactured from Secondary Processing Mill Residues. *Polymers* **2021**, *13*, 2487. https://doi.org/10.3390/polym13152487

Academic Editors: Antonio M. Borrero-López, Concepción Valencia-Barragán, Esperanza Cortés Triviño, Adrián Tenorio-Alfonso and Clara Delgado-Sánchez

Received: 1 July 2021
Accepted: 23 July 2021
Published: 28 July 2021

Publisher's Note: MDPI stays neutral with regard to jurisdictional claims in published maps and institutional affiliations.

Abstract: The generation of secondary processing mill residues from wood processing facilities is extensive in the United States. Wood flour can be manufactured utilizing these residues and an important application of wood flour is as a filler in the wood–plastic composites (WPCs). Scientific research on wood flour production from mill residues is limited. One of the greatest costs involved in the supply chain of WPCs manufacturing is the transportation cost. Wood flour, constrained by low bulk densities, is commonly transported by truck trailers without attaining allowable weight limits. Because of this, shipping costs often exceed the material costs, consequently increasing raw material costs for WPC manufacturers and the price of finished products. A bulk density study of wood flour (190–220 kg/m^3) and wood pellets (700–750 kg/m^3) shows that a tractor-trailer can carry more than three times the weight of pellets compared to flour. Thus, this study focuses on exploring the utilization of mill residues from four wood species in Maine to produce raw materials for manufacturing WPCs. Two types of raw materials for the manufacture of WPCs, i.e., wood flour and wood pellets, were produced and a study of their properties was performed. At the species level, red maple 40-mesh wood flour had the highest bulk density and lowest moisture content. Spruce-fir wood flour particles were the finest (d_{gw} of 0.18 mm). For all species, the 18–40 wood flour mesh size possessed the highest aspect ratio. Similarly, on average, wood pellets manufactured from 40-mesh particles had a lower moisture content, higher bulk density, and better durability than the pellets from unsieved wood flour. Red maple pellets had the lowest moisture content (0.12%) and the highest bulk density (738 kg/m^3). The results concluded that the processing of residues into wood flour and then into pellets reduced the moisture content by 76.8% and increased the bulk density by 747%. These material property parameters are an important attempt to provide information that can facilitate the more cost-efficient transport of wood residue feedstocks over longer distances.

Keywords: secondary processing mill residues; wood flour; wood pellets; wood–plastic composites; physical properties; transportation costs

1. Introduction

Many forest-based industries, including sawmills and turning mills, produce residues often termed "secondary processing mill residues". Secondary processing mill residues are the remnants after timber has been utilized in the wood mills for the processing of primary products such as lumber. Secondary processing residues are usually clean, uniform, on-site and low in moisture content and can be prepared for further utilization. Some of the examples include sawdust, planar shavings, small chips, etc. They have several direct or indirect applications such as the production of heat and energy, as the raw material for particleboard, pellets or char, for agricultural use (livestock bedding), and landscape applications. In the US, the volume of wood residues generated in 2019 was 15,288,915 m^3 and the volume of wood chips and particles was 46,355,003 m^3 [1]. The state of Maine generates around 1.6 million tons of green residues from the wood mills each year. Maine produces roughly 800 million board feet, from which roughly 800, 000 tons of clean chips and 800,000 tons of bark and sawdust are produced [2]. Kingsley [3] has mentioned, as a decent rule of thumb, that for every 1000 board feet of lumber produced, 2 tons of residues are produced, out of which 1 ton is clean chips (free from bark, adhesives, metals, microorganism attack, etc.).

One of the applications of mill residues can be grinding them into a powder form referred to as "wood flour" or "wood fiber". Wood flour is a finely ground wood cellulose. It is a highly comminuted wood material with a very fine particle size distribution typically produced from mill residues using several grinder types and sized by mechanical or air screening methods [4]. Wood flour is composed of fine particles passing through a screen with 850-micron openings or 20 US standard mesh [5]. Such reduced particle sizes can be achieved by fine grinders (e.g., hammermill, disc refiner, pin mill or chain mill). The reduction in biomass size changes the particle size and shape, increases bulk density, improves flow properties, increases porosity, and generates new surface area [6,7]. Compared to the other grinding mills, hammer mills have an advantage because of their ability to finely grind a greater variety of materials [8]. Nominal biomass particle sizes produced by hammer mill grinding depend on the processing characteristics of the mill. The two material variables that distinguish wood flour are the species and size [9]. High-quality wood flour can be made from hardwoods attributable to their durability and strength. Commercial production of wood flour started in 1906, and the first commercial product from wood flour was a gear shift knob for a Rolls Royce automobile in 1916 [10]. Wood flour uses can be grouped into absorbent, chemically reactive substances, chemically inert fillers, modifiers of physical properties, mildly abrasive materials, and decorative materials. Species mostly preferred for wood flour production include white pine (eastern, western and sugar pines), aspen, spruce, hemlock and, to some extent, balsam fir, paper birch, and the southern pines. In the US, 75% of wood flour is manufactured from white pine. The greatest application of wood flour is its utilization as lignocellulosic fiber for reinforcing plastics. Wood filler is compounded into the polymer matrix in order to enhance the properties of the polymer to behave more similarly to wood than a polymer. Wood–plastic composites (WPCs) have significant advantages from material, environmental, and economic perspectives [11–13].

Similarly, the other application of mill residues can be as a source of raw material for manufacturing pellets. Sources of raw materials for wood pellets include fallen branches, thinning and broken stems from the forest, and residues generated in sawmills such as sawdust, chips, etc. Wood pellets are 0.5–3 cm long and cylindrical (6–8 mm) compressed materials produced in pellet mills under high pressure (≈300 MPa) and high temperature (≈120 °C) [14]. Compression of the biomass or residues into pellets involves elastic and plastic deformation of particles and softening of natural binders such as starch, protein, lignin, fats and fibers for binding the particles together [15]. The type and amounts of extractives contribute to the fundamental difference between the softwood and hardwood pellets [16]. According to Calderon et al. [17], the utilization of wood pellets throughout the world has increased from approximately 12 million metric tons in 2008 to 56 million metric tons in 2018.

Transportation of wood flour is conducted using multi-walled paper bags (approximately 23 kg or 50 lbs) or bulk bags (typically 1.5 cubic meters or 55 cubic feet) or by bulk trailers [6]. Wood flour, being a low-density fluffy material, occupies less bulk weight for transportation. Considering shipment to the point of destination, the cost is often more than that for sawdust [5]. Wood flour transportation over longer distances can incur excess shipping costs as compared to the material price. Compared to other biomass fuels, pellets are easy to handle, store and transport [18]. Wood pellets, compared to other fibrous materials of wood and solid biofuel materials, have an increased energy output per unit volume [19]. The average bulk density of wood flour is 190–220 kg/m^3 or 12–14 $lb./ft^3$ [20] and the bulk density of wood pellets is 700–750 kg/m^3 or 43–47 $lb./ft^3$ [21]. This comparative study shows that the average bulk density of wood pellets is four times greater than wood flour, which suggests the storage footprint area for wood pellets is reduced by four times.

Very little information is available in the literature and technical sources regarding the secondary processing mill residues and their applications. On the other hand, there is no recent research on wood flour production from mill residues and its properties' characterization. Studies on wood flour production are quite outdated [5,21]. This suggests there is a wide gap to find current research on wood flour production. Nevertheless, a comparative study of the properties of wood flour based on different wood species and various mesh size level is still lacking. Literature regarding the applications of mm dimension residues on the pelletization of wood pellets is abundant. However, fine micron scale wood fibers being processed into pellets is limited in literature sources. There has been considerable research on WPC manufacturing using wood flour and studies on its properties, but only limited work has been reported on WPC manufacturing using wood pellets. Butylina et al. [22] reported on the comparative properties of WPCs using wood flour, wood pellets and heat-treated fibers in a polypropylene matrix. This study is focused on comparing wood flour and wood pellets with the overall goal of reducing the transportation costs of wood fillers for WPC manufacturing.

In addressing all the above-mentioned motives, one of the objectives of this study is to produce wood flour using the clean mill residues in Maine and characterize its properties. The characterization, i.e., the study of the morphology, moisture content, bulk density, particle size distribution and aspect ratio of the wood flour from each of the four Maine species (Northern White Cedar, Eastern White Pine, Spruce-Fir and Red Maple) can help better understand their specific properties. The second objective is the application of fine wood flour from each wood species in the manufacturing of wood pellets, and the studying of its properties. Finally, a comparative analysis of the mill residues, flour and pellets, focusing mostly on the parameters that affect production and transportation, i.e., in terms of moisture and density, is carried out. This study hopes to convey how each processing step causes a change in these material parameters. We expect this study will be a baseline study for future work on wood and polymer composite manufacturing from different raw materials in an attempt to gain maximum manufacturing efficiency.

2. Materials and Methods

2.1. Materials

Mill residues of around 100–150 kg typical planar shavings, sawdust, and small chips were obtained from local wood mills in Maine. Northern White Cedar (*Thuja occidentalis*) from (Katahdin Forest Products, Oakfield, ME, USA), Eastern White Pine (*Pinus strobus*) from (Hancock Lumber, Pittsfield, ME, USA), Eastern Spruce-Balsam Fir (*Picea rubens-Abies balsamea*) from (Pleasant River Lumber, Dover-Foxcroft, ME, USA) and Red Maple (*Acer rubrum*) from (Lumbra Hardwoods, Milo, ME, USA) were obtained through on-site visits. Figure 1 below presents the images of secondary processing processing mill residues of four different wood species in the study area. The scale bar is 3 cm in length.These residues were clean and free from bark, adhesives, metals, etc., and were either air-dried or kiln-dried. These residues were utilized to manufacture wood flour and then wood pellets. The raw material for manufacturing wood pellets, i.e., wood flour, was produced in the lab.

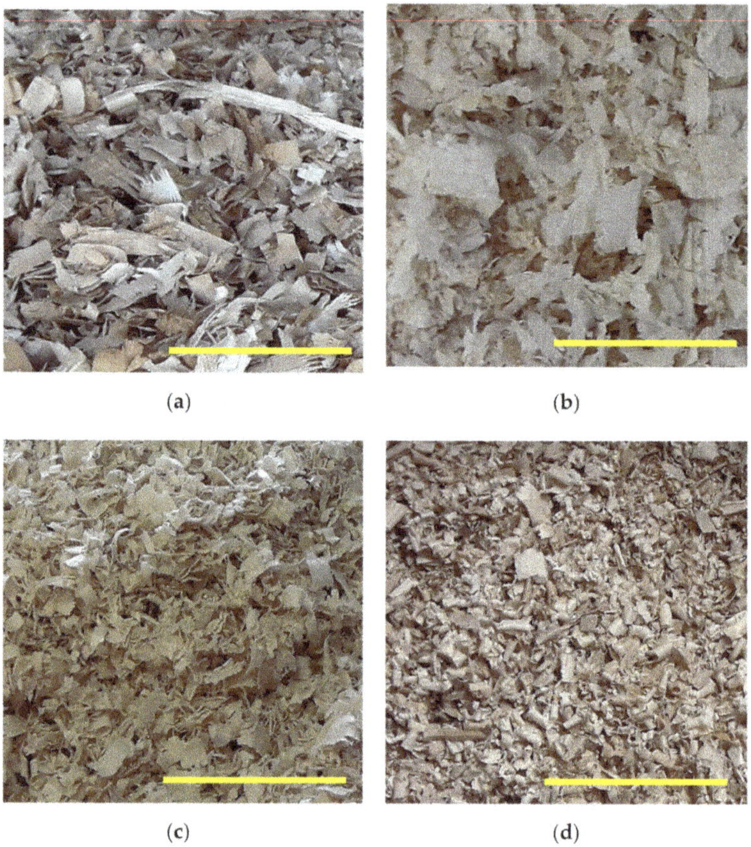

Figure 1. Mill residues of (a) White Cedar (b) White Pine (c) Spruce-Fir (d) Red Maple.

2.2. Manufacturing Process of Wood Flour and Wood Pellets

A Bliss Eliminator Hammermill (Bliss Industries LLC, Ponca City, OK, USA) with a screen size of 0.5 mm was used to produce wood flour. The residues were manually fed into the hopper of the hammermill separately for each species. The faces, edges, and corners of the hammers cut and shattered the material and threw it forcibly against the casing. Further size reduction took place in the layers of the material retained on the screen. The wood flour was then removed from the collection box. A Gilson screens haker (Gilson Company Inc., Worthington, OH, USA) was then used for screening the wood flour fractions into 20, 40, 60, 80, and 100 mesh sizes to study different characteristics of the wood flour, including morphology and other physical properties.

Usually, for the manufacturing of wood pellets that have other applications, for example, as biofuels, raw materials of up to 2 mm in size are preferred. However, in our experiments, the particle sizes were smaller (<500 μm). Several literature sources describe material characteristics, processing methods, and production parameters for wood pellet production [23–25]. For each species, two categories of wood pellets, one with unsieved flour and the other with a 40 mesh size fraction, were used for the manufacturing of pellets. The wood pellets were pelletized at the Technology Research Center (TRC), Old Town, ME, USA. A Lawson Mills Pellet Mill LM72A (Lawson Mills Biomass Solutions Ltd., North Wiltshire, PE, Canada) was used for the production of wood pellets. An integrated fines removal system captured fines for recycling. This machine allows up to two additives for binding. Even though the capacity of the pelletizer depends on the raw material, this machine processes as much as 160 kg of materials per hour. The ground wood flour from the hammer mill had a relatively low moisture content as compared to the requirement for pelletizing into pellets; therefore, water was added and mixed manually to ensure the equal distribution of moisture throughout the wood flour. Depending on the wood species, the moisture content of the wood flour was maintained between 10 and 15%. Maciejewska et al. [26] mentioned that the wood particles must be brought to the moisture content of 12–17% of weight by volume as required by the pellet press. Under high temperature and pressure, the wood flour was fed into a pellet mill and forced through a round opening called a "die" of quarter-inch thickness where the flour is compacted to form a solid mass of pellets. The high temperature in the machine allows the lignin in the wood to heat and this acts as a binder for the formation of pellets [27]. Van der Waals electrostatic or magnetic forces contribute to the process of pellet formation [28,29]. The wood pellets were then allowed to cool.

2.3. Characterization of Mill Residues, Wood Flour, and Wood Pellets

The moisture content of the mill residues, hammer mill grindings and screened wood flour, as well as of the wood pellets, was determined for each species. The American Society of Testing and Materials (ASTM) Standard D4442–20 Standard Test Methods for Direct Moisture Content Measurement of Wood and Wood-Based Materials (ASTM international, West Conshohocken, PA, USA) was followed. Moisture content, according to ASTM D9–20, (ASTM international, West Conshohocken, PA, USA) is the amount of water contained in the wood, usually expressed as a percentage of the mass of the oven-dry wood. The oven-dry method was followed where the samples were kept in the oven for 24 h at 103 ± 2 °C . The moisture content of the samples was calculated using the following formula:

$$\text{Moisture Content}(\%)\frac{\text{Original mass} - \text{Oven dry mass}}{\text{Oven dry mass}} \times 100\% \quad (1)$$

Similar to moisture content, the bulk density was also measured for the mill residues, hammer mill grindings, screened wood flour, and wood pellets. The ASTM E873–82 Standard Test Method for Bulk Density of Densified Particulate Biomass Fuels (ASTM International, West Conshohocken, PA, USA) was followed. Bulk density is the mass per unit volume of loose materials, powders and other divided "solids" at specified moisture content levels. A wooden cube of volume 1/4 ft^3 (7079 cm^3) was taken and weighed to record its weight. Then, the samples were poured with at least 5 droppings of the box from a height of 150 mm to ensure enough settling of the samples at the bottom, and filled up to a certain line as marked. The formula to determine the bulk density is:

$$\text{Bulk Density } (\text{kg}/\text{m}^3) = \frac{\text{Weight of cubical box and sample} - \text{Weight of cubical box}}{\text{Volume of cubical box}} \quad (2)$$

A Zeiss NVision 40 Scanning Electron Microscope (SEM) (Carl Zeiss Microscopy, LLC, White Plains, NY, USA) with a capacity of up to 1.2 nm resolution was used to observe the morphology of each mesh size fraction of wood flour for each species. The wood flour samples were sputter-coated with an Au/Pd conductive layer before the SEM observations. The thickness of the Au/ Pd coating was 6 nm. Since the particle size of wood flour is comparatively bigger for observation using SEM images, the magnification of the images was 50× and with a surface area of 100 μm in a high vacuum 3.56×10^{-6} Torr; the electron source voltage was 3 kV.

A Ro-Tap Shaker (Retsch Inc., Newton, PA, USA) was used for the particle size analysis of the wood flour. The American National Standards Institute/American Society of Agricultural Engineers (ASAE) S319.4 Method of Determining and Expressing Fineness of Feed Materials by Sieving (ASAE, St. Joseph, MI, USA) was followed. The geometric mean diameter or medium size of particles by the mass, geometric standard deviation of log-normal distribution by mass in ten-based logarithm, and geometric standard deviation of log-normal distribution by mass in natural logarithm were calculated based on the standard.

Similarly, the aspect ratio, which is the ratio of the images' width and height that describes the particles' shape, was calculated for each mesh size of each species. The small sample of each fraction/mesh size of the wood particles was placed on the black background with a scale and the image was taken through a digital camera. Image J software was used to express the average aspect ratio. The formula for the calculation of the aspect ratio is:

$$\text{Aspect Ratio of wood flour} = \frac{\text{Major axis}}{\text{Minor axis}} = \frac{\text{width}}{\text{height}} \tag{3}$$

The pellet dimensions were determined using a Vernier caliper (Master Gage and Tool Co., Danville, VA, USA). The diameter and length of the pellets were determined. For diameter, two types of measurement, one normal and the other with the angle of 90° of the previous measurement, was taken. Five wood pellets for each sample were taken randomly and the average value was calculated. Similarly, to determine the Pellets Durability Index (PDI), ASAE S269.4 (ASAE, St. Joseph, MI, USA) was followed. A durability tester/tumbler (Seedburo Equipment Company, Des Plaines, IL, USA) was used in the experiment and the formula to calculate PDI is:

$$\text{PDI} = \frac{\text{Mass of pellets retained on the } \frac{1}{8} \text{ inch sieve after tumbling}}{\text{mass of pellets before tumbling}} \times 100\% \tag{4}$$

To measure the ash content of the wood pellets, a thermo-gravimetric analyzer (TGA) 701 from LECO Corporation (St. Joseph, MI, USA), with an operating voltage of ~230 V, was used. To control the atmosphere inside the furnace, pneumatic gas supply/air of 3.10 bars was set up. The allowable temperature range in the TGA was 25–1000 °C. A maximum of 19 samples could be analyzed per batch. To be used in TGA, the wood pellet samples were powdered and loaded in a crucible. After 24 h, the value of the ash content was recorded on the computer system connected to the machine (1.2x TGA701, LECO Corporation, St. Joseph, MI, USA).

2.4. Statistical Analysis

A one-way Analysis of Variance (ANOVA) with a 0.05 significance level was used to determine the significant differences in the means of two or more variables. The statistical association of wood species and mesh size with the bulk density, moisture content and aspect ratio of the wood flour was analyzed. Here, wood species and mesh size are the independent variables whereas bulk density, moisture content and aspect ratio are the dependent variables. Tukey's test was performed as post-hoc analysis, whenever applicable, to figure out which groups in the sample differ more or less.

3. Results and Discussion

Figure 2 shows the SEM micrographs of the 40-mesh size wood flour for each of the four wood species in the study. Images were taken at 50× magnification and a scale bar of 100 µm. SEM images of 18, 35, 60, 80, and 100 mesh sizes of wood flour, for each of the species, are shown in Figure A1 in Appendix A. Compared to Cedar and Maple wood flour, the flour of Pine and Spruce-Fir was finer. Softwoods are flexible, whereas hardwoods are stiffer; furthermore, the chemical compositions between these species can contribute to the differences observed [30]. Despite being a softwood, wood flour fibers of Cedar appeared thicker and less fractured than the wood flour of Pine and Spruce-Fir. The presence of extractives in Cedar could cause some lubricity during processing, thus contributing to the observations of thicker fibers. The wood flour particles appear as layered tube wood-like structures [31,32]. The production of wood flour results in fiber bundles rather than individual fibers [33]. Depending on the species, differences can be observed in the appearance of the flour particles. Maple flour appears with a smoother surface than the other wood species particles. In general, softwood flour fibers are long and thin, which is attributable to the abundance of longitudinal tracheids. Hardwood fibers are short and thick, which is attributable to the abundance of radial and axial cellular components. Because of this, the softwood wood flour appears rougher compared to the maple flour in the SEM images. Even though sieving was performed for a considerable length of time, it was difficult to obtain the diameter of the particles equal to the mesh size [34]. This results in the discrepancies in the diameters of the particles for the same mesh size in the images.

Figure 2. SEM images of wood flour: (**a**) Cedar 40 mesh, (**b**) Pine 40 mesh, (**c**) Spruce-Fir 40 mesh, (**d**) Maple 40 mesh.

Two parameters of wood fillers addressed in the study of WPC material properties are the wood species and mesh size [35]. Figure 3 is a graphical representation of the relationship between mesh size and moisture content as well as mesh size and bulk density. Compared to other wood species, the moisture content of cedar flour was highest (8.4%), with spruce-fir being the lowest (5.4%), and the bulk density of maple flour was highest (268 kg/m^3), with pine flour (142 kg/m^3) being the lowest. In the case of softwoods, the moisture content of the flour is proportional to the moisture of the residue feedstock. Wood moisture content is an important controlling parameter in the manufacturing of WPCs as a moisture level above 1% can cause the composite to foam in the extruder, i.e., produce microvoids of irregular and heterogeneous shapes [36,37].

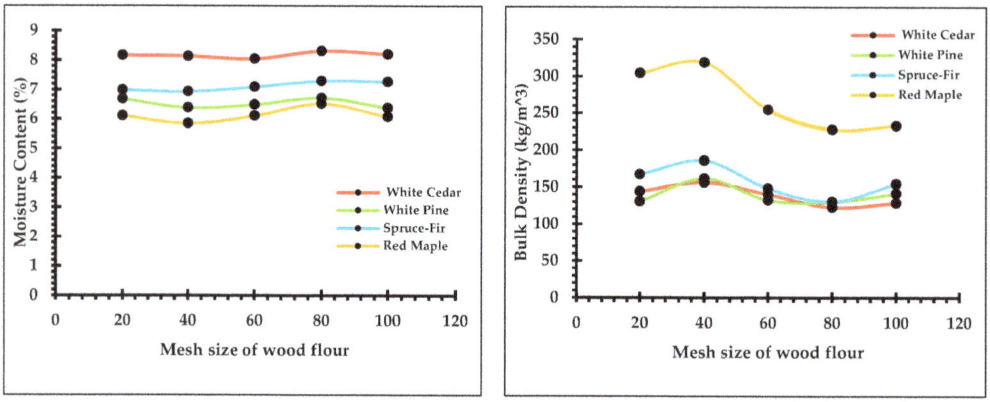

(**a**) (**b**)

Figure 3. Graphs of (**a**) change in moisture content with mesh size, and (**b**) change in bulk density with mesh size.

A one-way ANOVA test was run to determine the association of mesh size with moisture content and bulk density. From the statistical test, it was observed that the type of wood species had an association with the moisture content and bulk density of the wood flour. Similarly, the test showed there is no significant difference in the values of moisture content and bulk density with the change in mesh size of the wood flour. However, for all species, comparatively, the 80-mesh flour had the highest moisture content and 40-mesh the lowest moisture content. Similarly, the bulk density of 40-mesh wood flour was highest and of 80-mesh the lowest for all species. This shows there is an inverse relationship between moisture content and bulk density among different mesh sizes of wood flour. Patterson [38] mentioned that wood filler size between 40 and 80 mesh is the easiest to work with in the manufacturing of WPC products. The abnormality in the behavior of the moisture content and bulk density of wood flour larger and finer than 40 and 80 mesh, respectively, could also be a contributing reason.

Table 1 lists the results of the wood flour particle size distribution analysis. Wood particle size distribution has a great role in the properties of WPCs and the study of particle size distribution on properties has produced different conclusions by different researchers [29]. From Table 1, it is shown that 68% of the particles are within the range of d_{gw} for 16th and 84th percentile. Even though the same screen size of 0.5 mm was used in the hammer mill for grinding the residues, the mean geometric diameter of the wood flour particles was different among the different species. Spruce-fir wood flour had the smallest sized particles, with a geometric mean diameter of 0.18 mm, and pine flour had the largest sized particles (0.25 mm) compared to the other species. The geometric standard deviation of the diameter of pine flour particles was 0.10 mm, which was the lowest among the different wood species. This might be attributable to the utilization of commercial kiln-dried pine residues in the hammermilling process, which were highly uniform and clean. The fineness of the wood flour particles depends somewhat on the raw material and the manufacturing process used [6]. The resulting particle sizes obtained through the hammermill process vary widely, which is attributable to the hammer speed, the extent of wear on the hammer and screen, screen area, air flow, method of discharge, kind of raw material, moisture content of the raw material, etc. [39,40].

Figure 4 is the graph of aspect ratio for the different mesh sizes of wood flour for each species. Aspect ratio is an important characteristic of wood flour that affects the material properties of WPCs. Typically, the aspect ratio falls in the range of 1 to 5 for wood flour [21,41]. In this study, the aspect ratio was in the range of 2 to 3.5. The aspect ratio of commercial pine flour is 3.3 to 4.5 [42]. From a one-way ANOVA, a correlation among wood species and aspect ratio, as well as mesh size with the aspect ratio values, was not apparent. Similar to statistical results, the graph in Figure 4 shows certain variation in the aspect ratio values for different mesh sizes of different wood species. The aspect ratio, to some extent, is dependent on the grinding methods, moisture mass fraction, and wood species [43]. The same authors mentioned that, in general, there is decrease in aspect ratio with the decrease in the particle size of wood. On average, the 18–40 mesh size flour appears to have the highest values of aspect ratio for all species and 100 mesh has the lowest value. From the post-hoc analysis, it was observed that, compared to other mesh sizes, 100 mesh has the greatest association with aspect ratio. This implies that, for all species, 100 mesh size has the smallest value of aspect ratio. However, it should be noted that obtaining pure mesh size flour fractions from the sieving process is an arduous task.

Table 1. Particles size distribution analysis of wood flour for different wood species.

Species Type	Geometric Mean Diameter (d_{gw}) in mm	d_{gw} for 84th Percentile in mm	d_{gw} for 16th Percentile in mm	Geometric Standard Deviation (S_{gw}) in mm	Log-Normal Geometric Standard Deviation (S_{log}) in mm
White Cedar	0.22	0.03	1.84	0.12	0.23
White Pine	0.25	0.03	2.56	0.10	0.17
Spruce-Fir	0.18	0.02	1.57	0.11	0.26
Red Maple	0.21	0.03	1.52	0.14	0.27

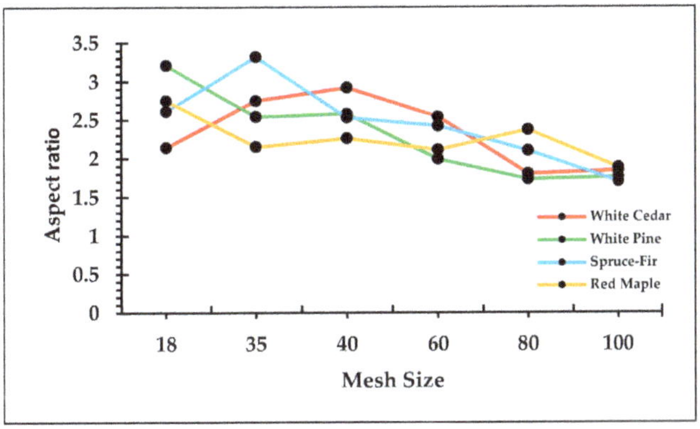

Figure 4. Relation of aspect ratio with different mesh sizes.

Figure 5 shows how the moisture content and bulk density changes with the processing steps of mill residues into flour and then into pellets. On average, when the residues are ground into wood flour, the moisture content is reduced by 54%, compressing wood flour into pellets reduced moisture content by 52.3%, and overall, when processing residues into pellets, moisture content is reduced by 76.8%. On average, the moisture content of residues was 2.2 times higher than the wood flour, and wood flour had 3.2 times higher moisture content than the pellets. The moisture content of residue feedstock was 7.4 times greater than that of the pellets. This change in moisture content was significant for cedar, contributing a reduction in moisture content of around 94% in the manufacture of pellets. This might be because of the higher extractives and lignin in cedar contributing to higher adhesion and bonding to form pellets. Bardfield and Levi [44] reported a decrease in wood pellet quality when lignin together with extractive content increases above a threshold level of 34%. Chen et al. [45] also observed the increase in bonding and overall pellet strength with the higher percentage of extractives and lignin. Converse to moisture content, the bulk density increases, which is attributable to processing from residues to flour to pellets.

On average, when grinding residues into wood flour, the bulk density increased by 119%, converting from flour to pellets, the bulk density increased by 276%, and, on converting residues into pellets, the bulk density increased by 747%. In other words, the bulk density of wood flour was 2.2 times greater than that of residues, that of pellets was 3.8 times greater than that of flour, and then that of pellets was 8.5 times greater than that of the residues. The change in bulk density was significant for pine, contributing an increase of around 1075% in the bulk density of pellets compared to the residue feedstock. This might be attributable to the utilization of fluffy compressed kiln dried commercial residues of pine with a lower bulk density than the other species. Thus, on average, the moisture content of pellets was 3.2 times less and the bulk density 3.8 times greater than for the wood flour. Mani et al. [46] suggested the size of raw materials has an inverse relation in the bulk density of pellets, i.e., smaller particles produce pellets with higher density because of the larger particle surface area, and vice-versa. However, Bergstrom et al. [47] showed almost equal densities of the pellets irrespective of the variation in particles sizes of the raw feedstock materials.

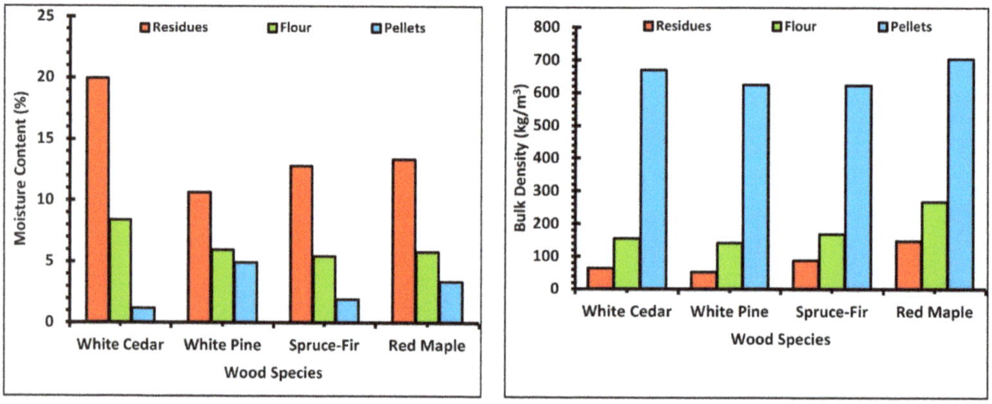

(a) (b)

Figure 5. Plots of (**a**) change in moisture content with wood processing steps, and (**b**) change in bulk density with different wood processing steps.

In Table 2, the physical properties of the wood pellets manufactured using 40-mesh wood flour and unsieved wood flour are presented. In most cases, the ash content of hardwood pellets is around three times greater than those manufactured from softwoods. However, we observed that cedar pellets have a greater ash content than the maple pellets, which is attributable to the presence of higher extractives in the heartwood of cedar. Since a quarter-inch die was used in the pelletizing process, the diameter of the pellets is near to the value. The length of the produced pellets had a greater range of sizes. Compared to the normal greater size of raw materials used in the pelletizing process, the particle size of the raw materials in this study was smaller, which might be the main reason influencing the variation in dimensions of the pellets, i.e., basically the length. Samuelsson et al. [48] observed a smaller correlation between the pellet length and the bulk density. However, they observed higher correlation in the effect of moisture content of raw materials to the pellet length. On the other hand, the durability of cedar and pine pellets was highest, which represents higher bonding of the materials in these pellets. The authors of [29,49] reported moisture in the biomass, to a certain limit (<23%), which acts as a binder during pelletization. Cedar and pine particles had higher moisture content than the rest of the two wood species, which might have contributed to higher bonding of the materials.

The images of wood pellets using 40-mesh size wood flour raw material are shown in Figure 6. Each of the scale bar is 3 cm in length. The production process of wood pellets from softwood wood flour was easier. For maple, during the pellet manufacturing from the wood flour, a large number of fine particles were generated in the waste collection chamber relative to the softwoods. Less desirable chemical composition of the particles is one of the important causes behind the higher generation of fines [29]. This might be attributed to the lower lignin content in maple that reduces binding of the particles and induces higher friction during pelletization [50]. Some of the past researchers [45,51] have mentioned that producing pellets from hardwood raw materials is arduous due to higher frictional forces in the compression channels of the die compared to the softwoods, leading to blockage of the pellet mill. On the other hand, researchers such as Thiffault et al. [51] have suggested that, under adequate conditioning and pelletizing, hardwood residues can still be processed into pellets. In our experiment, pellets from maple were manufactured successfully. However, there was significant production of fine particles during the manufacturing process. The type and amount of extractives contributes to the fundamental differences between hardwoods' and softwoods' pelletizing properties [52]. Extractive content influences the pelletizing process and pellets' quality [15,48]. The role of lignin plasticization when the biomass is processing has been studied by several authors. The authors of [53,54] are some of those who suggested that, as the temperature increases, lignin becomes more flexible, thus increasing the flow of molecules. This results in the enhancement of surface contact between particles, enabling the inter-penetration of polymer chain ends and segments between adjacent fibers. Consequently, new secondary bonds and entanglements are established once the polymer is cooled down below the glass-transition temperature of lignin. In a comparative study of the properties of pellets produced using unsieved wood flour and 40-mesh wood flour, it was observed that using the fractionated 40-mesh wood flour produced better quality pellets than the unsieved wood flour. This might be attributable to the greater uniformity of the particle size distribution of the fibers for the fractionated 40-mesh flour. The uniform size of the particles is also responsible for the smooth appearance of the wood pellets, which increases ease of storage and transportation.

(a)

(b)

Figure 6. *Cont.*

Figure 6. Wood pellets manufactured from 40 mesh wood flour: (a) White cedar, (b) White pine, (c) Spruce-fir, (d) Red maple.

Table 2. Various properties of wood pellets pelletized from unsieved and 40-mesh flour.

Wood Species	Moisture Content (%)	Bulk Density (kg/m^3)	Ash (gm)	Avg. Diameter (inches)	Avg. Length (inches)	Avg. Range of Length (inches)	Durability (%)
White Cedar (unsieved)	1.19	670	0.53	0.24	0.40	0.16–0.67	95
White Cedar (40 mesh)	2.67	699	0.45	0.24	0.40	0.14–0.73	97
White Pine (unsieved)	4.94	625	0.18	0.24	0.35	0.12–0.60	93.8
White Pine (40 mesh)	4.12	634	0.19	0.24	0.40	0.14–0.73	97
Spruce-Fir (unsieved)	1.91	624	0.32	0.24	0.48	0.19–0.93	75.8
Spruce-Fir (40 mesh)	2.57	671	0.28	0.23	0.46	0.17–0.84	82.5
Red Maple (unsieved)	3.39	705	0.41	0.24	0.49	0.17–0.93	67.2
Red Maple (40 mesh)	0.12	738	0.44	0.23	0.48	0.17–0.95	83.8

4. Conclusions

The following conclusions are drawn from the present study:

- On average, when grinding residues into wood flour, moisture is reduced by 54%, pressing wood flour into pellets reduces moisture by 52.3%, and overall, when processing residues to pellets, moisture content is reduced by 76.8%. A decrease in moisture content is a critical factor from a transportation perspective.
- On average, the bulk density increased by 119% on the comminution of residues into wood flour, it increased by 276% when converting wood flour to pellets, and, when converting residues into pellets, this value increased, on average, by 747%. An increase in bulk density is an important factor from a transportation point of view.
- Compressing fine wood flour into pellets produced pellets with a higher range in terms of dimension, and this value varied based on species in the study.
- It was found that 40-mesh fractionated wood flour produced better quality pellets than the unsieved wood flour.
- It is challenging to obtain the pure fractionated mesh sizes of the wood flour in terms of dimensions and aspect ratio even after sieving for a longer period of time. Because of the tendency of fibers with different mesh sizes to exhibit different physical properties, it is difficult to find direct relationships between material properties.

5. Future Research Work

This study presents a characterization of wood flour and wood pellets, mostly focusing on moisture content and bulk density, with the aim of reducing transportation costs. We expect that both of these raw materials have similar influence on the material properties of WPCs. The second part of this study will involve using these raw materials as fillers in plastics and comparing the physical and mechanical properties of the resulting WPCs. Future research work on processing flour into pellets along with using additives used in WPCs manufacturing, application with different polymers and different formulations, etc. for a better comparative study is highly recommended.

Author Contributions: Conceptualization, D.J.G., Y.H. and G.P.; methodology, D.J.G. and G.P.; software, G.P.; validation, D.J.G., Y.H. and G.P.; formal analysis, D.J.G. and G.P.; investigation, G.P.; resources, D.J.G.; data curation, G.P.; writing—original draft preparation, G.P.; writing—review and editing, D.J.G. and G.P.; visualization, D.J.G. and G.P.; supervision, D.J.G.; project administration, D.J.G.; funding acquisition, D.J.G. and Y.H. All authors have read and agreed to the published version of the manuscript.

Funding: This research was funded by United States Department of Agriculture (USDA) Agriculture Research Service (ARS) with Agreement Number/FAIN: 58–0204-9–166 and Project Number: 0204–41510-001–90S.

Institutional Review Board Statement: Not applicable.

Informed Consent Statement: Not applicable.

Data Availability Statement: All the necessary data are included within the article. Upon request, the data in the article can be made available from the corresponding authors.

Acknowledgments: The authors would like to acknowledge USDA ARS project for the financial support, School of Forest Resources at the University of Maine for the availability of laboratory equipment and the associated lab assistants for the training and guidance. This project was supported by the USDA McIntire-Stennis Project Number ME0–41809 through the Maine Agricultural & Forest Experiment Station. Maine Agricultural and Forest Experiment Station Publication Number 3832.

Conflicts of Interest: The authors declare no conflict of interest.

Appendix A

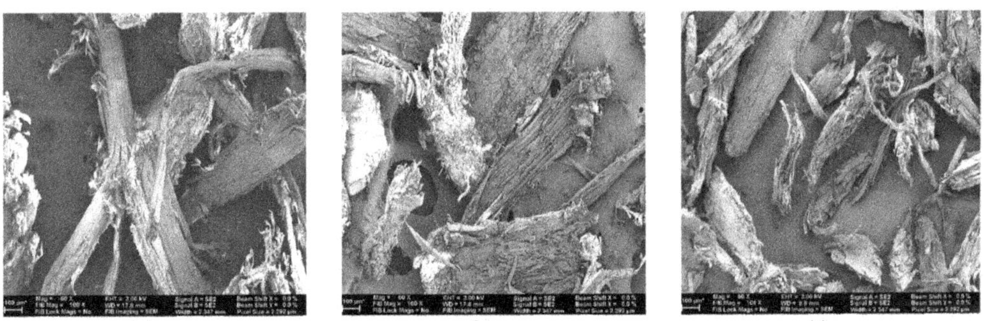

(a) (b) (c)

Figure A1. *Cont.*

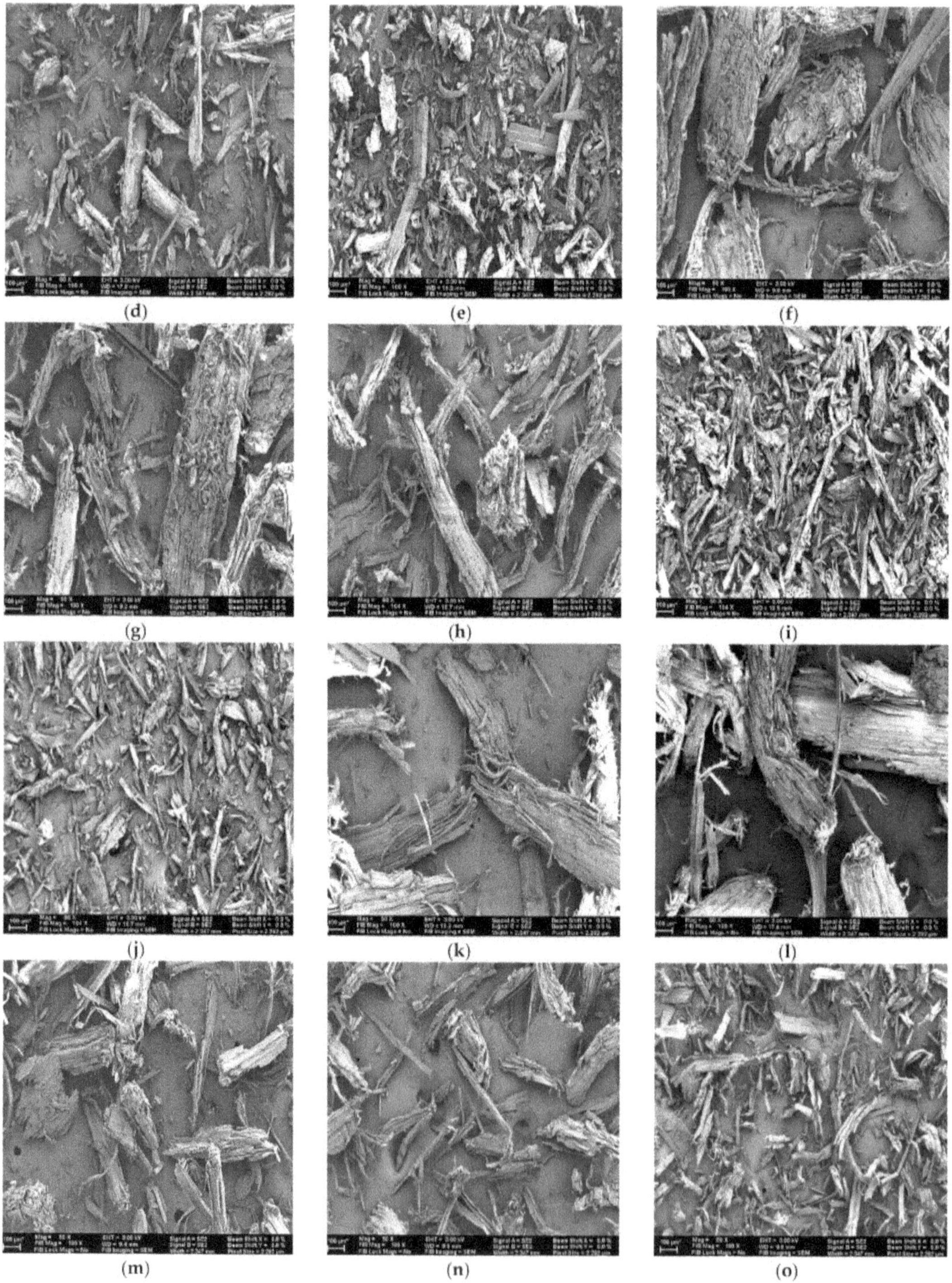

(d) (e) (f) (g) (h) (i) (j) (k) (l) (m) (n) (o)

Figure A1. *Cont.*

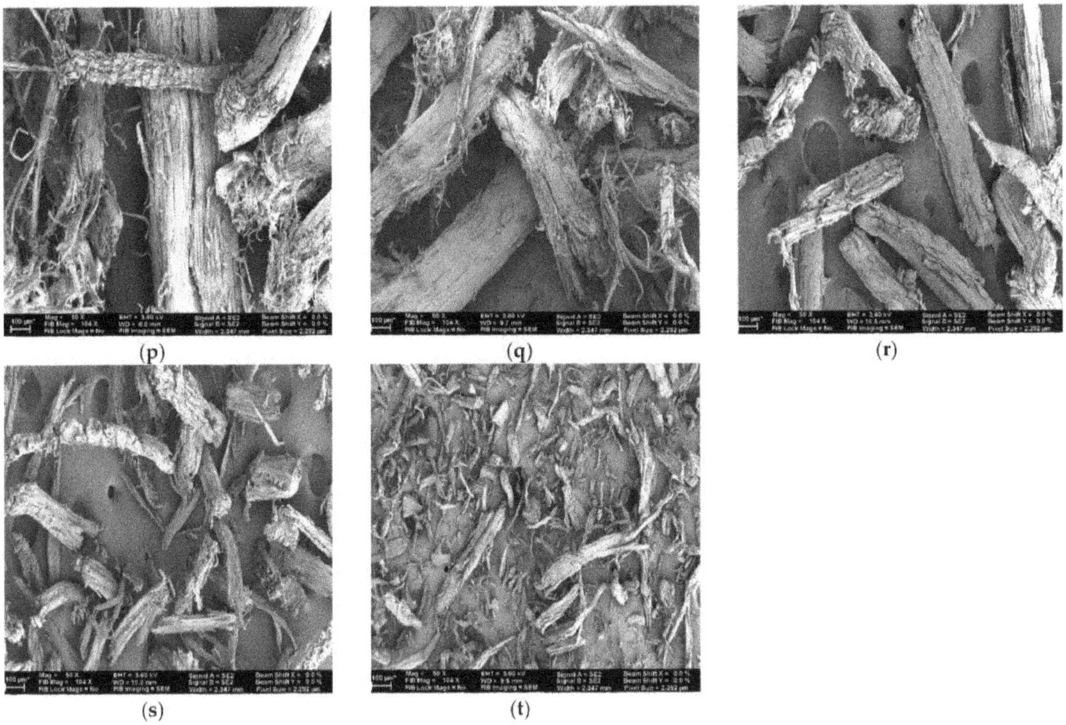

Figure A1. SEM images of wood flour: (**a**) Cedar 18 mesh, (**b**) Cedar 35 mesh, (**c**) Cedar 60 mesh, (**d**) Cedar 80 mesh, (**e**) Cedar 100 mesh, (**f**) Pine 18 mesh, (**g**) Pine 35 mesh, (**h**) Pine 60 mesh, (**i**) Pine 80 mesh, (**j**) Pine 100 mesh, (**k**) Spruce-Fir 18 mesh, (**l**) Spruce-Fir 35 mesh, (**m**) Spruce-Fir 60 mesh, (**n**) Spruce-Fir 80 mesh, (**o**) Spruce-Fir 100 mesh, (**p**) Maple 18 mesh, (**q**) Maple 35 mesh, (**r**) Maple 60 mesh, (**s**) Maple 80 mesh, (**t**) Maple 100 mesh.

References

1. Food and Agriculture Organization of the United Nations. *FAOSTAT Statistical Database*; FAO: Rome, Italy, 2019.
2. Butler, B.J. *Forests of Maine. Report to the U.S. Department of Agriculture*; Forest Service, Northern Research Station: Newton Square, PA, USA, 2018. [CrossRef]
3. Kingsley, E. Residues Becoming a Problem for Northeastern Mills. Available online: https://www.forest2market.com/blog/residues-becoming-a-problem-for-northeastern-mills (accessed on 21 September 2020).
4. Reineke, L.H. *Wood Flour. FPL Report No. 0113*; U.S. Department of Agriculture, Forest Service, Forest Products Laboratory: Madison, WI, USA, 1966.
5. Reineke, L.H. *Wood Flour. FPL Report No. R565 (Revised)*; U.S. Department of Agriculture, Forest Service, Forest Products Laboratory: Madison, WI, USA, 1945.
6. Drzymala, Z. Industrial briquetting: Fundamentals and methods. In *Studies in Mechanical Engineering*; PWN-Polish Scientific Publishers: Warsaw, Poland, 1993; Volume 13.
7. Bitra, V.S.P.; Womac, A.R.; Chevanan, N.; Miu, P.I.; Igathinathane, C.; Sokhansanj, S.; Smith, D.R. Direct mechanical energy measures of hammer mill comminution of switchgrass, wheat straw, and corn stover and analysis of their particle size distributions. *J. Powder Technol.* **2009**, *193*, 132–145. [CrossRef]
8. Scholten, R.L. The Effects of Prebreaking on the Efficiency of Hammermill Particle Size Reduction Systems in Feed Manufacturing. Master of Science, Kansas State University, Manhattan, KS, USA, 1985.
9. Stark, N.M.; Berger, M.J. Effect of particle size on properties of wood-flour reinforced polypropylene composites. In Proceedings of the Fourth International Conference on Woodfiber-Plastic Composites, Madison, WI, USA, 12–14 May 1997.
10. Gordon, J.E. *The New Science of Strong Materials (or Why You Don't Fall through the Floor)*; Princeton University Press: Princeton, NJ, USA, 1988; p. 179.
11. Kim, J.K.; Pal, K. Overview of wood-plastic composites and uses. In *Recent Advances in the Processing of Wood-Plastic Composites, Engineering Materials*; Kim, J.K., Pal, K., Eds.; Springer: Berlin/Heidelberg, Germany, 2010; Volume 32, pp. 1–22.

12. Kutnar, A.; Muthu, S.S. *Environmental Impacts of Traditional and Innovative Forest-Based Bioproducts*; Springer Science and Business Media: Singapore, 2016. [CrossRef]
13. Rangappa, S.M.; Parameswaranpillai, J.; Kumar, M.H.; Siengchin, S. *Wood Polymer Composites: Recent Advancements and Applications*; Springer Nature Singapore Pte Ltd: Singapore, 2021. [CrossRef]
14. Nielsen, N.P.K.; Gardner, D.J.; Felby, C. Effect of extractives and storage on the pelletizing process of sawdust. *Fuel* **2010**, *89*, 94–98. [CrossRef]
15. Gilvari, H.; Cutz, L.; Tiringer, U.; Mol, A.; Jong, W.; Schott, D. The effect of environmental conditions on the degradation behavior of biomass pellets. *Polym. J.* **2020**, *12*, 970. [CrossRef]
16. Nielsen, N.P.K.; Gardner, D.J.; Poulsen, T.; Felby, C. Importance of temperature, moisture content, and species for the con-version process of wood residues into fuel pellets. *Wood Fiber Sci.* **2009**, *41*, 414–425.
17. Calderón, C.; Gauthier, G.; Jossart, J.M. Bioenergy Europe Statistical Report 2019. Available online: https://bioenergyeurope.org/statistical-report.html (accessed on 12 September 2019).
18. Proskurina, S.; Rimppi, H.; Heinimö, J.; Hansson, J.; Orlov, A.; Raghu, K.; Vakkilainen, E. Logistical, economic, environmental and regulatory conditions for future wood pellet transportation by sea to Europe: The case of Northwest Russian seaports. *Renew. Sustain. Energy Rev.* **2016**, *56*, 38–50. [CrossRef]
19. Thraen, D.; Schaubach, K.; Mai-Moulin, T. *Global Wood Pellet Industry and Trade Study 2017*; IEA Bioenergy: Paris, France, 2017.
20. Clemons, C.M. Wood flour. In *Functional Fillers for Plastics*; Wiley-VCH: Weinheim, Germany, 2010; pp. 269–290.
21. Tumuluru, J.S.; Tabil, L.; Opoku, A.; Mosqueda, M.R.; Fadeyi, O. Effect of process variables on the quality characteristics of pelleted wheat distiller's dried grains with solubles. *Biosyst. Eng.* **2010**, *105*, 466–475. [CrossRef]
22. Butylina, S.; Martikka, O.; Kärki, T. Properties of wood fibre-polypropylene composites: Effect of wood fibre source. *Appl. Compos. Mater.* **2011**, *18*, 101–111. [CrossRef]
23. Wilson, T.; Buffington, D. Factors Affecting the Durability of Fuel Pellets. Master's Thesis, Pennsylvania State University, University Park, PA, USA, 2010.
24. Peksa-Blanchard, M.; Dolzan, P.; Grassi, A.; Heinimo, J.; Junginger, M.; Ranta, T.; Walter, A. *Global Wood Pellets Markets and Industry: Policy Drivers, Market Status and Raw Material Potential*; IEA Bioenergy Task 40 Technical Report 40; IEA Bioenergy: Paris, France, 2007.
25. Kaliyan, N.; Morey, R.V. Factors affecting strength and durability of densified products. In Proceedings of the 2006 American Society of Agricultural and Biological Engineers (ASABE) Annual Meeting, Portland, OR, USA, 9–12 July 2006. [CrossRef]
26. Maciejewska, A.; Veringa, H.; Sanders, J.; Peteves, S.D. *Co-firing of Biomass with Coal Constraints and Role of Biomass Pre-Treatment*; DG JRC Institute for Energy Publications Office: Luxembourg, 2006.
27. Guo, M.; Song, W.; Buhain, J. Bioenergy and biofuels: History, status, and perspective. *Renew. Sustain. Energy Rev.* **2015**, *42*, 712–725. [CrossRef]
28. Schineberger, L. *Understanding Adhesives*; Hitchcock Publishing, Co.: Wheaton, IL, USA, 1971.
29. Tumuluru, J.S.; Wright, C.T. *A Review on Biomass Densification Technology for Energy Application*; Idaho National Lab: Idaho Falls, ID, USA, 2010. [CrossRef]
30. Zazyczny, J.; Matuana, L.M. Fillers and reinforcing agents. In *PVC Handbook*; Hanser Publications: Munich, Germany, 2005; pp. 235–275.
31. Yin, B.; Hakkarainen, M. Green plasticizers from liquefied wood. *Waste Biomass Valoriz.* **2014**, *5*, 651–659. [CrossRef]
32. Pfaendner, R.; Melz, T. High-quality plastic additives made from natural raw materials benefit from the circular economy. *Biog. Plast. Addit.* **2020**, 161–178. [CrossRef]
33. Matuana, L.; Stark, N. *The Use of Wood Fibers as Reinforcements in Composites*; USDA Forest Service, Forest Products Laboratory: Madison, WI, USA; pp. 648–688. [CrossRef]
34. Chaudemanche, S.; Perrot, A.; Pimbert, S.; Lecompte, T.; Faure, F. Properties of an industrial extruded HDPE-WPC: The effect of the size distribution of wood flour particles. *Constr. Build. Mater.* **2018**, *162*, 543–552. [CrossRef]
35. Khonsari, A.; Taghiyari, H.R.; Karimi, A.; Tajvidi, M. Study on the effects of wood flour geometry on physical and mechanical properties of wood-plastic composites. *Maderas Cienc. Tecnol.* **2015**, *17*, 545–558. [CrossRef]
36. Rizvi, G.; Matuana, L.M.; Park, C.B. Foaming of PS/wood fiber composites using moisture as a blowing agent. *Polym. Eng. Sci.* **2000**, *40*, 2124–2132. [CrossRef]
37. Matuana, L.M.; Mengeloglu, F. Manufacture of rigid PVC/wood-flour composite foams using moisture contained in wood as foaming agent. *J. Vinyl Addit. Technol.* **2002**, *8*, 264–270. [CrossRef]
38. Patterson, J. New opportunities with wood-flour-foamed PVC. *J. Vinyl Addit. Technol.* **2001**, *7*, 138–141. [CrossRef]
39. Baker, R.J. Factors that Affect the Granulation and Capacity in Grinding of Corn, Oats, and Sorghum Grain with a Hammermill. Master's Thesis, Kansas State University of Agriculture and Applied Science, Manhattan, KS, USA, 1960.
40. Wilcox, R.A.; DeYoe, C.W.; Pfost, H.B. A method for determining and expressing the size of feed particles by sieving. *Poult. Sci.* **1970**, *49*, 9–13. [CrossRef]
41. Hietala, M.; Samuelsson, E.; Niinimäki, J.; Oksman, K. The effect of pre-softened wood chips on wood fibre aspect ratio and mechanical properties of wood-polymer composites. *Compos. Part A Appl. Sci. Manuf.* **2011**, *42*, 2110–2116. [CrossRef]
42. Stark, N.M.; Rowlands, R.E. Effects of wood fiber characteristics on mechanical properties of wood/polypropylene composites. *Wood Fiber Sci.* **2003**, *35*, 167–174.

43. Karinkanta, P.; Ammala, A.; Illikainen, M.; Niinimaki, J. Fine grinding of wood—Overview from wood breakage to applications. *Biomass Bioenergy* **2018**, *113*, 31–44. [CrossRef]
44. Bradfield, J.; Levi, M.P. Effect of species and wood to bark ratio on pelleting of southern woods. *For. Prod. J.* **1984**, *34*, 61–63.
45. Chen, P.Y.S.; Haygreen, J.G.; Graham, M.A. An evaluation of wood/coal pellets made in a laboratory pelletizer. *For. Prod. J.* **1989**, *39*, 53–58.
46. Mani, S.; Tabil, L.G.; Sokhansanj, S. Effects of compressive force, particle size and moisture content on mechanical properties of biomass pellets from grasses. *Biomass Bioenergy* **2006**, *30*, 648–654. [CrossRef]
47. Bergström, D.; Israelsson, S.; Öhman, M.; Dahlqvist, S.A.; Gref, R.; Boman, C.; Wasterlund, I. Effects of raw material particle size distribution on the characteristics of Scots pine sawdust fuel pellets. *Fuel Process. Technol.* **2008**, *89*, 1324–1329. [CrossRef]
48. Samuelsson, R.; Larsson, S.H.; Thyrel, M.; Lestander, T.A. Moisture content and storage time influence the binding mechanisms in biofuel wood pellets. *Appl. Energy* **2012**, *99*, 109–115. [CrossRef]
49. Kaliyan, N.; Morey, R.V. Factors affecting strength and durability of densified biomass products. *Biomass Bioenergy* **2009**, *33*, 337–359. [CrossRef]
50. Thiffault, E.; Barrette, J.; Blanchet, P.; Nguyen, Q.N.; Adjalle, K. Optimizing quality of wood pellets made of hardwood processing residues. *Forests* **2019**, *10*, 607. [CrossRef]
51. Holm, J.K.; Henriksen, U.B.; Hustad, J.E.; Sørensen, L.H. Toward an understanding of controlling parameters in softwood and hardwood pellets production. *Energy Fuels* **2006**, *20*, 2686–2694. [CrossRef]
52. Ekman, R. The chemistry of wood resin. In *Pitch Control, Wood Resin and Deresination*; TAPPI: Peachtree Corners, GA, USA, 2000.
53. Back, E.L. The bonding mechanism in hardboard manufacture review report. *Holzforschung* **1987**, *41*, 247–258. [CrossRef]
54. Bouajila, J.; Limare, A.; Joly, C.; Dole, P. Lignin plasticization to improve binderless fiberboard mechanical properties. *Polym. Eng. Sci.* **2005**, 809–816. [CrossRef]

polymers

MDPI

Article

The Effect of Cellulose Nanocrystal Suspension Treatment on Suspension Viscosity and Casted Film Property

Yucheng Peng * and Brian Via

School of Forestry and Wildlife Sciences, Auburn University, 520 Devall Drive, Auburn, AL 36849, USA; brianvia@auburn.edu
* Correspondence: yzp0027@auburn.edu

Abstract: Cellulose nanocrystals (CNCs) have attracted significant interest in different industrial sectors. Many applications have been developed and more are being explored. Pre-treatment of the suspension plays a critical role for different applications. In this study, different pre-treatment methods, including homogenization, ultrasonication, and mixing with a magnetic stirrer were applied to a CNC suspension. After treatment, the rheological behaviors of the treated CNC suspensions were characterized using a rotational viscometer. The treated suspensions were then used to cast films for characterization by ultraviolet-visible (UV-Vis) and Fourier transform near-infrared spectroscopy (FT-NIR). All the CNC suspensions demonstrated a shear thinning phenomena. Homogenization or ultrasonication significantly decreased the suspension viscosity compared with the suspension mixed by a magnetic stirrer. The viscosity of CNC suspension changed with time after treatment and settlement of treated CNC suspensions in room conditions increased the viscosity dramatically with time. Different UV and visible light interferences were observed for the CNC films generated from suspensions treated by different methods. The degree of crystallinity of the CNC films evaluated by FT-NIR showed that the film from suspension treated by homogenization and ultrasonication has the highest degree of crystallinity. Pre-treatments of CNC suspension affected the suspension viscosities and formed film properties.

Keywords: cellulose nanocrystal; viscosity; film; homogenization; ultrasonication

Citation: Peng, Y.; Via, B. The Effect of Cellulose Nanocrystal Suspension Treatment on Suspension Viscosity and Casted Film Property. *Polymers* **2021**, *13*, 2168. https://doi.org/10.3390/polym13132168

Academic Editors: Antonio M. Borrero-López, Concepción Valencia-Barragán, Esperanza Cortés Triviño, Adrián Tenorio-Alfonso and Clara Delgado-Sánchez

Received: 18 May 2021
Accepted: 28 June 2021
Published: 30 June 2021

1. Introduction

The interest in renewable nanomaterial of cellulose nanocrystals (CNCs) has been growing significantly for the past several decades. Many applications have been developed and more are being explored [1–6]. In woody biomass, CNC is the fundamental building block in crystalline structure to form cellulose elementary fibrils (microfibrils/nanofibrils). The cellulose microfibril/nanofibril consists of an additional cellulose amorphous region and the amorphous region arranges alternatively with the crystalline region along the fiber axis [7,8]. Production of CNCs from woody biomass is typically performed through a strong acid hydrolysis of cellulose materials, digesting the amorphous region. In this process, the cellulose source, acid reactant, and hydrolysis parameter impact the surface chemistry and particle morphology of the separated CNC particles. Using sulfuric acid, a stable colloidal suspension of CNCs in water can be prepared and the presence of negatively charged sulfate half-ester groups on the CNC surface prompts a well-dispersion aqueous suspension of CNCs [6,9–11].

CNC suspensions demonstrate unique properties due to the dispersion in water of rod or spindle shape particles with the nanometer dimensions in diameter and length [8,12]. At different concentrations, CNC suspensions are known to form either a chiral nematic ordered structure (anisotropic) or a gel-like (random dispersed or isotropic) material [13–17]. The formation of an ordered structure because of self-assembly depends on CNC particle morphology, surface chemistry, and temperature [17,18]. Ionic strength of the suspension can also be changed by adding different ions to control the phase transition between the chiral nematic ordered structure and the gel-like material [19,20]. After drying to form films, the effect of CNC suspension concentration, external shear force, and temperature on properties of casted CNC films were observed to be critical and complex [18,21–23]. Simultaneously, the self-assembled structure of CNC particles during the drying process of making film can be fine-tuned and customized to improve film quality, including the properties of optical, physical, mechanical, etc. Intensive research has been conducted in this regard for exploring the formation of multifunctional thin films from CNCs for a variety of applications [24–29]. However, the direct correlation between the CNC suspension properties and the formed film properties has not been clearly established. In this study effort was made to have a better understanding of the formation of CNC films through different pre-treatments of a CNC suspension. We treated the CNC suspension first with different methods, changing the suspension properties, mainly on viscosity. The films were then casted from the treated CNC suspensions and were characterized. The effect of CNC suspension viscosity change on the film characteristics was investigated.

Ultrasonication pre-treatment has been commonly employed to treat CNC suspensions for better particles dispersion and distribution [30–40]. The well-recognized functions of ultrasound treatment include better dispersion of individual CNC particles in the suspension [30,33,36], fragmentation of CNC particles in length [31,32,35], and decrease of CNC suspension viscosity [33]. Right after ultrasonication, the film casted from the CNC suspensions has also been investigated. Drying of the CNC suspension with self-assembly structure formed chiral nematic CNC films and the original CNC particle arrangements presented in the suspension maintained in the formed solid film [21,41]. The ultrasonication was observed to impact the CNC particles arrangement in the suspension and during the film drying process. Different ultrasonication treatment parameters shifted the visible light reflection band from the CNC films due to the change of the chiral nematic pitch. A higher ultrasound input energy to the CNC suspension induced a greater chiral nematic pitch for the solid film [42]. The structure change of the ordered chiral nematic domain in the CNC suspension also led to the rheological behavior change of the CNC suspensions [33]. However, the rheological behavior change along time after the ultrasonication treatment has not been recognized and studied. This is a critical factor to be considered when casting films from a CNC suspension if the suspension viscosity keeps changing. The CNC self-assembly nature would theoretically change or recover the effect from ultrasonication treatment. This is especially a concern under the circumstance of reduced CNC suspension viscosity caused by the ultrasonication treatment. The timing casting the CNC film from the suspension after the treatment would also be an important factor to be studied for controlling and optimizing the film quality.

The focus of this research was to characterize the rheological property change of a CNC suspension with time using a rotational viscometer after ultrasonication treatment. The effect of rheological property change of the CNC suspension on the film formation and film properties was also evaluated in this study. Two common mechanical treatment methods in addition to ultrasonication were also included in this study: mechanical stirring using a magnetic stirrer and high shear homogenization. For large-scale industrial applications, processing of suspensions using mechanical method is more widely practiced and can be more conveniently adopted. The rheological information provided in this research would offer significant insights to the industrial partners on processing CNC suspensions for a variety of applications, facilitating product development and market exploration. In this study, we used a rotational viscometer to analyze the effect of different suspension treatments on rheological behaviors of CNC suspensions due to its inexpensive nature and convenient operation in research laboratories and large-scale production plants. The change of viscosity of the CNC suspension was also researched with suspension settlement time. A novel observation was that the CNC suspension viscosity increased with time after each treatment and the viscosity change is closely related to the treatment method and history.

2. Materials and Methods

2.1. Materials

CNC was purchased in suspension format at 6 wt.% from CelluForce (Montreal, QC, Canada). The suspension was stored in a freezer at 6 °C after receiving.

2.2. CNC Suspension Preparation

The CNC suspension was sampled and settled in a sealed high-density polyethylene (HDPE) container at room temperature of 22 ± 1 °C and relative humidity (RH) of 50 ± 5%. After reaching the temperature equilibrium, the suspension was designated as the original CNC suspension (CNC_O). A magnetic stirrer (Corning, Glendale, AZ, USA) was then used to mix the original CNC suspension at room temperature with spin bar rotation at 870 revolution per minute (RPM) for one hour and the CNC suspension obtained was named CNC_M. Ultrasonication treatment of the CNC_M suspension was performed in a sonication bath using a VWR Ultrasonic cleaner (Radnor, PA, USA) for one hour with the container top loosely sealed to minimize the water evaporation from the suspension and the suspension after treatment was designated as CNC_U. Homogenization of the CNC_M suspension was conducted through an IKA disperser T18 digital Ultra-turrax (Wilmington, NC, USA) at 15,000 rpm for 2 min. After homogenization, air bubbles were introduced into the suspension and magnetic stirrer mixing for another hour at 870 RPM was used to eliminate the air bubbles. The obtained suspension was designated as CNC_H. The homogenized CNC suspension was also treated additionally with ultrasonication for another hour to eliminate the air bubbles and the suspension was then named as CNC_H+U. After the treatment, all the suspensions were settled in closed HDPE containers at room conditions for overnight for next step operation.

2.3. Rheological Property Characterization

The rheological behaviors of the CNC suspensions were characterized using an IKA Rotavisc lo-vi viscometer (Wilmington, NC, USA). A cylindrical spindle SP1 with a spindle guard leg was used for the apparent viscosity measurement in a 600 mL low form beaker at room temperature 22 ± 1 °C. The spindle SP1 has a dimension of 18.8 mm in diameter and 65.1 mm in length. The CNC suspension around 600 mL was required for each measurement. The suspension holding beaker has an inner diameter of 84.5 mm. The viscometer can rotate the spindle with a continuous speed change from 0 to 200 RPM. The measurement temperatures were recorded by the viscometer and were observed to be changing within 1 °C of the room temperature. During the measurement, spindle rotation speeds were controlled with the torque readings changing from around 10% to 90% of the full-scale torque of this viscometer which is 673.7 dyne-centimeters. The primary readings from this experiment include the torque percentage and the spindle rotation speed in RPM. The torque percentage were recorded when the reading from the viscometer became stabilized and it took less than 1 min to get one data point. Different spindle rotation speeds were generated during the measurement to obtain enough data points for a curve fitting in Excel to characterize the rheological behaviors of the CNC suspensions. Depending on the viscosities of different CNC suspensions, the spindle rotation speeds were in different ranges and the rheological properties characterized will be limited in that shear rate range. The applied torques were in the similar range from around 10% to 90%.

The rheological characterization was performed on all the CNC suspensions settled overnight at room temperature in closed HDPE containers. Prior to the apparent viscosity measurement, the suspension was poured into a 600 mL beaker carefully, avoiding introducing air bubbles into the suspension. After the initial viscosity measurement, the CNC suspensions treated by different methods were then allowed to settle at room temperature for different time periods for follow-up apparent viscosity measurements. The first viscosity measurement after setting the suspension overnight was designated as the data from day one. Then, the suspension viscosity measurement conducted on day five means the suspension has been settled for five days since the treatment. The stability of the CNC suspensions from the perspective of apparent viscosity was investigated.

2.4. CNC Film Preparation

The CNC_O suspension was cast directly into a polystyrene Petri dish with a diameter of 100 mm and drying of the suspension was performed at room conditions (22 ± 1 °C and 50 ± 5% RH) to obtain the solid film. The volume needed in the casting process to cover the whole surface of the Petri dish is around 10 mL. The solid CNC films for suspensions of CNC_M, CNC_H, CNC_U, and CNC_H+U were generated using the same method from the corresponding suspensions which have been settled in the room conditions for overnight after each of the specific treatments. The film moisture content was measured after drying the suspension at the room conditions to a constant weight.

2.5. CNC Film Characterization by Ultraviolet-Visible (UV-Vis) Spectrosocpy

The UV-Vis molecular absorption spectra of all the CNC film samples were collected between 190–800 nm at 2 nm spectral resolution using a Genesys 150 UV-Vis Spectrophotemeter from Thermo Fisher Scientific (Waltham, MA, USA).

2.6. CNC Film Characterization by Fourier Transform near Infrared (FT-NIR) Spectroscopy

The FT-NIR spectroscopy was employed to investigate the CNC films and the diffuse reflection spectra between 10,000 and 4000 cm^{-1} were collected using a PerkinElmer Spectrum 400 FT-NIR spectrometer (Waltham, MA, USA) at a 2 nm spectral resolution and 32 scans per sample. Prior to scanning, a reference check was performed on the system using a Spectralon reference standard. The CNC film specimens were then placed on the NIR window with a 0.8 cm diameter which was cleaned for each run using ethanol. The Spectralon reference material was put on top the CNC film specimen during the NIR spectrum acquirement.

Analysis of the FT-NIR spectra to investigate the degree of crystallinity of the CNC films was based on a method developed in the references in the FT-NIR spectra range of 7200–6000 cm^{-1} [43,44]. An assignment of the first overtone of O-H stretching vibration peaks was performed according to the second derivative of the NIR spectra with wavenumber in the range of 7200–6000 cm^{-1} [43,44]. Four regions are identified: amorphous (A_m) region at 7014 cm^{-1}, semicrystalline (S_c) region at 6750 cm^{-1}, and intramolecular hydrogen bonded crystalline regions (region C_I and region C_{II}) at 6474 and 6290 cm^{-1}. The method developed to evaluate the crystallinity of cellulose structure used the NIR absorption peaks of A_m, C_I and C_{II} [44] and the relative degree of crystallinity (CR_{NIR}) was calculated using the equation below:

$$CR_{NIR} = \frac{A(C_I) + A(C_{II})}{A(A_m) + A(C_I) + A(C_{II})} \quad (1)$$

where $A(A_m)$, $A(C_I)$, and $A(C_{II})$ are the area under the NIR spectra curve in the range of 7104–7200, 6454–6494, and 6270–6310 cm^{-1}.

3. Results and Discussion

The viscosity of the original CNC suspension (CNC_O) was also measured and the primary readings of spindle torque kept changing during the measurement. No consistent readings can be obtained and it is difficult to perform an accurate viscosity measurement. The results are not shown here. The possible explanation is that the random shaking of the suspension during the shipping process impacted the dispersion/distribution/agglomeration of the CNC particles in the liquid which in turn affected the viscosity measurement process. The uneven distribution of CNC particles in the suspension can be visually observed when pouring the CNC suspension into a container. Some locations in the suspension was observed to be more transparent than other locations, demonstrating the different concentrations of the CNC particles in different locations of the container. The viscosity measurement of the other four CNC suspensions treated by different methods generated consistent results and the primary readings from the viscometer for the measurement, including the spindle torque and rotation speed (RPM), are shown in Figure 1a. A power law relationship with a high coefficient of determination (R^2) between the spindle torque and the spindle rotation speed (RPM) was observed (Figure 1a). Rotating the spindle at the same RPM in different CNC suspensions required different torques. Based on the power law relationships shown in Figure 1a the CNC suspension after mixing by the magnetic stirrer required the highest torque to rotate the spindle when compared with other suspensions. Homogenization and ultrasonication treatments significantly decreased the torques required to rotate the spindle in the CNC suspensions at the same spindle rotation speed.

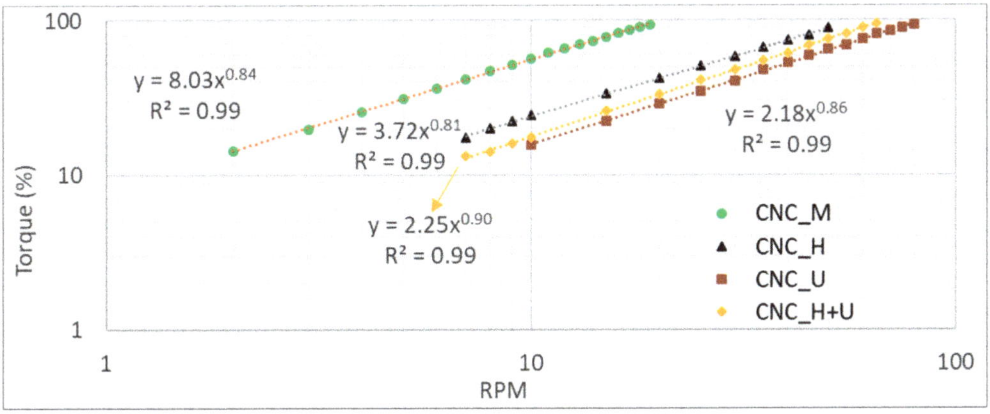

(**a**)

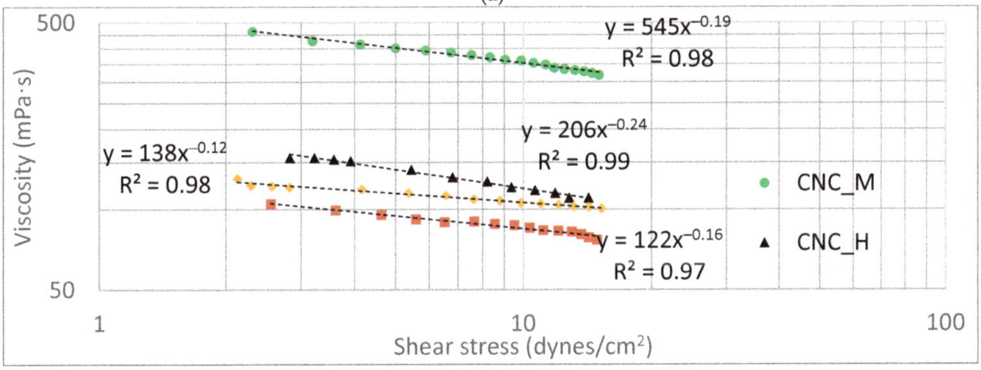

(**b**)

Figure 1. *Cont.*

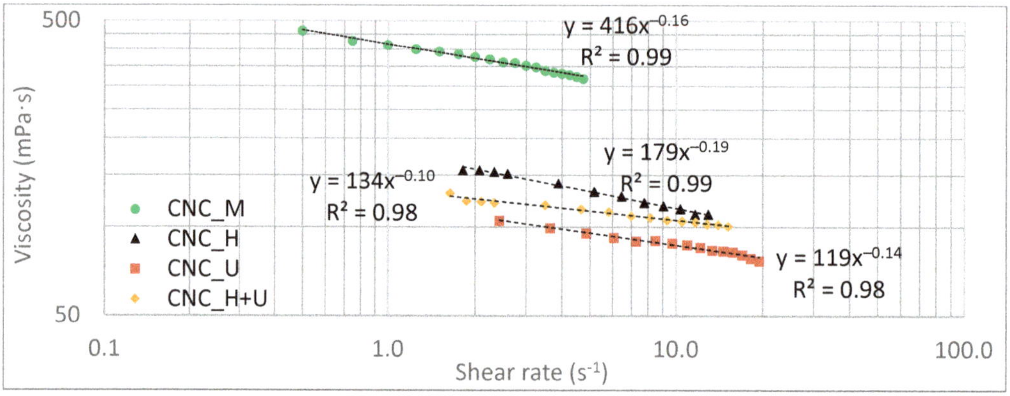

(c)

Figure 1. Viscosity measurement results of the CNC suspensions treated by different methods (**a**) torque versus RPM, (**b**) viscosity versus shear stress, and (**c**) viscosity versus shear rate.

The viscosity can be directly read from the rotational viscometer. However, the viscosity data from the viscometer is calculated using the primary readings of spindle rotation torque and speed assuming that the liquid is a Newtonian liquid. Obviously, the CNC suspensions are not Newtonian since the relationship between spindle rotational torque and RPM is not linear (Figure 1a). In addition, many research references also showed that the CNC suspensions are non-Newtonian [33–35]. Therefore, the viscosity data of the CNC suspensions in this study were calculated based on a template method [45–47].

The CNC suspension viscosities (mPa·s) were plotted against shear stress (dynes/cm^2) and shear rate (sec^{-1}) and the relationships are shown in Figure 1b,c, respectively. During the viscosity measurement, similar magnitudes of shear stress were applied (10–90% of the full capacity of the viscometer torque) to all the CNC suspensions. In this specific shear stress range, the viscosities of CNC suspensions decreased with increasing shear stress in a power law model. Under the same shear stress, the suspension of CNC_M has the highest viscosity, followed by the suspensions of CNC_H and CNC_H+U and the suspension of CNC_U has the lowest viscosity. When applying a same external shear stress to the CNC suspensions, the flow behaviors of these suspensions would be different even they have the same concentration of 6 wt.% and the same chemical components. Under the same shear stress of 5 dynes/cm^2, the viscosity of the suspension of CNC_M, CNC_H, CNC_U, and CNC_H+U are around 402, 141, 95, and 115 mPa·s based on the power law model calculation shown in Figure 1b. When compared with the viscosity of the suspension mixed by the magnetic stirrer, homogenization decreased the viscosity of CNC suspension by almost three times and ultrasonication decreased the viscosity by more than four times. The combination treatment of homogenization and ultrasonication, however, did not generate a direct addition effect on decreasing the suspension viscosity. In the measured shear stress range, the suspension viscosity treated by the combination of homogenization and ultrasonication locates between those treated by homogenization and ultrasonication individually. The relationships between the suspension viscosity and the shear stress are different as well for the suspensions treated by different treatment methods (Figure 1b). The viscosity results demonstrated that different treatments on the CNC suspension significantly impacted their flow behaviors under a same shear stress which needs to be considered in different applications, such as using CNC suspension for making films or coatings.

When a same shear rate is applied to the CNC suspension treated by the different methods significant differences in the apparent viscosities were also observed and this is demonstrated in Figure 1c with the plots of viscosity against shear rate. At the shear rate of 4 s^{-1}, for example, the apparent viscosities of the CNC suspensions are 334, 138, 98, and

116 mPa·s for CNC_M, CNC_H, CNC_U, and CNC_H+U. The plot of the viscosity against shear rate in Figure 1c also showed that all the CNC suspensions are a shear-thinning liquid with a fitted power law flow model (Oswald De Waele model) in the shear rate range studied here. The relationship between the viscosity and shear rate and the fitted power law models with high coefficients of determination are included in Figure 1c. In a power law model, the viscosity of the liquid is correlated with the shear rate in the Equation (2).

$$\eta = k\dot{\gamma}^{n-1} \tag{2}$$

where k is the consistency index and n is the flow behavior index. For the CNC suspension, homogenization and ultrasonication treatments significantly reduced the viscosity compared with the viscosity of CNC suspension mixed by the magnetic stirrer and this viscosity change can be represented by the consistency indices in the power law flow models. The consistency indices are much lower for CNC_H (179) and CNC_U (119) when compared with that for CNC_M (416) (Figure 1c). The consistency index in the power law flow model is directly correlated with the viscosity of the suspension [48]. In the shear rate range researched in this study the models obtained in Figure 1c can be used to predict the viscosity of the CNC suspensions. The shear mixing action introduced by the magnetic stirrer can uniformly distribute and/or disperse CNC particles, originally unevenly distributed in the original CNC suspension, to some degree and a consistent measurement using a rotational viscometer can be obtained. A more dramatical shear mixing in homogenization treatment could help to disperse and distribute the CNC particles in the suspension uniformly to a higher degree, resulting in less agglomeration of the CNC particles. According to the rheological theory of rigid particles in a Newtonian liquid [48], the rate of work creating shear of a volume of material is related to the suspension viscosity which is determined by the orientation of the rodlike particles in the suspension. When the long axis of the particle is aligned with the flow direction the lowest rate of work needs to be done to shear the material. On the other side, more unit energy will be needed to shear the suspension when the long axis of the rodlike particle aligns in an angle between 0 and 180 degrees with the flow direction and the highest rate of energy is required to rotate the particle when the long axis is perpendicular to the flow direction [48]. For the CNC suspension, homogenization decreased the viscosity significantly, which indicates that the treatment reduced the barrier to align the CNC particles with the flow direction, either by enhanced dispersion or increased alignment of the CNC particles directly in the suspension. Homogenization is a common practice used to improve particle dispersion in a suspension and the decreased viscosity of the CNC suspension studied here demonstrated that the rate of work required to align the dispersed CNC particles is lower than that needed for agglomerated CNC particles.

Ultrasonication decreased the CNC suspension viscosity even further compared with that treated by homogenization. However, there is no mechanical action to align the CNC particles during the ultrasonication treatment. The well-known function of ultrasonication is to enhance the dispersion [30,33,36] and decrease the particle size by rupture the rodlike particles into shorter ones [31,32,35]. Both actions help to align the CNC particles along the flow direction when the suspension is sheared by the spindle during the viscosity measurement, leading to lower viscosity readings. For the CNC suspension treated by homogenization first followed by ultrasonication the viscosity in the studied shear rate range is between the viscosities of the suspension treated by each individual method. After homogenization, good CNC particles dispersion and alignment can be achieved and the following ultrasonication could improve the dispersion more and shorten the CNC particles. The combined treatment, however, is not as effective as ultrasonication with respect to lowering the suspension viscosity. The CNC particle alignment caused by homogenization might change the suspension structure, offsetting or diverting the effect of ultrasonication.

The flow behavior index of the CNC suspension also changed after homogenization and ultrasonication, indicating the underlined dispersion, alignment, and rupture of the

CNC particles in the suspension. Water is a Newtonian liquid ($n = 1$) and adding rodlike CNC particles drives the flow behavior of the suspension away from the Newtonian flow ($n < 1$ or $n > 1$). The negative value of $n - 1$ ($n < 1$) in the power law models of Figure 1c demonstrated the shear thinning behaviors for all the CNC suspensions. A suspension having a smaller value of n (smaller value of $n - 1$) indicates that the flow of the suspension is relatively further away from the Newtonian flow than for the suspension having a larger n value. The suspension processed by combination treatment of homogenization and ultrasonication has the largest n value (0.90) followed by the suspension treated by ultrasonication (0.86) and homogenization (0.81). The suspension mixed by the magnetic stirrer has the flow behavior index of 0.84. During the viscosity measurement process, the rotation of the spindle sheared the suspension at a much lower shear rate than that applied during the mixing by the magnetic stirrer and the spindle rotation would not change the nature of the suspension during the viscosity measurement. The flow behavior index of 0.84 for the magnetic stirrer mixed suspension reflected the suspension flow property after mixing one hour at 870 RPM using the magnetic stirrer. Homogenization at a much higher rotational speed of 15,000 RPM significantly changed the CNC particle dispersion and alignment, changing the flow behavior index to 0.81. For the treatment of ultrasonication, no further mechanical alignment of the CNC particles occurred in the suspension. However, the CNC particle rupture changed the morphologies and dispersion of CNC [33,35,38], possibly facilitating the particle alignment during the viscosity measurement process with the rotation of spindle which in turn makes the suspension flow relatively more like a Newtonian fluid (0.86) compared with suspension treated by homogenization (0.81). When the CNC suspension was treated first by homogenization, the following ultrasonication increased the flow behavior index further to 0.90 The viscosity of the suspension processed by ultrasonication treatment did not decrease at the same rate with shear rate as that of homogenized suspension (Figure 1c). High shear mechanical mixing increases the viscosity decreasing rate with shear rate while ultrasonication decreases the viscosity decreasing rate with shear rate. The other conclusion may be drawn is that the ultrasonication drives the liquid toward a Newtonian flow while mechanical shearing moves the suspension toward non-Newtonian flow.

The apparent viscosity change of the CNC suspension with time after ultrasonication was studied. The CNC suspension settled in a sealed HDPE container was poured into a 600 mL beaker for apparent viscosity measurement and the measurement was conducted at day five, ten, and sixty after the ultrasonication treatment. The primary reading of torque in percentage was plotted against the spindle rotational speed in RPM and the curves are shown in Figure 2a in a logarithm scale. When a same torque was applied to the CNC suspension at different times after ultrasonication treatment, different spindle rotation speeds were obtained. The longer the CNC suspension settled, the lower spindle rotational speed was required to reach the same torque. An exponential relationship was established to represent the torque change with RPM for the CNC suspensions (Figure 2a) and it was observed that the fitted curves in the logarithm scale are parallel with each other up to ten days. There is a slight shifting for the curve at the day sixty and it is not parallel with other curves.

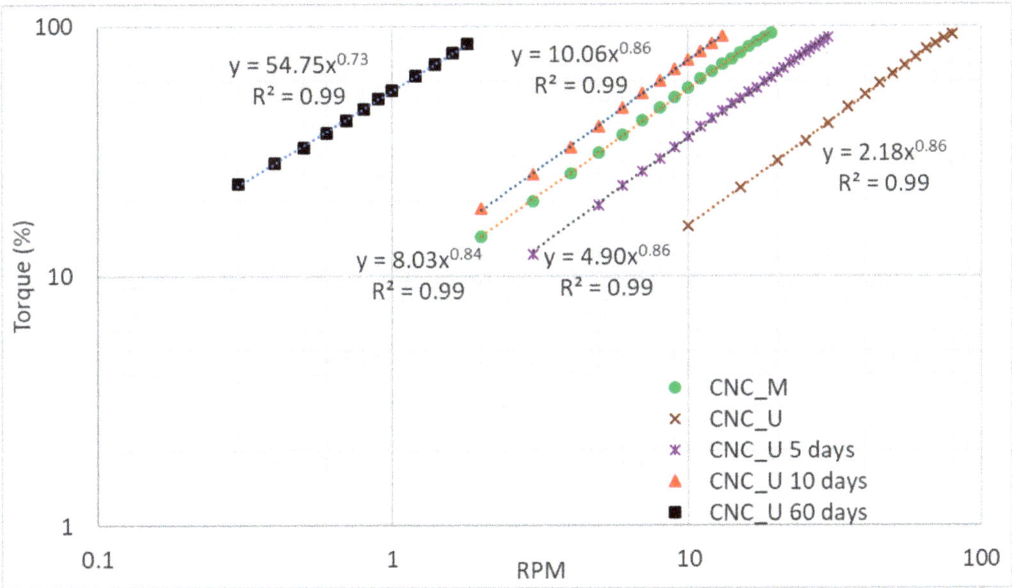

(a)

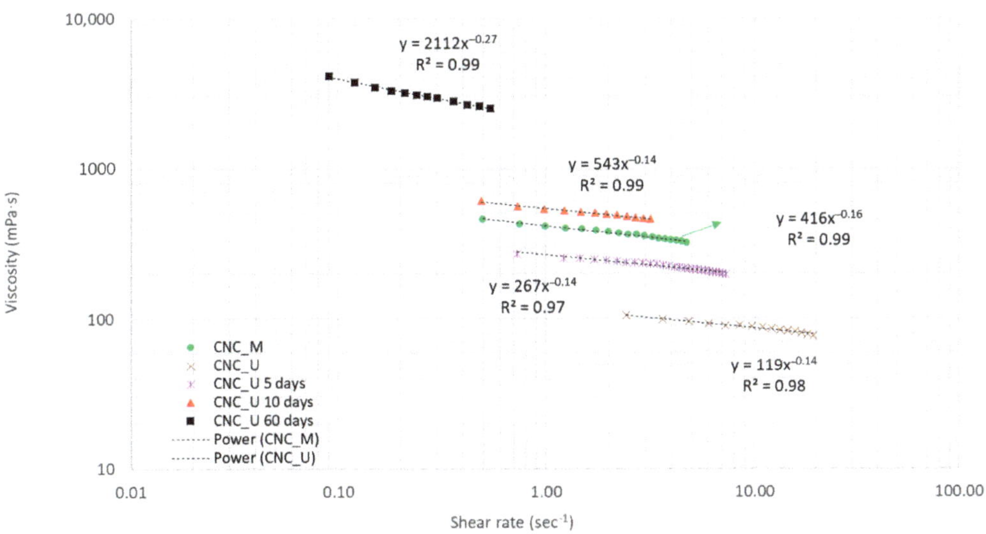

(b)

Figure 2. The change of viscosity with time for the CNC_U suspension (**a**) torque versus RPM and (**b**) viscosity versus shear rate.

The viscosity data change of CNC_U suspension with time was calculated and the relationship between viscosity and shear rate is shown in Figure 2b. Power law flow models were observed for all the relationships between viscosity and shear rate measured at different time periods. After sixty days, the viscosity of the suspension increased significantly, and it is more than ten time greater than the viscosity of the CNC_U suspension at day one based on the power law flow model calculation for the same shear rate. The viscosity of the CNC suspension is a dynamic variable, changing with time. After the ultrasonication, the dispersed and ruptured CNC particles could self-assemble into liquid crystalline or colloidal structures [25] and this self-assembly proceeded faster due to the lower viscosity of the suspension after ultrasonication. With the suspension settlement for different time periods various degrees of liquid crystalline or colloidal structures would be formed in the suspension. The strong interactions between the CNC particles in the liquid crystalline or colloidal structures would require much higher energy to rotate or to be sheared along the flow direction, resulting in the increase of the viscosity. The viscosity change can be demonstrated by the consistency indices for the power law flow models. The suspension at different times behaved differently. The flow behavior index did not change for up to ten days ($n = 0.86$) and then decreased to 0.73 at the day sixty. For up to ten days, the viscosity of CNC_U suspension decreases at the same rate with increasing shear rate while at the day sixty, the viscosity decreases at a faster rate with increasing shear rate. The liquid crystalline or colloidal structures formed by self-assembly of CNC particles did not significantly change the shear thinning flow behavior of the suspension for the first ten days and then the higher degree of self-assembly would decrease the flow behavior index gradually until reaching 0.73 at the day sixty.

A similar test was performed on the CNC_H+U suspension. The apparent viscosity measurement was conducted on day one, day three, day six, and day sixty. The results for the spindle torque versus RPM are shown in Figure 3a and the plots of apparent viscosity against shear rate are shown in Figure 3b. The results for CNC_M and CNC_H are added for comparison. The CNC_H+U suspension showed the same changing trends with time as demonstrated by the CNC_U suspension. The apparent viscosity of the suspension increased with settling time and at the day sixty, the apparent viscosity is significantly higher than that at day one. The consistency indices in the power law flow models increased from 134, 316, 412, to 2358 from day one, three, six, to sixty. However, the flow behavior indices for the CNC_H+U suspension in the power law flow mode did not keep the same for any time periods and slight change was observed even for day three. The effect of homogenization on suspension viscosity is different than that of ultrasonication and the combined treatment of homogenization and ultrasonication would alter the CNC particle nature existed in the suspension. Therefore, the formation of liquid crystalline or colloidal structures would proceed in a different way than that shown in the suspension treated by ultrasonication. After the treatment, the flow behavior of the CNC suspension kept changing with time and the viscosity of the suspension decreases at different rates with increasing shear rate at different times. Overall, the flow behavior index changed toward a lower number from 0.90 at day one to 0.68 at the day sixty, moving further away from a Newtonian liquid.

The CNC suspension stability after different treatments was also evaluated by characterizing the viscosity change with time. After a settlement for sixty days, the viscosities for the suspensions of CNC_M, CNC_H, CNC_U, and CNC_H+U were measured and the primary readings of spindle torque and RPM are shown in Figure 4a. The relationship between viscosity and shear rate are shown in Figure 4b. All the suspensions became significantly more viscous than before. After settlement for sixty days, suspensions of CNC_M, CNC_H, and CNC_U were mixed using a magnetic stirrer for one hour at 870 RPM in closed containers and then sitting at room temperature overnight for the viscosity measurement. The suspension of CNC_H+U was mixed overnight using the magnetic stirrer at 870 RPM for the viscosity measurement. The results for the viscosities are shown in Figure 5. After mixing, the viscosities of all the CNC suspensions were reduced significantly in the tested shear rate range. It can be seen from Figure 5 that one-hour magnetic stirrer mixing of the CNC suspensions decreased the suspension viscosities to the viscosity range near the initial viscosity of the suspension of CNC_M. Overnight mixing of the suspension of CNC_H+U did not lead to a significant difference on viscosity reduction compared with one-hour magnetic stirrer mixing did to the other suspensions. Magnetic stirrer can prepare a CNC suspension with a specific range of viscosity determined by the suspension treatment history.

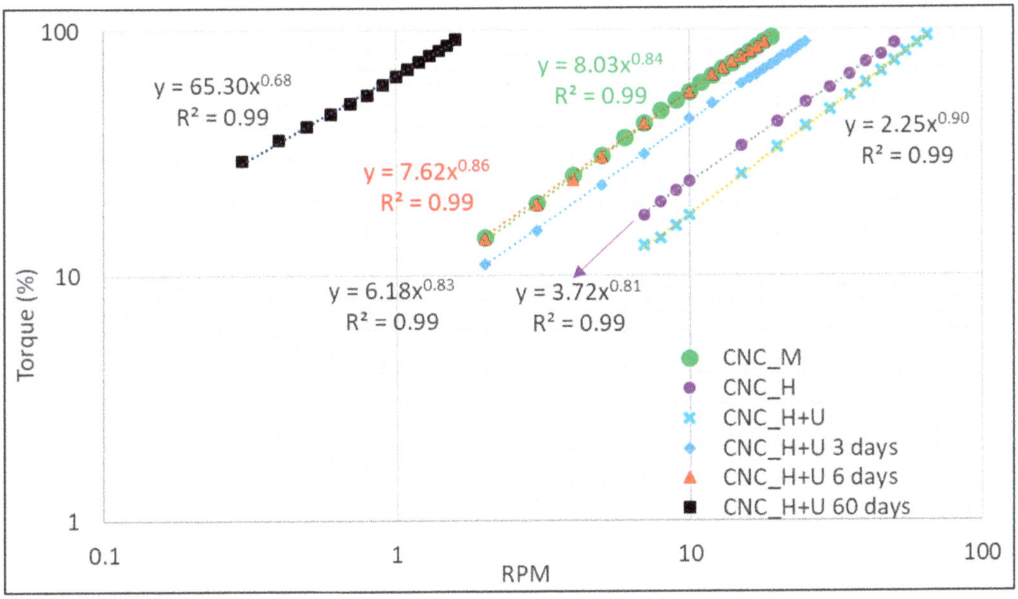

(a)

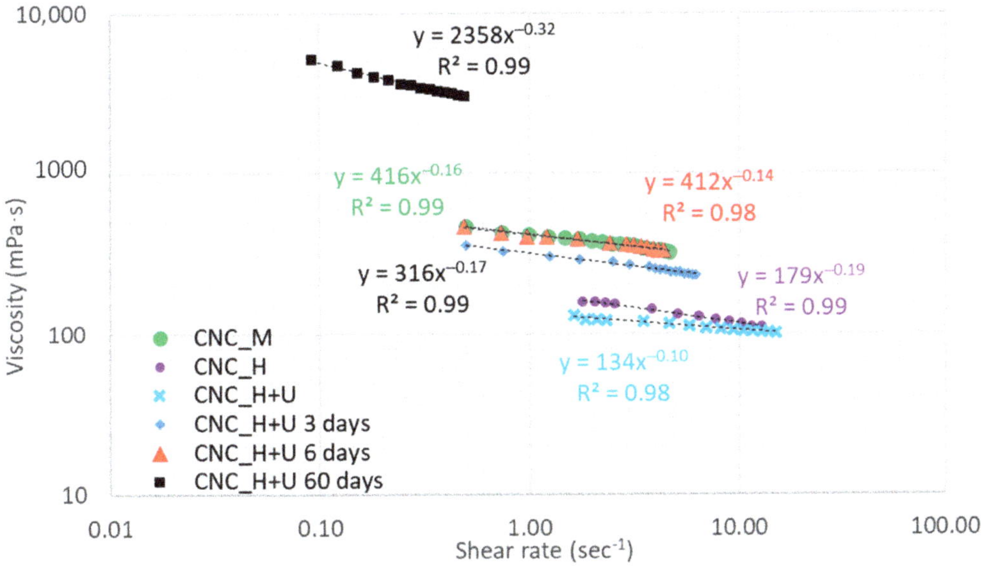

(b)

Figure 3. The change of viscosity with time for the CNC_H+U suspension (**a**) torque versus RPM and (**b**) viscosity versus shear rate.

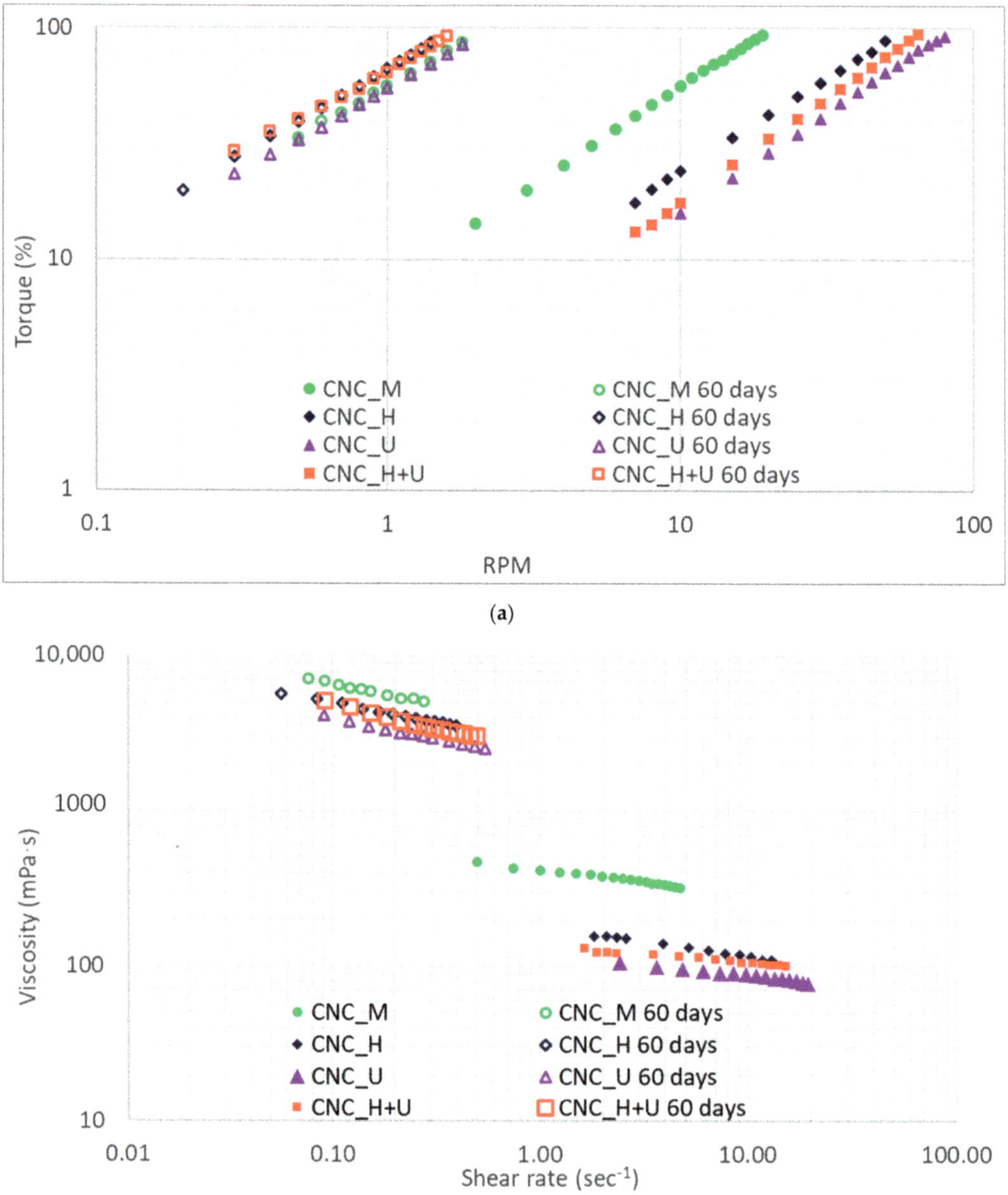

(a)

(b)

Figure 4. The viscosity of the CNC suspension after settling for sixty days (**a**) torque versus RPM and (**b**) viscosity versus shear rate.

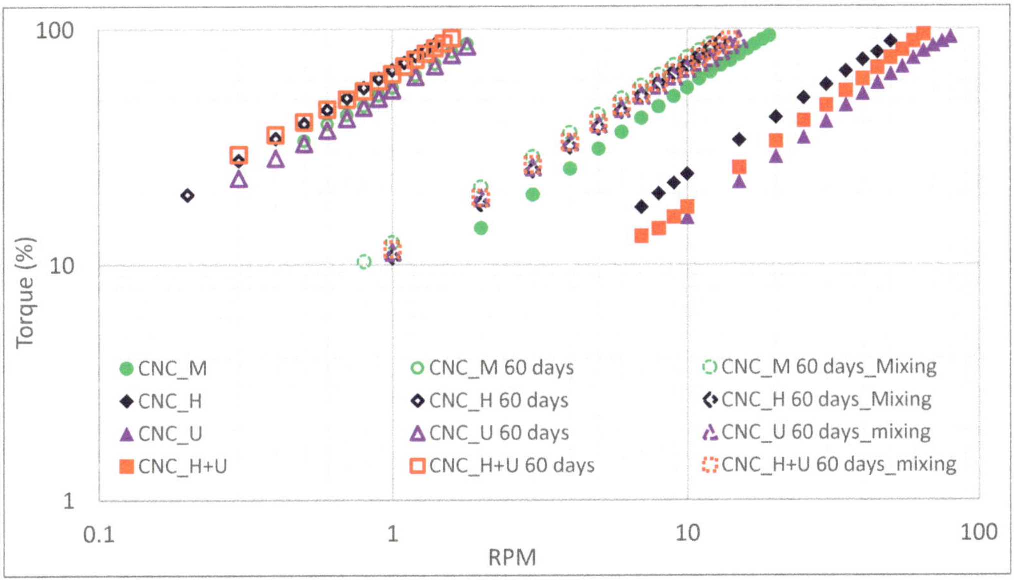

(a)

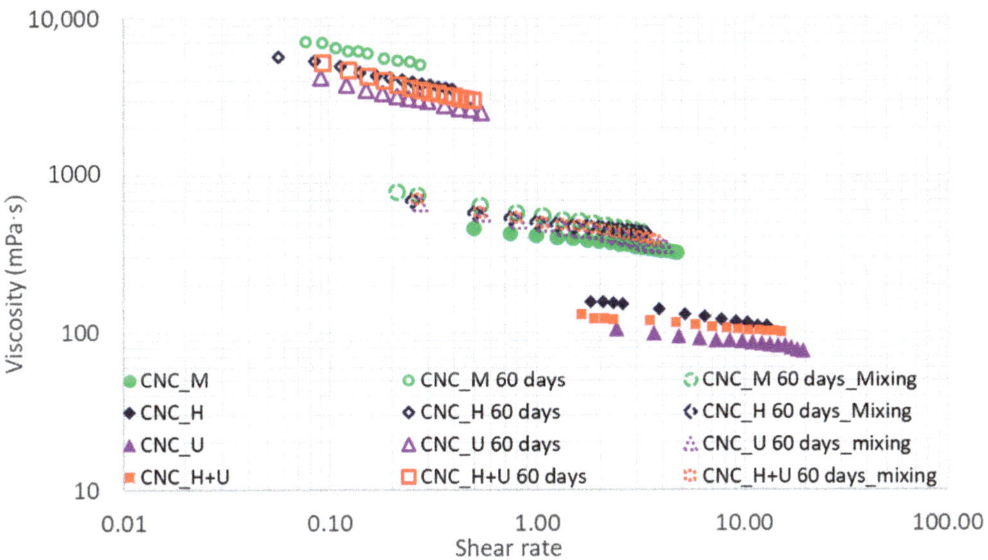

(b)

Figure 5. The effect of mixing by magnetic stirrer on the CNC suspension viscosity (**a**) torque versus RPM and (**b**) viscosity versus shear rate.

A quantitative analysis on the viscosity changes for settling and mixing are summarized in Table 1. All the plots of viscosity versus shear rate in Figure 5b for the CNC suspensions were fitted with power law flow models and model parameters are shown in Table 1. Settlement of the suspensions of CNC_M, CNC_H, CNC_U, and CNC_H+U for 60 days increased the consistency indices in the power law models from 416, 179, 119, and 134 to 3568, 2791, 2112, and 2358. The flow behavior indices for all the suspension have also decreased significantly with settlement (Table 1). After mixing with a magnetic stirrer, the consistency indices decreased while the flow behavior indices increased. The consistency indices decreased from 3568, 2791, 2112, and 2358 to 563, 524, 497, and 518 for the suspensions of CNC_M, CNC_H, CNC_U, and CNC_H+U, which are greater than the consistency indices for these suspensions at day one after the treatments. Magnetic stirrer mixing can reduce the suspension viscosity while it cannot reduce the viscosity down to the level that the homogenization and ultrasonication did to the suspensions. This may indicate that the magnetic stirrer cannot disassemble the formed liquid crystalline or colloidal structures as effective as homogenization or ultrasonication. When different treatment methods were applied to the CNC suspension, the CNC material, or the dispersion of the CNC particles in the suspension has been changed, resulting in different flow responses of the liquid when the same shear stress is applied. High shear action of homogenization and high energy sonication are capable of dispersing the CNC particles in suspension with the possibility of dissembling the formed liquid crystalline structures to a higher degree. After initial suspension treatment the consistency indices in the power law flow model of the suspensions are in the order of CNC_M > CNC_H > CNC_H+U > CNC_U and this order kept the same during the settlement and the second time magnetic stirrer mixing treatment (Table 1). This observation possibly indicates that the treatment history of the CNC suspension has an impact on the suspension viscosity.

Table 1. The power law flow model parameters for the CNC suspensions.

Suspension	Power Law Flow Model Parameter (Consistency Index—*k*; Flow Behavior Index—*n*)					
	Day One		Day 60		Mixing after Day 60	
	k	*n*	*k*	*n*	*k*	*n*
CNC_M	416	0.84	3568	0.73	563	0.79
CNC_H	179	0.81	2791	0.75	524	0.83
CNC_U	119	0.86	2112	0.73	497	0.77
CNC_H+U	134	0.90	2358	0.68	518	0.79

Solidified CNC films showed different morphologies when cast from the same suspension treated by different methods. After drying at the room condition to a constant weight, the moisture content of the films was measured to be 8.1%, 7.9%, 8.1%, 8.1% and 8.3% for the films of CNC_O, CNC_M, CNC_H, CNC_U, and CNC_H+U and no significant difference was observed for films generated from suspension treated by different methods. Casting original CNC suspension (CNC_O) formed warping films with deformation wrinkles after drying (Figure 6a). After treatment with different methods, flat films were formed for all the treated suspensions (Figure 6b) and no significant difference among different treatments was observed. The UV-Vis spectra of the solidified films were obtained in the range of 190–800 nm and the spectra normalized by film thickness are shown in Figure 7. Different UV-Vis spectra were obtained for the CNC films after different suspension treatment methods. In the visible light wavelength range from 380 to 740 nm the film of CNC_O showed the strongest absorbance of light while the films of CNC_U and CNC_H+U demonstrated the weakest interference with the visible light. The absorbance spectra in the measurement process are calculated using the transmittance measured by the spectrophotometer and the absorbance signal may include the reflected and scattered light. The film of CNC_O contains many wrinkles and would generate varied reflected and scattered light signals which makes it has the strongest visible light absorption. The absorbance of visible light for the CNC films cast from mechanically treated CNC suspension, including magnetic stirrer mixing and homogenization, is stronger than the films cast from ultrasonication treated suspensions (Figure 7). The CNC_M and CNC_H films have the same visible light absorbance. In the visible light range of 380 to 450 nm, the CNC_U film absorb less visible light than that of CNC_H+U film while in the wavelength range of 450 to 740 nm they showed the same visible light absorbance.

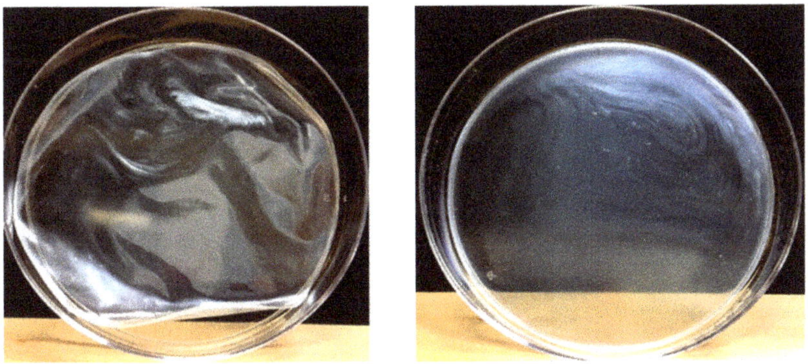

(a) (b)

Figure 6. Solidified films from casting solutions (**a**) CNC_O and (**b**) CNC_M.

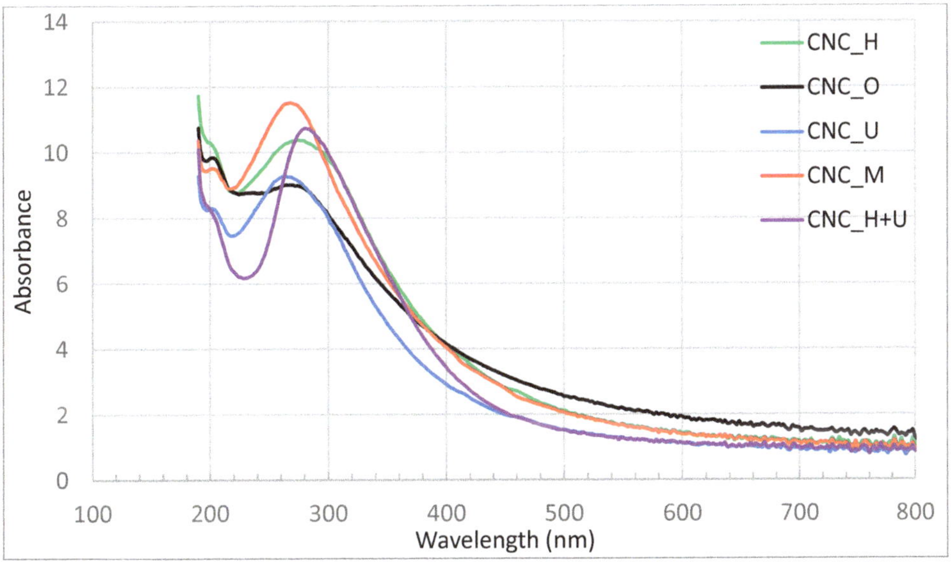

Figure 7. The UV-Vis spectra of CNC films.

In the UV light range of 190 to 380 nm, significantly different absorbances were observed for these films (Figure 7). Strong absorption below 200 nm is generally observed for cellulose [49] and the same strong absorption can be seen in Figure 7 for these CNC films. Two absorption peaks can be observed in Figure 7 at around 210 and 270 nm for the CNC films of CNC_O, CNC_M, CNC_H, and CNC_U. The relatively weak absorption at 210 nm is associated with the carbohydrate structure of CNC, which can be regarded as being composed of alcohol, ether, and hydrocarbon groups joined by aliphatic bonds [50]. The absorption peak at 210 nm can be identified to be at the similar magnitude in the films of CNC_O, CNC_M, and CNC_H while in the film of CNC_U the intensity of the peak is lower than the other three. The absorption peak at 270 nm in cellulose was explained in different ways from the literature. In acid treated cellulose and regenerated cellulose, an absorption peak at around 270 nm was attributed to the carboxyl groups formed during the treatment [51] while others attributed this absorption peak to the acetal groups in cellulose [52]. CNC is treated by sulfuric acid and oxidation is possible to form carboxyl group. In addition, sulfate ester group was introduced onto CNC surface and sulfate ion demonstrated similar absorption peak at around 270 nm [53]. Therefore, the absorption peak at 270 nm for the CNC films can be ascribed to the carboxyl and sulfate group attached to CNC particles. For the film of CNC_H+U, the UV absorption peak at 210 nm is overlapped with the film of CNC_U and the absorption peak at around 270 nm shifted to around 280 nm after homogenization and ultrasonication treatment of the suspension when compared with other CNC films. The high shear forces induced during the homogenization action followed by ultrasonication could change the structures of CNC differently than each single treatment which would change the absorption spectrum in the UV range, demonstrating the peak shifting. Further investigation is needed to figure out the behind mechanism of the UV absorption peak shifting for the film of CNC_H+U. The peak intensity absorbing at the wavelength of 270 nm can be ranked in the order CNC_M > CNC_H > CNC_U. For the CNC_H+U film, the intensity at the shifted peak is relatively higher than each individual homogenization and ultrasonication treatment. Pre-treatment of CNC suspension using different methods did change the UV and visible light absorption properties.

The original NIR diffuse-reflection spectra of absorbance versus wavenumber for all the CNC films are shown in Figure 8a. In this study, we specifically focused on the absorption bands in the range of 7200–6000 cm^{-1} and the detail spectra in this range is shown in Figure 8b. In this wavenumber range, the spectra are mainly dominated by the first overtone of O-H stretching vibrations [54]. In the cellulose polymer structure, three hydroxyl groups, one primary and two secondary, are attached to each glucose molecule and complex three dimensional structures, including amorphous and crystalline regions, are formed through inter- and intramolecular hydrogen bonds [55]. CNC particles after drying from suspension in air could aggregate into amorphous and crystalline structures due to the formation of hydrogen bonds among available hydroxyl groups. Different suspension treatment methods resulted in different suspension viscosities and are expected to change the hydrogen bonds formation process during drying, resulting in different rates to form amorphous and crystalline structures. The absorption intensity of each peak in the original spectra varied for different CNC films (Figure 8b). Based on the assignments of the four regions for different CNC films, including amorphous (A_m) region at 7014 cm^{-1}, semicrystalline (S_c) region at 6750 cm^{-1}, and intramolecular hydrogen bonded crystalline regions (C_I and C_{II}) at 6474 and 6290 cm^{-1}, the degree of crystallinity can be calculated using Equation (1). The assigned four peaks and the three shaded areas (A(A_m), A(C_I), and A(C_{II})) used for calculating the crystallinity of the CNC films are shown in Figure 8b. A baseline correction was performed using the Spectrum Quant quantitative analysis software and the area under the original NIR spectra was also calculated in the specified wavenumber range using the same software. According to the calculation based on the Equation (1), the degrees of crystallinity for the CNC films of CNC_O, CNC_M, CNC_U, and CNC_H are 57% while the degree of crystallinity of CNC_H+U film is 60%. The CNC suspension treatment through the combination of the homogenization and ultrasonication increased the crystallinity of the dried film and a change in hydrogen bond aggregation was possibly occurred during the film drying process. After the high shear treatment through homogenization better dispersion of CNC particles with a specific alignment could be generated in the suspension. The ultrasonication treatment would then change the dispersion of CNC again and created new surfaces for further aggregation under the preserved CNC particle alignment. Simultaneously, the ultrasonication is performed on the CNC suspension which has a much lower viscosity than suspension treated with magnetic stirrer. The structure change caused by the combination of homogenization and ultrasonication is different than that treated purely by homogenization or ultrasonication.

Other treatment showed almost the same crystallinity for the CNC film which may indicate that these treatments did not significantly change the hydrogen bond formation process during the film drying process.

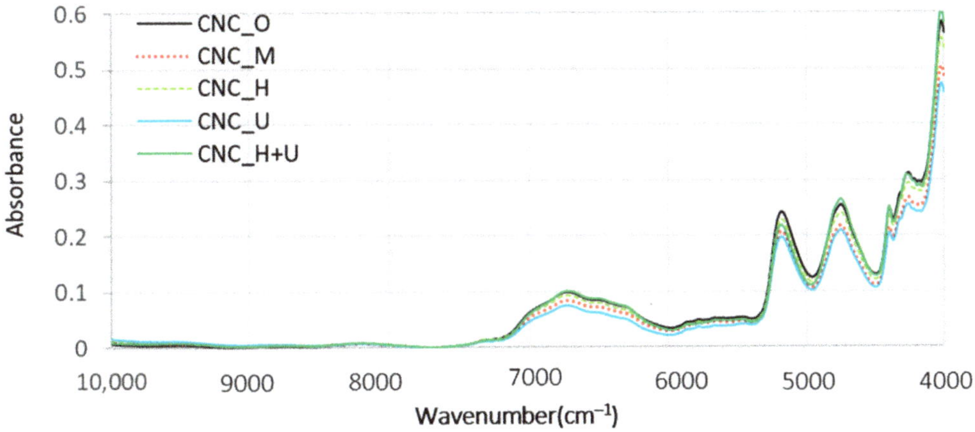

(a)

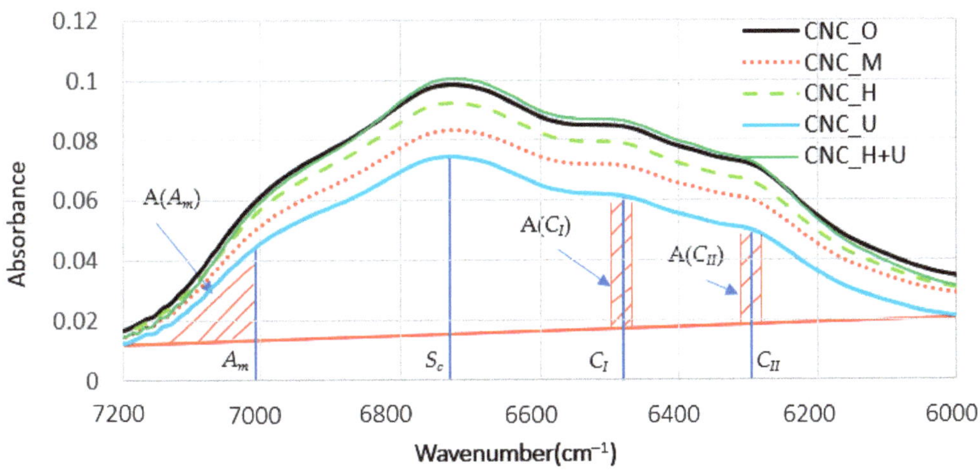

(b)

Figure 8. The original NIR spectra of the CNC films (**a**) the whole wavenumber range and (**b**) detail in the range of 7200–6000 cm^{-1}.

4. Conclusions

The rheological behavior of a CNC suspension at 6 wt.% was characterized using a rotational viscometer. In the researched shear rate range, the CNC suspensions treated by magnetic stirrer mixing, homogenization, ultrasonication, and combination of homogenization and ultrasonication demonstrated shear thinning phenomena which were fitted by power law flow models. Homogenization and ultrasonication can significantly decrease the viscosity of the CNC suspension compared with the suspension mixed by a magnetic stirrer. The viscosity of CNC suspension changed with time in a dynamic process after treatment and settlement of treated CNC suspensions in the room conditions increased the viscosity as a function of time. Mixing again with a magnetic stirrer decreased the suspension viscosity significantly to a different level. The flow behaviors of the suspension also changed with treatment and settlement.

CNC films cast from the suspensions treated by different methods were characterized using UV-Vis spectrophotometer and FT-NIR spectroscopy. Different UV and visible light interferences were observed for the CNC films. Mechanically mixed CNC suspension generated films with higher visible light absorption than films formed from ultrasonication treated suspension, especially in the lower range wavelengths of visible light. The crystallinity characterization using FT-NIR showed that the CNC film obtained from the treatment of homogenization and ultrasonication has the highest degree of crystallinity of 60% while other films have a same degree of crystallinity of 57%. In summary, pre-treatment of the CNC suspension did change the suspension properties and the properties of film formed thereof. The pre-treatment method should be specified for CNC suspension for any downstream application development, including coating, film manufacturing, viscosity modification, etc.

Author Contributions: Conceptualization, Y.P.; methodology, Y.P. and B.V.; investigation, Y.P.; resources, Y.P. and B.V.; writing—original draft preparation, Y.P.; writing—review and editing, Y.P. and B.V.; project administration, Y.P.; funding acquisition, Y.P. Both authors have read and agreed to the published version of the manuscript.

Funding: This work was supported by the USDA National Institute of Food and Agriculture, McIntire Stennis project ALAZ00079 and by the Alabama Agricultural Experiment Station.

Institutional Review Board Statement: Not applicable.

Informed Consent Statement: Not applicable.

Acknowledgments: Any opinions, findings, and conclusions, or recommendations expressed in this material are those of the author(s) and do not necessarily reflect those of the funding agencies. Y.P. also acknowledges the support from the School of Forestry and Wildlife Sciences and the Center for Bioenergy and Bioproducts at Auburn University.

Conflicts of Interest: The authors declare no conflict of interest.

References

1. Grishkewich, N.; Mohammed, N.; Tang, J.T.; Tam, K.C. Recent advances in the application of cellulose nanocrystals. *Curr. Opin. Colloid Interface Sci.* **2017**, *29*, 32–45. [CrossRef]
2. Mu, R.J.; Hong, X.; Ni, Y.S.; Li, Y.Z.; Pang, J.; Wang, Q.; Xiao, J.B.; Zheng, Y.F. Recent trends and applications of cellulose nanocrystals in food industry. *Trends Food Sci. Tech.* **2019**, *93*, 136–144. [CrossRef]
3. Huang, S.T.; Liu, X.H.; Chang, C.Y.; Wang, Y.X. Recent developments and prospective food-related applications of cellulose nanocrystals: A review. *Cellulose* **2020**, *27*, 2991–3011. [CrossRef]
4. Wijaya, C.J.; Ismadji, S.; Gunawan, S. A Review of Lignocellulosic-Derived Nanoparticles for Drug Delivery Applications: Lignin Nanoparticles, Xylan Nanoparticles, and Cellulose Nanocrystals. *Molecules* **2021**, *26*, 676. [CrossRef] [PubMed]
5. George, J.; Sabapathi, S.N. Cellulose nanocrystals: Synthesis, functional properties, and applications. *Nanotechnol. Sci. Appl.* **2015**, *8*, 45–54. [CrossRef] [PubMed]
6. Trache, D.; Tarchoun, A.F.; Derradji, M.; Hamidon, T.S.; Masruchin, N.; Brosse, N.; Hussin, M.H. Nanocellulose: From Fundamentals to Advanced Applications. *Front Chem.* **2020**, *8*. [CrossRef] [PubMed]
7. Fengel, D.; Wegener, G. *Wood—Chemistry, Ultrastructure, Reactions*; De Gruyter: Berlin, Germany, 1983.
8. Moon, R.J.; Martini, A.; Nairn, J.; Simonsen, J.; Youngblood, J. Cellulose nanomaterials review: Structure, properties and nanocomposites. *Chem. Soc. Rev.* **2011**, *40*, 3941–3994. [CrossRef]
9. Rånby, B.G. Fibrous macromolecular systems. cellulose and muscle. The Colloidal properties of cellulose micelles. *Discuss. Faraday Soc.* **1951**, *11*, 158–164. [CrossRef]
10. Mao, J.; Abushammala, H.; Brown, N.; Laborie, M.P. Comparative Assessment of Methods for Producing Cellulose I Nanocrystals from Cellulosic Sources. *Nanocellul. Their Prep. Prop. Appl.* **2017**, *1251*, 19–53.
11. Nascimento, D.M.; Nunes, Y.L.; Figueiredo, M.C.B.; de Azeredo, H.M.C.; Aouada, F.A.; Feitosa, J.P.A.; Rosa, M.F.; Dufresne, A. Nanocellulose nanocomposite hydrogels: Technological and environmental issues. *Green Chem.* **2018**, *20*, 2428–2448. [CrossRef]
12. Habibi, Y.; Lucia, L.A.; Rojas, O.J. Cellulose Nanocrystals: Chemistry, Self-Assembly, and Applications. *Chem. Rev.* **2010**, *110*, 3479–3500. [CrossRef]
13. Revol, J.F.; Bradford, H.; Giasson, J.; Marchessault, R.H.; Gray, D.G. Helicoidal Self-Ordering of Cellulose Microfibrils in Aqueous Suspension. *Int. J. Biol. Macromol.* **1992**, *14*, 170–172. [CrossRef]

14. Revol, J.F.; Godbout, L.; Dong, X.M.; Gray, D.G.; Chanzy, H.; Maret, G. Chiral Nematic Suspensions of Cellulose Crystallites—Phase-Separation and Magnetic-Field Orientation. *Liq. Cryst.* **1994**, *16*, 127–134. [CrossRef]
15. Beck-Candanedo, S.; Roman, M.; Gray, D.G. Effect of reaction conditions on the properties and behavior of wood cellulose nanocrystal suspensions. *Biomacromolecules* **2005**, *6*, 1048–1054. [CrossRef]
16. Liu, D.G.; Chen, X.Y.; Yue, Y.Y.; Chen, M.D.; Wu, Q.L. Structure and rheology of nanocrystalline cellulose. *Carbohyd. Polym.* **2011**, *84*, 316–322. [CrossRef]
17. Urena-Benavides, E.E.; Ao, G.Y.; Davis, V.A.; Kitchens, C.L. Rheology and Phase Behavior of Lyotropic Cellulose Nanocrystal Suspensions. *Macromolecules* **2011**, *44*, 8990–8998. [CrossRef]
18. Dong, X.M.; Revol, J.F.; Gray, D.G. Effect of microcrystallite preparation conditions on the formation of colloid crystals of cellulose. *Cellulose* **1998**, *5*, 19–32. [CrossRef]
19. Dong, X.M.; Kimura, T.; Revol, J.F.; Gray, D.G. Effects of ionic strength on the isotropic-chiral nematic phase transition of suspensions of cellulose crystallites. *Langmuir* **1996**, *12*, 2076–2082. [CrossRef]
20. Shafiei-Sabet, S.; Hamad, W.Y.; Hatzikiriakos, S.G. Ionic strength effects on the microstructure and shear rheology of cellulose nanocrystal suspensions. *Cellulose* **2014**, *21*, 3347–3359. [CrossRef]
21. Revol, J.F.; Godbout, L.; Gray, D.G. Solid self-assembled films of cellulose with chiral nematic order and optically variable properties. *J. Pulp Pap. Sci.* **1998**, *24*, 146–149.
22. Shopsowitz, K.E.; Qi, H.; Hamad, W.Y.; MacLachlan, M.J. Free-standing mesoporous silica films with tunable chiral nematic structures. *Nature* **2010**, *468*, 422–425. [CrossRef] [PubMed]
23. Pan, J.H.; Hamad, W.; Straus, S.K. Parameters Affecting the Chiral Nematic Phase of Nanocrystalline Cellulose Films. *Macromolecules* **2010**, *43*, 3851–3858. [CrossRef]
24. Hubbe, M.A.; Ferrer, A.; Tyagi, P.; Yin, Y.; Salas, C.; Pal, L.; Rojas, O.J. Nanocellulose in thin films, coatings, and plies for packaging applications: A review. *BioResources* **2017**, *12*, 2143–2233. [CrossRef]
25. Lagerwall, J.P.F.; Schutz, C.; Salajkova, M.; Noh, J.; Park, J.H.; Scalia, G.; Bergstrom, L. Cellulose nanocrystal-based materials: From liquid crystal self-assembly and glass formation to multifunctional thin films. *NPG Asia Mater.* **2014**, *6*. [CrossRef]
26. Bai, L.; Wang, Z.L.; He, Y.D.; Song, F.; Wang, X.L.; Wang, Y.Z. Flexible Photonic Cellulose Nanocrystal Films as a Platform with Multisensing Functions. *ACS Sustain. Chem. Eng.* **2020**, *8*, 18484–18491. [CrossRef]
27. Figueroa-Lopez, K.J.; Torres-Giner, S.; Angulo, I.; Pardo-Figuerez, M.; Escuin, J.M.; Bourbon, A.I.; Cabedo, L.; Nevo, Y.; Cerqueira, M.A.; Lagaron, J.M. Development of Active Barrier Multilayer Films Based on Electrospun Antimicrobial Hot-Tack Food Waste Derived Poly(3-hydroxybutyrate-co-3-hydroxyvalerate) and Cellulose Nanocrystal Interlayers. *Nanomaterials* **2020**, *10*, 2356. [CrossRef] [PubMed]
28. Kwon, G.; Lee, K.; Kim, D.; Jeon, Y.; Kim, U.J.; You, J. Cellulose nanocrystal-coated TEMPO-oxidized cellulose nanofiber films for high performance all-cellulose nanocomposites. *J. Hazard. Mater.* **2020**, *398*, 123100. [CrossRef]
29. Ling, Z.; Wang, K.L.; Liu, W.Y.; Tang, W.; Yong, Q. Tuning the cellulose nanocrystal alignments for supramolecular assembly of chiral nematic films with highly efficient UVB shielding capability. *J. Mater. Chem. C* **2020**, *8*, 8493–8501. [CrossRef]
30. van den Berg, O.; Capadona, J.R.; Weder, C. Preparation of homogeneous dispersions of tunicate cellulose whiskers in organic solvents. *Biomacromolecules* **2007**, *8*, 1353–1357. [CrossRef]
31. Maestri, C.A.; Abrami, M.; Hazan, S.; Chiste, E.; Golan, Y.; Rohrer, J.; Bernkop-Schnurch, A.; Grassi, M.; Scarpa, M.; Bettotti, P. Role of sonication pre-treatment and cation valence in the sol-gel transition of nano-cellulose suspensions. *Sci. Rep.* **2017**, *7*. [CrossRef]
32. Liao, J.S.; Pham, K.A.; Breedveld, V. Rheological characterization and modeling of cellulose nanocrystal and TEMPO-oxidized cellulose nanofibril suspensions. *Cellulose* **2020**, *27*, 3741–3757. [CrossRef]
33. Shafiei-Sabet, S.; Hamad, W.Y.; Hatzikiriakos, S.G. Rheology of Nanocrystalline Cellulose Aqueous Suspensions. *Langmuir* **2012**, *28*, 17124–17133. [CrossRef]
34. Xu, Y.; Atrens, A.; Stokes, J.R. Structure and rheology of liquid crystal hydroglass formed in aqueous nanocrystalline cellulose suspensions. *J. Colloid Interface Sci.* **2019**, *555*, 702–713. [CrossRef] [PubMed]
35. Xu, Y.; Atrens, A.D.; Stokes, J.R. Rheology and microstructure of aqueous suspensions of nanocrystalline cellulose rods. *J. Colloid Interface Sci.* **2017**, *496*, 130–140. [CrossRef]
36. Beck, S.; Bouchard, J.; Berry, R. Dispersibility in Water of Dried Nanocrystalline Cellulose. *Biomacromolecules* **2012**, *13*, 1486–1494. [CrossRef]
37. Cao, Y.Z.; Zavattieri, P.; Youngblood, J.; Moon, R.; Weiss, J. The relationship between cellulose nanocrystal dispersion and strength. *Constr. Build. Mater.* **2016**, *119*, 71–79. [CrossRef]
38. Chang, H.; Luo, J.; Davijani, A.A.B.; Chien, A.T.; Wang, P.H.; Liu, H.C.; Kumar, S. Individually dispersed wood-based cellulose nanocrystals. *ACS Appl. Mater. Interfaces* **2016**, *8*, 5768–5771. [CrossRef] [PubMed]
39. Girard, M.; Vidal, D.; Bertrand, F.; Tavares, J.R.; Heuzey, M.C. Evidence-based guidelines for the ultrasonic dispersion of cellulose nanocrystals. *Ultrason. Sonochem.* **2021**, *71*. [CrossRef]
40. Shojaeiarani, J.; Bajwa, D.; Holt, G. Sonication amplitude and processing time influence the cellulose nanocrystals morphology and dispersion. *Nanocomposites* **2020**, *6*, 41–46. [CrossRef]
41. Mu, X.Y.; Gray, D.G. Formation of Chiral Nematic Films from Cellulose Nanocrystal Suspensions Is a Two-Stage Process. *Langmuir* **2014**, *30*, 9256–9260. [CrossRef]

42. Beck, S.; Bouchard, J.; Berry, R. Controlling the Reflection Wavelength of Iridescent Solid Films of Nanocrystalline Cellulose. *Biomacromolecules* **2011**, *12*, 167–172. [CrossRef] [PubMed]
43. Tsuchikawa, S.; Siesler, H.W. Near-infrared spectroscopic monitoring of the diffusion process of deuterium-labeled molecules in wood. Part I: Softwood. *Appl. Spectrosc.* **2003**, *57*, 667–674. [CrossRef]
44. Inagaki, T.; Siesler, H.W.; Mitsui, K.; Tsuchikawa, S. Difference of the Crystal Structure of Cellulose in Wood after Hydrothermal and Aging Degradation: A NIR Spectroscopy and XRD Study. *Biomacromolecules* **2010**, *11*, 2300–2305. [CrossRef] [PubMed]
45. Rosen, M.R. A rheogram template for power law fluids II. rapid rheological characterization in a coaxial cylinder. *J. Colloid Interface Sci.* **1972**, *30*, 413–417. [CrossRef]
46. Rosen, M.R. A rheogram template for power law fluids: Technique for characterizing the rheological properties of emulsions and polymer solutions. *J. Colloid Interface Sci.* **1971**, *36*, 350–358. [CrossRef]
47. Brookfield Engineering Laboratories. *More Solutions to Sticky Problems: A Guide to Getting More from Your Brookfield Viscometer & Rheometer*; Brookfield Engineering Laboratories: Middleboro, MA, USA, 2017.
48. Mueller, S.; Llewellin, E.W.; Mader, H.M. The rheology of suspensions of solid particles. *Proc. R. Soc. A Math. Phys. Eng. Sci.* **2009**, *466*, 1201–1228. [CrossRef]
49. Bos, A. The UV spectra of cellulose and some model compounds. *J. Appl. Polym. Sci.* **1972**, *16*, 2567–2576. [CrossRef]
50. Perkampus, H.-H. *UV-Vis Atlas of Organic Compounds*, 2nd ed.; Wiley-VCH: Weinheim, Germany, 1992.
51. de la Orden, M.U.; Matias, M.C.; Sanchez, C.G.; Urreaga, J.M. Ultraviolet spectroscopic study of the cellulose functionalization with silanes. *Spectrosc. Lett.* **1999**, *32*, 993–1003. [CrossRef]
52. Beélik, A.; Hamilton, J.K. The ultraviolet irradiation of model compounds related to cellulose. *J. Org. Chem.* **1961**, *26*, 5074–5080. [CrossRef]
53. Karim, S.M.; Samuei, R. Absorption spectra of the sulphite and sulphate ions. *Proc. Indian Acad. Sci. Sect. A* **1934**, *1*, 398–406. [CrossRef]
54. Schwanninger, M.; Rodrigues, J.C.; Fackler, K. A review of band assignments in near infrared spectra of wood and wood components. *J. Near Infrared Spectrosc.* **2011**, *19*, 287–308. [CrossRef]
55. Jarvis, M. Chemistry—Cellulose stacks up. *Nature* **2003**, *426*, 611–612. [CrossRef] [PubMed]

polymers

MDPI

Review

Emergence of Polymeric Material Utilising Sustainable Radiation Curable Palm Oil-Based Products for Advanced Technology Applications

Rida Tajau [1,2,*], Rosiah Rohani [1,*], Mohd Sofian Alias [2], Nurul Huda Mudri [2], Khairul Azhar Abdul Halim [2], Mohd Hamzah Harun [2], Naurah Mat Isa [2], Rosley Che Ismail [2], Sharilla Muhammad Faisal [2], Marina Talib [2], Muhammad Rawi Mohamed Zin [2], Izzati Izni Yusoff [1], Nadiah Khairul Zaman [1] and Iqma Asyila Ilias [1]

1 Department of Chemical & Process Engineering, Faculty of Engineering and Built Environment, Universiti Kebangsaan Malaysia, UKM Bangi, Selangor 43600, Malaysia; nurul.izzati.izni@gmail.com (I.I.Y.); nadiahkz@gmail.com (N.K.Z.); p111889@siswa.ukm.edu.my (I.A.I.)

2 Radiation Processing Technology Division, Malaysian Nuclear Agency, Bangi, Kajang, Selangor 43000, Malaysia; sofian@nuclearmalaysia.gov.my (M.S.A.); nurul_huda@nuclearmalaysia.gov.my (N.H.M.); azharhalim@nuclearmalaysia.gov.my (K.A.A.H.); hamzah@nuclearmalaysia.gov.my (M.H.H.); naurah@nuclearmalaysia.gov.my (N.M.I.); rosley@nuclearmalaysia.gov.my (R.C.I.); sharilla@nuclearmalaysia.gov.my (S.M.F.); marina@nuclearmalaysia.gov.my (M.T.); muhammad_rawi@nuclearmalaysia.gov.my (M.R.M.Z.)

* Correspondence: rida@nuclearmalaysia.gov.my (R.T.); rosiah@ukm.edu.my (R.R.)

Abstract: In countries that are rich with oil palm, the use of palm oil to produce bio-based acrylates and polyol can be the most eminent raw materials used for developing new and advanced natural polymeric materials involving radiation technique, like coating resins, nanoparticles, scaffold, nanocomposites, and lithography for different branches of the industry. The presence of hydrocarbon chains, carbon double bonds, and ester bonds in palm oil allows it to open up the possibility of fine-tuning its unique structures in the development of novel materials. Cross-linking, reversible addition-fragmentation chain transfer (RAFT), polymerization, grafting, and degradation are among the radiation mechanisms triggered by gamma, electron beam, ultraviolet, or laser irradiation sources. These radiation techniques are widely used in the development of polymeric materials because they are considered as the most versatile, inexpensive, easy, and effective methods. Therefore, this review summarized and emphasized on several recent studies that have reported on emerging radiation processing technologies for the production of radiation curable palm oil-based polymeric materials with a promising future in certain industries and biomedical applications. This review also discusses the rich potential of biopolymeric materials for advanced technology applications.

Keywords: biomaterial; palm oil; radiation curing; radiation processing; surface coatings; VOC

Citation: Tajau, R.; Rohani, R.; Alias, M.S.; Mudri, N.H.; Abdul Halim, K.A.; Harun, M.H.; Mat Isa, N.; Che Ismail, R.; Muhammad Faisal, S.; Talib, M.; et al. Emergence of Polymeric Material Utilising Sustainable Radiation Curable Palm Oil-Based Products for Advanced Technology Applications. *Polymers* **2021**, *13*, 1865. https://doi.org/10.3390/polym13111865

Academic Editor: Antonio M. Borrero-López

Received: 21 April 2021
Accepted: 11 May 2021
Published: 4 June 2021

1. Introduction

The current evolution of radiation curable bio-based materials from various oil sources using radiation processing technologies has attracted the interest of scientists and technologists for developing novel biomaterials for diverse uses [1–6]. Synthetic polymers used in the processing of biomaterials have certain disadvantages, such as being highly toxic, non-biodegradable, and expensive. On the other hand, natural polymeric materials are simple to synthesise, low-cost, biodegradable, biocompatible, non-immunogenic, non-toxic, and biologically safety [7].

Natural vegetable oils and fatty acids that are mostly edible (e.g., palm oil, soya oil, jojoba oil, rapeseed oil, olive oil, and canola oil) have been widely used in the development of polymeric biomaterials, such as in coatings, lubricants, agrochemicals, and plasticisers [8–10]. The presence of carbon double bonding (-C=C-) in vegetable oils makes it easy to function with reactive groups, such as epoxy, hydroxyl, carboxyl, and acrylate, which allows the preparation of polyurethane, elastomers, plastics, and pressure-sensitive adhesives (PSA) [10–12]. Therefore, it has given attractive benefits to vegetable oils, unlike the toxic, costly, and non-environmentally friendly feedstocks taken from petroleum. Palm oil is among the most attractive renewable alternatives, and palm oil's inherent biodegradability is also a desirable feature in the context of rising environmental concerns compared to petroleum-based products concerning to minimising production costs and maximising inputs and outputs, such as food supply, water management, recovery of energy and products, and waste treatment [13]. Indonesia, Malaysia, and Thailand being the largest producers of palm oil in ASEAN countries will contribute to a significant field of palm oil industry for the oleochemical market.

Fatty acids, methyl esters, fatty alcohols, and fatty amines (Figure 1i) derived from natural oil are essential building blocks for the production of polyurethanes, surfactants, lubricants, and other products in the oleochemical industry (Figure 1ii). Therefore, the palm oil industry plays a crucial role in contributing to environmental sustainability and helps to reduce poverty of millions of palm oil workers and their families. Besides, these based sources have essential chemical compositions and ingredients for the development of food and non-food products.

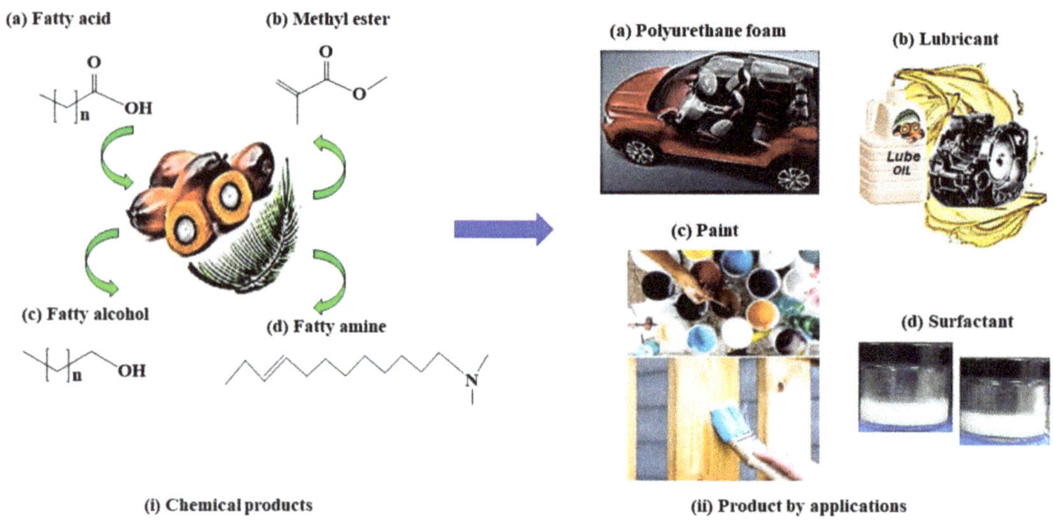

Figure 1. Palm oil used in oleochemical industry: (**i**) chemical products (**ii**) and formulated products by applications.

The development of radiation curable palm oil-based biomaterials seems to be one of the potential future applications of palm oil products, as this novel material has the potential to contribute positively to the palm oil industry. Expansion of the existing applications of palm oil products into the pharmaceutical industry is of interest because palm oil is an edible vegetable oil with excellent physicochemical properties, such as ability to modify chemical composition for biomaterials development. Likewise, other palm oil-based products, such as refined-bleached-deodorised (RBD) palm oil, palm oil, and palm stearin are abundantly available in Malaysia. They are known to be economical, sustainable, and environmentally biodegradable [3]. Palm oil is consumed worldwide as food products and has positive health properties, including antioxidants, anti-cancer, cholesterol-reducing effects, and is biologically active as pro-vitamin [14–16]. These palm oil-based products are promising prospects for manufacturing biomaterials that become alternative products to other polymers from synthetic/chemical-based.

Meanwhile, radiation processing technology has been reportedly used to modify the properties of materials in either bulk or surface structures with the aim to produce novel materials and products that pose desirable structures and properties [17–20]. Radiation processing can be properly controlled based on its delivery dosage, penetration depth, time of exposure, and raw material(s) used. This radiation method is known for its simplicity, rapid process, easy to control, and efficient for sterilisation. It is also a solvent-free process with low or no volatile organic compounds (VOC) and require fast-curing at room temperature only. These advantages allow the synthesis to be performed at low energy and require minimal working space. Radiation is also a particular energy source which can cause chemical reactions at any temperature (including ambient temperature), pressure, phase (gas, liquid or solid), and without assessment of a catalyst [21].

The general advantages of radiation processing techniques are simple process control, cost-effective, and safe for the environment. Therefore, these techniques are commonly used for modification of polymeric materials, polymer cross-linking, monomer and polymer curing, polymer grafting, and polymer degradation. These techniques are being utilised to develop new materials and advanced products, for example, nanomaterials and lithography (nanogels, metal nanoparticles, carbon nanotubes, nanocomposites), for different branches of industry [22,23]. Besides, radiation processing using gamma, electron beam, and UV sources have being explored and used for preservation and protecting cultural heritage artefacts, including desinsection, disinfection, restoration, and conservation to prevent further decay and damage [24].

Due to numerous benefits, this review is formulated to focus on the radiation curable palm oil-based materials prepared from radiation processing technologies. Critical discussions are presented in this review, especially on the palm oil-based polymeric sources, such as acrylate, polyol, and polyurethane, their properties and present use. The fundamental of radiation engineering for the production of biomaterials and their possible usage in various present and future applications are also discussed. This review is specifically aimed to strengthen the biomaterials benefits and potential applications in comparison to the well-known synthetic polymers. It is also especially aimed to reinforce radiation processing technology to be used for converting radiation curable palm oil-based products into novel biomaterials.

2. Fundamentals of Radiation Processing in Polymeric Biomaterial Design

Radiation-induced material changes are dependent on the source and form of the radiation and the energy deposited [25]. Radiation is classified into two primary classifications, which are non-ionising and ionising radiations. Ionising radiation is extraordinarily energetic and has sufficient energy as particles and electromagnetic waves to ionise a material molecule. Non-ionising radiation is the term used for radiation in the electromagnetic waves, where insufficient energy is required to induce the ionisation that requires a photo-initiator. The method of energy deposition relies on the source of radiation, such as particulate radiation (including electrons, positrons, protons, neutrons, and ions) or electromagnetic radiation (including infrared, visible and ultraviolet radiation, and X-rays and gamma rays).

Polymer irradiation is an industrial process that is very useful. It is an alternative to more conventional chemical processes in which the method may induce or modify the properties of the material. Primary interactions of polymers (either natural or synthetic polymers) with ionising radiation include ionisation, excitation of molecules and atoms, stabilisation of electrons by the generation of hot electrons, ion neutralisation, and free radical formation. These result in the formation of excited states of reactive intermediates, free radicals, ions, atoms, and molecules, as shown in Figure 2.

In non-ionising radiation, such as UV radiation, free radicals are produced typically by hydrogen abstraction when the photo-initiator is exposed to UV light (Figure 3a). Radicals formed during the initiation step (Figure 3b) propagate further via a free radical process (Figure 3c). Figure 3 demonstrates the mechanism of the reaction of the free radical formation by photo-initiators.

a. Ionization: $P \rightsquigarrow P^{+} + e^{-}$

b. Excitation: $P \rightsquigarrow P^{*}$

c. Thermal stabilization of electron: $e \rightarrow e^{-}_{th}$

d. Neutralization: $P^{+} + e^{-}_{th} \rightarrow P^{*}$

e. Free radicals formation: $P^{*} \rightarrow R_1^{\bullet} + R_2^{\bullet}$

f. Disproportionation: $\rightarrow R^{\bullet} + H^{\bullet}$

Figure 2. Primary processes occurring of polymer in ionising irradiation. * P is a polymer and $R^{\bullet}$ is a radical. Source: Adapted Vasile and Butnaru [26].

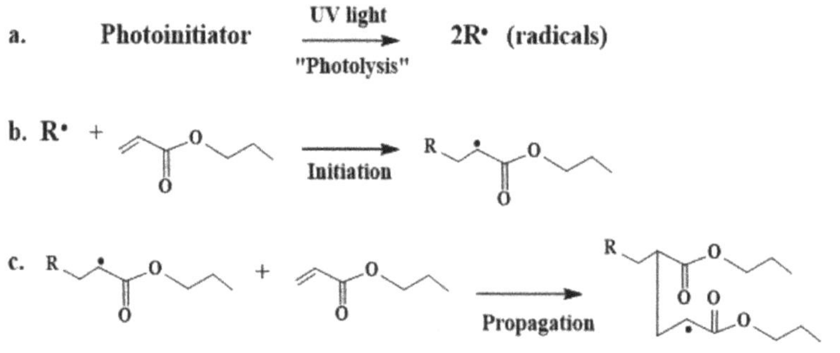

Figure 3. Primary mechanism of free radical polymerisation under UV radiation of (**a**) photolysis, (**b**) initiation, and (**c**) propagation.

In the case of polymer in a dilute aqueous solution, the primary reaction involves the radiolysis of water, where water molecules absorb the energy of ionising radiation. This reaction results in the creation of reactive intermediates: hydroxyl radicals ($^{\bullet}OH$), hydrogen atoms ($H^{\bullet}$), hydrated electrons (e^{-}_{aq}), hydrogen ion (H^{+}), hydrogen peroxide (H_2O_2), and hydrogen molecule (H_2) (Figure 4).

$$H_2O \xrightarrow{\text{ionizing radiation}} {}^{\bullet}OH,\ H^{\bullet},\ e^{-}_{aq},\ H^{+},\ H_2O_2,\ H_2$$

Figure 4. Primary process of radiolysis of water under irradiation. Source: Adapted Rosiak and Ulanski [27].

Furthermore, the subsequent interactions include either the transfer of electrons; the abstraction of hydrogen atoms; or the addition of C=C, C=N, C=S, aromatic rings, or an electron-rich functional group. This process contributes to the creation of radical sites in the polymer chain. It also allows the macroradicals to follow several different pathways of reaction, such as chain scission, hydrogen transfer, inter- or intramolecular recombination, and disproportion, which ultimately result in the formation of oxidised products, branching, grafts, scission/degradation, or cross-linking of the main chains of polymers, as shown in Figure 5 [28].

Chain branching forms a growing chain of polymers from the original chain of polymers, cross-linking forms an insoluble three-dimensional network of polymers, whereas degradation or scission involves the breaking of chemical bonds on the polymer backbone. These schemes of chains can be seen in Figure 5. Chain branching and cross-linking processes increase the polymeric molecular weight, while degradation or scission causes the initial molecular weight to be reduced. Overall, radiation processing technology is widely used to modify the material properties, and novel materials with desirable structures and properties can be produced as well.

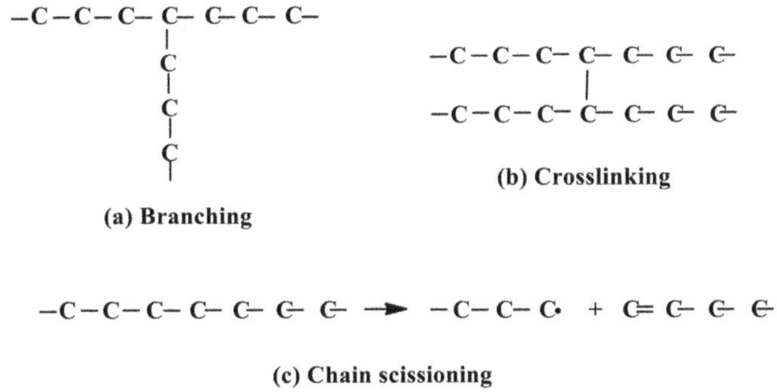

Figure 5. Scheme of chain branching, cross-linking, and chain scissioning.

3. Radiation Curable Palm Oil-Based Materials

Bio-based materials produced from vegetable oils via radiation curable process have received great attention for various applications in recent years. Palm oil, in particular, is directly obtained from the red-orange mesocarp of oil palm *Elaeis guineensis* fruit. It is edible and used worldwide as the main ingredient in many food products [29]. Palm oil has a complex mixture of triglyceride compounds, which can be seen in Figure 6. It is composed of fatty acids, including palmitic, stearic, myristic, oleic, and linoleic acids (Figure 7).

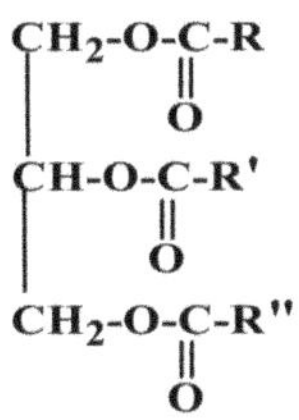

Figure 6. General chemical structure of palm oil (-R can be represented by palmitic, stearic, myristic, oleic, and linoleic).

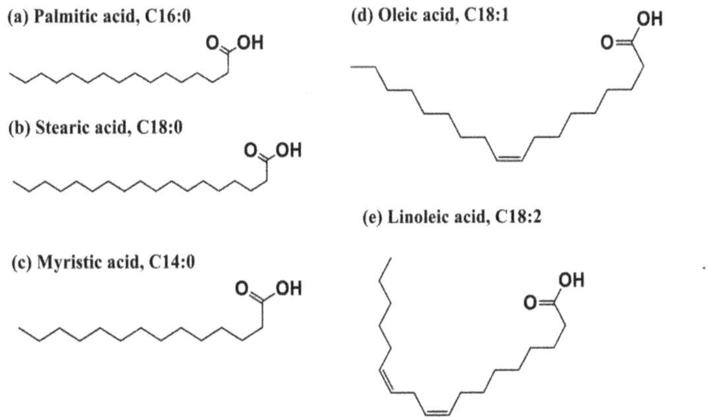

Figure 7. Chemical fatty acids structure in palm oil (C-C: saturation; C=C: unsaturation). It has high levels of saturated fatty acids, about 50% of which include palmitic (44.3%), stearic (4.6%), and myristic (1.0%). It also contains around 38.7% of oleic acid monounsaturated fatty acid and 10.5% of linoleic acid polyunsaturated fatty acid. Source: Adapted from Amir Reza, Suraya, and Azmi [30].

In the recent past, palm oil has gained significant interest and been extensively researched for nutritional, medicinal, and chemical applications, such as antioxidant activities, cholesterol-lowering benefits, anti-cancer effects, protection against atherosclerosis, emulsion and cosmetic formulations, and non-food products [31]. The palm oil products, such as polyol [32] and epoxidised palm oil acrylate (EPOLA®) [33], which are referred as Acrylated Palm Olein (APO) [34], have received considerable interest in many applications, such as surface finishing, packaging, automotive, and biomaterial (Figure 8). These products have shown remarkable physicochemical properties, as reported in many studies, which have been summarised in Table 1. The molecular weights and functional groups of polyol ester and APO play a vital role in the development and determination of the properties of the new products. They possess low molecular weight of below 5500 Daltons and hydrocarbon chains with ester compounds that could play a big role in hydrolytic degradation process. This inherent property could give additional biodegradability of APO and polyol ester compared to petrochemical-based polymers or synthetic polymers. Besides, their chemical structures, such as carbon double bonds (-C=C-) and hydroxyl groups (-OH), are also essentially useful in fine-tuning and developing new products with unique properties.

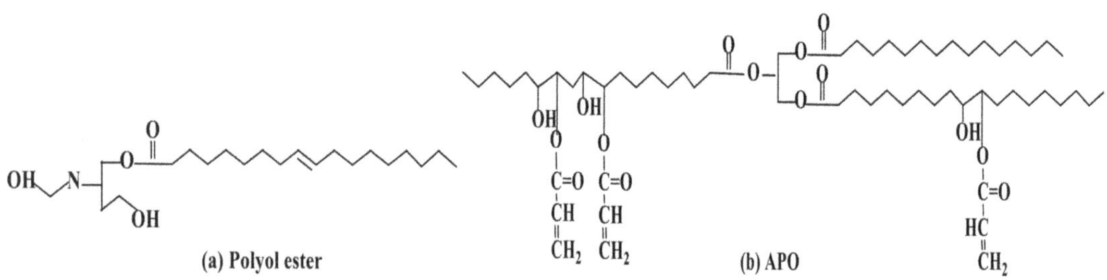

Figure 8. The general chemical structure and products of (**a**) polyol ester and (**b**) APO.

Table 1. The physicochemical properties of APO and polyol ester.

Type of Palm Oil-Based Products	APO [34]	Polyol Ester [32]
Acid value (mg KOH/g oil)	8.37	39.74
Saponification value (mg KOH/g oil)	221.89	297.00
Hydroxyl value (mg KOH/g oil)	46.41	182.51
Ester value	209.14	257.26
Viscosity (cps)	1774.67	1006.33
Molecular weight (Daltons)	1750.04	5004.00
Colour	Brownish	Dark

Many studies have demonstrated that EPOLA®, APO and polyol have shown promising characteristics in applications for radiation processing technology. For example, the presence of carbon double bonds in EPOLA® and/or APO are reactive sites to promote modification of materials through radiation technology, such as via UV and gamma radiation. Upon exposure to radiation, these bonds then form cross-linked structure that can give enormous properties to new products (refer to Figure 9a). Moreover, the polyol reactive hydroxyl group provides a medium for chemical modification and surface functionalisation. Palm oil-based polyols with functional hydroxyl groups, for instance, have been widely used to modify polyurethane products (the image can be seen in Figure 9b) [35].

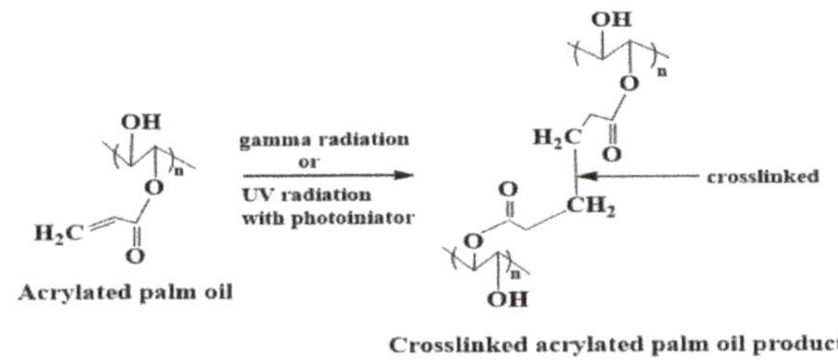

(a) Radiation cross-linking of the carbon double bonds of the acrylated palm oil products.

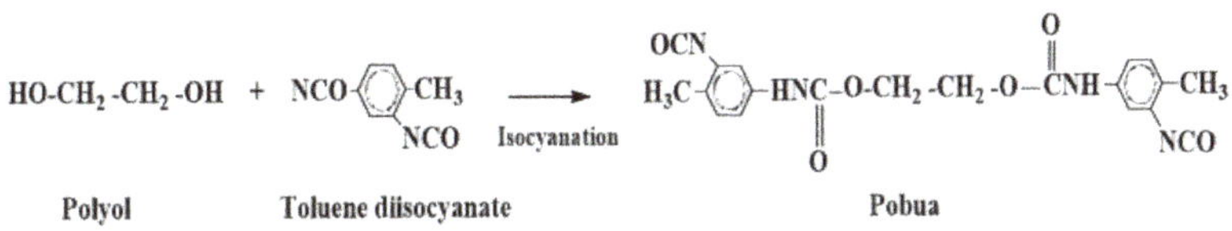

(b) Example of isocyanation reaction in the production of polyurethane based on palm oil.

Figure 9. Application of acrylated palm oil and polyol in (**a**) radiation polymerisation of EPOLA® or APO and (**b**) isocyanation of polyol in polyurethane production.

Concurrently, Figure 10 shows the recent applications of advanced radiation curing technology utilising palm oil-based materials. EPOLA® can been seen to produce radiation-curable products, such as coatings (refer to Figure 10a) [36], pressure-sensitive adhesives (refer to Figure 10b) [37], printing inks and overprint varnishes [38], while APO is used to develop drug delivery systems. Meanwhile, polyols based on palm oil are used to produce polyurethane for the manufacture of foams, elastomers, plastics, and cast and injection-moulded components. These show that radiation-curable products based on palm oil have many usages for different sectors, such as agricultural, military, automobile, manufacturing and medical [10,37]. Next, descriptions of the application of radiation processing for the production of radiation-cured palm oil products for usage in various future developments will be addressed in the following section.

(a) (b)

Figure 10. Radiation-curable of palm oil-based products with UV curing. (**a**) Acrylated and polyurethane resins for wood coating; (**b**) Polyurethane resins for PSA. Source: Unpublished data.

4. Current Application of Radiation Curable Palm Oil-Based Polymeric Materials

4.1. Surface Coating

The production of EPOLA® and polyol-type polymers is believed to ultimately eliminate a large portion of the supply of petroleum-based polymers. Since they are derived from natural resources, their usages seem to be potentially cost-effective for developing natural polymers compared to petroleum-based polymers. This is because palm oil is a sustainable resource and has a reliable manufacturing line for production of raw materials. Besides, a majority of acrylic and polyol-type polymers currently in the market are petrochemical-based. Therefore, these innovations could boost industries that utilise polymeric coatings to focus on using sustainable commodities, such as from palm oil as potentially cheaper feedstocks, rather than petrochemical-based substances that are usually costly and non-sustainable [3].

Radiation-curable materials, such as EPOLA®, polyol, and polyurethane have shown inherent properties for application in coating industries, which include its physicochemical, environment, and chemical resistance, thermal and mechanical properties [2,3,39–46]. Thermal stability and mechanical properties are mainly referred in the characterisation of the surface coating's thermal durability and reliability. Negligence in considering these properties would result in products cracking, delamination of the layers, and products deformation. These properties play a critical role in controlling the deposited coating elements in order to ensure that the quality of the substrates is maintained.

The thermal and mechanical properties of EPOLA® and polyol-based products and their most promising applications upon exposure to UV irradiation are summarised in Table 2. Based on Table 2, EPOLA® is seen to be widely used in various types of applications, such as surface coating resin, overprint varnish, adhesive, inks, and thermoplastic production. In contrast, polyol is used in miscellaneous polyurethane materials, such as ceiling panels, flora foams, polyurethane sheets, wall panels, roof insulators, laminated boards, adhesives, coatings, cushion, flexible foams, and automotive parts [3,47–50]. Materials based on EPOLA® and polyol are produced from products based on palm oil, such as epoxy palm olein and oleic acid. *They are specially developed as radiation curable resin for making surface coating products.* Next, the hardness test results presented in Table 2 shows that the acrylic and polyurethane resins based on EPOLA® have nearly identical ranges of deformation resistance properties as compared to polyol-based polyurethane. Meanwhile, the polyol-based polyurethane also displayed lower temperature degradation as compared to the resins-based EPOLA®. Different thermal and mechanical properties presented in Table 2 were attributed to various processing and formulation methods for ingredients, photo-initiators, and absorbed radiation dose of the resin products developed from EPOLA® and polyol.

The substitution of radiation-curable resins from palm oil is known to substitute petrochemical-based resins with multifunctional developmental coating resins. Multifunctional coatings can provide additional durability advantages, including surface resistance to mildew, fungus, bacteria, and microorganisms; anti-fouling without metals; superhydrophobicity; self-healing anti-corrosion; self-healing scratch repair; and thermochromic. Such features can open up modern, evolving applications of radiation-curable palm oil-based resin, such as in automobile, aerospace, and telecommunication devices; military and heavy-duty equipment; and agricultural machinery. The radiation-curable palm oil-based materials also gained strong attraction due to their variety of use, not only in coating, over print varnishes, ink, and adhesive industrial applications but also in the medical industry in particular to fulfil the needs for the production of biomaterial from natural resources relative to petrochemical source materials that tend to cause toxicity. Therefore, to ensure that the safety and effectiveness of these biomaterials are acceptable for the delivery of drugs throughout the human body, the production of a drug delivery system using radiation-curable palm oil-based materials was foreseen relevant.

Table 2. The thermal and mechanical properties of the EPOLA® and polyol-based products.

Developed Products	Thermal and Mechanical Properties	References
Acrylated resin from epoxidized oil (EPOLA®-based products)	Adhesion: 1.15 MPa Pencil hardness: H-2H	[51]
UV curable PSA of maleinated acrylated epoxidized palm oil (EPOLA®-based products)	Good and high adhesion properties	[52]
UV-curable polyurethane coating (EPOLA®-based products)	Adhesion: 0.5–1.5 MPa Pencil hardness: B and 2B Volatile organic compound: 5–15%	[53]
UV radiation-curable acrylic resin (EPOLA®-based products)	Degradation: 250–550 °C Pendulum hardness: 60% Tensile strength: 5.2–6.2 MPa	[42]
UV-curable acrylated coating (EPOLA®-based products)	Pendulum hardness: 55–75%	[43]
UV radiation-curable polyurethane resin (Polyol-based products)	Degradation: 200–455 °C Pencil hardness: B-2H	[48]
UV/EB-curable acrylated coating (EPOLA®-based products)	Scratch resistance: 0.1–0.9 N Pendulum hardness: 5–50% Pencil hardness: B-6H	[54]
UV curable acrylated polyol ester prepolymer (EPOLA®-based products)	Pendulum hardness: 49.4%	[55]

4.2. Drug Delivery Systems

In recent years, palm oil has received considerable research interest in the delivery of drugs. This is supported by the summary presented in Table 3 that outlines the latest production of micro- or nano-scale palm oil-based products used in drug delivery. The summary shows that the standard techniques for the development of the micro- or nano-scale palm oil-based products include double casting, emulsion, micro-emulsion, nano-emulsion, high pressure homogenisation (HPH), particles from gas saturated solutions (PGSS), free thaw, thin-film hydration, laser ablation, and radiation techniques to be used in drug carrier field. Table 3 also summarises the current strategies for designing pharmaceutical products that utilise palm oil-based materials using different methods.

The results summarised in Table 3 also shows that palm oil-based products developed using different techniques stated above are capable of producing less than 500 nm of particle size. The produced particles in this range of size are found to be highly useful for oral, transdermal, topical, parenteral, pulmonary, antibacterial, and drug delivery systems, especially for breast cancer. Microemulsion is found as the most commonly used technique in manufacturing pharmaceutical products due to its low-cost, simplicity, efficient method, and affordability. On the other hand, the combination of microemulsion and radiation processing techniques has been announced recently as a promising strategy for creating a new type of drug carrier [17,27,28,56–58]. Using these combined techniques, the three-dimensional (3D) network structure and nanoparticles from the palm oil-based products such as APO and polyol ester has been developed.

Table 3. Current development of palm oil-based products in drug delivery.

Product Development	Methods	Properties	Potential Application	References
Palm oil-based paracetamol suppositories	Double casting	More than 95% drug release	Oral delivery	[59]
Transparent microemulsion from palm oil-derived isopropyl palmitate (IPP) with Tween 80 and 1- butanol	Microemulsion	Particle size <150 nm pH: 6.76 to 7.80 Charge: −32 to −75 mV Storage stability for 4 weeks.	Transdermal and Topical	[60]
Palm oil based-organogels and microemulsions from palm oil, span 80/tween 80 mixture (organogel) and water	Microemulsion	Highly haemo-compatible Non-irritant	Topical delivery	[61]
Palm oil and soy oil-based organogels	Microemulsion	Good spreadability and vscosity profile highly biocompatible	Topical delivery	[62]
Palm Oil Esters (POEs)-based nanoemulsions	Microemulsion	Less than 500 nm	Topical delivery	[63]
Palm oil esters (POEs)-based nano-cream	Microemulsion	Less than 130 nm −20 mV Stable	Topical delivery	[64]
Microspheres encapsulating terbutaline sulphate nanoparticles using hydrogenated palm oil	Emulsion	3.9 um	Pulmonary delivery	[65]
Palm oil-based nanoparticle	Microemulsion and ionizing radiation	Particle size: 70–220 nm	Preliminary study for oral delivery	[66]
Solid lipid nanoparticle (SLN) using varieties of emulsifier of palm oil (S154) and lecithin (Lipoid 100)	High pressure homogenization (HPH)	Particle size: 140–300 nm Zeta potential of about −13 and −20 mV uniform size distribution	Colloidal drug delivery	[67]
Microcomposites theophylline/hydrogenated palm oil	Particles from gas saturated solutions (PGSS)	Mean particle size: 2.5–3.0 um Contain initially from 0.5 to 3.5% (weight) of theophylline uniform size distribution	Controlled drug delivery systems	[68]
Palm oil-based Liposomal Dox (Doxorubicin hydrochloride)	Freeze-thaw	(a) Distribution sizes of 438 and 453 nm (b) Zeta potential of about −31 and −32 mV Stability	Breast cancer delivery	[69]
Palm oil-based liposome	Thin film hydration	(a) Mean particle size: 340–450 nm (b) Zeta potential: −26 to −33 mV (c) Stable Biodegradability	Drug delivery device	[70]
Palm-oil-based Liposomal Dox (Doxorubicin hydrochloride)	Freeze-thaw	Distribution sizes of 438 and 453 nm Zeta potential of about −31 and −32 mV Stability	Breast cancer delivery	[71]
Palm oil-based liposome	Thin film hydration	Mean particle size: 340–450 nm Zeta potential: −26 to −33 mV Stable Biodegradability	Drug delivery device	[72]
Silver nanoparticles	Laser ablation	2.5–2 nm	Nanocomposite fabrications and antibacterial	[71]
Gold nanoparticles	Laser ablation	8–20 nm	Drug delivery device	[30]

The development of drug delivery systems via radiation of materials, such as polymeric micelle and nanoparticles has been preliminary initiated using APO. Gamma-induced cross-linking has been successfully conducted on different types of APO's microemulsion for developing the micro/nanoparticles [72]. These nanoparticles were dispersed in aqueous media at the critical micelle concentration of the surfactant and exposed to gamma irradiation to generate the nanoparticles. Figure 11 shows the TEM images of APO nanoparticles produced via gamma irradiation. The use of gamma-ray irradiation induces the hydroxyl radical forming in water to promote the double bond of acrylate carbon in APO that facilitates the formation of cross-link C-C bonds. The disappearance of the acrylate carbon double bond (-C=C-) at peaks 1641, 1620, 1409, 981, and 817 cm^{-1} analysed using FTIR indicate that the APO nanoparticles have undergone cross-linking after the irradiation, as shown in the CTAB microemulsion system in Figure 12 [66]. These findings showed that the irradiation technique could be used to produce nanoparticles for usage in drug carriers of below than 100 nm particle size, as confirmed in TEM images presented in Figure 11.

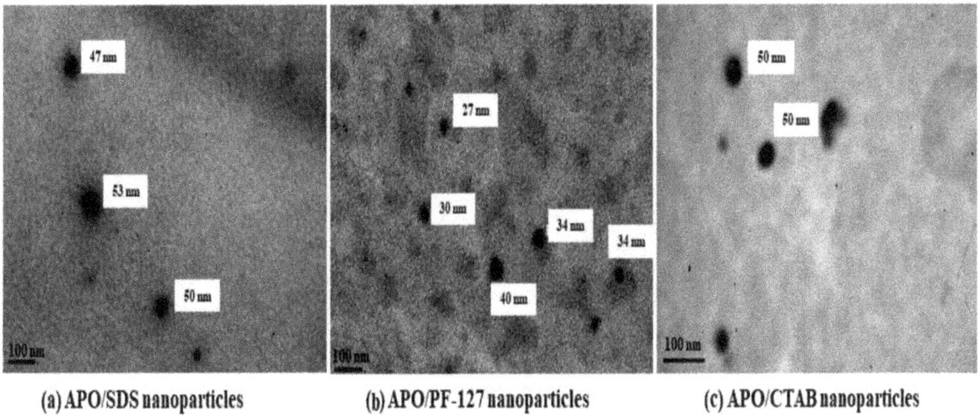

Figure 11. TEM images of the gamma-irradiated APO nanoparticles at different surfactants' microemulsion system. APO composition of 0.18% in the microemulsion produces a variety of sizes of nanoparticles, such as 47–53 nm in the sodium dodecyl sulphate (SDS), 27–40 nm in the Pluronic-127 (PF-127), and 50 nm in the cetyltrimethylammonium bromide (CTAB) microemulsions. Source: Unpublished data.

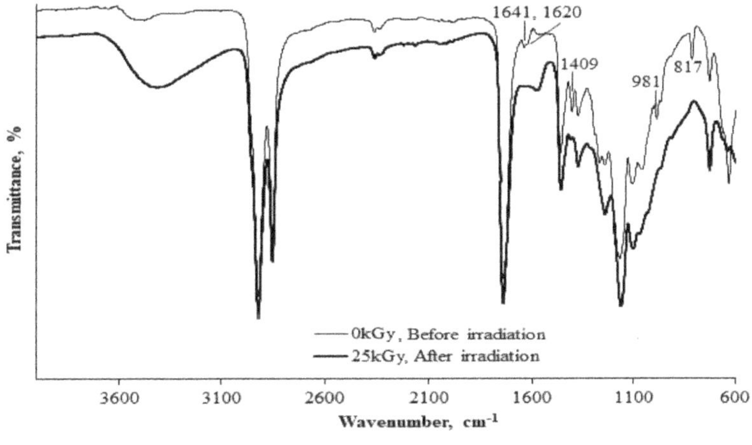

Figure 12. FTIR spectra of APO/CTAB matrix (**a**) non-irradiated; (**b**) irradiated at 25 kGy.

Furthermore, Sadrolhosseini et al. reported on gold nanoparticles (Au-NPs) synthesised from palm oil using green and simple laser irradiation techniques [30,73]. Using the electron transfer from the carboxylic group to Au-NPs, the tail of the carbonyl band of palm oil was used as a capping agent for Au-NP. Gold nanoparticles with a particle size ranging from 8.92 to 19.73 nm were formed in the spherical shape of palm oil. These Au-NPs have a wide range of applications in drug delivery, microelectronics, environments, photonics, electronics, photodynamic therapy, therapeutic agent delivery, biosensors, sensors, medical diagnosis, and catalysis.

In recent a study by Tajau et al., gamma irradiation-induced RAFT was used for the development of targeted copolymer nanoparticles using APO and polyol ester as reactants of starter polymer matrices (Figure 13) [74]. The poly(APO-b-polyol ester) nanoparticles are made from palm oil products using the RAFT technique induced by gamma radiation. These copolymer nanoparticles are potentially used in drug delivery systems for breast cancer.

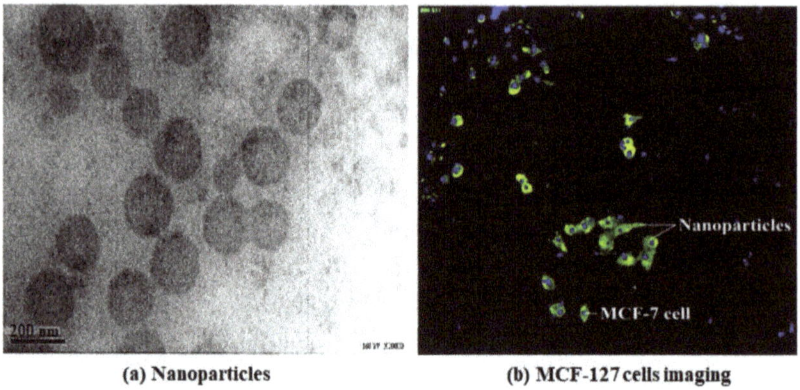

Figure 13. Peptide functionalised- and paclitaxel-loaded poly (APO-*b*-polyol ester) nanoparticles against MCF-7 cells. (**a**) Peptide was functionalised to poly(APO-*b*-polyol ester) nanoparticles using chemical conjugation technique and was loaded with paclitaxel; (**b**) Promising results were shown where the functionalised nanoparticles were found to be able to cause inhibition of human breast cancer cells. Source: unpublished data.

Reports also showed that various copolymers were developed using oleic acid (OA) by RAFT polymerization [75–77]. Previously, Zengin et al. developed hybrid silica-coated magnetic nanoparticles using oleic acid as a stabiliser via RAFT polymerization [75,76]. The surface-mediated RAFT polymerisation approach enables for synthesising less than 20 nm hybrid composite particles size. These particles have displayed superparamagnetic properties (saturation magnetisation = 35.4 emu g^{-1}) and precise identification of molecular cholesterol. These nanoparticles could become a single magnetic domain at such a small scale and perform like a single super spin that shows strong magnetic susceptibility, enabling these nanoparticles to have a better and faster magnetic reaction relative to bulk magnets because under the control of an applied magnetic magnet, these superparamagnetic nanoparticles imprinted with cholesterol would contribute to the recovery of spiked human serum, milk, yolk, and beef for use in cell sorting, biomolecule enrichment and isolation, and drug delivery applications [76].

Meanwhile, a pH-responsive supramolecular graft copolymer micelle (Dextran-g-OA) composed of dextran and poly (oleic acid) was produced by Karmakar et al. via RAFT polymerization [77]. The Dextran-g-OA micelle of less than 300 nm in particle size demonstrated good loading efficacy and excellent release features to nifedipine for the oral delivery system. It also showed its cyto-compatible nature towards MG-63 cells for gastro-retentive drug delivery system. Nifedipine is used for the treatment of high blood pressure, cardiovascular disease, and chest pain relief. Comparatively, short gastric emptying time in humans may result in an incomplete release of Nifedipine from the existing drug delivery system that leads to a decreased effectiveness of the administered dose. As a result, this discovery could solve existing concerns to improve the effectiveness of nifedipine delivery to specific organs [78].

The RAFT polymerisation is a useful tool in the development of engineered polymeric materials due to its ability to perform at room temperature, aqueous media, and a controllable initiation process. However, the synthesis of copolymeric nanoparticles of the radiation-curable palm oil-based materials has yet to be widely reported primarily through radiation and RAFT techniques. Therefore, the usage of radiation and RAFT techniques for the development and modification of palm oil-based materials is also known to have a significant potential for drug delivery application. This review is functioned to explain further on this technique for producing palm oil-based copolymers nanoparticles for cancer therapy.

Besides, in the following section, it is highlighted that APO is not only a drug nanocarrier, but it also can be used in the production of biopolymeric scaffolding materials for use in medicine for the reconstruction and replacement of tissues and organs. Therefore, the use of APO in the production of drug delivery system and scaffolding is found necessary to replace existing synthetic medical devices due to safety and local cost concerns.

4.3. Scaffolds

Radiation technologies play important roles in the development of tissue engineering, including the preparation and optimisation of instructive scaffolds. A new photopolymer resin prepared from APO resin has been formulated using a micro-stereo-lithography technique to be used in the manufacturing of tissue scaffold with a regulated microstructure and bioactivity, tailored porosity structure as well as inclusion phase shape and size [79]. The manufacturing of a tissue scaffold is carried out by an additive layer process in which the 3D object is sliced into a series of 2D layers with each of these layers produced until the 3D part is formed. Current calibration printing studies are performed using 100% poly(ethylene glycol) dimethacrylate (PEGDMA) with a photo-initiator present in each resin formulation using UV light. The appropriate design of the engineered tissue scaffold is carried out using CAD (computer-aided configuration) software. Figure 14 provides examples of scaffold products manufactured from APO using a micro-stereo-lithography technique. The generated tissue scaffolds, i.e., hearing aid and 3D denture, were predicted to have a controlled bioactivity due to starter scaffolding matrices made of radiation-curable palm-based materials that make it safe, biodegradable, and biocompatible as well as possess the desired porosity and mechanical properties [80].

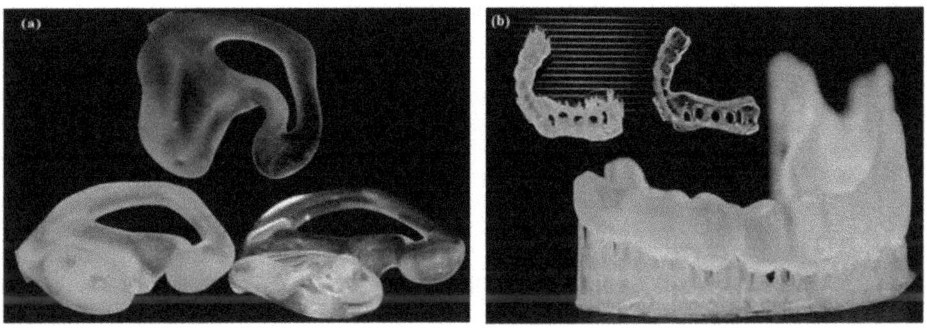

Figure 14. CAD model design using microstereolithography technique for (**a**) hearing aids and (**b**) 3D dentures. Source: Adapted from Tajau and Talib [80].

Meanwhile, acrylic-based polymers, such as polyacrylic acid (PAA), are well known for their excellent biocompatibility and efficiency in mechanical properties. This has benefitted their usage for various significant biomedical applications, such as contact lenses, corneal prosthesis, bone cement, tissue engineering, nanofibres, nanocomposites, and nanoparticles [81]. Modification of polyacrylic acid using irradiation is especially useful for improving the properties of scaffolds based on acrylic. Radiation techniques have been reported as reliable methods for acrylic-based hydrogel synthesis for manufacturing of engineered tissue scaffolds [81]. For example, a cross-linked PAA hydrogel synthesised by cross-linking ionising radiation has been developed by Jabari et al. [82] and Rosiak et al. [27] for therapeutic dressing. Lee et al. also worked on UV irradiation method in interpenetration polymer network (IPN) hydrogels synthesis that composed of chitosan and poly(acrylic acid) (PAAc). A significant improvement has been shown in its mechanical properties, even in a hydrated condition [83]. Another reported study by Fabbri et al. also showed that the successful preparation of graphene oxide/acrylic resins through UV radiation of electrically conductive acrylic resins containing graphene oxide (rGO) has resulted in improved properties of conductive acrylic composites [84]. It can be said that almost all these studies revealed that when reacting to radiation, acrylic-based polymeric materials have a promising characteristic, as it continues to evolve as more sophisticated materials for tissue engineering products.

In a different study reported nearly a decade ago by Rao et al., they developed a poly(3-hydroxybutyrate-co-4-hydroxybutyrate), P(3HB-co-4HB) that was obtained by biosynthesis of *Cupriavidus necator* from spent palm oil [85]. Their findings showed that inexpensive spent palm oil is a good source of carbon in the processing of polyhydroxyalkanoates (PHA) efficiently using *Cupriavidus necator* via biosynthesis technique. This biopolymer ultimately can become a modern absorbable biomaterial for medical usage [86]. Kamilah et al. also supported their finding whereby they demonstrated that the palm oil-based waste cooking oil (PO-WCO) was found suitable as a sustainable source of carbon for cell growth and poly(3-hydroxybutyrate) P(3HB) biosynthesis [86]. However, radiation technique was not applied in both reported studies, thus has opened up the gap for utilising the technique in similar studies.

The production of radiation curable palm oil-based acrylic scaffolds through radiation processing could be carried out in no time in order to identify new strategies for improving and displaying superior biomaterial properties, especially for medical applications, as this technique is versatile and can be easily adapted. In addition to the materials being the starter matrices for polymeric scaffolds, as discussed in the following section, radiation-curable palm oil-based materials are also introduced in the use of polymeric composites as dental filling materials.

4.4. Dentistry

A recent US patent granted in 2014 claimed on the polyurethane oligomer based on palm oil for usage as a dental restorative material [11]. Palm oil-based polyols possess more than two functional groups that induce branching of the polymer and tend to have higher crosslinking density as well as three-dimensional linked structures that provide greater mechanical strength and dimensional stability. The use of palm oil-based polyol can also improve the biocompatibility of the synthesised composition. The polyol and diisocyanate formulations have both soft and hard segments which are grafted by the polyol with isocyanate terminals with an unsaturated acrylic monomer to form a prepolymer with unsaturated terminals which is curable with free radical initiators. As a result, this can provide better dental filling materials that are fatigue resistance and have flexural properties. Figure 15 shows the UV light cure filling teeth dental process using curable materials, such as palm oil-based polyol dental filling materials. These composites of palm polyol were found susceptible to UV radiation, which were used as a coat for filling teeth cavities.

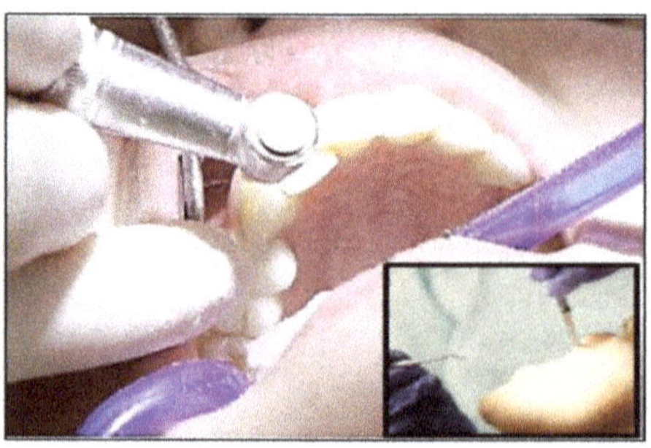

Figure 15. Ultraviolet light cure palm oil-based fillings.

For several decades, acrylic-based resins with application in dentistry made up of polymethylmethacrylate (PMMA) or polyethylmethacrylate (PEMA) have dominated denture technology [87,88]. However, some of the drawbacks of these materials are the toxicity of the residual monomer, the complex wrapping procedure, complicated processing, and low resistance. Furthermore, there are also several different types of resin compounds, such as styrene, polycarbonate, polyethylene, polyethylene glycol, polydimethylsiloxane, poly(e-caprolactone), polypyrrole, hexamethyldisilazane, epiminic, polyurethane, vinyl, polyamide, acetal, and polyglass available for dental use in the manufacture of dentistry resins, most of which come from petrochemical-based polymer [87,89]. A recent study by Ionta et al. revealed that palm oil has gained superior protective capacity against tooth erosion, which appears to be a promising alternative for the prevention of initial erosion of enamel compared to coconut oil, safflower oil, sunflower oil, and olive oil [90]. Therefore, new resin alternatives, such as radiation-curable palm oil-based resins with a natural, renewable source, inexpensive, biodegradable, biocompatible, and promising physicochemical properties compared to petrochemical-based acrylics, are thus more suitable for dental applications.

Other features were reported in a few studies that were conducted on surface coating, overprint varnish (OPV), printing inks, pressure-sensitive adhesives (PSA), and dentistry for radiation-curable palm oil-based product applications [1,91]. These applications appeared to have a little exploration, but the palm oil-based products do have more possibilities to fine-tune the functionality of the radiation-curable products to satisfy the need for the application, including agricultural for seed treatments and coatings [92,93], restoration and consolidation of various cultural heritage artefacts as well as the use of radiation technology to disinfect cultural heritage artefacts [24,94,95].

5. Emerging Applications of Radiation Processing for Modification of Palm Oil-Based Polymeric Biomaterials for Potential Use in Biomedical Application

Radiation-modified palm oil-based materials is an example that has great potential in promoting new products for use in biomedical application, such as medical surface coating, medical device adhesives, smart drug delivery system, and tissue engineering devices. Besides, their favourable properties, such as natural properties, non-toxic materials, and cost-effective production lines due to the low cost of raw material can make palm oil-based products a good option for developing novel and advanced technology for use as polymeric biomaterials in the biomedical application.

5.1. Medical Surface Coating

The development of radiation modification of palm oil-based materials in advanced applications for superhydrophobic biomaterials, such as temporally implants, contact lenses, controlled drug-release coatings medical device coatings, and antimicrobial coatings could be of great promise. Several authors have reported potential development on radiation modification of palm-oil based material coating, such as acrylate, polyol, and urethane [40,47,96,97]. For instance, Said et al. used the combination of biopolymer of an acrylate palm oil-based EPOLA®/PETIA with different amount of SiO_2 nanoparticles to produce good optical nanocomposites for numerous applications, including painting, sealants, adhesives, and coatings [54]. As a result, the increase of surface hardness and gel content values were achieved by increasing the silica loading in photocurable resin. In addition, the rapid decrease in water absorption in the presence of SiO_2 allows such nanocomposites particularly desirable for coating applications. The nanocomposites comprising nanosilica distinguish exceptional barrier properties against moisture and gases as well as excellent resistance to corrosion [98]. Salleh et al. reported on producing hyperbranched curable polyurethane acrylate (HBPUA) made from palm oil oleic acid for applications in surface curing technology [40]. HBPUA is described as green and/or eco-friendly curable resin products with excellent thermal stability, especially upon subjecting to UV/EB irradiations.

Alias et al. introduced zirconium acrylate in EPOLA® curable coating formulation in order to enhance EPOLA® properties [99]. The presence of zirconium acrylate increases the adhesion and hardness of curable coating due to increment of the roughness and polarity of the surface. Meanwhile, the zirconium structure and properties contributed to the enhancement of film hardness. Mudri et al. also used EPOLA® as a coating material on printed pages or known as overprint varnishes (OPV) [41]. The purpose of applying OPV is to enhance gloss, stain resistance, edge fusion resistance, resistance to burning and scuffing, and resistance to discolouration from environmental absorption of impurities. The EPOLA®-based OPV material possesses excellent properties for adhesion and glossiness.

Overall, the development of radiation-curable palm oil-based materials for surface coatings can reduce the risk of infection and prevent bacterial or microorganism attachment to the material surfaces. These materials must be chemically and physically non-reactive to the body as well as excellently comply when embedded in tissues. Therefore, the medical surface coating is necessary to reduce toxicity, side effects, and bacterial or microorganism infections without affecting public health.

5.2. Medical Light Curing Adhesives

Radiation processing of palm oil-based products could provide a wide variety of medical light curing adhesives for tape manufacture, pressure-sensitive adhesive (PSA), 3 M, and general adhesives. Radiation curable PSA was produced from different palm oil resins, such as EPOLA® and EPOMA (epoxidised palm oil methacrylate). Such resins have found to improve coating, curing, and adhesive performances. Mahmood et al. studied the EPOLA®-PSA developed with various low-transition temperature acrylate and methacrylate monomers before curing with UV irradiation [100]. The findings showed that EPOLA®-based PSA exhibited fast curing speed as well as appropriate compatibility formulations, as demonstrated by their cured film properties, such as surface tackiness, peel adhesion, and creep resistance.

In 2017, Tugiman et al. produced UV curable maleinated acrylic epoxy palm oil PSA using UV LED system, which resulted in a small percentage of double conversion rate that was below 50% due to monomers and maybe even light transmission. This work is just a few known UV curable PSAs that have been reported for palm oils. The PSA synthesis of palm oil-based materials from the studies shows promising results in the production of radiation-curable palm oil-based PSA for medical device application [52].

5.3. Nanoscale Radionuclides and Radiolabelled Nanomaterial

Nanoscale radionuclides and radiolabelled nanomaterial for imaging, *diagnosis, and treatment of* APO and polyol ester had promising properties used by radiation-induced RAFT polymerisation and cross-linking techniques to produce targeted nanoparticles utilising radiation techniques. New features for the design of smart nanoparticles for nuclear molecular imaging and therapeutic purposes include the surface functionalisation of these nanoparticles with biomolecules, radioactive materials and also with a magnetic, plasmonic or fluorescent substance. The use of radiation modification in palm oil-based materials in nuclear nanomedicine in these latest technologies is expected to bring a paradigm shift in multimodal molecular imagery, thus facilitating a new era in medical imaging agents for diagnosis.

5.4. Three-Dimensional (3D) Scaffold for Tissue Engineering Application

The production of a photopolymer resin-based palm oil scaffold for bone tissue engineering therapy is a significant field for providing supply of biomaterials relating to the osseointegration of bioresorbable scaffolds and the control of bone infection. Talib et al. investigated the construction of a scaffold using materials based on palm oil, such as APO, using a micro stereolithography technique [79]. The study obtained favourable properties of dynamics, composition, bioactivity, porosity, and density. More innovation on the scaffolding of APO may display more substantial interests in potential applications, such as for tissue engineering. Meanwhile, potential physicochemical, mechanical, biodegradable, and biocompatibility properties were discovered in a water-blown porous polyurethane scaffold made of palm glycerol monostearate and glutaric acid, which demonstrated their tremendous biomaterial potential for soft and rigid tissue engineering applications [101,102].

Furthermore, a new soft biomaterial made from aliphatic polyurethane based on palm oil has recently been developed by Yeol et al. for tissue engineering applications. A polyester polyol (PPP) obtained from epoxidised palm olein and glutaric acid was used to prepare the polyurethane [103]. The polyurethanes displayed superior water absorption, >90% porosity, biodegradable, good thermal stability, and mechanical properties that made the generated polyurethanes able to be used for tissue repair and tissue regeneration applications.

In addition to the palm-based oleochemicals used for the development of biomaterials for tissue engineering, the palm oil mill effluent (POME) generated from palm oil mills and the palm oil empty fruit bunch (EFB) developed from palm oil extraction have also generated significant interest in the production of modern scaffold polyhydroxyalkanoates [104,105]. These studies have shown a promising hope that palm oil-based biomaterials can be used in a wide variety of tissue engineering applications. The different forms of biopolymers derived from palm-based oleochemicals and oil palm waste materials as biomaterials have significantly contributed to the manufacture of scaffolding products. Finally, radiation processing technology could be used to obtain more ideal characteristics of biomaterials for modifying the scaffolding products.

6. Way Forward

Finally, radiation-curable bio-based products from palm oil have shown great interest in biomedical applications for surface coating, adhesive, nanoscale radionuclides and radiolabelled nanomaterials, and scaffolding. For example, antimicrobial and anticorrosion properties can be added to the radiation curable polymeric materials for use in medical surface coatings for an active approach to killing microbial and surface protection and also for use in *dental* restorative in a wide variety of nanocomposites. It also can act as core platforms for medical radionuclide attachment and radiolabelled for molecular imaging and targeted drug delivery and as light cure adhesives, curable palm oil formulations that can be assembled as medical and surgical adhesive tapes for skin, wound dressing, and wound bandages. Meanwhile, 3D bioprinting can facilitate the development of functional tissue structures with radiation-curable polymeric based on palm oil approaches to mimic human tissues and organs. Therefore, new coatings like radiation-curable polymeric materials based on palm oil can benefit the production of medical devices. They are natural, enable the delivery of drugs, are surface-modified, and can be used to form artificial tissue. Radiation-curable products derived from palm oil have a bright future in the development of biomaterials since they originally come from sustainable resources are a convenient and environmentally friendly technology for the processing of raw materials and non-toxic products, and because of the availability of much cheaper products that make them comparable in terms of biocompatibility, biodegradability, durability, and the reasonability production cost compared to petrochemical-based products. Furthermore, radiation processing technology provides a simple, reliable, and environmentally safe manufacturing facility and is a powerful tool to induce physical and chemical reactions for use in the production of advanced modern materials.

7. Conclusions

As a conclusion, polymeric-based triglycerides and monomers, such as radiation-curable palm oil-based polymeric materials, have become alternatives to synthetic polymers. Thus, they are potentially useful for a diverse range of industries including biomedical applications such as surface coatings, drug delivery systems, and scaffolds. These polymeric materials are highly favourable for the above applications, as they are derived from natural resources, non-toxic, inexpensive, biocompatible, and biodegradable and possess promising thermal and mechanical properties. Moreover, the radiation-curable palm oil products obtained can be used as polymeric biomaterials in imaging, diagnostics and therapeutic purposes, medical surface coating and adhesives, radionuclides and radiolabelled nanomaterial, and tissue engineering. Therefore, the development of polymeric biomaterials based on radiation-induced palm oil fatty acids is found to be significant for broadening the use and application of palm oil in biomedical applications. Last but not least, the most considerable attraction and valuable insight in this invention is that the primary raw material comes from palm oil, which can be obtained easily from palm oil-producing countries.

Author Contributions: R.T., R.R., M.S.A., N.H.M., K.A.A.H., M.H.H., N.M.I., R.C.I., S.M.F. and I.A.I. designed and performed the coating projects. R.T. and R.C.I. performed the PSA project. M.T. designed and directed the scaffold project. R.T., R.R., I.I.Y. and N.K.Z. designed and performed the drug delivery system project. R.T. wrote the paper with input from all authors in consultation with R.R. and M.R.M.Z. The first draft of the manuscript was written by R.T. All authors have read and agreed to the published version of the manuscript.

Funding: This research was funded by Malaysia Nuclear Agency, grant number NM-R&D-20-52, Ministry of Higher Education (Malaysia), grant number FRGS/1/2018/TK02/UKM/02/2, Universiti Kebangsaan Malaysia, grant number DIP/2019/012, and the Scholarship was funded by Department of Public Service (JPA, Malaysia), grant number HLP/2016.

Institutional Review Board Statement: Not applicable.

Informed Consent Statement: Not applicable.

Data Availability Statement: Not applicable.

Acknowledgments: The authors would like to extend their appreciation to the local inventors in Malaysian Nuclear Agency produced radiation modification palm oil-based materials, namely, the late senior research officer of Allahyarham Hj. Mohd Hilmi Mahmood, Khairul Zaman Mohd Dahlan, Mek Zah Salleh and Nik Ghazali Nik Salleh. Acknowledgements are owed to those who contribute to and promote this research directly and continuously, including all in the Radiation Curing and Synthesis Group (KSPS) staff at the Malaysian Nuclear Agency.

Conflicts of Interest: The authors declare no conflict of interest.

References

1. Salleh, M.Z.; Tajau, R.; Mudri, N.H. *The Evolution of Palm Oil Acrylates within 20 Years in Nuclear Malaysia*; INIS-MY–2017-072; IAEA: Vienna, Austria, 2016; pp. 1–6.
2. Hanzah, N.A.; Tuan Ismail, T.N.M.; Adnan, S.; Mohd Noor, N.; Sattar, M.N.; Mohd Zan, A.; Poo Palam, K.D.; Abu Bakar, Z.; Zolkarnain, N.; Hoong, S.S.; et al. Development of Palm oil-based polyols and polyurethane products in MPOB. *Palm Oil Dev.* **2019**, *71*, 13–17.
3. Cheong, M.Y.; Zafarizal Aldrin, A.H. Potential of palm oil as a renewable material for radiation curable coating. *Palm Oil Dev.* **2018**, *68*, 14–17.
4. Mohan, S.; Oluwafemi, O.S.; Kalarikkal, N.; Thomas, S.; Songca, S.P. Biopolymers-Application in Nanoscience and Nanotechnology. In *Recent Advances in Biopolymers*; Perveen, F.K., Ed.; Intech Open: London, UK, 2016.
5. Palanisamy, C.P.; Cui, B.; Zhang, H.; Jayaraman, S.; Muthukaliannan, G.K. A Comprehensive Review on Corn Starch-Based Nanomaterials: Properties, simulations, and applications. *Polymers* **2020**, *12*, 2161. [CrossRef]
6. Samrot, A.V.; Tan, C.S.; Bisyarah, U.; Mirarmandi, A.; Faradjeva, E.; Abu bakar, A.; Alia, H.H.; Angalene, J.L.A.; Kumar, S.S. Production, characterization and application of nanocarriers made of polysaccharides, proteins, bio-polyesters and other biopolymers: A review. *Int. J. Biol. Macromol.* **2020**, *165*, 3088–3105. [CrossRef]
7. Han, Y.; Duan, Q.; Li, Y.; Tian, J. Preparation and characterization of chitosan-based nanoparticles as protein delivery system. *Adv. Polym. Technol.* **2018**, *37*, 1–7. [CrossRef]
8. Wurm, F.R.; Weiss, C.K. Nanoparticles from renewable polymers. *Front. Chem.* **2014**, *2*, 49. [CrossRef] [PubMed]
9. Ab Latif, F.E.; Zainal Abidin, Z.; Cardona, F.; Awang Biak, D.R.; Abdan, K.; Mohd Tahir, P.; Liew, K.E. Bio-resin production through ethylene unsaturated carbon using vegetable oils. *Processes* **2020**, *8*, 48. [CrossRef]
10. Mustapha, R.; Rahmat, A.R.; Abdul Majid, R.; Mustapha, S.N.H. Vegetable oil-based epoxy resins and their composites with bio-based hardener: A short review. *Polym. Plast. Technol. Mater.* **2019**, *58*, 1311–1326. [CrossRef]
11. Gan, Y.; Jiang, X. Photo-cured materials from vegetable oils. In *Green Materials from Plant Oils*; Liu, Z., Kraus, G., Eds.; Royal Society of Chemistry: Cambridge, UK, 2014; Volume RSC Green Chemistry; pp. 1–27.

12. Mohd Noor, M.A.; Tuan Ismail, T.N.M.; Ghazali, R. Bio-based content of oligomers derived from palm oil: Sample combustion and liquid scintillation counting technique. *Malays. J. Anal. Sci.* **2020**, *24*, 906–917.
13. Hassana, N.; Idris, A.; Akhtarb, J. Overview on bio-refinery concept in Malaysia: Potential high value-added products from palm oil biomass. *J. Kejuruter.* **2019**, *2*, 113–124.
14. Idriss, M.; Hodroj, M.H.; Fakhoury, R.; Rizk, S. Beta-Tocotrienol Exhibits More Cytotoxic Effects than Gamma-Tocotrienol on Breast Cancer Cells by Promoting Apoptosis via a P53-Independent PI3-Kinase Dependent Pathway. *Biomolecules* **2020**, *10*, 577. [CrossRef] [PubMed]
15. Tomeo, A.C.; Geller, M.; Watkins, T.R.; Gapor, A.; Bierenbaum, M.L. Antioxidant effects of tocotrienols in patients with hyperlipidemia and carotid stenosis. *Lipids* **1995**, *30*, 1179–1183. [CrossRef]
16. Pignitter, M.; Hernler, N.; Zaunschirm, M.; Kienesberger, J.; Somoza, M.M.; Kraemer, K.; Somoza, V. Evaluation of Palm Oil as a Suitable Vegetable Oil for Vitamin A Fortification Programs. *Nutrients* **2016**, *8*, 378. [CrossRef] [PubMed]
17. Ashfaq, A.; Clochard, M.C.; Coqueret, X.; Dispenza, C.; Driscoll, M.S.; Ulanski, P.; Al-Sheikhly, A. Polymerization Reactions and Modifications of Polymers by Ionizing Radiation. *Polymers* **2020**, *12*, 2877. [CrossRef] [PubMed]
18. Freitas, L.F.D.; Varca, G.H.C.; Batista, J.G.D.S.; Lugão, A.B. An Overview of the Synthesis of Gold Nanoparticles Using Radiation Technologies. *Nanomaterials* **2018**, *8*, 939. [CrossRef] [PubMed]
19. Gerasimov, G.Y. Radiation methods in nanotechnology. *J. Eng. Phys. Thermophys.* **2011**, *84*, 947–963. [CrossRef]
20. Tajau, R.; Hashim, K.; Sharif, J.; Ratnam, C.T. *Teknologi Pemprosesan Sinaran Mengion (Ionizing Radiation Processing Technology)*; Dewan Bahasa dan Pustaka: Kuala Lumpur, Malaysia, 2017.
21. Sun, Y.; Chmielewski, A.G. *Applications of Ionizing Radiation in Materials Processing*; Institute of Nuclear Chemistry and Technology: Warszawa, Poland, 2017; Volume 1, pp. 1–244.
22. Quinn, J.F.; Davis, T.P.; Barner, L.; Barner-Kowollik, C. The application of ionizing radiation in reversible addition refragmentation chain transfer (RAFT) polymerization: Renaissance of a key synthetic and kinetic tool. *Polymer* **2007**, *48*, 6467–6480. [CrossRef]
23. Chmielewska, D. Radiation methods and uses in nanotechnology. In *Applications of Ionizing Radiation in Materials Processing*; Sun, Y., Chmielewski, A.G., Eds.; Institute of Nuclear Chemistry and Technology: Warszawa, Poland, 2017; Volume 1, pp. 395–414.
24. IAEA. *Uses of Ionizing Radiation for Tangible Cultural Heritage Conservation*; International Atomic Energy Agency: Vienna, Austria, 2017; pp. 1–264.
25. Adlienė, D. Basic radiation physics and sources of radiation. In *Applications of Ionizing Radiation in Materials Processing*; Sun, Y., Chmielewski, A.G., Eds.; Institute of Nuclear Chemistry and Technology: Warszawa, Poland, 2017; Volume 1, pp. 7–31.
26. Vasile, C.; Butnaru, E. Radiation chemistry of organic solids. In *Applications of Ionizing Radiation in Materials Processing*; Sun, Y., Chmielewski, A.G., Eds.; Institute of Nuclear Chemistry and Technology: Warszawa, Poland, 2017; Volume 1, pp. 117–141.
27. Rosiak, J.M.; Ulanski, P. Synthesis of hydrogels by irradiation of polymers in aqueous solution. *Radiat. Phys. Chem.* **1999**, *55*, 139–151. [CrossRef]
28. Ulanski, P.; Janik, I.; Rosiak, J.M. Radiation formation of polymeric nanogels. *Radiat. Phys. Chem.* **1998**, *52*, 289–294. [CrossRef]
29. Abdulkarim, M.F.; Abdullah, G.Z.; Sakeena, M.H.F.; Chitneni, M.; Yam, M.F.; Mahdi, E.S.; Salman, I.M.; Ameer, O.Z.; Sattar, M.A.; Basri, M.; et al. Study of pseudoternary phase diagram behaviour and the effect of several tweens and spans on palm oil esters characteristics. *Int. J. Drug Deliv.* **2011**, *3*, 95–100. [CrossRef]
30. Amir Reza, S.; Suraya, A.R.; Azmi, Z. Synthesis of gold nanoparticles dispersed in palm oil using laser ablation technique. *J. Nanomater.* **2017**, 1–5. [CrossRef]
31. Syafinaz, Z.; Mahiran, B.; Hamidon, B.; Ahmad Fuad, S.; Siti Salwa, A.G.; Roghayeh, A.K.; Emilia, A.M. Formulation optimization of a palm-based nanoemulsion system containing levodopa. *Int. J. Mol. Sci.* **2012**, *13*, 13049–13064.
32. Tajau, R.; Rohani, R.; Wan Isahak, W.N.R.; Salleh, M.Z. Palm oil-based precursors for development of polymeric delivery system. *Malays. J. Anal. Sci.* **2017**, *21*, 496–511.
33. Agensi Nuklear Malaysia. *EPOLA*; Trade Mark 2016074912; Intellectual Property Corporation of Malaysia (MyIPO): Kuala Lumpur, Malaysia, 2016.
34. Tajau, R.; Rohani, R.; Salleh, M.Z. Physicochemical and thermal properties of acrylated palm olein as promising biopolymer. *J. Polym. Environ.* **2020**, *28*, 2734–2748. [CrossRef]
35. Mahmood, M.H.; Tajau, R.; Mohd Dahlan, K.Z.; Salleh, M.Z.; Che Ismail, R. *Synthesis and Production of Palm Oil Based Urethane Acrylates (Pobua) for Use in Uv/Eb Curing of Coating Adhesives and Printing Inks*; Malaysia Patent No. MY-176255-A; Intellectual Property Corporation of Malaysia (MyIPO): Kuala Lumpur, Malaysia, 2020.
36. Tajau, R.; Ibrahim, M.I.; Mohd Yunus, N.; Mahmood, M.H.; Salleh, M.Z.; Nik Salleh, N.G. Development of palm oil-based UV-curable epoxy acrylate and urethane acrylate resins for wood coating application. In Proceedings of the International Conference on Nuclear Science Technology and Engineering 2013 (iNuSTEC2013), Kuala Lumpur, Malaysia, 30 September–2 October 2014; pp. 164–169.
37. Teo, K.T.; Aziz, H.; Gan, S.N. UV-curable urethane acrylate resin from palm fatty acid distillate. *Polymer* **2018**, *10*, 1374. [CrossRef]
38. Nik Salleh, N.G.; Salleh, M.Z.; Tajau, R.; Alias, M.S.; Harun, M.H.; Mudri, N.H.; Abdul Halim, K.A.; Muhammad Faisal, S.; Che Ismail, R. *Radiation Curable Resin for Overprint Varnish*; Malaysia Filing Patent No. PI 2016704912; Intellectual Property Corporation of Malaysia (MyIPO): Kuala Lumpur, Malaysia, 2016.
39. Mahmood, M.H.; Abdullah, Z.; Sakurai, Y.; Mohd Dahlan, K.Z. Effects of monomers on the properties of palm-oil-based radiation curable pressure sensitive adhesives (PSA)—A prepolymer method. *Radiat. Phys. Chem.* **2001**, *60*, 129–137. [CrossRef]

40. Salleh, M.Z.; Badri, K.; Tajau, R.; Nik Salleh, N.G. The production of green polymer-Hyperbranced curable palm oil oleic acid. *Adv. Mateials Res.* **2014**, *1024*, 197–200. [CrossRef]
41. Mudri, N.H.; Nik Salleh, N.G.; Salleh, M.Z.; Alias, M.S.; Tajau, R.; Che Ismail, R.; Abdurrahaman, N. Development of UV Curable Overprint Varnishes (OPV) Formulation from Epoxidized Palm Olein Acrylated (EPOLA). In Proceedings of the Seminar R&D Nuklear Malaysia 2016 Kajang, Selangor, Malaysia, 14–16 October 2014.
42. Asraf, M.S.; Ahmad, M.; Ibrahim, N.A.; Mohd Dahlan, K.Z.; Tajau, R.; Mahmood, M.H.; Wan Yunus, W.M.Z. Synthesis of radiation curable palm oil-based epoxy acrylate: NMR and FTIR spectroscopic investigations. *Molecules* **2015**, *20*, 14191–14211.
43. Nurliyana, A.R.; Khairiah, B.; Nik Ghazali, N.S. UV-curable acrylated coating from epoxidized palm oil. *Aip Conf. Proc.* **2014**, *1614*, 439–445.
44. Tajau, R.; Mahmood, M.H.; Salleh, M.Z.; Mohd Dahlan, K.Z.; Che Ismail, R.; Muhammad Faisal, S.; Sheikh Abdul Rahman, S.M.Z. Production of uv-curable palm oil resins/oligomers using laboratory scale and pilot scale systems. *Sains Malays.* **2013**, *42*, 459–467.
45. Mohd Azam, A.; Ooi, T.L.; Salmiah, A.; Umaru, S.I.; Mohd Ishak, Z.A. *New UV-Curable Acrylated Polyester Prepolymers from Palm Oil Based Products*; INIS-MY-034; International Atomic Energy Agency: Vienna, Austria, 1999; pp. 383–390.
46. Kosheela Devi, P.P.; Hazimah, A.H.; Tuan Noor Maznee, T.I.; Yeong, S.K.; Hoong, S.S.; Kushairi, A.; Choo, Y.M. Optimisation on synthesis of acrylated epoxidised palm olein using response surface methodology. *J. Oil Palm Res.* **2015**, *27*, 366–376.
47. Mek Zah, S.; Khairiah, H.B.; Mohd Hilmi, M.; Sahrim, A. Synthesis of UV-curable hyperbranched urethane acrylate from palm oil oleic acid. *J. Nucl. Relat. Technol.* **2010**, *7*, 38–48.
48. Salleh, M.Z.; Badri, K.; Ahmad, S.; Mahmood, M.H.; Tajau, R.; Nik Salleh, N.G.; Latif, M.L. Properties of radiation curable hyperbranched polyurethane acrylate from palm oil oleic acid. *Nucl. Sci. Tech.* **2013**, *24*, 1–6.
49. Hazimah, A.H.; Kosheela Devi, P.P.; Tuan Noor Maznee, T.I.; Mohd Norhisham, S.; Hoong, S.S.; Yeong, S.K.; Salmiah, A.; Ooi, T.L. Palm-based polyols and polyurethanes: Two decades of expriences. *Oil Palm Bull.* **2011**, *63*, 33–42.
50. Ooi, T.L.; Salmiah, A.; Hazimah, A.H.; Chong, Y.J. An overview of R&D in palm oil-based polyols and polyurethanes in MPOB. *Palm Oil Dev.* **2006**, *44*, 1–7.
51. Sharifah Nafisah, S.I.; Ahmad Faiza, M.; Rahmah, M. Synthesis and properties of acrylated epoxidized palm oil for UV cured coatings. *Solid State Sci. Technol.* **2017**, *25*, 86–95.
52. Tugiman, N.; Rahmat, A.R.; Jamaluddin, J.; Wan Tajulruddin, W.N.; Mustapha, R. Preliminary study on degree of conversion of uv curablemaleinated acrylated epoxidised palm oil pressure sensitive adhesives synthesised with isobornyl methacrylate monomer via FTIR-ATR analysis. *Chem. Eng. Trans.* **2017**, *56*, 223–228.
53. Onn, M.; Mohd, A.F.; Yhaya, M.F. The effect of reactives diluents to the physical properties of acrylated palm oil based polyurethane coatings. In Proceedings of the MATEC Web of Conferences, Melaka, Malaysia, 1–2 December 2015; pp. 1–7.
54. Said, H.M.; Nik Salleh, N.G.; Alias, M.S.; El-Naggar, A.W.M. Synthesis and characterization of hard materials based on radiation cured bio-polymer and nanoparticles. *J. Radiat. Res. Appl. Sci.* **2013**, *6*, 71–78. [CrossRef]
55. Cheong, M.Y.; Ooi, T.L.; Ahmad, S.; Wan Yunus, W.M.Z.; Kuang, D. Synthesis and characterization of palm-based resin for UV coating. *J. Appl. Polym. Sci.* **2009**, *111*, 2353–2361. [CrossRef]
56. Ulanski, P.; Rosiak, J.M. Polymeric nano/microgels. In *Encyclopedia of Nanoscience and Nanotechnology*; Nalwa, H.S., Ed.; American Scientific Publishers: Los Angeles, CA, USA, 2004; Volume 10, pp. 1–26.
57. Dispenza, C.; Alessi, S.; Spadaro, G. Radiation processing of polymers in aqueous media. In *Applications of Ionizing Radiation in Materials Processing*; Sun, Y., Chmielewski, A.G., Eds.; Institute of Nuclear Chemistry and Technology: Warszawa, Poland, 2017; Volume 2, pp. 291–326.
58. Hamzah, Y.; Wan Yunus, W.M.Z.; Mat Isa, N.; Tajau, R.; Hashim, K.; Mohd Dahlan, K.Z. Synthesis of polyethylene glycol diacrylate nanogel using irradiation of inverse micelles technique. *E-Polymer* **2012**, *12*, 1–6. [CrossRef]
59. Noordin, M.I.; Yong, C.L.; Mofat, I.; Zainuddin, Z.; Arya, A.; Nyamathulla, S. Evaluation of palm oil-based paracetamol suppositories by differential scanning calorimetry. *Trop. J. Pharm. Res.* **2014**, *13*, 23–29. [CrossRef]
60. Basheer, H.S.; Mohamed Ibrahim, N.; Ghareeb, M.M. Characterization of microemulsions prepared using isopropyl palmitate with various surfactants and cosurfactants. *Trop. J. Pharm. Res.* **2013**, *12*, 305–310.
61. Pradhan, S.; Sagiri, S.S.; Singh, V.K.; Pal, K.; Ray, S.S.; Pradhan, D.K. Palm oil-based organogels and microemulsions for delivery of antimicrobial drugs. *J. Appl. Polym. Sci.* **2014**, *131*, 39979. [CrossRef]
62. Baran, N.; Singh, V.K.; Pal, K.; Anis, A.; Pradhan, D.K.; Pramanika, K. Development and characterization of soy lecithin and palm oil-based organogels. *Polym. Plast. Technol. Eng.* **2014**, *53*, 865–879. [CrossRef]
63. Sakeena, M.H.; Muthanna, F.A.; Ghassan, Z.A.; Kanakal, M.M.; Elrashid, S.M.; Munavvar, A.S.; Azmin, M.N. Formulation and in vitro evaluation of ketoprofen in palm oil esters nanoemulsion for topical delivery. *J. Oleo Sci.* **2010**, *59*, 223–228. [CrossRef] [PubMed]
64. Musthanna, F.A.; Ghassan Zuhair, A.; Mallikarjun, C.; Elrashid Saleh, M.; Mun, F.Y.; Ahmed, F.; Ibrahim, M.S.; Omar Ziad, A.; Mohanad, N.S.; Munawar Zubaid, A.; et al. Formulation and characterization of palm oil esters based nanocream for topical delivery of piroxicam. *Int. J. Drug Deliv.* **2010**, *2*, 287–298.
65. Cook, R.O.; Pannu, R.K.; Kellaway, I.W. Novel sustained release microspheres for pulmonary drug delivery. *J. Control. Release* **2005**, *104*, 79–90. [CrossRef] [PubMed]

66. Tajau, R.; Mohd Dahlan, K.Z.; Mahmood, M.H.; Wan Yunus, W.M.Z.; Hashim, K. Radiation induced formation of acrylated palm oil (APO) nanoparticles using cetyltrimethylammonium bromide microemulsion system. *J. Adv. Mater. Res.* **2012**, *364*, 278–282. [CrossRef]
67. Abbasalipourkabir, R.; Salehzadeh, A.; Abdullah, R. Characterization and stability of nanostructured lipid carriers as drug delivery system. *Pak. J. Biol. Sci.* **2012**, *15*, 141–146. [CrossRef]
68. Rodrigues, M.; Peirico, N.; Matos, H.; Azevedo, E.G.; Lobato, M.R.; Almeida, A.J. Microcomposites theophylline/hydrogenated palm oil from a PGSS process for controlled drug delivery systems. *J. Supercrit. Fluids* **2004**, *29*, 175–184. [CrossRef]
69. Bahareh, S.; Mohamed Ibrahim, N.; Shaharuddin, M.; Rosnani, H.; Afendi, D.; Javar, H.A. Development and characterization of liposomal doxorubicin hydrochloride with palm oil. *Biomed Res. Int.* **2014**, *2014*, 765426.
70. Bahareh, S.; Mohamed Ibrahim, N.; Shaharuddin, M.; Rosnani, H.; Afendi, D.; Javar, H.A. Degradation study of liposome drug delivery prepared with palm oil; Using 1H NMR. *Int. J. Pharm. Sci. Res.* **2015**, *6*, 2741–2747.
71. Reza, Z.; Azmi, Z.; Hossein, A.A.; Amir Reza, S.; Mohd Adzir, M. Fabrication of silver nanoparticles dispersed in palm oil using laser ablation. *Int. J. Mol. Sci.* **2010**, *11*, 4764–4770.
72. Tajau, R.; Mohd Dahlan, K.Z.; Mahmood, M.H.; Salleh, M.Z. *A Composition Susceptible to Irradiation for Use as a Compound Carrier*; Malaysia Patent No. MY-182996-A; Intellectual Property Corporation of Malaysia (MyIPO): Kuala Lumpur, Malaysia, 2021.
73. Sadrolhosseini, A.R.; Mahdi, M.A.; Alizadeh, F.; Rashid, S.A. Laser Ablation Technique for Synthesis of Metal Nanoparticle in Liquid, Laser Technology and its Applications. In *Laser Technology and its Applications*; Ma, Y., Ed.; IntechOpen: London, UK, 2019; pp. 63–84.
74. Tajau, R.; Rohani, R.; Abdul Hamid, S.S.; Adam, Z.; Mohd Janib, S.N.; Salleh, M.Z. Surface functionalisation of poly-APO-b-polyol ester cross-linked copolymers as core-shell nanoparticles for targeted breast cancer therapy. *Sci. Rep.* **2020**, *10*, 1–17. [CrossRef] [PubMed]
75. Caykara, T. Polymer Brushes by surface-mediated raft polymerization for biological function. In *Polymer and Biopolymer Brushes: For Materials Science and Biotechnology*; Azzaroni, O., Szleife, I., Eds.; John Wiley & Sons: Hoboken, NJ, USA, 2017; Volume 1, pp. 97–122.
76. Zengin, A.; Yildirim, E.; Tamer, U.; Caykara, T. Molecularly imprinted superparamagnetic iron oxide nanoparticles for rapid enrichment and separation of cholesterol. *Analyst* **2013**, *138*, 7238–7245. [CrossRef] [PubMed]
77. Karmakar, P.D.; Seesala, V.S.; Pal, A.; Dhara, S.; Chatterjee, S.; Pal, S. Synthesis of raft-mediated amphiphilic graft copolymeric micelle using dextran and poly (oleic acid) toward oral delivery of nifedipine. *J. Polym. Sci. Part. A Polym. Chem.* **2018**, *56*, 2354–2363. [CrossRef]
78. Shaikh, N.; Payghan, S.A.; Desouza, J.I. Formulation of gastroretentive drug delivery system (floating tablets) of nifedipine. *Int. J. Pharm. Sci. Res.* **2011**, *2*, 2929–2933.
79. Talib, M. Formulation of photopolymer resin prepared from acrylated palm oil resin for tissue engineering application. In Proceedings of the IAEA Fourth Research Coordinating Meeting of the IAEA Coordinate Research Project (F23030-E31007) on Instructive Surfaces and Scaffolds for Tissue Engineering Using Radiation Technology, Sao Paulo, Brazil, 8–12 April 2019.
80. Tajau, R.; Talib, M. Biobahan (Biomaterial). In *Teknologi Pemprosesan Sinaran Mengion (Ionizing Radiation Processing Technology)*; Tajau, R., Hashim, K., Sharif, J., Ratnam, C.T., Eds.; Dewan Bahasa Dan Pustaka: Kuala Lumpur, Malaysia, 2017; pp. 199–222.
81. Serrano-Aroca, A.; Deb, S. Acrylic-based hydrogels as advanced biomaterials. In *Acrylate Polymers for Advanced Applications*; Serrano-Aroca, A., Ed.; Intechopen: London, UK, 2020.
82. Jabbari, E.; Nozari, S. Swelling behavior of acrylic acid hydrogels prepared by γ-radiation crosslinking of polyacrylic acid in aqueous solution. *Eur. Polym. J.* **2000**, *36*, 2685–2692. [CrossRef]
83. Lee, J.; Kim, S.; Kim, S.; Lee, Y.; Le, E.K. Synthesis and characteristics of interpenetrating polymer network hydrogel composed of chitosan and poly(acrylic acid). *Polymer* **1998**, *73*, 113–120. [CrossRef]
84. Fabbri, P.; Valentini, L.; Bittolo Bon, S.; Foix, D.; Pasquali, L.; Montecchi, M.; Sangermano, M. In-situ graphene oxide reduction during UV-photopolymerization of graphene oxide/acrylic resins mixtures. *Polymer* **2012**, *53*, 6039–6044. [CrossRef]
85. Rao, U.; Sridhar, R.; Sehgal, P.K. Biosynthesis and biocompatibility of poly(3-hydroxybutyrate-co-4-hydroxybutyrate) produced by Cupriavidus necator from spent palm oil. *Biochem. Eng. J.* **2010**, *49*, 13–20. [CrossRef]
86. Kamilah, H.; Al-Gheethi, A.; Yang, T.A.; Sudesh, K. The use of palm oil-based waste cooking oil to enhance the production of polyhydroxybutyrate [P(3HB)] by cupriavidus necator H16 strain. *Arab. J. Sci. Eng.* **2019**, *44*, 673–674. [CrossRef]
87. Ardelean, L.C.; Bortun, C.M.; Podariu, A.C.; Rusu, L.C. Acrylates and their alternatives in dental applications. In *Acrylic Polymers in Healthcare*; Reddy, B.S.R., Ed.; Intechopen: London, UK, 2017.
88. Tyliszczak, B.; Drabczyk, A.; Kudłacik-Kramarczyk, S. Acrylates in Dental Applications. In *Acrylic Polymers in Healthcare*; Reddy, B.S.R., Ed.; Intechopen: London, UK, 2017.
89. Rokaya, D.; Srimaneepong, V.; Sapkota, J.; Qin, J.; Siraleartmukul, K.; Siriwongrungson, V. Polymeric materials and films in dentistry: An overview. *J. Adv. Res.* **2018**, *14*, 25–34. [CrossRef] [PubMed]
90. Ionta, F.Q.; Alencar, C.R.B.; Val, P.P.; Boteon, A.P.; Jordão, M.C.; Honório, H.M.; Buzalaf, M.A.R.; Rios, D. Effect of vegetable oils applied over acquired enamel pellicle on initial erosion. *J. Appl. Oral Sci.* **2017**, *25*, 420–426. [CrossRef]
91. Gan, S.N.; Kasim, N.H.B.A.; Alsanabana, F.; Radzi, Z.B.; Yahya, N.A.B. Polyurethane Oligomers for Use in Restorative Dentistry. U.S. Patent No. 8,703,897, 22 April 2014.

92. Sharma, K.K.; Singh, U.S.; Sharma, P.; Kumar, A.; Sharma, L. Seed treatments for sustainable agriculture-A review. *J. Appl. Nat. Sci.* **2015**, *7*, 521–539. [CrossRef]
93. Ryan, J.W.; Juen, D.R. Crosslinked Silicone Coatings for Botanical Seeds. U.S. Patent No. 4,753,035, 28 June 1988.
94. Fierascu, R.C.; Doni, M.; Fierascu, I. Selected aspects regarding the restoration/conservation of traditional wood and masonry building materials: A short overview of the last decade findings. *Appl. Sci.* **2020**, *10*, 1164. [CrossRef]
95. Cao, Y.; Salvini, A.; Camaiti, M. Current status and future prospects of applying bioinspired superhydrophobic materials for conservation of stone artworks. *Coatings* **2020**, *10*, 353. [CrossRef]
96. Mohd Nor, H.; Mahmood, M.H.; Kifli, H.; Abdul Rahman, M. The use of epoxidised palm oil products for the synthesis of radiation curable resins: Several factors affecting the synthesis of epoxidised RBD palm olein acrylate (EPOLA). *Elaeis* **1992**, *4*, 60–64.
97. Mahmood, M.H.; Tajau, R.; Salleh, M.Z.; Che Ismail, R. *Method for Manufacturing Palm Oil Based Hydroxyl Containing Products for Use in Making Polyurethane Materials;* Malaysia Patent No. MY-142814-A; Intellectual Property Corporation of Malaysia (MyIPO): Kuala Lumpur, Malaysia, 2011.
98. Sangermano, M.; Malucelli, G.; Amerio, E.; Priola, A.; Billi, E.; Rizza, G. Photopolymerization of epoxy coatings containing silica nanoparticles. *Prog. Org. Coat.* **2005**, *54*, 134–138. [CrossRef]
99. Alias, M.S.; Nik Salleh, N.G.; Salleh, M.Z.; Harun, M.H.; Tajau, R.; Abdul Halim, K.A.; Mudri, N.H. The influence of zirconium acrylate on curable palm oil resin as coating material by using UV technique. In Proceedings of the Seminar R&D Nuklear Malaysia 2016, Malaysia Nuclear Agency, Kajang, Selangor, Malaysia, 7–9 November 2016.
100. Mahmood, M.H.; Mohd Nor, H.; Abd Raof, K.H.; Roslan, I.; Mohd Rosli, M.R.; Shariff, S.M.; Abdullah, Z. Development of pressure sensitive adhesives from palm oil acrylated resins by irradiation—A preliminary study. *J. Sains Nukl. Malays.* **1995**, *13*, 61–68.
101. Ng, W.S.; Lee, C.S.; Chuah, C.H.; Cheng, S.F. Preparation and modification of water-blown porous biodegradable polyurethane foams with palm oil-based polyester polyol. *Ind. Crop. Prod.* **2017**, *97*, 65–78. [CrossRef]
102. Ng, W.S.; Choy, S.L.; Cheng, S.F.; Cheng, H.C.; Shew, F.W. Biocompatible polyurethane scaffolds prepared from glycerol monostearate-derived polyester polyol. *J. Polym. Environ.* **2018**, *26*, 2881–2900. [CrossRef]
103. Yeoh, F.H.; Lee, C.S.; Kang, Y.B.; Wong, S.F.; Cheng, S.F.; Ng, W.S. Production of biodegradable palm oil-based polyurethane as potential biomaterial for biomedical applications. *Polymer* **2020**, *112*, 1842. [CrossRef]
104. Yusup, E.; Mahzan, S.; Mohammad, B.; Wan Daud, W.R. A Novel Approach for Bone Scaffold from Oil Palm Empty Fruit Bunch-Cellulose Phosphate/Glass Material. *Adv. Mater. Res.* **2013**, *748*, 180–183. [CrossRef]
105. Zubairi, S.I.; Mantalaris, A.; Bismarck, A.; Aizad, S. Polyhydroxyalkanoates (PHAs) for tissue engineering applications: Biotransformation of palm oil mill effluent (POME) to value-added polymers. *J. Teknol.* **2016**, *78*, 13–29. [CrossRef]

polymers MDPI

Article

Novel Mangosteen-Leaves-Based Marker Ink: Color Lightness, Viscosity, Optimized Composition, and Microstructural Analysis

Mohd Salahuddin Mohd Basri [1,2,3,*], Brenda Liew Min Ren [1], Rosnita A. Talib [1], Rabitah Zakaria [1] and Siti Hasnah Kamarudin [4]

1 Department of Process and Food Engineering, Faculty of Engineering, Universiti Putra Malaysia (UPM), Serdang 43400, Selangor, Malaysia; brendaliew96@gmail.com (B.L.M.R.); rosnita@upm.edu.my (R.A.T.); rabitah@upm.edu.my (R.Z.)
2 Laboratory of Halal Science Research, Halal Products Research Institute, Universiti Putra Malaysia (UPM), Serdang 43400, Selangor, Malaysia
3 Laboratory of Biopolymer and Derivatives, Institute of Tropical Forestry and Forest Products (INTROP), Universiti Putra Malaysia (UPM), Serdang 43400, Selangor, Malaysia
4 School of Industrial Technology, Faculty of Applied Sciences, Universiti Teknologi MARA (UiTM), Shah Alam 40450, Selangor, Malaysia; sitihasnahkam@uitm.edu.my
* Correspondence: salahuddin@upm.edu.my

Abstract: Dry mangosteen leaves are one of the raw materials used to produce marker ink. However, research using this free and abundant resource is rather limited. The less efficient one-factor-at-a-time (OFAT) approach was mostly used in past studies on plant-based marker ink. The use of statistical analysis and the regression coefficient model (mathematical model) was considered essential in predicting the best combination of factors in formulating mangosteen leaf-based marker ink. Ideally, ink should have maximum color lightness, minimum viscosity, and fast-drying speed. The objective of this study to study the effect of glycerol and carboxymethyl cellulose (CMC) on the color lightness and viscosity of mangosteen-leaves-based marker ink. The viscosity, color lightness, and drying properties of the ink were tested, the significant effect of glycerol and CMC (responses) on ink properties was identified and the prediction model on the optimum value of the responses was developed by using response surface methodology (RSM). The microstructure of mangosteen leaves was analyzed to study the surface morphology and cell structure during dye extraction. A low amount of glycerol used was found to increase the value of color lightness. A decrease in CMC amounts resulted in low viscosity of marker ink. The optimum formulation for the ink can be achieved when the weight percents of glycerol, benzalkonium chloride, ferrous sulphate, and CMC are set at 5, 5, 1, and 3, respectively. SEM micrographs showed the greatest amount of cell wall structure collapse on samples boiled with the lowest amount of glycerol.

Keywords: mangosteen leaves; marker ink; color lightness; viscosity; optimization; RSM

Citation: Mohd Basri, M.S.; Liew Min Ren, B.; A. Talib, R.; Zakaria, R.; Kamarudin, S.H. Novel Mangosteen-Leaves-Based Marker Ink: Color Lightness, Viscosity, Optimized Composition, and Microstructural Analysis. *Polymers* **2021**, *13*, 1581. https://doi.org/10.3390/polym13101581

Academic Editors: Antonio M. Borrero-López and Cédric Delattre

Received: 1 April 2021
Accepted: 7 May 2021
Published: 14 May 2021

1. Introduction

For centuries, ink is made from natural products such as berries, barks, and extracts of leaves. These have been used as raw materials to create various colors and to produce ink, dye, or paint when mixed with other substances. Early records showed that tea leaves had been used to make ink due to their shades that able to produce yellow, green, brown, or black ink [1]. Ink has been studied in a wide area of applications including inkjet printing [2,3] and bio-ink [4]. Inks are commonly made today from non-renewable synthetic resources such as petroleum- and chemical-based solvents, which are harmful to both users and the environment. Unprotected contact with ink may cause severe headaches, skin irritation, or damage to the nervous system likely due to the effects of solvents or pigment, such as p-anisidine, in the ink.

Noah et al. [5] conducted a preliminary study on the production of brown ink from *Gmelina arborea* fruit extract. The objective was to assess the potential of producing brown ink from the extract and compare the ink produced with commercial inks using an ink flotation test. Five different concentration levels (100%, 80%, 60%, 40%, and 20%) of ink were produced using ethanol as diluent and coconut vinegar as an additive to preserve the ink from biodegradation and to enhance its stability and permanence on paper once dried. The results showed that the production of brown ink from *Gmelina arborea* fruit extract was feasible.

Nwafulugo et al. [6] studied the marker ink production from a huckleberry (*Vaccinium membranaceum*) ink extract. The liquid was extracted by cooking and mashing the berries with the addition of apple cider (vinegar) as a solvent extractor. The extracted liquid was then mixed with 108 mL of gum arabic, 18 mL of methanol, and two teaspoons of dye, yielding an ink sample in each mixture. The physical properties of the sample were tested and compared to commercial ink. The results showed that the ink produced was of high quality, with a pH of 9.3, drying time of 2.3 s, and viscosity of 9.5×10^{-4} Ns/m^2.

Dagde et al. [7] studied the formulation of whiteboard marker ink using locally sourced raw materials such as natural wood from mango (*Mangifera* sp.), rubber (*Hevea brasiliensis*), and ugba or oil bean trees (*Pentaclethra macrophylla*). The pyrolysis process carried out on the different woods produced the final product called charcoal. In terms of odor, color, hazardous reaction, pH, density, and viscosity, the various ink samples were formulated and compared to international standards. The results showed a good match, indicating that the whiteboard marker ink produced is reliable and of high quality. The pH results for ugba (oil bean), rubber, and mango charcoal ink are 5.43, 6.79, and 7.41, respectively. Ugba (oil bean) charcoal ink showed the highest viscosity followed by rubber and mango charcoal ink.

Mangosteen (*Garcinia mangostana*) is planted in most of the countries of Southeast Asia—including Indonesia, Malaysia, the Philippines, and Thailand. Mature mangosteen trees can range from 6 to 25 m in height. The primary active components of the fruits are known as xanthones, which have several benefits, including anti-inflammatory properties, anti-allergic, and anti-convulsants [8]. The rind is reported to contain from 7% to 15% tannin and is used for tanning leather and dyeing in black color [9]. The rind of the mangosteen is also known for use as an ingredient in soap, shampoo, and conditioner [10].

One of the most important uses of mangosteen is as a source of natural dye for fabric and solar cells because of its prominent color. Fully ripened mangosteen pericarp contains anthocyanins, such as cyanidin-3-sophoroside and cyanidin-3-glucoside, which contribute to the dark purple or red coloration of its pericarp [11]. Anthocyanin can be used as a natural color extract to replace synthetic colorants because of its light color and higher solubility. Food scientists and horticulturists have meticulously examined anthocyanins because of their strong effect on the coloring quality of pre-harvested and post-harvested vegetables and fruits [12]. Anthocyanins are grouped under the family of flavonoid compounds, which contain solvable water pigments. These color pigments are also present in fruits and flowers and able to attract insects and animals [13].

Other dye compounds, such as tannin, can also be extracted from mangosteen for its brown coloring property [14]. These natural dyes exhibit excellent potential in the textile industry. They can be obtained easily from the leaves, pericarp fruit waste, and barks of the mangosteen tree. Due to their biodegradable and nontoxic properties, they are safe to the environment compared with synthetic dyes. Researchers for instance have successfully dyed cotton fabrics using mangosteen extract. In general, mangosteen sources have been used for a variety of purposes, ranging from industrial products to advanced technology and biomedical applications. The use of its fruits has recently been studied in postharvest biology, food science, and engineering fields.

Carboxymethyl cellulose (CMC) and glycerol are the additives used in the production of marker ink. CMC, also known as cellulose gum is a cellulose derivative with carboxymethyl groups ($-CH_2-COOH$) incorporated into some of the hydroxyl groups of the glucopyranose monomers that contribute to the cellulose backbone. It is often used as its sodium salt, sodium CMC. Habib and Khoda [15] studied the development of clay-based novel bio-ink by optimizing the composition of materials, which are montmorillonite (MMT), CMC, and sodium alginate. CMC is a water-soluble polysaccharide with a high molecular weight that is used as a viscosity modifier or thickener. It is reported that the binding of CMC's matrix protein assisted in cell migration and cell attachment. The incremental addition of CMC into alginate and MMT mixture makes the bio-ink more viscous. Bio-ink with very high viscosity requires high air pressure to suspend the filament that may limit the use of available bio-printer. The rheological properties of CMC were investigated by Ghannam and Esmail [16]. An increase in CMC concentration in the solutions changes the behavior from Newtonian to shear thinning. The higher the concentration, the lower the flow behavior index. The increase in CMC concentration is accompanied by stronger time-dependence of the rheological properties.

Glycerol (1,2,3-propanetriol) is a colorless, odorless, viscous liquid with a sweet taste. It is made from both natural and petrochemical substances [17]. Presently, the quantity of glycerol accounted for approximately 160,000 tonnes annually and is used for technical applications; this is expected to grow at an annual rate of 2.8%. Glycerol is typically used as a softener and plasticizer in alkyd resins and regenerated cellulose to determine properties in surface coatings and paints including flexibility, pliability, and toughness. It serves as binders in paints and inks [18]. Medeiros et al. [19] investigated the effects of glycerol concentration on thermophysical properties of the water-glycerol solution. The results showed that with an increase in the amount of glycerol, the density and dynamic viscosity were increased.

Although many studies on marker ink have been conducted based on plant material, research on the development and production of marker ink using dried mangosteen leaves is rather limited. The number of studies looking into the effects of various factors on the properties of marker ink is still minimal. Besides, the proper method in determining the optimum formulation using statistical optimization has not been applied in these studies.

Despite a large number of publications on biodegradable and environmentally friendly marker inks reported in the literature in recent years [7,20–22], most studies were carried out in the one-factor-a-time (OFAT) approach. For greater efficiency, the application of statistical analysis and regression coefficient (mathematical model) are necessary for better prediction on the optimum composition of marker ink with high color lightness and low viscosity. Compared to OFAT, the design of the experiment (DOE) provides many advantages, including low requirements for resources (experimental runs, time, material, and manpower), accurate measurement of main effects and interactions, and the ability to simultaneously analyze several variables [23]. Furthermore, the response surface methodology (RSM), which was originally coined by Box and Wilson [24], is used commonly as a mathematical model for enquiring into significant effects, interactions, and optimization studies. The central composite design (CCD) has been proven to be the best model for analysis [25].

Since the RSM approach, specifically CCD, has been widely used in polymer optimization [26–29], it will, therefore, be adopted in this study. The main objectives of this study are: (i) to identify the significant effects of different additives on the viscosity, color lightness, and drying speed of mangosteen-leaves-based marker ink; (ii) to determine the optimum composition of the mangosteen-leaves-based marker ink using RSM; and (iii) to study the microstructure of mangosteen leaves during the dye extraction process.

2. Materials and Methods

2.1. Factors and Levels of the Design of Experiment

In the study, CMC and glycerol—designated as V_1 and V_2 respectively—were chosen as factors. Other factors, such as ferrous sulphate and benzalkonium chloride, were kept constant at 1 and 5 wt %, respectively. Five weight percent (5 wt %) denoted as 5% of the total weight of the leave's boiled water. Factors and levels used in the DOE are shown in Table 1.

Table 1. Factors and levels.

Factor	Unit	Notation	Levels				
			−2	−1	0	1	2
CMC	wt %	V_1	1	2	3	4	5
Glycerol	wt %	V_2	5	10	15	20	25

2.2. Design of Experiment

RSM was used to design and conduct experiments due to its lower experimental runs as compared to the conventional experiment. Table 2 shows the difference in the total number of experimental runs between full-design RSM and a full factorial design (classical method) based on five-level factors. The result shows that for analyzing four factors, the full-design RSM only requires a minimum of 30 experimental runs (with one replication) as compared to 625 for the full factorial design.

Table 2. Total number of experimental runs for full factorial design and RSM based on five-level factors

Factors	Levels	Total Number of Experimental Runs	
		Full Factorial Design	RSM
2	5	25	14
3	5	125	20
4	5	625	30
5	5	3125	54

At each design stage, five levels and two factors were applied in the CCD and with three replications for a total of 39 experimental runs. The factors were selected based on preliminary lab work, their significant effect on the responses, and working range (workability). Table 3 displays the complete CCD with coded and uncoded levels of these factors. The value for the total block is 1 with the experiments carried out in randomized order. For analysis using RSM, $V_1 = V_2 = 0$ represent the center point. Replicating an entire DOE to increase the number of data points can be expensive and time-consuming. An alternative way to increase the accuracy of the analysis is to use center points. By increasing the number of center points in the DOE, the probability of analyzing data and perform prediction of the optimum value for the factors can be increased. Therefore, a total of 15 experimental runs has been designed for center points resulting in a total of 39 experimental runs.

The significance of the main factors and their interactions is calculated using the analysis of variance (ANOVA). The value of 95% was set as the significance level which reflected the p-value of 0.05. Based on the value of the correlation coefficient (R^2), the regression coefficient model (mathematical model) developed in the ANOVA table is used for optimization purposes. Experimental data are fitted with a second-order polynomial model to obtain the regression coefficient model. The general mathematical model obtained from the analysis is shown in Equation (1)

$$Y = \beta_0 + \sum_{t=1}^{3} \beta_i X_i + \sum_{i}^{3} \beta_{ii} X_i^2 + \sum_{i-1}^{2} \sum_{j=i+1}^{3} \beta_{ij} X_i X_j \quad (1)$$

where Y is the response, β_0, β_i, β_{ii}, and β_{ij} are regression coefficients for the intercept, linear, quadratic, and interaction terms, respectively. X_i and X_j are coded values for the independent variables [30].

Table 3. Design matrix.

Sample	Coded Factor		Uncoded Factor	
	V_1	V_2	V_1	V_2
S1	−2	0	5	3
S2	0	2	15	5
S3	1	−1	20	2
S4	−1	1	10	4
S5	−2	0	5	3
S6	0	0	15	3
S7	−1	−1	10	2
S8	0	0	15	3
S9	2	0	25	3
S10	0	0	15	3
S11	0	0	15	3
S12	0	−2	15	1
S13	1	−1	20	2
S14	0	0	15	3
S15	−2	0	5	3
S16	−1	−1	10	2
S17	−1	−1	10	2
S18	0	0	15	3
S19	1	1	20	4
S20	0	0	15	3
S21	0	0	15	3
S22	0	0	15	3
S23	1	−1	20	2
S24	1	1	20	4
S25	0	0	15	3
S26	−1	1	10	4
S27	0	−2	15	1
S28	0	2	15	5
S29	0	0	15	3
S30	0	0	15	3
S31	2	0	25	3
S32	0	−2	15	1
S33	−1	1	10	4
S34	0	0	15	3
S35	2	0	25	3
S36	1	1	20	4
S37	0	0	15	3
S38	0	2	15	5
S39	0	0	15	3

The desirability function was used to optimize the central composite design. The desirability function was created in this study to minimize the viscosity and color lightness of the ink. Validation was carried out by triplicate under optimal conditions.

2.3. Raw Materials and Sample Preparation

Dry mangosteen leaves, used as the main component in the marker ink production, were obtained from the Family Fruits Farm, Batang Kali, Selangor, Malaysia. The CMC, ferrous sulphate, benzalkonium chloride, and glycerol used in this study was purchased from Merck KGaA, Darmstadt, Germany. The specification for glycerol is colorless, and it contains 1.0% water, viscous, and clear liquid. CMC is an odorless white powder with a viscosity of 400 to 800 centipoise (cp), in a 2% solution in water at 25 °C. Ferrous sulphate has a density of 1.898 g/mL at 25 °C while benzalkonium chloride is a clear, colorless liquid with a concentration of 50 to 55 *w*/*w*%. The marker ink was prepared by following the steps, as shown in Figure 1.

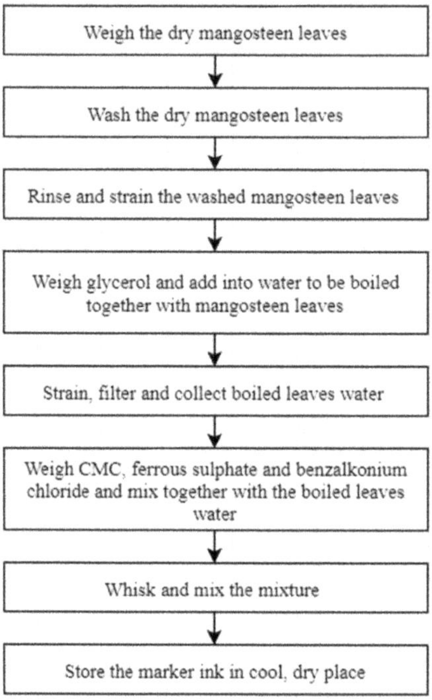

Figure 1. Fabrication process of mangosteen-leaves-based marker ink.

Approximately 75 g of dry mangosteen leaves were first weighed by using a digital weighing scale (Mettler Toledo, Columbus, OH, USA) which was able to read up to 0.001 g and for a sample weight up to 220 g. These were then washed thoroughly and rinsed. Glycerol ranging from 15 g to 62.5 g was weighed by using a digital weighing scale and 375 mL of water was measured using a measuring cylinder, and both were mixed in a beaker and thoroughly stirred for 30 s. The mixture was boiled together on a hot plate (Thermo Fisher Scientific, Waltham, MA, USA) at a heating degree of 8 (circa. 300 °C) for 30 min to increase the extraction rate of dye from the leaves. The hot plate can be heated up to 350 °C and the heating rate can be set from 0% to 100% at 1% intervals. The beaker was covered with plastic foil, to reduce vapor evaporation. Once boiled, the heating degree was reduced to 2 (circa. 100 °C). After boiling the hot plate was switched off, and the boiled mixture was collected into a container using a strainer.

CMC, ferrous sulphate, and benzalkonium chloride of 1–12.5 g, 3 g, and 15 g, respectively were weighed using a digital scale and added into the boiled mixture containing mangosteen leaves. To avoid clumping, the mixture was thoroughly stirred with a whisker, resulting in a smooth and homogeneous ink solution. The ink produced was stored in a glass container, sealed with an airtight cover, and kept in a cool, dry place for 24 h prior to testing.

2.4. Viscosity Test

The ink viscosity was investigated to develop a low viscous ink solution with a high drying rate. Ink samples were accordingly tested 24 h following preparation. The advanced AR-G2 rheometer (TA Instruments, New Castle, DE, USA) was used to conduct the viscosity test in accordance with the ASTM D-445 standard at 23 °C. A cone plate geometry with properties of 60 mm cone diameter, 1:0:0 cone angle ratio (deg:min:s), and 30 μm truncation was installed to the rheometer. The zero-gap was set for 30 μm and 5 g ink samples were placed on the Peltier plate by using a dropper. The experiment was initiated by clicking the 'run' icon, and the results were recorded.

2.5. Color Lightness Test

Color lightness was evaluated using a HunterLAB Ultrascan PRO d/8 Spectrophotometer (Hunter Associates Laboratory, Inc., Reston, VA, USA) operated in accordance with ASTM E1348. Approximately 50 mL of ink sample was used and placed in front of the porthole and the readings recorded. The colorimetric values were reported in terms of CIE color analysis for lightness (L*). The lightness, L*, represents the darkest black at L* = 0, and the brightest white at L* = 100.

2.6. Thermogravimetric Analysis

Thermogravimetric analysis (TGA) was conducted on marker ink samples using Mettler Toledo AG-TGA/SDTA (Model 851e, Mettler Toledo AG, Greifensee, Zürich, Switzerland). For specimen preparation, 20 mg of ink was used for each test. Nitrogen was used as a medium and the specimens were heated from 20 °C to 200 °C with a rate of 25 °C/min. Three ink samples that exhibited high, moderate, and low color lightness were tested to observe variation in their drying rates. As drying progressed, the weight of coating accordingly decreased due to solvent evaporation. The amount of solvent loss with time was monitored through mass change as recorded on the balance [31].

The drying rate of coating is the weight of solvent loss per time divided by the area of evaporation and this can be calculated using Equation (2) where R_{mass} is the evaporation rate, W is the weight of sample at a specific time, t is the time, Δt is the time interval between measurements, and A is the evaporation area.

$$R_{mass} = -\frac{dW}{Adt} \approx -\frac{W_{t+\Delta t} - W_t}{A\Delta t} \quad (2)$$

2.7. Microstructure of Rice Husk Ash

The scanning electron microscope (SEM) is one of the most versatile instruments available for the examination and analysis of microstructure morphology of plant materials [32]. Since the discovery that electrons can be deflected by a magnet, electron microscopy was developed by substituting the light source with a high-energy electron beam [33].

SEM analysis was carried out using the Hitachi S-3400N variable SEM. The microstructure images of mangosteen leaves, including raw samples, or those boiled without glycerol and with glycerol, and samples which exhibited high, moderate, and low color lightness and viscosity properties. Samples were prepared by placing on aluminum sample stubs with carbon conductive tape and were then sputter-coated with gold-palladium using Baltec SCD-005 Sputter Coater to produce high quality and bright image. The stub with coated samples was inserted into the sample chamber of the SEM for viewing.

3. Results and Discussion

The complete design matrix and response values of viscosity and color lightness are given in Table 4. Before designing an experiment using RSM, a preliminary study or screening process using a fractional factorial design (FrFD) method is required. The screening process is essential to identify the most significant factors and to reduce the number of less significant ones when designing an experiment using RSM. For the screening process, four factors including glycerol, ferrous sulphate, benzalkonium chloride, and CMC were studied. Glycerol and CMC showed the most significant effect on the responses and therefore were selected as factors in this study. Data were analyzed using MINITAB.

Table 4. Design matrix and response value for the viscosity and color lightness test.

Sample	Glycerol (V_1)	CMC (V_2)	Glycerol (V_1)	CMC (V_2)	Viscosity (Pa.s)	Color lightness (L*)
S1	−2	0	5	3	0.0221	24.70
S2	0	2	15	5	0.7278	25.83
S3	1	−1	20	2	0.0132	27.20
S4	−1	1	10	4	0.2320	25.11
S5	−2	0	5	3	0.0163	24.73
S6	0	0	15	3	0.1137	25.79
S7	−1	−1	10	2	0.0076	24.93
S8	0	0	15	3	0.1090	25.78
S9	2	0	25	3	0.0091	28.80
S10	0	0	15	3	0.1117	25.62
S11	0	0	15	3	0.1098	25.58
S12	0	−2	15	1	0.0045	25.83
S13	1	−1	20	2	0.0111	27.16
S14	0	0	15	3	0.1083	25.80
S15	−2	0	5	3	0.0203	24.78
S16	−1	−1	10	2	0.0109	25.14
S17	−1	−1	10	2	0.0084	24.99
S18	0	0	15	3	0.1042	25.61
S19	1	1	20	4	0.1595	27.40
S20	0	0	15	3	0.0948	25.78
S21	0	0	15	3	0.1077	25.84
S22	0	0	15	3	0.1064	25.85
S23	1	−1	20	2	0.0211	27.13
S24	1	1	20	4	0.1571	27.48
S25	0	0	15	3	0.1063	25.74
S26	−1	1	10	4	0.2245	25.09
S27	0	−2	15	1	0.0045	25.70
S28	0	2	15	5	0.7082	25.85
S29	0	0	15	3	0.1037	25.63
S30	0	0	15	3	0.1083	25.72
S31	2	0	25	3	0.0096	28.78
S32	0	−2	15	1	0.0045	25.62
S33	−1	1	10	4	0.2173	25.22
S34	0	0	15	3	0.0973	25.72
S35	2	0	25	3	0.0100	28.78
S36	1	1	20	4	0.1344	27.00
S37	0	0	15	3	0.0960	25.73
S38	0	2	15	5	0.6262	25.72
S39	0	0	15	3	0.1083	25.76

3.1. Statistical Analysis of Color Lightness and Viscosity Properties

A linear regression model was fitted to the experimental data using the least square technique. Several main parameters were considered in evaluating the statistical results, namely the coefficients of regression, the standard error of coefficient, and the *p*-value of each factor and its interactions for both responses which are color lightness and viscosity. The results in Table 5 indicated that all factors and interaction effects were highly significant ($p < 0.000$) except for V_2 with $p < 0.113$ and 'V_2* V_2' with $p < 0.529$. Values for $R^2 = 0.9873$ and R^2 (adjusted) = 0.9858 were considerably high, which indicated that 98.73% of sample variation in the response was attributed to the factors.

For viscosity, the *p*-value for both factors (color lightness and viscosity) and their interactions were considered significant at below the confidence level of 95% (*p* of 0.050). The results shown in Table 6 indicated that all factors and interaction effects were significant. The *p* values of all factors and their interactions were highly significant ($p < 0.000$) except for V_1 which did not affect the viscosity of the ink ($p < 0.382$). Values for $R^2 = 0.9292$

and R^2 (adjusted) = 0.9209 were considered very high, which indicated that 92.92% of the sample variation in the response was attributed to the independent variables.

Table 5. Estimated effects and coefficient for glycerol and CMC on the color lightness.

Term	Notation	Coefficient	Std. Error of Coefficient	p
Constant		25.7739	0.03071	0.000
Glycerol	V_1	1.0331	0.02135	0.000
CMC	V_2	0.0347	0.02135	0.113
Glycerol*Glycerol	V_1* V_1	0.2607	0.01545	0.000
CMC*CMC	V_2* V_2	0.0098	0.01545	0.529
		R^2 = 0.9873 R^2 (adj) = 0.9858		

Table 6. Estimated effects and coefficient for glycerol and CMC on the viscosity test

Term	Notation	Coefficient	Std. Error of Coefficient	p
Constant		0.0938	0.01191	0.000
Glycerol	V_1	−0.0073	0.00828	0.382
CMC	V_2	0.1431	0.00828	0.000
Glycerol*Glycerol	V_1* V_1	−0.0235	0.00599	0.000
CMC*CMC	V_2* V_2	0.0593	0.00599	0.000
		R^2 = 0.9292 R^2 (adj) = 0.9209		

Equations (3) and (4) represent the regression models for the color lightness and viscosity, respectively.

$$Y_{COL} = 25.7739 + 1.0331(V_1) + 0.0347(V_2) + 0.2607\left({V_1}^2\right) + 0.0098\left({V_2}^2\right) \quad (3)$$

$$Y_{VIS} = 0.0938 - 0.0073(V_1) + 0.1431(V_2) - 0.0235\left({V_1}^2\right) - 0.0593\left({V_2}^2\right) \quad (4)$$

where Y_{COL} and Y_{VIS} represent the responses which are color lightness and viscosity, respectively whereas V_1 and V_2 are the decoded values of glycerol and CMC, respectively. The regression models can be used to calculate and analyze the effect of factors on the properties of plant-based marker ink.

3.2. Effect of Factors on Color Lightness and Viscosity

ANOVA and regression models were used to analyze the effect of various factors on the properties of the plant-based marker ink. Contour plots were used for better illustration. Figures 2 and 3 illustrate the effect of glycerol (V_1) and CMC (V_2) on the responses, respectively. The findings show that higher V_1 resulted in higher color lightness, and lower V_2 resulted in lower viscosity. Based on Figure 2, the color lightness of mangosteen-leaves-based marker ink decreased when the ratio of glycerol concentration in the ink formulation is approaching 5. This indicates that a low concentration of glycerol reduces the value of color lightness in the ink.

An increase in the amount of glycerol in a liquid may increase the boiling temperature of the liquid [19]. Glycerol is alcohol with low vapor pressure. It changes its colligative properties when mixed with water, increasing the boiling point and lowering the melting point of the solution. According to Kowalska et al. [34], an increase in boiling temperature from 20 to 80 °C (glycerol content of 30% and below) may consequently decrease the total dye extraction yield. The study also discovered that increasing the glycerol concentration from 30% to 80%, while maintaining the same boiling temperature reduced the amount of extracted anthocyanins, which is in conformance with the results obtained in this study. Anthocyanin pigments are known as natural colorants due to their wide range of colors and high solubility in aqueous media [35].

The contour plot in Figure 3 shows the effect of glycerol and CMC concentration on the viscosity of plant-based marker ink. At a glycerol concentration of 5, the viscosity decreases as the CMC concentration decreases. Ghannam and Esmail [16] investigated the rheological properties of CMC solutions in a concentration ranging from 1% to 5%. When the concentration was lowest, they observed Newtonian behavior, while at higher concentrations, they observed non-Newtonian behavior such as pseudoplastic, thixotropic, and viscoelastic behavior.

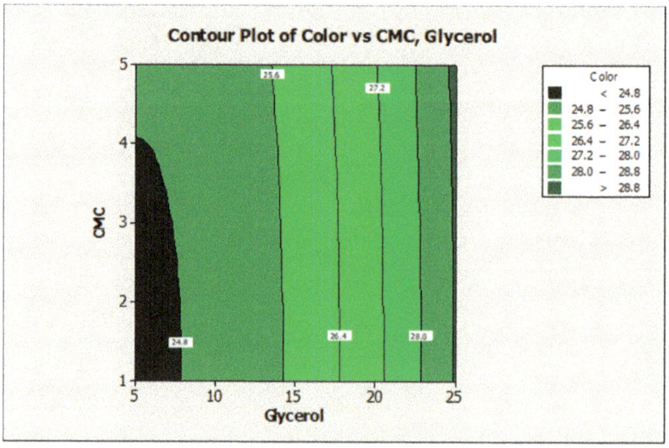

Figure 2. Contour plot for the effect of glycerol and CMC concentration on the color lightness of mangosteen-leaves-based marker ink.

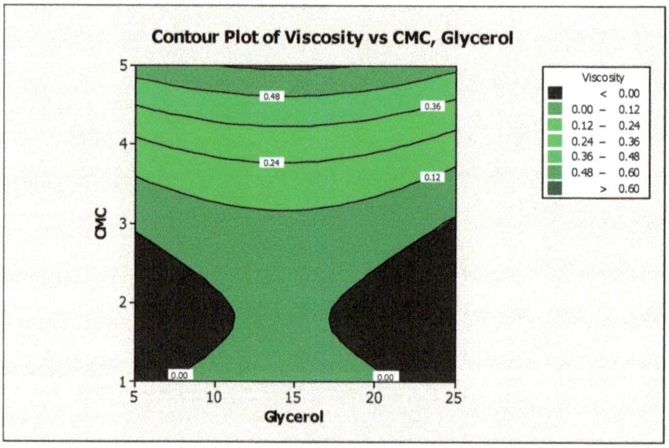

Figure 3. Contour plot for the effect of glycerol and CMC concentration on the viscosity of mangosteen-leaves-based marker ink.

Similar results were obtained by Edali et al. [36], who investigated the rheological behavior of CMC solutions also at higher concentrations. They found that the solution possessed both non-Newtonian and viscoelastic properties. Other studies also concluded that the apparent viscosity of CMC solution increased with rising concentrations. The results were obtained from the flow curves of the CMC solutions at different concentrations plotted over a log–log scale. This rise in apparent viscosity is due to the increase in the interactions between the CMC molecules [37].

3.3. Optimization of the Responses

The optimum formulation of marker ink can be achieved with the combination of glycerol ratio = 5 and CMC ratio = 3, as shown in Figure 4. The desirability of optimization was calculated as 0.98495, indicating that all parameters were within the target, which was to obtain desirable ink properties.

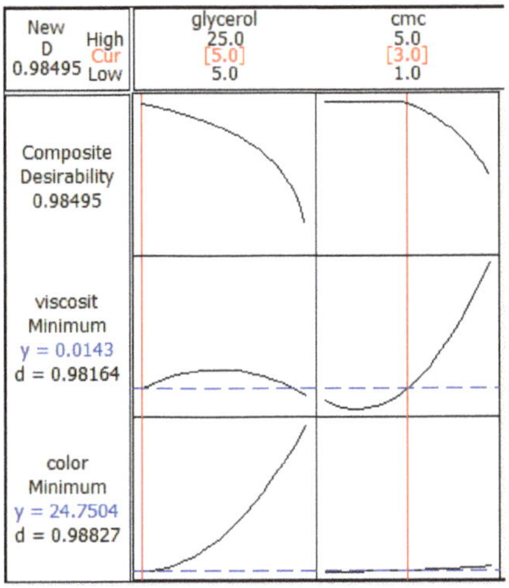

Figure 4. Optimization plot for glycerol and CMC.

The value of composite desirability was calculated at 0.98495, which is high and which proved that the prediction is accurate. It also indicates that all parameters were within the target of obtaining the desirable ink properties. The optimization was completed within a set of parameters. For color properties, the target and upper values were set at 24.7 and 29.0, respectively. The target and upper boundary values for viscosity properties were set at 0.001 and 0.7278, respectively.

3.4. Experimental Validation

From Table 7, it is clear that the average error for color lightness and viscosity were well below 15% at only 0.50% and 9.09%, respectively. It was concluded that the regression model established using this method was able to optimize accurately the values for the responses.

Table 7. Experimental validation for mangosteen-leaves-based marker ink properties.

Sample	Color Lightness (L*)			Viscosity (Pa.s)		
	Experimental Value	Predicted Value	Error (%)	Experimental Value	Predicted Value	Error (%)
SV1	24.99	24.75	0.97	0.0144	0.0143	0.70
SV2	24.80	24.75	0.20	0.0152	0.0143	6.29
SV3	24.83	24.75	0.32	0.0172	0.0143	20.28
	$\bar{x}$ Error		0.50	$\bar{x}$ Error		9.09

3.5. Microstructure Analysis of Mangosteen Leaves

Five samples were selected to determine the microstructure of mangosteen leaves and the drying properties of the marker ink. The samples were dried mangosteen leaves, boiled leaves without glycerol, and with glycerol showing low (sample S26), moderate (sample S19), and high (sample S35) value of color lightness. The sample was analyzed using an SEM.

Figure 5 illustrates SEM micrographs of dry mangosteen leaves at 100× and 500× magnifications. The former magnification shows an image with a relatively corrugated surface, in the form of wrinkles, developed on the leaf samples. The degree of surface corrugation was attributed to the drying process of the leaves at ambient temperature. The cell structure shrank slightly as a result of water evaporating from the leaves. The latter magnification shows no pores on the smooth surface of the dry leaves. This confirms that the leaves dried due to the evaporation of water at ambient temperature and not because of the influence of elevated temperature. In this stage, the drying shrinkage was closely related to the total volume of evaporated water [38].

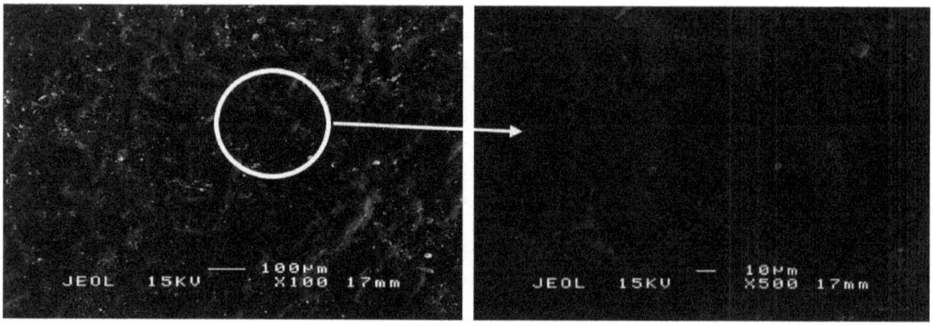

Figure 5. SEM micrographs of dry mangosteen leaves at 100× (**left**) and 500× (**right**) magnification.

Glycerol was found to exert the most significant effect on color lightness as compared to that of other additives. SEM scanning was conducted to investigate the effect of different amounts of glycerol on the microstructure of mangosteen leaves. Its addition will subsequently affect the amount of dye extracted from the leaves.

Based on Figure 6, the SEM micrographs of all samples showed corrugated leaf surface attributed to cell shrinkage. Figure 6a shows shrunken cells but with most of the cell wall structures remaining intact adjoining each other. Under high-temperature conditions, the leaves developed a folded cuticle on the adaxial epidermis layer [39]. Further magnification at 500× revealed a more definite structure of the leaf samples as compared to those under 100× magnification. Based on Figure 6b, boiled mangosteen leaves without glycerol showed good bonding between cell wall structures and considerably smooth surfaces. The dye was extracted from the leaves by passing it through the small pores between the cells. These pores were observed at the intercellular space, between neighboring epidermal cells. The moderate and large size of irregular oblong pores and disrupted orientation of membranes were formed due to heat stress during the boiling process.

A few recent studies described the effect of high temperature on the anatomical and ultrastructural changes which occurred in the leaf preparations including its organization, the protective outer layer of epidermis and cuticle as well as stomata [40]. The result is in agreement with that of Salem-Fnayou, Bouamama, Ghorbel and Mliki [39] which proved that a high-temperature condition can result in the folded cuticle and disrupted cell wall organization on the adaxial epidermis layer.

The microstructures of Figure 6a,c were found to be quite similar since the cell wall structures were still intact and adjoining each other. However, it can be seen that cell wall structures in sample S35 in Figure 6c,d had collapsed in certain areas and small pores had developed. The occurrence was due to the highest amount of glycerol used in boiling the sample. At high concentrations, glycerol is capable of altering plant cell structure. However, the collapse of cell wall structures is not as severe as samples S19 and S26. Glycerol is generally used as a solvent in water to extract natural pigments. The alcohol also acts as a pre-treatment agent and affects the final color produced through improving color performance [41].

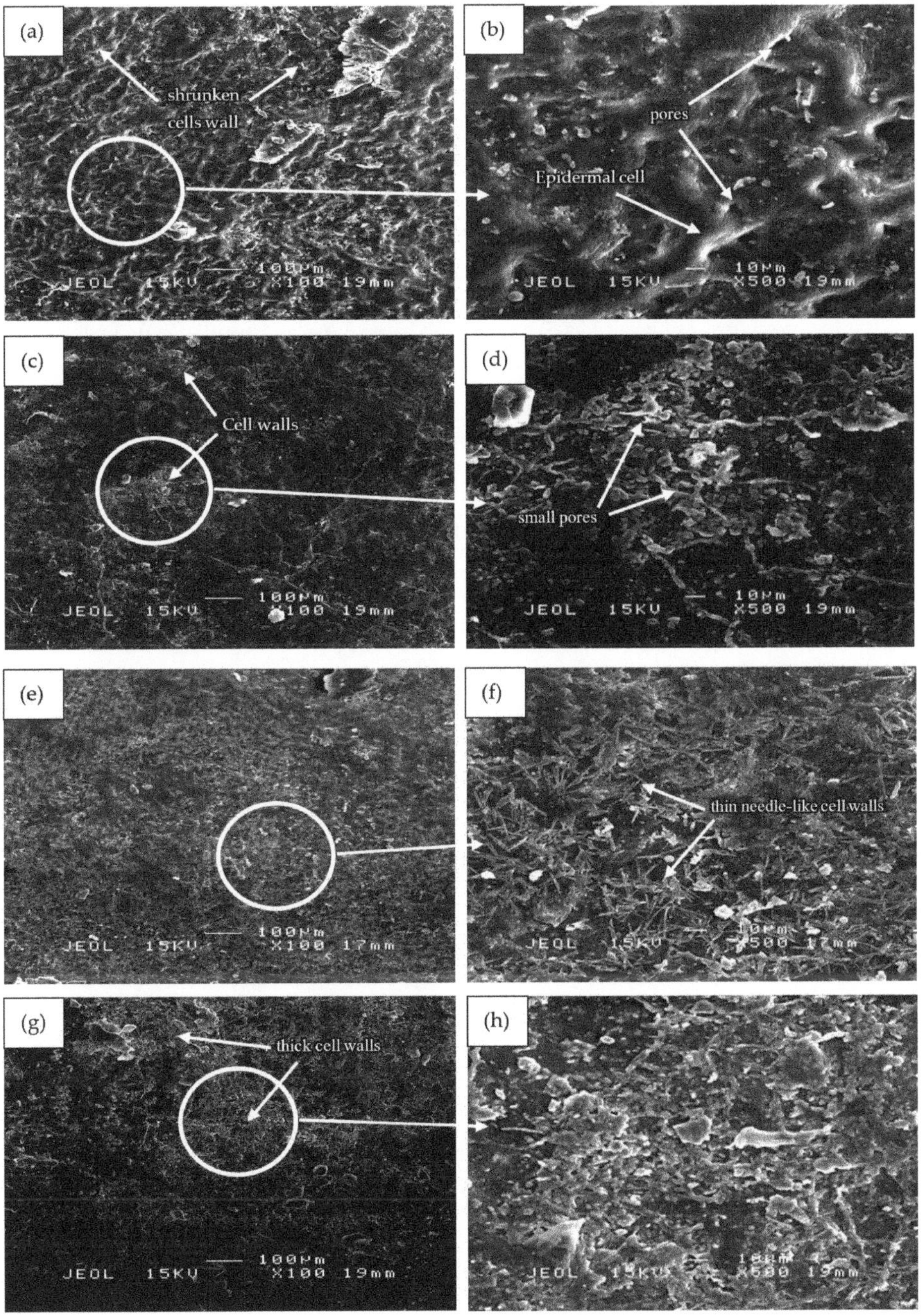

Figure 6. SEM micrographs of (**a**,**b**) boiled dry mangosteen leaves without glycerol, (**c**,**d**) sample S35, (**e**,**f**) sample S19, and (**g**,**h**) sample S26 at 100× and 500× magnification.

Figure 6e of sample S19 covered with a thin needle-like and small-sized collapsed cell wall structures which mostly covered the leaf surface. Sample S26, as seen in Figure 6g, showed the highest amount of collapsed cell wall when the least amount of glycerol was used in boiling. The sample, therefore, showed the most significant effect of glycerol and in agreement with Figure 2. The thicker cell wall structures in sample S26 as compared to that of sample S19 can be visualized based on the brightness of the aggregated structures. The formation of the large and thick cell wall structures was probably due to the agglomeration of the high amount of the collapsed small-sized cell wall structures. Due to the large surface area and strong attractive interaction between particles, the aggregation process must have occurred as similarly recorded by other researchers [42]. Aggregation caused the fracture mechanism to change, with the fracture initiating and developing from the aggregation center [43].

The findings suggest that a decrease in the amount of glycerol in boiled water may alter the cell wall structure of the leaves and subsequently increase the dye extracted. The role of glycerol as dye extraction solvent has two main effects: first, the weakening of the membrane structures of the cells in which the anthocyanin are stored in cell vacuoles, and second, the transformation of the extracted anthocyanin into their flavylium form, which is the most stable form [44]. This condition has been extensively used in technological processes to extract dye from other natural sources such as fresh berries and grape pomace [45].

3.6. *Drying Properties of Marker Ink*

The thermogravimetric analysis (TGA) is a suitable alternative method in determining the drying rate of a solution as measured through sample weight loss during a drying process [31]. The drying rate is affected by the viscosity of a fluid and the rate of evaporation of a solvent [46]. In this study, four ink samples tested exhibited low, moderate, high, and optimum color lightness and viscosity properties. To better visualize the weight loss of the ink samples during the TGA test, a graph of weight loss (in%) as a function of time was plotted as shown in Figure 7.

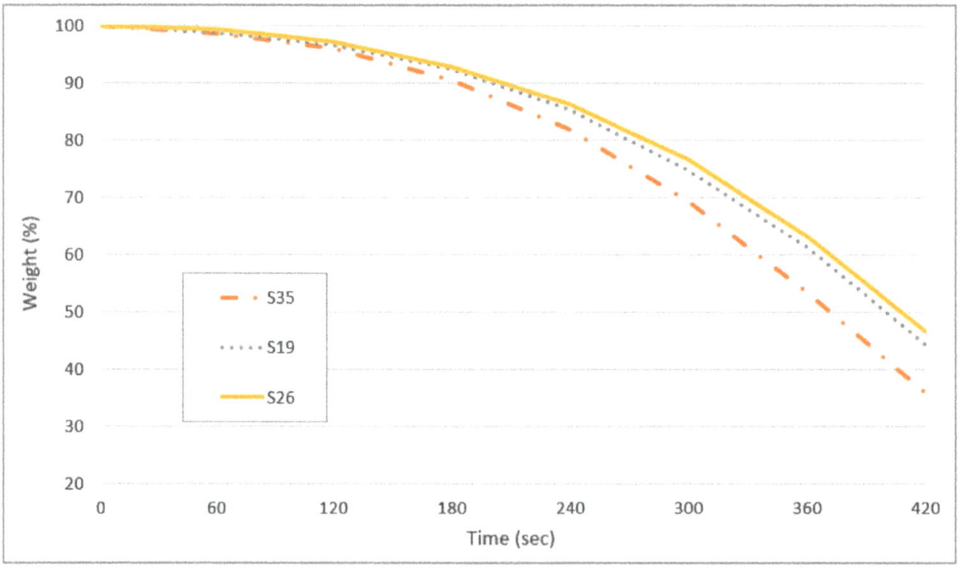

Figure 7. Weight loss as a function of time of sample S35, S19, and S26.

The weight loss for all samples for the first 120 s was very minimal, with a maximum of approximately 4% loss recorded for sample S35. As the test temperature was increased the rate of 25 °C for every 60 s, the initial drying rate was increased in the first 50 °C.

External factors of a volatile matter, including higher air velocity, lower air humidity, and higher air temperature contribute to the increase in drying rate [47]. The drying curve represents the loss of volatiles first by evaporation from a saturated surface above the ink film followed in turn by a period of evaporation from a saturated surface of gradually decreasing area and finally when the volatile compounds evaporate from the interior of the ink's solid [48]. Sample S35 recorded the highest percentage of weight loss at 64.1% followed by sample S19 (55.8%), and S26 (53.4%). It can be concluded that the sample S35 with the highest viscosity showed the fastest drying time.

Figure 8 shows the drying rate curves as a function of time in seconds. The drying rates for samples S19 and S26 were comparable at between 0.01% and 0.13% per second. Sample S35 shows the fastest drying rate of approximately 0.15% per second over 420 s. Due to the low viscosity of sample S35, the diffusion of solvent molecules (water) proceeds at a faster rate than a higher viscosity sample which thus has a slower ability to evaporate [49].

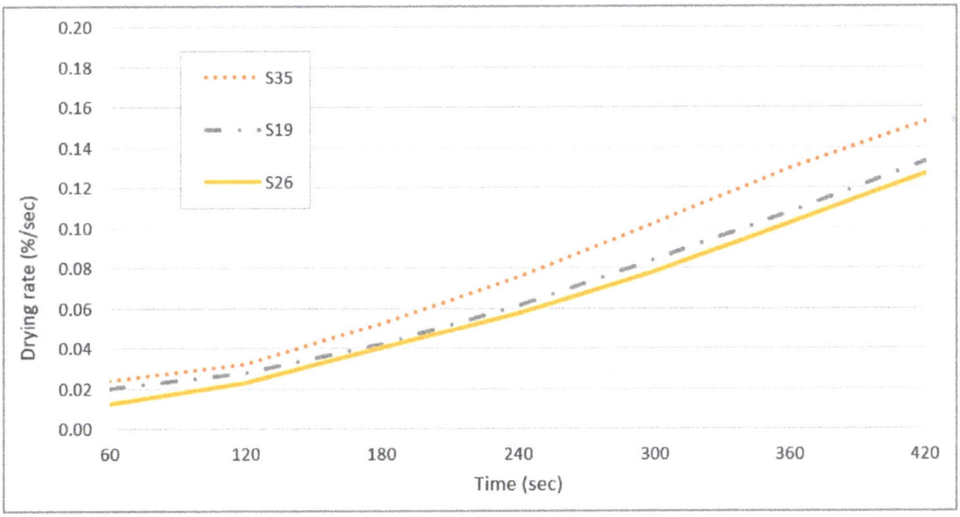

Figure 8. Weight loss as a function of time of sample S35, S19, and S26.

The slower drying rate of highly viscous samples is due to diffusional mass transport limitations in the particle phase, arising from the viscous phase of the particles or due to a combination of oligomer degradation and mass transfer limitations [49]. If a viscous phase is formed, the mixing within the particle bulk will be kinetically limited, and an equilibrium process cannot represent the gas-particle partitioning. The drying rate is mainly governed by the reversible decomposition of oligomers back to monomers [50].

4. Conclusions

For the color lightness test, all factors and interaction effects were highly significant ($p < 0.000$) except for CMC. For the viscosity test, all factors and their interactions were highly significant ($p < 0.000$) except for glycerol. The contour plot indicated that the lowest amount of glycerol results in the lowest value of color lightness. A low amount of CMC results in low viscosity. The optimum formulation of marker ink can be achieved when the ratio of glycerol, benzalkonium chloride, ferrous sulphate, and CMC is set at 5, 5, 1, and 3, respectively. SEM micrographs showed the largest amount of cell wall structure collapse in leaf samples boiled with the lowest amount of glycerol as compared to other samples boiled in the lower amount of glycerol. Sample S35 recorded the highest percentage of weight loss at 64.1% followed by sample S19 (55.8%) and S26 (53.4%). It can be concluded that the sample S35, which had the highest viscosity, showed the fastest drying time. Several potential studies have been identified for future investigation including color saturation, anthocyanin content, and color components.

Author Contributions: Conceptualization, M.S.M.B. and B.L.M.R.; Methodology, M.S.M.B.; Software, M.S.M.B. and B.L.M.R.; Validation, M.S.M.B.; Formal analysis, M.S.M.B. and S.H.K.; Investigation, R.A.T.; Resources, M.S.M.B.; Data curation, R.A.T. and R.Z.; Writing—original draft preparation, M.S.M.B. and B.L.M.R.; Writing—review and editing, B.L.M.R., R.A.T., and R.Z.; Visualization, S.H.K.; Supervision, M.S.M.B., R.A.T., and R.Z.; Project administration, M.S.M.B. and B.L.M.R.; Funding acquisition, M.S.M.B. All authors have read and agreed to the published version of the manuscript.

Funding: This research received no external funding.

Institutional Review Board Statement: Not applicable.

Informed Consent Statement: Not applicable.

Data Availability Statement: The data presented in this study are available on request from the corresponding author.

Acknowledgments: Appreciation is given to all assistant engineers in the department of process and food engineering for their guidance and laboratory assistance.

Conflicts of Interest: The authors declare no conflict of interest.

References

1. Piero, A.R.L.; Puglisi, I.; Rapisarda, A.P.; Petrone, G. Anthocyanins Accumulation and Related Gene Expression in Red Orange Fruit Induced by Low Temperature Storage. *J. Agric. Food Chem.* **2005**, *53*, 9083–9088. [CrossRef] [PubMed]
2. Sun, J.; Li, Y.; Liu, G.; Chen, S.; Zhang, Y.; Chen, C.; Chu, F.; Song, Y. Fabricating High-Resolution Metal Pattern with Inkjet Printed Water-Soluble Sacrificial Layer. *ACS Appl. Mater. Interfaces* **2020**, *12*, 22108–22114. [CrossRef]
3. Sun, J.; Bao, B.; He, M.; Zhou, H.; Song, Y. Recent Advances in Controlling the Depositing Morphologies of Inkjet Droplets. *ACS Appl. Mater. Interfaces* **2015**, *7*, 28086–28099. [CrossRef] [PubMed]
4. Puertas-Bartolomé, M.; Włodarczyk-Biegun, M.K.; Del Campo, A.; Vázquez-Lasa, B.; San Román, J. 3D Printing of a Reactive Hydrogel Bio-Ink Using a Static Mixing Tool. *Polymers* **2020**, *12*, 1986. [CrossRef] [PubMed]
5. Noah, A.; Usman, S.; Alao, O.; Omisakin, O.; Olawale, A. Preliminary Investigation on Production of Brown Ink from Gmelina arborea (ROXB) Fruit Extract. *Int. J. Sci. Basic Appl. Res.* **2014**, *18*, 297–303.
6. Nwafulugo, F.; Samuel, F.; Nyam, T.; Omale, S.; Nwosibe, P.O. Marker Ink Production from Berry Seed Extract. *Int. J. Sci. Eng. Res.* **2019**, *10*, 698–704.
7. Dagde, K.K.; Nwosa, G.; Ukpaka, C. Formulation of White Board Marker Ink Using Locally Sourced Raw Materials. *Eur. J. Eng. Res. Sci.* **2019**, *4*, 107–114. [CrossRef]
8. Shan, T.; Ma, Q.; Guo, K.; Liu, J.; Li, W.; Wang, F.; Wu, E. Xanthones from Mangosteen Extracts as Natural Chemopreventive Agents: Potential Anticancer Drugs. *Curr. Mol. Med.* **2011**, *11*, 666–677. [CrossRef] [PubMed]
9. Nakasone, H.Y.; Paull, R.E. *Tropical Fruits*; Cab International: Wallingford, UK, 1998; pp. 132–148.
10. Yapwattanaphun, C.; Subhadrabandhu, S.; Sugiura, A.; Yonemori, K.; Utsunomiya, N. Utilization of some Garcinia species in Thailand. In Proceedings of the 2000 International Symposium on Tropical and Subtropical Fruits, Cairns, Australia, 25 November–1 December 2000; Volume 575, pp. 563–570.
11. Abdul-Rahman, A.; Goh, H.-H.; Loke, K.-K.; Noor, N.M.; Aizat, W.M. RNA-seq analysis of mangosteen (*Garcinia mangostana* L.) fruit ripening. *Genom. Data* **2017**, *12*, 159–160. [CrossRef]
12. Wrolstad, R. Anthocyanin Pigments—Bioactivity and Coloring Properties. *J. Food Sci.* **2004**, *69*, C419–C425. [CrossRef]
13. Rasmussen, S.E.; Frederiksen, H.; Krogholm, K.S.; Poulsen, L. Dietary proanthocyanidins: Occurrence, dietary intake, bioavailability, and protection against cardiovascular disease. *Mol. Nutr. Food Res.* **2005**, *49*, 159–174. [CrossRef]
14. Kusumawati, N.; Santoso, A.B.; Sianita, M.M.; Muslim, S. Extraction, Characterization and Application of Natural Dyes from the Fresh Mangosteen (*Garcinia mangostana* L.) Peel. *Int. J. Adv. Sci. Eng. Inf. Technol.* **2017**, *7*, 878–885. [CrossRef]
15. Habib, M.A.; Khoda, B. Development of clay based novel bio-ink for 3D bio-printing process. *Procedia Manuf.* **2018**, *26*, 846–856. [CrossRef]
16. Ghannam, M.T.; Esmail, M.N. Rheological properties of carboxymethyl cellulose. *J. Appl. Polym. Sci.* **1997**, *64*, 289–301. [CrossRef]
17. Pagliaro, M.; Rossi, M. *The Future of Glycerol*; John Wiley & Sons: Hoboken, NJ, USA, 2008.
18. Bonnardeaux, J. *Glycerin Overview*; Western Australia Department of Agriculture and Food: Kensington, Australia, 2006.
19. Medeiros, P.S.G.; Barbosa, C.R.F.; Fontes, F.A.O. Effects of addition glycerol co-product of biodiesel in the thermophysical properties of water-glycerol solution applied as secondary coolant. In Proceedings of the 13th Brazilian Congress of Thermal Sciences and Engineering, Uberlandia, Brazil, 5–10 December 2010.

20. Cao, H.; Ai, L.; Yang, Z.; Zhu, Y. Application of Xanthan Gum as a Pre-Treatment and Sharpness Evaluation for Inkjet Printing on Polyester. *Polymers* **2019**, *11*, 1504. [CrossRef] [PubMed]
21. Almoiqli, M.; Aldalbahi, A.; Rahaman, M.; Govindasami, P.; Alzahly, S. Influence of Biopolymer Carrageenan and Glycerine on the Properties of Extrusion Printed Inks of Carbon Nanotubes. *Polymers* **2018**, *10*, 1148. [CrossRef]
22. Osemeahon, S.; John, M.; Dimas, B. Evaluation of the Use of Carbon Black from Waste Materials for the Production of Erasable White Board Ink. *Chem. Res. J.* **2020**, *5*, 97–104.
23. Czitrom, V. One-factor-at-a-time versus designed experiments. *Am. Stat.* **1999**, *53*, 126–131.
24. Box, G.E.; Wilson, K.B. On the experimental attainment of optimum conditions. In *Breakthroughs in Statistics*; Springer: Berlin/Heidelberg, Germany, 1992; pp. 270–310.
25. Czyrski, A.; Jarzębski, H. Response Surface Methodology as a Useful Tool for Evaluation of the Recovery of the Fluoroquinolones from Plasma—The Study on Applicability of Box-Behnken Design, Central Composite Design and Doehlert Design. *Processes* **2020**, *8*, 473. [CrossRef]
26. Flaifel, M.H. An Approach Towards Optimization Appraisal of Thermal Conductivity of Magnetic Thermoplastic Elastomeric Nanocomposites Using Response Surface Methodology. *Polymers* **2020**, *12*, 2030. [CrossRef] [PubMed]
27. Hassan, M.Z.; Roslan, S.A.; Sapuan, S.; Rasid, Z.A.; Mohd Nor, A.F.; Md Daud, M.Y.; Dolah, R.; Yusoff, M.Z.M. Mercerization Optimization of Bamboo (Bambusa vulgaris) Fiber-Reinforced Epoxy Composite Structures Using a Box–Behnken Design. *Polymers* **2020**, *12*, 1367. [CrossRef]
28. Azeem, B.; KuShaari, K.; Naqvi, M.; Keong, L.K.; AlMesfer, M.K.; Al-Qodah, Z.; Naqvi, S.R.; Elboughdiri, N. Production and Characterization of Controlled Release Urea Using Biopolymer and Geopolymer as Coating Materials. *Polymers* **2020**, *12*, 400. [CrossRef]
29. Chen, Y.; Wang, F.; Dong, L.; Li, Z.; Chen, L.; He, X.; Gong, J.; Zhang, J.; Li, Q. Design and Optimization of Flexible Polypyrrole/Bacterial Cellulose Conductive Nanocomposites Using Response Surface Methodology. *Polymers* **2019**, *11*, 960. [CrossRef] [PubMed]
30. Tabaraki, R.; Nateghi, A. Optimization of ultrasonic-assisted extraction of natural antioxidants from rice bran using response surface methodology. *Ultrason. Sonochem.* **2011**, *18*, 1279–1286. [CrossRef] [PubMed]
31. Kim, S.S.; Hyun, J.C. Drying of Coated Film. In *Handbook of Solvents*; ChemTec Publishing: Scarborough, ON, Canada, 2001; Volume 386.
32. Zhou, W.; Apkarian, R.; Wang, Z.L.; Joy, D. Fundamentals of Scanning Electron Microscopy (SEM). In *Scanning Microscopy for Nanotechnology*; Springer: Berlin/Heidelberg, Germany, 2006; pp. 1–40.
33. Wells, O.C.; Joy, D.C. The early history and future of the SEM. *Surf. Interface Anal.* **2006**, *38*, 1738–1742. [CrossRef]
34. Kowalska, G.; Wyrostek, J.; Kowalski, R.; Pankiewicz, U. Evaluation of glycerol usage for the extraction of anthocyanins from black chokeberry and elderberry fruits. *J. Appl. Res. Med. Aromat. Plants* **2021**, *22*, 100296. [CrossRef]
35. Teszlák, P.; Gaál, K.; Pour Nikfardjam, M.S. Influence of grapevine flower treatment with gibberellic acid (GA3) on polyphenol content of *Vitis vinifera* L. wine. *Anal. Chim. Acta* **2005**, *543*, 275–281. [CrossRef]
36. Edali, M.; Esmail, M.N.; Vatistas, G.H. Rheological properties of high concentrations of carboxymethyl cellulose solutions. *J. Appl. Polym. Sci.* **2001**, *79*, 1787–1801. [CrossRef]
37. Benchabane, A.; Bekkour, K. Rheological properties of carboxymethyl cellulose (CMC) solutions. *Colloid Polym. Sci.* **2008**, *286*, 1173–1180. [CrossRef]
38. Lewicki, P.P. Effect of pre-drying treatment, drying and rehydration on plant tissue properties: A review. *Int. J. Food Prop.* **1998**, *1*, 1–22. [CrossRef]
39. Ben Salem-Fnayou, A.; Bouamama, B.; Ghorbel, A.; Mliki, A. Investigations on the leaf anatomy and ultrastructure of grapevine (*Vitis vinifera*) under heat stress. *Microsc. Res. Tech.* **2011**, *74*, 756–762. [CrossRef]
40. Baniwal, S.K.; Bharti, K.; Chan, K.Y.; Fauth, M.; Ganguli, A.; Kotak, S.; Mishra, S.K.; Nover, L.; Port, M.; Scharf, K.-D.; et al. Heat stress response in plants: A complex game with chaperones and more than twenty heat stress transcription factors. *J. Biosci.* **2004**, *29*, 471–487. [CrossRef]
41. Hou, X.; Chen, G.; Xing, T.; Wei, Z. Reactive ink formulated with various alcohols for improved properties and printing quality onto cotton fabrics. *J. Eng. Fibers Fabr.* **2019**, *14*, 1558925019849242. [CrossRef]
42. Jouault, N.; Vallat, P.; Dalmas, F.; Said, S.; Jestin, J.; Boué, F. Well-Dispersed Fractal Aggregates as Filler in Polymer−Silica Nanocomposites: Long-Range Effects in Rheology. *Macromolecules* **2009**, *42*, 2031–2040. [CrossRef]
43. Esbati, A.H.; Irani, S. Effect of functionalized process and CNTs aggregation on fracture mechanism and mechanical properties of polymer nanocomposite. *Mech. Mater.* **2018**, *118*, 106–119. [CrossRef]
44. Ben Amor, B.; Allaf, K. Impact of texturing using instant pressure drop treatment prior to solvent extraction of anthocyanins from Malaysian Roselle (*Hibiscus sabdariffa*). *Food Chem.* **2009**, *115*, 820–825. [CrossRef]
45. Nabli, R.; Achour, S.; Jourdes, M.; Helal, A.N.; Ezzili, B.; Teissedre, P.-L. Anthocyanin composition and extraction from Grenache noir (*Vitis vinifera* L.) vine leaf using an experimental design. I—By ethanol or sulfur dioxide. *OENO One* **2012**, *46*, 295. [CrossRef]
46. Munekata, T.; Suzuki, T.; Yamakawa, S.; Asahi, R. Effects of viscosity, surface tension, and evaporation rate of solvent on dry colloidal structures: A lattice Boltzmann study. *Phys. Rev. E* **2013**, *88*, 052314. [CrossRef] [PubMed]
47. Berk, Z. *Food Process Engineering and Technology*; Academic Press: Cambridge, MA, USA, 2018.

48. Varadarajan, P. To Determine the Drying Rate and the Rewettability Characteristics of Water Based Flexographic News Inks. Master's Thesis, Rochester Institute of Technology, Rochester, NY, USA, 1990.
49. Vaden, T.D.; Imre, D.; Beránek, J.; Shrivastava, M.; Zelenyuk, A. Evaporation kinetics and phase of laboratory and ambient secondary organic aerosol. *Proc. Natl. Acad. Sci. USA* **2011**, *108*, 2190–2195. [CrossRef]
50. Roldin, P.; Eriksson, A.; Nordin, E.; Hermansson, E.; Mogensen, D.; Rusanen, A.; Boy, M.; Swietlicki, E.; Svenningsson, B.; Zelenyuk, A.; et al. Modelling non-equilibrium secondary organic aerosol formation and evaporation with the aerosol dynamics, gas- and particle-phase chemistry kinetic multilayer model ADCHAM. *Atmos. Chem. Phys.* **2014**, *14*, 7953–7993. [CrossRef]

MDPI
St. Alban-Anlage 66
4052 Basel
Switzerland
Tel. +41 61 683 77 34
Fax +41 61 302 89 18
www.mdpi.com

Polymers Editorial Office
E-mail: polymers@mdpi.com
www.mdpi.com/journal/polymers

www.ingramcontent.com/pod-product-compliance
Lightning Source LLC
Chambersburg PA
CBHW061453220926
51850CB00004B/82
* 9 7 8 3 0 3 6 5 6 5 0 6 4 *